Environmental Science and Engineering

The ultimate goal of this series is to contribute to the protection of our environment, which calls for both profound research and the ongoing development of solutions and measurements by experts in the field. Accordingly, the series promotes not only a deeper understanding of environmental processes and the evaluation of management strategies, but also design and technology aimed at improving environmental quality. Books focusing on the former are published in the subseries Environmental Science, those focusing on the latter in the subseries Environmental Engineering.

Tonni Agustiono Kurniawan · Abdelkader Anouzla
Editors

Algae as a Natural Solution for Challenges in Water-Food-Energy Nexus

Toward Carbon Neutrality

Springer

Editors
Tonni Agustiono Kurniawan
Universiti Teknologi Malaysia
Johor Bahru, Malaysia

Xiamen University
Xiamen, China

Abdelkader Anouzla
Faculty of Science and Technology
Mohammedia
University of Hassan II Casablanca
Mohammedia, Morocco

ISSN 1863-5520 ISSN 1863-5539 (electronic)
Environmental Science and Engineering
ISBN 978-981-97-2370-6 ISBN 978-981-97-2371-3 (eBook)
https://doi.org/10.1007/978-981-97-2371-3

This Springer imprint is published by the registered company Springer Nature Singapore Pte Ltd.
The registered company address is: 152 Beach Road, #21-01/04 Gateway East, Singapore 189721, Singapore

Paper in this product is recyclable.

Preface

Welcome to our book, *Algae as a Natural Solution for Challenges in the Water-Food-Energy Nexus: Toward Carbon Neutrality*. In the following pages, we embark on a journey through the world of algae, a group of organisms that hold future promise for addressing critical global challenges such as energy shortage, water pollution, and climate change.

As we turn these pages, let us recognize the urgency of our mission. Algae—our ancient allies—are not merely a curiosity but a lifeline. They offer hope for a sustainable future, where water, food, and energy coexist harmoniously. Join us on this voyage as we unlock the potential of algae and steer toward a carbon-neutral horizon.

Currently, the planet faces an intricate web of interconnected challenges as the confluence of water, food, and energy (Selvaraj et al. 2023). The delicate balance between water availability, food security, climate change, and sustainable energy production is at the heart of this nexus. As humanity navigates the intricate web of interdependencies within this nexus, so does the urgency to find new innovations that harmonize the essential elements.

For millennia, humans have interacted with algae—both macro (seaweed and kelp) and micro (unicellular) to meet their needs (Tavares et al. 2023). The organisms have been cultivated as food sources, providing sustenance to coastal communities and nourishing our bodies. The remarkable potential of algae emerges as a beacon of hope, offering a natural-based solution to multifaceted problems that the world encounters. Nevertheless, presently their importance extends far beyond traditional cuisine.

Although being often overlooked in mainstream discourse, algae possess a wealth of untapped potential. They have a unique ability to convert sunlight and CO_2 into a wide range of biochemical compounds. Despite being categorized as animals, they metabolize the same way as plants, producing O_2 to replenish what humans consume. This cycle acts as a carbon capture system, whereby harmful CO_2 in the atmosphere is converted to useful O_2(Yang et al. 2024). Micro-algae produce a wide range of other compounds found inside the cells, and this makes them useful at combating the effects of climate change on the environment.

What makes them stand out in a complex of water, food, and energy nexus? They can promote:

1. *Resource efficiency*: Algae have evolved to be highly efficient at utilizing resources. Their ability to thrive in diverse environments—whether freshwater, brackish water, or even seawater—sets them apart from other organisms. By harnessing non-arable land and non-potable water, algae complement traditional agriculture (Yamashita et al. 2009).
2. *Carbon sequestration*: Algae's remarkable capacity to sequester CO_2 during photosynthesis contributes to their sustainability. As the world strives for carbon neutrality, algae play a vital role in mitigating its carbon footprint (Ren, 2021).
3. *Nutrient-rich biomass*: Algae produce proteins, lipids, and carbohydrates that are highly digestible. They are rich in essential fatty acids, vitamins, and minerals—an ideal nutritional profile for human need (Diaz et al. 2023).

With their unique capabilities, this book delves into the intricacies of large-scale algae production for food. Our contributors explore breeding techniques, cultivation methods, and the quest for enhanced nutritional qualities. Algae's journey from biofuel research to mainstream food utilization is another path paved with innovation (Srivastava et al. 2023). We also uncover the role of algae in sustainable agriculture, where they serve as biofertilizers, livestock feed supplements, and sources of plant-based protein. By delving into the burgeoning field of algal biofuels, the organisms hold the key to unlocking renewable energy sources that can power the world without exacerbating climate change.

In recent years, there has been an increased interest in growing algae in a rapidly evolving field of renewable energy. Tremendous research on micro-algae has claimed that the tiny organisms have potential in generating clean energy such as biofuels, high-valuable products like biofertilizers, bioplastics, supplements, and aquafeed, while mitigating environmental-related issues such as bio-adsorbent, biochar, and soil-mediated agent (Varela Villarreal et al. 2020).

For this purpose, natural solutions such as algae have been explored and widely applied globally in recent years to deal with climate change. One of the algal types is micro-algae that can be used for biodiesel production. Micro-algae are tiny reservoirs of a plethora of biofuels. Biofuels are the need of today, and researchers around the globe have explored the options for biological fuel production. Algae are an optimal raw material because they occupy between 4% and 7% of the surface area needed to produce the same yield of a land-based crop, do not require fresh water, can be grown in arid zones near the coast, and avoid monocultivation of products to make fuel (Kim et al. 2012; Milledge and Heaven, 2013; Farrokheh et al. 2021). Algae have inherent with the high-lipid content found in some species being a fundamental edge.

Another technological benefit is algae's high per-acre productivity. Since micro-algae are a unique food source, algal cultivation for fuel does not interfere with food production at the levels that cultivation of other feedstock such as corn. Because algae grow in different environments, it could be produced on acreage that is not agriculturally productive. The use of algal-based fuel would result in a tiny fraction of the net GHG that can be traced to fuel use presently. Scaling up algae farming could result in yields of commercial products other than fuel (Wagener, 1983).

Additionally, the utilization of algae for wastewater treatment helps to minimize the amount of organic matter and capture inorganic contaminants such as heavy metals. Although excess growth of algae can poison drinking water and contaminate water sources, they can provide dissolved oxygen (DO) to other living organisms or reduce DO significantly that massive fish die-offs take place (Macusi, 2008; Paul et al. 2020; Ismael et al. 2021).

Conversely, finding carbon capture technologies is vital to minimize GHG emissions in the world. The utilization of micro-algae with a higher capture rate than trees that can be produced in reactors, represents an option for capturing CO_2 in industries and cities. In addition, the research aims to produce in an environmentally sustainable way by extracting a by-product from a waste that has already been produced by another anthropogenic process.

We also explore how algae can be harnessed to purify contaminated water sources, providing a lifeline to communities grappling with water scarcity and pollution (Erler et al. 2018). However, water pollution due to algae has so far received little attention outside scientific circles. Hence, not many scientific books addressing the emerging paradigm of algal management with respect to the problems of algal pollution have been published.

At the heart of this book lies a profound exploration of the role algae can play in steering the world toward carbon neutrality. As the specter of climate change grows ever more ominous, urgent action is vital to mitigate GHG emissions and transition toward a low-carbon future (Malyan et al. 2021). Through rigorous research and insightful analysis, the contributors of this book illuminate the myriad ways in which algae can play key roles to this endeavor.

The journey through these pages takes us on a comprehensive tour of algae's vast potential. We examine the intricacies of algal biology, their diverse habitats, and their remarkable ability to thrive in a wide range of environmental conditions (Tang et al. 2021). From microscopic diatoms to towering kelp forests, from freshwater ponds to vast oceanic expanses, algae inhabit a multitude of niches, each offering unique opportunities for exploration and exploitation. Therefore, this book delves into the practical applications of algae across the water-food-energy nexus.

This book meets the need of our societies, university students and policy makers on scientific approaches to deal with algal pollution partly by using it as biodiesel production and as a biosorbent for water treatment. For this reason, this book disseminates knowledge to readers on how algae may contribute to emerging understanding in climate change mitigation with respect to the relationship between algae as a low-cost biomaterial and climate change that paves the ways for a new direction of mitigation and adaptation in the future (Ansar et al. 2023).

It is expected that this book will inspire layman and other readers on how to contribute to the UN SDGs #6 'Clean Water' by utilizing algae for biodiesel production wastewater treatment, food application, and climate change mitigation. Optimizing the benefits of both algal water treatment and biofuel production demands maximization of total nutrient removal, biomass production, and lipid content of the biomass because algal species known for high nutrient removal and lipid production are easily suspended single-celled algae that are technically difficult to harvest efficiently by gravity alone.

For this reason, this book provides an overview of challenges and opportunities for algal management to mitigate climate change by offering new perspectives on how to control water pollution due to algae, while converting it to biosorbent and biodiesel that could be commercialized in market. The work also explores how to improve the performance of algae for such purposes (Guan et al. 2023). By identifying existing knowledge gap, this work unlocks new research directions for further development of algal management to address global environmental pollution.

As the editors of this volume, we are deeply honored to present this compendium of knowledge to readers, who share our passion for sustainability and innovation. We extend our heartfelt gratitude to the contributors, whose expertise and dedication have enriched the pages with invaluable insights. It is our sincere hope that this book can serve as a catalyst for dialogue, inspiration, and action, spurring renewed efforts to harness the power of algae in service of a sustainable and equitable world.

In closing, we invite readers to embark on a journey of discovery, as this book explores the boundless potential of algae as a natural-based solution for the challenges facing the water-food-energy nexus. Together, let us chart a course toward a future, where carbon neutrality is not merely a distant dream, but a tangible reality, powered by the remarkable resilience and ingenuity of the natural world.

Xiamen, China — Tonni Agustiono Kurniawan, Ph.D.
Mohammedia, Morocco — Abdelkader Anouzla, Ph.D.
February 2024

References

Ansar BSK, Kavusi E, Dehghanian Z et al (2023). Removal of organic and inorganic contaminants from the air, soil, and water by algae. Environ Sci Pollut Res 30:116538–116566. https://doi.org/10.1007/s11356-022-21283-x

Diaz CJ, Douglas KJ, Kang K, Kolarik AL, Malinovski R, Torres-Tiji Y, Molino JV, Badary A, Mayfield SP (2023) Developing algae as a sustainable food source. Front Nutr 9:1029841. https://doi.org/10.3389/fnut.2022.1029841

Erler DV, Nothdurft L, McNeil M et al (2018) Tracing nitrate sources using the isotopic composition of skeletal-bound organic matter from the calcareous green algae *Halimeda*. Coral Reefs 37:1003–1011. https://doi.org/10.1007/s00338-018-01742-z

Farrokheh A, Tahvildari K, Nozari M (2021). Comparison of biodiesel production using the oil of *Chlorella vulgaris* micro-algae by electrolysis and reflux methods using CaO/KOH-Fe_3O_4 and KF/KOH-Fe_3O_4 as magnetic nano catalysts. Waste Biomass Valor 12:3315–3329. https://doi.org/10.1007/s12649-020-01229-5

Guan B, Ning S, Ding X et al (2023) Comprehensive study of algal blooms variation in Jiaozhou Bay based on google earth engine and deep learning. Sci Rep 13:13930. https://doi.org/10.1038/s41598-023-41138-w

Ismael AA (2021) Climate change and its impact on harmful algae in the Egyptian Mediterranean Waters. In: La Rosa D, Privitera R (eds) Innovation in urban and regional planning. INPUT 2021. Lecture notes in civil engineering, vol 146. Springer, Cham. https://doi.org/10.1007/978-3-030-68824-0_44

Kim JK, Um BH, Kim TH (2012) Bioethanol production from micro-algae, *Schizocytrium* sp., using hydrothermal treatment and biological conversion. Korean J Chem Eng 29:209–214. https://doi.org/10.1007/s11814-011-0169-3

Macusi E (2008) Variability and community organization in moderately exposed tropical rocky shore algal communities as influenced by different consumer groups. Nat Prec. https://doi.org/10.1038/npre.2008.2267.1

Malyan SK, Bhatia A, Tomer R et al (2021) Mitigation of yield-scaled greenhouse gas emissions from irrigated rice through *Azolla*, blue-green algae, and plant growth–promoting bacteria. Environ Sci Pollut Res 28:51425–51439. https://doi.org/10.1007/s11356-021-14210-z

Milledge JJ, Heaven S (2013) A review of the harvesting of micro-algae for biofuel production. Rev Environ Sci Biotechnol 12:165–178. https://doi.org/10.1007/s11157-012-9301-z

Paul V, Shekharaiah PSC, Kushwaha S, Sapre A, Dasgupta S, Sanyal D (2020) Role of algae in CO_2 sequestration addressing climate change: a review. In: Deb D, Dixit A, Chandra L (eds) Renewable energy and climate change. Smart innovation, systems and technologies, vol 161. Springer, Singapore. https://doi.org/10.1007/978-981-32-9578-0_23

Ren W (2021) Study on the removable carbon sink estimation and decomposition of influencing factors of mariculture shellfish and algae in China—a two-dimensional perspective based on scale and structure. Environ Sci Pollut Res 28:21528–21539. https://doi.org/10.1007/s11356-020-11997-1

Selvaraj S, Bains A, Sharma M et al (2023). Freshwater edible algae polysaccharides: A recent overview of novel extraction technologies, characterization, and future food applications. J Polym Environ. https://doi.org/10.1007/s10924-023-03049-9

Srivastava N, Mishra PK (2023) Basic research advancement for algal biofuels production. https://doi.org/10.1007/978-981-19-6810-5

Tang Y, Ding S, Wu Y et al (2021). Mechanism of cobalt migration in lake sediments during algae blooms. J Soils Sediments 21:3415–3426. https://doi.org/10.1007/s11368-021-02917-y

Tavares JO, Cotas J, Valado A, Pereira L (2023) Algae food products as a healthcare solution. Mar. Drugs 21:578. https://doi.org/10.3390/md21110578

Villarreal VJ, Burgués C, Rösch C (2020). Acceptability of genetically engineered algae biofuels in Europe: opinions of experts and stakeholders. Biotechnol Biofuels 13:92. https://doi.org/10.1186/s13068-020-01730-y

Yamashita MK, Yokotani T, Hashimoto H, Sawaki N, Notoya M (2009). Sodium and potassium uptake of Ulva—Application of marine macro-algae for space agriculture. Adv Space Res. https://doi.org/10.1016/j.asr.2009.02.004

Wagener K (1983) Mass cultures of marine algae for energy farming in coastal deserts. Int J Biometeorol 27: 227–233. https://doi.org/10.1007/BF02184238

Yang Y, Tang SJ, Chen P (2024) Carbon capture and utilization by algae with high concentration CO_2 or bicarbonate as carbon source. Sci Total Environ. https://doi.org/10.1016/j.scitotenv.2024.170325

Contents

Part I
Towards a Sustainable Algal Management

Chapter 1
Micro-environment Establishment for Promoting Diverse Algal Growth

Alper Baran Sözmen

Abstract Algae offer sustainable sources for water, food, and energy production. They can remove pollutants and pathogens from wastewater and provide potable water. Also they can also be used in creating dietary supplements, functional foods, and biofuels. An optimal micro-environment is essential to promote their growth and productivity in specific applications, and for this purpose it is important to take various physical factors such as light, temperature, hydrodynamics, agitation, and shear stress and chemical factors such as nutrients, pH, and dissolved oxygen into consideration. These factors affect algae growth and productivity, both on their own and by a combinational affect. It is important to properly monitor and manage these factors to ensure optimal algae growth and productivity. This chapter provides an overview of physical and chemical factors that influence growth of algae, emphasizing the importance of micro-environment management in promoting sustainable algae-based industries. It also includes the strategies that can be applied for this purpose, which covers the cultivation techniques and methods to manage cultivation microenvironment in terms of pH, nutrients, and dissolved oxygen.

Keywords Microenvironment · Algal growth · Environmental factors

1.1 Introduction

A microenvironment is a smaller, more localized environment that exists within a larger surrounding one which refers to a specific and limited space where certain physical, chemical, and biological conditions may differ from those in the broader surroundings (Bhatia and Sharma 2015; Boonen et al. 2000). In microbiology, a microenvironment is a term used to refer to a small, localized niche within a larger ecosystem or habitat. The unique conditions found in microenvironments create

A. B. Sözmen (✉)
Bioengineering Department, İzmir Institute of Technology, İzmir, Turkey
e-mail: alpersozmen@iyte.edu.tr

T. A. Kurniawan and A. Anouzla (eds.), *Algae as a Natural Solution for Challenges in Water-Food-Energy Nexus*, Environmental Science and Engineering,
https://doi.org/10.1007/978-981-97-2371-3_1

habitats that support the growth and survival of resident microorganisms. It is important to note that microenvironments exist in various natural and artificial settings. These settings can include soil, water bodies, and even within industrial processes. Furthermore, the conditions found within these environments can vary significantly, including factors such as temperature, pH, oxygen levels, nutrient availability, and the presence of other organisms. Because of these differences, different types of microorganisms may thrive or be restricted to these specific microenvironments, leading to diverse microbial communities within a larger ecosystem.

In case of algae, since they generally inhabit aquatic environments, an algal microenvironment is also generally within a water body, though they share macro environment various algal microenvironments exist such as biofilms, microalgae mats, phytoplankton blooms, and coral symbioses. In some rarer cases microalgae can inhabit soil surfaces or form associations with other organisms, such as lichens. In any form, contents of an algal microenvironment include but is not limited to; algae, water, nutrients, light, and other microorganisms. It is important to understand algal microenvironments, as they significantly influence microbial interactions, metabolic activities, and the overall composition and functioning of algal communities in various environments. Additionally, understanding microenvironments can help developing strategies to manipulate these conditions for various purposes, such as bioremediation or the production of certain products. Overall, the study of microenvironments is a critical component of algal research.

Microenvironments are characterized by their distinct physical and chemical conditions in comparison to a wider environment surrounding them (Fig. 1.1). These conditions can be influenced by factors such as sunlight exposure, moisture levels, nutrient availability, and the presence of other organisms. As a result, microenvironments can create gradients of these factors, leading to diverse niches for different microorganisms to thrive or languish.

This chapter provides an overview of the factors that influence the microenvironment of algae, including light, temperature, hydrodynamics, nutrients, pH, and dissolved oxygen. The optimal range of each factor varies depending on the species of algae and the specific application. Therefore, carefully managing these factors is essential to promote diverse algal growth. Additionally, strategies to provide said conditions are discussed for nutrients, pH, and dissolved oxygen.

1.2 Physical Factors Influencing Algal Micro-environment

1.2.1 Light

There are many factors that affect the growth of algae, and among them, effect of the light level and quality are the main determinant factors of it (Wangpraseurt et al. 2015). Main light related parameters to consider for algae harvesting are light intensity, wavelength, exposure period, and light penetration (Bialevich et al. 2022;

Fig. 1.1 Algae cell in their microenvironment with nutrients, cellular waste, mutual shading, and dissolved oxygen

Ho Gim et al. 2016; Maltsev et al. 2021; Senge et al. 2014; Williams and Laurens 2010; Yan et al. 2013; Zarmi et al. n.d.).

Light saturation is the definition where the quality and intensity of incident light are at their optimum point for highest growth rate and it is specific for each algal strain. In laboratory cultures, the light intensity is usually too low to allow for logarithmic growth, while in nature, the intensity is often above saturation for most of the day and may even be high enough to cause photo-inhibition, also it may be too low during rainy days due to cloudy sky. This variation of light intensity and photoperiod in nature likely inhibit algal growth (Yan et al. 2013). Hence, light intensity and photo-period are integral parameters to control for algae cultivation (Mohsenpour et al. 2020; Termini et al. 2011). The range where algae can continue photosynthesis without inhibition or stratification does differ among the taxonomic groups, therefore the optimal growth conditions should be determined specifically for the relevant species which are to be cultivated (Kommareddy and Anderson 2003; McGlathery and Pedersen 1999). Furthermore, depending on the light intensity, photo-period would also take effect on algal growth. Alternating illumination cycles are detrimental for growth, as well even beyond light saturation it is possible to efficiently exploit light energy with the application of photo-period (Dubinsky et al. 1986). In photobioreactors, even though there are no visible multicellular structures of algae, the system can be imagined as possessing several layers where outermost layers absorbing most of the light energy that may lead to a decrease in photosynthesis efficiency (Sforza et al. 2012).

The effect of light on algal growth differs greatly depending on the imminent microenvironment, especially for multicellular structured algae such as algal biofilms and coral symbioses (Jo et al. 2022; Lyons and Kolter 2015). Whether it is a lake, a pond, or a photobioreactor, algae live in a dynamic environment which is subjected to wave motion, light scattering, and mutual shading; hence through the microenvironment a light gradient forms by the contribution of these environmental factors (Touloupakis et al. 2022; Wangpraseurt et al. 2015). Moreover, the photosynthesis rate, and accordingly, the growth rate of algae are greatly under influence of these factors too, to reach an optimal growth rate sufficient light penetration through the microenvironment must be met (Maltsev et al. 2021). Several methods are proposed to achieve such conditions such as; utilization of optical fibers as supplementary light sources (Wondraczek et al. 2019), limiting nutrition especially nitrogen sources (Wágner et al. 2018), applying "attached cultivation" techniques (Wang et al. 2015), and agitation (Sobczuk et al. 2006).

In addition to light intensity and penetration, wavelength of the incident light has major influence on photosynthesis rate since algae have various types of pigment specific to various wavelengths of light (Senge et al. 2014). Every algae cell contains chlorophyll a and may contain other pigments with various absorption characteristics (Mulders et al. 2014). Chlorophyll a has an absorption maxima wavelength of 420 nm (660 nm with different organic solutions), and other pigments have absorbance maxima values in the range between 420 and 707 nm (da Silva and Lombardi 2020). This difference in the absorption characteristics effects the ability of algae to exploit light energy with different spectral composition (Teo et al. 2014). It is crucial to provide the algae with light that is composed of relevant spectral composition both for growth of algae and also accumulation of specific compounds synthesized by algae (Sharma et al. 2020).

1.2.2 Temperature

Algal species have an optimal temperature value where they exhibit their highest growth rate and will typically grow best within a range that generally centralized at this optimum value, utilizing available nutrients and light more efficiently (Mei et al. 2022; Ras et al. 2013; Singh and Singh 2015). There are considerable differences between species for growth at low and high temperature, the growth and temperature relationship follows a skewed normal distribution curve like behavior (Sözmen et al. 2022). A higher rate of growth is possible if the temperature is held constant at the specific optimum value and sufficient heat transfer is provided for equal distribution (Ahlgren 1987; Moss 1973). On the other hand temperature stress might have a positive effect on accumulation of secondary metabolites such as fatty acids and phenolic compounds (Raven and Geider 1988). Fatty acid content of algae is especially important when the algal biomass is aimed to be used as animal feed, fertilizer, or biodiesel and it's been showed that short term heat-stress promotes accumulation of fatty acids (Nalley et al. 2018). Whereas for phenolic compounds there is no generalized effect of temperature, the effect of temperature is specific for specific phenolic compounds however total phenolic compound accumulation tends to increase with heat stress as well (Mannino et al. 2015; Newsted 2004; Sozmen et al. 2018). Depending on the application and desired end product, be it biomass or a specific phenolic compound; temperature optimization and sufficient distribution throughout the cultivation media is crucial for efficient production (El Shenawy et al. 2020; Knutson et al. 2018; Pawlita-Posmyk et al. n.d.; Ras et al. 2013).

1.2.3 Hydrodynamics

Agitation and aerating are crucial in the cultivation of algae. These processes ensure that every cell has access to the necessary resources for growth and reproduction by providing uniform distribution of nutrients (de la Noüe et al. 1984; Mittal and Ranade 2023). In addition, proper mixing and aeration enable the penetration and uniform distribution of light throughout the system, which is vital for photosynthesis and biomass production. Without proper mixing, microalgae cultures may also suffer from settling and aggregation, which can significantly lower the productivity (Reddy Ronda et al. 2012; Zhou et al. 2018). Algae cultures must be continuously mixed to keep all cells in suspension with sufficient access to light, for maintaining high levels of biomass productivity. A well-designed mixing system in a photo-bioreactor not only enables nutrient distribution and light penetration into the culture but also promotes efficient gaseous exchange, resulting in healthy growth and development of the algae (Legrand et al. 2021; Tham et al. 2022; Zhang et al. 2013). However excess aeration or agitation might lead to shear stress which refers to the mechanical force exerted on the surface of algae cells due to the movement of the surrounding liquid (Fadlallah et al. 2021).

Algae cells are susceptible to damage from high shear stress, which can cause cell rupture and breakage of cell walls or membranes (Wang and Lan 2018). This can lead to cell death and the release of cell contents into the culture medium. Shear stress can also alter the size and shape of algae cells, resulting in elongated, compressed, or distorted cells that may not efficiently absorb light for photosynthesis, and most probably do not function as regular cells in terms of productivity (Fanesi et al. 2021; Schriber et al. 2019; Xiao et al. 2016). Moderate shear stress can cause algae cells to cluster or form flocs, which can settle unevenly and reduce exposure to light due to mutual shading and nutrients due to hindered diffusion, ultimately impacting growth (Giraldo et al. 2021; Heyland et al. 2021; Lee and Palsson 1993; Rajendran et al. 2013; Sobczuk et al. n.d.). On the other hand, shear stress can also activate stress response mechanisms in algae, which can enhance resilience against mechanical forces. Additionally, shear stress can improve nutrient uptake rates and support cellular respiration by aiding in the transfer of oxygen in the culture efficiency (Michels et al. n.d.; Orefice et al. 2019; Wang and Lan 2018). The impact of shear stress on growth rate is complex and depends on the specific algal species, with some being more resistant to mechanical forces than others. Also, algae can adapt to specific shear stress levels over time, leading to the selection of strains better adapted to these conditions. It is important to find a balance between optimizing mixing, nutrient distribution, and gas exchange while minimizing harmful effects to manage shear stress. This involves adjusting agitation or aeration parameters, selecting appropriate methods, and monitoring shear stress levels to maintain a healthy and productive algal culture (Fadlallah et al. 2021; Legrand et al. 2021; Muhammad Orlando et al. 2018; Tham et al. 2022; Wang and Lan 2018; Zhang et al. 2013).

In summary, proper agitation and aeration are essential for the success of algae cultivation (de Souza Kirnev et al. 2022). They not only ensure uniform distribution of nutrients, air, and temperature in the culture but also facilitate light penetration and gaseous exchange in the microenvironment. Ultimately, this leads to higher biomass productivity and a healthier algae cells (Javanmardian and Palsson 1991; Suh et al. 1991; Suh and Lee 2003; Ugwu et al. 2008). Different types of systems can be used to proper agitation and aeration, depending on the scale of the operation. Where magnetic stirrers or air bubblers can be utilized for small-scale algae cultivation, larger-scale operations may require more advanced systems such as impellers, airlifts, or bubble columns. It is important to choose the appropriate mixing system for the specific needs of the culture to ensure optimal growth and productivity (de Souza Kirnev et al. 2022; Suh and Lee 2003; Wang and Lan 2018).

1.3 Chemical Factors Influencing Algal Micro-environment

1.3.1 Nutrients

Algae require a variety of nutrients to support their physiological processes, including photosynthesis, respiration, and cell division (Juneja et al. 2013; Raven and Beardall 2003).

Nitrogen (N): Nitrogen is one of the fundamental component of amino acids, proteins, nucleic acids, and chlorophyll. Adequate nitrogen is essential for cell growth and division (Gordillo et al. 1998; Juneja et al. 2013). Algae have different nitrogen preferences; some thrive in environments with high levels of nitrogen (eutrophic), while others prefer low-nitrogen conditions (oligotrophic) (Kumar and Bera 2020). Nitrogen limitation can lead to reduced growth and chlorosis (yellowing) of cells which is reversible via introduction of a readily available nitrogen source into culture medium of dormant cells (Alexander Klotz et al. 2016; Cointet et al. 2019).

Phosphorus (P): Phosphorus is a component of nucleic acids, ATP, and cell membranes; it comprises 1% of the algae dry biomass (Borchardt and Azad 1968). It is essential for energy transfer and storage. Like nitrogen, phosphorus availability influences algal growth and in natural environments it is mostly the primary limiting nutrient (Larned 1998). Phosphorus limitation can lead to slower growth rates and shifts in algal community, and lower lipid compositions (Xin et al. 2010; Zhuang et al. 2018).

Potassium (K): Potassium is involved in osmoregulation, enzyme activation, and photosynthesis. It also plays a role in maintaining cell turgor pressure (Takagi et al. 2000). Adequate potassium levels (although dependent on the strain concentrations between 15 and 350 mg/ml) promote healthy cell functioning and robust growth (Plato and Denovan 1974).

Magnesium (Mg): Magnesium is the central component of chlorophyll molecules, critical for photosynthesis (Salman et al. 2023). A deficiency in magnesium can lead to chlorosis and reduced photosynthetic efficiency (Wang et al. 2014).

Calcium (Ca): Calcium is important for maintaining cell wall integrity, stability, and cell division. It also participates in signaling pathways and enzyme activation (Tang et al. 2022). Calcium deficiency can result in weakened cell structures and reduced growth (Terauchi et al. 2016; Wang et al. 2014).

Sulfur (S): Sulfur is a component of amino acids, vitamins, and coenzymes. It is vital for protein synthesis and enzyme activity (Davidian and Kopriva 2010). Sulfur deficiency can lead to reduced protein production and growth inhibition (Zhang et al. 2004).

Trace Elements (Micronutrients): Micronutrients, such as iron (Fe), manganese (Mn), zinc (Zn), copper (Cu), molybdenum (Mo), and boron (B), are required in small amounts as the name indicates. They act as cofactors for enzymes and are involved in various metabolic processes. Deficiencies in trace elements can impact growth and lead to physiological disorders (Bruland et al. 1991).

Carbon (C): While carbon is primarily obtained from atmospheric CO_2, its availability influences algal growth. It is used in photosynthesis to produce organic compounds. Carbon limitation can slow down growth rates (Berman-Frank and Dubinsky 1999). On the other hand, an organic source of carbon may also be provided to the culture, which is called a mixotrophic culture, for some strains (Qian et al. 2023). In contrast to their autotrophic counterparts, mixotrophic algae exhibit a unique ability to adopt a dual nutritional strategy. These algae are versatile in their feeding habits, capable of both photosynthesis and heterotrophic feeding (Lin et al. 2018). This means that mixotrophic algae can utilize external sources of organic carbon, such as organic molecules or other microorganisms, to supplement their nutritional requirements. This adaptability enables them to flourish in environments where light and nutrient availability fluctuate (Jeong et al. 2010). Mixotrophy provides these algae with a distinct advantage, as they can switch between autotrophic and heterotrophic modes. This self-sufficiency in energy production and carbon assimilation allows autotrophic algae to thrive in environments where light and nutrients are available, such as sunlit water bodies, ponds, or controlled photobioreactors (Liu et al. 2022).

Vitamins and Growth Factors: Some algae require certain vitamins and growth factors for optimal growth. These compounds are involved in various metabolic pathways and can impact algal health and reproduction (Croft et al. 2006; Tandon et al. 2017).

Vitamin B1 (Thiamine): Thiamine is important for various metabolic reactions, including carbohydrate metabolism and the synthesis of certain amino acids. It serves as a cofactor for enzymes involved in energy production (Gobler et al. 2007; Paerl et al. 2015).

Vitamin B7 (Biotin): Biotin is a coenzyme involved in carboxylation reactions, which are important for fatty acid synthesis, glucose metabolism, and amino acid metabolism (Helliwell 2017; Sañudo-Wilhelmy et al. 2014).

Vitamin B12 (Cobalamin): Cobalamin is essential for the synthesis of certain amino acids and nucleotides. It plays a role in maintaining healthy cell structures and supports various metabolic processes (Gobler et al. 2007; Grant et al. 2014; Grossman 2016).

Vitamin C (Ascorbic Acid): As an antioxidant, vitamin C helps protect cells from oxidative stress caused by reactive oxygen species (ROS). It's involved in photosynthetic reactions and enhances stress tolerance (Gallie 2013; Messyasz et al. 2015).

Vitamin K: Vitamin K is involved in photosynthetic electron transport and helps transport electrons within the photosynthetic apparatus.

Nicotinamide Adenine Dinucleotide (NAD) and NADP: NAD and NADP are coenzymes involved in redox reactions and energy metabolism. They play a critical role in transferring electrons during various metabolic processes (Lin et al. 2010; Matsumura-Kadota et al. 1982).

Pantothenic Acid: Pantothenic acid is essential for the synthesis of coenzyme A (CoA), which is involved in many metabolic pathways, including the citric acid cycle and fatty acid synthesis (Avidan et al. 2015; Raven and Beardall 2003; Zhang and Bryant 2014).

Auxins: Auxins are plant growth regulators that influence cell elongation and differentiation. They can affect cell division and differentiation in some algae species (Davidson 1950; Han et al. 2018; Piotrowska-Niczyporuk et al. 2020; Zhang and Van Duijn 2014).

Cytokinins: Cytokinins are growth-promoting hormones that regulate cell division and growth in plants and algae (Piotrowska and Czerpak 2009; Piotrowska-Niczyporuk et al. 2020).

Gibberellins: Gibberellins are plant hormones that influence stem elongation and flowering. They can affect growth and development in certain algae (Stirk et al. 2013; Yu et al. 2016).

Ethylene: Ethylene is a plant hormone involved in fruit ripening and senescence. Although the data on algae is scarce, it might play a role in certain algae responses to stress or developmental changes (Ju et al. 2015; Pierik et al. 2006).

Abscisic Acid: Abscisic acid is a hormone that regulates responses to environmental stresses. It could impact the stress response of some algae species (Cowan and Rose 1991; Kobayashi et al. 1997; Yoshida et al. 2004).

Nutrient Ratios: The ratios of different nutrients can also affect algal growth. The Redfield ratio, for example, describes the ideal molar ratio of nitrogen, phosphorus, and carbon (106:16:1), which is close to the composition of algal biomass (Gao et al. 2019; Townsend et al. 2007, 2008; Zhao et al. 2023).

Balancing nutrient availability is crucial for promoting healthy algal growth and preventing issues like nutrient imbalance, eutrophication, and algal blooms. In controlled cultivation settings, providing the right concentrations of essential nutrients and monitoring nutrient levels regularly can help optimize algal culture conditions and enhance productivity for various applications, including biofuel production, bioremediation, and aquaculture (Juneja et al. 2013).

1.3.2 pH

The pH is a critical factor that significantly influences algae growth and overall aquatic ecosystem health. It impacts various biological and chemical processes in the water, particularly those related to nutrient availability, enzyme activity, and chemical speciation. Maintaining appropriate pH levels is crucial for optimizing algae growth and ensuring a stable environment (Qiu et al. 2017; Zhang et al. 2014). pH also affects the solubility and availability of nutrients in the water. Many essential nutrients, such as phosphorus and iron, have optimal solubility at specific pH ranges. If the pH is outside these ranges, nutrient availability to algae can be compromised, leading to nutrient deficiencies and hindered growth (Ebrahimzadeh et al. 2021; Schneider et al. 2013). Moreover, it influences the efficiency of photosynthesis and respiration. These processes are pH-sensitive, and deviations from the optimal pH range can impact carbon dioxide uptake, oxygen release, and resultantly energy production (Gehl and Colman 1985; Gerloff-Elias et al. 2005). pH can influence the toxicity of certain elements. For example, at higher pH levels, metals like aluminum become more soluble and toxic to algae. Maintaining appropriate pH can mitigate the harmful effects of such elements (Kinross et al. 2000; Lindemann et al. 1990). Lastly, pH affects the integrity of cell membranes. Drastic pH shifts can disrupt the lipid bilayer structure of cell membranes, leading to cellular stress and damage (Lavoie et al. 2012).

Buffering capacity refers to the ability of a solution to resist changes in pH when an acidic or alkaline compound is added. It is essential for maintaining stable pH conditions in aquatic systems, especially those subject to fluctuations due to environmental factors or metabolic activities. A sufficient buffering capacity can prevent aforementioned hinderances (Choi et al. 2017; Sorokin 1971).

In summary, maintaining the appropriate pH range and ensuring sufficient buffering capacity are critical for promoting healthy algae growth and maintaining a stable aquatic environment. Careful monitoring and management of pH conditions can contribute to the success of algae cultivation (Gehl and Colman 1985; Lívanský 1982; Qiu et al. 2017).

1.3.3 Dissolved Oxygen

Dissolved oxygen is another important factor in algal cultures because it affects growth and the overall health of aquatic ecosystems. The availability of oxygen in the dissolved form within the culture medium influences metabolic activities, which are intricately tied to algal growth (Kazbar et al. 2019). This oxygen content impacts many physiological processes, including growth rate, cellular functions, and interactions with other organisms (Masojídek et al. 2021). Photosynthesis and respiration are two pivotal processes central to algal metabolism. During photosynthesis, algae use light energy to synthesize oxygen, increasing the oxygen content within the medium. Conversely, respiration consumes oxygen, releasing energy from organic molecules. The balance between these two processes is crucial for overall growth and vitality (Franke et al. 2022; Rearte et al. 2021).

Oxygen saturation is a crucial concept in algal cultures. Dissolved oxygen capacity within water depends on temperature, atmospheric pressure, and salinity. Algal cultures should maintain dissolved oxygen levels at or near saturation levels to maintain optimal growth (Abiusi et al. 2020; Prakash Bhuyar et al. 2019; Ugwu et al. 2007). On the other hand, Oxygen limitation occurs when algae don't have sufficient oxygen for their metabolic demands (Haas et al. 2014). This limitation hampers growth rates, reduces photosynthetic efficiency, and can even provoke stress responses. In other respect, oxygen supersaturation, although rare, can yield oxidative stress and cellular impairment due to excessive oxygen exposure (Davisunjaya Sarian et al. 2016; Kazbar et al. 2019). Dissolved oxygen levels must be meticulously monitored for optimal growth rates and biomass yield. Factors such as species-specific oxygen requirements, temperature, light intensity, and nutrient availability all play a role in determining the optimal dissolved oxygen concentration (Titica et al. 2018). Regular monitoring of oxygen levels, along with strategies to maintain suitable aeration or oxygenation of the culture medium, are essential in creating an environment that fosters robust and healthy algal growth (Genkov and Spreitzer 2009; Kazbar et al. 2019).

1.4 Strategies for Managing Micro-environment for Optimal Algae Growth

1.4.1 Cultivation Techniques and Microenvironment Management

Cultivation techniques play an essential role in managing the microenvironment to optimize algae growth. Different techniques offer varying levels of control over factors such as light, temperature, nutrient availability, and pH. The choice of cultivation method can directly impact the microenvironment, influencing algal growth, diversity, and overall productivity (Goswami et al. 2022; Lane 2021; Molitor et al. 2023; Touloupakis et al. 2022; Yaashikaa et al. 2022).

Open Ponds

Open ponds are large, shallow basins in which microalgae are grown under natural sunlight. They provide natural sunlight and exposure to ambient conditions (Klein and Davis 2020). However, they rely on the surrounding environment for temperature regulation and nutrient input (Narala et al. 2016). The microenvironment in open ponds involves managing factors like water depth, temperature, light availability, and nutrient concentrations. Mixing mechanisms such as paddlewheels or air bubbling help maintain uniform conditions and prevent settling. However, challenges like contamination and susceptibility to environmental fluctuations need careful consideration (de Andrade et al. 2022; Deruyck et al. 2019; Pham et al. 2014).

Photobioreactors

Photobioreactors are closed systems that offer precise control over the microenvironment. They enable precise optimization of growth conditions, promoting specific algal species or high-value products (Touloupakis et al. 2022). Their designs can be tubular, flat-panel, or bubble column, which would be elected based on the cultivated strains characteristics such as shear stress sensitivity. Light intensity, temperature, pH, dissolved oxygen, and nutrient levels can be tightly regulated (Hulatt and Thomas 2011; Iluz and Abu-Ghosh 2016; Ojo et al. 2014; Tagliaferro et al. 2019; Zafar et al. 2023). Photobioreactors are well-suited for high-value products and research due to their controlled conditions, but they require more energy and investment (Narala et al. 2016).

Raceway Ponds

Raceway ponds are semi-circular or oval channels where water circulates continuously. They offer a compromise between open ponds and closed systems, providing some level of environmental and microenvironmental control via circulation that helps distribute nutrients (Narala et al. 2016; Ranganathan et al. 2017). The microenvironmental factors such as flow rate, nutrient concentrations, and temperature can be maintained uniformly in raceway ponds when compared to open ponds. Raceway ponds are cost-effective but can be susceptible to contamination and temperature fluctuations, like open ponds (Bell et al. 2016; Cunha et al. 2020; Prussi et al. 2014).

Vertical Algae Farming

Vertical algae farming involves cultivating algae on vertically arranged surfaces to optimize space utilization (Liu et al. 2013). The microenvironment can be managed by controlling light exposure, nutrient delivery, and temperature. This method enhances productivity in limited spaces, but efficient light distribution and prevention of shading are critical considerations (Carbone et al. 2017; Karimi et al. 2021; Meng et al. 2021).

Hanging Bags or Floating Systems

Algae can be cultivated in hanging bags or floating systems on natural water bodies. These systems utilize the natural microenvironment while providing some level of protection from contaminants. However, they offer limited control over factors like light intensity and temperature (Dogaris et al. 2015; Zhu et al. n.d.).

1.4.2 Microenvironmental Nutrient Management Strategies

Microenvironmental nutrient management strategies in algae cultivation involve controlling and optimizing nutrient availability to promote growth, productivity, and desired outcomes. These strategies aim to provide algae with essential nutrients while preventing imbalances or nutrient limitations that could hinder growth (Legrand et al. 2021). Regularly monitoring nutrient concentrations is the first step in choosing the right strategy to maintain optimal algal growth (Radzun et al. 2015). This helps to understand the nutrient dynamics and adjust supplementation accordingly. Nutrient ratios in the microenvironment of an algal culture depend on various factors, including the specific algal species, cultivation method, growth objectives, and desired outcomes (Slegers et al. 2015).

The N:P ratio is a crucial factor that influences algal growth and nutrient uptake. Typically, the N:P ratio is around 7:1, but this can vary widely depending on the algal species. Some algae might thrive at lower N:P ratios, while others might prefer higher ratios (Larned 1998; Xin et al. 2010; Zhuang et al. 2018). It's essential to research the nutrient preferences of the specific algal species you are cultivating. For diatom species that utilize silicon to build their cell walls, the N:Si ratio is also important. This ratio can vary widely among diatom species but often falls within the range of 15:1–30:1 (Gilpin et al. 2004; Kudo 2003). The C:N ratio is especially relevant when cultivating algae for lipid production or other carbon-rich compounds. A higher C:N ratio can promote the accumulation of carbon-rich molecules (Feng et al. 2011; Química et al. 2022). However, excessively high C:N ratios might lead to nitrogen limitation, so finding a balance is important (Cai et al. 2022). Considering the ratios of other essential nutrients such as potassium, magnesium, and trace elements, they can vary based on the nutritional requirements of the specific algal species. It's important to note that nutrient ratios can be adjusted based on the desired outcomes. For example, if the aim of cultivation is for lipid production (e.g., biofuel generation), nutrient ratios might be adjusted to promote lipid accumulation. Different algae species have evolved to thrive in diverse environments, and their nutrient preferences reflect these adaptations. Conducting preliminary experiments or referencing literature specific to the chosen algal species can provide valuable insights into the optimal nutrient ratios (Croft et al. 2006; Maliaka et al. 2018; Omar Faruque et al. 2020; Orefice et al. 2019).

In summary, nutrient ratios in algal cultures should be tailored to the specific species, growth objectives, and cultivation conditions. While general guidelines can provide a starting point, experimentation and careful observation of algal responses to different ratios are key to achieving optimal growth and desired outcomes (Xin et al. 2010; Yaakob et al. 2021).

1.4.3 pH Management Strategies for Microenvironment

Microenvironmental pH management strategies are crucial for maintaining optimal conditions in algae cultivation. pH plays a significant role in nutrient availability, metabolic processes, and overall algal growth as previously mentioned. Effective pH management ensures that the culture remains within a suitable range for the target algal species and for effective pH management regularly monitoring the pH levels in the culture medium is essential (Chen et al. 2021; Qiu et al. 2017; Titica et al. 2018). Monitoring of pH drift, particularly within open systems, emerges as a meticulous practice This allows the tracking of fluctuations, provides the ability to react and take corrective measures promptly which would augment the well-being of the algae and obviates potential stressors (Galès et al. 2020; Isiramen et al. 2022).

It is imperative to ascertain the optimal pH range that aligns with the growth behavior of the specific algal species. The inherent variation in pH preferences among different species underscores the necessity of determining the pH spectrum conducive to their thriving (Munir et al. 2015). A careful approach to pH stabilization involves the integration of buffering systems within the culture. Buffers, exemplified by bicarbonate and phosphate compounds, assume a pivotal role in amending abrupt pH perturbations. The selection of the most proper buffer is dependent on both the desired pH interval and the specific algae strain under consideration (Smith and Foy 2007; Scherholz et al. 2013). The advancement of pH control mechanisms is augmented through the adoption of automated systems orchestrating pH adjustments by means of judiciously dosed acid or base solutions. These automated systems proffer a steadfast maintenance of pH levels within the stipulated range, fostering a milieu harmonious with the requisites of algal proliferation (Khor et al. 2022; Yu et al. 2022).

A nuanced grasp of the nexus between pH and photosynthesis, as well as respiration, underscores the dynamic character of pH within the culture microenvironment. The photochemical process of photosynthesis engenders pH elevation, precipitated by oxygen liberation, whereas the respiratory act engenders pH attenuation (Gehl and Colman 1985; Haraguchi and Zheng 2021). Monitoring pH variances in response to light cycles offers insights into this dynamic interplay (Cheng et al. 2019). A similar phenomenon is encountered with aeration and carbon dioxide injection, where both hold sway over pH modulation. CO_2 injection tends to depress pH, while aeration exhibits an elevating influence. A purposeful calibration of aeration and CO_2 injection regimes facilitates the attainment of the aspired pH range. In the course of supplementing cultures with nutrients or other pertinent compounds other than CO_2 or oxygen, careful oversight of pH dynamics is required. Certain nutrient solutions have the potential to generate pH fluctuations, thereby necessitating pH rectification during the course of nutrient dosing (Cui et al. 2021; Goldman et al. 1982; Mouget et al. 1995; Raven et al. 2020; Rousch and Sommerfeld 1999).

The advancement of pH control mechanisms is augmented through the adoption of automated systems managing pH adjustments by means of adding well calculated doses acid or base solutions. These automated systems offer facile maintenance of pH levels within the stipulated range (Gao 2021; Philip Mayol et al. 2017). In cases wherein pH attenuation is necessary, utilization of organic acids, such as citric acid or acetic acid, holds potential (Bushell et al. 2019).

Akin to pH, the concept of alkalinity emerges as a noticeable aspect in pH management. Alkalinity, the capacity of a medium to withstand pH fluctuations, is intrinsically connected to buffering efficacy (Gerardi and Lytle 2015). Monitoring and adjusting alkalinity profiles are in parallel with the aim of preserving pH homeostasis. Conversely, for pH stabilization within alkaline bounds, the inclusion of bicarbonate serves as a suitable pH-buffering strategy, preventing volatile pH oscillations (Choi et al. 2019; Wang et al. 2019).

Overall, the central principle to manage pH lies in carefully maintaining pH levels within the ideal range suited to the specific algal species. This fundamental approach, upheld through careful monitoring, strategic utilization of buffering systems, and the integration of automated controls, serves as the foundation for achieving effective pH control in algal cultivation (Gao 2021; Titica et al. 2018).

1.4.4 Dissolved Oxygen Management Strategies

Effective management of dissolved oxygen is paramount in cultivating algae, as it profoundly influences their metabolic processes, growth kinetics, and overall physiological well-being (Kazbar et al. 2019; Ugwu et al. 2007). The maintenance of optimal dissolved oxygen levels within the microenvironment is essential to sustain vital respiratory and photosynthetic activities, while concurrently prohibiting the challenges posed by both oxygen insufficiency and supraphysiological concentrations (Gao et al. 2022; Rezayian et al. 2019; Sousa et al. 2013).

Firstly, the strategic deployment of aeration mechanisms, encompassing diffusers, fine bubble aerators, and paddlewheels, facilitates the infusion of atmospheric air into the culture. This aeration regimen facilitates the crucial exchange of gases, augmenting oxygen levels within the aqueous environment (Suh and Lee 2003; Touloupakis et al. 2022). Central to this pursuit is, similar to pH and nutrient management, the precise monitoring of oxygen saturation levels in the culture (Jiménez et al. 2003; Moheimani and Borowitzka 2007). Sensors engineered to gauge dissolved oxygen concentrations offer real-time data on the oxygen dynamics. This actionable data empowers operators to set up timely interventions as needed. The characterization of oxygen saturation—denoting the maximum oxygen concentration feasible at specific temperature and pressure conditions—constitutes a pivotal guideline too (Dormido et al. 2014; Nguyen and Rittmann 2018; Wei et al. 2019). In this context temperature dynamics merit due consideration, as dissolved oxygen solubility wanes with elevating temperatures (Ugwu et al. 2007). This necessitates proportional modulations in aeration rates to sustain appropriate oxygen levels. The control of aeration intensity predicates the growth phases at hand. During periods of accelerated growth, when oxygen demand attains prominence, intensified aeration serves as an operative approach. Resultantly, the synergistic integration of agitation and aeration strategies furthers the uniform distribution of oxygen, also curbing stagnation zones in the culture (de la Noüe et al. 1984). Moreover, mechanisms facilitating homogenization, such as mechanical stirrers or paddlewheels, have proven instrumental in propagating uniform oxygen distribution. This uniformly diffuses the essential gas throughout the culture, countering any tendencies toward stratification. These dynamic adjustments augment encompassing algal biomass density, growth kinetics, and prevailing environmental hindrances (Bautista-Monroy et al. 2022; Carvalho et al. 2006; Grubisic et al. 2019; Itano et al. 2019).

The relation between photosynthetic activity and dissolved oxygen levels emphasizes the need for nuanced adjustments. Photosynthesis generates oxygen, thereby causes oxygen concentration increments (Proppe and Harrel 2010). Cautious modulation of aeration during luminous phases averts undesired oxygen accumulation. Of utmost importance is the prevention of dissolved oxygen supersaturation, such states predispose algal entities to oxidative stress (Prasad et al. 2014). Prudent regulation of aeration prevents this phenomenon (Stanbury et al. 2017). Moreover, sensitivities to oxidative stress diverge among algal species, precipitating the need for discernment. Some species thrive with elevated oxygen concentrations, while others flourish in oxygen-deprived environments (Chiranjeevi and Patil 2020; Kazbar et al. 2019; Mendoza et al. 2013).

The implementation of feedback control systems, synchronized with real-time dissolved oxygen measurements, assumes an outstanding role in streamlining algae cultivation. This adaptive mechanisms ensure the perpetuation of oxygen levels within the desired spectrum (Fernández et al. 2014; Pawlowski et al. 2017; Titica et al. 2018). In summary, dissolved oxygen management within the microenvironment is essential for sustenance of algae cultivation. The pivotal duality of upholding oxygen sufficiency while forestalling supersaturation clears the nuanced aspects of this undertaking. Constant monitoring, calibrated aeration, and adaptive control systems, are key points of effective dissolved oxygen management of algal cultivation (Gao et al. 2022; Kazbar et al. 2019; Marquez et al. 1995; Mendoza et al. 2013).

1.5 Conclusions and Future Aspects

The establishment and management of the microenvironmental control play a pivotal role in fostering diverse and robust algal growth. Throughout this chapter, we have explored the interactions of factors such as light, temperature, nutrients, pH, and dissolved oxygen, each contributing to the balance required to cultivate a thriving algal population. By comprehending the unique requirements of different algal species and tailoring cultivation strategies accordingly, it is possible to optimize the microenvironment to promote their growth and productivity.

The significance of microenvironment management extends far beyond laboratory experiments, going into the domains of water, food, and energy resources. As global population increases and resource scarcity rises, algae-based industries offer a promising field to address these challenges sustainably. The successful manipulation of the microenvironment not only bolsters the production of biofuels, biofertilizers, and high-value compounds but also emerges as a potential solution for water remediation, aquaculture, and even as a nutritional supplement for human consumption.

Looking ahead, the future of algae-based industries is ripe with opportunities for innovation and research. Exploring advanced cultivation techniques, genetic engineering, bioprocess optimization, and novel bioreactor designs can further enhance the capacity to harness the potential of microalgae and macroalgae alike.

In conclusion, algae stand as a promising alternative to conventional resources for sustainable water, food, and energy. By improving the ability to create tailored microenvironments, it would be possible to make substantial contributions to a greener future. It is clear: the potential of algae as a sustainable source of water, food, and energy is boundless and transformative. The chapters that follow in this book delve deeper into these facets.

References

Abiusi F, Wijffels RH, Janssen M (2020) Doubling of microalgae productivity by oxygen balanced mixotrophy. https://doi.org/10.1021/acssuschemeng.0c00990

Ahlgren G (1987) Temperature functions in biology and their application to algal growth constants. Oikos 49(2):177. https://doi.org/10.2307/3566025

Alexander Klotz A, Georg J, Bucinská L, Jendrossek D, Hess WR, Forchhammer K (2016) Awakening of a dormant cyanobacterium from nitrogen chlorosis reveals a genetically determined program. Curr Biol 26. https://doi.org/10.1016/j.cub.2016.08.054

Avidan O, Brandis A, Rogachev I, Pick U (2015) Enhanced acetyl-CoA production is associated with increased triglyceride accumulation in the green alga *Chlorella desiccata*. J Exp Bot 66(13):3725–3735. https://doi.org/10.1093/jxb/erv166

Bautista-Monroy SS, Chávez-Urbiola EA, Ortega-Palacios R, González-Sánchez A, Gómez-Aldapa CA, Rodríguez-Nava O, Salgado-Ramírez JC, Cadena-Ramírez A (2022) Insights of raceway bioreactor scale-up: effect of agitation on microalgae culture and reduction of the liquid medium speed. Appl Sci 12(3):1513. https://doi.org/10.3390/APP12031513

Bell TAS, Prithiviraj B, Wahlen BD, Fields MW, Peyton BM (2016) A lipid-accumulating alga maintains growth in outdoor, alkaliphilic raceway pond with mixed microbial communities. Front Microbiol 6. https://doi.org/10.3389/FMICB.2015.01480

Berman-Frank I, Dubinsky Z (1999) Balanced growth in aquatic plants: myth or reality? Phytoplankton use the imbalance between carbon assimilation and biomass production to their strategic advantage. www.jstor.org

Bhatia S, Sharma K (2015) Microenvironmentation in micropropagation. In: Modern applications of plant biotechnology in pharmaceutical sciences, pp 345–360. https://doi.org/10.1016/B978-0-12-802221-4.00010-8

Bhuyar P, Hong DD, Mandia E, Rahim MHA, Maniam GP, Govindan N (2019) Desalination of polymer and chemical industrial wastewater by using green photosynthetic microalgae, Chlorella sp. Maejo Int J Energy Environ Commun 1(3):9–19. https://doi.org/10.54279/mijeec.v1i3.244924

Bialevich V, Zachleder V, Bišová K (2022) The effect of variable light source and light intensity on the growth of three algal species. Cells 11(8):1293. https://doi.org/10.3390/CELLS11081293/S1

Boonen C, Joniaux O, Janssens K, Berckmans D, Lemeur R, Kharoubi A, Pien H (2000) Modeling dynamic behavior of leaf temperature at three-dimensional positions to step variations in air temperature and light. Trans ASABE 43(6):1755–1766. https://doi.org/10.13031/2013.3078

Borchardt J, Azad HS (1968) Biological extraction of nutrients. J Water Pollut Control Fed. https://www.jstor.org/stable/25036132

Bruland KW, Donat JR, Hutchins DA (1991) Interactive influences of bioactive trace metals on biological production in oceanic waters. Limnol Oceanogr 36(8):1555–1577. https://doi.org/10.4319/LO.1991.36.8.1555

Bushell FML, Tonner PD, Jabbari S, Schmid AK, Lund PA (2019) Synergistic impacts of organic acids and pH on growth of *Pseudomonas aeruginosa*: a comparison of parametric and Bayesian non-parametric methods to model growth. Front Microbiol 10:418502. https://doi.org/10.3389/FMICB.2018.03196/BIBTEX

Cai Y, Zhai L, Fang X, Wu K, Liu Y, Cui X, Wang Y, Yu Z, Ruan R, Liu T, Zhang Q (2022) Effects of C/N ratio on the growth and protein accumulation of heterotrophic Chlorella in broken rice hydrolysate. Biotechnol Biofuels Bioprod 15(1). https://doi.org/10.1186/s13068-022-02204-z

Carbone DA, Olivieri G, Pollio A, Gabriele, Melkonian M (2017) Growth and biomass productivity of *Scenedesmus vacuolatus* on a twin layer system and a comparison with other types of cultivations. Appl Microbiol Biotechnol 101(23–24):8321–8329. https://doi.org/10.1007/S00253-017-8515-Y

Carvalho AP, Meireles LA, Malcata FX (2006) Microalgal reactors: a review of enclosed system designs and performances. Biotechnol Prog. https://doi.org/10.1021/bp060065r

Chen WH, Dillon WDN, Armstrong EA, Moratti SC, McGraw CM (2021) Self-referencing optical fiber pH sensor for marine microenvironments. Talanta 225:121969. https://doi.org/10.1016/J.TALANTA.2020.121969

Cheng Y, He D, He J, Niu G, Gao R (2019) Effect of light/dark cycle on photosynthetic pathway switching and CO_2 absorption in two Dendrobium species. Front Plant Sci 10:451620. https://doi.org/10.3389/FPLS.2019.00659/BIBTEX

Chiranjeevi P, Patil SA (2020) Microbial fuel cell coupled with microalgae cultivation for wastewater treatment and energy recovery. In: Integrated microbial fuel cells for wastewater treatment, pp 213–227. https://doi.org/10.1016/B978-0-12-817493-7.00010-2

Choi YY, Joun JM, Lee J, Hong ME, Pham HM, Chang WS, Sim SJ (2017) Development of large-scale and economic pH control system for outdoor cultivation of microalgae *Haematococcus pluvialis* using industrial flue gas. Biores Technol 244:1235–1244. https://doi.org/10.1016/J.BIORTECH.2017.05.147

Choi JY, Kinney KA, Katz LE (2019) Effect of $CaCO_3$(S) nucleation modes on algae removal from alkaline water. Environ Sci Technol 53(20):11694–11703. https://doi.org/10.1021/ACS.EST.5B05255

Cointet E, Wielgosz-Collin G, Bougaran G, Rabesaotra V, Gonçalves O, Méléder V (2019) Effects of light and nitrogen availability on photosynthetic efficiency and fatty acid content of three original benthic diatom strains. PLoS ONE 14(11). https://doi.org/10.1371/JOURNAL.PONE.0224701

Cowan AK, Rose PD (1991) Abscisic acid metabolism in salt-stressed cells of *Dunaliella salina* possible interrelationship with fB-carotene accumulation. Plant Physiol 97:798–803. https://academic.oup.com/plphys/article/97/2/798/6087137

Croft MT, Warren MJ, Smith AG (2006) Algae need their vitamins. Eukaryot Cell 5(8):1175–1183. https://doi.org/10.1128/EC.00097-06

Cui J, Jin Z, Wang Y, Gao S, Fu Z, Yang Y, Wang Y (2021) Mechanism of eutrophication process during algal decomposition at the water/sediment interface. J Clean Prod 309. https://doi.org/10.1016/J.JCLEPRO.2021.127175

Cunha P, Pereira H, Costa M, Pereira J, Silva JT, Fernandes N, Varela J, Silva J, Simões M (2020) *Nannochloropsis oceanica* cultivation in pilot-scale raceway ponds—from design to cultivation. Appl Sci 10(5). https://doi.org/10.3390/APP10051725

da Silva JC, Lombardi AT (2020) Chlorophylls in microalgae: occurrence, distribution, and biosynthesis. In: Pigments from microalgae handbook, pp 1–18. https://doi.org/10.1007/978-3-030-50971-2_1/FIGURES/5

Davidian J-C, Kopriva S (2010) Regulation of sulfate uptake and assimilation—the same or not the same? Mol Plant 3. https://doi.org/10.1093/mp/ssq001

Davidson FF (1950) The effects of auxins on the growth of marine algae. Am J Bot 37(7):502. https://doi.org/10.2307/2438025

Davisunjaya Sarian F, Yunita Rahman D, Schepers O, Jos Elise Cornelis van der Maarel M (2016) Effects of oxygen limitation on the biosynthesis of photo pigments in the red microalgae *Galdieria sulphuraria* strain 074G. https://doi.org/10.1371/journal.pone.0148358

de Andrade FP, de Farias Silva CE, Medeiros JA, Vieira RC, de Sá Filho MLF, Santos GKS (2022) Consortium between microalgae and other microbiological groups: a promising approach to emphasise the sustainability of open cultivation systems for wastewater treatment. J Water Process Eng 50:103211. https://doi.org/10.1016/J.JWPE.2022.103211

de la Noüe J, Cloutier-Mantha L, Walsh P, Picard G (1984) Influence of agitation and aeration modes on biomass production by Oocystis sp. grown on wastewaters. Biomass 4(1):43–58. https://doi.org/10.1016/0144-4565(84)90034-9

de Souza Kirnev PC, de Souza Vandenberghe LP, Soccol CR, de Carvalho JC (2022) Mixing and agitation in photobioreactors. In: Innovations in fermentation and phytopharmaceutical technologies, pp 13–35. https://doi.org/10.1016/B978-0-12-821877-8.00005-1

Deruyck B, Thi Nguyen KH, Decaestecker E, Muylaert K (2019) Modeling the impact of rotifer contamination on microalgal production in open pond, photobioreactor and thin layer cultivation systems. Algal Res 38. https://doi.org/10.1016/J.ALGAL.2018.101398

Dogaris I, Welch M, Meiser A, Walmsley L, Philippidis G (2015) A novel horizontal photobioreactor for high-density cultivation of microalgae. Biores Technol 198:316–324. https://doi.org/10.1016/J.BIORTECH.2015.09.030

Dormido R, Sánchez J, Duro N, Dormido-Canto S, Guinaldo M, Dormido S (2014) An interactive tool for outdoor computer controlled cultivation of microalgae in a tubular photobioreactor system. Sensors 14(3):4466–4483. https://doi.org/10.3390/S140304466

Dubinsky Z, Falkowski PG, Wyman K (1986) Light harvesting and utilization by phytoplankton. Plant Cell Physiol 27(7):1335–1349. https://doi.org/10.1093/OXFORDJOURNALS.PCP.A077232

Ebrahimzadeh G, Alimohammadi M, Kahkah MRR, Mahvi AH (2021) Relationship between algae diversity and water quality—a case study: Chah Niemeh reservoir Southeast of Iran. J Environ Health Sci Eng 19(1):437. https://doi.org/10.1007/S40201-021-00616-X

El Shenawy EA, Elkelawy M, Bastawissi HAE, Taha M, Panchal H, Sadasivuni KK, Thakar N (2020) Effect of cultivation parameters and heat management on the algae species growth conditions and biomass production in a continuous feedstock photobioreactor. Renew Energy 148:807–815. https://doi.org/10.1016/J.RENENE.2019.10.166

Fadlallah H, Peerhossaini H, De Groot C, Jarrahi M (2021) Motility response to hydrodynamic stress during the growth cycle in active fluid suspensions. J Fluids Eng Trans ASME 143(7). https://doi.org/10.1115/1.4050054

Fanesi A, Lavayssière M, Breton C, Bernard O, Briandet R, Lopes F (2021) Shear stress affects the architecture and cohesion of *Chlorella vulgaris* biofilms. Sci Rep 11:4002. https://doi.org/10.1038/s41598-021-83523-3

Feng Y, Li C, Zhang D (2011) Lipid production of *Chlorella vulgaris* cultured in artificial wastewater medium. Biores Technol 102. https://doi.org/10.1016/j.biortech.2010.06.016

Fernández I, Gabriel F, Acién A, Berenguel M, Joséluis J, Guzmán G (2014) First principles model of a tubular photobioreactor for microalgal production. https://doi.org/10.1021/ie501438r
Franke S, Steingröwer J, Walther T, Krujatz F (2022) The oxygen paradigm—quantitative impact of high concentrations of dissolved oxygen on kinetics and large-scale production of *Arthrospira platensis*. ChemEngineering 6(1). https://doi.org/10.3390/chemengineering6010014
Galès A, Triplet S, Geoffroy T, Roques C, Carré C, Le Floc'h E, Lanfranchi M, Simier M, Roque d'Orbcastel E, Przybyla C (2020) Control of the pH for marine microalgae polycultures: a key point for CO_2 fixation improvement in intensive cultures. J CO_2 Util 38. https://doi.org/10.1016/j.jcou.2020.01.019
Gallie DR (2013) L-ascorbic acid: a multifunctional molecule supporting plant growth and development. Scientifica 2013:1–24. https://doi.org/10.1155/2013/795964
Gao K (2021) Approaches and involved principles to control pH/pCO_2 stability in algal cultures. J Appl Phycol 33(6):3497–3505. https://doi.org/10.1007/S10811-021-02585-Y/FIGURES/6
Gao F, Yang HL, Li C, Peng YY, Lu MM, Jin WH, Bao JJ, Guo YM (2019) Effect of organic carbon to nitrogen ratio in wastewater on growth, nutrient uptake and lipid accumulation of a mixotrophic microalgae Chlorella sp. Biores Technol 282:118–124. https://doi.org/10.1016/J.BIORTECH.2019.03.011
Gao S, Edmundson S, Huesemann M (2022) Oxygen stress mitigation for microalgal biomass productivity improvement in outdoor raceway ponds. Algal Res 68:102901. https://doi.org/10.1016/J.ALGAL.2022.102901
Gehl KA, Colman B (1985) Effect of external pH on the internal pH of *Chlorella saccharophila*. Plant Physiol 77(4):917. https://doi.org/10.1104/PP.77.4.917
Genkov T, Spreitzer RJ (2009) Highly conserved small subunit residues influence Rubisco large subunit catalysis. J Biol Chem 284(44):30105–30112. https://doi.org/10.1074/jbc.M109.044081
Gerardi MH, Lytle B (2015) Algae, alkalinity, and pH. In: The biology and troubleshooting of facultative lagoons, pp 105–109. https://doi.org/10.1002/9781118981771.CH13
Gerloff-Elias A, Spijkerman E, Pröschold T (2005) Effect of external pH on the growth, photosynthesis and photosynthetic electron transport of *Chlamydomonas acidophila* Negoro, isolated from an extremely acidic lake (pH 2.6). Plant Cell Environ 28(10):1218–1229. https://doi.org/10.1111/J.1365-3040.2005.01357.X
Gilpin LC, Davidson K, Roberts E (2004) The influence of changes in nitrogen: silicon ratios on diatom growth dynamics. J Sea Res 51(1):21–35. https://doi.org/10.1016/J.SEARES.2003.05.005
Giraldo ND, Correa SM, Arbeláez A, Figueroa FL, Ríos-Estepa R, Atehortúa L (2021) Reducing self-shading effects in *Botryococcus braunii* cultures: effect of Mg^{2+} deficiency on optical and biochemical properties, photosynthesis and lipidomic profile. Bioresour Bioprocess 8(1). https://doi.org/10.1186/s40643-021-00389-z
Gobler CJ, Norman C, Panzeca C, Taylor GT, Sañudo-Wilhelmy SA (2007) Effect of B-vitamins (B1, B12) and inorganic nutrients on algal bloom dynamics in a coastal ecosystem. Aquat Microb Ecol 49(2):181–194. https://doi.org/10.3354/AME01132
Goldman JC, Dennett MR, Riley CB (1982) Effect of nitrogen-mediated changes in alkalinity on pH control and CO_2 supply in intensive microalgal cultures. Biotechnol Bioeng 24(3):619–631. https://doi.org/10.1002/BIT.260240308
Gordillo FJL, Goutx M, Figueroa FL, Niell FX (1998) Effects of light intensity, CO_2 and nitrogen supply on lipid class composition of *Dunaliella viridis*. J Appl Phycol 10(2):135. https://doi.org/10.1023/A:1008067022973
Goswami RK, Agrawal K, Upadhyaya HM, Gupta VK, Verma P (2022) Microalgae conversion to alternative energy, operating environment and economic footprint: an influential approach towards energy conversion, and management. Energy Convers Manage 269. https://doi.org/10.1016/j.enconman.2022.116118
Grant MA, Kazamia E, Cicuta P, Smith AG (2014) Direct exchange of vitamin B12 is demonstrated by modelling the growth dynamics of algal-bacterial cocultures. ISME J 8:1418–1427. https://doi.org/10.1038/ismej.2014.9

Grossman A (2016) Nutrient acquisition: the generation of bioactive vitamin B12 by microalgae. Curr Biol 26(8):R319–R321. https://doi.org/10.1016/J.CUB.2016.02.047

Grubisic M, Santek MI, Santek B (2019) Potential of microalgae for the production of different biotechnological products. Chem Biochem Eng Q 33(2):161–182. https://go.gale.com/ps/i.do?p=AONE&sw=w&issn=03529568&v=2.1&it=r&id=GALE%7CA624293620&sid=googleScholar&linkaccess=fulltext

Haas AF, Smith JE, Thompson M, Deheyn DD (2014) Effects of reduced dissolved oxygen concentrations on physiology and fluorescence of hermatypic corals and benthic algae. PeerJ. https://doi.org/10.7717/peerj.235

Han X, Zeng H, Bartocci P, Fantozzi F, Yan Y (2018) Phytohormones and effects on growth and metabolites of microalgae: a review. Fermentation 4(2). MDPI AG. https://doi.org/10.3390/fermentation4020025

Haraguchi A, Zheng J (2021) Effect of pH on photosynthesis of *Euglena mutabilis* Schmitz, an acidophilic benthic flagellate. Hydrobiology 1(1):2–9. https://doi.org/10.3390/HYDROBIOLOGY1010002

Helliwell KE (2017) The roles of B vitamins in phytoplankton nutrition: new perspectives and prospects. New Phytol 216(1):62–68. https://doi.org/10.1111/NPH.14669

Heyland A, Roszell J, Chau J, Chai K, Eaton A, Nolan K, Madden K, Ahmed WH (2021) Mass transfer and flow characterization of novel algae-based nutrient removal system. Biotechnol Biofuels 14(1). https://doi.org/10.1186/s13068-021-01951-9

Ho Gim G, Ryu J, Jong Kim M, Il Kim P, Wouk Kim S (2016) Effects of carbon source and light intensity on the growth and total lipid production of three microalgae under different culture conditions. J Ind Microbiol Biotechnol 43:605–616. https://doi.org/10.1007/s10295-016-1741-y

Hulatt CJ, Thomas DN (2011) Energy efficiency of an outdoor microalgal photobioreactor sited at mid-temperate latitude. Biores Technol 102(12):6687–6695. https://doi.org/10.1016/J.BIORTECH.2011.03.098

Iluz D, Abu-Ghosh S (2016) A novel photobioreactor creating fluctuating light from solar energy for a higher light-to-biomass conversion efficiency. Energy Convers Manage 126:767–773. https://doi.org/10.1016/J.ENCONMAN.2016.08.045

Isiramen OE, Bahri PA, Moheimani NR, Vadiveloo A, Shayesteh H, Parlevliet DA (2022) Improving pH control and carbon dioxide utilisation efficiency in microalgae cultivation systems with the use of a proportional-integral + dead-zone control strategy. Biores Technol Rep 17:100917. https://doi.org/10.1016/J.BITEB.2021.100917

Itano T, Inagaki T, Nakamura C, Hashimoto R, Negoro N, Hyodo J, Honda S (2019) Water circulation induced by mechanical aerators in a rectangular vessel for shrimp aquaculture. Aquacult Eng 85:106–113. https://doi.org/10.1016/J.AQUAENG.2019.03.006

Javanmardian M, Palsson BO (1991) High-density photoautotrophic algal cultures: design, construction, and operation of a novel photobioreactor system. Biotechnol Bioeng 38(10):1182–1189. https://doi.org/10.1002/BIT.260381010

Jeong H, Yoo J, Du Y, Kim JS, Seong KA, Kang NS, Kim TH (2010) Growth, feeding and ecological roles of the mixotrophic and heterotrophic dinoflagellates in marine planktonic food webs. Ocean Sci J 2:65–91. https://doi.org/10.1007/s12601-010-0007-2

Jiménez C, Cossío BR, Niell FX (2003) Relationship between physicochemical variables and productivity in open ponds for the production of Spirulina: a predictive model of algal yield. Aquaculture 221. https://doi.org/10.1016/S0044-8486(03)00123-6

Jo J, Price-Whelan A, Dietrich LEP (2022) Gradients and consequences of heterogeneity in biofilms. Nat Rev Microbiol 20(10):593. https://doi.org/10.1038/S41579-022-00692-2

Ju C, Van De Poel B, Cooper ED, Thierer JH, Gibbons TR, Delwiche CF, Chang C (2015) Conservation of ethylene as a plant hormone over 450 million years of evolution. https://doi.org/10.1038/NPLANTS.2014.4

Juneja A, Ceballos RM, Murthy GS (2013) Effects of environmental factors and nutrient availability on the biochemical composition of algae for biofuels production: a review. Energies 6(9):4607–4638. https://doi.org/10.3390/EN6094607

Karimi Z, Laughinghouse HD, Davis VA, Blersch DM (2021) Substrate properties as controlling parameters in attached algal cultivation. Appl Microbiol Biotechnol 105(5):1823–1835. https://doi.org/10.1007/S00253-021-11127-Y

Kazbar A, Cogne G, Urbain B, Marec H, Le-Gouic B, Tallec J, Takache H, Ismail A, Pruvost J (2019) Effect of dissolved oxygen concentration on microalgal culture in photobioreactors. Algal Res 39

Khor JG, Lim HR, Chia WY, Chew KW (2022) Automated cultivation system for microalgae: growth factors and control. Curr Nutr Food Sci 18(9):776–779. https://doi.org/10.2174/1573401318666220421132428

Kinross JH, Read PA, Christofi N (2000) The influence of pH and aluminium on the growth of filamentous algae in artificial streams. Arch Hydrobiol 149(1):67–86. https://doi.org/10.1127/ARCHIV-HYDROBIOL/149/2000/67

Klein B, Davis R (2020) Algal biomass production via open pond algae farm cultivation: 2021 state of technology and future research. https://www.nrel.gov/docs/fy22osti/82417.pdf

Knutson CM, McLaughlin EM, Barney BM (2018) Effect of temperature control on green algae grown under continuous culture. Algal Res 35:301–308. https://doi.org/10.1016/J.ALGAL.2018.08.020

Kobayashi M, Hirai N, Kurimura Y, Ohigashi H, Tsuji Y (1997) Abscisic acid-dependent algal morphogenesis in the unicellular green alga *Haematococcus pluvialis*. Plant Growth Regul 22:79–85

Kommareddy A, Anderson G (2003) Study of light as a parameter in the growth of algae in a photo-bio reactor (PBR). https://doi.org/10.13031/2013.13868

Kudo I (2003) Change in the uptake and cellular Si:N ratio in diatoms responding to the ambient Si:N ratio and growth phase. Mar Biol 143. https://doi.org/10.1007/s00227-003-1063-2

Kumar A, Bera S (2020) Revisiting nitrogen utilization in algae: a review on the process of regulation and assimilation. Biores Technol Rep 12:100584. https://doi.org/10.1016/J.BITEB.2020.100584

Lane TW (2021) Barriers to microalgal mass cultivation. Curr Opin Biotechnol 2022:323–328. https://doi.org/10.1016/j.copbio.2021.09.013

Larned ST (1998) Nitrogen-versus phosphorus-limited growth and sources of nutrients for coral reef macroalgae. Mar Biol 132

Lavoie M, Le Faucheur S, Boullemant A, Fortin C, Campbell PGC (2012) The influence of pH on algal cell membrane permeability and its implications for the uptake of lipophilic metal complexes 1. J Phycol 48(2):293–302. https://doi.org/10.1111/J.1529-8817.2012.01126.X

Lee C-G, Palsson BO (1993) High-density algal photobioreactors using light-emitting diodes

Legrand J, Artu A, Pruvost J, Pruvost J (2021) A review on photobioreactor design and modelling for microalgae production. React Chem Eng 6:1134. https://doi.org/10.1039/d0re00450b

Lin H, Kwan AL, Dutcher SK (2010) Synthesizing and salvaging NAD+: lessons learned from *Chlamydomonas reinhardtii*. PLoS Genet 6(9):1001105. https://doi.org/10.1371/journal.pgen.1001105

Lin CH, Flynn KJ, Mitra A, Glibert PM (2018) Algae with mixed feeding capabilities may have advantage forming blooms. Front Mar Sci 5. https://doi.org/10.3389/FMARS.2018.00320/FULL

Lindemann J, Holtkamp E, Herrmann R (1990) The impact of aluminium on green algae isolated from two hydrochemically different headwater streams, Bavaria, Germany. Environ Pollut 67(1):61–77. https://doi.org/10.1016/0269-7491(90)90172-9

Liu T, Wang J, Hu Q, Cheng P, Ji B, Liu J, Chen Y, Zhang W, Chen X, Chen L, Gao L, Ji C, Wang H (2013) Attached cultivation technology of microalgae for efficient biomass feedstock production. Biores Technol 127:216–222. https://doi.org/10.1016/J.BIORTECH.2012.09.100

Liu K, Ng HYT, Gao Z, Liu H (2022) Selective feeding of a mixotrophic dinoflagellate (Lepidodinium sp.) in response to experimental warming and inorganic nutrient imbalance. Front Microbiol 13:805306. https://doi.org/10.3389/FMICB.2022.805306/BIBTEX

Lívanský K (1982) Buffering capacity of the chief components of nutritive media for algae. Folia Microbiol 27(6):455–459. https://doi.org/10.1007/BF02876459/METRICS

Lyons NA, Kolter R (2015) On the evolution of bacterial multicellularity. Curr Opin Microbiol 24:21. https://doi.org/10.1016/J.MIB.2014.12.007

Maliaka V, Faassen EJ, Smolders AJP, Lürling M (2018) The impact of warming and nutrients on algae production and microcystins in seston from the iconic lake lesser Prespa, Greece. Toxins 10(4). https://doi.org/10.3390/TOXINS10040144

Maltsev Y, Maltseva K, Kulikovskiy M, Maltseva S (2021) Influence of light conditions on microalgae growth and content of lipids, carotenoids, and fatty acid composition. Biology 10(10). https://doi.org/10.3390/BIOLOGY10101060

Mannino AM, Vaglica V, Cammarata M, Oddo E (2015) Effects of temperature on total phenolic compounds in *Cystoseira amentacea* (C. Agardh) Bory (Fucales, Phaeophyceae) from southern Mediterranean Sea. Plant Biosyst 150(1):152–160. https://doi.org/10.1080/11263504.2014.941033

Marquez FJ, Sasaki K, Nishio N, Nagai S (1995) Inhibitory effect of oxygen accumulation on the growth of *Spirulina platensis*. Biotechnol Lett 17(2):225–228. https://doi.org/10.1007/BF00127993

Masojídek J, Ranglová K, Lakatos GE, Benavides AMS, Torzillo G (2021) Variables governing photosynthesis and growth in microalgae mass cultures. Processes 9(5):820. https://doi.org/10.3390/PR9050820

Matsumura-Kadota H, Muto S, Miyachi S (1982) Light-induced conversion of NAD+ to NADP+ in Chlorella cells. Biochim Biophys Acta (BBA) Bioenerg 679(2):300–307. https://doi.org/10.1016/0005-2728(82)90300-0

McGlathery KJ, Pedersen MF (1999) The effect of growth irradiance on the coupling of carbon and nitrogen metabolism in *Chaetomorpha linum* (Chlorophyta). J Phycol 35(4):721–731. https://doi.org/10.1046/J.1529-8817.1999.3540721.X

Mei X, Gao S, Liu Y, Hu J, Razlustkij V, Rudstam LG, Jeppesen E, Liu Z, Zhang X (2022) Effects of elevated temperature on resources competition of nutrient and light between benthic and planktonic algae. Front Environ Sci 10:624. https://doi.org/10.3389/FENVS.2022.908088/BIBTEX

Mendoza JL, Granados MR, de Godos I, Acién FG, Molina E, Heaven S, Banks CJ (2013) Oxygen transfer and evolution in microalgal culture in open raceways. Biores Technol 137:188–195. https://doi.org/10.1016/j.biortech.2013.03.127

Meng Y, Li A, Li H, Shen Z, Ma T, Liu J, Zhou Z, Feng Q, Sun Y (2021) Effect of membrane blocking on attached cultivation of microalgae. J Clean Prod 284. https://doi.org/10.1016/J.JCLEPRO.2020.124695

Messyasz B, Leska B, Fabrowska J, Pikosz M, Cieslak A, Schroeder G (2015) Effects of organic compounds on the macroalgae culture of *Aegagropila linnaei*. Open Chem 13(1):1040–1044. https://doi.org/10.1515/chem-2015-0119

Michels MHA, Van Der Goot AJ, Norsker N-H, Wijffels RH (n.d.) Effects of shear stress on the microalgae *Chaetoceros muelleri*. https://doi.org/10.1007/s00449-010-0415-9

Mittal R, Ranade V (2023) Bioactives from microalgae: a review on process intensification using hydrodynamic cavitation. J Appl Phycol 35(3):1129–1161. https://doi.org/10.1007/S10811-023-02945-W

Moheimani NR, Borowitzka MA (2007) Limits to productivity of the alga *Pleurochrysis carterae* (Haptophyta) grown in outdoor raceway ponds. Biotechnol Bioeng 96(1):27–36. https://doi.org/10.1002/BIT.21169

Mohsenpour SF, Hennige S, Willoughby N, Adeloye A, Gutierrez T (2020) Integrating micro-algae into wastewater treatment: a review. https://doi.org/10.1016/j.scitotenv.2020.142168

Molitor HR, Schaeffer AK, Schnoor JL (2023) Sustainably cultivating and harvesting microalgae through sedimentation and forward osmosis using wastes. ACS Omega 13:2. https://doi.org/10.1021/acsomega.1c01474

Moss B (1973) The influence of environmental factors on the distribution of freshwater algae: an experimental study: III. Effects of temperature, vitamin requirements and inorganic nitrogen compounds on growth. J Ecol 61(1):179. https://doi.org/10.2307/2258926

Mouget J-L, Dakhama A, Lavoie MC, Noüe J (1995) Algal growth enhancement by bacteria: is consumption of photosynthetic oxygen involved? FEMS Microbiol Ecol 18(1):35–43. https://doi.org/10.1111/J.1574-6941.1995.TB00159.X

Muhammad Orlando A, Rafii Ardiansyah S, Rahman A, Betawati Prihantini N (2018) Effects of aeration intensity as agitation in simple photobioreactors on leptolyngbya (cyanobacteria) growth as biofuel feedstock. E3S Web Conf. https://doi.org/10.1051/e3sconf/20186702011

Mulders KJM, Lamers PP, Martens DE, Wijffels RH (2014) Phototrophic pigment production with microalgae: biological constraints and opportunities. J Phycol 50(2):229–242. https://doi.org/10.1111/JPY.12173

Munir N, Naz S, Munir N, Imtiaz A, Sharif N, Naz S (2015) Optimization of growth conditions of different algal strains and determination of their lipid contents. Use of halophytic nanoparticles as a sustainable approach towards improvement of crop yield and nutritional quality in saline conditions. Emotions and their effects at work place. J Anim Plant Sci. https://www.researchgate.net/publication/275519543

Nalley JO, O'Donnell DR, Litchman E (2018) Temperature effects on growth rates and fatty acid content in freshwater algae and cyanobacteria. Algal Res 35:500–507. https://doi.org/10.1016/J.ALGAL.2018.09.018

Narala RR, Garg S, Sharma KK, Thomas-Hall SR, Deme M, Li Y, Schenk PM (2016) Comparison of microalgae cultivation in photobioreactor, open raceway pond, and a two-stage hybrid system. Front Energy Res 4:208116. https://doi.org/10.3389/FENRG.2016.00029/BIBTEX

Newsted JL (2004) Effect of light, temperature, and pH on the accumulation of phenol by *Selenastrum capricornutum*, a green alga. Ecotoxicol Environ Saf 59(2):237–243. https://doi.org/10.1016/j.ecoenv.2003.07.009

Nguyen BT, Rittmann BE (2018) Low-cost optical sensor to automatically monitor and control biomass concentration in microalgal cultivation. Algal Res 32:101–106. https://doi.org/10.1016/J.ALGAL.2018.03.013

Ojo EO, Auta H, Baganz F, Lye GJ (2014) Engineering characterisation of a shaken, single-use photobioreactor for early stage microalgae cultivation using *Chlorella sorokiniana*. Biores Technol 173:367–375. https://doi.org/10.1016/J.BIORTECH.2014.09.060

Omar Faruque M, Ilyas M, Mozahar Hossain M, Abdur Razzak S (2020) Influence of nitrogen to phosphorus ratio and CO_2 concentration on lipids accumulation of *Scenedesmus dimorphus* for bioenergy production and CO_2 biofixation. Chem Asian J 15(24):4307–4320. https://doi.org/10.1002/ASIA.202001063

Orefice I, Musella M, Smerilli A, Sansone C, Chandrasekaran R, Corato F, Brunet C (2019) Role of nutrient concentrations and water movement on diatom's productivity in culture. Sci Rep 9. https://doi.org/10.1038/s41598-018-37611-6

Paerl RW, Bertrand EM, Allen AE, Palenik B, Azam F (2015) Vitamin B1 ecophysiology of marine picoeukaryotic algae: strain-specific differences and a new role for bacteria in vitamin cycling. Limnol Oceanogr 60(1):215–228. https://doi.org/10.1002/LNO.10009

Pawlita-Posmyk M, Wzorek M, Płaczek M (n.d.) The influence of temperature on algal biomass growth for biogas production. https://doi.org/10.1051/matecconf/201824004008

Pawlowski A, Guzmán JL, Berenguel M, Acién FG, Dormido S (2017) Event-based control systems for microalgae culture in industrial reactors. In: Prospects and challenges in algal biotechnology, pp 1–48. https://doi.org/10.1007/978-981-10-1950-0_1/COVER

Pham DT, Everaert G, Janssens N, Alvarado A, Nopens I, Goethals PLM (2014) Algal community analysis in a waste stabilisation pond. Ecol Eng 73:302–306. https://doi.org/10.1016/J.ECOLENG.2014.09.046

Philip Mayol A, Ubando AT, Mandia E, Michael Mendoza DM, Sybingco E, Culaba AB, Dadios E (2017) Development of a microalgal automated cultivation system on *Tetradesmus obliquus*. J Comput Innov Eng Appl 2(1)

Pierik R, Tholen D, Poorter H, Visser EJW, Voesenek LACJ (2006) The Janus face of ethylene: growth inhibition and stimulation. Trends Plant Sci 11(4):176–183. https://doi.org/10.1016/j.tplants.2006.02.006

Piotrowska A, Czerpak R (2009) Cellular response of light/dark-grown green alga *Chlorella vulgaris* Beijerinck (Chlorophyceae) to exogenous adenine- and phenylurea-type cytokinins. Acta Physiol Plant 31(3):573–585. https://doi.org/10.1007/S11738-008-0267-Y/FIGURES/11

Piotrowska-Niczyporuk A, Bajguz A, Kotowska U, Zambrzycka-Szelewa E, Sienkiewicz A (2020) Auxins and cytokinins regulate phytohormone homeostasis and thiol-mediated detoxification in the green alga *Acutodesmus obliquus* exposed to lead stress. Sci Rep 10(1). https://doi.org/10.1038/s41598-020-67085-4

Plato P, Denovan JT (1974) The influence of potassium on the removal of 137Cs by live Chlorella from low level radioactive wastes. Radiat Bot 14(1):37–41. https://doi.org/10.1016/S0033-7560(74)90084-2

Prasad BSRV, Srinivasu PDN, Varma PS, Raman AV, Ray S (2014) Dynamics of dissolved oxygen in relation to saturation and health of an aquatic body: a case for Chilka Lagoon, India. J Ecosyst 2014:1–17. https://doi.org/10.1155/2014/526245

Proppe D, Harrel S (2010) The effects of photosynthesis and cellular respiration on dissolved oxygen concentration. Sci Act 44(1):10–15. https://doi.org/10.3200/SATS.44.1.10-15

Prussi M, Buffi M, Casini D, Chiaramonti D, Martelli F, Carnevale M, Tredici MR, Rodolfi L (2014) Experimental and numerical investigations of mixing in raceway ponds for algae cultivation. Biomass Bioenerg 67:390–400. https://doi.org/10.1016/J.BIOMBIOE.2014.05.024

Qian J, Fu S, Li J, Toda T, Li H, Sekine M, Takayama Y, Koga S, Shao S, Fan L, Xu P, Zhang X, Cheng J, Jin Z, Zhou W (2023) Effects of organic carbon sources on algal biofilm formation and insight into mechanism. Algal Res 71:103075. https://doi.org/10.1016/J.ALGAL.2023.103075

Qiu R, Gao S, Lopez PA, Ogden KL (2017) Effects of pH on cell growth, lipid production and CO_2 addition of microalgae *Chlorella sorokiniana*. Algal Res 28:192–199. https://doi.org/10.1016/J.ALGAL.2017.11.004

Química I, Fernando Barajas-Solano A, Guarin-Villegas E, Maciel Remolina-Páez L, Patricia Bermúdez-Castro J, Oriana Mogollón-Londoño S, Eduardo Contreras-Ropero J, Bibiana García-Martínez J (2022) Effect of de carbon/nitrogen ratio on the production of microalgae-based carotenoids [Efecto de la relación carbono/Nitrógeno en la producción de carotenoids en microalgas]. Biotechnol Biofuels Bioprod 15. https://doi.org/10.25100/iyc.v22i1.8686

Radzun KA, Wolf J, Jakob G, Zhang E, Stephens E, Ross I, Hankamer B (2015) Automated nutrient screening system enables high-throughput optimisation of microalgae production conditions. Biotechnol Biofuels 8(1). https://doi.org/10.1186/S13068-015-0238-7

Rajendran A, Anderson GA, Yan X, Gent S (2013) Light in a photobioreactor. In: 2013 Kansas City, Missouri, 21–24 July 2013. https://doi.org/10.13031/AIM.20131620672

Ranganathan P, Amal JC, Savithri S, Haridas A (2017) Experimental and modelling of *Arthrospira platensis* cultivation in open raceway ponds. Biores Technol 242:197–205. https://doi.org/10.1016/J.BIORTECH.2017.03.150

Ras M, Steyer JP, Bernard O (2013) Temperature effect on microalgae: a crucial factor for outdoor production. Rev Environ Sci Biotechnol 12(2):153–164. https://doi.org/10.1007/S11157-013-9310-6/FIGURES/5

Raven JA, Beardall J (2003) Carbohydrate metabolism and respiration in algae, pp 205–224. https://doi.org/10.1007/978-94-007-1038-2_10

Raven JA, Geider RJ (1988) Temperature and algal growth. New Phytol 110:441–461. https://doi.org/10.1111/j.1469-8137.1988.tb00282.x

Raven JA, Gobler CJ, Hansen PJ (2020) Dynamic CO_2 and pH levels in coastal, estuarine, and inland waters: theoretical and observed effects on harmful algal blooms. Harmful Algae 91. https://doi.org/10.1016/J.HAL.2019.03.012

Rearte TA, Celis-Plá PSM, Neori A, Masojídek J, Torzillo G, Gómez-Serrano C, Silva Benavides AM, Álvarez-Gómez F, Abdala-Díaz RT, Ranglová K, Caporgno M, Massocato TF, da Silva JC, Al Mahrouqui H, Atzmüller R, Figueroa FL (2021) Photosynthetic performance of *Chlorella vulgaris* R117 mass culture is moderated by diurnal oxygen gradients in an outdoor thin layer cascade. Algal Res 54:102176. https://doi.org/10.1016/J.ALGAL.2020.102176

Reddy Ronda S, Sekhar Bokka C, Ketineni C, Rijal B, Rao Allu P (2012) Aeration effect on *Spirulina platensis* growth and γ-linolenic acid production. Braz J Microbiol 12–20

Rezayian M, Niknam V, Ebrahimzadeh H (2019) Oxidative damage and antioxidative system in algae. Toxicol Rep 6:1309–1313. https://doi.org/10.1016/j.toxrep.2019.10.001

Rousch JM, Sommerfeld MR (1999) Effect of manganese and nickel on growth of selected algae in pH buffered medium. Water Res 33(10):2448–2454. https://doi.org/10.1016/S0043-1354(98)00474-6

Salman JM, Grmasha RA, Stenger-Kovács C, Lengyel E, Al-Sareji OJ, Al-Cheban AMAA, Meiczinger M (2023) Influence of magnesium concentrations on the biomass and biochemical variations in the freshwater algae, *Chlorella vulgaris*. Heliyon 9(1):e13072. https://doi.org/10.1016/J.HELIYON.2023.E13072

Sañudo-Wilhelmy SA, Gómez-Consarnau L, Suffridge C, Webb EA (2014) The role of B vitamins in marine biogeochemistry. Annu Rev Mar Sci 6:339–367. https://doi.org/10.1146/ANNUREV-MARINE-120710-100912

Scherholz C, Bmc C, Scherholz ML, Curtis WR (2013) Achieving pH control in microalgal cultures through fed-batch addition of stoichiometrically-balanced growth media. BMC Biotechnol 13:1. https://doi.org/10.1186/1472-6750-13-39

Schneider SC, Kahlert M, Kelly MG (2013) Interactions between pH and nutrients on benthic algae in streams and consequences for ecological status assessment and species richness patterns. Sci Total Environ 444:73–84. https://doi.org/10.1016/J.SCITOTENV.2012.11.034

Schriber CL, Venable ME, Schriber CL, Venable ME (2019) The effects of environmental stressors on aggregation in *Parachlorella kessleri*. J Environ Prot 10(12):1657–1676. https://doi.org/10.4236/JEP.2019.1012099

Senge MO, Ryan AA, Letchford KA, Macgowan SA, Mielke T (2014) Chlorophylls, symmetry, chirality, and photosynthesis. Symmetry 6:781–843. https://doi.org/10.3390/sym6030781

Sforza E, Simionato D, Giacometti GM, Bertucco A, Morosinotto T (2012) Adjusted light and dark cycles can optimize photosynthetic efficiency in algae growing in photobioreactors. PLoS ONE 7(6):38975. https://doi.org/10.1371/journal.pone.0038975

Sharma N, Fleurent G, Awwad F, Cheng M, Meddeb-Mouelhi F, Budge SM, Germain H, Desgagné-Penix I (2020) Red light variation an effective alternative to regulate biomass and lipid profiles in *Phaeodactylum tricornutum*. Appl Sci 10(7):2531. https://doi.org/10.3390/APP10072531

Singh SP, Singh P (2015) Effect of temperature and light on the growth of algae species: a review. Renew Sustain Energy Rev 50:431–444. https://doi.org/10.1016/J.RSER.2015.05.024

Slegers PM, Leduc S, Wijffels RH, van Straten G, van Boxtel AJB (2015) Logistic analysis of algae cultivation. Biores Technol 179:314–322. https://doi.org/10.1016/J.BIORTECH.2014.12.033

Smith RV, Foy RH (2007) Improved hydrogen ion buffering of media for the culture of freshwater algae. Br Phycol J. https://doi.org/10.1080/00071617400650271

Sobczuk TM, Camacho FG, Grima EM, Chisti Y (2006) Effects of agitation on the microalgae *Phaeodactylum tricornutum* and *Porphyridium cruentum*. Bioprocess Biosyst Eng 28(4):243–250. https://doi.org/10.1007/s00449-005-0030-3

Sobczuk TM, Garcıá AF, Molina CE, Ae G, Chisti Y (n.d.) Effects of agitation on the microalgae *Phaeodactylum tricornutum* and *Porphyridium cruentum*. https://doi.org/10.1007/s00449-005-0030-3

Sorokin C (1971) The capacity of growing algal cells to affect the pH and the buffering properties of the medium. Plant Cell Physiol 12(6):979–987. https://doi.org/10.1093/OXFORDJOURNALS.PCP.A074702

Sousa C, Compadre A, Vermuë MH, Wijffels RH (2013) Effect of oxygen at low and high light intensities on the growth of *Neochloris oleoabundans*. Algal Res 2(2):122–126. https://doi.org/10.1016/j.algal.2013.01.007

Sozmen AB, Canbay E, Sozmen EY, Ovez B (2018) The effect of temperature and light intensity during cultivation of *Chlorella miniata* on antioxidant, anti-inflammatory potentials and phenolic compound accumulation. Biocatal Agric Biotechnol 14:366–374. https://doi.org/10.1016/J.BCAB.2018.03.023

Sözmen AB, Ata A, Ovez B (2022) Optimization of the algal species *Chlorella miniata* growth: mathematical modelling and evaluation of temperature and light intensity effects. Biocatal Agric Biotechnol 39:102239. https://doi.org/10.1016/J.BCAB.2021.102239

Stanbury PF, Whitaker A, Hall SJ (2017) Aeration and agitation. In: Principles of fermentation technology, pp 537–618. https://doi.org/10.1016/B978-0-08-099953-1.00009-0

Stirk WA, Bálint P, Tarkowská D, Novák O, Strnad M, Ördög V, van Staden J (2013) Hormone profiles in microalgae: gibberellins and brassinosteroids. Plant Physiol Biochem PPB 70:348–353. https://doi.org/10.1016/J.PLAPHY.2013.05.037

Suh IS, Lee CG (2003) Photobioreactor engineering: design and performance. Biotechnol Bioprocess Eng 8(6):313–321. https://doi.org/10.1007/BF02949274/METRICS

Suh I-S, Schumpe A, Deckwer W-D, Kulicke W-M (1991) Gas-liquid mass transfer in the bubble column with viscoelastic liquid. Can J Chem Eng 69(2):506–512. https://doi.org/10.1002/CJCE.5450690215

Tagliaferro GV, Izário Filho HJ, Chandel AK, da Silva SS, Silva MB, dos Santos JC (2019) Continuous cultivation of *Chlorella minutissima* 26a in a tube-cylinder internal-loop airlift photobioreactor to support 3G biorefineries. Renew Energy 130:439–445. https://doi.org/10.1016/J.RENENE.2018.06.041

Takagi M, Watanabe K, Yamaberi K, Yoshida T (2000) Limited feeding of potassium nitrate for intracellular lipid and triglyceride accumulation of Nannochloris sp. UTEX LB1999. Appl Microbiol Biotechnol 54

Tandon P, Jin Q, Huang L (2017) A promising approach to enhance microalgae productivity by exogenous supply of vitamins. Microb Cell Fact 16(1). BioMed Central Ltd. https://doi.org/10.1186/s12934-017-0834-2

Tang C-C, Zhang X-Y, Wang R, Wang T-Y, He Z-W, Wang XC (2022) Calcium ions-effect on performance, growth and extracellular nature of microalgal-bacterial symbiosis system treating wastewater. Environ Res 207. https://doi.org/10.1016/j.envres.2021.112228

Teo CL, Atta M, Bukhari A, Taisir M, Yusuf AM, Idris A (2014) Enhancing growth and lipid production of marine microalgae for biodiesel production via the use of different LED wavelengths. Biores Technol 162:38–44. https://doi.org/10.1016/J.BIORTECH.2014.03.113

Terauchi M, Nagasato C, Inoue A, Ito T, Motomura T (2016) Distribution of alginate and cellulose and regulatory role of calcium in the cell wall of the brown alga *Ectocarpus siliculosus* (Ectocarpales, Phaeophyceae). Planta 244. https://doi.org/10.1007/s00425-016-2516-4

Termini ID, Prassone A, Cattaneo C, Rovatti M (2011) On the nitrogen and phosphorus removal in algal photobioreactors. Ecol Eng 37:976–980. https://doi.org/10.1016/j.ecoleng.2011.01.006

Tham PE, Ng YJ, Vadivelu N, Lim HR, Khoo KS, Chew KW, Show PL (2022) Sustainable smart photobioreactor for continuous cultivation of microalgae embedded with Internet of Things. Biores Technol 346:126558. https://doi.org/10.1016/J.BIORTECH.2021.126558

Titica M, Ifrim G, Barbu M, Kazbar A, Maree H, Caraman S, Pruvost J (2018) Simultaneous control of pH and dissolved oxygen in closed photobioreactor. ICSTCC

Touloupakis E, Faraloni C, Carlozzi P (2022) An outline of photosynthetic microorganism growth inside closed photobioreactor designs. Biores Technol Rep 18:101066. https://doi.org/10.1016/j.biteb.2022.101066

Townsend SA, Schult JH, Douglas MM, Skinner S (2007) Ratio of nitrogen to phosphorus in phytoplank. Freshw Biol. https://doi.org/10.1111/j.1365-2427.2007.01916.x

Townsend SA, Schult JH, Douglas MM, Skinner S (2008) Does the Redfield ratio infer nutrient limitation in the macroalga *Spirogyra fluviatilis*? Freshw Biol 53(3):509–520. https://doi.org/10.1111/J.1365-2427.2007.01916.X

Ugwu CU, Aoyagi H, Uchiyama H (2007) Influence of irradiance, dissolved oxygen concentration, and temperature on the growth of *Chlorella sorokiniana*. Photosynthetica 45(2)

Ugwu CU, Aoyagi H, Uchiyama H (2008) Photobioreactors for mass cultivation of algae. Biores Technol 99(10):4021–4028. https://doi.org/10.1016/J.BIORTECH.2007.01.046

Wágner DS, Valverde-Pérez B, Plósz BG (2018) Light attenuation in photobioreactors and algal pigmentation under different growth conditions—model identification and complexity assessment. https://doi.org/10.1016/j.algal.2018.08.019
Wang C, Lan CQ (2018) Effects of shear stress on microalgae—a review. Biotechnol Adv 36(4):986–1002. https://doi.org/10.1016/J.BIOTECHADV.2018.03.001
Wang S, Zhao S-X, Wei C-L, Yu S-Y, Shi J-P, Zhang B-G (2014) Effect of magnesium deficiency on photosynthetic physiology and triacylglyceride (TAG) accumulation of *Chlorella vulgaris*. Huan Jing Ke Xue 35(4):1462–1467
Wang J, Liu J, Liu T (2015) The difference in effective light penetration may explain the superiority in photosynthetic efficiency of attached cultivation over the conventional open pond for microalgae. Biotechnol Biofuels 8(1). https://doi.org/10.1186/s13068-015-0240-0
Wang B, Qin Y, Shen J, Wang G, Zhang Q, Liu M (2019) Experimental study on water sensitivity and salt sensitivity of lignite reservoir under different pH. J Petrol Sci Eng 172:1202–1214. https://doi.org/10.1016/J.PETROL.2018.09.036
Wangpraseurt D, Pernice M, Guagliardo P, Kilburn MR, Clode PL, Polerecky L, Kühl M (2015) Light microenvironment and single-cell gradients of carbon fixation in tissues of symbiont-bearing corals. ISME J 10(3):788–792. https://doi.org/10.1038/ismej.2015.133
Wei Y, Jiao Y, An D, Li D, Li W, Wei Q (2019) Review of dissolved oxygen detection technology: from laboratory analysis to online intelligent detection. Sensors 19(18):3995. https://doi.org/10.3390/S19183995
Williams PJLB, Laurens LML (2010) Microalgae as biodiesel & biomass feedstocks: review & analysis of the biochemistry, energetics & economics. Energy Environ Sci 3(5):554–590. https://doi.org/10.1039/b924978h
Wondraczek L, Gründler A, Reupert A, Wondraczek K, Schmidt MA, Pohnert G, Nolte S (2019) Biomimetic light dilution using side-emitting optical fiber for enhancing the productivity of microalgae reactors. Sci Rep 9(1). https://doi.org/10.1038/s41598-019-45955-w
Xiao Y, Li Z, Li C, Zhang Z, Guo J (2016) Effect of small-scale turbulence on the physiology and morphology of two bloom-forming cyanobacteria. PLoS ONE 11(12):168925. https://doi.org/10.1371/journal.pone.0168925
Xin L, Hong-ying H, Ke G, Ying-xue S (2010) Effects of different nitrogen and phosphorus concentrations on the growth, nutrient uptake, and lipid accumulation of a freshwater microalga Scenedesmus sp. Biores Technol 101(14):5494–5500. https://doi.org/10.1016/J.BIORTECH.2010.02.016
Yaakob MA, Mohamed RMSR, Al-Gheethi A, Ravishankar GA, Ambati RR (2021) Influence of nitrogen and phosphorus on microalgal growth, biomass, lipid, and fatty acid production: an overview. Cells 10(2):1–19. https://doi.org/10.3390/CELLS10020393
Yaashikaa PR, Keerthana Devi M, Senthil Kumar P, Rangasamy G, Rajendran S, Xiao L (2022) A review on pretreatment methods, photobioreactor design and metabolic engineering approaches of algal biomass for enhanced biohydrogen production. Int J Hydrogen Energy. https://doi.org/10.1016/J.IJHYDENE.2022.10.092
Yan C, Zhang L, Luo X, Zheng Z (2013) Effects of various LED light wavelengths and intensities on the performance of purifying synthetic domestic sewage by microalgae at different influent C/N ratios. Ecol Eng 51:24–32. https://doi.org/10.1016/j.ecoleng.2012.12.051
Yoshida K, Igarashi E, Wakatsuki E, Miyamoto K, Hirata K (2004) Mitigation of osmotic and salt stresses by abscisic acid through reduction of stress-derived oxidative damage in *Chlamydomonas reinhardtii*. Plant Sci 167(6):1335–1341. https://doi.org/10.1016/J.PLANTSCI.2004.07.002
Yu XJ, Sun J, Sun YQ, Zheng JY, Wang Z (2016) Metabolomics analysis of phytohormone gibberellin improving lipid and DHA accumulation in Aurantiochytrium sp. Biochem Eng J 112:258–268. https://doi.org/10.1016/J.BEJ.2016.05.002
Yu H, Kim J, Rhee C, Shin J, Shin SG, Lee C (2022) Effects of different pH control strategies on microalgae cultivation and nutrient removal from anaerobic digestion effluent. Microorganisms 10(2):357. https://doi.org/10.3390/MICROORGANISMS10020357/S1

Zafar SU, Mehra A, Nesamma AA, Jutur PP (2023) Innovations in algal biorefineries for production of sustainable value chain biochemicals from the photosynthetic cell factories. Algal Res 69. https://doi.org/10.1016/j.algal.2022.102949

Zarmi Y, Gordon JM, Mahulkar A, Khopkar AR, Patil SD, Banerjee A, Reddy G, Griffin TP (n.d.) Enhanced algal photosynthetic photon efficiency by pulsed light. https://doi.org/10.1016/j.isci.2020.101115

Zhang S, Bryant DA (2014) Learning new tricks from an old cycle: the TCA cycle in cyanobacteria, algae and plants. Perspect Phycol 1(2):73–86. https://doi.org/10.1127/PIP/2014/0016

Zhang S, Van Duijn B (2014) Cellular auxin transport in algae. Plants 3:58–69. https://doi.org/10.3390/plants3010058

Zhang Z, Shrager J, Jain M, Chang CW, Vallon O, Grossman AR (2004) Insights into the survival of *Chlamydomonas reinhardtii* during sulfur starvation based on microarray analysis of gene expression. Eukaryot Cell 3(5):1331–1348. https://doi.org/10.1128/EC.3.5.1331-1348.2004

Zhang QH, Wu X, Xue SZ, Wang ZH, Yan CH, Cong W (2013) Hydrodynamic characteristics and microalgae cultivation in a novel flat-plate photobioreactor. Biotechnol Prog 29(1):127–134. https://doi.org/10.1002/BTPR.1641

Zhang Q, Wang T, Hong Y (2014) Investigation of initial pH effects on growth of an oleaginous microalgae Chlorella sp. HQ for lipid production and nutrient uptake. Water Sci Technol 70(4):712–719. https://doi.org/10.2166/WST.2014.285

Zhao Y, Wang Z, Chen M, Huang X, Luo Z (2023) Effects of nitrogen to phosphorus ratios on algal growth and arsenate metabolism by *Microcystis aeruginosa* with dissolved organic phosphorus and nitrate as nutrients. Algal Res 69:102922. https://doi.org/10.1016/J.ALGAL.2022.102922

Zhou Y, Han L-R, He H-W, Sang B, Yu D-L, Feng J-T, Zhang X (2018) Effects of agitation, aeration and temperature on production of a novel glycoprotein GP-1 by *Streptomyces kanasenisi* ZX01 and scale-up based on volumetric oxygen transfer coefficient. Molecules 23(1). https://doi.org/10.3390/molecules23010125

Zhu C, Chi Z, Bi C, Zhao Y, Cai H (n.d.) Hydrodynamic performance of floating photobioreactors driven by wave energy. Biotechnol Biofuels. https://doi.org/10.1186/s13068-019-1396-9

Zhuang LL, Azimi Y, Yu D, Wu YH, Hu HY (2018) Effects of nitrogen and phosphorus concentrations on the growth of microalgae Scenedesmus. LX1 in suspended-solid phase photobioreactors (ssPBR). Biomass Bioenergy 109:47–53. https://doi.org/10.1016/J.BIOMBIOE.2017.12.017

Chapter 2
Microalgae: Production, Consumption and Challenges

Sadaf Gul, Laila Shahnaz, Sana Raiz, and Muhammad Farrakh Nawaz

Abstract Microalgae are the unicellular, autotrophic organisms present in marine or fresh water environment. They may be present as single cell or in the form of colonies. Microalgae are the source of various substances that are beneficial for health like proteins, lipids, lutine, precursors of vitamins, carotenoids and vitamin A, B1, B3, B6, B12, E. Furthermore, the use of microalgae for medicinal purposes can be traced back to ancient Chinese medicines due to their antibiotic, antimycobacterial, antiviral, anti-inflammatory, analgesic, antipyretic, antiproliferative, and cytotoxic activities. Recent advances in science and technology have revealed the drawback of using synthetic products for health and environment. This revelation has compelled people to use natural products. With the increasing trend of organic products in every sector of life and their benefits over synthetic products, utilization of microalgae has also gain popularity in various sectors viz., food, pharmaceuticals, cosmetics, cosmeceuticals, biofuel, biofertilizers, bioplastics, waste water treatment, aquaculture etc. Products from microalgae help preserving the environment and reduction of environmental pollution. Microalgae are also useful in bioremediation of heavy metals like; mercury, lead, chromium, cadmium and arsenic from water. One of the examples of utilization of microalgae as food is *Spirulina*, which is now commercially available in the form of powder and tablets. It is known as super food because of its high macro and micronutrient content. It is also used and feed for domestic animals and fish. Other commercially grown microalgae include *Chlorella*, *Skeletonema*, *Tetraselmis*, *Isochrysis*, *Pavlova*, *Phaeodactylum*, *Chaetoceros*, *Nannochloropsis*, *Oscillatoria* etc. For the commercial utilization, microalgae can be cultivated in open ponds (circular ponds or raceways), closed ponds (tubular or flat-plate photobioreactors) or in natural water bodies then harvested and processed

S. Gul (✉) · L. Shahnaz · S. Raiz
Department of Botany, University of Karachi, Karachi, Pakistan
e-mail: sadafgpk@yahoo.com

M. F. Nawaz
Institute of Environmental Studies, University of Karachi, Karachi, Pakistan

T. A. Kurniawan and A. Anouzla (eds.), *Algae as a Natural Solution for Challenges in Water-Food-Energy Nexus*, Environmental Science and Engineering,
https://doi.org/10.1007/978-981-97-2371-3_2

for respective use. Despite of their utilization in various products, microalgae cultivation faces many challenges like high production and maintenance cost, contamination by other microorganisms. To overcome these challenges further research and development of more cost effective and efficient methods are required.

Keywords Algae · Biofuels · Food · Heavy metals · Aqautic

2.1 Introduction

Algae are the diverse group of organisms that evolved from more than one ancestral lineage. They are a simple and primitive type of photosynthetic biota of this universe that are not only photoautotrophs but show a variety of modes of nutrition e.g. heterotrophy, mixotrophy, amphitrophy, phagotrophy, lithotrophy, organotrophy and chemo-organotrophy (Shameel 2003, 2012). They are capable to maintain their nutritional requirement according to their habitats. Their reproductive organs (sexual and asexual) are devoid of the sterile layer of tissues and they do not undergo true embryogenesis (Shameel 2008). They are classified due to their morphology, ultrastructures and phycochemical attributes. They are distributed in three kingdoms; 1—Monera; including two phyla (Cyanophycota and Prochlorophycota), 2—Protista; with twelve phyla (Glaucophycota, Volvophycota, Euglenophycota, Dinophycota, Cryptophycota, Chrysophycota, Prymnesiophycota, Bacillariophycota, Xanthophycota, Eustigmophycota, Raphidophycota and Porphyridiophycota), and 3—Phycota (comprises of Chlorophycota, Charophycota, Vaucheriophycota, Phaeophycota and Rhodophycota) (Shameel 2001, 2012) (Fig. 2.1).

Microalgae are also known as microphytes. They are the unicellular, photosynthetic microorganisms comprised of eukaryotic and prokaryotic organisms i.e., protists, cyanobacteria and blue-green alga. As they are microscopic, so, they are not visible by naked eye. They are present in both, marine and fresh water environment. Their cells are present individually or in the forms of colonies. They contribute approximately 50% of the oxygen we breathe. Some microalgae have both abilities of photosynthesizing and feeding on other organisms therefore they are called mixotrophs. Further, the members of Cyanophycota, Chlorophycota, Bacillariophycota and Chrysophycota are abundantly found in fresh, Brackish, marine water environments. As well as in deserts, snowfields, hot springs, and terrestrial habitats (Guschina and Harwood 2006).

2.2 Division of Microalgae

Description regarding the taxonomic position of microalgae and the specific classification to the class level is discussed hereunder with diverse references (Whitaker and Margulis 1978; Shameel 2001, 2003, 2008, 2012). The characteristics of phyla are

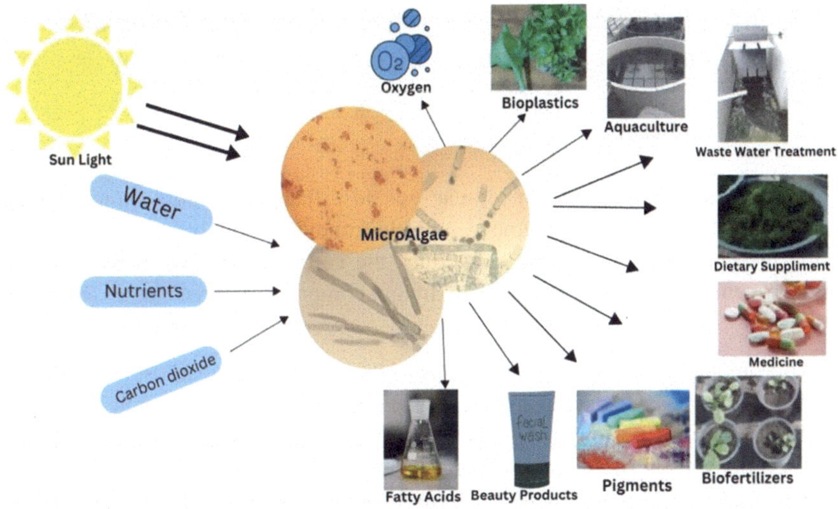

Fig. 2.1 Graphical illustration of services and products of microalgae

discussed in detail with the specific genera. Moreover, the attributes of the classes are elaborated where find necessary.

2.2.1 Kingdom: Monera (Prokaryotes, Unicellular, Colonial, Filamentous)

2.2.1.1 Cyanophycota

The organisms exhibited prokaryotic cellular organization. Individuals are unicellular, palmelloid, or colonial and filamentous forms of algae having chlorophyll a, phycobiliproteins, and xanthophylls (Echinenone and Myxoxanthophyll). Glycogen is reserved food material and the cell wall is made up of peptidoglycan. They do not show sexual reproduction. Representative genera are *Chroococcus, Nostoc, Spirulina, Scytonema, Lyngbya and Oscllatoria.*

2.2.2 *Kingdom: Protista (Eukaryotes, Solitary and Colonial Organisms, Coenobium, Unicellular Group)*

2.2.2.1 Volvophycota

Individuals are Eukaryotic, unicellular coccids, flagellates and colonial green microalgae. Photosynthetic pigments are chlorophyll a and b; starch is present inside the chloroplast; cellulose constitutes the cell wall; flagellates with two to four acronematic types of flagella. This phylum comprises of three classes (Shameel 2012):

a. Micromonadophyceae: The organisms of this class are the earliest green flagellates and scales present outside the cells. Flagellates exhibited spindle formation during mitotic division (e.g. *Micromonas*).
b. Volvcophyceae: These are unicellular colonial algae; microtubules break apart. Flagellates surrounded by theca, form by joining and merging of scales; flagella attached to the apical pit (*Tetraselmis*). Unicellular motile with cup-shaped chloroplast, an eyespot, two flagellated, contractile vacuoles and a pyrenoid (*Chlamydomonas*); Motile flagellated vegetative cells of *Volvox* with cup-shaped chloroplast; non-motile unicellular and colonial members with pseudocilia (*Tetraspora*) gametes and zoospores are formed by cell division (*Chlorococcum*). Non-motile unicellular and autocolony-forming organisms (*Chlorella*) and some individuals demonstrated desmoschisis (daughter cells retained within the parent cell; *Chlorosarcina*).
c. Desmidiophycee: They are commonly called desmids and occur in unicellular and filamentous form and no fragmentation takes place. Cell wall is with or without pores; sexual reproduction by conjugation. Semi-cells do not form in cell division (*Mesotaenium*, *Cosmarium* and *Closterium*).

2.2.2.2 Euglenophycota

They are single-celled eukaryotic flagellates also known as euglenoids without any special ingestion organelles; photosynthetic pigments are chlorophyll a and b having one or two emerging flagella rooted from two basal bodies, one flagellum with a row of coiled fibrilar hairs; inside the plasma membrane proteinous pellicles are arranged; mesokaryotic type of nucleus; amitotic cell division. Sexual reproduction does not take place (*Euglena*).

2.2.2.3 Dinophycota

They are predominantly unicellular flagellates. Chlorophyll a and c_2 are the main photosynthetic pigments; Two flagella are heterokontic type (longitudinal flagellum is small and smooth whereas, transverse flagellum is long, wavy, covered with hairs).

Starch is present in the cytoplasm; cell architecture is comprised of cellulosic plates with flattened vesicles; composing the epicone and hypocone; in some genera, a layer of acidic polysaccharides surrounds the plasma membrane termed as glycocalyx; scales are also present outside the plasma membrane (*Peridinium*, *Ceratium*, *Dinophysis*, Gonyaulax and *Gymnodinium*).

2.2.2.4 Cryptophycota

They are unicellular cyptoflagellates. Chlorophyll a and c_2 support photosynthesis along with phycobiliproteins; two apically and laterally inserted flagella with two rows of lateral hairs; two membranes of the chloroplast endoplasmic reticulum (CRE) surround the chloroplast envelope. Nucleomorph is located in between the inner and out membranes of the CRE. Periplastis located under the plasma membrane, composed of a sequence of plates the shape, number and arrangement of the plates are different in different genera. The special type of projectile is present in a groove near the base of flagella called ejectosomes (*Cryptosomonas*, *Thecomonas*, *Rhodomonas*).

2.2.2.5 Chrysophycota

They are golden brown unicellular and colonial flagellates. Flagella are two in number the anterior tinsel and posterior whiplash. Photosynthesis performs in the presence of chlorophyll a, c_1 and c_2 along with echinenone and fucoxanthin; chrysolaminarin is found as reserved food material; the cell wall is composed of cellulose and lorica; silica scales are radially and bilaterally arranged outside the plasma membrane; the cysts or resting spores are formed in unfavorable conditions.

a. Chrysophyceae: Cells contain two flagella perpendicular to each other with photoreceptors. The contractile vacuole and eyespot are also present. Silica scales are radially and biradially arranged. *Bicosoeca* in that genus chloroplast is completely absent. From the anterior region of the cell, two unequal flagella emerge in *Ochromonas*; single flagella but the second flagella present as a remnant of it (*Chromulina*); the cell wall is composed of eight or fine siliceous plates (*Tetraparma*); single-celled flagellates or non-motile small capsule-like cells; life cycle involves two distinct phases (*Hibberdia*); mucilaginous thallus shows capsoid or parenchymatous structure tightly attached to the substratum (*Hydrurus*); Motile naked cellular structure; eyespot is present; flagella are laterally inserted (*Chrysomeris*).
b. Dictyochphyceae: They are marine silicoflagellates (with external siliceous skeleton) and choanoflagellates. The amoeboid vegetative cells are characterized by the presence of tentacles and rhizopodial (*Dictyocha*, *Pedinella*, *Rhizochromulina*, and *Pseudopedinella*).
c. Synurophyceae: They are silicoflagellates closely related to the Chrysophyceae. Flagella are inserted parallel to one another; photoreceptors are located near the

base of flagella; a contractile vacuole is found at the posterior of the cell; scales are framed as a bilaterally symmetrical structure (*Synura*).

d. Pelagophyceae: Individuals are unicellular and extremely small in size, characterized as the ultraplankton with single tinsel flagellum in some members of this class. It is also recognized as the brown tide-forming algae (*Pelagomonas*).
e. Pinguiophyceae: They are marine phytoplankton with remarkably high amounts of polyunsaturated fatty acids; cells are biflagellated posterior whiplash and anterior tinsel flagellum with triplicate hairs (*Pinguiochrysis*).

2.2.2.6 Prymnesiophycota

The member of this phylum also categorized as haptophytes, are uninucleated marine plankton. The haptonema is a filamentous appendage thinner than the flagellum present between the two smooth flagella. Chlorophyll a, c_1 and c_2 and fucoxanthin; chrysolaminarin is found in the vesicles in the cytoplasm of the cell. Most of the members are covered by scales made up of calcium carbonate in the form of calcites; the aragonite form is uncommon. Organisms with deposition of $CaCO_3$ are called coccolithophorids (*Coccosphaera, Isochrysis, Prymnesium*).

2.2.2.7 Bacillariophycota

The members of this phylum are known as diatoms. Unicellular individuals sometimes form colonies; phototrophic autotrophs and colourless heterotrophs; chlorophyll-a, c_1 and c_2 and fucoxanthin that gives the brownish colour to them; single tinsel flagellum associated only with sperm; vesicles present in the cytoplasm contain storage food material chrysolaminarin; the siliceous wall, shell or capsule called a frustule. It has two overlapping halves, each called valve. The older and larger one is epitheca and the younger and smaller one is hypotheca.

2.2.2.8 Bacillariophyceae

Frustules incorporated with silica that forms different patterns. Centric diatoms (order: Biddulfiales), ornamented with radial and gonoid patterns (*Cosinodiscus, Triceratium*); cells contain many chloroplasts; cells do not have raphe; motile sperm have a single tinsel flagellum (Shameel 2012).

Pennate diatoms (order: Bacillariales), frustules ornate in pennate and trellisoid arrangement; no flagellation observed in spermatozoids; raphe present; sexual reproduction by conjugation (*Bacillaria*).

2.2.2.9 Xanthophycota

They are yellow-green flagellates with only chlorophyll-a-containing cells. Primarily, they are coccoid, capsoid, or filamentous; mostly in freshwater environments; only a few are found in marine water. Characterized by motile cells forwardly directed tinsel flagellum and posteriorly directed whiplash. The cell-wall is composed of cellulose with two overlapping halves and there are about 600 species. A-sexual reproduction by aplanospores, zoospores and fragmentation; a small number of genera reproduced sexually (*Heterogloea*, *Heterochloris*).

2.2.2.10 Eustigmophycota

The organisms included are spherical, small and pale-green also called eustigmophytes and are closely related to xanthophycota. They have chlorophyll-a along with β-carotene, violaxanthin and vaucherioxanthin; flagella are heterokontic type one or two apically inserted. The eyespot is positioned outside the chloroplast; t-shaped flagellar swelling is present at the base of the tinsel flagellum adjacent to the eyespot; pyrenoid is stalked structure. Most species produced single flagellar naked and rather amoeboid zoospores (*Vischeria*).

2.2.2.11 Raphidophycota

They are commonly named chloromonads; cells have muciferous bodies; an individual cell has anterior tinsel and posterior whiplash flagella, and some have a projectile. Chloroplast is small in size and numerous; Fat particles are reserved food material; the cell-wall is covered with a gelatinous matrix (*Fibrocapsa*).

2.2.2.12 Porphyridiophycota

The red microalgae are classified in this phylum (Shameel 2012). Unicellular and multicellular organisms contain chlorophyll-a and d, phycobiliproteins, antherxanthin and neoxanthin. Cells are characterized by the presence of a single large or a few minor plastids with pyrenoids. Cellulosic cell wall incorporated with amorphous mucilaginous material. The genus *Cyanidium* occurs in the volcanic area where acidic pH and temperature are high (56 °C). The cells of *Porphyridium* are enclosed in a common mucilaginous wall (Fig. 2.2).

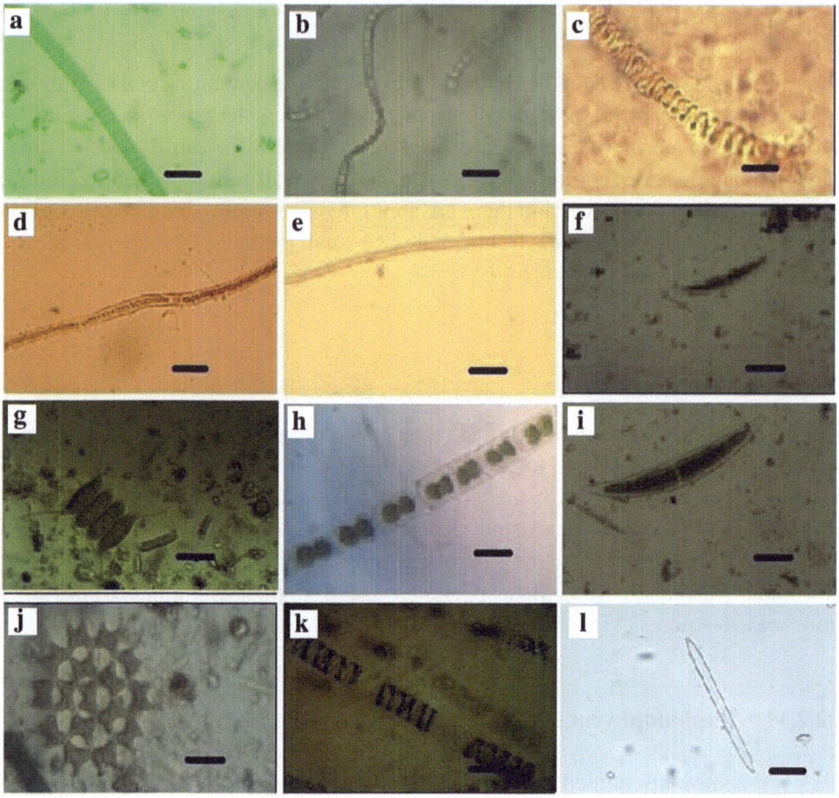

Fig. 2.2 Pictorial view of different microalgae. **a** *Oscillatoria* sp., **b** *Anabaena* sp., **c** *Spirulina* sp., **d** *Lyngbya* sp., **e** *Phormidium* sp., **f** *Microcoleus* sp., **g** *Scenedesmus* sp., **h** *Zygnema* sp., **i** *Closterium* sp., **j** *Pediastrum* sp., **k** *Spirogyra* sp., **l** *Syndra* sp. Scale = 10 μm

2.2.3 Kingdom: Phycota (Simple, Multicellular and Coenocytic)

2.2.3.1 Chlorophycota

The green algae classify under this phylum have chlorophyll-a and b; lutein and siphonoxanthin. Two to four isokontic types of terminal flagella; the cell-wall is made up of cellulose; Starch is present as the photosynthetic product.

a. Ulvophyceae: Simple individuals, filamentous, branched and unbranched or thalloid forms of thallus organization; cell division normal; naked zoospores are found in *Klebsormidium*. *Ctenocladus* is a filamentous marine alga; filaments with parietal chloroplast are in *Ulothrix* and an H-shaped cell wall is observed in *Microspora*. Simple branched or heterotrichous *Chaetophora*; discoid thallus structure and branched filaments and motile cells have scales in *Coleochaete*, reproduced by oogamous sexual reproduction.

b. Zygnemophyceae: Unbranched filaments mostly grow in freshwater; cap-cell formation or cell division represented by annual ring formation, sexual reproduction takes place by oogamy and stephanokontic type of zoospores are formed in *Oedogonium*; sexual reproduction by conjugation representative members are *Spirogyra* and *Zygnema.*
c. Siphonocladophyceae: This class comprises coenocytic, siphonaeous and filamentous thalli; branched, coenocytic, filamentous freshwater and marine form is *Cladophora* each cell has reticulate chloroplast; filaments with parietal chloroplast *Acrosiphonia.*

2.2.3.2 Phaeophycota

A variety of structures and sizes of thalli are found in this phylum ranging from the microscopic to the several meters high genera (*Laminaria*, *Macrocystis*, and *Ecklonia*) called Kelp. The large Kelp are generally found in the northern hemisphere (Atlantic and Pacific oceans), which are cold water areas. Pigments are chlorophyll-a, c_1 and c_2; fucoxanthin and violaxanthin; two heterokontic laterally inserted flagella; Primarily marine algae some filamentous genera occur in freshwater *Bodanell*, *Heribaudiella*, *Pleurocladia* but *Sphacelaria* and *Ectocarpus* both are fresh as well as marine water algae. Other marine filamentous and thread-like forms are *Tiloptris*, *Chordaria*, and *Sporochnus*.

2.3 Cultivation of Microalgae

Due to their single cell structure, microalgae grow rapidly with high accumulation of bioproducts within their cells (Randrianarison and Ashraf 2017). Unlike macroalga, microalgae cannot be collected from natural bodies for mass utilization. So, it is necessary to culture and harvest them. For culturing the microalgae, the conditions which are necessary include good and ample amount of light (as microalgae are all photosynthetic organisms, so light is the primary requirement for their cultivation), effective transportation (transportation of gases i.e., oxygen and carbon dioxide for proper function of photosynthesis and respiration, nutrients which are essential for optimum growth of algae and excretory material which is expelled out of the cells should be eliminated from the vessel for proper functioning of the system), prevention of contamination (by other microorganisms to obtain pure culture), procedure as simple as possible (simple processes are easy to maintain and cost effective as compare to complex methods), cost effectiveness (the production cost should be less enough to use the method for commercial purposes). Microalgae can be cultured in ponds and photobioreactors.

2.3.1 *Ponds*

2.3.1.1 Open Ponds

Microalga can be cultivated in large outdoor ponds. These ponds may have natural bed or lined with cement, PVC sheets or polyethylene. Agricultural-grade nutrient medium are used. Comparatively it is less expensive but suitable for fast growing species only for e.g., *Skeletonema costatum*. It is energy efficient and eases to scale up. The culture growth is not consistent due to possible fluctuation in environment i.e., temperature, sunlight etc. Another issue is the easy contamination by protozoans and bacteria. However, this problem can be overcome by cultivating algal species which can grow in high salinity and high pH (Stark and O'Gara 2012). Open ponds are of two types; circular ponds and raceway ponds.

2.3.1.2 Circular Ponds

As the name indicates they are circular in shape with a rotating agitator at the center to prevent settling the algal biomass in the bottom by mixing of water efficiently. They are 30–70 cm deep and diameter reaches up to 45 m. It is limited in size because of poor mixing ability. It has high energy consumption and high building cost. This system is widely used in Japan and Taiwan for *Chlorella* cultivation. Further, they can be used for the treatment of waste water by combining them with waste water channels.

2.3.1.3 Raceway Ponds

They are frequently and successfully used for culturing alga since 1950s. They have paddle wheels for mixing and preventing settlement of algae in bottom. Mixing helps in revolving the algal cells from bottom to upward in pond for better absorption of light and carbon dioxide (Molina et al. 2000). They are shaped like a race track with the approximate depth of 0.25–0.30 m, width may reach up to hectares. They are constructed as single channel or as a group of channels linked together. The material used for their building includes concrete, compacted soil, plastic etc. Its design helps to harvest maximum sunlight by alga. They receive water from a natural source or a well. These ponds are energy efficient but have low biomass production.

2.3.1.4 Cultivation in Natural Water Bodies

Another method of cultivation of microalgae in open water is in natural water bodies. Hutt Lagoon, Australia is one of the largest natural water bodies where microalgae are cultivated commercially (Borowitzka and Borowitzka 1990). It does not require land use and it can be done in waste water while no clean water is required. So, it does not take too much natural resources and helpful in bioremediation of waste water (Posadas et al. 2017). As they are photosynthetic organisms and they utilize carbon dioxide from environment, hence help in reducing greenhouse effect.

2.3.2 Photobioreactors

It is a closed system for the cultivation of microalgae. They have various advantages over open pond system such as; the chances of contamination are minimum, single strain culture is possible, it requires less land area, they have higher metabolic and nutrient efficiency resulting in the production of higher biomass (Posten 2009). On the other hand, they are expensive to maintain. Some of the widely used photobioreactor types are; tubular photobioreactor and flat-plate photobioreactor.

2.3.2.1 Tubular Photobioreactor

They are the most popular PBRs because of their easy construction, controlled gas transfer, large surface area that result in good biomass production (Ugwu et al. 2008). They consist of long transparent tubes of either glass or plastic for the maximum sunlight harvesting. Mechanical pump or air lift system is used to circulate the alga within the tubes (Xu 2007).

2.3.2.2 Flat-Plate Photobioreactor

It comprises of rectangular 1–5 cm wide compartments of glass or other transparent material with an air lift system to circulate the culture within the compartments for better exchange of gasses between air and culture. This is the most photosynthetically efficient type of photobioreactors (Yan et al. 2016).

2.4 Harvesting and Processing

It is evident from the preceding discussion that microalgae are essential for the production of various biomolecules and other components. The harvesting and processing sometimes become very energy consuming, time taken and expensive processes. So, it is mandatory to take vast observation of all the methods available to choose an energy efficient and less expensive method for product recovery. There are four methodological natures for microalgal harvesting present i.e., physical, chemical, biological and electrochemical based. They include filtration, centrifugation, flocculation and flotation etc. After successful harvesting, microalgal cells are processed for the extraction of the compound of interest. For extraction, algal cells are first broken down to release the intracellular compounds. This extraction is typically done by one of the four following methods; (1) mechanical extraction, (2) chemical extraction, (3) physical extraction and (4) breakdown of cell by enzymes. Among them, chemical breakdown by ionic liquids is one of the popular methods present (Tan et al. 2020).

2.5 Utilization of Microalgae

With the increasing trend of natural products instead of synthetic, microalgae also gain huge popularity. They are becoming popular among food, cosmetic as well as pharmaceutical industries. There are many benefits of using microalgae for commercial purpose. First and foremost, it helps in reducing various forms of environmental pollution like water pollution and plastic pollution. They also decrease air pollution. There are following potential commercial uses of microalgae.

2.5.1 Human Nutrition

A variety of microalgal products are used in the form of tablets, capsules and powder for human nutrition (Pulz and Gross 2004). The high-quality fats and fiber contents of microalgae make them an alternate source of food (Chen et al. 2022) and feed. Such as, *Spirulina* sp. and *Chlorella* sp. are rich in protein contents and widely used in the preparation of algal flour, drink sand other dietary items (Pulz and Gross 2004; Khan et al. 2018; Pereira et al. 2021; Wang et al. 2021).

2.5.1.1 Vitamins

Navicula sp., *Porphyridium cruentum* and *Dunaliella salina* are golden packages of vitamin C, E and β-carotene, which is the preliminary compound of vitamin A and multiple components of Vit B family likely B_2, B_3, B_6 and B_7 are essential micronutrients for humans and animals' nutrition. Vit B_7 is involved in metabolic processes. Whereas, Vit E specifically tocopherol present in every cell helps in regulating muscular function and enhances immunity (Bharathi 2021).

2.5.1.2 Phycophenols

Microalgal species are the best source of phenols that have antioxidant, chemotherapeutic, antimicrobial and anticancer properties (Jorgensen et al. 1999). Phenols also function as a shield against UV radiations that are carcinogenic (Huang and Ferraro 1992).

2.5.1.3 Lipids

The human body is not fully capable to produced ω-3 fatty acids and their most important derivatives i.e., eicosapentaenoic (EPA) and docosahexaenoic (DHA) acids. They are the common components of fish liver oil and are of diatom origin. It is interesting to point out that microalgae are the only source of commercially available polyunsaturated fatty acids (PUSFAs) (Spolaore et al. 2006). Both EPA and DHA have significant impacts on cardiovascular and inflammatory diseases and also support brain functioning (Ocean Nutrition Canada 2010). More than 14,000 studies have been approving the use of omega-3 fatty acids as a diet supplement for humans as well as animals (Khan et al. 2018). The global demand for microalgal DHA is swiftly moving with a 15 million US $ of market value.

2.5.2 Potential Uses of Blue-Green Algae in the Healthcare Sector

The polysaccharides extracted from several blue-green algae such as *Amorphonostoc puntiforme*, *Gloeocapsa turgidus*, *Sphaeronostoc coeruleum*, *Stratonosto clinckia* f. *spongiaeforme* and *Synechococcus cedrorum* are effective against rabies and herpes-1 virus, and extracellular mucilaginous sheath of *Aphanothece halophytica* showed inhibitory activity against H1N1 virus tested in mice (Mansour et al. 2011). Spirulan and Ca-spirulan polysaccharides of *Spirulina* and acidic polysaccharide of *Nostoc flagelliforme* encountered influenza, human immunodeficiency viruses 1, and 2 and type-1 (HSV) (Kanekiyo et al. 2005). *Anabaena*, *Calothrix*, *Lyngbya*, *Oscilltoria*, *Phormidium* and *Scytonema* contain sulfoglycolipids namely Sulfolipid-1 significantly inhibited the activity of the HIV-1 virus (Gustafson et al. 1989; Reshef et al. 1997). Cyanovrin-N of *Nostoc ellipsosorum*, Scytovirin isolated from *Scytonema varium* and microvirin of *Microcystis aeruginosa* have shown remarkable inhibitory activity to HIV, HIV-1 and Lentiviruses (Huskens et al. 2010). Additionally, Cyanovrin-N also fights against HSV-6 and measles (Dey et al. 2000). Different strains of the corona, influenza A and B, enteric and respiratory syncytial (RSV) viruses have also been significantly effective by Cyanovrin-N (Van et al. 2007).

Further, *Nostoc* sp. is a natural source of more than twenty cryptophycins that are derivatives of macrolide and depsipeptide and have anticancer properties (Chaganty et al. 2004). Particularly, cryptophycin 1 has inhibitory activity when tested against the nasopharyngeal, colorectal, breast and lung cancer cell lines. It was observed that it has more potential than regular cancer drugs. Curacin A extracted from *Lyngbya majuscula* showed growth inhibition of breast cancer cell lines but it was inactive in vivo and Apratoxin-D has efficacy against human lung cancer cells (Gutierrez et al. 2008). Borophycin and its associated compound Borophycin-8 obtained from *N. spongiaeforme* var. *tenue* and *N. linckia* have showed carcinoma cell growth inhibitory activities (Gupta et al. 2012). Similarly, *Calothrix* is responsible to produced Calothrixins-A and -B, very low doses of these compounds can kill the human HeLa cancer cell and induce apoptosis in human Jurkat cancer cells (Rickards et al. 1999; Chen et al. 2003). It is also reported that Calothrixins-A and -B exhibited antimalarial activity against the FAF6 strain of *Plasmodium falciparum* (Rickards et al. 1999). Somocystinamide-A is a disulfide dimer of *L. majuscule* that exhibited anticarcinogenic activity impeding angiogenesis, and cell proliferation and stimulating the process of apoptosis (Gupta et al. 2012).

Other microalgae, *Anabaena* extracts have significant antibacterial activity against *Staphylococcus aureus* resistant to vancomycin has been reported by Bhateja et al. (2006). *Lyngbya majuscule* has known for the isolation of the first antibiotic from blue-green algae Malyngolide showed antibacterial activity against *Mycobacterium smegmatis*, *Streptococcus pyogenes*, *S. aureus* and *Bacillus subtillus* (Cardellina et al. 1979). A variety of compounds have been isolated from *Fischerella*, *Hapalosiphon*, *Micrococcus*, *M. lacustris*, *Nostoc commune*, *N. sphaericum* and *Westiollopsis* and be effective against *B. anthracis*, *B. subtillus*, *B. cereus*, *E. coli*, *Salmonella gallinarum*, *S. typhi*, *Staphylococcus epidermidis*, *S. albus*, *S. aureus*, *Klebsiella pneumoniae* and *Vibrio cholerae* (Kim et al. 2012).

Some cyclic peptides and lipopeptides were isolated from *Microcystis aeruginosa* (Portmann et al. 2008), and *Oscillatoria* was found in marine habitats (Linington et al. 2007), *L. majuscula* (McPhail et al. 2007) were active against *Plasmodium falciparum* W2 and chloroquine-resistant strain of *P. falciparum*.

Several bioactive compounds isolated from *Anabaena laxa*, *Fischerella*, *Hapalosiphon*, *Hormothamnion enteromorphoides*, *L. confervoides*, *L. majuscula*, *Nostoc*, *Stigonema*, *Oscillatoria angustissima*, *O. subuliformis*, *Phormidium fragile* and *Tolypothrix panasensis* are active against *Acremonium*, *Aspergillus* sp., *Candida albicans*, *C. glabrata*, *C. krusei*, *C. kefyr*, *Cryptococcus neoformans*, *Penicillium*, *Trichoderma viride* (Senthil et al. 2013).

2.5.3 Microalgal Application as Stable-Isotope-Labeled Constituents

Microalgae are a possible source of such biochemical compounds that are used as stable isotope-labeled natural products. Their applications are in the clinical and immunological laboratories as well as in structure elucidation of terpenes, proteins and carbohydrates by advanced techniques of mass spectroscopy, NMR and IR radiometry (Pulz and Gross 2004; Spolaore et al. 2006).

2.5.4 Role of Microalgae in the Field of Nanotechnology

The combination of natural products of microalgae and nanostructures (nanocylindrical tubes, nanosurfaces and nanosphere) has been applied in the food packaging industry and the area of wearable technology for the manufacturing of band-aids and sensors to monitor health. This technology has the potential to enhance the rate of synthesis of biomass and ultimately the biochemical molecules. The biotechnology of microalgae delivers the integration of nanoparticles that develop the mechanism for wastewater treatment plants and regulates the activity of pathogens in the agriculture sector (Moreira et al. 2022).

2.5.5 Microalgae as Food

Microalgae have long been used as food or food supplements. They contain important dietary elements like; protein, lipids, bioactive compounds, vitamins and precursors of various vitamins. The evidences for the consumption of *Arthrospira* species are linked to sixteenth century where Aztecs harvested this alga from Texcoco Lake, Mexico city for bread making. Similarly, Lichen (an association between algae and fungi) locally known as "pathar ka phool" was used in royal cuisine of Sub-continent Asia. Large scale cultivation of microalga as food can be traced back to 1940s (Spoehr et al. 1949). *Chlorella* was the first microalga produced and sold commercially as food (Soong 1980). Many species of microalgae genera viz., *Chlorella, Tetraselmis, Isochrysis, Pavlova, Phaeodactylum, Chaetoceros, Nannochloropsis, Skeletonema,* and *Thalassiosira* have been assessed for their potential use as animal and human food (Spolaore et al. 2006). Besides their nutritional composition (carbohydrates, proteins, vitamins, lipids, antioxidants) they are also a great source of healthy fatty acids (Calijuri et al. 2022). Like spirulina is considered a rich source of linoleic acid and poly unsaturated fatty acids (Mahajan and Kamat 1995).

2.5.6 Waste Water Treatment

The three main types of waste water are municipal, industrial and agricultural waste waters. All of these can be treated by algae for the removal of nutrients from waste water and for the removal of heavy metals (in case of industrial waste) it's a promising way to reduce water pollution (Zhou et al. 2012). Microalgae not only remove the contaminants but the resultant cultivation can also be used as animal feed, biofuel production and fertilizer. For example, *Oscillatoria,* when used to treat diluted waste water in circular pond results in > 80% of ammonia and 50% of total organic carbon reduction (Sheehan et al. 1998). The cyanobacteria cultivated in mercury contaminated water showed the accumulation of mercury in their cells (Misra et al. 1985) which shows the potential of these organisms in bioremediation of heavy metals. Reports are also available on removal of copper, zinc and ferrous from waste water by the cultivation of algal culture (Mittal et al. 1992). They also reduce BOD (Biological oxygen demand) and COD (Chemical oxygen demand) in waste water. They also found to reduce magnesium, chloride, phosphorus, sodium, sulphate and nitrogenous compounds in waste waters (Abdelfattah et al. 2022).

Although traditional chemical treatment of waste water is still more in practice as compared to biological treatment but researchers are striving to develop advanced methods of algal treatment of waste water that are cost effective and need less man power (Abdelfattah et al. 2022). This practice is useful in two ways: firstly it cleans waste water and secondly the resultant algal culture could be used as biofertilizers or in other ways.

2.5.7 Aquaculture

Aquaculture is one of the large industries around the globe that involves molluscs, fishes, prawns and some crustaceans. Microalgae that have rapid growth rate with good nutritional profile, no toxin production and can have stable growth over the fluctuation of temperature are the choice of aqua culturists as live feed. There are around 20 species of microalgae that are used for this purpose which includes; *Isochrysis* sp., *Pavlova lutheri*, *Chaetoceros calcitrans*, *C. muelleri*, *C. gracilis*, *Thalassiosira pseudonana*, *Skeletonema* spp., *Tetraselmis suecica*, *Rhodomonas* spp., *Pyramimonas* spp., *Navicula* spp., *Nitzschia* spp. etc. (Wikfors and Ohno 2001). They can be used singly or in combinations with other diets. For this purpose, microalgae can be produced in outdoor or indoor ponds. In addition to this, some heterotrophic microalgae can be produced by fermentation technology. Using microalgae as a feed has single drawback i.e., high production cost.

2.5.8 Pharmaceuticals

Historically not many species of microalga were used medicinally but the utilization of Nostoc for medicinal purposes in ancient Chinese medicine was recorded in Materia Medica in 1596. Similarly, utilization of lichens for medicinal purposes has long history. Microalgae are gaining popularity due to their antibiotic, antimycobacterial, antiviral, anti-inflammatory, analgesic, antipyretic, antiproliferative, and cytotoxic effects (Müller 2001). Arthrospira is known for its antiviral and anti-inflammatory properties (Mao et al. 2005). Further the pigments (chlorophyll, carotenoids, phycobiliproteins) present in microalgae are considered as precursors of various vitamins. They also have antioxidant and anti-inflammatory properties. They are rich source of essential elements like zinc, iodine, selenium, iron, manganese, copper, phosphorus, sodium, nitrogen, magnesium, cobalt, molybdenum, sulfur, and

Table 2.1 Antibiotic obtained from different microalgae

S. no.	Microalgae	Antibiotic compounds	References
2	*Arthrospira*	Calcium spirulan	Lee et al. (1998)
1	*Chlorella*	Chlorellin	Pratt et al. (1944)
4	*Chlorococcum*	α-linolenic acid	Ohta et al. (1995)
5	*Dunaliella*	β-ionone and neophytadiene	Tinoco et al. (2016)
6	*Lyngbya mujuscula*	Aurilides B and C	Han et al. (2006)
7	*Moorea producens*	Malyngamide A	Sabry et al. (2017)
3	*Nostoc*	Cybastacines A (1) and B (2)	Cabanillas et al. (2018)
9	*Okeania* sp.	Odoamide	Sueyoshi et al. (2016)
11	*Porphyridium* sp.	Sulfated polysaccharides	Huheihel et al. (2001)
10	*Spirulina platensis*	Protein extract	Alzahrani (2018)
8	*Symploca* sp.	Symplostatin 1	Mooberry et al. (2003)

calcium. They contain vitamins of B group. They also contain omega 3 and omega 6, which are also the super foods for brain. *Spirulina platensis* is a good source of vitamins like; vitamin A, B1, B3, B6, B12, E and folic acid (Dianursanti et al. 2019). They have the potential to be used for extracting several types of antibiotic compounds (Table 2.1).

2.5.9 Natural Resources of Raw Materials for Cosmetic Products

Cosmetics are the products that can be applied on skin, nails or hair either to improve the quality or for beautification purpose. Another emerging trend is of cosmeceuticals, they are the combination of cosmetics and pharmaceutical products. They are used to increase health and beauty for long lasting effect in contrast to cosmetics which gives temporary results. Microalgae when exposed to the stress conditions such as UV rays. They produce valuable compounds, especially for sun protection, anti-aging and skin whitening. These compounds are valuable for cosmetic industry (Yarkent et al. 2020). *Nannochlosopsis* and *Anabaena* spp. are known to produce sunscreen effect. Similarly, many anti-aging compounds like carotenoids precursors of vitamin A can be obtained from many species of microalgae. Fucosterol which is naturally occurring sterol having anti-oxidant and anti-cancer properties can be isolated by *Olisthodiscus luteus*, *Nannochloropsis salina*. Squalene used as antistatic and emollient in moisturizing creams, also known to decrease hyper pigmentation can be obtained from *Botryococcus braunii*, *Schizochytrium mangrovei* and *Thraustochytrium* sp. (Ryu et al. 2015). Mycosporine-like amino acids can be obtained by *Nostoc commune*, *Anabaena variabilis*, *Aphanothece halophytica*. Retinol can be isolated by species like *Spirulina* sp. and *Dunaliela* sp. Skin whitening agents such as Oscillapeptin G from *Oscillatoria* agardhii, asthaxanthin from *Haematococcus pluvialis* and zeaxanthin from *N. oculata* showed whitening of skin by inhibiting tyrosinase in skin (Shen et al. 2011) *Spirulina* sp. and *Chlorella* sp. are the most reliable organisms that acquired a position in the cosmetic industry. Their chemical components have been used for skin and hair damage repair, help in sebum and have anti-inflammatory properties (Kim et al. 2008), anti-aging creams, moisturizers, sun blocks, face masks and skin softening creams (Spolaore et al. 2006). France has the largest market for algal beautifying products with a requirement of 5000 tonnes (Kim et al. 2008).

2.5.10 Biofertilizers

The algae based organic fertilizers are the best component to improve soil fertility, water holding capacity and physical parameters. The presence of significant nutrients makes them harmless as well as cost-effective and eco-friendly (Kawalekar 2013). Microalgal organic fertilizers are possibly plant growth promoters and can increase the yield of crops and reduce salt stress. The integrated micro-algal fertilizers enhance the biomass, uptake of nutrients and better crop production (Faheed and Fattah 2008). The blue-green algae are capable of eliminating the CO_2 from the atmosphere and reducing the damages caused by the greenhouses. Through the process of nitrogen fixation blue-green algae convert the atmospheric nitrogen into the amino acid and ultimately to protein. Their inoculum increases the yield of rice grains, and the presence of secondary metabolites, vitamins and carotenoids makes them beneficial for plant growth. The commercially available species of microalgae are *Arthrospira* spp. *Chaetoceros* spp., *Chlorella* spp., *Dunaliella* spp. and *Isochrysis* spp. (Priyadarshani and Rath 2012). But the most commonly available micro-algal genera are *Arthrospira* and *Chlorella* (Görs et al. 2010). They are widely used in food, animal feed, fuel, pharmaceutical industry and in the agriculture sector (Metting et al. 1990; Bulgari et al. 2015; Ertani et al. 2015).

The liquid extracts of *Chlorella vulgaris* and *Scenedesmus obliquus* function as plant growth hormones like auxin and cytokinins (Freudenberg et al. 2021). Shukla and Gupta (1967) investigated the effects of algal extract on rice grain revealing the earlier and higher growth rate. Another study in the hydroponic conditions exhibited the increased root and shoot lengths of tomato plants when subjected to microalgae (Kumari et al. 2011; Hernández-Herrera et al. 2014; Supraja et al. 2020). Microalgae can easily be cultured in domestic and industrial wastewater in the presence of atmospheric CO_2 (Behera and Balasubramanian 2019). Moreover, the obtained micro-algal biomass is a great source of lipids, protein and carbohydrates and is utilized for bio-diesel, bioethanol and biohydrogen production (Ronga et al. 2019; Behera et al. 2020). Proteins and amino acids are used in agriculture.

2.5.11 *Bioplastics/Biopolymers*

Usage of plastic-based products is increasing day by day due to huge global market and disposable packaging. The traditional petroleum based plastic is almost un-degradable and possess great hazard to the environment. To reduce environmental pollution and to reduce the burden of plastic on the Earth, scientists always search for natural and biodegradable forms of plastic. In their quest, bioplastic made from microalga is a breakthrough product. For the production of bioplastics, polyhydroxybutyrates (PHBs) and starch are produced inter-cellularly then extracted and further processed for the production of bioplastics. Bioplastics are used in two forms: 100% organic or with the blends of additives or polymers. Studies showed that blending was beneficial for commercial utilization. The two commonly used algae are *Spirulina* and *Chlorella vulgaris*. Among them plastic obtained by *Chlorella* sp. is more durable. Moreover, starch granules can be obtained from *C. sorokiniana* that is a great biopolymer. There is need of improvement in this sector for commercialization (Onen Cinar et al. 2020).

2.5.12 *Phytoremediation*

Microalgae have great potential to remove pollutants from the environment. The organic and inorganic chemical constitutions of industrial and domestic wastewater are a source of nutrients. The nitrogen and phosphorous responsible for increased biomass by utilizing the available nutrients in turn reduced the BOD and COD. They are also capable to reduce the heavy metals in wastewater. The application of various species of microalgae including *Phormidium, Oscillatoria, Synechococcus, Lyngbya, Chroococcuss, Chlorella vulgaris, Spirulina, Scenedesmus* and *Anabaena* in wastewater treatment plants has been used in the recent past (Munoz and Guieysse 2006).

2.6 Algal Pigments and Their Commercial Role in Industries

2.6.1 *Carotenoids and Xanthophylls*

They are naturally occurring lipid-soluble components of dark red, orange and yellow color pigments of fruits, vegetables and some aquatic animals (Del Campo et al. 2007; Naidoo et al. 2013). These pigments are widely distributed in terrestrial plants, microalgae and other prokaryotes being capable to support the function and structure of photosynthetic complexes as well as increasing the light spectrum to capture the light for photosynthesis (Pulz and Gross 2004; Naidoo et al. 2013; Pereira et al.

2021). The adverse growing conditions of microalgae can increase the production of astaxanthin, β-carotene and keto-carotene (Orosa et al. 2000).

Global market of carotenoids and their consumption in the edibles, drugs and aquaculture sector have been continuously increasing at the rate of 2.2% annually (Del Campo et al. 2007; Guedes et al. 2011). Due to their antioxidant properties, they can fight against carcinoma, obesity and high blood pressure, and hepatotoxins (Pereira et al. 2021). *Dunaliella salina* and *Haematococcus pluvialis* are the main producers of two widely demanded carotenoids i.e., β-carotene and astaxanthin (Pulz and Gross 2004; Spolaore et al. 2006; Khan et al. 2018; Rammuni et al. 2019). Leutin and zeaxanthin are also carotenoids found in microalgae are function as the shield against solar retinopathy and the formation of free radicals causes oxidative stress on the retina (Kotake-Nara and Nagao 2011; Pereira et al. 2021).

2.6.2 Phycobiliproteins

These are water-soluble accessory pigments that occur in the members of Cyanophycota and Rhodophycta. Their presence broadens the visible light spectrum and the light-harvesting ability of photosynthetic organisms has increased by them. They are of different types i.e., 1—Phycocyanin, 2—Phycoerythrin, 3—Allophycocyanin and 4—Phycoerythrocyanin. They contribute to the blue, bluish-green, purple and red-orange colors of the organisms and have high commercial demands. Phycocyanin and phycoerythrin are extracted from two profitable genera *Spirulina* and *Porphyridium* (Spolaore et al. 2006; Khan et al. 2018).

In medical technology and clinical and immunological research, these colored proteins are used in fluorescent dyes, for tracing the labeled antibodies and other biological molecules. They are also being used as a food color to chewing gums, candies and soft drinks. The blue color which is obtained from *Porphyridium aeruginem* added to a world-known brand of soft drinks namely Pepsi. The natural colors derivative of phycobiliproteins hit the international market with about 50 million US $ (Del Campo et al. 2007).

2.7 Market Values of Microalgae, Products, and Practical Applications

There are hundreds and thousands of microalgal species universally exist. It is a rough estimate that approximately 35,000 species of microalgae have been taxonomically identified (Cheng and Ogden 2011; Guiry 2012). But a few of them are being used for mass algal production and valued products. Microalgal biotechnology has become the most promising field of science concerning the various purposes of microalgal biomass in aquaculture, pharmaceutical, natural resources of energy and fuel and food supplements for human beings due to the high protein contents (Del Campo et al. 2007; Wang et al. 2021). About 10,000 tons y^{-1} dry-weight of microalgal biomass was required by various industries worldwide with 1.25 million US dollars in revenue generation annually (Becker 2007). At present 10^7 tonnes of algae are being harvested per annum, which is a source of food, feed, natural coloring and flavoring agents acquired by the non-transgenic blue biotechnology industry (Kumar et al. 2021). The economic worth of microalgal products was 32.60 billion US dollars in 2017, and it is predicted to increase by 53.43 billion US dollars by 2026 (Credence Research 2016). It was predicted by Allied Market Research that the international microalgal business compound annual growth rate will be 5.4% by 2021–2028 (Web1 2021).

2.8 Challenges

Microalgae have been used by humans for millennia as a food source. The vital role of microalgae in daily life cannot be overstated but the challenges in the production go side by side. Microalgae may be found practically anywhere be it on the earth, including water. Microalgal biomass is utilized for food and feed additives since it is the principal producer. Microalgae are major commercial sources.

The development of microalgae-based systems for biotechnological application is highly dependent on the success of the mass culture and extraction process. Because of disposal and classification of genotypic data the detection of new species is easier than before. However, the screening of bioactive compounds on a large scale is needed for the fast and explicit categorization of algal insolation in abundant quality. Preservation of microbial gene pools is a challenge as the bio-catalytic compounds has the greater contribution in the industry of biotechnology. Another is the extraction challenge which is because of size of microalgal cells and their listing is quite a hassle because they are smaller in size.

The situation is further worsened that the prices of equipment are sky-rocketing. Moreover, numerous biomasses harvesting techniques are available for the algae extraction but none of them can be applied universally as some are costly and others technically unfit for the environment. The development of microalgae-based systems for biotechnological application is highly dependent on the success of the mass culture and extraction process. Furthermore, microalgae have the potential to be utilized as a feedstock for biodiesel production. For effective production, it is essential to identify the most appropriate higher lipid producing algal strains. Through lipid analysis, we can evaluate the quality of the algal bio-fuel and compare them to classified biodiesel standards. In addition, it is possible to determine the suitability of the selected algae for biodiesel by examining its fatty acid composition and predicting the fuel properties.

There is a general agreement among the subject matter experts that the use of wastewater treatment HRAP is the most cost-effective way to cultivate algae for biofuel production. There are numerous advantages to using this method, including pocket friendly wastewater treatment and algal biomass production. Despite this, there is still a need to improve conditions to ensure efficient algal growth and nutrient removal in different climates. However, challenges remain there when it comes to large-scale lipid optimization and harvesting of algal biomass and overcoming these will help to ease the cost of algae production.

2.8.1 High Production Costs

Microalgae production worldwide is still in its early developmental stage, faced with challenges of high production cost. Additionally, if compared with normal protein sources the present price of algal based products still remains on top (Sui et al. 2020). Cultivation is the major cost contributor for microalgae products. However, harvesting and dewatering of microalgae biomass are also contributing equal part to the total costs. Various studies have stated that the harvesting costs are 20–30% of the total production costs (Fasaei et al. 2018). As reported by Ruiz et al. (2016), by enhancing production scales and selecting proper production location the commercial manufacturing cost of microalgae products can be remarkably reduced.

2.8.2 Contamination Problems and Collection

Algae can be cultivated in open ponds for large scale due to comparatively inexpensiveness, but they are easily contaminated (Khan et al. 2018). Additionally, predators such as rotifers can destroy the algal culture, and huge fluctuation in weather can make nutrients, light intensity, and CO_2 challenging. Apart from this, there are difficulties in algal collection and only small number of species can be successfully grown (Xu et al. 2009).

2.8.3 Light

As the high intensity of sunlight can lead to photoinhibition, producing toxic photo products including H_2O_2, O_2^-, $1O_2 \cdot OH$, and triplet chlorophyll, and also damage the Photosystem II reaction center. On the other hand, low light intensities, may limit photosynthetic activity and thus, biomass productivity whereas, the extent of sensitivity to light changes from species to species (Xu et al. 2009).

2.8.4 Temperature

The most difficult parameters are light and temperature to enhance in large-scale open culture systems. There is significant decline in productivity due to daily or annual fluctuations in temperature. Ideal temperatures for growth are generally between 20 and 30 °C (Chisti 2008). Further, many species with reduced growth rates can tolerate temperatures of up to 15 °C lower than their optimum, but a few degrees increase than optimal temperature can always leads to cell death (Mata et al. 2010).

2.8.5 Nutrients Stress

The main nutrient requirements for microalgal growth are nitrogen, phosphorous or silica. These requirements of nutrients, pH and osmolarity are species dependent. Biomass productivity may decline if there is any deviation from ideal levels. There must be sufficient supply of all essential nutrients for proper photosynthesis and growth, but insufficiency of key nutrients (e.g., nitrate, phosphate or silica) may form accumulation of desired products such as lipid (Griffiths et al. 2011). However, ammonia discharged from livestock and food industries stops the growth and physiological activities of microalgae at higher concentration (Tam and Wong 1996).

2.8.6 Salinity and pH

The pH of the culture media is another important factor affecting microalgae growth. Microalgae species have different pH requirements that mostly ranges from 6 to 8.5. Most algal species are pH sensitive while others can cope with wide range of pH. As pH increases there is increase in salinity of the culture media, which is very harmful for algal cells (Khan et al. 2018).

2.8.7 Cultivation System and Photobioreactor

The micro-algal cultivation methods are of two types in open systems e.g. in lakes and ponds and closed systems in photobioreactors (PBR). The natural culturing system is the most feasible and low cost as compared to the closed one but it requires more energy to regulate the nutrient accessibility and maintain 15 cm of water level for the availability of sunlight. However, open systems uncontrollable environmental conditions expose them to bacteria and other micro-organisms and loss of water due to evaporation (Xu et al. 2019). The photobioreactor conditions are favourable for contamination-free cell growth and reduce the loss of CO_2. These systems are challenging and high-cost to operate. The growing cells can be damaged by the physical and chemical factors involved in the PBR operations (Mata et al. 2010).

2.9 Conclusion

Microalgae are a vast and multifaceted group of organisms that have the potential to be used in various aspects; human food as a source of high-quality protein, vitamins and enzymes, lipids, polyunsaturated fatty acids (PUSFA), and animal feed. They have a variety of compounds that are being utilized in the pharmaceuticals, neutraceuticals, and cosmetics industries. The naturally occurring carotenoids, xanthophylls and phycobiliproteins are widely used in consumable items, drugs and in aquaculture. Phycobiliproteins are a highly demanded natural product used in medical technology and immunological research. They are capable to supply bioactive polysaccharides, antioxidants, chemotherapeutic, anticancer, antibacterial, antifungal and antiviral agents. Algae are an essential component of aquatic ecosystems and there is a need to adopt a mechanism for their growth and sustainability to maintain the future supply line. They are also the source of pigments, fatty acids, raw material for cosmetics and skin care products. They can be potentially utilized in aquaculture and in bioplastics as well as biofertilizers. All these qualities make them a potential candidate for industries and their demand is rising in international markets. But growing them is also challenging due to huge production cost, risks of contamination and other environmental factors. It is a need of an hour to device easy and low cost methods to cultivate microalgae so that maximum benefits can be harvested.

References

Abdelfattah A, Ali SS, Ramadan H, El-Aswar EI, Eltawab R, Ho SH, Sun J (2022) Microalgae-based wastewater treatment: mechanisms, challenges, recent advances, and future prospects. Environ Sci Ecotechnol 100205

Alzahrani M (2018) Proteins and their enzymatic hydrolysates from the marine diatom *Nitzschia laevis* and screening for their in vitro antioxidant, antihypertension, anti-inflammatory and antimicrobial activities. Ph.D. thesis, Research Space, Auckland

Becker EW (2007) Microalgae as a source of protein. Biotechnol Adv 25:207–210

Behera B, Balasubramanian P (2019) Natural plant extracts as an economical and ecofriendly alternative for harvesting microalgae. Biores Technol 283:45–52

Behera B, Patra S, Balasubramanian P (2020) Biological nutrient recovery from human urine by enriching mixed microalgal consortium for biodiesel production. J Environ Manag 260:110111

Bharathi MJ (2021) Bioactive compounds from algae. In: Sangeetha J, Thangadurai D, Elumalai S, Thimmappa SC (eds) Phycobiotechnology, 1st edn. Apple Academic Press Inc., USA, 474 pp

Bhateja P, Mathur T, Pandy M, Fatma T, Rattan A (2006) Activity of blue green microalgae extracts against in vitro generated *Staphylococcus aureus* with reduced susceptibility to vancomycin. Fitoterapia 77(3):233–235

Borowitzka LJ, Borowitzka MA (1990) Commercial production of β-carotene by *Dunaliella salina* in open ponds. Bull Mar Sci 47(1):244–252

Bulgari R, Cocetta G, Trivellini A, Vernieri P, Ferrante A (2015) Biostimulants and crop responses: a review. Biol Agric Hortic 31(1):1–17

Cabanillas AH, Tena Pérez V, Maderuelo Corral S, Rosero Valencia DF, Martel Quintana A, Ortega Doménech M, Rumbero Sanchez A (2018) Cybastacines A and B: antibiotic sesterterpenes from a Nostoc sp. cyanobacterium. J Nat Prod 81(2):410–413

Calijuri ML, Silva TA, Magalhães IB, de Paula Pereira ASA, Marangon BB, de Assis LR, Lorentz JF (2022) Bioproducts from microalgae biomass: technology, sustainability, challenges and opportunities. Chemosphere 305:135508

Cardellina JH II, Moore RE, Arnold EV, Clardy J (1979) Structure and absolute configuration of malyngolide, an antibiotic from the marine blue-green alga *Lyngbya majuscula* Gomont. J Org Chem 44(23):4039–4042

Chaganty S, Golakoti T, Heltzel C, Moore RE, Yoshida WY (2004) Isolation and structure determination of cryptophycins 38, 326, and 327 from the terrestrial cyanobacterium Nostoc sp. GSV 224. J Nat Prod 67(8):1403–1406

Chen X, Smith GD, Waring P (2003) Human cancer cell (Jurkat) killing by the cyanobacterium metabolite calothrixin A. J Appl Phycol 15(4):269–277

Chen C, Tang T, Shi Q, Zhou Z, Fan J (2022) The potential and challenge of microalgae as promising future food sources. Trends Food Sci Technol 129:99–112

Cheng KC, Ogden KL (2011) Algal biofuels. Chem Eng Prog 107:42–47

Chisti Y (2008) Biodiesel from microalgae beats bioethanol. Trends Biotechnol 25(3):126–131

Credence Research (2016) https://www.credenceresearch.com/report/algae-products-market

Del Campo JA, Garcia-Gonzälez M, Guerrero MG (2007) Outdoor cultivation of microalgae for carotenoid production: current state and perspectives. Appl Microbiol Biotechnol 74:1163–1174

Dey B, Lerner DL, Lusso P, Boyd MR, Elder JH, Berger EA (2000) Multiple antiviral activities of cyanovirin-N: blocking of human immunodeficiency virus type 1 gp120 interaction with CD4 and coreceptor and inhibition of diverse enveloped viruses. J Virol 74(10):4562–4569

Dianursanti D, Noviasari C, Windiani L, Gozan M (2019) Effect of compatibilizer addition in *Spirulina platensis* based bioplastic production. AIP Conf Proc 2092(1). AIP Publishing

Ertani A, Sambo P, Nicoletto C, Santagata S, Schiavon M, Nardi S (2015) The use of organic biostimulants in hot pepper plants to help low input sustainable agriculture. Chem Biol Technol Agric 2(1):1–10

Faheed FA, Fattah ZA (2008) Effect of *Chlorella vulgaris* as bio-fertilizer on growth parameters and metabolic aspects of lettuce plant. J Agric Soc Sci 4:165–169

Fasaei F, Bitter JH, Slegers PM, Van Boxtel AJB (2018) Techno-economic evaluation of microalgae harvesting and dewatering systems. Algal Res 31:347–362

Freudenberg RA, Baier T, Einhaus A, Wobbe L, Kruse O (2021) High cell density cultivation enables efficient and sustainable recombinant polyamine production in the microalga *Chlamydomonas reinhardtii*. Biores Technol 323:124542

Görs M, Schumann R, Hepperle D, Karsten U (2010) Quality analysis of commercial *Chlorella* products used as dietary supplement in human nutrition. J Appl Phycol 22:265–276

Griffiths MJ, Dicks RG, Richardson C, Harrison STL (2011) Advantages and challenges of microalgae as a source of oil for biodiesel. In: Stoytcheva M, Montero J (eds) Biodiesel—feedstocks and processing technologies. In Tech, Rijeka, pp 177–200

Guedes AC, Amaro HM, Malcata FX (2011) Microalgae as sources of carotenoids. Mar Drugs 9:625–644

Guiry MD (2012) How many species of algae are there? J Phycol 48(5):1057–1063

Gupta C, Dhan P, Amr P, Gupta S (2012) Whey proteins: a novel source of bioceuticals. Middle East J Sci Res 12(3):365–375

Guschina IA, Harwood JL (2006) Lipids and lipid metabolism in eukaryotic algae. Prog Lipid Res 45(2):160–186

Gustafson KR, Cardellina JH, Fuller RW, Weislow OS, Kiser RF, Snader KM, Patterson GML, Boyd MR (1989) AIDS-antiviral sulfolipids from cyanobacteria (blue-green algae). J Nat Cancer Inst 81(16):1254–1258

Gutierrez M, Suyama TL, Engene N, Wingerd JS, Matainaho T, Gerwick WH (2008) Apratoxin D, a potent cytotoxic cyclodepsipeptide from Papua New Guinea collections of the marine cyanobacteria *Lyngbya majuscula* and *Lyngbya sordida*. J Nat Prod 71(6):1099–1103

Han B, Gross H, Goeger DE, Mooberry SL, Gerwick WH (2006) Aurilides B and C, cancer cell toxins from a Papua New Guinea collection of the marine cyanobacterium *Lyngbya majuscula*. J Nat Prod 69:572–575

Hernández-Herrera RM, Santacruz-Ruvalcaba F, Ruiz-López MA, Norrie J, Hernández-Carmona G (2014) Effect of liquid seaweed extracts on growth of tomato seedlings (*Solanum lycopersicum* L.). J Appl Phycol 26:619–628

Huang MT, Ferraro T (1992) Phenolic compounds in food and cancer. In: Huang MT, Ho CT, Lee CY (eds) Phenolic compounds in food and their effects on health. American Chemical Society, Washington, DC, pp 8–34

Huheihel M, Ishanu V, Tal J, Arad SM (2001) Antiviral effect of red microalgal polysaccharides on *Herpes simplex* and *Varicella zoster* viruses. J Appl Phycol 13:127–134

Huskens D, Fèrir G, Vermeire K, Kehr J, Balzarini J, Dittmann E, Schols D (2010) Microvirin, a novel alpha (1,2)-mannose-specific lectin isolated from *Microcystis aeruginosa*, has anti-HIV-1 activity comparable with that of cyanovirin-N but a much higher safety profile. J Biol Chem 285(32):24845–24854

Jorgensen LV, Madsen HL, Thomsen MK, Dragsted LO, Skibsted LH (1999) Regulation of phenolic antioxidants from phenoxyl radicals: an ESR and electrochemical study of antioxidant hierarchy. Free Radical Res 30:207–220

Kanekiyo K, Lee JB, Hayashi K, Takenaka H, Hayakwa Y, Endo S, Hayashi T (2005) Isolation of an antiviral polysaccharide, Nostoflan, from a terrestrial cyanobacterium, *Nostoc flagelliforme*. J Nat Prod 68(7):1037–1041

Kawalekar JS (2013) Role of biofertilizers and biopesticides for sustainable agriculture. J Bio Innov 2(3):73–78

Khan MI, Shin JH, Kim JD (2018) The promising future of microalgae: current status, challenges, and optimization of a sustainable and renewable industry for biofuels, feed, and other products. Rev Micro Cell Fact 17(36):1–21

Kim S, Ravichandran YD, Khan SB, Kim YT (2008) Prospective of the cosmeceuticals derived from marine organisms. Biotechnol Bioprocess Eng 3:511–523

Kim H, Lantvit D, Hwang CH, Kroll DJ, Swanson SM, Franzblau SG, Orjala J (2012) Indole alkaloids from two cyanobacteria, *Westiellopsis* sp. and *Fischerella muscicola*. Bioorg Med Chem 20(17):5290–5295

Kotake-Nara E, Nagao A (2011) Absorption and metabolism of xanthophylls. Mar Drugs 9:1024–1037

Kumar R, Elumalai S, Pabbi S (2021) Prospecting microalgae for biomolecules and biofuels. In: Sangeetha J, Thangadurai D, Elumalai S, Thimmappa SC (eds) Phycobiotechnology, 1st edn. Apple Academic Press Inc., USA, 474 pp

Kumari R, Kaur I, Bhatnagar AK (2011) Effect of aqueous extract of *Sargassum johnstonii* Setchell & Gardner on growth, yield and quality of *Lycopersicon esculentum* Mill. J Appl Phycol 23:623–633

Lee JB, Hayashi T, Hayashi K, Sankawa U, Maeda M, Nemoto T, Nakanishi H (1998) Further purification and structural analysis of calcium spirulan from *Spirulina platensis*. J Nat Prod 61(9):1101–1104

Linington RG, Gonzalez J, Urena L, Romero LI, Ortega-Barria E, Gerwick WH (2007) Venturamides A and B: antimalarial constituents of the Panamanian marine cyanobacterium *Oscillatoria* sp. J Nat Prod 70:397–401

Mahajan G, Kamat M (1995) γ-Linolenic acid production from *Spirulina platensis*. Appl Microbiol Biotechnol 43(3):466–469

Mansour H, Shoman S, Kdodier M (2011) Antiviral effect of edaphic cyanophytes on rabies and herpes-1 viruses. Acta Biol Hung 62(2):194–203

Mao TK, Water JVD, Gershwin ME (2005) Effects of a *Spirulina*-based dietary supplement on cytokine production from allergic rhinitis patients. J Med Food 8:27–30

Mata TM, Martins AA, Caetano NS (2010) Microalgae for biodiesel production and other applications: a review. Renew Sustain Energy Rev 14(1):217–232

McPhail KL, Correa J, Linington RG, Gonzalez J, Ortega-Barria E, Capson TL (2007) Antimalarial linear lipopeptides from panamanian strain of marine cyanobacterium *Lyngbya majuscula*. J Nat Prod 70:984–988

Metting B, Zimmerman WJ, Crouch I, Van Staden J (1990) Agronomic uses of seaweed and microalgae. Intro Appl Phycol 589–627

Misra BB, Nanda DR, Mishra BN (1985) Reclamation with blue green algae, mercury uptake by algae cultured in solid waste of a chlor—alkali factory and its effect on growth and pigmentation. J Environ Biol 6(4):223–231

Mittal S, Sengar RMS, Kavshik BD (1992) Uptake and toxicity of heavy metals to algae. Ind J Microb Environ 32(1):51–55

Molina E, Acién Fernández FG, García Camacho F, Camacho Rubio F, Chisti Y (2000) Scale-up of tubular photobioreactors. J Appl Phycol 12:355–368

Mooberry SL, Leal RM, Tinley TL, Luesch H, Moore RE, Corbett TH (2003) The molecular pharmacology of symplostatin 1: a new antimitotic dolastatin 10 analog. Int J Cancer 104:512–521

Moreira JB, de Morais EG, de Almeida ACA, Pontes JF, Grenh A, Barreira L, Varela J, Costa JAV, de Morais MG (2022) Microalgal applications in nanotechnology: an outstanding tool for nanocompounds synthesis and bioproducts obtention. In: Prasad R (ed) Nanotechnology in the life sciences. Springer Nature, Switzerland

Müller K (2001) Pharmaceutically relevant metabolites from lichens. Appl Microbiol Biotechnol 56:9–16

Munoz R, Guieysse B (2006) Algal-bacterial processes for the treatment of hazardous contaminations: a review. Water Res 40:2799–2815

Naidoo T, Zulu N, Maharajh D, Lallo R (2013) Value-added products from microalgae. In: Bux F (ed) Biotechnological applications of microalgae. CRC Press, Taylor & Francis Group, Boca Raton, London, New York, pp 137–160

Ocean Nutrition Canada (2010) Market opportunities. http://www.ocean-nutrition.com/omeg3/market

Ohta S, Shiomi Y, Kawashima A, Aozasa O, Nakao T, Nagate T, Kitamura K, Miyata H (1995) Antibiotic effect of linolenic acid from *Chlorococcum* strain HS-101 and *Dunaliella primolecta* on methicillin-resistant *Staphylococcus aureus*. J Appl Phycol 7:121–127

Onen Cinar S, Chong ZK, Kucuker MA, Wieczorek N, Cengiz U, Kuchta K (2020) Bioplastic production from microalgae: a review. Int J Environ Res Public Health 17(11):3842

Orosa M, Torres E, Fidalgo P, Abalde J (2000) Production and analysis of secondary carotenoids in green algae. J Appl Phycol 12:553–556

Pereira AG, Otero P, Echave J, Carreira-Casais A, Chamorro F, Collazo N, Jaboui A, Lourenço-Lopes C, Simal-Gandara J, Prieto MA (2021) Xanthophylls from the sea: algae as source of bioactive carotenoids. Mar Drugs 19:188

Portmann C, Bloom JF, Kaiser M, Brun R, Jüttner F, Gademann K (2008) Isolation of aerucyclamides C and D and structure revision of microcyclamide 7806A, heterocyclic ribosomal peptides from *Microcystis aeruginosa* PCC 7806 and their antiparasite evaluation. J Nat Prod 71:1891–1896

Posadas E, Alcántara C, García-Encina PA, Gouveia L, Guieysse, B, Norvill Z, Acien FG, Markou G, Congestri R, Koreiviene J, Muñoz R (2017) Microalgae cultivation in wastewater. In: Microalgae-based biofuels and bioproducts. Woodhead Publishing, pp 67–91

Posten C (2009) Design principles of photobioreactors for cultivation of microalgae. Eng Life Sci 9(3):165–177

Pratt R, Daniels TC, Eiler JJ, Gunnison JB, Kumler WD, Oneto JF, Strait LA, Spoehr HA, Hardin GJ, Milner HW, Smith JHC, Strain HH (1944) Chlorellin, an antibacterial substance from *Chlorella*. Science 99(2574):351–352

Priyadarshani I, Rath B (2012) Commercial and industrial applications of micro algae—a review. J Algal Biomass Util 3(4):89–100

Pulz O, Gross W (2004) Valuable products from biotechnology of microalgae. Appl Microbiol Biotechnol 65:635–648

Rammuni MN, Ariyadsa TU, Nimarshana PHV, Attalage RA (2019) Comparative assessment on the extraction of carotenoids from microalgae sources: astaxanthin from *H. pluvialis* and β-carotene from *D. salina*. Food Chem 30:128–134

Randrianarison G, Ashraf MA (2017) Microalgae: a potential plant for energy production. Geol Ecol Landsc 1(2):104–120

Reshef V, Mizrachi E, Maretzki T, Silberstein C, Loya S, Hizi A, Carmeli S (1997) New acylated-sulfoglycolipids and digalactolipids and related known glycolipids from cyanobacteria with a potential to inhibit the reverse transcriptase of HIV-1. J Nat Prod 60(12):1251–1260

Rickards RW, Rothschild JM, Willis AC, De Chazal NM, Kirk J, Kirk K, Saliba KJ, Smith GD (1999) Calothrixins A and B, novel pentacyclic metabolites from *Calothrix* cyanobacterium with potent activity against malaria parasites and human cancer cells. Tetrahydron 55(47):13513–13520

Ronga D, Biazzi E, Parati K, Carminati D, Carminati E, Tava A (2019) Microalgal biostimulants and biofertilisers in crop productions. Agronomy 9(4):192

Ruiz J, Olivieri G, de Vree J, Bosma R, Willems P, Reith JH, Eppink MHM, Kleinegris DMM, Wijffels RH, Barbosa MJ (2016) Towards industrial products from microalgae. Energy Environ Sci 9:3036–3043

Ryu BM, Himaya SWA, Kim SK (2015) Applications of microalgae-derived active ingredients as cosmeceuticals. In: Handbook of marine microalgae: biotechnology advances. Elsevier Inc.

Sabry OM, Goeger DE, Gerwick WH (2017) Biologically active new metabolites from a Florida collection of *Moorea producens*. Nat Prod Res 31:555–561

Senthil KNS, Sivasubramanian V, Mukund S (2013) Antimicrobial and antifungal activity of extracts of *Phormidium fragile* Gomont. J Algal Biomass Util 4(3):66–71

Shameel M (2001) An approach to the classification of algae in the new millennium. Pak J Mar Biol 7(1&2):233–250

Shameel M (2003) Biodiversity and natural history of algae. J Nat Hist Wildl 2(1):1–9

Shameel M (2008) Change of divisional nomenclature in the Shameelian classification of algae. Int J Phycol Phycochem 4(2):225–232

Shameel M (2012) Nomenclature changes in the Shameelian classification of algae. Int J Phycol Phycochem 8(1):7–22

Sheehan J, Dunahay T, Benemann J, Roessler P (1998) Look back at the US Department of Energy's aquatic species program: biodiesel from algae; close-out report (No. NREL/TP-580-24190). National Renewable Energy Lab., Golden, CO

Shen CT, Chen PY, Wu JJ, Lee TM, Hsu SL, Chang CMJ, Young CC, Shieh CJ (2011) Purification of algal anti-tyrosinase zeaxanthin from *Nannochloropsis oculata* using supercritical anti-solvent precipitation. J Supercrit Fluids 55:955–962

Shukla AC, Gupta AB (1967) Influence of algal growth-promoting substances on growth, yield and protein contents of rice plants. Nature 213:744–744

Soong P (1980) Production and development of *Chlorella* and *Spirulina* in Taiwan. In: Shelef G, Soeder CJ (eds) Algae biomass. Elsevier/North Holland Biomedical Press, Amsterdam, pp 97–113

Spoehr HA, Smith JHC, Strain HH, Milner HW, Hardin GJ (1949) Fatty acid antibacterials from plants. Carnegie Institution of Washington Publication, Washington, DC, p 586

Spolaore P, Joannis-Cassan C, Duran E, Isambert A (2006) Commercial applications of microalgae. J Biosci Bioeng 101(2):87–96

Stark M, O'Gara I (2012) An introduction to photosynthetic microalgae. Disrupt Sci Technol 1(2):65–67

Sueyoshi K, Kaneda M, Sumimoto S, Oishi S, Fujii N, Suenaga K, Teruya T (2016) Odoamide, a cytotoxic cyclodepsipeptide from the marine cyanobacterium *Okeania* sp. Tetrahedron 72:5472–5478

Sui Y, Jiang Y, Moretti M, Vlaeminck S (2020) Harvesting time and biomass composition affect the economics of microalgae production. J Clean Prod 259:120782

Supraja KV, Behera B, Balasubramanian P (2020) Performance evaluation of hydroponic system for co-cultivation of microalgae and tomato plant. J Clean Prod 272:122823

Tam NFY, Wong YS (1996) Effect of ammonia concentration on growth of *Chlorella vulgaris* and nitrogen removal from media. Biores Technol 57:45–50

Tan JS, Lee SY, Chew KW, Lam MK, Lim JW, Ho S, Show PL (2020) A review on microalgae cultivation and harvesting, and their biomass extraction processing using ionic liquids. Bioengineered 11(11):116–129

Tinoco NA, Uekane TM, Tsukui A, Aguiar PFD, Teixeira CM, Rezende CM (2016) Generation of volatile compounds from carotenoids of *Dunaliella bardawil* algae by water bath heating and microwave irradiation. J Braz Chem Soc 27:1452–1458

Ugwu CU, Aoyagi H, Uchiyama H (2008) Photobioreactors for mass cultivation of algae. Biores Technol 99(10):4021–4028

Van DMFJUM, De Haan CAM, Schuurman NMP, Haijema BJ, Peumans WJ, Van DEJM, Delputte PL, Balzarini HF, Egberink HF (2007) Antiviral activity of carbohydrate-binding agents against *Nidovirales* in cell culture. Antiviral Res 76(1):21–29

Wang Y, Tibbetts SM, McGinn PJ (2021) Microalgae as sources of high-quality protein for human food and protein supplements. Foods 10(12):3002

Web1 (2021) https://www.alliedmarketresearch.com/microalga-market-A13419. Accessed 10/08/2023

Whitaker R, Margulis L (1978) Protist classification of the kingdoms of organisms. Biosystems 10:3–18

Wikfors GH, Ohno M (2001) Impact of algal research in aquaculture. J Phycol 37:968–974

Xu Z (2007) Biological production of hydrogen from renewable resources. In: Yang S-T (ed) Bioprocessing for value-added products from renewable resources. Elsevier, pp 527–557

Xu L, Weather PJ, Xiong XR, Liu CZ (2009) Microalgal bioreactors: challenges and opportunities. Rev Eng Life Sci 9(3):178–189

Xu X, Gu X, Wang Z, Shatner W, Wang Z (2019) Progress, challenges and solutions of research on photosynthetic carbon sequestration efficiency of microalgae. Renew Sustain Energy Rev 110:65–82

Yan C, Zhang Q, Xue S, Sun Z, Wu X, Wang Z, Lu Y, Cong W (2016) A novel low-cost thin-film flat plate photobioreactor for microalgae cultivation. Biotechnol Bioprocess Eng 21:103–109

Yarkent Ç, Gürlek C, Oncel SS (2020) Potential of microalgal compounds in trending natural cosmetics: a review. Sustain Chem Pharm 17:100304

Zhou W, Hu B, Li Y, Min M, Mohr M, Du Z, Chen P, Ruan R (2012) Mass cultivation of microalgae on animal wastewater: a sequential two-stage cultivation process for energy crop and omega-3-rich animal feed production. Appl Biochem Biotechnol 168:348–363

Chapter 3
Recent Advances in Algal Nexus for Circular Economy

Richard Luan Silva Machado, Darissa Alves Dutra, Adriane Terezinha Schneider, Rosangela Rodrigues Dias, Leila Queiroz Zepka, and Eduardo Jacob-Lopes

Abstract Microalgae are widespread microorganisms, considered conservative alternatives for the synthesis of various products with high and low added value. This unique feedstock can create nature-based products capable of replacing conventional materials with high carbon emissions. The microalgae-based biorefinery, following the cascade principle, is an approach that assumes great relevance in the circular economy, making microalgae-based processes less dystopian. This is because, in a microalgae biorefinery, the natural capital is renewable, recovered, and reused as much as possible, creating secondary products that can flow back into the biosphere and close the carbon cycle after use. Thus, in this chapter, we present a closer look at the concept of biorefinery applied to microalgae as a path to the sustainability and resilience of these systems. We also turn our attention to the environmental and economic life cycle analysis applied to the microalgae biorefinery. Ultimately, the mishaps and the long road ahead of the microalgae biorefinery are critically discussed.

Keywords Microalgae · Biorefinery · Circular economy · Sustainability · Renewable energy · Life cycle assessment · Renewable biomass

3.1 Introduction to the Biorefinery Concept

Microalgae are estimated as renewable bioresources as a result of their high growth rate and suitability in growing environments (Patel et al. 2022). The term biorefinery has grown universally, being intrinsically related to the similarities between an oil refinery that generates fossil fuels (Chandrasekhar et al. 2022). However, the most significant difference between these two refineries relates to the types of feedstocks

R. L. S. Machado · D. A. Dutra · A. T. Schneider · R. R. Dias · L. Q. Zepka · E. Jacob-Lopes (✉)
Bioprocess Intensification Group, Federal University of Santa Maria, UFSM, Roraima Avenue 1000, Santa Maria, RS 97105-900, Brazil
e-mail: ejacoblopes@gmail.com

T. A. Kurniawan and A. Anouzla (eds.), *Algae as a Natural Solution for Challenges in Water-Food-Energy Nexus*, Environmental Science and Engineering,
https://doi.org/10.1007/978-981-97-2371-3_3

used and their effect on sustainability. Microalgae, in particular, may have great potential as a starting point for application in a biorefinery (Barati et al. 2021). These microorganisms effectively convert solar energy and carbon dioxide (CO_2), in aquatic environments, into biomass composed of parts such as carbohydrates, lipids, and proteins (Zhang et al. 2023). Studies are constantly being developed to make this feedstock the next step for biotechnology, for example, Aketo et al. (2020) isolated the *NKY372003* strain of microalgae with a high carbohydrate index directed to biorefineries, making it an ideal solution for the production of green energy, such as biofuels (biohydrogen, biomethane, bioethanol, and biodiesel) and application in food, with high potential for mitigation in view of the energy crisis and food shortages (Khoo et al. 2023; Goswami et al. 2022; Jacob-Lopes et al. 2020).

The microalgae biorefinery converts them into products with high and low added value, following a production line associated with the cascade principle, integrating a system that uses microalgae to produce this high range of multi-products, which normally involves a series of steps (Okeke et al. 2022; Cabello-Galindo et al. 2023; Monlau et al. 2021). The first is the cultivation of microalgae, which can occur in closed systems, such as photobioreactors, or in open systems, such as ponds (Sharma et al. 2022; Tan et al. 2020; Depra et al. 2022). During cultivation, microalgae are exposed to sunlight, water, and nutrients, allowing exponential growth (Díaz et al. 2022). It is noteworthy that the development of these microalgae is based on pathways such as photoautotrophy, making use of inorganic carbon through photosynthesis (Yu et al. 2022; Manhaeghe et al. 2020) and also the heterotrophic pathway using organic carbon in the absence of light (Cao et al. 2023b). It can also be mixotrophic, not necessarily using any carbon source for its photosynthetic metabolism (Verma et al. 2020; Santos et al. 2021). Subsequently, this microalgae biomass is recovered through different methods, such as centrifugation, filtration, and flotation (Fal et al. 2023).

In the state of the art, one of the fractions of microalgae biomass is oil or lipid, extracted from residual biomass and used to obtain biodiesel, considered a sustainable alternative to its equivalent diesel oil, since, in addition to being produced in a renewable way, it has a lower carbon quota (Li et al. 2020; Elkelawy et al. 2021; Malode et al. 2021). In addition to oil, biomass also contains proteins, carbohydrates, and other bioactive compounds (Zhang et al. 2023). These fractions can be bases for use in various sectors. For example, microalgae proteins can be used as food ingredients, nutritional supplements, and in the animal feed industry (Kumar et al. 2022). Carbohydrates, in turn, can be transformed into bioplastics, fertilizers, and value-added chemicals (Lu and Xiao 2022; Chong et al. 2022; Cao et al. 2023a). Bioactive compounds produced by microalgae have been gaining commercial ground, mainly due to their applicability in functional foods, presenting new bioactive ingredients used by the food industry. These compounds make a range of scientifically supported claims for product functionality. Therefore, great importance has been given to these, such as carotenoids, chlorophylls, phenolic compounds and polyunsaturated fatty acids (PUFAs), which are closely associated with their positive effects against cardiovascular diseases, immunoprotection, anti-aging, photoprotection and modulation of oxidative stress (Depra et al. 2023; Jacob-Lopes et al. 2019; Nascimento et al. 2021).

Microalgae biorefineries also offer environmental benefits, as they are able to capture a high level of carbon dioxide (CO_2) during the photosynthesis process, helping to mitigate greenhouse gas (GHG) emissions (Cabello-Galindo et al. 2023; Koutra et al. 2023; Bhattacharya and Goswami 2020). In addition, microalgae cultivation can be carried out in areas not used for conventional agriculture, not competing with food resources. However, there are still challenges associated with microalgae biorefinery (Cheirsilp and Maneechote 2022). Large-scale cultivation and efficient recovery of microalgae biomass are areas that have encouraged continued research and development. The mass production of microalgae can be considered economically unfeasible, as the demand for water and nutrient resources is significant. However, the ability of these microorganisms to utilize key water pollutants, such as phosphorus and nitrogen, as nutrients is the key to allowing process optimization (Chhandama and Rai 2023). The idea of using this biomass in production cycles that positively impact the environment, such as the use of these microalgae in the treatment of effluents, opens up opportunities for reuse, recycling, and remanufacturing, conceptualized as the circular economy (Fathima and Chatterjee 2022; Ubando et al. 2020; Chandrasekhar et al. 2022).

In front of this scenario, this chapter seeks to expose, in a more attentive way, the concept of a microalgae biorefinery, demonstrating microalgae as a sustainable and resilient bioresource for the future, aiming at the gradual replacement and dependence on fossil resources, hopefully for the mitigation of changes climate change and the transition to a greener economy. Exposing more comprehensively the environmental and economic life cycle applied to microalgae biorefineries. Finally, obstacles and difficulties during the implementation of a microalgae biorefinery are presented and discussed.

3.2 Microalgae Biorefinery: Pathway to Sustainability and Resilience

Faced with the growing demand for energy, the high level of greenhouse gas (GHG) emissions, and the extinction of fossil resources, microalgae biomass is born as an essential source of feedstocks, seeking to meet the demands of humanity (Singh et al. 2019). Among the numerous systems applicable to the production of microalgae biomass, there is the circular economy plan aimed at the "closed loop" linked to the microalgae biorefinery (Choi et al. 2021). The circular economy is characterized and focused on the integration of the sustainable use of resources provided by nature, valuing the biomass obtained, with the use of processes and the multiple outputs of products through cascade application (Stegmann et al. 2020; Sherwood 2020; Monlau et al. 2021).

This biorefinery approach allows full use of microalgae biomass, minimizing waste and maximizing savings. This encompasses the use of products with high and low added value, from biomass, as an alternative source of polyunsaturated acids

(PUFAs) and production of biofuels, respectively, as shown in Fig. 3.1 (Khoo et al. 2023; Kumaran et al. 2023; Monlau et al. 2021). The microalgae biorefinery must play a prominent role in the twenty-first century. Challenges related to global climate change and over-reliance on traditional sources of fossil energy progressively fuel investments in innovative technologies with more environmental sustainability and resilience in this industrial segment (Sharma et al. 2022; Lu and Xiao 2022; Tsolakis et al. 2023; Cao et al. 2023a).

The current production of microalgae comprises cultivation systems, which can have different effects on the productivity and biochemical composition of the produced biomass, related to open or closed systems. The open cultivation method has beneficial characteristics such as the production of algae on a large scale, and can be operated with limited cost and capital, however, it is very conducive to contamination. The closed system comprises photobioreactors, which allow the action of direct light to carry out photosynthesis. It is noteworthy that in these systems it is possible to avoid contamination, making it possible to control the temperature and pH, causing a lower environmental impact and flexible technical designs. However, there are challenges to be overcome for the large-scale implementation of the microalgae biorefinery. Microalgae cultivation technology needs to be improved to maximize productivity and reduce production costs. In this sense, the production of microalgae can be intrinsically linked to the cultivation provided by wastewater, acting beneficially in the treatment of water through bioremediation mediated by these microorganisms (Oliveira et al. 2022). It is noteworthy that these are the main precursors

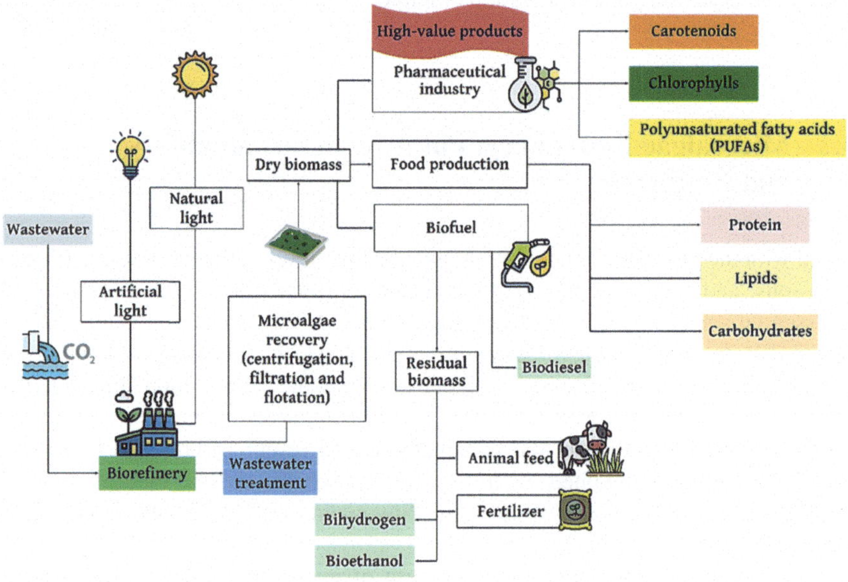

Fig. 3.1 Microalgae biorefinery in synergy with water treatment and multi-product production through the application of the cascade principle

of liquid oxygen on earth, in addition, almost 88% of carbon fixation in the aquatic environment is through the photosynthetic activity of microalgae (Abdelfattah et al. 2022; Duerschlag et al. 2022).

Typically, the cellular content of microalgae comprises more than 50% provided by carbon dioxide (CO_2), which is responsible for about 79.5% of emissions, which generate global warming and result in climate change, allocating only 11.4%, 6.1%, and 3% for methane, nitrous oxide and fluorinated gases, respectively. Anthropogenic CO_2 emissions are the main cause of climate change. About 1.83 tons of this gas are fixed to produce 1 kg of microalgae biomass, for example (Nath et al. 2023). The main lever for the use of microalgae is related to its great capacity to absorb CO_2, often found in wastewater, which indicates its unparalleled capacity to remove this gas, taking advantage of it as a nutrient, in the biological carbon cycle (Fathima and Chatterjee 2022). Currently, CO_2 biosequestration by microalgae cells has gained much prominence, making use of these high-emission gases to form high- and low-value biomass within a biorefinery (Mustafa et al. 2021).

Biomass is considered a bioresource, which unquestionably has contributed to the development of the bioeconomy. From this, microalgae, in particular, have received a lot of attention in recent years due to their accelerated growth and the production of high and low-added value resources, such as the conversion of biomass produced into biofuels, food and feed (Sarwer et al. 2022). One of the renewable fuels obtained from the oil of the residual biomass of microalgae is biodiesel, which is considered an alternative for the success of green energy. The biodiesel production process from microalgae uses methods such as transesterification. However, the generation of biofuels from microalgae still does not offer resilience in the market, due to its high production costs, thus preventing commercial competition (Jacob-Lopes et al. 2020). Estimates show an average of 2–2.59 US$ for biodiesel production, considering the process stages (cultivation, harvesting, biomass treatment, extraction, and transesterification) (Nazloo et al. 2022). For example, in one study, *Chlamydomonas malina* at low temperatures (8 °C) and lux intensity around 250 $\mu mol/m^2/s$ produced a number of lipids and PUFAs around 161.3 mg/L/d and 85.5 mg/L/d, respectively (Morales-Sánchez et al. 2020). In another experiment performed, *Chlorella pyrenoidosa C. ZTY4* demonstrated a total fatty acid of about 66.4% (Zhang et al. 2019). Demonstrating that the use of microalgae for the use of fatty acids can become a sustainable alternative for future applications.

3.2.1 Environmental Life Cycle Analysis Applied to Microalgae Biorefinery

Achieving sustainability alone is one of the challenges currently faced, especially when applied to large-scale production, such as a microalgae biorefinery, where several points still need to be analyzed and applied. Assessing the environmental impacts of processes and products based on microalgae is the central point for biorefineries to become a sustainable and ecologically correct platform (Dias et al. 2023). In this context, the life cycle analysis tool through sustainability metrics and indicators, helps us and directs us toward the decision-making path for eco-friendly development in the production chains (Depra et al. 2023). In turn, it is important to emphasize that the production chain, when constituted in a cascade, collaborates with advantages in different spheres at the same time, like for example with the circular bioeconomy and sustainability (Karan et al. 2023).

Regarding the life cycle analysis regarding the carbon footprint, the impact category analyzed is the global warming potential, which measures the emission of greenhouse gases such as carbon dioxide from processes and products. As an example, we can highlight the values for this category of environmental impact for the production of 1 kg of β-carotene through the microalgae *Dunaliella salina*, astaxanthin from *Haematococcus pluvialis* and phycobiliprotein from *Arthrospira platensis* the values of 852.49, 2804.52 and 1177.18 kg CO_2eq, in this order (Depra et al. 2020). About this, the literature suggests that 98% of ecosystem impacts during production are a consequence of the need for energy in the upstream and downstream stages (Depra et al. 2020).

In light of this, the transition energy matrix with high greenhouse gas emissions to clean and renewable energy is one of the essential factors that will make the microalgae biorefinery progressively move towards successful results in terms of environmental impact. In the literature, studies on the reduction of the carbon footprint by replacing the coal-fired electric matrix with energy from nuclear, hydroelectric, wind, biomass, photovoltaic, biogas, natural gas, and waste sources demonstrate savings in carbon dioxide emissions that revolve around approximately 45–99%. Evidently, given the demand for the mitigation of greenhouse gas emissions, the carbon credits market drives the implementation of the green label. However, the purchase of carbon credits may be masking the real emission of greenhouse gases from products and processes along the entire chain (Dias et al. 2022).

Finally, the microalgae biorefinery can achieve clean, sustainable, and economical production, since the analysis of the life cycle of processes and products shows which metrics and sustainability indicators most need attention. Achieving this feat in a future that is increasingly close and inevitable is essential.

3.2.2 *Mishaps and Economic Challenges Applied to the Microalgae Biorefinery*

Economically, a microalgae biorefinery is a facility that would integrate a vast amount of conversion technologies, which makes it unfeasible in economic terms, such as thermochemical and biochemical routes for the growth of microorganisms and their subsequent conversion into biomass in a sustainable flow of products biobased (Ubando et al. 2020; Lim et al. 2022; Pereira et al. 2021). However, microalgae have shown great interest on the part of researchers and industries in recent years, as potentially sustainable sources for the production of precious bioproducts used in the areas of food, renewable energy, nutraceuticals and pharmaceuticals (Katiyar et al. 2021; Derossi et al. 2021). The use of an integrated biorefinery with sequential extraction in order to close the biological carbon cycle can help to face the challenges posed by the circular bioeconomy, producing high and low value products at the same time (Thangam et al. 2021; Fal et al. 2023; Stegmann et al. 2020).

In the state of the art, microalgae can produce an extensive range of high-value bioactive compounds, such as carotenoids, proteins, lipids and some other bioproducts (Fu et al. 2019). Among the vast alternatives for applying biomass, biofuels stand out for their promising advantages in reducing the carbon dioxide (CO_2) footprint and mitigating the production of renewable energy. However, its high production cost makes the process unfeasible, being more expensive than petroleum-derived fuels (Aslam et al. 2020; Ummalyma et al. 2020). For example, the cost of 10,000 tons of microalgae biomass comprising 30% fatty acids is mathematically converted to 2.80 USD/L. Compared to the cost of oil, for example (1.10 USD/L) in the United States, the costs of producing biofuels from microalgae are higher. Therefore, the search for innovative and low-cost treatments for the efficient generation of biofuels from microalgae can be the initial substitution strategy in favor of the environment. Notoriously, the adoption of this strategy would not leverage economic growth and the viability of the biofuel production process, the same refers to the production of value-added bioresources, such as nutrients, food supplements, pigments and the adoption of bioremediation. Furthermore, research is also dedicated to looking for microalgae that allow and have a profile for the development of the large-scale biorefinery concept, although there are on average 0.2–0.8 million species of microalgae, only a small portion of these is intended for potential uses, some of them are discussed along with their applications in Table 3.1 (Chandrasekhar et al. 2022; Barati et al. 2021).

Globally, the microalgae biorefinery continues to face a challenge in its commercialization. Resources and cost allocation involving their "upstream processing", strategies to reacquire algal biomass, prerequisite for "downstream processing", are still trying to cross the valley of death. Different studies related to the low cost of cultures, as a source of effluents with the implementation of multiple processing (pre-treatment and biomass recovery) in synergy with the biorefinery, are promising alternatives to overcome the high cost (Okeke et al. 2022; Singh et al. 2019; Ummalyma et al. 2023). Some studies (Saral et al. 2022; Roles et al. 2021; Wiatrowski et al. 2022) sought to compare the costs of different products and consider their process optimizations, seeking to reduce costs in unit operations associated with the microalgae biorefinery. According to Slegers et al. (2020) the microalgae biorefinery based on the cascade concept increases the use of bioresources from this biomass from 25 to 95%, considering that the multiproduct biorefinery is more advantageous in the value chains when compared to the production of products individuals (Monlau et al. 2021).

Table 3.1 Disparity of microalgae and their applications within a biorefinery

Species of microalgae	Applications	References
Scenedesmus obliquus	Production of biohydrogen and biodiesel	Trivedi et al. (2022) and Satheesh et al. (2023)
Neochloris sp.	Production of biodiesel and biogas	Catalina et al. (2020) and Ummalyma et al. (2023)
Nannochloropsis sp.	Lipid extraction, production of bio-oil, biodiesel (PUFAs)	Huang et al. (2022), Aguilar-Ruiz et al. (2020), Ahmad et al. (2022) and Resdi et al. (2021)
Chlorella sp.	Extraction of lipids and production biodiesel	Danilović et al. (2021) and Ahmad et al. (2022)
Chlorella pyrenoidosa	Production of biodiesel, biomethane and biohydrogen	Ngatcha et al. (2022) and Satheesh et al. (2023)
Arthrospira sp.	Production of bio-oil, cosmetics (regulation of skin microbiota or preservation), food additives, and nutraceuticals	Wils et al. (2021), Barkia et al. (2019) and Kandasamy et al. (2019)
Scenedesmus sp.	Lutein recovery for food, bioplastic, bioethanol, biogas, bio-oil industries	Fan et al. (2022), Thangam et al. (2021), Ahmad et al. (2022) and Kalwani et al. (2022)
Chlorella vulgaris	Extraction of natural polyphenolic antioxidants, biodiesel production, bioethanol, bioremediation, food additives, animal feed ingredients	Mahmood et al. (2019), Nematian et al. (2020), Mustafa et al. (2021) and Kumaran et al. (2023)
Yarrowia lipolytica	*Production of biomethane and lipids*	Yu et al. (2019)
Chlorella sorokiniana	Production of biodiesel and biohydrogen	Constantino et al. (2021) and Satheesh et al. (2023)
Dunaliela sp.	Production bio-oil, biogas (PUFAs)	Kaushik and Raza (2019) and Elleuch et al. (2021)
Desmodesmus sp.	Production biodiesel, bioethanol, bio-oil, bioplastic (PUFAs)	Abinandan et al. (2020), Sarfraz et al. (2021) and Jothibasu et al. (2022)
Tribonema sp.	Production biodiesel, bioethanol, unsaturated fatty acids (USFA), polyunsaturated fatty acids (PUFAs)	Huo et al. (2020), Qi et al. (2023) and Sengupta et al. (2023)

Given this scenario, the biorefinery has become the target of great economic efforts, however, its profitable expansion is mediated by commitments such as technological innovations, which aim to reduce operating costs, the elaboration of the concept of multiproduct biorefinery, valuing biomass through the cascade principle and the improvement of a new line of special products for application in food. The application of the cascade concept can generate a higher cost for the biorefinery, but it is offset by the amount of biomass generated in the process. Therefore, the application of these conditions can generate very viable multiproduct recipes, which can make the microalgae biorefinery economically competitive in the future (Slegers et al. 2020).

3.3 The Setbacks and the Long Road Ahead of the Microalgae Biorefinery

Despite the significant technological advances in microalgae biorefineries, from an economic point of view, it is difficult to establish a multi-product biorefinery capable of handling large volumes of commodities. Current projections indicate that a conventional microalgae-based multiproduct biorefinery is too high a value. Typical downstream post-treatment processes in large-scale biotechnological processes represent about 30% of the total costs, while in existing microalgae biorefineries, this proportion ranges from 50 to 60%. To reduce costs, it will be necessary to develop simplified after-treatment approaches involving fewer operational steps (Vermuë et al. 2018).

Although the microalgae biorefinery is not yet a utopian system, it is encouraging to note that there is a growing awareness of the importance of seeking sustainable alternatives. The microalgae biorefinery has significant potential, both in terms of producing biofuels and high-value-added products. It is essential to establish effective collaboration between different actors in society to transform this potential into a reality in line with nature. As mentioned, it is necessary to connect the dots between investors, companies, academia, governmental and non-governmental organizations. This cooperation will allow the sharing of knowledge, resources, and experiences, boosting the development of the microalgae biorefinery (Banu et al. 2020; Okeke et al. 2022).

Investors play a crucial role in providing the necessary financial resources for the research, development, and construction of the facilities. The academy plays a key role in research and technological advancement, seeking innovative solutions and improving biomass cultivation and inheritance processes. Government organizations have the responsibility to establish adequate regulations, encourage supportive policies, and promote actions that favor sustainable development. Non-governmental organizations play an important role in raising public awareness, promoting sustainability, and protecting the environment (Rosa et al. 2023).

Currently, only a few companies are present in the algae biorefinery sector, establishing themselves and gaining a significant position in the global market. Companies such as Algatechnologies, and BioReal Inc., among others, are involved in the manufacture and sale of pigments derived from microalgae, which are used as biotherapeutics. Mera Pharmaceuticals Inc., a marine biotechnology company, has successfully produced astaxanthin from the species *Haematococcus pluvialis*. Cyanotech Corporation in Hawaii is recognized as the world's leading manufacturer of Spirulina, a food component approved by the FDA (Food and Drugs Administration), which is marketed in over 30 countries, including the United States. The company Algeon

Inc., based in the USA, was founded in 2011 and became known to produce Spirulina and Chlorella. Many of their products are made from the photosynthetic protist Euglena Graciosa, including b-glucan and whole-cell extracts (Chandra et al. 2019).

There are many government-supported algae processing experimentation facilities are in operation in the Americas and Europe. These include Algenol, a pilot-scale integrated test facility for processing algae in the United States, Algae PARC in the Netherlands, the ALL-Gas Project in Spain, the Cellana algae biorefinery demonstration in the United States, Global Algae Innovation, also in the United States, focused on the production of low-cost food and fuel, and the Muradel test factory, located in Australia. The Food and Agriculture Organization of the United Nations (FAO), as an integral part of the United Nations, provides comprehensive information and statistics on algal biomass on a global scale. However, this data is often fragmented and scattered due to relatively lower production than other biomass forms. Although certain regions such as Europe and North America have databases operated by the European Joint Research Center (EU-JRC) and the National Renewable Energy Laboratory (NREL) respectively, many developing countries lack regulations and legal obligations to report algae data comprehensively (Araújo et al. 2021; Behera and Paramasivan 2022).

With the lack of data, the investment that is necessary for the implementation of a microalgae biorefinery ends up being impeded due to the scarcity of these accurate and true data, making its commercialization and progression in the market a bottleneck. In order to reinforce the viability of the microalgae biorefineries process, the application of the LCA tool has been explored in studies, the intention is to try to solve the bottlenecks related to the lack of information, which generates speculation about the long-term sustainability, promoting a closing of doors in the stakeholder market (Behera and Paramasivan 2022).

Some studies have addressed the aforementioned challenges and suggested the commercial viability of microalgae biorefineries through Life Cycle Analysis (LCA). However, most zero-waste biorefinery approaches are still limited to laboratory-scale studies, which raises uncertainty about sustainability on a commercial scale. The LCA is used to assess the environmental impact of the biorefinery supply chain, demonstrating that microalgae biorefinery setups align with circular bioeconomy principles (Ubando et al. 2022). This academic production serves as a starting point for validating microalgae biorefineries. If all sectors of society work together, in the future this green gold can be further refined, and the economic aspects can be compared to those of biorefineries that use conventional feedstocks, such as the soy biorefinery (Taelman et al. 2015).

Despite the promising potential of microalgae biorefinery, there are several setbacks and challenges to be overcome to make it a viable commercial reality. It is essential to invest in research and development to reduce production costs, making microalgae biorefinery more economically feasible. Collaboration between academia, industry, governments, and investors is crucial to drive research and development in this field. The implementation of sustainable cultivation practices, such as the use of recycled water and the development of efficient waste treatment systems, is essential to minimize the negative effects on the environment. Although there are challenges ahead, perseverance and collaborative work can lead to the success of microalgae biorefinery, paving the way for sustainable production and a significant contribution to a greener future.

3.4 Conclusion

The microalgae biorefinery is a striking sustainable alternative, aimed at various sectors, such as the production of biofuels, food, water treatment. In addition, the energy transition, mediated by microalgae biofuels, is something promising that aligns the biorefinery in a prosperous direction in its impact on sustainability. It is noteworthy that synergistic bioremediation with microalgae biorefinery (adhering to the cascade concept) would make this a promising strategy against the elimination of fossil subsidies and the reuse of water resources. However, despite recent advances and the enormous potential of microalgae biorefineries, there are still challenges to be faced in their implementation, mainly in the economic field. It is necessary to develop more efficient and economically viable technologies for the cultivation, processing and extraction of microalgae products. More efficient ecological techniques still need to be promoted for microalgae biorefining, such as the development of more research on biodiesel production on a pilot scale. Furthermore, it is important to ensure the environmental sustainability of these practices, avoiding negative impacts on marine ecosystems and the availability of resources used to produce this green gold.

References

Abdelfattah A, Ali SS, Ramadan H, El-Aswar EI, Eltawab R, Ho SH, Elsamahy T, Li S, El-Sheekh MM, Schagerl M, Kornaros M, Sun J (2022) Microalgae-based wastewater treatment: mechanisms, challenges, recent advances, and future prospects. Environ Sci Ecotechnol 100205

Abinandan S, Subashchandrabose SR, Venkateswarlu K, Megharaj M (2020) Sustainable iron recovery and biodiesel yield by acid-adapted microalgae, *Desmodesmus* sp. MAS1 and *Heterochlorella* sp. MAS3, grown in synthetic acid mine drainage. ACS Omega 5(12):6888–6894

Aguilar-Ruiz RJ, Martínez-Macias MDR, Sánchez-Machado DI, López-Cervantes J, Dévora-Isiordia GE, Nateras-Ramírez O (2020) Removal of copper improves the lipid content in *Nannochloropsis oculata* culture. Environ Sci Pollut Res 27:44195–44204

Ahmad A, Banat F, Alsafar H, Hasan SW (2022) Algae biotechnology for industrial wastewater treatment, bioenergy production, and high-value bioproducts. Sci Total Environ 806:150585

Aketo T, Hashizume R, Yabu Y, Hoshikawa Y, Nojima D, Maeda Y, Yoshino T, Takano H, Tanaka T (2020) Characterization of a novel marine unicellular alga, *Pseudoneochloris* sp. strain NKY372003 as a high carbohydrate producer. J Biosci Bioeng 129(6):687–692

Araújo R, Vázquez Calderón F, Sánchez López J, Azevedo IC, Bruhn A, Fluch S, Tesende MG, Ghaderiardakani F, Ilmjärv T, Laurans M, Monagail MM, Mangini S, Peteiro C, Rebours C, Stefansson T, Ullmann J (2021) Current status of the algae production industry in Europe: an emerging sector of the blue bioeconomy. Front Mar Sci 7:626389

Aslam A, Fazal T, uz Zaman Q, Shan A, Rehman F, Iqbal J, Rashid A, Rehman MSU (2020) Biorefinery of microalgae for nonfuel products. In: Microalgae cultivation for biofuels production. Academic Press, pp 197–209

Banu JR, Kavitha S, Gunasekaran M, Kumar G (2020) Microalgae based biorefinery promoting circular bioeconomy-techno economic and life-cycle analysis. Biores Technol 302:122822

Barati B, Zafar FF, Rupani PF, Wang S (2021) Bacterial pretreatment of microalgae and the potential of novel nature hydrolytic sources. Environ Technol Innov 21:101362

Barkia I, Saari N, Manning SR (2019) Microalgae for high-value products towards human health and nutrition. Mar Drugs 17(5):304

Behera B, Paramasivan B (2022) Research trends and market opportunities of microalgal biorefinery technologies from circular bioeconomy perspectives. Biores Technol 351:127038

Bhattacharya M, Goswami S (2020) Microalgae—a green multi-product biorefinery for future industrial prospects. Biocatal Agric Biotechnol 25:101580

Cabello-Galindo A, Rodríguez-Jasso RM, Ruiz HA (2023) Microalgae biomass: a model of a sustainable third-generation of biorefinery concept. In: Green approach to alternative fuel for a sustainable future. Elsevier, pp 369–379

Cao TND, Mukhtar H, Le LT, Tran DPH, Ngo MTT, Nguyen TB, Bui XT (2023a) Roles of microalgae-based biofertilizer in sustainability of green agriculture and food-water-energy security nexus. Sci Total Environ 870:161927

Cao Y, Yang S, Wang J, Kong W, Guo B, Xi Y, Zhang A, Yue B (2023b) Metabolomic exploration of the physiological regulatory mechanism of the growth and metabolism characteristics of *Chlorella vulgaris* under photoautotrophic, mixotrophic, and heterotrophic cultivation conditions. Biomass Bioenergy 173:106775

Catalina ASR, Kwaijtaal J, Peinado OC, Van Den Berg C, Wijffels RH, Eppink MH (2020) Multistep fractionation of microalgal biomolecules using selective aqueous two-phase systems. ACS Sustain Chem Eng 8(6):2441–2452

Chandra R, Iqbal HM, Vishal G, Lee HS, Nagra S (2019) Algal biorefinery: a sustainable approach to valorize algal-based biomass towards multiple product recovery. Biores Technol 278:346–359

Chandrasekhar K, Raj T, Ramanaiah SV, Kumar G, Banu JR, Varjani S, Sharma P, Pandey A, Kumar S, Kim SH (2022) Algae biorefinery: a promising approach to promote microalgae industry and waste utilization. J Biotechnol 345:1–16

Cheirsilp B, Maneechote W (2022) Insight on zero waste approach for sustainable microalgae biorefinery: sequential fractionation, conversion and applications for high-to-low value-added products. Biores Technol Rep 101003

Chhandama MVL, Rai PK (2023) Coupling bioremediation and biorefinery prospects of microalgae for circular economy. Biores Technol Rep 101479

Choi HI, Hwang SW, Patel AK, Sim SJ (2021) Microalgal biorefinery: a sustainable technology toward circular bioeconomy and microalgal biomass valorization. In: Biomass, biofuels, biochemicals. Elsevier, pp 323–350

Chong JWR, Tan X, Khoo KS, Ng HS, Jonglertjunya W, Yew GY, Show PL (2022) Microalgae-based bioplastics: future solution towards mitigation of plastic wastes. Environ Res 206:112620

Constantino A, Rodrigues B, Leon R, Barros R, Raposo S (2021) Alternative chemo-enzymatic hydrolysis strategy applied to different microalgae species for bioethanol production. Algal Res 56:102329

Danilović BR, Đorđević NG, Karabegović IT, Troter DZ, Savić DS, Veljković VB (2021) Enhancing lipid extraction from green microalgae *Chlorella* sp. using a deep eutectic solvent pretreatment. Chem Ind Chem Eng Q 27(4):313–317

Depra MC, Severo IA, dos Santos AM, Zepka LQ, Jacob-Lopes E (2020) Environmental impacts on commercial microalgae-based products: sustainability metrics and indicators. Algal Res 51:102056

Depra MC, da Motta Pacheco LFC, Rossi R, Nunes M, Goldemberg S, Suzuki S, Menezes M, Jacob-Lopes E (2022) Hybrid photobioreactors: the success-to-failure experiences on pilot scale. In: 3rd generation biofuels. Woodhead Publishing, pp 1019–1035

Depra MC, Zepka LQ, Jacob-Lopes E (2023) Microalgae-based ingredients and products: toward the global sustainable development goals. In: Handbook of food and feed from microalgae. Academic Press, pp 603–610

Derossi A, Francavilla M, Monteleone M, Caporizzi R, Severini C (2021) From biorefinery of microalgal biomass to vacuum impregnation of fruit. A multidisciplinary strategy to develop innovative food with increased nutritional properties. Innov Food Sci Emerg Technol 70:102677

Dias RR, Deprá MC, Zepka LQ, Jacob-Lopes E (2022) Roadmap to net-zero carbon emissions in commercial microalgae-based products: environmental sustainability and carbon offset costs. J Appl Phycol 34(3):1255–1268

Dias RR, Schneider AT, Fagundes MB (2023) Challenges and opportunities for microalgae biotechnology development. Microalgae Based Sys Process Integr Process Intensification Approaches 41

Díaz V, Leyva-Díaz JC, Almécija MC, Poyatos JM, del Mar Muñío M, Martín-Pascual J (2022) Microalgae bioreactor for nutrient removal and resource recovery from wastewater in the paradigm of circular economy. Biores Technol 127968

Duerschlag J, Mohr W, Ferdelman TG, LaRoche J, Desai D, Croot PL, Voß D, Zielinski O, Lavik G, Littmann S, Tschitschko B, Bartlau N, Osterholz H, Dittmar T, Kuypers MM (2022) Niche partitioning by photosynthetic plankton as a driver of CO_2-fixation across the oligotrophic South Pacific Subtropical Ocean. ISME J 16(2):465–476

Elkelawy M, Bastawissi HAE, El Shenawy EA, Taha M, Panchal H, Sadasivuni KK (2021) Study of performance, combustion, and emissions parameters of DI-diesel engine fueled with algae biodiesel/diesel/n-pentane blends. Energy Convers Manage X 10:100058

Elleuch J, Amor FB, Chaaben Z, Frikha F, Michaud P, Fendri I, Abdelkafi S (2021) Zinc biosorption by *Dunaliella* sp. AL-1: mechanism and effects on cell metabolism. Sci Total Environ 773:145024

Fal S, Smouni A, El Arroussi H (2023) Integrated microalgae-based biorefinery for wastewater treatment, industrial CO_2 sequestration and microalgal biomass valorization: a circular bioeconomy approach. Environ Adv 12:100365

Fan C, Liu Y, Shan Y, Cao X (2022) A priori design of new natural deep eutectic solvent for lutein recovery from microalgae. Food Chem 376:131930

Fathima J, Chatterjee P (2022) A techno-economic assessment of nutrient recovery from wastewater using microalgae: scenario in India collected from published literature. Water Sci Technol 86(6):1325–1341

Fu W, Nelson DR, Mystikou A, Daakour S, Salehi-Ashtiani K (2019) Advances in microalgal research and engineering development. Curr Opin Biotechnol 59:157–164

Goswami RK, Agrawal K, Upadhyaya HM, Gupta VK, Verma P (2022) Microalgae conversion to alternative energy, operating environment and economic footprint: an influential approach towards energy conversion, and management. Energy Convers Manage 269:116118

Huang R, He Y, Yao X, Yu Y, Song W, Yang W, Cheng J (2022) Disintegration of wet microalgae biomass with deep-eutectic-solvent-assisted hydrothermal treatment for sustainable lipid extraction. Green Chem 24(4):1615 1626

Huo S, Basheer S, Liu F, Elshobary M, Zhang C, Qian J, Xu L, Arslan M, Cui F, Zan X, Zhu F, Zou B, Ding Q, Ma H (2020) Bacterial intervention on the growth, nutrient removal and lipid production of filamentous oleaginous microalgae *Tribonema* sp. Algal Res 52:102088

Jacob-Lopes E, Maroneze MM, Deprá MC, Sartori RB, Dias RR, Zepka LQ (2019) Bioactive food compounds from microalgae: an innovative framework on industrial biorefineries. Curr Opin Food Sci 25:1–7

Jacob-Lopes E, Santos AB, Severo IA, Deprá MC, Maroneze MM, Zepka LQ (2020) Dual production of bioenergy in heterotrophic cultures of cyanobacteria: process performance, carbon balance, biofuel quality and sustainability metrics. Biomass Bioenergy 142:105756

Jothibasu K, Muniraj I, Jayakumar T, Ray B, Dhar DW, Karthikeyan S, Rakesh S (2022) Impact of microalgal cell wall biology on downstream processing and nutrient removal for fuels and value-added products. Biochem Eng J 187:108642

Kalwani M, Devi A, Patil K, Kumari A, Dalvi V, Malik A, Tyagi A, Shukla P, Pabbi S (2022) Microalgae-mediated wastewater treatment and enrichment of wastewater-cultivated biomass for biofuel production. In: Expanding horizon of cyanobacterial biology. Academic Press, pp 259–281

Kandasamy S, Zhang B, He Z, Chen H, Feng H, Wang Q, Bhuvanendran N, Esakkimuthu S, Ashokkumar V, Krishnamoorthi M (2019) Hydrothermal liquefaction of microalgae using Fe_3O_4 nanostructures as efficient catalyst for the production of bio-oil: optimization of reaction parameters by response surface methodology. Biomass Bioenergy 131:105417

Karan H, Roles J, Hankamer B, Ross IL (2023) Targeting greens and yellows: a solar biorefinery analysis for the microalgae-based co-production of pigments, proteins, and fuel. Algal Res 74:103187

Katiyar R, Banerjee S, Arora A (2021) Recent advances in the integrated biorefinery concept for the valorization of algal biomass through sustainable routes. Biofuels Bioprod Biorefin 15(3):879–898

Kaushik G, Raza K (2019) Potential of novel *Dunaliella salina* from sambhar salt lake, India, for bioremediation of hexavalent chromium from aqueous effluents: an optimized green approach. Ecotoxicol Environ Saf 180:430–438

Khoo KS, Ahmad I, Chew KW, Iwamoto K, Bhatnagar A, Show PL (2023) Enhanced microalgal lipid production for biofuel using different strategies including genetic modification of microalgae: a review. Prog Energy Combust Sci 96:101071

Koutra E, Ali SS, Sakarika M, Kornaros M (2023) Biorefinery and bioremediation potential of microalgae. In: Valorization of microalgal biomass and wastewater treatment. Elsevier, pp 197–217

Kumar R, Hegde AS, Sharma K, Parmar P, Srivatsan V (2022) Microalgae as a sustainable source of edible proteins and bioactive peptides—current trends and future prospects. Food Res Int 111338

Kumaran M, Palanisamy KM, Bhuyar P, Maniam GP, Rahim MHA, Govindan N (2023) Agriculture of microalgae *Chlorella vulgaris* for polyunsaturated fatty acids (PUFAs) production employing palm oil mill effluents (POME) for future food, wastewater, and energy nexus. Energy Nexus 9:100169

Li G, Lu Z, Zhang J, Li H, Zhou Y, Zayan AMI, Huang Z (2020) Life cycle assessment of biofuel production from microalgae cultivated in anaerobic digested wastewater. Int J Agric Biol Eng 13(1):241–246

Lim JY, Teng SY, How BS, Nam K, Heo S, Máša V, Stehlík P, Yoo CK (2022) From microalgae to bioenergy: identifying optimally integrated biorefinery pathways and harvest scheduling under uncertainties in predicted climate. Renew Sustain Energy Rev 168:112865

Lu Q, Xiao Y (2022) From manure to high-value fertilizer: the employment of microalgae as a nutrient carrier for sustainable agriculture. Algal Res 67:102855

Mahmood WMAW, Lorwirachsutee A, Theodoropoulos C, Gonzalez-Miquel M (2019) Polyol-based deep eutectic solvents for extraction of natural polyphenolic antioxidants from *Chlorella vulgaris*. ACS Sustain Chem Eng 7(5):5018–5026

Malode SJ, Prabhu KK, Mascarenhas RJ, Shetti NP, Aminabhavi TM (2021) Recent advances and viability in biofuel production. Energy Convers Manage X 10:100070

Manhaeghe D, Blomme T, Van Hulle SWH, Rousseau DPL (2020) Experimental assessment and mathematical modelling of the growth of *Chlorella vulgaris* under photoautotrophic, heterotrophic and mixotrophic conditions. Water Res 184:116152

Monlau F, Suarez-Alvarez S, Lallement A, Vaca-Medina G, Giacinti G, Munarriz M, Urreta I, Raynaud C, Castañón S (2021) A cascade biorefinery for the valorization of microalgal biomass: biodiesel, biogas, fertilizers and high valuable compounds. Algal Res 59:102433

Morales-Sánchez D, Schulze PS, Kiron V, Wijffels RH (2020) Production of carbohydrates, lipids, and polyunsaturated fatty acids (PUFA) by the polar marine microalga Chlamydomonas malina RCC2488. Algal Res 50:102016

Mustafa S, Bhatti HN, Maqbool M, Iqbal M (2021) Microalgae biosorption, bioaccumulation and biodegradation efficiency for the remediation of wastewater and carbon dioxide mitigation: prospects, challenges and opportunities. J Water Process Eng 41:102009

Nascimento TC, Pinheiro PN, Fernandes AS, Murador DC, Neves BV, de Menezes CR, Rosso VV, Jacob-Lopes E, Zepka LQ (2021) Bioaccessibility and intestinal uptake of carotenoids from microalgae *Scenedesmus obliquus*. LWT 140:110780

Nath PC, Bhunia B, Bandyopadhyay TK (2023) Carbon dioxide capture and its enhanced utilization using microalgae. In: Green sustainable process for chemical and environmental engineering and science. Elsevier, pp 531–546

Nazloo EK, Moheimani NR, Ennaceri H (2022) Biodiesel production from wet microalgae: progress and challenges. Algal Res 102902

Nematian T, Shakeri A, Salehi Z, Saboury AA (2020) Lipase immobilized on functionalized superparamagnetic few-layer graphene oxide as an efficient nanobiocatalyst for biodiesel production from *Chlorella vulgaris* bio-oil. Biotechnol Biofuels 13:1–15

Ngatcha ADP, Muhammad G, Lv Y, Xiong W, Zhao A, Xu J, Alam MA (2022) Microalgae biomass pre-treatment with deep eutectic solvent to optimize lipid isolation in biodiesel production. Biomass Convers Bioref 12(Suppl 1):133–143

Okeke ES, Ejeromedoghene O, Okoye CO, Ezeorba TPC, Nyaruaba R, Ikechukwu CK, Oladipo A, Orege JI (2022) Microalgae biorefinery: an integrated route for the sustainable production of high-value-added products. Energy Convers Manage X 100323

Oliveira CYB, Jacob A, Nader C, Oliveira CDL, Matos ÂP, Araújo ES, Shabnam N, Ashok B, Gálvez AO (2022) An overview on microalgae as renewable resources for meeting sustainable development goals. J Environ Manage 320:115897

Patel A, Krikigianni E, Rova U, Christakopoulos P, Matsakas L (2022) Bioprocessing of volatile fatty acids by oleaginous freshwater microalgae and their potential for biofuel and protein production. Chem Eng J 438:135529

Pereira JF, Papaiconomou N, Ventura SP (2021) Envisioning the future of industrial bioprocesses through biorefinery. Front Bioeng Biotechnol 9:136

Qi S, Wang Z, Hu Y, Lei J, Zhan X, Stengel DB (2023) Selective enrichment of auto-floating microalgae for wastewater bioremediation and biofuel/bioproduct production. Algal Res 69:102911

Resdi R, Lim JS, Idris A (2021) Batch kinetics of nutrients removal for palm oil mill effluent and recovery of lipid by *Nannochloropsis* sp. J Water Process Eng 40:101767

Roles J, Yarnold J, Hussey K, Hankamer B (2021) Techno-economic evaluation of microalgae high-density liquid fuel production at 12 international locations. Biotechnol Biofuels 14(1):1–19

Rosa MDH, Alves CJ, dos Santos FN, de Souza AO, Zavareze EDR, Pinto E, Noseda MD, Ramos D, de Pereira CMP (2023) Macroalgae and microalgae biomass as feedstock for products applied to bioenergy and food industry: a brief review. Energies 16(4):1820

Santos VZD, Vieira KR, Nass PP, Zepka LQ, Jacob-Lopes E (2021) Application of microalgae consortia/cocultures in wastewater treatment. In: Recent advances in microbial degradation, pp 131–154

Saral JS, Satheesh AR, Ranganathan P (2022) Economic and environmental analysis of algal biorefinery for the production of renewable fuels and co-product. Energy Convers Manage X 14:100189

Sarfraz R, Taneez M, Sardar S, Danish L, Hameed A (2021) Evaluation of *Desmodesmus subspicatus* for the treatment of wastewater. Int J Environ Anal Chem 1–12

Sarwer A, Hamed SM, Osman AI, Jamil F, Al-Muhtaseb AAH, Alhajeri NS, Rooney DW (2022) Algal biomass valorization for biofuel production and carbon sequestration: a review. Environ Chem Lett 20(5):2797–2851

Satheesh S, Pugazhendi A, Al-Mur BA, Balasubramani R (2023) Biohydrogen production coupled with wastewater treatment using selected microalgae. Chemosphere 334:138932

Sengupta SL, Chaudhuri RG, Dutta S (2023) A critical review on phycoremediation of pollutants from wastewater—a novel algae based secondary treatment with the opportunities of production of value-added products
Sharma P, Gujjala LKS, Varjani S, Kumar S (2022) Emerging microalgae-based technologies in biorefinery and risk assessment issues: bioeconomy for sustainable development. Sci Total Environ 813:152417
Sherwood J (2020) The significance of biomass in a circular economy. Biores Technol 300:122755
Singh A, Jasso RMR, Gonzalez-Gloria KD, Rosales M, Cerda RB, Aguilar CN, Singhania RR, Ruiz HA (2019) The enzyme biorefinery platform for advanced biofuels production. Biores Technol Rep 7:100257
Slegers PM, Olivieri G, Breitmayer E, Sijtsma L, Eppink MH, Wijffels RH, Reith JH (2020) Design of value chains for microalgal biorefinery at industrial scale: process integration and techno-economic analysis. Front Bioeng Biotechnol 1048
Stegmann P, Londo M, Junginger M (2020) The circular bioeconomy: its elements and role in European bioeconomy clusters. Resour Conserv Recycl X 6:100029
Taelman SE, De Meester S, Van Dijk W, Da Silva V, Dewulf J (2015) Environmental sustainability analysis of a protein-rich livestock feed ingredient in The Netherlands: microalgae production versus soybean import. Resour Conserv Recycl 101:61–72
Tan JS, Lee SY, Chew KW, Lam MK, Lim JW, Ho SH, Show PL (2020) A review on microalgae cultivation and harvesting, and their biomass extraction processing using ionic liquids. Bioengineered 11(1):116–129
Thangam KR, Santhiya A, Sri SA, MubarakAli D, Karthikumar S, Kumar RS, Thajudin N, Soosai MR, Varalakshmi P, Moorthy IG, Pugazhendhi A (2021) Bio-refinery approaches based concomitant microalgal biofuel production and wastewater treatment. Sci Total Environ 785:147267
Trivedi J, Agrawal D, Atray N, Ray A (2022) Enhanced lipid production in *Scenedesmus obliquus* via nitrogen starvation in a two-stage cultivation process and evaluation for biodiesel production. Fuel 316:123418
Tsolakis N, Goldsmith AT, Aivazidou E, Kumar M (2023) Microalgae-based circular supply chain configurations using Industry 4.0 technologies for pharmaceuticals. J Clean Prod 395:136397
Ubando AT, Felix CB, Chen WH (2020) Biorefineries in circular bioeconomy: a comprehensive review. Biores Technol 299:122585
Ubando AT, Ng EAS, Chen WH, Culaba AB, Kwon EE (2022) Life cycle assessment of microalgal biorefinery: a state-of-the-art review. Biores Technol 360:127615
Ummalyma SB, Sahoo D, Pandey A (2020) Microalgal biorefineries for industrial products. In: Microalgae cultivation for biofuels production. Academic Press, pp 187–195
Ummalyma SB, Chiang A, Herojit N, Arumugam M (2023) Sustainable microalgal cultivation in poultry slaughterhouse wastewater for biorefinery products and pollutant removal. Biores Technol 374:128790
Verma R, Kumari KK, Srivastava A, Kumar A (2020) Photoautotrophic, mixotrophic, and heterotrophic culture media optimization for enhanced microalgae production. J Environ Chem Eng 8(5):104149
Vermuë MH, Eppink MHM, Wijffels RH, Van Den Berg C (2018) Multi-product microalgae biorefineries: from concept towards reality. Trends Biotechnol 36(2):216–227
Wiatrowski M, Klein BC, Davis RW, Quiroz-Arita C, Tan EC, Hunt RW, Davis RE (2022) Techno-economic assessment for the production of algal fuels and value-added products: opportunities for high-protein microalgae conversion. Biotechnol Biofuels Bioprod 15(1):1–14
Wils L, Leman-Loubière C, Bellin N, Clément-Larosière B, Pinault M, Chevalier S, Gueiffier-Enguehard C, Bodet C, Boudesocque-Delaye L (2021) Natural deep eutectic solvent formulations for Spirulina: preparation, intensification, and skin impact. Algal Res 56:102317
Yu Q, Qin L, Liu Y, Sun Y, Xu H, Wang Z, Yuan Z (2019) In situ deep eutectic solvent pretreatment to improve lignin removal from garden wastes and enhance production of bio-methane and microbial lipids. Biores Technol 271:210–217

Yu L, Li T, Ma J, Zhao Q, Wensel P, Lian J, Chen S (2022) A kinetic model of heterotrophic and mixotrophic cultivation of the potential biofuel organism microalgae *Chlorella sorokiniana.* Algal Res 64:102701

Zhang TY, Wu YH, Wang JH, Wang XX, Deantes-Espinosa VM, Dao GH, Hua-Guo D, Xin T, Hu HY (2019) Heterotrophic cultivation of microalgae in straw lignocellulose hydrolysate for production of high-value biomass rich in polyunsaturated fatty acids (PUFA). Chem Eng J 367:37–44

Zhang C, Ming J, Sun X, Zhu Y, An G, Chen G, Yang Y (2023) Development of a green and efficient photocatalytic mesh microalgae biorefinery (PMMB) system for sustainable biomass conversion under real solar light. Chem Eng J 143260

Chapter 4
Appraisal and Identification of Algal Bloom Region, Prevention and Management Approaches

Anuj Sharma, Praveen Sharma, and Sharma Mona

Abstract Algae is a photosynthetic ubiquitous organism that can be grown in numerous forms. Being photosynthetic in nature algae serve as a source of oxygen, and food and also act as indicators of pollution. So, algae seem to be quite beneficial for the aquatic ecosystem but—excess of everything is bad. Exaggerated growth of algae leads to the formation of blooms known as algal blooms. The high inflow of nutrients in a water body is the primary cause of algal blooms. Algal blooms lead to various socio-economic impacts such as toxicity to humans and aquatic biodiversity and thus these blooms are also termed as harmful algal blooms (HABs). For proper management of algal bloom regions, the first step is to identify the cause and then methods like aeration, biological augmentation, and proper disposal of organic wastes that can be used to prevent algal bloom formation. The scientific community should explore more about finding a sustainable way to prevent algal bloom. In the present book chapter, the basic concept of algal blooms is discussed along with methods for appraisal and identification of algal bloom regions. Moreover, the prevention and management strategies for algal blooms are also highlighted. This chapter also emphasizes various impacts of algal bloom on water, aquatic diversity, and humans. The algal bloom is also discussed through the lens of sustainability.

Keywords Harmful algal blooms (HABs) · Wastewater · Algae · Impacts · Management · Prevention

A. Sharma · P. Sharma
Department of Environmental Science and Engineering, Guru Jambheshwar University of Science and Technology, Hisar, Haryana 125001, India
e-mail: anujvashist2511@gmail.com

P. Sharma
e-mail: parveensharma@gjust.org

S. Mona (✉)
Department of Environmental Studies, School of Interdisciplinary and Applied Sciences, Central University of Haryana, Mahendergarh, Haryana 123031, India
e-mail: mona@cuh.ac.in

T. A. Kurniawan and A. Anouzla (eds.), *Algae as a Natural Solution for Challenges in Water-Food-Energy Nexus*, Environmental Science and Engineering,
https://doi.org/10.1007/978-981-97-2371-3_4

4.1 Introduction

Algae are a group of organisms that are majorly aquatic in nature and synthesize their own food using photosynthesis. Being photosynthetic in nature algae is a good source of oxygen and helps in the sequestration of carbon from the atmosphere thus helping in controlling climate change. Algae is an important source of crude oil, along with these algae has numerous applications in the food industry, nutraceuticals, cosmetics, energy production, animal husbandry, bioremediation, and pharmaceutical industry. Because of these vast applications of algae in different fields, algae are believed to be an organism, that not only helps the natural ecosystem but also helpful for mankind in terms of economy, medicines, and food. So, for the betterment of humans and the natural ecosystem growth of algae is important. But exaggerated growth of some species of algae like diatoms leads to the formation of blooms of algae that are term as algal blooms (David and Kidwell 2015; Sanseverino et al. 2016; Sharma et al. 2023).

The algal bloom is an exaggerated growth of algae in marine or freshwater, that can be easily recognized by its color. Microalgae and macroalgae both can lead to the formation of algal bloom. Cyanobacteria, diatoms, and dinoflagellates are some of the major sets of organisms that lead to algal bloom. Generally, dinoflagellates and diatoms are responsible for the formation of algal bloom in marine water and cyanobacteria in freshwater. As algae is a photosynthetic organism its growth revolves around exposure to the sun, the concentration of carbon dioxides in the atmosphere, and accessibility of nutrients such as nitrogen and phosphorous (Ballah et al. 2019; Mittal et al. 2023). Along with these, some factors like the concentration of salt in water, temperature, and pH of water, stability of water, and human-induced factors like excessive discharge of nutrients in the water body. Excessive load of nutrients like nitrogen and phosphorous in the water body leads to the condition of eutrophication. Eutrophication favours the formation of algal blooms. Eutrophication leads to the formation of visible algal blooms and visibility of the water body is also reduced (Ballah et al. 2019; Bhukal et al. 2022; Kumar et al. 2023).

Algal blooms produce various toxins that are harmful to aquatic organisms along with humans. Thus, algal blooms are also known by the name harmful algal blooms (HABs). Toxins produced by algal blooms result in a foul smell and odour of water. Water affected by algal bloom is unfit for human consumption because it can lead to serious health issues in the human body. Algal blooms because of their toxin-producing ability become the cause of death of aquatic organisms. Along with health, algal blooms also have some socio-economic impacts on humans (David and Kidwell 2015; Ballah et al. 2019; Sharma et al. 2023). Because of algal blooms fishes die and their population along with other organisms drastically decreased thus detrimentally affecting the fisheries industry. Tourism is also reduced because of algal bloom as it leads to foul smell and odour of water. As day by day, we are heading toward industrial growth of our countries, which leads to the enhancement of nutrients in the water body because of discharge from various industries. Thus, cases of HABs are increases as development proceeds. There is a need to explore more identification

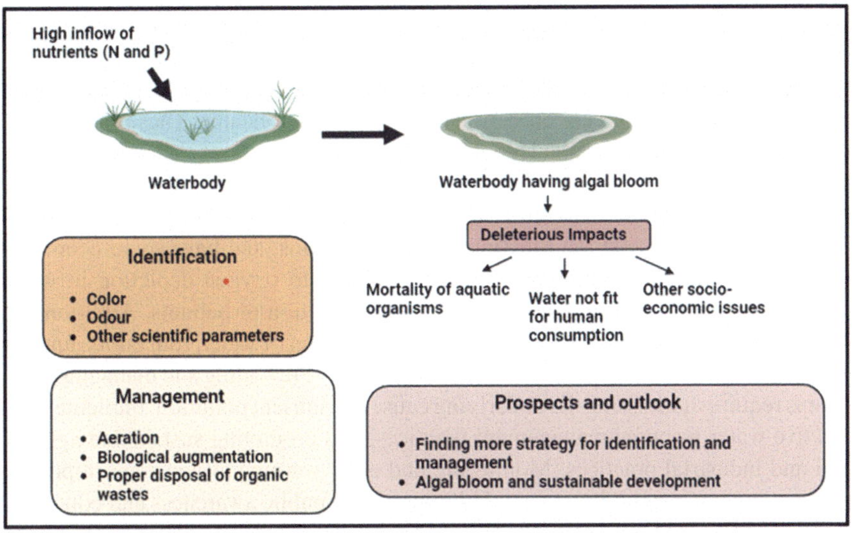

Fig. 4.1 Basic concept of harmful algal blooms (HABs)

and management techniques by the scientific community to tackle the menace of HABs (Sanseverino et al. 2016; Guo et al. 2021; Li et al. 2022a).

In the present chapter, the basic concept of algal blooms is discussed along with methods for appraisal and identification of algal bloom regions. Moreover, the methods to prevent and manage algal blooms are also discussed. The Socio-economic impacts of algal bloom on humans and aquatic body also highlighted in this chapter. Algal bloom is also discussed by interlinking it with sustainable development. Figure 4.1 represents the basic concept of HABs.

4.2 Algae and Algal Bloom

Algae are a diverse group of photosynthetic organisms that are found in various aquatic environments, including oceans, lakes, rivers, and ponds. They play a crucial role in ecosystems as primary producers, converting sunlight, carbon dioxide, and nutrients into organic matter through photosynthesis. Algae encompass a wide range of species, including microscopic phytoplankton and larger macroalgae (seaweeds). Under certain conditions, algae can undergo rapid and excessive growth, leading to the formation of algal blooms (Mona et al. 2011, 2013, 2020). Algal blooms occur when nutrient levels, particularly phosphorus and nitrogen, are elevated, creating an environment that favors the growth of algae. Factors that contribute to algal bloom formation include nutrient pollution from agricultural runoff, urban stormwater runoff, wastewater discharges, and industrial activities. Algal blooms can be composed of different types of algae, including cyanobacteria (blue-green algae), diatoms, dinoflagellates, and green algae. The specific composition of a bloom depends on factors such as water temperature, nutrient availability, light levels, and the surrounding ecosystem. While not all algal blooms are harmful, certain blooms can have detrimental effects on ecosystems and human health. These blooms are often referred to as harmful algal blooms (HABs) (Sanseverino et al. 2016; Kaur et al. 2019a, b, 2020). Harmful algal blooms can produce toxins that are harmful to aquatic organisms, including fish, shellfish, marine mammals, and even humans. The toxins can cause illness, impact the functioning of ecosystems, and have socio-economic consequences. Harmful algal blooms can also lead to oxygen depletion in water bodies, which can result in fish kills and the degradation of habitats. Additionally, blooms can create aesthetic issues, such as discoloration of water, foul odors, and the formation of thick mats or scums on the water surface. Preventing and managing algal blooms requires addressing the underlying causes of nutrient pollution, implementing effective water resource management practices, and promoting sustainable agricultural and industrial practices. Monitoring and early warning systems are important for detecting and responding to algal blooms, while public awareness and education play a crucial role in promoting responsible practices and minimizing the risks associated with algal blooms (David and Kidwell 2015; Ballah et al. 2019; Guo et al. 2021; Mittal et al. 2023).

4.3 Appraisal and Identification of Algal Bloom Region

Single finger test: In this test instead of hand, a stick can also be used. Simply after wearing gloves lift a part of the bloom in hand and as the water is drained from the hand and only algal biomass is left. Then closely examine algal biomass if it is long stringy masses then this algal bloom is of a filamentous algae and if the biomass is in the form of small bits, then this algal bloom must be because of planktonic algae.

To appraise and identify algal bloom regions, several methods and tools can be utilized. Algal blooms are the rapid growth of algae in aquatic ecosystems, often causing harmful effects on the environment and other organisms. Here are some approaches commonly used for appraisal and identification: (i) Satellite Imagery: Remote sensing using satellites can provide valuable information about algal blooms on a large scale. Satellites equipped with sensors capable of detecting specific wavelengths of light can identify and monitor the presence of algal blooms. Chlorophyll-a, a pigment present in algae, can be measured to estimate the biomass and distribution of algal blooms. Various satellite missions, such as MODIS and SeaWiFS, have been used for this purpose (Li et al. 2022a; Binding et al. 2023; Zhu et al. 2023; Gernez et al. 2023). (ii) Water Sampling and Analysis: Conducting regular water sampling in affected areas allows for direct measurement and analysis of algal bloom parameters. These parameters may include chlorophyll-a concentration, nutrient levels (e.g., nitrogen and phosphorus), dissolved oxygen, and pH. By analysing the water samples, the presence of algal blooms and their characteristics can be determined (Ballah et al.

2019; Liu et al. 2022a; Summers and Ryder 2023; Sharma et al. 2023). (iii) Environmental Monitoring Stations: Deploying monitoring stations in areas prone to algal blooms provides continuous data collection. These stations can measure water quality parameters, such as temperature, dissolved oxygen, turbidity, and nutrient concentrations. Real-time monitoring can help identify and track the development of algal blooms over time (Ballah et al. 2019; Guo et al. 2021; Kumar et al. 2023). (iv) Remote Sensing Aircraft: Like satellite imagery, remote sensing aircraft equipped with specialized sensors can capture high-resolution images of algal bloom regions. This method allows for more detailed analysis and can provide additional information beyond what satellites can offer (Li et al. 2022a; Binding et al. 2023; Zhu et al. 2023; Gernez et al. 2023). (v) Citizen Science Initiatives: Engaging citizen scientists in monitoring and reporting algal blooms can help increase coverage and data collection. Mobile applications or websites can be developed to allow individuals to report and document algal bloom occurrences, providing valuable information for identification and appraisal. It is important to note that algal blooms can vary in their causes, severity, and impact. Some blooms may be harmful and toxic, while others may be non-toxic. The specific identification and appraisal methods may differ depending on the type and characteristics of the algal bloom in question. Consulting with local environmental agencies, research institutions, or experts in the field can provide further guidance on specific approaches for your region of interest (David and Kidwell 2015; Coyne et al. 2022; Liu et al. 2022a; Summers and Ryder 2023) (Table 4.1).

4.4 Some Examples of Algal Bloom Incidents Around the World

Algal bloom incidents have occurred in various regions around the world, impacting both freshwater and marine ecosystems. Here are a few notable examples of algal bloom incidents: (i) Lake Erie, United States, and Canada: Lake Erie has experienced significant algal bloom incidents, primarily caused by excessive nutrient pollution, particularly phosphorus from agricultural runoff. These blooms are predominantly composed of blue-green algae (cyanobacteria) and have resulted in the release of toxins, leading to water quality issues and harm to aquatic life. The largest recorded bloom in Lake Erie occurred in 2011, covering approximately 5000 km^2 (Martin et al. 2021). (ii) Red Tide in Florida, United States: The Gulf of Mexico and the Atlantic coast of Florida have witnessed recurring occurrences of harmful algal blooms known as red tide. Red tides are caused by the rapid growth of a marine dinoflagellate called Karenia brevis. These blooms can produce toxins that harm marine organisms and cause respiratory issues in humans. They have led to fish kills, marine mammal deaths, and economic losses in coastal communities (Liu et al. 2022b; Kurtz et al. 2023). (iii) Lake Victoria, East Africa: Lake Victoria, the largest freshwater lake in Africa, has experienced significant algal blooms, particularly blue-green algae. These

Table 4.1 Characteristic features of algae and HABs caused by them

Algae	Features of algal blooms	References
Cladophora	• Color of bloom ranges from green to yellow-brown to orange • Bloom found in the form of filamentous mats • Bloom has a specific odour	Sahoo and Baweja (2015)
Euglena	• Green to red color • Bloom is found in areas having high loads of organic wastes or organic matter • Bloom has a very dense population of algae	Sahoo and Baweja (2015)
Chlamydomonas	• Bloom is of red color • Have a fishy smell when algae are present in a large number • Can be found in a variety of habitats including alpine snow	Sahoo and Baweja (2015)
Volvox	• Usually yellowish-brown to light green in color • Colonies are surrounded by a gelatinous matrix • Individuals join together to form complex colonies, about 1 mm in diameter	Sahoo and Baweja (2015)
Spirogyra	• Vibrant green or yellowish-brown in color • A dense form of bloom can deplete oxygen, and cause the mortality of fishes • Provides breeding sites for mosquitos	Sahoo and Baweja (2015)

blooms are linked to nutrient pollution from agricultural runoff, sewage discharge, and invasive species. Algal blooms have led to the decline of native fish populations, reduced water quality, and negative impacts on local communities dependent on the lake for their livelihoods (Simiyu et al. 2022; Omara et al. 2023). (iv) Indian Scenario: Several incidents of algal blooms were reported from the Malabar coast, off Goa, off Gujarat, and off Kochi (D'Silva et al. 2012; Oyeku and Mandal 2021). These examples highlight the global nature of algal bloom incidents and the need for effective monitoring, mitigation, and management strategies to address the underlying causes and minimize their ecological and socio-economic impacts. Figure 4.2 represents a waterbody affected by algal blooms.

4.5 Socio-economic Impacts of the Algal Boom

Algal blooms can have significant socio-economic impacts on affected regions. Here are some of the key socio-economic consequences associated with algal blooms: (i) Economic Losses in Fisheries, (ii) Impacts on Tourism and Recreation, (iii) Water Supply and Treatment Costs, (iv) Human Health Concerns, (v) Impact on Agricultural Practices, (vi) Damage to Ecosystem Services (Rolton et al. 2022; Kourantidou et al.

Fig. 4.2 Waterbody affected by HABs in the rural area (28.990410, 76.983740) of Haryana, India

2022). Addressing the socio-economic impacts of algal blooms requires a combination of preventive measures, such as reducing nutrient pollution, implementing effective water management practices, and promoting sustainable agricultural and industrial practices (Sanseverino et al. 2016; Weir et al. 2022). Early warning systems, monitoring programs, and public awareness campaigns can also help mitigate the impacts and promote adaptive strategies for affected communities.

4.5.1 Impacts on Humans and Aquatic Biodiversity

Algal blooms, particularly harmful algal blooms (HABs), can have various impacts on human health. These impacts can occur through direct contact with contaminated water, inhalation of aerosolized toxins, or consumption of seafood tainted with algal toxins. Here are some key impacts of algal blooms on human health: (i) Gastrointestinal Issues, (ii) Skin Irritation and Rashes, (iii) Respiratory Problems, (iv) Neurological Effects, (v) Allergies and Sensitization. It is important to note that not all algal blooms produce toxins, and not all algal species are harmful to humans. However, it is difficult to visually determine the toxicity of an algal bloom, so it is advisable to exercise caution and follow any local advisories or warnings regarding water use and consumption of seafood during algal bloom events. If you suspect exposure to algal toxins or experience symptoms after contact with algal bloom-affected water or seafood, it is recommended to seek medical attention and inform healthcare professionals about the potential algal bloom exposure (Young et al. 2020; Sultana et al. 2022; Lad et al. 2022). Algal blooms can have significant impacts on aquatic biodiversity, affecting various organisms and ecosystems. Here are some key impacts of algal blooms on aquatic biodiversity: (i) Oxygen Depletion, (ii) Habitat Modification, (iii) Disruption of Food Webs, (iv) Harmful Toxins, (v) Biodiversity Loss, (vi) Coral Reef Degradation. Efforts to mitigate the impacts of algal blooms on aquatic biodiversity include nutrient management strategies to reduce nutrient pollution, improving water quality through proper land use practices, and implementing ecosystem-based approaches for the conservation and restoration of affected ecosystems (Amorim and do Nascimento Moura 2021; Ahmed et al. 2022; Xue et al. 2023).

4.6 Prevention and Management Approaches

Preventing and managing algal blooms require a combination of strategies that target the underlying causes of bloom formation, such as excessive nutrient pollution and other environmental factors. Here are some key prevention and management approaches for algal blooms: (i) Nutrient Management, (ii) Watershed Management, (iii) Advanced Wastewater Treatment, (iv) Reducing Stormwater Runoff, (v) Monitoring and Early Warning Systems, (vi) Integrated Pest Management (IPM), (vii) Public Awareness and Education, (viii) Research and Innovation (Balaji-Prasath et al. 2022; ElFar et al. 2022; Tian et al. 2022; Summers and Ryder 2023). Prevention and management strategies are crucial to mitigate their adverse effects. Prevention measures focus on reducing nutrient inputs into water bodies. This involves controlling agricultural runoff, industrial discharges, and sewage treatment to limit the amount of nitrogen and phosphorus, the primary nutrients responsible for algal growth, entering the water. Additionally, promoting sustainable land use practices and watershed management can help prevent nutrient pollution. In terms of management, early detection, and monitoring systems are vital to identify and predict algal blooms. Implementing a timely response can prevent their escalation. Physical measures like aeration and water circulation can disrupt bloom formation, while the use of chemical agents can mitigate their impact in localized areas. Biological methods such as introducing natural competitors or grazers of algae can help maintain a balance in the ecosystem. Integrated approaches that combine various methods often yield better results. Public awareness and education play an essential role in preventing algal blooms. Encouraging responsible behavior, such as avoiding excessive use of fertilizers and proper waste disposal, can contribute to long-term prevention efforts (Bramburger et al. 2022; Li et al. 2022b; Tian et al. 2022; Summers and Ryder 2023; Molares-Ulloa et al. 2023). Overall, effective prevention and management of algal blooms require a collaborative effort from governments, communities, industries, and environmental organizations to safeguard the health and sustainability of aquatic ecosystems. Addressing algal blooms requires a collaborative effort among governments, communities, industries, and scientific communities. Integrated approaches that combine multiple strategies are most effective in preventing and managing algal blooms and safeguarding the health and ecological integrity of aquatic ecosystems.

4.7 Algal Bloom and Sustainability

Algal blooms have significant implications for sustainability, as they can disrupt the balance and functioning of aquatic ecosystems, impact human health, and have socio-economic consequences. Here's a closer look at the relationship between algal blooms and sustainability: (i) Environmental Impact, (ii) Water Resource Management, (iii) Biodiversity Conservation, (iv) Human Health and Well-being, (v) Economic Sustainability, (vi) Climate Change (Anderson et al. 2019; Brown et al. 2020; Gokul et al. 2020; Wang et al. 2021; Trottet et al. 2022). Overall, achieving sustainability requires effective prevention and management of algal blooms. This involves adopting practices that minimize nutrient pollution, promote sustainable water resource management, conserve biodiversity, protect human health, and address the underlying drivers of algal bloom formation. By integrating algal bloom management into broader sustainability efforts, we can work towards resilient and healthy aquatic ecosystems for present and future generations.

4.8 Conclusions and Future Outlooks

In conclusion, the appraisal and identification of algal bloom regions are crucial for effective prevention and management strategies. Algal blooms, characterized by rapid and excessive growth of algae, have significant ecological and economic impacts. By employing advanced remote sensing techniques, such as satellite imagery and hyperspectral analysis, scientists can accurately identify and monitor algal bloom regions. This enables early warning systems and prompt response measures to be implemented. Moving forward, the future outlook for algal bloom prevention and management lies in a multidisciplinary approach. Improved understanding of the underlying causes of algal blooms, including nutrient pollution and climate change, will inform targeted prevention strategies. Innovative technologies, such as autonomous underwater vehicles and drones, can facilitate real-time monitoring and data collection, enabling more precise management interventions. Furthermore, collaboration between researchers, policymakers, and stakeholders is essential. Encouraging sustainable agricultural practices, implementing effective wastewater treatment methods, and developing eco-friendly algaecides are some potential strategies. Embracing natural solutions, like biomanipulation and the use of beneficial microorganisms, holds promise for long-term algal bloom control. In conclusion, through continued research, technological advancements, and collaborative efforts, it is possible to mitigate the negative impacts of algal blooms and safeguard aquatic ecosystems, human health, and economic activities that rely on clean water resources.

References

Ahmed SF, Kumar PS, Kabir M et al (2022) Threats, challenges and sustainable conservation strategies for freshwater biodiversity. Environ Res 214. https://doi.org/10.1016/j.envres.2022.113808

Amorim CA, do Nascimento Moura A (2021) Ecological impacts of freshwater algal blooms on water quality, plankton biodiversity, structure, and ecosystem functioning. Sci Total Environ 758. https://doi.org/10.1016/j.scitotenv.2020.143605

Anderson CR, Berdalet E, Kudela RM et al (2019) Scaling up from regional case studies to a global harmful algal bloom observing system. Front Mar Sci 6. https://doi.org/10.3389/fmars.2019.00250

Balaji-Prasath B, Wang Y, Su YP et al (2022) Methods to control harmful algal blooms: a review. Environ Chem Lett. https://doi.org/10.1007/s10311-022-01457-2

Ballah M, Bhoyroo V, Neetoo H (2019) Assessment of the physico-chemical quality and extent of algal proliferation in water from an impounding reservoir prone to eutrophication. J Ecol Environ 43. https://doi.org/10.1186/s41610-018-0094-z

Bhukal S, Sharma A, Rishi et al (2022) Spirulina based iron oxide nanoparticles for adsorptive removal of crystal violet dye. Top Catal 65:1675–1685. https://doi.org/10.1007/s11244-022-01640-3

Binding CE, Zeng C, Pizzolato L et al (2023) Reporting on the status, trends, and drivers of algal blooms on Lake of the Woods using satellite-derived bloom indices (2002–2021). J Great Lakes Res 49:32–43. https://doi.org/10.1016/j.jglr.2022.12.007

Bramburger AJ, Filstrup CT, Reavie ED et al (2022) Paradox versus paradigm: a disconnect between understanding and management of freshwater cyanobacterial harmful algae blooms. Freshw Biol. https://doi.org/10.1111/fwb.14019

Brown AR, Lilley M, Shutler J et al (2020) Assessing risks and mitigating impacts of harmful algal blooms on mariculture and marine fisheries. Rev Aquac 12:1663–1688. https://doi.org/10.1111/raq.12403

Coyne KJ, Wang Y, Johnson G (2022) Algicidal bacteria: a review of current knowledge and applications to control harmful algal blooms. Front Microbiol 13. https://doi.org/10.3389/fmicb.2022.871177

David MR, Kidwell M (2015) Programmatic environmental assessment for the prevention, control, and mitigation of harmful algal blooms program

D'Silva MS, Anil AC, Naik RK, D'Costa PM (2012) Algal blooms: a perspective from the coasts of India. Nat Hazards 63:1225–1253. https://doi.org/10.1007/s11069-012-0190-9

ElFar OA, Aron NSM, Chew KW, Show PL (2022) Sustainable management of algal blooms in ponds and rivers. In: Biomass, biofuels, biochemicals: circular bioeconomy: technologies for waste remediation. Elsevier, pp 431–444

Gernez P, Zoffoli ML, Lacour T et al (2023) The many shades of red tides: Sentinel-2 optical types of highly-concentrated harmful algal blooms. Remote Sens Environ 287. https://doi.org/10.1016/j.rse.2023.113486

Gokul EA, Raitsos DE, Gittings JA, Hoteit I (2020) Developing an atlas of harmful algal blooms in the Red Sea: linkages to local aquaculture. Remote Sens (Basel) 12:1–14. https://doi.org/10.3390/rs12223695

Guo J, Ma Y, Lee JHW (2021) Real-time automated identification of algal bloom species for fisheries management in subtropical coastal waters. J Hydro-Environ Res 36:1–32. https://doi.org/10.1016/j.jher.2021.03.002

Kaur M, Sharma P, Kumari S (2019a) Equilibrium studies for copper removal from aqueous solution using nanoadsorbent synthesized from rice husk. SN Appl Sci 1. https://doi.org/10.1007/s42452-019-1024-0

Kaur M, Sharma P, Kumari S (2019b) Response surface methodology approach for optimization of Cu^{2+} and Pb^{2+} removal using nanoadsorbent developed from rice husk. Mater Today Commun 21. https://doi.org/10.1016/j.mtcomm.2019.100623

Kaur M, Kumari S, Sharma P (2020) Removal of Pb(II) from aqueous solution using nanoadsorbent of *Oryza sativa* husk: isotherm, kinetic and thermodynamic studies. Biotechnol Rep 25. https://doi.org/10.1016/j.btre.2019.e00410

Kourantidou M, Jin D, Schumacker EJ (2022) Socioeconomic disruptions of harmful algal blooms in indigenous communities: the case of Quinault Indian nation. Harmful Algae 118. https://doi.org/10.1016/j.hal.2022.102316

Kumar S, Saini N, Sujata et al (2023) Algae for wastewater treatment: an integrated approach for circular biorefinery. In: Integrated waste management: the circular economy, pp 185–195

Kurtz BE, Landmeyer JE, Culter JK (2023) Precipitation, submarine groundwater discharge of nitrogen, and red tides along the southwest Florida Gulf coast. Heliyon 9. https://doi.org/10.1016/j.heliyon.2023.e16046

Lad A, Breidenbach JD, Su RC et al (2022) As we drink and breathe: adverse health effects of microcystins and other harmful algal bloom toxins in the liver, gut, lungs and beyond. Life 12. https://doi.org/10.3390/life12030418

Lake Erie algae blooms. Alliance for the Great Lakes. https://greatlakes.org/campaigns/lake-erie-algae-blooms/. Accessed 20 July 2023

Li C, Tao B, Liu Y et al (2022a) Assessment of VIIRS on the identification of harmful algal bloom types in the coasts of the East China Sea. Remote Sens (Basel) 14. https://doi.org/10.3390/rs14092089

Li Y, Huang Y, Ji D et al (2022b) Storm and floods increase the duration and extent of phosphorus limitation on algal blooms in a tributary of the Three Gorges Reservoir, China. J Hydrol (Amst) 607. https://doi.org/10.1016/j.jhydrol.2022.127562

Liu J, Xia C, Xie H et al (2022a) Accurate monitoring of algal blooms in key nearshore zones of lakes and reservoirs using binocular video surveillance system. Water (Switzerland) 14. https://doi.org/10.3390/w14223728

Liu Y, Weisberg RH, Zheng L et al (2022b) Termination of the 2018 Florida red tide event: a tracer model perspective. Estuar Coast Shelf Sci 272. https://doi.org/10.1016/j.ecss.2022.107901

Martin JF, Kalcic MM, Aloysius N et al (2021) Evaluating management options to reduce Lake Erie algal blooms using an ensemble of watershed models. J Environ Manage 280. https://doi.org/10.1016/j.jenvman.2020.111710

Mittal R, Sharma A, Bhardwaj AK et al (2023) Removal of chromium(VI) using spirulina assisted synthesized mesoporous iron oxide nanoparticles. Inorg Chem Commun 154:110881. https://doi.org/10.1016/j.inoche.2023.110881

Molares-Ulloa A, Rivero D, Ruiz JG et al (2023) Hybrid machine learning techniques in the management of harmful algal blooms impact. Comput Electron Agric 211:107988. https://doi.org/10.1016/j.compag.2023.107988

Mona S, Kaushik A, Kaushik CP (2011) Hydrogen production and metal-dye bioremoval by a *Nostoc linckia* strain isolated from textile mill oxidation pond. Bioresour Technol 102:3200–3205. https://doi.org/10.1016/j.biortech.2010.11.005

Mona S, Kaushik A, Kaushik CP (2013) Prolonged hydrogen production by Nostoc in photobioreactor and multi-stage use of the biological waste for column biosorption of some dyes and metals. Biomass Bioenergy 54:27–35. https://doi.org/10.1016/j.biombioe.2013.03.001

Mona S, Kumar SS, Kumar V et al (2020) Green technology for sustainable biohydrogen production (waste to energy): a review. Sci Total Environ 728. https://doi.org/10.1016/j.scitotenv.2020.138481

Omara T, Nagawa CB, Kyarimpa C et al (2023) Lacustrine cyanobacteria, algal blooms and cyanotoxins in East Africa: implications for human and ecological health protection. Phycology 3:147–167. https://doi.org/10.3390/phycology3010010

Oyeku OG, Mandal SK (2021) Historical occurrences of marine microalgal blooms in Indian peninsula: probable causes and implications. Oceanologia 63:51–70. https://doi.org/10.1016/j.oceano.2020.08.008

Rolton A, Rhodes L, Hutson KS et al (2022) Effects of harmful algal blooms on fish and shellfish species: a case study of New Zealand in a changing environment. Toxins (Basel) 14. https://doi.org/10.3390/toxins14050341

Sahoo D, Baweja P (2015) General characteristics of algae. In: The algae world, pp 3–29

Sanseverino I, Conduto D, Pozzoli L et al (2016) Algal bloom and its economic impact

Sharma A, Pal K, Saini N et al (2023) Remediation of contaminants from wastewater using algal nanoparticles via green chemistry approach: an organized review. Nanotechnology 34:352001. https://doi.org/10.1088/1361-6528/acd45a

Simiyu BM, Amukhuma HS, Sitoki L et al (2022) Interannual variability of water quality conditions in the Nyanza Gulf of Lake Victoria, Kenya. J Great Lakes Res 48:97–109. https://doi.org/10.1016/j.jglr.2021.10.017

Sultana S, Awal S, Shaika NA, Khan S (2022) Cyanobacterial blooms in earthen aquaculture ponds and their impact on fisheries and human health in Bangladesh. Aquac Res. https://doi.org/10.1111/are.16011

Summers EJ, Ryder JL (2023) A critical review of operational strategies for the management of harmful algal blooms (HABs) in Inland reservoirs. J Environ Manage 330. https://doi.org/10.1016/j.jenvman.2022.117141

Tian J, Guo S, Wang J et al (2022) Preemptive warning and control strategies for algal blooms in the downstream of Han River, China. Ecol Indic 142. https://doi.org/10.1016/j.ecolind.2022.109190

Trottet A, George C, Drillet G, Lauro FM (2022) Aquaculture in coastal urbanized areas: a comparative review of the challenges posed by harmful algal blooms. Crit Rev Environ Sci Technol 52:2888–2929. https://doi.org/10.1080/10643389.2021.1897372

Wang D, Zhang S, Zhang H, Lin S (2021) Omics study of harmful algal blooms in China: current status, challenges, and future perspectives. Harmful Algae 107. https://doi.org/10.1016/j.hal.2021.102079

Weir MJ, Kourantidou M, Jin D (2022) Economic impacts of harmful algal blooms on fishery-dependent communities. Harmful Algae 118. https://doi.org/10.1016/j.hal.2022.102321

Xue L, Hu J, Wang Z, et al (2023) Assessing risks of algal blooms in water transfer based on algal growth potential. Environ Monit Assess 195. https://doi.org/10.1007/s10661-023-11514-0

Young N, Sharpe RA, Barciela R et al (2020) Marine harmful algal blooms and human health: a systematic scoping review. Harmful Algae 98. https://doi.org/10.1016/j.hal.2020.101901

Zhu S, Wu Y, Ma X (2023) Deep learning-based algal bloom identification method from remote sensing images—take China's Chaohu Lake as an example. Sustainability (Switzerland) 15. https://doi.org/10.3390/su15054545

Chapter 5
Toward the Establishment of Nature-Based Solution (NbS) Using Seagrasses and Macroalgae to Control Harmful Algal Bloom

Nobuharu Inaba

Abstract The global population is estimated to grow to around 8.5 billion in 2030 and 9.7 billion in 2050, an increase of approximately 2 billion people worldwide in the next 30 years, as the United Nations World Population Prospect 2022 report. To improve global food security, fisheries and aquaculture are crucial in responding to increased seafood demand. However, expanding anthropogenic activities will likely worsen the impact on aquatic ecosystems. Harmful algal blooms (HABs) are a global environmental issue that indiscriminately kill and/or contaminate seafoods. In extreme cases, humans die from exposure to toxic seafood products. Over the past few decades, there has been growing evidence of an acceleration in geographic expansion and an increase in the severity of HABs related to climate change. Therefore, establishing sustainable control measures against HAB outbreaks is urgently needed. Biological control using naturally-living bacteria that kill and/or inhibit the growth of HAB species has been extensively studied worldwide as an environmentally friendly measure. This chapter introduces and discusses the relationship between climate change and HAB outbreaks and countermeasures, focusing on the possibility of controlling HABs using bacteria inhabiting macroalgal and seagrass beds as a Nature-based Solution (NbS), adding value to efforts to establish a carbon-neutral society.

Keywords Nature-based solution (NbS) · Harmful algal bloom · Climate change · Biological control · Raphidophytes · Dinoflagellates · Growth-limiting bacteria (GLB) · Macroalgae · Seagrass · Sustainable

---

N. Inaba (✉)
Civil Engineering Research Institute for Cold Region, Public Works Research Institute, National Research and Development Agency, Hiragishi 1-3-1-34, Toyohira-Ku, Sapporo 062-8602, Hokkaido, Japan
e-mail: inaba-n@ceri.go.jp

School of Biological, Earth and Environmental Sciences, Faculty of Science, University of New South Wales, Sydney, NSW 2052, Australia

T. A. Kurniawan and A. Anouzla (eds.), *Algae as a Natural Solution for Challenges in Water-Food-Energy Nexus*, Environmental Science and Engineering,
https://doi.org/10.1007/978-981-97-2371-3_5

Abbreviations

HAB Harmful algal bloom
GLB Growth-limiting bacteria

5.1 Introduction

Microalgae are an essential group of microorganisms that support the food chain as the primary producers in aquatic environments. On the other hand, blooms formed by harmful and toxic algae, so-called harmful algal blooms (HABs), cause mass die-offs of fish, shellfish, and poisonous bivalves, resulting in severe economic damage, especially in the fisheries and aquaculture industries (Adams et al. 2018). Furthermore, the risk of HABs has been considered elevated due to environmental changes associated with accelerating climate change (Wells et al. 2020), and a special report on oceans and cryosphere released by the Intergovernmental Panel on Climate Change (IPCC) in September 2019 clearly stated for the first time that there is a direct link between ocean warming and the increased occurrence of HABs in coastal areas (IPCC 2019). As the demand for marine products continues to increase owing to the global spread of fish consumption and population growth (OECD and FAO 2021; United Nations 2022), it is imperative to establish effective and practical measures to control HAB outbreaks, which threaten food security and are increasingly at risk due to warming. In addition to ocean warming, various anthropogenic stresses in coastal areas, such as the release of pollutants and reclamation, have also continuously increased owing to population expansion and industrialization (Jha et al. 2023; Sengupta et al. 2023), resulting in the degradation of the healthy coastal environment, such as losses of seagrass and macroalgal beds (Waycott et al. 2009; Filbee-Dexter and Wernberg 2018), which may worsen HAB occurrences (Abdenadher et al. 2012; Inaba et al. 2017). Thus, environmentally friendly and sustainable measures are indeed demanded and surely a priority matter in the context of controlling HABs (Imai et al. 2021).

5.2 Climate Change and HABs

Reports on the increased risk of HABs due to various environmental changes associated with accelerating climate change have been published one after another (Moore et al. 2011; Sun et al. 2011; Gobler et al. 2017; Brandenburg et al. 2019; Mardones et al. 2023). The reported effects of ocean warming caused by climate change on HAB events include growth rate and toxic production, prolonged blooms due to extended periods of high water temperature, and the expansion of distribution toward polar regions (Wells et al. 2020). Increasing concentration of CO_2 induced anthropogenically in the atmosphere is also absorbed by the ocean, causing the decrease of pH; so-called "Ocean acidification" (Dore et al. 2009), threatening numerous marine organisms with calcium carbonate ($CaCO_3$) shells and skeletons (Orr et al. 2005), such as shellfish, sea urchins, corals, and even some phytoplankton species called coccolithophores belonging to haptophytes, the major primary producers of the surface ocean (de Vargas et al. 2007). As for the effects of increased CO_2 concentration in seawater on HAB species, Sun et al.(2011) demonstrated increased production of domoic acid, which is known to be responsible for amnesic shellfish poisoning by *Pseudo-nitzschia* and a meta-analysis of 26 studies conducted on the relationship between phytoplankton growth and pCO_2 concentrations by Brandenburg et al. (2019), found that increased pCO_2 enhanced the growth rate of HAB species (mainly dinoflagellates), but had no effect on the growth of other phytoplankton species. Gobler et al. (2017) used high-resolution sea surface temperature records to evaluate trends in the growth rates and duration of blooms for two toxic dinoflagellates, *Alexandrium fundyense* and *Dinophysis acuminata*, and concluded that increasing ocean temperatures facilitated the intensification of HAB events by significantly increasing the growth rates and duration of blooms caused by these two toxic species. In the context of the expansion of distribution and extended bloom period, the formation of blooms by warm water HAB species has been confirmed in northern Japan (Shimada 2021), and in 2015, the first fish and invertebrate (salmon, Japanese common squid, and ezo-abalone) deaths occurred because of the red tide of the warm water HAB species "*Karenia mikimotoi*" in Hokkaido (Shimada et al. 2016). A red tide caused catastrophic economic damage (approximately 9 billion yen *as of September 2022), mainly along the Pacific coast of East Hokkaido in 2021, which was revealed to contain *K. mikimotoi* (Iwataki et al. 2022). In U.S. water, the geographic distribution of tropical dinoflagellates in the genus *Gambierdiscus*, known to cause Ciguatera Poisoning (Chinain et al. 2021), has been shown to expand into higher latitudes, associated with ocean warming (Anderson et al. 2021). Moore et al. (2011) evaluated bloom scenarios for *Alexandrium catenella*, the causative dinoflagellate of paralytic shellfish poisoning (PSP), in Puget Sound, WA, USA, using climate change projections, and predicted a significant expansion of the bloom season. The unique relationship between marine macrophytes and climate change may also be attributed to the occurrence of HAB. Inaba et al. (2017) documented that the cysts of fish-killing raphidophyte *Heterosigma akashiwo* were only detected where a seagrass bed disappeared. In contrast, no cysts were found at other locations where seagrasses were thriving and with high-density bacteria that showed growth-limiting activity against target HAB species in Puget Sound. Jacob-Palmer et al. (2020) also demonstrated a similar phenomenon in that the abundance of dinoflagellates was consistently low within the seagrass beds, emphasizing the allelopathic effect of *Zostera marina*. This phenomenon is also thought to occur in macroalgal beds (Inaba et al. 2020; Sylvers and Gobler 2021). These results indicate the mitigating effect of macrophytes on HABs. However, climate change and multiple anthropogenic stressors, such as the release of pollutants, reclamation, and eutrophication, have resulted in the degradation of seagrass and macroalgal beds globally (Waycott et al. 2009; Filbee-Dexter and Wernberg 2018), which may worsen HAB occurrences in coastal areas (Abdenadher et al. 2012).

5.3 Direct Countermeasures Against HABs

5.3.1 The Recent Research on Physicochemical and Biological Approaches

Various physicochemical methods have been tested as direct countermeasures against HABs (Anderson et al. 2017), such as ultrasonication (Lürling and Tolman 2014), ultraviolet radiation (Li et al. 2020), clay flocculation (Seger et al. 2017; Yu et al. 2017), ozone (Ng et al. 2023), chlorine (Dixon et al. 2017), and algicides (Pokrzywinski et al. 2012, 2017). However, its negative impact on the surrounding environment, cost, and scale of control remain a concern (Imai et al. 2021). To date, clay spraying is the only practical method implemented in the natural environment, and it is suggested to play a key role as an "Emergency measure" with a high density of HAB species in water (Imai et al. 2021). Biological controls using naturally living organisms are expected to be the most environmentally friendly approach and have been widely applied in agriculture for decades, such as to protect crops from pests, improve soil, and clean polluted areas (FAO and WHO 2017; Kenis et al. 2019). Biological methods using organisms living in aquatic environments have also attracted attention (Imai et al. 2021; Kibuye et al. 2021; Balaji-Prasath et al. 2022). Candidates are among the diverse groups of microorganisms, including viruses (Nakayama et al. 2020), bacteria (Coyne et al. 2022), protistan grazers (Lim et al. 2017), parasitic protists (Nishitani et al. 2021), and diatoms (Imai et al. 2021). Nakayama et al. (2020) documented the first time field trial to test the effectiveness of the sediment containing RNA virus (HcRNAV) during the early-stage bloom of *Heterocapsa circularisquama* and successfully reduced the density of *H. circularisquama.* Bacteria that kill or inhibit the growth of various microalgae, including major HAB species (Fig. 5.1), so-called algicidal bacteria, and growth-inhibiting bacteria (hereafter collectively referred to as growth-limiting bacteria (GLB)), have always been at the center of the topic (Mayali and Azam 2004; Meyer et al. 2017; Coyne et al. 2022), and a number of studies have demonstrated promising results for controlling blooms at laboratory scales (Shi et al. 2018; Zheng et al. 2018). Recent findings show that GLB are abundant in seagrass and macroalgal beds (Inaba and Imai 2023), providing new perspectives for establishing an environmentally friendly sustainable approach (see Sect. 5.3.2). Imai et al. (2021) described the successful induction to increase diatoms and suppression of *A. catenella* (Group I) in Osaka Bay by the bottom perturbation, aiming at competitive overwhelming HAB species. Although promising methods are being developed worldwide, they also face challenges, such as biosafety, cost, scaling up, time to effect, regulations, and local understanding (Imai et al. 2021; Coyne et al. 2022; Inaba and Imai 2023).

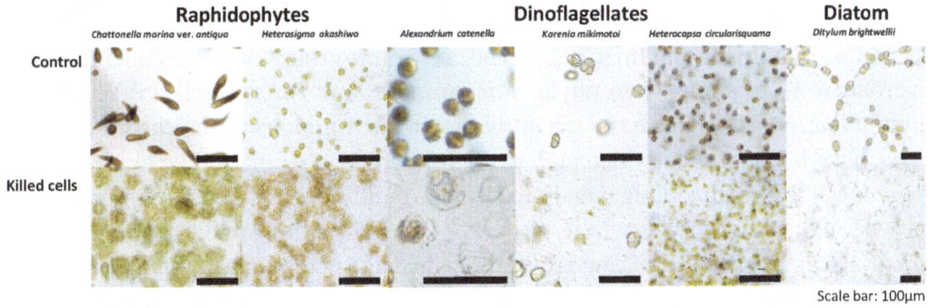

Fig. 5.1 Healthy vegetative cells of major HAB species and a diatom (*Ditylum brightwellii*) (upper), and lysed cells after adding microalgal growth-limiting bacteria (bottom)

5.3.2 Microalgal Growth-Limiting Bacteria (GLB) Associated with Seagrass and Macroalgal Beds and Their Application

In recent years, several studies conducted in the coastal areas of Japan and the USA have revealed that microalgal growth-limiting bacteria (GLB) inhabit high densities in seagrass and macroalgal beds (Imai 2015; Inaba et al. 2017, 2020; Imai et al. 2020). In fact, when eelgrass (*Z. marina*) bed seawater was filtered and added to artificially created *Chattonella* blooms, the suppression and termination of the blooms were confirmed even under nutrient-rich conditions, and a dramatic increase in GLB in the seawater during the process was observed, demonstrating the effectiveness of eelgrass bed seawater containing GLB in suppressing *Chattonella* blooms (Inaba et al. 2019). Table 5.1 lists the GLB isolated from seagrass beds in previous studies. Most studies investigating GLB associated with seagrasses were conducted at the eelgrass "*Z. marina*" beds (Fig. 5.2), and detected densities of GLB from the *Z. marina* biofilm ranged between 10^5 and 10^8 CFU g^{-1} wet leaf (Imai 2015; Inaba et al. 2017, 2018; Onishi et al. 2021). A similar GLB density (7.5×10^7 CFU g^{-1}) against *A. catenella* was detected in *Z. japonica* collected in Puget Sound, WA, USA (Inaba et al. 2017). The taxonomic groups of GLB associated with seagrasses included four phyla: *Actinobacteria*, *Bacteroidetes*, *Firmicutes*, and *Proteobacteria*. Most notably, many genera of bacteria, such as *Labrenzia*, *Sphingobium*, and *Thalassospira*, the class *Alphaproteobacteria*, *Marinomonas* and *Teredinibacter*, the class *Gammaproteobacteria*, and *Cellulophaga* (eater of cellulose), the class *Flavobacteriia*, detected as GLB, were previously reported to be associated with the degradation of lignocellulosic compounds (Masai et al. 2007; Espinosa et al. 2010; Inaba et al. 2019; Pang et al. 2022), which are most likely to be abundant throughout the year at the seagrass bed. GLB isolated from macroalgal beds in previous studies is listed in Table 5.2. Types of macroalgae reported as sources of GLB include green algae (*Ulva* sp., *U. lactuca*, *U. australis*, and *Cladophora ohkuboana*), red algae (*Gelidium* sp., *Corallina pilulifera*, *G. elegans*, *Chondrus ocellatus*, and *Gloiopeltis furcata*), and brown algae (*Saccharina japonica*, *Sargassum thunbergii*, and *S. dupulicatum*). The densities of GLB detected from the macroalgal biofilm ranged between 10^4 and 10^7 CFU (or MPN) g^{-1} wet weight (Inaba and Imai 2023). The taxonomic groups of GLB associated with macroalgae were among two phyla: *Bacteroidetes* and *Proteobacteria*. Within the phylum *Bacteroidetes*, bacterial strains belonging to *Aquimarina*, the class *Flavobacteriia*, have been frequently detected as GLB in green, red, and brown algae, as well as in seawater from macroalgal beds and the surrounding environment. *Aquimarina* is known to exhibit agarolytic activity (Lin et al. 2012), bleaching of the red alga *Delisea pulchra* (Kumar et al. 2016), and is also described as a potential pathogen of the sugar kelp *Saccharina latissima* (Liu et al. 2022). However, pathogenic activity against *D. pulchra* has only been detected under stress caused by elevated seawater temperatures (Kumar et al. 2016). *Octadecabacter*, *Roseobacter*, *Roseovarius*, *Phaeobacter*, and *Sulfitobacter*, the class *Alphaproteobacteria*, were found as GLB within the phylum *Proteobacteria*. Most are members of the *Roseobacter* clade, known to be closely associated to macroalgae (Luo and Moran 2014). For instance, morphogenetic compounds required for algal morphogenesis are released by a strain of the genus *Roseovarius* (Kessler et al. 2018). Bacteria belonging to the genus *Phaeobacter* are biofilm-forming marine bacteria associated with the green alga *Ulva australis* (Rao et al. 2006), which produces antibiotic tropodithietic acid (Rasmussen et al. 2018), and have also been studied as candidates for probiotic bacteria against several fish pathogens (D'Alvise et al. 2013; Rasmussen et al. 2018; Tesdorpf et al. 2022). Among *Gammaproteobacteria*, *Alteromonas* and *Pseudoalteromonas* are often detected as GLB. They are well-known as common GLB in coastal areas (Sakami et al. 2017; Lyu et al. 2019). *Alteromonas* and *Pseudoalteromonas* have also been isolated from brown algae as bacteria producing alginolytic enzymes (Sawabe et al. 1992; Matsushima et al. 2009). Since dinoflagellates such as *Alexandrium* and *Karenia*, which are major HAB species, are known to be composed of cellulose, and raphidophytes such as *Chattonella* and *Heterosigma* are known to be composed of polysaccharides called glycocalyx on their cell surface (Wang et al. 2020; Yokote and Honjo 1985), abundant bacteria in their original role as decomposers and symbionts at seagrass and macroalgal beds contribute to the suppression of HAB species. New findings regarding GLB associated with seagrasses and macroalgae may open new paths using bacteria amplified by nature to control HABs instead of artificial approaches, such as direct spraying of effective bacteria, bacterial algicides, and immobilized or encapsulated bacteria (Coyne et al. 2022), in terms of overcoming many of the challenges described for other measures, and could also add value to efforts to establish a carbon-neutral society using seagrasses and macroalgae as carbon sinks.

This chapter introduces and discusses the relationship between climate change and HAB outbreaks and countermeasures, focusing on the possibility of controlling HABs using bacteria inhabiting macroalgal and seagrass beds as a Nature-based Solution (NbS), adding value to efforts to establish a carbon-neutral society.

Table 5.1 A list of GLB isolated from seagrass beds in previous studies (modified from Inaba et al. 2018)

Source of isolation	Identified GLB	Killed or inhibited microalgal species	Sampling location	Reference
Seagrass biofilm (*Zostera marina*)	Flavobacteriaceae bacterium	*Chaetoceros mitra* *Heterocapsa circularisquama* *Chattonella marina* var. *antiqua* *Heterosigma akashiwo*	Usujiri Fishing Port, Hokkaido, Japan	Onishi et al. (2014)
Seagrass biofilm (*Z. marina*)	*Arthrobacter, Erythrobacter, Gaetbulibacter, Teredinibacter*	*H. akashiwo* *Alexandrium catenella*	Puget Sound, WA, USA	Inaba et al. (2017)
Seagrass biofilm (*Z. japonica*)	*Alteromonas, Gaetbulibacter, Halomonas, Kordia, Shewanella*	*H. akashiwo* *A. catenella*		
Seawater at seagrass bed (free-living)	*Alteromonas, Arthrobacter, Bacillus, Hydrogenophaga, Pseudoalteromonas, Shewanella*	*H. akashiwo* *A. catenella*		
Seawater at seagrass bed (particle-associated)	*Alteromonas, Erythrobacter, Litorimonas, Olleya, Pseudoalteromonas, Thalassospira*	*H. akashiwo* *A. catenella*		
Seagrass biofilm (*Z. marina*)	*Alteromonas, Marinomonas, Vibrio*	*C. marina* var. *antiqua*	Ariake Sea, Kumamoto, Japan	Inaba et al. (2018)
Seawater at seagrass bed	*Alteromonas, Nautella, Pseudoalteromonas, Vibrio*			
Seagrass biofilm (*Z. marina*)	*Alteromonas, Cellulophaga, Erythrobacter, Phaeobacter, Rhodobacter, Roseobacter, Sphingobium, Teredinibacter, Vibrio*	*C. marina* var. *antiqua*	Genji Bay, Okayama, Japan	Inaba et al. (2019)
Seawater at seagrass bed	*Alteromonas, Erythrobacter, Flavobacterium, Marinomonas, Phaeobacter, Pseudovibrio, Rhodobacter, Roseobacter, Shewanella, Silicibacter, Sphingobium, Thalassolituus, Vibrio*			

(continued)

Table 5.1 (continued)

Source of isolation	Identified GLB	Killed or inhibited microalgal species	Sampling location	Reference
Seagrass biofilm (*Z. marina*)	*Alteromonas,* Bacteroidetes bacterium, *Kordia, Krokinobacter, Sulfitobacter, Zobellia*	*A. catenella*	Akkeshi-ko Estuary, Hokkaido, Japan	Onishi et al. (2021)
Seawater at seagrass bed (free-living)	*Alteromonas, Sulfitobacter, Vibrio*			
Seawater at seagrass bed (particle-associated)	*Alteromonas, Labrenzia, Leifsonia, Marinobacter, Sphingomonas, Vibrio*			

Fig. 5.2 Eelgrass "*Zostera marina*" bed in Puget Sound, WA, USA

5.4 Prospective

Global declines in seagrass and macroalgal beds, vital habitats for diverse organisms supporting healthy coastal ecosystems and food security and playing a critical role in climate mitigation, have been accelerating (Waycott et al. 2009; Cullen-Unsworth and Unsworth 2018; Duarte et al. 2022). Recent findings indicate that bacteria that kill and/or inhibit the growth of various HAB species are abundant in seagrass and

Table 5.2 A list of GLB isolated from macroalgal beds in previous studies (modified from Inaba and Imai 2023)

Source of isolation	Identified GLB	Killed or inhibited microalgal species	Sampling location	Reference
Macroalgal biofilm (*Gelidium* sp.)	*Alteromonas*	*Chattonella marina*	Osaka Bay, Osaka, Japan	Imai et al. (2002, 2006)
		C. ovata		
		Heterosigma akashiwo		
	Vibrio	*Karenia mikimotoi*		
		C. ovata		
		H. akashiwo		
Macroalgal biofilm (*Ulva* sp.)	*Octadecabacter*	*C. marina* var. *antiqua*		
	Pseudoalteromonas	*K. mikimotoi*		
		Fibrocapsa japonica		
		H. akashiwo		
	Vibrio	*K. mikimotoi*		
		C. ovata		
		H. akashiwo		
Seawater at macroalgal bed	Rhodobacteraceae bacterium	*K. mikimotoi*		
		C. ovata		
		F. japonica		
		H. akashiwo		
Macroalgal biofilm (*U. lactuca*)	*Altererythrobacter*, *Halomonas*, *Shewanella*	*H. akashiwo*	Puget Sound, WA, USA	Inaba et al. (2017)
		Alexandrium catenella		
Seawater at macroalgal bed	*Thalassospira*, *Vibrio*, *Agarivorans*, *Aquimarina*, *Tenacibaculum*	*A. catenella*	Shinori-cho, Hokkaido, Japan	Imai et al. (2020)
		Cochlodinium polykrikoides		
		Heterocapsa circularisquama		
		C. marina var. *antiqua*		
		H. akashiwo		
		Ditylum brightwellii		
Macroalgal biofilm (*Saccharina japonica*)	Rhodobacteraceae bacterium, *Vibrio*	*A. catenella*		
		D. brightwellii		
Macroalgal biofilm (*Sargassum thunbergii*)	Bacterium WHC1-4	*A. catenella*		

(continued)

Table 5.2 (continued)

Source of isolation	Identified GLB	Killed or inhibited microalgal species	Sampling location	Reference
Macroalgal biofilm (*Corallina pilulifera*)	Flavobacteriaceae bacterium	*A. catenella*		
Macroalgal biofilm (*U. pertusa*)	*Tenacibaculum*	*A. catenella*		
		D. brightwellii		
Seawater at introduced *Ulva* bed	*Nautella, Alteromonas, Pseudoalteromonas, Tenacibaculum*	*C. marina* var. *antiqua*	Kushimoto-cho, Wakayama, Japan	Inaba et al. (2020)
		K. mikimotoi		
		H. circularisquama		
Introduced *U. pertusa* biofilm	*Aquimarina, Pseudoalteromonas, Vibrio*	*C. marina* var. *antiqua*		
		H. circularisquama		
Seawater at macroalgal bed	*Roseobacter, Alteromonas, Pseudoalteromonas, Vibrio, Aquimarina, Winogradskyella*	*C. marina* var. *antiqua*	Taiji-cho, Wakayama, Japan	
		H. akashiwo		
		K. mikimotoi		
		H. circularisquama		
Macroalgal biofilm (*S. dupulicatum*)	*Pseudoalteromonas, Vibrio, Aquimarina, Dokdonia, Winogradskyella*	*A. catenella*		
		K. mikimotoi		
		H. circularisquama		
Macroalgal biofilm (*G. elegans*)	*Phaeobacter, Alteromonas, Aquimarina*	*C. marina* var. *antiqua*		
		H. circularisquama		
Macroalgal biofilm (*Cladophora ohkuboana*)	*Roseovarius, Aquimarina, Winogradskyella*	*C. marina* var. *antiqua*		
		H. akashiwo		
		A. catenella		
		H. circularisquama		
Macroalgal biofilm (*U. pertusa*)	*Aquimarina, Dokdonia, Winogradskyella*	*A. catenella*		
		K. mikimotoi		
		H. circularisquama		
Seawater at macroalgal bed	*Sphingorhabdus, Aquimarina*	*H. akashiwo*	Shinori-cho, Hokkaido, Japan	Inaba and Imai (2023)
		A. catenella		
		K. mikimotoi		
Sediment at macroalgal bed	*Loktanella, Algibacter*	*A. catenella*		

(continued)

Table 5.2 (continued)

Source of isolation	Identified GLB	Killed or inhibited microalgal species	Sampling location	Reference
Pebble at macroalgal bed	*Aquimarina, Granulosicoccus*	*A. catenella*		
Macroalgal biofilm (*U. pertusa*)	*Sulfitobacter, Olleya*	*A. catenella*		
Macroalgal biofilm (*Chondrus ocellatus*)	*Aquimarina, Kordia*	*A. catenella*		
Macroalgal biofilm (*Gloiopeltis furcata*)	*Litoreibacter, Stappia*	*A. catenella*		

macroalgal beds, whereas HAB species are scarce (Inaba et al. 2017; Inaba and Imai 2023; Jacob-Palmer et al. 2020); future efforts to restore, protect, and adequately introduce seagrass and macroalgal beds could enhance coastal resilience against HAB events (Fig. 5.3). Moreover, these efforts may not only reduce coastal HAB events but also prevent anthropogenically intensified hypoxia and/or anoxia on the seafloor (Rabalais et al. 2010) by moderating the fluctuation of microalgal biomass in the coastal area (Imai et al. 2017). However, the mechanism by which GLB is supplied from these beds to the surrounding environment through physical diffusion and physiological-biological processes and the amount of GLB required to effectively express its inhibitory capacity is still unclear (Inaba and Imai 2023). Thus, further understanding of these mechanisms centered on repeated field trials is required to establish a practically feasible and sustainable Nature-based Solution (NbS) using natural macrophyte communities. Furthermore, promoting active global conservation efforts in partnership with the industry, government, academia, and private sector is imperative.

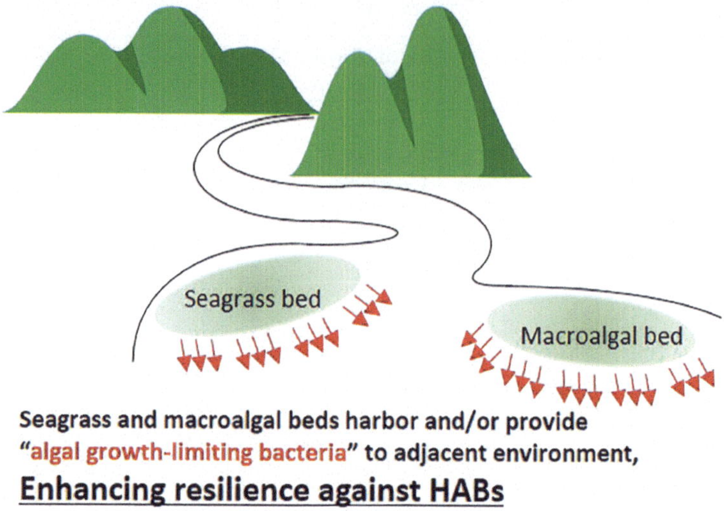

Fig. 5.3 Efforts to restore, protect, and adequately introduce seagrass and macroalgal beds to enhance coastal resilience against HAB events (Inaba 2021)

References

Abdenadher M et al (2012) Factors determining the dynamics of toxic blooms of *Alexandrium minutum* during a 10-year study along the shallow southwestern Mediterranean coasts. Estuar Coast Shelf Sci 106:102–111

Adams CM et al (2018) Assessing the economic consequences of harmful algal blooms. In: Shumway SE, Burkholder JM, Morton SL (eds) Harmful algal blooms: a compendium desk reference. Wiley, Hoboken, pp 337–354

Anderson CR, Sellner KG, Anderson DM (2017) Bloom prevention and control. In: Anderson DM, Boerlage SFE, Dixon MB (eds) Harmful algal blooms (HABs) and desalination: a guide to impacts, monitoring and management. Intergovernmental Oceanographic Commission of UNESCO, Paris, pp 205–215

Anderson DM et al (2021) Marine harmful algal blooms (HABs) in the United States: history, current status and future trends. Harmful Algae 102:101975

Balaji-Prasath B et al (2022) Methods to control harmful algal blooms: a review. Environ Chem Lett 20:3133–3152

Brandenburg KM, Velthuis M, Van de Waal DB (2019) Meta-analysis reveals enhanced growth of marine harmful algae from temperate regions with warming and elevated CO_2 levels. Glob Change Biol 25(8):2607–2618

Chinain M et al (2021) Ciguatera poisonings: a global review of occurrences and trends. Harmful Algae 102:101873

Coyne K, Wang Y, Johnson G (2022) Algicidal Bacteria: a review of current knowledge and applications to control harmful algal blooms. Front Microbiol 13:871177

Cullen-Unsworth LC, Unsworth R (2018) A call for seagrass protection. Science 361(6401):446–448

D'Alvise PW et al (2013) Protection of cod larvae from vibriosis by *Phaeobacter* spp.: a comparison of strains and introduction times. Aquaculture 384–387:82–86 de Vargas C et al (2007) Origin and evolution of coccolithophores: from coastal hunters to oceanic farmers. In: Falkowski PG, Knoll AH (eds) Evolution of primary producers in the sea. Elsevier Academic Press, pp 251–285

Dixon MB et al (2017) Removal of algal toxins and taste and odor compounds during desalination. In: Anderson DM, Boerlage SFE, Dixon MB (eds) Harmful algal blooms (HABs) and desalination: a guide to impacts, monitoring and management. Intergovernmental Oceanographic Commission of UNESCO, Paris, pp 315–332

Dore JE et al (2009) Physical and biogeochemical modulation of ocean acidification in the central North Pacific. Proc Natl Acad Sci USA 106(30):12235–12240

Duarte CM et al (2022) Global estimates of the extent and production of macroalgal forests. Glob Ecol Biogeogr 31(7):1422–1439

Espinosa E et al (2010) Taxonomic study of *Marinomonas* strains isolated from the seagrass *Posidonia oceanica*, with descriptions of *Marinomonas balearica* sp. Nov. and *Marinomonas pollencensis* sp. Nov. Int J Syst Evol Microbiol 60:93–98

FAO and WHO (2017) International code of conduct on pesticide management: guidelines for the registration of microbial, botanical and semiochemical pest control agents for plant protection and public health uses. Food and Agriculture Organization of the United Nations (FAO) and World Health Organization (WHO), Rome, p 86

Filbee-Dexter K, Wernberg T (2018) Rise of turfs: a new battlefront for globally declining kelp forests. Bioscience 68:64–76

Gobler CJ et al (2017) Ocean warming since 1982 has expanded the niche of toxic algal blooms in the North Atlantic and North Pacific oceans. Proc Natl Acad Sci USA 114(19):4975–4980

Imai I et al (2006) Algicidal bacteria isolated from the surface of seaweeds from the coast of Osaka Bay in the Seto Inland Sea, Japan. Afr J Mar Sci 28:319–323

Imai I (2015) Interactions between harmful algae and algicidal and growth-inhibiting bacteria associated with seaweeds and seagrasses. In: Ohtsuka S et al (eds) Marine protists. Springer Japan, Tokyo, pp 597–619

Imai I, Fujimaru D, Nishigaki T (2002) Co-culture of fish with macroalgae and associated bacteria: a possible mitigation strategy for noxious red tides in enclosed coastal sea. Fish Sci 68:493–496

Imai I, Inaba N, Sakami T (2017) Environmentally friendly strategies for the prevention of harmful algal blooms using algicidal bacteria associated with seagrass beds. In: Proenca LAO, Hallegraeff GM (eds) Marine and fresh-water harmful algae. Proceedings of the 17th international conference on harmful algae. ISSHA, pp 160–165

Imai I et al (2020) Biological control of toxic and noxious algal blooms with environment-friendly strategies. Kaiyo Monthly 52(5):217–226

Imai I, Inaba N, Yamamoto K (2021) Harmful algal blooms and environmentally friendly control strategies in Japan. Fish Sci 87:437–464

Inaba N (2021) Seeking a sustainable way to control harmful algal bloom. Open Access Government July, pp 438–439

Inaba N, Imai I (2023) Using macroalgal beds to create a low-risk environment for harmful algal blooms. Kaiyo Monthly 55(10):532–538

Inaba N et al (2017) Algicidal and growth-inhibiting bacteria associated with seagrass and macroalgae beds in Puget Sound, WA, USA. Harmful Algae 62:136–147

Inaba N et al (2018) Temporal variability of algicidal and growth-inhibiting bacteria at an eelgrass bed in the Ariake Sea, Japan. Bioremediat J 22:1–14

Inaba N et al (2019) Dynamics of seagrass bed microbial communities in artificial *Chattonella* blooms: a laboratory microcosm study. Harmful Algae 84:139–150

Inaba N et al (2020) Distribution of harmful algal growth-limiting bacteria on artificially introduced *Ulva* and natural macroalgal beds. Appl Sci 10(16):5658

IPCC (2019) IPCC special report on the ocean and cryosphere in a changing climate (eds: Pörtner H-O et al). Cambridge University Press, Cambridge, p 755

Iwataki M et al (2022) Morphological variation and phylogeny of *Karenia selliformis* (Gymnodiniales, Dinophyceae) in an intensive cold-water algal bloom in eastern Hokkaido, Japan. Harmful Algae 114:102204
Jacobs-Palmer E et al (2020) A halo of reduced dinoflagellate abundances in and around eelgrass beds. PeerJ 8:e8869
Jha DK et al (2023) Editorial: coastal and marine environmental quality assessments. Front Mar Sci 10:1141278
Kenis M et al (2019) Guide to the classical biological control of insect pests in planted and natural forests. Food and Agriculture Organization of the United Nations (FAO), Rome, p 113
Kessler RW et al (2018) Macroalgal–bacterial interactions: role of dimethylsulfoniopropionate in microbial gardening by *Ulva* (Chlorophyta). Mol Ecol 27(8):1808–1819
Kibuye FA, Zamyadi A, Wert EC (2021) A critical review on operation and performance of source water control strategies for cyanobacterial blooms: part II-mechanical and biological control methods. Harmful Algae 109:102119
Kumar V et al (2016) Multiple opportunistic pathogens can cause a bleaching disease in the red seaweed *Delisea pulchra*. Environ Microbiol 18(11):3962–3975
Li S et al (2020) UV-C irradiation for harmful algal blooms control: a literature review on effectiveness, mechanisms, influencing factors and facilities. Sci Total Environ 723:137986
Lim AS et al (2017) Ichthyotoxic *Cochlodinium polykrikoides* red tides offshore in the South Sea, Korea in 2014: II. Heterotrophic protists and their grazing impacts on red-tide organisms. Algae 32:199–222
Lin B et al (2012) *Aquimarina agarilytica* sp. nov., an agarolytic species isolated from a red alga. Int J Syst Evol Microbiol 62:869–873
Liu Y et al (2022) A deep dive into the epibiotic communities on aquacultured sugar kelp *Saccharina latissima* in Southern New England. Algal Res 63:102654
Luo H, Moran MA (2014) Evolutionary ecology of the marine *Roseobacter* clade. Microbiol Mol Biol Rev 78(4):573–587
Lürling M, Tolman Y (2014) Beating the blues: is there any music in fighting cyanobacteria with ultrasound? Water Res 66:361–373
Lyu YH et al (2019) Optimized culturing conditions for an algicidal bacterium *Pseudoalteromonas* sp. SP48 on harmful algal blooms caused by *Alexandrium tamarense*. MicrobiologyOpen 8(8):e00803
Mardones JI et al (2023) Extreme harmful algal blooms, climate change, and potential risk of eutrophication in Patagonian fjords: Insights from an exceptional *Heterosigma akashiwo* fish-killing event. Prog Oceanogr 210:102921
Masai E, Katayama Y, Fukuda M (2007) Genetic and biochemical investigations on bacterial catabolic pathways for lignin-derived aromatic compounds. Biosci Biotechnol Biochem 71(1):1–15
Matsushima R et al (2009) Analysis of extracellular alginate lyase and its gene from a marine bacterial strain, *Pseudoalteromonas atlantica* AR06. Appl Microbiol Biotechnol 86:567–576
Mayali X, Azam F (2004) Algicidal bacteria in the sea and their impact on algal blooms. J Eukaryot Microbiol 51:139–144
Meyer N et al (2017) Strategies and ecological roles of algicidal bacteria. FEMS Microbiol Rev 41:880–899
Moore SK, Mantua NJ, Salathé Jr EP (2011) Past trends and future scenarios for environmental conditions favoring the accumulation of paralytic shellfish toxins in Puget Sound shellfish. Harmful Algae 10(5):521–529
Nakayama N et al (2020) Evaluation of a virus-based control method to protect cultured oysters from the harmful dinoflagellate *Heterocapsa circularisquama*. Aquaculture 529(4):735625
Ng PH et al (2023) Assessment of ozone nanobubble technology to reduce freshwater algae. Aquac Res 2023:9539102

Nishitani G et al (2021) A novel parasite strain of *Amoebophrya* sp. infecting the toxic dinoflagellate *Alexandrium catenella* (Group I) and its effect on the host bloom in Osaka Bay, Japan. Harmful Algae 110:102123

OECD, FAO (2021) OECD-FAO agricultural outlook 2021–2030. OECD Publishing, Paris, p 337

Onishi Y et al (2014) The seagrass Zostera marina harbors growth-inhibiting bacteria against the toxic dinoflagellate Alexandrium tamarense. Fish Sci 80:353–362

Onishi Y et al (2021) Distribution of growth-inhibiting bacteria against the toxic dinoflagellate *Alexandrium catenella* (Group I) in Akkeshi-ko estuary and Akkeshi Bay, Hokkaido, Japan. Appl Sci 11:172

Orr JC et al (2005) Anthropogenic ocean acidification over the twenty-first century and its impact on calcifying organisms. Nature 437(7059):681–686

Pang M et al (2022) Seasonal succession of bacterial communities in cultured *Caulerpa lentillifera* detected by high-throughput sequencing. Open Life Sci 17:10–21

Pokrzywinski KL et al (2012) Investigation of the algicidal exudate produced by *Shewanella* sp IRI-160 and its effect on dinoflagellates. Harmful Algae 19:23–29

Pokrzywinski KL et al (2017) Effects of the bacterial algicide IRI-160AA on cellular morphology of harmful dinoflagellates. Harmful Algae 62:127–135

Rabalais NN et al (2010) Dynamics and distribution of natural and human-caused hypoxia. Biogeosciences 7:585–619

Rao D, Webb JS, Kjelleberg S (2006) Microbial colonization and competition on the marine alga *Ulva australis*. Appl Environ Microbiol 72:5547–5555

Rasmussen BB et al (2018) Effect of TDA-producing *Phaeobacter inhibens* on the fish pathogen *Vibrio anguillarum* in non-axenic algae and copepod systems. Microb Biotechnol 11(6):1070–1079

Sakami T et al (2017) Distribution of three algicidal *Alteromonas* sp. strains in seagrass beds and surrounding areas in the Seto Inland Sea, Japan. Fish Sci 83:113–121

Sawabe T, Ezura Y, Kimura T (1992) Characterization of an alginolytic marine bacterium from decaying rishiri-kombu *Laminaria japonica* var. *ochotensis*. Nippon Suisan Gakkaishi 58:141–145

Seger A, Park T-G, Hallegraeff G (2017) Assessment of the efficacy of clay flocculation in Korean fish farm waters: *Cochlodinium* cell removal and mitigation of ichthyotoxicity. Harmful Algae 61:46–55

Sengupta D et al (2023) Mapping 21st century global coastal land reclamation. Earth's Future 11:e2022EF002927

Shi XG et al (2018) Isolation of an algicidal bacterium and its effects against the harmful algal bloom dinoflagellate *Prorocentrum donghaiense* (Dinophyceae). Harmful Algae 80:72–79

Shimada H (2021) Long-term fluctuation of red tide and shellfish toxin along the coast of Hokkaido. Sci Rep Hokkaido Fish Res Inst 100:1–12

Shimada H et al (2016) First record of red tide due to the harmful dinoflagellate *Karenia mikimotoi* in Hakodate Bay, southern Hokkaido, in autumn 2015. Nippon Suisan Gakkaishi 82(6):934–938

Sun J et al (2011) Effects of changing pCO_2 and phosphate availability on domoic acid production and physiology of the marine harmful bloom diatom *Pseudo-nitzschia multiseries*. Limnol Oceanogr 56(3):829–840

Sylvers PH, Gobler CJ (2021) Mitigation of harmful algal blooms caused by *Alexandrium catenella* and reduction in saxitoxin accumulation in bivalves using cultivable seaweeds. Harmful Algae 105:102056

Tesdorpf JE et al (2022) *Roseobacter* group probiotics exhibit differential killing of fish pathogenic *Tenacibaculum* species. Appl Environ Microbiol 88(6):e0241821

United Nations (2022) World population prospects 2022: summary of results. United Nations Department of Economic and Social Affairs, New York, p 38

Wang M et al (2020) Algal cell lysis by bacteria: a review and comparison to conventional methods. Algal Res 46:101794

Waycott M et al (2009) Accelerating loss of seagrasses across the globe threatens coastal ecosystems. Proc Natl Acad Sci USA 106(30):12377–12381

Wells ML et al (2020) Future HAB science: directions and challenges in a changing climate. Harmful Algae 91:101632

Yokote M, Honjo T (1985) Morphological and histochemical demonstration of glycocalyx on the cell surface of *Chattonella antiqua*, a "naked flagellate." Experientia 41:1143–1145

Yu Z et al (2017) Mitigation of harmful algal blooms using modified clays: theory, mechanisms, and applications. Harmful Algae 69:48–64

Zheng N et al (2018) Diverse algicidal bacteria associated with harmful bloom-forming *Karenia mikimotoi* in estuarine soil and seawater. Sci Total Environ 631–632:1415–1420

Part II
Algal Management in Wastewater Treatment

Chapter 6
Water Remediation to Water Mining: Cradle to Cradle in Wastewater Treatment Using Algae

Manali Date, Deepali Kulkarni, and Dipika Jaspal

Abstract Algal boom is a direct consequence of rampant eutrophication of water bodies. Presence of algae is considered objectionable as it leads to low dissolved oxygen level and decreased sunlight, thereby deteriorating the quality of water and adversely affecting aquatic flora and fauna. There has been a meticulous rate of research and development in the areas of application of algae, for the purpose of wastewater remediation and reuse. These efforts on the utilisation of algae have successfully brought algae into positive shades of limelight. Different technologies are being developed to utilize this biomass to the maximum extent in waste water remediation to solve the problems of algal disposal. Around 90–99% of removal has been achieved for toxic metals, harmful organic wastes and other inorganic matter in different types of wastewaters by using algae alone or a bio consortium of algae with other useful microspecies of bacteria and other plants. Algae in addition to proving as efficient substances in waste water remediation are also being explored for the recovery and recycling of several useful components. Along with remediation of wastewater; algae have also been found to be a suitable medium for mining or recovery of various metals and nutrients like phosphates and nitrates. Some recent studies have even shown adsorption of 25–30 mg/g for gold and uranium from wastewater by algal species like *Galdieria sulphuraria* and *Hydrodictyon reticulatum.* This chapter is an attempt to present the research carried out on the use of algae in the domain of wastewater remediation and mining.

Keywords Algae · Wastewater · Remediation · Reuse · Water Mining · Metal Recovery

M. Date
Department of Civil Engineering, Symbiosis Institute of Technology (SIT), Symbiosis International (Deemed University) (SIU), Lavale, Pune, Maharashtra, India

D. Kulkarni · D. Jaspal (✉)
Department of Applied Science, Symbiosis Institute of Technology (SIT), Symbiosis International (Deemed University) (SIU), Lavale, Pune, Maharashtra, India
e-mail: dipikajaspal@gmail.com

T. A. Kurniawan and A. Anouzla (eds.), *Algae as a Natural Solution for Challenges in Water-Food-Energy Nexus*, Environmental Science and Engineering,
https://doi.org/10.1007/978-981-97-2371-3_6

6.1 Introduction

Phycoremediation, which is the use of algae in wastewater remediation is currently an emerging technique for the treatment of contaminated water, as a result of domestic, industrial, or agricultural effluents, proving to be a promising approach to eliminate the omnipresent problem of wastewater remediation. Such a kind of bio treatment either alone or merged with different types of treatments can be implemented for enhancing the overall quality of the wastewater by reducing and or removing its detrimental characteristics. Some of the primary characteristics of wastewater which need to be addressed are chemical oxygen demand (COD), biochemical oxygen demand (BOD), total suspended solids (TSS), total dissolved solids (TDS), heavy metals, hazardous and toxic constituents.

The removal of nitrogen, which is in the form of inorganic nitrates and nitrites, and phosphorus, as phosphates, is a must since their presence is harmful for health of living beings. The norms related to the removal of the mentioned are becoming due to the imparted hazardous traits (Fewtrell 2004; Wang and Turner 2013) Algae have a tendency to consume these micronutrients like nitrates and phosphates, and carbon dioxide and thereby leading to the destruction of polluting, unwanted substances present in wastewater. Application of microalgae is a green solution for remediation of wastewater (Goswami et al. 2023).

Adequate treatment of wastewater using bioremediation by algae is generally carried out either for safe disposal into fresh water resources or for reuse of treated wastewater. However, this can be extended further to the mining of vital nutrients and elements especially phosphates (forms of phosphorus), nitrates (forms of nitrogen), elements like iron, copper, zinc, lead, manganese etc. which can be used for different kinds of applications.

It has been found that algae-based remediation processes produce enormous amounts of algae biomass sludge. Unlike the sludge produced by conventional physical, chemical and other biological methods, this sludge has no disposal issues and can be sold as a bio-fertilizer having good nutritional value (Bhatt et al. 2022). This chapter ardently explains the existing and under explored systems in employing algae in wastewater remediation and water mining activities. Discussion on cost efficiency involved in these operations is also been presented. Figure 6.1 Bioremediation flow diagram of wastewater using algae.

6.2 Algae and Wastewater Remediation

A review on the treatment of emerging compounds in wastewater by algal biosorption was carried out by Ramesh et al. (2023). The application of algae for remediation of toxic heavy metals, dyes and pharmaceutical components was studied. It was inferred that biosorption of toxic and hazardous components of wastewater has great

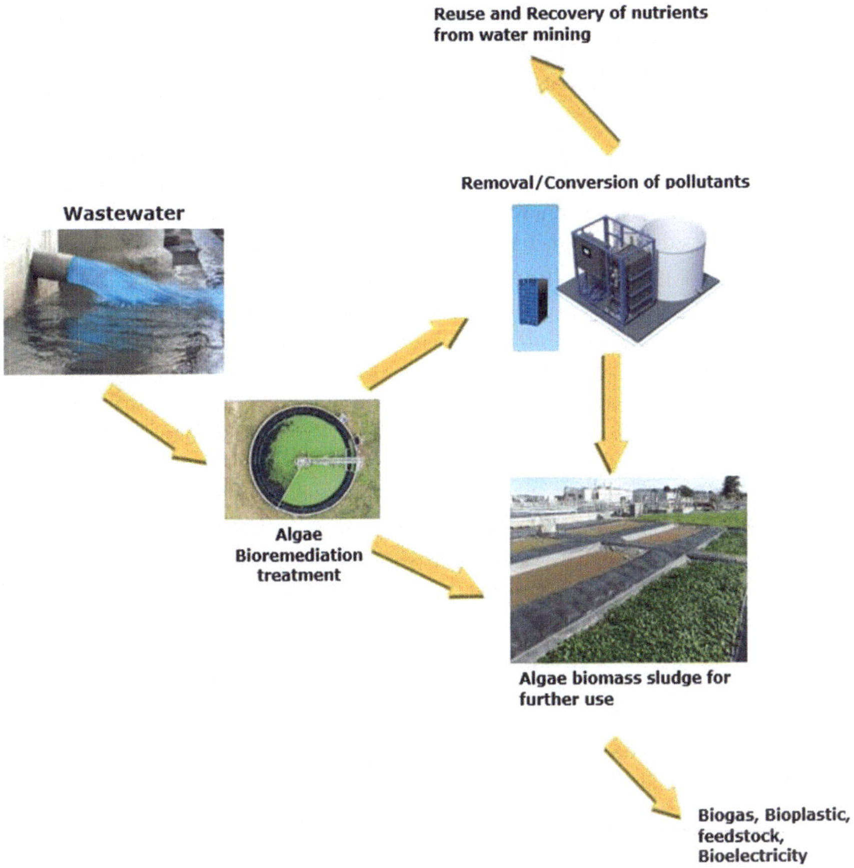

Fig. 6.1 Bioremediation flow diagram of wastewater using algae

future potential since it is environmentally friendly, feasible, and practical and more economic when compared to other bio adsorbents.

6.2.1 Remediation of Organic and Inorganic Compounds

In a research study by Uma et al. (2023) marine microalgae species were used. They were activated using bacterial sludge from a customised sequential batch reactor. These were used for removal of nitrate with high concentrations of 500 mg/L in 32 h period. There was reduction observed in turbidity, dissolved solids, total nitrogen and total phosphorus levels of the simulated wastewater. The settling velocity of wastewater increased, which concluded that algae could be harvested as precipitates.

The type of algae species applied included: *Chroococcus minor, Spirulina platensis, Synechococcus elongates, Chroococcus* and *Phormidium.*

In an experimental trial conducted by Wu et al. in 2023, sixteen different types of micropollutants were targeted for removal by microalgae *Chlorella sorokiniana.* The adversities brought about by different wastewater characteristics was analysed, for which the microalgae species were harvested in three categories of wastewater namely municipal wastewater, wastewater which was anaerobically digested and secondary clarified effluent in batch and continuous types of reactor. Elimination of eleven organic pollutants from domestic and anaerobically digested wastewater was observed in batch mode. In the continuous mode having 0.8 days as hydraulic retention time, the removal of all organic pollutants decreased. This concluded that longer retention time is required for targeting efficient removal of organic pollutants in such a case. There was not much removal observed in the case of secondary clarified effluent because of the limited amount of required nutrients for microalgal growth.

Zhang et al. in 2020 expounded the treatment of phenol and phenolic compounds by adsorption and degradation, using algae, was cost-effective and eco-friendly, but higher concentrations of phenol can definitely impart ecotoxicity. The structure of algae may get hampered due to this. Li et al. (2022a, b) discussed the various possible mechanisms of algae bioremediation like bioaccumulation, bio adsorption and biodegradation. Another application of algae which was focused was its use as a predecessor agent for manufacturing biochar to improve its properties substantially. This type of algae modified biochar was useful in removal of antibiotics and for advanced oxidation processes. Bio electrochemical processes was also noted as a new arena of treatment of wastewater integration of biological with electrochemical methods (Li et al. 2022a, b).

6.2.1.1 Industrial Wastewater Remediation

Dairy wastewater has high organic matter and requires pre-treatment before biological treatment. One such dairy wastewater was treated using microalgae in a study by Ma et al. (2023). Ten strains of microalgae were screened out of which only these were able to sustain such high Chemical oxygen demand: *Chlorella pyrenoidosa, Scenedesmus,* and *Chlorella vulgaris.* A photo bioreactor was constructed which enabled the removal of organic matter as well as build-up of biomass thereby capturing carbon dioxide.

Treatment of aquaculture wastewater is a challenging task as it contains a plethora of nutrients which if released directly causes eutrophication of water bodies. It consumes the nutrients and assimilates them into microbial biomass which can serve as food for larger aquatic creatures.

Wastewater valorisation was applied by Gorzelnik et al. in 2023 for unsterilized type of aquaculture wastewater using microalgae species *Haematococcus pluvialis, Chlorella vulgaris,* and *Scenedesmus dimorphus.* Removal of phosphorus was 100%, while there was only about 50–70% removal of nitrogen. This was improved by reducing the hydraulic retention time and dose of phosphorus added for cultivation of microalgae. Daphnia magna fish, a microalgae eater consumed the nutrient assimilated microalgae.

In another research work for treatment of marine and freshwater, classes of shrimp aquaculture wastewater, using six different species of *Chlorella* microalgae, were tested, from zero to 60% concentration (volume: volume). Out of these, 30 and 40% concentrations (volume: volume) were found to be most optimum. More than 80–90% of reduction of nitrite, ammonia and orthophosphate was observed. The cell density of applied microalgae increased ten times the original cell density due to absorption and assimilation of nutrients in the shrimp aquaculture wastewater sample. Moreover, it was concluded that this technique was a sustainable and novel technique suitable for treating aquaculture wastewater in large amounts (Nasir et al. 2023).

6.2.2 Remediation of Pharmaceutical Compounds

Microalgae catalysed by zero valent iron in continuous mode run microbial fuel cell was tested for removal of antibiotics Chlortetracycline by Zhang et al. (2023a, b). Research was extended to decipher if problems of contamination in the electrolyte of cathode chamber, and acidification of anode chamber occurred. Moreover, the build-up of antibiotic resistant genes (ARG) and reduction in the efficiency of removal of nitrogen and phosphorus was also checked. Additional removal by about 14.24%, in COD, 20.27% ammonical nitrogen and 32.27% total nitrogen was obtained in the cathode compartment due to the microalgae. Zero valent iron contributed in the removal of Chlortetracycline, led to anaerobic type of ammonia oxidation and enhanced denitrification. It reduced transmission of and increased removal of nitrogen and phosphorus.

Ali et al. in 2018 devised a chemisorption category of technique by utilizing modified form of *Scenedesmus obliquus* microalgae for the removal of tramadol, ibuprofen and other pharmaceutical compounds by adsorption. A second order pseudo-order reaction and Freundlich isotherm with more than 90% of remediation of tramadol was observed in 45 min time at natural pH, 0.5 g/L of basic microalgae adsorbent. It was found that this modified mass of microalgae was reusable for at least three runs, with effectively same removal.

6.2.3 Remediation of Dyes

Targeting the removal of dyes using microalgae is a very essential technique which needs to be developed to large scale due its sustainability. In a study by Abdulhameed et al. (2022), removal of two dyes: crystal violet and methylene blue using dosages of 20–100 mg were tried at pH 4–10 for 2–6 min of adsorption. Pseudo-second order kinetics with Langmuir type adsorption was observed and it was an endothermic type of reaction. Adsorption by microalgae was due to H-bonding, n-π stacking and electrostatic forces. Quantity of adsorbed dye was 243 mg/g of crystal violet and 297.1 mg/g methylene blue which was an indicative of the future of application of microalgae adsorption in dye removal and decolourization of wastewater.

6.2.4 Remediation of Heavy Metals

Acidic microalgae adsorbent was used by Almomani and Bhosale (2021) for the removal of toxic metals like aluminium, nickel and copper. Algae strains of *Spirulina platensis* and *Chlorella vulgar* were used in dosage range of 0.2–2.5 g, pH 5–8 and time of contact 5–10 min. Maximum removal of aluminium was observed and overall removal of all metals was above 50% at acidic to near neutral pH range. Reaction kinetics were of second order and followed Langmuir adsorption isotherms. The mechanism of adsorption was spontaneous, endothermic in nature and based on forces of electrostatic attraction.

The efficacy of filamentous freshwater algae as a bio adsorbent was examined by Singh et al., in 2021 in a series of experiments using phytoremediation for synthetic wastewater in batch mode. Removal of chromium (VI) and methylene blue dye was targeted. It was found to be of pseudo second order with D-R isotherm and more than 90% of removal of the dye and heavy metal were found. Wastewater remediation through algae is thus a low cost, evolving technology having huge scope in future. Another investigation for four different types of algae *Vaucheria debaryana, Cladophora glomerata, Oedogonium westii*, and *Zygnema insigne* was carried out by Khan et al. (2017).

Remediation of heavy metals like cadmium, nickel, lead and chromium was investigated from a real sample of industrial wastewater. There was substantial reduction in values of total conductivity, chemical oxygen demand, bio chemical oxygen demand, total dissolved solids, nitrate and increase in amounts of dissolved oxygen. Substantial removal (more than 50%) of heavy metals with preference of a particular algae towards a particular heavy metal was elucidated. Biosorption of heavy metal chromium by use of algae *Spirulina platensis* immobilized in sodium alginate beads was studied by Sharmila (2020). Several parameters like pH, initial concentration, dose of adsorbent and depth of bead were optimized. pH 2 with algae and pH 6 without algae were found to be most optimum for removal above 80%.

6.2.5 Algae Consortium for Wastewater Remediation

Initial growth of micro algae is faster than bacteria, after which it attains a steady state to equalize the growth of bacteria (Zhang et al. 2023a, b). A comparison was made in the study of municipal wastewater by using algae and bacteria. In comparison to the lone treatment by bacteria or algae, the consortium of bacteria and algae proved to give better outcomes in settling, removal efficiency and quantity of extra cellular polymeric substances. However, the synergy disrupted the growth of denitrifying bacteria. Dominating eukaryotic algae species in the algae bacteria consortium were *Trebouxiophyceae* (Wang et al. 2023a).

A merger of microalgae with fungi in symbiotic form of microalgae and fungi was found to be promising by Wang et al. (2023b) in removal of heavy metals like cadmium and chromium from wastewater. The interaction of fungi with microalgae imparted synergy to the overall system and fungi microalgae mass used was *Aspergillus funigatus* and *Synechocystis.* Research is still lacking in the area of how the immobilization mechanism occurs of adsorption of heavy metals occurs using such a fungi microalgae mass. Removal of cadmium (II) was 90% and of chromium (IV) was around 80%. Both extracellular and intracellular adsorptions were observed with extracellular adsorption in forefront. The exact mechanism between fungi and microalgae mass for elimination of cadmium and chromium was based theoretically on concept of optimization of research parameters and molecular response mechanism approach.

Another consortium of micro-algae and bacteria was applied by (Wang et al. 2023a) for remediating domestic wastewater. It was found to be successful in removal of carbon, nitrogen and phosphorus. Different types of algae were tried and gave excellent results in terms of performance, quantity of extracellular polymeric matter and settling characteristic and overall nutrient removal. Dominant algae species and other organisms were *Trebouxiophyceae, Saccharimonadales, Propionibacteriaceae, Propioniciclava,* and *Micropruina.* Combination of algae with bacteria in the remediation process is economic due to lower requirement of aeration for cultivation of algae in comparison to mono culture of algae (Papadopoulos et al. 2023).

A continuous mode photo reactor was studied by Zhang et al. in 2023 with bio films of bacteria and micro algae, developed inside the photo reactor. A hydraulic retention time (HRT) of 9 h was applied for the remediation of municipal wastewater. At the end of retention time, the biofilms in aerated conditions successfully removed 71.4% of total nitrogen, almost equivalent percentage of phosphorus and 90% of total organic carbon (TOC). The dominant eukaryotic algal type in this case was found to be *Chlorophyta* (Zhang et al. 2023a, b).

6.3 Need for Water Mining

One of the biggest issues the society facing today is the access to water in terms of its utility, energy, environmental, and financial implications. The worldwide water demand is anticipated to outgo the water availability by 2030, according to the World Bank's Water Resources Group. Wastewater with metals like lead, mercury and organic compounds like dyes and paints is a great environmental threat. Many governing bodies have designed guidelines for their prohibition (Patel et al. 2021) but due to their numerous applications; their discharge in surrounding environment becomes inevitable. This can be modified by investigating alternate water sources and creating creative approaches for sustainable water management. For example, utilising industrial and urban wastewater and saltwater desalination. These types of projects contribute to ensuring access to pure water and over all sanitization. Water mining will make it easier to scale up the manufacturing of attractive biobased products from wastewater treatment by-products. To achieve this; development for an innovative spin-off from granular sludge produced in wastewater treatment at a large scale is necessary. Cost-effective and eco-friendly recovery of minerals, phosphates, nitrates, metals, organic and inorganic molecules from wastewater treatment effluents or surface water will not only facilitate to improve water resources but in turn the surrounding environment and will also avoid their wastage. Water mining thus assures for pollutant free wastewater and their effective extraction. Studies have shown that algae-based treatments are more effective than chemical ones at eliminating several heavy metals and nutrients from wastewater. The use of microalgae in such a recovery and remediation of medicines, metals and organic pollutants in waste-water effluent is thus gaining popularity (Maryjoseph and Ketheesan 2020).

6.3.1 Algae and Water Mining

This method has harnessed noteworthy role as one of the green bioremediation methodologies for heavily polluted wastewater treatment because of effective recovery of nutrients, organic compounds, and inorganic compounds from wastewater (Topal et al. 2022). Microalgal metabolic activities facilitates for the production of a number of significant products. Microalgae utilise light energy to grow mixotrophically with carbohydrates and photo-autotrophically with CO_2. They can be harvested heterotrophically in deficiency of light if organic sugar is available in the form of energy and carbon. Thus, they can be easily cultivated in presence of merely required CO_2 and water; which is easily available in even wastewater. Microalgae can undergo genetic alterations that cause them to produce a range of chemicals.

Due to the availability of a complete sequencing database, genetic engineering is frequently studied for microalgae like *Chlorella, Dunaliella salina, Phaeodactylum, Chlamydomonas,* and *Synechocystis* species (Sivaramakrishnan et al. 2022). Bioaccumulation is a two-step mechanism. The first step is biosorption, here, heavy metals are adsorbed on the cell surface; the second step is metabolic process, wherein heavy metals are sent to the cells.

Algae has a tendency to resist toxicity of heavy metals, which makes it suitable for functioning in presence of heavy metals. This is due to the precipitation and complexation on the cell surface of algae, secretion to keep out cells, efflux, changing the heavy metal oxides state, chelating action of algae with intracellular polyphosphate bodies, polysaccharide and polypeptide, and lastly sub-organelles sequestration and categorisation (Zhang and Luo 2022).

In order to prevent metal toxicity, microalgae cope with heavy metal stress in a variety of ways which includes exclusion, complex formation, compartmentalization, and the binding proteins synthesis like metallothionein's (MTs) and phytochelatins (PCs). Algae even allows transfer of location of metal into cellular vacuoles thereby avoiding metal toxicity. Through a variety of active ligands and metal defence mechanisms, *Chlorella pyrenoidosa* can grow even under high concentration of metal like zinc, lead, copper, manganese, cadmium and nickel from wastewater.

The structure of algal cells is made up of thin (-SH) pools, which begin to decrease when metals are bound to them. As a result, the number of antioxidants in cells put them at a risk of oxidative stress state. Algal cells are exposed to a disruption of larger concentration of metal-induced oxygen. This amplifies internal lipid peroxidation and causing algal cell membrane rupture. Consequently, the lifespan of algal cells is shortened. Thus, phycoremediation of heavy metals is achieved showing a good association between biochemical compound and photosynthetic pigments (Kothari et al. 2022).

6.3.2 *Components from Water Mining*

6.3.2.1 Mining of Pharmaceuticals, Nutrients and Cosmetics

The nutritional, pharmaceutical, and cosmetic industries can be benefitted from the abundance of biomolecules found in microalgae, including proteins, lipids, pigments, vitamins, and antioxidants. Pharma by-products are one of the most emerging containments in wastewater. The most often reported and in-depth investigated species in pharma wastewater treatment are *Scenedesmus, Chlorella*, and *Chlamydomonas*. This is owing to their resilience and adaptability under adverse environmental conditions (Sivaramakrishnan et al. 2022).

6.3.2.2 Mining of Metals

The adsorption of cadmium metal (II) on algal biochar was found to be better compared to pomelo peel and corncob biochar. This was attributed to the presence of abundant mesoporous structure and presence of surface oxygen, nitrogen as functional groups on the algal biochar (Song et al. 2023). Zhang et al. (2023a, b) showed that microalgae *Chlorella vulgaris* in presence of fungi *Aspergillus flavus* in the form of pellets is an effective biomass for recovery of copper (II) from wastewater. The process exhibited a good adsorption capacity of 80.42 mg/g of copper. The uniform dispersion and suspension of some unicellular microalgae make it difficult to retrieve them after they have adsorbed metals from wastewater.

Photosynthesis of microalgae like *Scenedesmus* and *Chlorella* in swine water reduced the concentration of copper by 95%. Presence of copper from 0.1 to 3.0 mg/L affected the bioaccumulation of other components phosphorus, ammonia and even carbohydrates (de Sousa Oliveira et al. 2023).

The wastewater produced during the extraction of gold had a concentration of 10–40 mg/L and a total acid concentration of 4 M. After 30-min exposure, it was demonstrated that *Galdieria sulphuraria* algal cells can directly adsorb gold from such simulated effluent, with an adsorption capacity of 35 ± 2.5 mg/g gold. These algal processes have selectivity greater than that of an ion-exchange resin and are comparable to activated carbon (Miyashita et al. 2022). Another algae species *H. reticulatum* showed a great potential to bioaccumulate cadmium, chromium, lead, manganese, copper, and other elements. The highest uranium bioaccumulation capacity was able to achieve 24.47 ± 0.86 mg/g by dry biomass. This was due to the bioaccumulation features and acute toxicity of uranium to *Hydrodictyon reticulatum* (Zhang and Luo 2022).

6.3.2.3 Mining of Organic and Inorganic Components

Algal bioremediation treats waste streams and recover nutrients simultaneously. Wastewater stands to be a cheap feedstock for biosynthesis of organic components like proteins, unsaturated carboxylic acids etc. *Chlorella* and *Scenedesmus* due to their adaptive nature have demonstrated their utility in enhancing wastewater and flue gases (Leong and Chang 2023).

Phosphorus recovery through absorption in the growth of algae is a low-tech, low-cost ecological technique. Phosphorus recovery depends on simple biological phosphorus intake for algal growth, the precipitation of phosphorus salts with available ions, and the creation of oxygen to promote bacterial action. Even in temperate climates, microalgae have demonstrated their capacity to efficiently extract ammonium ion (NH_4^+) and phosphates from wastewater (Peng et al. 2018). Algae coupled constructed wetland is a good method for removal and recovery of not only phosphorus, nitrogen but even chemical oxygen (Chauhan and Prajapati 2022).

6.3.3 Methods and Technologies of Water Mining

The effectiveness of marine algae in chelating heavy metals from contaminated wastes has been the subject of numerous investigations. Both living and dead algae can be used as adsorbents for heavy metals, but recent research has shown that latter may be more efficient for metal chelation and less expensive than living cells. This is because it is neither affected by the toxicity of metal ions, nor it needs food or other essential elements for biological growth (Nowwar et al. 2022). Highly acidic conditions found in metal-containing wastewater from aqua regia are generally incompatible with algal biomass. Hence their tolerance towards acidic conditions must be raised to make them possible to recover metals from such wastewater practically.

The ability of biomaterials ability to adsorb substances is known to enhance by lyophilization. Lyophilization (process of preserving by quick freezing and removing ice by vacuum evaporation) has benefits in terms of cost and environmental impact as well. It increases the amount of solid matter in the adsorbent and reduces the need for chemicals, which makes it easier to transport, store, and manipulate. It has been successfully demonstrated to use lyophilized *G. sulphuraria* cells for the direct recovery of gold from its diluted aqua regia solutions (Miyashita et al. 2022).

6.3.4 Governing Factors for Water Mining

6.3.4.1 Concentration of Metal

Though the concentration of iron is low in the mine waters, it gets mainly accumulated by metabolic and photosynthetic processes inside the algae via absorption rather than adsorption on the algae surface. Similarly, zinc being an essential micronutrient, its accumulation is found in algal growth of mining waterbodies. The physical and chemical processes involved in algae growth can be adversely affected by high zinc concentrations in algae. The bioconcentration parameters of several metals decreased on average, in the following order: iron >> lead >> copper > cadmium > zinc (Du et al. 2022). The optimum concentration for algal growth of *Chlorella pyrenoidosa* was low concentration of heavy metals; where outcome showed highest biomass growth in 50% concentration of common effluent treatment plant wastewater (Kothari et al. 2022). When the copper concentration was raised, the biomass of *Scenedesmus dimorphus* was found to decrease (de Sousa Oliveira et al. 2023).

6.3.4.2 Temperature and Season

Temperature and seasons play an important role in the growth of algal species and in turn their capacity of bioaccumulation. A case study on the two most frequently found filamentous green algal species viz. *Ulothrix* and *Oedogonium* in the Frongoch (UK) acid mine drainage was done. It was discovered that iron and lead were primarily bioaccumulated in these algae, while cadmium, copper and zinc were adsorbed on the algae surface along with bioaccumulation. The collections of samples in summer showed more accumulation of the metals compared to other seasons which can be attributed to increased concentration of metals in waterbodies due to enhanced evaporation (Du et al. 2022).

6.3.4.3 Type of Species

Different species of microalgae have been used for remediation and recovery of various nutrients based on their capacity to extract a specific nutrient. *Chlorella* and *spirulina* are the most popular species utilised for nutrition and other algae-based food items. Microalgae can be differed genetically to manufacture a wide range of chemicals, including foreign substances.

Due to the availability of a complete sequencing database, many microalgae, including *Chlorella, Dunaliella salina, Phaeodactylum, Chlamydomonas,* and *Synechocystis* are being explored (Sivaramakrishnan et al. 2022). Peng et al. (2018) has mentioned various species of algae applicable for phosphorus recovery.

6.3.4.4 Ore Mines, Irrigation, Industrialisation and Other Human Activities

Ore mining activities by humans releases a large number of metals in the surrounding water bodies thus making it important to be recovered (Topal et al. 2022). Industrial processes including pharmaceuticals and drugs, cosmetics, paints and dyes, textile, pesticides, fertilizers etc. releases not only metals but also organic and many inorganic components in the nearby water reservoirs. A fresh algal species *C. fracta* showed a high accumulation capacity of toxic pollutants like chromium, iron, barium, lead, cadmium and manganese metals from gallery waters from mining areas (Topal et al. 2022).

6.4 Algal Bioremediation Wastes as Bioproducts

In the last few decades, these systems based have been investigated as an alternative to traditional municipal wastewater treatment systems. In comparison to conventional treatments like activated sludge, these photosynthetic perform wastewater remediation with larger percentage of removal efficiency and low costs.

Furthermore, the microalgal biomass that is produced during the treatment process can be used for making bioproducts like bioplastics, pigments, biogas and biofuel (Álvarez-González et al. 2023). An attempt for green concepts of biofuel production from macroalgae *Eucheuma denticulatum* biomass on an industrial scale has been found thermodynamically feasible (Loh et al. 2023).

Algal biorefinery platform offers a good enhancement to wastewater utility through biosynthesis of biofuels like biogas, ethanol and biodiesel (Leong and Chang 2023). The synthesis of numerous value-added products like biogas, biodiesel, pharmaceuticals, cosmetics and bio-adsorbents onsite can be done using algal biomass produced in common effluent treatment plant of wastewater (Kothari et al. 2022).

Bioethanol can be produced from algae after pre-treatment, hydrolysis, and fermentation. Carbohydrate content is higher in cell walls and starch-based cytoplasm, which can be used as a viable feedstock for the synthesis of bioethanol. These carbohydrates undergo enzymatic hydrolysis to yield monomeric sugars, which are then fermented to produce bioethanol. When compared to conventional crop feedstocks, using algae as a feedstock for the production of biofuel has benefits like quick harvesting, inexpensive farming, CO_2 mitigation, high removal, faster biomass growth rate, and less water use (Sivaramakrishnan et al. 2022).

6.5 Economic and Environmental Aspects of Algae Utilization

There are two methods for cultivation of algae. Either in an open pond or in the form of a closed type of photo-bioreactor (Bhatt et al. 2022). Owing to its operational characteristics and control in terms of nutrients build up is better obtained in a closed type of bio reactor. However, constructing a photo bio-reactor is more costly than that of an open pond construction (Harun et al. 2011).Growth of algae requires fully aerated conditions which increases the cost of cultivation due to long term aerator costs.

As suggested by Liu et al. (2020), factors under consideration while choosing right category of microalgae for bioremediation include supply of inorganic carbon, supply of nitrogen, phosphorus and other vital nutrients, type of reactor pond, its depth, location constraints, interaction of bacteria with algae, mixing and cultivating conditions and schedule. Amongst the mono culture and polyculture categories of filamentous algae, the former is more suitable for treatment plant-based applications than latter as it is less challenging in terms of maintaining its efficiency and quality control.

A case study of Brazil was presented by Vassalle et al. (2023) to decode how treatment of wastewater through use of algae has environmental footprint. Efficiency of an up-flow anaerobic sludge blanket (UASB) reactor merged with high-rate algal ponds was compared with that of UASB merged with trickling filters, wetlands and polishing ponds.

Algae consumes carbon, thereby reducing global carbon footprint. The integrated system of UASB with high-rate algal ponds was found to have the least environmental impacts. These were found to be feasible for small community-based treatment plants since high capital cost was compensated due to recovered heat and electricity, gave low operation and maintenance costs.

6.6 Future Perspectives

The merit and scope of microalgae is due to the fact that it can be easily and rapidly grown in different types of wastewaters. Abinandan et al. in 2018 inferred that the current practices of wastewater treatment under application have high costs and huge energy consumption which paves the need for newer approaches of treatment. Biological treatment modes are evergreen in terms of proficiency of application, cost effectiveness and energy factor. Even after significant removal of contaminants using microalgae, there is still limited acceptance of this technology among people (Li et al. 2022a, b).

Another parameter affecting the future scope of algae was emphasized by Mohsenpour et al. (2021). They discussed about the enhancement of quality of water bodies which were deteriorating lately due to eutrophocation. This could be brought down by reducing nitrogen and phosphorus in water bodies. Mixotrophic breed of algae is an emerging trend of algae in this regard. Conventional methods have been in use for a long time period and hence, perception of the users should be diverted towards this novel technique.

Abdelfattah et al. (2023) stated in his review that the attributes of algae are such that they have immense future prospects of bio-remediation of pollutants in domestic and industrial wastewater. Major merits of microalgae cultivation for wastewater treatment are its energy efficiency, highly transformative and reactive and adaptability of micro-algae in all extreme conditions.

Additionally, algae can utilise a number of mechanisms at once that work in concert to remove pharmaceuticals and other harmful substances from the environment. Among the most thoroughly studied remediation methods are sorption and accumulation by biological processes (Hejna et al. 2022).

6.7 Conclusions

The obligatory occurrence of huge urbanization, industrialization and agricultural events is responsible for adding high levels of toxic and hazard causing pollutants in wastewater. Algae based treatment systems are energy efficient as they utilize less amount of energy, light and resources in comparison to other systems, particularly those incorporating bacteria since they require dense oxygen environment.

The overall parameters responsible for affecting the cost of the technology are bioreactor configuration and possibility of merging conventional biological treatment with microalgae bioremediation to bring about reduction in overall cost and energy areas. It should be highlighted that despite the enormous diversity of microalgae species, only a few numbers of species have received enough research into their capacity for bioremediating pollutants. Consequently, screening programmes are required to verify the selective removal of these pollutants by a variety of microalgae species. It is also necessary to research effective algal-based remediation methods. The need of the hour is to categorise various algae production systems and to explain the benefits and drawbacks of algal phyco-remediation.

Many of the conventional wastewater treatments are not well-designed to effectively remove these contaminants. The phyco-remediation method, which appears to be a potential option as a cheap and environmentally benign wastewater treatment. The removal or biotransformation of contaminants using macro- or microalgae is a method that is constantly being rectified to address the problem of wastewater contamination.

References

Abdelfattah A, Ali SS, Ramadan H, El-Aswar EI, Eltawab R, Ho S-H, Elsamahy T, Li S, El- Sheekh MM, Schagerl M, Kornaros M, Sun J (2023) Microalgae-based wastewater treatment: mechanisms, challenges, recent advances, and future prospects. Environ Sci Ecotechnol 13:100205. https://doi.org/10.1016/j.ese.2022.100205

Abdulhameed AS, Jawad AH, Kashi E, Radzun KA, Alothman ZA, Wilson LD (2022) Insight into adsorption mechanism, modeling, and desirability function of crystal violet and methylene blue dyes by microalgae: Box-Behnken design application. Algal Res 67:102864. https://doi.org/10.1016/j.algal.2022.102864

Abinandan S, Subashchandrabose SR, Venkateswarlu K, Megharaj M (2018) Nutrient removal and biomass production: advances in microalgal biotechnology for wastewater treatment. Crit Rev Biotechnol 38(8):1244–1260. https://doi.org/10.1080/07388551.2018.1472066

Ali MEM, Abd El-Aty AM, Badawy MI, Ali RK (2018) Removal of pharmaceutical pollutants from synthetic wastewater using chemically modified biomass of green alga *Scenedesmus obliquus*. Ecotoxicol Environ Saf 151:144–152. https://doi.org/10.1016/j.ecoenv.2018.01.012

Almomani F, Bhosale RR (2021) Bio-sorption of toxic metals from industrial wastewater by algae strains *Spirulina platensis* and *Chlorella vulgaris*: application of isotherm, kinetic models and process optimization. Sci Total Environ 755:142654. https://doi.org/10.1016/j.scitotenv.2020.142654

Álvarez-González A, Uggetti E, Serrano L, Gorchs G, Escolà Casas M, Matamoros V, Gonzalez-Flo E, Díez-Montero R (2023) The potential of wastewater grown microalgae for agricultural purposes: contaminants of emerging concern, heavy metals and pathogens assessment. Environ Pollut 324:121399. https://doi.org/10.1016/j.envpol.2023.121399

Bhatt P, Bhandari G, Turco RF, Aminikhoei Z, Bhatt K, Simsek H (2022) Algae in wastewater treatment, mechanism, and application of biomass for production of value-added product. Environ Pollut 309:119688. https://doi.org/10.1016/j.envpol.2022.119688

Chauhan A, Prajapati SK (2022) Algae coupled constructed wetland system for wastewater treatment. In: Integrated environmental technologies for wastewater treatment and sustainable development, pp 191–216. https://doi.org/10.1016/B978-0-323-91180-1.00003-X

de Sousa Oliveira AP, Assemany P, Covell L, Calijuri ML (2023) Copper multifaceted interferences during swine wastewater treatment in high-rate algal ponds: alterations on nutrient removal, biomass composition and resource recovery. Environ Pollut 324:121364. https://doi.org/10.1016/j.envpol.2023.121364

Du T, Bogush A, Edwards P, Stanley P, Lombardi AT, Campos LC (2022) Bioaccumulation of metals by algae from acid mine drainage: a case study of Frongoch Mine (UK). Environ Sci Pollut Res 29(21):32261–32270. https://doi.org/10.1007/s11356-022-19604-1

Fewtrell L (2004) Drinking-water nitrate, methemoglobinemia, and global burden of disease: a discussion. Environ Health Perspect 112(14):1371–1374. https://doi.org/10.1289/ehp.7216

Gorzelnik SA, Zhu X, Angelidaki I, Koski M, Valverde-Pérez B (2023) Daphnia magna as biological harvesters for green microalgae grown on recirculated aquaculture system effluents. Sci Total Environ 873. https://doi.org/10.1016/j.scitotenv.2023.162247

Goswami RK, Agrawal K, Verma P (2023) Multifaceted role of microalgae for municipal wastewater treatment: a futuristic outlook toward wastewater management. CLEAN—Soil, Air, Water 51(3):2100286. https://doi.org/10.1002/clen.202100286

Harun R, Davidson M, Doyle M, Gopiraj R, Danquah M, Forde G (2011) Technoeconomic analysis of an integrated microalgae photobioreactor, biodiesel and biogas production facility. Biomass Bioenerg 35(1):741–747. https://doi.org/10.1016/j.biombioe.2010.10.007

Hejna M, Kapuścińska D, Aksmann A (2022) Pharmaceuticals in the aquatic environment: a review on eco-toxicology and the remediation potential of algae. Int J Environ Res Public Health 19(13). https://doi.org/10.3390/ijerph19137717

Khan S, Shamshad I, Waqas M, Nawab J, Ming L (2017) Remediating industrial wastewater containing potentially toxic elements with four freshwater algae. Ecol Eng 102:536–541. https://doi.org/10.1016/j.ecoleng.2017.02.038

Kothari R, Pandey A, Ahmad S, Singh HM, Pathak VV, Tyagi VV, Kumar K, Sari A (2022) Utilization of *Chlorella pyrenoidosa* for remediation of common effluent treatment plant wastewater in coupling with co-relational study: an experimental approach. Bull Environ Contam Toxicol 108(3):507–517. https://doi.org/10.1007/s00128-021-03292-7

Leong YK, Chang J-S (2023) Waste stream valorization-based low-carbon bioeconomy utilizing algae as a biorefinery platform. Renew Sustain Energy Rev 178:113245. https://doi.org/10.1016/j.rser.2023.113245

Li M, Zamyadi A, Zhang W, Dumée LF, Gao L (2022a) Algae-based water treatment: a promising and sustainable approach. J Water Process Eng 46:102630. https://doi.org/10.1016/j.jwpe.2022.102630

Li S, Show PL, Ngo HH, Ho S-H (2022b) Algae-mediated antibiotic wastewater treatment: a critical review. Environ Sci Ecotechnol 9:100145. https://doi.org/10.1016/j.ese.2022.100145

Liu J, Pemberton B, Lewis J, Scales PJ, Martin GJO (2020) Wastewater treatment using filamentous algae—a review. Biores Technol 298:122556. https://doi.org/10.1016/j.biortech.2019.122556

Loh SR, Tan IS, Foo HCY, Tan YH, Lam MK, Lim S (2023) Energy analysis of a holistic zero waste macroalgae-based third-generation bioethanol biorefinery approach: biowaste to bioenergy. Environ Technol Innov 30:103089. https://doi.org/10.1016/j.eti.2023.103089

Ma M, Yu Z, Jiang L, Hou Q, Xie Z, Liu M, Yu S, Pei H (2023) Alga-based dairy wastewater treatment scheme: candidates screening, process advancement, and economic analysis. J Clean Prod 390:136105. https://doi.org/10.1016/J.jclepro2023.136105

Maryjoseph S, Ketheesan B (2020) Microalgae based wastewater treatment for the removal of emerging contaminants: a review of challenges and opportunities. Case Stud Chem Environ Eng 2. https://doi.org/10.1016/j.cscee.2020.100046

Miyashita SI, Ogura T, Kondo T, Fujii SI, Inagaki K, Takahashi Y, Minoda A (2022) Recovery of Au from dilute aqua regia solutions via adsorption on the lyophilized cells of a unicellular red alga *Galdieria sulphuraria*: a mechanism study. J Hazard Mater 425:127982. https://doi.org/10.1016/j.jhazmat.2021.127982

Mohsenpour SF, Hennige S, Willoughby N, Adeloye A, Gutierrez T (2021) Integrating micro-algae into wastewater treatment: a review. Sci Total Environ 752:142168. https://doi.org/10.1016/j.scitotenv.2020.142168

Nasir NM, Jusoh A, Harun R, Ibrahim NNLN, Rasit N, Ghani WAWAK, Kurniawan SB (2023) Nutrient consumption of green microalgae, Chlorella sp. during the bioremediation of shrimp aquaculture wastewater. Algal Res 72. https://doi.org/10.1016/j.algal.2023.103110

Nowwar AI, Farghal II, Ismail MA, Amin MA (2022) Impact of irrigation with wastewater on accumulation of heavy metals in *Phaseolus vulgaris* L. and its remediation. J Soil Sci Plant Nutr. https://doi.org/10.1007/s42729-022-01080-8

Papadopoulos KP, Economou CN, Stefanidou N, Moustaka-Gouni M, Genitsaris S, Aggelis G, Tekerlekopoulou AG, Vayenas DV (2023) A semi-continuous algal-bacterial wastewater treatment process coupled with bioethanol production. J Environ Manage 326:116717. https://doi.org/10.1016/j.jenvman.2022.116717

Patel A, Enman J, Gulkova A, Guntoro PI, Dutkiewicz A, Ghorbani Y, Rova U, Christakopoulos P, Matsakas L (2021) Integrating biometallurgical recovery of metals with biogenic synthesis of nanoparticles. Chemosphere 263. https://doi.org/10.1016/j.chemosphere.2020.128306

Peng L, Dai H, Wu Y, Peng Y, Lu X (2018) A comprehensive review of the available media and approaches for phosphorus recovery from wastewater. Water Air Soil Pollut 229(4). https://doi.org/10.1007/s11270-018-3706-4

Ramesh B, Saravanan A, Senthil Kumar P, Yaashikaa PR, Thamarai P, Shaji A, Rangasamy G (2023) A review on algae biosorption for the removal of hazardous pollutants from wastewater: limiting factors, prospects and recommendations. Environ Pollut 327:121572. https://doi.org/10.1016/j.envpol.2023.121572

Sharmila A (2020) Cr (VI) removal from industrial wastewater using immobilized algae. Biosci Biotechnol Res Commun 13(15):108–114. https://doi.org/10.21786/bbrc/13.15/19

Singh A, Pal DB, Kumar S, Srivastva N, Syed A, Elgorban AM, Singh R, Gupta VK (2021) Studies on zero-cost algae-based phytoremediation of dye and heavy metal from simulated wastewater. Biores Technol 342:125971. https://doi.org/10.1016/j.biortech.2021.125971

Sivaramakrishnan R, Suresh S, Kanwal S, Ramadoss G, Ramprakash B, Incharoensakdi A (2022) Microalgal biorefinery concepts' developments for biofuel and bioproducts: current perspective and bottlenecks. Int J Mol Sci 23(5). https://doi.org/10.3390/ijms23052623

Song Z, Li L, Chen Y, Duan X, Ren N (2023) Efficient removal and recovery of Cd^{2+} from aqueous solutions by capacitive deionization (CDI) method using biochars. J Mater Sci Technol 148:10–18. https://doi.org/10.1016/j.jmst.2022.11.016

Topal M, Arslan Topal EI, Öbek E (2022) A green algae *Cladophora fracta* for accumulation of toxic/harmful pollutants causing environmental pollution in mine gallery waters. Int J Environ Sci Technol 19(5):4481–4490. https://doi.org/10.1007/s13762-021-03479-9

Uma VS, Panigrahi S, Chandrasekaran S, Srinivas CV, Venkatraman B (2023) Development of marine activated algal-bacterial granule: a novel replacement to the conventional algal remediation processes. Algal Res 69. https://doi.org/10.1016/j.algal.2022.102914

Vassalle L, Ferrer I, Passos F, Filho CRM, Garfí M (2023) Nature-based solutions for wastewater treatment and bioenergy recovery: a comparative life cycle assessment. Sci Total Environ 880:163291. https://doi.org/10.1016/j.scitotenv.2023.163291

Wang H, Turner JA (2013) Photoelectrochemical reduction of nitrates at the illuminated p-$GaInP_2$ photoelectrode. Energy Environ Sci 6(6):1802–1805. https://doi.org/10.1039/c3ee40745d

Wang J, Deng Y, Chen W, Song H, Zan F, Wu X, Ren Y (2023a) Revealing the role of algae in algae enhanced bacteria consortia for municipal wastewater treatment: performance, characteristics, and microbial pathways. J Water Process Eng 53:103640. https://doi.org/10.1016/j.jwpe.2023.103640

Wang J, Tian Q, Cui L, Cheng J, Zhou H, Zhang Y, Peng A, Shen L (2023) Bioimmobilization and transformation of chromium and cadmium in the fungi-microalgae symbiotic system. J Hazardous Mater 445. https://doi.org/10.1016/j.jhazmat.2022.130507

Wu K, Atasoy M, Zweers H, Rijnaarts H, Langenhoff A, Fernandes TV (2023) Impact of wastewater characteristics on the removal of organic micropollutants by *Chlorella sorokiniana*. J Hazardous Mater 453. https://doi.org/10.1016/j.jhazmat.2023.131451

Zhang J, Luo X (2022) Bioaccumulation characteristics and acute toxicity of uranium in *Hydrodictyon reticulatum*: an algae with potential for wastewater remediation. Chemosphere 289:133189. https://doi.org/10.1016/j.chemosphere.2021.133189

Zhang C, Wang X, Ma Z, Luan Z, Wang Y, Wang Z, Wang L (2020) Removal of phenolic substances from wastewater by algae. A review. Environ Chem Lett 18(2):377–392. https://doi.org/10.1007/s10311-019-00953-2

Zhang C, Laipan M, Zhang L, Yu S, Li Y, Guo J (2023a) Capturing effects of filamentous fungi *Aspergillus flavus* ZJ-1 on microalgae *Chlorella vulgaris* WZ-1 and the application of their co-integrated fungi-algae pellets for Cu (II) adsorption. J Hazard Mater 442:130105. https://doi.org/10.1016/j.jhazmat.2022.130105

Zhang X, Ji B, Tian J, Liu Y (2023b) Development, performance and microbial community analysis of a continuous-flow microalgal-bacterial biofilm photoreactor for municipal wastewater treatment. J Environ Manage 338:117770. https://doi.org/10.1016/j.jenvman.2023.117770

Chapter 7
Algae Technologies for Environmental Management and Bioremediation

Andrés F. Barajas-Solano, Janet B. Garcia-Martínez, Jefferson E. Contreras Ropero, and Antonio Zuorro

Abstract The application of algal-based bioremediation technologies hold the promise to recycle high content of N, P, and micronutrients already loaded in different wastewater streams which can potentially reduce environmental impact of the industry, while generating an interesting biomass which can be transformed into renewable metabolites with different grades of usability (from low to high-end products), including biopesticides, biofertilizers, animal feed and high-value metabolites such as colorants and food additives. However, different challenges such as the optimization of upstream and downstream processes must be overcome to deliver the true potential of algal-based bioremediation processes. The following chapter summarizes state-of-the-art and the challenges in the usage of microalgal and cyanobacterial strains in the recovery of nutrients and the synthesis of bio-based products from different wastewaters.

Keywords Wastewater treatment · Microalgal strains · Nutrient removal · Biomass production

7.1 Introduction

The emergence of algae-based technologies in the field of environmental management and bioremediation has offered a promising solution to address the challenges posed by the increasing discharge of wastewater and its detrimental effects on the

A. F. Barajas-Solano (✉) · J. B. Garcia-Martínez · J. E. Contreras Ropero
Department of Environmental Sciences, Universidad Francisco de Paula Santander, Av. Gran Colombia No. 12E-96, 540003 Cúcuta, Colombia
e-mail: andresfernandobs@ufps.edu.co

A. Zuorro
Department of Chemical Engineering, Materials, and Environment, Sapienza University, Via Eudossiana 18, 00184 Roma, Italy
e-mail: antonio.zuorro@uniroma1.it

T. A. Kurniawan and A. Anouzla (eds.), *Algae as a Natural Solution for Challenges in Water-Food-Energy Nexus*, Environmental Science and Engineering,
https://doi.org/10.1007/978-981-97-2371-3_7

environment (Arun et al. 2020). Wastewater is a complex liquid with varying biological and physicochemical compositions, often characterized by elevated levels of organic compounds, microorganisms, suspended solids, and heavy metals, among other pollutants. The presence of these contaminants in effluents poses significant environmental risks to ecosystems, disrupts vital ecological services, and directly impacts water availability, exacerbating the strain caused by water stress induced by climate change (Haddeland et al. 2014; IPCC 2018). In this context, the development and implementation of efficient wastewater treatment technologies have become of paramount importance to mitigate adverse environmental consequences.

Conventional approaches for wastewater treatment consist of a combination of biological and physicochemical methods tailored to the specific composition and disposal needs of the effluent. In recent times, there has been a concentrated effort in research to enhance these conventional systems and introduce hybrid treatment techniques that can provide more effective water quality improvement and reduce environmental impact. Techniques such as coagulation-flocculation (Mohamad et al. 2021; Zhao et al. 2021) and adsorption/ion exchange (Tomonaga et al. 2021; López-Maldonado and Oropeza-Guzmán 2021), falling under the umbrella of physicochemical treatments, have demonstrated their efficacy in treating wastewater from various industries, including palm oil production, laundry wastewater, and industrial and municipal wastewater.

On the other hand, the primary aim of biological treatments is to maintain a sufficient level of dissolved oxygen by removing organic and inorganic contaminants (Sodhi et al. 2021; Xu et al. 2021) that have the potential to directly harm aquatic ecosystems and the organisms depending on them. These biological treatment processes have shown promise in the removal of various organic and inorganic compounds, including the elimination of genetic material from severe acute respiratory syndrome coronavirus (SARS-CoV-2) through the use of traditional activated sludge systems (Kumar et al. 2021). Additionally, these treatments have the added benefit of simultaneous removal of phosphorus and nitrogen, as well as the production of volatile fatty acids, methane, and solubilization of chemical oxygen demand, volatile suspended solids, and complex macromolecular compounds (He et al. 2021; Mohamad et al. 2021).

Over the years, Various methods have been implemented to improve the quality of wastewater from both livestock facilities and domestic sources. One of these methods is the utilization of artificial wetlands (Karpisek et al. 2021), which have demonstrated their effectiveness in treating effluents from coffee processing (Said et al. 2021), textile wastewater (Tasneem et al. 2021), and the removal of contaminants, including heavy metals (Masinire et al. 2021), pharmaceuticals, personal care products, and pesticides (Al-Baldawi et al. 2021; Brunhoferova et al. 2021). These treatment systems make use of the photosynthetic capabilities of macrophytic plants to facilitate bioremediation techniques. On the other hand, microorganisms, particularly microalgae, have shown their potential as powerful tools for environmental management and bioremediation in various industrial sectors worldwide (Cardoso et al. 2021; Wu et al. 2020; Oliveira et al. 2019; Fazal et al. 2021). These organisms have emerged as a promising solution to address these challenges. Their rapid growth and ability to thrive in nutrient-rich wastewater make them highly suitable for effectively removing undesirable elements (Ahmad et al. 2022). Due to their photosynthetic nature, microalgae have gained significant attention in the field of phytoremediation. This is primarily because of their remarkable ability to remove contaminants from various sources of wastewater, including those associated with pig farming, poultry farming, textile industries, mining operations, and landfill leachates (Viegas et al. 2021; Zuhra et al. 2021). These microorganisms have been efficient in adsorbing heavy metals and transforming them into less harmful forms (Liu et al. 2021; Plöhn et al. 2021). Moreover, microalgae provide the additional advantage of alleviating the burden of recalcitrant pollutants, which are notoriously difficult to degrade (Mohamad et al. 2021).

7.2 Microalgae Based Remediation Systems

Various bioremediation systems have emerged with a dual purpose of removing contaminants and generating biomass for the extraction of metabolites (Dogaris et al. 2019; Hernández-García et al. 2019; Li et al. 2022; Su 2021). Another area in development is the production of biofuels from microalgae. The biomass of these microorganisms is rich in lipids and other compounds suitable for obtaining renewable biofuels, thereby contributing to reducing dependence on fossil fuels and greenhouse gas emissions (Chew et al. 2018; Hu et al. 2019). These microorganisms not only thrive in diverse soil and water conditions but also have a faster energy conversion rate from solar energy to biomass compared to plants (Chia et al. 2018). However, it is important to note that the production costs of microalgae-based biodiesel are still relatively high compared to traditional fossil fuels (Chew et al. 2018; Hu et al. 2019). Furthermore, their biomass holds significant value and can be used to produce a range of metabolites with industrial applications (Dayana Priyadharshini et al. 2021; Kholssi et al. 2021). A clear example of this is the application of technified microalgae cultures in post-consumer aquaculture waters, which offers the possibility of obtaining economically viable products for the national aquaculture sector. These products include protein and lipid-rich concentrates that serve as high-quality feed for fish and other aquatic animals, enhancing overall efficiency and sustainability in the farming process (Ahmed et al. 2021). Additionally, microalgae biomass resulting from nutrient removal can be processed into biofertilizers, providing essential nutrients for terrestrial crops and reducing dependence on chemical fertilizers (Ansari et al. 2017; Ahmad et al. 2022). This not only benefits agriculture but also promotes a circular and sustainable nutrient cycle between aquaculture and agriculture. Moreover, these organisms have a wide range of applications in various industries. In the food industry, they are recognized for their potential as emulsifiers and nutritional supplements due to their rich content of proteins, amino acids, minerals, polyunsaturated fatty acids, and vitamins (Silva et al. 2021; Siahbalaei et al. 2021; Bai et al. 2021). Additionally, they have been used as feedstock for livestock, poultry, and fish production (Silva et al. 2021). Beyond the food industry, the pharmaceutical and cosmetic sectors have shown interest in other microalgae-derived bioproducts. These bioproducts include heavy-metal-free polysaccharides that promote the generation of collagen I and possess potential properties such as antiallergic, wound-healing, antioxidant, and antitumor effects (Miguel et al. 2021; Tseng et al. 2021). Moreover, microalgae can produce biostimulants that enhance plant defenses against biotic stress (Rachidi et al. 2021).

The introduction of algae-based biorefineries has opened up new possibilities for improving cultivation methods and increasing the yield of essential cellular elements such as proteins, lipids, carbohydrates, and bioactive compounds (Figueroa-Torres et al. 2021). These advances not only increase production but also contribute to minimizing environmental damage (Ortiz et al. 2021). The key steps of microalgae-based biorefinery include cultivation and harvesting, cell disruption, and extraction of value-added compounds (García-Martínez et al. 2022a, b). Suitable microalgae species must be selected based on their carbohydrate, lipid, and protein content to ensure the production of high-quality biofuels and value-added products (Converti et al. 2009). The integration of upstream and downstream processes can make microalgae's value-added products more economically viable (García-Martínez et al. 2022a, b). Therefore, a systematic approach combining genome sequencing and process-scale engineering is crucial for the development of future microalgae biorefineries (Converti et al. 2009). Microalgae-based biorefineries offer an innovative solution to the growing need for sustainable compounds while remediating water pollution associated with conventional industrial techniques (Sivaramakrishnan and Incharoensakdi 2021).

In wastewater bioremediation, various types of algae are used, each with its own characteristics and unique capabilities. For instance, microalgae have phycoremediation capabilities, involving the removal or biotransformation of contaminants from liquid or gaseous media (Foglia et al. 2021; Giri et al. 2021). These microorganisms can absorb and metabolize contaminants, thereby reducing their concentration in the water (Touliabah et al. 2022). On the other hand, macroalgae are larger, more visible algae that can be used to remove heavy metals and nutrients from wastewater (Giri et al. 2021). By selecting the appropriate type of algae for specific contaminants, wastewater treatment systems can effectively target and remove pollutants, resulting in improved water quality and environmental sustainability (Al-Anbari et al. 2022; Glover et al. 2022).

However, cultivating and harvesting algal biomass for wastewater bioremediation requires careful consideration of several parameters. Factors such as nutrient availability, light intensity, temperature, and pH levels play a crucial role in algae growth and productivity (de Benedetti et al. 2015; Raghuvanshi et al. 2018). Understanding these parameters and optimizing cultivation conditions can improve the efficiency and effectiveness of algae-based wastewater treatment systems (Raghuvanshi et al. 2018). Additionally, selecting appropriate harvesting methods is critical to ensure efficient extraction of algae biomass from the water. Various harvesting techniques, such as centrifugation, filtration, and flocculation, have been explored to maximize algae biomass recovery. A critical analysis of algae cultivation and harvesting methods is essential to enhance overall performance and sustainability of algae technologies in environmental management and wastewater bioremediation (Li et al. 2020; Chen et al. 2020). The future of algae technologies in environmental management and wastewater bioremediation is very promising for a wide range of applications. With continued research and development efforts, these technologies are expected to become more efficient, cost-effective, and scalable. In this context, several key areas of research are focused on obtaining sustainable products.

7.3 Algal-Based Biofuels Production

Biofuels derived from algal biomass have been proposed as an interesting player in transitioning towards renewable energy sources (Li et al. 2020). Ongoing research aims to enhance the efficiency of biofuel production processes, including lipid extraction and conversion technologies. Integrating algae cultivation with other renewable energy systems, such as solar and wind, can further enhance the sustainability of biofuel production (Catone et al. 2021; Ahmad et al. 2022). Concerning biofuel production, algae-based technology has demonstrated high potential. According to a study conducted by Ahmad et al. (2022), algae-based biofuel production can be a green and cost-effective process. However, it is important to consider that the production cost of biofuels from algae is still higher compared to fossil fuels. Judd et al. (2017) reported that the selling price of algae oil exceeds \$2 per liter, while fossil fuel prices are below \$1 per liter. This poses a challenge in terms of competitiveness in the biofuel market. Additionally, algae-based biofuel production faces technical challenges. For instance, Cerón et al. (2008) found that the transesterification process yield using wet algae biomass was 20% lower compared to other processes. This indicates that further research and development are required to improve the efficiency of algae-based biofuel production processes. Moreover, the bioremediation of wastewater and algae-based biofuel production raise environmental concerns. For example, algae overgrowth in water bodies can have adverse effects on aquatic ecosystems. Dourou et al. (2018) reported that excessive algae growth can deplete oxygen in the water, leading to the death of fish and other aquatic life forms. Additionally, the decomposition of dead algae can release toxic compounds and contribute to water eutrophication. It is crucial to note that the efficiency of nutrient removal can vary depending on specific system conditions and the algae species used. For example, Satayavibul and Ratanatamskul (2021) found that algae-based wastewater ponds achieved efficient nitrogen removal but had lower performance in phosphorus removal. Furthermore, wastewater bioremediation using algae can also address the elimination of organic and inorganic contaminants. Yang et al. (2018) demonstrated that algae can efficiently remove organic contaminants, such as pesticides and phenolic compounds, through adsorption and biodegradation mechanisms. However, the efficiency of wastewater bioremediation using algae may be affected by various factors, such as the initial contaminant concentration, wastewater composition, and algae cultivation conditions. For example, Raghuvanshi et al. (2018) found that the efficiency of organic contaminant removal by algae was higher in anaerobically digested kitchen wastewater compared to dairy wastewater. This indicates that wastewater composition can influence the ability of algae to remove specific contaminants.

Wastewater bioremediation using algae may have limitations concerning scalability and applicability to different types of liquid wastes. For instance, Xu et al. (2014) discovered that algae nitrogen removal efficiency was higher in pig wastewater compared to cattle wastewater. This suggests that specific wastewater characteristics can influence the capacity of algae to remove contaminants.

7.4 Carbon Capture Coupled with Wastewater Treatment

Algae possess the capability to capture and utilize carbon dioxide (CO_2) from various sources, including industrial emissions and combustion gases. Advances in algae cultivation techniques and photobioreactors can enhance CO_2 sequestration efficiency. Moreover, the captured CO_2 can be used as a raw material to produce valuable products such as biofuels, bioplastics, and biochemicals (Olguín et al. 2022). The utilization of algae for carbon capture and utilization has been extensively studied and validated by numerous researchers. For instance, Olguín (2003) highlighted key issues for cost-effective nutrient removal processes in phytoremediation. The author emphasized the importance of selecting appropriate algae species for specific phytoremediation applications, as different species have varying capacities for carbon capture and utilization and differing levels of tolerance to pollutants (Pinto et al. 2002).

The potential of algae for carbon capture and utilization has also been explored in the context of reducing greenhouse gas emissions. Algae can capture carbon dioxide from various sources, including industrial emissions and wastewater treatment plants. This not only helps mitigate climate change but also provides a valuable raw material for high-value product production. A study by Chisti (2007) highlighted the potential of algae-based biofuels as a sustainable alternative to fossil fuels, with the additional benefit of carbon capture and utilization (Mulbry et al. 2008). These microorganisms have demonstrated the ability to degrade these compounds through a combination of exogenously supplied energy and carbon sources, as well as the effects of induction and resonance of the substituents (Olguín 2003). For example, in a study by Papazi and Kotzabasis (2007), enhanced phenolic compound removal from wastewater was achieved by improving the input and flow of inorganic carbon sources. However, it is essential to note that challenges and limitations exist regarding the implementation of algae technologies for environmental management and bioremediation. One of the main challenges is the scalability of algae cultivation. While laboratory-scale studies have shown promising results, scaling up algae production for large-scale applications can be a technical and economic challenge. Factors such as the availability of suitable land and water resources, as well as energy and nutrient requirements for algae cultivation, contribute to this challenge (Mulbry et al. 2008). Additionally, the economic viability of algae technologies for carbon capture and utilization remains a topic of debate. While algae can capture carbon dioxide and convert it into biomass, the cost of algae cultivation and processing can be high. Furthermore, the market for algae-based products, such as biofuels, is still developing and may not yet support large-scale implementation of algae technologies (Chisti 2007).

7.5 Nutrient Recovery and Circular Economy

Algae technologies offer the potential to recover valuable nutrients such as nitrogen and phosphorus from wastewater and agricultural runoff. These nutrients can be recycled and used as fertilizers in agriculture, reducing the reliance on synthetic fertilizers and minimizing environmental pollution. Integrating algae-based systems with wastewater treatment plants and agricultural practices can create a more sustainable and circular economy (Ashokkumar et al. 2022; Zhou et al. 2013). Current microalgae cultivation techniques have a high demand for nutrients and result in negative energy production, posing challenges for the competitiveness and sustainability of microalgae-based compound production (Zhou et al. 2013). An approach to optimize processes and enhance the sustainability of microalgae production is nutrient recirculation and reuse (Gahane et al. 2022). The closed-loop production process of microalgae offers several advantages in terms of nutrient recovery and energy generation. By using anaerobic digestion and microalgae cultivation, it is possible to recycle nutrients and reduce dependence on external inputs such as nitrogen and phosphorus fertilizers. A study focused on using digested liquid (DL) from anaerobic digestion as a nutrient source for microalgae cultivation (Yang et al. 2018). Researchers found that the DL contained high levels of ammonia, phosphorus, and potassium, which are essential nutrients for microalgae growth. This indicates that DL can serve as a suitable medium for microalgae cultivation, reducing the need for synthetic fertilizers. Furthermore, the study demonstrated that microalgae cultured with DL had a higher biomass concentration compared to those grown in a standard medium (Yang et al. 2018). This suggests that the recycled nutrients in DL can effectively support microalgae growth and biomass production. By utilizing DL, the nutrient costs associated with microalgae-based biofuel production can be significantly reduced.

Several studies have investigated the potential of using microalgae for the simultaneous removal of nutrients from wastewater and lipid production. For example, Yoo et al. (2010) selected freshwater microalgae for their ability to remove nutrients from wastewater while producing lipids. They found that certain microalgae strains were effective in removing nitrogen and phosphorus from wastewater, indicating their nutrient recovery potential. Another study explored the potential use of lipid-extracted microalgae (LEM) for methane production and nutrient recycling (Cai et al. 2013). It was found that LEM had a higher crude lipid content compared to raw microalgae (RM), making them a promising feedstock for biofuel production. The researchers discovered that the methane production potential of LEM could offset the cost of microalgae-based biofuel production. This highlights the energy recovery potential through anaerobic digestion in the closed-loop process.

7.6 Value-Added Products

In addition to biofuels, algae can be used to produce a wide range of value-added products, including high-value chemicals, nutraceuticals, and animal feed. Research efforts are directed toward exploring the potential of algae to produce these products and developing cost-effective methods for extraction and purification (Miguel et al. 2021; Tseng et al. 2021). The use of microalgae for medical purposes, such as a source of proteins and antibiotics, was first explored during World War II (Chisti 2007). In the 1970s, microalgae gained increased interest as a renewable energy source during the energy crisis, and research focused on hydrogen and methane production (Chisti 2007). Since then, extensive research has been conducted on the use of microalgae for wastewater treatment, carbon sequestration, and the production of economically valuable products (Chisti 2007; García-Martínez et al. 2022a, b). In addition to biofuels, microalgae can be used to produce a wide range of value-added products, including high-value chemicals, nutraceuticals, and animal feed (Converti et al. 2009). The content of carbohydrates, proteins, lipids, and pigments in microalgae makes them suitable for the production of these valuable products (Khan and Shin 2018; Sivaramakrishnan et al. 2022). Carbohydrates, proteins, and lipids can be found in microalgae species in the range of 8–69.7%, 5–74%, and 7–65%, respectively (Converti et al. 2009). This versatility allows microalgae to be used in multiple industries, including agriculture, livestock, medicine, culinary, and cosmetics (García-Martínez et al. 2022a, b).

Regarding environmental management and bioremediation, microalgae hold a crucial position, as they can capture carbon dioxide during biomass production, mitigating greenhouse gas impact. This capability makes them a potential tool for bioremediation. Additionally, microalgae can be used for wastewater treatment as they can remove nutrients and contaminants from water bodies (Ortiz-Betancur et al. 2022; García-Martínez et al. 2022a, b). This makes them a sustainable and environmentally friendly solution for water pollution control. Recent breakthroughs in algae biotechnology have unveiled exciting prospects for environmental management and bioremediation. One area that has seen remarkable progress is the field of genetic engineering and algae strain enhancement. Researchers have successfully manipulated algae metabolism to bolster their bioremediation capabilities, enabling efficient removal of contaminants and heavy metals from water bodies (Ummalyma et al. 2017). These advancements have significant implications for wastewater treatment and environmental restoration (Pacheco et al. 2015; Ummalyma et al. 2017). Moreover, strides have been made in developing high-performance algae strains for biofuel production (Khan and Shin 2018). These genetically modified strains hold promise in providing a sustainable and cost-effective source of biofuels (Fabris et al. 2020). The progress in genetic engineering and strain improvement techniques has paved the way for more effective and efficient algae-based solutions for environmental management (Fabris et al. 2020). Another area witnessing remarkable progress in algae biotechnology is the potential integration of algae with other sustainable technologies. Algae cultivation has proven to be an efficient option for wastewater bioremediation, with microalgae particularly effective in nutrient recovery from wastewater (Pacheco et al. 2015). Such integration with wastewater treatment systems can improve nutrient removal efficiency and mitigate the environmental impact of wastewater discharge (Ortiz-Betancur et al. 2022). Furthermore, researchers have explored the technological advancement of photosynthetic microalgae in microbial fuel cells for energy production and wastewater treatment (Khan and Shin 2018; Fabris et al. 2020).

The headway made in algae biotechnology has also spurred research and investment opportunities in various fields. Algae's potential in nutrition, aquaculture, pharmaceuticals, and biofuels has gained recognition (Sivaramakrishnan et al. 2022; Rosenberg et al. 2008). As a result, increased research and investment are underway to further explore algae's potential and develop new products and applications (Fabris et al. 2020). These endeavors not only advance algae biotechnology but also hold promise in addressing pressing environmental issues and promoting sustainable development (Pacheco et al. 2015).

The integration of algae technologies with other sustainable solutions presents intriguing avenues for research and investment. By combining algae with other renewable energy sources like solar and wind, energy production efficiency and sustainability can be enhanced (Nhat et al. 2018). This integration can also lead to a circular economy approach, where wastewater serves as a nutrient source for algae growth, aiding in wastewater treatment while generating biomass for diverse applications, including biofuel production (Rahman et al. 2020). Exploring synergies between algae technologies and other sustainable solutions can lead to integrated systems with multiple environmental and economic benefits. Investing in research and development in this area can contribute to the advancement of efficient technologies for environmental management. However, for widespread adoption of sustainable algae solutions, political support and funding are vital. Governments and regulatory bodies play a crucial role in creating an enabling environment for the development and implementation of algae technologies. This includes providing financial incentives, grants, and subsidies for algae-related projects and initiatives. Additionally, policies can be implemented to encourage algae use in diverse industries such as agriculture, energy production, and pollution control (Fabris et al. 2020; Roy et al. 2022). Active support and funding of algae- based solutions can expedite the transition towards a more sustainable and environmentally friendly future. Similarly, collaboration among industries, stakeholders, and research institutions is essential for successful algae solution implementation. Working together allows for the sharing of knowledge and expertise, leading to the development of more effective algae technologies. Industries play a pivotal role in adopting and integrating algae-based solutions into their operations, fostering more sustainable and ecological business practices. Partnerships between academia, government agencies, and private sectors facilitate the exchange of ideas, resources, and funding, promoting innovation and widespread adoption of sustainable algae solutions (Fabris et al. 2020; Roy et al.

2022). Through collaboration, we can harness the full potential of algae technologies and make a significant impact on environmental management and bioremediation. To fully embrace sustainable algae solutions, continuous research and development in algae technologies are crucial. While algae have demonstrated immense potential in various fields, including biofuels, food ingredients, and medicinal products (Khan and Shin 2018), challenges remain, such as improving microalgae growth rates and optimizing product synthesis (Khan and Shin 2018). Thus, supporting ongoing research efforts and investing in algae technology development will pave the way for new breakthroughs and innovative solutions in environmental management and bioremediation.

References

Ahmad A, Banat F, Alsafar H, Hasan SW (2022) Algae biotechnology for industrial wastewater treatment, bioenergy production, and high-value bioproducts. Sci Total Environ 806:150585. https://doi.org/10.1016/j.scitotenv.2021.150585

Ahmed W, Núñez-Delgado A, Mehmood S, Ali S, Qaswar M, Shakoor A, Chen D-Y (2021) Highly efficient uranium (VI) capture from aqueous solution by means of a hydroxyapatite-biochar nanocomposite: adsorption behavior and mechanism. Environ Res 201. https://doi.org/10.1016/j.envres.2021.111518

Al-Anbari MA, Altaee SA, Kareem SL (2022) Using life cycle assessment (LCA) in appraisal sustainability indicators of Najaf wastewater treatment plant. Egypt J Chem 65(9):513–519. https://doi.org/10.21608/EJCHEM.2022.113093.5139

Al-Baldawi IA, Mohammed AA, Mutar ZH, Abdullah SRS, Jasim SS, Almansoory AF, Ismail N (2021) Application of phytotechnology in alleviating pharmaceuticals and personal care products (PPCPs) in wastewater: source, impacts, treatment, mechanisms, fate, and SWOT analysis. J Clean Prod 319. https://doi.org/10.1016/j.jclepro.2021.128584

Ansari FA, Singh P, Guldhe A, Bux F (2017) Microalgal cultivation using aquaculture wastewater: integrated biomass generation and nutrient remediation. Algal Res 21:169–177. https://doi.org/10.1016/j.algal.2016.11.015

Arun J, Gopinath KP, Sundar Rajan P, Felix V, Joselyn Monica M, Malolan R (2020) A conceptual review on microalgae biorefinery through thermochemical and biological pathways: bio-circular approach on carbon capture and wastewater treatment. Bioresour Technol Rep 11. https://doi.org/10.1016/j.biteb.2020.100477

Chen W-H, Thomas J, Ngamcharussrivichai C (2022) Advanced technologies on thesustainable approaches for conversion of organic waste to valuable bioproducts: emerging circular bioeconomy perspective. Fuel 324. https://doi.org/10.1016/j.fuel.2022.124313

Bai R, Su P, Guo Z, Nguyen TT, Diao Y, Heimann K, Zhang W (2021) An efficient protein isolation process for use in *Limnospira maxima*: a biorefinery approach. J Food Compos Anal 104. https://doi.org/10.1016/j.jfca.2021.104173

BrunhoferovaH, Venditti S, Schlienz M, Hansen J (2021) Removal of 27 micropollutants by selected wetland macrophytes in hydroponic conditions. Chemosphere 281. https://doi.org/10.1016/j.chemosphere.2021.130980

Cai T, Park SY, Li Y (2013) Nutrient recovery from wastewater streams by microalgae: status and prospects. Renew Sustain Energy Rev 19:360–369. https://doi.org/10.1016/j.rser.2012.11.030

Cardoso LG, Lombardi AT, de Jesus Silva JS, Lemos PVF, Costa JAV, de Souza CO, Druzian JI, Chinalia FA (2021) Scaling-up production of Spirulina sp. LEB18 grown in aquaculture wastewater. Aquaculture 544. https://doi.org/10.1016/j.aquaculture.2021.737045

Catone CM, Ripa M, Geremia E, Ulgiati S (2021) Bio-products from algae-based biorefinery on wastewater: a review. J Environ Manage 293:112792. https://doi.org/10.1016/j.jenvman.2021.112792

Cerón M, Campos I, Sánchez JFP, Acién F, Molina E, Sevilla JMF (2008) Recovery of lutein from microalgae biomass: development of a process for *Scenedesmus almeriensis* biomass. J Agric Food Chem 56(24):11761–11766. https://doi.org/10.1021/jf8025875

Chen Y-D, Liu F, Ren N-Q, Ho S-H (2020) Revolutions in algal biochar for different applications: state-of-the-art techniques and future scenarios. Chin Chem Lett 31(10):2591–2602. https://doi.org/10.1016/j.cclet.2020.08.019

Chew KW, Chia SR, Show PL, Ling TC, Arya SS, Chang J-S (2018) Food waste compost as an organic nutrient source for the cultivation of *Chlorella vulgaris*. Biores Technol 267:356–362. https://doi.org/10.1016/j.biortech.2018.07.069

Chia SR, Chew KW, Show PL, Yap YJ, Ong HC, Ling TC, Chang J-S (2018) Analysis of economic and environmental aspects of microalgae biorefinery for biofuels production: a review. Biotechnol J 13(6). https://doi.org/10.1002/biot.201700618

Chisti Y (2007) Biodiesel from microalgae. Biotechnol Adv 25(3):294–306. https://doi.org/10.1016/j.biotechadv.2007.02.001

Converti A, Casazza AA, Ortiz EY, Perego P, Del Borghi A (2009) Effect of temperature and nitrogen concentration on the growth and lipid content of *Nannochloropsis oculata* and *Chlorella vulgaris* for biodiesel production. Chem Eng Process 48(6):1146–1151. https://doi.org/10.1016/j.cep.2009.03.006

Dayana Priyadharshini S, Suresh Babu P, Manikandan S, Subbaiya R, Govarthanan M, Karmegam N (2021) Phycoremediation of wastewater for pollutant removal: a green approach to environmental protection and long-term remediation. Environ Pollut 290:117989. https://doi.org/10.1016/j.envpol.2021.117989

de Benedetti B, Barbera AC, Freni P, Tecchio P (2015) Wastewater valorization adopting the microalgae accelerated growth. Desalin Water Treat 53(4):1001–1011. https://doi.org/10.1080/19443994.2013.848412

Dogaris I, Loya B, Cox J, Philippidis G (2019) Study of landfill leachate as a sustainable source of water and nutrients for algal biofuels and bioproducts using the microalga *Picochlorum oculatum* in a novel scalable bioreactor. Biores Technol 282:18–27. https://doi.org/10.1016/j.biortech.2019.03.003

Dourou M, Aggeli D, Papanikolaou S, Aggelis G (2018) Critical steps in carbon metabolism affecting lipid accumulation and their regulation in oleaginous microorganisms. Appl Microbiol Biotechnol 102:2509–2523. https://doi.org/10.1007/s00253-018-8813-z

Fabris M, Abbriano RM, Pernice M, Sutherland DL, Commault AS, Hall C, Labeeuw L, McCauley JI, Kuzhiuparambil U, Ray P, Ralph PJ (2020) Emerging technologies in algal biotechnology: toward the establishment of a sustainable, algae-based bioeconomy. Front Plant Sci 11. https://doi.org/10.3389/fpls.2020.00279

Fazal T, Rehman MSU, Javed F, Akhtar M, Mushtaq A, Hafeez A, Alaud Din A, Iqbal J, Rashid N, Rehman F (2021) Integrating bioremediation of textile wastewater with biodiesel production using microalgae (*Chlorella vulgaris*). Chemosphere 281. https://doi.org/10.1016/j.chemosphere.2021.130758

Figueroa-Torres GM, Pittman JK, Theodoropoulos C (2021) Optimisation of microalgal cultivation via nutrient-enhanced strategies: the biorefinery paradigm. Biotechnol Biofuels 14(1). https://doi.org/10.1186/s13068-021-01912-2

Foglia A, Andreola C, Cipolletta G, Radini S, Akyol Ç, Eusebi AL, Stanchev P, Katsou E, Fatone F (2021) Comparative life cycle environmental and economic assessment of anaerobic membrane bioreactor and disinfection for reclaimed water reuse in agricultural irrigation: a case study in Italy. J Clean Prod 293. https://doi.org/10.1016/j.jclepro.2021.126201

Gahane D, Biswal D, Mandavgane SA (2022) Life cycle assessment of biomass pyrolysis. Bioenergy Res 15(3):1387–1406. https://doi.org/10.1007/s12155-022-10390-9

García-Martínez JB, Contreras-Ropero JE, Urbina-Suarez NA, López-Barrera GL, Barajas- Solano AF, Kafarov V, Barajas-Ferreira C, Ibarra-Mojica DM, Zuorro A (2022a) A simulation analysis of a microalgal-production plant for the transformation of inland-fisheries wastewater in sustainable feed. Water 14(2):250. https://doi.org/10.3390/w14020250

García-Martínez JB, Sanchez-Tobos LP, Carvajal-Albarracín NA, Barajas-Solano AF, Barajas-Ferreira C, Kafarov V, Zuorro A (2022b) The circular economy approach to improving CNP ratio in inland fishery wastewater for increasing algal biomass production. Water 14(5):749. https://doi.org/10.3390/w14050749

Giri A, Khandayataray P, Murthy MK, Samal D (2021) Biochemical and molecular identification of lipolytic bacteria isolated from beverage industrial wastewater and optimization of lipase-secreting bacteria. Biomass Convers Biorefinery. https://doi.org/10.1007/s13399-021-01890-3

Glover CJ, Phillips JA, Marchand EA, Hiibel SR (2022) Modeling and life cycle assessment of a membrane bioreactor–membrane distillation wastewater treatment system for potable reuse. Separations 9(6). https://doi.org/10.3390/separations9060151

Haddeland I, Heinke J, Biemans H, Eisner S, Flörke M, Hanasaki N, Konzmann M, Ludwig F, Masaki Y, Schewe J, Stacke T, Tessler ZD, Wada Y, Wisser D (2014) Global water resources affected by human interventions and climate change. Proc Natl Acad Sci 111(9):3251–3256. https://doi.org/10.1073/pnas.1222475110

He Z-W, Jin H-Y, Ren Y-X, Yang W-J, Tang C-C, Yang C-X, Zhou A-J, Liu W-Z, Wang A-J (2021) Stepwise alkaline treatment coupled with ammonia stripping to enhance short-chain fatty acids production from waste activated sludge. Bioresour Technol 341. https://doi.org/10.1016/j.biortech.2021.125824

Hernández-García A, Velásquez-Orta SB, Novelo E, Yáñez-Noguez I, Monje-Ramírez I, Orta Ledesma MT (2019) Wastewater-leachate treatment by microalgae: biomass, carbohydrate and lipid production. Ecotoxicol Environ Saf 174:435–444. https://doi.org/10.1016/j.ecoenv.2019.02.052

Hu Z, Qi Y, Zhao L, Chen G (2019) Interactions between microalgae and microorganisms for wastewater remediation and biofuel production. Waste and Biomass Valorization 10(12):3907–3919. https://doi.org/10.1007/s12649-018-0325-7

IPCC (2018) Global warming of 1.5 °C. An IPCC Special Report on the impacts of global warming of 1.5 °C above pre-industrial levels and related global greenhouse gas emission pathways, in the context of strengthening the global response to the threat of climate change. In: IPCC—Sr15, vol 2, Issue October. Retrieved from www.environmentalgraphiti.org

Judd S, Momani FA, Znad H, Ketife AMA (2017) The cost benefit of Algal technology for combined CO_2 mitigation and nutrient abatement. Renew Sustain Energy Rev 71:379–387. https://doi.org/10.1016/j.rser.2016.12.068

Karpisek V, Hasnine MDT, Cicek N, Yuan Q (2021) Natural wetlands contribution on phosphorus removal in small northern communities in Canada. Environ Technol (United Kingdom). https://doi.org/10.1080/09593330.2021.1965221

Khan MI, Shin JH (2018) The promising future of microalgae: current status, challenges, and optimization of a sustainable and renewable industry for biofuels, feed, and other products. Microb Cell Fact 17(1). https://doi.org/10.1186/s12934-018-0879-x

Kholssi R, Ramos PV, Marks EAN, Montero O, Rad C (2021) Biotechnological uses of microalgae: a review on the state of the art and challenges for the circular economy. Biocatalysis Agric Biotechnol 36. https://doi.org/10.1016/j.bcab.2021.102114

KumarM, Kuroda K, Joshi M, Bhattacharya P, Barcelo D (2021) First comparison of conventional activated sludge versus root-zone treatment for SARS-CoV-2 RNA removal from wastewaters: statistical and temporal significance. Chem Eng J 425. https://doi.org/10.1016/j.cej.2021.130635

Li S, Li X, Ho S-H (2022) Microalgae as a solution of third world energy crisis for biofuels production from wastewater toward carbon neutrality: an updated review. Chemosphere 291:132863. https://doi.org/10.1016/j.chemosphere.2021.132863

Li P, Yuan X, Luo Y (2020) Life cycle assessment for carbon balance of a wastewater treatment integrated microalgae biofuel production process. In: Computer aided chemical engineering, vol 48, pp 1699–1704. https://doi.org/10.1016/B978-0-12-823377-1.50284-6

Liu W, Iordan CM, Cherubini F, Hu X, Fu D (2021) Environmental impacts assessment of wastewater treatment and sludge disposal systems under two sewage discharge standards: a case study in Kunshan, China. J Clean Prod 287. https://doi.org/10.1016/j.jclepro.2020.125046

López-Maldonado EA, Oropeza-Guzmán MT (2021) Nejayote biopolyelectrolytes multifunctionality (glucurono ferulauted arabinoxylans) in the separation of hazardous metal ions from industrial wastewater. Chem Eng J 423. https://doi.org/10.1016/j.cej.2021.130210

Masinire F, Adenuga DO, Tichapondwa SM, Chirwa EMN (2021) Phytoremediation of Cr(VI) in wastewater using the vetiver grass (*Chrysopogon zizanioides*). Miner Eng 172. https://doi.org/10.1016/j.mineng.2021.107141

Miguel SP, Ribeiro MP, Otero A, Coutinho P (2021) Application of microalgae and microalgal bioactive compounds in skin regeneration. Algal Res 58. https://doi.org/10.1016/j.algal.2021.102395

Mohamad NA, Hamzah S, Che Harun MH, Ali A, Rasit N, Awang M, Rahman WRWA, Azmi AAAR, Abu Habib AA, Amri Zahid MS, Fahmi Mustofa AA, Latfi SA, Aripin SM, Saad R (2021) Integration of copperas and calcium hydroxide as a chemical coagulant and coagulant aid for efficient treatment of palm oil mill effluent. Chemosphere 281. https://doi.org/10.1016/j.chemosphere.2021.130873

Mulbry W, Kondrad SL, Buyer JS (2008) Treatment of dairy and swine manure effluents using freshwater algae: fatty acid content and composition of algal biomass at different manure loading rates. J Appl Phycol 20(6):1079–1085. https://doi.org/10.1007/s10811-008-9314-8

Nhat PVH, Ngo HH, Guo W, Chang SW, Nguyen DD, Nguyen P, Bui X, Zhang X, Guo J (2018) Can algae-based technologies be an affordable green process for biofuel production and wastewater remediation? Bioresour Technol 256:491–501. https://doi.org/10.1016/j.biortech.2018.02.031

Olguín EJ (2003) Phycoremediation: key issues for cost-effective nutrient removal processes. Biotechnol Adv 22(1–2):81–91. https://doi.org/10.1016/s0734-9750(03)00130-7

Olguín EJ, Sánchez-Galván G, Arias-Olguín II, Melo FS, González-Portela RE, Cruz LJ, De Philippis R, Adessi A (2022) Microalgae-based biorefineries: challenges and future trends to produce carbohydrate enriched biomass, high-added value products and bioactive compounds. Biology 11(8):1146. https://doi.org/10.3390/biology11081146

Oliveira AC, Barata A, Batista AP, Gouveia L (2019) *Scenedesmus obliquus* in poultry wastewater bioremediation. Environ Technol (United Kingdom) 40(28):3735–3744. https://doi.org/10.1080/09593330.2018.1488003

Ortiz A, García-Galán MJ, García J, Díez-Montero R (2021) Optimization and operation of ademonstrative full scale microalgae harvesting unit based on coagulation, flocculation and sedimentation. Sep Purif Technol 259. https://doi.org/10.1016/j.seppur.2020.118171

Ortiz-Betancur JJ, Herrera-Ochoa MS, García-Martínez JB, Urbina-Suarez NA, López-Barrera G L, Barajas-Solano AF, Bryan SJ, Zuorro A (2022) Application of Chlorella Sp. and scenedesmus Sp. in the bioconversion of urban leachates into industrially relevant metabolites. Appl Sci 12(5):2462. https://doi.org/10.3390/app12052462

Pacheco MM, Hoeltz M, Moraes MSA, Schneider RCS (2015) Microalgae: cultivation techniques and wastewater phycoremediation. J Environ Sci Health Part A 50(6):585–601. https://doi.org/10.1080/10934529.2015.994951

Papazi A, Kotzabasis K (2007) Bioenergetic strategy of microalgae for the biodegradation of phenolic compounds—exogenously supplied energy and carbon sources adjust the level of biodegradation. J Biotechnol 129(4):706–716. https://doi.org/10.1016/j.jbiotec.2007.02.021

Pinto G, Pollio A, Previtera L, Temussi F (2002) Bioenergetic strategy of microalgae for the biodegradation of phenolic compounds—exogenously supplied energy and carbon sources adjust the level of biodegradation. Biotech Lett 24(24):2047–2051. https://doi.org/10.1023/a:1021367304315

PlöhnM, Escudero-Oñate C, Funk C (2021) Biosorption of Cd(II) by Nordic microalgae: tolerance, kinetics and equilibrium studies. Algal Res 59. https://doi.org/10.1016/j.algal.2021.102471
Rachidi F, Benhima R, Kasmi Y, Sbabou L, Arroussi HE (2021) Evaluation of microalgae polysaccharides as biostimulants of tomato plant defense using metabolomics and biochemical approaches. Sci Rep 11(1). https://doi.org/10.1038/s41598-020-78820-2
Raghuvanshi S, Bhakar V, Chava R, Sangwan KS (2018) Comparative study using life cycle approach for the biodiesel production from microalgae grown in wastewater and fresh water. Procedia CIRP 69:568–572. https://doi.org/10.1016/j.procir.2017.11.030
Rahman A, Agrawal S, Nawaz T, Pan S, Selvaratnam T (2020) A review of algae-based produced water treatment for biomass and biofuel production. Water 12(9):2351. https://doi.org/10.3390/w12092351
Rosenberg JN, Oyler GA, Wilkinson L, Betenbaugh MJ (2008) A green light for engineered algae: redirecting metabolism to fuel a biotechnology revolution. Curr Opin Biotechnol 19(5):430–436. https://doi.org/10.1016/j.copbio.2008.07.008
Roy AK, Gogoi N, Yasmin F, Farooq MO (2022) The use of algae for environmental sustainability: trends and future prospects. Environ Sci Pollut Res 29(27):40373–40383. https://doi.org/10.1007/s11356-022-19636-7
Said NSM, Kurniawan SB, Abdullah SRS, Hasan HA, Othman AR, Ismail N (2021) Competence of *Lepironia articulata* in eradicating chemical oxygen demand and ammoniacal nitrogen in coffee processing mill effluent and its potential as green straw. Sci Total Environ 799. https://doi.org/10.1016/j.scitotenv.2021.149315
Satayavibul A, Ratanatamskul C (2021) Life cycle assessment of a novel zero organic-waste model using the integrated anaerobic digester and oxidation-ditch membrane bioreactor for high-rise building application. Waste and Biomass Valorization 12(10):5425–5436. https://doi.org/10.1007/s12649-021-01418-w
SiahbalaeiR, Kavoosi G, Noroozi M (2021) Protein nutritional quality, amino acid profile, anti-amylase and anti-glucosidase properties of microalgae: inhibition and mechanisms of action through in vitro and in silico studies. LWT 150. https://doi.org/10.1016/j.lwt.2021.112023
Silva METD, Leal MA, Resende MDO, Martins MA, Coimbra JSDR (2021) *Scenedesmus obliquus* protein concentrate: a sustainable alternative emulsifier for the food industry. Algal Res 59. https://doi.org/10.1016/j.algal.2021.102468
Sivaramakrishnan R, Suresh S, Kanwal S, Ramadoss G, Ramprakash B, Incharoensakdi A (2022) Microalgal biorefinery concepts' developments for biofuel and bioproducts: current perspective and bottlenecks. Int J Mol Sci 23(5):2623. https://doi.org/10.3390/ijms23052623
Sivaramakrishnan R, Incharoensakdi A (2021) Higher efficiency of microalgal biorefinery is achieved with integrated than one-way method. Fuel 300. https://doi.org/10.1016/j.fuel.2021.120988
Sodhi V, Singh C, Pal Singh Cheema P, Sharma R, Bansal A, Kumar Jha M (2021) Simultaneous sludge minimization, pollutant and nitrogen removal using integrated MBBR configuration for tannery wastewater treatment. Bioresour Technol 341. https://doi.org/10.1016/j.biortech.2021.125748
Su Y (2021) Revisiting carbon, nitrogen, and phosphorus metabolisms in microalgae for wastewater treatment. Sci Total Environ 762:144590. https://doi.org/10.1016/j.scitotenv.2020.144590
Tasneem A, Sarker P, Akter S, Mouna SSP, Rahaman MS, Mohinuzzaman M, Uddin MK, Kabir MM (2021) Textile wastewater treatment by combination of chemical and phytoremediation processes. Pollution 7(1):43–54. https://doi.org/10.22059/poll.2020.304569.835
Tomonaga H, Tanigaki Y, Hayashi K, Matsuyama T, Ida J (2021) Adsorption properties of poly(NIPAM-co-AA) immobilized on silica-coated magnetite nanoparticles prepared with different acrylic acid content for various heavy metal ions. Chem Eng Res Des 171:213–224. https://doi.org/10.1016/j.cherd.2021.05.005
Touliabah HE, El-Sheekh MM, Ismail MM, El-Kassas HY (2022) A review of microalgae- and cyanobacteria-based biodegradation of organic pollutants. Molecules 27(3):1141. https://doi.org/10.3390/molecules27031141

Tseng C-C, Yeh H-Y, Liao Z-H, Hung S-W, Chen B, Lee P-T, Nan F-H, Shih W-L, Chang C-C, Lee M-C (2021) An in vitro study shows the potential of *Nostoc commune* (Cyanobacteria) polysaccharides extract for wound-healing and anti-allergic use in the cosmetics industry. J Funct Foods 87. https://doi.org/10.1016/j.jff.2021.104754

Ummalyma SB, Pandey A, Sukumaran R, Sahoo D (2017) Bioremediation by microalgae: current and emerging trends for effluents treatments for value addition of waste streams. Energy, environment, and sustainability, pp 355–375. https://doi.org/10.1007/978-981-10-7434-9_19

Viegas C, Nobre C, Mota A, Vilarinho C, Gouveia L, Gonçalves M (2021) A circular approach for landfill leachate treatment: chemical precipitation with biomass ash followed by bioremediation through microalgae. J Environ Chem Eng 9(3). https://doi.org/10.1016/j.jece.2021.105187

Wu W, Cheng L-C, Chang J-S (2020) Environmental life cycle comparisons of pig farming integrated with anaerobic digestion and algae-based wastewater treatment. J Environ Manage 264. https://doi.org/10.1016/j.jenvman.2020.110512

Xu M, Bernards MT, Hu Z (2014) Algae-facilitated chemical phosphorus removal during high-density *Chlorella emersonii* cultivation in a membrane bioreactor. Biores Technol 153:383–387. https://doi.org/10.1016/j.biortech.2013.12.026

Xu X, Liu G-H, Li Q, Wang H, Sun X, Shao Y, Zhang J, Liu S, Luo F, Wei Q, Sun W, Li Y, Qi L (2021) Optimization nutrient removal at different volume ratio of anoxic-to-aerobic zone in integrated fixed-film activated sludge (IFAS) system. Sci Total Environ 795. https://doi.org/10.1016/j.scitotenv.2021.148824

Yang L, Tan X, Si B, Zhao F, Chu H, Zhou X, Zhang Y (2018) Nutrients recycling and energy evaluation in a closed microalgal biofuel production system. Algal Res-Biomass Biofuels Bioprod 33:399–405. https://doi.org/10.1016/j.algal.2018.06.009

Yoo CY, Jun SY, Lee JY, Ahn C, Oh H (2010) Selection of microalgae for lipid production under high levels carbon dioxide. Biores Technol 101(1):S71–S74. https://doi.org/10.1016/j.biortech.2009.03.030

Zhao C, Zhou J, Yan Y, Yang L, Xing G, Li H, Wu P, Wang M, Zheng H (2021) Application of coagulation/flocculation in oily wastewater treatment: a review. Sci Total Environ 765. https://doi.org/10.1016/j.scitotenv.2020.142795

Zhou Y, Schideman L, Yu G, Zhang Y (2013) A synergistic combination of algal wastewater treatment and hydrothermal biofuel production maximized by nutrient and carbon recycling. Energy Environ Sci 6(12):3765. https://doi.org/10.1039/c3ee24241b

Zuhra, Malahayati N, Emalya N, Rinaldi W, Yunardi Y (2021) Microalgae for simultaneous removal of organic and nitrogenous compounds from wastewater. IOP Conf Ser Earth Environ Sci 667(1). https://doi.org/10.1088/1755-1315/667/1/012044

Chapter 8
Laboratory and Field Studies of Microalgae in Wastewater Treatment in the Removal of Heavy Metals

Ojeaga Evans Imanah, Blessing Edidiong Akachukwu, and Omolola Valentina Imanah

Abstract In this review article, a comprehensive report on the potential of microalgae in wastewater treatment for heavy metal removal is presented. It explores technologies for wastewater treatment, various microalgae types, both modified and unmodified, detailing their unique properties in metal binding. Through laboratory and field studies, the study evaluates their efficiency in real-world conditions. While showcasing microalgae's potential, the research emphasizes the need for further field studies to solidify their application for heavy metal removal.

8.1 Introduction

Water, an indispensable yet limited resource, is fundamental for the survival of all living beings. Globally, fresh water serves diverse purposes such as household and industrial use, irrigation, aquaculture, mining, and power generation. Regrettably, the growing global population, rapid social changes, and economic advancements have led to a scarcity of accessible fresh water and a surge in global wastewater production. Surprisingly, unsafe water claims more lives annually than all forms of violence, including wars (Salman 2005). Water pollution stands as a critical global issue affecting both developed and developing nations, endangering the economic

The original version of the chapter has been revised: The chapter author name has been removed. A correction to this chapter can be found at
https://doi.org/10.1007/978-981-97-2371-3_35

O. E. Imanah (✉) · B. E. Akachukwu
Department of Chemistry, University of Benin, Benin City, Nigeria
e-mail: ojeaga.imanah@uniben.edu

O. V. Imanah
Department of Biology, Brigham Young University, Utah, USA
e-mail: imanaho@byu.edu

T. A. Kurniawan and A. Anouzla (eds.), *Algae as a Natural Solution for Challenges in Water-Food-Energy Nexus*, Environmental Science and Engineering,
https://doi.org/10.1007/978-981-97-2371-3_8

and physical well-being of billions (Mateo et al. 2017). The improper disposal of wastewater from homes, businesses, and industries significantly contributes to environmental pollution. Furthermore, agricultural activities like pesticide and fertilizer applications contaminate groundwater and surface water, especially when untreated wastewater is used for irrigation. Among numerous contaminants, heavy metals pose the most significant threat. Alarmingly, 80 percent of municipal wastewater is discharged untreated into water bodies globally. Industries release millions of tons of heavy metals, solvents, toxic sludge, and other waste materials into water bodies annually (WWAP 2017). These pollutants result in severe health conditions, including cancer, hormone disruption, and impaired brain function. Urgent action is vital to address this issue.

Heavy metals constitute a group of metals and metalloids with atomic weights and densities exceeding 6 g cm−3 (Laureano-Anzaldo et al. 2021). They tend to form complexes, exhibit high reactivity, and persist in the environment due to increased biochemical activity (Hussain et al. 2021). While these toxic metallic elements are naturally present in the earth's crust, human and industrial activities introduce them into public water supplies. Industrial sources include pesticide and petrochemical applications, mining, plating, smelting, paper manufacturing, and the textile industry (Qasem et al. 2021). Natural sources encompass volcanic processes, rock weathering, and soil erosion. Heavy metals are non-biodegradable and can persist in the human body for extended periods (Hahladakis et al. 2013). Particularly concerning heavy metals include mercury, lead, cadmium, arsenic, nickel, chromium, and zinc.

Mercury, highly toxic, affects the kidney, circulatory system, and central nervous system in high concentrations (Azimi et al. 2017). Lead exposure damages the brain, leading to coma and convulsions. It can cause high blood pressure and kidney damage in adults and poses severe risks to pregnant women and young children (Naseem and Tahir 2001). Cadmium poisoning leads to kidney, bone, and lung diseases and even cancer with prolonged exposure (Rahimzadeh et al. 2017). Inorganic arsenic, a confirmed carcinogen, induces symptoms such as vomiting, skin lesions, and infant mortality (Naujokas et al. 2013). Nickel, considered a potential carcinogen, affects the lungs and prostate and can cause birth disorders (IARC 1990). Cr(VI) affects multiple body systems and can lead to respiratory problems and altered genetic material (IARC 1990). Zinc, beneficial in small quantities, becomes toxic in larger amounts, causing nausea and skin irritations (ATSDR 2005).

Efficient wastewater management is crucial, and the potential of safely reused wastewater remains underappreciated. Removing heavy metals from wastewater is essential for enhancing water quality. Fortunately, various technologies such as membrane filtration, ozonation, ion- exchange, electrochemical methods, coagulation-flocculation, and adsorption are in active use. This study specifically focuses on the use of microalgae in removing heavy metals from wastewater, exploring both laboratory and field applications.

8.2 Technologies for Wastewater Treatment

Researchers worldwide have actively engaged in developing innovative wastewater treatment methods, many of which have been integrated into urban water systems. The fundamental principle guiding these methods is the sustainable reuse of wastewater, conserving the rapidly depleting natural water resources (Shah et al. 2020). These approaches primarily employ physicochemical, electrochemical, or advanced oxidation processes. This review explores various techniques, including membrane filtration, ozonation, ion-exchange, electrochemical methods, coagulation-flocculation, and adsorption.

8.2.1 Membrane Filtration

Membrane filtration stands out as a highly efficient, pollution-free, and seamless wastewater treatment technology. It finds extensive applications in beverage industries, biotechnology, and wastewater treatment. Techniques like ultrafiltration, reverse osmosis, microfiltration, nanofiltration, and electrodialysis are pivotal membrane processes, differing mainly in their pore structures, permeability, and operating pressures (Murthy and Chaudhari 2009).

8.2.2 Ozonation

Ozonation, a form of advanced oxidation, excels in eliminating hazardous, resistant, and non- biodegradable organic and inorganic pollutants, including heavy metals. By generating potent oxidants like hydroxyl radicals (OH−), it rapidly oxidizes organic compounds (Kolosov et al. 2018). Catalytic ozonation processes using materials like activated carbon have been successful in removing heavy metals. Studies from China and Russia demonstrated the effective removal of heavy metals using ozonation coupled with calcium peroxide (Zhai and Jiang 2014; Massalimova et al. 2016).

8.2.3 Ion-Exchange

Ion-exchange, owing to its rapid kinetics and high selectivity, has gained widespread acceptance for heavy metal removal from wastewater. This method operates through the reversible interchange of ions between solid and liquid phases. Utilizing ion-exchange resins, both synthetic and natural, cations are exchanged with metals in wastewater. Natural zeolites have proven as effective as synthetic resins, demonstrating excellent cation exchange capacities (Ostroski et al. 2009). This method successfully removes specific heavy metal ions such as Pb^{2+}, Hg^{2+}, Cd^{2+}, Ni^{2+}, V^{4+}, V^{5+}, Cr^{3+}, Cr^{4+}, Cu^{2+}, and Zn^{2+} (Dąbrowski et al. 2004).

8.2.4 Electrochemical Method

Electrochemical treatment involves redox reactions, where oxidation occurs at the anode and reduction at the cathode, leading to metal removal. While effective, this method demands substantial electricity and significant capital investment. Variants like electrocoagulation (EC), electro floatation (EF), and electrodeposition are notable electrochemical treatment technologies (Wang et al. 2007).

8.2.5 Coagulation and Flocculation

Coagulation destabilizes colloids, forming microflocs, which are subsequently encouraged to clump through flocculation. These mechanisms, combining charge neutralization, adsorption, entrapment, and complexation, efficiently remove metals, turbidity, color, and bacteria from wastewater. Specific metals like copper and lead are efficiently removed through physical separation, while nickel and zinc require adsorption or ion exchange processes (Karvelas et al. 2003; Martin et al. 1995).

8.2.6 Adsorption

Adsorption, a potent physicochemical method, reduces heavy metal concentration in wastewater. Utilizing adsorbents from agricultural waste, industrial by-products, or natural materials, this method involves mass transfer between the liquid phase and a solid adsorbent. Activated carbon, carbon nanotubes, and sawdust are common adsorbents. Among these, activated carbon is widely used, although its high production and regeneration costs pose challenges (Hussain et al. 2021). Adsorption stands out as a feasible, cost-effective, and environmentally friendly method, producing high-quality effluent (Demirbas 2008).

While these methods have shown success over the years, each has its limitations. For instance, membrane filtration and ion exchange methods incur high operational costs due to expensive equipment and substrates. Conventional methods like coagulation-flocculation are ineffective at low metal concentrations and generate sludge pollutants. Ozonation poses challenges due to chemical costs and handling, while electrochemical methods require regular replacement of anodes and cathodes. Recognizing the need for cost-effective and eco-friendly alternatives has driven scientists to explore biological methods for wastewater treatment, specifically microalgae- based systems. Recent studies have highlighted the remarkable efficiency of microalgae in removing heavy metals and other pollutants from wastewater (Atif et al. 2023). This review critically examines both laboratory and field research on microalgae's potential in heavy metal removal, providing valuable insights for policymakers and stakeholders considering microalgae- based water treatment facilities (Atif et al. 2023).

8.3 Microalgae

8.3.1 Types and Characteristics

There are two primary types of algae based on their cellular structures: microalgae and macroalgae. This review focuses on microalgae, which are unicellular algal species living singly or in colonies, ranging in size from 1 to 400 mm (Raheem et al. 2018). These photosynthetic microorganisms can be found in both saline and freshwater environments, converting sunlight, water, and carbon dioxide into biomass (John et al. 2010). Microalgae encompass various types, including eukaryotic protists, prokaryotic cyanobacteria, and blue-green algae. Algae are classified based on their predominant pigments, resulting in green, red, or brown varieties. Some microalgae lack photosynthetic pigments and are heterotrophic, relying on other organisms for sustenance, while others, known as mixotrophs, display both photosynthetic and heterotrophic behavior.

Microalgae are considered promising candidates for capturing CO_2 due to their exceptional biomass productivity and the ability to produce various value-added products. They can sequester CO_2 from both the atmosphere and industrial emissions (Brennan and Owende 2010). The high productivity of microalgae ensures rapid generation of desired products and facilitates efficient CO_2 fixation (Masoud et al. 2021). Additionally, photosynthetic microalgae significantly contribute to atmospheric oxygen levels. Integrating microalgae into wastewater treatment processes provides the benefit of oxygen (O_2) generation. Heterotrophic bacteria biodegrade carbonaceous materials and produce O_2 through photosynthesis (Mohsenpour et al. 2021). Microalgae can utilize both inorganic and organic nitrogen, such as ammonia, nitrite, and nitrates (Ross et al. 2018). This method can also be employed to remove heavy metals and emerging contaminants (Hena et al. 2021). Certain freshwater microalgae have demonstrated the ability to extract organic toxins under tropical conditions (Hossain et al. 2019). Studies have shown that green microalgae effectively removed 30–70% of pharmaceuticals and personal care products (PPCPs) from domestic wastewater (Jimenez et al. 2017). Utilizing microalgae in wastewater treatment offers several advantages, including lower energy requirements and reduced operational costs (Renuka et al. 2015). Moreover, microalgae serve as renewable, sustainable, and economical sources of biofuels, making them increasingly attractive for research and development.

In the context of heavy metal removal from wastewater, microalgae act as effective adsorbents due to their cell wall's binding affinity for heavy metals (Hwang et al. 2016). The removal mechanism involves absorption, detoxification by metallothionein's, complexation by chelates and polysaccharides, alkaline precipitation, and adsorption (Wang et al. 2017). Metal ions are passively taken up by cells through biosorption, where heavy metal ions are trapped at binding sites in the cell structure (Kumar et al. 2015). Active uptake involves the cellular metabolic cycle in absorbing heavy metals, a process known as bioaccumulation (Ahmed et al. 2022). A study by El-Sheekh et al. (2016) demonstrated the effectiveness of marine and freshwater algae (Chlorella salina and Chlorella Vulgaris, respectively) in reducing various pollutants in wastewater. Chlorella Vulgaris exhibited higher efficiency in removing heavy metals compared to Chlorella salina. The removal efficiency depends on the physicochemical properties of the algal biomass, such as pore structures and surface chemistry, and the characteristics of the adsorbates in the wastewater.

Regulatory bodies emphasize controlling toxic compounds in water and wastewater due to their significant environmental and health risks. Table 8.2 provides an overview of discharge limits for industrial effluents and permissible limits for drinking water set by regulatory bodies such as Malaysia's Department of Environment (DOE), the United States Environmental Protection Agency (USEPA), and the World Health Organization (WHO), all converted into mg $L{-1}$ (ppm) for ease of comparison.

8.4 Modification of Microalgae

Over the years, gene modification of animals, plants, fungi and even microalgae have led to the production of improved food products and other consumables. For decades, it has been possible to alter the genetic makeup of living organisms by introducing a specific piece of DNA to give it a new property, hence producing an improved version of its original form. This process is done using biotechnology-based techniques called genetic engineering (Spicer and Molnar 2018). Modifications on algal biomass to produce engineered activated carbon is necessary in improving adsorption capacities towards heavy metals. One way to enhance the adsorption properties of materials involves two main techniques: physical activation and chemical activation (Sizmur et al. 2017). Physical activation, often seen in steam treatment, focuses on altering the material's structure. On the other hand, chemical activation, exemplified by methods like KOH treatment, modifies the material's properties through chemical means. While both methods are expected to increase the surface area of the adsorbents, chemical activation using KOH might pose a drawback. This method eliminates beneficial exchangeable cations (Mg^{2+}, Na^{2+}, K^{+}) by converting them into its basic solution, potentially leading to disadvantages (Kim et al. 2016).

The genetic engineering of microalgae can help to overcome the recurrent limitation of metabolic capacity for higher accumulation of desired biomolecules, thus eventually improving the economic feasibility of the production process (Kumar

Table 8.1 Metal Pollutants, their MCL (Maximum Contaminant Level) and toxicities (US EPA et al. 2024) (Rules et al 1987)

Contaminants	MCL [mg L^{-1}]	Potential health effects from long-term exposure above the MCL (unless specified as short term)
Antimony	0.006	Rise in cholesterol, drop in blood sugar
Arsenic	0.010	Skin issues, circulatory problems, heightened cancer risk
Cadmium	0.005	Impaired kidney function
Chromium (total)	0.1	Skin rash due to allergy
Copper	TT action level = 1.3	Prolonged exposure may lead to liver or kidney damage
Lead	TT action level = 0.015	Youngsters might face developmental delays affecting attention and learning, while adults could suffer kidney problems and high blood pressure
Manganese	0.05	Insomnia, weakness, muscle spasms, reproductive system damage
Mercury (inorganic)	0.002	Impaired kidney function
Nickel	0.10	Skin irritation, feeling sick, ongoing asthma, coughing, known to cause cancer in humans
Zinc	0.80	Depression, fatigue, neurological symptoms, excessive thirst

[a] Maximum Contaminant Level (MCL): The highest allowed contaminant level in drinking water
[b] Treatment Technique (TT): A necessary method to lower contaminant levels in drinking water

Table 8.2 Permissible limits/Standard for heavy metal concentration in drinking water and industrial effluent discharge (Mahmud et al. 2016; Pandey et al. 2016)

Heavy metals	Maximum permissible limits (ppm) in drinking water		Maximum permissible limits in industrial effluents in India	
	WHO	US EPA	Inland Sewers- IS2490	Public Sewers- IS3306
As	0.05	0.01	0.2	0.2
Pb	0.05	0.015	0.1	1.0
Cd	0.005	0.005	2.0	1.0
Cr	0.05	0.05	0.1	2.0
Hg	0.001	0.002	0.01	0.01
Zn	5.0	5.0	5.0	15.0
Cu	1.5	1.3	3.0	3.0
Co	0.01	–	–	–

et al. 2020). Though genetic engineering of microalgae has proven to boost laboratory production performance, achievable success in the industry so far is limited. With genetic engineering technologies developing rapidly and intensive investigations going on, more bioproducts are expected to be produced by genetically modified microalgae and even much more to be prospected (Grama et al. 2022). The figures below show the Fourier transform infrared (FTIR) spectra, Scanning electron microscopy (SEM) and X-ray diffraction (XRD) patterns of modified and unmodified algae biomass. Although pore formation is important for microalgae adsorption, (Cheng et al. 2023), result by Joo et al. (2021) showed that the presence of pores was not the critical factor in heavy metal sequestration but the zeta potentials. The SEM microgram of C. vulgaris treated with deionized water, HCl and NaOH is presented in Fig. 8.1. Although, NaOH treated microalgae had higher pores, DI treated microalgae had better sorption capacity for heavy metal because of significantly higher zeta potential. In the same vein, Joo et al. (2021) showed that microalgae bio sorbent is significantly affected by the different growth media of ammonium phosphate (AP), ammonium sulphate (AS), glucose (GL) BG-11, and BG-11 free medium (MM) as shown if Fig. 8.2. Out of all the modified C. vulgaris biosorbents, the ones cultivated in a nitrogen- and phosphorus-rich medium (L-AP) exhibited the highest qe ($Ce = 0.1$ mM) for removing five heavy metals. This performance was statistically superior compared to others ($p \leq 0.05$, two-way ANOVA).

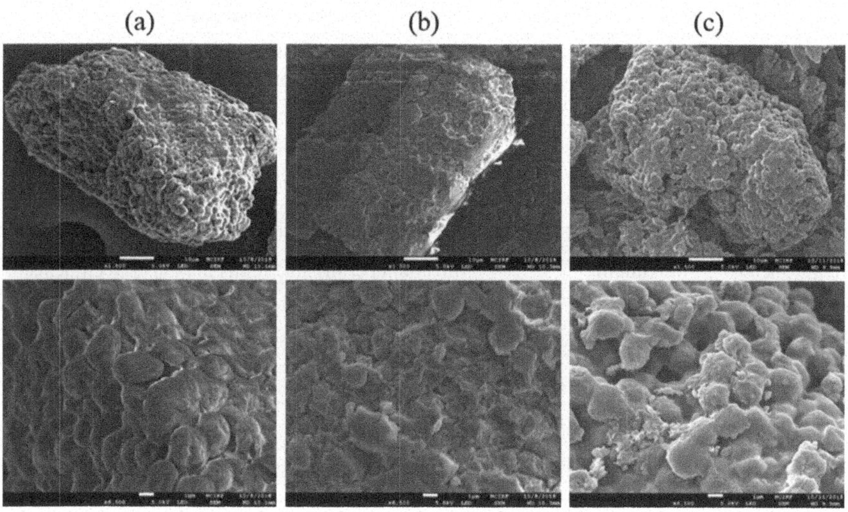

Fig. 8.1 SEM photos of unmodified and chemically modified biosorbents: **a** C- DI, **b** C-HCl, and **c** C-NaOH (Joo et al. 2021)

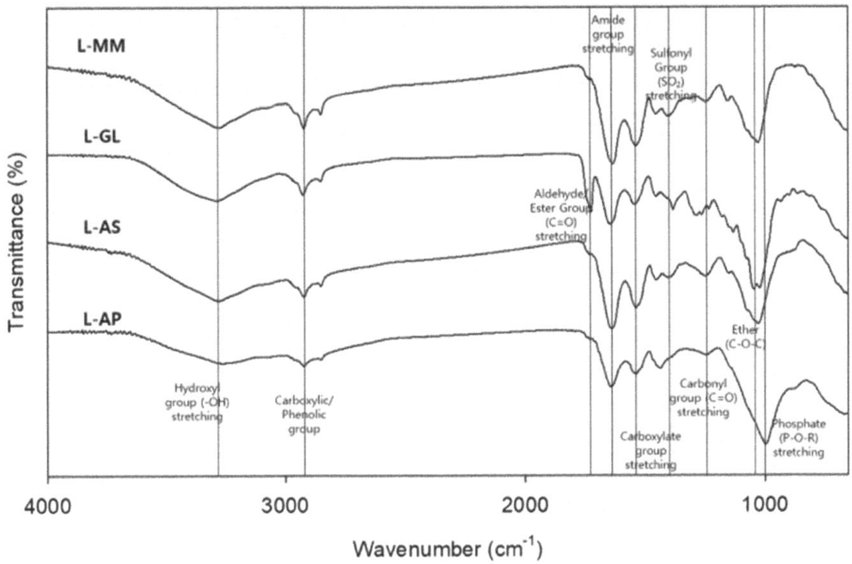

Fig. 8.2 FTIR spectrum of FTIR spectra for algal biosorbents prepared using different growth media (Joo et al. 2021)

8.5 Laboratory Studies

Over the years, extensive research has been conducted on the efficacy of microalgae in eliminating heavy metals from wastewater. These heavy metals are typically present in their ionic forms such as cations of Cd^{2+}/Cd (II), Cu^{+}/Cu(I), Cu^{2+}/Cu (II), and Cr^{3+}/Cr (III), as well as anions of As^{3-} and $CrO_4{}^{2-}$ (Sizmur et al. 2017). The algal adsorbent's ability to attract these heavy metals depends on factors like ion charges, atomic weights, and radius.

In a study by Chan et al. (2014), three microalgae strains, namely Chlorella vulgaris, Spirulina maxima, and a naturally occurring algae sample from wastewater treatment plant effluent, were examined for their capacity to absorb copper and zinc in both untreated and autoclaved secondary effluent substrates. The results revealed that these microalgae could remove up to 81.7% of copper and 94.1% of zinc in untreated secondary effluent, with concentrations as low as 7.8 ppb and 0.6 ppb, respectively, after 10 days. However, the efficiency varied among microalgae strains. Additionally, the autoclaved secondary effluent showed higher heavy metal removal rates, suggesting that existing microorganisms hindered the growth and metal uptake of microalgae in untreated effluent.

Another study by Sultana et al. (2019) employed Chlorella kessleri microalgae to biosorb heavy metals from synthetic wastewater. By optimizing parameters such as initial solution pH, temperature, and microalgae biomass dose, a maximum lead (II) removal efficiency of 99.54% was achieved at specific conditions. The study also explored the simultaneous removal of multiple heavy metals, with the removal efficiency ranking as Pb (II) > Co (II) > Cu (II) > Cd (II) > Cr (II). Their findings supported the pseudo-second order kinetic model for lead (II) biosorption.

In a recent investigation, green algae were cultivated and tested for their ability to remove heavy metals from coal mine wastewater (Makhanya et al. 2021). Results demonstrated significant removal percentages, notably 85% for Fe, 95% for Zn, and 99% for Cd. Copper and aluminum were completely removed. The study also highlighted correlations between physicochemical parameters and heavy metals, indicating moderate to strong positive relationships for temperature, weak correlations for pH, and strong positive correlations for COD.

Table 8.3 provides a summary of the adsorption capacities of different algae biomasses used to remove heavy metal ions from wastewater. The table includes information on the type of algae, the specific heavy metals targeted, initial metal concentrations (C0, in mg L^{-1}), the quantity of algae used (in grams), pH levels (dimensionless), contact duration (in minutes), and both the experimental (Qexp) and theoretical adsorption capacities (Qmax) (in mg g^{-1}) of the respective microalgae.

These research findings highlight the significant potential of microalgae in efficiently eliminating heavy metals from wastewater. This progress opens avenues for the development of eco-friendly and highly effective methods for wastewater treatment.

When compared to macroalgae, microalgae biosorbents demonstrate superior adsorption capacities for heavy metal pollutants, exceeding 100 mg g^{-1}, indicating the influence of particle size in wastewater metal adsorption (Abdel-Raouf et al. 2012). The predominant adsorption model for all microalgae biomass is the Langmuir isotherm. Notably, Chlorella sp. QB-102, cultivated in a high-phosphorus environment (265 mg L^{-1}), exhibited remarkable Pb (II) removal efficiency with a high Q_{max} value of 635.8 mg g^{-1} (Li et al. 2019). Similarly, Scenedesmus obliquus demonstrated exceptional adsorption performance for Zn (II) with Q_{max} = 330 mg g^{-1} and Q_{exp} = 836.5 mg g^{-1} (Monteiro et al. 2011), while Scenedesmus quadricauda exhibited significant Pb (II) adsorption capacities (Q_{max} = 333.3 mg g^{-1}; Q_{exp} = 311.5 mg g^{-1}) (Mirghaffari et al. 2015). This data highlights the microalgae's favorable affinity for heavy metal removal, including Cu (II), Cr (III), Cd (II), Pb (II), and others. In another study, the potential of microalgae in reducing heavy metal concentrations in water and wastewater, focusing on cobalt ion removal using a rotary biofilm reactor named BIOALGA was explored (Travieso et al. 2002). Scenedesmus obliquus was immobilized in the reactor and operated in batch mode for 20 days. A maximum cobalt ion removal efficiency of 94.5% was achieved within 10 days. Experimental results were analyzed using a pseudo first-order kinetic model, accurately predicting reactor behavior with less than 7% deviation from experimental values.

In summary, laboratory investigations demonstrate the immense potential of algae biomass- derived adsorbents in effectively treating diverse wastewater types (Lee et al. 2022).

Table 8.3 Heavy metal removal efficiency of various microalgae species

Microalgae	Pollutants	C_0, mg $_L$ $^{-1}$	dosage (g $_L$ $^{-1}$)	Temparature (^OC)	pH	Time (mins)	Q ($_{max}$) (mg g^{-1})	Q (exp)	References
High Phosphorus Cultivated Chlorella sp.	Pb (II)	280	0.2	25	6	30	635	636.0	Li et al. (2019)
Scenedesmus obliqus	Zn (II)	75	0.02	25	6.0–7.0	90	330	836.5	Monteiro et al. (2011)
Scenedesmus quadricauda	Cd (II)	10	0.20	RT	5.0	60	135 1	130.0	Mirghaff ari et al. (2015)
Chlamydomonas reinhardtii	Pb (II) Cr (VI) Pb (II)	10 100 100	0.20 0.10 0.20	RT RT 25	5.0 2.0 5.0 –6.0	60 120 60	333 3 24.9 384.4	311.5 18.2 35.9	Arıca et al. (2005) Bayramo glu et al. (2006)
Chlorella vulgaris	Cd (II) Ni (II)	150 150	1.0 1.0	25 25	4.0 4.0	120 120	86.6 58.4	–	Aksu and Donmez (2006)
Chara aculeolata	Pb (II) Cd (II) Zn (II)	2 0.3 6	2 2 2	25 25 25	4.0 4.0 4.0	15 90 30	105 26 23.0 4 15.2	54.05 3.69 4.48	Sooksaw at et al. (2016)

(continued)

Table 8.3 (continued)

Microalgae	Pollutants	C_0, mg $_L$ $^{-1}$	dosage (g $_L$ $^{-1}$)	Temparature (^{O}C)	pH	Time (mins)	Q ($_{max}$) (mg g^{-1})	Q (exp)	References
Cystoseira indicia	Cu(II) Co(II)	100 100	1 1	25 25	5.0 5.0	240 240	103 09 59.5 2	76.11 54.59	Akbari et al. (2015)
Eucheuma denticulatum	Pb(II) Cu(II) Fe(II)	10 10 10	4 4 4	25–50 25–50 25–50	5.0 5.0 5.0	120 120 120	81.9 7 66.2 51.0	2.31 1.71 1.55	Rahman and Sathasivam (2016)

8.6 Field Studies

In the realm of wastewater treatment, microalgae find their application in open systems, dwelling in natural water bodies like rivers, ponds, and organic plants, as well as artificial water reservoirs and containers (Nizami et al. 2017). The potential of algae-based systems in removing toxic minerals, including lead, cadmium, mercury, scandium, tin, arsenic, and bromine, is currently under exploration (Soeder et al. 1978). Noteworthy reports highlight algae's remarkable ability to accumulate various metal ions such as Cu^{2+}, Pb^{2+}, Cr^{3+}, Ni^{2+}, Cd^{2+}, Co^{2+}, Fe^{2+}, and Mn^{2+} (Chen et al. 2008). Field experiments conducted by Gale (1986) unveiled the pivotal role of live photosynthetic microalgae in detoxifying mine wastewater. Utilizing cyanobacteria in artificial pools and meanders, an impressive 99% removal of dissolved and particulate metals was achieved.

Specifically, Soeder et al. (1978) demonstrated Coelastrum proboscideum's capacity to absorb 100% of lead from a 1.0 ppm solution within 20 hours at 23 °C and approximately 90% after merely 1.5 h at 30 °C. The sorption dynamics of lead were found to be significantly influenced by temperature, with lead content in the biomass being higher at lower suspension density, indicating a gradual dilution effect by newly formed biomass.

Conversely, cadmium exhibited a lower sorption potential. In a 24-hour period, cadmium was absorbed less efficiently, with a maximal concentration of about 25 ppm absorbed from a 40 ppb solution. The sorption of cadmium displayed a strong dependence on pH, with sorption rates increasing in acidic conditions, highlighting the role of hydrogen ion concentration in the sorption process. The study emphasized that environmental contamination in industrial areas is primarily dominated by lead, with lesser and less toxicologically significant levels of antimony, cadmium, and selenium.

In a study by McHardy and George (1990) focusing on Cladophora glomerata in artificial freshwater channels, the algae demonstrated exceptional proficiency in accumulating zinc. Experiments involving zinc concentrations ranging from 0 to 4.0 mg per liter at current velocities of 20–33 cm per second for up to 3 hours revealed a correlation between increased zinc concentration in the test solution and enhanced zinc uptake by Cladophora. Concentration factors observed in the experiments ranged from 1.9 to 5.2×103, closely aligning with the range of 1.1 to 9.9×103 obtained in field studies conducted earlier (McHardy and George 1985). This close correspondence hints at the potential use of Cladophora glomerata as a biological monitor for zinc contamination.

8.7 Conclusion and Future Perspectives

The existing body of literature pertaining to the application of microalgae in heavy metal removal from wastewater has been thoroughly examined, encompassing both laboratory and field studies. This comprehensive review delves into the adsorption capabilities of various microalgae-based adsorbents, analyzing the pertinent parameters involved. In the realm of wastewater management, the global focus on efficient treatment methods has prompted the exploration of biological solutions, notably the utilization of microalgae for heavy metal elimination. Microalgae present a promising alternative for effluent treatment when compared to other technologies due to their cost-effectiveness, minimal energy consumption, reduced sludge generation, and high selectivity for metal ions.

Nevertheless, the next frontier in microalgae-based water treatment lies in conducting extensive experimental studies that simulate real-world conditions encountered in ponds, lakes, and other natural water bodies. Most existing research primarily focuses on laboratory processes, lacking the significance of replicating the actual interactions with wastewater in natural environments. This gap necessitates in-depth exploration to accumulate precise data essential for the commercialization and industrial implementation of wastewater treatment methods.

References

Abdel-Raouf N, Al-Homaidan AA, Ibraheem IBM (2012) Microalgae and wastewater treatment. Saudi J Biol Sci 19:257–275

Agency for Toxic Substances and Disease Registry [ATSDR] (2005) Toxicological profile for zinc. U.S. Department of Health and Human Services, Public Health Service, Agency for Toxic Substances and Disease Registry, Division of Toxicology and Environmental Medicine/ Applied Toxicology Branch, Atlanta. Retrieved from http://www.atsdr.cdc.gov/toxprofiles/tp60.pdf. Accessed 21 Apr 2024

Ahmed SF, Mofijur M, Parisa TA, Islam N, Kusumo F, Inayat A, Ong HC (2022) Progress and challenges of contaminate removal from wastewater using microalgae biomass. Chemosphere 286:131656

Akbari M, Hallajisani A, Keshtkar AR, Shahbeig H, Ali S (2015) Ghorbanian, Equilibrium and kinetic study and modeling of Cu (II) and Co (II) synergistic biosorption from Cu (II)-Co (II) single and binary mixtures on brown algae C. Indica. J Environ Chem Eng 3:140–149

Aksu Z, Donmez G (2006) Binary biosorption of cadmium (II) and nickel (II) onto dried Chlorella vulgaris: Co-ion effect on mono-component isotherm parameters. Process Biochemi 41:860–868

Arıca MY, Tüzün I, Yalçın E, Ince O, Bayramoglu G (2005) Utilisation of native, heat and acid-treated microalgae Chlamydomonas reinhardtii preparations for biosorption of Cr (VI) ions. Process Biochem 40:2351–2358

Atif A, Zunera K, Allam AA, Jamaan SA (2023) Wastewater treatment by using microalgae:insights into fate, transport, and associated challenges. Chemosphere 338:2023, 139501, ISSN 0045-6535

Azimi A, Azari A, Rezakazemi M, Ansarpour M (2017) Removal of heavy metals from industrial wastewaters: a review. ChemBioEng Rev 4(1):37–59

Bayramoglu G, Tuzun I, Celik G, Yilmaz M, Arica MY (2006) Biosorption of mercury (II), cadmium (II) and lead (II) ions from aqueous system by microalgae. Chlamydomonas reinhardtii immobilized in alginate beads. Int J Miner Process 81:35–43

Brennan L, Owende P (2010) Biofuels from microalgae—a review of technologies for production, processing, and extractions of biofuels and co-products. Renew Sustain Energy Rev 14(2):557–577

Chan A, Salsali H, McBean E (2014) Heavy metal removal (copper and zinc) in secondary effluent from wastewater treatment plants by microalgae. ACS Sustain Chem Eng 2(2):130–137

Chen Z, Ma W, Han M (2008) Biosorption of nickel and copper onto treated alga (Undaria pinnatifida): application of isotherm and kinetic models. J Hazard Mater 2008(155):327–333

Cheng X, Ling P, Iqbal MS, Liu F, Xu J, Wang X (2023) Water adsorption properties of microalgae powders: thermodynamic analysis and structural characteristics. J Stored Prod Res 101:102093

Dąbrowski A, Hubicki Z, Podkościelny P, Robens E (2004) Selective removal of the heavy metal ions from waters and industrial wastewaters by ion-exchange method. Chemosphere 56(2):91–106

Demirbas A (2008) Heavy metal adsorption onto agro-based waste materials: a review. J Hazard Mater 2008 Sep 15;157(2–3):220–229

El-Sheekh MM, Farghl AA, Galal HR, Bayoumi HS (2016) Bioremediation of different types of polluted water using microalgae. Rendiconti Lincei 27(2):401–410

Gale NL (1986) The role of algae and other microorganisms in metal detoxification and environmental clean-up. Biotechnol Bioeng Symp 16:171–180

Grama SB, Liu Z, Li J (2022) Emerging trends in genetic engineering of microalgae for commercial applications. Mar Drugs 20(5):285

Hahladakis J, Smaragdaki E, Vasilaki G, Gidarakos E (2013) Use of sediment quality guidelines and pollution indicators for the assessment of heavy metal and PAH contamination in Greek surficial sea and lake sediments. Environ Monit Assess 185:2843–2853

Hena S, Gutierrez L, Croue J (2021) Removal of pharmaceutical and personal care products (PPCPs) from wastewater using microalgae: a review. J Hazard Mater 403. https://doiorg/101016/jjhazmat2020124041

Hossain N, Zaini J, Mahlia TMI, Azad, AK (2019) Elemental, morphological and thermal analysis of mixed microalgae species from drain water Renew Energy 131:617624. https://doi.org/10.1016/j.renene.2018.07.082

Hussain A, Madan S, Madan R (2021) Removal of heavy metals from wastewater by adsorption. IntechOpen. https://doi.org/10.5772/intechopen.95841

Hwang J, Church J, Lee S, Park J, Hyoung LW (2016) Use of microalgae for advanced wastewater treatment and sustainable bioenergy generation. Environ. Eng Sci 33(11):882–897. https://doi.org/10.1089/ees2016.0132

IARC (1990) Chromium, nickel and welding. IARC monographs on the evaluation of carcinogenic risks to humans, vol 49. ISBN-13: 978-92-832-1249-2

Jimenez S, Mico MM, Arnaldos M, Ferrero E, Malfeito JJ, Medina F, Contreras S (2017)Integrated processes for produced water polishing: enhanced flotation/sedimentation combined with advanced oxidation processes. Chemosphere 168:309317

John R, Andrea S, Astrid A (2010) Bioenergy and the potential contribution of agricultural biotechnologies in developing countries. Biomass Bioenergy 34(10):1427–1439, ISSN 0961-9534

Joo G, Lee W, Choi Y (2021) Heavy metal adsorption capacity of powdered *Chlorella vulgaris* biosorbent: effect of chemical modification and growth media. Environ Sci Pollut Res 28:25390–25399. https://doi.org/10.1007/s11356-021-12396-w

Karvelas M, Katsoyiannis A, Samara C (2003) Occurrence and fate of heavy metals in the wastewater treatment process. Chemosphere 53:1201–1210

Kim BS, Lee HW, Park SH, Baek K, Jeon JK, Cho HJ, Jung SC, Kim SC, Park YK (2016) Removal of Cu^{2+} by biochars derived from green macroalgae. Environ Sci Pollut Res Int 23:985–994

KolosovML, Peyot V, Yargeau (2018) Novel materials for catalytic ozonation of wastewater for disinfection and removal of micro pollutants. Sci Total Environ 644:1207–1218

Kumar KS, Dahms H, Won E, Lee J, Shin K (2015) Ecotoxicology and environmental safety microalgae a promising tool for heavy metal remediation. Ecotoxicol Environ Saf 113:329352

Kumar G, Shekh A, Jakhu S, Sharma Y, Kapoor R, Sharma TR (2020) Bioengineering of microalgae: recent advances, perspectives, and regulatory challenges for industrial application. Front Bioeng Biotechnol 8. https://www.frontiersin.org/articles/10.3389/fbioe.2020.0091410.3389/fbioe.2020.00914. ISSN2296-4185

Kurniawan TA, Chan GY, Lo WH, Babel S (2006) Physico–chemical treatment techniques for wastewater laden with heavy metals. Chem Eng J 118(1–2):83–98

Kurniawan TS, Babel (2003) Second international conferences on energy technology towards a clean environment (RCETE), Phuket, Feb 2003

Laureano-Anzaldo CM, González-López ME, Pérez-Fonseca AA, Cruz-Barba LE, Robledo- Ortíz JR (2021) Plasma-enhanced modification of polysaccharides for wastewater treatment: a review. Carbohyd Polym 252:117195

Lee XJ, Ong HC, Ooi J, Yu KL, Tham TC, Chen WH, Ok YS (2022) Engineered macroalgal and microalgal adsorbents: Synthesis routes and adsorptive performance on hazardous water contaminants. J Hazard Mater 423:126921

Li Y, Song S, Xia L, Yin H, Garcia-Meza JV, Ju W (2019) Enhanced Pb (II) removal by algal based biosorbent cultivated in high phosphorus cultures. Chem Eng J 361:167–179

Mahmud HNM, Obidul AKM, Yahya R (2016) Removal of heavy metal ions from wasterwater/aqueous solution by polypyrrole-based adsorbents: a review. RSC Adv 6.https://doi.org/10.1039/C5RA24358K

MakhanyaBN, Nyandeni, NSF, Ndulini MS, Mthembu (2021) Application of green microalgae biofilms for heavy metals removal from mine effluent. Phys Chem Earth Parts A/B/C 124(Part 1):103079, ISSN 1474-7065. https://doi.org/10.1016/j.pce.2021.103079

Maliutina K, Tahmasebi A, Yu J (2018) Effects of pressure on morphology and structure of bio-char from pressurized entrained-flow pyrolysis of microalgae. Data in Brief. 18. https://doi.org/10.1016/j.dib.2018.03.048

Martin J, Dai MH, Cauwet G (1995) Significance of colloids in the biogeochemical cycling of organic carbon and trace metals in the Venice Lagoon (Italy). Limnol Oceanogr 40:119–131

Masoud H, Farhad K, Alireza B, Mohammad RG, Sajjad H (2021) Chapter 24—Carbon dioxide as a main source of air pollution: prospective and current trends to control. In: Avelino NÃºÃ±ez-Delgado (ed) Sorbents materials for controlling environmental pollution. Elsevier, pp 623–688, ISBN 9780128200421

Massalimova IA, Il'yasova RR, Samsonov MR, Mustafi AG, Medvedev YA (2016) Recovery of heavy metal ions with calcium peroxide microparticles. Russ J Appl Chem 89:360–366

Mateo-Sagasta J, Zadeh SM, Turral H, Burke J (2017) Water pollution from agriculture: a global review. Food and Agriculture Organization of the United Nations, Rome and the International Water Management Institute on behalf of the Water Land and Ecosystems Research Program, Colombo

McHardy BM, George JJ (1985) The uptake of selected heavy metals by the green alga Cladophora glomerata. Symp Biol Hung 29:3–19

McHardy BM, George JJ (1990) Bioaccumulation and toxicity of zinc in green alga Cladophora glomerata. Environ Pollut 1990(66):55–66

Mirghaffari N, Moeini E, Farhadian O (2015) Biosorption of Cd and Pb ions from aqueous solutions by biomass of the green microalga, Scenedesmus quadricauda. J Appl Phycol 27:311–320

Mohsenpour S, Fatemeh H, Sebastian WN, Adeloye A, Gutierrez T (2021) Integrating micro- algae into wastewater treatment: a review. Sci Total Environ 752 (September 2020):142168

Monteiro CM, Castro PML, Xavier MF (2011) Biosorption of zinc ions from aqueous solution by the microalga Scenedesmus obliquus. Environ Chem Lett 9:169–176

Murthy ZVP, Chaudhari LB (2009) Separation of binary heavy metals from aqueous solutions by nanofiltration and characterization of the membrane using Spieglere Kedemmodel. Chem Eng J 150:181–187

Naseem R, Tahir SS (2001) Removal of Pb(II) from aqueous solution by using bentonite as an adsorbent. Water Res 35:3982–3986

Naujokas MF, Anderson B, Ahsan H, Aposhian VH, Graziano JH, Thompson C, Suk WA (2013) The broad scope of health effects from chronic arsenic exposure: update on a worldwide public health problem. Environ Health Perspect 121:295–302

Nizami A, Rehan M, Waqas M, Naqvi M, Ouda O, Shahzad K, Miandad R, Khan M, Syamsiro M, Ismail I (2017) Waste bio refineries enabling circular economies in developing countries. Biores Technol 241:1101–1117. https://doi.org/10.1016/j.biortech.2017.05.097

Ostroski IC, Barros MASD, Silvab EA, Dantas JH, Arroyo PA, Lima OCM (2009) A comparative study for the ion exchange of Fe (III) and Zn (II) on zeolite Na Y. J Hazard Mater 161:1404–1412
Oyaro N, Ogendi J, Murago EN, Gitonga E (2007) The contents of Pb, Cu, Zn and Cd in meat in nairobi, Kenya
Pandey SK, Singh P, Singh J, Sachan S, Srivastava S, Singh SK (2016) Nanocarbon-based electrochemical detection of heavy metals. Electroanalysis 28(10):2472–2488
Qasem NAA, Mohammed RH, Lawal DU (2021) Removal of heavy metal ions from wastewater: a comprehensive and critical review. npj Clean Water 4:36
Raheem A, Prinsen P, Vuppaladadiyam AK, Zhao M, Luque R (2018) A review on sustainable microalgae based biofuel and bioenergy production: recent developments. J Clean Prod 181:4259
Rahimzadeh MR, Rahimzadeh MR, Kazemi S, Moghadamnia AA (2017) Cadmium toxicity and treatment: an update. Caspian J Intern Med 8(3):135
Rahman MS, Sathasivam KV (2016) Heavy metal biosorption potential of a Malaysian Rhodophyte (Eucheuma denticulatum) from aqueous solutions. Int J Environ Sci Technol 13:1973–1988
Renuka N, Sood A, Prasanna R, Ahluwalia AS (2015) Phytoremediation of wastewaters a synergistic approach using microalgae for bioremediation and biomass generation. Int J Environ Sci Technol 12(4):1443–1460. 101007/s13762-014-0700-2
Ross ME, Davis K, McColl R, Stanley MS, Day JG, Semiao AJC (2018) Nitrogen uptake by the macro-algae cladophora coelothrix and cladophora parriaudii: influence on growth, nitrogen preference and biochemical composition. Algal Res 30:110
Rules SWT, Rule GW, Rule RTC, Rule TC, Above E (1987) National primary drinking water regulations. Synth Organic Chem
Salman SM (2005) United Nations general assembly resolution: international decade for action, water for life, 2005–2015: a water forum contribution. Water Int 30(3):415–418
Shah AI, Dar MU, Bhat RA, Singh JP, Singh K, Bhat SA (2020) Prospectives and challenges of wastewater treatment technologies to combat contaminants of emerging concerns. Ecological Eng 152:105882
Sizmur T, Fresno T, Akgül G, Frost H, Moreno-Jimenez E (2017) Biochar modification to enhance sorption of inorganics from water. Bioresour Technol 246:34–47
Soeder CJ, Payer HD, Runkel KH, Beine J, Briele E (1978) Sorption and concentration of toxic minerals by mass cultures of Chlorococcales. Mitt Internat Verein Limnol 1978(21):575–584
Sooksawat N, Meetam M, Kruatrachue M, Pokethitiyook P, Inthorn D (2016) Equilibrium and kinetic studies on biosorption potential of charophyte biomass to remove heavy metals from synthetic metal solution and municipal wastewater. Bioremediat J 20:240–251
Spicer A, Molnar A (2018) Gene editing of microalgae: scientific progress and regulatory challenges in Europe. Biology 7(1):21
Sultana N, Hossain SZ, Mohammed ME, Irfan MF, Haq B, Faruque MO, Razzak SA, Hossain MM (2020) Experimental study and parameters optimization of microalgae based heavy metals removal process using a hybrid response surface methodology-crow search algorithm. Sci Rep 10(1):15068
Tamaddon F, Nasiri A, Yazdanpanah G (2020) Photocatalytic degradation of ciprofloxacin using CuFe2O4@methyl cellulose based magnetic nanobiocomposite. MethodsX 7:74–81
Tchobanoglous G, Burton FL, Stensel HD (2003) Wastewater engineering: treatment and reuse. McGraw-Hill, Boston
Travieso L, Pellón A, Benítez F, Sánchez E, Borja R, O'farrill N, Weiland P (2002) BIOALGA reactor: preliminary studies for heavy metals removal. Biochem Eng J 12(2):87–91
Wang LK, Hung YT, Shammas NK (2007) Advanced physicochemical treatment technologies. In: Handb Environ Eng 5. Humana, New Jersey
Wang JH, Tian YZ, Guo HD, Xue QX, Xiao XW, Hong YH (2017) Microalgae based advanced municipal wastewater treatment for reuse in water bodies Appl Microbiol Biotechnol 101(7):26592675. https://doiorg/101007/s002530178184x
Water UN (2015) Water, food and energy. United Nations. [Online] https://www.unwater.org/water-facts/water-food-and-energy

WWAP (2017) The United Nations World Water Development Report 2017: Wastewater, the untapped resource. United Nations World Water Assessment Programme (WWAP). Paris, United Nations Educational, Scientific and Cultural Organization

Zhai J, Jiang CH (2014) Synthesis of calcium peroxide microparticles in aqueous at room temperature and its application in heavy metal ions removal from waste liquid of COD determining. Adv Mater Res 864–867:648–653

Chapter 9
Strategies for Removal of Emerging Compounds of Concern Through Algal Niche Adaptation

Wafa Hassen, Bilel Hassen, Marwa El Ouaer, and Abdennaceur Hassen

Abstract The modern world currently and in the future faces increasing challenges related to the many problems of environmental pollution of air, water, and soil, climate change, chemical exposures, etc., and their negative impact on animal, plant, and human health. The attenuation of this pollution and its effect requires more and more innovative, efficient, and very consistent financial resources. New biotechnical applications have been the pillar of this emerging pollution remediation to guarantee a healthy and friendly living environment for all living organisms. Bioremediation has received considerable attention as an effective, cost-effective, and environmentally friendly method of detoxifying air, water, contaminated soil, and natural sites. Among the different organisms of the globe, such as algae, fungi, bacteria, and some plant species used for bioremediation, bioremediation based on microalgae are a more sustainable and environmental alternative for different mitigations of residues and their valorization through the production of biofuels and useful compounds like enzymes. It is a renewable energy source because of its excellent properties for the elimination of pollutants. The various residues generated owing to different human activities have many so-called emerging compounds, of toxic quality and nature that are harmful to humans and their environment. These compounds of mineral, biological, organic, radioactive nature, etc., causing pollution and biological, physical, or chemical alteration, occur at low concentrations in water, air, or soil, in the order of micrograms or nanograms per liter. They mainly have a wide range of pharmaceutical residues, hormones, pesticides, cosmetics, various industrial and agricultural wastes, etc., and can exert a toxic or ecotoxic action on all or part of the organisms or the ecosystem. Also, they are hard to eliminate and require new techniques

W. Hassen · M. El Ouaer · A. Hassen (✉)
Laboratory of Treatment and Wastewater Valorization, Center of Research and Water Technologies (CERTE), Tourist Road of Soliman, P.O. Box 273, 8020 Soliman, Tunisia
e-mail: abdohas@gmail.com

B. Hassen
Laboratories of Microbiology and Pathology of Aquatic Organisms, National Institute of Navy Sciences and Technologies (INSTM), Street, March 2nd, 1934, 2025, Salammbô, Tunisia

T. A. Kurniawan and A. Anouzla (eds.), *Algae as a Natural Solution for Challenges in Water-Food-Energy Nexus*, Environmental Science and Engineering,
https://doi.org/10.1007/978-981-97-2371-3_9

and increased financial means to fight them. Microalgae as a bioremediation operator is an interesting technique and process from the viewpoint of the treatment of these micro-pollutants and the costs of implementation. This chapter aims to present the techniques of bioremediation operated by microalgae and their advantages for treating emerging micro-pollutants to provide new efficient and inexpensive alternative solutions for large-scale pollution problems.

Keywords Emerging micro-pollutants · Bioremediation · Microalgae · Environmental and health protection

9.1 Introduction

Residual products of any natural or human activity are often considered emerging substances and may present risks to health, the environment, or even treatment processes. These substances, which are many and from diverse sources, can include nanoparticles, micro-pollutants, GMOs, pesticides, and drug residues and are often considered harmful and generate various problems. This environmental pollution, and according to the World Health Organization, about 12.6 million people died during 2012 because they lived in unhealthy environments resulting from air, water, and soil pollution, climate change, chemical exposures, etc. The removal of residual polluting compounds from the natural environment is an important issue because of their potentially harmful effects on the environment, humans, and living beings (Saravanan et al. 2018). So the treatment of these residual substances is ultimate. The population explosion and modern industrial activities contribute to the generation of very many new wastes or residues of all kinds of solids, liquids, and gases. These pollutant wastes are released by urban dwellings and generated by industrial units and usually undergo a complex chemical and biological transformation in the environment, which can form more toxic and persistent by-products known as micro-pollutants, as mentioned above (Wang et al. 2019). This set of conventional treatment pathways primarily targets the attenuation or total removal of these gaseous and liquid pollutant streams, and this attenuation or removal has been approached through physical, chemical, and other methods. Most of these techniques are costly and have varying performances (Hassen et al. 2003; Leong and Chang 2020; Yadav and Sen 2017). Faced with emerging concerns about ecological and biodiversity preservation and strict government regulations on discharge safety, the need to develop innovative and sustainable processes for waste mitigation has gained global interest (Saravanan et al. 2018).

Thus, all conventional treatment pathways concerned with these new pollutants, encompassing liquid or solid residues such as landfill leachate, wastewater treatment plant (WWTP) sludge, etc., seem more particularly sensitive to these new emerging compounds, the micro-pollutants.

Bioremediation has received considerable attention as an efficient, cost-effective, and environmentally friendly method of detoxifying contaminated water and soil.

All around the globe, different organisms, such as algae, fungi, bacteria, and some plant species, are utilized for bioremediation. Microalgae-based bioremediation is a greener and more sustainable alternative for different types of waste mitigation and renewable biofuel production because of its excellent pollutant removal properties (Mathimani et al. 2019; Mathimani and Pugazhendhi 2019; Zada et al. 2018). In recent years, new effective and inexpensive techniques have emerged to address the nuisances and threats of these emerging substances. Among these novel treatment techniques, bioremediation has received considerable attention as an efficient, cost-effective, and environmentally friendly method of detoxification of contaminated water and soil and is proving to be a promising, low-cost, and easy-to-implement technique. It is defined as a process using biological operators such as microorganisms (bacteria, algae, and fungi), plants, or enzymes to remediate a site or an environment altered by contaminants and especially emerging substances and return it to its native, healthy state. It thus leads to depollution and site rehabilitation. Among the biological operators of bioremediation, we can mention microscopic algae or microalgae.

So, microalgae have been applied to reduce toxic and recalcitrant pollutants in municipal wastewater, especially wastewater from industrial sources. Microalgae can remove different contaminants through different methods, such as biosorption, bioaccumulation, and biodegradation. Microalgae showed exceptional biological characteristics as they can grow in extreme environmental conditions and have been widely used as alternatives to conventional methods. So, a new emerging trend developed in recent years for using microalgae for the bioremediation of toxic pollutants because of several advantages: a simple and robust process; a lack of constraints related to toxicity, a rapid growth rate compared to higher plants; and the formation of added value products and bio-energy or biofuels.

But it is commonly known that the bioremediation of a polluted site goes through three required steps: bio-stimulation (introduction of nutrients stimulating the operator), bio-augmentation (introduction of a specific, selected, and performing operator), and intrinsic bioremediation (reduction of the concentration of contaminants by natural biological processes in the absence of any human intervention). Faced with the development of our modern societies and in search of new processes and innovative active molecules, microalgae present themselves as formidable and excellent potential operators of bioremediation, and very many species with wide biodiversity have made up a reservoir of metabolites and biochemical properties almost unexplored, allowing the development and implementation of new biotechnological applications.

Microalgae are photosynthetic unicellular organisms at the base of the food chain in the aquatic environment and are characterized by their extraordinary biodiversity (from 40,000 to 70,000 species) and their great adaptability. They are ubiquitous in the aquatic environment and can be found on all surfaces of the globe: from the oceans (i.e., 2/3 of the Earth's surface) to the Arctic ice, including hyper-salty lakes, eternal snow, wet forests, and the walls of houses (Bougaran and Saint-Jean 2014). Phytoplankton, which includes microalgae and phototrophic bacteria, engages in 50% of the world's primary production and plays a major role in the sequestration of atmospheric carbon dioxide. To date, about twenty species of microalgae are cultivated for use, mainly in aquaculture and food processing. Some microalgae belong to the "green lineage," which is at the origin of all terrestrial plants and is characterized by a cellular organelle, the green plast. Other original lineages (such as diatoms, rhodophytes, and dinoflagellates) have developed only in an aquatic environment and present a diversity that has remained unexploited until now regarding their biotechnological potential. Thus, in comparison with terrestrial plants and bacteria, microalgae are at the dawn of biotechnology, whose applications are as varied as they are promising in the fields of food, health, energy, or bioremediation.

Among the algal species with added value in biotechnology and bioremediation, we can mention Chlorella, which has proven over the last twenty years to be a good bioremediation operator, particularly in wastewater treatment and bio-energy or biofuel production. Chlorella is a unicellular microalga that is non-mobile, spherical, or ellipsoidal. The size of the cells is barely 2–12 μm in diameter, and this chlorella lives in freshwater streams, lakes, and oceans. They are deep green because of their chlorophyll content, and their name means "the greenest of green foods".

The goal of our study is to present the techniques of bioremediation by microalgae and their advantages for treating emerging micro-pollutants and for providing new efficient and low-cost alternative solutions for large-scale pollution problems. The manuscript included a general introduction, two chapters, and a conclusion. The first chapter will be devoted to the presentation of the different emerging micro-pollutants, their dangers to the environment and health, as well as the conventional methods of depollution. In the second chapter, we will present generalities on microalgae and their application in bioremediation, considering their advantages.

9.2 Emerging Micro-pollutants

The contamination of the environment, particularly the aquatic environment, by many new varieties of pollutants, known as emerging micro-pollutants because of their small quantities and their new appearance, is currently a major environmental concern in all countries. Various pollutants resulting from human activity, particularly in urban areas, are released more or less long-term into the soil, oceans, and surface water bodies, degrading their quality and disrupting the proper ecological functioning of these ecosystems. Thus, a wide range of micropollutants is often found in soil, surface water, and ocean water (Luo et al. 2014). So stringent regulations have been put in place to combat, limit, and remediate their presence in the environment. One of the main sources of these micro-pollutants is the agrarian sector and urban and industrial sanitation via wastewater treatment plant (WWTP) effluents and landfill leachate. These sources of discharge have been considered for some time as an important source of introducing these compounds of toxic nature into the natural environment, and their contribution to the contamination of this environment has been shown and confirmed. In these discharges, a wide range of residual compounds are recorded at trace levels, i.e., ng to μg per liter or per kg of dry matter, such as pharmaceutical residues and hormones (PPHs), pesticides, phthalates, artificial sweeteners, personal care products, etc.

Most of these compounds are endocrine disruptors and/or toxic to humans and the environment (Werheni et al. 2021). In this context, the limitation of the contribution of these micro-pollutants to the environment and their remediation is a key issue that managers, politicians, and environmentalists have taken up. In this perspective, three main strategies have been considered: (1) reduction at the source of these micro-pollutants by limiting/prohibiting the use or consumption of certain molecules, (2) optimization of conventional treatment processes to improve abatement in WWTPs and the application of new efficient processes; and (3) the implementation of specific processes via a combined tertiary bio-physical–chemical treatment of reactivation, adsorption, and oxidation of WWTP effluents and controlled landfills, such as UV, photoreactivation, ozone, activated carbon, etc.

The physical–chemical conditions of the environment, the physical–chemical properties of the micro-pollutant such as pH, alkalinity, temperature, etc., and mainly the hydrophobicity of the micro-pollutant molecules are crucial and ultimate parameters in the abatement of micro-pollutants. Some hydrophilic molecules are well abated by adsorption, such as betaxolol, while hydrophobic molecules are poorly abated, such as those of bezafibrate (Alexander et al. 2012). Among the many emerging micro-pollutants are pharmaceutical residues such as acetaminophen, diclofenac, minocycline, meclocycline sulfosalicylate, chlortetracycline, demeclocycline, doxazosin, betaxolol, chlordiazepoxide, zolpidem, bromazepam, etc.

Table 9.1 lists the main pollutants recognized so far by the environmental and health science communities. But algae, especially microalgae, are often used as bio-operators for the remediation of these micropollutants, the key objective of this chapter.

9.3 Environmental Challenges Related to Micro-pollutants and Health Risks Associated with These Emerging Compounds

Micro-pollutants are toxic substances likely to provoke disruption and alteration of the functions of a living organism, leading to harmful effects and even death. Thus, numerous cases of acute and accidental pollution worldwide have drawn attention to the health and environmental impacts of these micropollutants, which have effects even at low concentrations. The influence of the duration of exposure or chronic exposure of the individual to micro-pollutants is strongly suspected, but knowledge is still fragmentary and not yet conclusive. The impact of a substance on an organism would depend not only on the intrinsic toxicity, mutagenesis, and carcinogenesis of the micropollutant but also on the duration and mode of exposure of the organism to this substance. The risk assessment after this exposure is complex, especially considering the multiplicity of micro-pollutants and their residual degradation products

Table 9.1 List of emerging micro-pollutant compounds and their subcategory (Marcoux et al. 2012 modified)

Category	Subcategory
Algal toxins	Cyanotoxins
Anticorrosives	–
Anti-foaming agents	–
Anti-fouling agents	Organotin or organostannan
Antioxidants	Some compounds
Biocides	Biocides, pesticides, and antimicrobial agents
Bio-terrorism/sabotage agents	
Complex agents	Various microbial toxins
Detergents	Aromatic sulfonates, linear alkylbenzene sulfonates (LAS) ethoxylates/carboxylates of octyl/nonyl phenols, and others
Drinking water disinfection by-products	Trihalomethanes (THMs), haloacetic acids (HAAs)
Drugs	Various antimicrobial residues
Flame retardants	Brominated compounds polybromo diphenyl ether (PBDEs) organophosphates, chlorinated kerosene
Food additives	Various residual food dyes
Fragrances	Nitrated musks polycyclic musks, others
Petroleum additives	Dialkyl ether, HAP
Industrial chemicals	Common heavy metals like copper, zinc, cobalt, chromium, etc.
Nanoparticles (NP)	Fullerenes and carbon nanotubes Carbon black silicon-based NP calcium-based NP aluminum-based NP Cellulose gypsum limestone magnesite magnesium oxide marble mica titanium dioxide; copper, zinc diester
Others	
Perfluoroalkylated substances	Perfluoro-alkyl and metabolites
Body care products	Sunscreen insect repellents parabens (hydroxyl-benzoic acid esters)
Pesticides and their metabolites	Chloro-phenoxy-acetic acid acetanilide phenyl urea and sulfonylurea herbicides triazines Nitro phenols organophosphate pesticides carbamates pyrethroids others

(continued)

Table 9.1 (continued)

Category	Subcategory
Pharmaceutical products	Analgesics anorectics anthelmintics antibacterials anticonvulsants antidepressants antidiabetics antiemetics antihistamines antihypertensives anti-inflammatories antimicrobial agents Antineoplastics anxiolytics beta-blockers Blood viscosity agents bronchodilators diuretics lipid regulators sedatives, hypnotics, steroids, and hormones psychiatric drugs X-ray contrast media
Additives for plastics	Phthalates others
Corrosion inhibitors, anti-scaling agents, trace of heavy metals	Methyl-benzo-triazoles (MBT) tolyl-triazoles (TT) Mercury, Chromium, Copper, etc.
Wood preservatives	Phenols, PCP

present in an environment and the exposure routes of organisms, such as inhalation, skin application, and ingestion (Table 9.2). The regulatory approach is based on laboratory tests carried out on animal or plant species. The results are extrapolated, hence the precautionary factors to be considered in reaching health standards. For the environment, the environmental quality standards defined by each country in the form of the Water Framework Directive are the only existing tools.

In urban areas, the discharges of wastewater treatment plants, the combined overflows of rainy weather, and the stormwater of separate networks are recognized as being responsible for the diffuse contribution of micro-pollutants to the natural environment. Generally, the chemical risk of contamination of the environment and natural resources by micro-pollutants can be linked either to the contamination of water resources by phytosanitary products, nitrates, solvents, hydrocarbons, etc., or to the treatment of water intended for human consumption such as aluminum and iron derivatives used as coagulants, disinfection by-products essentially resulting from chlorination, or to potential alterations of water during distribution or storage by humic by-products, lead, polycyclic aromatic hydrocarbons, etc. Another aspect that has progressed in many countries in recent years concerns the dangers resulting from the non-rigorous agricultural reuse of wastewater, especially sludge from wastewater treatment plants, regarding toxic metals and various organic molecules that are still poorly identified. Besides the many micro-pollutants of chemical origin, there are also those of biological origin, mainly from microorganisms such as diverse secondary metabolites, toxins, antibiotics, etc.

Biological pollution is fortunately mitigated by natural processes of self-purification following the destruction of pathogenic microorganisms by the action of solar ultraviolet rays, by the vital competition of saprophytic microorganisms, bacteriophages, etc. It is important to point out that this self-purification is increasingly hindered by a large number of physical–chemical discharges, particularly synthetic

Table 9.2 Main families of emerging micro-pollutants and their potential toxicities

Family/subfamily	Types of toxicity
PAH	Carcinogens (especially benzo/o/pyrene), suspected endocrine disruptors
PCB	Endocrine disruptors and carcinogens
PDDE	Developmental disorders, effects on the liver and thyroid (experimental studies in animals), effects on humans (human and animal studies), and human effects are not well known
Chlorinated solvents	Intoxications, neurological and/or carcinogenic effects on the liver and blood
Benzene solvents	Highly toxic to blood cells, bone marrow, carcinogenic, especially benzene)
Chloroalkanes	Carcinogens
Chlorophenols	Toxic to aquatic organisms. Irritation (skin and mucous membranes), damage to the digestive and respiratory systems, damage to the nervous system, and effects on the heart, liver, and kidneys (accidental intoxication in the workplace)
Chloroanilines	Very toxic to aquatic organisms. Toxic to humans: respiratory damage, methemoglobinemia
Alkyl phenols	Potential endocrine disruptors (mimic the action of estrogen)
Phthalates	Low direct toxicity to the aquatic ecosystem (low solubility in water). Disruptive effects were seen on fish and carcinogenic (DEHP)
Bisphenol A	Potential endocrine disruptors in aquatic fauna and humans
MTBE (methyl-tert-butyl ether)	Damage to the nervous, respiratory, and digestive systems of mice has effects on humans that are not well known
Pesticides	Carcinogenic (leukemia, glioma), immune, neurological, or reproductive effects, allergic reactions (depending on the mode of penetration into the body)
Biocides	Strong suspicion of carcinogenicity (formalin), allergies, or endocrine-disrupting effects
Pharmaceutical products	Possible long-term effects, especially in the case of synergy

detergents, and hydrocarbons, which disrupt the biological balance and prevent the re-oxygenation of water.

The impacts of micro-pollutants on human health, living beings, and the natural environment are variable depending on the toxicity of the compounds and the duration and manner of exposure. Ecotoxicological analyses are provided to detect micropollutants by observing and measuring the changes they cause in a living organism, known as biological indicators. The impact is estimated at the molecular, biochemical, cellular, physiological, or behavioral level, and one could thus obtain an estimate of pollution and its possible effects on animal and plant health and the general natural environment. The detection, depollution, or remediation of these compounds is the ultimate and urgent way to delimit their fate in the natural environment, both terrestrial and aquatic, and in humans, and to avoid the dissemination of these micro-pollutants in the environment, leading to general contamination that is difficult to control. Priority is placed on water pollution caused by antibiotics, steroidal estrogens, cholesterol-lowering drugs, and psychotropic drugs. The role of microalgae as an agent promoting the bioremediation of these micro-pollutants, which are usually recognized as harmful to man and his environment, will be detailed in the following.

9.4 Classification of Algae

Algae are organisms that, independently from animals and plants, belong to a third group of organisms called protists. They are unicellular or multicellular but do not present differentiation in tissues and organs. They consist of two groups: the lower protists, or prokaryotes, and the higher protists, or eukaryotes.

Prokaryotes have no real nucleus and have a simple DNA molecule, often circular or so-called "coiled." They include bacteria, actinomycetes, or filamentous bacteria, and Cyanophyceae, or blue-green algae.

The eukaryotes have a nucleus with a nuclear membrane, well-developed cell organelles, and individualized chromosomes. These eukaryotes include fungi, algae, and protozoa, with the latter two groups representing smaller biomass in soil and water. The name "microalgae," sometimes called "microphytes," refers to microscopic algae. They are unicellular organisms, microalgae, or undifferentiated multicellular organisms, "macroalgae," which are often photosynthetic eukaryotes or prokaryotes, whose cells can be identified by the presence or absence of organelles and a nucleus, respectively.

Thus, algae is a photosynthetic organism that grows in a variety of habitats, mostly aquatic environments, and can convert light energy and carbon dioxide ("CO_2") into a set of organic materials, or "biomass." There are two main categories of algae: macroalgae and microalgae. Macroalgae are multicellular algae close to a centimeter in size and grow most often in natural freshwater or saltwater basins (Wen and Johnson 2009). These organisms are required for coastal ecosystems in terms of composition, function, and productivity (Dawes and Mathiesen 2008). These macroalgae are mainly classified into three groups differentiated by their pigmentation: red, green, and brown. Microalgae are micrometer-sized, single-celled algae that grow in suspension, mainly in aqueous solutions (Wen and Johnson 2009). Microalgae are characterized by high- and low-moisture habitats and colonize mainly various aquatic water bodies, soils, and especially hydromorphone soils. They colonize all biotopes exposed to light and can be mobile via flagella. The vast majority of microalgae can also grow and be active at night through osmotrophy, so they are typical mixotrophic (autotrophic and osmotrophic simultaneously) unicellular organisms.

Thus, algae are one of the most primitive forms of plants and belong to diverse groups of photosynthetic microorganisms ranging from unicellular (microalgae or phytoplankton) to multicellular (macroalgae or filamentous) with sizes less than

400 μm and diameters normally between 1 and 30 μm (Raheem et al. 2015, 2018; Sambusiti et al. 2015). Microalgae are classified based on their life cycle, basic cell structure, and variation in photosynthetic pigmentation as green, blue-green, red, brown, and golden algae. The most important classes of microalgae in abundance are shown in Fig. 9.1: (a) diatoms (Bacillariophyceae); (b) green (Chlorophyceae); (c) blue and blue-green cyanobacteria (Cyanophyceae); (d) golden (Chrysophyceae); and (e) red algae (Rhodophyceae) (Vassilis et al. 2016). Diatoms represent the most dominant life form and the most biomass-producing group on Earth (Demirbas and Demirbas 2011). Microalgae, diatoms, and cyanobacteria have cell sizes ranging from 0.5 to 200 μm and adopt different shapes: elongated, filamentous, or spherical (Gerardo et al. 2015).

These microorganisms can be divided into photosynthetic, non-photosynthetic, autotrophic, heterotrophic, mixotrophic, prokaryotic, and eukaryotic (Sambusiti et al. 2015; Bharathiraja et al. 2015). So these simple chlorophyll-containing microorganisms can capture CO_2 emitted from various sources and convert it into algal biomass during photosynthesis in the presence of light, water, nutrients, and the usual temperature (Sambusiti et al. 2015; Bharathiraja et al. 2015). It is estimated that about 55,000 species and over 100,000 strains of algae are available in diverse marine, brackish, freshwater, and terrestrial ecosystems (Bharathiraja et al. 2015). Recently, the focus

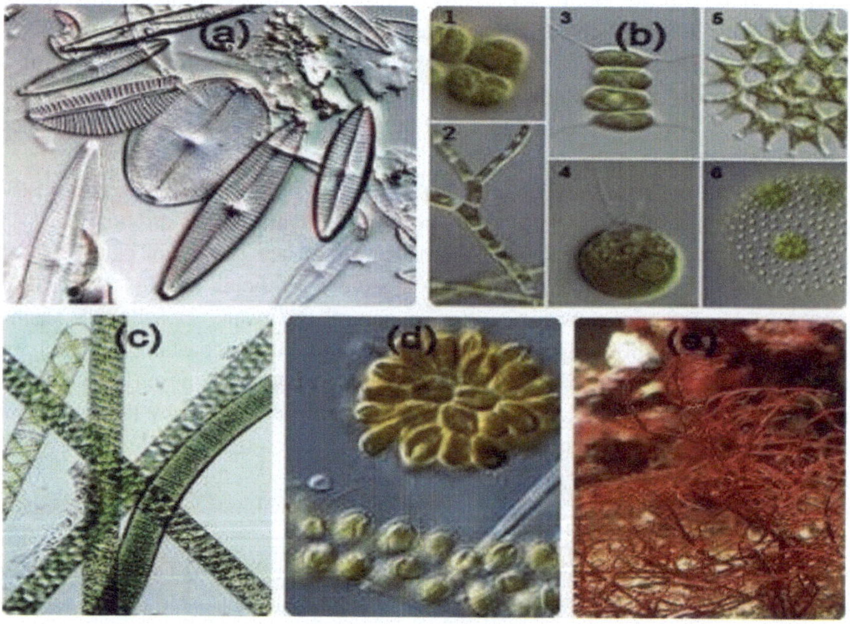

Fig. 9.1 Classification of microalgae: **a** diatoms (Bacillariophyceae); **b** green (Chlorophyceae): (1. Desmotetra, 2. Stigeoclonium, 3. Desmodesmus, 4. Chlamydomonas, 5. Pediastrum, 6. Volvox); **c** blue and blue-green cyanobacteria (Cyanophyceae); **d** golden (Chrysophyceae); and **e** red algae (Rhodophyceae). Adapted from this link: http://bioweb.uwlax.edu/bio203/2011/fedor_kara/gallery.htm

has been on algal biomass for numerous beneficial applications, including wastewater treatment, leachate treatment, micro-pollutant treatment, bio-energy (biofuel) production, etc. (Rosenberg et al. 2008).

The most important classes of microalgae in abundance are shown in Fig. 9.1: (a) diatoms (Bacillariophyceae); (b) green (Chlorophyceae); (c) blue and blue-green cyanobacteria (Cyanophyceae); (d) golden (Chrysophyceae); and (e) red algae (Rhodophyceae) (Vassilis et al. 2016). Diatoms represent the most dominant life form and biomass-producing group on Earth (Demirbas and Demirbas 2011). Microalgae, diatoms, and cyanobacteria have cell sizes ranging from 0.5 to 200 μm and adopt different shapes: elongated, filamentous, or spherical (Gerardo et al. 2015). These microorganisms can be divided into photosynthetic, non-photosynthetic, autotrophic, heterotrophic, mixotrophic, prokaryotic, and eukaryotic (Sambusiti et al. 2015; Bharathiraja et al. 2015).

To summarize this paragraph, macroalgae are multicellular algae close to a centimeter in size and grow most often in natural freshwater or saltwater basins (Wen and Johnson 2009). These organisms are necessary for coastal ecosystems in terms of composition, function, and productivity (Littler and Littler 2000; Dawes and Mathiesen 2008), and they are divided mainly into three groups of macroalgae differentiated by pigmentation: red, green, and brown. Microalgae are micrometers in size and are single-celled algae that grow in suspension, mainly in aqueous solutions (Wen and Johnson 2009). Both macro and microalgae are photosynthetic organisms capable of growing in a variety of habitats, mostly aquatic environments, and can convert light energy and carbon dioxide, or "CO_2," into an array of organic matter, or "biomass." These photosynthetic microorganisms are considered the foremost producers of oxygen, which is essential for the respiration of common living organisms, via the fixation of atmospheric CO_2 and its transformation into O_2. Photosynthesis is affected by chlorophyll, mainly, as well as other photosynthetic pigments.

In this chapter, we will only talk about the role of microalgae in the purification of the main residual compounds in the natural environment, in particular the emerging polluting compounds. These include liquid residual products of solid waste or leachates, the chemical industry, pesticides, metals, surfactants, pharmaceuticals and petrochemicals, antibiotics, cosmetics, including many hormones, poly- and perfluorinated alkyl substances, etc. (Tables 9.1 and 9.2; Fig. 9.2).

9.4.1 Habitats and Adaptation Capacities of Microalgae

The capacities of microalgues for adaptation and survival are confirmed by their high character of colonization and survival in many various environments. They can be found in thermal waters and ice, in acidic or hyper-saline waters, in caves, in symbiosis with any living organism, and even as parasites, among others, of man. They can also develop on rigid surfaces, such as walls or tree trunks, and even on submerged structures. Some species can withstand low or, paradoxically, extreme

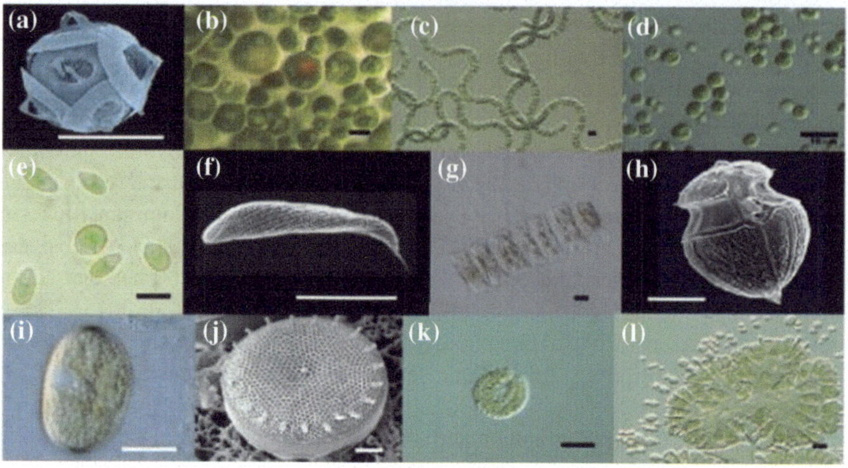

Fig. 9.2 Morphological diversity of microalgae (Sumi 2009). Sustainable seabed mining: guidelines and a new concept for Atlantis II Deep, LRET Collegium 2012 book series—scientific figures on ResearchGate. Available from: https://www.researchgate.net/figure/Application-Areas-of-Microalgae-SourceSumi_2009_fig13_260984355 [accessed 12 May 2023]

temperatures. This adaptability results from their morphological properties as well as their ability to synthesize different varieties of secondary metabolites (Becerra Celis 2009).

9.4.2 Main Microalgae Characteristics

These microorganisms synthesize, through photosynthesis, oxygen and primary organic metabolites such as carbohydrates, lipids, and proteins. The microalgae cell structure included a nucleus, a plasmic membrane, and organelles necessary for its survival, mainly mitochondria, amyloplasts, chloroplasts, and osteoblasts, with three main types of pigments: chlorophylls, carotenoids, and phycobilin proteins. Also, microalgae showed variable shapes, often spherical (Porphyritic), crescent-shaped (Clostridium), spiral (Arthrosporic), droplet-shaped (Chlamydomonas), and even star-shaped (Staurastrum) (Fig. 9.1).

9.4.3 Main Roles of Microalgae

Algae play an important role in the carbon cycle and, in a more general way, in all biogeochemical cycles involving water bodies, lakes, and oceans. Some play an important role in the phenomena of bioaccumulation and bioconcentration in the food chain. It is important to mention that the first uni-algal culture was carried out by Beijerinck (1890) with the species *Chlorella vulgaris*. These photosynthetic microorganisms have characteristics largely equivalent to those of micro-factories.

The sensitivity of some microalgae to certain pollutants, such as copper and hydrocarbons, can give them a bio-indicator value. Some algae are already used as a test for measuring cytotoxicity in a specific environment. Gilbert and Galgani (1990) developed a new method to evaluate the effects of different contaminants on marine microalgae using *Tetraselmis suesica* and *Skeletonema costatum*. This test is based on the measurement of the cell esterase activity by using a fluorescence marker, fluorescein diacetate (FDA). The manipulations will be performed on microplates, and the measurements will be performed by a fluorescence plate reader. The authors have piloted and handled the instruments and data processing using computer software developed specifically for this test. Applications to rapid eco-toxicological tests via the calculation of the median lethal concentration (LC 50) and an evaluation of the quality of the environment by studying water samples are adopted. The median lethal concentration (LC50) will often be determined using these formulas:

$$\text{LC50} = \text{LC100} - \sum \text{cont. Diff. X mean\% mortality/\% control.}$$

This testing is now largely commercialized and used as a reliable ecotoxicological test.

9.5 Different Biotechnological Applications of Microalgae

The production of energy and many molecules of various interests via numerous species of microalgae is undoubtedly very promising, and since the beginning of the twenty-first century, increasing efforts and research activities have been mobilized in this direction and around the potential and power offered by these microorganisms. In addition to the treatment and bioremediation of polluted sites due to unsound industrial, urban, and agricultural activities, other avenues remain to be explored, such as the direct production of hydrogen, vitamins, fodder, etc., and the use of these organisms in microbial batteries.

Agro-food applications: Microalgae have been cultivated for over 30 years for animal nutrition in aquaculture, but also for cosmetology and more recently for their interest in the pharmaceutical, bioremediation, and energy sectors. Microalgae are also a food product, considered in human and animal nutrition in some countries, and their consumption comprises the whole consumption of the microalgae or even the extraction, processing, and packaging of some molecules known for their bioactive properties. Only some species of microalgae are approved for human consumption. Thus, the green microalgae *Chlorella* sp., rich in starch and a source of polyunsaturated fatty acids ($\omega 3$ and $\omega 6$), has been exploited as a food supplement for its tonic and detoxifying properties. The market for these by-products is in progress on a worldwide scale. Besides, other microalgae rich in original pigments, such as ß-carotene, are also used as food supplements (Fig. 9.3).

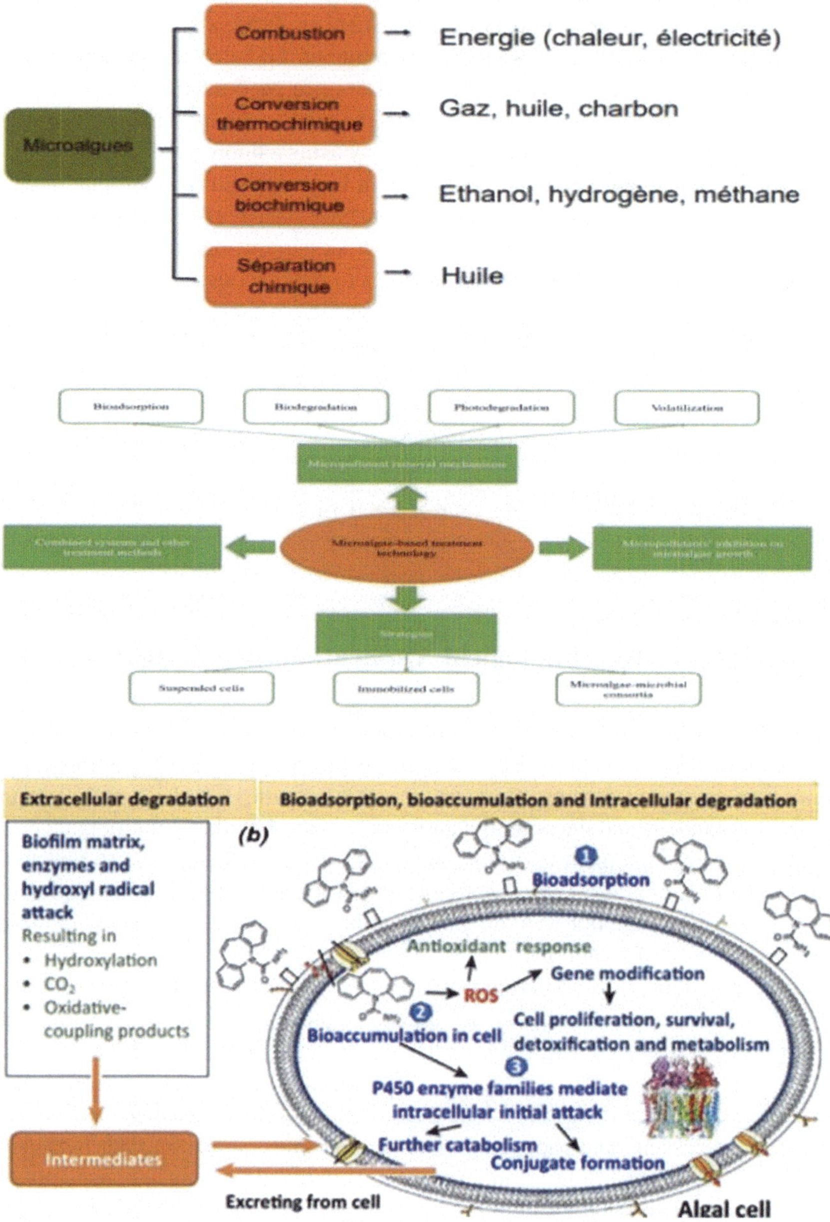

Fig. 9.3 The main recent biotechnological approaches and applications of microalgae

9.5.1 Pharmaceutical and Medical Applications

The biodiversity of microalgae, like that of terrestrial plants, represents a real reservoir of original molecules of medical interest. Especially since microalgae belong to evolutionary lineages sometimes very distant from each other, but terrestrial plants are derived from the single green microalgal lineage, this underlying chemical diversity offers the opportunity to develop the pharmacopeia of tomorrow through original compounds (Hosseini et al. 2013). Recently, special attention has been paid to marine-origin sub-compounds such as carbohydrates, peptides, lipids, and carotenoids, which are extracted from microalgae and have anti-cancer, anti-inflammatory, antimicrobial (e.g., anti-COVID-19 activity), and antioxidant properties in biomedicine and pharmaceutical biotechnology.

In parallel, these photosynthetic marine microalgae with several biotechnological applications were suitable hosts for the production of recombinant proteins/peptides, such as monoclonal antibodies and vaccines. The study of Khavari et al. (2021) has reported results concerning the use of silica-based nanoparticles originating from diatoms as drug delivery carriers owing to their biodegradability, easy functionalization, low cost, and simple features compared to synthetics, which make these agents proper alternatives for synthetic silica nanoparticles. Therefore, diatom-based nanoparticles were revealed as a viable option for the delivery of anti-cancer drugs and for reducing the side effects of cancer chemotherapy. In the same way, these residual microalgal sub-compounds can include various other known metabolites, like ω3 (EPA, DHA) and ω6 (γ-linolenic acid, GLA, and arachidonic acid, AA) polyunsaturated fatty acids. These lipids would have protective effects against cardiovascular diseases and some cancers. Pigments, which may have antioxidant or oxidant activities, are also the subject of much research into the prevention and treatment of cancer. Some exopolysaccharides secreted by microalgae have shown antiviral capacities in the Herpes simplex virus type 1, responsible for oral infections, the measles virus, the MV, responsible for measles; or the influenza virus A, responsible for influenza; and anti-proliferative properties on cancerous cell lines. In the same way, the study of Hosseini et al. (2013) has shown that Spirulina spp. and its processing products were employed in agriculture, the food industry, pharmaceutics, perfumery, and medicine. These Spirulina microalgae had several pharmacological activities such as antimicrobial (including antiviral and antibacterial), anti-cancer, metal-protective (prevention of heavy metal poisoning against Cd, Pb, Fe, and Hg), as well as immuno-stimulant and antioxidant effects related to their rich content of protein, polysaccharides, lipids, essential amino and fatty acids, dietary minerals, and vitamins.

9.5.2 A Renewable Energy Sources

The scarcity of fossil energy resources and the anthropic pressure on the global climate have resulted in the exploration of renewable energy sources. Recent studies have shown that microalgal biomass is one of the most promising sources of renewable biodiesel capable of meeting the big demand for transport fuels. The production of useful compounds from microalgae culturing allowed for the enhancement and diversification of food-borne sources for humans, animals, and several other by-products of medical and pharmaceutical interest. Microalgal biomass is known for its high component content in proteins, carbohydrates, and lipids (Um and Kim 2009; Sidney et al. 2010). The biomass composition of various microalgae in terms of those main components is shown in Table 9.3.

Thus, with their high photosynthetic activity, microalgae can capture and convert solar energy into versatile biomass that can be an energy source. The microalgae biomass will be valued in the production of methane by anaerobic digestion or ethanol via an alcoholic fermentation process and in the biosynthesis of high-energy molecules such as lipids and hydrogen.

Table 9.3 Biomass composition of microalgae expressed on a dry matter basis (Um and Kim 2009; Sidney et al. 2010)

Microalgal strain	Protein	Carbohydrates	Lipid
		Rates dry weight (%)	
Anabaena cylindrica	43–56	25–30	4–7
Botryococcus braunii	40	2	33
Chlamydomonas rheinhardii	48	17	21
Chlorella pyrenoidosa	57	26	2
Chlorella vulgaris	41–58	12–17	10–22
Dunaliella bioculata	49	4	8
Dunaliella salina	57	32	6
Dunaliella tertiolecta	29	14	11
Euglena gracilis	39–61	14–18	14–20
Porphyridium cruentum	28–39	40–57	9–14
Prymnesium parvum	28–45	25–33	22–39
Scenedesmus dimorphus	8–18	21–52	16–40
Scenedesmus obliquus	50–56	10–17	12–14
Scenedesmus quadricauda	47	–	1.9
Spirogyra sp.	6–20	33–64	11–21
Spirulina maxima	60–71	13–16	6–7
Spirulina platensis	42–63	8–14	4–11
Synechoccus sp.	63	15	11
Tetraselmis maculata	52	15	3

Because of their high lipid content, which represents 30–60% of dry weight for some algal species like *Chlorella* sp., *Neochloris oleoabondans, and Phaeodactylum tricornutum*, these microalgae are proving to be good agents for the production of third-generation biofuels, i.e., independent of agricultural products, waste, or conversion of other forms of residual biomass (Song et al. 2013). Different cultivation technologies can be implemented and started, ranging from paddle-wheel agitated tanks, which are not highly productive and relatively economical, to highly efficient but much more expensive photo-bioreactors. The estimated lipid yields for agitated ponds are a function of the climatic factors, in particular the temperature, sunshine, etc.

9.5.3 Agents for Bioremediation

Because of their large specific cell surface, microalgae have interesting absorption capacities for the various polluted sites' decontamination. They act as very good operators for some remediation and cleanup applications owing to their incredible capability as biocatalysts in environmental appliances and their robustness and simple nutritional demand (Manikandan et al. 2022). Also, microalgae can variously collect bio-based chemical compounds useful for diversified medicinal, nutritional, and environmental applications, especially as cost-effective and beneficial green alternatives. As mentioned above, the emergence of several serious environmental problems related to the large propagation of various new polluting agents had imposed a new, low-cost, effective, and friendly process implementation. Among these various polluting agents, heavy metals were looked at as the principal polluting and toxic agents because of their carcinogenicity and toxicity, even in low quantities, for the natural environmental ecosystem, and human and animal health systems.

On the other side, using microalgae for the remediation of sites contaminated by industrial effluents and heavy metals has also been widely considered and has led to potentially higher performance and lower cost than conventional coagulation-flocculation-settling treatment technologies. This ability to concentrate metallic elements by biosorption and bioaccumulation leads not only to applications in the detoxification of polluted environments but also to processes to recover metals of interest such as gold, and even the sequestration of radioactive elements. Microalgae as a bioremediation agent was regarded as prominent because of its vast benefits, including great accessibility, cost-effectiveness, excellent toxic metal-eliminating efficiency, and being friendly to the ecosystem.

Thus, the high growth rates of microalgae, associated with their capacity to absorb nitrogen and phosphorus, can be used for urban or industrial wastewater purification, livestock effluents, and landfill leachate (Gayosso-Morales et al. 2023; El Ouaer et al. 2022). They have been used for a long time in natural lagoon wastewater processes associated with aquatic plants or macrophytes, as efficient purifying agents. In these complex wastewater-polluted environments, microalgae of diverse species can maintain cooperative and even synergistic relationships with bacteria, which increases the kinetics and efficiency of the purification system. In activated sludge systems, the oxygen produced by the microalgae favors a reduction in energy spending related to its continuous supply and speeds up the general purification processes (Nguyen et al. 2020a, b).

The study of Nguyen et al. (2020a, b) showed a mixed culture of microalgae and commonly activated sludge microbes by using a photo-bioreactor system a high total biomass generation, and a dominant role in COD and nitrogen removal via microalgal assimilation. The clear symbiosis between microalgae and autochthon bacteria in activated sludge led to an increase in COD removal by 37.5–45.7%

Using microalgae for leachate remediation from controlled landfills has been successfully applied for leachate treatment and recovery in some countries (El Ouaer et al. 2022). Because these leachates were a dangerous source of bad odors and contamination of deep-water resources, their efficient and low-cost-effective treatment was regarded as the ultimate, and the idea to use the microalgae as an operator agent for the leachate treatment was exploited and associated with the valorization of the high algal biomass generated for biofuels or biodiesel production.

The various micro-pollutant removal mechanisms in microalgae-based treatment systems are concerned mainly with biodegradation and molecule transformation, photo-degradation via UV solar radiation, volatilization via micro-pollutant transformation as volatile gases, and sorption by ants at the biomass cell surfaces. All the most relevant and known contaminant-removal processes occurring in microalgae are illustrated in Fig. 9.3. As earlier underlined by Nguyen et al. (2020a, b), the most important pathways for the removal of micropollutants looked to be biodegradation and photodegradation, but sorption and volatilization are significant only for hydrophobic compounds. In addition, we provide an overview of treatment methods for micro-pollutant removal technology to investigate the advantages and limitations of specific Chlorella microalgae-based methods compared to each other.

9.6 Various Scientific Aspects of Interest in Using Chlorella

As the above evokes, several properties and benefits have made chlorella a suitable and interesting field for scientific research in the fundamental study of photosynthesis and applied levels in the medicinal, food, and especially environmental fields (Jin et al. 2011; Cheng et al. 2016). These properties include low trophic requirements (Vilchez et al. 1997); ease of cultivation (Chevallier 1991); rapid growth: it divides every 24 h (Cheng et al. 2016), and protein-richness (60%) (Bhatnagar et al. 2011).

The similarity of photosynthetic pigments and end products to those of angiosperms (Liu et al. 2016). Given their richness in amino acids (60%), chlorophylls, carotenoids, carbohydrates, and vitamins A, B, and C, chlorella is a feed additive to improve nutrition in humans and fish (Merchant et al. 2000; Enzing et al. 2000; Ursu et al. 2014). Tanaka et al. (1997) showed that oral administration of *Chlorella vulgaris* can reduce or even prevent ulcer formation.

Therapeutically, it has been proven that chlorella has bactericidal power. They can synthesize an antibiotic called chlorellin, which helps fight certain bacteria. Konishi et al. (1990) showed antibacterial power against *Escherichia coli* in the genus Chlorella.

Besides therapeutic and nutritional capacities, a third part is not lacking in importance: the purifying power of wastewater (Chong et al. 2000; Vilchez et al. 1997). This power has been verified in most macroscopic algae, but especially in microalgae and especially in chlorella, which has shown enormous potential for purification of water and effluents of different harmful compounds to the ecosystem, such as heavy metals, various aromatic and organic compounds, pesticides, biocides, and herbicides (Vilchez et al. 1997).

9.6.1 Use of Chlorella in Various Environmental Processes

Some species of Chlorella are well known for their capacity to complex and break down various residual mineral and organic micropollutants through chelation agents or enzymatic activities (Chouteau 2004). These important Chlorella characteristics have led to some new biological applications to treat industrial or municipal effluents based on the properties of the selected Chlorella cultured either in conventional tanks, in batches (Lim et al. 2010), or in more intensive processes such as bioreactors (De-Bashan et al. 2008).

9.6.2 Photo-Degradation

Light is the universal source of solar energy and organic compound production for all living organisms on the planet via photosynthesis, and as photosynthetic organisms, all microalgae-based treatment systems require illumination. During their growth, microalgae, like all living organisms, had to carry out and ensure the synthesis and degradation of various compounds for photosynthesis in a microalgae-based treatment system, and among the degradation processes known in microalgae is photo-degradation. So, photo-degradation can play an important role in the removal of micro-pollutants in wastewater treatment systems using microalgae-based technologies, like natural or aerated lagoons, which are largely universally used. The photosynthesis phenomenon will induce and orient a pathway for micro-pollutant removal through direct or indirect UV photo-degradation by generating complex chemical reactions leading to bond breaking and degradation (Gruchlik et al. 2018).

The availability of oxygen resulting from microalgal photosynthesis also helps to generate various dissolved organic substances and reactive oxygen species (e.g., hydroxyl radicals, singlet oxygen), following the reactions of oxidation and photolysis. The micro-pollutant structure, particularly the presence of double and triple bonds, plays an important role in these complex reactions of degradation and involves a great dependence on the chemical properties of contaminants. On the other hand, these phenomena of photodegradation can occur directly outside or indirectly inside the microalgal living cells. In this context, Mathon et al. (2019) studied the solar photo-degradation of 23 organic micro-pollutants at different water column depths, and they showed that photo-degradation was most effective in the first 10 cm of the water column; among these studied 23 micro-pollutants, ketoprofen, diclofenac, fenofibric acid, and metronidazole looked like the fast-photodegradable compounds. Meanwhile, six slow-photodegradable micropollutants were registered: diazepam, atrazine, dimethoate, diuron, simazine, and cyclophosphamide (Mathon et al. 2019). Among the disadvantages of the natural application of the microalgal photo-degradation system, the change in solar irradiance intensity is largely affected both by the geographic location and season and by the great dependence on the chemical properties of contaminants.

In addition to the ultimate presence of light, the presence of microalgae in an aquatic water system conditions the general processes of bio- and photo-degradation of the various pollutants available in the water via oxidation and photo-oxidation. These degradation processes, due to the complex bio-physical-chemical nature of the contaminants in aquatic environments, are untimely related to the structure of the compound in question and essentially to its chemical properties. This degradation generated various transformed compounds and some intermediate by-products that may be more toxic than the parent compounds. That is why it is important to identify these residual by-products. For example, the study of Rivera-Utrilla et al. (2013) showed that acridine, a photo-degradation product of carbamazepine, is mutagenic and carcinogenic. Photodegradation products of naproxen are also more toxic than the parent compound. Also, in some circumstances, photodegradation could not be the principal removal mechanism for micropollutants but an additional one.

Volatilization The process of volatilization is known as when gas-transformed contaminants are evacuated from the water matrix and released into the atmosphere, generating intense air pollution. These compounds include volatile and semi-volatile MPs once in an open microalgae-based treatment system. The elimination of trace pollutants by volatilization during the microalgae-based process depends on Henry's law constants for the analyzed trace pollutants, and volatilization is usually considered negligible and deeply affected by the air stripping intensity and temperature in the system. In any case, volatilization is not treatment but the passage and change of state of any compound from a liquid to a gaseous state. These gases will be converted into pollutants from a liquid phase and released into the atmosphere instead of transforming and degrading them into smaller molecules and compounds. Lastly, the physicochemical characteristics of microalgae and the complex nature of their structures have usually and particularly resulted in low micropollutant removal (Wang et al. 2017).

9.6.3 Bioaccumulation and Bioadsorption of Various Micro-pollutants (MPs) by Microalgae

Bioaccumulation of biosorption is a physical–chemical process that can be defined as substances' removal from any solution using biological materials, like microbial cells, plants as macrophytes, macro or microalgae, etc. The biosorption process is involved in the whole external and inner biological cell accumulation, and these cells serve as sorbent elements or matrices (Hassen et al. 2003). Due to dominant functional groups, e.g., carboxyl, phosphoryl, amides, amines, etc., the microalgae cell wall is negatively charged, and important complex electrostatic interactions with dissolved contaminant cationic groups in water can occur. These cationic groups will be effectively attached to the microalgal surface, or stored inside the cell if they cross the cytoplasmic membrane barrier and result in contaminant biosorption. Because adsorption is extracellular, the process depends largely on hydrophobicity, structure, and functional groups of different MPs and microalgae species. The inner microalgal cell storage depended largely on the microalgal species and the chemical structure of MPs.

In this way, Cho et al. (2019) have investigated the environmental fate and transport of some pharmaceutical wastes by examining and measuring the adsorptive interaction of 30 MPs with the microalgae *Chlorella vulgaris*. Their main findings showed that all pharmaceutical compounds were adsorbed to *Chlorella vulgaris* through dispersive and hydrophobic contacts, but not by polar, hydrogen-bonding, or anionic interactions (Cho et al. 2019). The study of Wilt et al. (2016) has shown that the MPs adsorbed by Chlorella, including diclofenac, ibuprofen, paracetamol, metoprolol, trimethoprim, carbamazepine, estrone, b-estradiol, and ethinyl-sterol, varied from a lower limit of detection to 16.7% of the removed one. This result is related to the content of algal cell walls, which often have several polymers like cellulose, pectin, chitin, alginate, glycan, lignin, etc.

Another important factor that can influence and affect the phenomenon of biosorption and the microalgae's ability to absorb, catch, and trap the MPs may be the pH value and the amount/composition of extracellular polymeric substances in the medium (Cheng et al. 2019). Thus, large extracellular polymeric substances availability must limit or even inhibit the microalgal process of biosorption; on the other hand, the extreme pH values of the medium, acid, or basic, make the exchanges dynamic and transfer the outside toward the inside of the microalgal cell and vice versa; thus, values around the neutrality tend to slow down these exchanges between the cell and its external medium despite a clear and important cell growth around the neutrality.

On the other hand, many other significant studies have shown that algae and chlorella can fix a wide range of concentrated heavy metals from aquatic environments, such as Cu, Zn, Pb, and Cd. (Priya et al. 2014; Richards and Mullins 2013). In freshwater, a stimulating and nutritive effect of soluble petroleum by-products has been registered in various algae, like *Chlorella vulgaris*, *Anabaena flosaquae*,

Ankistrodesmus convultus, and *Tabellaria fenestrata*. *Chlorella vulgaris* can metabolize alkanes (such as n-tridecane, n-tetradecane, and n-pentadecane) in the presence of light and CO_2. The degradation is so thorough that the products of oxidation, such as alcohol and ketones, are only found in trace amounts in the culture medium (Mouchet 1986).

Very stable organic MPs such as polychlorinated biphenyls (PCBs) can be absorbed and fixed without algae metabolism, and some species can achieve a concentration factor of 1000 as compared to the ambient medium. The capacity of fixation varies according to the species; for example, *Chlorella geitleri* absorbs more PCBs than *Scenedesmus acutus*. PCBs are removed from water even by dead cells, and the mechanism of PCB fixation results in dissolution by the cellular lipids or even simple physical adsorption (Mouchet 1986).

The disruptive effect of some MPs on the enzymatic activity in Chlorella has been applied to determine these pollutants in the natural environment. *Chlorella vulgaris* showed two families of particularly interesting enzymes, such as phosphatases and esterases. The localization of these enzymes is membranous, allowing and ensuring fast enzymatic reactions.

Phosphatases are enzymes whose function consists of managing the phosphorus level in the cell, an essential element of growth during some deficiency periods. These enzymes hydrolyze the phosphorus molecules to release the hot phosphate. This phosphate can then diffuse passively through the cell membrane to establish intracellular stocks. In microalgae, three types of phosphatases, acid, neutral, or alkaline, can exist depending on the pH (Chouteau 2004). The alkaline phosphatase is inhibited in the presence of some heavy metals like Pb^{2+}, Cd^{2+}, and Zn^{2+}.

The esterases ensure the hydrolysis of the molecules' ester bonds and are inhibited by carbamates and organophosphates and the high conductivity of the milieu (Nistor and Emneus 1999; Regel et al. 2002). In this context, they have initiated the development of bioassays or biosensors based on the detection of these enzymatic activities. The work of Chouteau in 2004 allowed the development of a bi-enzymatic conductimetric biosensor using *Chlorella vulgaris* cells immobilized in a bovine serum albumin-based membrane cross-linked with glutaraldehyde vapors. This biosensor can highlight a family of pollutants such as heavy metals and certain organophosphate or carbamate molecules at low concentrations (concentrations of 0.1 ppb for Cd^{2+}, 1 ppb for Pb^{2+}, and 10 ppb for Zn^{2+} and carbamates).

The study of Priya et al. (2014) has shown that chlorella can fix mineral substances like NH^{4+}, NO_3^-, and PO_4^-, considered common pollutants in sewage effluents, and can serve as important nutrients for algae, hence the interest in applying chlorella in the general water treatment process and new biological wastewater treatments based on chlorella strains. These new biological processes based on Chlorella strains as the main operator will be started either in classical lagoons or in more intensive processes such as immobilized algal cultures in agar beads (De-Bashan and Bashan 2010) or on rotating disks (Abdel-Raouf et al. 2012).

9.6.4 Wastewater Treatment

The ability of Chlorella to remove textile dye from textile wastewater through an HRAP (high algal pool) system was studied (Table 9.4). The biosorption of textile dye by Chlorella biomass was also studied. The mechanism of color removal by *C. vulgaris* is biosorption, and it follows the kinetic model of Langmuir and Freundlich. HRAP using *C. vulgaris* offers a good system to treat wastewater from textile mills before final disposal (Lim et al. 2010).

In the case of the biocide of tributyltin (TBT), a toxic chemical used for various industrial purposes such as slim control in paper mills, disinfection of circulating industrial cooling waters, antifouling agents, and the preservation of wood, the study of Luan et al. (2006) has shown that Chlorella degrades this biocide into DBT (dibutyltin) and MBT (monobutyltin). Chlorella could grow in residual textile water with no dilution, and its purifying power decreased with an increase in pollutant content. TBT was rapidly eliminated in the first 5 min with an average elimination rate of 85% by dead chlorella, and no difference in elimination was registered between the first 5 min and the treatment's three days. The percentages of TBT eliminated by living Chlorella cells were significantly lower than those by dead cells of the same species. After three days of treatment, the percentages obtained for TBT removal by living Chlorella cells were only half that of dead cells (Nora et al. 2002).

In a controlled photo-bioreactor, *C. sorokiniana* can grow as a high-density population in wastewater at 40–42 °C and a light intensity of 2500 μmol m^2 s^{-1} for 5 h daily, and effectively remove ammonium from this water better than at a low temperature of 28 °C and a low light intensity of 60 μmol m^2 s^{-1}. When co-immobilized with the bacterium *Azospirillum brasilense,* which promoted algal growth (De-Bashan

Table 9.4 Various investigations using Chlorella for wastewater treatment (El Ouaer et al. 2018)

Pollutant	Chlorella strain	Living cell	Immobilized in alginate	Reduction rate	Algal growth rate	References
Textile dye: Supranol Red 3BW		+	–	50% 3BW 45.1% NH_4–N 33.3% PO_4–P 62.3% DCO	0.17 à 2.26 mg Chla/l	(Lim et al. 2010)
Biocide (Tributyltin: TBT)	*C. vulgaris*	+	+	90% TBT	1 à 2.7 et 3.3 ($\times 10^6$ cells)	(Luan et al. 2006)
	C. miniata *C. sorokiniana*	–/+ –/+	– –	85% (–)/ 40% (+)		(Nora et al. 2002)
Ammonium (NH_4^+)	*C. srokiniana*	–	+	+		(De-Bashan et al. 2008)

et al. 2004), the Chlorella population grows rapidly and allows more ammonium removal (Gonzalez and Bashan 2000).

9.6.5 Leachate Treatment

Important studies about using microalgae as wastewater treatment operators are readily available, while work on the treatment of landfill leachate with chlorella is almost nonexistent. Leachate is a complex colloidal solution generated principally by precipitation and percolating through solid waste layers deposited in a controlled landfill. The leachates were known for their extra-complex content and their bad odor. Their composition depended largely on the age of the landfill and the landfilled waste. They can load highly environmentally harmful and emergent compounds that may then enter the environment via air, soil, water tables, etc.

The staff of Lin et al. (2007) in China has started their research on the leachate treatment by chlorella. So, two microalgae, identified as *Chlorella pyrenoidosa* and *Chlamydomonas snowiae*, were isolated from a high-ammonia leachate pond in the Li Keng landfill in Guangzhou, China. Their growth and nutrient removal rates were determined in a dilution series of landfill leachate under laboratory conditions, and their growth rates were compared with a *C. pyrenoidosa* strain isolated from an unpolluted river. Their main results showed that the densities of *C. pyrenoidosa* (LK) and *C. pyrenoidosa* growth increased by 19.4 and 1.7 times, respectively, in the 30% leachate. For *C. snowiae*, negative growth was confirmed by decreased cell density and observed in crude leachates, 30, 50, and 80% leachate rates. In the same way, relative removal rates of NH_4^+, ortho-P, and COD in 10 and 30% leachate rates were higher than in 50, 80, and 100% leachate rates, respectively. Chlorella-treated 10 and 30% leachates allowed germination of *Brassica chinensis* seeds at 65 and 50%, respectively, while being lower than the germination rate recorded in distilled water (97%). The massive cultivation of chlorella in the wastewater or leachate treatment sector can be a first step toward the valorization of the biomass produced for energy production as biofuels.

The contribution of our research team on this aspect concerns the treatment of leachates from the controlled landfill of Djebel-Chakir, a western suburb of Greater Tunis, by a species of microalgae "*Chlorella* sp." isolated from the site and replacing the conventional technique of coagulation-flocculation-settling set-up on the site, recognized as very costly in energy, not ecological, and generating recalcitrant residual by-products. In addition to the treatment, the valorization of the collected algal biomass for the production of third-generation biofuels. All the results obtained have been conclusive and very interesting (El Ouaer 2017a, b, 2020a, b, 2022).

9.7 Biofuel's Production

9.7.1 Biofuel Production Interests

Biofuels cover all solid, liquid, or gaseous fuels produced from biomass and are intended for energy recovery in transport or other uses. Biomass, a carbon-based renewable energy source for liquid fuels, has the potential to mitigate fossil fuel dependency and environmental problems (Thangalazhy-Gopakumar et al. 2012). Energy needs to be related to technological development and industrialization, and the growing population is increasing more and more; however, fossil energy deposits are estimated to be exhausted within 50 years at the present consumption rate (Beisson et al. 2013). Burning oil releases trapped fossil carbon and increases atmospheric CO_2, one of the major greenhouse gases responsible for climate change. Biofuels have an advantage over fossil fuels because they limit greenhouse gas emissions since their CO_2 emission balance is neutral (Gavrilescu and Chisti 2005), and the CO_2 emitted is compensated by the CO_2 fixed via photosynthesis, making them a friendly and greener alternative energy source. To materialize and consolidate this initiative, the European Union will reduce its dependence on fossil energy resources monopolized by the Organization of the Petroleum Exporting Countries (OPEC) through the diversification of other new energy sources (Demirbas and Balat 2006).

9.7.2 Biofuel Sources and Production

Biofuels are used as additives or complements to these fossil fuels: diesel (incorporated as biodiesel), gasoline (incorporated as ethanol), kerosene, and gaseous fuels.

The classification of biofuel generation is based on the biomass used, the process, and the final product (Bélair 2012). A substantial distinction is made between first-, second-, and third-generation biofuels (Fig. 9.4).

9.7.2.1 First-Generation Biofuels

The first-generation biofuels are based on crop reserve organs such as seeds of cereals (wheat, maize) or oilseeds (rapeseed, sunflowers, jatropha), roots of beet or sugarcane, and fruits of the oil palm (Drabik and Venus 2017; Saladini et al. 2016). These plants reserve organs that store sugar (beet and cane), starch (wheat, maize), or oil (rapeseed, sunflower, palm, and jatropha). As these reserve organs are also used for human food, the production of biofuels competes with food production. In addition, first-generation biofuels require a large area of cultivable land, significant amounts of water, agricultural machinery, fertilizer, and pesticides. Biodiesel can be produced

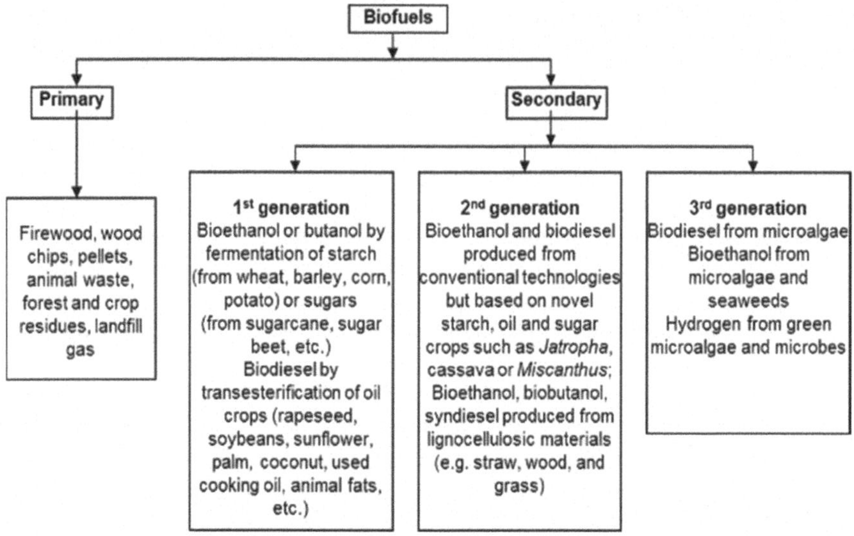

Fig. 9.4 Organization of different biofuel categories (modified and reported by Dragone et al. 2010)

by the transesterification of vegetable oil or ethanol via the fermentation of synthetic plant sugars (Fig. 9.4).

9.7.2.2 Third-Generation Biofuels

Third-generation biofuels were obtained from microalgal biomass such as biodiesel and kerosene (Fig. 9.4). The use of microalgae to produce biodiesel will not compromise the production of food, feed, and other crop-derived products (Chisti 2007). As largely evoked earlier, these microalgae can be used for wastewater treatment, and the oil content of some microalgae can exceed 80% of their dry weight as compared to oil plants such as soybeans and oil palms, for which the oil yield does not exceed 5% of their weight. This biofuel has not yet reached the industrial stage and is in the research and development stage (Fig. 9.5).

9.7.2.3 Production of Biofuels from Microalgae

Through photosynthesis, the microalgae use solar energy, water, and carbon dioxide (CO_2) to elaborate biomass and are responsible for 40–50% of CO_2 fixation on Earth. Some species accumulate up to 60% of their dry weight in oils as lipid reserves made up essentially of triglycerides and complex compounds containing long carbon chains, which represent an interesting resource for biodiesel synthesis. For example, in 2016, the demand for oil in the US was around 19.6 million barrels per day,

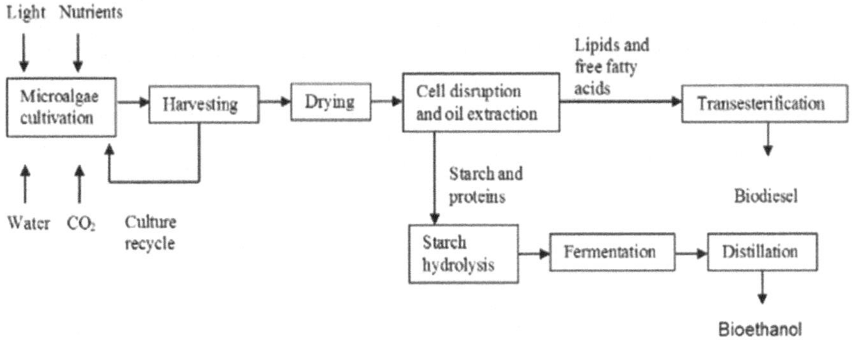

Fig. 9.5 General diagram of the process for biodiesel and bioethanol production from microalgae

and scientific communities estimated that with one hectare of algae, it would be possible to produce the equivalent of half a barrel of oil in algae fuel. This algal fuel production would therefore take nearly 40 million hectares to cover the U.S. demand for hydrocarbons or about 400,000 km^2. This is an area a little less extensive than France, a little more than Germany, and close to a country like Iraq. This is about 4% of the surface of the United States (EESI) (2016).

In recent years, using algae as a bio-indicator of environmental status, the quality of the environment worldwide has suffered from a great deterioration because of uncontrolled industrial discharges, intensive use of chemical fertilizers in agriculture, and abusive and irrational exploitation of natural resources. Globally, the pollution of water and the natural environment has become a current problem that affects all countries. Its consequences are, on the one hand, often irreversible ecosystem degradation and the rarefaction of precious and vital natural resources. So, it is useful and judicious to use living organisms whose sensitivity and ecological requirements allow for judging the state of the environment and its disturbances. It is especially for soil, air, and water pollution that the method of biological indicators can prove valuable and ultimate.

In this respect, many microalgae can survive in polluted waters and environments, which makes them indicators or bio-indicators of pollution. Bio-indicators are living organisms that, through their observation and monitoring, can provide important information on the natural state of the environment. Their monitoring and observation often replace heavier and more expensive methods of analysis and examination and will let the different interventions be done quickly and more efficiently. Among these indicator microalgae, *Chlorella vulgaris* is widely used as a bio-indicator of ecosystem contamination. As we have mentioned several times, *Chlorella vulgaris* is unicellular eukaryotic and green freshwater microalgae belonging to the class Chlorophyceae, discovered by a Dutch researcher, M. W. Beijerinck, in 1890, ranging in size from 3 to 5 μm and in two round or ellipsoidal forms, and rich in carotenoids, chlorophyll a and b, and starch grains (Clément-La Rosière 2012; Rowley 2010).

These organisms' characteristics make them interesting for multiple applications:

- A sensitivity to pollutants, which lets them be used as bio-indicators in standardized bioassays (growth tests) or not (tests on chlorophyll fluorescence, enzymatic activities, etc.).
- Chlorella is often used for wastewater detoxification by the absorption of heavy metals such as arsenic (Karadjova et al. 2008), thallium (Turner and Furniss 2013), chromium (Rai et al. 2013), copper, zinc, cobalt, and manganese (Chojnacka 2007).
- *Chlorella vulgaris* has two families of enzymes of particular interest: phosphatases and esterases, with membrane localization of some of these phosphatases and esterases.

Microalgae culture reactors are often organized and controlled systems to obtain high productivity, and they can be in open ponds or closed systems known as photobioreactors. The average productivity of dry biomass is 10–25 g m^{-2} day^{-1}, with an average content of the order of 0.5 g L^{-1} (Becker 1994; Lee 2001).

Microalgae, as a mixotrophic organism, combining autotrophy and heterotrophy metabolism, are grown in an intensive photobioreactor culturing, in closed containers made of transparent materials to allow light scattering and penetration. The design of the photo-bioreactors is optimized to ensure good light exposure and biomass production and to allow control of all growth parameters like pH, temperature, oxygen, agitation, etc. (Fig. 9.6).

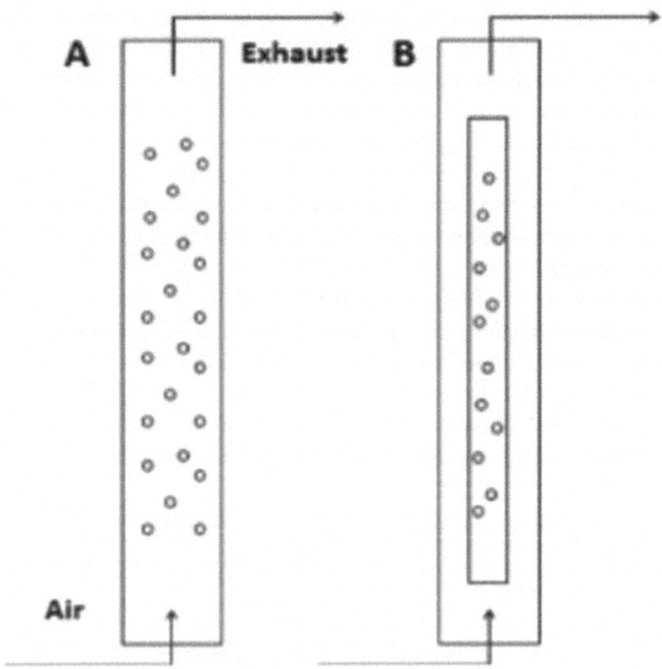

Fig. 9.6 Diagram of a tubular photo-bioreactor

9.8 Bioaccumulation of MPs, Heavy Metals, and Residual Toxic Compounds in Microalgae

In ecotoxicology, the bioaccumulation of some MPs, heavy metals, and various emergent and deleterious compounds in microalgae is known as the process of biosorption, by which certain organisms can accumulate or store pollutants in their bodies. This bioaccumulation included biosorption on the surface by contact and biosorption inside the organism by ingestion. Bioaccumulation, or bio-concentration, is defined as the process by which substances are present in a living organism at a higher concentration than in its surrounding environment. Thus, the concept of bioaccumulation or biosorption results from the net balance of an organism's mechanisms for material accumulation, storage, and excretion.

Bioaccumulation, a key phenomenon at the organism level, is expressed as the difference between residual compounds that penetrate through biological barriers and the amount eliminated to the outside environment, known as the excretion process. Penetration, storage in target organs, and elimination will depend on some specific abiotic environmental factors, the nature of the residual contaminant, and the physiological and biochemical characteristics of the organism or species considered. For example, *C. vulgaris* can build over 70% of chromium in its wall when grown in the presence of this metal in solution (0.1–100 ppm) (Rai et al. 2013).

The speciation of the residual compound, its concentration, and its possible interactions with the bio-physicochemical parameters prevailing in the environment, in particular the presence of organic and inorganic ligands that could form complexes with the residual compounds, conditioned the bioaccumulation process. The use of powdered *Chlorella vulgaris* algal cells for copper biosorption was studied. The effects of contact time, biosorbent dose, equilibrium pH, and lead or zinc ions on the biosorption of copper were studied. Kinetic studies showed that copper biosorption by algal cells was rapid, with equilibrium reached within 15 min, and this kinetics followed pseudo-second-order (Abu al-Rub et al. 2006). The study of Wang et al. (2021) about cadmium binding capacity and mechanisms of cadmium biosorption by *Chlorella vulgaris* K-01 cells showed that the mechanism of ion binding is based on chemo-sorption with a high first sorption rate, and the capacity of Cd^{2+} removal by living biomass of *Chlorella vulgaris* K-01 was independent of the initial pH over a broad pH range (6.0–9.0). They conclude that *Chlorella vulgaris* K-01 has substantial potential for cadmium adsorption and removal, and is suitable for industrial-scale applications. In the same sense of concepts, the study of Ahmad et al. (2019), about freely suspended and Ca-alginate immobilized C. vulgaris cells for the biosorption of Fe (II), Mn (II), and Zn (II) ions from the aqueous solution, has shown that the biosorption capacity of algal cells depends strongly on the operational conditions, such as pH, first metal ion concentration, dosages, contact time, and temperature. The maximum biosorption of Fe (II) 43.43, Mn (II) 40.98, and Zn (II) 37.43 mg/g was achieved with Ca-alginate immobilized algal cells at an ideal pH of 6.0, algal cell dosage of 0.6 g/L, and contact time of 450 min at room temperature.

Also, the study of Hubbe et al. (2014) has shown that the use of algal biomass as a biosorption matrix is also a promising, environmentally friendly, and low-cost option for removing the dissolved residual fractions of petroleum. In addition, some recent studies have proposed the use of biosorption to eliminate crude oil and polycyclic aromatic hydrocarbons (PAHs) from contaminated environments (Olivella et al. 2013; Flores-Chaparro et al. 2017). PAHs from naturally occurring fossil fuels or incomplete combustion of organic materials have been found in polluted water and soil. The study conducted by Pathak et al. (2018) regarding the biosorption of PAHs on different microalgae has shown the removal of PAHs, including fluoranthene (1.0 mg/L), pyrene (1.0 mg/L), and a mixture of fluoranthene and pyrene (each at 0.5 mg/L), by four microalgal biomasses: *Chlorella vulgaris, Scenedesmus platydiscus, Scenedesmus quadricauda,* and *Selenastrum capricornutum.* Their study showed that *Selenastrum capricornutum* was the most efficient species for PAH removal/transformation (78%), but C. vulgaris was the least efficient in removing and transforming PAHs (48%). All the species under investigation, except for *S. platydiscus*, showed better removal efficiency of fluoranthene than of pyrene. The removal rate of a mixture of fluoranthene and pyrene was comparable to or greater than the removal of each compound taken alone, showing that one PAH stimulates the removal of the other PAH.

The removal and transformation of seven high-molecular-weight polycyclic aromatic hydrocarbons (PAHs) from water, namely benzene [a] anthracene, benzo [b] fluoranthene, benzo [k] fluoranthene, benzo [a] pyrene, dibenzo [a, h] anthracene, indenol [1, 2, 3-c, d] pyrene, and benzo [g, h, i] perylene, were studied by a freshwater microalga *Selenastrum capricornutum* under gold and white light irradiation. The removal efficiency of seven PAHs and the difference between living and dead microalgal cells were PAH compound-dependent. Benz [a] anthracene and benzo [a] pyrene were highly transformed into live and dead algal cells, and dead cells displayed greater transformation levels than live cells. Likewise, the study of Aksu and Tezer (2005) about the biosorption of three vinyl sulfone-type reactive dyes [Remazol Black B (RB), Remazol Red RR (RR), and Remazol Golden Yellow RNL (RGY)] onto dried *Chlorella vulgaris* was investigated in a batch system and showed the highest dye uptake capacity at the initial pH value of 2.0 and temperature between 25 and 35 °C for all dyes. The biosorption capacity of the algae increased with initial dye concentrations, up to 800 mg l^{-1} for the RB and RR dyes and up to 200 mg l^{-1} for the RGY dye. Among the three dyes, RB was adsorbed most effectively by the biosorbent to a maximum of about 419.5 mg g^{-1}.

9.9 Perspectives on the Development and Implementation of Microalgal Remediation Technologies

Since microcode has obvious biotechnological potential, phycoculture or algoculture, which refers to the mass cultivation of algae for industrial and commercial purposes, is a young, recent, and promising activity. Its development must come with a research effort to reduce production costs and make the yields of the cultures on an industrial scale more reliable.

The cultivation technologies to be used are decisive for the economic profitability of these microalgae productions. Using large volumes of water leads to high-energy expenditure, which is materialized by the necessity of the agitation of the cultures, thus ensuring the regular exposure of the algal cells to light and their access to the nutritive elements. These process conditions are very energy-consuming, and the harvesting of a few grams of dry matter per liter of water is also costly. The development and implementation of new technologies, which reduce these costs, are promising ways for larger-scale industrialization.

Similarly today, microalgae producers use wild species from the natural environment. Strategies for the selection and improvement of species, notably through directed evolution, are emerging. They should lead to substantial gains in productivity, as in agriculture, and contribute to developing the field.

The stability of crop performance is the final challenge for algal industrial farming. It requires deepening our knowledge of the physiology of the cultivated species to better understand their needs for mineral elements, such as nitrogen and phosphorus in particular, and studying their various possible beneficial interactions with the autochthonous bacterial consortia of the environment and the prevailing environmental factors.

The technical and scientific advances, in their current dynamics, let us envisage rapid progress on these biological and technological axes, which will lead to the development of industrial cultures of microalgae in the next few years. The improvement in biomass yields and molecules of interest is undoubtedly very important and will contribute to merging the economies of industrial crops.

The improvement of these strains by genetic engineering and getting genetically modified organisms (GMOs) has significantly increased their productivity for molecules of interest like lipids, pigments, etc. The use of algae as a bioremediation operator, notably for emerging toxic and recalcitrant residual compounds, should be encouraged in the future.

The selective diversity and the efficient originality of these photosynthetic and very bioactive organisms associated with their high productivity, the variety of the envisaged fields of application related to the concept of health, and the bioremediation of man and his environment are all so many insurances and guarantees of a great future for these microalgae.

Finally, bioremediation is an ideal solution to reduce and prevent residual and recalcitrant pollutants. It is an innovative, low-cost, clean, and efficient technology whose main operator is microorganisms, and the microalgae are in the front line with exceptional biological characteristics, guaranteeing a great development in very difficult and even extreme environmental conditions, and which can largely replace conventional methods, often very expensive and associated to hard implementation. So, a new development is emerging for the future use of microalgae for the bioremediation of toxic waste compounds due to several advantages: a simple and robust process, the absence of toxicity constraints, a rapid growth rate compared to higher plants as well as the formation of value-added products and bio-energy.

9.10 Conclusion

Microalgae, recognized as one of the oldest living organisms, are thallophytic and belong to a class of photosynthetic single-celled organisms that produce sugar and oxygen from water, carbon dioxide (CO_2), and solar energy. Although their main photosynthesis activity, they are typically more effective solar energy converters as compared to higher plants because of their basic cellular structure, their ability to grow in an aqueous solution, and their easier access to dissolved nutrients, like water, CO_2, and various minerals.

These green microalgae, endowed with an easy and undemanding culture in fresh or saltwater and with rapid growth kinetics on average 100 times faster than those of terrestrial plants, have been used as a purification agent for urban, agro-food, and industrial wastewater. These waters are most of the time loaded with a multitude of new compounds, often harmful to the environment and toxic to humans; these new compounds are called emergent. The effects, positive or negative, of these new compounds on man and his environment are still poorly known. In addition, these microalgae, at the end of much serious recent scientific research in this field, prove to be good operators of bioremediation, enhancing most of these emerging compounds, such as the production of biodiesel from sugars and triglycerides they produce, and many pharmaceutical molecules with added value, such as vitamins, antibiotics, etc. So, microalgae offer great potential as a sustainable feedstock for the production of third-generation biofuels, such as biodiesel and bioethanol, to synthesize several useful and cost-effective substances, and to bioremediate various recalcitrant emergent compounds. These green microalgal cells can also be a safe place to store synthesized energy or substances. At a time of global warming and energy transition, microalgae represent an interesting avenue for sustainable energy production and a reliable tool for wastewater treatment, cleaning, and bioremediation of emerging compounds whose fate in the environment is still poorly known and natural sites contaminated by man and his industries.

However, several important scientific and technical barriers still need to be overcome before microalgal bioremediation field usage, the production of bioactive substances of interest, and large-scale biofuel fabrication become commercial realities. Technological developments, including advances in photo-bioreactor design, selection of efficient microalgae strains for bioremediation amenable to easy and rapid cultivation and harvesting, and biomass processing, are important areas that can lead to improved profitability and, consequently, efficient commercial production and exploitation of substances of interest generated via the microalgae usage strategy.

References

Abdel-Raouf N, Al-Homaidan AA, Ibraheem IBM (2012) Microalgae and wastewater treatment. Saudi J Biol Sci 19(3):257–275

Abu Al-Rub FA, El-Naas MH, Ashour I, Al-Marzouqi M (2006) Biosorption of copper on *Chlorella vulgaris* from single, binary, and ternary metal aqueous solutions. Process Biochem 41(2):457–464. ISSN 1359-5113. https://doi.org/10.1016/j.procbio.2005.07.018

Ahmad A, Bhat AH, Buang A (2019) Enhanced biosorption of transition metals by living *Chlorella vulgaris* immobilized in Ca-alginate beads. Environ Technol 40(14):1793–1809. https://doi.org/10.1080/09593330.2018.1430171. Epub 1 Feb 2018. PMID: 29,345,546

Aksu Z, Tezer S (2005) Biosorption of reactive dyes on the Green Alga *Chlorella vulgaris*. Process Biochem 40(3):1347–1361. https://doi.org/10.1016/j.procbio.2004.06.007

Alexander JT, Hai FI, Al-aboud TM (2012) Chemical coagulation-based processes for trace organic contaminant removal: current state and future potential. J Environ Manage 111(0):195–207

Beisson F, Li-Beisson Y, Peltier G (2013) Des microalgues pour la production de biocarburants. Clefs CEA 61:50–52

Becker EW (1994) Microalgae: biotechnology and microbiology. Cambridge Studies in Biotechnology, vol 10. Cambridge University Press, Cambridge, 293 pp

Beijerinck MW (1890) Culturversuche mit Zoochlorellen, Lichenengonidien und anderen niederen Algen. Botanische Zeitung, vol 47. pp 725–739, 741–754, 757–768, 781–785

Bélair V (2012) Développement de nouvelles techniques d'extraction des lipides à partir des microalgues en vue de leur valorisation en biocarburant. Thèse Université Laval, Québec, Canada

Bharathiraja B, Chakravarthy M, Ranjith Kumar R, Yogendran D, Yuvaraj D, Jayamuthunagai J, Palani S et al (2015) Aquatic biomass (algae) as a future feedstock for bio-refineries: a review on cultivation, processing, and products. Renew Sustain Energy Rev 47:634–653

Bhatnagar A, Chinnasamy S, Singh M, Das KC (2011) Renewable biomass production by mixotrophic algae in the presence of various carbon sources and wastewater. Appl Energy 88(10):3425–3431

Bougaran G, Saint-Jean B (2014) Microalgues: de petits végétaux aux grandes promesses! Biofutur 360:28–31. Open Access Version: https://archimer.ifremer.fr/doc/00252/36321/

Becerra-Celis GBC (2009) Proposition de stratégies de commande pour la culture de microalgue dans un photobioréacteur continu. Autre. École Centrale Paris. Français. ⟨NNT: 2009ECAP0014⟩. ⟨tel-00453307⟩

Cheng J, Ye Q, Xu J, Yang Z, Zhou J, Cen K (2016) Improving pollutant removal by microalgae Chlorella PY-ZU1 with 15% CO_2 from undiluted anaerobic digestion effluent of food wastes with ozonation pretreatment. Biores Technol 216:273–279

Cheng DL, Ngo HH, Guo WS, Chang SW, Nguyen DD, Kumar SM (2019) Microalgae biomass from swine wastewater and its conversion to bioenergy. Biores Technol 275:109–122. https://doi.org/10.1016/j.biortech.2018.12.019

Chevallier F (1991b) Essais de cryoconservation d'algues unicellulaires d'intérêts aquacoles. IFREMER, France

Chisti Y (2007) Biodiesel from microalgae. Biotechnol Adv 25(3):294–306

Cho C, Zhao Y, Yun YS (2019) QSAR modeling for predicting adsorption of neutral, cationic, and anionic pharmaceuticals and other neutral compounds to microalgae *Chlorella vulgaris* in the aquatic environment. Water Res 151:288–295. https://doi.org/10.1016/j.watres.2018.12.033

Chong AM, Wong YS, Tam NF (2000) Performance of different microalgal species in removing nickel and zinc from industrial wastewater. Chemosphere 41(1–2):251–257

Chouteau C (2004) Développement d'un biocapteur conductimétrique bi-enzymatique à cellules algales. Thèse Université Claude Bernard-Lyon 1, France

Dawes CJ, Mathieson AC (2008) The seaweeds of Florida. University Press of Florida, Gainesville, FL, 8 + 592 p. Am J Plant Sci 10(10)

de Wilt A, Butkovskyi A, Tuantet K, Leal LH, Fernandes TV, Langenhoff A, Zeeman G (2016) Micropollutant removal in an algal treatment system fed with source-separated wastewater streams. J Hazard Mater 304:84–92. https://doi.org/10.1016/j.jhazmat.2015.10.033

De-Bashan LE, Bashan Y (2004) Recent advances in removing phosphorus from wastewater and its future use as fertilizer (1997–2003). Water Res 38(19):4222–4246

De-Bashan LE, Bashan Y (2010) Immobilized microalgae for removing pollutants: review of practical aspects. Biores Technol 101(6):1611–1627

De-Bashan LE, Hernandez JP, Morey T, Bashan Y (2004) Microalgae growth-promoting bacteria as 'helpers' for microalgae: a novel approach for removing ammonium and phosphorus from municipal wastewater. Water Res 38(2):466–474

De-Bashan LE, Trejo A, Huss VAR, Hernandez JP, Bashan Y (2008) *Chlorella sorokiniana* UTEX 2805, a heat and intense, sunlight-tolerant microalga with the potential for removing ammonium from wastewater. Biores Technol 99(11):4980–4989

Demirbas MF, Balat M (2006) Recent advances on the production and utilization trends of bio-fuels: a global perspective. Energy Convers Manage 47(15–16):2371–2381

Demirbas M, Demirbas F (2011) Importance of algae oil as a source of biodiesel. Energy Convers Manage 52(1):163–170. ISSN 0196-8904. https://doi.org/10.1016/j.enconman.2010.06.055

Drabik D, Venus TJ (2017) Chapter 2.2.2—water use for biofuels in Europe A2. In: Ziolkowska JR, Peterson JM (eds) Competition for water resources. Elsevier, pp 144–159

Dragone G, Fernandes B, Vicente António A, Teixeira José A (2010) Third-generation biofuels from microalgae. Report on projects INNOVALGAE (PTDC/AAC-AMB/108,511/2008) and ALGANOL

El Ouaer M, Kallel A, Kasmi M, Hassen A, Trabelsi I (2017a) Tunisian landfill leachate treatment using *Chlorella* sp.: effective factors and microalgae strain performance. Arab J Geosci 10(20):1–9. https://doi.org/10.1007/s12517-017-3241-4

El Ouaer M, Halaoui M, Trabelsi I, Hassen A (2017b) Adaptation and use of *Chlorella* sp. for landfill leachate treatment. Int J Innov Appl Stud 14(3):807–818

El Ouaer M, Trabelsi I, Turki N, Amjad Kallel, Halaoui M, Hassen A (2018) Recovery of landfill leachate as culture medium for two microalgae: Chlorella sp., and Scenedesmus sp. Environ Dev Sustain. https://doi.org/10.1007/s10668-019-00314-7

El Ouaer M, Turki N, Kallel A, Trabelsi I, Hassen A (2020a) Recovery of landfill leachate as a cultural medium for two microalgae: *Chlorella* sp. and *Scenedesmus* sp. Projects: Renewable energy, energy recovery, and compaction soil. Wastewater treatment in rural and remote areas. Environ Dev Sustain 22(3). https://doi.org/10.1007/s10668-019-00314-7

El Ouaer M, Turki N, Kallel A, Trabelsi I, Hassen A (2020b) Recovery of landfill leachate as a cultural medium for two microalgae: *Chlorella* sp. and *Scenedesmus* sp. Arab J Geosci 10(20). https://doi.org/10.1007/s12517-017-3241-4

El Ouaer M, Turki N, Ibrahim C, Hussain Chaabene S, Kallel A, Trabelsi I, Hassen A (2022) Insights into the use of landfill leachate to grow *Chlorella* sp. for lipid and fatty acids production. Clean Technol Environ Policy. https://doi.org/10.1007/s10098-022-02460-5

Environmental and Energy Study Institute (EESI) (2016) Marine microalgae: the future of sustainable biofuel newsletters climate change solutions EESI impact. https://www.eesi.org/articles/view/marine-microalgae-the-future-of-sustainable-biofuel

Enzing C, Ploeg M, Barbosa M, Sijtsma L, Vigani M, Parisi C, Rodríguez Cerezo E (2000) Microalgae-based products for the food and feed sector: an outlook for Europe. Publications Office of the Institute for Prospective Technological Studies, Luxembourg

Flores-Chaparro CE, Ruiz LFC, de la Torre MCA, Huerta-Diaz MA, Rangel-Mendez JR (2017) Biosorption removal of benzene and toluene by three dried macroalgae at different ionic strengths and temperatures: algae biochemical composition and kinetics. J Environ Manage 193:126–135. https://doi.org/10.1016/j.jenvman.2017.02.005

Gavrilescu M, Chisti Y (2005) Biotechnology—a sustainable alternative for the chemical industry. Biotechnol Adv 23(7–8):471–499

Gayosso-Morales MA, Rivas-Castillo AM, Lucas-Gómez I, López-Fernández A, Calderón AV, Fernández-Martínez E, Bernal JO, González-Pérez BK (2023) Microalgae, a current option for the bioremediation of pharmaceuticals: a review. Folia Microbiol (Praha) 68(2):167–179. https://doi.org/10.1007/s12223-022-01013-z. Epub 11 Nov 2022. PMID: 36,367,638

Gerardo Michael L, Van Den S, Hende B, Vervaeren H, Thea C, Skill Stephen C (2015) Harvesting of microalgae within a biorefinery approach: a review of the developments and case studies from pilot plants. Algal Res 11:248–262. https://doi.org/10.1016/j.algal.2015.06.019

Gilbert F, Galgani F (1990) Test éco toxicologique sur microalgues et tirage NB lecteur de plaques: mesures d'activités enzymatiques sur cellules in vivo. Application à l'évaluation de la qualité du milieu. Institut Français de Recherche pour l'Exploitation de la Mer, IFREMER, Centre de Nantes, Direction des Recherches Océaniques. DRO-90-06-MR, p 39

Gonzalez LE, Bashan Y (2000) Increased growth of the microalga *Chlorella vulgaris* when co-immobilized and co-cultured in alginate beads with the plant-growth-promoting bacterium *Azospirillum brasilense*. Appl Environ Microbiol 66(4):1527–1531

Gruchlik Y, Linge K, Joll C (2018) Removal of organic micropollutants in waste stabilization ponds: a review. J Environ Manage 206:202–214. https://doi.org/10.1016/j.jenvman.2017.10.020

Hassen A, Jamoussi F, Saidi N, Mabrouki Z, Fakhfakh E (2003) Microbial and copper adsorption by smectitic clay-an experimental study. Environ Technol 24(9):1117–1127. https://doi.org/10.1080/095933303093856527

Hosseini SM, Khosravi-Darani K, Mozafari MR (2013) Nutritional and medical applications of spirulina microalgae. Mini Rev Med Chem 13(8):1231–1237. https://doi.org/10.2174/1389557511313080009. PMID: 23,544,470

Hubbe MA, Park J, Park S (2014) Cellulosic substrate for the removal of pollutants from aqueous systems: a review. Part 4. Dissolved petrochemical compounds. Bioresour Technol 9:7782–7925

Jin J, Yang L, Chan SMN, Luan T, Li Y, Tam NFY (2011) Effect of nutrients on the biodegradation of tributyltin (TBT) by alginate immobilized microalga, *Chlorella vulgaris*, in natural river water. J Hazard Mater 185(2–3):1582–1586

Johnson MB, Wen Z (2009) Production of biodiesel fuel from the microalga *Schizochytrium limacinum* by direct transesterification of algal biomass. Energy Fuels 23(10):5179–5183. https://doi.org/10.1021/ef900704h

Karadjova IB, Slaveykova VI, Tsalev DL (2008) The bio uptake and toxicity of arsenic species on the green microalga Chlorella salina in seawater. Aquat. Toxicol 87(4):264–271. ISSN 0166-445X, https://doi.org/10.1016/j.aquatox.2008.02.006

Khavari F, Saidijam M, Taheri M et al (2021) Microalgae: therapeutic potential and applications. Mol Biol Rep 48:4757–4765. https://doi.org/10.1007/s11033-021-06422-w

Konishi F, Tanaka K, Kumamoto S, Hasegawa T, Okuda M, Yano I, Yoshikai Y, Nomoto K (1990) Enhanced resistance against *Escherichia coli* infection by subcutaneous administration of the hot-water extract of *Chlorella vulgaris* in cyclophosphamide-treated mice. Cancer Immunol Immunother CII 32(1):1–7

Leong YK, Chang JS (2020) Bioremediation of heavy metals using microalgae: recent advances and mechanisms. Bioresour Technol 303:122886. https://doi.org/10.1016/j.biortech.2020.122886. Epub 23 Jan 2020. PMID: 32,046,940

Lee Lees HR (2001) The demography and interactions of Ecklonia radiata in southern New Zealand. Thesis. University of Canterbury. New Zealand. https://doi.org/10.26021/6593

Lim S-L, Chu W-L, Phang SM (2010) Use of *Chlorella vulgaris* for bioremediation of textile wastewater. Biores Technol 101(19):7314–7322

Lin L, Chan GYS, Jiang BL, Lan CY (2007) Use of ammoniacal nitrogen tolerance microalgae in landfill leachate treatment. Waste Manage 27(10):1376–1382

Littler DS, Littler MM (2000) Caribbean reef plants: an identification guide to the reef plants of the Caribbean, Bahamas, Florida and the Gulf of Mexico. Offshore Graphics, Washington DC. Am J Plant Sci 6(19)

Liu J, Mao X, Zhou W, Guarnieri MT (2016) Simultaneous production of triacylglycerol and high-value carotenoids by the astaxanthin-producing oleaginous green microalga *Chlorella zofingiensis*. Biores Technol 214:319–327

Luan TG, Jin J, Chan SMN, Wong YS, Tam NFY (2006) Biosorption and biodegradation of tributyltin (TBT) by alginate immobilized *Chlorella vulgaris* beads in several treatment cycles. Process Biochem 41(7):1560–1565

Luo L, Wang P, Lin L, Luan T, Ke L, Tam NFY (2014) Removal and transformation of high-molecular-weight polycyclic aromatic hydrocarbons in water by live and dead microalgae. Process Biochem 49(10):1723–1732

Manikandan A, Suresh Babu P, Shyamalagowri S, Kamaraj M, Muthukumaran P, Aravind J (2022) The emerging role of microalgae in heavy metal bioremediation. J Basic Microbiol 62(3–4):330–347. https://doi.org/10.1002/jobm.202100363. Epub 1 Nov 2021. PMID: 34,724,223

Marcoux M-A, Matias M, Olivier F (2012) Substances émergentes, polluants émergents dans les déchets: analyze et prospective. ECOGEOS Rapport Final. Étude N° 10-0143/1A

Mathimani T, Pugazhendhi A (2019) Utilization of algae for biofuel, bio-products, and bio-remediation. Biocatal Agric Biotechnol. Corpus ID: 134,991,195. https://doi.org/10.1016/J.BCAB.2018.12.007

Mathimani T, Baldinelli A, Rajendran K, Prabakar D, Matheswaran M, Leeuwen RPV, Pugazhendhi A (2019) Review on cultivation and thermochemical conversion of microalgae to fuels and chemicals: process evaluation and knowledge gaps. J Clean Product 208:1053–1064. https://doi.org/10.1016/j.jclepro.2018.10.096

Mathon B, Coquery M, Miège C, Vandycke A, Choubert JM (2019) Influence of water depth and season on the photodegradation of micropollutants in a free-water surface constructed wetland receiving treated wastewater. Chemosphere 235:260–270. https://doi.org/10.1016/j.chemosphere.2019.06.140

Merchant RE, Carmack CA, Wise CM (2000) Nutritional supplementation with *Chlorella pyrenoidosa* for patients with fibromyalgia syndrome: a pilot study. Phytother Res PTR 14(3):167–173

Mouchet P (1986) Review of algal reactions to mineral and organic micropollutants; ecological consequences and possible industrial applications. Water Res 20(4):399–412

Nguyen HT, Yoon Y, Ngo HH, Jang A (2020a) The application of microalgae in removing organic micropollutants in wastewater. Crit Rev Environ Sci Technol. https://doi.org/10.1080/10643389.2020.1753633

Nguyen TT, An Binh Q, Bui XT, Ngo HH, Vo HNP, Andrew Lin KY, Vo TD, Guo W, Lin C, Breider F (2020b) Co-culture of microalgae-activated sludge for wastewater treatment and biomass production: exploring their role under different inoculation ratios. Bioresour Technol 314:123754. https://doi.org/10.1016/j.biortech.2020.123754. Epub 2 July 2020. PMID: 32,650,264

Nistor C, Emneus J (1999) Bioanalytical tools for monitoring polar pollutants. Waste Manage 19(2):147–170

Nora NF, Chong AMY, Wong YS (2002) Removal of tributyltin (TBT) by live and dead microalgal cells. Mar Pollut Bull 45(1):362–371

Olivella M, Jové P, Bianchi A, Bazzicalupi C, Cano L (2013) An integrated approach to understanding the sorption mechanism of phenanthrene by cork. Chemosphere 90:1939–1944. https://doi.org/10.1016/j.chemosphere.2012.10.035

Pathak B, Gupta S, Verma R (2018) Biosorption and biodegradation of polycyclic aromatic hydrocarbons (PAHs) by microalgae. In: Green adsorbents for pollutant removal. Springer, Heidelberg

Priya M, Gurung N, Mukherjee K, Bose S (2014) 23—microalgae in removal of heavy metal and organic pollutants from soil A2—Das, Surajit. In: Microbial biodegradation and bioremediation. Elsevier, Oxford, pp 519–537

Purev O, Park C, Kim H, Myung E, Choi N, Cho K (2023) *Spirulina platensis* immobilized alginate beads for removal of Pb (II) from aqueous solutions. Int J Environ Res Publ Health 20(2):1106. https://doi.org/10.3390/ijerph20021106. PMID: 36,673,865

Raheem A, Azlina WAKGW, Tau YH, Danquah MK (2015) Thermochemical conversion of microalgal biomass for biofuel production.vol 49:990–999. https://doi.org/10.1016/j.rser.2015.04.186

Raheem A, Prinsen P, Vuppaladadiyam Arun K, Zhao M, Luque R (2018) A review on sustainable microalgae-based biofuel and bioenergy production: recent developments. J Clean Product 181:42–59. ISSN 0959-6526. https://doi.org/10.1016/j.jclepro.2018.01.125. https://www.sciencedirect.com/science/article/pii/S0959652618301471

Rai UN, Singh NK, Upadhyay AK, Verma S (2013) Chromate tolerance and accumulation in Chlorella vulgaris L.: role of antioxidant enzymes and biochemical changes indetoxification of metals. Bioresour Technol 136: 604–609

Regel RH, Ferris JM, Ganf GG, Brookes JD (2002) Algal esterase activity as a bio-measure of environmental degradation in a freshwater creek. Aquat Toxicol (Amsterdam Neth) 59(3–4):209–223

Richards RG, Mullins BJ (2013) "Using microalgae for combined lipid production and heavy metal removal from leachate. Ecol Model 249:59–67

Rivera-Utrilla J, Sánchez-Polo M, Ferro-García MA, Prados-Joya G, Ocampo-Pérez R (2013) Pharmaceuticals as emerging contaminants and their removal from water. A review. Chemosphere 93(7):1268–1287. Corpus ID: 19,892,631. https://doi.org/10.1016/j.chemosphere.2013.07.059

Rosenberg JN, Oyler GA, Wilkinson L, Betenbaugh MJ (2008) A green light for engineered algae: redirecting metabolism to fuel a biotechnology revolution. Curr Opin Biotechnol 19(5):430–436. https://doi.org/10.1016/j.copbio.2008.07.008

Rowley WM (2010) Nitrogen and phosphorus biomass-kinetic model for Chlorella vulgaris in a biofuel production scheme

Saladini F, Patrizi N, Pulselli FM, Marchettini N, Bastianoni S (2016) Guidelines for energy evaluation of first, second, and third generation biofuels. Renew Sustain Energy Rev 66:221–227

Sambusiti C, Monlau F, Ficara E, Musatti A, Rollini M, Barakat A, Malpei F (2015) Comparison of various post-treatments for recovering methane from agricultural digestation. Fuel Process Technol 137:359–365. https://doi.org/10.1016/j.fuproc.2015.04.028

Saravanan K, Duraisamy S, Ramasamy G, Kumarasamy A, Balakrishnan S (2018) Evaluation of the saccharification and fermentation process of two different seaweeds for eco-friendly bioethanol production". Biocatal Agric Biotechnol 14:444–449. https://doi.org/10.1016/j.bcab.2018.03.017

Song M, Haiyan P, Wenrong H, Guixia M (2013) Evaluation of the potential of 10 microalgal strains for biodiesel production. Bioresource Technol 141:245–251. ISSN 0960-8524. https://doi.org/10.1016/j.biortech.2013.02.024

Sumi Y (2009) Microalgae, pioneering the future: application and utilization. Life Sci Res unit Q Rev No. 34

Sydney EB, Sturm W, de Carvalho JC, Thomaz-Soccol V, Larroche C, Pandey A, Soccol CR (2010) Potential carbon dioxide fixation by industrially important microalgae. Biores Technol 101:5892–5896

Tanaka K, Yamada A, Noda K, Shoyama Y, Kubo C, Nomoto K (1997) Oral administration of unicellular green algae, *Chlorella vulgaris*, prevents stress-induced ulcers. Planta Med 63(5):465–466

Thangalazhy-Gopakumar S, Adhikari S, Chattanathan SA, Gupta RB (2012) Catalytic pyrolysis of green algae for hydrocarbon production using H+ZSM-5 catalyst. Biores Technol 118:150–157

Turner A, Furniss O (2012) An evaluation of the toxicity and bioaccumulation of thallium in the coastal marine environment using the macro alga, Ulva lactuca. Mar Pollut Bull 64:2720–2724

Um B-H, Kim Y-S (2009) Review: a chance for Korea to advance algal-biodiesel technology. J Ind Eng Chem 15:1–7

Ursu A-V, Marcati A, Sayd T, Sante-Lhoutellier V, Djelveh G, Michaud P (2014) Extraction, fractionation and functional properties of proteins from the microalgae *Chlorella vulgaris*. Biores Technol 157:134–139

Vassilis SV, Vassileva CG (2016) Composition, properties, and challenges of algae biomass for biofuel application: an overview, Fuel 181:1–3. ISSN 0016-2361, https://doi.org/10.1016/j.fuel.2016.04.106

Vilchez C, Garbayo I, Lobato MV, Vega J (1997) Microalgae-mediated chemical production and waste removal. Enzyme Microb Technol 20(8):562–572

Wang Y, Liu J, Kang D, Wu C, Wu Y (2017) Removal of pharmaceuticals and personal care products from wastewater using algae-based technologies: a review. Rev Environ Sci Bio/technol 16(4):717–735. https://doi.org/10.1007/s11157-017-9446-x

Wang H, Yang B, Zhang Q, Zhu W (2019) Catalytic routes for the conversion of lignocellulosic biomass to aviation fuel range hydrocarbons. Renew Sustain Energy Rev 120. Article ID 109,612. https://doi.org/10.1016/j.rser.2019.109612

Wang L, Liu J, Filipiak M, Mungunkhuyag K, Jedynak P, Burczyk J, Fu P, Malec P, (2021) Fast and efficient cadmium biosorption by *Chlorella vulgaris* K-01 strain: the role of cell walls in metal sequestration. Algal Res 60:102, 497. ISSN 2211-9264. https://doi.org/10.1016/j.algal.2021.102497

Werheni AR, Hidri Y, Bagues M, Hassen A (2021) Pentachlorophenol attenuation and biodegradation process in Tunisian forest soil. Arab J Geosci 14(24):1–14. https://doi.org/10.1007/s12517-021-09218-1

Zada B, Wang C, Park J-B, Jeong SH, Park JE, Hawaibam BS, Seon-Won K (2018) Metabolic engineering of *Escherichia coli* for production of mixed isoprenoid alcohols and their derivatives. Biotechnol. Biofuels 11:210. https://doi.org/10.1186/s13068-018-1210-0

Chapter 10
Implementation of Algal Approach in Techno-socio-economical Aspect of Wastewater Treatment

Tazkiaturrizki, Astri Rinanti, Melati Ferianita Fachrul, Diana Irvindiaty Hendrawan, Sarah Aphirta, Sheilla Megagupita Putri Marendra, and Naomi Oshin Laurensa Sipahutar

Abstract The implementation of the algae approach in the technological, social, and economic aspects of wastewater treatment offers an environmentally friendly solution to address the problem of water pollution. Algae play an important role in absorbing pollutants, recovering nutrients, reducing greenhouse gas emissions, and producing biomass for renewable energy. Knowledge of their classification, characteristics, growth, reproduction, and interaction with the environment will help increase the effectiveness of wastewater treatment. The process of wastewater treatment by microalgae metabolic activity involves certain principles, mechanisms, systems, and technologies that enable the improvement of treatment parameters and efficiency. Integrating microalgae technology with other wastewater treatment technologies, such as hybrid systems and technology combinations, can improve overall system synergy and efficiency. Economic aspects and cost analysis need to

Tazkiaturrizki · A. Rinanti (✉) · M. F. Fachrul · D. I. Hendrawan · S. Aphirta · S. M. P. Marendra · N. O. L. Sipahutar
Environmental Engineering Department, Faculty of Landscape Architecture and Environmental Technology, Universitas Trisakti, Jakarta, Indonesia
e-mail: astririnanti@trisakti.ac.id

Tazkiaturrizki
e-mail: tazkiaturrizki13@trisakti.ac.id

M. F. Fachrul
e-mail: melati@trisakti.ac.id

D. I. Hendrawan
e-mail: dianahendrawan@trisakti.ac.id

S. Aphirta
e-mail: aphirtasarah@trisakti.ac.id

S. M. P. Marendra
e-mail: sheilla@trisakti.ac.id

N. O. L. Sipahutar
e-mail: naomi082001900055@std.trisakti.ac.id

T. A. Kurniawan and A. Anouzla (eds.), *Algae as a Natural Solution for Challenges in Water-Food-Energy Nexus*, Environmental Science and Engineering,
https://doi.org/10.1007/978-981-97-2371-3_10

be carefully considered, including investment costs, operations, economic studies, processing scale, and product marketing potential and sources of income. Meanwhile, the social and policy aspects discussed include public awareness and acceptance, policies and regulations that support the use of algae, as well as partnerships and collaboration between government, industry, and society. Case studies and real applications show experiences and lessons from implementation in various countries, as well as success factors and constraints in these implementations. In facing future challenges and prospects, it is necessary to carry out further research and development of algae technology to increase efficiency and innovation in wastewater treatment and overcome various obstacles that may be encountered during implementation.

Keywords Wastewater treatment · Microalgae technology · Water pollution · Techno-socio-economical aspect · Environmentally friendly · Renewable energy

10.1 Introduction

10.1.1 Background on the Problem of Wastewater Treatment

Issues related to the implementation of the algae approach in the technological, social with regard to public awareness, and economic aspects of wastewater treatment involve various factors. The background of this issue encompasses several related aspects, including the increasing industrial activities and population growth, which have led to increasingly severe water pollution problems. Domestic and industrial wastewater contains pollutants such as excessive nutrients, heavy metals, and hazardous chemicals that can endanger the environment and human health. Conventional wastewater treatment methods—such as sedimentation, filtration, and biological treatment—have limitations in addressing various types of pollutants. Some methods also require high operational and maintenance costs and generate secondary waste that needs further treatment (Mitra et al. 2022).

Algae offer an environmentally friendly and sustainable alternative solution for wastewater treatment. Algae can absorb pollutants like excessive nutrients, heavy metals, and hazardous chemicals through bioaccumulation and biodegradation processes, often eliminating the need for chemical additives as shown in Fig. 10.1 (Valchev and Ribarova 2022). Furthermore, algae can be utilized for the production of biogas, biofuel, and other high-value products, thereby enhancing the economic aspects of wastewater treatment.

Wastewater treatment technology using algae is still relatively new and not yet fully developed. It creates challenges in adapting and integrating this technology into existing wastewater treatment systems. Additionally, the process efficiency, scalability, and maintenance of algae systems are also issues that need to be addressed. The implementation of the algae approach in wastewater treatment requires specific technologies and equipment, such as photobioreactors and environmental parameter

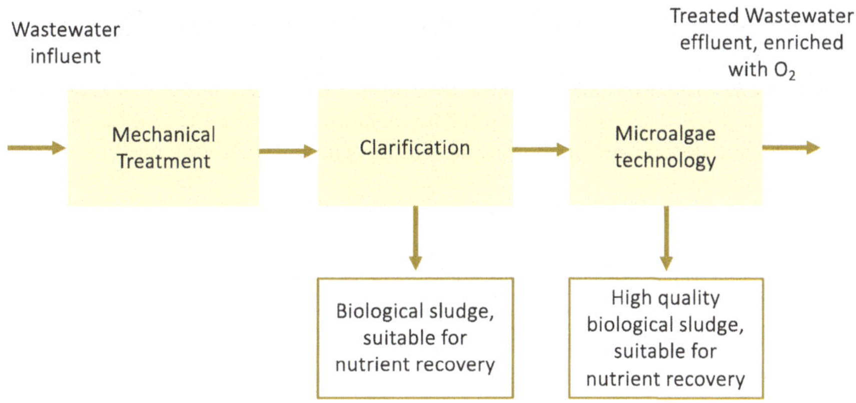

Fig. 10.1 Non-chemical usable

control systems as shown in Fig. 10.2 (You et al. 2023). Research and development of efficient and economical technologies are still needed to maximize the potential of algae in wastewater treatment.

The public may be unfamiliar with or uncertain about algae-based wastewater treatment technology; hence, effective education and socialization are necessary. Wastewater treatment with algae requires awareness and support from the public. Some individuals may not realize the importance of wastewater treatment or feel skeptical about the effectiveness of algae-based approaches. Public education and outreach are needed to enhance the understanding of the benefits and potential of this technology. From an economic perspective, implementing the algae approach in wastewater treatment requires significant initial investment, including installation,

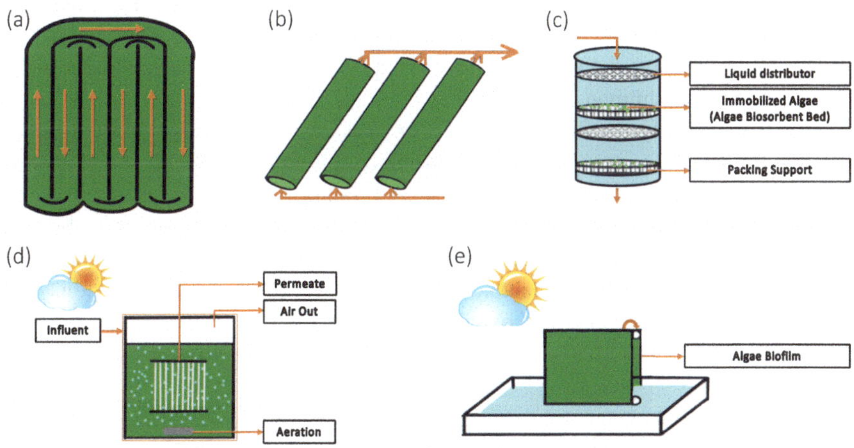

Fig. 10.2 Photobioreactor types. **a** Suspended open systems, **b** suspended closed systems, **c** immobilized algae system, **d** membrane photobioreactors, and **e** algae biofilm

operation, and maintenance costs. Additionally, high initial financing, regulations, and government support also present obstacles to the implementation of this technology. Further research and development to improve the efficiency and effectiveness of this technology also require substantial financial support. In many cases, collaboration between the government and private sectors is necessary to provide the required funding and incentives.

The existing regulations and policies may not yet facilitate the implementation of the algae approach in wastewater treatment. The government needs to review and update the existing policies and regulations to create a conducive environment for the development and application of this technology. The implementation of the algae approach in wastewater treatment requires collaboration between the industrial sector, the government, and society (Obaideen et al. 2022). Creating supportive policies, increasing public awareness, and investing in research and technology development are crucial steps to address these challenges. The hurdles in coordination and integration of efforts from various stakeholders can hinder the implementation of this technology.

10.1.2 The Importance of Environmentally-Friendly Technology in Wastewater Treatment

Eco-friendly technology in wastewater treatment is pivotal for several reasons:

(1) Wastewater treatment with environmentally-friendly technology helps maintain the quality of water resources such as rivers, lakes, and groundwater. By reducing contamination and pollution, this technology ensures that water resources remain safe and suitable for various purposes, e.g., consumption, irrigation, and industrial use. Wastewater treatment with environmentally-friendly methods, like the algae approach, helps reduce water pollution and protect existing water resources. Preserving water resources supports aquatic life, agricultural needs, industries, and human consumption.

(2) Environmentally-friendly technology in wastewater treatment reduces negative impacts on ecosystems and biodiversity. Poorly treated water pollution can threaten the lives of flora and fauna and disrupt the ecosystem balance. By employing environmentally-friendly technology, we contribute to environmental protection and the preservation of biodiversity.

(3) Environmentally-friendly technology reduces greenhouse gas emissions and air pollution often generated by conventional wastewater treatment methods. It contributes to global efforts to mitigate climate change impacts and maintain air quality.

(4) Effective wastewater treatment with environmentally-friendly technology reduces the dissemination of waterborne diseases such as diarrhea and cholera, thereby improving public health and well-being. Poorly treated wastewater

can contaminate drinking water sources and contribute to disease transmission. Environmentally-friendly wastewater treatment technology helps reduce the risk of waterborne diseases and enhances the quality of life.

(5) Environmentally-friendly technology in wastewater treatment often reduces energy and hazardous chemical consumption. Eco-friendly methods like the algae approach utilize natural and renewable resources, such as sunlight, to produce the energy needed in the treatment process. It decreases reliance on fossil energy sources and enhances the sustainability of the system. Algae-based approaches, for instance, leverage natural biological processes that produce minimal residue and have lower environmental impact compared to conventional methods. Environmentally-friendly wastewater treatment technology also includes resource recovery methods, such as nutrient and energy extraction from wastewater. For example, microorganisms in the algae approach can convert nutrients in wastewater into biomass, which can be used as feedstock for energy production or other high-value products.

(6) Implementing environmentally-friendly technology in wastewater treatment creates new economic opportunities, such as the development of innovative products and technologies, job creation, and cost savings. The development and implementation of environmentally-friendly wastewater treatment technology foster new business opportunities and employment in the green industry. Furthermore, it supports sustainable economic growth and encourages a transition towards a circular economy. For instance, in the algae-based approach, algae grown in the wastewater treatment process can be used as raw material sources for fisheries, animal feed, bioenergy, or pharmaceutical products.

(7) Implementing environmentally-friendly technology in wastewater treatment can serve as a means to raise public awareness regarding the importance of water resources and environmental preservation. It will encourage more environmentally-friendly behaviors and support for sustainable environmental policies.

(8) The use of environmentally-friendly technology in wastewater treatment enables companies and governments to comply with increasingly strict environmental regulations and avoid penalties or fines that may arise from violations.

To ensure environmental sustainability and societal well-being, it is essential to adopt environmentally-friendly technology in wastewater treatment, such as the algae-based approach, which combines technological, social, and economic advantages (Zapata-Mendoza et al. 2022; Siwi et al. 2017).

10.1.3 The Role of Algae in Wastewater Treatment

Microalgae have several advantages compared to physical and chemical treatment methods in heavy metal absorption from wastewater. The significant role of microalgae in wastewater treatment lies in their ability to remove pollutants, produce high-value biomass, and potentially reduce greenhouse gas emissions. Some roles of microalgae in wastewater treatment are as follows:

(1) Biofiltration and nutrient absorption: Microalgae can form biofilms on various surfaces, such as biological disks, during the biofiltration process. These biofilms enable microalgae to absorb pollutants from wastewater, including heavy metals and organic compounds. This process helps reduce the concentration of pollutants in wastewater and improves water quality, involving the absorption, biodegradation, and bioaccumulation of pollutants by microalgae. For instance, microalgae can remove nitrogen and phosphorus from wastewater by assimilating the pollutants as nutrients for growth. Microalgae require nutrients like nitrogen and phosphorus for growth and reproduction (Yaakob et al. 2021). Wastewater rich in these essential elements serves as a good nutrient source for microalgae. When microalgae absorb these nutrients from wastewater, the nutrient content in the wastewater decreases, thereby reducing the potential for environmental pollution and eutrophication. Therefore, algae are considered efficient in removing excess nutrients from wastewater, such as phosphate and nitrogen. These nutrients are major pollutants responsible for water eutrophication. Algae convert these nutrients into biomass through photosynthesis and cell growth.
(2) Biodegradability: Microalgae are living organisms that can easily degrade in the environment after accumulating heavy metals. This means that algae-based methods are more environmentally friendly compared to physical and chemical methods, which often result in secondary waste that is difficult to degrade or requires further management (Touliabah et al. 2022; Kautsar et al. 2021).
(3) Bioremediation and phytoremediation: Microalgae as living organisms play a crucial role in bioremediation of wastewater due to their ability to absorb and accumulate pollutants in water. Specifically, since microalgae are producers or similar to plants (phyto), the process of removing heavy metals and other pollutants from wastewater is called phytoremediation. Algae can absorb, accumulate, and transform heavy metals into less hazardous and stable forms, thereby reducing the toxicity of wastewater. This process involves the absorption of heavy metals, excess nutrients, and organic compounds by algae through bioadsorption, biosorption, and biodegradation processes.
(4) Reduction of pathogenic microorganisms: Some types of algae, especially microalgae, can reduce the number of pathogens in wastewater through predation, competition, and production of antimicrobial compounds. It helps decrease the spread of water-related diseases.
(5) Biomitigation: Microalgae photosynthesize to convert carbon dioxide (CO_2) from the air into biomass (Onyeaka et al. 2021). This process helps reduce CO_2 emissions produced by industrial sectors and power plants, thus mitigating the impact of climate change. Additionally, using algae in wastewater treatment reduces the need for conventional aerobic processes that generate greenhouse gas emissions such as methane and nitrous oxide. Through photosynthesis, algae also produce oxygen as a byproduct, which enhances water quality by

increasing dissolved oxygen levels. This condition supports aquatic life and other biological processes in water bodies.

(6) Bioproduction of high-value biomass: Microalgae can produce biomass that can be used for various applications, such as raw materials for bioenergy (biofuel), animal feed, fertilizers, and high-value products like pigments, omega-3 fatty acids, and proteins. Thus, algae not only treat wastewater but also create valuable commercial products, promoting new business opportunities and supports the circular economy.

(7) Anaerobic wastewater treatment: Microalgae can be combined with bacteria in anaerobic wastewater treatment systems to produce biogas (Kazimierowicz et al. 2023). Microalgae supply oxygen through photosynthesis to support bacterial growth, which subsequently generates biogas that can be used as a renewable energy source.

Implementing the utilization of algae in wastewater treatment based on technological, social, and economic aspects requires consideration of the above roles of algae. Integrating algae-based technology into existing wastewater treatment systems and developing new methods involving algae will contribute to more efficient, sustainable, and environmentally-friendly wastewater treatment (Maryjoseph and Ketheesan 2020). To implement the algae approach in wastewater treatment, systems such as photobioreactors, algae ponds, and hybrid systems with conventional technologies can be used. The success of implementing this technology will depend on the type of algae used, environmental conditions, and technological, social, and economic support.

(8) Operational conditions: The use of microalgae in wastewater treatment generally requires lighter operational conditions (e.g., temperature, pressure, and pH) compared to physical and chemical methods. This reduces energy consumption and operational costs in wastewater treatment systems and decreases the risk of accidents and environmental damage that may occur due to extreme operational conditions.

10.2 Algae Biology and Ecology

10.2.1 Classification and Characteristics of Algae

The ecological aspects of algae classification and characteristics are crucial for understanding how they can be utilized in the implementation of the algae approach in the technological, social, and economic aspects of wastewater treatment. Several characteristics and classifications of algae can be categorized based on various criteria, such as photosynthetic pigments, cell shape, and cell wall composition. Generally, algae are divided into several main groups, including:

Cyanobacteria, also known as blue-green algae, are prokaryotes capable of performing oxygenic photosynthesis found in diverse environments, ranging from fresh to saline water, soil, and even extreme environments such as hot springs and polar regions. Through oxygenic photosynthesis, oxygen is produced as a byproduct of water splitting. This process has a significant impact on the Earth's atmosphere, transforming the early carbon dioxide-rich atmosphere into one with higher oxygen content. This transformation enabled the development of life as we know it today. *Cyanobacteria* play a crucial role in nutrient and oxygen cycles on Earth and represent one of the earliest groups of photosynthetic organisms. Their photosynthetic ability is essential in the history of life on Earth.

Cyanobacteria contain chlorophyll a, the primary pigment found in all photosynthetic organisms. Moreover, these microalgae contain additional pigments known as phycobiliproteins, such as allophycocyanin and phycocyanin, which give *Cyanobacteria* their characteristic blue-green color. These phycobiliproteins help absorb light of various wavelengths and enhance photosynthetic efficiency. *Cyanobacteria* employ two photosystems (Photosystem I and Photosystem II) in the photosynthesis process, similar to green plants and algae. These photosystems work together to produce energy (ATP) and electron carriers (NADPH) required for converting carbon dioxide into carbohydrates.

Some *Cyanobacteria* are also capable of nitrogen fixation, converting atmospheric nitrogen gas (N_2) into ammonia (NH_3) usable by other organisms. This process contributes to the nitrogen cycle and enables *Cyanobacteria* to thrive in environments with low nitrogen availability. Certain *Cyanobacteria* possess specialized structures called heterocysts, allowing them to separate photosynthetic activity and nitrogen fixation within a single organism. This separation is essential as the enzyme nitrogenase, used for nitrogen fixation, is highly sensitive to oxygen. By separating these processes, *Cyanobacteria* can maintain efficient photosynthesis while facilitating nitrogen fixation.

Cyanobacteria play a crucial role in wastewater treatment due to their ability to remove pollutants, absorb excess nutrients, and produce oxygen through photosynthesis. Some common cyanobacteria species used in wastewater treatment are as follows:

(a) *Anabaena*, a filamentous *Cyanobacteria* genus capable of nitrogen fixation. This species has been employed in wastewater treatment to remove excess nitrogen and phosphorus and aid in controlling harmful algae growth.
(b) Like *Anabaena*, *Nostoc* is also a filamentous *Cyanobacteria* capable of nitrogen fixation (Kollmen and Strieth 2022). This species has been utilized in wastewater treatment to reduce excess nutrients, particularly nitrogen and phosphorus.
(c) *Microcystis* is a colonial *Cyanobacteria* found in freshwater. Although known for producing the microcystin toxin, this species has also been used in wastewater treatment to remove excess nutrients and certain pollutants (Xiao et al. 2018).
(d) *Oscillatoria* is a freshwater blue-green alga that can be harnessed in wastewater treatment to remove pollutants such as heavy metals and absorb excess nutrients (Ankit et al. 2022).

(e) *Phormidium* is commonly found in freshwater environments. This species is reported to reduce excess nutrients and remove some pollutants, such as heavy metals, from wastewater (Lakmali et al. 2022).
(f) *Spirulina* (now popular as *Arthrospira*) is a filamentous *Cyanobacteria* known as a food supplement due to its high protein, vitamin, and mineral content. Additionally, *Spirulina* is used in wastewater treatment to remove excess nutrients and pollutants, such as heavy metals (Nege et al. 2020).

Chlorophyta, or green algae, is a large group of photosynthetic organisms found in various environments such as freshwater, saline water, soil, and even extreme environments like snow and ice (Cho et al. 2022). Green algae exhibit a wide diversity of forms and sizes, ranging from single-celled organisms to colonial and filamentous structures, as well as larger macroalgae known as *Ulvophyceae*. One of the distinguishing characteristics of *Chlorophyta* from other algal groups is the presence of chlorophyll a and chlorophyll b as the primary pigments used in the photosynthesis process. These pigments give them their characteristic green color. Besides chlorophyll, green algae contain carotenoids and xanthophylls, which play roles in light absorption and protect cells from oxidative stress.

The cell walls of green algae are typically made of cellulose, providing structure and mechanical support. In some cases, the cell walls may also contain other polysaccharides such as pectin and hemicellulose. Similar to *Cyanobacteria* and green plants, green algae utilize two photosystems (Photosystem I and Photosystem II) in the photosynthesis. The outcome of this process is the production of oxygen, energy (ATP), and electron carriers (NADPH) used to convert carbon dioxide into carbohydrates. Green algae exhibit a highly diverse reproductive strategy, including asexual reproduction through cell division, fragmentation, and the formation of asexual spores, as well as sexual reproduction involving gametes and zygospores. Traditionally, green algae are classified into three main classes, namely *Chlorophyceae*, *Charophyceae*, and *Ulvophyceae*, based on cell form and structure, pigments, and reproductive patterns. However, this classification continues to evolve with further genetic research. Green algae play vital roles in ecosystems, serving as primary producers in food chains, oxygen producers, and carbon fixers.

Green algae play a significant role in wastewater treatment due to their ability to absorb excess nutrients, remove pollutants, and produce oxygen through the photosynthesis process. Some common genera of green algae used in wastewater treatment are as follows:

(a) *Chlorella* is a genus of microscopic green algae, known for its rapid growth and high photosynthetic efficiency. *Chlorella* has been employed in wastewater treatment to remove excess nutrients such as nitrogen and phosphorus, as well as some pollutants like heavy metals (Karima et al. 2018).
(b) *Scenedesmus* is a colonial genus of green algae commonly found in freshwater. This species has been utilized in wastewater treatment to reduce excessive nutrient content, particularly nitrogen and phosphorus (Qader and Shek 2023), and to remove pollutants such as heavy metals.

(c) *Haematococcus* is a microscopic green algae known as a producer of astaxanthin, an antioxidant pigment with various applications in the nutraceutical and cosmetic industries (Oslan et al. 2021). *Haematococcus* is also used in wastewater treatment to remove excess nutrients and some pollutants.
(d) *Spirogyra* is a filamentous genus of green algae frequently found in freshwater. This species has been employed in wastewater treatment to eliminate excess nutrients, especially nitrogen and phosphorus (Muriuki Githaiga et al. 2020), and to reduce pollutant content, including heavy metals.
(e) *Volvox* is a colonial green algae composed of cells arranged in a spherical shape. This species has been used in wastewater treatment to reduce excessive nutrient content and remove pollutants, such as heavy metals (Diaconu et al. 2023).
(f) *Dunaliella* is a microscopic green algae known as a producer of beta-carotene, a pigment used in food and supplement industries. While the following studies do not specifically discuss *Dunaliella's* application in wastewater treatment containing heavy metals and excess nutrients, they describe the algae's potential in wastewater treatment through phototrophy, mixotrophy, and production of extracellular polymeric substances that can function as biosorbents (Mega et al. 2021).

Phaeophyta, also known as brown algae, is a group of algae primarily found in marine and some freshwater environments. Brown algae encompass a wide range of forms, from microscopic filamentous types to larger macroalgae known as seaweeds (Hakim and Patel 2020). Brown algae contain both chlorophyll a and chlorophyll c, as well as an additional pigment called fucoxanthin, which imparts the characteristic brown color. These pigments are used in the process of photosynthesis. Additionally, brown algae contain various carotenoids and xanthophylls.

The cell walls of brown algae are composed of cellulose and a specific polysaccharide called alginate. Alginate is commercially used as a thickening and binding agent in various food and pharmaceutical products. Brown algae employ two photosystems (Photosystem I and Photosystem II) in the photosynthesis, similar to green algae and *Cyanobacteria*. The result of this process is the production of oxygen, energy (ATP), and electron carrier molecules (NADPH), which are utilized to convert carbon dioxide into carbohydrates. Reproduction in *Phaeophyta* involves a life cycle with three different generations: the haploid gametophyte, the diploid sporophyte, and the tetrasporophyte. Sexual and asexual reproduction occur at different stages in this life cycle. Traditionally, *Phaeophyta* is classified into several orders based on morphology and reproductive structures, such as *Fucales*, *Laminariales*, *Ectocarpales*, and *Sphacelariales*. However, this classification continues to evolve with further genetic research.

Brown algae play a crucial role in ecosystems, particularly as primary producers in marine food chains and as habitats for many marine species. They also have various commercial applications, such as in the production of alginates, foods, and nutritional supplements (e.g., fucoidan and laminarin). While brown algae are less commonly used in wastewater treatment compared to green algae and *Cyanobacteria*, some species have been studied for their ability to remove pollutants and excess nutrients from wastewater. Several potentially beneficial brown algae species in wastewater treatment include:

(a) *Ascophyllum nodosum*, also known as knotted wrack, is a brown algae species found on rocky shores in cold marine regions. This species has been studied for its ability to remove heavy metals, such as copper, lead, and zinc, from wastewater.
(b) *Sargassum muticum* and *Sargassum vulgare* have been studied for their ability to remove excess nutrients, such as phosphate and nitrate, as well as heavy metals like lead and mercury, from wastewater (Hansen et al. 2023).
(c) *Fucus vesiculosus* and *Fucus serratus* have been studied for their ability to remove heavy metals, such as cadmium, copper, and nickel, from wastewater (Hansen et al. 2023).
(d) *Laminaria*, also known as kelp, including *Laminaria digitata* and *Laminaria hyperborea*, has been researched for its ability to remove heavy metals, such as lead and zinc, from wastewater (Venardou et al. 2023).

Selecting the appropriate brown algae species and managing environmental conditions are crucial to optimize the wastewater treatment process using brown algae.

Rhodophyta, or red algae, is a group of algae primarily found in marine waters, although some species are also found in freshwater and wet terrestrial environments (Cotas et al. 2020). Red algae exhibit diverse forms and sizes, ranging from microscopic filaments to more complex macroalgae known as seaweeds (Hakim and Patel 2020). Red algae contain chlorophyll a and chlorophyll d as well as additional pigments called phycobiliproteins, which give them their distinctive red color. These pigments are involved in the photosynthesis process. Red algae also contain carotenoids and other xanthophylls. Their cell walls consist of cellulose, special polysaccharides called agar and carrageenan, as well as proteins. Agar and carrageenan are commercially used as thickeners and binders in various food and pharmaceutical products.

Red algae utilize two photosystems (Photosystem I and Photosystem II) in the process of photosynthesis, similar to green algae and *Cyanobacteria*. However, due to the presence of chlorophyll d and phycobiliproteins, red algae can utilize a different spectrum of light for photosynthesis compared to other algal groups. Reproduction in *Rhodophyta* involves a life cycle with three different generations: haploid gametophyte, diploid sporophyte, and carposporophyte. Sexual and asexual reproduction occur at different stages in this life cycle (Vieira et al. 2018). Traditionally, *Rhodophyta* is classified into several classes based on their morphology and reproductive structures, namely *Florideophyceae*, *Bangiophyceae*, and *Cyanidiophyceae*. However, this classification continues to evolve with further genetic research.

Red algae play a vital role in ecosystems, particularly as primary producers in marine food chains and as habitats for numerous marine species. They also have diverse commercial applications, including the production of agar, carrageenan, and nutritional supplements (e.g., proteins and pigments). Although the use of

Rhodophyta in wastewater treatment has not been extensively explored, some studies have shown the potential of red algae in removing pollutants such as heavy metals and excess nutrients from wastewater. Several *Rhodophyta* species that may be beneficial in wastewater treatment are as follows:

(a) *Gracilaria verrucosa* has the ability to remove excess nutrients, such as phosphates and nitrates, as well as heavy metals, such as lead and mercury, from wastewater.
(b) Gelidium species like *Gelidium amansii* and *Gelidium elegans* have the ability to absorb heavy metals, such as cadmium, copper, and nickel, from wastewater.
(c) *Palmaria palmata* and *Palmaria decipiens* have been reported to effectively remove excess nutrients, such as phosphates and nitrates, from wastewater.
(d) *Chondrus crispus* and *Chondrus ocellatus* possess the ability to remove heavy metals, such as lead and zinc, from wastewater.
(e) *Porphyra* is a genus of red algae that includes many species used in the production of nori, a traditional Japanese food. Some studies have shown that *Porphyra yezoensis* is effective in removing excess nutrients, such as phosphates and nitrates, from wastewater.

When used in wastewater treatment, red algae, such as the aforementioned species, can help reduce contaminants through biosorption and bioaccumulation processes—pollutants bind to the cell walls or are internalized into the algae cells, thereby reducing their concentration in wastewater.

Diatom (*Bacillariophyta*) is a group of microscopic algae found in various freshwater and marine habitats (Fu et al. 2022). Diatoms are known to play a crucial role in aquatic ecosystems and the carbon cycle, acting as primary producers in aquatic food chains. Diatoms have a unique cell wall called a frustule, made of biogenic silica (SiO_2). The frustule has a complex layered structure with repeating patterns and pores, giving diatoms their distinctive shapes and strength. Diatoms contain chlorophyll a and chlorophyll c, as well as additional pigments called fucoxanthin, which impart the characteristic golden-brown color. These pigments are used in the photosynthesis process. Additionally, diatoms also contain carotenoids and other xanthophylls. Diatoms utilize two photosystems (Photosystem I and Photosystem II) in the photosynthesis, similar to green algae and cyanobacteria. However, due to the presence of chlorophyll c and fucoxanthin, diatoms can use a different spectrum of light for photosynthesis compared to other algal groups.

Diatoms mainly reproduce asexually through binary fission. As the silica cell wall cannot expand, diatoms produce smaller daughter cells from the parent cell. Sexual reproduction also occurs in some diatom species, although it is less common. Diatoms are traditionally classified into two main groups based on frustule morphology: centric diatoms (Centrales), which have radial symmetry, and pennate diatoms (Pennales), which have bilateral symmetry. However, this classification continues to evolve with further genetic research. Diatoms have various commercial applications, such as the production of biosilica, pigments, and nutritional supplements (e.g., omega-3 fatty acids) (Sharma et al. 2021). Diatoms have also been studied for their potential in wastewater treatment, particularly in removing excess nutrients such as phosphates and nitrates, as well as organic pollutants and heavy metals. Diatoms can accumulate these pollutants within their cells or remove them through biodegradation and biosorption processes, thereby reducing pollutant concentrations in wastewater and improving water quality. The following are some diatoms that have potential benefits in wastewater treatment:

(a) *Phaeodactylum tricornutum* is a marine diatom that has been studied for its ability to remove excess nutrients such as phosphates and nitrates, as well as heavy metals like copper, cadmium, and lead from wastewater.
(b) *Nitzschia palea* is a freshwater diatom that has been studied for its ability to remove organic pollutants such as phenol and volatile organic chemicals from wastewater (Sudarshan et al. 2019).
(c) *Navicula* is capable of removing excess nutrients such as phosphates and nitrates, as well as heavy metals like chromium, nickel, and zinc from wastewater (Chugh et al. 2022).
(d) Pinnularia is capable of removing pollutants like heavy metals and excess nutrients from wastewater (Bwapwa et al. 2017).
(e) Thalassiosira can remove excess nutrients and organic pollutants from wastewater (Al-Jabri et al. 2021).

Algae have specific characteristics that can be described as follows (Babich et al. 2022):

(1) Algae obtain energy through photosynthesis, converting sunlight energy into chemical energy stored in chemical bonds. The ability of algae to photosynthesize depends on the photosynthetic pigments they possess, such as chlorophyll, carotenoids, and phycoerythrin.
(2) Algae exhibit diverse tolerance to environmental conditions, such as temperature, salinity, pH, and nutrient concentration. This characteristic is important in selecting the most suitable algae for use in specific wastewater treatment systems.
(3) The growth rate and reproduction of algae significantly influence the efficiency of wastewater treatment processes. Some types of algae, like microalgae, have rapid growth rates, enabling them to treat wastewater in a short time.
(4) Algae's characteristics in absorbing pollutants, such as excess nutrients, heavy metals, and organic compounds, are crucial in the context of wastewater treatment. Some algae have high affinities for specific pollutants, making them effective bioremediation agents.
(5) Some types of algae, especially microalgae, have the potential to produce high-value products such as lipids, proteins, pigments, and polysaccharides. These products can be further processed into biofuels, animal feed, fertilizers, and chemicals.

10.2.2 Growth and Reproduction of Algae

Algae growth and reproduction have significant implications for the implementation of algae utilization in wastewater treatment. The selection of algae species with optimal growth and reproduction rates becomes a crucial factor in achieving efficient and sustainable outcomes, as described below (Sharma et al. 2021; Ziganshina et al. 2022):

(1) Processing efficiency: Algae with rapid growth rates can treat wastewater in a shorter time, leading to higher processing efficiency and enabling the system to handle larger wastewater loads. Additionally, fast algal reproduction helps maintain a sufficient algal concentration in the system, ensuring effective wastewater treatment.
(2) System scalability: Fast growth and reproduction of algae allow algae-based wastewater treatment systems to be easily scalable. Systems capable of handling larger wastewater loads are more appealing to industries and governments for adoption, as they offer more efficient and sustainable treatment solutions.
(3) Process stability: Algae that can adapt and reproduce quickly under various environmental conditions contribute to higher process stability. Stable systems are easier to operate and maintain and are more resilient to changes in operational and environmental conditions.
(4) Production of high-value biomass: Algae with rapid growth rates yield a larger biomass. This biomass can be further processed into high-value products such as biofuels, animal feed, fertilizers, and chemicals. High biomass production enhances the economic value of algae-based wastewater treatment systems and supports the transition to a circular economy.
(5) Social and community support: Algae with accelerated growth and reproduction rates can deliver a more immediate positive impact on the environment and water quality. This will increase community support for algae-based wastewater treatment technology and promote its adoption on a larger scale.
(6) Operational and maintenance costs: Rapid growth and reproduction of algae can reduce operational and maintenance costs of wastewater treatment systems. Algae that grow and reproduce quickly require less maintenance intervention and are more resilient to disturbances caused by external factors, such as changes in temperature and light conditions.

10.2.3 Algae Interaction with the Environment

The interaction of algae with the environment is crucial in influencing the implementation of algae utilization for wastewater treatment. Environmental factors significantly impact algae growth, explained as follows:

Availability of Light

Light influences the rate of photosynthesis and the growth of algae, thereby affecting the efficiency of algae-based wastewater treatment systems (Katam et al. 2022). Systems designed considering optimal light availability will yield better results in wastewater treatment. Light availability can be controlled by adjusting the pond depth, providing additional illumination, and considering factors such as light intensity, duration, and spectrum. Through the photosynthesis process, algae produce oxygen as a byproduct, enhancing water quality by increasing dissolved oxygen levels. This condition supports aquatic life and other biological processes occurring in water bodies. In the context of photosynthesis, algae also play a role in carbon dioxide (CO_2) sequestration from the environment. By means of photosynthesis, algae convert CO_2 into biomass, thereby reducing greenhouse gas emissions and contributing to mitigating the impacts of climate change.

Availability of Nutrients in Wastewater

The concentration of nutrients in wastewater, such as nitrogen and phosphorus, affects algae growth (Alazaiza et al. 2023). Algae utilize these nutrients as a food source, converting them into biomass. Optimal nutrient conditions support rapid and efficient algae growth, thereby enhancing the effectiveness of wastewater treatment systems. The optimal nitrogen and phosphorus concentrations for algae growth depend significantly on the algae species. Generally, nitrogen concentrations ranging from 10 to 60 mg/L and phosphorus level of 1–8 mg/L are considered suitable, although these requirements may vary depending on the physiological conditions of the microalgae involved.

Temperature

Algae's ability to grow and reproduce under various environmental conditions, such as temperature and pH, impacts the effectiveness of algae-based wastewater treatment systems (Oruganti et al. 2022). Algae that tolerate environmental changes are easier to manage and can adapt to different types of wastewater. Water temperature and pH affect the rate of algae growth and metabolic activity. Each algae species has an optimal range of temperature and pH for its growth. In designing algae-based wastewater treatment systems, it is essential to consider the appropriate temperature and pH conditions suitable for the utilized algae. Generally, temperatures ranging from 20 to 30 °C are considered optimal for the growth of many microalgae species. This optimal temperature range can increase the rate of algae growth and the efficiency of algae-based wastewater treatment systems.

Temperatures above 30 °C are generally unsuitable for microalgae growth for several reasons listed below (Barten et al. 2020):

(a) Protein denaturation: High temperatures can cause protein denaturation, resulting in the alteration of protein structure and loss of function. Proteins are vital components in the metabolism and cell growth of algae, so protein denaturation will inhibit algae growth.

(b) Oxidative stress: High temperatures can increase the production of reactive oxygen species (ROS) within algae cells. Excessive ROS can cause cell damage, including structural changes to membranes, DNA damage, and oxidation of proteins and lipids. Algae have antioxidant systems to combat ROS, but if ROS production exceeds the antioxidant capacity, oxidative stress will hinder algae growth.
(c) Enzyme reaction rate changes: High temperatures can affect the rate of enzymatic reactions within algae cells. Each enzyme has an optimal temperature at which it functions most efficiently. If the temperature exceeds the optimal range, enzyme activity decreases, reducing the rate of metabolism and algae growth.
(d) Evaporation: High temperatures can increase the rate of evaporation in algae cultivation systems, leading to a reduction in water volume and nutrient concentration available to algae. Decreased water and nutrient availability will impact algae growth.

However, some references discuss the existence of thermophilic microalgae that thrive better at temperatures above 30 °C, thus potentially enhancing the efficiency of algae-based wastewater treatment systems.

Environmental pH

The optimum pH for algae growth in the context of wastewater treatment generally ranges from 7 to 9, although this may vary depending on the physiological characteristics of the algae species and the type of wastewater (Ma and Jian 2023). Environmental pH significantly influences the growth of microalgae in wastewater treatment for several reasons:

(a) The level of pH affects the solubility and availability of essential nutrients such as phosphorus, nitrogen, and metal ions in water. Some nutrients are more readily utilized by algae at specific pH levels. If the pH is unsuitable, these nutrients become less soluble and challenging for algae to absorb, thereby inhibiting their growth.
(b) Enzymes within algae cells are highly sensitive to pH changes. Each enzyme has an optimum pH at which it functions most efficiently. If the environmental pH falls outside the optimal range, enzyme activity decreases, leading to a reduction in metabolic rate and algae growth.
(c) Extreme environmental pH can cause changes in the structure and stability of algal cell membranes, which can result in cell leakage and cell death. Cell membranes also play a role in nutrient and ion transport across the membrane, so changes in membrane stability will affect algae's ability to absorb nutrients.
(d) Some algae species are more tolerant of pH fluctuations than others. Tolerant algae species may still grow under less optimal pH conditions, while more sensitive species may experience reduced growth or even die (Masojídek et al. 2021). Therefore, the suitable environmental pH for algae growth depends on the species used in the wastewater treatment system.

Chlorella vulgaris shows better growth at pH 6 compared to higher pH levels (7–9). Additionally, lipid production increases at pH 6, potentially improving the quality of algal biomass for bioenergy purposes. Generally, microalgae exhibit better growth rates at higher pH levels (approximately 7–9) in the context of wastewater treatment. However, this discovery suggests that certain microalgae species may thrive well at pH 5 and pH 6, which could enhance the efficiency of algae-based wastewater treatment systems at these pH levels.

Salinity of Wastewater

The salinity of wastewater can affect algae growth and activity (Al-Enazi 2020). Some algae species are more tolerant of high salt concentrations, while others are better suited to freshwater conditions. Some microalgae species grow better at low salinity levels, such as freshwater, while others are more tolerant of higher salinity levels, such as seawater or brackish water. Generally, low to moderate salinity (around 0.1–3% based on total dissolved solids content) is advantageous for microalgae growth in the context of algae-based wastewater treatment. Selecting the appropriate algae species that can thrive under the salinity level of the wastewater will enhance treatment efficiency.

Presence of Heavy Metals in the Environment

Algae growth is generally inhibited by high concentrations of heavy metals (Wang et al. 2022), as these metals can be toxic to algae. However, some studies indicate that low concentrations of heavy metals can act as nutrients and may affect algae growth rates. Therefore, it is essential to consider algae species that are tolerant of wastewater toxicity conditions. The ability of algae to absorb pollutants, such as excess nutrients, heavy metals, and organic compounds, is crucial in wastewater treatment. The interaction of algae with these pollutants through bioadsorption, biosorption, and biodegradation processes helps to remove pollutants from the water and improve water quality.

Interactions with Other Microorganisms or Synecology

In wastewater treatment systems, algae interact with various other microorganisms, such as bacteria, protozoa, and invertebrates. These interactions can be symbiotic, competitive, or predator–prey relationships. Some types of algae, particularly microalgae, can reduce the number of pathogens in wastewater through predation, competition, and the production of antimicrobial compounds. This helps to decrease the spread of waterborne diseases. By understanding and managing these interactions effectively, the efficiency of wastewater treatment can be enhanced (Młyński et al. 2020).

(a) Predation: Microalgae serve as a food source for other microscopic organisms known as protozoa and metazoa. These organisms prey on microalgae and pathogens present in wastewater. In this process, the number of pathogens decreases as they become prey for protozoa and metazoa. Thus, by increasing the population of microalgae in wastewater, the number of pathogens can be regulated through the microscopic food chain.

(b) Competition: Microalgae compete with pathogens and other microorganisms in wastewater for limited resources, including phosphate, nitrate, and ammonium. Microalgae can utilize these nutrients more efficiently than pathogens, limiting their growth. Moreover, microalgae can absorb heavy metals and harmful chemicals, further hindering pathogen growth.

(c) Production of antimicrobial compounds: Some microalgae species are known to produce antimicrobial compounds that inhibit pathogen growth. These compounds include aldehydes, unsaturated fatty acids, polysaccharides, and peptides. These compounds exhibit effective antimicrobial activities against various pathogens, such as bacteria, viruses, protozoa, and fungi. The production of antimicrobial compounds by microalgae can enhance the efficiency of wastewater treatment systems in reducing the number of pathogens and maintaining water quality.

Consequently, through predation, competition, and the production of antimicrobial compounds, microalgae play a significant role in reducing the number of pathogens in wastewater. The use of microalgae in wastewater treatment is not only environmentally friendly but also enhances treatment efficiency and effectiveness while reducing operational costs.

10.3 Wastewater Treatment Process with Algae

10.3.1 Principles and Mechanisms

Wastewater treatment through harnessing the metabolic activity of microalgae involves several key principles and mechanisms (Abdelfattah et al. 2023). This process is commonly known as phytoremediation or algae-based bioremediation. The principles and mechanisms involved in wastewater treatment using algae can be explained as follows: Algae are photosynthetic organisms capable of converting solar energy into chemical energy through photosynthesis. During this process, algae absorb carbon dioxide (CO_2), thereby reducing the CO_2 levels in wastewater, while releasing oxygen (O_2) into the water to aid in the oxidation and degradation of organic matter in the wastewater. High levels of dissolved oxygen are crucial for supporting the life of other aquatic organisms.

Algae require nutrients such as nitrogen (N) and phosphorus (P) for their growth and development. These nutrients are often found in high concentrations in industrial wastewater, typically present in the form of ammonia, nitrate, and phosphate. The uptake of nutrients by algae helps to reduce the excessive nutrient concentration in wastewater, thereby mitigating the potential for eutrophication upon wastewater discharge into the environment. Algae are capable of degrading and absorbing various dissolved pollutants in wastewater, such as heavy metals, pesticides, and organic compounds. This process is known as biodegradation (for pollutants broken down into simpler compounds) and bioaccumulation (for pollutants that accumulate within algae cells). Certain microalgae can degrade organic compounds present in wastewater, such as pesticides, hydrocarbons, and phenols. This process involves enzymes produced by algae that break down the organic compounds into simpler and less toxic components.

Microalgae can absorb and accumulate heavy metals, such as mercury, cadmium, and lead, from wastewater (Ankit et al. 2022). This process involves adsorption, precipitation, and complexation mechanisms. The absorption of heavy metals by algae helps to reduce the concentration of these metals in wastewater, thus lowering the risk of environmental pollution. During growth and reproduction, algae form larger and denser aggregates called flocs. This process is referred to as coagulation and flocculation. Some microalgae produce extracellular polysaccharides that act as natural coagulants and flocculants. Coagulation and flocculation processes aid in the agglomeration of solid particles and colloids dissolved in wastewater, facilitating the separation of solids from the treated water.

The phase of dense algae growth helps to aggregate particles and pollutants in the wastewater, thereby facilitating solid–liquid separation. After coagulation and flocculation, algal solids and flocs containing pollutants can be separated from the wastewater through solid–liquid separation processes. This can be achieved through gravity-based methods (such as sedimentation), filtration, or centrifugation. Following solid–liquid separation, the algal biomass that has absorbed pollutants can be utilized for various purposes, such as bioenergy production (e.g., biogas or biodiesel), animal feed, fertilizers, or chemicals. The utilization of algal biomass helps to reduce environmental impacts and creates additional economic value from the wastewater treatment process.

During the wastewater treatment process, microalgae grow and produce biomass. This biomass can be further processed into high-value products, such as biofuels, animal feed, fertilizers, and chemicals. The production of biomass by algae also enhances the economic value of algae-based wastewater treatment systems. Overall, industrial wastewater treatment using algae is an environmentally-friendly approach that combines natural mechanisms with wastewater treatment technologies. In practice, microalgae-based wastewater treatment systems may involve the utilization of single algae species or consortia of algae with diverse metabolic capabilities.

10.3.2 Processing Parameters and Efficiency

Monitoring and controlling parameters are crucial to ensure the efficiency of microalgae-based wastewater treatment and to reduce negative environmental impacts. Microalgae can grow under various environmental conditions, such as changes in temperature, salinity, or water quality, and can adapt to different types of wastewaters. Several parameters affect the efficiency of microalgae-based wastewater treatment, including physical, chemical, and biological factors as listed below:

(1) Light availability: Light serves as an energy source for microalgae photosynthesis. Adequate light intensity, spectrum, and duration affect the rate of photosynthesis and algae growth, which, in turn, influences the efficiency of wastewater treatment.
(2) Temperature: Temperature affects the rate of metabolism and growth of microalgae. Each microalgae species has an optimal temperature range for growth and metabolic activity. Optimal temperature enhances the efficiency of the treatment process, while temperatures that are too high or too low can inhibit algae growth and reduce treatment efficiency. Controlling the temperature in the wastewater treatment system helps achieve higher efficiency.
(3) Nutrient availability: Nutrients including nitrogen, phosphorus, carbon, and micronutrients affect the growth and metabolic activity of microalgae. A balanced nutrient concentration ensures optimal algae growth and efficient pollutant uptake.
(4) pH of wastewater: The pH of wastewater affects the activity of enzymes and the permeability of microalgae cell membranes. Each microalgae species has an optimal pH range for growth and metabolic activity. Optimal pH ensures proper enzyme function and efficient pollutant uptake. Controlling the pH in the wastewater treatment system contributes to higher efficiency.
(5) Selection of appropriate microalgae species with favorable growth characteristics and high pollutant absorption capacity enhances the efficiency of microalgae-based wastewater treatment.
(6) Microalgae concentration in the treatment system affects the system's ability to treat pollutants (Amaro et al. 2023). An optimal concentration ensures efficient pollutant uptake and robust algae growth. The optimal microalgae concentration for wastewater treatment, ensuring efficient nutrient and heavy metal removal, falls within the range of 0.5–2 g/L. Some studies have shown that a microalgae concentration of 1 g/L can achieve high efficiency in nutrient and heavy metal uptake. Higher microalgae concentrations may enhance heavy metal and nutrient removal, but increased microalgae growth requires more energy and nutrients. Therefore, finding a balance between removal efficiency and operational costs is essential.

    Using microalgae concentrations higher than 1 g/L in wastewater treatment can offer advantages and benefits, although some considerations need to be taken into account to determine the most optimal concentration. The advantages of using microalgae concentrations higher than 1 g/L (Abdelfattah et al. 2023) include higher biomass production, which can be utilized for various applications, such as bioenergy production (biodiesel and biogas), animal feed, and biologically-based chemical products.

    However, prior to increasing microalgae concentration in wastewater treatment, population density considerations should be factored in. High population density can increase competition among microalgae for resources (nutrients and light), thereby reducing growth and removal efficiency.
(7) Adequate aeration and agitation in microalgae-based wastewater treatment systems ensure sufficient dissolved oxygen for biological oxidation processes

and maintain a homogenous distribution of algae in the system. In other words, aeration and agitation affect gas and nutrient transfer between wastewater and microalgae, as well as prevent microalgae settling in the system. Adequate aeration and agitation improve the efficiency of microalgae-based wastewater treatment (Srimongkol et al. 2022).

Thus, optimal aeration and agitation may need to be adjusted based on the physiological needs of microalgae species, the design of the photobioreactor or conventional pond system, and the specific characteristics of the wastewater. Additionally, the optimal conditions for aeration and agitation should also consider energy costs and their impact on microalgae life in the system (Valchev and Ribarova 2022).

(8) Hydraulic Retention Time (HRT) refers to the time required for wastewater to pass through the treatment system. An optimal HRT ensures sufficient contact between microalgae and pollutants, thereby enhancing removal efficiency.

10.3.3 Systems and Technology Used

Various systems and technologies can be implemented for wastewater treatment utilizing microalgae. The selection of appropriate systems and technologies for microalgae-based wastewater treatment depends on various factors (Goh et al. 2023), including wastewater characteristics, treatment objectives, climatic and geographic conditions, as well as land availability and resources. Some common alternative systems and technologies are as follows:

Ponds and Raceway Pond Systems

Ponds are open systems used for microalgae cultivation. Open ponds are the simplest cultivation system, consisting of shallow channels or ponds with water circulation to maintain even algal biomass distribution. Shallow and wide open ponds allow for good light penetration and sufficient water circulation. Generally, the optimal pond design for microalgae-based wastewater treatment will include efficient aeration and agitation systems, proper lighting, as well as optimal temperature and pH conditions. However, more detailed specifications will depend on the type of microalgae used, wastewater characteristics, and the final objectives of wastewater treatment (e.g., biofuel production, nutrient processing, or heavy metal absorption).

Raceway Pond Systems are a variation of open ponds designed to enhance water circulation and gas exchange. This system consists of U-shaped channels with paddle wheels or other mechanical devices continuously stirring the water. The raceway pond system allows for more efficient and uniform microalgae growth, enhancing the wastewater treatment capability. In general, the optimal design of ponds or raceway pond systems for microalgae-based wastewater treatment should consider aeration and agitation efficiency, light utilization, as well as ease of microalgae biomass maintenance and collection.

Although open ponds are cheaper and easier to operate and maintain compared to photobioreactor (PBR) systems, they have limited control over the environment (especially temperature and pH), making them more vulnerable to contamination and changes in environmental conditions (Li et al. 2023). Open ponds are generally used for large-scale wastewater treatment due to relatively lower construction and maintenance costs. Therefore, the design and operation of raceway pond systems need to be tailored based on the specific needs of the wastewater treatment system being used.

High-Rate Algal Pond (HRAP)

HRAP is a specialized open pond system designed for high microalgae growth rates. This system features shallow channels or ponds with strong water circulation to enhance the contact between microalgae and nutrients in the wastewater (Li et al. 2023). HRAP is commonly used for nutrient-rich wastewater treatment and biomass production. In general, the optimal design of HRAP for microalgae-based wastewater treatment should consider aeration efficiency, pond depth, light utilization, temperature control, as well as appropriate pH and nutrient concentration. The selection of suitable microalgae species and operational conditions are also crucial for achieving optimal outcomes in HRAP systems.

Algal Turf Scrubber (ATS) and Biofilm-Based System

Algal Turf Scrubber (ATS) and biofilm-based systems are two different approaches for utilizing microalgae in wastewater treatment. They share similarities and differences in terms of working principles, system design, and treatment efficiency (Reinecke et al. 2023). Both systems utilize microalgae for wastewater treatment and the removal of pollutants such as nutrients and heavy metals; both harness the microalgae's photosynthesis to convert available nutrients in wastewater into biomass, which can be further processed for various purposes such as biogas or biofuel production. In addition, both systems enable more efficient pollutant uptake compared to conventional wastewater treatment systems.

The primary differences between both systems can be explained as follows:

(1) System design: The Algal Turf Scrubber (ATS) is a system that utilizes coarse and porous substrates (such as mesh or fibers) on which microalgae grow in the form of "turf" or a thin carpet. Wastewater flows over the substrate, and microalgae absorb pollutants as they grow on it. On the other hand, biofilm-based systems involve microalgae growing as a thin layer or biofilm on suitable surfaces such as plates, rods, or spheres.

(2) Biomass maintenance and recovery: In ATS, microalgae biomass can be easily removed from the substrate through brushing or scraping, allowing periodic biomass recovery and preventing competition for space and nutrients among microalgae. In biofilm-based systems, biomass recovery may be more challenging and require more complex separation techniques, such as peeling or using enzymes to detach the biofilm from the surface.

(3) Light efficiency: ATS allows better light penetration into microalgae growing on the substrate, optimizing photosynthesis and microalgae growth. Biofilm-based systems may have limitations in light penetration, especially when the biofilm becomes thick, which can affect treatment efficiency.
(4) Resilience to disturbances: ATS tends to be more tolerant to changes in environmental conditions such as temperature, pH, and pollutant concentrations. Biofilm-based systems may be more vulnerable to such changes, which can affect microalgae growth and activity within the biofilm.
(5) Space requirement: ATS usually requires a larger area compared to biofilm-based systems due to the flowing water over the substrate. Biofilm-based systems can save space through vertical layouts or using media with complex shapes to increase the available surface area for biofilm growth.

Overall, the optimal design of ATS for microalgae-based wastewater treatment needs to consider the appropriate substrate selection, microalgae species, and operational conditions (such as lighting and aeration) (Valchev and Ribarova 2022). Wastewater treatment efficiency can also be enhanced with proper nutrient control and regular system maintenance. In the context of microalgae-based wastewater treatment, biofilm plays a significant role in pollutant removal from wastewater. Microalgae in the biofilm can take up nutrients such as phosphate and nitrogen, as well as organic compounds from wastewater and convert them into biomass. This process is known as bioremediation and can help reduce contaminants in wastewater before it is discharged into the environment or reused in industrial or agricultural processes (Ojha et al. 2021). Moreover, microalgae can produce oxygen through photosynthesis, which aids in the biological oxidation process to further break down organic compounds.

To build an optimal biofilm-based system for microalgae-based wastewater treatment, considerations include selecting appropriate substrates, effective microalgae species, optimal lighting, and proper operational conditions (such as aeration and nutrient control) need to be accounted for. The biofilm-based system offers advantages in space utilization and easier biomass separation compared to systems without biofilm. Biofilms are formed when microorganisms, like microalgae, attach to a surface and start producing extracellular polymeric substances (EPS), including proteins, polysaccharides, and nucleic acids. These EPS form a matrix that protects and supports the associated microorganisms within the biofilm. The mechanism of biofilm formation in microalgae-based wastewater treatment occurs sequentially as follows (Srimongkol et al. 2022):

(1) Initiation: Microalgae initially attach to available surfaces, such as plant stems or reactor surfaces. This initial attachment is reversible and influenced by interactions between microalgae and the surface, such as van der Waals forces, electrostatic repulsion, and hydrophobic interactions.
(2) Irreversible attachment: After the initial attachment, microalgae begin to produce EPS. This process makes the attachment irreversible and aids in biofilm growth and development.

(3) Growth and maturation: During this phase, the biofilm starts to grow and develop more complex structures. Microalgae in the biofilm acquire nutrients from wastewater and continue to reproduce. Formation of chemical and physical gradients within the biofilm allows various microorganism species, including bacteria and protozoa, to coexist and contribute to the wastewater treatment process.
(4) Dispersion: At some point, parts of the biofilm will detach from the matrix and enter the wastewater. This dispersion allows the spread of microalgae and other microorganisms, maintaining the cycle of biofilm growth.

Photobioreactor (PBR)

The photobioreactor is a closed system specifically designed for the growth of microalgae with strict environmental controls, such as temperature, pH, and illumination. In the context of wastewater treatment using microalgae, the photobioreactor provides an optimal environment for microalgae growth and maximizes their ability to remove contaminants from wastewater through photosynthesis and bioremediation, resulting in higher wastewater treatment efficiency compared to open systems.

Some advantages of photobioreactors include the ability to control environmental conditions—e.g., temperature, pH, nutrients, and light intensity—all of which are crucial for microalgae growth and activity. The concentration of microalgae can be maintained at higher levels compared to open systems, thereby enhancing wastewater treatment efficiency. Being closed and isolated from the surrounding environment, photobioreactors reduce the risk of contamination by pathogens or unwanted microorganisms. Furthermore, photobioreactors typically have smaller sizes and require less land compared to open ponds, making them more suitable for areas with limited space.

However, wastewater treatment using photobioreactors also has some disadvantages as follows:

(1) Photobioreactor requires higher initial investment costs compared to open systems (such as open ponds or flow-through systems) due to the need for more sophisticated equipment and technology.
(2) Photobioreactors generally have higher operational costs because of the strict control of environmental conditions, including temperature, pH, and light intensity. Additionally, the energy required to operate equipment, such as pumps and control systems, can increase operational costs.
(3) Although photobioreactors can achieve higher biomass concentrations, scaling up photobioreactor systems to an industrial level can be challenging. The design and operation of large-scale photobioreactors are more complex and require more resources.
(4) In photobioreactors, the risk of fouling or formation of deposits on the reactor surface is higher, subsequently reducing light efficiency and hinder microalgae growth. Regular cleaning and maintenance are required to maintain photobioreactor performance.

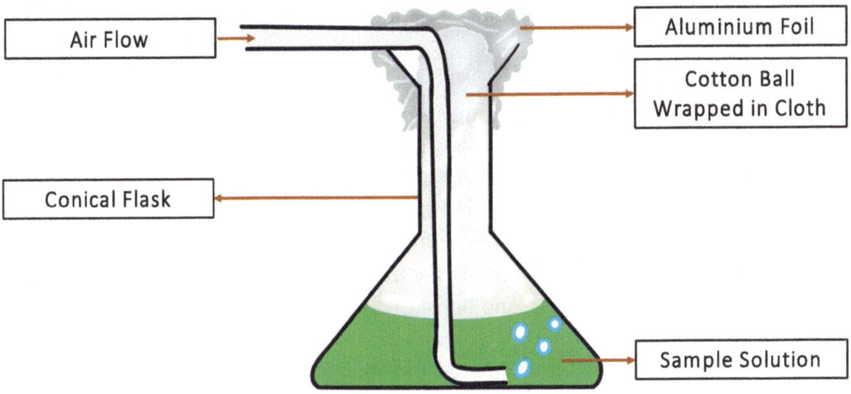

Fig. 10.3 Conical flask prototype

(5) Closed photobioreactors have limitations in gas exchange, such as oxygen and carbon dioxide, which are important for microalgae growth. Inadequate design can lead to gas accumulation, negatively affecting the microalgae activity.

Photobioreactors can be designed in the form of tubes, panels, or transparent plastic bags (Huang et al. 2017). However, the construction, investment, and maintenance costs of photobioreactors are generally higher compared to open pond systems. Various types of photobioreactors have been developed, listed as follows:

(1) Tubular photobioreactor, which is commonly used for microalgae cultivation and wastewater treatment (Egbo et al. 2018). The tubular photobioreactor consists of transparent tubes typically made of glass or plastic, in which microalgae are grown. These tubes are connected in series or parallel configurations and can have straight, spiral, or helical shapes, or even in prototype forms, as shown in Fig. 10.3 (Alazaiza et al. 2023).

Tube photobioreactor allows for efficient light penetration into the microalgae culture, which is essential for the photosynthesis process. Efficient light penetration affects the growth rate of microalgae and the efficiency of wastewater treatment. Tube photobioreactor often employs an airlift system or pumps to create circulation within the reactor. This mixing ensures even distribution of microalgae and sufficient access to nutrients, light, and CO_2 required for photosynthesis. On top of that, mixing aids in gas transfer, such as oxygen and CO_2, between the microalgae culture and the environment. In the tube photobioreactor, microalgae grow in high concentrations, enabling them to efficiently treat wastewater and reduce pollution loads.

Environmental conditions such as temperature, pH, and nutrient concentration can be easily regulated and monitored compared to open systems like open ponds. Such preferable environmental control allows for optimal microalgae growth and higher efficiency in wastewater treatment. Tube photobioreactors can be scaled up from laboratory to industrial scale, although construction and maintenance costs may be higher compared to open systems. However, the increased biomass productivity and wastewater treatment efficiency can help balance these additional costs.

(2) Flat panel photobioreactor is another type of photobioreactor used for cultivating microalgae in the context of wastewater treatment (Benner et al. 2022). This photobioreactor utilizes transparent flat plates or slightly tilted plates, allowing microalgae to grow between them. Flat panel photobioreactor provides a larger surface area for light penetration, which is crucial for microalgae photosynthesis Effective light penetration in the flat panel photobioreactor enables more efficient microalgae growth and boosts wastewater treatment efficiency.

In the flat panel photobioreactor, mixing and gas transfer between microalgae and the environment are regulated through an airlift system or gas bubble arrangement as shown in Fig. 10.4 (Benner et al. 2022). Proper mixing ensures sufficient access to nutrients, light, and CO_2 required for photosynthesis, while removing by-products such as oxygen yielded during this process.

Microalgae in the flat panel photobioreactor effectively assimilate pollutants in wastewater, including nutrients (nitrogen and phosphorus), heavy metals, and organic compounds. The high concentration of microalgae biomass in the flat panel photobioreactor enhances wastewater treatment efficiency and reduces pollution loads. Better control over environmental conditions—such as temperature, pH, and nutrient concentration—can be achieved compared to

Fig. 10.4 Gas bubble photobioreactor (Benner et al. 2022)

open systems like open ponds. This improved environmental control allows for optimal microalgae growth and higher efficiency in wastewater treatment. Flat panel photobioreactors can be scaled up from laboratory to industrial scale. However, construction and maintenance costs may be higher compared to open systems or tube photobioreactors. Scaling down and increased biomass productivity can help balance these additional costs.

Overall, flat panel photobioreactor is a suitable choice for microalgae cultivation and wastewater treatment, as it offers proper environmental control, efficient light penetration, and high treatment efficiency (Goh et al. 2023). However, it also has some limitations, such as higher costs and more complex installation compared to tube photobioreactors.

(3) Airlift photobioreactor is a system that utilizes air or gas to create circulation within the reactor. Airlift photobioreactors can take the form of tubes or flat panels. This photobioreactor operates on the principle of airlift—gas (usually air or CO_2) is injected into the bottom of the reactor, which then rises through the liquid column, generating natural circulation within the reactor. Microalgae are cultivated in a column consisting of one or several separate compartments. Airlift promotes effective mixing, gas transfer, and reduces the risk of contamination.

In the airlift photobioreactor, mixing and gas transfer between microalgae and the environment occur efficiently due to the natural circulation generated by gas injection (Cui et al. 2020). Proper mixing ensures sufficient access to nutrients, light, and CO_2 required for photosynthesis, as well as the removal of by-products such as oxygen yielded during this process.

Light penetration in the airlift photobioreactor depends on the reactor design, such as the shape and dimensions of the tube or panel. Airlift photobioreactors are generally designed with smaller diameters or thicknesses to maximize light penetration, which is crucial for microalgae growth and wastewater treatment efficiency. Microalgae in the airlift photobioreactor effectively assimilate pollutants in wastewater, including nutrients (nitrogen and phosphorus), heavy metals, and organic compounds. The high concentration of microalgae biomass in the airlift photobioreactor enhances wastewater treatment efficiency and reduces pollution loads.

Airlift photobioreactor allows for better control over environmental conditions, such as temperature, pH, and nutrient concentration, compared to open systems like open ponds. This improved environmental control allows for optimal microalgae growth and higher efficiency in wastewater treatment. Airlift photobioreactors can be scaled up from laboratory to industrial scale. Construction and maintenance costs may be lower compared to tube or flat panel photobioreactors, primarily due to the natural circulation generated by gas injection, which reduces the need for mechanical mixing equipment. Overall, airlift photobioreactor is a viable choice for microalgae cultivation and wastewater treatment, as it offers convenient environmental control, high mixing and gas transfer efficiency, as well as high treatment efficiency.

(4) Bubble column photobioreactor: This photobioreactor utilizes a transparent column filled with the culture medium (Benner et al. 2022), and air or gas is

bubbled through the bottom of the column. The resulting gas bubbles create circulation and mixing within the medium, promoting efficient microalgae growth. Bubble column photobioreactor is easy to design and has relatively low construction and maintenance costs.

Bubble column photobioreactor achieves effective mixing through the rising gas bubbles. This mixing ensures sufficient access to nutrients, light, and CO_2 required for photosynthesis; and eliminates by-products such as oxygen. Light penetration in the bubble column photobioreactor depends on the reactor design, its size, and density of the gas bubbles. In some cases, large or excessive gas bubbles may reduce light penetration, thus decreasing photosynthesis efficiency and microalgae growth.

Microalgae in the bubble column photobioreactor effectively assimilate pollutants in wastewater, including nutrients (nitrogen and phosphorus), heavy metals, and organic compounds. The high concentration of microalgae biomass in the bubble column photobioreactor enhances wastewater treatment efficiency and reduces pollution loads. Better environmental control allows for optimal microalgae growth and higher efficiency in wastewater treatment. Like other types of photobioreactors, bubble column photobioreactors can be scaled up from laboratory to industrial scale. Construction and maintenance costs may be lower compared to some other types of photobioreactors, such as tube or flat panel photobioreactors.

Airlift photobioreactor and bubble column photobioreactor are two different types of photobioreactors but share some similarities (Ding et al. 2021), which are elaborated in Table 10.1.

While both types of photobioreactors can be used for microalgae cultivation and wastewater treatment, the choice between them will depend on specific requirements, costs, and expected efficiencies in a particular application (Rinanti et al. 2014).

Table 10.1 Similarities and differences between airlift and bubble column photobioreactors

Similarities	
Both belong to photobioreactors based on the principle of mixing through gas flow	
Differences	
Airlift photobioreactor	Bubble column photobioreactor
• This photobioreactor utilizes the airlift principle, where gas is injected into the bottom of the riser (upward column) and liquid is drawn upwards by the gas bubbles	• Bubble column photobioreactor involves gas injection into the bottom of a single liquid column, creating rising bubbles that mix its contents
• In the airlift photobioreactor, fluid circulation occurs between the riser and downcomer (downward column) due to the density difference between them	• Mixing in the bubble column photobioreactor occurs as gas bubbles rise within a single liquid column
• Airlift photobioreactor generally has a more complex design, involving multiple components such as riser, downcomer, and gas injector	• Bubble column photobioreactor has a simpler design as it only involves a single liquid column and gas injection system

Biofilm Photobioreactor

A biofilm photobioreactor is a type of photobioreactor designed to cultivate microalgae in the form of biofilms on solid surfaces within the reactor (Moreno Osorio et al. 2021). Microalgae adhere to a substrate, typically in the form of plates, threads, or other materials with a large surface area for microalgae growth. The biofilm photobioreactor offers several advantages compared to suspended liquid photobioreactors, as explained below:

(a) Allowing microalgae to grow at higher cell densities, thereby enhancing system productivity and efficiency.
(b) Providing protection for microalgae from unfavorable environmental conditions, such as temperature changes, pH variations, or fluctuations in nutrient concentrations.
(c) Enabling better control of fluid flow, as it can be independently regulated from microalgae growth.
(d) Facilitating easier biomass separation from the wastewater since microalgae grow on solid substrates rather than being dispersed in suspension.
(e) Demonstrating effectiveness in wastewater treatment, as microalgae growing in biofilms deliver higher removal efficiency of nutrients and pollutants from wastewater compared to suspended liquid systems.

However, biofilm photobioreactors also present several challenges (Guzzon et al. 2019), including:

(a) Requiring a uniform light distribution to all microalgae cells growing in the biofilm photobioreactor.
(b) Entailing more intensive maintenance compared to suspended liquid photobioreactors, including cleaning or replacing substrates covered with biofilm.
(c) Involving biofilm thickness monitoring to ensure efficient photosynthesis, as an excessively thick biofilm can hinder light penetration and lead to anaerobic conditions within the biofilm.

10.3.4 Integration of Algae Technology with Other Wastewater Treatment Technologies

The integration of algae technology with other wastewater treatment technologies can be achieved through various methods, depending on the specific objectives, characteristics of the wastewater, and operational conditions (Gururani et al. 2022). This integration allows for improved efficiency and effectiveness in wastewater treatment. Several examples of such integrations are elaborated below:

Hybrid Reactor System

The hybrid reactor system is a system that combines features and advantages from two or more different types of reactors to achieve higher efficiency and overcome the limitations of each individual system. In the context of microalgae-based wastewater treatment, the hybrid reactor system typically combines features from open ponds (algae ponds) and photobioreactors.

Open ponds are a relatively simple and cost-effective system commonly used for large-scale cultivation of microalgae. However, this system has lower productivity than photobioreactors and is susceptible to contamination and environmental changes. On the other hand, photobioreactors are closed systems that offer more controlled growth conditions for microalgae, resulting in higher biomass productivity, reduced contamination risk, and better control over growth parameters. Nevertheless, photobioreactors generally have higher operational and maintenance costs compared to open ponds.

A hybrid reactor system that combines features from open ponds and photobioreactors can achieve higher efficiency in microalgae-based wastewater treatment by leveraging the advantages of both systems. For instance, microalgae can initially grow in open ponds to reduce the initial pollution load of wastewater, then be transferred to photobioreactors for further growth and more efficient nutrient processing. Hybrid reactors can also be used to produce high-value products, such as pigments or lipids for bioenergy production.

Hybrid systems also integrate microalgae-based approaches with conventional wastewater treatment technologies, such as active sludge, anaerobic, or aerobic systems. In this hybrid method, microalgae cooperate with bacteria and other microorganisms to enhance wastewater treatment efficiency. Bacteria degrade organic compounds and produce CO_2, which microalgae utilize for photosynthesis. As a result, microalgae release oxygen, which bacteria use for oxidation. This system improves wastewater treatment efficiency and generates a synergistic effect between microalgae and bacteria. Hybrid systems allow better environmental control during the initial stages and take advantage of the low maintenance costs of open ponds during the final stages, reducing operational costs and improving wastewater treatment efficiency.

In this system, microalgae are utilized alongside aerobic and anaerobic processes for wastewater treatment (Amenorfenyo et al. 2019). Microalgae can be grown in photobioreactors or open ponds and combined with conventional wastewater treatment systems, such as aerobic and anaerobic systems, after initial processing stages. In the advanced treatment stage following aerobic or anaerobic processes, the pretreated wastewater serves as a nutrient source for microalgae growth. The remaining nutrients are consumed, lowering pathogen levels. Microalgae take up the remaining nutrients and pollutants, while their photosynthesis produces oxygen to facilitate the aerobic process and reduce energy requirements for aeration. This process enhances treatment efficiency, including the reduction of excess nutrients, heavy metals, and hazardous organic compounds; leading to improved wastewater quality and enabling safe recycling or environmentally safe disposal.

Generally, literature focusing on hybrid reactor systems for microalgae-based wastewater treatment is still limited. However, by studying various existing systems (e.g., open ponds, photobioreactors, and other wastewater treatment technologies), it is expected that innovative and efficient hybrid reactor systems can be developed for microalgae-based wastewater treatment applications.

Integration with Phytoremediation Systems

The integration of a photobioreactor with a phytoremediation system can create a more efficient and effective wastewater treatment by harnessing the capabilities of microalgae and plants in removing pollutants (Cui et al. 2020). In this integrated system, wastewater is first processed through the photobioreactor containing microalgae. The microalgae will consume nutrients such as phosphates and nitrogen, as well as degrade organic compounds and produce oxygen through photosynthesis. After passing through the photobioreactor, the wastewater—which has already seen a reduction in contaminants—will be directed to the phytoremediation system.

The phytoremediation system involves the use of plants (usually aquatic plants) to remove or reduce remaining pollutants in the wastewater. The plants absorb contaminants through their roots and accumulate them in their tissues or convert them into less harmful compounds through various mechanisms, such as phytodegradation, phytostabilization, or phytovolatilization. Additionally, plants can enhance the activity of microorganisms in the root zone, known as the rhizosphere, aiding in contaminant degradation.

The integration of a photobioreactor and phytoremediation offers several advantages, including:

(1) Higher treatment efficiency: The combination of microalgae and plants allows for more efficient and effective treatment, as they can target various types of contaminants that may not be removed by a single system alone.
(2) Nutrient load reduction: By combining photobioreactors and phytoremediation, the system can be more effective in reducing nutrient concentrations such as phosphates and nitrogen, which are major contributors to eutrophication and water pollution.
(3) Sustainability: The integrated system utilizes natural processes to remove pollutants and produces biomass that can be used as an energy source or raw material for industries, such as animal feed, fertilizer, or chemicals.
(4) Flexibility: Integrated systems can be tailored to specific wastewater treatment needs and can be upgraded or modified according to changes in wastewater quality or treatment objectives.

Despite the numerous advantages, the design and operation of this integrated system require a thorough understanding of microalgae and plant biology, as well as photobioreactor and phytoremediation technologies (Benner et al. 2022). While wastewater treatment in an integrated photobioreactor and phytoremediation system has its benefits, there are also several drawbacks to consider:

(1) Investment and operational costs: Integrating photobioreactors and phytoremediation systems generally involves higher investment and operational costs compared to operating each system separately. Additional costs may include equipment, maintenance, and more complicated environmental controls.

(2) Design and operation challenges: Designing, constructing, and operating this integrated system entail a deeper understanding of both systems and how they can work efficiently and effectively together, potentially requiring more human resources and technical expertise.
(3) The potential of lower contaminant availability: In this integrated system, wastewater that has passed through the photobioreactor may already have seen significant contaminant reduction. As a result, plants in the phytoremediation system may have limited access to contaminants, diminishing the effectiveness of phytoremediation.
(4) Intensive maintenance: Involving a combination of two methods, this integrated system naturally requires more intensive maintenance. For instance, regular cleaning and upkeep are critical to sustain photobioreactor performance, while pruning plants and managing biomass may be necessary for the phytoremediation system.
(5) Scalability: Scaling up this integrated system for large-scale wastewater treatment may pose challenges due to its complexity of design and operations.

Although there are some disadvantages in utilizing an integrated photobioreactor and phytoremediation system for wastewater treatment, this technology still holds great potential in reducing pollutants and restoring water quality. In many cases, the advantages of this integrated system may outweigh its drawbacks, depending on specific wastewater treatment needs and objectives.

Integration with Coagulation-Flocculation Systems

Microalgae-based wastewater treatment integrated with coagulation and flocculation is an approach that combines biological and chemical methods to treat wastewater and remove various pollutants (Al-Jabri et al. 2021). The integrated process of wastewater treatment with coagulation and flocculation occurs as follows:

(1) In the initial stage, wastewater is directed through a photobioreactor or an open system containing microalgae. Microalgae consume nutrients (such as phosphates and nitrogen) and degrade organic compounds. During this initial process, microalgae also produce oxygen through photosynthesis, which can aid in the biological oxidation process to further break down organic compounds.
(2) After undergoing microalgae-based treatment, the wastewater is then directed to the coagulation system. In this stage, chemical coagulants (e.g., aluminum sulfate, ferric chloride, or polymers) are added to the wastewater. These coagulants react with suspended and colloidal particles in the wastewater, causing these particles to clump together into larger flocs.
(3) Flocculation follows, where the flocs formed during coagulation are agglomerated into larger and denser aggregates. This process usually involves gentle stirring or agitation of the wastewater to facilitate contact between the flocs and aid their coalescence.
(4) After flocculation, the larger and denser flocs are separated from the wastewater, either through sedimentation, flotation, or filtration. As a result, the generated outflow will have lower particle and pollutant content.

The integration of microalgae-based wastewater treatment with coagulation and flocculation offers several advantages (You et al. 2023):

(1) Higher treatment efficiency: The combination of biological and chemical processes can achieve higher treatment efficiency compared to operating each method separately, as it can remove various pollutants through different mechanisms.
(2) Nutrient load reduction: Microalgae-based treatment is effective in reducing nutrients such as phosphates and nitrogen, while coagulation and flocculation can help remove suspended and colloidal particles that may contain additional pollutants.
(3) Process stability: Combining both methods can enhance the stability of the wastewater treatment process, as the deficiency or failure of one system can be compensated by the other.

However, this integrated system still has some disadvantages, including:

(1) Investment and operational costs: Integrating microalgae-based treatment with coagulation and flocculation may involve higher investment and operational costs compared to operating each system separately. Additional costs may include equipment, chemicals, maintenance, and more complicated environmental controls.
(2) System complexity: Integrated systems are more complex and require a deeper understanding of both treatment methods, as well as knowledge of how they can work efficiently and effectively together. This may require more human resources and technical expertise.
(3) Use of chemicals: The use of chemical coagulants in the coagulation and flocculation processes can pose environmental and health concerns if not handled properly. Careful selection of chemical coagulants is vital, and appropriate management and monitoring measures are required to reduce potential negative impacts on the environment and human health.
(4) Sludge management: The coagulation and flocculation processes yield sludge as a byproduct that needs further treatment or safe disposal. Managing the sludge generated from the integrated system may be more complex due to the involvement of microalgae biomass and chemical coagulants.
(5) Scalability: Scaling up this integrated system for large-scale wastewater treatment may pose challenges due to its complexity of design and operations. Moreover, meeting the energy and chemical requirements on a larger scale may be an issue that needs to be addressed (Fudge et al. 2021).

Integration with Membrane Systems

Integrating microalgae-based approaches with membrane technologies like ultrafiltration and nanofiltration creates a more efficient and effective wastewater treatment system by harnessing the capabilities of microalgae in removing pollutants (Amenorfenyo et al. 2019), followed by further separation and purification through membrane apparatus to segregate microalgae from the treated wastewater. This process enables the collection of nutrient-rich algal biomass, which can be utilized for other purposes such as bioenergy production or animal feed.

In this integrated system, wastewater is first processed by microalgae in a photobioreactor or open system. Microalgae consume nutrients (including phosphates and nitrogen) and degrade organic compounds. After undergoing microalgae-based treatment, the wastewater with reduced contaminants is directed to the membrane system. Ultrafiltration (UF) is a filtration process that uses membranes with pore sizes ranging from 0.01 to 0.1 μm, that can effectively take out suspended particles, bacteria, and most viruses. Nanofiltration (NF) uses membranes with smaller pores, about 0.001–0.01 μm, allowing the removal of dissolved ions, small organic molecules, and most viruses.

The integration of microalgae-based wastewater treatment with membrane technologies offers several advantages:

(1) Higher treatment efficiency: The combination of microalgae and membrane technologies allows for more efficient and effective treatment, as they can target various types of contaminants that may not be removed by a single system alone.
(2) Improved water quality: Membrane technologies such as ultrafiltration and nanofiltration can produce higher-grade water quality as they effectively remove particles, bacteria, viruses, as well as most organic and inorganic contaminants.
(3) Potential for water recycling: The water generated by this integrated system can reach a sufficiently high quality for recycling and reuse in various applications, such as irrigation, cooling, or even drinking water, depending on water quality standards and additional treatments that may be required.
(4) Sustainability: The integrated system utilizes natural processes to remove pollutants and produce biomass that can be used as a source of energy or raw materials for industries, such as animal feed, fertilizers, or chemicals.

However, the integration of microalgae-based wastewater treatment with membrane technologies also has some disadvantages, including higher investment and operational costs, as well as the potential for membrane fouling by microalgae biomass or other particles. In this regard, pre-separation or additional treatment may be necessary to reduce particle load before the wastewater enters the membrane system (Goh et al. 2023).

10.4 Benefits and Potential of Algae in Wastewater Treatment

10.4.1 Absorption and Bioaccumulation of Pollutants

Utilization of microalgae offers an efficient, sustainable, and environmentally friendly solution to address water and soil pollution issues. Overall, incorporating microalgae in wastewater treatment provides various benefits, encompassing environmental, energy, and economic aspects (Goh et al. 2023). The absorption and accumulation mechanisms involved in microalgae-based wastewater treatment entail several steps, including nutrient absorption followed by biodegradation of organic pollutants, and subsequent bioaccumulation. Throughout this process, microalgae undergo photosynthesis.

(a) Nutrient absorption: Microalgae need nutrients like nitrogen and phosphorus for their growth. These nutrients are typically found in high amounts in wastewater, and microalgae have the ability to absorb them from the wastewater to support their growth and reproduction. This process aids in reducing nutrient concentrations in the wastewater, thus preventing eutrophication when the water is discharged into the environment.
(b) Biodegradation: Microalgae can also degrade some organic pollutants in wastewater, such as aromatic compounds, pesticides, and hydrocarbons. This process involves enzymes produced by microalgae to break down pollutants into simpler and less hazardous components.
(c) Bioaccumulation: Besides biodegradation, microalgae can accumulate pollutants within their cells, including heavy metals and radionuclides. This process involves the transportation of pollutants across the microalgae cell membrane and their accumulation in the cytoplasm or cell organelles. Bioaccumulation helps lower pollutant levels in wastewater but can lead to these pollutants being concentrated in microalgae biomass.
(d) Carbon dioxide fixation: Photosynthesis, occurring in microalgae, involves the utilization of carbon dioxide (CO_2) as a carbon source for biomass synthesis. Microalgae can absorb and fix CO_2 from the environment, either from the atmosphere or industrial exhaust gases, thereby reducing greenhouse gas emissions. Wastewater treatment using microalgae can help decrease CO_2 emissions while simultaneously generating biomass for various purposes, such as bioenergy production, animal feed, and fertilizer.

Several advantages and potential uses of microalgae in wastewater treatment, related to pollutant absorption and bioaccumulation in water and soil, are as follows (Abdelfattah et al. 2023):

(1) Microalgae can absorb and accumulate heavy metals, excess nutrients, and other organic pollutants found in wastewater. This process, called bioremediation, helps mitigate the negative impacts of pollution on aquatic and soil ecosystems.

(2) Certain microalgae species can degrade hazardous organic compounds and pathogens (such as pesticides and hydrocarbons) into simpler and less harmful compounds, improving the overall efficiency of the wastewater treatment system and produce cleaner effluent.
(3) Microalgae can accumulate rare metals like gold, silver, and platinum from wastewater. These metals can be recovered from microalgae biomass for industrial use.
(4) Microalgae-based wastewater treatment requires significantly less space compared to conventional methods like aerobic or anaerobic processes. This reduces pressure on land and enhances the sustainability of wastewater treatment systems. The use of microalgae in wastewater treatment is an environmentally friendly solution, as it does not produce hazardous waste and helps decrease water, soil, and air pollution. Additionally, microalgae can be cultivated in infertile or degraded land, thus avoiding competition with productive agricultural areas.

10.4.2 Production of Biomass and Renewable Energy

Microalgae-based wastewater treatment requires lower energy compared to conventional methods such as aerobic and anaerobic processes. Microalgae can grow rapidly and produce biomass rich in proteins, lipids, and carbohydrates that can be further processed for various purposes, including animal feed, organic fertilizer, cosmetics, pharmaceuticals, and nutraceuticals. This provides greater economic opportunities for industries related to microalgae. Moreover, microalgae can produce biomass as a raw material for bioenergy production (e.g., biogas, bioethanol, or biodiesel), hence reducing dependence on fossil fuels.

10.4.3 Nutrition Restoration and Reduction of Greenhouse Gas Emissions

Wastewater is often rich in nutrients derived from various sources, such as domestic, industrial, and agricultural waste. Nutrient recovery processes occur during wastewater treatment by microalgae for the following reasons (Valchev and Ribarova 2022):

(a) Nutrient absorption: Microalgae actively absorb available nutrients in wastewater, including ammonia (NH_4^+), nitrate (NO_3^-), and phosphate (PO_4^{3-}), to support their growth and diverse metabolic activities as shown in Fig. 10.5 (Gururani et al. 2022) process helps reduce nutrient concentrations in wastewater.
(b) Microalgae biomass: The nutrients taken up by microalgae from wastewater are stored in the form of cellular biomass. This microalgae biomass can be

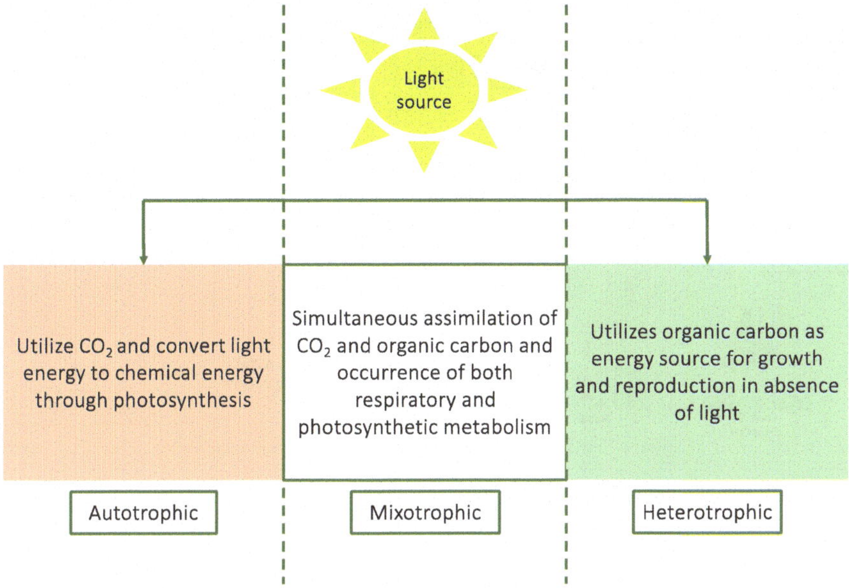

Fig. 10.5 Microalgae metabolism

utilized as nutrient-rich organic fertilizer, allowing the nutrients extracted from wastewater to be returned to the environment in a more eco-friendly form.

(c) Resource recovery: The use of microalgae in wastewater treatment not only helps reduce pollutant concentrations but also enables the recovery and recycling of valuable resources, such as nutrients. This can lessen reliance on non-renewable natural resources and enhance the sustainability of wastewater treatment systems.

(d) Mitigation of eutrophication: Eutrophication is a process where surface water bodies, such as lakes or rivers, become excessively rich in nutrients, leading to overgrowth of plants and disruptions in aquatic ecosystems. By diminishing nutrient levels in the discharged wastewater, microalgae-based wastewater treatment can help prevent eutrophication and its negative impacts on aquatic ecosystems.

Overall, nutrition restoration occurs during the wastewater treatment by microalgae due to their ability to uptake and accumulate nutrients in their biomass, thus lowering nutrient concentrations in wastewater and enabling the recovery and recycling of these valuable resources. Microalgae also play a significant role in reducing greenhouse gas (GHG) emissions during the wastewater treatment process, involving the following activities:

- Carbon dioxide (CO_2) fixation: One of the major components of greenhouse gases is CO_2. Microalgae utilize CO_2 as a carbon source during photosynthesis to form biomass. In wastewater treatment, microalgae can absorb CO_2 from both

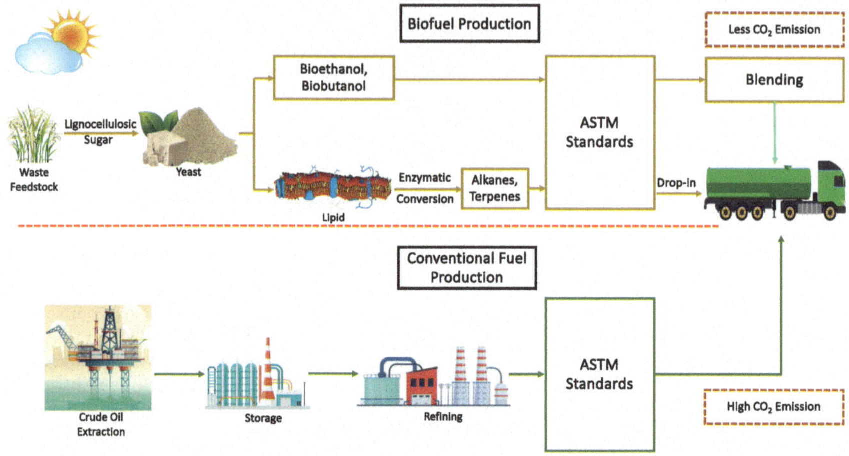

Fig. 10.6 Generation of crude oil and biofuels

the atmosphere and industrial exhaust gases. The process of CO_2 fixation by microalgae reduces the amount of CO_2 released into the atmosphere.

- Substitution of fossil energy sources: The microalgae biomass produced during wastewater treatment can be used as a renewable energy source, such as biofuel. By replacing some of the use of fossil fuels with microalgae-based biofuel, greenhouse gas emissions from fossil fuel combustion can be cut down, as portrayed in Fig. 10.6 (Malode et al. 2021).
- Methane (CH_4) emission reduction: Microalgae-based approaches can lower methane emissions yielded during conventional wastewater treatment processes, such as anaerobic digestion. Methane is a more potent greenhouse gas than CO_2. By decreasing methane production during wastewater treatment, microalgae can contribute to an overall reduction in greenhouse gas emissions.
- Utilization of microalgae biomass as fertilizer: The biomass generated from wastewater treatment with microalgae can be utilized as nutrient-rich organic fertilizer. The use of organic fertilizers can lessen reliance on synthetic chemical fertilizers—the production of which involves significant greenhouse gas emissions.

Overall, the mechanisms for decreasing greenhouse gas emissions during the wastewater treatment by microalgae involve CO_2 fixation, substitution of fossil energy sources, methane emission reduction, and the utilization of microalgae biomass as organic fertilizer. These processes help lower the carbon footprint of wastewater treatment and contribute to climate change mitigation efforts. With further research and development, microalgae can become an innovative and sustainable solution to address water and soil pollution issues. With diverse advantages and potential applications, the use of microalgae in wastewater treatment could become one of the promising future technologies to maintain ecological balance.

10.5 Economic Aspects and Cost Analysis

10.5.1 Investment and Operational Costs

The investment costs for implementing microalgae-based wastewater treatment systems can vary significantly depending on several factors and are highly determined by the scale and complexity of the project, local conditions, final objectives, and regulatory requirements, as elaborated below (Obaideen et al. 2022):

(1) Project scale: The investment costs will differ for small, medium, and large-scale projects. Small-scale projects, such as those in laboratories or research facilities, normally have lower costs compared to large-scale industrial wastewater treatment systems.
(2) Technology used: The choice of technology, such as open ponds, closed ponds, or photobioreactors, will affect the investment costs. Photobioreactors typically have higher initial costs compared to open or closed pond systems due to more advanced apparatus and more convenient environmental control.
(3) Integration with other wastewater treatment systems: If the microalgae-based approach is integrated with other wastewater treatment technologies (e.g., aerobic, anaerobic, phytoremediation, coagulation-flocculation, or membrane systems), the investment costs will vary depending on the selected methods and the level of integration.
(4) Required infrastructure: The investment costs will also depend on the infrastructure needed, such as pond construction, water treatment systems, pumping systems, and control systems.
(5) Project location: Investment costs can vary depending on the project location, including land prices, construction costs, and labor costs.

The cost components to consider when calculating investments costs are identified as follows:

(1) Initial costs required for designing and planning the microalgae-based wastewater treatment system, including feasibility studies, technical design, and environmental permits.
(2) Costs for constructing and installing the facilities for the microalgae-based wastewater treatment, such as ponds, photobioreactors, aeration systems, and other supporting infrastructure.
(3) Costs for acquiring and maintaining the microalgae used in the wastewater treatment system, including procurement, transportation, and inoculum preparation.
(4) Monitoring and evaluation: Costs needed for monitoring the performance of the microalgae-based wastewater treatment system and evaluating the achievement of environmental and economic goals, including sampling, laboratory analysis, and reporting.

In addition to the initial investment costs, it is essential to consider operational and maintenance costs. These ongoing costs are necessary to operate and maintain the microalgae-based wastewater treatment system, including energy costs, chemical expenses, equipment maintenance, and labor, as explained below:

(1) Energy costs required to operate equipment in the microalgae-based wastewater treatment system, such as pumps, aeration systems, and biomass processing apparatus—whether for living biomass only or both dead and living biomass, as depicted in Fig. 10.7 (Ojha et al. 2021). The energy required for microalgae-based systems is generally lower than conventional methods.
(2) The expenses incurred for the chemical substances needed in the wastewater treatment process, such as coagulants, flocculants, and additional nutrients that may be required to support microalgae growth.
(3) The costs associated with the labor force required to operate and maintain the microalgae-based wastewater treatment system, including salary expenses, training, and personnel management.
(4) The expenditures needed for the maintenance and repair of equipment utilized in the microalgae-based wastewater treatment system, such as photobioreactors, ponds, and solid–liquid separation devices.
(5) The expenses related to the collection, processing, and utilization of microalgae biomass generated from the wastewater treatment system, including costs for drying equipment and converting biomass into value-added products such as bioenergy or animal feed.
(6) The costs involved in periodically monitoring the performance of the microalgae-based wastewater treatment system and evaluating the achievement of environmental and economic objectives, including expenses for sample collection, laboratory analysis, and reporting.

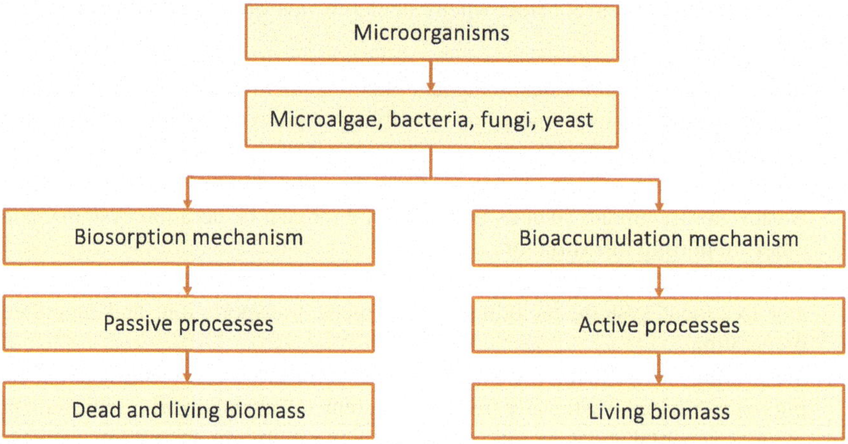

Fig. 10.7 Dead and living biomass

Due to the multitude of factors affecting investment costs, it is challenging to provide precise figures for the implementation expenses of microalgae-based wastewater treatment systems. However, some studies indicate that the cost of treating wastewater with microalgae ranges between \$0.10 and \$0.50 per cubic meter of water processed. These costs will vary depending on the aspects mentioned above. To obtain a more accurate cost estimate, it is advisable to conduct a feasibility study that takes into account the specific factors of the planned project and to consult with experts and suppliers of microalgae-based wastewater treatment technology.

10.5.2 Assessment of Economics and Processing Scale

The economic study related to the project scale when implementing microalgae-based approaches involves analyzing the costs and benefits associated with various aspects of the wastewater treatment system (Srimongkol et al. 2022), including investment, operational, and maintenance costs, as well as the potential revenue from the products generated. Several factors that need to be considered in the economic study are as follows:

(1) Treatment scale: The scope of microalgae-based wastewater treatment can range from small-scale systems (e.g., domestic or small industrial wastewater treatment) to large-scale systems (e.g., urban or large industrial wastewater treatment). The investment and operational costs will vary depending on the project magnitude, and economies of scale may apply, where the cost per unit of treatment decreases along with the increasing processing capacity. Small-scale systems may have lower investment costs, but the operational cost per unit volume of treated wastewater may be higher compared to large-scale systems. Large-scale systems can achieve economies of scale in operational costs but may require a higher initial investment.

(2) Technology used: The technology employed in the microalgae-based wastewater treatment system, such as open ponds, photobioreactors, or a combination with other treatment technologies, will determine the investment, operational, and maintenance costs. Photobioreactors are generally more expensive in terms of investment but can yield higher microalgae growth rates and removal efficiency. Advanced technologies may offer higher efficiency and better control over environmental conditions, but they may also have higher initial costs.

(3) Wastewater quality: The quality of the influent will affect the operational costs and efficiency of the microalgae-based treatment system (Al-Jabri et al. 2021). Wastewater with higher pollutant concentrations may require higher processing costs, but it can also provide a richer nutrient source for microalgae—which, in turn, can increase biomass productivity and the revenue potential from the products generated.

(4) Treatment objectives: The aims of the wastewater treatment system, such as achieving emission standards or utilizing the resources contained in the wastewater (e.g., nutrients or energy), will impact the economic costs and benefits of the microalgae-based wastewater treatment system.
(5) Products generated: The commodities obtained from the microalgae-based wastewater treatment system, such as biomass, bioenergy, or animal feed, have economic value that can help cover the investment and operational costs of the treatment system. The economic study should consider the revenue potential from the products generated, as well as the market and prices for these products.
(6) Government subsidies and incentives: In some cases, the government may offer incentives or subsidies for microalgae-based wastewater treatment, especially if the system helps achieve environmental or sustainability targets. These incentives can help reduce investment and operational costs, thus enhancing the economic feasibility of the treatment system.

A comprehensive economic study will aid in assessing the feasibility and potential success of the microalgae-based wastewater treatment system within a specific context. Cost–benefit analysis, sensitivity analysis, and feasibility studies can help identify the most economical and sustainable treatment solutions for a particular project, as portrayed in Fig. 10.8 (Srimongkol et al. 2022).

(1) Cost–benefit analysis is a systematic process used to compare the costs and benefits associated with a microalgae-based wastewater treatment project. Its purpose is to determine whether the project provides positive value to investors and/or the community. In other words, cost–benefit analysis aims to identify solutions that offer the highest benefit relative to the incurred costs. The analyzed expenses

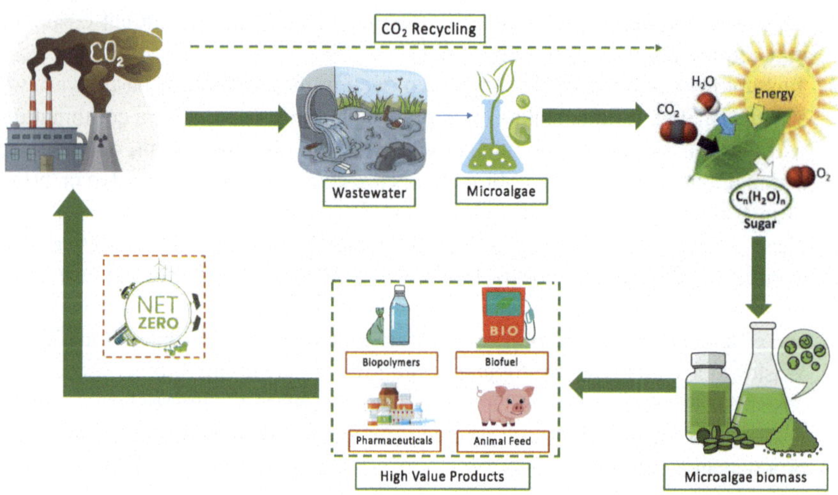

Fig. 10.8 Biorefinery economic model

include investment, operational, and maintenance costs, while the benefits encompass improvements in water quality, greenhouse gas emission reduction, and the potential utilization of by-products such as microalgae biomass. Cost–benefit analysis helps identify the most economical treatment options and indicates whether the project is economically viable.

(2) Sensitivity analysis is a process to test how changes in variables or assumptions used in the cost–benefit analysis can influence the results. These variables may include, for example, investment costs, energy prices, by-product prices, and processing efficiency. Sensitivity analysis assists in determining critical variables that affect the success of the microalgae-based wastewater treatment and can help optimize the system's design and operation. It also identifies the key factors with the most significant influence on the project's economic outcomes and demonstrates the project's susceptibility to changes in market conditions or technological advancements.

(3) Feasibility study is a comprehensive evaluation that covers the technical, economic, environmental, and social aspects of a microalgae-based wastewater treatment project. It aims to determine whether the project is feasible and can be implemented under specific conditions. The feasibility study includes cost–benefit analysis and sensitivity analysis, as well as an evaluation of the technology used, wastewater quality, regulatory requirements, financial analysis, and the project's environmental and social impacts. The feasibility study helps identify the most sustainable and viable treatment solutions for the specific needs of a project.

By employing these three tools, decision-makers can better understand the potential and risks associated with the microalgae-based wastewater treatment project, as well as identify the most economical and sustainable treatment solutions for specific requirements (Srimongkol et al. 2022). Furthermore, these analyses can assist in proposing the project to investors or other supporting parties, and aid in planning and implementing the wastewater treatment project.

10.5.3 Product Marketing Potential and Sources of Income

The implementation of microalgae utilization for wastewater treatment not only improves water quality but also generates diverse products and sources of income. The potential products and revenue sources that can be derived from microalgae-based wastewater treatment are as follows:

Bioenergy Production

Microalgae biomass generated from wastewater treatment can be converted into various types of bioenergy, e.g., biogas, biodiesel, or bioethanol. In this context, the produced bioenergy can be sold in the energy market or used internally to reduce energy costs of the treatment system (Malode et al. 2021).

The choice between using fossil energy (such as kerosene) and biogas for household needs depends on several factors, such as energy availability, cost, environmental impact, and practicality. These aspects affect the revenue potential of biogas produced from microalgae feedstock. The following considerations need to be taken into account when choosing between kerosene and biogas for household needs:

(a) Availability and sustainability: Kerosene is a non-renewable fossil energy source, hence its availability is limited. In contrast, biogas is a renewable energy source produced from the degradation of organic materials (e.g., household waste, animal manure, and plants).
(b) Cost: Fossil fuel costs (such as kerosene) often tend to be more stable and cheaper in certain regions, especially in oil-producing countries. However, biogas can be a cheaper alternative if there is existing infrastructure for biogas production and distribution, particularly in rural areas or places where kerosene is difficult to access or expensive.
(c) Environmental impact: Biogas has a lower environmental impact compared to kerosene. It produces less greenhouse gas emissions and reduces reliance on fossil energy. Moreover, biogas produced from organic waste helps to manage waste effectively and decrease environmental pollution.
(d) Practicality: The use of kerosene for household needs is generally easier and more practical due to the already established infrastructure in many areas. On the other hand, the use of biogas entails more complex production and storage systems. However, in some countries, especially in rural areas, small-scale biogas systems have become practical and efficient solutions to meet household energy needs.

The relative benefits of using kerosene or biogas for household purposes depend on local situations and conditions. Over the long haul, transitioning to biogas may be more advantageous in terms of environmental sustainability and resource management. However, in the short run, kerosene might be more practical and cost-effective in certain regions. Therefore, it is crucial to consider these factors before deciding on the most beneficial energy source for household needs (Bušić et al. 2018). Similarly, the choice between fossil fuels (such as gasoline and diesel) and biodiesel for transportation fuel depends on several factors, including energy availability, cost, environmental impact, and vehicle performance. Some considerations regarding the selection of gasoline, diesel, or biodiesel—affecting the potential income from biodiesel produced by microalgae—are as follows:

(a) Availability and sustainability: Gasoline and diesel are non-renewable fossil fuel sources, leading to limited availability. On the other hand, biodiesel is derived from renewable microalgae, making it a sustainable energy source.
(b) Cost: The cost of gasoline and diesel fuel is often more stable and can be cheaper in some regions, particularly in oil-producing countries. However, biodiesel prices may vary depending on feedstock prices and production scale. In certain countries, government incentives can make biodiesel prices more competitive.
(c) Environmental impact: Biodiesel has a lower environmental impact compared to gasoline and diesel. It yields less greenhouse gas emissions and reduces

dependence on fossil fuels. Moreover, using biodiesel can help mitigate air pollution from particulate emissions and volatile organic compounds.

(d) Vehicle performance: Biodiesel has better lubricating properties than gasoline and diesel, decreasing friction and wear in engines. However, biodiesel has a slightly lower energy value than gasoline and diesel, which may result in a tad higher fuel consumption. Most modern vehicles can operate with a biodiesel blend of up to 20% (B20) without significant engine modifications. Some vehicles are even specifically designed to run on pure biodiesel (B100).

Based on the above description, the potential income from bioenergy production using microalgae biomass derived from wastewater treatment relies on several factors, such as production scale, conversion efficiency, and the market price of bioenergy products. The following are some factors influencing the potential income from microalgae-based bioenergy production:

(a) Microalgae biomass concentration: The productivity of microalgae in wastewater treatment systems affects the amount of biomass available for conversion into bioenergy. Higher biomass concentration leads to greater bioenergy production potential.
(b) Conversion efficiency: The process of converting microalgae biomass into bioenergy has varying efficiencies depending on the technology used. Higher efficiency results in more bioenergy per unit biomass, thus increasing the income potential.
(c) Market price of bioenergy: The income potential from bioenergy production heavily relies on the market price of bioenergy products (e.g., biogas, biodiesel, or bioethanol). This price is affected by economic factors and policies, including demand and supply, government incentives, and environmental regulations.
(d) Production costs: When calculating the income potential, it is crucial to consider bioenergy production costs, including operational and maintenance costs of wastewater treatment systems, biomass conversion costs, and byproduct processing costs.

Given the variability in these factors, it is difficult to provide specific income figures for bioenergy production derived from microalgae-based wastewater treatment. Nonetheless, the potential for income growth can be fostered by improving system efficiency, conducting research and development of innovative conversion technologies, and implementing supportive and favorable policies.

Livestock Feed

Microalgae biomass is rich in protein, fatty acids, and other beneficial nutrients for livestock (Bušić et al. 2018; Lamminen 2021). After undergoing drying and processing, microalgae biomass can be sold as livestock feed for the farming industry. The income derived from microalgae-based livestock feed business in the farming industry depends significantly on various factors as listed below:

(1) Production scale: Larger microalgae production scale can decrease production costs per unit and increase revenue. Production efficiency also influences the costs and the generated income.
(2) Selling price: The selling price of microalgae-based livestock feed depends on market conditions, demand, and competition with other livestock feed products. Competitive pricing can help boost sales and revenue.
(3) Production costs: Microalgae production costs include energy, water, nutrient inputs, maintenance, and equipment expenses. Lowering production costs will raise the profits from the livestock feed business (Chanana et al. 2023). Energy is required to operate photobioreactors, pumps, heating or cooling systems, lighting systems (if used), and other instruments. Energy costs will depend on local electricity tariffs and equipment efficiency. Water is a crucial component in microalgae cultivation, and its cost will be determined by the water source, required treatment, and local water prices. Equipment and infrastructure costs include photobioreactors, lighting systems, pumps, control systems, tanks, and other facilities requisite for *Chlorella* production. These expenses will vary depending on the production scale and technology used. Labor is required to operate and maintain the system, perform quality control, and carry out other tasks related to *Chlorella* production. These costs will depend on the number of workers and wage rates at the location. Following the *Chlorella* growth, processing costs are indispensable, including separation, drying, and packaging costs to manufacture the final sellable product.
(4) Nutritional value: The nutrient content of microalgae (e.g., protein, essential fatty acids, and vitamins) can affect product appeal and boost selling prices.

- Microalgae typically contain a high amount of protein, around 40–70% of their dry weight, making them valuable as a protein source in livestock feed. Products with higher protein content are usually priced higher.
- Some microalgae types, such as *Nannochloropsis* and *Schizochytrium*, are rich in omega-3 fatty acids like DHA and EPA, which are highly valuable in the fish and shrimp feed industries.
- Certain microalgae species, including *Spirulina* and *Haematococcus*, contain pigments like phycocyanin and astaxanthin, which have high commercial value and can elevate the selling price of products.

Estimated selling prices of livestock feed products based on their nutritional value are as follows:

- *Chlorella* is one of the most commonly used microalgae in livestock feed due to its high protein content (around 50–60%) and essential fatty acids. Its selling price ranges from $10 to $25 per kilogram, depending on quality and nutrient content.
- *Spirulina* has an incredibly high protein content (around 60–70%) and is rich in vitamins, minerals, and essential fatty acids. Its selling price ranges from $15 to $40 per kilogram, depending on quality and nutrient content.

- *Haematococcus pluvialis* contains astaxanthin, a powerful antioxidant with numerous health and nutritional benefits. The selling price of *Haematococcus pluvialis* is usually higher than other microalgae types and can reach $100 per kilogram or more, depending on the astaxanthin content and product quality.
- *Nannochloropsis* has a high protein content (around 50%) and is rich in omega-3 fatty acids. Its selling price ranges from $10 to $25 per kilogram, depending on quality and nutrient content.

(5) Marketing and distribution: Effective marketing strategies and a wide distribution network can help increase sales and revenue for microalgae-based livestock feed businesses.

(6) Regulation and quality standards: Complying with existing regulations and quality standards for livestock feed can impact manufacturing costs and product acceptance in the market. In general, the selling price of microalgae-based livestock feed ranges from $5 to $15 per kilogram or more, depending on quality and nutritional composition of the product.

Nutritional Supplements

Microalgae contain abundant essential nutrients, such as omega-3 fatty acids, proteins, vitamins, and minerals (Koyande et al. 2019). Following the extraction and purification processes, these nutrients can be marketed as dietary supplements or used as raw materials in the food and beverage industry.

Raw Materials for Industries

Microalgae serve as a source of various bioactive compounds with commercial value, as they have significant potential as raw materials in the pharmaceutical, cosmetic, and chemical industries. Some commercially valuable bioactive compounds are as follows:

(a) Astaxanthin is a carotenoid pigment found in microalgae like *Haematococcus pluvialis*. It possesses powerful antioxidant properties and has been utilized in dietary supplements, skincare products, and cosmetics.

(b) Phycocyanin is a blue-green pigment found in *Spirulina* (*Arthrospira platensis*). It exhibits antioxidant, anti-inflammatory, and immunomodulatory properties and finds applications in the pharmaceutical, food, and cosmetic industries.

(c) Phycobiliproteins are pigments found in *Cyanobacteria* and *Rhodophyta* (red algae), such as phycoerythrin and allophycocyanin. They are used as natural colorants, fluorophores, and diagnostic markers in the pharmaceutical and biotechnology industries.

(d) Certain microalgae produce polysaccharides that act as antioxidants, antitumor, antiviral, and anti-inflammatory agents. Examples include polysaccharides from *Porphyridium* sp., *Chlorella* sp., and *Dunaliella* sp.

(e) Some microalgae types, such as *Nannochloropsis* and *Schizochytrium*, yield omega-3 fatty acids like DHA (docosahexaenoic acid) and EPA (eicosapentaenoic acid). These fatty acids offer wide-ranging health benefits and are used in the pharmaceutical, nutraceutical, and food industries.
(f) Certain microalgae, like *Dinoflagellates*, generate toxins that can be used in pharmaceutical and biomedical research. Examples of these toxins include okadaic acid and brevetoxins.
(g) Several microalgae produce peptides that function as antimicrobial, antitumor, and antioxidant agents. These peptides can be isolated and utilized in pharmaceutical and biotechnology research.

Water Quality Improvement

Microalgae-based wastewater treatment can produce cleaner water that meets the water quality standards set by the government. In some cases (Goh et al. 2023), the treated water can be reused for irrigation, cooling water, or other industrial processes, thus decreasing operational costs in an industry.

(a) Irrigation: Water treated by microalgae typically contains fewer pollutants and excess nutrients, making it safe for agricultural irrigation. The use of treated water can alleviate pressure on freshwater sources and contribute to the preservation of water resources.
(b) Cooling water: In industrial processes, cooling water is used to regulate the temperature of machines or manufacturing operations. Water treated by microalgae can be used as cooling water due to its reduced pollutant content and improved quality.
(c) Other industrial processes: Water treated by microalgae can be employed in various industrial processes, such as chemical, oil and gas, food and beverage, or pharmaceutical productions. The use of treated water can abate the demand for fresh water and help cut down natural resource consumption.
(d) Fisheries and aquaculture: Water treated by microalgae can also be used in fisheries and aquaculture systems. Cleaner water can elevate the environmental quality for fish and other organisms, hence improving productivity and animal welfare.
(e) Water storage and sustainable water resources: Water treated by microalgae can be stored in reservoirs or artificial lakes to ensure water resource availability in the future. Treated water can also replenish groundwater or reduce pressure on existing freshwater sources.

The use of microalgae-treated wastewater in various applications not only helps lower the consumption of freshwater resources but also generates additional income for wastewater treatment system stakeholders.

Carbon Emission Credit Sales

The use of microalgae in wastewater treatment can diminish greenhouse gas emissions through CO_2 absorption and eco-friendly bioenergy production. In this context, companies implementing microalgae-based wastewater treatment systems can earn carbon emission credits that can be sold in the carbon market or used to meet emission reduction targets. To maximize income from microalgae utilization in wastewater treatment, it is crucial to identify the most promising products and markets, as well as optimize the manufacturing and treatment processes of microalgae biomass. Effective marketing strategies and collaboration with related industries can also help enhance the economic value of microalgae-based wastewater treatment systems.

10.6 Social and Policy Aspects

10.6.1 Community Awareness and Acceptance

The implementation of microalgae utilization for wastewater treatment indeed requires awareness, acceptance, and active participation from the community. Society needs to comprehend the negative impacts of wastewater on the environment and human health. By increasing awareness of the importance of wastewater treatment (You et al. 2023), the community will be more inclined to support microalgae-based solutions. For widespread adoption of microalgae-based approaches, the public must understand and embrace this technology. Educating the community about the benefits and sustainability of this solution will help them accept it as a viable alternative to conventional wastewater treatment methods. The community also needs to actively engage in the wastewater treatment process using microalgae, e.g., participating in local wastewater treatment initiatives, supporting research and development of microalgae technology, and adopting environmentally friendly practices to reduce water pollution.

The application of microalgae-based approaches requires collaboration between the government, industry, research institutions, and the community. Society is obliged to support government policies that promote the use of microalgae in wastewater treatment and collaborate with industries to develop effective and economical solutions.

To ensure the successful implementation of microalgae technology, education and training must be provided to the community. This will help them understand the mechanism and how to manage microalgae-based wastewater treatment systems. The public can participate in small-scale wastewater treatment projects that use microalgae, such as in residential areas, offices, or agriculture. Such involvement will promote this technology at the local level and create more sustainable solutions for waste management. With community awareness, acceptance, and active participation in the use of microalgae for wastewater treatment, the implementation of this technology will be more effective and efficient, thereby reducing the negative impact of wastewater on the environment and human health.

10.6.2 Policies and Regulations Supporting the Use of Algae

Policies and regulations supporting the use of algae as agents in wastewater treatment vary between different countries and regions. Some common policies and regulations frequently encountered include (Amaro et al. 2023):

(1) Environmental policies: These policies encourage the use of environmentally friendly technologies to reduce water pollution and protect water resources. The use of algae as agents in wastewater treatment aligns with environmental policies.
(2) Wastewater regulations: These regulations establish standards for wastewater quality that industries and wastewater treatment facilities must meet. Algae can help achieve these standards as they are effective in absorbing and removing specific pollutants in wastewater.
(3) Subsidies and incentives: Governments often provide subsidies and incentives to industries and wastewater treatment facilities that employ environmentally friendly technologies, including algae utilization. These incentives may include tax reductions, financial assistance, or technical support. The Government of the Republic of Indonesia has taken steps to support industries and wastewater treatment facilities that adopt eco-friendly methods, including algae usage. Examples of incentives and subsidies provided are:

    (a) Tax reductions on:

        - Value-added tax (VAT) covered by the government for environmentally friendly equipment and technologies, including algae-based wastewater treatment systems.
        - Corporate income tax exemption or abatement for companies investing in environmentally friendly technologies, such as algae-based wastewater treatment.

    (b) Financial assistance:

        - Soft loans with low-interest rates for companies interested in adopting environmentally friendly technologies in wastewater treatment.
        - Provision of research grants and development assistance for projects using algae technology in wastewater treatment.
        - Subsidies for companies generating renewable energy from algae biomass yielded during wastewater treatment.

    (c) Technical support:

        - Training and workshops for industry workers on the implementation of environmentally friendly technologies, including algae-based wastewater treatment.
        - Consultation facilities and technical support from the government or research institutions to assist companies in adopting and optimizing algae-based technologies.

- Collaboration among companies, government, and research institutions in the development and enhancement of algae-based wastewater treatment technologies.

These policies and incentives aim to encourage companies and wastewater treatment facilities to adopt eco-friendly technologies, including algae utilization, to achieve desired environmental and sustainability goals. However, it is essential to note that the provided policies and incentives may change over time, depending on government priorities and economic conditions. Therefore, it is crucial to continually monitor the latest developments in policies and incentives offered by the Government of the Republic of Indonesia.

(4) Research and development policies: Governments and research institutions in various countries may have policies to support studies and development of innovative technologies in wastewater treatment, including the use of algae. These policies may include research funding allocations, government-industry partnerships, and international collaborations.

(5) Education and training policies: The algae-based approach to wastewater treatment requires specialized knowledge and skills. Hence, policies on education and training supporting the development of human resources in this field will be a vital aspect to promote algae utilization.

(6) Industrial regulations and standards: Some industries that generate wastewater may be required to comply with specific standards and regulations encompassing the use of environmentally friendly technologies, including algae-based wastewater treatment.

10.6.3 Partnerships and Collaborations Between Government, Industry, and Society

By fostering effective partnerships and collaborations between the government, industry, and society, the implementation of microalgae utilization in wastewater treatment can be facilitated, leading to broader environmental, economic, and social benefits (Srimongkol et al. 2022). Several strategies to promote effective partnerships and collaborations include the following:

(1) The government can develop policies and regulations that support the use of microalgae in wastewater treatment, such as stringent water quality standards, tax incentives or subsidies, as well as research and development support for eco-friendly technologies.

(2) The government needs to take an active role in providing the necessary infrastructure and technical support to facilitate the adoption of microalgae-based technologies in wastewater treatment. This may involve constructing wastewater treatment facilities, providing equipment, and offering technical support for system maintenance and operations.

(3) Collaboration between the government, industry, and research institutions is pivotal to develop efficient, effective, and economical microalgae-based wastewater treatment methods. Research efforts can focus on optimizing microalgae growth, improving photobioreactor or pond technologies, as well as processing and utilizing by-products. Government and industry can support research and development through funding, resources, and technical assistance.
(4) The government, industry, and educational institutions can collaborate to devise training and education programs in order to amplify the grasp of the benefits and technologies of microalgae-based approaches. Such training can help build local capacity in designing, operating, and maintaining microalgae-based wastewater treatment systems. Educating the public and industries involved in microalgae-based technologies is essential to improve understanding and technical capacity for implementing wastewater treatment systems. The government and educational institutions can cooperate to provide training programs, facilities, resources, and technical support.
(5) The government, industry, and society need to work together to promote the use of microalgae-based technologies in wastewater treatment. Marketing and promotional strategies may include public awareness campaigns, supporting the marketing of by-products, and recognizing best practices. Partnerships between the government, industry, and society can help identify business opportunities, reduce risks, and create synergies in resource management (Srimongkol et al. 2022).
(6) The government and financial institutions can provide financial support in the form of loans, grants, or incentives to companies implementing microalgae-based wastewater treatment systems. This financial aid can help lower initial investment costs and accelerate the adoption of this technology.
(7) Society can play an active role in supporting the implementation of microalgae utilization in wastewater treatment, for example, by participating in water resource management programs or local environmental initiatives. Increased awareness and support from the public for this technology will help hasten its adoption and success in wastewater treatment.

Collaboration between diverse industries (e.g., wastewater treatment, agriculture, energy, and pharmaceuticals) can help create synergies and efficiencies in resource utilization and usage of by-products derived from microalgae-based wastewater treatment systems.

10.7 Case Studies and Real Applications

10.7.1 Experience and Lessons from Implementation in Different Countries

Various countries in Asia have implemented the utilization of microalgae in wastewater treatment. Many lessons and experiences can be drawn from the use of microalgae-based approaches, as described below (Geremia et al. 2021):

(1) India: Microalgae utilization in wastewater treatment has been a subject of research in India. Several studies have demonstrated the effectiveness of microalgae in reducing pollutants such as nitrogen, phosphorus, and heavy metals. Lessons learned from India include the importance of research and development to optimize microalgae-based technologies and identify competent local microalgae species for wastewater treatment.
(2) China: As a country with large-scale industries, China has experienced increased water pollution due to industrial and agricultural activities. Microalgae utilization for wastewater treatment has been a focus of research and implementation in various regions. China has invested significant resources in the research and development of microalgae-based technologies for wastewater treatment and bioenergy production (Goh et al. 2023). Lessons learned from China highlight the prominence of collaboration between the government, industry, and research institutions in developing and adopting microalgae-based wastewater treatment technologies.
(3) Taiwan: Taiwan has employed microalgae for wastewater treatment in the fisheries sector, producing clean water for reuse in aquaculture and mitigating environmental impacts. Lessons learned from Taiwan emphasize the utilization of microalgae in specific sectors, such as fisheries, and the integration of wastewater treatment technologies with relevant industries.
(4) Indonesia: Several small-scale studies and implementations have been conducted in Indonesia to utilize microalgae in domestic and industrial wastewater treatment. Lessons learned from Indonesia underscore the importance of developing technologies suitable for local conditions and specific needs in wastewater treatment (Gómez-Sanabria et al. 2020).

In Europe, the utilization of microalgae in wastewater treatment has also been adopted and implemented in various research and pilot projects in the following countries:

(1) Spain: Spain has carried out several studies and projects related to microalgae use in wastewater treatment, particularly in the fisheries and agriculture sectors. One of the paramount lessons learned from Spain is the importance of cross-sector collaboration in developing innovative solutions harnessing microalgae to address water pollution issues.

(2) Netherlands: The Netherlands has numerous projects and initiatives focusing on using microalgae for wastewater treatment and producing high-value products like bioenergy and nutritional supplements (Watanabe and Isdepsky 2021). Lessons from the Netherlands emphasize the significance of research and development to enhance the efficiency and economics of microalgae-based approaches.
(3) Germany: Germany has invested considerable resources in research and development of microalgae-based technologies, including their use in wastewater treatment (You et al. 2023). Lessons learned from Germany highlight the importance of government support in the form of policies, regulations, and incentives to promote the adoption of environmentally-friendly technologies.
(4) France: France has implemented various research and pilot projects focusing on microalgae use in wastewater treatment, especially in agriculture and industry. Lessons from France emphasize the integration of microalgae-based approaches with conventional wastewater treatment methods to create efficient and sustainable solutions.
(5) United Kingdom (UK): The UK has undertaken several research and projects exploring microalgae utilization in wastewater treatment. One of the notable projects involves using photobioreactors harnessing microalgae to reduce greenhouse gas emissions and produce high-value products from wastewater.
(6) Italy: Italy has conducted various research and pilot projects related to microalgae use in wastewater treatment, particularly in the fisheries and industry sectors. Italy has experimented with integrated wastewater treatment systems combining microalgae and conventional technologies to create efficient and sustainable solutions.
(7) Sweden: Sweden has invested in research and development of microalgae-based methods for wastewater treatment and bioenergy production. One of the noteworthy projects involves using microalgae to treat wastewater from the pulp and paper industry.
(8) Norway: Norway has also adopted microalgae-based technologies in wastewater treatment, especially in the fisheries and aquaculture sectors. One of the eminent projects involves using microalgae to treat wastewater from the fisheries industry and produce high-value products such as protein and omega-3 fatty acids.
(9) Denmark: Denmark has carried out research and projects exploring microalgae use in wastewater treatment, particularly in agriculture and industry sectors. Examples of renowned projects include using microalgae to treat wastewater from farms and produce high-value products such as animal feed and bioenergy.

Some countries in Africa and America have also adopted and implemented microalgae utilization in wastewater treatment (Mohsenpour et al. 2021). Here are a couple of instances:

(1) South Africa: South Africa has undertaken several research and projects exploring microalgae use in wastewater treatment, particularly in agriculture and industry sectors. Microalgae are used to treat wastewater from farms and industries to produce high-value products such as animal feed and bioenergy.

(2) Nigeria: Nigeria has experimented with microalgae-based technologies in wastewater treatment, especially in the fisheries and aquaculture sectors. Microalgae are used to treat wastewater from the fisheries industry and produce high-value products such as protein and omega-3 fatty acids.
(3) Egypt: Egypt has conducted research on the utilization of microalgae for wastewater treatment in the agriculture and industry sectors. One of the acclaimed projects involves using microalgae to treat wastewater from the textile industry and produce high-value products such as bioenergy.
(4) United States (USA): The USA has invested in research and development of microalgae-based technologies for wastewater treatment and bioenergy production. Some projects have incorporated microalgae to treat wastewater from farms and industries to generate high-value products such as animal feed and bioenergy.
(5) Canada: Canada has also adopted microalgae-based approaches in wastewater treatment, principally in the fisheries and aquaculture sectors. One of the remarkable projects involves using microalgae to treat wastewater from the fisheries industry and yield high-value products such as protein and omega-3 fatty acids.
(6) Brazil: Brazil has carried out research and projects exploring microalgae use in wastewater treatment, particularly in agriculture and industry sectors. Examples of projects include using microalgae to treat wastewater from farms and produce high-value products such as animal feed and bioenergy.

From the experiences of these countries, several common lessons can be drawn, including:

(1) The importance of research and development to identify effective local microalgae species and optimize microalgae-based approaches in wastewater treatment.
(2) The significance of policies and regulations that support the use of microalgae-based methods in wastewater treatment, including government incentives to promote environmentally-friendly technology adoption.
(3) The importance of collaboration between the government, industry, and research institutions in developing and adopting microalgae-based wastewater treatment systems.
(4) The utilization of microalgae in specific sectors, such as fisheries and agriculture, to mitigate environmental impacts and create business opportunities.
(5) The development of technologies tailored to local conditions and specific needs in wastewater treatment.

10.7.2 Success Factors and Constraints in Implementation

Aspects contributing to the successful implementation of microalgae utilization in wastewater treatment include (Mohsenpour et al. 2021):

(1) Research and development: Continuously advancing knowledge and technology through research and development efforts will optimize microalgae-based operations, boost efficiency, and lessen costs.
(2) Government support: Policymaking, regulations, and incentives aid from the government play a crucial role in encouraging the adoption of microalgae-based technologies in wastewater treatment.
(3) Cross-sector collaboration: Collaboration among the government, industry, and research institutions facilitates the dissemination of technology, funding, and development of the necessary infrastructure.
(4) Education and training: Raising awareness and understanding of the benefits and potential of microalgae in wastewater treatment will help promote this technology across various sectors.
(5) Technology integration: Integrating microalgae-based approaches with conventional wastewater treatment methods can create more efficient and sustainable solutions.

Furthermore, the following primary challenges have been identified in implementing microalgae utilization in wastewater treatment:

(1) Investment costs: The expenses entailed in establishing infrastructure and acquiring necessary equipment for microalgae-based wastewater treatment systems can be a hindrance, particularly for developing countries or small and medium-sized industries.
(2) Operational costs: Although operational costs may diminish over time along with improved efficiency, they can still be a constraint for some management parties.
(3) Limited knowledge and advanced technology: The lack of understanding and the modern instruments required to utilize microalgae in wastewater treatment may restrict the adoption and effectiveness of this method.
(4) Variability in wastewater quality: The considerable variation in wastewater quality, particularly concerning nutrient content and pollutants, can influence the efficiency and effectiveness of microalgae-based wastewater treatment.
(5) Scaling up the technology: Developing and optimizing microalgae-based wastewater treatment systems for larger scales can be a challenge, notably in controlling microalgae growth conditions and managing the waste.

10.7.3 Recommendations and Strategies for Further Development

The following strategies are recommended to further develop the implementation of microalgae-based wastewater (Srimongkol et al. 2022):

(1) Focus on research and development: Research efforts should prioritize improving efficiency, lowering costs, and optimizing microalgae-based

approaches, including studies on the most effective microalgae species, optimal growth conditions, and high-value product extraction methods.

(2) Government support: The government needs to provide favorable policy support, such as fiscal incentives, subsidies, or regulations encouraging the adoption of microalgae-based technologies in wastewater treatment.

(3) Enhanced collaboration: Strengthen the partnership and cooperation between the government, industry, and research institutions to expedite technology dissemination, financing, and necessary infrastructure development.

(4) Community empowerment: Conduct education and training campaigns to raise awareness and understanding of the benefits and potential of microalgae utilization in wastewater treatment.

(5) Technology integration: Combine microalgae-based approaches with conventional wastewater treatment methods to create more efficient and sustainable solutions, including integrated systems incorporating wastewater treatment with renewable energy production and by-product utilization.

(6) Scalability and adaptability: Develop and optimize microalgae-based wastewater treatment systems for larger scales and different types of waste water, including designing modular and flexible schemes adaptable to various conditions and needs.

(7) Market and product development: Increase the added value of products derived from microalgae-based wastewater treatment (e.g., animal feed, bioenergy, and chemicals) by expanding new markets and applications.

(8) Regular evaluation and monitoring: Periodically assess the performance of microalgae-based wastewater treatment systems to identify areas for improvement and ensure the technology's long-term sustainability.

(9) Knowledge exchange: Facilitate the exchange of knowledge and experiences between countries and sectors that have successfully implemented microalgae utilization in wastewater treatment, thus enhancing technology effectiveness and accelerating global adoption.

(10) Community engagement: Involve the community in decision-making processes and the implementation of microalgae-based wastewater treatment to foster social support and ensure project sustainability in the long run.

10.8 Challenges and Future Prospects

The following challenges and prospects in several studies regarding the development of microalgae approaches need to be addressed to enhance efficiency and innovation in wastewater treatment (Srimongkol et al. 2022):

(1) Microalgae species selection: Research should focus on identifying the most efficient microalgae species for pollutant and nutrient removal from wastewater, with high potential for biomass production and valuable product generation.

(2) Growth conditions optimization: Investigate optimal growth conditions for microalgae (e.g., temperature, pH, light intensity, and nutrient concentration) to increase wastewater treatment efficiency and biomass production.
(3) Photobioreactor design optimization: Improve photobioreactor design to enhance light utilization, microalgae growth, as well as ease control of growth conditions and wastewater treatment.
(4) Microalgae separation and product recovery: Research and develop efficient and energy-saving methods for microalgae separation from wastewater and recovery of high-value products (e.g., proteins, fatty acids, pigments, and bioenergy).
(5) Co-cultivation technology: Combine microalgae with other organisms, such as bacteria or fungi, to boost wastewater treatment efficiency and high-value product generation.
(6) By-products valorization: Develop methods to convert by-products from microalgae-based wastewater treatment into valuable commodities like bioenergy, fertilizers, or chemicals.
(7) Integration with other technologies: Incorporate microalgae-based approaches with conventional wastewater treatment or renewable energy systems to create more efficient and sustainable solutions.
(8) Mathematical modeling: Employ mathematical modeling and computer simulations to optimize the design of microalgae-based wastewater treatment systems and predict its performance under various operating conditions.
(9) Monitoring and control technology: Devise real-time monitoring and control technologies for microalgae-based wastewater treatment systems to maximize efficiency and lower operational costs.
(10) Scalability and adaptability: Inspect and construct microalgae-based water treatment systems that can be adapted and scaled up for various types of wastewater and operational scales to allow broad applications across different sectors and environments.

Through research and development in microalgae technology, we can improve efficiency and innovation in wastewater treatment (You et al. 2023), opening up opportunities for new applications and high-value products derived from microalgae.

References

Abdelfattah A, Ali SS, Ramadan H, El-Aswar EI, Eltawab R, Ho SH, Elsamahy T, Li S, El-Sheekh MM, Schagerl M, Kornaros M, Sun J (2023) Microalgae-based wastewater treatment: mechanisms, challenges, recent advances, and future prospects. Environ Sci Ecotechnol 13. Editorial Board, Research of Environmental Sciences. https://doi.org/10.1016/j.ese.2022.100205

Alazaiza MYD, He S, Su D, Abu Amr SS, Toh PY, Bashir MJK (2023) Sewage water treatment using *Chlorella vulgaris* microalgae for simultaneous nutrient separation and biomass production. Separations 10(4). https://doi.org/10.3390/separations10040229

Al-Enazi NM (2020) Salinization and wastewater effects on the growth and some cell contents of *Scenedesmus bijugatus*. Saudi J Biol Sci 27(7):1773–1780. https://doi.org/10.1016/j.sjbs.2020.05.021

Al-Jabri H, Das P, Khan S, Thaher M, Abdulquadir M (2021) Treatment of wastewaters by microalgae and the potential applications of the produced biomass—a review. Water (Switzerland) 13(1). MDPI AG. https://doi.org/10.3390/w13010027

Amaro HM, Salgado EM, Nunes OC, Pires JCM, Esteves AF (2023) Microalgae systems—environmental agents for wastewater treatment and further potential biomass valorisation. J Environ Manag 337. Academic Press. https://doi.org/10.1016/j.jenvman.2023.117678

Amenorfenyo DK, Huang X, Zhang Y, Zeng Q, Zhang N, Ren J, Huang Q (2019) Microalgae brewery wastewater treatment: potentials, benefits and the challenges. Int J Environ Res Public Health 16(11). https://doi.org/10.3390/ijerph16111910

Ankit, Bauddh K, Korstad J (2022) Phycoremediation: use of algae to sequester heavy metals. Hydrobiology 1(3):288–303. https://doi.org/10.3390/hydrobiology1030021

Babich O, Sukhikh S, Larina V, Kalashnikova O, Kashirskikh E, Prosekov A, Noskova S, Ivanova S, Fendri I, Smaoui S, Abdelkafi S, Michaud P, Dolganyuk V (2022) Algae: study of edible and biologically active fractions, their properties and applications. Plants 11(6). MDPI. https://doi.org/10.3390/plants11060780

Barten RJP, Wijffels RH, Barbosa MJ (2020) Bioprospecting and characterization of temperature tolerant microalgae from Bonaire. Algal Res 50. https://doi.org/10.1016/j.algal.2020.102008

Benner P, Meier L, Pfeffer A, Krüger K, Oropeza Vargas JE, Weuster-Botz D (2022) Lab-scale photobioreactor systems: principles, applications, and scalability. Bioprocess Biosyst Eng 45(5):791–813. Springer Science and Business Media Deutschland GmbH. https://doi.org/10.1007/s00449-022-02711-1

Bušić A, Kundas S, Morzak G, Belskaya H, Mardetko N, Šantek MI, Komes D, Novak S, Šantek B (2018) Recent trends in biodiesel and biogas production. Food Technol Biotechnol 56(2):152–173. University of Zagreb. https://doi.org/10.17113/ftb.56.02.18.5547

Bwapwa JK, Jaiyeola AT, Chetty R (2017) Bioremediation of acid mine drainage using algae strains: a review. S Afr J Chem Eng 24:62–70. Elsevier B.V. https://doi.org/10.1016/j.sajce.2017.06.005

Chanana I, Kaur P, Kumar L, Kumar P, Kulshreshtha S (2023) Advancements in microalgal biorefinery technologies and their economic analysis and positioning in energy resource market. Fermentation 9(3). MDPI. https://doi.org/10.3390/fermentation9030202

Cho KH, Wolny J, Kase JA, Unno T, Pachepsky Y (2022) Interactions of *E. coli* with algae and aquatic vegetation in natural waters. Water Res 209. Elsevier Ltd. https://doi.org/10.1016/j.watres.2021.117952

Chugh M, Kumar L, Shah MP, Bharadvaja N (2022) Algal bioremediation of heavy metals: an insight into removal mechanisms, recovery of by-products, challenges, and future opportunities. Energy Nexus 7. Elsevier Ltd. https://doi.org/10.1016/j.nexus.2022.100129

Cotas J, Leandro A, Pacheco D, Gonçalves AMM, Pereira L (2020) A comprehensive review of the nutraceutical and therapeutic applications of red seaweeds (Rhodophyta). Life 10(3). MDPI AG. https://doi.org/10.3390/life10030019

Cui X, Yang J, Feng Y, Zhang W (2020) Simulation of a novel tubular microalgae photobioreactor with aerated tangent inner tubes: improvements in mixing performance and flashing-light effects. Archaea 2020. https://doi.org/10.1155/2020/8815263

Diaconu LI, Covaliu-Mierlă CI, Păunescu O, Covaliu LD, Iovu H, Paraschiv G (2023) Phytoremediation of wastewater containing lead and manganese ions using algae. Biology 12(6):773. https://doi.org/10.3390/biology12060773

Ding N, Li C, Wang T, Guo M, Mohsin A, Zhang S (2021) Evaluation of an enclosed air-lift photobioreactor (ALPBR) for biomass and lipid biosynthesis of microalgal cells grown under fluid-induced shear stress. Biotechnol Biotechnol Equip 35(1):139–149. https://doi.org/10.1080/13102818.2020.1856717

Egbo MK, Okoani AO, Okoh IE (2018) Photobioreactors for microalgae cultivation—an overview. http://www.ijser.org

Fu W, Shu Y, Yi Z, Su Y, Pan Y, Zhang F, Brynjolfsson S (2022) Diatom morphology and adaptation: current progress and potentials for sustainable development. Sustain Horiz 2:100015. https://doi.org/10.1016/j.horiz.2022.100015

Fudge T, Bulmer I, Bowman K, Pathmakanthan S, Gambier W, Dehouche Z, Al-Salem SM, Constantinou A (2021) Microbial electrolysis cells for decentralized wastewater treatment: the next steps. Water (Switzerland) 13(4). https://doi.org/10.3390/w13040445

Geremia E, Ripa M, Catone CM, Ulgiati S (2021) A review about microalgae wastewater treatment for bioremediation and biomass production—a new challenge for Europe. Environments 8(12). MDPI. https://doi.org/10.3390/environments8120136

Goh PS, Lau WJ, Ismail AF, Samawati Z, Liang YY, Kanakaraju D (2023) Microalgae-enabled wastewater treatment: a sustainable strategy for bioremediation of pesticides. Water (Switzerland) 15(1). MDPI. https://doi.org/10.3390/w15010070

Gómez-Sanabria A, Zusman E, Höglund-Isaksson L, Klimont Z, Lee SY, Akahoshi K, Farzaneh H, Chairunnisa (2020) Sustainable wastewater management in Indonesia's fish processing industry: bringing governance into scenario analysis. J Environ Manag 275. https://doi.org/10.1016/j.jenvman.2020.111241

Gururani P, Bhatnagar P, Kumar V, Vlaskin MS, Grigorenko AV (2022) Algal consortiums: a novel and integrated approach for wastewater treatment. Water (Switzerland) 14(22). MDPI. https://doi.org/10.3390/w14223784

Guzzon A, Di Pippo F, Congestri R (2019) Wastewater biofilm photosynthesis in photobioreactors. Microorganisms 7(8). https://doi.org/10.3390/microorganisms7080252

Hakim MM, Patel IC (2020) A review on phytoconstituents of marine brown algae. Futur J Pharm Sci 6(1). https://doi.org/10.1186/s43094-020-00147-6

Hansen HK, Gutiérrez C, Valencia N, Gotschlich C, Lazo A, Lazo P, Ortiz-Soto R (2023) Selection of operation conditions for a batch brown seaweed biosorption system for removal of copper from aqueous solutions. Metals 13(6). https://doi.org/10.3390/met13061008

Huang Q, Jiang F, Wang L, Yang C (2017) Design of photobioreactors for mass cultivation of photosynthetic organisms. Engineering 3(3):318–329. https://doi.org/10.1016/J.ENG.2017.03.020

Karima A, Silalahi MDS, Rinanti A (2018) Increasing content of lipid in tropical microalgae Chlorella sorokiniana and Closterium sp. with variation of nitrogen content and extraction temperature. Matec Web Conf 197(1). https://doi.org/10.1051/matecconf/201819713019

Katam K, Ananthula R, Anumala S, Sriariyanun M, Bhattacharyya D (2022) The impact of light intensity and wavelength on the performance of algal-bacterial culture treating domestic wastewater. E3S Web Conf 355. https://doi.org/10.1051/e3sconf/202235502003

Kautsar, ML, Rinanti, A (2021) Performance evaluation of domestic wastewater treatment: Case study "X building at Jakarta". IOP Conf Ser Earth Environ Sci 802(1).

Kazimierowicz J, Dębowski M, Zieliński M (2023) Taxonomic structure evolution, chemical composition and anaerobic digestibility of microalgae-bacterial granular sludge (M-BGS) grown during treatment of digestate. Appl Sci (Switzerland) 13(2). https://doi.org/10.3390/app13021098

Kollmen J, Strieth D (2022) The beneficial effects of cyanobacterial co-culture on plant growth. Life 12(2). MDPI. https://doi.org/10.3390/life12020223

Koyande AK, Chew KW, Rambabu K, Tao Y, Chu DT, Show PL (2019) Microalgae: a potential alternative to health supplementation for humans. Food Sci Hum Wellness 8(1):16–24. Elsevier B.V. https://doi.org/10.1016/j.fshw.2019.03.001

Lakmali WGM, Athukorala ADSNP, Jayasundera KB (2022) Investigation of Pb(II) bioremediation potential of algae and cyanobacteria strains isolated from polluted water. Water Sci Eng 15(3):237–246. https://doi.org/10.1016/j.wse.2022.04.003

Lamminen M (2021) Nutritional value of microalgae for ruminants and implications from microalgae production. CAB Rev Perspect Agric Vet Sci Nutr Nat Resour 2021. https://doi.org/10.1079/PAVSNNR202116054

Li M, Wang Y, Zhang J, Liu B, Xue H, Wu L, Li Z (2023) Knowledge mapping of high-rate algal ponds research. Water (Switzerland) 15(10). https://doi.org/10.3390/w15101916

Ma X, Jian W (2023) Growth conditions and growth kinetics of *Chlorella vulgaris* cultured in domestic sewage. Sustainability (Switzerland) 15(3). https://doi.org/10.3390/su15032162

Malode SJ, Prabhu KK, Mascarenhas RJ, Shetti NP, Aminabhavi TM (2021) Recent advances and viability in biofuel production. Energy Convers Manag X 10. https://doi.org/10.1016/j.ecmx.2020.100070

Maryjoseph S, Ketheesan B (2020) Microalgae based wastewater treatment for the removal of emerging contaminants: a review of challenges and opportunities. Case Stud Chem Environ Eng 2. https://doi.org/10.1016/j.cscee.2020.100046

Masojídek J, Ranglová K, Lakatos GE, Benavides AMS, Torzillo G (2021) Variables governing photosynthesis and growth in microalgae mass cultures. Processes 9(5). MDPI AG. https://doi.org/10.3390/pr9050820

Mega SJS, Helena S, Safitri I (2021) Optimization of β-carotene production in *Dunaliella salina* using led and different culture media. J Ilm Kelaut 7(1):36–41

Mitra S, Chakraborty AJ, Tareq AM, Emran TB, Nainu F, Khusro A, Idris AM, Khandaker MU, Osman H, Alhumaydhi FA, Simal-Gandara J (2022) Impact of heavy metals on the environment and human health: novel therapeutic insights to counter the toxicity. J King Saud Univ Sci 34(3). https://doi.org/10.1016/j.jksus.2022.101865

Młyński D, Młyńska A, Chmielowski K, Pawełek J (2020) Investigation of the wastewater treatment plant processes efficiency using statistical tools. Sustainability (switzerland) 12(24):1–16. https://doi.org/10.3390/su122410522

Mohsenpour SF, Hennige S, Willoughby N, Adeloye A, Gutierrez T (2021). Integrating microalgae into wastewater treatment: a review. Sci Total Environ 752. Elsevier B.V. https://doi.org/10.1016/j.scitotenv.2020.142168

Moreno Osorio JH, Pollio A, Frunzo L, Lens PNL, Esposito G (2021) A review of microalgal biofilm technologies: definition, applications, settings and analysis. Front Chem Eng 3. https://doi.org/10.3389/fceng.2021.737710

Muriuki Githaiga B, Waithaka P, Withaka PN, George OO (2020) Nutrient removal efficiency of *Spirogyra* sp. and *Oedogonium* sp. in wastewater from Egerton University, Kenya. Isolation and screening for antimicrobials from actinomycetes. View project plant biocides view project. Int J Adv Biotechnol Res 11(2):18–25. https://www.researchgate.net/publication/356978216

Nege AS, Masithah ED, Khotib J (2020) Trends in the uses of *Spirulina* microalga: a mini-review. J Ilm Perikan Kelaut 12(1):149–166. https://doi.org/10.20473/jipk.v12i1.17506

Obaideen K, Shehata N, Sayed ET, Abdelkareem MA, Mahmoud MS, Olabi AG (2022) The role of wastewater treatment in achieving sustainable development goals (SDGs) and sustainability guideline. Energy Nexus 7. https://doi.org/10.1016/j.nexus.2022.100112

Ojha N, Karn R, Abbas S, Bhugra S (2021) Bioremediation of industrial wastewater: a review. IOP Conf Ser Earth Environ Sci 796(1). https://doi.org/10.1088/1755-1315/796/1/012012

Onyeaka H, Miri T, Obileke KC, Hart A, Anumudu C, Al-Sharify ZT (2021) Minimizing carbon footprint via microalgae as a biological capture. Carbon Capture Sci Technol 1. Elsevier Ltd. https://doi.org/10.1016/j.ccst.2021.100007

Oruganti RK, Katam K, Show PL, Gadhamshetty V, Upadhyayula VKK, Bhattacharyya D (2022) A comprehensive review on the use of algal-bacterial systems for wastewater treatment with emphasis on nutrient and micropollutant removal. Bioengineered 13(4):10412–10453. Taylor and Francis Ltd. https://doi.org/10.1080/21655979.2022.2056823

Oslan SNH, Tan JS, Oslan SN, Matanjun P, Mokhtar RAM, Shapawi R, Huda N (2021) *Haematococcus pluvialis* as a potential source of astaxanthin with diverse applications in industrial sectors: current research and future directions. Molecules 26(21). MDPI. https://doi.org/10.3390/molecules26216470

Qader MQ, Shek YA (2023) Using microalga *Scenedesmus quadricauda* for the improvement of municipal wastewater quality. Iraqi J Sci 2178–2188. https://doi.org/10.24996/ijs.2023.64.5.7

Reinecke D, Bischoff LS, Klassen V, Blifernez-Klassen O, Grimm P, Kruse O, Klose H, Schurr U (2023) Nutrient recovery from wastewaters by algal biofilm for fertilizer production part 1: case study on the techno-economical aspects at pilot-scale. Sep Purif Technol 305. https://doi.org/10.1016/j.seppur.2022.122471

Rinanti, A, Kardena, E, Astuti, DI, Dewi, K (2014) Improvement of carbon dioxide removal through artificial light intensity and temperature by constructed green microalgae consortium in a vertical bubble column photobioreactor. Malays J Microbiol 10(1):29–37

Sharma N, Simon DP, Diaz-Garza AM, Fantino E, Messaabi A, Meddeb-Mouelhi F, Germain H, Desgagné-Penix I (2021) Diatoms biotechnology: various industrial applications for a greener tomorrow. Front Mar Sci 8. Frontiers Media S.A. https://doi.org/10.3389/fmars.2021.636613

Siwi, WP, Rinanti, A, Silalahi, MDS, Hadisoebroto, R, Fachrul, MF (2017) Effect of immobilized biosorbents on the heavy metals (Cu 2+) biosorption with variations of temperature and initial concentration of waste. IOP Conf Ser Earth Environ Sci 106(1).

Srimongkol P, Sangtanoo P, Songserm P, Watsuntorn W, Karnchanatat A (2022) Microalgae-based wastewater treatment for developing economic and environmental sustainability: current status and future prospects. Front Bioeng Biotechnol 10. Frontiers Media S.A. https://doi.org/10.3389/fbioe.2022.904046

Sudarshan B, Maheshwar P, Priya P, Sanjay K (2019) Volatile and phenolic compounds in freshwater diatom *Nitzschia palea* as a potential oxidative damage protective and anti-inflammatory source. Pharmacogn Mag 15(64):228. https://doi.org/10.4103/pm.pm_649_18

Touliabah HES, El-Sheekh MM, Ismail MM, El-Kassas H (2022) A review of microalgae and cyanobacteria-based biodegradation of organic pollutants. Molecules 27(3). MDPI. https://doi.org/10.3390/molecules27031141

Valchev D, Ribarova I (2022) A review on the reliability and the readiness level of microalgae-based nutrient recovery technologies for secondary treated effluent in municipal wastewater treatment plants. Processes 10(2). MDPI. https://doi.org/10.3390/pr10020399

Venardou B, O'Doherty JV, Garcia-Vaquero M, Kiely C, Rajauria G, McDonnell MJ, Ryan MT, Sweeney T (2023) In vitro evaluation of brown seaweed *Laminaria* spp. as a source of antibacterial and prebiotic extracts that could modulate the gastrointestinal microbiota of weaned pigs. Animals 13(5). https://doi.org/10.3390/ani13050823

Vieira VMNCS, Engelen AH, Huanel OR, Guillemin ML (2018) Differentiation of haploid and diploid fertilities in *Gracilaria chilensis* affect ploidy ratio. BMC Evol Biol 18(1). https://doi.org/10.1186/s12862-018-1287-x

Wang J, Yan B, Zhang H, Huang L, Wang H, Lan Q, Yin M, Zhu Z, Yan X, Zhu A, Fu C, Wu Y (2022) Heavy metals exacerbate the effect of temperature on the growth of *Chlorella* sp.: implications on algal blooms and management. Processes 10(12). https://doi.org/10.3390/pr10122638

Watanabe MM, Isdepsky A (2021) Biocrude oil production by integrating microalgae polyculture and wastewater treatment: novel proposal on the use of deep water-depth polyculture of mixotrophic microalgae. Energies 14(21). MDPI. https://doi.org/10.3390/en14216992

Xiao M, Li M, Reynolds CS (2018) Colony formation in the cyanobacterium *Microcystis*. Biol Rev 93(3):1399–1420. https://doi.org/10.1111/brv.12401

Yaakob MA, Mohamed RMSR, Al-Gheethi A, Ravishankar GA, Ambati RR (2021) Influence of nitrogen and phosphorus on microalgal growth, biomass, lipid, and fatty acid production: an overview. Cells 10(2):1–19. MDPI. https://doi.org/10.3390/cells10020393

You N, Deng S, Wang C, Ngo HH, Wang X, Yu H, Tang L, Han J (2023) Review and opinions on the research, development and application of microalgae culture technologies for resource recovery from wastewater. Water (Switzerland) 15(6). MDPI. https://doi.org/10.3390/w15061192

Zapata-Mendoza PCO, Berrios-Tauccaya OJ, Tirado-Kulieva VA, Gonzales-Malca JA, Ricse-Reyes DR, Berrios-Zevallos AA, Seminario-Sanz RS (2022) Environmentally friendly technologies for wastewater treatment in food processing plants: a bibliometric analysis. Sustainability (Switzerland) 14(22). MDPI. https://doi.org/10.3390/su142214698

Ziganshina EE, Bulynina SS, Yureva KA, Ziganshin AM (2022) Growth parameters of various green microalgae species in effluent from biogas reactors: the importance of effluent concentration. Plants 11(24). https://doi.org/10.3390/plants11243583

Chapter 11
The Application of the Algal Approach in the Technological, Socioeconomic, and Wastewater Treatment Domains

Dina M. El-Sherif, Alaa El Din Mahmoud, and Ayman N. Saber

Abstract The use of algae as a natural and sustainable method for treating wastewater has attracted considerable attention in recent years because of its potential to minimize the environmental impact of conventional wastewater treatment procedures. The chapter overviews the most cutting-edge algal-based wastewater treatment technologies currently available, including open pond systems, photobioreactors, and hybrid systems. This can potentially provide sustainable solutions for water pollution control while creating new revenue streams. Additionally, it covers the benefit and drawbacks of each technology as well as its economic viability. The chapter emphasizes the significance of considering socioeconomic factors when putting algal-based wastewater treatment systems into place. These elements include stakeholder participation, public perception, and community acceptance. Furthermore, this chapter covers the possibility of using algae-based wastewater treatment to provide new revenue opportunities by producing added-value biofuels, bioplastics, and other products. In addition, we carried out a bibliometric analysis with a particular emphasis on applying algal technologies in wastewater treatment. It seeks to pinpoint gaps, trends, and patterns in the literature on this issue. An additional study is necessary for the optimal use of these technologies and to guarantee their long-term survival in various socio-economic factors.

D. M. El-Sherif (✉)
Collaborative Innovation Center of Food Safety and Quality Control in Jiangsu Province, School of Food Science and Technology, Jiangnan University, Wuxi 214122, China
e-mail: dina.moh55@outlook.com

National Institute of Oceanography and Fisheries (NIOF), Cairo, Egypt

A. E. D. Mahmoud
Environmental Sciences Department, Faculty of Science, Alexandria University, Alexandria 21511, Egypt
e-mail: alaa-mahmoud@alexu.edu.eg

Green Technology Group, Faculty of Science, Alexandria University, Alexandria 21511, Egypt

A. N. Saber
Department of Pesticide Residues and Environmental Pollution, Central Agricultural Pesticide Laboratory, Agriculture Research Center, Giza, Egypt

T. A. Kurniawan and A. Anouzla (eds.), *Algae as a Natural Solution for Challenges in Water-Food-Energy Nexus*, Environmental Science and Engineering,
https://doi.org/10.1007/978-981-97-2371-3_11

Keywords Algae · Sustainability · Wastewater treatment · Bibliometric · Socioeconomic · Design systems

11.1 Introduction

Wastewater refers to the discharged water from the commercial, pharmaceutical and industrial production process, which contains heavy metals (Mahmoud et al. 2016, 2021a, b; Nasr et al. 2017), antibiotics (Abdelfatah et al. 2022; Mahmoud 2022), dyes (Mahmoud et al. 2018, 2020), and organics (Mahmoud 2020; Mahmoud et al. 2022a), with complex components. Wastewater treatment is essential in modern society to protect the environment and human health by removing pollutants before releasing the treated water back into the environment (Mahmoud et al. 2021b). Various conventional and advanced treatment technologies have been developed to treat wastewater, but they often require high energy consumption and generate significant amounts of waste (Din Mahmoud and Fawzy 2016; Nagy et al. 2023).

Microalgae have outstanding potential in removing heavy metals. This is because the cell wall of microalgae is composed of negatively charged carbohydrates and polysaccharides, which can bind tightly to positively charged metals (Wang and Hong 2022). Algae are tiny photosynthetic creatures that can thrive in various conditions, including sewage. They can eliminate these toxins from water and wastewater by using the nutrients and organic substances in the wastewater as a food supply. Algae-based treatment systems are highly efficient in removing carbon, nutrients (nitrogen and phosphorus, organic compounds, and heavy metals from wastewater and aquatic environments. Gonçalves et al. (2017) reported that algae biofilms can achieve high nutrient removal efficiencies, with nitrogen removal efficiencies ranging from 58 to 98% and phosphorus removal efficiencies ranging from 56 to 99%. Lutzu et al. (2021) showed that algae biofilm can remove PAHs from wastewater with removal efficiencies ranging from 45 to 100%, depending on the specific algal species and the growth conditions. In addition, they can be utilized to recover nutrients and water for reuse (Li et al. 2019; Priya et al. 2022). Chlorella, Nitzschia, and Scenedesmus, other species of microalgae like *Cyanophyta coccal, Dactylococcopsis* sp., and Microspora are used to treat fish waste in the water, and *Teilingia* sp., *Anabaena* sp., and *Pinnularia* sp. assist in treating wastewater from slaughterhouses (Mishra et al. 2023).

The recent literature has reported various aspects of circular water treatment using an algal approach (Lutzu et al. 2021). Investigation using various algal strains and their optimal growth conditions for better treatment efficiency; temperature, inoculum ratio, pH, reactor type, light intensity, CO_2, and nutrition levels are just a few of the growth characteristics that are highly reliant on and unique to the type of microalgae being used (Coşgun et al. 2021). In order to obtain maximal treatment effectiveness and high biomass yield, it is crucial to supply the strain with the ideal growth circumstances. Machine learning models can be used to choose the right strains for wastewater treatment and the best growth conditions (Mahmoud et al. 2023). Due to the intermittent nature of microalgal development, optimizing its cultivation is a non-direct process that frequently calls for dynamic optimization techniques. Microalgae cultivation is a difficult biological process to comprehend despite having numerous applications. Numcrous uncontrollable elements, including reactor design and culture conditions, could have an impact on biomass productivity (Coşgun et al. 2021). Table 11.1 summarizes categorical and continuous variables. The Microalgal integration with wastewater treatment greatly advances both direct and indirect Sustainable Development Goals (SDG) progress. SDG-2 (zero hunger), SDG-6 (clean water and sanitation), SDG-7 (cheap and clean energy), SDG-9 (industry, innovation, and infrastructure), SDG-12 (responsible consumption and production), SDG-14 (life below water), and SDG-15 (life on land) are all directly impacted by microalgae. The integration and accomplishment of many SDGs are made possible by microalgal biotechnology (Olabi et al. 2023; Oruganti et al. 2023). Algae biofilm technology offers a sustainable and efficient solution for wastewater treatment while also producing valuable biomass (Barnharst et al. 2023). The technology works through a combination of physical, chemical, and biological processes and can be used for a variety of applications in a circular economy. As research continues, algae biofilm technology will likely become increasingly common for wastewater treatment and biomass production (Mahmoud and Krasucka 2023).

When microalgae production is combined with wastewater treatment, biological flocculation using fungi or bacteria holds many potentials as a low-cost method for harvesting microalgae for bulk biomass production. This is because wastewater can provide the necessary carbon source for flocculating microorganisms. In accordance

Table 11.1 Types of variables controlling the algal approach in wastewater treatment

Type of variables	Examples
Categorical variables	
Microalgae class	*Bacillariophyceae, Coccolithophyceae Chlorodendrophyceae, Eustigmatophyceae, Cryptophyceae, Pavlovophyceae, Trebouxiophyceae, Florideophyceae, Mediophyceae, Porphyridiophyceae, Ulvophyceae*
Cultivation type	Heterotrophic, photoautotrophic, photoheterotrophic, mixotrophic
Pretreatment method	Grounding, lyophilization, sonication, bead milling, liquid nitrogen, autoclaving, microwave, steam explosion
Continuous variables	
Growth medium and contents	C-inorganic (3.3 mol L^{-1}), PO_4 (0.012 mol L^{-1}), N (0.24 mol L^{-1}), C-orgainc (0.20 mol L^{-1}), Tris-base (0.040 mol L^{-1}), K (0.054 mol L^{-1}), Ca (0.012 mol L^{-1}), Fe (0.00018 mol L^{-1}), Si (0.00011 mol L^{-1}), Co (0.000017 mol L^{-1}), Cu (0.000063 mol L^{-1}), B (0.0018 mol L^{-1}), Vitamin B_{12} (3 mol L^{-1}), Vitamin B_1 (3 mol L^{-1}), Biotin (0.080 mol L^{-1})
pH control	3–12
Temperature	10–50 °C
Light intensity	0–700 (μmol m^{-2} s^{-1})
CO_2 content	0–30%

with the number of fatty acids present, around one kg of algae can convert 20% of its biomass into biodiesel (Murthy and Kumar 2023). Proteins and carbohydrates, also found in the harvested algae-like components, are used to make biogas. Abdelzaher et al. (2023) demonstrate that Wadi-Dara on the Red Sea Coast, Egypt produces 5% of its biofuel from algae.

This chapter aims to provide a holistic view on examining the technical feasibility of using algal approaches for wastewater treatment and evaluate the social and economic impacts of implementing algal wastewater treatment. This may involve public perception and acceptance of algal wastewater treatment, cost considerations compared to conventional wastewater treatment methods, and potential economic benefits from byproducts such as algal biomass for biofuels. In this chapter, we also conduct a bibliometric analysis of the existing research on implementing algal approaches for wastewater treatment. Bibliometric analysis involves the statistical analysis of publications to gain insights into research trends and patterns, mapping the literature's growth over time, and analyzing the most prolific authors, institutions, and countries. Moreover, identifying highly cited publications would help identify major gaps and opportunities to guide future research. This chapter also gives recommendations based on the findings from studying the above aspects.

11.2 Bibliometric Analysis

11.2.1 Data Source

On June 3, 2023, we conducted a quantitative bibliometric analysis and network visualization of original research articles published on the implementation of algae in wastewater treatment. On the Scopus database, we performed the following query:

(TITLE-ABS-KEY (alga*) AND TITLE-ABS-KEY ("wastewater treat*")) was used to retrieve highly relevant research published between 1969 and 2022 on the implementation of algae in wastewater treatment from the SCOPUS electronic database. The search yielded a total of 4067 publications.

11.2.2 Data Extraction

The retrieved articles' titles, keywords and abstracts were then screened based on the selection criteria. The data concerning citation, bibliographical information, author keywords, and funding details were exported into a "csv" (comma separated values) format for analysis. Inclusion and exclusion criteria were developed, as shown in Table 11.2.

Table 11.2 Inclusion and exclusion criteria of the retrieved articles

Inclusion criteria	Exclusion criteria
Original research articles	Other publications that are not original research articles
Journals only	Books, book series, trade journal
Articles that were highly relevant to algae and wastewater treatment in the title, keywords and/or abstract)	Articles that did not mention algae or wastewater treatment in the title, keywords and/or abstract)
Articles were in the English language	Other languages

11.2.3 Data Analysis

11.2.3.1 Top Ten Authors

As shown in Fig. 11.1, Muñoz, R. had the highest number (n = 45) of articles related to the implementation of algae in wastewater treatment research, followed by Ruan, R. (44 publications), and Craggs, R.J. (37 publications).

11.2.3.2 Top Fifteen Journals

All included papers were published in 646 journals; Table 11.3 includes the top fifteen contributing journals, which account for 46.84% of the publications on the implementation of algae in wastewater treatment research.

Journal of Bioresource Technology published the most papers (440/4067, 10.82%), followed by Water Science and Technology (227/4067, 5.58%) and Water Research (215/4067, 5.29%).

11.2.3.3 Top Ten Institutions

Figure 11.2 shows the list of the top ten institutions having the highest number of these articles. Harbin Institute of Technology in China was the institution with the highest number (n = 90) of such articles, followed by New Mexico State University in the United States (n = 70), then the National Institute of Water and Atmospheric Research in New Zealand (n = 58).

11.2.3.4 Top Thirteen Countries

A total of 113 countries were identified as the country affiliations of the authors of the original articles on the implementation of algae in wastewater treatment. Figure 11.3

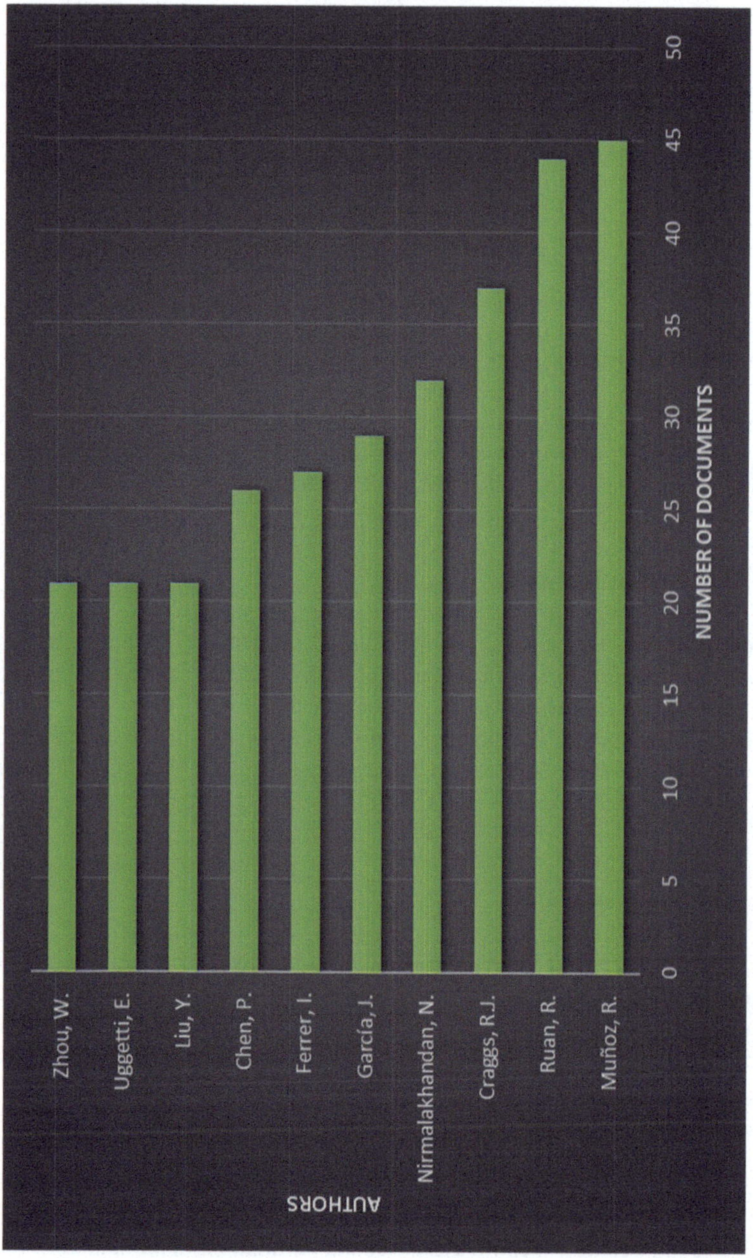

Fig. 11.1 Top 10 most productive authors in implementation of algae in wastewater treatment's research

Table 11.3 Top 15 journals publishing research on the implementation of algae in wastewater treatment research, 1969–2022

Rank	CiteScore 2021	SNIP 2021	SJR 2021	Journal	Country	Categories	Publications, n	Percentage[a], %
1	17.4	2.060	2.354	Bioresource Technology	United Kingdom	Renewable Energy, Sustainability and the Environment	440	10.82
2	3.4	0.648	0.447	Water Science and Technology	United Kingdom	Water Science and Technology	227	5.58
3	18.0	2.499	2.806	Water Research	United Kingdom	Water Science and Technology	215	5.29
4	14.1	2.175	1.806	Science of the Total Environment	Netherlands	Environmental Engineering Pollution	189	4.65
5	11.7	1.603	1.505	Chemosphere	United Kingdom	Environmental Chemistry	137	3.37
6	7.4	1.178	0.878	Algal Research	United Kingdom	Agricultural and Biological Sciences	106	2.61
7	6.6	1.154	0.831	Environmental Science and Pollution Research	Germany	Environmental Chemistry	71	1.75
8	6.0	1.353	0.567	Journal of Environmental Management	United States	Waste Management and Disposal	71	1.75
9	–	0.580	–	Journal of Hazardous Materials	Netherlands	Waste Management and Disposal	70	1.72
10	1.7	0.327	0.240	Desalination and Water Treatment	United States	Water Science and Technology	67	1.65
11	15.8	2.444	1.921	Journal of Cleaner Production	United Kingdom	Renewable Energy, Sustainability and the Environment	67	1.65

(continued)

Table 11.3 (continued)

Rank	CiteScore 2021	SNIP 2021	SJR 2021	Journal	Country	Categories	Publications, n	Percentage[a], %
12	5.7	1.604	1.076	Environmental Technology	United States	Environmental Engineering	63	1.55
13	7.7	1.359	1.015	Ecological Engineering	Netherlands	Environmental Engineering	61	1.50
14	4.8	1.128	0.716	Water (Switzerland)	Switzerland	Water Science and Technology	61	1.50
15	14.8	2.049	2.635	Environmental Science and Technology	United States	Environmental Chemistry	59	1.45

[a] The total number of retrieved papers on implementation of algae in wastewater treatment's research from 1969 to 2022 (N = 4067) was used as the denominator

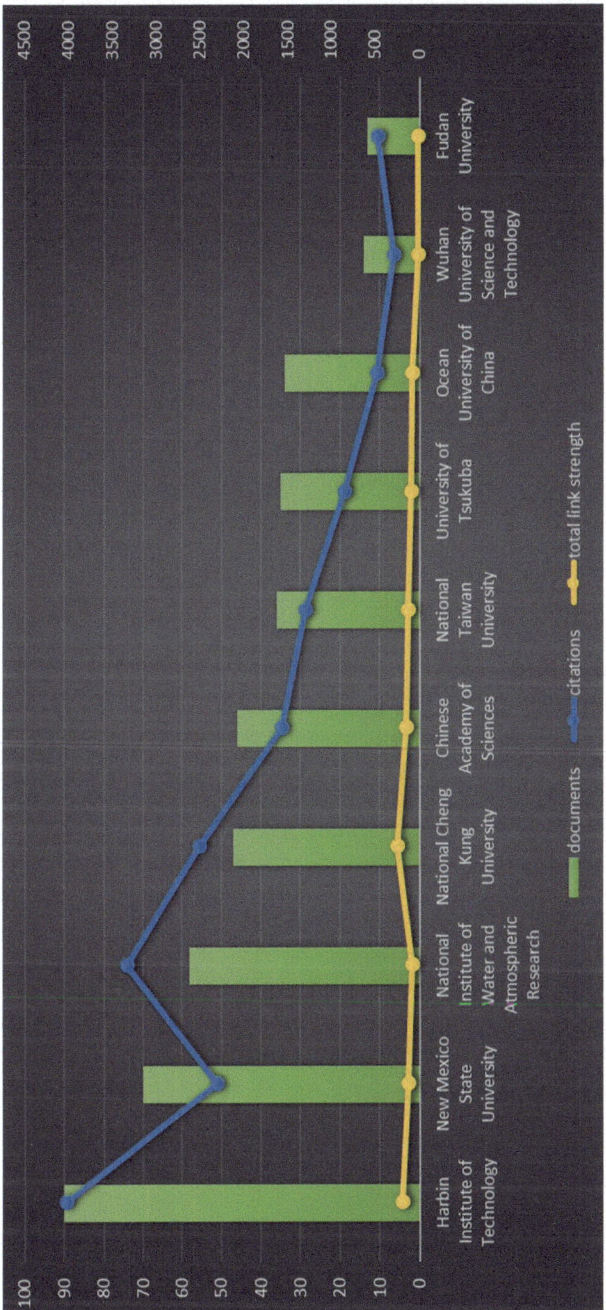

Fig. 11.2 Top 10 organization publishing implementation of algae in wastewater treatment's research

represents the top thirteen countries that publish research articles on the implementation of algae in wastewater treatment, 1969 to 2022 and its citations. Egypt was among the top 30 countries. China has the most publications (873/4067, 21.47%), followed by the USA (707/4067, 17.38%), India (395/4067, 09.71%), and Spain (279/4067, 6.86%). Figure 11.4 shows the collaboration of the top 74 productive countries, which are classified into nine distinct clusters.

However, research from a few African countries (South Africa, Saudi Arabia, and Egypt) has been noted since 2011, contributing less than 5%. Therefore, there is a greater need for research on the implementation of algae in wastewater treatment from non-Western countries' viewpoints. The findings also reveal that the nodes of South Africa, the USA, Sweden, Canada, the United Kingdom and Hong Kong are dark blue, indicating that early research was conducted in these countries.

11.2.3.5 Articles

Figure 11.5, which represents the development of the literature through time and illustrates the expansion of interest in the study field, indicates that between 2010 and 2022, researchers' interest increased. Figure 11.5 shows the evolution and growth of publication production during the previous few years.

11.2.3.6 Research Funders

Out of the 4067 articles on the implementation of algae in wastewater treatment research, only 3012 (74.06%) were externally funded. Figure 11.6 shows all the funding organisations' names (n = 15). Out of these, four were in China (National Natural Science Foundation of China, Fundamental Research Funds for the Central Universities, National Key Research and Development Program of China, and China Postdoctoral Science Foundation), and there were in Europe (Horizon 2020 Framework Programme, European Regional Development Fund, and European Commission).

All of the funding organizations funded at least five original research articles. The National Natural Science Foundation of China was at the top, which funded 456 original research articles, followed by the National Science Foundation in the US-funded 108 research articles.

11.2.3.7 Hotspots

Figure 11.7 shows the bibliometric map of author keywords obtained through co-occurrence analysis, according to which six groups of interrelated keywords were identified. The grouped themes are associated with wastewater treatment and bioremediation (red, green, and violet, nodes). Similarly, terms associated with energy products such as biofuel, biogas, biorefinery, and lipids, which show the approach

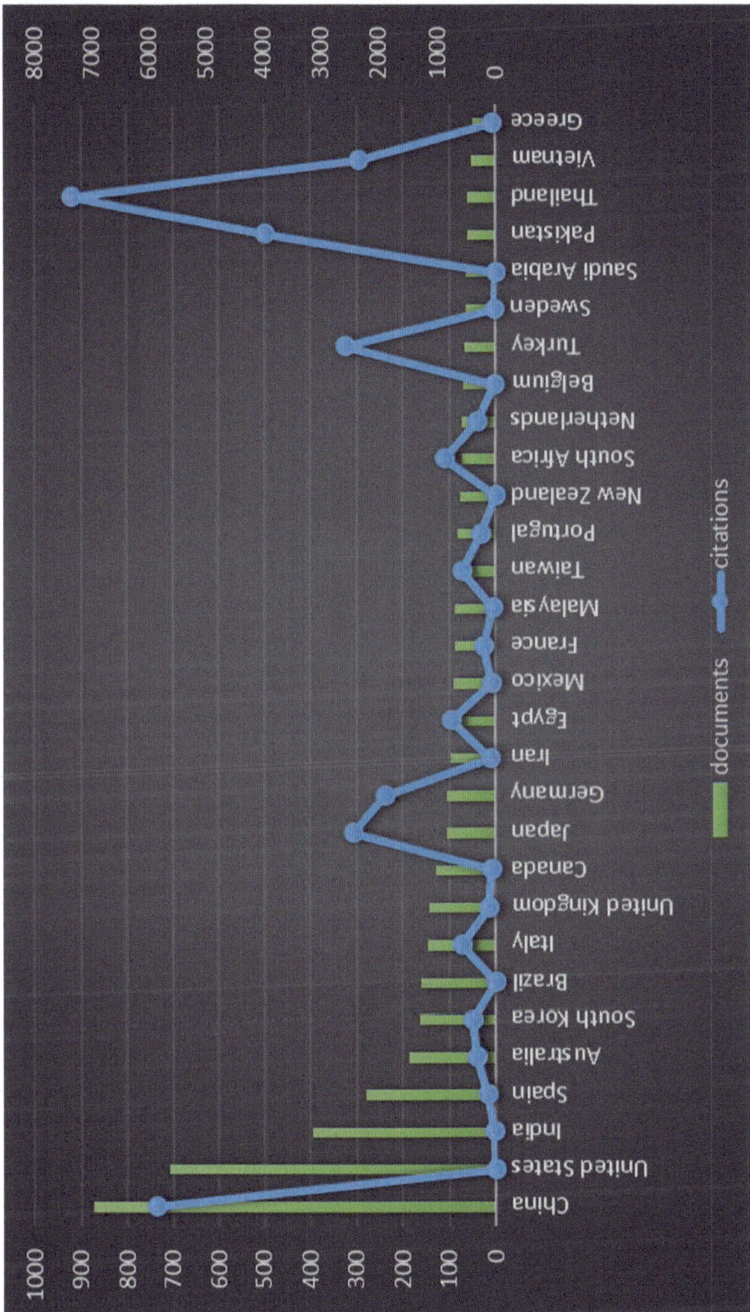

Fig. 11.3 Top 30 countries publishing implementation of algae in wastewater treatment's research

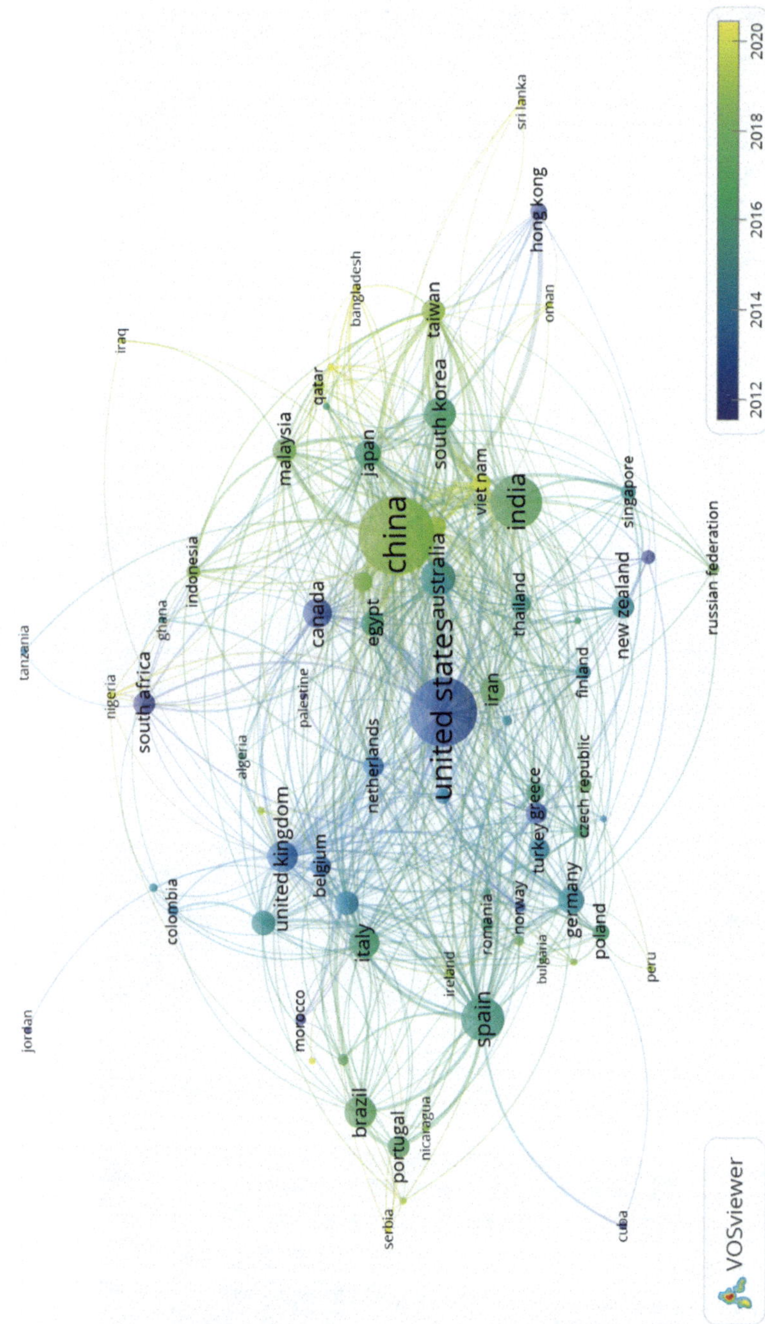

Fig. 11.4 Overlay visualization of countries that contributed to implementation of algae in wastewater treatment's research

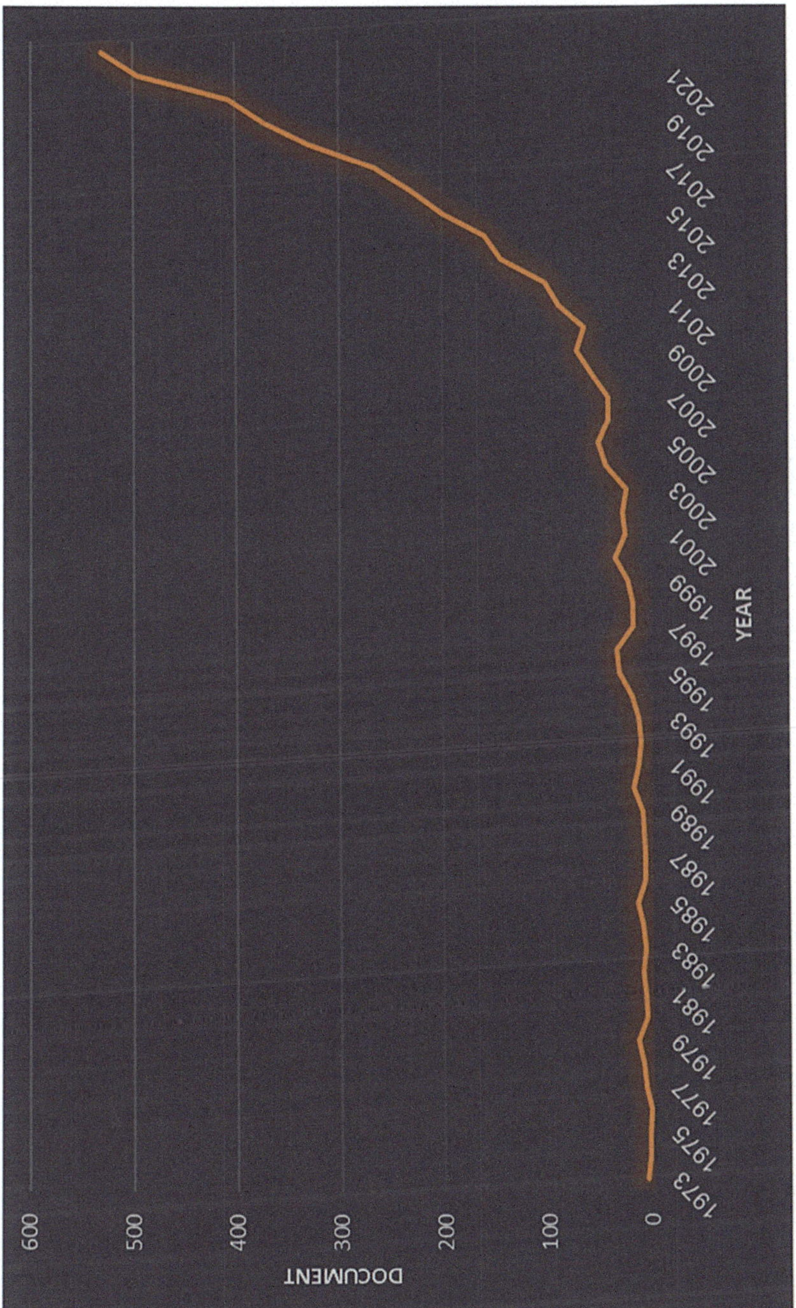

Fig. 11.5 Number of publications on implementation of algae in wastewater treatment's research

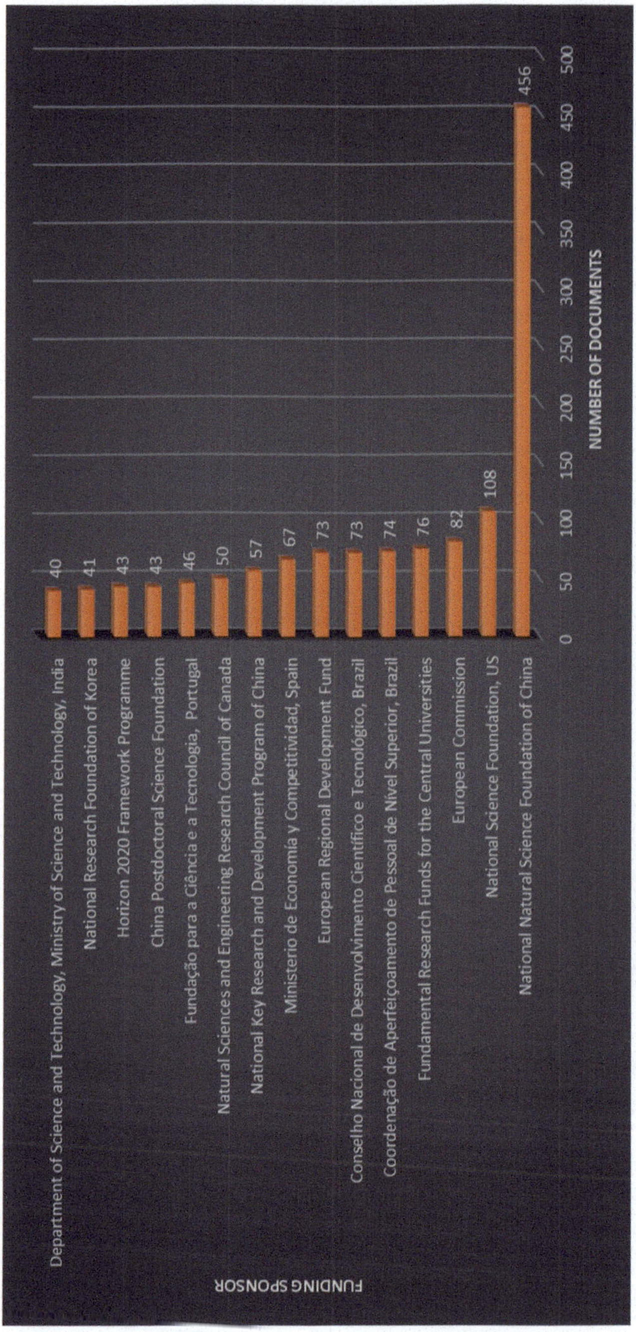

Fig. 11.6 Funders of original research on implementation of algae in wastewater treatment's research

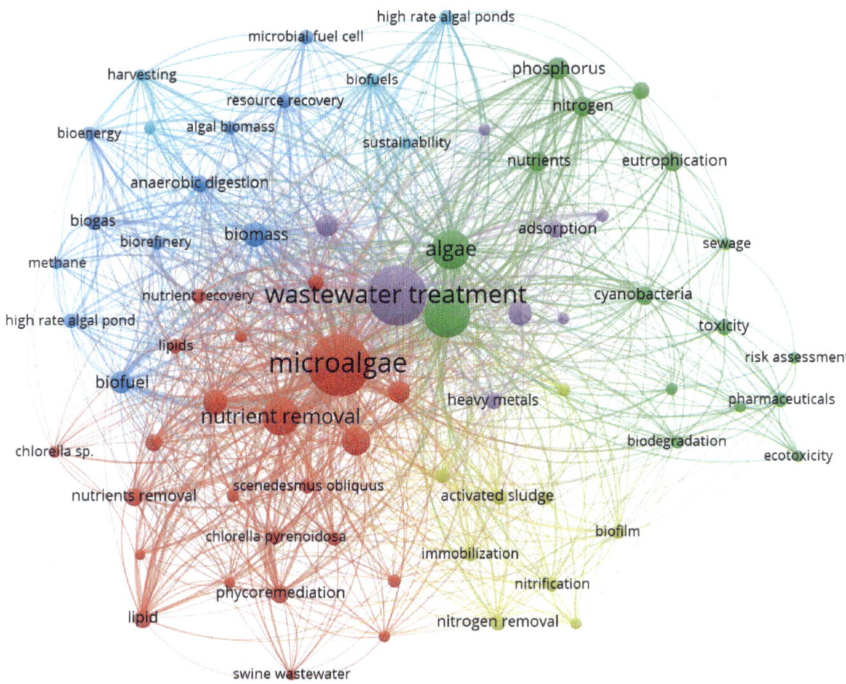

Fig. 11.7 Bibliometric map of author keywords in publications on implementation of algae in wastewater treatment's research. Six theme groups: red, green, blue, yellow, violet, and sky blue

toward the use of microalgal biomass as an alternative to meet energy demands worldwide (blue node). From a broader perspective, it was possible to establish the main keywords reported based on their co-occurrence: microalgae (750), wastewater treatment (675), wastewater (424), algae (300), nutrient removal (265), Chlorella vulgaris (159), biodiesel (121), and biomass (115) were the words used with greater proximity in the published documents according to the data from Scopus reports.

11.3 Circular Wastewater Treatment Using the Algal Approach

Due to the potential of the algal approach to offer a sustainable and affordable alternative for wastewater treatment, circular wastewater treatment employing an algal method has recently received much interest (Ahmad et al. 2022). In addition to being very effective in removing nutrients, organic matter, and heavy metals from wastewater, algae-based treatment systems can also be utilized to recover nutrients and water for reuse. Matamoros et al. (Matamoros et al. 2015) tested the removal of 26 organic microcontaminants from urban wastewater in two pilot high-rate algal ponds. Microalgae-based wastewater treatment systems enable removal efficiency up to 90–99%. It is found that the hydraulic retention time (HRT) of algae-based wastewater treatment is significantly long (> 10 days) compared to bacteria based, just a few hours (Nguyen et al. 2022) (Fig. 11.8).

Selecting the right strain is crucial for growing algae in wastewater because wastewaters with an improper N/P ratio, such as secondary effluent, may prevent algal growth. This N/P ratio has repeatedly led to faster uptake of nitrogen than phosphorus for the growth of algae in wastewater (Nguyen et al. 2022). According to the empirical formula for microalgae (C106H263O110N16P), more nitrogen is required for cell formation.

Due to their rapid growth rates and excellent adaptability to various wastewater streams, Chlorella sp. and Scenedesmus sp. strains have been extensively employed and discovered to be ideal for cultivation in wastewater (Li et al. 2019).

Algae biofilm has emerged as a promising alternative to suspended algae for circular wastewater treatment due to its enhanced stability and higher biomass productivity (Leong et al. 2021). Algae biofilm can be grown on various support materials, including natural fibres, synthetic polymers, and ceramic materials, which provide a surface for algal attachment and growth (Khosravizadeh et al. 2023; Wang et al. 2023).

The use of algae biofilm in circular wastewater treatment offers many advantages, including improved stability, higher biomass productivity, and increased production of value-added products (Barlow et al. 2016; Magalhães et al. 2021).

One of the key advantages of algae biofilm in circular wastewater treatment is its ability to maintain stable treatment performance over a longer period of time compared to suspended algae. Algae biofilm is less susceptible to fluctuations in environmental conditions such as light intensity, pH, and temperature, which can reduce the risk of treatment failure and improve process stability (Valchev and Ribarova 2022). Moreover, algae biofilm can be easily harvested and processed for various applications, making it a valuable resource for circular economy approaches (Alazaiza et al. 2022).

Using support materials for algae biofilm growth in circular wastewater treatment can also improve the efficiency of the treatment process. Support materials provide a surface for algal attachment and growth, which can increase the system's biomass productivity and nutrient removal efficiency (de Assis et al. 2019). The choice of support material can also affect the quality of the treated water, as different materials can have different surface properties and chemical compositions that may affect the adsorption of pollutants and nutrients (Wang and Hong 2022).

Furthermore, using algae biofilm in circular wastewater treatment can also reduce energy consumption and improve the sustainability of the treatment process. Algae biofilm can be grown on various support materials made from renewable resources, such as bamboo and sugarcane bagasse, which can reduce the carbon footprint of the treatment process (Pandey et al. 2014). In addition, the use of algae biofilm can also reduce the need for aeration, which is a significant source of energy consumption in traditional suspended algae systems (Yang et al. 2020).

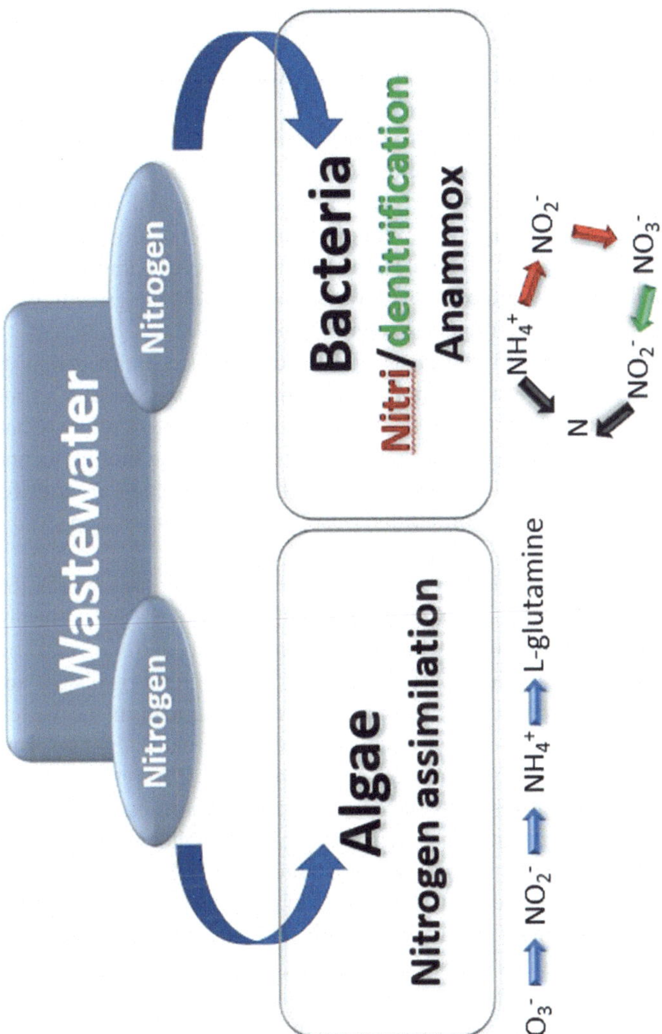

Fig. 11.8 Difference between algae and bacteria for nitrogen removal from wastewater

Algae biofilm involves using algae to form a biofilm on a surface, such as a membrane or a rock bed, to treat wastewater while also producing biomass that can be used for various applications (Murshid et al. 2023). Here are the details and mechanism of this process:

i. **Formation of algae biofilm**: Algae are introduced into the wastewater and allowed to grow on the surface of a membrane or a rock bed. The algae form a biofilm, which is a thin layer of microorganisms that adheres to the surface. The biofilm provides a habitat for the algae to grow and helps remove contaminants from the wastewater (Wang et al. 2022).
ii. **Wastewater treatment**: The algae biofilm removes contaminants from the wastewater through a combination of physical, chemical, and biological processes. The algae absorb nutrients such as nitrogen and phosphorus from the wastewater, which helps to prevent eutrophication in receiving water bodies (El-Maghrabi et al. 2022). The algae also produce oxygen through photosynthesis, which helps to aerate the wastewater and promote the growth of aerobic bacteria that break down organic matter (Chan et al. 2022).
iii. **Biomass production**: As the algae grow, they produce biomass that can be harvested and used for various applications. Biomass can be used as a source of biofuel, as a feedstock for animal feed, or as a fertilizer for agriculture (Chaukura et al. 2022). By producing biomass from wastewater, algae biofilm technology can help to create a circular economy in which waste products are transformed into valuable resources (Mahmoud and Desokey 2022; Mahmoud and Mahmoud 2023).
iv. **Mechanisms of algae biofilm**: The algae biofilm technology works through several mechanisms. Firstly, the biofilm provides a surface area for the algae to grow and also helps to protect them from predators and environmental stresses. Secondly, the biofilm provides a habitat for other microorganisms, such as bacteria and fungi, which can help to break down contaminants in the wastewater. Thirdly, biofilm can help to create a stable environment for the algae by providing a buffer against changes in water quality or temperature (Zhou et al. 2018). While algae- based treatment systems offer many advantages as a sustainable and cost-effective solution for wastewater treatment, several challenges also need to be addressed to implement these systems on a larger scale (El Din Mahmoud 2020). One of the main challenges is the variability of algal growth and the difficulty in controlling algal biomass production (Singh et al. 2011). Algae require specific environmental conditions such as temperature, light, and nutrients to grow, and these conditions can be difficult to maintain in a wastewater treatment system that is subject to fluctuations in influent quality and quantity. As a result, algal biomass production can be unpredictable, affecting the treatment process's efficiency. Another challenge is the cost of implementing and maintaining an algae-based treatment system (Bwapwa et al. 2017). These systems require significant capital investment and ongoing maintenance to ensure optimal performance. In addition, the cost of harvesting and processing algal biomass for various applications can be high, limiting the system's economic feasibility.

Furthermore, regulatory and public acceptance challenges are associated with using algae for wastewater treatment (Chrispim et al. 2019). There may be concerns about the safety of using algae to treat wastewater, especially if the treated water is intended for reuse. Regulatory challenges may also be related to using algal biomass for different applications, which may require additional permits and approvals. Finally, technical challenges are associated with integrating algae-based treatment systems with other technologies (Yadav et al. 2021). For example, the use of membrane filtration to improve the quality of treated water can be complicated by the presence of algae, which can clog the membranes and reduce filtration efficiency (Yadav et al. 2021).

To optimize algae biofilm for economic feasibility in wastewater treatment, several strategies can be considered:

i. **Choice of Algal Species**: The selection of algal species is a critical factor for the economic feasibility of algae-based wastewater treatment. Different algal species have different growth rates, nutrient uptake capacities, and biomass production yields. By selecting fast-growing algal species with high biomass productivity, the treatment process can be optimized for economic feasibility (Chai et al. 2021; Rahman et al. 2020).
ii. **Growth Conditions**: The growth conditions of the algae biofilm can also affect the economic feasibility of the treatment process. By optimizing growth conditions such as light intensity, temperature, and nutrient availability, the biomass productivity and nutrient removal efficiency of the algae biofilm can be increased, which can reduce the overall cost of the treatment process (Abinandan et al. 2018).
iii. **Support Materials**: The choice of support material for algae biofilm growth can also affect the economic feasibility of the treatment process. Support materials made from renewable resources such as natural fibres and agricultural waste products can reduce the treatment process cost while providing a sustainable solution for wastewater treatment (Rosli et al. 2020).
iv. **Harvesting and Processing**: Harvesting and processing algae biomass for various applications can significantly cost algae-based wastewater treatment. By optimizing the harvesting and processing techniques, the cost of producing algal biomass for various applications can be reduced, which can increase the economic feasibility of the treatment process (Singh et al. 2023).
v. **Integration with Other Technologies**: Integration of algae-based treatment systems with other technologies, such as membrane filtration and anaerobic digestion, can also improve the economic feasibility of the treatment process. Membrane filtration can improve the quality of the treated water, while anaerobic digestion can generate biogas and reduce the overall cost of the treatment process (Muthu and Khadir 2022; Sankaran et al. 2014).

Overall, optimizing algae biofilm for economic feasibility in wastewater treatment requires a holistic approach that considers the selection of algal species, growth conditions, support materials, harvesting and processing techniques, and integration with other technologies. Algae biofilm can be a sustainable and cost-effective solution for wastewater treatment by optimising these factors while generating valuable products and mitigating climate change (Huang et al. 2023).

11.4 Techno-economic Analysis of Algal Approaches

Microalgae technology undoubtedly offers various outstanding advantages in terms of its efficacy in energy consumption, nutrition and the environment; nevertheless it can only be advantageous when economically feasible procedures with positive net energy balance are established. Sheehan et al. (1998) mentioned in the closeout report of the National Renewable Energy Laboratory's famous Aquatic Species Programme in the late 1990s, "When the time is accurate, we fully supposed to realize the renewed concern in algae as a source of fuels and other chemicals". However, a comprehensive method to develop microalgae in integrated bio-refinery must reach such impressive results. Aside from biodiesel production from microalgae, a new scenario has evolved recently that examines several value-added products (VAPs) derived from microalgae with a market potential of up to \$600/kg, primarily β carotones (Cezare-Gomes et al. 2019) and astaxanthin (Panis and Carreon 2016). Furthermore, microalgae are expected to have a wide range of applications, including water treatment (bioremediation) (Delrue et al. 2012; Olguín 2012). In recent years, the potential of microalgae to metabolize hazardous substances has been used for the bioremediation of industrial effluents (Kumar et al. 2020).

As a result, there is a significant opportunity for building a sustainable industry where microalgae production comprises a variety of VAPs and may be made independent of freshwater use through effluent remediation. In addition, renewable energy may be generated from cellulose-enriched microalgae biomass following VAP extraction by anaerobic digestion and bio-manure generation (Kumar et al. 2020). As a result, these technologies may be easily scaled up to deliver various high-value goods while minimizing the strain on freshwater supplies. But before any new technology can be predicted to be commercially viable, its economic potential must be carefully examined. For instance, microalgae-based techniques are presently in the early stages of commercialization in several developing countries. Since their future is unknown, it is necessary to conduct industrial feasibility analyses and sensitivity evaluations based on peer-reviewed data and sector expertise. The seeming persuasive advantages may be confirmed, which will aid policy decisions for future investments in research and development of sustainable and affordable microalgal technologies can be developed (Nagarajan et al. 2022).

But not only social and environmental considerations and commercial ones must be made while creating and promoting microalgae biofuels. To evaluate the viability of numerous microalgae- based biofuel methods from an economic standpoint, several techno-economic assessments (TEA) were carried out (Table 11.4).

According to this research, the fuel cost varies significantly according to many assumptions, such as goods, oil contents, biomass productivity, farming methods, production capacity, and conversion technologies (Xin et al. 2016).

11.4.1 Baseline Economics

11.4.1.1 Capital Cost

The system's most significant capital investment was photo bioreactor PBR ($127,755/year), representing 47.7% of the overall project cost. Two centrifuges cost $2555 per year each, and the harvesting system included a flocculation tank for $1278. Since the PBR had four levels and significantly reduced land use, the land's capital investment was at its lowest. 0.58% of the project's overall facility spending was spent on land (1552 dollars annually) (Ansari et al. 2021).

11.4.1.2 Operating Cost

Wastewater provided 100% of the nourishment for microalgae production. As a result, electricity was required, costing $5438 per year. Chemical and electrical costs for microalgae harvesting were $396,985 and $13,372 per year, respectively. The annual depreciation was $26,7633. Overhead expenditures were $110,000 per year (50% of total labor costs), whereas maintenance costs were $161,457 per year (2% of installation costs) (Ansari et al. 2021; Kumar et al. 2020; Xin et al. 2016).

Among the operational costs, the expenses for flocculants were the most significant (33.8%). Equipment depreciation came in second with a cost of 22.8%, followed by operator and manager wages at about 19% of overall operating costs. In addition, maintenance, insurance, and taxes made up 13.7% of all operational expenses (Sathinathan et al. 2023).

11.4.1.3 Revenue

Due to lower hazardous emissions, wastewater treatment costs were reduced in both operating and cover costs. Centrate contributes around 15–20% of the nitrogen load (Fux et al. 2002) and 17% of the phosphorus load (Wang et al. 2010) in the influent wastewater for the aeration basin. Cost reductions from central treatment were determined based on the amount of nitrogen load in the wastewater.

Research has highlighted the use of digestate or wastewater for microalgae cultivation to recycle nutrients (mainly nitrogen and phosphorus). In contrast to a theoretical calculation, the nutrients were eliminated at a rate of 0.058 kg TN/kg dry algae, 1.51 kg COD/kg dry algae and 0.13 kg TP/kg dry algae. The high TP elimination rate was primarily reason by the rain that fell during cultivation. Once this removed P

Table 11.4 Techno-economic findings for the bio-fuel production

Product	Culture system	Oil content (%)	Biomass productivity	Capacity (/year)	Biofuel selling price	Conversion technology	References
TAG	Open pond	25	25 g m^{-2} day^{-1}	1950.58 ha	$8.52/gallon		Davis et al. (2014)
TAG	PBR	25	1.25 kg m^{-3} day^{-1}	1950.58	$18.1/gallon		
Green diesel	Open pond	25	25 gm^{-2} day^{-1}	1950.58	$9.84/gallon	Hydro-treating	
Green diesel	PBR	25	1.25 kg m^{-3} day^{-1}	1950.58	$20.53/gallon	Hydro-treating	
Algae	PBR		48 g m^{-2} day^{-1}	1 M kg	$2.95/kg		Chisti (2007)
Renewable diesel	Open pond	25	25 g m^{-2} day^{-1}	1955 ha	$10.66/– $12.15/ gallon	Hydro-treating	Xin et al. (2016)
Algae	Pond		35 gm^{-2} day^{-1}	1 M kg	$3.8/kg		Chisti (2007)
Renewable diesel				5 billion gallons	$4.35/GGE	HTL	(Davis et al. (2014)
Biodiesel	Solar lit PBR			500 ha	$21.72/L	Solvent extraction	Amer et al. (2011)
Biodiesel	Open POND			500 ha	$3.55/L		
Biodiesel	PBR	20–50	0.65–1.9 kg m^{-3} day^{-1}		$2.8/L	Hydro-treating	Delrue et al. (2012)
Biodiesel	Open pond	20–50	25 g m^{-2} day^{-1}	333.3 ha	$1.68/L	Hydro-treating	
Biodiesel	Raceway + PBR	20–50			$2.69/L	Hydro-treating	
Biodiesel	Open pond	50	60 g m^{-2} day^{-1}		$0.42/L	Oil extraction and transesterification	Nagarajan et al. (2022)
Biodiesel	PBR	25	25 g m^{-2} day^{-1}		$9.89/L		Richardson et al. (2012)
Biodiesel	Open pond	25	1.25 kg m^{-3} day^{-1}		$3.91/L		
Biodiesel	Open pond	50	30 g m^{-2} day^{-1}		$0.97/L		Nagarajan et al. (2022)

(continued)

Table 11.4 (continued)

Product	Culture system	Oil content (%)	Biomass productivity	Capacity (/year)	Biofuel selling price	Conversion technology	References
Biofuel	Algae remnant				\$1.49–1.80/L	Catalytic pyrolysis	Thilakaratne et al. (2014)
Bio-oil using FWC		20–30			\$0.61–3.45/L	Oil extraction	Orfield et al. (2014)
Bio-oil	Race pond	25	22 g m^{-2} day^{-1}	100 or 400 ha	\$7.19–9.93/gal a \$0.67–7.89/gal b	Hexane for extraction	Lundquist et al. (2010)

TAG: intermediate oil; FWC: co-utilization of flue gas and wastewater; ha: hectare (Xin et al. 2016) had been converted to biochar and burned, the P-rich ash could be used as fertilizer. The effectiveness of STP removal was about 77.5%. The STP was 201.5 mg/L at the sludge centrifuge center. It was predicated on the idea that P-rich ash with phosphorite phosphate content would cost similar. A 35% income tax rate was predicted.

The economic study used two indicators: the internal rate of return (IRR) and the net present value (NPV) (Xin et al. 2016).

$$NPV = R \times \frac{1-(1+i)^{-n}}{i} - \text{Initial Investment}$$

where R is the projected net cash inflow per period, i is the needed rate of return per period; and n is the number of periods throughout which the project is expected to generate cash inflows. If the NPV equals zero, the IRR is obtained by solving for i. The cash flows generated by the project over its useful life were used to construct these two economic metrics. Depending on the general economic conditions, both consider the time worth of money through an interest rate.

11.4.1.4 Economic Evaluation Indicators

To exclude the influence of short-term price changes, four average crude oil equivalent prices were chosen: 12 months in 2014 ($93.26), 20 months from January 2014 to August 2015 ($76.6), 24 months from September 2013 to August 2015 ($80.43), and 36 months from September 2012 to August 2015 ($85.15). The IRRs were estimated as 5.41%, 3.07%, 3.63%, and 4.3% for the four different crude oil prices, demonstrating that the alternative energy price significantly influenced IRRs. If P-rich ash had not been employed as P-fertilizer in the system, the IRRs in the selected prices would be 2.13%, 6.08%, 5.05%, and 3.89% (Xin et al. 2016).

11.4.1.5 Sensitivity Analysis

According to the system's sensitivity analysis, IRR is sensitive to six factors: algae biomass concentration (1.33%, 5.41%, 8.11%), chemical cost (1.94%, 5.41%, 8.45%), bio-oil yield ratio (0.78%, 5.41%, 9.60%), LHV ratio (6.61%, 5.41%, 4.14%), PBR investment (3.08%, 5.41%, 8.72%), and crude oil price (6.67%, 5.41%, 5.08%). Per the uncertainty analysis, the system's IRR was most sensitive to PBR cost, bio-oil yield ratio (BYR), chemical cost (Cc), and algal concentration (Sathinathan et al. 2023; Xin et al. 2016).

11.4.2 The Effect of Crucial Technology Improvement

The cost of producing biomass, which was previously calculated to be about $0.33/kg, has been significantly reduced by microalgae grown on citrate. The results of Ventura et al. indicate that nutrients make up 29.51–31.46% of operational costs for the desired production of 1000 dry tonnes of algae annually, or $145.5–$149.3 in nutrients per tonne of dry algae. Generating algae using wastewater can reduce annual expenditures for central treatment ($564,7768) and fertilizer ($550,00). However, the project's IRR (3–5.4%) was still subpar compared to crude oil. It was essential to raise the IRR and earnings (Xin et al. 2016).

The results of the sensitivity analysis show that by reducing the capital cost of the PBR, raising the productivity of algae, improving harvest technology to minimize operating costs, and improving bio-oil conversion efficiency to obtain higher fuel yields, the performance of the wastewater-based algal biofuel production system could be improved (Ventura et al. 2013).

To evaluate the computation results and estimate the impact of significant technology upgrading, four examples were created: (i) Base Case: Actual technology; (ii) Case 1: developing PBR technology; (iii) Case 2: developing harvest technology; and (iv) Case 3: increasing conversion technology. The Base Case was built on the data from the pilot scale. The premise of Cases 1–3 is that technology has advanced. In four scenarios P-rich ash was used as a P-fertilizer to investigate the effectiveness of co-product use (Sathinathan et al. 2023; Xin et al. 2016).

11.4.2.1 Upgrading the PBR

It can be demonstrated that the production of bio-oil has a considerable impact on the IRR due to the high capital cost. If the machinery itself could be improved and tailored, particularly for the production of microalgae, significant cost reductions would be realized. The cost of PBR might be reduced by 10–20% if crude oil average prices were $93.26/bbl and $80.43/bbl, respectively, boosting the IRR from 5.41% to 6.71% and 8.25%, or from 3.63% to 4.83% and 6.26% (Xin et al. 2016).

11.4.2.2 Innovating Harvesting Technology

According to Gao et al. (2010), flocculation can be produced utilizing organic, inorganic, or polymeric flocculants or methods like auto-flocculation, microbiological flocculation, or electro-flocculation. In the baseline phase, conventional flocculation techniques (chemical flocculation) were applied. Contrarily, the cost of the chemicals came to $396,985, or 33.8% of the total. The IRR would increase from 5.41% to 10.1% and 10.7%, or from 3.63% to 8.84% and 9.72%, if chemical expenses could be reduced by 50% and 60%, respectively, and the average price of crude oil was $93.26/bbl and $80.43/bbl (Xin et al. 2016).

Although many techniques for harvesting have been documented, the electro-coagulation-flotation (EC) approach was adopted because of its simplicity and process scale-up capacity. Because it is less sensitive to pH variations, more versatile, and less expensive, EC is a suitable replacement for CC (chemical coagulation). The EC was therefore promoted as a more effective and promising method of harvesting microalgae on a big scale than CC technology (Golzary et al. 2015).

The cost of energy and the replacement of aluminium electrodes were the main operating costs for electro-flocculation. According to calculations made by Lee et al. (2013), the operational cost (electrical and aluminium dissolving) is $0.04 per cubic metre. With this cutting-edge method, the operating cost for a harvesting rate of 5420 m^3 d^1 was $216.8 d^1 or $79,132 per year; the IRR would be 11.8% and significantly improve project viability (Lee et al. 2013; Xin et al. 2016).

Algal harvesting will also become more appealing thanks to additional cutting-edge techniques, such as combining electro-flocculation with dispersed air flotation and microbial flocculation, greatly enhancing project viability. Of course, the characteristics and desired products of the microalgae greatly influence the best way to collect them (Golzary et al. 2015).

11.4.2.3 Improving Productivity

In order to treat wastewater and recover nutrients in the form of microalgal biomass, high rate algal ponds (HRAP) are a part of an advanced pond system that was initially created in the 1950s. The basic idea underlying HRAP wastewater treatment is that aerobic bacterial breakdown of organic substances is driven by microalgal photosynthesis, which in turn produces the oxygen needed for photosynthesis (Oswald 1988). Microalgae are able to directly ingest soluble organic molecules, allowing them to reduce the chemical oxygen requirement when combined with bacterial degradation of organic compounds. Microalgae can develop rapidly in HRAPs due to its architecture, which improves nutrient removal through assimilation into their biomass. As a result, the HRAP treats wastewater in a combined secondary and partial tertiary technique (Goh et al. 2022).

With a variety of wastewaters, including municipal, industrial, and agricultural wastewaters, and at a variety of scales, including full-scale, pilot scale and mesocosm (Cai et al. 2013; Craggs et al. 2012a; De Godos et al. 2009; García et al. 2006), the efficacy of HRAPs for wastewater treatment has been demonstrated. According to study (Craggs et al. 2014), they offer better and more reliable wastewater treatment than facultative ponds and have higher microalgal productivity. HRAP systems have the additional advantage of recovering nutrients from wastewater via algal biomass for use as fertiliser, feed, or a feedstock for biofuel production in addition to the improved wastewater treatment (Craggs et al. 2014). The possibility of using wastewater algae as a feedstock for the production of biofuels has recently drawn a lot of interest (Craggs et al. 2012b; Park et al. 2011; Rawat et al. 2011). Due to their rapid growth rates, year-round growth, and low water and land requirements compared to crop plants, microalgae are thought to be preferable to conventional agricultural crops (Rawat et al. 2011).

The entire algal biomass can be used to produce a variety of goods without further processing. These include food (such as *Spirulina supplements*), energy (such as syngas or bio-oil), soil additives (such as biostimulants), and anaerobic digestion feedstock (such as biostimulants) (Choi et al. 2019; Sutherland et al. 2021). The biomass produced during the algae-based wastewater treatment process can be utilised to make biocrude, biofuel, bioplastics, and animal feed (Goswami et al. 2022; Nguyen et al. 2023), among other products. By replacing components now made from carbon-intensive processes using fossil fuels, these applications could offset or reduce carbon emissions in wastewater treatment facilities. As an illustration, Naaz et al. (2019) showed that two algae species, *Chlorella pyrenoidosa* and *Phormidium*, could be grown in municipal wastewater using a 100 L outdoor attached biofilm system. The generated biomass remained steady at 3.5 g/m^2day with a high lipid content (35% of total solid), while the elimination of nutrients was between 53 and 87% after 6 days of HRT. The generation of biocrude had a higher net energy ratio when compared to biomethane, which is produced by the anaerobic digestion of the biomass that was produced, and biocrude, which is produced by the hydrothermal liquefaction of the biomass (Naaz et al. 2019). Despite having a high energy ratio, the synthesis of biomethane from WWT based on algae may have multiple advantages. For instance, the biogas produced by the anaerobic digester can be used to supply CO_2 for the growth of algae. Nutrients can be added using the digestate. Converti et al. (2009), who paired a mixed sludge anaerobic digester with a photobioreactor to produce biogas with a CH_4 level exceeding 70%, developed this arrangement initially. Additionally, this method enables the direct exploitation of algal biomass (Nguyen et al. 2022).

11.4.2.4 The Combined Effects of Several Technologies Improvement

The influence of various significant technologies, all improving, was calculated based on the 12-month average price in 2014 ($93.26), which relates to the benefits of numerous technological advancements. For instance, the IRR could rise to 18.7% if breakthroughs were made in three essential cultivation, harvesting, and downstream conversion technologies. Specifically, the TPI could be reduced by 20% by upgrading the PBR, harvesting could be done using electricity flocculation, and the yield rate of bio-oil could be increased by 40% by improving downstream conversion technology. The IRR would increase to 15.67% if TPI was decreased by 20% and electrical flocculation was used. The cumulative effects of numerous technology improvements may boost economic viability even in challenging market conditions (Xin et al. 2016).

11.4.3 Environmental Factors

11.4.3.1 Reduce Toxic and Waste

Natural reservoirs and the environment are under tremendous stress due to urbanization and industrialization. Worldwide, 380 trillion L/year of wastewater is produced due to inefficient and excessive water use in commercial, agricultural, and domestic activities. Wastewater is a better option for the growth of algae because it contains many nutrients. Thus, using algae to recover wastewater nutrients is a valuable strategy for sustaining aquatic environments (Goswami et al. 2021). Resource recovery and sustainable development can be accomplished quickly, safely, and affordably using an algal wastewater treatment system. Several types of algae have been successfully used to recover resources from contaminated wastewater and to achieve long-term development objectives. Researchers have investigated the bioremediation of dangerous toxins from the biomass of algae species such as *Nannochloropsis* sp., *Scenedesmus* sp., *Chlorella* sp., *C. sorokiniana*, *Tetraselmis* sp., and *C. pyrenoidosa*, and the wastewater from these algal cultures has been used to create bioenergy (Goswami et al. 2021).

Resources are recovered, and wastewater is treated in HRAPs (high-rate algal ponds). In this system, algae and biomass are valorized for biogas production (as well as for the recovery of nutrients for biofertilizers) (Arashiro et al. 2018). Recent studies have shown that an integrated algal system may successfully recover N (> 42%), P (> 69%), and energy from the treatment of sewage (Abeysiriwardana-Arachchige et al. 2022).

11.4.3.2 Land Use

Algal biomass-based effluent treatment has become a significant replacement with increased effectiveness and long-term growth. Numerous studies on the use of algae in treating municipal, agricultural, and industrial wastewater have shown improved nutrient recovery and pollutant removal efficiency (Maurya et al. 2022). Using algal biomass for wastewater treatment is practical, followed by nutrient recovery and biofuel production (Renuka et al. 2015). Additionally, a symbiotic interaction between bacteria and algae has been discovered, which enhances the removal of dangerous toxins. Algal strains can treat wastewater more effectively and affordably while adhering to stringent recycling and discharge rules (Whitton et al. 2015).

Additionally, after wastewater treatment, algal biomass can be converted into valuable products such as biogas, biofuels, biofertilizers, antiviral and antibacterial chemicals, and animal feed (Raheem et al. 2015). Previous studies have found that algae may recover resources from contaminated water while using biomass to produce bioenergy. Algal strains are crucial ecological players in the long-term evolution of the ecosystem (Bhatt et al. 2022).

11.4.3.3 Greenhouse Gas Emissions

Microalgae could greatly reduce excess energy demand (mostly from aeration units), reducing total energy requirements and indirect CO_2 emissions from wastewater treatment plants by providing oxygen through photosynthesis to heterotrophs or nitrifying bacteria. Traditional nutrient recovery techniques including nitrification–denitrification, phosphorus recovery, activated sludge, and sedimentation must overcome a number of challenges in order to more efficiently and affordably meet strict nutrient release regulations (Kumar and Pal 2015; Whitton et al. 2015). Additionally, the high standards for the circular economy, which include lower carbon emission, lower energy demand, and higher nutrient recovery, are challenging to meet with these traditional methods due to their high energy demand, time requirement, higher carbon and sludge release, and recyclable resource waste (Jin et al. 2014; Sun et al. 2016).

11.4.4 Social Factors

11.4.4.1 Increase in Quality of Life

Safety is a significant concern, both socially and technologically. There are concerns regarding microalgal cultures in wastewater treatment, where chemical processes frequently cause Secondary contamination, loss of valuable potential nutrients (N, P), and the essential for sludge disposal (Bhatt et al. 2022).

11.4.4.2 Employment

Microalgae wastewater treatment generates employment. Due to bioremediation and biomass production advancements, investors are motivated to build infrastructure in Microalgae WWTPs. This might also lead to the establishment of new job possibilities (Bhatt et al. 2022).

11.5 Conclusion and Recommendations

Algal systems may effectively remove nutrients such as nitrogen and phosphorus from wastewater, assisting in meeting discharge limitations. This benefits the environment by lowering chemical burdens in waterways. Algal cultivation combined with wastewater treatment can provide a source of biomass for uses such as biofuels and animal feed, increasing the economic sustainability of algal systems. However, extra processing is frequently required to make the biomass appropriate for these applications. The high capital and ongoing costs of algal wastewater treatment continue to be a barrier, while economics can improve with system scale and connection with existing wastewater infrastructure. More research and optimization of algal strains, cultivating approaches, and harvesting/processing technologies are required to increase the efficiency and cost- effectiveness of algal wastewater treatment. For assessing the true sustainability and advantages of algal solutions, a careful assessment of land requirements, water use, and life cycle assessment factors is required. Still, Future research should focus on optimizing the growth conditions and support materials for algae biofilm to improve treatment efficiency and economic feasibility.

References

Abdelfatah AM, El-Maghrabi N, Mahmoud AED, Fawzy M (2022) Synergetic effect of green synthesized reduced graphene oxide and nano-zero valent iron composite for the removal of doxycycline antibiotic from water. Sci Rep 12(1):1–19

Abdelzaher MA, Farahat EM, Abdel-Ghafar HM, Balboul BAA, Awad MM (2023) Environmental policy to develop a conceptual design for the water–energy–food nexus: a case study in Wadi-Dara on the Red Sea Coast, Egypt. Water 15(4):780

Abeysiriwardana-Arachchige ISA, Delanka-Pedige HMK, Munasinghe-Arachchige SP, Brewer CE, Nirmalakhandan N (2022) Modeling and optimization of an algal-based sewage treatment and resource recovery (STaRR) system. J Environ Chem Eng 10(2):107139

AbinandanS, Subashchandrabose SR, Venkateswarlu K, Megharaj M. 2018. Nutrient removal and biomass production: advances in microalgal biotechnology for wastewater treatment. Crit Rev Biotechnol 38(8):1244–1260. https://doi.org/10.1080/07388551.2018.1472066

Ahmad A, Banat F, Alsafar H, Hasan SW (2022) Algae biotechnology for industrial wastewater treatment, bioenergy production, and high-value bioproducts. Sci Total Environ 806:150585

Alazaiza MYD, Albahnasawi A, Ahmad Z, Bashir MJK, Al-Wahaibi T et al (2022) Potential use of algae for the bioremediation of different types of wastewater and contaminants: Production of bioproducts and biofuel for green circular economy. J Environ Manage 324:116415

Amer L, Adhikari B, Pellegrino J (2011) Technoeconomic analysis of five microalgae-to-biofuels processes of varying complexity. Biores Technol 102(20):9350–9359

Ansari FA, Nasr M, Rawat I, Bux F (2021) Artificial neural network and techno-economic estimation with algae-based tertiary wastewater treatment. J Water Process Eng 40:101761

Arashiro LT, Montero N, Ferrer I, Acién FG, Gómez C, Garfí M (2018) Life cycle assessment of high rate algal ponds for wastewater treatment and resource recovery. Sci Total Environ 622–623:1118–1130

Barlow J, Sims RC, Quinn JC (2016) Techno-economic and life-cycle assessment of an attached growth algal biorefinery. Biores Technol 220:360–368

Barnharst T, Rajendran A, Sun X, Hu B (2023) Process optimization of aquaculture wastewater treatment using a mycoalgae biofilm. Algal Res 70:103020

Bhatt P, Bhandari G, Turco RF, Aminikhoei Z, Bhatt K, Simsek H (2022) Algae in wastewater treatment, mechanism, and application of biomass for production of value-added product. Environ Pollut 309:119688

Bwapwa JK, Jaiyeola AT, Chetty R (2017) Bioremediation of acid mine drainage using algae strains: a review. S Afr J Chem Eng 24:62–70

Cai T, Park SY, Li Y (2013) Nutrient recovery from wastewater streams by microalgae: status and prospects. Renew Sustain Energy Rev 19:360–369

Cezare-Gomes EA, Mejia-da-Silva L del C, Pérez-Mora LS, Matsudo MC, Ferreira-Camargo LS et al (2019) Potential of microalgae carotenoids for industrial application. Appl Biochem Biotechnol 188(3):602–634

Chai WS, Tan WG, Halimatul Munawaroh HS, Gupta VK, Ho SH, Show PL (2021) Multifaceted roles of microalgae in the application of wastewater biotreatment: a review. Environ Pollut 269:116236
Chan SS, Khoo KS, Chew KW, Ling TC, Show PL (2022) Recent advances biodegradation and biosorption of organic compounds from wastewater: microalgae-bacteria consortium—a review. Biores Technol 344:126159
Chaukura N, Muzawazi ES, Katengeza G, Mahmoud AED (2022) Remediation technologies for contaminated soil systems. In: Emerging contaminants in the terrestrial-aquatic-atmosphere continuum: occurrence, health risks and mitigation, pp 353–365
Chisti Y (2007) Biodiesel from microalgae. Biotechnol Adv 25(3):294–306
Choi YY, Patel AK, Hong ME, Chang WS, Sim SJ (2019) Microalgae bioenergy with carbon capture and storage (BECCS): an emerging sustainable bioprocess for reduced CO_2 emission and biofuel production. Bioresour Technol Rep 7:100270
Chrispim MC, Scholz M, Nolasco MA (2019) Phosphorus recovery from municipal wastewater treatment: critical review of challenges and opportunities for developing countries. J Environ Manage 248:109268
Converti A, Oliveira RPS, Torres BR, Lodi A, Zilli M (2009) Biogas production and valorization by means of a two-step biological process. Biores Technol 100(23):5771–5776
Coşgun A, Günay ME, Yıldırım R (2021) Exploring the critical factors of algal biomass and lipid production for renewable fuel production by machine learning. Renew Energy 163:1299–1317
Craggs R, Sutherland D, Campbell H (2012a) Hectare-scale demonstration of high rate algal ponds for enhanced wastewater treatment and biofuel production. J Appl Phycol 24:329–337
Craggs RJ, Lundquist TJ, Benemann JR (2012b) Wastewater treatment and algal biofuel production. In: Algae for biofuels and energy. Springer, pp 153–163
Craggs R, Park J, Heubeck S, Sutherland D (2014) High rate algal pond systems for low-energy wastewater treatment, nutrient recovery and energy production. NZ J Bot 52(1):60–73
Davis RE, Fishman DB, Frank ED, Johnson MC, Jones SB et al (2014) Integrated evaluation of cost, emissions, and resource potential for algal biofuels at the national scale. Environ Sci Technol 48(10):6035–6042
de Assis LR, Calijuri ML, Assemany PP, Berg EC, Febroni LV, Bartolomeu TA (2019) Evaluation of the performance of different materials to support the attached growth of algal biomass. Algal Res 39:101440
De Godos I, Blanco S, García-Encina PA, Becares E, Muñoz R (2009) Long-term operation of high rate algal ponds for the bioremediation of piggery wastewaters at high loading rates. Biores Technol 100(19):4332–4339
Delrue F, Setier P-A, Sahut C, Cournac L, Roubaud A et al (2012) An economic, sustainability, and energetic model of biodiesel production from microalgae. Biores Technol 111:191–200
Din Mahmoud AE, Fawzy M (2016) Bio-based methods for wastewater treatment: green sorbents. In: Phytoremediation: Management of Environmental Contaminants, Volume 3, 209–38
El Din Mahmoud A (2020) Nanomaterials: green synthesis for water applications
El-Maghrabi N, Fawzy M, Mahmoud AED (2022) Efficient removal of phosphate from wastewater by a novel phyto-graphene composite derived from palm byproducts. ACS Omega 7(49):45386–45402
Fux C, Boehler M, Huber P, Brunner I, Siegrist H (2002) Biological treatment of ammonium-rich wastewater by partial nitritation and subsequent anaerobic ammonium oxidation (anammox) in a pilot plant. J Biotechnol 99(3):295–306
Gao S, Yang J, Tian J, Ma F, Tu G, Du M (2010) Electro-coagulation–flotation process for algae removal. J Hazard Mater 177(1–3):336–343
García J, Green BF, Lundquist T, Mujeriego R, Hernández-Mariné M, Oswald WJ (2006) Long term diurnal variations in contaminant removal in high rate ponds treating urban wastewater. Biores Technol 97(14):1709–1715
Goh PS, Lau WJ, Ismail AF, Samawati Z, Liang YY, Kanakaraju D (2022) Microalgae-enabled wastewater treatment: a sustainable strategy for bioremediation of pesticides. Water 15(1):70

Golzary A, Imanian S, Abdoli MA, Khodadadi A, Karbassi A (2015) A cost-effective strategy for marine microalgae separation by electro-coagulation–flotation process aimed at bio-crude oil production: optimization and evaluation study. Sep Purif Technol 147:156–165

Gonçalves AL, Pires JCM, Simões M (2017) A review on the use of microalgal consortia for wastewater treatment. Algal Res 24:403–415

Goswami RK, Mehariya S, Verma P, Lavecchia R, Zuorro A (2021) Microalgae-based biorefineries for sustainable resource recovery from wastewater. J Water Process Eng 40:101747

Goswami RK, Mehariya S, Karthikeyan OP, Gupta VK, Verma P (2022) Multifaceted application of microalgal biomass integrated with carbon dioxide reduction and wastewater remediation: a flexible concept for sustainable environment. J Clean Prod 339:130654

Huang KX, Vadiveloo A, Zhou JL, Yang L, Chen DZ, Gao F (2023) Integrated culture and harvest systems for improved microalgal biomass production and wastewater treatment. Biores Technol 376:128941

Jin L, Zhang G, Tian H (2014) Current state of sewage treatment in China. Water Res 66:85–98

Khosravizadeh N, Lu D, Liao Y, Liao B, Fatehi P (2023) Simulation and experimental analysis of microalgae and membrane surface interaction. Colloids and Interfaces 7(1):24

Kumar R, Pal P (2015) Assessing the feasibility of N and P recovery by struvite precipitation from nutrient-rich wastewater: a review. Environ Sci Pollut Res 22(22):17453–17464

Kumar AK, Sharma S, Dixit G, Shah E, Patel A (2020) Techno-economic analysis of microalgae production with simultaneous dairy effluent treatment using a pilot-scale high volume V-shape pond system. Renew Energy 145:1620–1632

Lee AK, Lewis DM, Ashman PJ (2013) Harvesting of marine microalgae by electroflocculation: the energetics, plant design, and economics. Appl Energy 108:45–53

Leong YK, Huang CY, Chang JS (2021) Pollution prevention and waste phycoremediation by algal-based wastewater treatment technologies: the applications of high-rate algal ponds (HRAPs) and algal turf scrubber (ATS). J Environ Manage 296:113193

Li K, Liu Q, Fang F, Luo R, Lu Q et al (2019) Microalgae-based wastewater treatment for nutrients recovery: a review. Biores Technol 291:121934

Lundquist TJ, Woertz IC, Quinn NWT, Benemann JR (2010) A realistic technology and engineering assessment of algae biofuel production

Lutzu GA, Ciurli A, Chiellini C, Di Caprio F, Concas A, Dunford NT (2021) Latest developments in wastewater treatment and biopolymer production by microalgae. J Environ Chem Eng 9(1):104926

Magalhães IB, Ferreira J, de Siqueira CJ, de Assis LR, Calijuri ML (2021) Technologies for improving microalgae biomass production coupled to effluent treatment: a life cycle approach. Algal Res 57:102346

Mahmoud AED (2020) Graphene-based nanomaterials for the removal of organic pollutants: insights into linear versus nonlinear mathematical models. J Environ Manage 270:110911

Mahmoud AED (2022) Recent advances of TiO_2 nanocomposites for photocatalytic degradation of water contaminants and rechargeable sodium ion batteries. Eng Mater, 757–770

Mahmoud AED, Desokey O (2022) Artificial intelligence-based smart waste management for the circular economy. Environ Manage Technol, 341–358

Mahmoud AED, Krasucka P (2023) Global water challenges and sustainability. Artif Intell Model Water Sustain Glob Challenges, 1–12

Mahmoud AED, Mahmoud E (2023) Circular economy models in water and wastewater sectors. Artif Intell Model Water Sustainab Glob Challenges, 253–272

Mahmoud AED, Stolle A, Stelter M (2018) Sustainable synthesis of high-surface-area graphite oxide via dry ball milling. ACS Sustain Chem Eng 6(5):6358–6369

Mahmoud AED, Fawzy M, Radwan A (2016) Optimization of cadmium (CD^{2+}) removal from aqueous solutions by novel biosorbent. Int J Phytoremediation 18(6):619–625. https://doi.org/10.1080/15226514.2015.1086305

Mahmoud AED, Franke M, Stelter M, Braeutigam P (2020) Mechanochemical versus chemical routes for graphitic precursors and their performance in micropollutants removal in water. Powder Technol 366:629–640

Mahmoud AED, Fawzy M, Hosny G, Obaid A (2021a) Equilibrium, kinetic, and diffusion models of chromium(VI) removal using *Phragmites australis* and *Ziziphus spina-christi* biomass. Int J Environ Sci Technol 18(8):2125–2136

Mahmoud AED, Umachandran K, Sawicka B, Mtewa TK (2021b) Water resources security and management for sustainable communities. Phytochem Military Health: Phytotoxins Nat Defenses, 509–522

Mahmoud AED, Franke M, Braeutigam P (2022a) Experimental and modeling of fixed-bed column study for phenolic compounds removal by graphite oxide. J Water Process Eng 49:103085

Mahmoud AED, Hosny M, El-Maghrabi N, Fawzy M (2022b) Facile synthesis of reduced graphene oxide by *Tecoma stans* extracts for efficient removal of Ni (II) from water: batch experiments and response surface methodology. Sustain Environ Res 32(1):1–16

Mahmoud AED, Khan NA, Hung YT (2023) Advances in artificial intelligence applications in sustainable water remediation. Artif Intell Model Water Sustain Glob Challenges, 53–70

Mahmoud AED, Fawzy M, Khan NA, Artif Intell Model Water Sustain Glob Challenges

Matamoros V, Gutiérrez R, Ferrer I, García J, Bayona JM (2015) Capability of microalgae-based wastewater treatment systems to remove emerging organic contaminants: a pilot-scale study. J Hazard Mater 288:34–42

Maurya R, Zhu X, Valverde-Pérez B, Ravi Kiran B, General T et al (2022) Advances in microalgal research for valorization of industrial wastewater. Biores Technol 343:126128

Mishra B, Tiwari A, Mahmoud AED (2023) Microalgal potential for sustainable aquaculture applications: bioremediation, biocontrol, aquafeed. Clean Technol Environ Policy 25(2):675–687

Murshid S, Antonysamy AJ, Dhakshinamoorthy GP, Jayaseelan A, Pugazhendhi A (2023) A review on biofilm-based reactors for wastewater treatment: recent advancements in biofilm carriers, kinetics, reactors, economics, and future perspectives. Sci Total Environ 892:164796

Murthy GP, Kumar R (2023) Chemical extraction of biodiesel and parametric analysis of DI-diesel engine using algae esterified oil-gasoline blends. Int J Chem Technol 7(1):6–24

Muthu SS, Khadir A (eds) (2022) Membrane based methods for dye containing wastewater

Naaz F, Bhattacharya A, Pant KK, Malik A (2019) Investigations on energy efficiency of biomethane/biocrude production from pilot scale wastewater grown algal biomass. Appl Energy 254:113656

Nagarajan D, Lee D-J, Varjani S, Lam SS, Allakhverdiev SI, Chang J-S (2022) Microalgae-based wastewater treatment—microalgae-bacteria consortia, multi-omics approaches and algal stress response. Sci Total Environ 845:157110

Nagy H, Fawzy M, Hafez E, Mahmoud AED (2023) Potentials of mono- and multi-metal ion removal from water with cotton stalks and date palm stone residuals. Environ Sci Pollut Res 1:1–17

Nasr M, Mahmoud AED, Fawzy M, Radwan A (2017) Artificial intelligence modeling of cadmium(II) biosorption using rice straw. Appl Water Sci 7(2):823–831

Nguyen LN, Aditya L, Vu HP, Johir AH, Bennar L et al (2022) Nutrient removal by algae-based wastewater treatment. Curr Pollut Rep 8(4):369–383

Nguyen LN, Vu MT, Vu HP, Johir MdAH, Labeeuw L et al (2023) Microalgae-based carbon capture and utilization: a critical review on current system developments and biomass utilization. Crit Rev Environ Sci Technol 53(2):216–238

Olabi AG, Shehata N, Sayed ET, Rodriguez C, Anyanwu RC et al (2023) Role of microalgae in achieving sustainable development goals and circular economy. Sci Total Environ 854:158689

Olguín EJ (2012) Dual purpose microalgae–bacteria-based systems that treat wastewater and produce biodiesel and chemical products within a biorefinery. Biotechnol Adv 30(5):1031–1046

Orfield ND, Keoleian GA, Love NG (2014) A GIS based national assessment of algal bio-oil production potential through flue gas and wastewater co-utilization. Biomass Bioenerg 63:76–85

Oruganti RK, Biji AP, Lanuyanger T, Show PL, Sriariyanun M et al (2023) Artificial intelligence and machine learning tools for high-performance microalgal wastewater treatment and algal biorefinery: a critical review. Sci Total Environ 876:162797

Oswald WJ (1988) Micro-algal and waste-water treatment. Micro-algal Biotechnol, 305–328

Pandey PK, Bharti V, Kumar K (2014) Biofilm in aquaculture production. Afr J Microbiol Res 8(13):1434–1443

Panis G, Carreon JR (2016) Commercial astaxanthin production derived by green alga *Haematococcus pluvialis* : a microalgae process model and a techno-economic assessment all through production line. Algal Res 18:175–190

Park JBK, Craggs RJ, Shilton AN (2011) Wastewater treatment high rate algal ponds for biofuel production. Biores Technol 102(1):35–42

Priya AK, Jalil AA, Vadivel S, Dutta K, Rajendran S et al (2022) Heavy metal remediation from wastewater using microalgae: recent advances and future trends. Chemosphere 305:135375

Raheem A, Wan Azlina WAKG, Taufiq Yap YH, Danquah MK, Harun R (2015) Thermochemical conversion of microalgal biomass for biofuel production. Renew Sustain Energy Rev 49:990–999

Rahman A, Agrawal S, Nawaz T, Pan S, Selvaratnam T (2020) A review of algae-based produced water treatment for biomass and biofuel production. Water 12(9):2351

Rawat I, Ranjith Kumar R, Mutanda T, Bux F (2011) Dual role of microalgae: phycoremediation of domestic wastewater and biomass production for sustainable biofuels production. Appl Energy 88(10):3411–3424

Renuka N, Sood A, Prasanna R, Ahluwalia AS (2015) Phycoremediation of wastewaters: a synergistic approach using microalgae for bioremediation and biomass generation. Int J Environ Sci Technol 12(4):1443–1460

Richardson JW, Johnson MD, Outlaw JL (2012) Economic comparison of open pond raceways to photo bio-reactors for profitable production of algae for transportation fuels in the Southwest. Algal Res 1(1):93–100

Rosli SS, Amalina Kadir WN, Wong CY, Han FY, Lim JW et al (2020) Insight review of attached microalgae growth focusing on support material packed in photobioreactor for sustainable biodiesel production and wastewater bioremediation. Renew Sustain Energy Rev 134:110306

Sankaran K, Premalatha M, Vijayasekaran M, Somasundaram VT (2014) DEPHY project: distillery wastewater treatment through anaerobic digestion and phycoremediation—a green industrial approach. Renew Sustain Energy Rev 37:634–643

Sathinathan P, Parab HM, Yusoff R, Ibrahim S, Vello V, Ngoh GC (2023) Photobioreactor design and parameters essential for algal cultivation using industrial wastewater: a review. Renew Sustain Energy Rev 173:113096

Sheehan J, Dunahay T, Benemann J, Roessler P (1998) Look back at the U.S. Department of Energy's Aquatic Species Program: biodiesel from algae; close-out report. Golden, CO

Singh A, Nigam PS, Murphy JD (2011) Mechanism and challenges in commercialisation of algal biofuels. Biores Technol 102(1):26–34

Singh HM, Sharma M, Tyagi VV, Goria K, Buddhi D et al (2023) Potential of biogenic and non-biogenic waste materials as flocculant for algal biomass harvesting: mechanism, parameters, challenges and future prospects. J Environ Manage 337:117591

Sun Y, Chen Z, Wu G, Wu Q, Zhang F et al (2016) Characteristics of water quality of municipal wastewater treatment plants in China: implications for resources utilization and management. J Clean Prod 131:1–9

Sutherland DL, McCauley J, Labeeuw L, Ray P, Kuzhiumparambil U et al (2021) How microalgal biotechnology can assist with the UN Sustainable Development Goals for natural resource management. Curr Res Environ Sustain 3:100050

Thilakaratne R, Wright MM, Brown RC (2014) A techno-economic analysis of microalgae remnant catalytic pyrolysis and upgrading to fuels. Fuel 128:104–112

Valchev D, Ribarova I (2022) A review on the reliability and the readiness level of microalgae-based nutrient recovery technologies for secondary treated effluent in municipal wastewater treatment plants. Processes 10(2):399

Ventura J-RS, Yang B, Lee Y-W, Lee K, Jahng D (2013) Life cycle analyses of CO_2, energy, and cost for four different routes of microalgal bioenergy conversion. Biores Technol 137:302–310

Wang X, Hong Y (2022) Microalgae biofilm and bacteria symbiosis in nutrient removal and carbon fixation from wastewater: a review. Curr Pollution Rep 8(2):128–146

Wang C, Tan Y, Zhu L, Zhou C, Yan X et al (2022) The intrinsic characteristics of microalgae biofilm and their potential applications in pollutants removal—a review. Algal Res 68:102849

Wang Y, Li L, Zhao D, Zhou W, Chen L et al (2023) Surface patterns of mortar plates influence Spirulina platensis biofilm attached cultivation: experiment and modeling. Algal Res 71:103079

Wang L, Min M, Li Y, Chen P, Chen Y et al (2010) Cultivation of green algae Chlorella sp. in different wastewaters from municipal wastewater treatment plant. Appl Biochem Biotechnol 162(4):1174–1186

Whitton R, Ometto F, Pidou M, Jarvis P, Villa R, Jefferson B (2015) Microalgae for municipal wastewater nutrient remediation: mechanisms, reactors and outlook for tertiary treatment. Environ Technol Rev 4(1):133–148

Xin C, Addy MM, Zhao J, Cheng Y, Cheng S et al (2016) Comprehensive techno-economic analysis of wastewater-based algal biofuel production: a case study. Biores Technol 211:584–593

Yadav G, Shanmugam S, Sivaramakrishnan R, Kumar D, Mathimani T et al (2021) Mechanism and challenges behind algae as a wastewater treatment choice for bioenergy production and beyond. Fuel 285:119093

Yang J, Shi W, Fang F, Guo J, Lu L et al (2020) Exploring the feasibility of sewage treatment by algal– bacterial consortia. https://doi.org/10.1080/07388551.2019.1709796

Zhou H, Sheng Y, Zhao X, Gross M, Wen Z (2018) Treatment of acidic sulfate-containing wastewater using revolving algae biofilm reactors: sulfur removal performance and microbial community characterization. Biores Technol 264:24–34948

Chapter 12
Algae Application for Treating Wastewater Contaminated with Heavy Metal Ions

Ali Aghababai Beni, Mina Haghmohammadi, Soheila Delnabi Asl, Seyyed Mostafa Hakimzadeh, and Arman Nezarat

Abstract Contamination of soil with lead and zinc is a widespread environmental problem associated with mining industry plants. Mining activities, production and processing activities, waste disposal and management practices, and atmospheric deposition are the main sources of lead and zinc contamination. Soil pollution and degradation, water pollution and its effects on aquatic ecosystems, plant uptake and crop contamination, health risks, and hazards associated with lead and zinc exposure, and effects on wildlife and biodiversity are the main environmental and health effects of lead and zinc contamination. The article discusses various sampling and analytical techniques, geographic information systems, and remote sensing techniques for monitoring and assessing lead and zinc contamination in soil. Soil remediation techniques, phytoremediation, and other plant-based approaches, as well as human health protection and risk management strategies, are the main mitigation and remediation strategies for lead and zinc contamination. Finally, the new techniques and approaches for mining waste management and remediation, green chemistry and sustainable practices in the mining industry, and the use of artificial intelligence for lead and zinc pollution control.

A. Aghababai Beni (✉)
Department of Chemical Engineering, Shahrekord Branch, Islamic Azad University, Shahrekord, Iran
e-mail: aliaghababai@yahoo.com

M. Haghmohammadi
Department of Organic Chemistry, Faculty of Chemistry, University of Semnan, Semnan, Iran

S. Delnabi Asl
Department of Analytical Chemistry, Faculty of Chemistry, Shahid Madani University, Azarbaijan, Iran

S. M. Hakimzadeh
Department of Chemical Engineering, Faculty of Chemical Engineer, University of Amirkabir, Tehran, Iran

A. Nezarat
Department of Chemical Engineering, Mahshahr Branch, Islamic Azad University, Mahshahr, Iran

T. A. Kurniawan and A. Anouzla (eds.), *Algae as a Natural Solution for Challenges in Water-Food-Energy Nexus*, Environmental Science and Engineering,
https://doi.org/10.1007/978-981-97-2371-3_12

Keywords Mining industry · Lead · Zinc · Soil pollution · Health effects · Monitoring · Pollution control · Sustainable practices

12.1 Introduction

12.1.1 Background and Significance of Wastewater Treatment

Wastewater treatment plays a critical role in addressing the escalating challenges of water pollution and ensuring the sustainable management of our precious water resources. With rapid industrialization, urbanization, and population growth, the generation of wastewater containing various contaminants, including heavy metal ions, has become a significant environmental concern (Cao et al. 2010).

Wastewater refers to any water that has been used and discarded, carrying with it a wide range of impurities, pollutants, and harmful substances. If left untreated, these contaminants can have devastating consequences for the environment and human health. Wastewater often contains heavy metal ions such as lead, cadmium, mercury, and chromium, which are known to be toxic and pose serious risks to ecosystems and public health (Abdollahi et al. 2022). The significance of wastewater treatment cannot be overstated. It aims to remove or reduce the concentration of pollutants and contaminants present in wastewater to safe levels before it is discharged into natural water bodies or reused for various purposes. The primary goals of wastewater treatment are to protect the environment, safeguard public health, and promote sustainable water resource management (Esmaeili and Aghababai Beni 2015).

The untreated discharge of wastewater containing heavy metal ions into rivers, lakes, or oceans can have detrimental effects on aquatic ecosystems. Heavy metal contamination can accumulate in sediments, bioaccumulate in aquatic organisms, and disrupt the delicate balance of aquatic ecosystems. This can lead to the decline of fish populations, damage to aquatic plants, and overall ecological degradation. By treating wastewater contaminated with heavy metal ions, we can minimize the negative impacts on ecosystems and preserve biodiversity (Rodrigues et al. 2012; Kayan et al. 2021).

Untreated wastewater contaminated with heavy metal ions can pose severe risks to human health, both directly and indirectly. Direct exposure to heavy metals through contaminated water sources or consumption of contaminated food can lead to a range of health problems, including organ damage, developmental disorders, and various acute and chronic illnesses. Additionally, heavy metals can enter the food chain, bioaccumulate in crops, and eventually reach human populations, posing long-term health risks. Proper wastewater treatment ensures that heavy metal ions are removed or reduced to safe levels, safeguarding public health and well-being (Winters et al. 2017). Water scarcity is an increasingly pressing global issue, making the effective management and conservation of water resources paramount. Wastewater treatment plays a crucial role in this context by facilitating the safe reuse of treated wastewater for various purposes, such as irrigation, industrial processes, and even potable water supply in water-stressed regions. By treating wastewater contaminated with heavy metal ions, we can recover valuable water resources and alleviate the pressure on freshwater sources (Peng et al. 2022).

12.1.2 Overview of Heavy Metal Contamination in Wastewater

Heavy metal contamination in wastewater poses significant environmental and public health concerns worldwide. Wastewater generated from various industrial activities, municipal sources, and agricultural runoff can contain elevated levels of heavy metal ions. These heavy metals, including lead (Pb), cadmium (Cd), mercury (Hg), and chromium (Cr), among others, are known for their persistence, toxicity, and potential for bioaccumulation in ecosystems (Es-Sahbany et al. 2019).

The sources of heavy metal contamination in wastewater are diverse. Industrial discharges, such as those from metal plating, mining operations, and manufacturing processes, are major contributors. Municipal wastewater can also carry heavy metals from domestic sources, including household cleaning products, personal care items, and corroded plumbing systems. In addition, agricultural activities involving the use of fertilizers and pesticides can lead to the runoff of heavy metals into water bodies (Orfanidis et al. 2015).

The presence of heavy metal ions in wastewater poses serious risks to both the environment and human health. Heavy metals are non-biodegradable and can persist in water bodies, soil, and sediments for extended periods. When released into the environment, they can accumulate in aquatic organisms, enter the food chain, and cause toxic effects in various species, including fish, birds, and mammals. Furthermore, the discharge of heavy metal-contaminated wastewater into water bodies can contaminate drinking water sources, leading to potential health risks for humans (Song et al. 2023).

Addressing the challenge of heavy metal contamination in wastewater requires effective treatment methods that can remove or reduce the concentration of these pollutants to safe levels. Algae-based applications have gained attention as a promising approach due to the remarkable ability of algae to accumulate heavy metals through processes such as bioaccumulation and biosorption. These algae-based systems offer the potential for sustainable and cost-effective wastewater treatment, simultaneously addressing heavy metal pollution and promoting resource recovery (Al Ketife et al. 2020).

12.1.3 *Importance of Algae in Wastewater Treatment*

Algae, a diverse group of photosynthetic organisms, have emerged as a promising tool in wastewater treatment due to their unique ability to remove contaminants, including heavy metal ions. The use of algae in wastewater treatment holds significant importance, offering numerous advantages over conventional treatment methods. These microorganisms possess exceptional capabilities for nutrient uptake, carbon fixation, and the biosorption or bioaccumulation of heavy metals, making them an effective and sustainable solution for treating wastewater contaminated with heavy metal ions (Al Ketife et al. 2020; Almomani and Bohsale 2021).

One of the primary advantages of using algae in wastewater treatment is their remarkable capacity for nutrient removal. Algae can assimilate nutrients such as nitrogen and phosphorus from wastewater, which are the key culprits behind eutrophication in natural water bodies. Through a process known as nutrient uptake, algae effectively reduce nutrient concentrations, thereby mitigating the harmful effects of eutrophication and promoting the restoration of water quality (Alazaiza et al. 2022).

Moreover, algae have demonstrated a unique affinity for heavy metal ions, making them efficient bioaccumulators and biosorbents of these contaminants. Through various mechanisms, such as ion exchange, complexation, and adsorption onto algal cell surfaces, algae can remove heavy metals from wastewater. This ability not only reduces heavy metal concentrations but also helps prevent their release into the environment, safeguarding both aquatic ecosystems and human health (Fazal et al. 2018).

Additionally, algae-based wastewater treatment systems offer several environmental benefits. Algae utilize carbon dioxide during photosynthesis, mitigating greenhouse gas emissions and aiding in carbon sequestration. Furthermore, they produce oxygen as a byproduct, contributing to the oxygenation of water bodies and supporting aquatic life (Mustafa et al. 2021).

The importance of algae in wastewater treatment extends beyond their efficacy in pollutant removal. Algae cultivation can be relatively simple and cost-effective, particularly in the case of non-potable wastewater treatment applications. Algae can thrive in a range of environments, including open ponds or closed photobioreactors, and can be cultivated using wastewater as a nutrient source, thereby reducing the need for additional inputs. This aspect contributes to the potential for sustainable and economically viable wastewater treatment solutions (Saúco et al. 2021).

12.2 Algae Species and Their Characteristics

12.2.1 Overview of Different Algae Species Used in Wastewater Treatment

Algae encompass a wide range of species with varying characteristics, and several of them have shown potential in wastewater treatment applications. Here, we provide a general overview of different algae species commonly used or studied in the context of wastewater treatment, based on commonly known information.

12.2.1.1 Chlorella Vulgaris

Chlorella vulgaris is a widely studied green microalga that has gained significant attention in the field of wastewater treatment, particularly for its remarkable ability to absorb and accumulate heavy metal ions. Chlorella vulgaris algae are photosynthetic organisms and primarily grow through a process called autotrophic growth (Oh et al. 2022). Chlorella vulgaris requires several essential nutrients to support its growth, including carbon dioxide (CO_2), water, sunlight, and various inorganic nutrients such as nitrogen, phosphorus, and trace elements (Li et al. 2023). These nutrients are obtained from the surrounding environment. Chlorella vulgaris utilizes photosynthesis to convert sunlight energy into chemical energy. Chlorophyll, the pigment present in the algae, captures light energy, which is then used to convert carbon dioxide and water into glucose and oxygen (Al-Khiat et al. 2023). This process occurs in specialized structures called chloroplasts within the algal cells. Chlorella vulgaris can reproduce through both asexual and sexual methods (Wang et al. 2021). In asexual reproduction, the algae undergoes cell division, producing genetically identical daughter cells. This process allows for rapid population growth under favorable conditions. Sexual reproduction involves the fusion of gametes from different individuals, leading to the formation of zygotes that can develop into new individuals (Alagawany et al. 2021). The growth of Chlorella vulgaris is influenced by various environmental factors. Adequate sunlight is essential for photosynthesis and biomass production (Matula and Nabity 2021). Temperature, pH level, nutrient availability, and light intensity are critical factors that can affect the growth rate and overall productivity of the algae. Chlorella vulgaris can be cultivated in various systems, including open ponds, closed photobioreactors, and raceway ponds. The cultivation method depends on factors such as scale of production, nutrient management, and the desired end-use of the algae (Ostera et al. 2020).

The wall structure of Chlorella vulgaris is composed of several layers that provide protection and support to the cell (Azam et al. 2022). Here's a breakdown of the wall structure: The outermost layer of the Chlorella vulgaris cell wall is composed of glycoproteins. These proteins have sugar molecules attached to them, which help in cell recognition and adhesion. Beneath the outer glycoprotein layer, there is a microfibrillar layer composed of cellulose (Barghchi et al. 2023). This layer provides rigidity and strength to the cell wall. Below the microfibrillar layer, there is a layer of sporopollenin. Sporopollenin is a chemically resistant polymer that provides protection against environmental stressors, such as UV radiation, desiccation, and chemical attacks. The innermost layer of the Chlorella vulgaris cell wall is another layer of microfibrils. These microfibrils are made up of cellulose and provide additional structural support (Xaaldi Kalhor et al. 2016).

Chlorella vulgaris algae exhibit distinctive functional groups in FTIR analysis, with specific peaks indicating their presence. The hydroxyl groups (–OH) appear as a broad peak in the range of 3200–3600 cm^{-1}, carbonyl groups (C=O) are observed around 1650–1750 cm^{-1}, amino groups ($-NH_2$) exhibit absorption around 3200–3500 cm^{-1}, phosphate groups (PO^{3-}) display peaks in the region of 900–1200 cm^{-1}, and alkene/alkyne groups (C=C and C≡C) manifest absorption bands at approximately 1600–1680 cm^{-1} and 2100–2260 cm^{-1}, respectively (Brar et al. 2022). These distinct peaks allow for the identification and analysis of the functional groups present in Chlorella vulgaris algae, providing insights into their biochemical composition and potential applications (Wang et al. 2021).

The surface morphology of Chlorella vulgaris algae appears as a textured and uneven surface. The algal cells typically display a spherical or oval shape, with a smooth or slightly rough surface texture. The cell wall can appear as a distinct outer layer surrounding the cell, often exhibiting microfibrils or fine striations. The surface may also feature small pores or openings, through which nutrients and gases can exchange (Bhuyar et al. 2020; Wanderley et al. 2020).

Adsorption of heavy metal ions by microalgae, particularly Chlorella vulgaris and other related species, has emerged as a promising and eco-friendly approach for heavy metal removal from aqueous environments. Several research articles discussed the efficient adsorption capabilities of microalgae in different contexts, shedding light on their potential as effective bioremediation agents.

The study by Sayadi et al. (2019) investigated the adsorption mechanism of Cd(II) using modified Spirulina platensis and Chlorella vulgaris powder. The modified biosorbents demonstrated a high adsorption efficiency for heavy metals, particularly cadmium. The results indicated that these algae-based adsorbents have the potential to effectively remove heavy metal ions from aqueous solutions.

In the research conducted by Yousefi et al. (2023), the bioremediation potential of Chlorella vulgaris and Scendesmus obliquus for heavy metals (copper and iron) and nutrients (nitrate and phosphate) removal from aqueous media was evaluated. The microalgae species exhibited impressive removal percentages for heavy metals, with Chlorella vulgaris and Scendesmus obliquus removing up to 99.9% and 99.19% of copper and iron, respectively. The presence of heavy metals also led to an increase in nutrient uptake by the microalgae, demonstrating their ability to simultaneously address multiple water quality issues.

In the study by Luo et al. (2023), the stress responses of Chlorella vulgaris to copper and oxytetracycline (OTC) were examined. Interestingly, OTC was found to mitigate the toxicity of copper in the combined stress condition, showcasing the potential of microalgae to adapt and alleviate heavy metal stress.

Andrzej Bajguz's research (Bajguz 2010) highlighted the concentration-dependent increase in antioxidant enzyme activities in Chlorella vulgaris following exposure to heavy metals (copper, lead, and cadmium). Moreover, the application of exogenous brassinolide further activated enzymatic and non-enzymatic systems, enhancing the microalgae's resistance to heavy metal stress. This suggests that microalgae can harness their antioxidant defense mechanisms to counteract heavy metal-induced oxidative stress.

Saavedra et al. (2019) explored the influence of organic matter (OM) and CO_2 addition on the bioremediation potential of Chlorella vulgaris and Scenedesmus almeriensis in a multimetallic matrix. The presence of OM was found to impact the biosorption capacity of heavy metals, while CO_2 addition inhibited Chlorella vulgaris growth. However, metal recovery using different eluents (HCl, NaOH, and $CaCO_2$) showed promise, with HCl 0.1 M providing the highest recovery efficiencies for various heavy metals.

These research articles collectively demonstrate the strong adsorption capabilities of Chlorella vulgaris and related microalgae for heavy metal removal from aqueous environments. The findings underscore the potential of microalgae-based bioremediation as a sustainable and environmentally friendly solution to address heavy metal pollution in various water bodies. The efficient removal of heavy metals by these microorganisms holds great promise for their application in industrial wastewater treatment and environmental remediation projects.

12.2.1.2 Spirulina Platensis

Spirulina platensis, commonly known as spirulina, is a type of cyanobacteria or blue-green algae. It is a filamentous organism that forms long, spiral-shaped chains of cells. While it does not have a true cell wall like plants or fungi, it has a distinct outer covering called the sheath (Alagawany et al. 2021). The sheath of Spirulina platensis is primarily composed of a complex mixture of polysaccharides, proteins, and lipids. It serves as a protective layer surrounding the filamentous chains of cells. The sheath provides mechanical support to the cells and helps maintain their structural integrity (Nakamoto et al. 2023). The polysaccharides present in the sheath of Spirulina platensis are mainly composed of glucose and other sugar molecules. These polysaccharides contribute to the rigidity of the cell wall and protect the organism from environmental stresses such as desiccation, high salinity, and temperature fluctuations (Rahnama et al. 2023).

Proteins are another essential component of the sheath. They are involved in various functions, including adhesion of the cells to each other, as well as to surfaces or substrates. The proteins also play a role in cell-to-cell communication and signal transduction processes within the organism (Abdel-Latif et al. 2022). Lipids, particularly glycolipids and phospholipids, are found in the sheath of Spirulina platensis. These lipids contribute to the fluidity and flexibility of the sheath, allowing the organism to adapt to changes in its environment. Lipids also play a role in energy storage and membrane stability (Alagawany et al. 2021).

The sheath of Spirulina platensis can vary in thickness and composition depending on environmental conditions. Factors such as light intensity, nutrient availability, and temperature can influence the structure and properties of the sheath. Changes in these conditions can lead to alterations in the cell wall composition, affecting the overall physiology and growth of Spirulina platensis (Ai et al. 2023). Spirulina platensis algae are photosynthetic organisms that primarily grow in freshwater environments. They have unique characteristics that enable them to thrive and multiply under specific conditions. Spirulina platensis algae prefer warm and alkaline environments. They can be found in natural bodies of water such as lakes, ponds, and alkaline lakes, as well as in human-made cultivation systems like open ponds or closed photobioreactors. Spirulina platensis requires a balanced supply of nutrients for optimal growth. The primary nutrients needed include carbon dioxide (CO_2), nitrogen, phosphorus, potassium, iron, and trace minerals. These nutrients can be obtained from the surrounding water or supplemented in the cultivation system (Adamia et al. 2018). Spirulina platensis algae require abundant sunlight or artificial light for photosynthesis. They grow best in temperatures ranging from 25 to 35 °C (77–95 °F). However, they can tolerate a wide range of temperatures, including both higher and lower extremes (Alagawany et al. 2021; Lafarga et al. 2021).

Spirulina platensis can be cultivated using different methods, including open ponds and closed photobioreactors. In open ponds, the algae are grown in shallow basins or raceways, where they are exposed to sunlight and agitated to prevent sedimentation and promote mixing. In closed systems, such as photobioreactors, the growth conditions can be more controlled, including light intensity, temperature, and nutrient supply (Nakamoto et al. 2023). Spirulina platensis reproduces through a process called binary fission, where one cell divides into two. Under favorable conditions, the algae multiply rapidly, forming long, spiral-shaped chains or filaments. These filaments can intertwine and create a dense biomass within the cultivation system.

Once the Spirulina platensis biomass has reached a desirable density, it is harvested for various purposes. Harvesting methods include filtration, sedimentation, or centrifugation. After harvesting, the algae can be processed into different forms, such as powder or flakes, for various applications.

The amide I and amide II peaks are indicative of protein content and are typically observed around 1650- 1670 cm^{-1} and 1540–1560 cm^{-1}, respectively. The peak around 2920–2960 cm^{-1} corresponds to the stretching vibrations of C-H bonds in aliphatic compounds such as lipids and carbohydrates (Dotto et al. 2013). The band at approximately 1040–1070 cm^{-1} corresponds to the stretching vibrations of C–O and C–C bonds in polysaccharides, while the peak around 1230–1260 cm^{-1} represents the stretching vibrations of C–O and C–N bonds. Additionally, the peak at approximately 3400–3600 cm^{-1} is associated with the stretching vibrations of O–H bonds in hydroxyl groups of carbohydrates and water (Çelekli et al. 2010). These functional groups and their corresponding peaks in the FTIR spectrum provide valuable information about the chemical composition of Spirulina platensis algae.

In SEM images, Spirulina platensis typically exhibits a distinctive filamentous or cylindrical shape. The filaments appear long and slender, resembling coiled or spiraled threads. The surface of the filaments often appears smooth and may have a slight wrinkled or undulating texture. The individual cells that make up the filament can be discerned, appearing as small, elongated structures arranged in a chain-like formation. The dimensions of the Spirulina platensis cells can vary, but they are generally in the range of a few micrometers in length and a few tenths of a micrometer in diameter. The surface of the cells may show fine surface features or structures, such as ridges or grooves, which can be observed at higher magnifications in SEM images (Dutta et al. 2021).

The diverse algae species, Spirulina platensis has attracted significant attention as a potential candidate for phycoremediation due to its high efficiency and versatility in metal ion removal (Cepoi et al. 2020). Sayadi et al. (2019) investigated the adsorption rate of heavy metal ions using modified Spirulina platensis and Chlorella vulgaris algae biosorbents. Their research highlighted the effectiveness of the modified biosorbents in removing heavy metals from aqueous solutions. They explored kinetic and isotherm models governing the adsorption process and assessed the thermodynamics parameters, indicating the feasibility of using Spirulina platensis for heavy metal removal. Rady et al. (2023) explored the potential of Spirulina platensis biomass in accumulating heavy metals from multicomponent copper-containing synthetic effluents. Their study revealed the high capacity of Spirulina platensis in accumulating copper, iron, nickel, and zinc from the synthetic effluents, making it a promising candidate for bioremediation in metal-contaminated sites. Their research also highlighted the importance of understanding the biochemical changes in Spirulina platensis during metal exposure, which can enhance its resistance capability to heavy metals. Ai et al. (2023) conducted a comprehensive review on the extraction methods, structural features, bioactivities, and potential toxicities of Spirulina platensis polysaccharides (SPPs). Their review showcased the various health-promoting activities of SPPs, including antioxidant, anticancer, immunomodulatory, and hypoglycemic effects. They also explored the influence of salinity fluctuations on the phycoremediation capacity of Pb by Spirulina platensis. Their findings revealed morphological changes in Spirulina platensis under different salinity levels, affecting the removal efficiency of Pb.

12.2.2 Factors Influencing Algae Growth and Metal Uptake

There are several factors that can influence algae growth and metal uptake. Some of the key factors include:

Nutrient Availability: Algae require nutrients such as nitrogen and phosphorus for growth. Increased availability of these nutrients, often due to runoff from agricultural or urban areas, can lead to excessive algal growth, a phenomenon known as eutrophication. Algae can take up metals along with these nutrients, leading to metal accumulation in their tissues (Rizzo et al. 2013; Kharraz et al. 2022).

Light: Algae are photosynthetic organisms that require light for photosynthesis, which is essential for their growth. Adequate light availability promotes algae growth.

However, excessive light or prolonged exposure to direct sunlight can be detrimental to algae, leading to photoinhibition and reduced growth (Nodeh et al. 2023).

Temperature: Algae exhibit different growth rates and metabolic activities at different temperatures. While some algae thrive in warmer waters, others prefer cooler environments. Temperature influences the rate of metabolic processes, including metal uptake and assimilation within algae cells. Extreme temperature conditions can negatively impact algae growth and metal uptake (Esmaeili and Aghababai Beni 2015).

pH: The pH of the water plays a crucial role in the growth and survival of algae. Different algae species have specific pH requirements for optimal growth. pH affects metal speciation and availability in aquatic environments, consequently influencing metal uptake by algae (Esmaeili et al. 2010).

Metal Concentrations: The concentration of metals in the water directly affects the metal uptake by algae. High metal concentrations can inhibit algal growth and may even be toxic to them. However, some algae species have developed mechanisms to tolerate or accumulate metals, enabling them to grow in metal-contaminated environments (Beni and Esmaeili 2020).

Water Chemistry: Various chemical factors in water, such as dissolved oxygen levels, salinity, and organic matter content, can influence algae growth and metal uptake. Changes in water chemistry can affect the availability and toxicity of metals, thereby impacting their uptake and accumulation by algae (Bunio et al. 2010).

Competition and Predation: Algae may face competition from other algae species or aquatic organisms for resources, including nutrients and light. Additionally, grazers and predators can consume algae, reducing their population and altering metal uptake dynamics within the ecosystem (Esmaeili and Darvish 2014).

12.3 Mechanisms of Heavy Metal Uptake by Algae

12.3.1 Bioaccumulation and Biosorption Processes

Algae can take up heavy metals through two primary mechanisms: bioaccumulation and biosorption (Liliana et al. 2021). **Bioaccumulation**: Bioaccumulation refers to the active uptake and accumulation of heavy metals within the cellular structures of algae (Sedlakova-Kadukova et al. 2019). Algae possess transport proteins and ion channels on their cell membranes that enable the selective uptake of metals. These metals can then be stored in vacuoles or bound to intracellular ligands such as proteins and peptides. Bioaccumulation is an energy-dependent process and can vary depending on the metal species and the physiological characteristics of the algae species involved (Chojnacka 2010).

Biosorption: Biosorption involves the passive binding or adsorption of heavy metals onto the surface of algae cells or extracellular components (Zinicovscaia et al. 2020). Algae possess negatively charged functional groups, such as carboxyl, sulfate, and phosphate groups, on their cell walls and extracellular polymeric substances (EPS) (Gupta and Diwan 2017; Mustafa et al. 2021). These functional groups can attract and bind heavy metal ions through electrostatic interactions, complexation, or ion exchange processes. Biosorption is a rapid process and can contribute significantly to the overall removal of heavy metals from the surrounding water (Winters et al. 2017; He et al. 2023).

Lozano-Bilbao et al. (2023) delve into the potential role of structuring algae as buffers against heavy metals and trace elements in the intertidal habitat of the Canary Islands. Through meticulous analysis of various species, including Anemonia sulcata, Palaemon elegans, Gongolaria abies-marina, and Jania virgata, using Inductively Coupled Plasma Optical Emission Spectrometry (ICP-OES), they unveil higher metal concentrations in structuring algae, particularly J. virgata. These findings offer valuable insights into the capacity of seaweeds to mitigate pollution in the intertidal region.

Al Ketife et al. (20202020) propose an innovative bioassimilation and bioaccumulation model to efficiently remove heavy metals from wastewater using algae. Employing Response Surface and Box Benken Methodology (BBM) alongside best-fit simulation, the researchers achieve remarkable removals of Cu^{2+} and Pb^{2+}, with the bioaccumulation process proving substantially more effective for Cu^{2+}. This model introduces a promising new strategy for heavy metal removal and recovery through algae-based processes.

In a study by Mao et al. (2023), the bioaccumulation and toxicity of perfluorooctanoic acid (PFOA) and perfluorooctane sulfonate (PFOS) in marine algae Chlorella sp. are thoroughly investigated. Their experiments reveal that exposure to PFOA and PFOS induces cell membrane damage, oxidative stress, and inhibition of algal growth. Surprisingly, the algae display a hormetic growth pattern, with chlorophyll a content increasing despite reduced photosynthesis. This research sheds light on the complex biological responses of marine microalgae to perfluoroalkyl acids (PFAAs).

Peng and et al. (2022) conduct a species-specific bioaccumulation and health risk assessment of heavy metals in seaweeds from the tropic coasts of the South China Sea. Their findings unveil significant variations in bioaccumulation levels among different seaweed species. Certain seaweed species, such as Sargassum oligocystum, Turbinaria ornata, Sargassum polycystum, and Sargassum thunbergii, pose a moderate to high health risk for children. This study underscores the need for meticulous evaluation of heavy metal contamination in seaweeds and its potential implications for human health.

12.3.2 Factors Affecting Heavy Metal Uptake by Algae

Metal Concentration: The concentration of heavy metals in the surrounding water affects their uptake by algae. Higher metal concentrations generally lead to increased uptake until saturation is reached, beyond which toxicity can occur (Javee et al. 2022).

pH: The pH of the water influences heavy metal speciation and availability. Different metal species have different affinities for algae uptake under different pH conditions. Generally, metal uptake increases at lower pH values, as metals tend to be more soluble and available for uptake (Khoshraftar et al. 2023).

Complexation with Ligands: Heavy metals in water may form complexes with ligands, such as organic matter or inorganic ligands. These complexes can influence the bioavailability and uptake of metals by algae. Stronger metal–ligand complexes may reduce metal uptake, while weaker complexes or free metal ions may be more readily taken up by algae (Jaafari and Yaghmaeian 2019).

Nutrient Interactions: Nutrient concentrations, particularly those of essential metals like iron, can affect the uptake of heavy metals by algae. Competition between different metal ions for uptake mechanisms can occur, and the presence of high nutrient levels can either enhance or inhibit heavy metal uptake (Djinni and Djoudi 2022; Swar et al. 2023).

Temperature: Temperature influences algae metabolism and growth rates, which, in turn, can affect metal uptake. Higher temperatures generally increase the metabolic activity of algae, potentially enhancing metal uptake. However, extreme temperatures can be detrimental to algae growth and metal uptake processes (Kaynar et al. 2021).

12.3.3 Role of Algae Cell Wall and Extracellular Polymeric Substances (EPS) in Metal Removal

The cell wall and extracellular polymeric substances (EPS) of algae play a significant role in metal removal processes. These components have several functions related to metal uptake and sequestration: **Adsorption Sites**: The cell wall and EPS contain functional groups, such as carboxyl, sulfate, and phosphate groups, which can serve as adsorption sites for heavy metals. These functional groups attract and bind metal ions, facilitating their removal from the surrounding water (Sayadi et al. 2019).

Complexation and Ion Exchange: The cell wall and EPS can complex with heavy metal ions through electrostatic interactions or form ion exchange sites. This binding can immobilize the metals, preventing their re-release into the water column (Dutta et al. 2021).

Physical Barrier: The cell wall acts as a physical barrier, restricting the entry of larger metal complexes or particles into the algae cell. This mechanism helps protect the cellular machinery from potential metal toxicity (Oh et al. 2022).

Metal Sequestration: Algae can produce EPS that can encapsulate heavy metals, sequestering them within the extracellular matrix. This sequestration process immobilizes the metals and reduces their bioavailability and potential toxicity (Atoku et al. 2021).

Overall, the cell wall and EPS of algae contribute to the removal and sequestration of heavy metals from the environment, playing an important role in metal detoxification and tolerance mechanisms.

12.4 Algae Cultivation for Wastewater Treatment

12.4.1 Different Cultivation Methods

Algae can be cultivated for wastewater treatment using various methods, including:

Open Ponds: Open ponds are shallow, earthen or lined basins where algae can be grown. They are cost- effective and relatively simple to set up and maintain. However, open ponds are susceptible to contamination, evaporation, and temperature fluctuations (Saúco et al. 2021; Brar et al. 2022).

Closed Photobioreactors: Photobioreactors are closed systems that allow for more controlled cultivation conditions. They can be tubular, flat-panel, or cylindrical in design. Photobioreactors offer advantages such as higher biomass productivity, better control over environmental factors (e.g., light, temperature, pH), and reduced contamination risks. However, they are typically more expensive to construct and operate compared to open ponds (Saúco et al. 2021; Bhandari et al. 2023).

12.4.2 Optimization of Growth Conditions for Maximum Metal Uptake

To maximize metal uptake by algae during wastewater treatment, several growth conditions can be optimized:

Nutrient Supply: Adjusting nutrient concentrations, particularly nitrogen and phosphorus, can optimize algal growth and metal uptake. However, excessive nutrient supply should be avoided to prevent eutrophication and the proliferation of undesirable algae species (Almomani and Bohsale 2021).

Light Intensity: Providing optimal light intensity promotes photosynthesis and enhances algal growth. Light levels should be balanced to prevent photoinhibition or excessive growth of algae, which may lead to self-shading and reduced metal uptake efficiency (Islam et al. 2022).

pH and Temperature: Maintaining suitable pH and temperature ranges specific to the algae species being cultivated can enhance their growth and metal uptake capacity. Different algae species have different pH and temperature preferences, and optimal conditions should be maintained accordingly (Klin et al. 2023).

Metal Concentrations: It is crucial to optimize the metal concentrations in the wastewater to ensure effective metal uptake by algae. Metal concentrations should be within a range that allows efficient uptake without causing toxicity or inhibiting algal growth (Casagli et al. 2022).

12.4.3 Harvesting and Biomass Recovery Techniques

Harvesting algae biomass from wastewater after metal uptake is an important step in the wastewater treatment process. Various techniques can be employed for harvesting and biomass recovery: **Sedimentation**: Allowing algae to settle by gravity in settling tanks or ponds, followed by decanting or mechanical separation, is a common method for biomass recovery. This technique is suitable for larger algae species with relatively high settling rates (Bhatt et al. 2022).

Filtration: Using various filtration methods, such as microfiltration or ultrafiltration, to separate algae from the wastewater can be effective. Filtration can be combined with other techniques to improve biomass recovery (González Fernández et al. 2023).

Centrifugation: Centrifuges are used to spin down algae cells, separating them from the liquid phase. This method is suitable for higher biomass concentrations and can provide efficient biomass recovery (Alazaiza et al. 2022).

Flocculation: The addition of chemicals, known as flocculants, can induce the clumping or aggregation of algae cells, making them easier to separate from the water. Flocculation is often combined with sedimentation or filtration for biomass recovery (Valenzuela-Elgueta et al. 2021).

After harvesting, the algae biomass can be further processed for various applications, including bioenergy production, bioremediation, or value-added products (Muthuraman and Kasianantham 2023). It's important to note that the specific choice of cultivation method, growth conditions, and biomass recovery techniques may depend on the specific goals, resources, and constraints of the wastewater treatment system.

12.5 Performance Evaluation of Algae-Based Wastewater Treatment Systems

12.5.1 Experimental Setups and Monitoring Parameters

When evaluating the performance of algae-based wastewater treatment systems, several experimental setups and monitoring parameters can be considered (Esmaeili and Aghababai Beni 2015):

Experimental Setups: The experimental setups can include small-scale laboratory-scale systems, pilot- scale systems, or full-scale systems. These setups can vary in terms of reactor types (e.g., open ponds, closed photobioreactors), wastewater flow rates, and treatment configurations.

Monitoring Parameters: Monitoring parameters can include various parameters related to wastewater quality, algae growth, and metal removal efficiency. Some key parameters to monitor include: Wastewater Characteristics: pH, temperature, organic matter content, nutrient concentrations (e.g., nitrogen, phosphorus), heavy metal concentrations, chemical oxygen demand (COD), biochemical oxygen demand (BOD), etc.

Algae Growth: Biomass concentration, chlorophyll-a content, specific growth rate, productivity, etc. **Metal Removal Efficiency**: Metal concentrations in influent and effluent wastewater, metal uptake by algae, metal speciation analysis, etc.

Other Parameters: Dissolved oxygen levels, light intensity, pH stability, turbidity, etc.

12.5.2 Removal Efficiency of Different Heavy Metals

The removal efficiency of different heavy metals can vary depending on several factors, including the algae species, metal species, cultivation conditions, and initial metal concentrations. It is important to assess the removal efficiency of specific heavy metals in the wastewater treatment system under consideration.

The removal efficiency of different metals can be evaluated individually, and a comprehensive analysis can provide insights into the overall effectiveness of the algae-based wastewater treatment system in removing heavy metals.

Atoku et al. (2021) conducted a study to assess the removal efficiency of heavy metals from industrial wastewater using microalgae. The results indicate that the three microalgae species showed significant potential in improving the quality of wastewater through phycoremediation. The removal efficiency for heavy metals, such as Cu, Fe, Ni, Pb, and Cd, ranged from 10% (Cu) to 90% (Pb) after 45 days of treatment. While the microalgae demonstrated appreciable metal removal capabilities, the treated wastewater still contained metal levels above the permissible limits set by the National Environmental Standards and Regulations Enforcement Agency (NESREA), with the exception of Zn. Nevertheless, the study highlights the promising potential of Nostoc commune, Oscillatoria limosa, and Chlorella vulgaris in the treatment of wastewater contaminated with heavy metals, making it a relevant contribution to eco-friendly technologies for industrial wastewater management. Yousefi er al. (2023) present a comprehensive study on the removal efficiency of heavy metals and nutrients from aqueous media using microalgae, specifically Chlorella vulgaris and Scendesmus obliquus, along with their biofilms. The results demonstrate the impressive bioremediation capabilities of these microalgae species, indicating their potential as environmentally friendly and cost-effective solutions for heavy metal treatment issues. The study reveals that both C. vulgaris and S. obliquus, as well as their mix culture, exhibit high removal efficiencies for copper and iron, ranging from 98.25% to 99.9% and 90.22% to 99.19%, respectively. The mix culture showed particular promise in iron uptake, while S. obliquus demonstrated greater resistance to copper and iron stress situations. Notably, the authors observed that copper exhibited more toxicity effects compared to iron, with C. vulgaris being the most vulnerable among the cultures. Moreover, the study highlights the importance of biofilm structure, showing its effective role in enhancing heavy metal resistance. Jaafari et al. (2019) investigated the optimization of heavy metal biosorption using

Chlorella coloniales algae from aqueous solutions. They employed the Box-Behnken design to study the interaction of initial metal concentration, reaction time, and algae dose for the biosorption of Cr, Cd, Co, Fe, and As. The results demonstrated that Chlorella coloniales algae exhibited excellent potential as an alternative for the removal of heavy metals, with removal efficiencies of more than 95% for Cr, Cd, Co, Fe, and As. The researchers found that heavy metal biosorption increased with an increase in the reaction time up to 100 h, after which it started to decrease. Moreover, the biosorption capacity of Chlorella coloniales algae increased with higher initial Cd concentrations up to 12 mg L^{-1}, beyond which no further increase was observed. At low concentrations of heavy metals, the algae showed effective bioaccumulation for Cr, Co, Fe, and As, but at higher concentrations, the efficiency of bioaccumulation decreased. Under the optimized process parameters, Chlorella coloniales algae exhibited impressive removal efficiencies, with values of 97.8%, 97.05%, 95.15%, 98.6%, and 96.5% for Cr, Cd, Co, Fe, and As, respectively. The study concluded that the use of C. coloniales algae could be a promising and cost-effective alternative to current expensive methods for removing heavy metals from aqueous solutions. Their findings contribute valuable insights into the potential application of Chlorella coloniales algae in heavy metal bioremediation efforts. Brar et al. (2022) present a novel approach for managing acid mine drainage (AMD) that not only addresses environmental sustainability but also explores the potential for bioenergy production. The study highlights a two-stage process involving AMD remediation and simultaneous lipid production using the microalga Chlorella vulgaris, which proves to be highly efficient with zero-waste generation. Through alkaline treatment at different pH values (8.0, 10.0, and 12.0), the authors achieved significant reductions in metal loads, including copper, cobalt, chromium, cadmium, nickel, and zinc, compared to the acidic pH range. Additionally, the researchers observed substantial reductions in sulfates, iron, aluminum, and manganese concentrations when AMD was treated at a pH value of 8.0, which was later chosen as the ideal medium to promote lipid accumulation by Chlorella vulgaris. The study demonstrates the remarkable efficiency of C. vulgaris in AMD treatment, making it a promising candidate for bioenergy production. Moreover, the researchers found that the presence of metals in the AMD positively impacts lipid production in microalgae, achieving a lipid yield of 0.18 g g^{-1} of glucose and a high lipid content of 0.35 g g^{-1} of biomass. The findings not only contribute to the field of environmental biotechnology but also hold great promise for sustainable biofuel production, emphasizing the importance of this novel approach in managing AMD and utilizing Chlorella vulgaris as a valuable resource for potential bioenergy applications. Sayadi et al. (2019), they have demonstrated the impressive removal efficiency of heavy metal ions, namely cadmium, lead, and copper, from aqueous solutions using biosorbents derived from modified Spirulina platensis and Chlorella vulgaris algae. The research focused on assessing the adsorption rate of these metal ions under various conditions, including different pH levels, contact times, adsorbent contents, and metal concentrations. The results clearly indicate that both modified biosorbents exhibit excellent capabilities in removing heavy metals, with variations observed in the removal percentages for different metals. At optimized conditions, the adsorbents showed remarkable removal efficiencies, with

Spirulina platensis achieving 92.76%, 94.09%, and 80.75% removal for Cd, Pb, and Cu, respectively, while Chlorella vulgaris reached removal percentages of 87.52%, 90.09%, and 84.75%, respectively. Notably, the lead removal percentage was found to be higher for both modified biosorbents compared to cadmium and copper. The use of Longmuir and second-order kinetics equations provided valuable insights into the adsorption kinetics, while the isotherm models highlighted the higher capacity of modified biosorbents for lead adsorption. Moreover, the thermodynamics assessment indicated that the adsorption process for Cd, Pb, and Cu ions on the modified biosorbents was favorable, spontaneous, and physico-chemical. Overall, the study underscores the high adsorption efficiency of Spirulina platensis and Chlorella vulgaris as effective and safe options for biosorption of heavy metals from aqueous solutions, thus offering significant potential for environmental remediation and water treatment applications.

12.5.3 Comparison with Other Conventional Wastewater Treatment Methods

Comparing algae-based wastewater treatment systems with other conventional methods can help assess their performance and advantages. Some conventional wastewater treatment methods commonly compared with algae-based systems include (Beni and Esmaeili 2020):

Activated Sludge Process: This process involves the use of microorganisms in an aerobic environment to treat wastewater. Algae-based systems can offer advantages such as higher nutrient and metal removal efficiencies, lower sludge production, and potential for bioenergy production.

Chemical Precipitation: Chemical precipitation involves adding chemicals to wastewater to form insoluble precipitates, which can be removed by sedimentation. Algae-based systems can be more sustainable and cost-effective for metal removal, as they can recover and concentrate metals in algal biomass for potential resource recovery. Adsorption: Adsorption processes utilize materials such as activated carbon or ion exchange resins to remove heavy metals from wastewater. Algae-based systems can provide an alternative adsorption mechanism through biosorption, potentially offering lower costs and environmental impacts.

12.6 Challenges and Limitations of Algae-Based Systems

12.6.1 Toxicity and Tolerance Limits of Algae to Heavy Metals

One challenge of algae-based systems is the toxicity of heavy metals to algae themselves. While some algae species have developed mechanisms to tolerate and accumulate heavy metals, high metal concentrations can still be toxic and inhibit their growth. The toxicity and tolerance limits vary among different algae species and different metals. It is crucial to understand these limits to ensure that the selected algae species can effectively tolerate and remove heavy metals without being negatively impacted.

Kim et al. (2023), the toxicity and tolerance limits of boron to three green algal species, Chlamydomonas reinhardtii, Chlorella sorokiniana, and Chlorella vulgaris, were investigated in the presence of ethylenediaminetetraacetic acid (EDTA). The research aimed to understand how the presence of EDTA, a common substance in aquatic ecosystems, affects the toxicity of boron to these algae. The findings revealed that C. vulgaris displayed the highest tolerance among the three algae, as its toxicity to boron remained relatively unchanged in the presence of EDTA compared to the other species. In contrast, both C. reinhardtii and C. sorokiniana exhibited heightened sensitivity to boron in the presence of EDTA. The study further utilized Visual MINTEQ modeling to demonstrate that EDTA combined with micronutrients in the water, forming chelates that reduced their availability for uptake by the algae, consequently increasing their sensitivity to boron toxicity. These valuable results shed light on the complex interactions between metals and chelating agents like EDTA, providing crucial insights for the future risk assessment and management of boron in aquatic ecosystems. Understanding the toxicity and tolerance limits of algae to heavy metals, as explored in this study, is crucial for addressing environmental concerns and developing effective strategies to safeguard aquatic ecosystems. Piotrowska-Niczyporuk et al. (2012) delve into the fascinating topic of phytohormones as regulators of heavy metal biosorption and toxicity in the green alga Chlorella vulgaris. The study investigates the influence of exogenously applied phytohormones, such as auxins, cytokinins, gibberellin, jasmonic acid, and polyamine (spermidine), on the growth and metabolism of Chlorella vulgaris under heavy metal stress (Cd, Cu, Pb). The results reveal that heavy metals exert varying degrees of inhibitory effects on algal growth and metabolism, with Cd displaying the highest phytotoxicity, followed by Pb and Cu. However, the presence of exogenously applied phytohormones modifies the toxic effects of heavy metals. Auxins, cytokinins, gibberellin, and spermidine show the ability to mitigate stress symptoms by inhibiting heavy metal biosorption, thus restoring algal growth and primary metabolite levels. Additionally, these phytohormones and polyamine enhance the activities of antioxidant enzymes and the accumulation of antioxidants in the cells, leading to increased antioxidant capacity and reduced oxidative stress. However, jasmonic acid exhibits a contrasting effect, enhancing heavy metal toxicity by increasing metal biosorption and ROS generation. Understanding the role of phytohormones and polyamine in the algal response to heavy metal stress is crucial for comprehending metal-specific mechanisms of toxicity and predicting toxicity accurately. The findings suggest that phytohormones and polyamine play significant roles in Chlorella vulgaris' response to abiotic stressors and its ability to adapt to metal contamination in aquatic environments.

These research findings underline the importance of selecting suitable algal species and understanding their tolerance limits when designing algae-based systems for heavy metal remediation. By choosing species that can effectively tolerate and remove heavy metals without being negatively impacted, we can develop more efficient and sustainable methods for water remediation and environmental protection. Additionally, the insights gained from the studies on phytohormones' role in regulating heavy metal toxicity provide a promising avenue for further research in developing novel strategies to enhance algal resistance to heavy metal stress. As we continue to address environmental concerns and strive to safeguard our ecosystems, this knowledge will undoubtedly contribute to the development of innovative and effective approaches to tackle heavy metal pollution and ensure the health and sustainability of our aquatic environments.

12.6.2 Cost Considerations and Scalability Issues

The cost of implementing and operating algae-based wastewater treatment systems can be a significant challenge (Du et al. 2016). Factors contributing to costs include infrastructure construction, energy requirements for mixing and aeration, nutrient supplementation, monitoring and control systems, and biomass harvesting and processing. Additionally, scalability can be an issue, as the efficient cultivation of algae on a large scale may require substantial space and resources. Balancing the cost-effectiveness and scalability of algae-based systems is essential for their practical application (Oh et al. 2022).

12.6.3 Potential Ecological Impacts and Unintended Consequences

The introduction of algae-based systems for wastewater treatment can have potential ecological impacts and unintended consequences. For example:

Algal blooms: If the system is not well-managed, the excessive growth of algae can lead to algal blooms, which can deplete dissolved oxygen levels in the water, harm aquatic organisms, and disrupt ecosystem balance (Bhuyar et al. 2020; Bhandari et al. 2023).

Species competition and biodiversity: Introducing non-native or invasive algae species for wastewater treatment can potentially lead to competition with native species and alter the ecological composition and biodiversity of water bodies (Alazaiza et al. 2022; Lozano-Bilbao et al. 2023).

Nutrient release: The release of excess nutrients from treated wastewater, particularly if not properly managed, can contribute to eutrophication and further degradation of water quality (Bhatt et al. 2022).

To mitigate these potential impacts, careful selection of native or non-invasive algae species, appropriate system design and operation, and integration with complementary ecological management strategies are necessary. Monitoring and regular assessment of the ecological impacts are also important to ensure the sustainability of algae-based systems.

12.7 Future Directions and Research Perspectives

12.7.1 Advancements in Algae-Based Wastewater Treatment Technologies

Future advancements in algae-based wastewater treatment technologies may focus on improving system efficiency, cost-effectiveness, and scalability. Some potential advancements include:

Genetic engineering of algae: Genetic modification of algae to enhance their metal uptake capabilities, improve biomass productivity, and optimize nutrient utilization.

Development of novel cultivation systems: Innovation in photobioreactor designs, including improved light distribution, enhanced mixing, and better control of environmental factors.

Algal biofilm systems: Exploring the use of algal biofilms for wastewater treatment, which can provide higher biomass retention, improved metal uptake efficiency, and reduced energy requirements.

Algal-bacterial symbiotic systems: Harnessing the symbiotic relationship between algae and bacteria for enhanced wastewater treatment capabilities, including nutrient removal and metal sequestration.

12.7.2 Integration of Algae Systems with Other Treatment Approaches

Integrating algae-based systems with other treatment approaches can enhance overall wastewater treatment efficiency and address specific challenges. Some potential integration options include:

Algae-bacterial consortia: Combining algae and specific bacterial strains to create synergistic effects in wastewater treatment, such as improved nutrient removal and metal transformation.

Hybrid systems: Integrating algae-based systems with conventional treatment methods, such as activated sludge processes or chemical precipitation, to optimize metal removal and nutrient recovery.

Algae-based post-treatment: Using algae systems as a polishing step after primary treatment processes to further remove nutrients, residual metals, and organic compounds.

12.7.3 Emerging Trends and Areas for Further Research

There are several emerging trends and areas for further research in algae-based wastewater treatment: Exploring efficient methods for recovering nutrients, such as nitrogen and phosphorus, from algal biomass for potential reuse in agriculture or other applications. Investigating the potential for converting algal biomass into valuable products, such as biofuels, bioplastics, or high-value chemicals, to enhance the economic viability of algae-based systems. Assessing the capacity of algae-based systems to remove microplastics from wastewater and exploring the mechanisms and optimization strategies for effective microplastic removal. Researching the feasibility and performance of algae-based wastewater treatment systems in decentralized settings, such as rural communities or small-scale industries, to address localized wastewater treatment needs.

12.8 Conclusion

Algae-based systems hold great promise as a highly effective approach for the treatment of heavy metal- contaminated wastewater. Through extensive research, valuable findings and contributions have been made in this field. One crucial aspect is the understanding of various factors that influence algae growth and metal uptake in aquatic environments. These factors include nutrient availability, light intensity, temperature, pH levels, metal concentrations, water chemistry, and the presence of competition or predation. The mechanisms of heavy metal uptake by algae have been explored in depth. Algae demonstrate the ability to accumulate metals through two primary processes - bioaccumulation and biosorption. In bioaccumulation, algae actively take up and store metals within their cells, while in biosorption, the passive binding of metals occurs onto the algae cell walls and extracellular polymeric substances. However, algae-based systems are not without challenges and limitations. It is vital to consider the toxicity and tolerance limits of algae to heavy metals. Additionally, cost factors, scalability issues, and potential ecological impacts must be carefully managed and optimized to ensure the sustainability and efficacy of algae-based wastewater treatment. Despite these challenges, algae-based systems offer numerous advantages in treating heavy metal-contaminated wastewater. One significant advantage lies in their high metal uptake efficiency, making them a natural and cost-effective method for metal removal from wastewater. Moreover, algae have demonstrated effectiveness in removing nutrients such as nitrogen and phosphorus, contributing to overall water quality improvement.

Resource recovery is another notable benefit of algae-based systems. Valuable metals can be extracted, and bioenergy or bioproducts can be produced from algal biomass. These systems also have a smaller ecological footprint compared to conventional treatment methods, utilizing natural processes that contribute to the restoration of ecosystems. The future of algae-based systems for treating heavy metal-contaminated wastewater is promising, with several prospects and potential impacts. Technological advancements, including genetic engineering, cultivation systems, and biomass harvesting techniques, can further enhance efficiency and cost-effectiveness in algae-based wastewater treatment. Integration with other technologies, such as bacteria-based systems or hybrid approaches, presents an opportunity to optimize metal removal and nutrient recovery. As a result, algae-based systems have the potential to play a significant role in sustainable water management, contributing to improved water quality, resource recovery, and reduced environmental impact. To continue advancing algae-based systems, ongoing research, knowledge sharing, and collaboration among researchers, practitioners, and policymakers are crucial. These efforts will help address challenges effectively and pave the way for more efficient and sustainable solutions in treating heavy metal-contaminated wastewater.

References

Abdel-Latif, H.M.R. et al. (2022) The effectiveness of Arthrospira platensis and microalgae in relieving stressful conditions affecting finfish and shellfish species: an overview. Aquaculture Rep 24. https://doi.org/10.1016/j.aqrep.2022.101135

Abdollahi SA, Mokhtariyan N, Ameri E (2022) Design, synthesis and application of a sponge-like nanocomposite ceramic for the treatment of Ni(II) and Co(II) wastewater in the zinc ingot industry. Arab J Chem 15(1):103477. https://doi.org/10.1016/j.arabjc.2021.103477

Adamia G et al (2018) About possibility of alga Spirulina application for phytoremediation of water polluted with 2,4,6-trinitrotoluene. Ann Agrarian Sci 16(3):348–351. https://doi.org/10.1016/j.aasci.2018.07.004

Ai X et al (2023) Polysaccharides from Spirulina platensis: extraction methods, structural features and bioactivities diversity. Int J Biol Macromolecules 231(1166):123211. https://doi.org/10.1016/j.ijbiomac.2023.123211

Al-Khiat SH et al (2023) Comparison of the microalgae Phormidium tenue and Chlorella vulgaris as biosorbents of Cd and Zn from aqueous environments. Environ Res 235:116675. https://doi.org/10.1016/j.envres.2023.116675

Alagawany M et al (2021) Nutritional applications of species of Spirulina and Chlorella in farmed fish: a review. Aquaculture 542:736841. https://doi.org/10.1016/j.aquaculture.2021.736841

Alazaiza MYD et al (2022) Potential use of algae for the bioremediation of different types of wastewater and contaminants: production of bioproducts and biofuel for green circular economy. J Environ Manage 324. https://doi.org/10.1016/j.jenvman.2022.116415

Almomani F, Bohsale RR (2021) Bio-sorption of toxic metals from industrial wastewater by algae strains Spirulina platensis and Chlorella vulgaris: application of isotherm, kinetic models and process optimization. Sci Total Environ 755. https://doi.org/10.1016/j.scitotenv.2020.142654

Atoku DI et al (2021) Evaluating the efficiency of Nostoc commune, Oscillatoria limosa and Chlorella vulgaris in a phycoremediation of heavy metals contaminated industrial wastewater. Sci Afr 12:e00817. https://doi.org/10.1016/j.sciaf.2021.e00817

Azam R et al (2022) Cultivation of two Chlorella species in Open sewage contaminated channel wastewater for biomass and biochemical profiles: comparative lab-scale approach. J Biotechnol 344:24–31. https://doi.org/10.1016/j.jbiotec.2021.11.006

Bajguz A (2010) An enhancing effect of exogenous brassinolide on the growth and antioxidant activity in Chlorella vulgaris cultures under heavy metals stress. Environ Experimental Botany 68(2):175–179. https://doi.org/10.1016/j.envexpbot.2009.11.003

Barghchi H et al (2023) The effects of Chlorella vulgaris on cardiovascular risk factors: a comprehensive review on putative molecular mechanisms. Biomed Pharmacother 162:114624. https://doi.org/10.1016/j.biopha.2023.114624

Beni AA, Esmaeili A (2020) Environmental technology & innovation biosorption, an efficient method for removing heavy metals from industrial effluents: a review. Environ Technol Innov 17:100503. https://doi.org/10.1016/j.eti.2019.100503

Bhandari M et al (2023) An integration of algae-mediated wastewater treatment and resource recovery through anaerobic digestion. J Environ Manage 342. https://doi.org/10.1016/j.jenvman.2023.118159

Bhatt P et al (2022) Algae in wastewater treatment, mechanism, and application of biomass for production of value-added product. Environ Pollution 309. https://doi.org/10.1016/j.envpol.2022.119688

Bhuyar P et al (2020) Salinity reduction from poly-chem-industrial wastewater by using microalgae (*Chlorella* sp.) collected from coastal region of peninsular Malaysia Spirulina project view project Algal biorefinery view project salinity reduction from poly-chem-industrial was. J Biol Med 1(1)

Brar KK et al (2022) Novel approach for the management of acid mine drainage (AMD) for the recovery of heavy metals along with lipid production by *Chlorella vulgaris*. J Environ Manage 308:114507. https://doi.org/10.1016/j.jenvman.2022.114507

Brar PK, Örmeci B, Dhir A (2022) Algae: a cohesive tool for biodiesel production alongwith wastewater treatment. Sustain Chem Pharmacy 28. https://doi.org/10.1016/j.scp.2022.100730

Bunio A et al (2010) The content of elements in rainwater and its relation to the frequency of hospitalization for diabetes and obesity in Opole Voivodship, Poland, during 2000–2002. Biol Trace Element Res 136(2):149–156. https://doi.org/10.1007/s12011-009-8531-6

Cao DW et al (2010) Characteristics of the biodiatomite dynamic membrane (cake layer) for municipal wastewater treatment. Desalination 250(2):544–547. https://doi.org/10.1016/j.desal.2009.09.020

Casagli F et al (2022) Optimizing resource recovery from wastewater with algae-bacteria membrane reactors. Chem Eng J 451.https://doi.org/10.1016/j.cej.2022.138488

Çelekli A, Yavuzatmaca M, Bozkurt H (2010) An eco-friendly process: predictive modelling of copper adsorption from aqueous solution on *Spirulina platensis*. J Hazardous Mater 173(1–3):123–129. https://doi.org/10.1016/j.jhazmat.2009.08.057

Cepoi L et al (2020) Growth and heavy metals accumulation by *Spirulina platensis* biomass from multicomponent copper containing synthetic effluents during repeated cultivation cycles. Ecol Eng 142:105637. https://doi.org/10.1016/j.ecoleng.2019.105637

Chojnacka K (2010) Biosorption and bioaccumulation—the prospects for practical applications. Environ Int 36(3):299–307. https://doi.org/10.1016/j.envint.2009.12.001

Djinni I, Djoudi W (2022) Streptomyces sp. WR1L1S8 a potent endophytic marine strain for heavy metal resistance and copper removal enhanced by RSM modeling. Acta Ecologica Sinica 42(2):80–89. https://doi.org/10.1016/j.chnaes.2021.04.004

Dotto GL et al (2013) Equilibrium and thermodynamics of azo dyes biosorption onto Spirulina platensis. Brazilian J Chem Eng 30(1):13–21. https://doi.org/10.1590/S0104-66322013000100003

Du Z et al (2016) Fast microwave-assisted preparation of a low-cost and recyclable carboxyl modified lignocellulose-biomass jute fiber for enhanced heavy metal removal from water. Bioresource Technol 201:41–49. https://doi.org/10.1016/j.biortech.2015.11.009

Dutta S et al (2021) Recent advances on the removal of dyes from wastewater using various adsorbents: a critical review. Mater Adv 2(14):4497–4531. https://doi.org/10.1039/d1ma00354b

Es-Sahbany H et al (2019) Removal of heavy metals (nickel) contained in wastewater-models by the adsorption technique on natural clay. Mater Today Proc 13:866–875. https://doi.org/10.1016/j.matpr.2019.04.050

Esmaeili A, Aghababai Beni A (2015) Novel membrane reactor design for heavy-metal removal byalginate nanoparticles. J Ind Eng Chem [Preprint]. https://doi.org/10.1016/j.jiec.2014.11.023

Esmaeili A, Darvish M (2014) Evaluation of the marine alga sargassum glaucescens for the adsorption of Zn(II) from aqueous solutions. Water Qual Res J Canada 49(4):339–345. https://doi.org/10.2166/wqrjc.2014.006

Esmaeili A, Ghasemi S, Sohrabipour J (2010) Biosorption of copper from wastewater by activated carbon preparation from alga *Sargassum* sp. Nat Product Res 24(4):341–348. https://doi.org/10.1080/14786410903064915

Fazal T et al (2018) Bioremediation of textile wastewater and successive biodiesel production using microalgae. Renew Sustain Energy Rev 82:3107–3126. https://doi.org/10.1016/j.rser.2017.10.029

González Fernández LA et al (2023) Fundamentals in applications of algae biomass: a review. J Environ Manage 338. https://doi.org/10.1016/j.jenvman.2023.117830

Gupta P, Diwan B (2017) Bacterial Exopolysaccharide mediated heavy metal removal: a Review on biosynthesis, mechanism and remediation strategies. Biotechnol Rep 13:58–71. https://doi.org/10.1016/j.btre.2016.12.006

He X et al (2023) Detoxification and removal of arsenite by *Pseudomonas* sp. SMS11: oxidation, biosorption and bioaccumulation. J Environ Manage 336:117641. https://doi.org/10.1016/j.jenvman.2023.117641

Islam MS et al (2022) Mathematical modelling and statistical optimization of fast cultivation of *Agardhiella subulata*: response surface methodology. Energy Nexus 7. https://doi.org/10.1016/j.nexus.2022.100115

Jaafari J, Yaghmaeian K (2019) Optimization of heavy metal biosorption onto freshwater algae (*Chlorella coloniales*) using response surface methodology (RSM). Chemosphere 217:447–455. https://doi.org/10.1016/j.chemosphere.2018.10.205

Javee A, Malairaj S, Subramani N (2022) Biosynthesis, protease optimization and purification of alkaline serine from Shewanella algae and its potential application as silver recovery. Sustain Chem Pharmacy 25. https://doi.org/10.1016/j.scp.2021.100595

Kayan I, Oz NA, Kantar C (2021) Comparison of treatability of four different chlorophenol-containing wastewater by pyrite-Fenton process combined with aerobic biodegradation: role of sludge acclimation. J Environ Manage 279:111781. https://doi.org/10.1016/j.jenvman.2020.111781

Kaynar UH et al (2021) Adsorption of thorium (IV) ions by metal ion doped ZnO nanomaterial prepared with combustion synthesis: empirical modelling and process optimization by response surface methodology (RSM). Appl Radiation Isotopes 178. https://doi.org/10.1016/j.apradiso.2021.109955

Al Ketife AMD, Al Momani F, Judd S (2020) A bioassimilation and bioaccumulation model for the removal of heavy metals from wastewater using algae: new strategy. Process Safety Environ Protection 144:52–64. https://doi.org/10.1016/j.psep.2020.07.018

Kharraz JA et al (2022) Membrane distillation bioreactor (MDBR) for wastewater treatment, water reuse, and resource recovery: a review. J Water Process Eng 47. https://doi.org/10.1016/j.jwpe.2022.102687

Khoshraftar Z, Masoumi H, Ghaemi A (2023) Experimental, response surface methodology (RSM) and mass transfer modeling of heavy metals elimination using dolomite powder as an economical adsorbent. Case Stud Chem Environ Eng 7. https://doi.org/10.1016/j.cscee.2023.100329

Kim D et al (2023) Effects of EDTA on boron toxicity to the green algae *Chlamydomonas reinhardtii*, *Chlorella sorokiniana*, and *Chlorella vulgaris*. J Water Process Eng 53:103823. https://doi.org/10.1016/j.jwpe.2023.103823

Klin M, Pniewski F, Latała A (2023) Multistage optimization of growth and physiological condition of brackish green microalgae with the use of natural waters. Aquaculture 562. https://doi.org/10.1016/j.aquaculture.2022.738820

Lafarga T et al (2021) Industrial production of spirulina as a protein source for bioactive peptide generation. Trends Food Sci Technol 116:176–185. https://doi.org/10.1016/j.tifs.2021.07.018

Li Y et al (2023) Atrazine biodegradation in water by co-immobilized *Citricoccus* sp. strain TT3 with Chlorella vulgaris under a harsh environment. Algal Res 70:102994. https://doi.org/10.1016/j.algal.2023.102994

Liliana C et al (2021) Biomass of Arthrospira platensis enriched with lithium by bioaccumulation and biosorption process. Food Biosci 41:100950. https://doi.org/10.1016/j.fbio.2021.100950

Lozano-Bilbao E et al (2023) Structuring algae as buffers for heavy metals and trace elements in the Canary Islands intertidal habitat. Mar Pollution Bull 190. https://doi.org/10.1016/j.marpolbul.2023.114890

Luo Y et al (2023) Stress of cupric ion and oxytetracycline in *Chlorella vulgaris* cultured in swine wastewater. Sci Total Environ 895:165120. https://doi.org/10.1016/j.scitotenv.2023.165120

Mao W et al (2023) Bioaccumulation and toxicity of perfluorooctanoic acid and perfluorooctane sulfonate in marine algae *Chlorella* sp. Sci Total Environ 870. https://doi.org/10.1016/j.scitotenv.2023.161882

Matula EE, Nabity DJA (2021) Metabolic response of *Chlorella vulgaris* to a transient thermal environment for supporting simultaneous air revitalization and thermal control in a crewed habitat. Acta Astronautica 187:406–415. https://doi.org/10.1016/j.actaastro.2021.07.003

Mustafa S et al (2021) Microalgae biosorption, bioaccumulation and biodegradation efficiency for the remediation of wastewater and carbon dioxide mitigation: prospects, challenges and opportunities. J Water Process Eng 41. https://doi.org/10.1016/j.jwpe.2021.102009

Muthuraman VS, Kasianantham N (2023) Valorization opportunities and adaptability assessment of algae based biofuels for futuristic sustainability—a review. Process Safety Environ Protection 174:694–721. https://doi.org/10.1016/j.psep.2023.04.043

Nakamoto MM et al (2023) Spirulina application in food packaging: gaps of knowledge and future trends. Trends Food Sci Technol 133:138–147. https://doi.org/10.1016/j.tifs.2023.02.001

Nodeh ZP, Beni AA, Moghadam AJ (2023) Development of evaporation technique for concentrating lead acid wastewater from the battery recycling plant, by nanocomposite ceramic substrates and solar/wind energy. J Environ Manage 328:116980. https://doi.org/10.1016/j.jenvman.2022.116980

Oh YK et al (2022) Cell disruption and lipid extraction from Chlorella species for biorefinery applications: recent advances. Bioresource Technol 366:128183. https://doi.org/10.1016/j.biortech.2022.128183

Orfanidis S et al (2015) Study of Cu(II) removal by *Cystoseira crinitophylla* biomass in batch and continuous flow biosorption. Chem Eng J 277:334–340. https://doi.org/10.1016/j.cej.2015.04.138

Ostera JM, Malanga G, Puntarulo S (2020) Assessment of oxidative balance in hydrophilic cellular environment in Chlorella vulgaris exposed to glyphosate. Chemosphere 248:125955. https://doi.org/10.1016/j.chemosphere.2020.125955

Peng Z et al (2022) Species-specific bioaccumulation and health risk assessment of heavy metal in seaweeds in tropic coasts of South China Sea. Sci Total Environ 832. https://doi.org/10.1016/j.scitotenv.2022.155031

Piotrowska-Niczyporuk A et al (2012) Phytohormones as regulators of heavy metal biosorption and toxicity in green alga *Chlorella vulgaris* (Chlorophyceae). Plant Physiol Biochem 52:52–65. https://doi.org/10.1016/j.plaphy.2011.11.009

Rady MM et al (2023) Spirulina platensis extract improves the production and defenses of the common bean grown in a heavy metals-contaminated saline soil. J Environ Sci (China) 129: 240–257. https://doi.org/10.1016/j.jes.2022.09.011

Rahnama I et al (2023) The effect of Spirulina supplementation on lipid profile: GRADE-assessed systematic review and dose-response meta-analysis of data from randomized controlled trials. Pharmacol Res 193. https://doi.org/10.1016/j.phrs.2023.106802

Rizzo AM et al (2013) Characterization of microalga Chlorella as a fuel and its thermogravimetric behavior. Applied Energy 102:24–31. https://doi.org/10.1016/j.apenergy.2012.08.039

Rodrigues MS et al (2012) Metal biosorption onto dry biomass of Arthrospira (Spirulina) platensis and *Chlorella vulgaris*: multi-metal systems. J Hazardous Mater 217–218:246–255. https://doi.org/10.1016/j.jhazmat.2012.03.022

Saavedra R et al (2019) Influence of organic matter and CO_2 supply on bioremediation of heavy metals by *Chlorella vulgaris* and *Scenedesmus almeriensis* in a multimetallic matrix. Ecotoxicol Environ Safety 182:109393. https://doi.org/10.1016/j.ecoenv.2019.109393

Saúco C et al (2021) Hybrid wastewater treatment system based in a combination of high rate algae pond and vertical constructed wetland system at large scale. J Water Process Eng 43. https://doi.org/10.1016/j.jwpe.2021.102311

Sayadi MH, Rashki O, Shahri E (2019) Application of modified *Spirulina platensis* and *Chlorella vulgaris* powder on the adsorption of heavy metals from aqueous solutions. J Environ Chem Eng 7(3):103169. https://doi.org/10.1016/j.jece.2019.103169

Sedlakova-Kadukova J et al (2019) Bioaccumulation and biosorption of zinc by a novel Streptomyces K11 strain isolated from highly alkaline aluminium brown mud disposal site. Ecotoxicol Environ Safety 167:204–211. https://doi.org/10.1016/j.ecoenv.2018.09.123

Song Q et al (2023) Multi-metal electrochemical response mechanism for direct copper recovery from waste printed circuit boards via sulfate- and chloride-system electrolysis. Resources Conserv Recycling 190:106804. https://doi.org/10.1016/j.resconrec.2022.106804

Swar SS, Boonnorat J, Ghimire A (2023) Algae-based treatment of a landfill leachate pretreated by coagulation-flocculation. J Environ Manage 342. https://doi.org/10.1016/j.jenvman.2023.118223

Valenzuela-Elgueta J, Delgado AV, Ahualli S (2021) Effect of cationic surfactant addition on the electrokinetics and stability of silica/kaolinite suspensions in copper hydrometallurgy conditions. Minerals Eng 169:106958. https://doi.org/10.1016/j.mineng.2021.106958

Wanderley MCA et al (2020) First report on Chlorella vulgaris collagenase production and purification by aqueous two-phase system. Sustain Chem Pharmacy 15. https://doi.org/10.1016/j.scp.2019.100202

Wang Z et al (2021) Aged microplastics decrease the bioavailability of coexisting heavy metals to microalga *Chlorella vulgaris*. Ecotoxicol Environ Safety 217:112199. https://doi.org/10.1016/j.ecoenv.2021.112199

Winters C, Guéguen C, Noble A (2017) Equilibrium and kinetic studies of Cu(II) and Ni(II) sorption on living *Euglena gracilis*. J Appl Phycol 29(3):1391–1398. https://doi.org/10.1007/s10811-016-1040-z

Xaaldi Kalhor A et al (2016) Biodiesel production in crude oil contaminated environment using *Chlorella vulgaris*. Bioresource Technol 222:190–194. https://doi.org/10.1016/j.biortech.2016.09.110

Yousefi Y et al (2023) Heavy metals (copper and iron) and nutrients (nitrate and phosphate) removal from aqueous medium by microalgae *Chlorella vulgaris* and *Scendesmus obliquus*, and their biofilms. Mar Environ Res 188:105989. https://doi.org/10.1016/j.marenvres.2023.105989

Zinicovscaia I et al (2020) Evaluation of biosorption and bioaccumulation capacity of cyanobacteria Arthrospira (spirulina) platensis for radionuclides. Algal Res 51. https://doi.org/10.1016/j.algal.2020.102075

Chapter 13
The Role of Microalgae as Bioindicators of Aquatic Contamination

Walter José Martínez-Burgos, Roberta Pozzan, Júlio Cesar de Carvalho, Matheus Cavali, André B. Mariano, José V. C. Vargas, Juan Ordonez, Ihana A. Severo, and Carlos Ricardo Soccol

Abstract Physical and chemical parameters, if used in an isolated way, may not reflect the actual state of water quality, nor the effect of contaminants on aquatic biota. Thus, biological tools have been developed and applied to assess the water quality and the effects of the contaminants on various aspects of the biology of living beings exposed to these substances. The presence, quantity, and distribution of some species, called bio indicators, can help to determine the water conditions, as well as the environmental impacts provoked by pollutants dumped in water bodies. One of the most used indicators of water quality is microalgae. These single-celled organisms can be found in freshwater or saline aquatic environments, playing fundamental importance as primary producers in the trophic chain. These organisms respond quickly to physical and chemical changes in the water, as well as to the presence of pollutants and enrichment of organic matter, which makes them widely used as bio indicators of aquatic contamination in several studies. Currently, there are internationally standardized eco toxicological tests that use microalgae as bio indicators, which,

W. J. Martínez-Burgos (✉) · J. C. de Carvalho · C. R. Soccol
Department of Bioprocess Engineering and Biotechnology, Federal University of Paraná, Centro Politécnico, Curitiba, PR 81531-990, Brazil
e-mail: ing.wjmartinez@gmail.com

R. Pozzan
Cellular Biology Department, Cellular Toxicology Laboratory, Federal University of Paraná, Centro Politécnico P.O, Curitiba, PR, Brazil

M. Cavali
Department of Sanitary and Environmental Engineering, Federal University of Santa Catarina, Florianópolis, Santa Catarina 88040-970, Brazil

A. B. Mariano · J. V. C. Vargas · I. A. Severo
Sustainable Energy Research and Development Center (NPDEAS), Federal University of Paraná, Curitiba, Paraná 81531-980, Brazil

J. Ordonez · I. A. Severo
Department of Mechanical Engineering, FAMU-FSU College of Engineering, Energy and Sustainability Center, Center for Advanced Power Systems (CAPS), Florida A&M University, Florida State University, Tallahassee, FL 32310-6046, USA

T. A. Kurniawan and A. Anouzla (eds.), *Algae as a Natural Solution for Challenges in Water-Food-Energy Nexus*, Environmental Science and Engineering,
https://doi.org/10.1007/978-981-97-2371-3_13

associated with the assessment of the presence of a contaminant, help to establish water quality criteria and policies to protect the environment. In this context, this chapter aims to present how microalgae species are used as bio indicators of water quality in the most diverse aquatic ecosystems and discuss their potential as an early sentinel in contaminated water bodies.

Keywords Water quality · Algae · Biological indicators · Biomarkers · Environmental impact

13.1 Introduction

Government agencies and the general public have been increasingly concerned with maintaining the quality of water resources. This is because population growth has caused the generation of high loads of waste due to industrial, agricultural, and domestic activities. These, in turn, can be effluents with substantial amounts of contaminants of different natures, both organic and inorganic, conventional or emerging pollutants (Anwer et al. 2019; Martinez-Burgos et al. 2021).

Nowadays, almost all aquatic ecosystems have been directly or indirectly influenced by anthropic actions across the planet, such as contamination. These negative impacts can disrupt the physical–chemical environment and thus alter the natural dynamics of biological communities. However, most conventional treatment systems are unable to completely remove contaminants from the water. Some traces of those compounds remain in treated water and can reach water bodies, once the effluents are disposed of in the environment (Vasilachi et al. 2021). Therefore, to assess the degree of contamination and the impacts that pollutants can provoke in aquatic environments, techniques are needed to determine the presence of these substances. The investigation of certain biological communities as bioindicators to carry out an assessment of the ecological status of water bodies is used for that purpose (Yadav et al. 2023).

The aquatic bioindicators mainly used are phytoplankton, macroalgae, microalgae, macroinvertebrates, and fish. For instance, algal communities are considered good indicators ofenvironmental changes. Indeed, species distribution and resource dynamics, in aquatic or terrestrial environments, have been affected by climate change and indiscriminate human activity (Cid et al. 2023).

Microalgae are key components of aquatic ecosystems, capable of occupying all environments that provide them with sufficient light and moisture, temporarily or permanently. They are primary food chain producers present in freshwater and saltwater bodies worldwide. Thus, microalgae are considered potential bioindicators because they can provide relevant information for both human and ecosystem health. Due to their rapid reproduction rates and very short life cycles, microalgae respond quickly to environmental changes, and any effect on them can impact the higher trophic levels. For those reasons, microalgae are suitable for water quality assessment (Omar 2010). Furthermore, as microalgae are easily grown in the laboratory (e.g., in photobioreactors under controlled conditions) (Deprá et al. 2021) or in outdoor systems, and sampling is simple, they are considered a model for rapid and low-cost testing, creating minimal impact on the resident biota. Due to these characteristics, the use of microalgae-based toxicity tests is growing. Currently, these tests are often required by authorities for chemical notifications and are also increasingly applied to manage pollutant discharges (Cid et al. 2023).

However, each microalgal species has its own characteristics, which may indicate in different ways the impacts that pollutants can cause to water bodies. Also, their growing depends on the environmental condition to which they are submitted (Cid et al. 2023; Deprá et al. 2021). In this sense, this chapter aims to present an overview of the role of microalgae as bioindicators of aquatic contamination. Hence, it is addressed the following topics: (i) types of bioindicators, (ii) use of microalgae as bioindicators of water quality and aquatic contamination, (iii) main microalgae species used as bioindicators and their specific applications, (iv) international standardized tests and evaluated parameters, and (v) molecular and biochemical biomarkers in microalgae.

13.2 Bioindicators

Bioindicators can be biological processes, species, or microbial communities that are used to assess the quality of the environment and its changes over time (Holt and Miller 2011). In general, environmental impacts are attributed to anthropogenic processes, such as pollution and changes in water and land utilization, although changes can also be triggered by natural stressors—e.g., dry, rainy, or freezing periods (Holt and Miller 2011; Rizo-Patrón et al. 2013). Therefore, monitoring the state of ecosystems requires the use of a set of bioindicators that are biologically, methodologically, and socially relevant and can be used effectively over time to assess trends and provide early warnings (Burger 2006b). The main characteristics that a good indicator must meet are presented in Fig. 13.1. Recently, bioindicators have gained more relevance because they can be used to evaluate ecological risks and meet regulatory requirements. Sometimes bioindicators can provide information for both human and ecosystem health (Burger 2006a), including the sustainability assessment or the elucidation of adverse effects mechanisms of xenobiotic products or compounds in the environment.

Among the different bioindicators, plants are highly sensitive to changes due to environmental stresses. For example, marine plants provide significant information on the state of the ocean environment because they are not mobile (Parmar et al. 2016). Animals are also bioindicators since variations in their populations may suggest alterations in ecosystems generated by pollution. Microorganisms are another category of bioindicators, and they are widely used to monitor the state of aquatic and terrestrial ecosystems due to their abundance, rapid growth, and simple utilization. For instance, when some microorganisms are exposed to polluting compounds (e.g., cadmium and

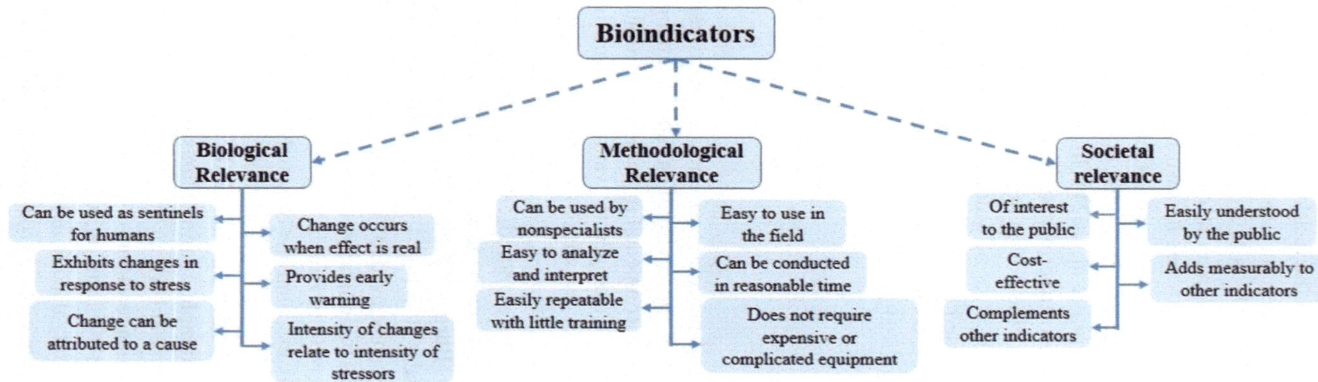

Fig. 13.1 Main characteristics that an environmental bioindicator must ful fill. Adapted from: (Bartell 2016; Burger 2006b; Parmar et al. 2016)

benzene) they can develop some biomarkers, such as stress proteins, that are considered early warning signs. The contamination by mercury, for example, can be determined through the enzymes mercuric reductase and organomercurial lyase, which are produced by the bacteria *Pseudomonas* spp. in the presence of this metal (Burger 2006b). Phytoplankton are also widely used as bioindicators microorganisms, among which microalgae stand out. Microalgae are frequently used as bioindicators of the health of aquatic and terrestrial ecosystems. Due to their abundance, they are easy to test and readily available.

13.3 Microalgae as Bioindicators of Water Quality and Aquatic Contamination

Microalgae are unicellular organisms that are part of marine phytoplankton. They are found floating in fresh and salt waters, mainly on the surface of water bodies where the incidence of light is greater (Abada and Segev 2018). These microorganisms are extremely important for maintaining the carbon balance of the planet because they fix a large part of the CO2 generated and produce the greatest amount of oxygen on the planet (Saifullah et al. 2014). Phytoplankton and microalgae are one of the most critical bioindicators of water quality in rivers, seas, and lakes. Small changes in the physical–chemical characteristics of the water can directly affect the marine phytoplankton, and these in turn can give a warning voice of the condition of the water body. According to (Cid et al. 2013), microalgae constitute the first level of aquatic trophic chains. Due to their microscopic size, it is possible to obtain samples at population and community levels. They can be grown in photobioreactors quickly under controlled conditions due to their short generation times (Peng et al. 2020). Accordingly, these microorganisms respond quickly to environmental changes, and any effect on them can affect the highest trophic levels. In addition, microalgae are highly sensitive to a wide variety of compounds, both organic and inorganic (Jasniet al. 2020), offering the possibility of studying the transgenerational effects of exposure to pollutants and being a model of choice to evaluate the effects of long-term exposure to pollutants at the population level. Therefore, microalgae are considered one of the ideal bioindicators (Galloet al. 2020; O'Neill and Rowan 2022).

Microalgae toxicity tests are often required by European environmental authorities for chemical notifications and are also increasingly being used to manage chemical discharges (Cidet al. 2013). The parameters that are generally evaluated in toxicity studies are cell growth and viability. The first is the most studied parameter and reflects the physiological state of the cells. The second parameter is a test that analyzes microalgae in a medium, aiming the evaluation of their activity qualitatively and quantitatively (An et al. 2021; Qi et al. 2022).

13.4 Main Species of Microalgae Used as Bioindicators and Their Specific Applications

Most microalgae toxicity tests have been carried out with freshwater microalgae (Wan and Lee 2021). Although microalgae are considered ideal bioindicators, there are few bioassays that can be classified as standards for marine environments (Cid et al. 2013). The main genera of microalgae that have been used in bioassays of different contaminants include *Chlamydomonas*, *Chlorella*, *Dunaliella*, *Scenedesmus*, and *Selenastrum*.

For instance, the microalgae *D. tertiolecta* has been shown to be more sensitive than other microalgae and cyanobacteria when exposed to different types of metals. For example, the IC50 of *D. tertiolecta* exposed to copper was 0.530 mg/L, while that of *Phaeodactylum tricornutum* was 1.19 mg/L at 72 h (Horvatić and Peršić 2007; Levy et al. 2007). When *D. tertiolecta* was exposed to silver nanoparticles (nAg) it showed an IC50 of 0.9 mg/L while the IC50 for *S. costatum* was 3.4 times higher (Gambardella et al. 2015). The microalgae *D. tertiolecta* also exhibited hypersensitivity to metolachlor herbicide with an EC50 of 0.011×10^{-3} mg/L, while the EC50 for *S. costatum* was 0.061 mg/L (Machado and Soares 2019). Gallo et al. (2020) also showed that the microalgae *D. tertiolecta* is more sensitive than *P. tricornutum* and *S. costatum* exposed to sediment elutriates. Therefore, it can be concluded that the microalgae *D. tertiolecta* could be a good indicator for heavy metals and herbicides.

In contrast, the microalgae of the *Charophyta* is commonly used as a bioindicator of water quality, whereas microalgae of the genus *Selanastrum* have been used as bioindicators of water contamination (Parmar et al. 2016). Hairunnadawiah et al. (2022) used microalgae to assess water quality from the Batujai Dam in Central Lombok. The results showed that the identified microalgae consisted of 5 classes, 16 orders, and 40 species. The 5 classes found consisted of Bacillariophycea, Chlorophyceae, Coscinodiscophyceae, Cyanophyceae, and Euglenoidae. The discovery of microalgae of the Euglenophyceae class indicates that large amounts of organic matter are entering the Batujai Dam. Therefore, the waters of this dam are classified as slightly polluted and then can be used in water fish farming, livestock, and agricultural irrigation activities. However, it is not recommended for human consumption or recreation.

Isnaini (2012) also highlighted that Euglenophyceae species live in fresh waters thatcontain high contamination of organic matter. On the other hand, (Arsad et al. 2021) used theepiphytic microalgae community as an aquatic bioindicator in the Brantas River, East Java, as away to monitor water quality. The results show that the periphyton community structure is inbalance, with six divisions of 59 different genera. The six divisions found are Bacillariophyta, Cyanophyta, Chlorophyta, Charophyta, Ochrophyta, and Rhodophyta. The most identified genusis the division Bacillariophyta with 28 genera, while the least identified genus is the division Rhodophyta with 1 genus. Based on the saprobic value obtained, the water quality of the river canbe classified as water with light to moderate levels of pollution. Other examples are presented in Table 13.1.

Table 13.1 Some studies using microalgae species as bioindicators and their applications

Microalgae	Application	References
Chlorella pyrenoidosa	Study of the acute and chronic effects of roxithromycin and its mechanism of removal by microalgae	Li et al. (2023)
Chlorella vulgaris	Microalgae cultivation in different proportions of activated sludge extracts from the treatment of synthetic effluent containing 2,4,6-trichlorophenol (2,4,6-TCP) for removal and use of toxicity	Wang et al. (2017)
Dunaliella tertiolecta	Chemically characterized municipal wastewater samples	Harbi et al. (2017)
Emiliania huxleyi and Chlorella vulgaris	Effect of waters contaminated by oil sands and the action of naphthenic acid, one of the most toxic organic pollutants present in oil sands process waters, both on the photosynthetic process and on plant growth. Both species could indicate the existence of pollution in aquatic ecosystems	Beddow et al. (2016)
Pseudokirchneriella subcapitata	Evaluation of the biochemical and structural modifications of the microalgae after exposure for 72 h to three high concentrations of Cd(II), Cr(VI), Cu(II) and Zn(II)	Machado et al. (2015)
Chlorella vulgaris and Chlamydomonas mexicana	Study of bisphenol toxicity, cellular stress, and biodegradation/bioaccumulation in both species. *C. mexicana* was more tolerant to the pollutant and could be used for the treatment of aqueous systems contaminated with the same	Ji et al. (2014)
Phaeodactylum tricornutum and Dunaliella tertiolecta	Interaction of nanoparticles CdSe/ZnS quantum dots (QDs) with the marine diatom and the green alga as biological models in the marine environment	Morelli et al. (2013)
Pseudokirchneriella subcapitata	Proteomic analysis induced by microalgae chromate in relation to changes in protein expression profile caused by exposure to chromium (Cr), more specifically, 0.2 and 1 ppm of potassium dichromate	Vannini et al. (2009)
Dunaliella primolecta	Develop assay for herbicides in marine environments	Santín-Montanyá et al. (2007)
Dunaliella tertiolecta	Study of algae growth responses to effluents from an aluminum coating plant and wastewater from a pharmaceutical plant producing analgesic/anti-inflammatory drugs	Türker Saçan and Akmehmet Balcioğlu (2006)
Dunaliella tertiolecta	Examination of the toxicity of Irgarol, both alone and in mixtures with other pesticides, as binary mixtures with three other pesticides, the fungicide chlorothalonil and the herbicides atrazine and 2,4-D	DeLorenzo and Serrano (2006)

(continued)

Table 13.1 (continued)

Microalgae	Application	References
Scenedesmus subspicatus	Evaluation of the toxicity of Cu in the presence of ethylenediaminetetraacetic acid (EDTA), a synthetic organic binder, and fulvic acid (FA), a typically natural organic binder. The extracellular Cu concentration level was a good indicator to measure the toxic effects of Cu on algal growth	Ma et al. (2003)
Selenastrum capricornutum	Monitoring river phytoplankton and several hydrological, physical, and chemical variables, in combination with bioassays for the characterization of nutrient, dissolved heavy metals and chlorophyll contents	O'Farrell et al. (2002)

13.5 International Standardized Tests and Evaluated Parameters

In the aquatic environment, microalgae present the role of primary producers in the food chain. It makes these microorganisms special to assess the adverse effects of xenobiotic substances through toxicity bioassays since a negative impact on microalgae can disturb higher trophic-level organisms (e.g., zooplankton and fish). In other words, a toxic effect on microalgae can interfere with the whole ecosystem balance, reducing water quality. Therefore, microalgae are considered suitable as bioindicators of aquatic contamination by being easily cultured and presenting strong sensitivity to inorganic and organic pollutants. Moreover, given their high surface-to-volume ratio and fast growth rate, microalgae provide short response times in the presence of toxic substances. Thence, microalgal tests and bioassays have been applied as useful tools to evaluate the potential toxicity of different compounds aiming at elucidating their impact on the aquatic environment (Lee et al. 2023; Tato and Beiras 2019).

Tests and bioassays using microalgae must be performed following some rules. For that reason, they are standardized by some organizations, such as International Organization for Standardization (ISO), Organization for Economic Cooperation and Development (OECD), and the American Society for Testing and Materials (ASTM). Accordingly, ISO 8692 (Water quality—Fresh water algal growth inhibition test with unicellular green algae) standardizes the evaluation of the growth inhibition of freshwater unicellular algae, whereas ISO 10253 (Water quality—Marine algal growth inhibition test with *Skeletonema* sp. and *Phaeodactylum tricornutum*) standardizes the assessment of the growth inhibition of marine unicellular algae (ISO 8692:2012) (Water Quality—Fresh Water Algal Growth Inhibition Test with Unicellular Green Algae), 2012; ISO 10253:2016 (Water Quality—Marine Algal Growth Inhibition Test with Skeletonema Sp. and Phaeodactylum Tricornutum), 2016). Test No. 201 from OECD (Freshwater Alga and Cyanobacteria, Growth Inhibition Test) provides guidelines to analyze the growth inhibition of freshwater microalgae and/or cyanobacteria exposed to a test substance (Test No. 201: Freshwater Alga and

Cyanobacteria, Growth Inhibition Test, 2006). The test ASTM E1218-21 (Standard Guide for Conducting Static Toxicity Tests with Microalgae), whose principles are similar to ISO and OECD tests, supplies the procedures to perform a toxicity test with both freshwater and saltwater microalgae (ASTM E1218-21: Standard Guide for Conducting Static Toxicity Tests with Microalgae, 2021). These three tests provide responses about the test substance's toxicity based on the microalgae's growth rate. Besides their short duration, they allow the examination of the effects of the evaluated substance upon multiple generations of an algal population (Boros and Ostafe2020; Tato and Beiras 2019).

The principle of the procedures standardized by ISO (8692 and 10253) is to cultivate exponentially growing microalgae cells and assess their growth in a medium containing the evaluated substance, which is tested in a range of concentrations. ISO 8692 recommends *Desmodesmus subspicatus* and *Pseudokirchneriella subcapitata* to test the effect of the target substance in freshwater environments, and ISO 10253 recommends the *Skeletonema* sp. And *Phaeodactylum tricornutum* to examine the effect of the target substance in marine environments. The response is the growth inhibition that is obtained by measuring the reduction in specific growth rate relative to a culture grown in identical conditions but not exposed to the evaluated substance (control). The test lasts 72 h, and the cell density is measured in intervals of at least 24 h. From both tests, the effective concentration is obtained, which means the concentration that caused a particular percentage reduction in the specific growth rate relative to the control (ISO 8692:2012 (Water Quality—Fresh Water Algal Growth Inhibition Test with Unicellular Green Algae), 2012; ISO 10253:2016 (Water Quality—Marine Algal Growth Inhibition Test with Skeletonema Sp. and Phaeodactylum Tricornutum), 2016). Thereby, several studies utilize both methodologies. For instance, ISO 8692 has been employed to evaluate the impact in fresh waters of heavy metals and polycyclic aromatic hydrocarbons (Baścik-Remisiewicz et al. 2011; Kreutzer et al. 2022), disinfectants and their byproducts (Chhetri et al. 2017), nanoparticles (Hund-Rinke and Simon 2006; Sorensen and Baun 2015), herbicides and their mixtures (Bisewska et al. 2012), and antibiotics (Halling-Sùrensen 2000). Similarly, ISO 10253 has been used to determine the impact in salt waters of some substances, such as microplastics (Piccardoet al. 2020), nanoparticles (Prosposito et al. 2019), and heavy metals (Renzi et al. 2014).

Regarding test No. 201 from OECD, batch cultures of exponentially growing organisms are prepared and exposed to the test substance within a period of usually 72 h. The strains suitable for the test are *Pseudokirchneriella subcapitata* and *Desmodesmus subspicatus*, but others can also be employed. The parameter measured is the growth decrease of the culture for each concentration of the evaluated substance, and the value is then compared with the control assay composed of unexposed cultures. Both growth and growth inhibition are determined by measuring the algal biomass (dry weight per volume) as a function of time. But since dry weight can be difficult to measure, other parameters, such as cell counts, cell volume, optical density, and fluorescence, can be used. At the end of the test, it is possible to obtain the values of ErCx and EyCx; the former expresses the concentration that promoted a specific percentage inhibition of the growth rate, and the latter indicates the concentration that promoted a specific percentage inhibition of the yield—biomass at the end minus biomass at the start of the test (Test No. 201: Freshwater Alga and Cyanobacteria, Growth Inhibition Test, 2006). This methodology has been useful for many studies that evaluated different types of substances—dyes (Croce et al. 2017), nanoparticles (Aruoja et al. 2009), pharmaceuticals and chlorophenols (Geiger et al. 2016), veterinary products (Coelho et al. 2011), mycotoxin (Eagles et al. 2019), and degradation products of flame retardants (Koch and Sures 2019)—or verified the ecotoxicity of engineered nanomaterials proposed for environmental applications (Fiorati et al. 2020).

Thus, according to ISO, OECD, and ASTM protocols, the use of the inhibition of microalgae growth rate is recommended as the most representative response to environmental stress caused by different xenobiotics to these organisms. By using these tests, the adverse effects of xenobiotic substances can be evaluated based on the results of the growth inhibition of microalgae. It enables the establishment of public policies for the conservation of the water quality and remediation of sites that have already been impacted. Moreover, standardized tests are essential for comparing the results of different research centers.

13.6 Molecular and Biochemical Biomarkers in Microalgae

Aquatic environments continuously receive polluting compounds and contaminants from human activities. The effects of these xenobiotics are found at the end of a chain of events that begins with the release of the pollutant by the emitting source, followed by its dispersion in the environment (which determines its environmental concentrations), and finally ends with the effects of the pollutant on organisms (Amoatey and Baawain 2019; Estêvão 2023).

The potential risk that these compounds represent for aquatic ecosystems can be evaluated through several parameters used as biological indicators. In this sense, biomarkers or biological markers are changes at the molecular, biochemical, histological or physiological level that can be measured experimentally and can indicate the occurrence of a certain normal, abnormal or pathological function of an organism or even a response of this organism to a pharmacological or xenobiotic agent exposure (Lionetto et al. 2019). The use of these biomarkers can help to determine the conditions of environmental contamination and pollution that are affecting living beings in a certain ecosystem, and can even be used to identify the severity or progression of determined contamination (Lomartire et al. 2021). Thus, biomarkers consist of the presence, absence or alteration of specific cells, molecules, genes, enzymes, or hormones. Biochemical and molecular biomarkers are of great relevance in environmental and ecotoxicological studies. Since they are very sensitive to the presence of certain xenobiotics and have high specificity, they are the first signs or effects of environmental contaminants and pollutants, including in aquatic organisms, such as microalgae (Pikula et al. 2019).

In order to eliminate toxic compounds from their bodies, living beings have developeddefense strategies. The biotransformation of xenobiotics basically consists of neutralizing the toxic potential of the chemical compound, removing toxic radicals or molecules by non-enzymatic reactions (scavenging) or enzymatic transformation of the compound so that it is easily eliminated contamination (Beiras 2018). In some cases, biotransformation produces a metabolite that is more toxic than the original compound, a process called bioactivation. If these metabolites accumulate and overcome the organism's defenses, then they can produce an injury that is manifested by a toxic response (Chen 2020).

Biotransformation is the most common mechanism that organisms use to detoxify andeliminate xenobiotics, and it is usually divided into two principal phases, which do not necessarily occur in sequence or in totality, and which depend on different biochemical processes (Fig. 13.2). Phase I (functionalization or oxidation) biotransforms xenobiotics by converting them intosubstrates of Phase-II enzymes; that is, Phase I consists of a set of reactions (generally oxidations, reductions, and hydrolysis) that make the compounds more hydrophilic and/or reactive, preparing them to be transformed by the reactions of Phase II. Phase II (conjugation) comprises conjugation reactions (via acetylation, glucuronidation, glutathionylation, methylation, and sulphation), in which the metabolite oxidized in phase I, with high energy binding, is conjugated with a non-toxic endogenous metabolite to form a hydrophilic product that can be eliminated from the organism. The elimination of the organism is called for some authors as Phase III (Beiras 2018).

As the synthesis of enzymes and other molecules involved in detoxification systems and xenobiotic metabolism is induced by the presence of these compounds in the environment in which living beings are found, the presence or depletion of these biomolecules has been widely used as biomarkers of aquatic environmental contamination (Beiras 2018). In microalgae, biochemical and molecular biomarkers of biotransformation that are widely used are the presenceor activity of Glutathione S-transferases (GSTs) and Glutathione Peroxidase (GPX), as enzymes active in Phase II, and cytochrome P450, the most common enzyme in Phase I (Chu et al. 2022; Hamed et al. 2022). Details on its action are presented in Table 13.2.

During the biotransformation process, especially in Phase I, free radicals are produced in the oxidation and reduction reactions promoted by cytochrome P450, just as they are produced naturally in the process of cellular respiration, the cellular machinery responsible for the constant flow of electrons in the respiratory chain (Chen 2020). These free radicals, such as reactive oxygen species (ROS) and reactive nitrogen species (RNS), can be used in various biological processes (e.g., cell signaling and carbon and nitrogen fixation in photosynthetic beings) or eliminated by mechanisms of the organism's antioxidant system (Pikula et al. 2019). However, when there is an imbalance between the use and neutralization of ROS and RNS and their production, causing an exacerbated increase of ROS/RNS intracellularly, there may be a series of damages to cellular structures and biomolecules, especially DNA, proteins, and lipids, what is called oxidative stress. These damaged molecules can be removed or restored by enzymatic systems, which can be modulated according to the oxidative state of the cell (Beiras 2018).

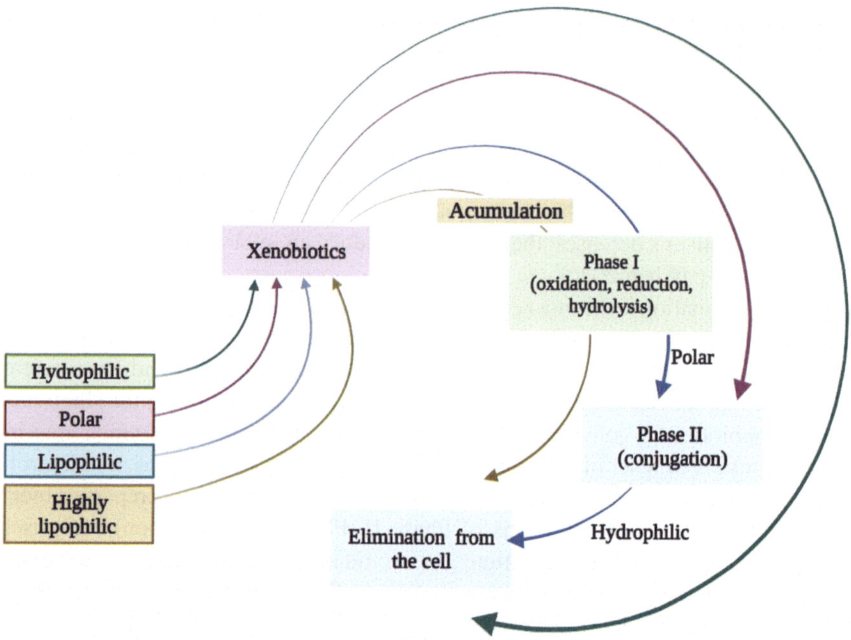

Fig. 13.2 Principal phases of xenobiotic biotransformation

Thus, one of the causes of the increase and acceleration of ROS production and, consequently, of oxidative stress, is the exposure of organisms to xenobiotics, such as organic pollutants and toxic metals. These compounds stimulate the generation of ROS and RNS in cells through disturbance of various metabolic processes, and it can be intensified in photosynthetic organisms, such as microalgae, that have many polyunsaturated lipids in their biological membranes, especially in mitochondria and chloroplasts, which is precisely where the electron transport chains are located in cells (Beiras 2018; Pikula et al. 2019).

The antioxidant system can act only for an initial period of contact with these substances, until the specific antioxidant mechanisms are depleted. Thus, changes in the antioxidant system at the biochemical and molecular level are one of the first signs of toxicity in organisms and allow mechanistically understanding some of the many actions of these substances in living beings, and therefore are considered excellent biomarkers of exposure to xenobiotics (Almeida et al. 2019; Pikula et al. 2019). In this way, it is possible to use the presence, absence, increase and decrease of ROS and RNS, as well as the activation or inhibition of antioxidant systems (enzymes and products of enzymatic action), as biomarkers of exposure to xenobiotics, which allows the detection of pollutants and their interferences in the antioxidant system of living beings in the first stages of environmental impacts (Beiras 2018; Pikula et al. 2019).

Table 13.2 Function and mechanism of action of different biochemical and molecular biomarkers in microalgae

Biomarker	Molecular classification	Function	Mechanism of action	Main actuation system	References
Nitrate reductase	Enzyme	Catalyst	Reduces nitrate to nitrite, the substrate for nitrite reductase; nitrite reduction products are used in amino acid biosynthesis	Antioxidant defense system	Wang et al. (2023)
Peroxiredoxins	Enzyme	Catalyst	Catalyzes the formation of H_2O from H_2O_2, corresponding alcohols from ROOH and NO_2 from $ONOO^-$	Biotransformation (phase II/ conjugation enzyme)	
Vanadium-bromoperoxidases (VBPx)	Enzyme	Catalyst	Catalyzes the production of volatile halocarbons, using H_2O_2 as substrate	Antioxidant defense system	Pikula et al. (2019)
Polyphenols (e.g., flavanoids e anthocyanins)	Polyphenols	Scavenger	They combine with non-specific free radicals (ROS and RNS) and alter their kinetics and reactions	Antioxidant defense system	
Tocopherols and Tocotrienols (vitamin E)	Vitamin	Scavenger	They donate phenolic hydrogen to lipid free radicals and donate electrons to the O_2^- singlet anion	Antioxidant defense system	

(continued)

Table 13.2 (continued)

Biomarker	Molecular classification	Function	Mechanism of action	Main actuation system	References
Glutathione	Tripeptide	Reducing agent	It donates electrons (non-specific donation) and acts as a substrate for enzymes, in addition to help repairing disulfide bonds in proteins; its oxidized form is glutathione disulfide (GSSH)	Antioxidant defense system	
Glutathione S-transferases	Enzyme	Catalyst	Conjugates glutathione (GSH) to electrophilic compounds such as free radicals, including ROS	Biotransformation (phase II / conjugation enzyme)	Abassi et al. (2019)
Glutathione Peroxidase	Enzyme	Catalyst	Promotes the conversion of H2O2 to H2O, or ROOH to alcohols	Biotransformation (phase II / conjugation enzyme)	Zhao et al. (2017)
Glutathione Reductase	Enzyme	Catalyst	Reduces GSSH to two GSH molecules using NADPH as a cofactor	Antioxidant defense system	Couto et al. (2016)
Cytochrome P450	Enzymatic system	Catalyst	Transfers electrons from NADPH + H^+ to oxygen, and transfers the OH group to the xenobiotic	Biotransformation (Phase I/ oxidation enzyme)	Torres et al. (2008)
Ascorbate Peroxidase	Enzyme	Catalyst	Catalyzes the reduction of peroxides (such as H_2O_2 and organic hydroperoxides) to water or alcohols, using ascorbate as an electron donor	Antioxidant defense system	

(continued)

Table 13.2 (continued)

Biomarker	Molecular classification	Function	Mechanism of action	Main actuation system	References
Superoxide Dismutase	Enzyme	Catalyst	Catalyzes the dismutation of superoxide anion (O_2^-) into oxygen (O_2) and hydrogen peroxide (H_2O_2)	Antioxidant defense system	
Catalase	Enzyme	Catalyst	Catalyzes the production of water using hydrogen peroxide (H_2O_2) as substrate	Antioxidant defense system	

In microalgae, the most used biochemical and molecular biomarkers of oxidative stress are Glutathione S-transferases (GST), Glutathione Reductase (GR), Glutathione Peroxidase (GPX), Ascorbate Peroxidase (APX), Superoxide Dismutase (SOD), Catalase (CAT) and Nitrate Reductase (NR), which fluctuate their activity according to the increase in oxidative stress caused by the xenobiotic, in addition to the generation of radical scavengers, such as Flavonoids, Polyphenols, Anthocyanins, Tocopherols, Tocotrienols and Glutathione (Pikula et al. 2019; Torres et al. 2008). The functions and mechanisms of action of each of these biomarkers are also shown in Table 13.2.

In addition to the increase or inhibition of these enzymes and proteins, lipid peroxidation—the oxidative degradation of lipids (especially polyunsaturated fatty acids in biological membranes) caused by free radicals—can also be used as a molecular biomarker of oxidative stress, including in microalgae. Malondialdehyde (MDA) is one of the main products of lipid peroxidation, and its cellular concentration after exposure to xenobiotics may indicate toxicity and effects on the lipid composition of cell membranes (Guo et al. 2020).

Another molecular biomarker used in microalgae is the class of proteins called metallothioneins. As their name suggests, metallothioneins respond to the presence of metals in the environment and, due to the high content of thiol groups (-SH) from cysteine residues in their structure, they can make connections to several metals (Beiras 2018; Torres et al. 2008).

There are three main classes of metallothioneins, present in different living beings. Class III metallothioneins (MTIII) are found in microalgae, with phytochelatins being the most representative elements of this class (Perales-Vela et al. 2006). Living beings have a basal rate of metallothionein, which acts in cellular metalloregulation, functioning as a reserve of essential metals and helping to maintain cellular homeostasis (Nordberg and Nordberg 2022). However, its synthesis is increased in the presence of metals in the environment, with cadmium, lead, zinc and copper, in that order, being the most potent inducers of its expression (Perales-Vela et al. 2006).

When heavy metals undergo cellular uptake, they compete with essential metals for intracellular ligands, and the metallothioneins then act as detoxifying agents, forming protein-metal complexes with the toxic metals. In addition, metallothioneins also act on the antioxidant system, since free thiols groups can scavenge ROS (Poluri and Mohan 2021). In the case of microalgae, the complex formed between MTIII and the metal ends up going to the cell vacuole by transporters capable of internalizing this complex in this organelle and also to the interior of chloroplasts and mitochondria (Fig. 13.3). The internalization of the MTIII-metal complex constitutes a heavy metal detoxification mechanism for these organisms, which may also explain why photosynthetic organisms, such as microalgae, have great tolerance and bioaccumulation of these inorganic pollutants (Perales-Vela et al. 2006).

The heat shock proteins (HSPs), in turn, are a group of proteins involved in maintaining protein homeostasis, where they assist in protein folding, aggregation, and trafficking processes (Reinle et al. 2022). They are divided into different families according to their molecular weight, the main ones being 90, 70, 60 and 1624 kDa, known as HSP90, HSP 70, HSP60 and low molecular weight, respectively (Bierkens et al. 1998). Each family of HSPs has different cellular functions and substrate specificities, but they generally work together, with each other and with other proteins and enzymes, and are generally very sensitive to heat (hence their name), which is a major causative agent of protein denaturation (Bierkens et al. 1998; Reinle et al. 2022).

As they are involved in the processes of conformational changes in proteins in general, some of the HSPs can be upregulated in a directly proportional way to the

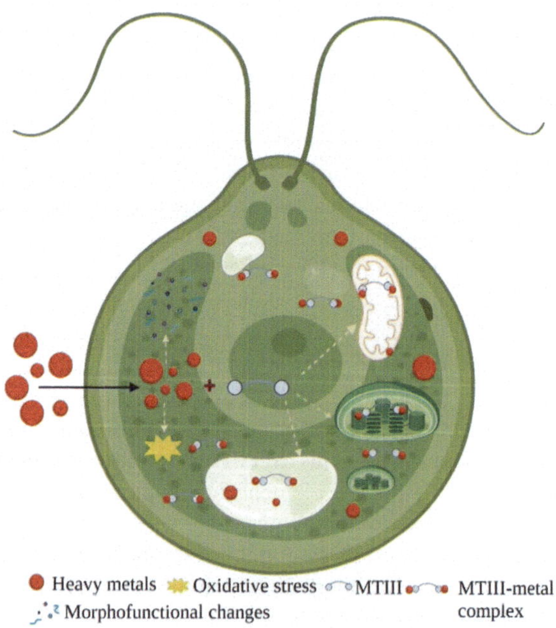

Fig. 13.3 MTIII-metal complexes and their cellular destination in microalgae

damage caused to these proteins. In this sense, HSPs are expressed to help prevent or repair proteotoxic effects, such as denaturing, unfolding and irreversible aggregation of proteins (Reinle et al. 2022). Therefore, as they are expressed in greater amounts when the cell is exposed to xenobiotics that cause such protein damage, HSPs are also considered important molecular biomarkers of contamination for environmental toxicology studies (Fig. 13.4) (Abassi et al. 2019; Bierkens et al. 1998).

HSPs seem to be very sensitive molecular biomarkers in microalgae, since they are highly expressed and activated very early in exposure to toxic agents, even at concentrations lower than doses considered lethal for organisms (Bierkens et al. 1998). Some studies have already demonstrated gene and protein expression of HSPs in a dose-dependent manner in microalgae exposed to non-lethal concentrations of xenobiotics (Bierkens et al. 1998; Tukaj and Tukaj 2010).

Table 13.3 presents examples of studies in which biochemical and molecular biomarkers were used to understand the effects and how substances or chemical compounds affect different species of microalgae, demonstrating how these changes at the molecular and biochemical level can indicate the occurrence of abnormal physiological conditions in these organisms as responses to exposure to pharmacological or xenobiotic agents.

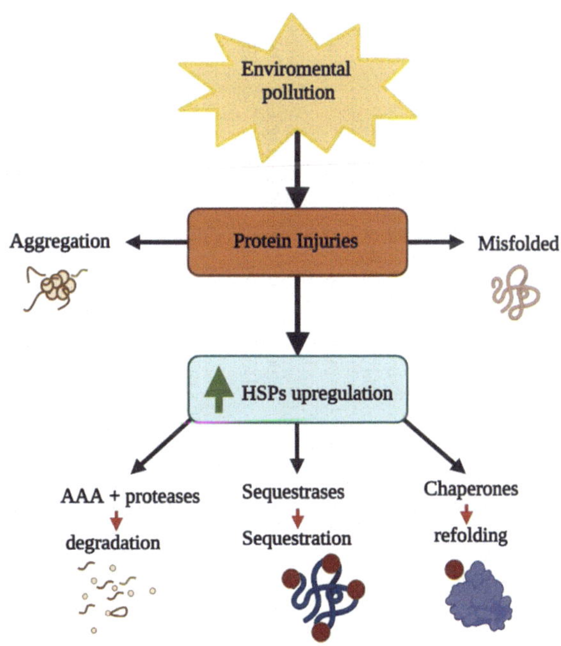

Fig. 13.4 HSPs response for protein injury. Environmental stress caused by pollutants or contaminants can induce protein injuries, such as aggregation or misfolding. These kinds of damages need to be repaired, and this is mostly provided by HSPs. Misfolded proteins can be refolded by chaperones, degraded by AAA + proteases or sequestrated by sequestrases. Adapted from (Reinle et al. 2022)

Table 13.3 Examples of studies that reported changes in biochemical and molecular levels in different species of microalgae exposed to xenobiotics

Xenobiotic	Microalgae species	Experimental conditions	Main observed results	References
Clarithromycin	Raphidocelis subcapitata	Exposure for 7 days, concentration levels of 0, 5, 20, 40, 80 $\mu g.L^{-1}$	Elevation of MDA, carotenoids and GSH; elevation of SOD, CAT, GP, and GST enzymatic activity	Guo et al. (2020
Benzophenone	Chlamydomonas reinhardtii	24h exposure to benzophenone 4 and 3 (97% purity)	Increased lipid peroxidation	Anido- Varela et al. (2022)
Sulfamethoxazole	Chlorella sorokiniana	Exposure to 0.1 $mg.L^{-1}$ and 5 $mg.L^{-1}$ for 10 days	Increased cytochrome p450 activity	Chu et al. (2022)
Polybrominated diphenyl ethers (PBDEs)	Alexandrium minutum	Exposure to 0.1, 0.5, and 1 $mg.L^{-1}$ BDE-47, for 96 h	Increase in ROS, consequently in SOD, CAT, GR and GPX activity	Zhao et al. (2017)
Cadmium, lead and chromium	Dunaliella salina, Amphora coffaeiformis and Navicula salinicola	Microalgae were grown for 14 days in the media supplemented with the metal solution	All metals induced up- regulation of CAT and SOD genes in *A. coffaeiformis* and *D. salina*; phytochelatin was up- regulated in *N. salinicola, A. coffaeiformis* and in *D. salina* after Cr exposure and also in *A. coffaeiformis* and *D. salina* after Cd exposure; Pb induced overexpression of phytochelatin for *D. salina*	Elleuch et al. (2021

(continued)

Table 13.3 (continued)

Xenobiotic	Microalgae species	Experimental conditions	Main observed results	References
Anthracene (PAH), cadmium and chloridazone (herbicide)	Desmodesmus subspicatus	Exposure to 0.08 and 0.44 $mg.dm^{-3}$ of anthracene, to 16.43 and 20.16 $mg.dm^{-3}$ of cadmium, and to 2.29 and 5.88 $mg.dm^{-3}$ of chloridazone for 6h and 24h; another treatment used the same concentrations, but with a thermal pretreatment of 40 °C for 1h	Increased cytoplasmic HSP70; heat treatment increased cadmium resistance	Tukaj and Tukaj (2010)
Crude oil components	Scenedesmus dimorphus	Exposure to crude oil for 21d at different concentrations (0, 0.01, 0.02, 0.04, 0.08, 0.1, 0.2, 0.4, 0.8, and 1% v/v)	Decreased content of chlorophylls, carotenoids, flavonoids, anthocyanins, and phenols; increased activity of SOD and increased activity of CAT and APX at lower concentrations	Ghodrati et al. (2022)
Atrazine	Chlamydomonas reinhardti	Exposure to 0.1, 0.25, 0.5, 1.0 and 2.0 mM atrazine for 96 h	Decrease in NR activity proportionally to the increase in concentrations; NR activity was completely depleted at the two highest concentrations	Fernández- Naveira et al. (2016)
Bisphenol A	Chlamydomonas reinhardti	6 h of exposure to several concentrations of BPA (from 10 to 50 $mg.L^{-1}$)	Elevation of O_2^- inside the cell	Esperanza et al. (2020)

13.7 Conclusions

The ever-increasing environmental pollution resulting mainly from water bodies, and the growing search for the identification of contaminants and their recovery, has become a critical object of study nowadays among researchers and other interested parties. Microalgae fit as an interesting tool since they play a role as bioindicators of aquatic contamination, and their metabolic response to xenobiotics can point to important biomarkers when exposed to pollution. Additionally, the utilization of microalgae as bioindicators is relatively accessible because, as among the various existing species, they are easily found and handled, serving as a resource for assessing ecological integrity. They often change their species composition or diversity, which can range from species-rich communities to smaller communities. Microalgae can purify water because they can control the concentration of dissolved oxygen and can remove nutrients from the environment by storing them in their biomass. Thus, when the aquatic environment is highly polluted, microalgae have direct or indirect mechanisms that can reveal such unregulated environmental factors.

Although some species have been widely known for several years, studies using-microalgae as bioindicators of contamination are recent and have been increasingly improved inresearch. The use of microalgae-based bioindicators is considered an operative means of assessing the quality of water bodies. The diverse information found in the literature proves the effectiveness of different species in indicating the main pollutants found in effluents, for example, in different environments, as well as in the removal of certain compounds. This function favors cost reductions by avoiding more complex analyses and the use of other tools, which can be as accurate as other methods and standardized parameters. The greatest advantage of using microalgae as bioindicators of aquatic contamination compared to conventional physicochemical methods is that they offer information on effects over time, while other typically applied methods only allow for punctual knowledge.

Finally, microalgae are increasingly used to monitor the ecological quality and health of the aquatic ecosystem and to define the effectiveness of management programs, restoration, or regulatory actions as well. A biological bioindicator based on microalgae is therefore of great value to reflect on the concept of environmental sustainability for water management.

References

Abada A, Segev E (2018) Multicellular features of phytoplankton. Front Mar Sci 5:1–7. https://doi.org/10.3389/fmars.2018.00144

Abassi S, Wang H, Ponmani T, Ki JS (2019) Small heat shock protein genes of the green algae Closterium ehrenbergii: cloning and differential expression under heat and heavy metal stresses. Environ Toxicol 34(9):1013–1024. https://doi.org/10.1002/tox.22772

Almeida AC, Gomes T, Habuda-Stanić M, Lomba JAB, Romić Ž, Turkalj JV, Lillicrap A (2019) Characterization of multiple biomarker responses using flow cytometry to improve environmental hazard assessment with the green microalgae Raphidocelis subcapitata. Sci Total

Environ 687:827–838. https://doi.org/10.1016/j.scitotenv.2019.06.124 ASTM E1218–21: Standard Guide for Conducting Static Toxicity Tests with Microalgae, (2021). https://www.astm.org/e1218-21.html

Amoatey P, Baawain MS (2019) Effects of pollution on freshwater aquatic organisms. Water Environ Res 91(10):1272–1287. https://doi.org/10.1002/wer.1221

An S, Hong S, Lee J, Cha J, Lee S, Moon H, Giesy JP, Seong J (2021) Identification of potential toxicants in sediments from an industrialized area in Pohang, South Korea: application of a cell viability assay of microalgae using flow cytometry. J Hazard Mater 405:124230. https://doi.org/10.1016/j.jhazmat.2020.124230

Anido-Varela L, Seoane M, Esperanza M, Cid Á, Rioboo C (2022) Cytotoxicity of BP-3 and BP-4: Blockage of extrusion pumps, oxidative damage and programmed cell death on Chlamydomonas reinhardtii. Aquat Toxicol 251(1907). https://doi.org/10.1016/j.aquatox.2022.106285

Anwer H, Mahmood A, Lee J, Kim KH, Park JW, Yip ACK (2019) Photocatalysts for degradation of dyes in industrial effluents: Opportunities and challenges. Nano Res 12(5):955–972. https://doi.org/10.1007/s12274-019-2287-0

Arsad S, Putra KT, Latifah N, Kadim MK, Mus M (2021) Epiphytic microalgae community as aquatic bioindicator in Brantas River, East Java, Indonesia. Biodiversitas 22:2961–2971

Aruoja V, Dubourguier HC, Kasemets K, Kahru A (2009) Toxicity of nanoparticles of CuO, ZnO and TiO_2 to microalgae Pseudokirchneriella subcapitata. Sci Total Environ 407(4):1461–1468. https://doi.org/10.1016/j.scitotenv.2008.10.053

Bartell SM (2016) Biomarkers, bioindicators, and ecological risk assessment—A brief review and evaluation assessment—A brief review and evaluation. Environ Bioindic 5275:60–73. https://doi.org/10.1080/15555270591004920

Baścik-Remisiewicz A, Aksmann A, Zak A, Kowalska M, Tukaj Z (2011) Toxicity of cadmium, anthracene, and their mixture to Desmodesmus subspicatus estimated by algal growth-inhibition ISO standard test. Arch Environ Contam Toxicol 60(4):610–617. https://doi.org/10.1007/s00244-010-9585-3

Beddow J, Johnson RJ, Lawson T, Breckels MN, Webster RJ, Smith BE, Rowland SJ, Whitby C (2016) The effect of oil sands process-affected water and model naphthenic acids on photosynthesis and growth in Emiliania huxleyi and Chlorella vulgaris. Chemosphere 145:416–423. https://doi.org/10.1016/j.chemosphere.2015.11.046

Beiras R (2018) Marine pollution : Sources, fate and effects of pollutants in coastal ecosystems. 2023.

Bierkens J, Maes J, Vander Plaetse F (1998) Dose-dependent induction of heat shock protein 70 synthesis in Raphidocelis subcapitata following exposure environmental pollutants. Environ Pollut 101(1):91–97. https://doi.org/10.1016/S0269-7491(98)00010-4

Bisewska J, Sarnowska EI, Tukaj ZH (2012) Phytotoxicity and antioxidative enzymes of green microalga (Desmodesmus subspicatus) and duckweed (Lemna minor) exposed to herbicides MCPA, chloridazon and their mixtures. J Environ Sci Health Part B Pesticides Food Contam Agric Wastes 47(8):814–822. https://doi.org/10.1080/03601234.2012.676443

Boros BV, Ostafe V (2020) Evaluation of ecotoxicology assessment methods of nanomaterials and their effects. In: Nanomaterials, vol. 10, Issue 4. MDPI AG. https://doi.org/10.3390/nano10040610

Burger J (2006a) Bioindicators: a review of their use in the environmental literature 1970–2005. Environ Bioindic 1(2):136–144. https://doi.org/10.1080/15555270600701540

Burger J (2006b) Bioindicators: types, development, and use in ecological assessment and research. Environ Bioindic 1(1):22–39. https://doi.org/10.1080/15555270590966483

Chen C-H (2020) Xenobiotic metabolic enzymes: bioactivation and antioxidant defense, S. Nature, Ed.; Vol. 4, Issue 1

Chhetri RK, Baun A, Andersen HR (2017) Algal toxicity of the alternative disinfectants performic acid (PFA), peracetic acid (PAA), chlorine dioxide (ClO_2) and their by-products hydrogen peroxide (H_2O_2) and chlorite (ClO_2-). Int J Hyg Environ Health 220(3):570–574. https://doi.org/10.1016/j.ijheh.2016.11.011

Chu Y, Zhang C, Wang R, Chen X, Ren N, Ho SH (2022) Biotransformation of sulfamethoxazole by microalgae: removal efficiency, pathways, and mechanisms. Water Res 221:118834. https://doi.org/10.1016/j.watres.2022.118834

Cid Á, Herrero C, Rioboo C, Esperanza M, Seoane M (2023) Chapter 19—Microalgae as whole-cell biosensors in the prospective assessment of toxic effects of emerging contaminants. In: Emerging contaminants in the environment, vol. 4, Issue 1, pp. 495–510

Cid Á, Prado R, Rioboo C, Suárez-Bregua P, Herrero C (2013) Use of microalgae as biological indicators of pollution: looking for new relevant cytotoxicity endpoints. Microalgae Biotechnol Microbiol Energy 311–324

Coelho S, Oliveira R, Pereira S, Musso C, Domingues I, Bhujel RC, Soares AMVM, Nogueira AJA (2011) Assessing lethal and sub-lethal effects of trichlorfon on different trophic levels. Aquat Toxicol 103(3–4):191–198. https://doi.org/10.1016/j.aquatox.2011.03.003

Couto N, Wood J, Barber J (2016) The role of glutathione reductase and related enzymes on cellular redox homoeostasis network. Free Radical Biol Med 95:27–42. https://doi.org/10.1016/j.freeradbiomed.2016.02.028

Croce R, Cinà F, Lombardo A, Crispeyn G, Cappelli CI, Vian M, Maiorana S, Benfenati E, Baderna D (2017) Aquatic toxicity of several textile dye formulations: acute and chronic assays with Daphnia magna and Raphidocelis subcapitata. Ecotoxicol Environ Saf 144:79–87. https://doi.org/10.1016/j.ecoenv.2017

DeLorenzo ME, Serrano L (2006) Mixture toxicity of the antifouling compound irgarol to the marine phytoplankton species Dunaliella tertiolecta. J Environ Sci Health Part B Pesticides Food Contam Agric Wastes 41(8):1349–1360. https://doi.org/10.1080/03601230600964100

Deprá MC, Severo IA, Dias RR, Zepka LQ, Jacob-Lopes E (2021) Chapter 2—Photobioreactor design for microalgae culture. In: Microalgae cultivation, recovery of compounds and applications, Vol. 4, Issue 1, pp 88–100

Eagles EJ, Benstead R, MacDonald S, Handy R, Hutchinson TH (2019) Impacts of the mycotoxin zearalenone on growth and photosynthetic responses in laboratory populations of freshwater macrophytes (Lemna minor) and microalgae (Pseudokirchneriella subcapitata). Ecotoxicol Environ Saf 169:225–231. https://doi.org/10.1016/j.ecoenv.2018.10.101

Elleuch J, Hmani R, Drira M, Michaud P, Fendri I, Abdelkafi S (2021) Potential of three local marine microalgae from Tunisian coasts for cadmium, lead and chromium removals. Sci Total Environ 799:149464. https://doi.org/10.1016/j.scitotenv.2021.149464

Esperanza M, Seoane M, Servia MJ, Cid Á (2020) Effects of bisphenol a on the microalga Chlamydomonas reinhardtii and the clam Corbicula fluminea. Ecotoxicol Environ Saf 197:110609. https://doi.org/10.1016/j.ecoenv.2020.110609

Estêvão MD (2023) Aquatic pollutants: risks, consequences, possible solutions and novel testing approaches. Fishes 8(2):8–10. https://doi.org/10.3390/fishes8020097

Fernández-Naveira A, Rioboo C, Cid A, Herrero C (2016) Atrazine induced changes in elemental and biochemical composition and nitrate reductase activity in Chlamydomonas reinhardtii. Eur J Phycol 51(3):338–345. https://doi.org/10.1080/09670262.2016.1163737

Fiorati A, Grassi G, Graziano A, Liberatori G, Pastori N, Melone L, Bonciani L, Pontorno L, Punta C, Corsi I (2020) Eco-design of nanostructured cellulose sponges for sea-water decontamination from heavy metal ions. J Cleaner Prod 246. https://doi.org/10.1016/j.jclepro.2019.119009

Gallo A, Guida M, Armiento G, Siciliano A, Mormile N, Carraturo F, Pellegrini D, Morroni L, Tosti E, Ferrante MI, Montresor M, Molisso F, Sacchi M, Danovaro R, Lofrano G, Libralato G (2020) Species-specific sensitivity of three microalgae tosediment elutriates. Mar Environ Res 156:104901. https://doi.org/10.1016/j.marenvres.2020.104901

Gambardella C, Costa E, Piazza V, Fabbrocini A, Magi E, Faimali M, Garaventa F (2015) Effect of silver nanoparticles on marine organisms belonging to different trophic levels. Mar Environ Res 111:41–49. https://doi.org/10.1016/j.marenvres.2015.06.001

Geiger E, Hornek-Gausterer R, Saçan MT (2016) Single and mixture toxicity of pharmaceuticals and chlorophenols to freshwater algae Chlorella vulgaris. Ecotoxicol Environ Saf 129:189–198. https://doi.org/10.1016/j.ecoenv.2016.03.032

Ghodrati M, Zarrini G, Kosari-Nasab M, Movafeghi A (2022) Capability of the microalga scenedesmus dimorphus for biodegradation of crude oil components: biological responses and catabolic intermediates. Clean: Soil, Air, Water 50(11):1–15. https://doi.org/10.1002/clen.202200207

Guo J, Peng J, Lei Y, Kanerva M, Li Q, Song J, Guo J, Sun H (2020) Comparison of oxidative stress induced by clarithromycin in two freshwater microalgae Raphidocelis subcapitata and Chlorella vulgaris. Aquat Toxicol 219:105376. https://doi.org/10.1016/j.aquatox.2019.105376

Hairunnadawiah, Khairuddin, Zulkifli L (2022) Microalgae diversity as a bioindicator of water quality in Batujai Dam, Central Lombok. J Appl Sci Environ Manage 315–322

Halling-Sùrensen B (2000) Algal toxicity of antibacterial agents used in intensive farming. Chemosphere 40:731–739

Hamed SM, Al-Nuaemi IJ, Korany SM, Alsherif EA, Mohamed HS, Abdelgawad H (2022) Hazard assessment and environmental fate of propiconazole degradation by microalgae: differential tolerance, antioxidant and detoxification pathway. J Environ Chem Eng 10(4):108170. https://doi.org/10.1016/j.jece.2022.108170

Harbi K, Makridis P, Koukoumis C, Papadionysiou M, Vgenis T, Kornaros M, Ntaikou I, Giokas S, Dailianis S (2017) Evaluation of a battery of marine species-based bioassays against raw and treated municipal wastewaters. J Hazard Mater 321:537–546. https://doi.org/10.1016/j.jhazmat.2016.09.036

Holt EA, Miller SW (2011) Bioindicators: using organisms to measure environmental impacts. Nat Educ Knowl 3(10):1–10

Horvatić J, Peršić V (2007) The effect of Ni 2+, Co 2+, Zn 2+, Cd 2+ and Hg 2+ on the growth rate of marine diatom Phaeodactylum tricornutum bohlin: microplate growth inhibition test. Bull Environ Contam Toxicol 79(5):494–498. https://doi.org/10.1007/s00128-007-9291-7

Hund-Rinke K, Simon M (2006) Ecotoxic effect of photocatalytic active nanoparticles (TiO_2) on algae and daphnids. Environ Sci Pollut Res 13(4):225–232. https://doi.org/10.1065/espr2006.06.311

Isnaini (2012) Phytoplankton community structure in the waters of the banyuasin river estuary, Banyuasin Regency, South Sumatra. Maspari J 04:58–68

Jasni J, Narayanan S, Haiza N, Yasin M, Mohamed P, Lin S, Liu C, Wu S, Jahim J, Sobri M (2020) Comparative toxicity effect of organic and inorganic substances in palm oil mill effluent (POME) using native microalgae species. J Water Proc Eng 34:101165. https://doi.org/10.1016/j.jwpe.2020.101165

Ji MK, Kabra AN, Choi J, Hwang JH, Kim JR, Abou-Shanab RAI, Oh YK, Jeon BH (2014) Biodegradation of bisphenol a by the freshwater microalgae Chlamydomonas mexicana and Chlorella vulgaris. Ecol Eng 73:260–269. https://doi.org/10.1016/j.ecoleng.2014.09.070

Koch C, Sures B (2019) Ecotoxicological characterization of possible degradation products of the polymeric flame retardant "Polymeric FR" using algae and Daphnia OECD tests. Sci Total Environ 656:101–107. https://doi.org/10.1016/j.scitotenv.2018.11.207

Kreutzer A, Faetsch S, Heise S, Hollert H, Witt G (2022) Passive dosing: assessing the toxicity of individual PAHs and recreated mixtures to the microalgae Raphidocelis subcapitata. Aquat Toxicol 249. https://doi.org/10.1016/j.aquatox.2022.106220

Lee J, Hong S, An SA, Khim JS (2023) Methodological advances and future directions of microalgal bioassays for evaluation of potential toxicity in environmental samples: a review. Environ Int 173. https://doi.org/10.1016/j.envint.2023.107869

Levy JL, Stauber JL, Jolley DF (2007) Sensitivity of marine microalgae to copper: the effect of biotic factors on copper adsorption and toxicity. Sci Total Environ 387(1–3):141–154. https://doi.org/10.1016/j.scitotenv.2007.07.016

Li C, Li P, Fu H, Chen J, Ye M, Zhai S, Hu F, Zhang C, Ge Y, Fortin C (2023) A comparative study of the accumulation and detoxification of copper and zinc in Chlamydomonas reinhardtii: the role of extracellular polymeric substances. Sci Total Environ 871:161995. https://doi.org/10.1016/j.scitotenv.2023.161995

Lionetto MG, Caricato R, Giordano ME (2019) Pollution biomarkers in environmental and human biomonitoring. Open Biomarkers J 9(1):1–9. https://doi.org/10.2174/1875318301909010001

Lomartire S, Marques JC, Gonçalves AMM (2021) Biomarkers based tools to assess environmental and chemical stressors in aquatic systems. Ecol Indicators 122:107207. https://doi.org/10.1016/j.ecolind.2020.107207

Ma M, Zhu W, Wang Z, Witkamp GJ (2003) Accumulation, assimilation and growth inhibition of copper on freshwater alga (Scenedesmus subspicatus 86.81 SAG) in the presence of EDTA and fulvic acid. Aquat Toxicol 63(3):221–228. https://doi.org/10.1016/S0166-445X(02)00179-0

Machado MD, Lopes AR, Soares EV (2015) Responses of the alga Pseudokirchneriella subcapitata to long-term exposure to metal stress. J Hazard Mater 296:82–92. https://doi.org/10.1016/j.jhazmat.2015.04.022

Machado MD, Soares EV (2019) Sensitivity of freshwater and marine green algae to three compounds of emerging concern. J Appl Phycol 31:399–408

Martinez-Burgos WJ, Bittencourt Sydney E, Bianchi Pedroni Medeiros A, Magalhães AI, de Carvalho JC, Karp SG, Porto de Souza Vandenberghe L, Junior Letti LA, Thomaz Soccol V, de Melo Pereira GV, Rodrigues C, Lorenci Woiciechowski A, Soccol CR (2021).Agro-industrial wastewater in a circular economy: characteristics, impacts and applications for bioenergy and biochemicals. Bioresour Technol 341(125795). https://doi.org/10.1016/j.biortech.2021.125795

Morelli E, Salvadori E, Bizzarri R, Cioni P, Gabellieri E (2013) Interaction ofCdSe/ZnS quantum dots with the marine diatom Phaeodactylum tricornutum and the green alga Dunaliella tertiolecta: a biophysical approach. Biophys Chem 182, 4–10. https://doi.org/10.1016/j.bpc.2013.06.007

Nordberg M, Nordberg GF (2022) Metallothionein and cadmium toxicology—Historical review and commentary. Biomolecules 12(3):1–15. https://doi.org/10.3390/biom12030360

O'Farrell I, Lombardo RJ, De Tezanos Pinto P, Loez C (2002) The assessment of water quality in the Lower Luján River (Buenos Aires, Argentina): phytoplankton and algal bioassays. Environ Pollut 120(2):207–218. https://doi.org/10.1016/S0269-7491(02)00136-7

Omar WMW (2010) Perspectives on the use of algae as biological indicators for monitoring and protecting aquatic environments, with special reference to Malaysian freshwater ecosystems. Trop Life Sci Res 21(2):51–67

O'Neill, E. A., & Rowan, N. J. (2022). Microalgae as a natural ecological bioindicator for the simple real-time monitoring of aquaculture wastewater quality including provision for assessing impact of extremes in climate variance – A comparative case study from the Republic of Ireland. *Science of the Total Environment*, *802*, 149800. https://doi.org/10.1016/j.scitotenv.2021.149800 Test No. 201: Freshwater Alga and Cyanobacteria, Growth Inhibition Test, (2006).

Parmar TK, Rawtani D, Agrawal YK (2016) Bioindicators: the natural indicator of environmental pollution. Front Life Sci 9(2):110–118. https://doi.org/10.1080/21553769.2016.1162753

Peng L, Fu D, Chu H, Wang Z, Qi H (2020) Biofuel production from microalgae: a review. Environ Chem Lett 18(2):285–297. https://doi.org/10.1007/s10311-019-00939-0

Perales-Vela HV, Peña-Castro JM, Cañizares-Villanueva RO (2006) Heavy metal detoxification in eukaryotic microalgae. Chemosphere 64(1):1–10. https://doi.org/10.1016/j.chemosphere.2005.11.024

Piccardo M, Provenza F, Grazioli E, Cavallo A, Terlizzi A, Renzi M (2020) PET microplastics toxicity on marine key species is influenced by pH, particle size and food variations. Sci Total Environ 715. https://doi.org/10.1016/j.scitotenv.2020.136947

Pikula KS, Zakharenko AM, Aruoja V, Golokhvast KS, Tsatsakis AM (2019) Oxidative stress and its biomarkers in microalgal ecotoxicology. Curr Opin Toxicol 13:8–15. https://doi.org/10.1016/j.cotox.2018.12.006

Poluri ST, Mohan K (2021) Metallothionein-and phytochelatin- assisted mechanism of heavy metal detoxification in microalgae. In: Springer (Ed.), Approaches to the remediation of inorganic pollutants p 2021

Prosposito P, Burratti L, Bellingeri A, Protano G, Faleri C, Corsi I, Battocchio C, Iucci G, Tortora L, Secchi V, Franchi S, Venditti I (2019) Bifunctionalized silver nanoparticles as Hg2+ plasmonic sensor in water: Synthesis, characterizations, and ecosafety. Nanomaterials 9(10). https://doi.org/10.3390/nano9101353

Qi X, Ru S, Xiong J (2022) Environmental technology and innovation ecotoxicological effects of sulfacetamide on a green microalga, Desmodesmus quadricauda: cell viability, antioxidant system, and biotransformation. Environ Technol Innov 26:102278. https://doi.org/10.1016/j.eti.2022.102278

Reinle K, Mogk A, Bukau B (2022) The diverse functions of small heat shock proteins in the proteostasis network: functions and mechanisms of sHsps. J Mol Biol 434(1). https://doi.org/10.1016/j.jmb.2021.167157

Renzi M, Roselli L, Giovani A, Focardi SE, Basset A (2014) Early warning tools for ecotoxicity assessment based on Phaeodactylum tricornutum. Ecotoxicology 23(6):1055–1072. https://doi.org/10.1007/s10646-014-1249-z

Rizo-Patrón VF, Kumar A, McCoy Colton MB, Springer M, Trama FA (2013) Macroinvertebrate communities as bioindicators of water quality in conventional and organic irrigated rice fields in Guanacaste, Costa Rica. Ecol Ind 29:68–78. https://doi.org/10.1016/j.ecolind.2012.12.013

Saifullah AZA, Karim A, Ahmad-yazid A (2014) Microalgae : an alternative source of renewable energy. Am J Eng Res 03(03):330–338

Santín-Montanyá I, Sandín-España P, García Baudín JM, Coll-Morales J (2007) Optimal growth of Dunaliella primolecta in axenic conditions to assay herbicides. Chemosphere 66(7):1315–1322. https://doi.org/10.1016/j.chemosphere.2006.07.019

Sorensen SN, Baun A (2015) Controlling silver nanoparticle exposure in algal toxicity testing-A matter of timing. Nanotoxicology 9(2):201–209. https://doi.org/10.3109/17435390.2014.913728

Tato T, Beiras R (2019) The use of the marine microalga Tisochrysis lutea (T-iso) in standard toxicity tests; comparative sensitivity with other test species. Front Mar Sci 6(JUL). https://doi.org/10.3389/fmars.2019.00488

Torres MA, Barros MP, Campos SCG, Pinto E, Rajamani S, Sayre RT, Colepicolo P (2008) Biochemical biomarkers in algae and marine pollution: a review. Ecotoxicol Environ Saf 71(1):1–15. https://doi.org/10.1016/j.ecoenv.2008.05.009

Tukaj S, Tukaj Z (2010) Distinct chemical contaminants induce the synthesis of Hsp70 proteins in green microalgae Desmodesmus subspicatus: heat pretreatment increases cadmium resistance. J Thermal Biol 35(5), 239–244. https://doi.org/10.1016/j.jtherbio.2010.05.007

Türker Saçan M, Akmehmet Balcioğlu I (2006) A case study on algal response to raw and treated effluents from an aluminum plating plant and a pharmaceutical plant. Ecotoxicol Environ Saf 64(2):234–243. https://doi.org/10.1016/j.ecoenv.2005.03.017

Vannini C, Marsoni M, Domingo G, Antognoni F, Biondi S, Bracale M (2009) Proteomic analysis of chromate-induced modifications in Pseudokirchneriella subcapitata. Chemosphere 76(10):1372–1379. https://doi.org/10.1016/j.chemosphere.2009.06.022

Vasilachi IC, Asiminicesei DM, Fertu DI, Gavrilescu M (2021) Occurrence and fate of emerging pollutants in water environment and options for their removal. Water 13(2):1–34. https://doi.org/10.3390/w13020181

Wan J, Lee C (2021) Influence of polystyrene microplastic and nanoplastic on copper toxicity in two freshwater microalgae. Environ Sci Pollut Res 28:33649–33668

Wang L, Chen X, Wang H, Zhang Y, Tang Q, Li J (2017) Chlorella vulgaris cultivation in sludge extracts from 2,4,6-TCP wastewater treatment for toxicity removal and utilization. J Environ Manage 187:146–153. https://doi.org/10.1016/j.jenvman.2016.11.020

Wang Y, Johnson GI, Postles A, Coyne KJ (2023) Nitrate reductase enzymes in alga Chattonella subsalsa are regulated by environmental cues at the translational and post-translational levels. Front Microbiol 14:1–10. https://doi.org/10.3389/fmicb.2023.1059074

Yadav S, Singh AK, Verma E (2023) Cyanobacteria as bioindicator of water pollution. In: Cyanobacterial biotechnology in the 21st Century, p 2023

Zhao Y, Wang Y, Li Y, Santschi PH, Quigg A (2017) Response of photosynthesis and the antioxidant defense system of two microalgal species (Alexandrium minutum and Dunaliella salina) to the toxicity of BDE-47. Mar Pollut Bull 124(1):459–469. https://doi.org/10.1016/j.marpolbul.2017.07.038

Part III
Microalgae for Biodiesel Production

Chapter 14
Energy Production from Algal-Bacterial and Other Biosolids for Electricity

Ramón Piloto-Rodríguez and Sven Pohl

Abstract Sustainable production of energy is a main issue in any society. In this respect, the production of zero emission biofuels and green electricity are in the spotlights today. The production of energy for algae is not new, but new technologies, which are more efficient, scale-up and with cost reduction are under development. In this line, the production of power and electricity from algae and different biosolids are alternatives. The culture of microalgae-bacteria consortia in sewage sludge contribute to clean energy production and to sustainability. A greater insight on the control of the interactions between microalgae and other microorganisms, particularly bacteria, are being useful for enhancing the efficiency of microalgal biomass production and co-products. Attention must be paid to the controlled utilization of microalgae-bacteria consortia. Multi-anode sediment microbial fuel cells have the potential for real-scale implementation in order to produce electricity. The production of hydrogen and biogas are also main targets for the implementing of such technologies. This chapter includes several issues related to power and electricity production from algae, including several bio solids. The chapter explores from the basic fundamentals, to fuel cell technologies, type of electrodes, reactors, energy outputs and issues related to the scale-up and practical applications of these technologies, trends and future.

Keywords Algae · Bacteria · Consortia · Biofuels · Electricity

R. Piloto-Rodríguez (✉)
Center for the Study of Renewable Energies, Technical University of Havana, Calle 114, No. 11901 e/119 y 127, Marianao 15, La Habana, Cuba
e-mail: rpiloto.rodriguez@gmail.com

S. Pohl
Institute of Thermodynamics, Energy Process Engineering and System Engineering, University of Applied Sciences of Mittelhessen (THM), Wiesenstrasse 14, 35390 Gießen, Germany
e-mail: sven.pohl@me.thm.de

T. A. Kurniawan and A. Anouzla (eds.), *Algae as a Natural Solution for Challenges in Water-Food-Energy Nexus*, Environmental Science and Engineering,
https://doi.org/10.1007/978-981-97-2371-3_14

14.1 Introduction

Energy production from algal-bacterial and biosolids represents a cutting-edge and sustainable approach to generating electricity. This innovative method harnesses the natural processes of photosynthesis and microbial activity to convert organic matter, such as algae and wastewater solids, into electrical energy. As we face growing concerns about energy security, environmental sustainability, and waste management, this emerging field offers a promising solution at the intersection of renewable energy and wastewater treatment. Algae, bacteria, and biosolids are readily available and renewable resources. Algae, in particular, are highly efficient at converting sunlight and carbon dioxide into biomass, making them an ideal candidate for energy production. The process involves two main steps: (I) Algae capture solar energy to produce organic matter through photosynthesis. (II) Specialized bacteria metabolize this organic matter, breaking it down into biogas (methane) or other biofuels, which can be used to generate electricity. This approach mitigates several environmental challenges simultaneously. It reduces greenhouse gas emissions by capturing CO_2 during photosynthesis and prevents the release of methane, a potent greenhouse gas, from decomposing organic waste. It also addresses the issue of wastewater treatment, transforming a potential pollutant into a valuable energy source. The integrated approach of algal-bacterial energy production optimizes resource use. It utilizes non-arable land for algae cultivation, reduces the need for synthetic fertilizers, and maximizes the energy potential of wastewater solids that would otherwise require costly disposal methods.

Energy production from algal-bacterial and biosolids can be tailored to various scales, from small municipal wastewater treatment plants to large industrial facilities. Its adaptability makes it suitable for both urban and rural settings. While promising, this technology is still in the research and development stage. Challenges include improving biomass yields, optimizing microbial processes, and refining system integration for higher energy conversion efficiencies.

As society seeks sustainable energy alternatives, this technology is gaining traction. It has the potential to contribute to decentralized energy production and address the energy needs of remote or off-grid areas. Further research and development will likely unlock its full potential and contribute to a cleaner, more sustainable energy future.

14.2 General Use of Algae for Energy Production

The development of new technologies for sustainable energy production, and the energy efficiency of the current ones, is a main issue in the actual energy scenario worldwide. In this respect, the production of energy from algae is still a cutting edge subject. Energy production from microalgae and macroalgae has several advantages.

Among them, high biomass yield in the case of microalgae, and because they represent an abundant natural resource, in the case of macroalgae are very important features. Especially, in the case of microalgae, they represent the living biomass source with the highest growing rate, by far (Sánchez-Borroto et al. 2018; Mohamed 2018; Pandey et al. 2014).

Algal biomass could be used to generate various energy outputs, by different conversion methods, depending on feedstock characteristics and demand of energy. Liquid and gaseous renewable fuels can be produced through processes such as transesterification, pyrolysis, hydrothermal liquefaction, gasification, and anaerobic digestion (Tobío-Pérez et al. 2021). This way, biodiesel, biogas, hydrogen and terpene based fuels; among others can be produced (Lapuerta et al. 2023; Tetali 2019; Tsolakis et al. 2019).

Microalgae are prokaryotic or eukaryotic photosynthetic organisms. They grow fast in fresh or salt water due to their simple cellular structure. Because of this, they are natural converters of solar energy. Microalgae use carbon dioxide as primary carbon source, and additionally they need water and nutrients. They are one of the oldest living organisms and energy sources in our planet and are represented by more than 300,000 species (Tobío-Pérez et al. 2021). An additional feature is that they can grow in waste water, getting their nutrients from this medium and being an efficient and clean water regeneration system. Because as nutrients they need nitrogen and phosphorus, they efficiently reduce the amount of these elements in water. Microalgae culture takes place in open or covered ponds, or closed photobioreactors, based on tubular, flat plate or other designs (Mohamed 2018; Pandey et al. 2014; Riveiro and Da Silva 2013).

Figure 14.1 represents a general squeme where a huge CO_2 producer (power plant) can be coupled to a green algae farm (open pound system in this case). This way, the CO_2 is captured and used as substrate for algae growing, and the nutrient and aqueous medium can be pumped from the sea, which is often in contact with the power plant. Normally, power plants are settled just beside the sea in order to use seawater as cooling system. The farm can be established beside the power plant, which is logic to be surrounded by poor quality soils, not competing this way with crops plantation. During the downstream process, fresh water and oxygen are delivered, and the algal biomass can be used for bioenergy production.

On the other hand, sewage sludge (also known as biosolids) is defined as the residual, semi-solid produced as a by-product during the sewage treatment of industrial, households or municipal wastewater (Kumar et al. 2017). Generally, biosolids contain 20% content of fats, 50% of carbohydrates, 30–40% of organic matter, 3% total nitrogen, 1.5% total phosphorus, 0.7% total potassium content, and certain levels of copper, zinc, iron, chromium, manganese, mercury, lead, nickel and cadmium. The heating value of biosolids is 12.000 kJ/kg, and they represent a nutrient-rich organic material. Among other well-known uses, it can be used as raw material for energy production.

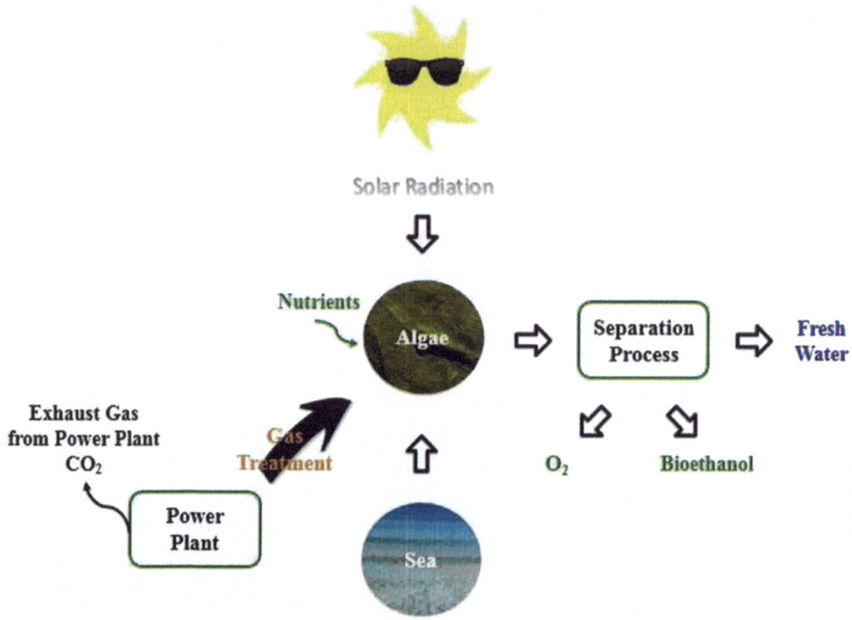

Fig. 14.1 Squeme of algae bioenergy production from an industrial CO_2 source

The use of algae and microalgae for energy production is a current line of scientific and technical development. It is a system full of new ideas and business opportunities, within the concepts of biorefinery and circular economy. Inside this field, the implementation of algal-bacterial consortia is an attractive pathway.

14.3 Electricity and Power Production from Algal-Bacterial and Other Biosolids

Anaerobic digestion is a bacterial decomposition process in the absence of oxygen that stabilizes organic wastes and produces biogas, which is a mixture of methane and carbon dioxide. It has been receiving more attention for a further energy use of biosolids. In this respect, biogas technology is one of the ways for deployment in conversion of organic residues to bioenergy and valuable fertilizers. The biogas can be produced with anaerobic bacteria or fermentation of biodegradable materials. Its heating value is 21–25 MJ/m^3, with an average concentration of 65% CH_4 (Mohamed 2018), and is considered similar to natural gas concerning energy content with the advantage of being renewable. Four key biochemical stages take place during anaerobic digestion, which are hydrolysis, acidogenesis, acetogenesis and methanogenesis; all enrolling bacterial activity.

The sewage sludge may be utilized for energy generation and recovery, not only by anaerobic digestion. In this respect, a dry preprocess should be implemented. Pyrolysis or Gasification, are thermochemical conversion processes that generate biofuels such as bio-oil and syngas, and can be applied for thermochemical conversion of bio feed such as algae, but these are not considered ideal feedstocks for such processes due to particular chemical and physical properties such as high water and ash content (Pfeil et al. 2020; Tobío-Pérez et al. 2021).

14.3.1 Substrates. The Role of Microorganisms

Substrates play a key role in the cultivation of algae for energy production. To harness the full potential of algae as an energy source, it is essential to understand the role of substrates in their growth and how microorganisms contribute to this process.

Algae require essential nutrients, including nitrogen, phosphorus, and micronutrients, for growth. These nutrients are often supplied through substrates such as wastewater, agricultural runoff, or synthetic nutrient solutions. The choice of substrate can significantly impact the growth rate and biomass yield of algae cultures. The composition and availability of nutrients must be carefully balanced to maximize productivity and prevent nutrient pollution. Algae primarily capture carbon dioxide from the atmosphere during photosynthesis. However, to enhance their growth and energy content, additional carbon sources can be supplied through substrates like flue gas from industrial processes or organic carbon-rich materials. This not only accelerates algae growth but also helps in carbon capture and utilization, mitigating greenhouse gas emissions. On the other hand, some substrates may naturally have a high or low pH, affecting the viability of the algae culture. Adjusting the pH using substrates or chemical additives is necessary to maintain an optimal growth environment.

Microbes can also aid in the degradation of organic substrates used in algae cultivation. For example, in wastewater-based systems, bacteria break down organic matter, releasing nutrients that are subsequently taken up by algae. This not only helps recycle nutrients but also improves the overall efficiency of the cultivation process.

Substrates are fundamental to algae cultivation for energy production, serving as sources of essential nutrients, carbon, and pH regulation. Their careful selection and management are critical for maximizing biomass yield and biofuel production while minimizing environmental impacts. Furthermore, the role of microorganisms in supporting algae growth cannot be understated. Beneficial symbiotic relationships and waste degradation by bacteria contribute significantly to the success of algae-based energy systems. As research continues to advance, optimizing substrate utilization and microbial interactions will be key factor to the sustainable development of algae-based biofuels.

14.3.2 Algae-Bacteria Consortia

The algae-bacteria consortium has been studied since more than seventy years (Vasseur et al. 2012). The description highlights the synergistic relationship between green algae and bacteria in the context of bioenergy processes. The symbiosis between these organisms allows for the enhancement of bioenergy production through natural collaboration and coexistence.

Photosynthetic bacteria play a role in the process, utilizing various reduced carbon or sulfur compounds as reductant for photosynthesis. Interestingly, in the absence of combined or free nitrogen, these bacteria have the ability to convert carbonaceous compounds into hydrogen and carbon dioxide. However, a limitation arises as these bacteria cannot use water as the substrate for hydrogen evolution.

On the other hand, algae possess an effective water-splitting complex. This complex enables algae to use water as an electron donor for photosynthesis, and it can be coupled with hydrogenase, an enzyme that facilitates hydrogen production. This connection allows for the production of hydrogen with a maximum potential net energy conversion efficiency exceeding 10%.

The combined capabilities of photosynthetic bacteria and algae create a system where solar radiation energy can be converted and stored as chemical free energy, contributing to bioenergy processes. Despite being a cutting-edge technology, it holds promise for sustainable and efficient energy production.

The Microbial Fuel Cell (MFC) is a fascinating technology that leverages the anaerobic respiration of microorganisms to directly convert organic waste into electricity. In this process, microorganisms act as biocatalysts, facilitating the conversion of fuel into electrical energy. The use of photosynthetic organisms, such as algae, in the cathode half-cell is particularly interesting.

In MFCs, the cathode half-cell requires oxygen for the reduction reaction. Photosynthetic organisms, such as algae, can play a dual role in this context. They not only contribute to meeting the oxygen demand for the reduction reaction but also produce biomass during photosynthesis. This biomass can serve as a direct fuel source for the microbial fuel cell.

Algal powder, in particular, has been identified as a suitable feedstock for MFCs. The photosynthetic activity of algae allows them to generate biomass, and the organic components of this biomass can be utilized as the fuel for microbial respiration, contributing to the generation of electricity.

Overall, MFCs offer a promising avenue for sustainable and direct electricity generation from organic waste, with the added benefit of incorporating photosynthetic organisms like algae into the process (Gajda et al. 2015).

The isolation of a consortium involving *Nannochloris* spp. along with a diverse bacterial community from naturally highly-productive marine coastal ecosystems is an interesting approach. In this consortium, *Nannochloris* spp. plays a role in photosynthesis, utilizing phosphate, ammonia, and acetate as sources of phosphorus, nitrogen, and carbon, respectively. It's noteworthy that this consortium was developed in a batch culture using artificial seawater, emphasizing its potential for sustainable and nutrient-efficient growth.

The involvement of heterotrophic bacteria in the consortium is significant. These bacteria assimilate acetate, contributing to their own metabolism, and may act as probiotics. Probiotic bacteria can play a beneficial role by providing essential compounds such as vitamins and carbon dioxide, which are crucial for the growth of microalgae like *Nannochloris* spp. This synergistic interaction between microalgae and bacteria can enhance the overall productivity and stability of the consortium.

The use of naturally occurring and locally adapted consortia offers advantages in terms of ecosystem compatibility and adaptability to specific environmental conditions. It also aligns with the principles of sustainable and eco-friendly approaches in biotechnology, where microorganisms are harnessed in consortia to achieve specific goals, such as efficient nutrient utilization and enhanced biomass production (Vasseur et al. 2012).

Negative effects of bacteria are however possible when acting as parasites on microalgae leading to their death and because they can also compete for nutrients. In natural conditions, the growth of the algal-bacterial consortium and resulting biomass will be determined by physical and chemical factors such as temperature, light and pH, with different optimal growths in both cases.

Vasseur et al. (2012) reported an algae-bacteria consortium controlled by *Nannochloris* spp. growing on a synthetic anaerobic digester effluent reaching a maximum carbon conversion efficiency of 3.6%. When bacteria were included in the medium, 6.3% of efficiency is reached.

Bacteria associated with microalgal cultures can be very useful for their growth, providing vitamins for better microalgal growth which could reduce production cost of microalgal biomass and therefore higher efficiency.

Micronutrients like vitamins and macronutrients like nitrogen, oxygen and carbon usually exchange between algae and bacteria. Photosynthetic oxygen can be consumed by bacteria creating a suitable environment for algal hydrogen production. Bacteria supply vitamin B_{12} to the algae in exchange for fixed carbon. Antibiotics can be produced by the bacteria for algal protection against other microorganisms (mutualism/commensalism) or for control of algal blooms. Acyl-homoserine lactone is produced by bacteria and is involved in biofilm formation between bacteria and algae cells (wastewater treatment and biomass harvesting improvement) (Fuentes et al. 2016). Bacterial communities associated with microalgae cultures can also regenerate or fix inorganic nutrients that otherwise would not be bioavailable to them.

The interaction between algae and bacteria may occur in sea, but in several different environments. Examples of peer algae-bacteria can be consulted in Fuentes et al. (2016). The bacteria associated with microalgae play a key role not only in the growth but also in the composition of the microalgal biomass. The chemical composition of the microalgal biomass is a key factor in aquaculture.

Hydrogen production by microalgae depends on a hydrogenase enzyme activity that is highly sensitive to oxygen. Anaerobic conditions are required for an efficient production of hydrogen by microalgae. It has been reported that such anaerobic environment suitable for algal hydrogen production might be conferred by the bacteria which can consume the oxygen generated photosynthetically by the algae, without damaging the photosynthetic apparatus. With the help of the bacteria that consume the oxygen evolved, the algae can capture light energy and produce hydrogen at the same time without further manipulation of the system (Fuentes et al. 2016; Melis and Happe 2001; White et al. 2009).

14.3.3 Electroactive Bacteria

Electroactive bacteria, such as *Geobacter sulfurreducens* and *Shewanella oneidensis*, have exclusive metabolic capabilities that allow them to transfer electrons extracellularly to solid substrates or other microbial partners. These bacteria are known for their ability to respire using extracellular electron transfer (EET) mechanisms, making them ideal candidates for bioelectrochemical applications. EET is achieved through specialized protein complexes like c-type cytochromes, and conductive nanowires, which facilitate the transfer of electrons from the bacterial cell to external electron acceptors.

The synergy between electroactive bacteria and algae in biofuel production relies on their complementary metabolic processes. Algae photosynthesize to produce organic matter through the fixation of carbon dioxide and sunlight. However, excess photosynthetic energy can lead to the accumulation of reducing equivalents such as NADH, which limits algal growth and productivity. Electroactive bacteria can alleviate this limitation by accepting excess electrons, effectively acting as electron sinks for algae. This interaction enhances algal growth and biomass production. Furthermore, electroactive bacteria can produce molecular hydrogen through fermentation processes.

The electroactive bacteria consume organic matter produced by algae, generating hydrogen as a byproduct. This process, known as photofermentation, not only boosts hydrogen production, but also reduces the buildup of organic substrates in the culture medium, preventing their inhibition of algal growth.

The combination of algae and electroactive bacteria in biofuel production has yielded promising results in recent years. The most notable biofuel products include hydrogen production through photofermentation, utilizing algal-derived organic matter as a substrate; biodiesel production, through lipids production, that can be converted into biodiesel. Electroactive bacteria enhance lipid accumulation by promoting algal growth and biomass productivity. This kind of bacteria can also help to reduce the oxygen content in algal cultures, making it more suitable for ethanol fermentation by yeast. This increases bioethanol yields from algal biomass.

Recent advances in the field include the engineering of electroactive bacteria and algae to optimize their interactions and increase biofuel yields. Synthetic biology approaches have been employed to enhance the efficiency of EET pathways, improve electron transfer rates, and tailor microbial consortia for specific biofuel production.

The integration of electroactive bacteria into algae-based biofuel production represents a promising road for sustainable energy generation. Future research should focus on addressing several challenges, developing robust and scalable bioprocesses, and conducting comprehensive environmental impact assessments. Additionally, exploring the potential of other electroactive microorganisms and algae species can further advance this technology. This symbiotic relationship leverages the unique metabolic capabilities of each organism to improve overall biofuel yields while mitigating the environmental impact of fossil fuel consumption. Continued research and development in this field are crucial for realizing the full potential of electroactive bacteria in the transition to a sustainable energy future.

14.4 Reactors

Open ponds are very popular and less expensive compared to tubular photobioreactors. They are constructed based on closed loops and recirculation channels. *Chlorella* and *Spirulina* are the most cultivated in open ponds. Nevertheless, a main drawback is the risk of culture contamination due to the direct contact with the environment. A general comparison of open ponds and tubular photobioreactors is shown in Table 14.1.

Turbulent flow is essential to avoid stratification of the algal biomass during culture. In raceway systems, 5 cm/s are enough to keep algae in suspension on the

Table 14.1 Comparison between two main reactors type

Item	Open pond	Photobioreactor
Design and building	Very simple	Diverse and more complex
Growth rates (g/m^2 day)	10–25 (Mohamed 2018)	1–500 (Mohamed 2018)
Biomass concentration (g/L)	0.1–0.2 (Pulz 2001)	2–8 (Pulz 2001)
Temperature control	Difficult	Easy
Contamination	High risk	Low risk
Light regime	Low	High
CO_2 to atmosphere	High	Almost null
Land use	High	Very low
Mixing cost (EUR/kg (db))	0.08 (Mohamed 2018)	1.27 (Mohamed 2018)
Flowing procedure	Paddlewheels	Airlift
Flexibility of production	Non-flexible to change between varieties	Flexible
Reproduction of production parameters	Possible	Possible into a certain range
Standardization	Non possible	Possible
Time interval until net production (weeks)	6–8 (Mohamed 2018)	2–4 (Mohamed 2018)

culture medium (Tobío-Pérez et al. 2021). Highlighting the importance is essential, as an overly intense turbulent flow not only jeopardizes cell integrity through stress but also concurrently escalates energy consumption. Remarkably, the generation of biodiesel from microalgae holds the potential to capture 1.83 kg of CO_2 for every kilogram of algae, establishing an efficient process for carbon dioxide sequestration.

The sustainable production of biogas, achieved by combining extensive algal cultivation with anaerobic digestion, emerges as a remarkably efficient technology for bioenergy provision. This method presents notable benefits, with a key feature being the sustainable expansion of photosynthetic biomass. Moreover, it leads to considerable cost savings in both the culture medium and the end effluent, rendering it a promising and economically feasible solution for enhancing production rates (Pulz 2001).

Photosynthetic bacteria display significantly elevated rates of hydrogen production when compared to other photosynthetic microorganisms. Moreover, these bacteria showcase remarkable resilience in challenging environments, rapid growth, and the capacity to harness both visible and near-infrared light for photosynthesis. Their genetic manipulation is relatively straightforward, allowing for the possibility of modifying them to metabolize readily available biomass waste materials as cost-effective alternatives to defined organic substrates (Weaver et al. 1979). Moreover, the potential for hydrogen production through it is promising. *Chlamydomonas reinhardtii* is the most studied unicellular green algae capable of hydrogen photoproduction. *C. reinhardtii* can produce hydrogen through two light-dependent pathways and through dark fermentation.

The efficiency of photobioreactors is intricately tied to their design, particularly in effectively managing light capture, transportation, distribution, and utilization for the photosynthesis process. The surface-to-volume ratio stands out as a pivotal factor influencing light capture. Reactor surfaces are typically crafted from materials like glass, glass fibers, PVC, or Plexiglas, and these material choices play a significant role in optimizing the performance of the photobioreactor, thereby facilitating the photosynthetic process.

The light regime serves as a crucial determinant of biomass growth, stress stages, and, ultimately, production costs. Open ponds depend on natural sunlight, yet the intermittent availability of sunlight due to natural factors can diminish process efficiency. Conversely, artificial light can be employed consistently, providing a more stable light source. However, the utilization of artificial light entails higher production costs for algal biomass, presenting a trade-off between process efficiency and overall production expenses (Chen et al. 2011). The influence of the light regime on algal culture can be classified into three primary phases: limitation, saturation, and inhibition. The light saturation phase is especially vital as it guarantees the highest specific growth rate. Nevertheless, the utilization of open ponds presents a challenge in achieving maximum growth rates. When the light source has a narrow spectrum that aligns with the photosynthesis absorption spectrum of microalgae, overall energy conversion is improved. Hence, aligning the light source spectrum with the photosynthetic absorption spectrum of microalgae becomes a critical factor for optimizing energy conversion in the cultivation process (Pandey et al. 2014). For this reason, the use of light-emitting diodes is the best choice as light source for microalgae cultures.

Turbulent flow is essential to warranty the maximal biomass yield. Depending on the type of reactor employed, impellers, airlift systems or paddlewheels are used. The higher risk of culture stress is by the use of paddlewheels.

Temperature is a crucial parameter. For outdoor culture, the medium temperature is related to the daily and environmental conditions. The optimal temperature for growing will depend on the algal specie or strain. The range is wide (22–50 °C (Lackner 2015)) but specific to each algae strain or specie.

pH plays a pivotal role as a key medium parameter affecting algae growth by influencing biochemical reactions. The ideal pH range for algae growth generally lies between 7 and 9. Beyond this range, especially at higher pH levels, organisms may undergo stress, potentially resulting in culture collapse. Moreover, heightened pH conditions can facilitate the conversion of a portion of the introduced CO_2 into bicarbonates, diminishing its availability as a nutrient source for the organisms. Thus, maintaining an appropriate pH level is crucial to uphold optimal conditions and foster the health of algae cultivation.

Irrespective of the reactor type or scale, several essential factors need consideration in algae cultivation, encompassing light intensity and wavelength, temperature, pH, conversion efficiency, and specific reactor-related process parameters. Maintaining a high cell concentration in the culture medium poses a significant challenge, leading many large-scale open pond reactors to operate in a semibatch mode to strike a balance between cultivated algae output and the input of fresh medium.

Various designs of column photobioreactors are available, falling into vertical and horizontal configurations. In vertical photobioreactors, a gas sparger is typically positioned at the bottom, transforming the inner gas into small bubbles that facilitate mixing, prevent agglomeration, and supply CO_2 while eliminating oxygen produced during photosynthesis. Vertical reactors are further categorized as bubble or airlifting reactors, with the sparger design playing a pivotal role in biomass production. Airlifting reactors, depending on their configuration, can be either internal or external loop systems.

Another type of reactor is the flat plate, comprising various rectangular plates configured to be illuminated on both sides. Coil-type systems are also widely used, designed to optimize space efficiency compared to other reactor designs. They feature a high surface area-to-volume ratio for effective illumination, along with ease of temperature and nutrient control. The scalability of coil-type systems is considered more straightforward compared to other photobioreactor types.

Biomass harvesting is a significant challenge that directly impacts the production cost of bioenergy from algae. Particularly in the case of microalgae, the biomass is composed by microorganisms. In such cases, harvesting methods play a crucial role in achieving an industrial production process in a cost-effective manner. The harvesting process typically involves flocculation, centrifugation, filtration, or sedimentation (Mohamed 2018).

Single-chamber microbial fuel cells typically have a cubical shape, though cylindrical configurations are also observed. In these systems, algae serve as terminal electron acceptors in the cathode, facilitating the reduction of oxygen produced through photosynthesis. To enhance oxygen reduction, cathodes may be exposed to the air.

On the other hand, double-chamber microbial fuel cells feature two separate chambers with the organism, divided by a proton exchange membrane. In some designs, the anode, membrane, and cathode are enclosed within a single tank, while others adopt an H-shaped configuration with the membrane embedded in one arm. The H-shaped setup, despite drawbacks such as increased fouling and reduced membrane area leading to lower power output, is advantageous for small-scale production due to its cost-effectiveness in terms of capital expenditure (Kannan and Donnellan 2021).

The sediment microbial fuel cell represents an innovative design wherein the anode is integrated into sedimental wastes containing anaerobic exo-electrogenic bacteria. Above this, the cathodic chamber facilitates the reduction of oxygen with the assistance of electron acceptors. Electron transfer from the sediment to the anode takes place through a potential difference via an electric wire. This system contributes to an increase in the population density of both photosynthetic organisms and exo-electrogenic bacteria (Kannan and Donnellan 2021; Song et al. 2020).

Figure 14.2 shows a reactor design based on two chambers with a multiple array anode system, which brings several advantages. The advantage of MFCs mainly lies in the use of microorganisms as the biocatalysts at the anode and cathode chambers of MFCs. Microorganisms thrive in the anode chamber, where organic matter is oxidized by microbes to produce electrons (Fakhimi et al. 2020). The anodic operating conditions play a crucial role in the overall performance of the MFC. Among these operating conditions, the choice of anodic pH value is particularly significant. This selection ensures not only the optimal growth and reproduction of microorganisms but also the efficient movement of protons through the membrane. The MFC achieves its maximum current at an anodic pH value between 7 and 8, with the current decreasing when the anodic pH value exceeds 9 or falls below 6 (Jia and Yuan 2018).

14.4.1 Electrodes and Electrochemical Reactions

Bacterial cells can also be applied as an energy source in microbial fuel cells. Microbial fuel cells employ the anaerobic respiration of microorganisms to convert organic waste into electricity. This promising approach integrates bacteria with electrochemical cells (Shlosberg et al. 2022a). In this system, electron transfer is facilitated by protein complexes that contain electron carriers, thylakoids, or living cells to the electrode. The process of electron transfer involves intracellular protein complexes capable of reducing electroactive metabolites, which can then exit the cells and reduce the anode, generating an electric current. Among the bacteria known for producing the highest rates of external electron transport are *Geobacter sulfurreducens* and *Shewanella oneidensis*.

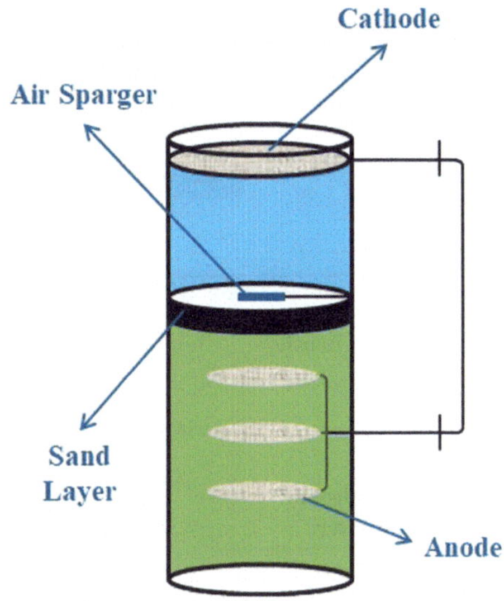

Fig. 14.2 Two-chamber tubular fuel cell reactor

Microbial fuel cells are bioreactors that convert organic matter into electricity. They are designed by having two chambers for anode and cathode separated by a proton exchange membrane (PEM). On the other hand, sediment microbial fuel cells (SMFC) are a type of MFC where the anode is embedded in sediment and cathode is submerged in water. Sediment Microbial Fuel Cells boast a relatively lower cost compared to MFCs due to the absence of a Proton Exchange Membrane (PEM) between the anode and cathode sides, eliminating the need for expensive reactor designs. PEMs, present in MFCs, require periodic replacement due to a decrease in proton transfer capacity over time. This aspect enhances the potential for real-scale implementation of SMFCs compared to MFCs. Furthermore, while the anode in MFCs is a fragile component that significantly affects operational conditions, SMFCs feature a robust anode. Despite these advantages, it is important to note that SMFC systems tend to produce lower power density than MFCs, primarily because of the lower concentration of organic matter in the sediment medium (Taskan et al. 2020). To overcome this, different low-cost organic substrate sources (or organic wastes) can be used in the sediment medium. In this respect, algae biomass is selected as one of the promising alternatives.

The electrode material holds significant importance in bioelectrochemical systems, playing a fundamental role in the process of electricity generation. An array of materials and configurations has been employed as anodes and cathodes in Microbial Fuel Cells and Sediment Microbial Fuel Cells. Carbon-based materials are frequently chosen as electrode materials due to their cost-effectiveness, biocompatibility with electroactive microorganisms, chemical stability, and neutral abundance, among other factors. However, the practical application of MFC systems is hindered by the low electrical conductivity and specific surface area of carbon electrodes (Taskan et al. 2020).

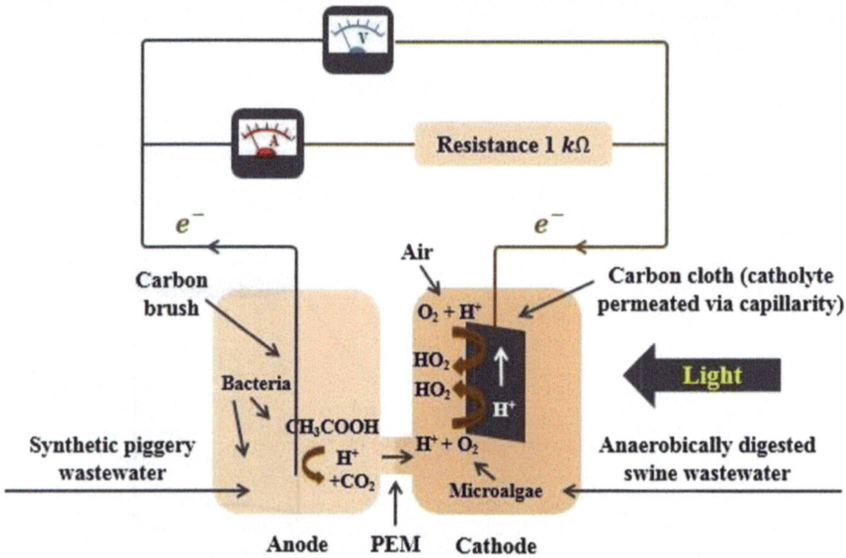

Fig. 14.3 General processes in a two chamber electrochemical reactor

Tin-coated copper mesh is reported as an anode material in MFC, and it demonstrated good chemical stability, electrochemical performance, and excellent biocompatibility (Taskan et al. 2016). Copper is not directly used as anode since easy oxidation. A general scheme representing the main chemical reactions taking place in each electrode in a two-chamber electrochemical reactor is shown in Fig. 14.3. A compilation of several experimental conditions for algae growing in such systems using different microorganisms is shown in Table 14.2.

Taskan et al. (2020) used algae biomass as a raw material to improve the power generation of multi-anode SMFC. The highest power production in SMFC was 2965 mW/m^2. The generation of biogas bubbles in the anode medium has the potential to disrupt the stability of the anode medium and its separation. Another drawback associated with Sediment Microbial Fuel Cells is the formation of biofilm on the surface of the cathode electrode, which restricts the influx of oxygen to the electrode. To overcome this limitation, regular cleaning of the cathode is necessary to remove the biofilm and maintain optimal performance. The microscopic observations confirmed that the biocompatibility of the anode material enhanced the bacterial adhesion and enrichment of electroactive bacterial groups (Gammaproteo, Deltaproteo and Alphaproteo bacteria). The mixed algae were previously cultured in a photobioreactor until obtaining enough biomass, and the cells were dewatered by centrifugation. After that, algae slurry solid content of 20% was obtained and used in the cell.

An essential advancement in microbial fuel cell technology is the development of bio-photoelectrochemical cells (BPEC). In BPECs, the process of photosynthesis

Table 14.2 Experimental conditions for current production from algae

Organism/ component	Anode	Applied bias potential (V)	Mediator	Photocurrent density (with mediator)	References
Thylakoids	Graphite	0.5	$Fe(CN)_6$	nd (100 A/cm^2)	Pinhassi et al. (2016)
Synechocystis sp.	Graphite	0.5	NADP + NAD	5 μA/cm^2 (20 A/ cm^2)	Shlosberg et al. (2021a)
Trichodesmium erythraeum	Graphite	0.5	NADP + thiamine cytochrome C $Fe(CN)_6$	35 μA/cm^2 (90–550 A/cm^2)	Shlosberg et al. (2022b)
Dunaliella salina	Graphite	0.5	NADP + NAD + thiamine	25 μA/cm^2 (100–130 A/cm^2)	Shlosberg et al. (2021b)
Ulva lactuca	Stainless steel	0.5	$Fe(CN)_6$	40 mA/cm^2 (80 mA/cm^2)	
Nymphaeaceae	Stainless steel	0.5	–	6 mA/cm^2	Shlosberg et al. (2022c)
Corpuscularia lehmannii	Iron	None	–	20 μA/cm^2	

nd not defined

is utilized to convert light energy into electricity or to generate high-energy chemical compounds (Shlosberg et al. 2022a). In contrast to non-photosynthetic bacteria, photosynthetic organisms possess the mechanisms necessary to convert sunlight into electricity. During the photosynthetic process, electrons are extracted from water and transferred through membrane protein-pigment complexes, leading to the creation of a proton gradient essential for synthesizing high-energy ATP molecules.

A very important and well-known technical factor that can enhance electricity generation is the ionic strength of the electrolyte solution. Increased salinity can increase the conductivity of the BPEC, representing a significant advantage for BPECs that are based on marine cyanobacteria. Their tolerance to salinity contributes to increase three times the generated by cyanobacteria (Shlosberg et al. 2022a). Nevertheless, the type of cyanobacteria selected could represent a significant difference in current generated. Factors such as type of chlorophyll contained, light intensities allowed and tolerance to salinity are determinant on their suitability for electricity generation.

On the other hand, the generation of electricity does not reduce the viability of the organisms in the BPEC. This may economically compensate for the low current production, enabling the BPEC technology to be integrated into industrial cultivation facilities without making any reduction to the yield of the crops (Shlosberg et al. 2022a). Moreover, Cyanobacteria species have the capability to form biofilms. The cellular architecture within a biofilm is compact and denser compared to a suspension, enhancing the number of cells in contact with the anode of the BPEC.

In contrast to photosynthetic microorganisms, seaweeds are large multicellular systems with diverse geometries and do not exist in suspension. Consequently, the standard configuration of electrochemical systems, where electrodes are immersed into the solution, is not suitable for such measurements. Three main types of configurations are employed to position the electrodes in the biological medium.

14.5 Energy Production

14.5.1 Hydrogen Production for Power and Electricity

Eukaryotic green algae are capable of absorbing sunlight and directly converting solar energy into hydrogen using Fe–Fe hydrogenases. This process takes advantage of the photosynthetic system of the algae, which links water splitting to hydrogen production (Hupp et al. 2022).

Two green algae species, *Chlamydomonas* and *Chlorella*, were utilized in conjunction with a starch-degrading bacterium from the Bacillus genus in a consortium. Continuous photoheterotrophic biohydrogen production was successfully achieved by optimizing the algal-bacterial ratio and adjusting culture conditions. The associations between algae and bacteria resulted in increased biomass and biohydrogen yield compared to normal conditions. *Chlorella* sp., in particular, exhibited a significantly higher biohydrogen production when both the bacterial partner and starch were introduced into the media. Importantly, the generation of electricity did not compromise the viability of organisms in the BPEC, allowing seamless integration of BPEC technology into industrial cultivation facilities without affecting crop yields (Hupp et al. 2022).

Photocurrent in such systems is influenced by several factors: interaction of the electrolytes with the electrodes obviously can change the electrochemical processes, and therefore the conversion performance; the electrode surface can be blocked by molecules' adherence, some proteins may actually precipitate due to high ionic strength conditions, non-metallic electrodes such as graphite or fluorine tin-oxide are used to increase interaction enhancing electrons transfer. Nonetheless, under certain conditions these anodes are less conductive or less compatible with the presence of photosynthetic organisms, leading to significantly reduction of the obtained current. Metals such as iron, stainless steel or aluminum are very good anode material. The current production can be improved by adding to the system certain vitamins of quinones, helping in the electrons transfer. Using proper current enhancers, up to 50–100 times can be increased (Hupp et al. 2022).

Pseudomonas sp. Strain D has been demonstrated to enhance photolysis-mediated hydrogen production in *C. reinhardtii* and to stimulate hydrogen evolution in the green algae *Chlorella* and *Scenedesmus*. In strain D, bacterial growth quickly depletes the relative oxygen content, creating a fully anaerobic environment conducive to activating algal hydrogenase. The collaboration between algae and bacteria effectively slows down the reduction of chlorophyll, boosts starch accumulation, and maintains protein content. These factors represent potential avenues for improving algal hydrogen production through the cooperative efforts of algae and bacteria (Ban et al. 2018).

14.5.2 Biodiesel and Bioethanol Generation

Biodiesel is typically produced through the transesterification of triglycerides found in straight vegetable oils, fats, or waste cooking oil. Common sources include palm, soybean, rapeseed, coconut, Jatropha, and Moringa. During transesterification, triglycerides react with methanol or ethanol in the presence of a catalyst, yielding fatty acid methyl esters (biodiesel) and glycerol.

Efforts to produce biodiesel from algae have gained attention. However, it's crucial to note that the attractive process for obtaining biodiesel from algae is primarily based on microalgae cultivation, which generally contains around 50% oil content. In the case of marine macroalgae, regardless of the algae species, the oil content is typically very low (1), and therefore, biodiesel production from it represents very low yields.

Microalgal produced triacylglycerol and other lipids can be transesterified into fatty acid methyl esters and are excellent as biodiesel precursors. Many studies have identified microalgae as feedstock for biodiesel production based on their high lipid contents but on their biomass growing rate. *Chlorella vulgaris* has a high lipid contents and is considered as a major species in the scale-up of algal biodiesel production. Different species of *Pseudomonas* are associated with *C. vulgaris*. The correlation between *Pseudomonas* populations with *Chlorella* may suggest the symbiotic association.

Co-culturing of the green microalga, *Auxenochlorella protothecoides*, with *Escherichia coli* leads to an increase in algal growth and a doubling of the neutral lipid content as compared to axenic growth, indicating a symbiotic relationship between them.

Bacterial strains, *Pelagibaca bermudensis* and *Stappia* sp., isolated from mass cultivation of *Tetraselmis striata*, shown growth-promoting effects on microalgae *T. striata*, with a high potential for industrial scale production (Yao et al. 2018).

Bioethanol can be derived from various carbohydrates, with starch and sugar fermentation being commonly employed in industrial processes. Alternatively, bioethanol production through fermentation using glycerol as a carbon source can be achieved with *E. coli* bacteria. This approach is particularly intriguing as glycerol is a by-product of biodiesel production. Bioethanol serves as a widely used renewable fuel, either used alone or blended with gasoline. While it has a lower energy density compared to gasoline and is hygroscopic, its renewable nature and higher octane number make it a valuable alternative.

In the case of microalgae residual biomass, incubation with anaerobic microbes stimulates methanogenesis, leading to the production of biogas. The quantity of biogas generated depends on the microalgal species utilized, as the relative proportions of proteins, carbohydrates, and lipids in the cells can influence the activities of anaerobic bacteria and Archaea. Another factor influencing the methanogenic potential of microalgae is the protease resistance of their cell walls, which can limit the effectiveness of primary microbial degradation in anaerobic digesters (Yao et al. 2018).

Many of the MFC studies have been performed in a batch mode analysis. To have any practical application in wastewater treatment, it is imperative for the process to be either continuous or semi-continuous.

14.6 Scale-Up and Practical Applications

Designing and sustaining microbial consortia poses challenges due to the intricacies of metabolic interactions, resource competition, and potential uncertainties over time. The shift from laboratory-scale experiments to large-scale biofuel production presents a considerable obstacle, encompassing concerns related to system stability, contamination, and cost-effectiveness.

The potential synergy between electroactive bacteria and algae in biofuel production has garnered significant attention in laboratory research. However, the transition from small-scale experiments to large-scale industrial applications introduces a multitude of practical challenges that need to be addressed for the successful implementation of such systems.

The selection of a bioreactor design is pivotal in the scale-up process of biofuel production. Different types of bioreactors, such as photobioreactors and open ponds, have been utilized for algae cultivation. The incorporation of electroactive bacteria adds a layer of complexity to reactor design, necessitating efficient mixing, control of oxygen levels, and maintenance of optimal pH conditions. It is crucial to optimize bioreactor configurations to accommodate both algae and electroactive bacteria, ensuring uniform growth and efficient electron transfer for a successful biofuel production process (Ghasemi and Gonçalves 2017; Choudhary and Schenk 2012).

Maintaining nutrient availability is crucial for algal growth. In large-scale operations, efficient nutrient delivery systems are necessary to prevent nutrient limitations or excesses, which can negatively impact biomass productivity and composition. Electroactive bacteria also require specific nutrient requirements, adding complexity to nutrient management strategies (Richmond 2017).

Large-scale systems are more vulnerable to contamination by unwanted microorganisms, posing a threat to the stability and performance of algae-bacteria consortia. Contaminants can outcompete the desired organisms, disrupting the consortium's functionality. Implementing robust sterilization and contamination control measures is crucial to maintain a consistent and reliable biofuel production process.

The scale-up process often demands increased energy input for activities like mixing, aeration, temperature control, and harvesting. Balancing the energy demand, aiming for sustainable biofuel production is a significant challenge. Optimizing energy efficiency and exploring renewable sources, such as solar and wind, can help mitigate the environmental impact of increased energy requirements.

Consistent and high biomass productivity and biofuel yields at the industrial scale present a challenge. Factors like light availability, nutrient fluctuations, and variations in environmental conditions can impact performance. Research efforts should focus on strain improvement, optimizing growth conditions, and real-time monitoring to maximize productivity.

Addressing scale-up and practical issues in biofuel production from algae-bacteria consortia requires interdisciplinary solutions. Overcoming these challenges is essential to unlock the full potential of this technology for sustainable bioenergy production. Continuous research, innovation, and collaboration between academia and industry are crucial to establish scalable and economically viable biofuel production processes.

14.7 Conclusions

Energy production from algal-bacterial consortia and biosolids is a cutting-edge and sustainable approach to electricity generation. This technology can be customized for various scales, from small municipal wastewater treatment plants to large industrial facilities, making it adaptable for both urban and rural settings. However, despite its promise, it is still in the research and development stage. Challenges include enhancing biomass yields, optimizing microbial processes, and refining system integration for higher energy conversion efficiencies.

Biofuels production from algae can be based on different substrates, which are crucial for algae cultivation as they provide essential nutrients, carbon, and regulate pH. Careful substrate selection and management are critical for maximizing biomass yield and biofuel production while minimizing environmental impacts.

The use of algae-bacteria consortia for biofuels production is an innovative technology in energy production. Photosynthetic bacteria, while unable to use water as an electron donor for photosynthesis, utilize various reduced carbon or sulfur compounds as reductant.

Electroactive bacteria, such as *Geobacter sulfurreducens* and *Shewanella oneidensis*, possess unique metabolic capabilities that enable them to transfer electrons extracellularly to solid substrates or other microbial partners. This makes them ideal candidates for bioelectrochemical applications, such as microbial fuel cells. These bioreactors convert organic matter into electricity and are designed with two chambers for the anode and cathode separated by a proton exchange membrane. Sediment microbial fuel cells, which lack a PEM between the anode and cathode sides, have a relatively lower cost compared to membrane fuel cells and do not require expensive reactor designs. However, they produce lower power density due to the low organic matter concentration in the sediment medium.

The production of biofuels like biogas, biodiesel, biohydrogen, and bioethanol is technically feasible from algae cultures. The collaboration of algae with bacteria opens up new possibilities for cutting edge technologies and applications, aiming for higher biofuels yield, increased energy efficiency, cleaner production, and lower production costs.

References

Ban H, Lin W, Wu F, Luo J (2018) Algal-bacterial cooperation improves algal photolysis-mediated hydrogen production. Biores Technol 251:350–357. https://doi.org/10.1016/j.biortech.2017.12.072

Chen C, Yeh K, Aisyah R, Lee DH, Chang J (2011) Cultivation, photobioreactor design and harvesting of microalgae for biodiesel production: a critical review. Biores Technol 102:71–81

Choudhary P, Schenk PM (2012) Bioprocess engineering of microalgae to produce a variety of consumer products. Renew Sustain Energy Rev 16(7):5217–5227

Fakhimi N, Gonzalez-Ballester D, Fernández E, Galván A, Dubini A (2020) Algae-bacteria consortia as a strategy to enhance H_2 production. Cell 9:1353. https://doi.org/10.3390/cells9061353

Fuentes JL, Garbayo I, Cuaresma M, Montero Z, González-del-Valle M, Vílchez C (2016) Impact of microalgae-bacteria interactions on the production of algal biomass and associated compounds. Mar Drugs 14:100. https://doi.org/10.3390/md14050100

Gajda I, Greenman J, Melhuish C, Ieropoulos I (2015) Self-sustainable electricity production from algae grown in a microbial fuel cell system. Biomass Bioenergy. https://doi.org/10.1016/j.biombioe.2015.05.017

Ghasemi NF, Gonçalves AL (2017) Bioreactor technology for microalgal cultivation. In: Microalgae-based biofuels and bioproducts. Academic Press, pp 167–188

Hupp B, Pap B, Farkas A, Maróti G (2022) Development of a microalgae-based continuous starch-to-hydrogen conversion approach. Fermentation 8:294. https://doi.org/10.3390/fermentation8070294

Jia H, Yuan Q (2018) Ammonium removal using algae-bacteria consortia: the effect of ammonium concentration, algae biomass, and light. Biodegradation 29(2):105–115. https://doi.org/10.1007/s10532-017-9816-7

Kannan N, Donnellan P (2021) Algae-assisted microbial fuel cells: a practical overview. Biores Technol Rep. https://doi.org/10.1016/j.biteb.2021.100747

Kumar V, Chopra AK, Kumar A (2017) Review on sewage sludge (biosolids) a resource for sustainable agriculture. Arch Agric Environ Sci 2(4):340–347. https://doi.org/10.26832/24566632.2017.020417

Lackner M (2015) 3rd-generation biofuels: bacteria and algae as sustainable producers and converters. In: Handbook of climate change mitigation and adaptation. Springer, New York. https://doi.org/10.1007/978-1-4614-6431-0_90-1

Lapuerta M, Tobío-Pérez I, Ortiz-Alvarez M, Donoso D, Canoira L, Piloto-Rodríguez R (2023) Heterogeneous catalytic conversion of terpenes into biofuels: an open pathway to sustainable fuels. Energies 16:2526. https://doi.org/10.3390/en16062526

Melis A, Happe T (2001) Hydrogen production. Green algae as a source of energy. Plant Physiol 127:740–748

Mohamed AE (2018) The role of microalgae in renewable energy production: challenges and opportunities. https://doi.org/10.5772/intechopen.73573

Pandey A, Lee DJ, Chisti Y, Soccol CR (2014) Biofuels from algae. Elsevier, Great Britain

Pfeil M, Piloto-Rodríguez R, Díaz Y, Sanchez-Borroto Y, Melo-Espinosa EA, Denfeld D (2020) Data on the thermochemical potential of six Cuban biomasses as bioenergy sources. Data Brief 105207

Pinhassi RI, Kallmann D, Saper G, Dotan H, Linkov A, Kay A (2016) Hybrid bio-photo-electro-chemical cells for solar water splitting. Nat Commun 7:12552. https://doi.org/10.1038/ncomms

Pulz O (2001) Photobioreactors: production systems for phototrophic microorganisms. Appl Microbiol Biotechnol 57:287–293

Richmond A (2017) Principles for attaining maximal microalgal productivity in photobioreactors: an overview. In: Handbook of microalgal culture. Wiley, pp 357–372

Riveiro LA, Da Silva PP (2013) Surveying techno-economic indicators of microalgae biofuel technologies. Renew Sustain Energy Rev 25:89–96

Sánchez-Borroto Y, Lapuerta M, Melo-Espinosa EA, Bolonio D, Tobío-Pérez I, Piloto-Rodríguez R (2018) Green-filamentous macroalgae *Chaetomorpha* cf. *gracilis* from Cuban wetlands as a feedstock to produce alternative fuel: a physicochemical characterization. Energy Sources Part A Recov Util Environ Effects. https://doi.org/10.1080/15567036.2018.1476931

Shlosberg Y, Eichenbaum B, Tóth TN, Levin G, Liveanu V, Schuster G (2021a) NADPH performs mediated electron transfer in cyanobacterial-driven bio-photoelectrochemical cells. iScience 24:101892. https://doi.org/10.1016/j.isci.2020

Shlosberg Y, Tóth TN, Eichenbaum B, Keysar L, Schuster G, Adir N (2021b) Electron mediation and photocurrent enhancement in *Dunaliella salina* driven bio-photo electrochemical cells. Catalysts 11:1–11. https://doi.org/10.3390/catal11101220

Shlosberg Y, Schuster G, Adir N (2022a) Harnessing photosynthesis to produce electricity using cyanobacteria, green algae, seaweeds and plants. Front Plant Sci 13:955843. https://doi.org/10.3389/fpls.2022

Shlosberg Y, Spungin D, Holland N, Schuster G, Frank I-B, Adir N (2022b) *Trichodesmium erythraeum* produces a higher photocurrent than other cyanobacterial species in bio-photo electrochemical cells. Biochim Biophys Acta (BBA) Bioenergetics 1863(8):148910. https://doi.org/10.1016/j.bbabio.2022

Shlosberg Y, Meirovich M, Yehezkeli O, Schuster G, Adir N (2022c) Production of photocurrent and hydrogen gas from intact plant leaves. Biosens Bioelect 215(1):114558. https://doi.org/10.1016/j.bios.2022

Song X, Wang W, Cao X, Wang Y, Zou L, Ge X, Zhao Y, Si Z, Wang Y (2020) *Chlorella vulgaris* on the cathode promoted the performance of sediment microbial fuel cells for electrogenesis and pollutant removal. Sci Total Environ 728:138011. https://doi.org/10.1016/j.scitotenv.2020.138011

Taskan E, Özkaya B, Hasar H (2016) Comprehensive evaluation of two different inoculums in MFC with a new tin-coated copper mesh anode electrode for producing electricity from a cottonseed oil industry effluent. Environ Prog Sustain Energy 35(1):110–116

Taskan B, Bakır M, Taskan E (2020) Enhanced power generation from algal biomass using multi-anode membrane-less sediment microbial fuel cell. Int J Energy Res 1–12. https://doi.org/10.1002/er.5894

Tetali SD (2019) Terpenes and isoprenoids: a wealth of compounds for global use. Planta. https://doi.org/10.1007/s00425-018-3056-x

Tobío-Pérez I, Alfonso-Cardero A, Díaz-Domínguez Y, Pohl S, Piloto-Rodríguez R, Lapuerta M (2021) Thermochemical conversion of *Sargassum* for energy production: a comprehensive review. Bioenergy Res. https://doi.org/10.1007/s12155-021-10382-1

Tsolakis N, Bam W, Srai JS, Kumar M (2019) Renewable chemical feedstock supply network design: the case of terpenes. J Clean Prod 222:802–822. https://doi.org/10.1016/j.jclepro.2019.02.108

Vasseur C, Bougaran G, Garnier M, Hamelin J, Leboulanger C, Le Chevanton M (2012) Carbon conversion efficiency and population dynamics of a marine algae-bacteria consortium growing on simplified synthetic digestate: first step in a bioprocess coupling algal production and anaerobic digestion. Biores Technol 119:79–87. https://doi.org/10.1016/j.biortech.2012.05.128

Weaver PP, Lien S, Seibert M (1979) Improved hydrogen photoproduction from photosynthetic bacteria and green algae. US Solar Energy Research Institute

White HK, Reimers CE, Cordes EE, Dilly GF, Girguis PR (2009) Quantitative population dynamics of microbial communities in plankton-fed microbial fuel cells. ISME J 3:635–646

Yao S, Lyu S, An Y, Lu J, Gjermansen C, Schramm A (2018) Microalgae-bacteria symbiosis in microalgal growth and biofuel production: a review. J Appl Microbiol 126:359–368

Chapter 15
Addressing Algal Bloom and Other Ecological Issues Caused by Microalgae Biomass Conversion Technology

Diana Irvindiaty Hendrawan, Astri Rinanti, Melati Ferianita Fachrul, Tazkiaturrizki, Astari Minarti, Sheilla Megagupita Putri Marendra, and Luthfia Aqilah Zahra

Abstract While biomass conversion technology for microalgae has provided economic value for humans, be they in the form of biofuels, food, animal feed, and chemicals, it may also bring environmental issues through an ecological phenomenon called algal bloom. An algal bloom is the growth of microalgae that occurs very quickly and excessively in waters. This phenomenon can cause environmental problems such as decreased water quality, eutrophication, and disruption of ecosystem balance. Algal bloom can produce poisons (toxins) that are harmful to humans, animals, and ecosystems. When an algal bloom occurs, food chains are disrupted and environmental quality declines. Algal bloom can also cause an increase in nutrients, especially nitrogen and phosphorus, in the waters. Along with this increase, there is a decrease in dissolved oxygen levels, resulting in the death of fish and other aquatic organisms. Therefore, controlling the amount of nutrients that enter the waters can

D. I. Hendrawan · A. Rinanti (✉) · M. F. Fachrul · Tazkiaturrizki · A. Minarti · S. M. P. Marendra · L. A. Zahra
Environmental Engineering Department, Faculty of Landscape Architecture and Environmental Technology, Universitas Trisakti, Jakarta, Indonesia
e-mail: astririnanti@trisakti.ac.id

D. I. Hendrawan
e-mail: dianahendrawan@trisakti.ac.id

M. F. Fachrul
e-mail: melati@trisakti.ac.id

Tazkiaturrizki
e-mail: tazkiaturrizki13@trisakti.ac.id

A. Minarti
e-mail: astari.minarti@trisakti.ac.id

S. M. P. Marendra
e-mail: sheilla@trisakti.ac.id

L. A. Zahra
e-mail: luthfia082001900042@std.trisakti.ac.id

T. A. Kurniawan and A. Anouzla (eds.), *Algae as a Natural Solution for Challenges in Water-Food-Energy Nexus*, Environmental Science and Engineering,
https://doi.org/10.1007/978-981-97-2371-3_15

help prevent eutrophication and algal bloom. Regular monitoring of water quality and the use of advanced technology to detect signs of impending algal blooms can help reduce their impact. In addition, it is necessary to develop and apply conversion technologies that are more environmentally friendly, for example, by using algae strains that are not invasive or have less impact on the environment. Research and education are needed to understand better the mechanism of algal bloom and its impact on ecosystems and to increase public awareness about the importance of sustainable water resource management. A multidisciplinary approach involving research, technology, and collaboration between government, industry, and society is necessary to overcome algal bloom and other ecological problems caused by microalgae biomass conversion technology.

Keywords Algal bloom · Biomass conversion · Biofuels · Nutrients · Eutrophication

15.1 Introduction

15.1.1 The Potential of Microalgae Biomass as a Raw Material for Commercial Products and Its Problems

Microalgae are microscopic single-celled photosynthetic organisms living in different water types, such as freshwater, seawater, and other wet environments (Khan et al. 2018). Microalgae, also called phytoplankton, are an essential part of the food chain in aquatic ecosystems. As primary producers in the aquatic food chain, microalgae can photosynthesize, significantly contributing to the Earth's carbon cycle (Praharyawan 2021). Microalgae belong to the protists group, widely varied in shape, size, and chemical composition (Maia et al. 2020). Microalgal biomass is produced through the process of photosynthesis, which starts from the absorption of sunlight, carbon dioxide, and nutrients from water to produce energy and organic matter (Fig. 15.1). It also consists of various organic matter such as proteins, carbohydrates, lipids, pigments, and other bioactive compounds (Martins et al. 2023). Due to its rich composition and organic matter content, and high growth rate, microalgal biomass has a high potential to be used in various biotechnology applications (Khan et al. 2018). These applications include the conversion of biomass into bioenergy in solid (Osman et al. 2021), liquid, and gas forms (e.g., biodiesel, bioethanol, and biogas). Besides that, microalgae are also used in commercial products such as food ingredients, animal feed, *food supplements*, and raw materials for the cosmetic industry (Anggraeni et al. 2019).

Microalgae can be utilized as commercial feedstock products for various technical and non-technical reasons. The following are some of these technical reasons:

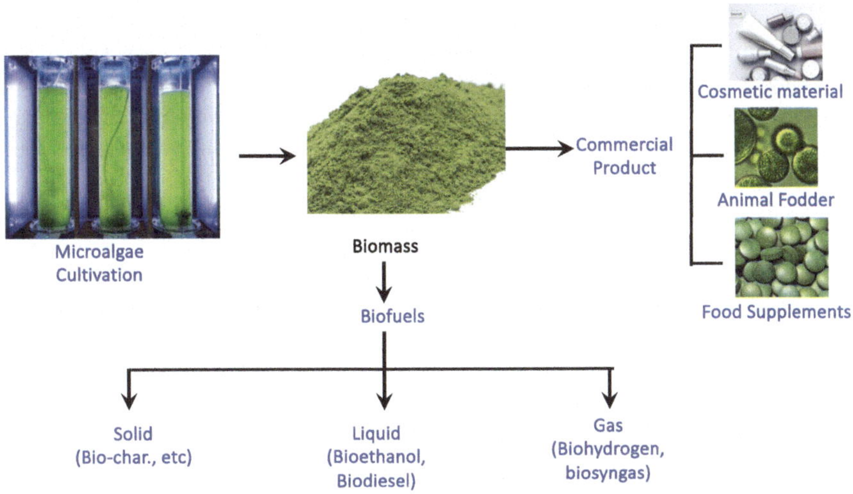

Fig. 15.1 Microalgae conversion process into biomass utilized as bioenergy and commercial products

1. Fast growth. Microalgae have a short life cycle and can reproduce rapidly. This speedy growth rate enables high biomass yields in a short time, making it an efficient and sustainable source of feedstock (Khan et al. 2018).
2. High nutrient content. Microalgae are rich in essential nutrients such as protein, omega-3 fatty acids, antioxidants, vitamins, and minerals. This high nutrient content makes them ideal for food products, beverages, and nutritional supplements (Raja et al. 2018).
3. Efficient photosynthesis. With its high photosynthetic efficiency, microalgae can produce more energy and biomass per unit area than terrestrial plants (Arenas et al. 2017). For this reason, microalgae can help reduce carbon dioxide emissions by absorbing CO_2 during photosynthesis.
4. Flexibility in habitat selection. Microalgae can grow in various environmental conditions, including freshwater, seawater, and wastewater, allowing for more diverse and sustainable use of resources.
5. Ability to produce bioactive compounds. Microalgae can produce various bioactive compounds such as carotenoids, phycobilin, fatty acids, polysaccharides, vitamins, sterols, and other biologically active molecules (Bhattacharjee 2016). These compounds have various applications in pharmaceutical products, cosmetics, and industrial needs.

In addition to the technical reasons outlined above, researchers agree that microalgae can serve as an environmentally friendly commercial feedstock. These non-technical reasons allow the utilization of microalgae to help reduce the carbon footprint and unsustainable use of natural resources. For example, using microalgae as an alternative fuel source such as biodiesel (Hossain et al. 2020) can reduce greenhouse gas emissions and dependence on fossil fuels.

Utilizing microalgae as a source of protein and nutrients can improve food security by helping to address hunger and malnutrition worldwide (Leandro et al. 2020). Microalgae can be an alternative source of nutrients that can reach areas with limitations in traditional crop production (Martins et al. 2023). Developing microalgae-based industries can create jobs and boost economies in previously underdeveloped areas, helping to reduce poverty and improve the life quality of local communities. Using microalgae in commercial products can encourage research and development of new technologies and increase public awareness of the potential for sustainable use of natural resources.

Through increasingly sophisticated conversion technologies, microalgal biomass used as feedstock can be converted into renewable energy source products, such as biogas, bioethanol, and biodiesel (Khan et al. 2018). It can also be utilized as other economically valuable products such as food, medicine, animal feed, and other chemicals. However, the development and application of biomass conversion technologies can cause several ecological problems, e.g.,

1. Large-scale microalgae cultivation that can fulfill biomass conversion needs requires enormous amounts of water and nutrients (de Carvalho et al. 2018), leading to resource competition with other human and environmental needs, such as agriculture and drinking water supply. Inefficient or unsustainable use of water resources can result in environmental problems, such as decreased water availability and changes in aquatic habitats.
2. Converting microalgal biomass to energy can produce greenhouse gas emissions, such as methane and carbon dioxide (Anwar et al. 2020). Although some of these emissions can be compensated by carbon sequestration during algal growth, the potential negative impact on climate change still needs to be considered.
3. Using fertilizers and excess nutrients in microalgae cultivation can pollute surrounding waters and trigger eutrophication and algal blooms detrimental to aquatic ecosystems (Tsikoti and Genitsaris 2021).
4. Spread of invasive microalgae. Without proper control, cultivated microalgae can spread to surrounding ecosystems, taking the place of indigenous microalgae species and potentially causing changes in community structure and ecosystem function.

15.1.2 Microalgae Biomass Conversion Process and Technology (Biofuel, Food, Animal Feed, and Chemicals)

Microalgae are microscopic photosynthetic organisms that have great potential in renewable fuel production due to their high content of lipids, carbohydrates, and proteins that can be converted into various energy products such as biodiesel, bioethanol, and biogas (Udayan et al. 2022). These fuels are considered more environmentally friendly as they reduce greenhouse gas emissions. However, the cost of

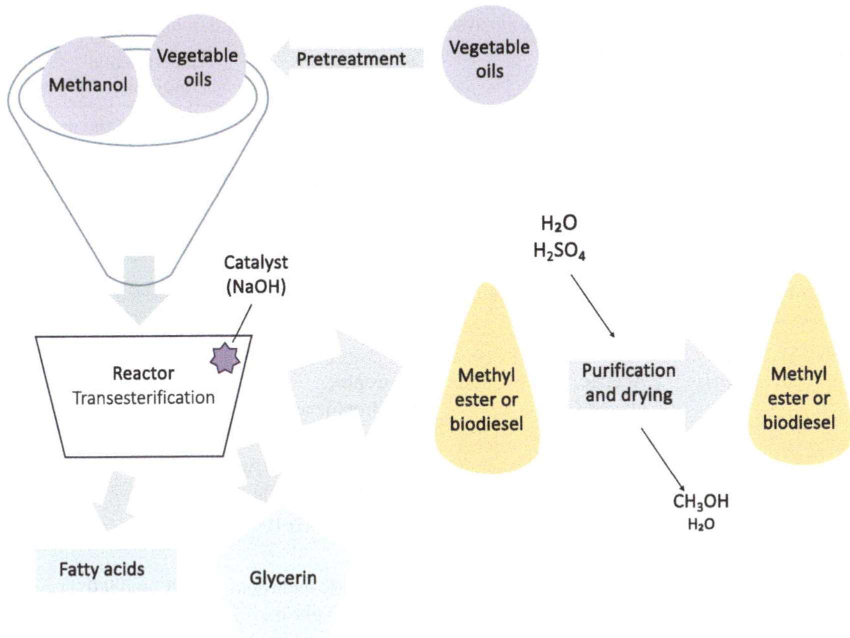

Fig. 15.2 Biodiesel production process

producing biofuels from microalgae is still high, and efficient processing technologies must be further developed. The process of converting microalgal biomass into bioenergy products involves several different steps and technologies.

Biodiesel Production

Microalgae-based biodiesel refers to biofuel produced from the conversion of lipids extracted from microalgae biomass such as (Fig. 15.2).

The microalgae used in biodiesel production are a group of microscopic photosynthetic organisms such as *Chlorella, Spirulina*, and *Nannochloropsis.* Microalgae-based biodiesel offers several advantages over biodiesel from traditional feedstocks such as palm, soybean, or castor oil. Several studies summarized in (Table 15.1) (Pal et al. 2019) show the growth media and cultivation systems used for oil-producing microalgae.

During growth, microalgae can accumulate significant amounts of lipids, reaching up to 20–50% of the algal dry weight depending on the species and cultivation conditions, as shown in (Table 15.2).

The conversion of microalgal biomass to biodiesel involves several key steps and technologies, as shown in (Fig. 15.3).

1. Microalgae cultivation. This process involves the growth and propagation of microalgae in controlled systems such as photobioreactors or open ponds.

Table 15.1 Growth media and cultivation systems used for oleaginous microalgae

Microalgae used	Media	Culture systems	Experimental conditions
Chlorella vulgaris	Modified basal medium, soy whey and thin stillage with 100 mM buffer strength	Heterotrophic growth, heterotrophic and mixotrophic batch cultivation, and mixotrophic batch cultivation	The pH of the media maintained was 5, 6, 6.8, 7.5 and 8 28 °C for 5 days
Scenedesmus obtusiusculus	Chitosan and tannin (CFL-PT) as flocculent		Salt-based medium
Chlorella vulgaris	Acetate, glucose, and glycerol	Autotrophic, heterotrophic and mixotrophic growth	150 rpm air flow rate of 200 mL/min
Scenedesmus obliquus	Secondary treated municipal wastewater and Bold's basal medium	75% wastewater (WW) + 25% Bold's Basal Medium (BBM) 100% BBM	
Chlorella pyrenoidosa II-H6	BG11, 13 days	Atmospheric and room temperature plasma technique	33 °C and 9.0
Nannochloropsis sp.	Chu 13 medium	Immobilized alginate beads in PBR	30 ± 2 °C for 7 days
Tribonema minus *Scenedesmus* sp.	BG11 medium Secondary effluent	Heterotrophic condition Heterotrophic	23 ± 1 °C
Scenedesmus obliquus SIT06	BG11	Photoautotrophic	28 °C for 10 days

Table 15.2 Several oil-producing algae and their oil content (Pal et al. 2019)

Algae	Oil content (% dry cell weight)
Botryococcus braunii	25–75
Chlorella sp.	28–31
Dunaliella sp.	23
Neochloris	35–54
Nannochloropsis sp.	31–68
Oleoabundans phaeodactylum	20–30

Microalgae will produce biomass with high lipid content suitable for biodiesel production.

2. Microalgae harvesting. After reaching the desired growth phase, the microalgal biomass is harvested by separating it from water through separation processes such as centrifugation, flocculation, or filtration (Ogbonna and Nwoba 2021).

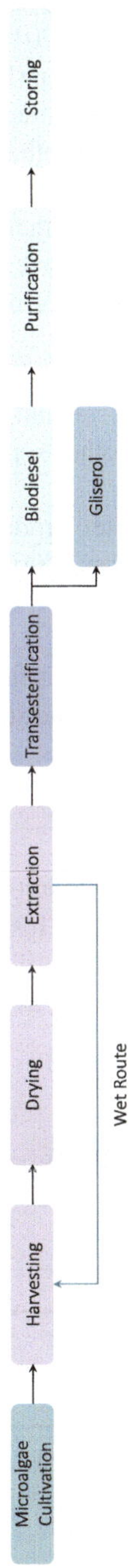

Fig. 15.3 General process of microalgae conversion to biodiesel

3. Biomass drying. The harvested microalgae biomass is then dried to reduce the moisture content. The drying process can be done by heat drying, freeze drying, or other suitable methods (Shekarabi et al. 2019).
4. Lipid extraction. The process is illustrated in (Fig. 15.4).

   After the biomass is dried, lipids are extracted from the microalgal biomass using organic solvents such as hexane, ethanol, or other methods such as supercritical or ultrasonic-assisted extraction (Ghasemi Naghdi et al. 2016).
5. Transesterification, The extracted lipids are then reacted with an alcohol (usually methanol or ethanol) in a catalyst (acidic, basic, or enzymatic) to produce methyl or ethyl esters, known as biodiesel and glycerol as a by-product. The process is illustrated in (Fig. 15.5).
6. Separation. After transesterification, the reaction mixture will contain biodiesel, glycerol, and other materials. Biodiesel and glycerol are separated through gravity separation, centrifugation, or other separation methods. The process is illustrated in (Fig. 15.6.).
7. Purification of biodiesel. The biodiesel produced may contain contaminants or residues from the production process. Purification processes such as water washing, filtration, or vacuum distillation remove contaminants and improve biodiesel quality.
8. Drying and storage of biodiesel. After purification, the biodiesel is dried to remove any moisture that may be present and stored under suitable conditions to prevent oxidation or quality changes.

Bioethanol Production

Microalgae-based bioethanol is a fuel that results from converting carbohydrates contained in microalgal biomass into ethanol (Hossain et al. 2020). During growth, microalgae can accumulate significant amounts of carbohydrates, such as cellulose, hemicellulose, and amylum, which can be used as feedstock sources for bioethanol

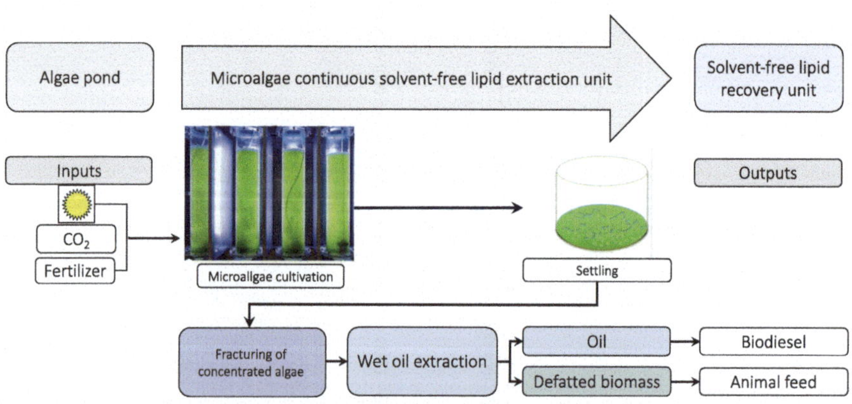

Fig. 15.4 Lipid extraction process

Fig. 15.5 Enzymatic transesterification to produce biodiesel and glycerol

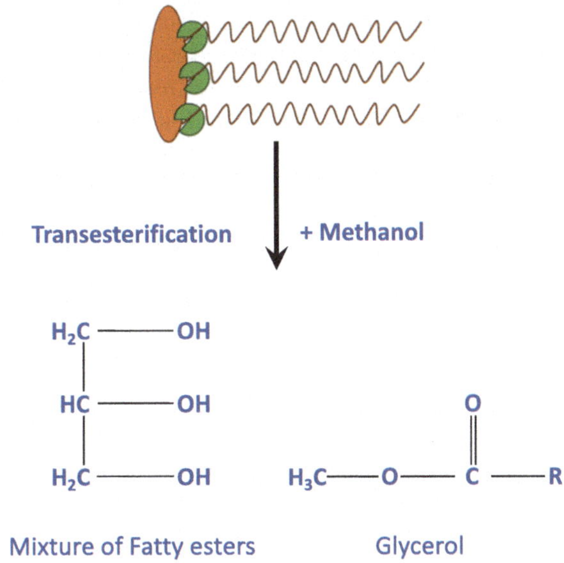

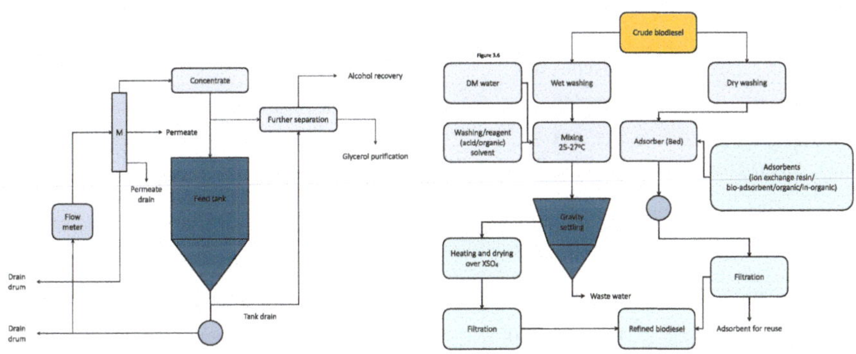

Fig. 15.6 Wet and dry biodiesel purification process

production (Khan et al. 2018). Microalgae-based bioethanol offers several advantages over bioethanol from traditional feedstocks such as sugarcane, corn, or wheat. The resulting bioethanol can be used as an environmentally friendly alternative fuel, blended with gasoline, or used purely in conventional gasoline engines (Bhan et al. 2020). Successively, the conversion of microalgal biomass into bioethanol consists of the following process steps.

1. Microalgae cultivation. Microalgae growth and propagation are carried out in controlled systems such as photobioreactors or open ponds. During growth, microalgae produce biomass with high carbohydrate content, which can be converted to bioethanol.

2. Microalgae harvesting. After reaching the desired growth phase, microalgal biomass is harvested through separation processes such as centrifugation, flocculation, or filtration to separate microalgal biomass from water and algal concentrate (Singh and Patidar 2018). Microalgal cell mass and concentration can be estimated by measuring chlorophyll content, microalgal cell number, absorbance (optical density) (Table 15.3) (Maltsev et al. 2021), dry weight, and ash-free dry weight. The most widely applied methods include flocculation, coagulation, gravity sedimentation, electricity-based processes, filtration, and centrifugation. Gravity sedimentation, flocculation, and centrifugation usually have high yields and lower costs. The harvesting process is further explained in (Fig. 15.7).
3. Pretreatment. Microalgae biomass undergoes pretreatment to destroy the cell structure and release the polysaccharides. Pretreatment methods include thermal, mechanical, chemical, or a combination of several methods, as seen in (Fig. 15.8).
4. Hydrolysis. The released polysaccharides are hydrolyzed into monosaccharides (e.g., glucose) using enzymes or chemical reagents. Enzymes that are commonly used in this process are amylase, cellulase, and xylanase (Córdova et al. 2018).

Table 15.3 Microalgae cell mass and concentration estimated by measuring chlorophyll content, microalgae cell number, absorbance (optical density)

Pigments	Absorption maxima in organic solvents, nm	Representatives
Chlorophyll *a*	420, 660	All algae
Chlorophyll *b*	435, 643	Green algae
Chlorophyll *c*	445, 625	Heterokontophyta, haptophyta, dinophyta, cryptophyta
Chlorophyll *d*	450, 690	Rodophyta, some cyanobacteria
Chlorophyll *f*	707	Some cyanobacteria
β-carotene	425, 450, 480	Most algae
α-carotene	420, 440, 470	Some algae (cryptophyta, haptophyta, dinophyta, chyrysophyceae), some cyanobacteria
Fucoxanthin	425, 450, 475	Heterokontophyta (bacillariophyceae phaeophyceae, chyrysophyceae), haptophyta
Phycoerythrin	490, 546, 576	Rodophyta, cryptophyta, cyanobacteria
Phycocyanin	618	Rodophyta, cryptophyta, cyanobacteria
Allophycocynanin	650	Rodophyta, cryptophyta, cyanobacteria

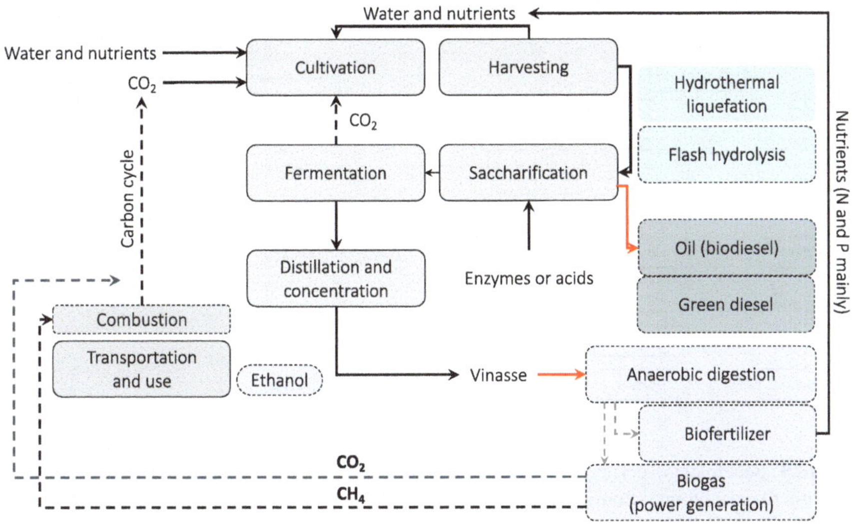

Fig. 15.7 The conversion process of microalgae to ethanol

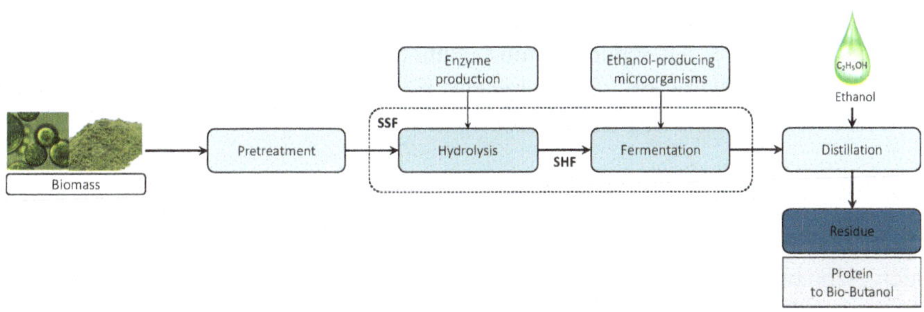

Fig. 15.8 Process flow of separate hydrolysis and fermentation (SHF) and simultaneous saccharification and fermentation (SSF)

5. Fermentation. Monosaccharides produced from the hydrolysis process are then fermented by microorganisms (such as *Saccharomyces cerevisiae*) to produce ethanol. Fermentation can be aerobic or anaerobic, depending on the type of microorganism used (Wongsurakul et al. 2022). The process is illustrated in (Fig. 15.9).
6. Ethanol separation. After fermentation, the resulting ethanol needs to be separated from the fermentation mixture. Separation processes include distillation, pervaporation, or extraction with solvents, which separate ethanol from water and other by-products (El-Dalatony et al. 2017).
7. Ethanol purification. Ethanol from the separation process may contain residues or contaminants (Wongsurakul et al. 2022). Purification processes such as

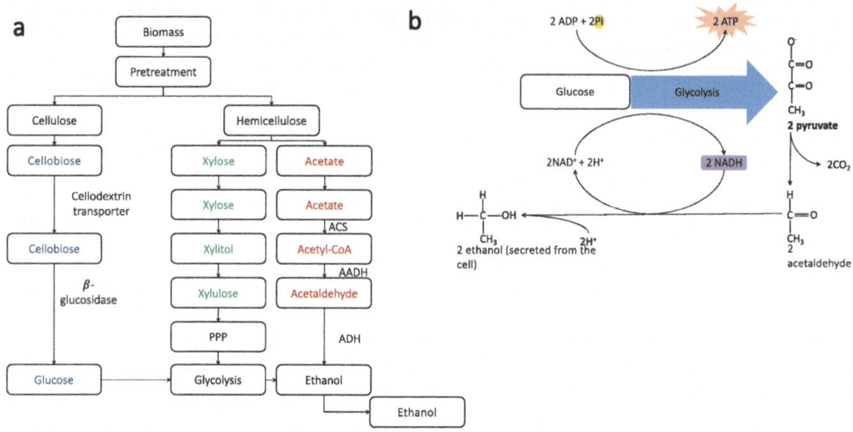

Fig. 15.9 Schematic overview of bioethanol production using cellobiose, xylose and acetic acid from microalgae biomass converted by yeast (left) and the metabolic pathways, i.e., glycolysis and fermentation (right)

azeotropic distillation, adsorption, or extraction with solvents are used to improve the quality of ethanol to the desired standard.

8. Ethanol storage. After purification, ethanol is stored under suitable conditions to prevent oxidation or quality changes. Ethanol should be stored in an airtight container and protected from direct light.

Biogas Production

Microalgae-based biogas refers to the gas produced from the anaerobic degradation of microalgae biomass (Feng et al. 2019). During growth, microalgae produce biomass containing carbohydrates, proteins, and lipids, which can be used as substrates for biogas production (Klin et al. 2020). The biogas produced consists mainly of methane (CH_4) and carbon dioxide (CO_2), as well as small amounts of other gases such as hydrogen sulfide (H_2S) and nitrogen (N_2) (Andriani et al. 2020). Biogas can be used as a renewable and environmentally friendly energy source for heating and power generation (Abanades et al. 2022), and as a vehicle fuel (Ullah Khan et al. 2017).

Microalgae-based biogas has many advantages over biogas produced from traditional feedstocks such as agricultural waste, food waste, or animal manure (Atelge et al. 2020) following are the technological steps and process of converting microalgal biomass into biogas:

1. Cultivation of microalgae. The growth and propagation of microalgae are carried out in controlled systems such as photobioreactors or open ponds (Zieliński et al. 2022). During growth, microalgae produce biomass containing carbohydrates, proteins, and lipids, which can be used as a substrate for biogas production (Sivaramakrishnan et al. 2022).

2. Microalgae harvesting. After reaching the desired growth phase, the microalgae biomass is harvested through separation processes such as centrifugation, flocculation, or filtration to separate the microalgae biomass from the water and algae concentrate (Najjar and Abu-Shamleh 2020).
3. Pretreatment. Microalgal biomass undergoes pretreatment to destroy cell structures and release intracellular components. Pretreatment methods include mechanical, physical, thermal, chemical treatments, and combination methods. The classification of pretreatment methods is shown in (Fig. 15.10).
4. Anaerobic digestion. The pretreated microalgae biomass is then fed into a biogas reactor and anaerobically degraded by microorganisms (such as methanogenic bacteria). This process produces biogas consisting of methane (CH_4), carbon dioxide (CO_2), and small amounts of other gases such as hydrogen sulfide (H_2S), nitrogen (N_2), and (NH_3) in Iron (II) oxide (Fe_2O_3) (Thanakunpaisit et al. 2017).
5. Biogas separation and purification. After anaerobic digestion, the biogas produced must be separated from the reactor mixture and cleaned from contaminants such as H_2S, CO_2, and water vapor. Separation and purification processes include absorption, scrubbing, or membrane technology, as illustrated in (Fig. 15.11).
6. Storage and use of biogas. The cleaned biogas is then stored under suitable conditions or used directly for various purposes, such as heating, power generation, and vehicle fuel (Ullah Khan et al. 2017).
7. Waste management. After the anaerobic digestion, the remaining biomass (digestate) must be managed appropriately. Digestate can be used as organic fertilizer, an additive in animal feed, or can be further processed to produce other products such as bioethanol or compost (Logan and Visvanathan 2019).

In addition, the remaining microalgal biomass can also be converted into economically valuable products such as food, medicine, animal feed, and various other chemicals.

Food and Other Chemicals

Microalgae biomass can be utilized as food ingredients. Some microalgae species, such as Spirulina (*Arthrospira platensis*) and *Chlorella*, have long been used as dietary supplements and rich sources of nutrients. Microalgae contain proteins, essential fatty acids, vitamins, minerals, and antioxidant pigments, making them nutritious food ingredients beneficial for health. Some examples of microalgae as a food source are described below:

1. Dried microalgal biomass processed into powder or tablet can be consumed as a dietary supplement to increase nutrient intake. *Spirulina* and *Chlorella* are examples of popular dietary supplements made from microalgae (Khan et al. 2018).
2. Microalgae biomass can be used as a raw material to manufacture various food products, such as bread, pasta, beverages, and snacks. Microalgae can add nutrition, color, and flavor to these food products (Pal et al. 2019).

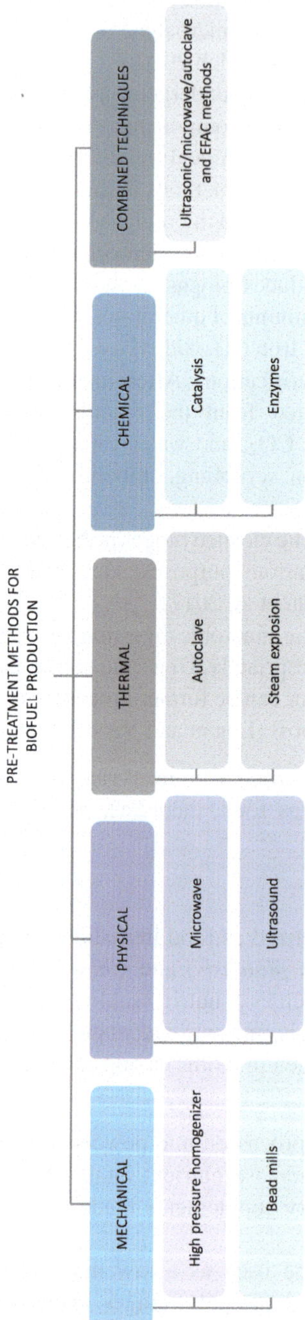

Fig. 15.10 Classification of cell pre-treatment methods

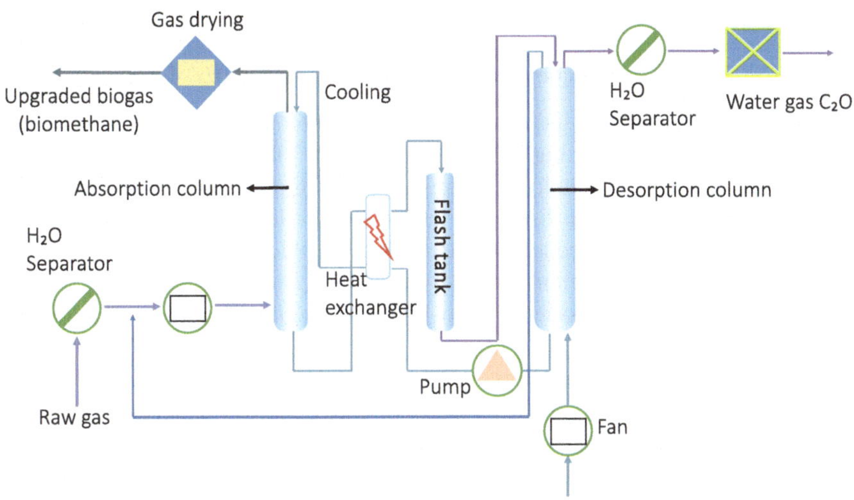

Fig. 15.11 Biogas purification by water scrubbing and organic solvent scrubbing to remove CO_2

3. The protein contained in microalgal biomass is of good quality with complete essential amino acids. Therefore, microalgae can substitute animal protein in vegetarian and vegan diets and reduce meat and animal product consumption.
4. Some microalgae species, such as *Nannochloropsis* and *Schizochytrium*, contain high omega-3 fatty acids (e.g., DHA and EPA). These fatty acids are essential for heart, brain, and immune system health. Thus, microalgal biomass can be used as an environmentally friendly and sustainable source of omega-3 fatty acids.

As a food ingredient, microalgal biomass offers advantages in terms of nutrition and sustainability, as well as the potential to reduce pressure on natural resources and the environment due to conventional food production (Liu et al. 2022). However, further research is needed for technological development supported by appropriate regulations to ensure the safety and quality of microalgae-containing food products. In addition, problems that may arise in this context include difficulties in maintaining product quality and safety and limitations in production scale (Albrektsen et al. 2022).

In addition to their potential as food raw materials, microalgae contain bioactive compounds such as carotenoids and polysaccharides that promotes skin health and beauty (Martínez-Ruiz et al. 2022). The use of microalgae for skin care products will require the development of stable and effective formulations and product safety and efficacy trials. Microalgae can also be processed into bioplastics, chemicals, and pigments. Problems using microalgae as industrial feedstock are the high production costs and complexities associated with scaling up mass-scale production.

Animal Feed

Microalgae biomass can be converted into feed for various types of livestock, e.g.,

- Poultry such as chickens, turkeys, ducks, and other birds can consume microalgae feed mainly due to their high protein and omega-3 fatty acid content can help improve meat and egg quality and support poultry growth and overall health.
- The use of microalgae in fish feed has become an important area of research, especially to replace or reduce the use of unsustainable fishmeal (Nagappan et al. 2021). Microalgae are rich in protein, omega-3 fatty acids, and pigments such as astaxanthin, which are essential for fish growth and health, as well as attractive colors in ornamental fish.
- Microalgae can also be used as feed for shrimp and mollusks, such as clams and oysters. Microalgae contain nutrients essential for these aquaculture species' growth and reproduction (Maizatul et al. 2017).
- Cattle, goats, sheep, and other ruminants can consume microalgae feed. Although microalgae will not replace grass or silage completely, they can be used as a supplement to protein and other nutrients to improve livestock health and productivity (Souza et al. 2021).
- Microalgae can be used as an additional source of protein and nutrients in pig feed. Using microalgae in pig feed can help improve meat quality and overall animal health.
- Rabbits and other herbivorous animals can also consume feed containing microalgae. The protein and other nutrients found in microalgae can support rabbit growth and health.

Overall, microalgal biomass has the potential to be a sustainable and nutritious feed source for various livestock species. However, proper feed formulation and further research on the optimal dosage and long-term effects of microalgae in animal feed are still needed to maximize its benefits (Nagappan et al. 2021). The technology and process of converting microalgal biomass into animal feed involve several important steps, starting from microalgae cultivation as in (Fig. 15.12) to the final processing into ready-to-use feed. The following are the general steps used in the process.

1. Microalgae cultivation. Microalgae are cultivated in closed systems (e.g., photobioreactors) or open systems (e.g., ponds or raceway ponds) under optimal conditions, such as sufficient lighting, temperature, pH, and nutrients. The selection of suitable microalgae species is also important, especially species that have a high content of protein, fatty acids, and other nutrients and are suitable for livestock needs.
2. Harvesting microalgae. After reaching the desired density, microalgae are harvested from the cultivation system using centrifugation, filtration, or flocculation methods. The aim is to separate the microalgae cells from the water and concentrate the microalgae biomass.
3. Biomass drying. The harvested microalgae biomass is then dried to reduce the moisture content. Hot air drying, spray drying, or freeze drying are commonly

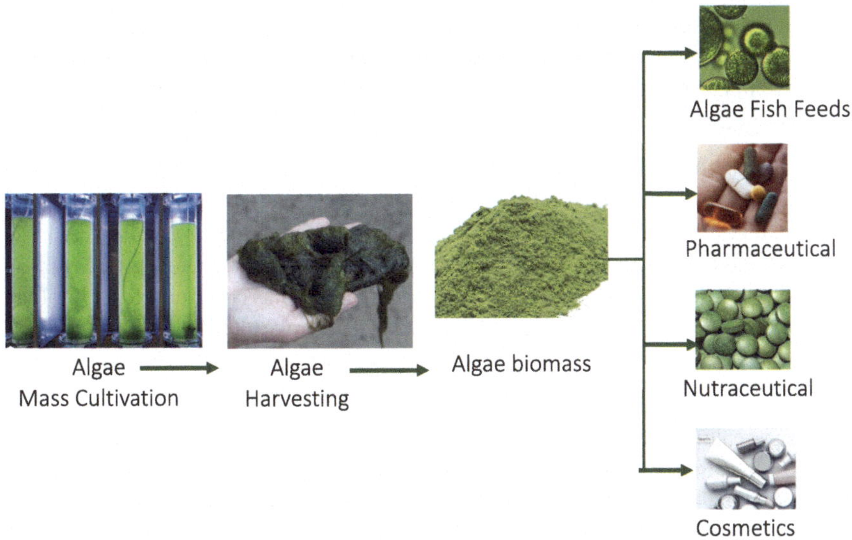

Fig. 15.12 Schematic diagram of the microalgae cultivation process (Yaakob et al. 2021)

used methods. Drying will help preserve the nutrients in the microalgae and facilitate storage and transportation (Vilatte et al. 2023).

4. Processing and nutrient extraction. Dried microalgal biomass can be further processed to extract desired nutrients, such as proteins, fatty acids, and pigments. The extraction process involves mechanical methods, such as solvent extraction or enzymatic methods. However, this stage may not always be necessary, depending on the animal feed application and the microalgae species used.
5. Animal feed formulation. After processing, the microalgal biomass or nutrient extracts can be mixed with other feed ingredients to create a balanced feed formulation that meets the nutritional needs of the livestock. This formulation will be customized according to the livestock species, age, and production goals (e.g., growth, reproduction, or egg production).
6. Pelletization. Feed formulations can be converted into pellets through pelletization to facilitate feeding. Pelletization helps to reduce nutrient losses during storage and feeding and improves feed consumption and digestion by livestock (Mores et al. 2020).

These technologies and processes can convert microalgal biomass into nutritious and sustainable animal feed. In general, the utilization of microalgae biomass into products of economic value still faces the following obstacles:

1. High production cost. Developing more efficient and energy-saving technologies is needed to reduce the cost of microalgae biomass production.
2. Limited availability of resources. Factors such as sunlight, temperature, and nutrients affect microalgae growth. The availability of optimal conditions for large-scale microalgae growth can be challenging (Chowdury et al. 2020).

3. Regulatory barriers. The use of microalgae in food, cosmetic or pharmaceutical products requires approval from a competent authority, such as the FDA or EFSA. This approval process can require significant time and resources (Ferreira de Oliveira and Bragotto 2022).

In addition to the three major problems mentioned above, the underlying problem of massive microalgal biomass production raises serious environmental issues, such as the potential for algal bloom, which needs to be addressed in an integrated and comprehensive manner because it has the potential to cause prolonged ecosystem imbalance (Sarker and Kaparaju 2023).Microalgae biomass conversion technology does not directly cause algal bloom. However, increasing microalgae production on a large scale can increase the risk of algal bloom if not appropriately managed. Here are some factors that may contribute to the risk of algal bloom:

1. Large-scale microalgae cultivation requires adequate intake of nutrients, such as phosphate and nitrogen, for optimal growth. If the nutrients provided exceed the microalgae's needs, the excess nutrients can pollute the surrounding waters and cause algal blooms, negatively affecting the aquatic ecosystem (Yaakob et al. 2021).
2. If not appropriately managed, Microalgae cultivation in open conditions, such as ponds or lakes, can increase the risk of algal blooms. For example, if an open cultivation system develops a leak or is contaminated by wild microalgae, this can lead to an overgrowth of algae in the surrounding waters.
3. Various microalgae species can be used in biofuel production. Some species may be more tolerant of extreme environmental conditions or more adaptable. If such species "leak" into the natural environment, they may dominate the local ecosystem and cause algal blooms (Patwardhan et al. 2022).

15.2 Algal Bloom and Its Impact on the Environment

15.2.1 Causes and Mechanism of Algal Bloom

An algal bloom is a phenomenon that occurs when the microalgae population in an aquatic ecosystem increases rapidly, resulting in a significant increase in phytoplankton biomass. The increase in biomass can alter the physical and chemical characteristics of the ecosystem (Song and Lv 2023). Algal blooms can result from natural or anthropogenic factors, such as increased nutrients (e.g., phosphorus and nitrogen), favorable temperatures, optimal light conditions, and water flow. Some factors causing algal bloom are:

1. Nutrient overload. High concentrations of nutrients, especially phosphorus and nitrogen in the water, can trigger massive microalgae growth. The source of these nutrients can come from industrial, agricultural, and urban waste discharges (e.g., fertilizers, livestock waste, and detergents) that pollute waters through surface

runoff, soil erosion, and infiltration into groundwater (Irfeey et al. 2023). Excess nutrients provide food for microalgae, which allows microalgae to thrive.

2. Water temperature. Warmer water temperatures, especially in spring and summer, often favor microalgae growth. Hot summers or temperature changes due to climate change can lead to increased water temperatures, which in turn can increase the risk of algal blooms.
3. Light conditions. Microalgae require light for photosynthesis. Availability of sufficient light or increased light intensity can trigger microalgal growth (Irfeey et al. 2023).
4. Hydrodynamic conditions. Changes in currents, flow velocity, and stagnant water layers can affect the distribution of nutrients and oxygen in the water (Nzayisenga et al. 2020). Slow or stagnant water flow can result in nutrient accumulation, creating favorable conditions for microalgae growth. In addition, reduced water flow due to water use or weather changes can also affect nutrient concentrations and microalgal growth.
5. The process that causes water layers to separate based on temperature and density, known as stratification, can also affect microalgae growth. When water layers separate, nutrients trapped in the lower layers may not be able to mix with the oxygenated upper layers, leading to excessive microalgae growth in the upper layers.

The mechanism of algal bloom involves an interaction between factors that create optimal conditions for microalgal growth. The process includes:

1. Increased nutrients. Algal blooms usually start when environmental conditions favor microalgal growth, such as excess nutrients (primarily nitrogen and phosphorus), optimal temperature and light, and changes in water flow. The high availability of nutrients in water provides an abundant food source for microalgae, promoting rapid microalgae growth and reproduction.
2. Microalgae growth. Under favorable conditions, microalgae begin to reproduce faster, too much to be controlled by their natural consumers (e.g., zooplankton) (Nagappan et al. 2021). This condition results in exponentially increasing microalgae populations.
3. Biomass accumulation. As the microalgae population continues to grow, the microalgae biomass also increases. Microalgae begin to form algal bloom, indicated by discoloration of the water surface, reduced water clarity, or the formation of a thick layer of floating microalgae. The algal bloom causes an increase in turbidity, which affects light penetration and photosynthetic activity (Udayan et al. 2022).
4. Ecosystem disruption. Prolonged and intensive algal blooms can result in chemical and biological changes in waters, such as decreased dissolved oxygen, toxin production, and changes in the composition of the aquatic organism community.
5. Algal blooms usually end when nutrients that support microalgal growth are depleted, or environmental conditions become less favorable, such as a drop in temperature or a change in water flow. At this point, the microalgae population will decline, and the aquatic ecosystem will return to its previous state. However,

microalgae that die and are decomposed by bacteria can cause hypoxia or oxygen deficiency in the water (Khan et al. 2018). This condition significantly affects the survival of other aquatic organisms, causing fish mortality which has a long-term adverse effect on the balance of the ecosystem.

15.2.2 The Impact of Algal Blooms on Water Quality, Ecosystems, and Human Health

Algal bloom can negatively impact water quality, which in turn can affect ecosystems, aquatic life, and human water use. The negative impacts of algal bloom on water quality are described below:

1. Decrease in dissolved oxygen. When the microalgae in an algal bloom die and begin to decompose, the decomposition process consumes dissolved oxygen in the water. This condition can lead to hypoxia or oxygen deprivation, harming the lives of aquatic organisms such as fish and invertebrates.
2. Increased turbidity. Algal blooms can cause an increase in water turbidity, which reduces light penetration and affects photosynthetic activity. High turbidity can also impair visibility for aquatic organisms and reduce water aesthetics (Nasution et al. 2021).
3. Toxin production. Some microalgae involved in algal blooms can produce harmful toxins to aquatic life, humans, and animals. These toxins can contaminate waters and cause poisoning if accumulated in the food chain or exposed through direct contact with contaminated water.
4. Water discoloration. Algal blooms often change the color of the water to green, red, brown, or even purplish-blue. This discoloration is caused by pigments in the microalgae cells and can detract from the water's aesthetics and signal a decline in water quality (Sun et al. 2023).
5. Odor and taste. Algal blooms can cause changes in the odor and taste of water, especially when microalgae begin to die and decompose (Ballah et al. 2019). These unpleasant odors and tastes can affect the quality of drinking water, swimming pools, and recreational waters and reduce the aesthetic value and attractiveness of water.
6. Water treatment disruption. Algal blooms can disrupt drinking water treatment processes, such as coagulation, filtration, and disinfection. High microalgae concentrations and toxin production can cause disruption to water treatment and increase treatment costs.

Algal blooms also pose many negative impacts on human health, especially when the blooms involve toxin-producing types of microalgae, namely:

1. Food poisoning. Toxins produced by some microalgae, such as cyanobacteria, can accumulate in the tissues of aquatic organisms such as fish and shellfish. Consumption of seafood contaminated with these toxins can cause food

poisoning with symptoms such as nausea, vomiting, diarrhea, headaches, and in severe cases, liver or kidney damage (Vilariño et al. 2018).

2. Respiratory distress. Some toxins produced by microalgae, such as brevetoxin produced by Karenia brevis, can irritate the human respiratory tract when inhaled (Ransom Hardison et al. 2019). This can cause symptoms such as coughing, sneezing, nasal congestion, and difficulty breathing, especially in individuals with respiratory diseases such as asthma or chronic obstructive pulmonary disease (COPD).
3. Skin and eye irritation. Direct contact with water containing toxins from algal blooms can irritate the skin, eyes, and mucous membranes. Possible symptoms include rash, itching, redness, swelling, and red or watery eyes.
4. Nervous system disorders. Some toxins produced by microalgae, such as anatoxin-a (Plata-Calzado et al. 2022) produced by some species of Cyanobacteria, can affect the human nervous system. Exposure to these toxins can cause symptoms such as muscle weakness, tingling, numbness, seizures, and in severe cases, paralysis or death.

The risk of negative impacts of algal bloom on human health can be prevented by doing several things, for example:

1. Avoid direct contact with water containing algal blooms, such as swimming, diving, or other water activities.
2. Do not consume seafood from waters experiencing algal blooms or suspected of containing toxins (Costa et al. 2021).
3. Follow the advice and warnings of local health authorities regarding the presence of algal bloom and the associated potential health risks.
4. Report the presence of algal bloom to local environmental authorities so that they can take necessary prevention and control measures.

15.2.3 Case Studies of Algal Bloom

The following are several case studies of algal blooms that occurred around the world in the past 10 years and their negative impacts:

- Galicia, Spain (2020). Algal blooms involving dinoflagellates in Galician coastal waters resulted in the closure of the shellfish fishing industry and significant economic impacts (Karlson et al. 2021).
- Lake St. Clair, Michigan, USA (2020). An algal bloom involving cyanobacteria in Lake St. Clair resulted in the closure of several recreational areas and a drinking water quality advisory. Negative impacts include health threats to residents and visitors and disruptions to tourism.
- India (2019). Algal blooms in India's coastal waters, especially in the Kerala region, have resulted in declining fish populations and negative impacts on the fishing industry and ecosystem.

- Utah, USA (2019). An algal bloom involving cyanobacteria in Utah Lake led to health warnings and the closure of several recreational areas. Negative impacts include health threats to residents and visitors and disruptions to tourism and water activities.
- New South Wales, Australia (2019). The algal bloom in the Darling River resulted in a mass die-off of fish, particularly Murray cod, an endangered species. Negative impacts include economic losses to the fishing industry and threats to the survival of fish species.
- Oregon, USA (2018). An algal bloom involving cyanobacteria in Detroit Lake, Oregon, led to the closure of several recreational areas and warnings regarding drinking water quality. Negative impacts include health threats to residents and visitors and disruptions to tourism.
- Lake Okeechobee, Florida, USA (2018). An algal bloom involving cyanobacteria in Lake Okeechobee caused a decline in water quality and negative impacts on the ecosystem and the fishing and tourism industries.
- Florida, USA (2016 and 2018). Algal blooms involving cyanobacteria and *Karenia brevis* (red tide) resulted in mass deaths of fish and other marine life, including dolphins, sea turtles, and manatees. These events also negatively impact tourism and the fishing industry.
- Lake Victoria, Africa (2017). An algal bloom involving cyanobacteria in Lake Victoria resulted in health threats to residents who depend on the lake for drinking water and fishing, as well as negative impacts on fish life and the ecosystem.

Meanwhile, several water bodies in Indonesia have experienced algal blooms, e.g.,

- Jatiluhur Reservoir in West Java, a source of raw water for Jakarta and surrounding areas, has experienced algal blooms resulting in water quality problems, including unpleasant odor and taste in drinking water.
- Kedung Ombo Reservoir in Central Java has also experienced algal bloom, resulting in a decline in water quality and negative impacts on ecosystems and fisheries.
- Rawa Pening in Central Java is one of the natural lakes that has experienced algal bloom, causing water discoloration and quality degradation, directly impacting fish life and ecosystems.
- Lake Maninjau in West Sumatra has experienced algal bloom, which resulted in a decline in water quality and negative impacts on fish life and ecosystems as well as tourism.
- Lake Matano in South Sulawesi has experienced an algal bloom that resulted in a decline in water quality and negative impacts on fish life and ecosystems.
- Jakarta Bay in DKI Jakarta has caused a decline in water quality and disruption to ecosystem life and fisheries.

15.3 Other Ecological Problems Caused by Microalgae Biomass Conversion Technology

15.3.1 Eutrophication

Microalgae biomass conversion technology involves using algae as an energy and fuel source. When microalgal biomass is mass-produced for conversion into such energy or fuel, it usually utilizes a growth medium containing high nutrients, such as nitrogen and phosphate fertilizers. When mass production is done in an open system (e.g., pond) and the growth medium is not adequately controlled, unused nutrients can flow out of the system and enter the surrounding environment (van der Wiel et al. 2019). This situation further caused increasing nutrient concentrations in the water, triggering undesirable algal blooms and eventually leading to eutrophication.

Another example is microalgae biomass treated in a closed system that uses wastewater as a growth medium. Suppose the wastewater contains high nutrients, such as nitrogen and phosphorus. In that case, these nutrients will enter the microalgae growth system and trigger unwanted algae growth, which can cause eutrophication. Based on this description, eutrophication can be defined as the process of increasing nutrient concentration, especially phosphorus and nitrogen, in a water body such as a lake, river, or ocean. These nutrients come from various sources, such as agricultural and domestic waste and industrial pollutants. This nutrient increase leads to excessive growth of phytoplankton, mainly algae, which can cause drastic changes in the aquatic ecosystem.

Eutrophication can cause various negative impacts, such as algal bloom, the formation of hypoxia (oxygen deprivation conditions) or dead zones (Sidabutar et al. 2021) and food chain disruption that causes ecological imbalance. Microalgae biomass conversion technology should be used carefully to avoid eutrophication and use appropriate and effective growth systems. In addition, waste treatment must also be appropriately managed to minimize negative impacts on the surrounding environment.

15.3.2 Disruption to Ecosystems and Food Chains

Algal blooms or explosions of wild algal populations can cause disruptions in food chains and imbalances in aquatic ecosystems (Kazmi et al. 2022). The stages of food chain disruption due to algal bloom in waters include the following:

1. An algal bloom occurs when the number of wild algae populations increases significantly over a short period. This can be triggered by increased levels of nutrients in the water, such as nitrogen and phosphorus, the primary food sources for algae.

2. Furthermore, algal blooms can cause a decrease in water quality because wild algae require large amounts of oxygen for growth. As a result, oxygen levels in the water will decrease and cause the death of aquatic organisms that need oxygen to survive. Excessive algae growth can damage aquatic ecosystems and decrease the dissolved oxygen that eventually leads to the death of fish and other organisms.
3. Some types of wild algae can produce harmful toxins to animals and humans. When algal blooms occur, the amount of wild algae that produce toxins also increases. This can threaten the health of fish, shrimp, other aquatic organisms, and humans who consume such seafood. Fish or birds that consume organisms contaminated by wild algal toxins can experience nervous system disorders, organ damage, and death. Because of this, the population of organisms at higher levels of the food chain will be reduced.

   Some types of toxin-producing algae can produce compounds that can kill zooplankton. These zooplankton are the first consumers in the aquatic food chain (Talib et al. 2022) and if their numbers are reduced, then populations of organisms at higher levels of the food chain will be affected. Some types of toxin-producing algae that can kill zooplankton include:

   (a) Dinoflagellates are a group of marine algae known for their ability to produce saxitoxin toxins that can affect aquatic organisms and humans. The toxin can kill zooplankton and affect human and animal health.
   (b) Diatoms. Some species of diatoms produce the toxin domoic acid, which can cause suffocation in fish and seabirds that consume contaminated organisms.
   (c) Cyanobacteria, or blue-green algae, are a group of algae commonly found in freshwater and seawater. Some species of cyanobacteria produce a toxin called microcystin, which can affect human and animal health and kill zooplankton.
   (d) Prymnesiophyta is a group of unicellular algae that produce bioluminescent toxins, which can cause injury to the cells of affected organisms (Núñez-Pons et al. 2018). *Prymnesium parvum* is an algal species found in fresh and saltwater. *Prymnesiophyceae* produce prorocentrolid and prymnesin toxins that can kill zooplankton and other aquatic organisms.
   (e) Coccolithophore. Some coccolithophore species produce toxic emitters that can affect human and animal health and kill zooplankton (Moore 2021).
   (f) *Alexandrium* spp. is a group of marine algae commonly found in warm waters. Some species of Alexandrium spp. produce toxins known as saxitoxins, which can affect zooplankton and other aquatic organisms.
   (g) *Karenia brevis* is a species of marine algae that can produce a toxin known as brevetoxin. This toxin can affect zooplankton and other aquatic organisms and cause poisoning in humans who consume contaminated fish or shellfish.

      If the zooplankton population decreases, the food available to fish as the second consumer in the food chain will also decrease. This can affect the fish population and cause an imbalance in the food chain. Ultimately, algal blooms can disrupt the aquatic food chain because increasing wild algal populations can affect the number and diversity of other organisms.

4. In addition, algal blooms can trigger the growth of invasive organisms, such as sea moss or wild corals, which can disrupt food chains and damage ecosystems (Fig. 15.13).

Invasive organisms are "new" species that did not exist before and quickly establish themselves in an algal bloom environment. Some examples of invasive organisms related to algal blooms include:

- *Caulerpa taxifolia* is an invasive sea lichen native to the Caribbean. The species has a rapid growth rate and outcompete the indigenous species, spreading into the Mediterranean Sea, and is now disrupting ecosystems and inhibiting coral reef growth.

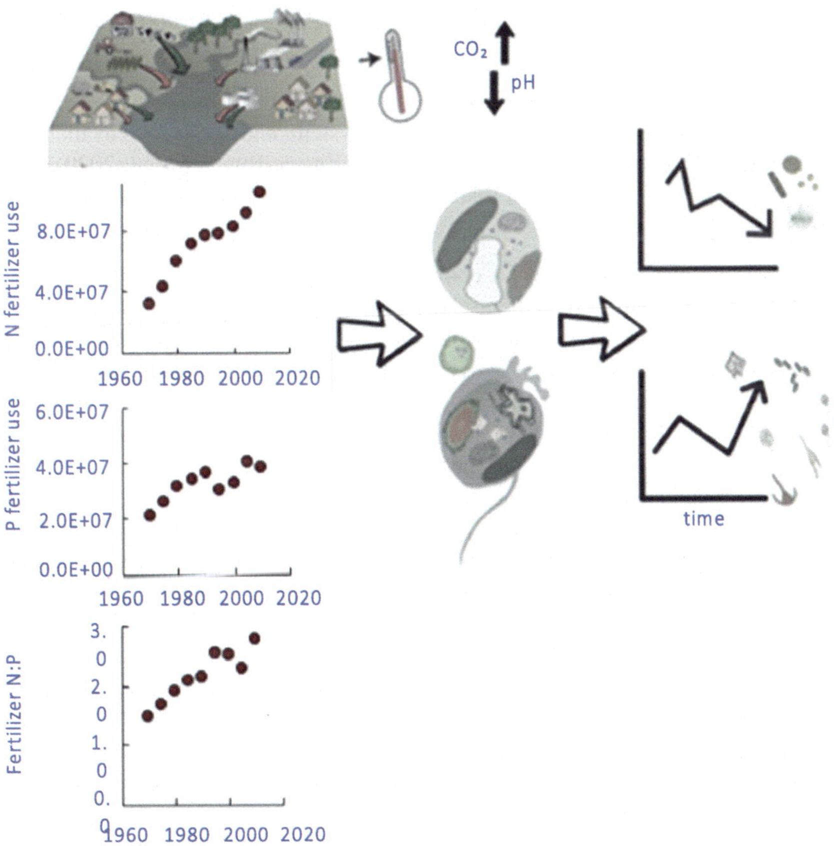

Fig. 15.13 The concept of aquatic ecosystem disturbance caused by algae blooming

- *Sargassum muticum* is a type of invasive brown algae that can grow in various conditions and disrupt coastal ecosystems, including coral reefs.
- Zebra Mussel *(Dreissena polymorpha)*: Although not algae, the zebra mussel is highly invasive and disrupts the aquatic food chain. It filters food particles from water that would otherwise be a food source for indigenous species.
- *Ostreopsis ovata* is a type of dinoflagellate that can cause toxic algal blooms. The toxin produced by this organism can harm marine life and humans.
- *Pfiesteria piscicida* is a type of dinoflagellate that can produce harmful toxins that affect fish and other organisms in the food chain.
- *Didymosphenia geminata* (Didymo) is also known as "rock slime": This freshwater alga has become invasive in many places, including New Zealand and North America forming thick algal blooms and disrupting the habitat of fish and other organisms in rivers and lakes.
- Lionfish (*Pterois volitans*), a venomous fish as an indigenous predator from the Pacific Ocean, has spread to the Caribbean and Atlantic Seas rapidly and massively. Lionfish feed on various fish and invertebrates, disrupting local food chains.

The growth of invasive organisms can disrupt food chains, damage ecosystems by competing with indigenous organisms in the ecosystem, and affect food availability for consumers at higher levels of the food chain (Emery-Butcher et al. 2020). The growth of invasive organisms can disrupt food chains and damage ecosystems as follows:

- Invasive organisms often have a competitive advantage in a new environment. These organisms may defend themselves through faster growth or have better self-defense systems than native organisms (Parr and Bishop 2022). This can result in a decrease in the population of native organisms, affecting the food chain.
- Invasive organisms can disrupt food chains by potentially feeding on foods generally eaten by indigenous organisms. This can reduce food availability for consumers at higher levels of the food chain.
- Invasive organisms can become prey for new predators that have no natural enemies in the new environment. On the other hand, invasive organisms can prey on indigenous species in the ecosystem. This can lead to a decline in the population of indigenous species and affect the entire food chain above due to predator–prey changes.
- Some invasive organisms can produce toxic or chemical substances that harm native organisms and other animals in the environment. These toxic substances cause new diseases in the ecosystem, which can affect the population of indigenous organisms, disrupting the balance of the food chain and destroying the balance of the ecosystem.
- Some invasive organisms, such as sea moss or wild coral, can rapidly grow and cover large areas. This can damage the physical structure of ecosystems, such as coral reefs or mangrove forests, which are habitats for many organisms and play a crucial role in the food chain.

15.3.3 Pollution of Water and Soil Due to Production Waste

As explained earlier, some algae species can produce toxins that are harmful to humans and animals. If wild algae grow excessively, the increased toxin levels can pollute waters, ultimately threatening the health of humans and living creatures along the food chain.

In addition to water pollution, microalgae biomass production has the potential to pollute the soil. Land pollution due to microalgae biomass production waste can occur when the waste generated from the microalgae biomass conversion process is not adequately managed. Microalgae biomass conversion technology uses microalgae as a renewable energy source, such as biofuels, chemicals, and other products. Some commonly used processes in microalgal biomass conversion include lipid extraction, fermentation, pyrolysis, and gasification. Solid and liquid wastes can be generated during these processes, which can contaminate the soil if not appropriately managed.

There are several ways that microalgae biomass production waste can contaminate soil:

1. Solid waste. After lipid extraction or other conversion processes, the remaining biomass is usually rich in nutrients and organic matter. However, if these solid wastes are carelessly disposed of on the ground, the nutrients can contaminate the soil and groundwater, causing pollution and eutrophication.
2. Liquid waste. The microalgae biomass conversion process often generates liquid waste containing chemicals and nutrients. If these wastes are not adequately treated before being discharged into the environment, they can pollute soil and groundwater and damage the surrounding ecosystem.
3. Chemicals. During the microalgae biomass conversion process, chemicals such as solvents, catalysts, and acids may be used. If these chemicals are not handled and disposed of properly, they can contaminate soil and groundwater and threaten human health and the environment.

Several management strategies to reduce the impact of soil pollution due to microalgae biomass production waste should be taken, for example:

1. Properly treat solid and liquid waste before disposal into the environment, for example, by composting, drying, or utilizing waste as fertilizer or animal feed (Ayilara et al. 2020).
2. Utilizing environmentally friendly and biodegradable chemicals in microalgae biomass conversion processes can reduce the potential for soil pollution. Selecting appropriate chemicals will depend on the specific microalgae biomass conversion process and operational conditions (Rinanti and Purwadi, 2019).

   Some recommended chemicals in the microalgae biomass conversion process are:

   (a) Enzymes such as cellulase, hemicellulase, and protease can be used to pretest and hydrolyze microalgae biomass. These enzymes are efficient and environmentally friendly biological catalysts and are biodegradable.

(b) Citric acid is a weak organic acid that can be used in lipid extraction and pretreatment of microalgae biomass. Citric acid is biodegradable and has a lower environmental impact than strong acids such as sulfuric acid or hydrochloric acid.
(c) Acetic acid is an organic acid that can be used in the pretreatment process of microalgae biomass. It is biodegradable and more environmentally friendly than strong inorganic acids.
(d) Oxalic acid is an organic acid that can be used to extract lipids from microalgal biomass. Oxalic acid is biodegradable and has a lower environmental impact than other strong acids.
(e) Sodium hydroxide (NaOH) is a base that can be used in the pretreatment process of microalgae biomass. Although NaOH is not a completely environmentally friendly chemical, its environmental impact can be minimized in the right concentration and with good waste management.
(f) Sodium bicarbonate ($NaHCO_3$) is the sodium salt of carbonic acid, which can be used in microalgal biomass's precipitation and flocculation process. Sodium bicarbonate is a biodegradable chemical and has a low environmental impact.

3. Regularly monitoring and controlling waste and chemical emissions in the microalgae biomass production process can help identify and address potential pollution problems before they become significant (Ali et al. 2022). Some of the things that can be done include:

(a) Monitor the quality of water used in the microalgae production process, including nutrient and pollutant content. Water containing excess nutrients or pollutants can increase the risk of algal blooms and environmental pollution.
(b) Monitor and control chemical consumption in the production process. Prioritize using environmentally friendly and biodegradable chemicals to reduce environmental impact and the risk of algal blooms.
(c) Ensure that the wastewater treatment system from the microalgae production process effectively reduces emissions of waste and pollutants to the environment, including regular monitoring of wastewater quality and treatment effectiveness.
(d) Regularly monitor the density and quality of microalgae biomass in the production system to ensure the process is running optimally and reduce the risk of algae bloom.
(e) Monitor the contamination in microalgae production systems caused by pathogenic microorganisms or potentially toxic algae species. Contamination can increase the risk of algal blooms and environmental pollution.
(f) Implement good management and operating practices to reduce the risk of algal bloom and environmental pollution, including optimizing microalgae growth conditions, controlling environmental factors such as temperature and lighting, and conducting regular equipment maintenance and sanitation.

(g) Ensure personnel in the microalgae biomass production process have adequate knowledge and training on algal bloom and environmental pollution control.
(h) Cooperate with authorities and communities in algal bloom and environmental pollution control efforts, including reporting algal bloom incidents or potential environmental problems to the appropriate authorities.

4. Optimize the production and conversion of microalgal biomass to reduce the amount of waste generated and ensure resources are used efficiently and sustainably.

15.4 Strategies to Deal with Algal Blooms and Other Ecological Problems

15.4.1 Management of Nutrients and Reduction of Pollutants in the Waters

Nutrient management and pollutant reduction in water is one of the crucial strategies for addressing algal blooms and other ecological problems. Excess nutrients, such as nitrogen and phosphorus, are often the leading cause of algal blooms and eutrophication. Pollution by other pollutants, such as chemicals and heavy metals, can also disrupt aquatic ecosystems. Here are some ways to manage nutrients and reduce pollutants in water:

1. Source control of pollutants by reducing the introduction of nutrients and pollutants into waters through regulation and control of industry, agriculture, and domestic waste disposal. This may include more efficient use of fertilizers, treatment of wastewater before discharge, and application of green technologies in industry. Some regions have established nutrient management, including imposing restrictions on using phosphate in detergents and fertilizers, as phosphate is a key nutrient promoting algae growth.
2. *Wetland* restoration and establishing vegetative buffer zones along the banks of rivers or lakes can help absorb nutrients and pollutants to avoid algal blooms in waterways. Restoring buffer zones involves maintaining riparian plants, swamps, and mangroves, which can absorb nutrients and pollutants before they reach the water. Buffer zones can also reduce erosion and sedimentation, which can affect water quality.
3. Implementation of Bioremediation Technology, specifically phytoremediation utilizing plants and/or microorganisms, to reduce and even remove pollutants from the environment. For example, some types of algae and plants can be used to absorb excess nutrients from water, preventing algal blooms.
4. Water flow management by using water flow management techniques, such as retention ponds, dams, and infiltration wells, to reduce the flow of water containing nutrients and pollutants into the waters.

5. Water quality monitoring by regularly monitoring water quality to identify sources of nutrients and pollutants and measuring the effectiveness of the management strategies implemented.
6. Educate the public to be more aware of the impacts of nutrients and pollutants on waters and ways to reduce these impacts, such as reducing fertilizer and pesticide use, managing household waste, and reporting pollution. This includes raising awareness about the importance of good waste management and sustainable agricultural practices.
7. Encourage research and development of new technologies or generate innovations to reduce nutrients and pollutants in waters and improve understanding of the relationship between nutrients, pollutants, and algal blooms.
8. Promote cooperation between government, industry, environmental groups, and communities to address water nutrient and pollutant issues.
9. Encourage the central and local governments to develop and implement effective regulations and policies to control pollution and reduce algal blooms.

By implementing these strategies, hopefully, the water's nutrients and pollutants can be controlled, which will help prevent algal blooms and address other ecological concerns. Effective and sustainable management of water bodies is essential for maintaining ecosystem balance and human well-being. Effective management strategies usually involve a combination of the various approaches mentioned above. The appropriate strategy usually varies depending on the specific ecosystem and pollution source (Imai et al. 2021; Rohmah et al. 2018).

15.4.2 Monitoring and Early Detection of Algal Blooms

Monitoring and early detection of algal blooms are essential in managing and addressing associated ecological issues. Early detection of algal blooms allows ecosystem authorities and managers to respond faster and reduce environmental and human health impacts. The following are some methods and strategies that can be used for monitoring and early detection of algal blooms (Binding et al. 2021):

1. Water Sampling and Analysis. Periodic sampling of water from aquatic systems and analysis of nutrient content, such as nitrogen and phosphorus, as well as algae with the potential to form blooms, is essential. Laboratory methods such as morphological analysis with various microscopes, liquid chromatography analysis, and spectroscopic techniques can be used to identify and quantify the concentration of algae in water samples.
2. Satellite Monitoring and Aerial Imagery (*Remote Sensing*). Using *remote sensing* technologies, such as satellite or *drone* sensing, can help monitor algal blooms from a distance. This technology is beneficial for monitoring large or inaccessible areas. Satellite imagery can identify changes in water color that indicate the presence of algal blooms. The images and data collected can be used to identify

areas of high algal activity and monitor the changing waters and the distribution of algal blooms over time.

3. Biosensors. A biosensor is a device that uses organisms or biological components to detect the presence of a specific material. Biosensors are designed to detect certain types of algae or toxins produced by algae and thus can be used for early detection of algal blooms. Biosensors can also be devices to measure chlorophyll fluorescence, aiding in the early detection of algal blooms. Biosensors can be placed in aquatic systems and provide real-time data on conditions that may favor algal blooms.
4. Monitor the presence of vulnerable species. Susceptible species to algal blooms, such as some fish and mollusks, can be monitored as indicators of ecosystem health and the potential for algal blooms. Different types of fish can be affected by algal blooms, and the specific susceptibility of fish can depend on several factors, including the type of algae involved, the type of toxin produced, and the characteristics of the fish species. Some fish species that are more susceptible to algal blooms and can be used as indicators of potential blooms are:

    (a) Prey fish, such as pike, bass, and tuna, can be particularly vulnerable to algal blooms because they are at the top of the food chain. If toxin-producing algae cause algal blooms, the toxins can build up in the food chain and reach high concentrations in predatory fish.
    (b) Fish that live in shallow waters, such as trout and salmon, can be particularly vulnerable to algal blooms. Salmon (*Salmo salar* and *Oncorhynchus* sp.) are sensitive to changes in water quality and the presence of toxins produced by some algae. Decreases in salmon larval populations or survival can indicate algal blooms. Trout (*Salmo trutta* and *Oncorhynchus mykiss*) are also susceptible to changes in water quality and dissolved oxygen levels. Trout can indicate algal blooms that are occurring or will occur, especially in fresh waters such as lakes and rivers.
    (c) Plankton-eating fish, such as herring and anchovy, can be vulnerable to algal blooms as toxin-producing algae can form part of their diet. Herring (*Clupea harengus*) are particularly sensitive to changes in water quality, especially low dissolved oxygen, which is often associated with algal blooms. Mass mortality of herring can indicate algal blooms that are occurring or about to occur.
    (d) Menhaden (*Brevoortia* sp.) is a small pelagic species highly susceptible to hypoxic conditions (low dissolved oxygen) caused by algal blooms. High mortality of menhaden can be an early sign of algal blooms.
    (e) Mullet (*Mugil* sp.) is a species sensitive to changes in water quality, including hypoxic conditions and the presence of toxins some algae produce. Declines in mullet populations or mass mortality can be indicative of algal blooms.
    (f) Monitoring these fish species and other species susceptible to algal blooms can help identify changes in water quality and provide early signs of blooms. However, note that mass fish mortality or population declines can be caused by other factors such as disease, pollution, or habitat changes. Therefore,

observations of vulnerable fish species should be combined with other detection and monitoring methods to identify algal blooms effectively.

5. Predictive Models. The development of predictive models that consider environmental factors such as temperature, humidity, rainfall, and nutrient concentrations can assist in identifying conditions that favor algal blooms. Based on historical data and current environmental conditions, computer models can be used to predict when and where algal blooms might occur. For example, models can use water temperature, salinity, and nutrient concentration data to predict the risk of algal blooms, allowing ecosystem managers to take action before algal blooms occur.
6. Inter-Agency Cooperation. Collaboration between research institutions, governments, and environmental organizations enables the exchange of data and information on algal blooms and the conditions that support blooms. This cooperation also enables effective coordination of actions to reduce the impact of algal blooms on the environment and human health.
7. Community Participation. Involving the community in monitoring algal blooms through citizen watch programs or reporting incidents of algal blooms is expected to aid in the early detection and response to this problem. Communities are trained to identify changes in water properties such as color, texture, or unusual odors. In addition, communities can also play a role in preventing algal blooms by reducing pollution and keeping the environment clean.

In order to manage and address algal blooms and other ecological issues, a combination of these multiple monitoring and early detection methods, along with nutrient management and pollutant reduction strategies, will be essential.

15.4.3 Use of Environmentally Friendly Technology in Microalgae Biomass Conversion

Using environmentally friendly technologies in microalgae biomass conversion can help reduce the negative environmental impacts that trigger algal blooms. Although green technology in microalgae biomass conversion does not directly play a role in monitoring and early detection of algal blooms, it can provide benefits in reducing the risk of algal blooms (Zabochnicka et al. 2022). Such green technologies in microalgae biomass conversion, such as using more efficient extraction methods, separation technologies that minimize the use of hazardous chemicals, and using biodegradable green chemicals, can help reduce the amount of waste and pollutants produced. Proper waste management is essential to prevent environmental pollution in the microalgae biomass conversion process. Environmentally friendly technologies can involve the treatment of solid and liquid wastes before disposal, such as composting, drying, or utilizing wastes as fertilizer or animal feed.

Efficient energy use in microalgal biomass conversion can reduce energy consumption and greenhouse gas emissions. This can help reduce broader environmental impacts and, indirectly, the risk of algal blooms associated with climate change (Zieliński et al. 2023). Sustainable resource use can be pursued by optimizing the production and conversion of microalgal biomass to reduce the amount of waste produced, which can help prevent environmental problems such as algal blooms. Overall, although green technologies in microalgal biomass conversion do not directly play a role in monitoring and early detection of algal blooms, they can help reduce the environmental impacts that trigger algal blooms. By reducing risks and negative environmental impacts, these green technologies can contribute to the prevention and management of algal blooms.

15.4.4 Restoration of Ecosystems Affected by Algal Blooms

Restoration of ecosystems affected by algal blooms involves a series of actions to restore environmental conditions and ecological balance (Sukenik & Kaplan 2021).

1. Eliminate algal blooms. The first step in ecosystem restoration is to eliminate existing algal blooms. This can be done by using environmentally friendly algaecide exposures, mechanical controls such as screening or physically removing algae, or using organisms that prey on algae, such as phytoplankton-eating fish or mussels.
2. Increase oxygen. Algal blooms can cause a decrease in oxygen levels in water, which negatively affects aquatic life. Aeration or other technologies that increase oxygen content in water can help restore optimal environmental conditions.
3. Habitat restoration by replacing dead vegetation or removing toxin-laden sediments to help restore damaged ecosystems. This includes restoration of wetlands, vegetative buffer zones, and damaged river or lake bed substrates.
4. Nutrient management by reducing nutrient inputs into aquatic systems affected by algal blooms is a significant effort to prevent recurrence. It is the control of pollutant sources such as industrial, agricultural, and domestic effluents, as well as the use of improved sewage treatment technologies and more sustainable agricultural practices.
5. Once restoration measures are taken, ongoing monitoring and evaluation are required to ensure the success of the restoration efforts. This includes monitoring water quality, algae population density, and overall ecosystem health.
6. In some cases, restoring fish populations affected by algal blooms may be necessary, for example, by releasing fish caught from elsewhere or a hatchery program to help increase the hard-hit fish population.
7. Community education and engagement about algal blooms and their impacts on ecosystems can help prevent blooms as a long-term restoration effort. Education

programs can include information campaigns, training, and developing awareness of the importance of good waste management and sustainable agricultural practices.

15.5 Circular Economy and Wastewater Management

15.5.1 The Circular Economy Concept and Its Application

The Circular Economy concept is an approach that aims to reduce resource waste and pollution and promote environmental sustainability through the efficient design of products, processes, and systems. It encompasses the principles of "reduce, reuse, recycle" and replaces the "take, make, dispose" paradigm commonly found in linear economies (Kirchherr et al. 2023). In the context of wastewater management and microalgae biomass conversion, the application of the circular economy concept can include various strategies, such as using wastewater as a nutrient source for microalgae growth: Wastewater rich in nutrients, such as nitrogen and phosphorus, can be used to grow microalgae. This way, wastewater is treated and simultaneously provides the necessary nutrient source for microalgae, reducing the need for artificial fertilizers.

After the microalgae biomass conversion process, the remaining water can be further treated to recover nutrients or other chemical compounds that can be used in other processes, such as agriculture or industry. In order to optimize resource use and reduce the environmental footprint, wastewater treatment systems can be combined with microalgae biomass conversion technologies. This enables pollution reduction, resource recovery, and the production of valuable raw materials, such as bioenergy or other high-value products. Microalgae biomass conversion processes often produce by-products that can be used for other purposes, such as animal feed, fertilizer, or raw materials for the chemical industry. Utilizing these by-products reduces resource wastage and creates added value from the production process.

In applying the circular economy concept, developing technologies that can be adapted and improved to meet changing needs and incorporate innovations must be considered. The application of the circular economy concept in wastewater management and microalgae biomass conversion can help create a system that is more environmentally friendly and efficient in resource use. This will support environmental and economic sustainability in the long run.

15.5.2 The Use of Microalgae to Treat Wastewater

Microalgae biomass conversion technology produces commercial products and leaves liquid and solid waste from the production process. Utilizing microalgae to treat wastewater is one example of applying the circular economy concept in wastewater management. Microalgae are photosynthetic microorganisms that can absorb nutrients, such as nitrogen and phosphorus, and organic chemicals from wastewater. This process helps reduce the levels of pollutants in wastewater and simultaneously supports the growth of microalgae (Abdelfattah et al. 2023). The following are some aspects related to the utilization of microalgae in wastewater treatment:

1. Bioremediation. Microalgae play a role in wastewater bioremediation by absorbing excess nutrients (such as nitrates and phosphates) and reducing organic matter levels that can cause pollution. This process helps prevent eutrophication and algal blooms in natural waters.
2. Biomass production. Microalgae growth supported by nutrients from wastewater produces biomass that can be converted into various high-value products, such as bioenergy (biodiesel, biogas, or bioethanol), animal feed, fertilizer, and industrial chemicals.
3. Integrated wastewater treatment. The utilization of microalgae in wastewater treatment can be combined with other treatment technologies, such as aerobic or anaerobic systems, to achieve more efficient and effective treatment levels. This integration allows for more optimized resource use and reduced environmental footprint.
4. Resource recovery. After a wastewater treatment process with microalgae, the remaining water can be further treated to recover resources such as nutrients, water, or other chemical compounds that can be used in other processes, such as agriculture or industry.
5. Education and community engagement. Utilizing microalgae in wastewater treatment can also be used as an educational tool and community involvement in environmental sustainability. The community can be involved in monitoring water quality and maintaining treatment systems that use microalgae.

15.5.3 Case Studies and Best Practices in Wastewater Treatment Using Microalgae

Several projects in Europe have successfully applied wastewater treatment technology using microalgae to produce renewable energy, high-value products, and recover resources (You et al. 2023), namely:

1. The All-gas project in Spain is one of the industrial-scale projects developed by Aqualia, a leading water utility company in Spain. The project focuses on using microalgae to treat wastewater and produce biogas as a renewable energy source. The project successfully combines wastewater treatment with renewable energy production and reduces greenhouse gas emissions. The All-gas project: The most significant wastewater to biofuel facility using algae.

2. The EU INCOVER project is a research initiative funded by the European Union under the Horizon 2020 program. The project includes developing and demonstrating three innovative microalgae-based solutions for wastewater treatment, including biomass production as a renewable energy source, nutrient recovery, and wastewater treatment for irrigation. The INCOVER project demonstrates the potential use of microalgae in integrated and sustainable wastewater treatment systems.
3. The BIOFAT (Biorefinery for Food, Fuel, and Materials) project in the European Union is a research project funded by the European Union under the FP7 program. The project aims to develop an integrated microalgae biorefinery process to produce various high-value products from wastewater, such as biodiesel, chemicals, and fertilizers. The project successfully demonstrated pilot and commercial-scale wastewater treatment using microalgae as a renewable source of energy and chemicals.

15.6 Policy, Regulation, and International Cooperation

15.6.1 Policies and Regulations that Support Algal Bloom Control and the Use of Environmentally Friendly Technologies

Policies and regulations that support the control of algal blooms and the use of environmentally friendly technologies are essential to prevent and manage the negative impacts of algal blooms on ecosystems and human health. Below are some policies and regulations that can help achieve this goal (Rashidi et al. 2021):

1. **Wastewater Management Policies** include policies to limit the amount of nutrients entering aquatic systems through human activities such as agriculture, industry, and domestic sewage, in an effort to reduce the risk of algal blooms. Regulations governing the use of fertilizers, implementation of sustainable agricultural practices, and adequate treatment of effluents can contribute to the control of algal blooms. Here are some international and national regulations governing nutrient restrictions:

    (a) The European Union's Water Framework Directive (WFD) (2000/60/EC), is the legal framework governing the management of water resources in EU member states. The WFD includes restrictions on nutrients, including nitrogen and phosphorus, coming from agriculture, industry, and domestic sewage. Other aspects addressed in the WFD include assessment methodologies, implementation challenges, and integration with other environmental frameworks (de Carvalho et al. 2018).

(b) The European Union's Urban Wastewater Treatment Directive (UWWTD) (91/271/EEC) regulates the collection, treatment, and discharge of municipal wastewater, including restrictions on pollutants that can cause algal blooms.
(c) The United States Clean Water Act (CWA) is a law that aims to restore and maintain the physical, chemical, and biological integrity of the nation's waters. The CWA includes requirements to reduce the amount of nutrients entering waters through restrictions on Total Maximum Daily Loads (TMDLs). In addition, the CWA also regulates the National Pollutant Discharge Elimination System (NPDES) which requires facilities that discharge wastewater to be licensed and meet certain emission standards.
(d) Nutrient Management Regulations in Canada are federal and provincial regulations that govern the management of nutrients in agricultural, industrial, and domestic waste activities. Some provinces such as Ontario have specific regulations called the Nutrient Management Act, which govern the use and handling of nutrients in the agricultural and industrial sectors.
(e) Australia's National Environment Protection (Assessment of Site Contamination) Measure is a regulation governing the assessment and remediation of site contamination, including the limitation of nutrients from human activities such as agriculture, industry, and domestic waste. Australia also has The Australian Guidelines for Water Recycling (AGWR), Water Act 2007 and Australia's National Water Quality Management Strategy (NWQMS). These three regulations generally provide a national framework for efficient and environmentally friendly wastewater management, including the reduction of pollutants that can trigger algal blooms.
(f) Regulation of the Minister of Environment and Forestry of the Republic of Indonesia No. 68 Year 2016 on Wastewater Quality Standard is a regulation governing the limitation of nutrients in wastewater originating from various sources, including agriculture, industry, and domestic waste.
(g) Hong Kong's Water Pollution Control Ordinance (WPCO) regulates wastewater management to reduce water pollution and potential algae blooms. The WPCO covers technical requirements for wastewater treatment facilities, water pollution control, such as the implementation of water quality standards, discharge permits, and enforcement. These regulations aim to reduce the negative impacts of water pollution on the environment, human health, and aquatic life.
(h) China's Environmental Protection Law is a law that regulates wastewater management and the discharge of pollutants into waters, including requirements for environmental licensing and efficient and environmentally friendly wastewater management.
(i) Various water quality standards have been developed around the world to help prevent and control algal blooms. Environmental Quality Standards (EQS) in various countries regulate the amount of nutrients allowed in surface water and wastewater to prevent and control algal blooms. These standards can be used as a basis for monitoring and enforcement.

The standards normally include the parameters of nutrient concentration, algae density, and toxicity levels. EQS vary depending on the country and local regulations (Leverett et al. 2021). The following are some examples of water quality standards adopted by several countries and international organizations:

i. The European Union—Water Framework Directive (WFD) and Environmental Quality Standards Directive (EQSD) are standards covering the parameters of nutrient concentration, phytoplankton density (including algae), and toxicity levels set to achieve the goal of good ecological status in surface waters.
ii. The US Environmental Protection Agency (EPA) established the United States—National Recommended Water Quality Criteria (NRWQC) which contains criteria for the concentration of nutrients such as nitrogen and phosphorus, as well as the toxicity of chemicals that may contribute to causing algal blooms.
iii. Australia—The Australian and New Zealand Guidelines for Fresh and Marine Water Quality (ANZG) are guidelines that provide information on nutrient concentrations, algal densities, and toxicity that are considered safe for various water uses, such as drinking water supply, ecosystem protection, and recreation.
iv. Canada—The Canadian Water Quality Guidelines for the Protection of Aquatic Life (CWQG) contain criteria for nutrient concentrations, algal densities, and toxicity that are considered safe to protect aquatic life.
v. World Health Organization (WHO)—Guidelines for Drinking-water Quality (GDWQ) is a WHO Guideline that contains information on drinking water quality parameters, including nutrient concentrations, algae density, and toxicity levels considered safe for human health.

Each country may have its own water quality standards that include the parameters of nutrient concentration, algae density, and toxicity levels. These standards are usually implemented by national or local environmental authorities. In Indonesia, water quality standards covering the parameters of nutrient concentration, algal density (Leverett et al. 2021) and toxicity levels to help prevent and control algal blooms can be found in the Regulation of the Minister of Environment and Forestry of the Republic of Indonesia No. 68 Year 2016 on Wastewater Quality Standards (Permen LHK 68/2016). It regulates the quality of wastewater that is allowed to be discharged into the environment, including limits on nutrient concentrations and toxicity levels that are considered safe for ecosystems and human health. While algae density may not specifically be regulated in the above regulation, water quality criteria that include nutrient concentrations and toxicity levels indirectly help prevent and control algae blooms by reducing the influx of nutrients and other pollutants that can trigger algae growth.

2. Energy and Industrial Policies are expected to support the development and use of environmentally friendly technologies in the energy and industrial sectors, such as tax incentives and research funds to help reduce negative environmental impacts that can trigger algal blooms. Here are some examples of relevant energy and industrial policies:

   (a) The European Union—Renewable Energy Directive (RED) and Energy Efficiency Directive (EED), is the European Union's Green Energy Policy that sets renewable energy and energy efficiency targets for EU member states. It supports the development and adoption of green technologies in the energy and industrial sectors through tax incentives for companies using green energy technologies, financial support schemes, and research funds for the development of new technologies.
   (b) The Renewable Portfolio Standard (RPS) in the United States is a policy that requires power companies to supply a portion of their energy from renewable sources. It encourages the use of green technologies and can help reduce negative impacts on the environment.
   (c) United States—The Production Tax Credit (PTC) and Investment Tax Credit (ITC) are federal tax incentives aimed at supporting the development and adoption of renewable energy and energy efficiency technologies in the energy and industrial sectors.
   (d) China—The 13th Five-Year Plan for Energy Development is a plan that includes targets for the development and use of renewable energy, energy efficiency, and green technologies in the energy and industrial sectors. It includes tax incentives, research funds, and other policy support.
   (e) India—National Solar Mission and National Wind Energy Mission, are part of India's energy policy to reduce dependence on fossil energy and promote the use of renewable energy. The policy includes tax incentives, financial support schemes, and research funds for the development of green technologies in the energy and industrial sectors.
   (f) Germany—Energiewende (Energy Transition) is Germany's highly ambitious energy policy, which aims to reduce greenhouse gas emissions and increase the use of renewable energy. It includes tax incentives, research funds, and other policy support for the development of green technologies in the energy and industrial sectors.
   (g) The Australian Renewable Energy Target (RET) is an Australian government policy designed to increase energy production from renewable sources. The policy includes tax incentives and research funding for companies that develop and use green technologies.
   (h) Feed-in Tariffs (FiTs) and Power Purchase Agreements (PPAs) in various countries are policy instruments used to encourage renewable energy development. These policies guarantee certain payments to renewable energy producers, which helps make the technology more economical and encourages its implementation.

(i) Indonesia—The National Energy General Plan (RUEN) is Indonesia's energy policy that includes targets for the development and use of renewable energy and energy efficiency in the energy and industrial sectors. The policy includes tax incentives, research funding, and other policy support. Indonesia's Renewable Energy Law No. 30 of 2007 establishes the legal framework for renewable energy development in Indonesia, which includes tax incentives and research funds for companies using environmentally friendly technologies.

3. Effective monitoring and enforcement are necessary to ensure regulatory compliance, as well as strict enforcement of violations to control algal blooms.
4. Community education and engagement is expected to increase public awareness about algal blooms, their impacts, and how to prevent them, helping to encourage more environmentally friendly behavior and support existing policies and regulations.
5. As the sources of pollutants and algal blooms are often transboundary, there is a need for international cooperation in developing and implementing effective policies and regulations. This cooperation can include information and knowledge sharing, technical support, and policy coordination between countries.

By combining these various policies and regulations, governments and international organizations can collaborate to control algal blooms and support the use of green technologies in various sectors. This will help protect the ecosystem.

15.6.2 International Initiatives and Cooperation in Tackling Algal Blooms and Ecological Problems

The following are some of the international initiatives and cooperation in addressing algal blooms and ecological problems (Kumar et al. 2021):

1. The Global Water Partnership (GWP) is an international network established in 1996 by the World Bank, Sweden, and the United Nations through the United Nations Development Programme (UNDP) (https://www.gwp.org/en/About-GWP/). GWP aims to support sustainable water resources management and create partnership-based solutions worldwide. GWP works with governments, international organizations, academic institutions, and the private sector to develop and promote Integrated Water Resources Management (IWRM) principles to address water-related challenges at global, regional, and national levels. GWP does not specifically targ*et al*gal blooms but, through initiatives and activities in sustainable water management, indirectly helps reduce the risk of algal blooms. Here are some examples of GWP activities related to water management and their impact on algal blooms:

   (a) Integrated Water Resources Management (IWRM). GWP strongly supports the concept of IWRM, which is a holistic approach to managing water resources. Through IWRM, GWP helps countries develop policies and practices to reduce human activities' negative impacts on water resources, including pollution that can cause algal blooms.

(b) The GWP Water, Climate, and Development Program (WACDEP) is a program designed to integrate climate change in water resources planning and management in Africa. Through WACDEP, GWP works with governments and local partners to reduce the impacts of climate change on water resources, including algal blooms exacerbated by climate change.

(c) GWP-Caribbean is a regional partner of GWP that aims to address water management challenges in the Caribbean. GWP-Caribbean has taken steps to reduce the risk of algal blooms, such as through the development of early warning and monitoring systems.

(d) Water Stewardship and Industry Partnerships is GWP's private sector partner to promote sustainable water management, including efforts to reduce industrial pollution that can cause algal blooms.

(e) GWP supports training and capacity building for water management professionals, including those working in water pollution control and algal blooms.

Although GWP does not explicitly focus on algal blooms, their activities in sustainable water management and mitigation of climate change impacts indirectly help reduce the risk of algal blooms. By working closely with government, private sector, and local partners, GWP helps build awareness and capacity to effectively manage water resources and reduce negative impacts on the environment.

2. The UNESCO Intergovernmental Oceanographic Commission (IOC) (https://ioc.unesco.org/about-ioc) is a commission under UNESCO that was established in 1960. The IOC aims to enhance international coordination and cooperation in research and management of marine resources. Through programs and projects focusing on ocean science, water management and disaster mitigation, the IOC contributes to the development of national knowledge and capacities to address challenges related to climate change, water sustainability, and food and water security. The IOC also plays an important role in early warning systems for tsunamis and other maritime disasters, and in the development and implementation of international policies and strategies for sustainable management of marine resources and disaster risk reduction. The IOC has developed programs such as the Global Harmful Algal Bloom (HAB) which aims to improve global understanding of the harmful algal bloom phenomenon and help countries identify, monitor, and mitigate the impacts of algal blooms on ecosystems and human health. Here are some of the main activities carried out by the Global HAB Program:

(a) The Global HAB Program supports research and monitoring of harmful algal blooms around the world. The program helps countries develop and implement effective monitoring technologies, such as the use of satellites, remote sensing, and early warning systems. This enables early detection of algal blooms and a faster response to mitigate their impacts.

(b) The Global HAB Program organizes various trainings, workshops, and courses to increase countries' capacity to manage and mitigate algal blooms. The program helps water management professionals, scientists and policy makers improve their knowledge and skills in tackling harmful algal blooms.

(c) The Global HAB Program facilitates the exchange of information and knowledge between countries affected by algal blooms. The program supports the development of global and regional networks, such as the Harmful Algal Bloom Information System (HABISYS) and the Northwest Pacific Action Plan (NOWPAP), which enable collaboration and information exchange between experts and institutions working on algal blooms.

(d) The Global HAB Program recognizes the importance of community engagement in reducing the impacts of algal blooms. The program works with local communities, coastal groups, and fishers to raise awareness about algal blooms and develop effective adaptation and mitigation strategies.

(e) The Global HAB Program helps countries develop effective policies to reduce the impacts of algal blooms on ecosystems and human health. The program provides guidance and recommendations to policymakers on how to reduce inputs of nutrients and other pollutants that can cause algal blooms and how to manage and mitigate the impacts of algal blooms that have already occurred.

2. The United Nations Environment Programme (UNEP) is a UN agency that focuses on environmental protection and sustainable development (https://www.unep.org/about-unep). The organization, founded in 1972, focuses on protecting the environment and empowering people to implement sustainable development. UNEP works to provide leadership and foster partnerships in caring for the environment by inspiring, informing, and empowering nations and communities to improve their quality of life without compromising future generations. UNEP has implemented various initiatives and programs to address global environmental issues, including water pollution and algal blooms. Here are some examples of activities undertaken by UNEP:

   (a) The Global Programme of Action for the Protection of the Marine Environment from Land-based Activities (GPA) is a global initiative that aims to address marine pollution from land-based activities, including pollution that can cause algal blooms. UNEP works with governments, international organizations, and the private sector to develop and implement pollution reduction strategies.

   (b) UNEP Freshwater Strategy. The strategy is designed to improve the management of freshwater resources and reduce water pollution. Through this strategy, UNEP helps countries develop effective policies and practices to reduce water pollution and its impacts on ecosystems, including algal blooms.

(c) The World Water Assessment Programme (WWAP) is a program led by UNESCO and implemented by UNEP, which specifically evaluates the state of water resources around the world. WWAP produces the World Water Development Report, which includes information on water pollution and algal blooms and strategies to address them.
(d) UNEP has developed guidelines and a framework for sustainable nutrient management, which aims to reduce the impact of excess nutrients, such as nitrogen and phosphorus, on the environment. These excess nutrients can cause algal blooms and result in negative impacts on ecosystems and human health.
(e) The Partnership for Clean Fuels and Vehicles (PCFV) is a global initiative led by UNEP that aims to reduce emissions and air pollution from vehicles. While the main focus of PCFV is air pollution, reducing emissions can also reduce the deposition of nutrients and pollutants from the atmosphere into waters, thereby helping to reduce the risk of algal blooms.
(f) UNEP works with various international organizations, governments, and the private sector to address environmental issues, including water pollution and algal blooms. UNEP also supports the establishment of networks and platforms for countries to share knowledge and experience in managing environmental issues.

Through these activities, UNEP helps countries identify and address water pollution and algal blooms and other environmental problems. UNEP emphasizes an integrated, multi-sectoral approach to achieve sustainable environmental management and reduce the risk of algal blooms.

4. HELCOM (Baltic Marine Environment Protection Commission) https://helcom.fi/about-us/

HELCOM (Baltic Marine Environment Protection Commission), or Helsinki Commission, is a regional environmental organization formed by the countries around the Baltic Sea and the European Union. The main objective of HELCOM is to protect the Baltic marine environment from all sources of pollution by coordinating the necessary actions by the member states. HELCOM was established in 1974 under the Convention on the Protection of the Marine Environment of the Baltic Region (Helsinki Convention). The Convention was revised in 1992 to accommodate political and environmental changes in the region. HELCOM member states cooperate in the monitoring, assessing, and managing the Baltic marine environment, and develop recommendations and environmental protection measures, including efforts to control algal blooms through nutrient load reduction and environmental monitoring.

5. OSPAR Commission (Oslo-Paris Commission) https://www.ospar.org/about is a regional environmental organization established in 1992 to protect and preserve the marine environment of the Northeast Atlantic and North Sea. The OSPAR Commission is the result of the merger of the Oslo Convention of 1972, which deals with waste disposal at sea, and the Paris Convention of 1974, which deals with marine pollution from land-based sources. OSPAR has 15 member states and

the European Union, which work together to control and reduce marine pollution, protect marine biodiversity and ecosystems, and monitor and assess the quality of the marine environment. The Commission develops strategies, recommendations and environmental management measures that are adopted by member states. Cooperation among member states within the OSPAR Commission plays an important role in improving the quality of the marine environment and ensuring the sustainability of ecosystems in the Northeast Atlantic and North Sea. OSPAR is an intergovernmental cooperation mechanism that aims to protect the marine environment of the Northeast Atlantic and North Sea including reducing pollution from various sources, which can cause algal blooms.

6. The International Nitrogen Initiative (INI) https://www.initrogen.org/about is a global program focused on understanding and addressing the impacts of the nitrogen cycle on the environment, economy, and human health. INI was established in 2003 as a result of recommendations from the Scientific Committee on Problems of the Environment (SCOPE) and operates under the auspices of Future Earth and the International Union of Nutritional Sciences (IUNS). INI's goal is to optimize nitrogen use in agricultural and industrial systems and reduce unwanted nitrogen emissions that can lead to climate change, air pollution, water quality degradation and biodiversity loss. INI achieves this goal through research, education, and policy development involving scientists, policy makers, and other stakeholders. INI has several regional centers responsible for coordinating activities in different regions around the world, including North America, Latin America, Europe, Africa, and Asia. The initiative supports international cooperation and multidisciplinary research aimed at effectively and sustainably addressing nitrogen issues is an international program that aims to optimize the use of nitrogen in agricultural and industrial systems, while reducing the negative impacts of nitrogen on the environment and human health by reducing the excessive influx of nutrients into aquatic systems, which can lead to algal blooms (McLellan et al. 2018). Here are some of the efforts made by THIS to prevent algal blooms:

   (a) Research and understanding. INI supports and facilitates research on the nitrogen cycle and its impact on the environment, including algal blooms. This research helps improve understanding of how excess nitrogen triggers algal blooms and how to reduce their impact.
   (b) Sustainable nutrient management. INI develops and promotes sustainable nutrient management practices, such as efficient fertilizer use and precision farming systems. These practices help reduce excess nutrient inputs to waters, thereby reducing the risk of algal blooms.
   (c) Policy and recommendations. INI provides guidance and recommendations to policy and decision makers on how to reduce the impact of excess nitrogen on the environment, including algal blooms. INI seeks to influence policy and practice at national and international levels to reduce nitrogen pollution.
   (d) Building Capacity and Education. INI seeks to increase the capacity and knowledge of experts, practitioners, and policymakers in effectively

managing nitrogen and reducing environmental impacts, including algal blooms. INI organizes workshops, trainings, and conferences to promote the exchange of knowledge and experience in the field of nitrogen management.

(e) Partnerships and collaboration. INI works closely with various international organizations, governments, the private sector, and research institutions to address the issue of excess nitrogen and its impact on the environment. INI helps coordinate and integrate efforts across different sectors and countries to reduce the impacts of excess nitrogen, including algal blooms.

Through these efforts, the International Nitrogen Initiative (INI) seeks to reduce the negative impact of excess nitrogen on the environment and prevent algal blooms.

7. Ramsar Convention on Wetlands. This convention is an intergovernmental agreement that aims to conserve and use wetlands wisely. The convention includes efforts to control algal blooms through nutrient source management and environmental monitoring (Xu et al. 2019). Although the main focus of the convention is wetland conservation, some efforts have been made to control algal blooms in the context of wetlands. Below are some examples of the Ramsar Convention on Wetlands' efforts to control algal blooms:

   (a) Sustainable wetland management. The Convention encourages sustainable and wise management of wetlands, which can help reduce the flow of nutrients into waters and reduce the risk of algal blooms. This management includes erosion control, ecosystem restoration, and efficient use of water resources.
   (b) The Ramsar Convention supports research and monitoring of wetland ecosystems, including water quality and the occurrence of algal blooms. A better understanding of the processes and factors that influence algal blooms in wetland ecosystems can help inform effective management strategies.
   (c) The convention seeks to raise public awareness about the importance of wetlands and the impact of algal blooms on ecosystems and human health. This includes extension and information campaigns on how to reduce the flow of nutrients and pollutants into wetlands and waters.
   (d) The Ramsar Convention provides guidelines and recommendations for member states on effective and sustainable wetland management. This includes policies and practices that can reduce the risk of algal blooms, such as pollutant control and nutrient management.
   (e) The Convention promotes international cooperation in the conservation and wise use of wetlands. Through the exchange of information, knowledge and best practices, countries can learn from each other on how to control algal blooms and protect wetland ecosystems.

International cooperation on algal blooms and other ecological issues is an important aspect of global efforts to protect the environment and biodiversity. Through these initiatives, countries can share information, resources, and best practices to address common environmental problems.

15.6.3 The Role of Government, Industry, and Society in Overcoming the Algal Bloom Problem

Addressing algal blooms requires an integrated and collaborative effort between government, industry, and society (Stauffer et al. 2019). Governments are urged to develop and implement regulations and policies that limit the release of nutrients and pollutants into waters; increase monitoring and research on algal blooms to better understand their causes, impacts and control. In addition, the government is expected to provide support and resources for efficient and environmentally friendly wastewater management and treatment. Collaboration and coordination between various government agencies in addressing the problem of algal blooms is also highly desirable. Public education and outreach campaigns on the importance of protecting water resources and reducing pollution need to be initiated by the government.

Industry is expected to adopt environmentally friendly technologies and practices that reduce the release of nutrients and pollutants into waters; implement effective environmental management systems to reduce operational impacts on aquatic ecosystems; invest in research and development of technologies that can help reduce algal blooms and their impacts. Industry needs to actively cooperate with governments and communities in a joint effort to address algal blooms and conduct transparent and accountable reporting on company environmental impacts and algal bloom reduction efforts.

In addition to the government and industry, the public is also expected to increase awareness and knowledge about algal blooms and their impact on the environment and human health. Other things that communities need to do are to reduce the use of chemical fertilizers and pesticides in agriculture and plantations and implement sustainable agricultural practices; reduce water pollution from domestic waste, such as ensuring good septic systems and reducing the use of phosphate detergents. In addition, the community needs to monitor and report algal blooms to the authorities and encourage effective control measures. Finally, communities need to participate in initiatives and activities that aim to protect and maintain the quality of water resources.

15.7 Research and Innovation of Environmentally Friendly Conversion Technology

15.7.1 Recent Research on Treating Algal Blooms and Other Ecological Problems

Here are some examples of collaborative research initiatives between universities, research centers, and industry that focus on the circular economy and wastewater management using microalgae:

1. AlgaePARC (Algae Production and Research Center) is managed by Wageningen University & Research, the Netherlands. AlgaePARC is a research center that combines expertise from academia, industry, and government to develop efficient and sustainable microalgae biomass production technologies. Researchers at AlgaePARC work on the development and optimization of microalgae cultivation systems, such as photobioreactors, open ponds, and systems that combine various technologies. The goal is to increase the efficiency and reduce the cost of microalgae production. AlgaePARC also focuses on improving photosynthetic efficiency, nutrient use, and microalgae growth by optimizing environmental conditions such as Light (Oostlander et al. 2020), temperature, and nutrients. Researchers at AlgaePARC study various microalgae species to determine the most suitable species for specific applications, such as the production of chemicals, feedstuffs, and bioenergy. To improve the productivity and resilience of microalgae species, researchers at AlgaePARC use conventional breeding techniques and genetic engineering. AlgaePARC also focuses on developing efficient processing technologies for the extraction of high-value products from microalgae, such as lipids, proteins, and carbohydrates. In addition, they are also examining ways to recover nutrients from waste streams and use them for microalgae growth. The development of mathematical models and life cycle analyses were also conducted to assess the environmental and economic impacts of microalgae production systems and identify opportunities for improvement.
2. Bielefeld University and Karlsruhe Institute of Technology, Germany—Project "MicroAlgae". This project is a collaboration between Bielefeld University, Karlsruhe Institute of Technology, and industrial partners to develop an innovative photobioreactor technology capable of treating wastewater and producing microalgae biomass for biodiesel production. The aim of the project is to reduce greenhouse gas emissions and produce high-value by-products from microalgae biomass.
3. Arizona State University—Arizona Center for Algae Technology and Innovation (AzCATI) is a collaboration between Arizona State University and industry partners to develop environmentally friendly and economical microalgae biomass production technologies and processes. AzCATI collaborates with industry and academic partners to accelerate research and development of microalgae technology and facilitate the transfer of technology and knowledge to the industrial sector. AzCATI provides education and training to students, researchers, and industry professionals on microalgae technology and its applications in various sectors (Khavari et al. 2021). AzCATI focuses on wastewater management, increasing microalgae productivity, and converting biomass into biodiesel and other value-added products. AzCATI also provides support and services to startups and companies interested in developing microalgae-based technologies or products, including access to research facilities, equipment, and technical expertise.
4. National University of Singapore (NUS)—Environmental Research Institute (NERI). NERI collaborates with industry and academic partners to develop wastewater treatment technologies using microalgae, with the aim of producing

microalgal biomass that can be converted into biodiesel and high-value by-products. The initiative aims to reduce water pollution and optimally utilize resources.

5. BioGAS + Project (European Union). The BioGAS + project is a consortium of universities, research centers and companies working together to develop innovative technologies for producing biogas from wastewater enriched with microalgae. In addition to producing biogas, the project also creates high-value by-products from the microalgae biomass, such as animal feed and fertilizer.

By involving various stakeholders in collaborative research initiatives, the process of discovering and developing new technologies becomes more efficient, and research results can be applied more quickly on an industrial scale.

15.7.2 Microalgae Biomass Conversion Technology Innovations That are Environmentally Friendly (Microwave Pyrolysis, Fermentation Processes for Bioethanol Production, Integration of Photobioreactor Technology (PBR) With Wastewater Treatment Systems)

Some environmentally friendly microalgae biomass conversion technology innovations include:

Lipid Extraction Technology with Green Solvents

The use of green solvents, such as peracetic acid or terpenes, can replace conventional, potentially hazardous organic solvents in the process of lipid extraction from microalgae. This technology enables a more environmentally friendly and safe lipid extraction. Lipid extraction technology with green solvents, such as peracetic acid or terpenes, is an innovative microalgal biomass conversion technology that is environmentally friendly for several reasons

1. Green solvents such as peracetic acid or terpenes are more environmentally friendly compared to conventional solvents, such as hexane or chloroform. Green solvents have lower environmental impact, are biodegradable, and have low potential toxicity.
2. Lipid extraction technology using green solvents offers higher extraction efficiency compared to conventional methods. This helps to increase production yield and reduce resource consumption in the microalgae biomass conversion process.
3. The lipid extraction process with green solvents produces less waste than conventional methods. This waste reduction contributes to the sustainability of the industry and reduces environmental impact.

4. The use of green solvents in the lipid extraction process has the potential to produce better quality products, such as lipids with higher purity or products with more stable characteristics. This can increase the commercial value of the product and expand its applications.
5. Green solvents are generally safer to use in industrial processes and have lower risks to human health. This can reduce the risk of occupational accidents and improve safety for workers involved in the lipid extraction process.

The process of lipid extraction with green solvents proceeds in stages as follows (Zieliński et al. 2022):

1. Microalgae cells are crushed or treated with mechanical, chemical, or enzymatic methods to break the cell wall and release the lipids inside.
2. After cell destruction, the cellular solids are separated from the solution to isolate the desired lipids.
3. Extraction of lipids with green solvents involves mixing a lipid solution with a green solvent, such as peracetic acid or terpenes, which will solubilize the lipids and separate them from other components in the solution.
4. After extraction, the lipid-solvent mixture is separated from the aqueous phase and other insoluble components through processes such as centrifugation, decantation, or filtration.
5. The extracted lipids are then separated from the solvent through evaporation, distillation, or other techniques, resulting in pure lipids ready for use.

Lipid Extraction Technology by Supercritical Method

This process has advantages such as low use of chemical solvents and shorter extraction time. Supercritical lipid extraction technology is an innovative microalgae biomass conversion technology that is environmentally friendly for several reasons (Saini et al. 2021).

1. Use of supercritical solvents. Supercritical methods often use supercritical CO_2 ($scCO_2$) as a solvent. CO_2 is a relatively non-toxic, non-flammable, and non-reactive solvent, making it safer and environmentally friendly compared to conventional organic solvents.
2. Extraction efficiency. The lipid extraction process by supercritical methods offers high extraction efficiency and good selectivity. This means that more lipids can be extracted from microalgal biomass by this method compared to conventional methods.
3. Waste reduction. In the supercritical extraction process, the amount of waste generated is usually lower than in conventional methods. In addition, the CO2 used as a solvent can be recycled and reused in the extraction process, reducing emissions and environmental impact.
4. Product quality: Supercritical methods produce lipids with higher purity and better quality, as the process is more selective in removing lipids from microalgal biomass. This increases the commercial value of the product and expands its applications.

5. Safety and health: Compared to conventional organic solvents, supercritical CO_2 is safer to use in industrial processes and has a lower risk to human health.

How does supercritical lipid extraction technology work? Here are some steps to perform lipid extraction by supercritical method

1. Biomass preparation. The dried and crushed microalgae biomass was placed in an extraction chamber.
2. Supercritical condition setting. CO_2 is compressed and heated to reach supercritical conditions (pressure and temperature above the critical point of CO_2). At this condition, CO_2 has unique solvent properties and is effective for dissolving lipids.
3. Lipid extraction. Supercritical CO_2 is flowed through an extraction chamber containing microalgae biomass. Lipids dissolve in the supercritical CO_2 and are carried out of the extraction chamber.
4. Lipid separation. After going through the extraction chamber, the mixture of supercritical CO_2 and lipids is flowed into the separation chamber, and then the pressure and temperature are reduced. The CO_2 will return to gas and the lipids will separate, resulting in pure lipids ready for use.
5. CO_2 recycling. The CO_2 separated from the lipids is then recycled and reused in the extraction process, reducing emissions and environmental impact.

Hydrothermal Technology

Hydrothermal conversion processes such as hydrothermal liquefaction and hydrothermal gasification are innovative microalgal biomass conversion technologies that are environmentally friendly (Grande et al. 2021) due to the following:

1. Through a hydrothermal conversion process microalgal biomass is converted into liquid or gaseous fuels that are biodegradable and sustainable. This means that in the event of a spill or leak, the environmental impact will be much lower than with fossil fuels. Hydrothermal technology also has the potential to reduce negative environmental impacts by reducing the amount of solid residue, producing cleaner energy, and promoting energy sustainability.
2. Hydrothermal liquefaction and hydrothermal gasification are efficient processes that convert microalgal biomass under high pressure and temperature conditions without the need for a pre-drying step. This reduces energy costs and makes the process more environmentally friendly.
3. The use of fuels produced from microalgal biomass can help diversify energy sources and reduce dependence on fossil fuels, which will have a positive impact on the environment and energy sustainability.

Utilization of Microalgal Biomass for Biogas Production

Microalgal biomass can be used as a substrate in anaerobic processes, such as anaerobic fermentation or anaerobic digestion, to produce biogas. The biogas produced can be used as fuel or converted into electricity and heat. The utilization of anaerobic digestion of microalgae is an important step to make microalgae-based biodiesel production more sustainable. In this study, the authors evaluated the anaerobic digestion of microalgae and assessed its potential to improve the sustainability of microalgae-based biodiesel production. This study focused on exploring the energy potential in the microalgal biomass remaining after lipid extraction for biodiesel production. Anaerobic digestion of microalgal biomass can produce biogas, which is a mixture of methane (CH_4) and carbon dioxide (CO_2) and other components in small amounts. This process can improve the sustainability of biodiesel production from microalgae by reducing environmental impacts and creating an additional energy source. The authors also discuss various factors that affect the efficiency of the anaerobic digestion process, such as microalgal biomass composition, process conditions, and pre-processing techniques used before anaerobic digestion. In addition, the authors identify future research areas that can improve the efficiency and sustainability of microalgae-based biodiesel production.

Microwave Pyrolysis

Microwave pyrolysis technology is a thermochemical process that converts microalgae biomass into solid, liquid, and gaseous fuels under low oxygen conditions (Zhou and Hu 2020). Microwave pyrolysis has advantages, such as higher energy efficiency and shorter process time. This process is considered an environmentally friendly microalgae biomass conversion technological innovation for several reasons:

1. Pyrolysis is a thermochemical process that occurs under low oxygen conditions, thereby reducing emissions of greenhouse gases such as carbon dioxide (CO_2) and oxides of nitrogen (NO_x) generated during the conversion process.
2. Microwave pyrolysis uses microwaves to generate heat, which improves energy efficiency and reduces energy consumption during the conversion process. This reduces the carbon footprint derived from this pyrolysis process and makes it more environmentally friendly.
3. Just like the hydrothermal conversion process, the use of microalgae biomass in microwave pyrolysis technology allows the utilization of a sustainable and fast-growing biomass source, without competing with agricultural land or food needs.
4. Microwave pyrolysis technology produces solid (biochar), liquid (bio-oil), and gas (biogas) products that can be used as fuel or feedstock for the chemical industry. These products are of high quality and can replace fossil fuels, thereby reducing environmental impact.
5. As with hydrothermal conversion, microalgal biomass absorbs CO_2 during growth, thus helping to reduce greenhouse gas emissions. The use of fuel produced from microwave pyrolysis technology will also reduce CO_2 emissions resulting from fossil fuel combustion (Megia et al. 2021; Rinanti and Purwadi 2018).
6. By producing solid, liquid, and gaseous fuels, microwave pyrolysis technology helps diversify energy sources and reduce dependence on fossil fuels.

Fermentation Process for Bioethanol Production

Microalgae can be fermented to produce bioethanol, an environmentally friendly alternative fuel. Carbohydrates in microalgae biomass are broken down or converted into simple sugars which are then fermented into ethanol. The fermentation process for bioethanol production from microalgae biomass is considered an environmentally friendly biomass conversion technology innovation for several reasons:

1. A sustainable source of biomass. Microalgae is a fast-growing and renewable biomass source. Compared to terrestrial crops, microalgae have higher productivity and do not require large areas of agricultural land, so they do not compete with land requirements for food production.
2. CO_2 sequestration. During growth, microalgae absorb CO_2 from the atmosphere, thus helping to reduce greenhouse gas emissions. In addition, bioethanol produced from the fermentation of microalgae biomass will reduce CO_2 emissions resulting from the combustion of fossil fuels.
3. Renewable fuel. Bioethanol is a renewable fuel that can be used as an alternative or blend with gasoline, thereby reducing dependence on fossil fuels and the environmental impacts associated with them.
4. Biodegradability. Bioethanol is more biodegradable in the environment than fossil fuels, so it has a lower environmental impact in the event of a spill or leak.
5. Reduction of pollutant emissions. Bioethanol produces lower emissions of some air pollutants, such as sulfur dioxide (SO_2) and particulates, compared to gasoline. This can help reduce air pollution and improve air quality.
6. Waste utilization. Fermentation processes for bioethanol production from microalgae can utilize microalgae biomass waste generated from other processes, such as lipid extraction for biodiesel production. By utilizing this waste, the fermentation process helps to reduce waste and make it more environmentally friendly.

Therefore, the fermentation process for bioethanol production is considered an innovative microalgal biomass conversion technology that is environmentally friendly because it utilizes a sustainable and fast-growing biomass source (Rozina et al. 2022), produces renewable fuels with lower environmental impact, and reduces greenhouse gas emissions. The use of bioethanol can also help diversify energy sources and reduce dependence on fossil fuels (Juliarnita et al. 2018).

Integration of Photobioreactor (PBR) Technology with Wastewater Treatment Systems

Specially designed photobioreactors can be used to treat wastewater by utilizing microalgae. This technology helps to reduce water pollution and utilize the nutrients contained in wastewater for microalgae growth and biomass production as in (Fig. 15.14).

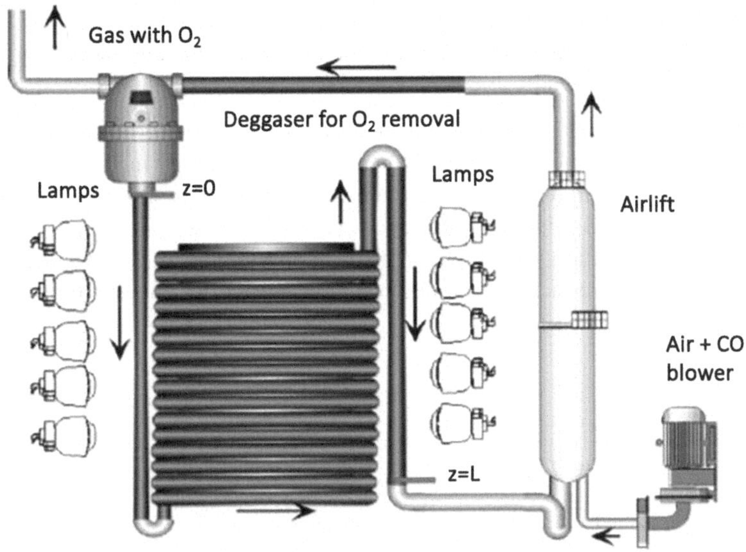

Fig. 15.14 The tubular photobioreactor (PBR)

The integration of photobioreactor (PBR) technology with wastewater treatment systems is considered an environmentally friendly conversion technology innovation for several reasons:

1. Photobioreactors integrated with wastewater treatment systems allow microalgae to grow in wastewater, helping to treat the water and reduce pollution. Microalgae can remove excess nutrients such as phosphate and nitrogen from wastewater, thereby reducing eutrophication and disruption of aquatic ecosystems.
2. During growth, microalgae in the photobioreactor absorb CO_2 from the atmosphere and from industrial emissions associated with wastewater treatment systems. This helps reduce greenhouse gas emissions and associated environmental impacts.
3. Microalgae grown in photobioreactors are a fast-growing and renewable source of biomass (Egbo et al. 2018). The use of microalgae as feedstock to produce fuels and other products may reduce dependence on non-renewable natural resources and help realizing a circular economy.
4. Photobioreactors can be placed in various scales and shapes, allowing flexible placement and efficient space utilization. In addition, photobioreactors integrated with wastewater treatment systems utilize existing resources, such as wastewater and CO_2 emissions, to produce microalgal biomass.
5. The microalgal biomass produced from the photobioreactor can be converted into various products, such as bioethanol, biodiesel, biogas, or other chemicals, which can replace fossil fuels and help diversify energy sources.

6. By utilizing microalgae to remove nutrients and contaminants from wastewater, the cost of wastewater treatment can be reduced, making the system more environmentally friendly and economical.

In summary, integrating photobioreactor technology with wastewater treatment systems is an environmentally friendly conversion technology innovation because it helps reduce water pollution and greenhouse gas emissions, utilize sustainable biomass sources, and improve space and resource efficiency. By developing and adopting environmentally friendly microalgae biomass conversion technology innovations, industries can actively reduce negative impacts on the environment and create more sustainable processes.

References

Abanades S, Abbaspour H, Ahmadi A, Das B, Ehyaei MA, Esmaeilion F, El Haj Assad M, Hajilounezhad T, Hmida A, Rosen MA, Safari S, Shabi MA, Silveira JL (2022) A conceptual review of sustainable electrical power generation from biogas. Energy Sci Eng 10(2):630–655. Wiley. https://doi.org/10.1002/ese3.1030

Abdelfattah A, Ali SS, Ramadan H, El-Aswar EI, Eltawab R, Ho SH, Elsamahy T, Li S, El-Sheekh MM, Schagerl M, Kornaros M, Sun J (2023) Microalgae-based wastewater treatment: Mechanisms, challenges, recent advances, and future prospects. Environ Sci Ecotechnol 13. https://doi.org/10.1016/j.ese.2022.100205

Albrektsen S, Kortet R, Skov PV, Ytteborg E, Gitlesen S, Kleinegris D, Mydland LT, Hansen JØ, Lock EJ, Mørkøre T, James P, Wang X, Whitaker RD, Vang B, Hatlen B, Daneshvar E, Bhatnagar A, Jensen LB, Øverland M (2022) Future feed resources in sustainable salmonid production: a review. Rev Aquaculture 14(4):1790–1812. https://doi.org/10.1111/raq.12673

Ali SS, Mastropetros SG, Schagerl M, Sakarika M, Elsamahy T, El-Sheekh M, Sun J, Kornaros M (2022) Recent advances in wastewater microalgae-based biofuels production: a state-of-the-art review. Energy Rep 8:13253–13280. https://doi.org/10.1016/j.egyr.2022.09.143

Andriani D, Rajani A, Kusnadi, Santosa A, Saepudin A, Wresta A, Atmaja TD (2020) A review on biogas purification through hydrogen sulphide removal. IOP Conf Ser Earth Environ Sci 483(1). https://doi.org/10.1088/1755-1315/483/1/012034

Anggraeni VJ, Wahyu S, Kusriani H, Kurnia D, Tinggi S, Bandung F (2019) Aktivitas antibakteri ekstrak mikroalga thalassiosira sp terhadap bakteri staphylococcus aureus, staphylococcus epidermidis dan propionibacterium acne. Jurnal Kimia Riset 4(1)

Anwar MN, Fayyaz A, Sohail NF, Khokhar MF, Baqar M, Yasar A, Rasool K, Nazir A, Raja MUF, Rehan M, Aghbashlo M, Tabatabaei M, Nizami AS (2020) CO_2 utilization: turning greenhouse gas into fuels and valuable products. J Environ Manage 260. https://doi.org/10.1016/j.jenvman.2019.110059

Arenas EG, Rodriguez Palacio MC, Juantorena AU, Fernando SE L, Sebastian PJ (2017) Microalgae as a potential source for biodiesel production: techniques, methods, and other challenges. Int J Energy Res 41(6):761–789. John Wiley and Sons Ltd. https://doi.org/10.1002/er.3663

Atelge MR, Krisa D, Kumar G, Eskicioglu C, Nguyen DD, Chang SW, Atabani AE, Al-Muhtaseb AH, Unalan S (2020) Biogas production from organic waste: recent progress and perspectives. Waste and Biomass Valorization 11(3):1019–1040. https://doi.org/10.1007/s12649-018-00546-0

Ayilara MS, Olanrewaju OS, Babalola OO, Odeyemi O (2020) Waste management through composting: challenges and potentials. Sustainability (Switzerland) 12(11). https://doi.org/10.3390/su12114456

Ballah M, Bhoyroo V, Neetoo H (2019) Assessment of the physico-chemical quality and extent of algal proliferation in water from an impounding reservoir prone to eutrophication. J Ecol Environ 43(1). https://doi.org/10.1186/s41610-018-0094-z

Bhan C, Verma L, Singh J (2020) Alternative fuels for sustainable development. In: Environmental concerns and sustainable development. Springer Singapore, 317–331. https://doi.org/10.1007/978-981-13-5889-0_16

Bhattacharjee M (2016) Pharmaceutically valuable bioactive compounds of algae. Asian J Pharm Clin Res 9(6):43–47. https://doi.org/10.22159/ajpcr.2016.v9i6.14507

Binding CE, Pizzolato L, Zeng C (2021) EO LakeWatch: delivering a comprehensive suite of remote sensing algal bloom indices for enhanced monitoring of Canadian eutrophic lakes. Ecol Indicators121. https://doi.org/10.1016/j.ecolind.2020.106999

Chowdury KH, Nahar N, Deb UK (2020) The growth factors involved in microalgae cultivation for biofuel production: a review. Comput Water Energy Environ Eng 09(04):185–215. https://doi.org/10.4236/cweee.2020.94012

Córdova O, Santis J, Ruiz-Fillipi G, Zuñiga ME, Fermoso FG, Chamy R (2018) Microalgae digestive pretreatment for increasing biogas production. Renew Sustain Energy Rev 82:2806–2813. https://doi.org/10.1016/j.rser.2017.10.005

Costa A, Alio V, Sciortino S, Nicastro L, Cangini M, Pino F, Servadei I, La Vignera A, Fortino G, Monaco S, Dall'ara S (2021) Algal blooms of Alexandrium spp. and paralytic shellfish poisoning toxicity events in mussels farmed in Sicily. Italian J Food Saf 10(1). https://doi.org/10.4081/ijfs.2021.9062

da Maia JL, Cardoso JS, da S. Mastrantonio DJ, Bierhals CK, Moreira JB, Costa JAV, de Morais MG (2020) Microalgae starch: a promising raw material for the bioethanol production. Int J Biol Macromol 165:2739–2749. https://doi.org/10.1016/j.ijbiomac.2020.10.159

de Carvalho JC, Sydney EB, Assú Tessari LF, Soccol CR (2018) Culture media for mass production of microalgae. In: Biomass, biofuels, biochemicals: biofuels from algae, 2nd edn. Elsevier, pp 33–50. https://doi.org/10.1016/B978-0-444-64192-2.00002-0

Egbo MK, Okoani OO, Okoh IE (2018) Photobioreactors for microalgae cultivation—an overview. Int J Sci Eng Res 9(11):65–74

El-Dalatony MM, Salama ES, Kurade MB, Hassan SHA, Oh SE, Kim S, Jeon BH (2017) Utilization of microalgal biofractions for bioethanol, higher alcohols, and biodiesel production: a review. Energies 10(12). https://doi.org/10.3390/en10122110

Emery-Butcher HE, Beatty SJ, Robson BJ (2020) The impacts of invasive ecosystem engineers in freshwaters: a review. Freshwater Biol 65(5):999–1015. https://doi.org/10.1111/fwb.13479

Feng RZ, Zaidi AA, Zhang K, Shi Y (2019) Optimisation of microwave pretreatment for biogas enhancement through anaerobic digestion of microalgal biomass. Periodica Polytech, Chem Eng 63(1):65–72. https://doi.org/10.3311/PPch.12334

Ferreira de Oliveira AP, Bragotto APA (2022) Microalgae-based products: food and public health. Future Foods6. https://doi.org/10.1016/j.fufo.2022.100157

Ghasemi Naghdi F, González González LM, Chan W, Schenk PM (2016) Progress on lipid extraction from wet algal biomass for biodiesel production. Microbial Biotechnol 9(6):718–726. https://doi.org/10.1111/1751-7915.12360

Grande L, Pedroarena I, Korili SA, Gil A (2021) Hydrothermal liquefaction of biomass as one of the most promising alternatives for the synthesis of advanced liquid biofuels: a review. Materials 14(18). https://doi.org/10.3390/ma14185286

Hossain N, Hasan MH, Mahlia TMI, Shamsuddin AH, Silitonga AS (2020) Feasibility of microalgae as feedstock for alternative fuel in Malaysia: a review. Energy Strategy Rev 32. https://doi.org/10.1016/j.esr.2020.100536

Imai I, Inaba N, Yamamoto K (2021) Harmful algal blooms and environmentally friendly control strategies in Japan. Fisheries Sci 87(4):437–464. https://doi.org/10.1007/s12562-021-01524-7

Irfeey AMM, Najim MMM, Alotaibi BA, Traore A (2023) Groundwater pollution impact on food security. Sustainability (Switzerland) 15(5). https://doi.org/10.3390/su15054202

Juliarnita IGA, Hadisoebroto R, Rinanti, A (2018) Bioethanol production from mixed culture microalgae biomass with temperature hydrolysis variation. Matec Web Conf 197(1).

Klemm K, Kobos J, Lehtinen S, Lundholm N, Mazur-Marzec H, Naustvoll L, Poelman M, Provoost P, De Rijcke M, Suikkanen S (2021) Harmful algal blooms and their effects in coastal seas of Northern Europe. Harmful Algae102. https://doi.org/10.1016/j.hal.2021.101989

Kazmi SSUH, Yapa N, Karunarathna SC, Suwannarach N (2022) Perceived intensification in harmful algal blooms is a wave of cumulative threat to the aquatic ecosystems. Biology 11(6). https://doi.org/10.3390/biology11060852

Khan MI, Shin JH, Kim JD (2018) The promising future of microalgae: current status, challenges, and optimization of a sustainable and renewable industry for biofuels, feed, and other products. Microbial Cell Factories 17(1). https://doi.org/10.1186/s12934-018-0879-x

Khavari F, Saidijam M, Taheri M, Nouri F (2021) Microalgae: therapeutic potentials and applications. Mol Biol Rep 48(5):4757–4765. https://doi.org/10.1007/s11033-021-06422-w

Kirchherr J, Yang NHN, Schulze-Spüntrup F, Heerink MJ, Hartley K (2023) Conceptualizing the circular economy (revisited): an analysis of 221 definitions. Resour Conserv Recycl 194. https://doi.org/10.1016/j.resconrec.2023.107001

Klin M, Pniewski F, Latała A (2020) Growth phase-dependent biochemical composition of green microalgae: theoretical considerations for biogas production. Bioresour Technol 303. https://doi.org/10.1016/j.biortech.2020.122875

Kumar R, Verma A, Shome A, Sinha R, Sinha S, Jha PK, Kumar R, Kumar P, Shubham, Das S, Sharma P, Prasad PVV (2021) Impacts of plastic pollution on ecosystem services, sustainable development goals, and need to focus on circular economy and policy interventions. Sustainability (Switzerland) 13(17). https://doi.org/10.3390/su13179963

Leandro A, Pacheco D, Cotas J, Marques JC, Pereira L, Gonçalves AMM (2020) Seaweed's bioactive candidate compounds to food industry and global food security. Life 10(8):1–37. https://doi.org/10.3390/life10080140

Leverett D, Merrington G, Crane M, Ryan J, Wilson I (2021) Environmental quality standards for diclofenac derived under the European Water Framework Directive: 1. Aquatic organisms. Environ Sci Eur 33(1). https://doi.org/10.1186/s12302-021-00574-z

Liu Y, Ren X, Fan C, Wu W, Zhang W, Wang Y (2022) Health benefits, food applications, and sustainability of microalgae-derived N-3 pufa. Foods 11(13). https://doi.org/10.3390/foods11131883

Logan M, Visvanathan C (2019) Management strategies for anaerobic digestate of organic fraction of municipal solid waste: current status and future prospects. Waste Manage Res 37(1_suppl):27–39. https://doi.org/10.1177/0734242X18816793

Maizatul AY, Radin Mohamed RMS, Al-Gheethi AA, Hashim MKA (2017) An overview of the utilisation of microalgae biomass derived from nutrient recycling of wet market wastewater and slaughterhouse wastewater. Int Aquatic Res 9(3):177–193. https://doi.org/10.1007/s40071-017-0168-z

Maltsev Y, Maltseva K, Kulikovskiy M, Maltseva S (2021) Influence of light conditions on microalgae growth and content of lipids, carotenoids, and fatty acid composition. Biology 10(10). https://doi.org/10.3390/biology10101060

Martínez-Ruiz M, Martínez-González CA, Kim DH, Santiesteban-Romero B, Reyes-Pardo H, Villaseñor-Zepeda KR, Meléndez-Sánchez ER, Ramírez-Gamboa D, Díaz-Zamorano AL, Sosa-Hernández JE, Coronado-Apodaca KG, Gámez-Méndez AM, Iqbal HMN, Parra-Saldivar R (2022) Microalgae bioactive compounds to topical applications products—a review. Molecules 27(11). https://doi.org/10.3390/molecules27113512

Martins R, Sales H, Pontes R, Nunes J, Gouveia I (2023) Food wastes and microalgae as sources of bioactive compounds and pigments in a modern biorefinery: a review. Antioxidants 12(2). https://doi.org/10.3390/antiox12020328

McLellan EL, Cassman KG, Eagle AJ, Woodbury PB, Sela S, Tonitto C, Marjerison RD, Van Es HM (2018) The nitrogen balancing act: tracking the environmental performance of food production. Bioscience 68(3):194–203. https://doi.org/10.1093/biosci/bix164

Megia PJ, Vizcaino AJ, Calles JA, Carrero A (2021) Hydrogen production technologies: from fossil fuels toward renewable sources. A mini review. Energy and Fuels 35(20):16403–16415. https://doi.org/10.1021/acs.energyfuels.1c02501

Moore D (2021) Saving the planet with appropriate biotechnology: coccolithophore cultivation and deployment. Mexican J Biotechnol 6(1):129–155. https://doi.org/10.29267/MXJB.2021.6.1.129

Mores ICV, Muramatsu K, Maiorka A, Orlando UAD, da Silva JMS, de Paulo LM, Gouveia ABVS, da Silva WJ, Sousa JG, e Rezende CSM, Minafra CS (2020). Pelleting on the nutritional quality of broiler feeds. J Agric Stud 8(3):193. https://doi.org/10.5296/jas.v8i3.16072

Nagappan S, Das P, AbdulQuadir M, Thaher M, Khan S, Mahata C, Al-Jabri H, Vatland AK, Kumar G (2021) Potential of microalgae as a sustainable feed ingredient for aquaculture. J Biotechnol 341:1–20. https://doi.org/10.1016/j.jbiotec.2021.09.003

Najjar YSH, Abu-Shamleh A (2020) Harvesting of microalgae by centrifugation for biodiesel production: a review. Algal Res 51. https://doi.org/10.1016/j.algal.2020.102046

Nasution AK, Takarina ND, Thoha H (2021) The presence and abundance of harmful dinoflagellate algae related to water quality in Jakarta bay, Indonesia. Biodiversitas 22(5):2909–2917. https://doi.org/10.13057/biodiv/d220556

Núñez-Pons L, Avila C, Romano G, Verde C, Giordano D (2018) UV-protective compounds in marine organisms from the southern ocean. Marine Drugs 16(9). https://doi.org/10.3390/md16090336

Nzayisenga JC, Farge X, Groll SL, Sellstedt A (2020) Effects of light intensity on growth and lipid production in microalgae grown in wastewater. Biotechnol Biofuels 13(1). https://doi.org/10.1186/s13068-019-1646-x

Ogbonna CN, Nwoba EG (2021) Bio-based flocculants for sustainable harvesting of microalgae for biofuel production. A review. Renew Sustain Energy Rev 139. https://doi.org/10.1016/j.rser.2020.110690

Oostlander PC, van Houcke J, Wijffels RH, Barbosa MJ (2020) Microalgae production cost in aquaculture hatcheries. Aquaculture525. https://doi.org/10.1016/j.aquaculture.2020.735310

Osman AI, Mehta N, Elgarahy AM, Al-Hinai A, Al-Muhtaseb AH, Rooney DW (2021). Conversion of biomass to biofuels and life cycle assessment: a review. Environ Chem Lett 19(6):4075–4118. https://doi.org/10.1007/s10311-021-01273-0

Pal P, Chew KW, Yen HW, Lim JW, Lam MK, Show PL (2019) Cultivation of oily microalgae for the production of third-generation biofuels. Sustainability (Switzerland) 11(19). https://doi.org/10.3390/su11195424

Parr CL, Bishop TR (2022) The response of ants to climate change. Global Change Biol 28(10):3188–3205. https://doi.org/10.1111/gcb.16140

Patwardhan SB, Pandit S, Ghosh D, Dhar DW, Banerjee S, Joshi S, Gupta PK, Lahiri D, Nag M, Ruokolainen J, Ray RR, Kumar Kesari K (2022) A concise review on the cultivation of microalgal biofilms for biofuel feedstock production. Biomass Convers Biorefinery. https://doi.org/10.1007/s13399-022-02783-9

Plata-Calzado C, Prieto AI, Cameán AM, Jos A (2022) Toxic effects produced by anatoxin-a under laboratory conditions: a review. Toxins 14(12). https://doi.org/10.3390/toxins14120861

Praharyawan S (2021) Bioteknologi & biosains indonesia peningkatan produksi biomassa sebagai strategi jitu dalam mempercepat produksi biodiesel berbasis mikroalga di Indonesia. Jurnal Bioteknologi & Biosains Indonesia 8(2):294–320. http://ejurnal.bppt.go.id/index.php/JBBI

Raja R, Coelho A, Hemaiswarya S, Kumar P, Carvalho IS, Alagarsamy A (2018) Applications of microalgal paste and powder as food and feed: an update using text mining tool. Beni-Suef Univ J Basic Appl Sci 7(4):740–747. https://doi.org/10.1016/j.bjbas.2018.10.004

Ransom Hardison D, Holland WC, Currier RD, Kirkpatrick B, Stumpf R, Fanara T, Burris D, Reich A, Kirkpatrick GJ, Wayne Litaker R (2019) Habscope: a tool for use by citizen scientists to facilitate early warning of respiratory irritation caused by toxic blooms of *Karenia brevis*. PLoS ONE 14(6). https://doi.org/10.1371/journal.pone.0218489

Rashidi H, Baulch H, Gill A, Bharadwaj L, Bradford L (2021) Monitoring, managing, and communicating risk of Harmful Algal Blooms (HABs) in recreational resources across Canada. Environ Health Insights15. https://doi.org/10.1177/11786302211014401

Rinanti A, Purwadi R (2018) Harvesting of freshwater microalgae biomass by Scenedesmus sp. as bioflocculant. IOP Conf Ser Earth Environ Sci 106(1).

Rinanti A, Purwadi, R (2019) Increasing carbohydrate and lipid productivity in tropical microalgae biomass as a sustainable biofuel feed stock. Energy Procedia 158:1215–1222.

Rohmah, Y, Rinanti, A, Hendrawan, DI (2018) The determination of ground water quality based on the presence of Escherichia coli on populated area (A case study: Pasar Minggu, South Jakarta). IOP Conf Ser Earth Environ Sci 106(1).

Rozina, Ahmad M, Zafar M (2022). Biomass as sustainable material for bioethanol production. In: Handbook of smart materials, technologies, and devices. Springer International Publishing, pp 1–24. https://doi.org/10.1007/978-3-030-58675-1_19-2

Saini RK, Prasad P, Shang X, Keum YS (2021) Advances in lipid extraction methods—a review. Int J Mol Sci 22(24). https://doi.org/10.3390/ijms222413643

Sarker NK, Kaparaj P (2023) A critical review on the status and progress of microalgae cultivation in outdoor photobioreactors conducted over 35 years (1986–2021). Energies 16(7). https://doi.org/10.3390/en16073105

Shekarabi SPH, Mehrgan MS, Razi N, Sabzi S (2019). Biochemical composition and fatty acid profile of the marine microalga *Isochrysis galbana* dried with different methods. J Microbiol Biotechnol Food Sci 9(3):521–524. https://doi.org/10.15414/jmbfs.2019/20.9.3.521-524

Sidabutar T, Srimariana ES, Cappenberg H, Wouthuyzen S (2021) An overview of harmful algal blooms and eutrophication in Jakarta Bay, Indonesia. IOP Conf Ser Earth Environ Sci 869(1). https://doi.org/10.1088/1755-1315/869/1/012039

Singh G, Patidar SK (2018) Microalgae harvesting techniques: a review. J Environ Manage 217:499–508. https://doi.org/10.1016/j.jenvman.2018.04.010

Sivaramakrishnan R, Suresh S, Kanwal S, Ramadoss G, Ramprakash B, Incharoensakdi A (2022) Microalgal biorefinery concepts' developments for biofuel and bioproducts: current perspective and bottlenecks. Int J Mol Sci 23(5). https://doi.org/10.3390/ijms23052623

Song Y, Lv X (2023. Study of phytoplankton biomass and environmental drivers in and around the Ross Sea Marine Protected Area. J Mar Sci Eng 11(4). https://doi.org/10.3390/jmse11040747

Souza CMM, Bastos TS, dos Santos MC (2021) Microalgae use in animal nutrition. Res Soc Dev 10(16):e53101622986. https://doi.org/10.33448/rsd-v10i16.22986

Stauffer BA, Bowers HA, Buckley E, Davis TW, Johengen TH, Kudela R, McManus MA, Purcell H, Smith GJ, Vander WA, Tamburri MN (2019) Considerations in harmful algal bloom research and monitoring: Perspectives from a consensus-building workshop and technology testing. Front Mar Sci 6(JUL). https://doi.org/10.3389/fmars.2019.00399

Sukenik A, Kaplan A (2021) Cyanobacterial harmful algal blooms in aquatic ecosystems: a comprehensive outlook on current and emerging mitigation and control approaches. Microorganisms 9(7). https://doi.org/10.3390/microorganisms9071472

Sun H, Wang Y, He Y, Liu B, Mou H, Chen F, Yang S (2023) Microalgae-derived pigments for the food industry. Marine Drugs 21(2). https://doi.org/10.3390/md21020082

Talib RH, Helal MM, Naji RK (2022) The dynamics of the aquatic food chain system in the contaminated environment. Iraqi J Sci 63(5):2173–2193. https://doi.org/10.24996/ijs.2022.63.5.31

Thanakunpaisit N, Jantarachat N, Onthong U (2017) Removal of hydrogen sulfide from biogas using laterite materials as an adsorbent. Energy Procedia 138:1134–1139. https://doi.org/10.1016/j.egypro.2017.10.215

Tsikoti C, Genitsaris S (2021) Review of harmful algal blooms in the coastal Mediterranean Sea, with a focus on Greek Waters. Diversity 13(8). https://doi.org/10.3390/d13080396

Udayan A, Pandey AK, Sirohi R, Sreekumar N, Sang BI, Sim SJ, Kim SH, Pandey A (2022) Production of microalgae with high lipid content and their potential as sources of nutraceuticals. Phytochem Rev. https://doi.org/10.1007/s11101-021-09784-y

Ullah Khan I, Hafiz Dzarfan Othman M, Hashim H, Matsuura T, Ismail AF, Rezaei-DashtArzhandi M, Wan Azelee I (2017) Biogas as a renewable energy fuel—a review of biogas upgrading, utilisation and storage. Energy Convers Manage 150:277–294. https://doi.org/10.1016/j.enconman.2017.08.035

van der Wiel BZ, Weijma J, van Middelaar CE, Kleinke M, Buisman CJN, Wichern F (2019) Restoring nutrient circularity: a review of nutrient stock and flow analyses of local agro-food-waste systems. Resour Conserv Recycl X 3. https://doi.org/10.1016/j.rcrx.2019.100014

Vilariño N, Louzao MC, Abal P, Cagide E, Carrera C, Vieytes MR, Botana LM (2018) Human poisoning from marine toxins: unknowns for optimal consumer protection. Toxins 10(8). https://doi.org/10.3390/toxins10080324

Vilatte A, Spencer-Milnes X, Jackson HO, Purton S, Parker B (2023) Spray drying is a viable technology for the preservation of recombinant proteins in microalgae. Microorganisms 11(2). https://doi.org/10.3390/microorganisms11020512

Wongsurakul P, Termtanun M, Kiatkittipong W, Lim JW, Kiatkittipong K, Pavasant P, Kumakiri I, Assabumrungrat S (2022) Comprehensive review on potential contamination in fuel ethanol production with proposed specific guideline criteria. Energies 15(9). https://doi.org/10.3390/en15092986

Xu T, Weng B, Yan D, Wang K, Li X, Bi W, Li M, Cheng X, Liu Y (2019) Wetlands of international importance: status, threats, and future protection. Int J Environ Res Public Health 16(10). https://doi.org/10.3390/ijerph16101818

Yaakob MA, Mohamed RMSR, Al-Gheethi A, Ravishankar GA, Ambati RR (2021) Influence of nitrogen and phosphorus on microalgal growth, biomass, lipid, and fatty acid production: an overview. Cells 10(2):1–19. https://doi.org/10.3390/cells10020393

You N, Deng S, Wang C, Ngo HH, Wang X, Yu H, Tang L, Han J (2023) Review and opinions on the research, development and application of microalgae culture technologies for resource recovery from wastewater. Water (Switzerland) 15(6). https://doi.org/10.3390/w15061192

Zabochnicka M, Krzywonos M, Romanowska-Duda Z, Szufa S, Darkalt A, Mubashar M (2022) Algal biomass utilization toward circular economy. Life 12(10). https://doi.org/10.3390/life12101480

Zhou Y, Hu C (2020) Catalytic thermochemical conversion of algae and upgrading of algal oil for the production of high-grade liquid fuel: a review. Catalysts 10(2). https://doi.org/10.3390/catal10020145

Zieliński M, Dębowski M, Kazimierowicz J (2022) Outflow from a biogas plant as a medium for microalgae biomass cultivation—pilot scale study and technical concept of a large-scale installation. Energies 15(8). https://doi.org/10.3390/en15082912

Zieliński M, Dębowski M, Kazimierowicz J, Świca I (2023) Microalgal carbon dioxide (CO_2) capture and utilization from the European Union perspective. Energies 16(3). https://doi.org/10.3390/en16031446

Chapter 16
Circular Bioeconomy Transition-Based Studies in Biorefineries of Microalgae Biomass

Renato Barbosa Pagnano, Thais Suzane Milessi, Arthur Santos Longati, Luísa Pereira Pinheiro, and Andreza Aparecida Longati

Abstract Anthropogenic activities that cause large emissions of greenhouse gases into the atmosphere have caused worldwide effects such as climate change and global warming, leading to growing global concern about the environmental impact of traditional industries. In this context, the promotion of microalgae biorefineries for renewable energy and chemical production plays an important role. These facilities are known as a way to maximize the value of biomass, reduce waste and produce different products (chemicals, biomaterials and bioenergy). This chapter provides a literature review on the conversion of microalgal biomass into energy, fuels and value-added products in a biorefinery and identifies the main circular bioeconomy policies around the world to promote this technology.

R. B. Pagnano
Department of Materials and Bioprocess Engineering, School of Chemical Engineering, University of Campinas, Campinas 13083-852, Brazil

T. S. Milessi · A. A. Longati (✉)
Chemical Engineering Department, Federal University of São Carlos, São Carlos 13565-905, Brazil
e-mail: andrezalongati@gmail.com

T. S. Milessi
Graduate Program of Chemical Engineering (PPGEQ), Federal University of São Carlos, São Carlos 13565-905, Brazil

A. S. Longati
Graduate Programs Mechanical Engineering, Federal University of São João del-Rei, Praça Frei Orlando, 170, Centro, São João del-Rei, Minas Gerais, CEP 36307-352, Brazil

L. P. Pinheiro
Doctoral Program in Electrical Engineering, Federal University of Itajubá, Avenida B.P.S, 1303, Pinheirinho, Itajubá, Minas Gerais 37500-903, Brazil

T. A. Kurniawan and A. Anouzla (eds.), *Algae as a Natural Solution for Challenges in Water-Food-Energy Nexus*, Environmental Science and Engineering,
https://doi.org/10.1007/978-981-97-2371-3_16

16.1 Introduction

In recent years, the effort to develop technologies to distance the dependence of energy based on fossil fuels has been a major concern in several society sectors. According to the International Energy Agency (IEA 2021), oil still corresponds to more than 40% of the world's total energy consumption, while biofuels and waste share only 10.4%. At the same time, environmental issues derived from using unrenewable energy sources continue to be present. The annual greenhouse gas emissions (GHG)—such as carbon dioxide (CO_2), methane (CH4), and nitrous oxide (N2O)—were higher in the last decade than in any other, as shown by the Intergovernmental Panel on Climate Change on the Sixth Assessment Report (IPCC). This growth has produced tangible effects, as GHG emissions from human activities are linked to approximately 1.1 °C of warming in global temperature since 1850–1900 and are expected to reach 1.5 °C of warming in the next 20 years (IPCC 2022).

The appeal to energy transition has engaged many countries, with more than 146 implementing renewable energy support policies by the end of 2015 (REN21 2016). In this scenario, biorefinery development could play a key role in the transition to a low- carbon economy. The biorefinery can be defined as a "*facility or network of facilities that convert biomass into a variety of chemical, biomaterials and energy*" (REN21 2016). These integrated facilities, designed to maximize biomass value and minimize waste, have been developed and are growing as a viable option in various parts of the world, and are expected to increase in the future. In Europe, according to the Joint Biorefinery Vision for 2030, a major share of the total demand for chemicals, energy and materials will be met by biomass as a feedstock (Clark and Deswarte 2015).

Progress in biorefinery studies has led to the search for different feedstocks. In this context, microalgae, prokaryotic unicellular phototrophs, appear as an advantageous possibility. The benefits of using this type of feedstock are intrinsically linked to its biomass characteristics. A major advantage of microalgae is related to their cultivation conditions. Firstly, the reduced demand for resources associated with this process: although it requires a culture medium enriched with nutrients and salts (Baicha et al. 2016), the other components involved in the process are essentially water and atmospheric CO_2, which minimizes the costs. However, it is relevant to state the large amount of water that can be involved in its cultivation. One way of bioremediating is to diversify its source, for example, using seawater in the case of marine algae and, even more advantageous, using wastewater rich in nutrients (Chia et al. 2018).

Secondly, the ability of microalgae to grow on degraded and infertile land is another advantage as it reduces competition with land and food crops (Yen et al. 2013; Chew et al. 2017). Finally, microalgae grow at an accelerated rate, typically doubling their biomass within 24 h, and further decreasing to 3.5 h during exponential growth. This is significantly faster than the biomass doubling time of other crops (Chisti 2007). However, the technologies used in biorefineries to transform microalgal biomass into bioenergy and chemicals are still under development, as is discussed below. Nevertheless, microalgae-based products are expected to be implemented on an industrial scale in the medium and long term (Cuevas-Castillo et al. 2020).

In this sense, it is expected that the microalgae biomass biorefineries will have sustainable gains and development in next years. Their advance will play an important role in the transition from a linear to a circular bioeconomy, since the implementation of these facilities are crucial to address global sustainability challenges and provide a more secure, sustainable and affordable energy system. Therefore, this work aims to consolidate the general conclusions available in the scientific literature regarding the different technologies for the conversion of microalgal biomass into energy, fuels and value-added products.

16.2 Circular Economy, Bioeconomy and Circular Bioeconomy

Kirchherr et al. (2023) provide a meta-definition of circular economy (CE), based on the analysis of 221 circular economy definitions: "*The CE is a regenerative economic system which necessitates a paradigm shift to replace the 'end of life' concept with reducing, alternatively reducing, recycling and recovering materials throughout the supply chain, with the aim to promote value maintenance and sustainable development, creating environmental quality, economic development, and social equity, to the benefit of current and future generations. It is enabled by an alliance of stakeholders (industry, consumers, policymakers, academia) and their technological innovations and capabilities*". Therefore, CE is a model of economic, social and environmental production and consumption that aims to build an autonomous and sustainable society in harmony with the issue of environmental resources.

According to McCormick and Kautto (2013) "*the bioeconomy can be defined as an economy where the basic building blocks for materials, chemicals and energy are derived from renewable biological resources*". Thus, bioeconomy is economic activity involving the use of biotechnology and biomass in the production of goods, services, or energy.

From the integration of CE and the bioeconomy, two sustainability-focused concepts that aim to transform the current linear, fossil-based economy into a more sustainable, efficient, waste-recycling one (D'Amato et al. 2020), it is obtained the circular bioeconomy (Fig. 16.1), also known as bio-based circular economy. The circular bioeconomy refers to using biomass derived from biological resources in a systemic approach to economic development (Leong et al. 2021).

According to D'Amato et al. (2020), the emerging concept of circular bioeconomy concept aims to address the limitations of individual concepts of CE and the bioeconomy. Thus, the circular bioeconomy is not only about adopting circularity principles such as biomass cascading, waste hierarchy, and efficiency in the use of biomass resources, but should also include elements such as sustainable sourcing of biomass,

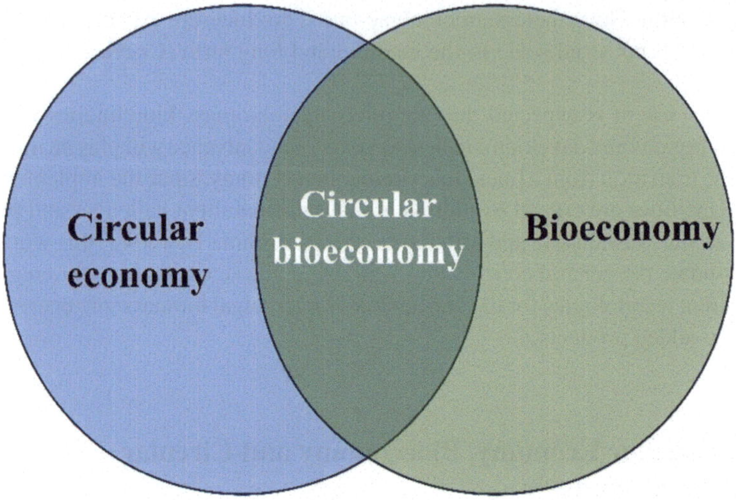

Fig. 16.1 Circular bioeconomy: the integration of circular economy and the bioeconomy

as well as aiming at use rather than ownership (e.g. sharing, renting, pay-per-use). For Tan and Lamers (2021), the circular bioeconomy is truly a bio-based circular carbon economy, representing, thus, a framework to increase circularity within the economic system.

16.3 Biomass from Microalgae: A Versatile Resource for Sustainable Biorefinery

The composition of microalgae is an important factor in understanding the importance of using microalgae, as it explains the variety of products derived from these organisms. Lipids represent 30–50% of their content by weight, which can be increased by modifying the culture conditions (Yen et al. 2013). A way to achieve this is to elevate the carbon to nitrogen (C/N) ratio in the medium, which will lead to lipid accumulation due to the exhaustion of the nitrogen source and its low levels during the cultivation phase (Chew et al. 2017). Stress conditions can also produce this effect, such as high temperature, pH shift, and high salt concentration (Chew et al. 2017). The high levels of lipid productivity, in comparison to other lipid-based energy crops, present microalgae as a promising raw material. Table 16.1 presents the species from which these macromolecules can be obtained and the extraction methods used in this process. Although bio-toxicity, chemical cost, and lipid degradation can be problems involved in the process, chemical-based extraction methods offer a low energy demand and scale-up potential (Chew et al. 2017).

Table 16.1 Microalgae species and their lipid extraction methods (adapted from Chew et al. 2017)

Species	Extraction/Purification methods	Yield/ Extraction efficiency	Comments	References
S. almeriensis	Solvent extraction [Kochert method: methanol–chloroform 1:2 (v/v)]	N/A	Need for organic solvents	Hernández et al. (2014)
S. almeriensis	Solvent extraction [Soxhlet method: methanol–chloroform 2:1 (v/v)]	8.0 DW% (dry weight)	Need for organic solvents	Hernández et al. (2014)
N. oculata	Ultrasonic extraction	25.0 DW%	Poor product quality	Adam et al. (2012)
P. ellipsoidea	Microwave extraction (140 °C)	37.5 wt.% (lipid)	Energy intensive. High extraction yield	Biller et al. (2013)
S. almeriensis	SC-CO_2	10.1 DW%	High operating condition. High extraction yield	Hernández et al. (2014)
S. almeriensis	Microwave SC-CO_2 (80 °C)	3.1 DW%	Low extraction yield	Hernández et al. (2014)
Chlorella sp.	Isotonic extraction	19 wt.% (lipid)	Energy intensive. High capital cost	Vanthoor-Koopmans et al. (2013)
C. vulgaris	Enzyme extraction	5.27%	High materials (enzyme) cost	Liang et al. (2012)

Microalgae lipids are divided into categories, each one with specific applications. Polyunsaturated fatty acids (PUFAs) are long-chain (over 20 carbons) molecules with high-value components in the health food market and are considered essential nutrients, not capable of being synthesized by higher eukaryotes. The main microalgae capable of producing the PUFAs are: *Nannochloropsis*, *Tetraselmis*, *Isochrysis*, *Thalassiosira*, and *Chaetoceros*. Eicosapentaenoic acid (EPA) and docosahexaenoic acid (DHA) are two of the main kinds of PUFAs commercially available, which is explained by their elevated levels of bioactivity (Yen et al. 2013).

On the other hand, short-chain lipid molecules (containing 14–20 carbons), in particular triacylglycerol (TAG), are suitable for biodiesel production (Yen et al. 2013). In the seek to produce carbon–neutral renewable liquid fuels that can substitute up to a certain level of petroleum-derived transport fuels, biodiesel from microalgae oil presents itself as a contender, especially when compared to other sources of biofuels. While the oil content of some microalgae species can reach up to 80% of the dry weight in biomass, agricultural oil crops currently used to produce biodiesel (such as soybean and oil palm) presents a far inferior value, with less than 5% of total biomass basis (Chisti 2007). That disparity translates into an inferior capacity to provide the amount of biodiesel necessary to overcome the need for petroleum-based transport fuels.

To demonstrate the productivity difference between the two bio-sources, a hypothetical scenario can be analyzed in which a country like the United States was to replace all petroleum-derived transport fuel with biodiesel. In the case of oil palm, considering a productive oil crop, the amount of oil needed to produce biodiesel would require more than 60% of all agricultural cropping land in the country (Chisti 2007). However, considering an average productivity of microalgae biomass in a production system in a tropical zone and assuming an average content of 30% dry weight in the biomass, the land needed would represent only 3% of the U.S. cropping area (Chisti 2007). This shows the potential of biodiesel produced from microalgae oil, which diminishes competition with food and other agricultural products, therefore being a far more realistic way to biodiesel in the near future.

Another critical component in microalgae is carbohydrates, usually more than 50% of their dry weight (Ho et al. 2012). Algal carbohydrates include starch, glucose, cellulose/hemicelluloses, and different kinds of polysaccharides (Yen et al. 2013). Algal carbohydrates present a diverse range of applications, with starch and glucose being used for biofuel production, more importantly, bioethanol (John et al. 2011) and hydrogen (Chochois et al. 2009).

As in other organisms, polysaccharides perform crucial biological functions related to storage, protection, and structural integrity in algae cells. Most notably coming from red microalgae, the application of these carbohydrates is possible in different fields, such as food, cosmetics, textiles, stabilizers, emulsifiers, lubricants, thickening agents, and clinical drugs (Yen et al. 2013). The multiplicity of applications is partly explained by their unique rheological property (viscoelasticity), stability in the body (greater than hyaluronic acid), and adsorption properties, which make them suitable to be used as water-soluble lubricants or as additives in other bio-lubricating fluids (Arad and Levy-Ontman 2010).

Among red unicells available, *Porphyridium* sp., found in seawater habitats, has been used in pharmaceutical applications due to its polysaccharides' antitumor and antiviral activities. The material non-toxicity compared to other sugars of a similar kind is an advantageous property that can lead to further pharmacological developments (Arad and Levy-Ontman 2010). Other kinds of red microalgae that have been studied include *P. cruentum* (seawater), *P. aerugienuem* (freshwater), and *Dixoniella grisea* (brackish water) (Arad and Levy-Ontman 2010). A standard method to obtain microalgae carbohydrates is chemical pretreatments, such as hydrogen sulfide and sodium hydroxide, followed by chemical hydrolysis. That is the case for *Chlorococcum infusionum*, which can present up to 89.6% in sugar yield (Karemore and Sen 2016).

Although generally undervalued in comparison to other microalgae products, proteins also compose an important fraction of these organisms, in a proportion that can vary from 30 to 80% in mass, depending on the species. The use of microalgae as a sustainable source of protein for food and feed has been studied and implemented. Table 16.2 shows the proportion proteins play in different microalgae species and other factors related to their use in human nutrition.

Table 16.2 Microalgae species and their protein compositions (adapted from Chew et al. 2017)

Species	Protein composition (%)	Nutritional quality (PDCAAS)[a]
C. vulgaris	51–81	Known for some strains – 0.54–64%
A. platensis	60–71	48%
Tetraselmis chuii	31–46	50–60%
Nannochloropsis oceanica	35–44	35.5%
Dunaliella salina	50–80	–
Galdieria sulphuraria	62	–

[a]PDCAAS—Protein Digestibility Corrected Amino Acid Score

Their potential as an alternative to conventional high-protein crops in animal feeding can be explained by their environmental advantages, such as the lack of necessity in using large terrestrial areas and the fact that their cultivation does not imply the use of fertilizers that could potentially pollute drinking water or marine ecosystems, unlike agriculture (Eilam et al. 2023). Regarding human health, using microalgae-derived protein can produce positive effects due to antioxidant, anti-inflammatory, and anti- cancer properties.

Finally, microalgae produce other valuable compounds, such as pigments, which have been used as vitamin precursors in food and animal feed, additives, cosmetics, biomaterials, and pharmaceutical industries. The main classes of pigments present in these organisms are carotenoids, chlorophylls, and phycobiliproteins. High levels of vitamins are also found in microalgae when compared to commodity feeds, for example, riboflavin, an essential vitamin in the case of maricultured animals (Chew et al. 2017).

The analysis of each of the compounds of microalgae biomass and their application shows the versatility of these organisms and their many advantages from an environmental and economic point of view. This wide range of possibilities makes microalgae a promising resource to be applied to sustainable biorefineries.

16.4 Microalgae in Biorefinery Context

Considering all the advantages of microalgae biomass already discussed in the previous topics, such as lack of competition with foods, fast growth and the adaptability to a wide range of substrates, the microalgae is seen as a high potential raw material in the development of biorefineries once it can captures CO_2 from atmosphere while simultaneously produce low and high-value-added bioproducts of industrial interest (Kashyap et al. 2023). In addition, the valorization of waste through microalgae bioprocessing is a greener and circular bioeconomy approach (Leong et al. 2021).

In general, biorefineries can produce products with low commercial value and in high quantities (Low Value High Volume—LVHV) such as bioethanol, biodiesel and biogas; or products with high added value and in small quantities (High Value Low Volume—HVLV) such as biochemicals, bionutrients and biomaterials (Budzianowski 2017), the choice and combination of products to be integrated in the plant is generally based on technical–economic and environmental analysis, seeking to find the plant with the greatest viability and sustainability. Considering the challenge faced by renewable biofuels in achieving cost competitive values with the relatively cheap fossil fuels, the coproduction of high-values products can help the biorefinery to achieve feasibility (Elias et al. 2021). Based on this concept, biorefineries can be divided into two main groups: biorefineries for energy production, where biomass is mainly used for the production of biofuels, energy and heat, and small amounts of differentiated bioproducts are produced to supplement the plant's income and achieve economic viability; and biorefineries for bioproducts production, where biomass is used mainly in the production of high-value bioproducts and the waste from biomass processing is used to generate energy (Takkellapati et al. 2018).

Taking this into consideration, microalgae is a particularly suitable and versatile raw material for HVLV products due to its composition. As mentioned above, microalgal are mainly composed of protein, carbohydrates, and lipids, which composition will vary according to the cultivation conditions (Parakh et al. 2023). These three main components are important precursors for the production of biodiesel (lipids); bioethanol, biobutanol and acetone (carbohydrates); and alternative food proteins (Siddiki et al. 2022). However, the microalgae potential as raw material goes beyond the existing technologies and biofuel production, once it can be used as raw material for a wide range of HVLP with high potential in food and pharmaceutical sectors, such as pigments, nutraceuticals, omega-3 fatty acids, vitamins, carbohydrates, starch, pesticides, and phycobiliproteins (Fig. 16.2) (Mutanda et al. 2020; Siddiki et al. 2022). These products are non-nutritive secondary metabolites, which are generally produced by the microalgae as a defense against environmental conditions (Ampofo and Abbey 2022).

In general, bioproducts obtaining from microalgae can be performed through direct extraction, such as for primary and secondary metabolites like lipids, proteins and vitamins; or obtained through the chemical or biochemical transformation of the microalgae metabolites, such as bioethanol, biodiesel and butanol. As previously mentioned, microalgae receive special attention as a sustainable resource for lipid production due to their high lipid accumulation and phototrophic growth rate (Xu et al. 2023). The oil extracted from microalgae generally is applied on the production of biodiesel through esterification or transesterification processes (Table 16.3). However, PUFAs are important compounds that are produced in the primary metabolism of microalgae and are not produced by the human body, such as omega-3 (including EPA and DHA) (Ampofo and Abbey 2022). These compounds can be directed extracted from microalgae and are characterized as HVLV bioproducts. Aussant et al. (2018) observed that *Nannochloropsis oculata* and *Isochrysis galbana* were able to produce EPA (2.52 mg/L) and DHA (1.08 mg/L), respectively, with their investigated in vitro nutritional indices (i.e., hypocholesterolemic, atherogenic and thrombogenic indices) falling within accepted health ranges. Bernaerts

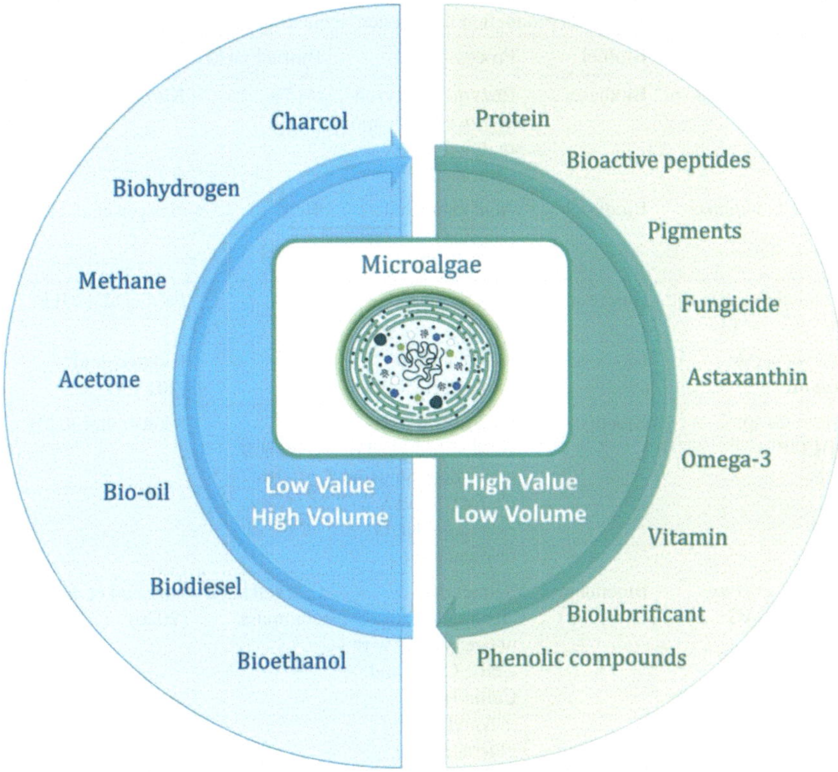

Fig. 16.2 Potential microalgal High Value Low Volume (HVLV) and Low Value High Volume (LVHV) bioproducts for biorefineries application

et al. (2020) achieved the significant amount of 139 mg of omega-3 fatty acids per gram of extracted oil by performing the cell disruption of *Nannochloropsis* sp.

Microalgae are also a feasible protein source, being rich in essential amino acids. Due to the high protein content, *Spirulina*, *Chlorella*, and *Dunaliella* are the most common microalgal species used as additives to food products such as pasta, noodles, cookies, ice cream, and bread to increase the nutritive values these foods (Siddiki et al. 2022). The variety of microalgae applied in animal feed is much bigger once it faces lower legal restrictions. Some microalgae are currently commercialized as dietary supplements and are mainly sold as capsules or tablets. However, *Arthrospira* and *Chlorella* species, are also found for human consumption without any kind of processing (Lafarga 2019; Geada et al. 2021). The protein fraction of microalgae can also be a source of bioactive peptides, which can be applied as nutraceutical ingredients in foods and pharmaceuticals, and can be obtained through cell lysis and protein extraction, followed by controlled enzymatic hydrolysis of algae protein (Ovando et al. 2018).

Table 16.3 Summary of biofuels production using microalgae as raw material

Microalgae Species	Biofuel	Processing	Biofuel yield	References
Aurantiochytrium sp. KRS101	Biodiesel	Enzyme catalyzed *in-situ* esterification in dimethyl carbonate	89.5%	Kim et al. (2016)
Chlorella vulgaris	Biodiesel	Acid catalyzed (H2SO4) transesterification	96%	Nguyen et al. (2020)
Nannochloropsis sp.	Biodiesel	Supercritical method	85.8%	Patil et al. (2011)
Botryococcus braunii	Biodiesel	Acid catalyzed transesterification	95%	Hidalgo et al. (2015)
Chlorella sp. ABC001	Bioethanol	Hydrolysis with H2SO4 followed by neutralization using Ca(OH)2 and *S. cerevisiae* KL17 fermentation	0.43 gethanol/ gsugars	Seon et al. (2020)
Haematococcus pluvialis SAG 192.80	Bioethanol	Sequential enzymatic pretreatments with Cellic Ctec2 and Cellic Htec2, α-Amylase and glucoamylase followed by *Saccharomyces cerevisiae* CCUG 53,310 fermentation	0.21gethanol/ gbiomass	Hosseini et al. (2020)
Porphyridium cruemtum KMMCC-1061	Bioethanol	Enzymatic hydrolysis with pectinase and cellulasefollowed by *Saccharomyces cerevisiae* KCTC 7906 fermentation	70.3%	Kim et al. (2017)

(continued)

Table 16.3 (continued)

Microalgae Species	Biofuel	Processing	Biofuel yield	References
Scenedesmus obliquus SAG 276.7	Bioethanol	Acid treatment followed by enzymatic hydrolysis with Viscozyme® L, AMG 300 L, and Pectinex Ultra SP-L followed by fermentation with 75% *S. cerevisiae.* and 25% *Pichia stipitis* ATCC 58, 785	87%	Farias Silva et al. (2018)
Galdieria sulphuraria	Biohydrogen	Hydrothermal liquefaction and pyrolysis in a membrane reactor	45.70 mol% H2	Ibrahim et al. (2020)
Chlorella PTCC 6010	Biohydrogen	Supercritical water gasification	41.46 mol% H2	Samiee-Zafarghandi et al. (2018)
Coccomyxa chodatii SAG 216– 2	Biohydrogen	Direct photolysis	200 mL/10^6 cells	Danial et al. (2023)
Chlorella sp	Biomethane	CO_2 assisted hydrothermal treatment and anaerobic digestion	194 mL CH4/ gVS	Wu et al. (2020)
Scenedesmus sp.	Biomethane	Rumen inoculated continuous stirred tank reactor for anaerobic digestion	0.185 mL CH4/gVS	Zamorano-López et al. (2019)
Porphyridium cruentum	Biomethane	Enzyme pretreated biomass in anaerobic reactor (500 m^3)	179 ml CH4/ gVS	Kendir Çakmak and Ugurlu (2020)

For the production of biofuels as bioethanol and biodiesel, a pretreatment step to break microalgae biomass structure and release the fermentation substrate is crucial for process success such as acid and enzymatic pretreatments (Parakh et al. 2023). Once the sugars are released, an alcoholic fermentation using the traditional yeast *Saccharomyces cerevisiae*, the most used microorganism in ethanol mills (Ramos et al. 2022), can be performed to produce ethanol. Some applications of microalgae for biofuel production are also summarized in Table 16.3.

In concern of HVLV directed extracted bioproducts, microalgae are excellent sources of vitamins, once it can accumulate vitamin and precursors as a result of its metabolism (Ampofo and Abbey 2022). Ljubic et al. (2020) observed that

Nannochloropsis oceanica accumulated (1 μ/gdryweight) of vitamin D3 (cholecalciferol) upon exposure to an UV-B dose of 36 kJ/m^2/day. *Dunaliella salina* is a good producer of provitamin A (β-carotene), being able to accumulate from 6 to 14% of its cell dry weight as provitamin A, depending on growth conditions, which corresponds to an yield of 100 g/Kg DCW (Arora and Philippidis 2023). Bernaerts et al. (2020) achieved the significant amount of 10 mg of β-carotene per gram of microalgae extracted oil by performing the cell disruption of *Nannochloropsis* sp. Additionally, *Dunaliella tertiolecta* and *Tetraselmis suecica* microalgae may be used to produce vitamin E which is a powerful antioxidant that is commonly used in cosmetics industries (Siddiki et al. 2022).

Lutein is a pigment that has gained attention in recent years due to its food coloring property and nutraceutical value, especially as antioxidant and anticancer agents, with *Chlorella protothecoides* being the microalgae specie dominant producer (about 4.6% on dry weight) (Ampofo and Abbey 2022). Other interesting HVLV are phenolic compounds and astaxanthin. Widowati et al. (2017) observed total phenolic levels of 16.87 and 17.80 mg/g with *Tetraselmis chuii* and *I. galbana*, respectively. According to Levasseur et al. (2020), the astaxanthin content in *Haematococcus pluvialis* can reach about 81% of its total carotenoids (7–20% of dry weight), which is interesting once this compound has an antioxidant capacity 10 times higher than β-carotene (Ampofo and Abbey 2022).

Considering all of this exposed, a wide range of microalgae derived products can be currently found in the market, such as BioAstin®, which is a astaxanthin carotenoid with high antioxidant activity (Cyanotech Corporation 2018); Omega-3 for animal nutrition (Veramaris 2023); and alternative proteins for food applications (ProVeg Incubator 2020; Sophie's Bionutrients 2023). However, with the advancement of biotechnology and the engineering of microorganisms that can become real cell factories, the range of bioproducts obtained from algae biomass fractions can increase substantially (De Vries et al. 2021). For exemple, Huo et al. (2022) developed an industrial *S. cerevisiae* strain capable to produce 2,3-butanediol from glucose. Wang et al. (2020) built a *S. Cerevisiae* yeast capable to produce muconic acid. Both components, 2,3-butanediol and muconic acid, are important build blocks that could be produced from microalgae carbohydrates to integrate biorefineries portfolios.

Most recently, microalgae have been seen as a natural choice in the implementation of microbial CO_2 capture and conversion into HVLV products. According to Akash et al. (2023), *Chlorella vulgaris* and Cyanobacteria are able to capture more than 90% of carbon dioxide in their cells and yield about 0.45 g/L/day of biomass in a typical photobioreactor, which is a very promising strategy to simultaneously product HVLV products and reduce the carbon footprint of the biorefinery.

16.5 Technological Advances in Microalgae in Biorefineries

More than a long prospect into the future, the employment of microalgae in biorefineries is already a short-term reality, and its development has been the focus of several studies in recent periods. Its application can be sectorized into two main categories upstream (UPS) and downstream processing (DPS) (Behera et al. 2022). UPS focuses on cultivation and nutrient recovery. Regarding cultivation, key-role parameters are the selection of "microalgae strains, supply of carbon dioxide (CO_2), nutrient source, and source of illumination" (Chew et al. 2017). Research focusing on understanding and improving UPS has tested these operating conditions, as shown in Aravantinou and Manariotis (2016) where initial biomass concentration, air supply, and light exposure (type and duration) were analyzed for *Chlorococcum* sp. Among other results, the research concluded that artificial light promoted faster growth and more CO_2 biofixation than sunlight exposure, thus demonstrating the importance of these factors in implementing microalgae cultivations (Aravantinou and Manariotis 2016).

Other recent developments have microalgae strain selection as their research object. The field for finding innovative algal species is wide as there around 0.2–0.8 million of them, even though only a limited number are currently used in biorefinery process (Okeke et al. 2022).

Aketo et al. (2020) manage to isolate *Pseudoneochloris sp.* strain NKY372003, considered a high carbohydrate producer, and different cultivation conditions were tested to find an optimal condition for biomass and carbohydrates productivities. The strain was found to produce considerably more biomass and carbohydrate than other current microalgal species, including *Pseudoneochloris marina*, from their own gender. In fact, the levels achieved by the strain of $67.8 \pm 7.0\%$ in carbohydrate content and 5.5 ± 0.2 g/L in carbohydrate production were considered to be the highest production of these macromolecules to ever be presented by microalgae (at the time the studied was conducted). Starch granules accumulated intracellularly by the strain can be converted into fermentable monosaccharides (e. g., glucose), which can be subsequently transformed into bioethanol and biobutanol via a fermentation process. Moreover, the starch produced can be used to synthesize useful compounds for fine chemicals, such as succinate, methyl levulinate and methyl lactate, obtained by the successfully fermentation. All these possible contributions to biofuels and fine chemicals productions place *Pseudoneochloris* sp. NKY372003 strain as a promising option for carbohydrate production in biorefinery applications.

Progress have also been made concerning the development or improvement of the capability of existing microalgal strains by biotechnology advances. For this purpose, genetic engineering stands out with multiple techniques to achieve the modification of algal strains in RNA level (Okeke et al. 2022). For example, Clurestered Regularly Interspaced Short Palindromic Repeats (CRISPR), RNA interference (RNAi), and Transcription activator-like effector nucleases (TALENs). These techniques aim the generation of valuable enzymes and genes capable of improving microalgae characteristics, whether related to increasing their production of high-value compounds, in particular lipid and pigmentation, or other properties as stress tolerance (Okeke et al. 2022). Either way, current and yet to be investigated methodologies seek enhancing the performance of these microorganism, turning them into even more valuable components of a bio-economy integration.

DPS are the processes conducted within the photobioreactor (PBR) and extraction and purification methods that ultimately aim to obtain valuable compounds from microalgae. Technology development in this sector has focused on mild separation techniques that can overcome the high costs involved in traditional ones, such as high-pressure heating and chemicals, bead beating, and homogenizers. These methods avoid using high pressure and solvents, which enables one to obtain the desired product without damaging the others (Chew et al. 2017). This is attained by a mild breakage in the process of cell wall disruption, for which alternative methods are flow techniques, in particular Pulsed Electric Field, Supersonic Flow Fluid Processing, and ultrasound (Vanthoor-Koopmans et al. 2013). Enzymes are also another option to achieve this goal.

One of the possible uses for microalgae-based biorefineries is its combination with produced water (PW). Commonly found in offshore oil fields and mature reservoirs, this wasterwater is generated along the oil extraction process. The amount by which this effluent is produced is rather concerning once it is approximately in a 3:1 ratio compared to oil extraction and is expected to increase to 12:1 by 2025 due to oil wells aging (Pessôa et al. 2022). Current methods to treat PW can involve many steps until being able to remove the oil and impurities from the water, as shown in Fig. 16.3.

These conventional treatment methods present a series of disadvantages, mainly regarding their expensiveness, sludge formation, power consumption, and toxic side products. All these downsides have shown the need for different technologies to emerge, for which bioremediation promoted by microalgae-based refineries appears as a decisive alternative, as it could use not only the effluents hydrocarbons and inorganic salts but also consume atmospheric CO_2 in the process (Pessôa et al. 2022), promoting a ecological process.

Moreover, the microalgae refinery approach using PW as a cultivation medium implicates in stress-generating factors for the microorganism, promoting lipidic accumulation, from which biodiesel production would certainly benefit. Brasil et al. (2017) described the inclusion of this PW use in microalgae biorefinery, which can lead to an interesting model of circular economy. Adapted to the context of the water-energy- food nexus, this integration aims the use of microalgae biomass generated in PW-based medium to produce different bioproducts. Apart from biofuels, this

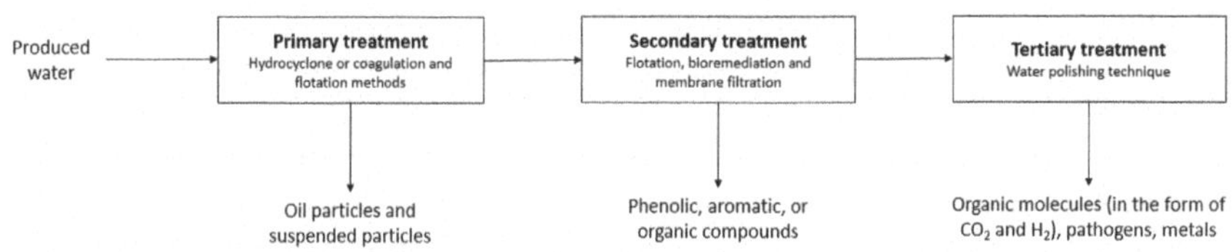

Fig. 16.3 Produced water conventional treatment process

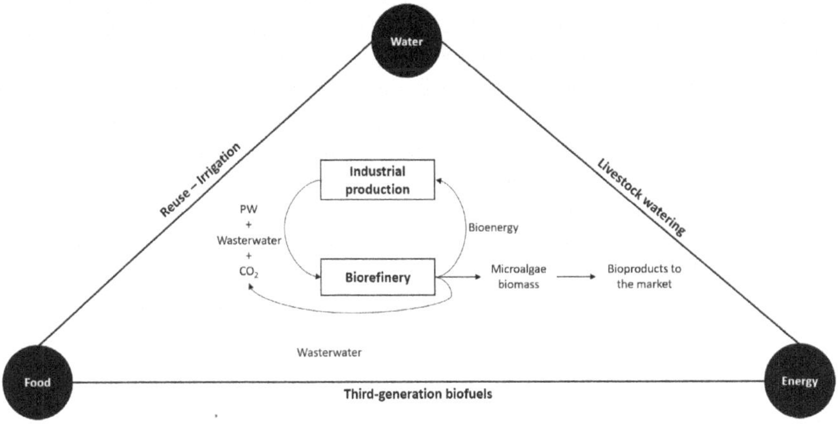

Fig. 16.4 Microalgae-based biorefinery approach in conformation to circular economy (water-energy-food nexus)

includes electricity, biomaterials and biologically based chemicals, as presented in Fig. 16.4. The implementation of that sustainable bioeconomy business model could also promote economic advantages. According to Brasil et al. (2017) in case of obtaining nutrients, water, and microalgal cultivation at low cost, a cost reduction of more than 50% could be achieved. Therefore, this possibility would represent a "synergistic integration between biorefining and other industries in Brazil" (Brasil et al. 2017), which could be replicated to other countries.

Nevertheless, adopting the growth of these organisms in PW still demands a combination of pretreatments, especially concerning the lack of nitrogenous and phosphorous in the medium, restraining the achievement of optimal constitutionality for the organism to develop. This problem could be tackled by mixing wastewater from other sources to produce water, which also mitigates the problem of dilution presented by Javed et al. (2019), in which case an additional cost of water supply would be needed. This synergistic combination could involve the participation of more than one company, contributing to the creation of a circular economy that would result in reducing effluent treatment costs (Pessôa et al. 2022).

Choosing effluents to be added to PW demand considering geographic, economic, and environmental criteria (Javed et al. 2019). In this context, a possibility presented by Kamyab et al. (2019) is palm oil mill effluent (POME), which was tested along with three other effluents in cultivating *C. pyrenoidosa*: piggery, mixed-kitchen, and domestic wastes. POME presented the most remarkable growth and lipid productivity among these four possibilities tested. Other studies of mixing effluents have been conducted with the same gender of *Chlorella* microalgae.

Further attempts to develop a circular bio-economy have been made focusing on different aspects of biorefineries. The integration of microalgae in the production of biofuels is an exciting possibility. This market is expected to grow at a compound annual growth rate (CAGR) of 8.8% until 2025 when it could reach a value of U$

10.7 billion (GVR 2020). This industry growth has been driven by a demand for reliable and sustainable alternatives for petroleum-based fuels, such as diesel and gasoline (Behera et al. 2022). A large part of the market is concentrated in the U.S., with the country presenting 30% of the market share, which is seized by companies like Sapphire Energy, ExxonMobil, and many others (Behera et al. 2022).

As mentioned before, biodiesel is the main product derived from microalgae in the energy sector. Khoo et al. (2011) studied the possibility of a hypothetical integration of PBR-raceway microalgae-to-biodiesel production in Singapore. This process comprises four main processes: cultivation, harvesting, lipid extraction, and oil conversion in biodiesel (and other coproducts). In the presented study, the route for obtaining this biofuel is transesterification, whose global reaction is represented in Eq. 16.1.

$$\text{Lipid} + \text{Methanol} \rightarrow \text{Biodisel (90\%)} + \text{Glycerol (10\%)} \tag{16.1}$$

The feasibility of this whole chain dramatically relies upon each process's energy efficiency. In order to study the final environmental benefits or drawbacks, life cycle assessment (LCA) was chosen, a "systematic environmental management tool" used for measuring input and output streams through each stage, as well as projecting their environmental performance. The methodology is based on defining the product's functional value (or functional unity), for which all cases analyzed were 1 MJ as the high calorific value of biodiesel. Energy balances were performed for each LCA stage, as well as the CO_2 balance of the system. The system boundary was modeled based on a cradle- to-gate model, beginning with microalgae cultivation, and ending with biodiesel as the main product. A total energy demand of 4.4.4 MJ was found, from which the majority was consumed during lipid extraction (85%), followed by biomass production (13%) and biodiesel production (2%).

A sensibility analysis was performed, allowing to find bottlenecks points that could be improved in the process in future research. The most significant improvement was found to be by reducing energy consumption in the extraction stage: decreasing it from 3.8 MJ to 1.3 MJ per MJ biodiesel would represent a 56% drop in the total life cycle (MJ/MJ). A final analysis was done comparing the LCA to other case studies.

16.6 Microalgae Biorefinery—Market and Challenges

The currently microalgae market is estimated in about US$ 11.8 billion in 2023, which the prediction to reach US$ to reach 25.4 billion in 2033 due to a Compound Annual Growth Rate (CAGR) of 8% (FMI 2023). The main producer is the North America, and the main products available in the market are based/derived from *Spirulina* and *Chlorella* microalgae, however *Spirulina* is the largest contributor to the market growth, which is currently mainly composed of food and feed products (AMR 2023). The rise in awareness of nutritional values of microalgae is the key factor driving its market growth, which includes the application in cosmetic, dietary supplements and food/feed industries (AMR 2023).

The main challenges for the scale up of microalgae biorefineries are the high cost of microalgae processing and photobioreactors, the need of sustainable techniques for extraction and purification of its metabolites, and the functional structure preservation during and after extraction procedure, which makes it mandatory the reduction of the costs of this biomass cultivation and bioproducts production (Ampofo and Abbey 2022; Kashyap et al. 2023). In addition, microbial contamination during the process is a problematic issue that can seriously lower biomass yields, which makes the need of the development of efficient suitable and cost-effective harvesting strategies (Mutanda et al. 2020). The use of wastewater as substrate is one of the strategies being studied once wastewater are rich in nitrates, phosphates and other micro, macro and trace elements (Mohsenpour et al. 2021).

There is a lot of alternatives being researched to overcome the challenges faced by the microalgae industries, such as the bioprospection of superior microalgal strains, isolation and purification of strains, strain selection for specific purposes, cultivation systems, biomass harvesting, microalgal disruption techniques and intracellular metabolite extraction, biochemical transformations and product recovery (Mutanda et al. 2020). High-throughput screening is an important and emerging technique where small-scale screening experiments are performed as a reliable and efficient way to investigate different cultivation conditions at once with several repetitions, besides providing a replicable culture environment and saving experimental time and material (Bonfiglio et al. 2021). This strategy is of major use in the identification of microalgal that produces high amounts of the desired HVLV metabolite. In addition, the optimization of microalgae cultivation strategies, such as temperature, pH, light quality and intensity, agitation, and CO_2 supply, can be optimized to lead to a maximum production of the target metabolite (Mutanda et al. 2020).

The selection and genetic engineering modifications of strains to hyper-accumulate the HVLV desired product is a highly performed approach too increase the cultivation step efficiency, however the acceptance and biosafety concerns about genetic modified microorganisms has brought attention to the improvement of microalgal yield via non-genetic techniques, such as the induction of oxidative stress, a biotechnological approach known as elicitation (Ampofo and Abbey 2022). In fact, another limitation faced by microalgae-derived bioproducts is related to strict food safety regulations. The safety of microalgae must be assessed before their commercialization, but restrictive regulatory requirements can delay their access to the market due to long and expensive authorization processes (Geada et al. 2021).

The downstream processing operations are mainly impaired by the dilute microalgae culture and the complex nature of its cell wall (Kashyap et al. 2023), which requires to be disrupted as stated previously. The requirement for multiple process steps can seriously impair the economic feasibility of the process, especially in concern of the lack of costly biomass pretreatments and the dependence of high cost enzymes to release the carbohydrate fraction from microalgae (Kim et al. 2022). Most recently, the consolidated bioprocessing of biomass is an emerging technology to obtain HVLV products without the use of exogenous enzymes, once the enzyme production, polysaccharides hydrolysis, and fermentation of the released sugars into HVLV products happens in a single reactor (Perez et al. 2022; Kim et al. 2022). In this sense, a lot of effort has being made in the development of a microorganism capable of efficiently secret enzymes (Claes et al. 2020; Cunha et al. 2020), however there still the need to increase enzymes production and reduce the CBP time to achieve the feasibility. With respect to the preservation of bioactive compounds properties during extraction and purification operations, there is the need to develop methods that optimize extraction yield while protecting the biochemical structures of less-stable bioactive compounds (Ampofo and Abbey 2022).

Considering all the above, to increase the industrial utilization of microalgae products, it is imperative to develop sustainable and environmentally friendly innovative techniques that are efficient for the transformation of raw microalgae biomass into HVLV products without compromising their nutritional and environmental effects (Ampofo and Abbey 2022). Through the integration and co-optimization of all operations of the microalgae biorefineries, the feasibility of the plant can be achieved.

16.7 Case Study and Plants in Industrial Level

As previously mentioned, some companies already introduced the production of biodiesel derived from microalgae as part of their current business. With more technological advances in their application in biorefinery models, the feasibility of these projects increases, making the conception of these units economically possible. The following pages are dedicated to analyzing implemented plants at the industrial level and their projects and initiatives related to the use of microalgae biomass and how it can impact circular transition.

Although the large-scale implementation of microalgae biorefineries is still restricted, plenty of international companies currently have operating industrial plants, including Sapphire Energy Inc, TerraVia Inc, Algenol Biotech, Cellana Inc, Solix Biofuels, Seambiotic, Algae. Tec., CyanoTech Corp., and Muradel Pty Ltd., among others (Bhattacharya and Goswami 2020). Originally from countries like the United States, Israel, Australia, and India, these companies have a production range from 100 million to 1 billion gallons per annum, which generates revenue from USD 93,000 to 16.8 million (Chanana et al. 2023).

One of the reasons for the economic feasibility of these companies comes from the constant improvement of the production process, remarkably in stages linked to extraction and pathways to use the microalgae, for which aqueous extraction techniques and mild liquid-based extraction can be mentioned. By enhancing output results, they can attenuate the high costs of the biorefinery production chain. The seeking of technological improvement can explain the relevant number of patents that have been filled in the last years, as seen in Table 16.4 (Chanana et al. 2023).

Sapphires Energy, Inc. (SEI) is a relevant example to analyze. As described in a U.S. Department of Energy (2015) final report on their unit production, SEI is a "leader in large-scale photosynthetic algal biomass production," presenting a business

Table 16.4 Submitted patents for microalgae biorefinery process improvement

Company	Patent theme	Patent code
Terravia Holdings	Tailored oils	US9719114B2
Yeda Research and Development Co. Ltd	Methods of producing algal oil and biofuel	WO2017033188A1
Solazyme Roquette Nutritionals, LLC	Microalgal flour	US10098371B2
Corbion Biotech, Inc	Protein-rich microalgal biomass compositions of optimized sensory quality	US10119947B2

model combining lab-based strain selection, cultivation, and harvest and production scale, with a production of Green Crude oil, an effective substitute for traditional crude oil. Following the biorefinery concept, their production aims not only for biofuel but feed and other high- value products (White and Tyler 2015).

Since 2012, SEI has been operating its Integrated Algal Biorefinery (IABR) in Luna County, next to Columbus, New Mexico (NM), U.S, while their extraction unit is located in Las Cruces, NM, less than 100 miles from the IABR. As shown in the scheme of Fig. 16.5, IABR holds microalgae cultivation in open ponds. The microorganism selected was Arthrospira sp, after acquiring its license from Earthrise Nutritionals, LLC (Earthrise). Earthside has produced these species for more than 30 years, holding cultivation on a large scale, which provided SEI with algal strains commercially proved and adapted to be grown in the expected large areas.

After cultivated, biomass is harvested using SEI's propriety technology, including steps of flocculating and de-watering in their dissolved air flotation (DAF) unit. This method, previously only associated with wastewater treatment, potable water production, and the paper industry, constitutes an integrated innovation. Another crucial step

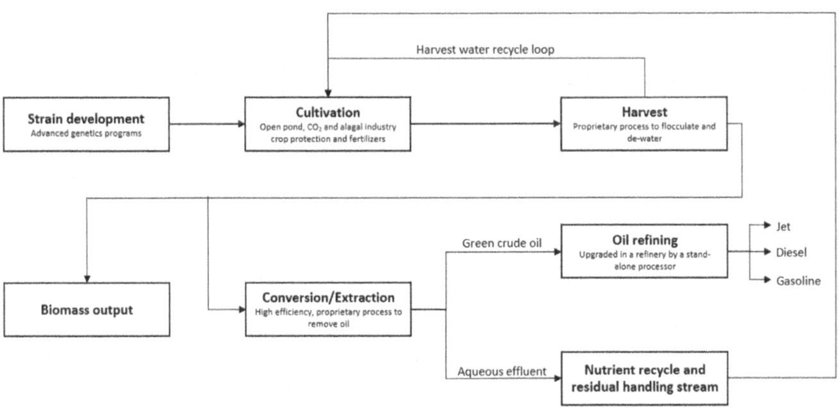

Fig. 16.5 SEI's commercial chain, from upstream to downstream processing and product generation (adapted from White and Tyler 2015)

that must guide other microalgae-producing industrial facilities is recycling water used in the cultivation and post-harvest. In the case of the IABR unit, over 97% of water in the harvesting system returns to production ponds, promoting a critical cost reduction and contributing to a sustainable biomass process (White and Tyler 2015).

Biomass produced is then sent to the extraction facility to be converted into Green Crude Oil. In fact, from June 2012 to November 2014, these facilities produced 524 metric tons (MT) of biomass (d.w. basis) and 2,587 gal of Green Crude oil (White and Tyler 2015). Constant improvements in the unit since the operation started, thus reducing costs in biomass production, can explain the success of producing biofuels from this source, positioning the IABR as a critical piece in SEI's integrated production chain.

Another example of a player developing into the biorefinery business is Algenol, an industrial biotechnology company with operations based in agal research and production. The company is sieged in Fort Myers, U.S., with a Research & Development facility in Berlin, Germany, and has been operating since 2006. Concerning the research for new microalgae, Algenol has already been responsible for developing more than 2000 strains, combined with 44 approved patent technology (and 38 pending applications by 2017). Key areas in which the company is developing include commercial algae production aiming for high-value products, innovating photobioreactor systems for their cultivation, and enhancing algae metabolism (Legere et al. 2017).

One of the company's recent ventures was the construction of the Integrated Biorefinery (IBR) in the same city as its Headquarters. This pilot-scale project received USD 25 million in financing from the U.S. Department of Energy under the "Demonstrating Integrated Biorefinery Operations" opportunity (Legere et al. 2017). The Dow Chenical Company (Dow) also supported the initiative as a partner, alongside the National Renewable Energy Laboratory (NREL), Georgia Institute of Technology (GaTech), and Membrane Technology and Research (MTR).

The project presented bioethanol production via their proprietary integrated process Direct-to-Ethanol®, which uses the proprietary algae strain AB1, which belongs to the cyanobacteria group (blue-green algae). One of the steps that differentiate this technology from the most consists of over-expressing genes for fermentation pathway enzymes in the microorganism, more specifically pyruvate decarboxylase (PDC) and alcohol dehydrogenase (ADH), thus enabling the generation of biomass and ethanol concomitantly and by the same organism. With the subsequent production of bio-crude oil from the biomass generated, the final result is a conversion of more than 85% of the introduced carbon into transportation fuel (Legere et al. 2017).

The transition of the laboratory-developed technology to a larger scale was not entirely economically viable, as shown by TEA analysis. However, the identification achieved of limiting factors for this production enables the company to keep developing the process to advance toward being completely financially stable (Legere et al. 2017). When seeking this objective, by the presented case and other industrial processes, diversification of high-value products generated and technical improvements that reduce costs in different stages can play critical roles.

One of the most recent projects in the microalgae biorefinery industry is being conducted by Microphyt, bioactive ingredients producer founded in 2007 and based in Baillargues, near Montpellier in the South of France. The biotechnology company was awarded in 2021 a €15 million grant from the "Bio-based Industries Joint Undertaking" (BBI-JU), partnership established between the European Commission and Bio-industries Consortium, to the initiate SCALE. The project's objective is to "build and operate the world's first fully integrated microalgae biorefinery", which focus on the food, food supplements, feed, and cosmetic sectors, by producing high nutritional natural active ingredients (Microphyt 2021).

Considered as a Flagship project, which is defined as a research and development project that presents a considerable amount of financial and scientific volume and number of partners (FFG 2013), the initiative aims for an annual capacity of over 100 tons in products. Production will be based in the cutting-edge technology photobioreactor CAMARGUE, developed and patented by Microphyt. Currently operating in pilot plants, this reactor proposes the microalgae cultivation in a two-phase flow tubular system, which differentiates from the traditional single-phase flow current technologies and from open systems, as raceway ponds (Microphyt 2020). This specific kind of closed system presents a wide range of advantages, such as: optimization in gas–liquid transference; higher control of conditions, allowing even more fragile strains to reproduce; high productivity, as seen in other photobioreactors, but not in ponds (Microphyt 2020). These and other characteristics result in a high scalability capacity of this technology, thus contributing to a dilution of the capital expenditure (CAPEX) demand and ultimately enabling that an industrial production, as the one SCALE propose, can be conceived.

Besides the continental financial aid, 11 partners from four European countries are supporting SCALE, including Chanel Parfums Beauté (France), Evolvys (Norway), Tecnalia (Spain) and Nutrasteward (United Kingdom) (Microphyt 2021), which shows the collective effort and importance in conceiving a microalgae-based enterprise and strengthening the value chain linked to this market. The biorefinery construction started in June 2021 and is set to be finished by May 2025, totaling 48 months, which will have a final cost of €24.5 million (SCALE 2022). A first report was generated in 2022, one year after its kick-off date, presenting the advances made up to this date. This included the new ingredients tests for targeted application by the partners, such as cosmetics by Chanel, and the diagnosis of the thermal and power demand in collaboration with Evolvys and Sunti (France) (SCALE 2022) demonstrating the value of an integrated partnership venture. Such project reinforces the relevance of microalgae biorefinery and its possible role in natural and sustainable production for multiple products, including the ones for nutrition and wellness ingredients.

The current scenario for microalgae biorefinery shows successful examples of conceiving and maintaining industrial-scale productions. Nevertheless, they still present a limited range of participation in the market, particularly when visualizing biofuel production as a possible contender to substitute fossil fuel, which continues to present a considerable economic advantage. Points yet to be improved include the high costs involved in different stages of this process, such as downstream processing, estimated to account for 40% of total production costs (Davis et al. 2016). Moreover, wastewater use can also create irregularities in biomass composition, resulting in more processing expenses (Chanana et al. 2023).

The development of new methodologies to tackle this bottleneck can benefit from techno-economic analysis (TEA), which focuses on analyzing the process's economic feasibility and potential to return the investment. LCA can also be advantageous, focusing on the environmental impact of the studied process. Combining these tools enables tracking possible innovations provided by the company's Research & Development department, which ultimately contributes to narrowing down research findings and choosing which improvements or newly developed processes present the highest benefit to be implemented (Chanana et al. 2023). As a whole, it can be foreseen that microalgae biorefinery has yet some steps of development to pass through but has presented a potential to be an economical and environmentally sustainable business model if well applied.

16.8 Policy Recommendation of Circular Bioeconomy in Microalgae Context

As described above, the idea of circularity is fundamental, with the aim of reducing waste, reusing and maximizing the potential for the use of inputs, based on a well-defined concept of the circular bioeconomy, inserted in the context of a green economy (Rossi et al. 2020). The focus of the sustainable circular bioeconomy is the creation of new opportunities for economic growth in sectors that are bio-based, taking into account the achievement of sustainable development goals, the challenges of raw material supply insecurity and the increase in environmental restrictions at different levels of jurisdiction (Denny 2021).

At this point, the transition to a sustainable circular bioeconomy will depend on technological progress, cost competitiveness, financing, facilitating investment, as well as on the availability of sustainable, standardized and certified biomass and, above all, on promoting coordinated public policies. Multilateral negotiations in the World Trade Organization (WTO) can contribute to many of these aspects (Denny 2021).

Although the discussions on the topic of sustainable development are not of recent origin, the current moment is characterized by the need for a cognitive change with regard to sustainability. There are elements that call for this urgency: on the one hand, the impending climate catastrophe and, on the other, the need to boost the economy. Of course, the only way to overcome this dichotomy is to identify points of convergence that will lead to economic development at the same time that socio-environmental benefits are generated. For resource-rich countries (UNEP 2011) such as Brazil, sustainability should be seen as a business opportunity, not an obstacle.

Since 2012, the European Commission has adopted the Bioeconomy Strategy, which aims at a more innovative economy, that reconciles food security, sustainable agriculture and fishing, and the sustainable and efficient use of resources for industrial purposes, ensuring biodiversity and environmental protection (European Comission 2018). In addition, the European Union has different programs, political instruments, wide range of financing mechanisms in order to guarantee the implementation and development of circular bioeconomy activities, the main ones being: Waste Framework Directive, FOOD2030 initiative, European Green Deal, Farm to fork Strategy, EU Forest Strategy, Circular Economy Action Plan, Common Agricultural Policy, Renewable energy directive, Blue Growth, EU Biodiversity strategy for 2030, Horizon Europe, Erasmus + (Naudet and Marrazzo 2021).

The European Commission has recently redefined its long-term goals with Fit55, which aims to develop a competitive, efficient and low-carbon economy by 2050. This document provides a large range of details about the pathway to achieve this commitment (EU 2020). These new targets can be potential barriers for Brazilian products or create a market for sustainable products of the country and for various nature-based climate solutions (Seddon et al. 2020). Thus, a sustainable circular bioeconomy can play an important role in achieving these goals.

Outside the WTO context, but related to trade in bio-inputs, Brazil joined the Nagoya Protocol (UN 2011), which aims to establish a multilateral framework to regulate biodiversity resources and to ensure the conservation of biological diversity, the sustainable use of its components, and the fair and equitable sharing of benefits arising from the use of genetic resources. The central idea of the Protocol is that access to and use of a genetic resource originating in a particular country creates an obligation to share benefits in order to contribute to the conservation of biological diversity in the country of origin. Its main objective is to promote access to genetic resources and associated traditional knowledge and to share the benefits arising from products developed from biodiversity resources (Denny 2021).

In December 2017, the Brazilian government created the National Biofuels Policy (RenovaBio) (Law N° 13.576). The main objectives are: to achieve the goals and commitments assumed in the Paris Agreement, promoting an energy transition in the Brazilian energy matrix, contribute to energy efficiency and reduction of greenhouse gas emissions, promote the expansion of production and use of biofuels in the national energy matrix, ensure predictability for the competitive participation of biofuels in the national fuel market (Brazil 2017; EPE 2017).

Also recently, Brazil established the National Policy of Payments for Environmental Services to create a way to reward sustainability initiatives, and its objective IX is precisely to stimulate scientific research related to "the development of methodologies for the implementation, monitoring, verification and certification of projects that pay for environmental services" (Brazil 2021). The regulation of a carbon market is also likely to contribute to the creation of ways of pricing externalities, thus encouraging the development of initiatives with lower impacts (Denny 2021).

In addition, there is another global policy that supports the circular bioeconomy, the 2030 Agenda formulated by the United Nations (UN) for Sustainable Development. This policy consists of an international document that proposes the coordination of actions by governments, businesses, academia, organized civil society and individuals to achieve 17 Sustainable Development Goals (SDGs) and their 169 related targets by 2030. These are guidelines without a legally binding character, soft law, but intended to promote a complex, multilevel, interdisciplinary, interdependent and among the various interested publics (Nasser 2008).

In this context, global policies encourage research and technological advances in the field of biorefinery, from biomass processing technologies to the development of new strains of microalgae through biotechnology and bioengineering (Okeke et al. 2022). In 2014, Brazil had the first industrial plant going into operation producing microalgae oil for chemical products (food and lubricants). Using sugarcane juice as a carbon source for microalgae cultivation. Despite this process, the country has great potential to attach the production and processing of microalgae to fuel industries (Brasil et al. 2017). Other countries use microalgae for the production of high added value products such as Japan, United States, China, Taiwan, Thailand, India, Australia (Olguín et al. 2022), however the commercialization of microalgae biofuels has been commercialized mainly in the United States, Australia and certain countries Europeans (Chia et al. 2018).

16.9 Conclusions

The microalgae biorefinery plays an important role in the circular bioeconomy, offering a transformative solution in the context of the world facing many environmental challenges and issues such as climate change, population growth and resource depletion. These facilities have several advantages such as the efficient use of resource (minimizing inputs and maximizing outputs), waste minimization (includes using by-products or waste streams from one stage as inputs for another, e.g. integration of different sectors such as agriculture, aquaculture and energy), environmental benefits (reducing environmental impact by reducing resource consumption, greenhouse gas emissions and pollution, e.g. sequestering carbon dioxide during growth, using minimal land and water resources compared to traditional crops), economic feasibility (production of multiple products in the biorefinery, increasing the revenue stream of the farm), collaborative innovation (Circular Bioeconomy principles drive innovation by encouraging the development of new technologies, processes and business models that support resource efficiency and waste reduction. This often requires collaboration between different stakeholders, including researchers, industry, policy makers and consumers, to create a thriving ecosystem for sustainable development). In addition, global policies play a key role in the development of microalgae biorefineries. Through these policies, various studies, researches and technologies are being developed and improved in order to achieve sustainable development goals, generate energy security, improve the characteristics of microalgae through biotechnology and bioengineering, reduce the operational and processing costs of microalgae.

Acknowledgements The authors thanks to the National Council for Scientific and Technological Development (CNPq) [#152289/2022-4 and #407716/2021-1], and São Paulo Research Foundation (FAPESP) [#2019/15851-2 and #2016/10636-8]. This work was also supported in part by Coordenação de Aperfeiçoamento de Pessoal de Nível Superior—Brasil (CAPES, Finance Code 001).

References

Adam F, Abert-Vian M, Peltier G, Chemat F (2012) "Solvent-free" ultrasound-assisted extraction of lipids from fresh microalgae cells: a green, clean and scalable process. Bioresour Technol 114:457–465. https://doi.org/10.1016/j.biortech.2012.02.096

Akash S, Sivaprakash B, Rajamohan N, Vo DVN (2023) Biotechnology to convert carbon dioxide into biogas, bioethanol, bioplastic and succinic acid using algae, bacteria and yeast: a review. Environ Chem Lett 21:1477–1497

Aketo T, Hashizume R, Yabu Y et al (2020) Characterization of a novel marine unicellular alga, *Pseudoneochloris* sp. strain NKY372003 as a high carbohydrate producer. J Biosci Bioeng 129:687–692. https://doi.org/10.1016/j.jbiosc.2019.12.010

AmpofoJ, Abbey, Lord (2022) Microalgae: bioactive composition, health benefits, safety and prospects as potential high-value ingredients for the functional food industry. Foods 11. https://doi.org/10.3390/foods11121744

AMR (2023) Microalgae market by type (Spirulina, Chlorella, Dunaliella Salina, and Aphanizomenon Flos-Aquae), Application (dietary supplements, food/feed, pharmaceutical, cosmetic, and others): global opportunity analysis and industry forecast 2021–2028. In: Allied market research. https://www.alliedmarketresearch.com/microalgae-market-A13419. Accessed 23 Aug 2023

Arad S, Levy-Ontman O (2010) Red microalgal cell-wall polysaccharides: biotechnological aspects. Curr Opin Biotechnol 21:358–364. https://doi.org/10.1016/j.copbio.2010.02.008

Aravantinou AF, Manariotis ID (2016) Effect of operating conditions on *Chlorococcum* sp. growth and lipid production. J Environ Chem Eng 4:1217–1223. https://doi.org/10.1016/j.jece.2016.01.028

AroraN, Philippidis GP (2023) The prospects of algae-derived vitamins and their precursors for sustainable cosmeceuticals. Processes 11. https://doi.org/10.3390/pr11020587

Aussant J, Guihéneuf F, Stengel DB (2018) Impact of temperature on fatty acid composition and nutritional value in eight species of microalgae. Appl Microbiol Biotechnol 102:5279–5297. https://doi.org/10.1007/s00253-018-9001-x

Baicha Z, Salar-García MJ, Ortiz-Martínez VM et al (2016) A critical review on microalgae as an alternative source for bioenergy production: a promising low cost substrate for microbial fuel cells. Fuel Process Technol 154:104–116. https://doi.org/10.1016/j.fuproc.2016.08.017

Behera B, Selvam S M, Paramasivan B (2022) Research trends and market opportunitiesof microalgal biorefinery technologies from circular bioeconomy perspectives. Bioresour Technol 351. https://doi.org/10.1016/j.biortech.2022.127038

Bernaerts TMM, Verstreken H, Dejonghe C et al (2020) Cell disruption of *Nannochloropsis* sp. improves in vitro bioaccessibility of carotenoids and ω3-LC- PUFA. J Funct Foods 65. https://doi.org/10.1016/j.jff.2019.103770

BhattacharyaM, Goswami S (2020) Microalgae—a green multi-product biorefinery for future industrial prospects. Biocatal Agric Biotechnol 25. https://doi.org/10.1016/j.bcab.2020.101580

Biller P, Friedman C, Ross AB (2013) Hydrothermal microwave processing of microalgae as a pre-treatment and extraction technique for bio-fuels and bio-products. Bioresour Technol 136:188–195. https://doi.org/10.1016/j.biortech.2013.02.088

BonfiglioF, Cagno M, Yamakawa CK, Mussatto SI (2021) Production of xylitol and carotenoids from switchgrass and Eucalyptus globulus hydrolysates obtained by intensified steam explosion pretreatment. Ind Crops Prod 170. https://doi.org/10.1016/j.indcrop.2021.113800

Brasil BSAF, Silva FCP, Siqueira FG (2017) Microalgae biorefineries: the Brazilian scenario in perspective. N Biotechnol 39:90–98. https://doi.org/10.1016/j.nbt.2016.04.007

Brazil (2017) Lei N° 13.576, de 26 de Dezembro de 2017 (Law N° 13.576, of December 26, 2017). Presidência da República, Brasília

Brazil (2021) Lei N° 14.119, de 13 de Janeiro de 2021 (Law N° 14.119, of January 13, 2021). Presidência da República, Brasil

Budzianowski WM (2017) High-value low-volume bioproducts coupled to bioenergies with potential to enhance business development of sustainable biorefineries. Renew Sustain Energy Rev 70:793–804. https://doi.org/10.1016/j.rser.2016.11.260

ChananaI, Kaur P, Kumar L et al (2023) Advancements in microalgal biorefinery technologies and their economic analysis and positioning in energy resource market. Fermentation 9. https://doi.org/10.3390/fermentation9030202

Chew KW, Yap JY, Show PL et al (2017) Microalgae biorefinery: high value products perspectives. Bioresour Technol 229:53–62. https://doi.org/10.1016/j.biortech.2017.01.006

ChiaSR, Chew KW, Show PL et al (2018) Analysis of economic and environmental aspects of microalgae biorefinery for biofuels production: a review. Biotechnol J 13. https://doi.org/10.1002/biot.201700618

Chisti Y (2007) Biodiesel from microalgae. Biotechnol Adv 25:294–306. https://doi.org/10.1016/j.biotechadv.2007.02.001

Chochois V, Dauvillée D, Beyly A et al (2009) Hydrogen production in Chlamydomonas: photosystem II-dependent and-independent pathways differ in their requirement for starch metabolism. Plant Physiol 151:631–640. https://doi.org/10.1104/pp.109.144576

Claes A, Deparis Q, Foulquié-Moreno MR, Thevelein JM (2020) Simultaneous secretion of seven lignocellulolytic enzymes by an industrial second-generation yeast strain enables efficient ethanol production from multiple polymeric substrates. Metab Eng 59:131–141. https://doi.org/10.1016/j.ymben.2020.02.004

Clark J, Deswarte F (2015) The biorefinery concept: an integrated approach. In: Introduction to chemicals from biomass. John Wiley & Sons, Ltd, pp 1–29

Cuevas-CastilloGA, Navarro-Pineda FS, Baz Rodríguez SA, Sacramento Rivero JC (2020) Advances on the processing of microalgal biomass for energy-driven biorefineries. Renew Sustain Energy Rev 125. https://doi.org/10.1016/j.rser.2019.109606

CunhaJT, RomaníA, Inokuma K, et al (2020) Consolidated bioprocessing of corn cob-derived hemicellulose: engineered industrial Saccharomyces cerevisiae as efficient whole cell biocatalysts. Biotechnol Biofuels 13. https://doi.org/10.1186/s13068-020-01780-2

Cyanotech Corporation (2018) Fulfilling the promise of WHOLE HEALTH through Hawaiian microalgae. In: Cyanotech. https://www.cyanotech.com/. Accessed 23 Aug 2023

D'AmatoD, Veijonaho S, Toppinen A (2020) Towards sustainability? Forest-based circular bioeconomy business models in Finnish SMEs. For Policy Econ 110. https://doi.org/10.1016/j.forpol.2018.12.004

Danial AW, Abdel-Basset R, Abdel-Kader HAA (2023) Tuning photosynthetic oxygen for hydrogen evolution in synergistically integrated, sulfur deprived consortia of *Coccomyxa chodatii* and *Rhodobium gokarnense* at dim and high light. Photosynth Res 155:203–218. https://doi.org/10.1007/s11120-022-00961-4

Davis R, Markham J, Kinchin C et al (2016) Process design and economics for the production of algal biomass: algal biomass production in open pond systems and processing through dewatering for downstream conversion. United States

DeVries L, Guevara-Rozo S, Cho MJ et al (2021) Tailoring renewable materials via plant biotechnology. Biotechnol Biofuels 14. https://doi.org/10.1186/s13068-021-02010-z
Denny DMT (2021) Bioeconomia e a Organização Mundial do Comércio: Um comentário inicial (Bioeconomy and the World Trade Organization: An initial comment). In: XI Simpósio Brasileiro sobre cortes e tribunais internacionais. NETI, pp 1–19
EilamY, Khattib H, Pintel N, Avni D (2023) Microalgae—sustainable source for alternative proteins and functional ingredients promoting gut and liver health. Glob Challenges 7. https://doi.org/10.1002/gch2.202200177
EliasAM, Longati AA, de Campos Giordano R, Furlan FF (2021) Retro-techno- economic-environmental analysis improves the operation efficiency of 1G-2G bioethanol and bioelectricity facilities. Appl Energy 282. https://doi.org/10.1016/j.apenergy.2020.116133
EPE (2017) RenovaBio: Biocombustíveis 2030—Nota Técnica: Sustentabilidade (RenovaBio: Biofuels 2030—Technical Note: Sustainability). Rio de Janeiro
EU(2020) Regulation (EU) 2020/852 of the European Parliament and of the Council of 18 June 2020 on the establishment of a framework to facilitate sustainable investment, and amending Regulation (EU) 2019/2088. European Parliament and Council of the European Union, Brussels
European Comission (2018) A sustainable bioeconomy for Europe: strengthening the connection between economy, society and the environment. Brussels
Farias Silva CE, Meneghello D, Bertucco A (2018) A systematic study regarding hydrolysis and ethanol fermentation from microalgal biomass. Biocatal Agric Biotechnol 14:172–182. https://doi.org/10.1016/j.bcab.2018.02.016FFG(2013)TechnicalGuidelinesforFlagshipProjects
FMI (2023) Microalgae market outlook (2023 to 2033). In: Future market insights, Inc. https://www.futuremarketinsights.com/reports/microalgae-market. Accessed 23 Aug 2023
GeadaP, Moreira C, Silva M et al (2021) Algal proteins: production strategies and nutritional and functional properties. Bioresour Technol 332. https://doi.org/10.1016/j.biortech.2021.125125
GVR (2020) Algae Biofuel Market Size, Share & Trend Analysis By Application (Transportation, Others), By Region (North America, Europe, Asia Pacific, ROW), By Country, And Segment Forecasts, 2020—2025. In: Grand view research. https://www.grandviewresearch.com/industry-analysis/algae-biofuel-market. Accessed 23 Aug 2023
Hernández D, Solana M, Riaño B et al (2014) Biofuels from microalgae: lipid extraction and methane production from the residual biomass in a biorefinery approach. Bioresour Technol 170:370–378. https://doi.org/10.1016/j.biortech.2014.07.109
Hidalgo P, Ciudad G, Schober S et al (2015) Biodiesel synthesis by direct transesterification of microalga Botryococcus braunii with continuous methanol reflux. Bioresour Technol 181:32–39. https://doi.org/10.1016/j.biortech.2015.01.047
Ho SH, Chen CY, Chang JS (2012) Effect of light intensity and nitrogen starvation on CO_2 fixation and lipid/carbohydrate production of an indigenous microalga Scenedesmus obliquus CNW-N. Bioresour Technol 113:244–252. https://doi.org/10.1016/j.biortech.2011.11.133
HosseiniA, Jazini M, Mahdieh M, Karimi K (2020) Efficient superantioxidant and biofuel production from microalga Haematococcus pluvialis via a biorefinery approach. Bioresour Technol 306. https://doi.org/10.1016/j.biortech.2020.123100
HuoG, Foulquié-Moreno MR, Thevelein JM (2022) Development of an industrial yeast strain for efficient production of 2,3-butanediol. Microb Cell Fact 21. https://doi.org/10.1186/s12934-022-01924-z
IbrahimAFM, Dandamudi KPR, Deng S, Lin YS (2020) Pyrolysis of hydrothermal liquefaction algal biochar for hydrogen production in a membrane reactor. Fuel 265. https://doi.org/10.1016/j.fuel.2019.116935
IEA (2021) Key world energy Statistics 2021
IPCC (2022) Climate change 2022: mitigation of climate change
JavedF, Aslam M, Rashid N, et al (2019) Microalgae-based biofuels, resource recovery and wastewater treatment: A pathway towards sustainable biorefinery. Fuel 255. https://doi.org/10.1016/j.fuel.2019.115826

John RP, Anisha GS, Nampoothiri KM, Pandey A (2011) Micro and macroalgal biomass: a renewable source for bioethanol. Bioresour Technol 102:186–193. https://doi.org/10.1016/j.biortech.2010.06.139

Kamyab H, Chelliapan S, Lee CT et al (2019) Microalgae cultivation using various sources of organic substrate for high lipid content. In: Green energy and technology. Springer, Heidelberg, pp 893–898

Karemore A, Sen R (2016) Downstream processing of microalgal feedstock for lipid and carbohydrate in a biorefinery concept: a holistic approach for biofuel applications. RSC Adv 6:29486–29496. https://doi.org/10.1039/c6ra01477a

KashyapM, Chakraborty S, Kumari A et al (2023) Strategies and challenges to enhance commercial viability of algal biorefineries for biofuel production. Bioresour Technol 387. https://doi.org/10.1016/j.biortech.2023.129551

KendirÇakmak E, Ugurlu A (2020) Enhanced biogas production of red microalgae via enzymatic pretreatment and preliminary economic assessment. Algal Res 50. https://doi.org/10.1016/j.algal.2020.101979

Khoo HH, Sharratt PN, Das P et al (2011) Life cycle energy and CO_2 analysis of microalgae-to-biodiesel: preliminary results and comparisons. Bioresour Technol 102:5800–5807. https://doi.org/10.1016/j.biortech.2011.02.055

Kim HM, Oh CH, Bae HJ (2017) Comparison of red microalgae (*Porphyridium cruentum*) culture conditions for bioethanol production. Bioresour Technol 233:44–50. https://doi.org/10.1016/j.biortech.2017.02.040

Kim KH, Lee OK, Kim CH et al (2016) Lipase-catalyzed in-situ biosynthesis of glycerol-free biodiesel from heterotrophic microalgae, *Aurantiochytrium* sp. KRS101 biomass. Bioresour Technol 211:472–477. https://doi.org/10.1016/j.biortech.2016.03.092

KimSR, Eckert CA, Mazzoli R (2022) Editorial: microorganisms for consolidated 2nd generation biorefining. Front Microbiol 13. https://doi.org/10.3389/fmicb.2022.940610

KirchherrJ, Yang NHN, Schulze-Spüntrup F et al (2023) Conceptualizing the circular economy (revisited): an analysis of 221 definitions. Resour Conserv Recycl 194. https://doi.org/10.1016/j.resconrec.2023.107001

LafargaT (2019) Effect of microalgal biomass incorporation into foods: Nutritional and sensorial attributes of the end products. Algal Res 41. https://doi.org/10.1016/j.algal.2019.101566

Legere E, Roessler P, Miller H et al (2017) Recovery act—integrated pilot-scale biorefinery for producing ethanol from hybrid algae. MTR

LeongHY, Chang CK, Khoo KS et al (2021) Waste biorefinery towards a sustainable circular bioeconomy: a solution to global issues. Biotechnol Biofuels 14. https://doi.org/10.1186/s13068-021-01939-5

LevasseurW, Perré P, Pozzobon V (2020) A review of high value-added molecules production by microalgae in light of the classification. Biotechnol Adv 41. https://doi.org/10.1016/j.biotechadv.2020.107545

Liang K, Zhang Q, Cong W (2012) Enzyme-assisted aqueous extraction of lipid from microalgae. J Agric Food Chem 60:11771–11776. https://doi.org/10.1021/jf302836v

LjubicA, Jacobsen C, Holdt SL, Jakobsen J (2020) Microalgae Nannochloropsis oceanica as a future new natural source of vitamin D3. Food Chem 320. https://doi.org/10.1016/j.foodchem.2020.126627

McCormick K, Kautto N (2013) The bioeconomy in Europe: an overview. Sustainability 5:2589–2608. https://doi.org/10.3390/su5062589

Microphyt (2021) Microphyt Announces Launch of SCALE, the world's first fully- integrated microalgae biorefinery. In: Microphyt. https://microphyt.eu/microphyt-announces-launch-of-scale-the-worlds-first-fully-integrated-microalgae-biorefinery/. Accessed 23 Aug 2023

Microphyt (2020) Tout ce que vous devez savoir sur la culture des microalgues. In: Microphyt. https://microphyt.eu/fr/tout-ce-que-vous-devez-savoir-sur-la-culture-des-microalgues/. Accessed 23 Aug 2023

MohsenpourSF, Hennige S, Willoughby N et al (2021) Integrating micro-algae into wastewater treatment: a review. Sci Total Environ 752. https://doi.org/10.1016/j.scitotenv.2020.142168

MutandaT, Naidoo D, Bwapwa JK, Anandraj A (2020) Biotechnological applications of microalgal oleaginous compounds: current trends on microalgal bioprocessing of products. Front Energy Res 8. https://doi.org/10.3389/fenrg.2020.598803

Nasser SH (2008) Sources and norms of international law: a study on Softlaw, 1 ed. Galda + Wilch Verlag

Naudet PM, Marrazzo G (2021) The governance of circular bioeconomy: practices and lessons learnt from European regions. Brussels

NguyenTT, Lam MK, Uemura Y, et al (2020) High biodiesel yield from wet microalgae paste via in-situ transesterification: effect of reaction parameters towards the selectivity of fatty acid esters. Fuel 272. https://doi.org/10.1016/j.fuel.2020.117718

Okeke ES, Ejeromedoghene O, Okoye CO et al (2022) Microalgae biorefinery:an integrated route for the sustainable production of high-value-added products. Energy Convers Manage X 16. https://doi.org/10.1016/j.ecmx.2022.100323

Olguín EJ, Sánchez-Galván G,Arias-Olguín II et al (2022) Microalgae-based biorefineries: challenges and future trends to produce carbohydrate enriched biomass, high-added value products and bioactive compounds. Biology (Basel) 11. https://doi.org/10.3390/biology11081146

Ovando CA, de Carvalho JC, de Melo V, Pereira G et al (2018) Functional properties and health benefits of bioactive peptides derived from Spirulina: a review. Food Rev Intl 34:34–51. https://doi.org/10.1080/87559129.2016.1210632

ParakhSK, Tian Z, Wong JZE, Tong YW (2023) From microalgae to bioenergy: recent advances in biochemical conversion processes. Fermentation 9. https://doi.org/10.3390/fermentation9060529

Patil PD, Gude VG, Mannarswamy A et al (2011) Optimization of direct conversion of wet algae to biodiesel under supercritical methanol conditions. Bioresour Technol 102:118–122. https://doi.org/10.1016/j.biortech.2010.06.031

PerezCL, Pereira LPR da C, Milessi TS et al (2022) Towards a practical industrial 2G ethanol production process based on immobilized recombinant *S. cerevisiae*: Medium and strain selection for robust integrated fixed-bed reactor operation. Renew Energy 185:363–375. https://doi.org/10.1016/j.renene.2021.12.044

PessôaLC, Cruz EP, Deamici KM et al (2022) A review of microalgae-based biorefineries approach for produced water treatment: Barriers, pretreatments, supplementation, and perspectives. J Environ Chem Eng 10. https://doi.org/10.1016/j.jece.2022.108096

ProVeg Incubator (2020) Naka foods: the future of meet in India. In: ProVeg Incubator. https://provegincubator.com/naka-foods-the-future-of-meat-in-india/. Accessed 23 Aug 2023

Ramos MDN, Milessi TS, Candido RG et al (2022) Enzymatic catalysis as a tool in biofuels production in Brazil: current status and perspectives. Energy Sustain Dev 68:103–119. https://doi.org/10.1016/j.esd.2022.03.007

REN21 (2016) Renewables 2016: global status report. Renewable Energy Policy Network for the 21st Century, Paris

RossiE, Bertassini AC, dos Ferreira CS et al (2020) Circular economy indicators for organizations considering sustainability and business models: plastic, textile and electro-electronic cases. J Clean Prod 247. https://doi.org/10.1016/j.jclepro.2019.119137

Samiee-Zafarghandi R, Karimi-Sabet J, Abdoli MA, Karbassi A (2018) Supercritical water gasification of microalga Chlorella PTCC 6010 for hydrogen production: Box-Behnken optimization and evaluating catalytic effect of MnO_2/SiO_2 and NiO/SiO_2. Renew Energy 126:189–201. https://doi.org/10.1016/j.renene.2018.03.043

SCALE (2022) First year update: scale program delivers significant achievements

SeddonN, Chausson A, Berry P et al (2020) Understanding the value and limits of nature-based solutions to climate change and other global challenges. Philos Trans R Soc B 375. https://doi.org/10.1098/rstb.2019.0120

Seon G, Kim HS, Cho JM et al (2020) Effect of post-treatment process of microalgal hydrolysateon bioethanol production. Sci Rep 10. https://doi.org/10.1038/s41598-020-73816-4

SiddikiSYA, Mofijur M, Kumar PS et al (2022) Microalgae biomass as a sustainable source for biofuel, biochemical and biobased value-added products: an integrated biorefinery concept. Fuel 307. https://doi.org/10.1016/j.fuel.2021.121782

Sophie's Bionutrients (2023) Sustainable urban protein. In: Sophie's Bionutrients B.V. https://sophiesbionutrients.com/. Accessed 23 Aug 2023

Takkellapati S, Li T, Gonzalez MA (2018) An overview of biorefinery-derived platform chemicals from a cellulose and hemicellulose biorefinery. Clean Technol Environ Policy 20:1615–1630. https://doi.org/10.1007/s10098-018-1568-5

TanECD, Lamers P (2021) Circular bioeconomy concepts—a perspective. Front Sustain 2. https://doi.org/10.3389/frsus.2021.701509

UNEP (2011) Towards a green economy: pathways to sustainable development and poverty eradication—a synthesis for policy makers

Vanthoor-Koopmans M, Wijffels RH, Barbosa MJ, Eppink MHM (2013) Biorefinery of microalgae for food and fuel. Bioresour Technol 135:142–149. https://doi.org/10.1016/j.biortech.2012.10.135

Wang G, Øzmerih S, Guerreiro R et al (2020) Improvement of cis, cis-muconic acid production in saccharomyces cerevisiae through biosensor-aided genome engineering. ACS Synth Biol 9:634–646. https://doi.org/10.1021/acssynbio.9b00477

White R, Tyler M (2015) Sapphire energy—integrated algal biorefinery. United States

Widowati I, Zainuri M, Kusumaningrum HP et al (2017) Antioxidant activity of three microalgae *Dunaliella salina, Tetraselmis chuii* and *Isochrysis galbana* clone Tahiti. In: IOP conference series: earth and environmental science. Institute of Physics Publishing

Wu H, Li J, Wang C et al (2020) Sequent production of proteins and biogas from Chlorella sp. via CO_2 assisted hydrothermal treatment and anaerobic digestion. J Clean Prod 277. https://doi.org/10.1016/j.jclepro.2020.123563

Yen HW, Hu IC, Chen CY et al (2013) Microalgae-based biorefinery—from biofuels to natural products. Bioresour Technol 135:166–174. https://doi.org/10.1016/j.biortech.2012.10.099

Zamorano-López N, Borrás L, Giménez JB et al (2019) Acclimatised rumen culture for raw microalgae conversion into biogas: linking microbial community structure and operational parameters in anaerobic membrane bioreactors (AnMBR). Bioresour Technol 290. https://doi.org/10.1016/j.biortech.2019.121787

Chapter 17
Circular Economy for Biodiesel Production by Managing Wastewater Using Microalgae

Astari Minarti, Astri Rinanti, Melati Ferianita Fachrul, Tazkiaturrizki, and Ranadiya Fadhila

Abstract A circular economy is an economic concept that aims to reduce waste and pollution while maximizing the use of resources through sustainable cycles of production, consumption, and recycling. In biodiesel production, a circular economy can be applied by managing liquid waste (wastewater) produced from the production process through microalgal metabolic activities. A circular economy begins with utilizing wastewater generated from biodiesel production, which contains nutrients and other organic matter for microalgae growth. Microalgae can be cultured in closed (such as photobioreactors) or open (such as ponds) systems filled with wastewater. Naturally, microalgae will absorb nutrients such as nitrogen, phosphorus, and carbon in wastewater. This process not only cleans the wastewater, but also reduces the risk of eutrophication and algal bloom, particularly if the wastewater is discharged without treatment. By absorbing nutrients from wastewater, microalgae will grow and reproduce, producing energy-rich biomass. This biomass can then be collected and further processed to produce high-value products such as biodiesel. Microalgae biomass contains lipids (fats) that can be converted into biodiesel through extraction, transesterification, and purification. The extracted lipids are then converted into methyl esters (biodiesel) with the help of a catalyst and alcohol. After lipid extraction, the remaining biomass rich in protein and carbohydrates can be used as animal

A. Minarti · A. Rinanti (✉) · M. F. Fachrul · Tazkiaturrizki · R. Fadhila
Environmental Engineering Department, Faculty of Landscape Architecture and Environmental Technology, Universitas Trisakti, Jakarta, Indonesia
e-mail: astririnanti@trisakti.ac.id

A. Minarti
e-mail: astari.minarti@trisakti.ac.id

M. F. Fachrul
e-mail: melati@trisakti.ac.id

Tazkiaturrizki
e-mail: tazkiaturrizki13@trisakti.ac.id

R. Fadhila
e-mail: ranadiya082001900059@std.trisakti.ac.id

T. A. Kurniawan and A. Anouzla (eds.), *Algae as a Natural Solution for Challenges in Water-Food-Energy Nexus*, Environmental Science and Engineering,
https://doi.org/10.1007/978-981-97-2371-3_17

feed, fertilizer, or feedstock for biogas production, thereby reducing the wastage of resources. Water treated with microalgae can be further processed to reduce the remaining pollutant content, if necessary, before being discharged into the environment or reused in industrial processes. A more efficient and environmentally friendly system can be created by implementing a microalgae-based circular economy at the production and waste treatment level. This can help reduce the negative impact of biodiesel production on the environment while creating added value from the waste generated.

17.1 Introduction to the Circular Economy

17.1.1 Definition and Principles of Circular Economy

The circular economy is an economic system that aims to reduce resource wastage and minimize environmental impact by redesigning, reusing, and recycling products and materials (Morseletto 2020). The concept is an alternative to the currently dominant linear economic model based on the principle of "take, make, dispose". The circular economy combines economic, environmental, and social principles to create a more sustainable and inclusive system. Some of the basic principles in a circular economy are as follows:

(1) Design environmentally friendly products and processes by developing products and processes that can be recycled, easily separated, and minimize the use of natural resources. The goal is to reduce waste and pollution from the beginning to the end of the product life cycle.
(2) Maintenance, repair, and refurbishment concept are implemented by increasing product life with good maintenance, repair when needed, and refurbishment to maintain product quality and function.
(3) Reuse and redistribution are reusing products or their components in other processes or applications or redistributing products no longer useful to individuals or organizations in need.
(4) Recycling and upcycling involve processing waste materials to produce new materials of higher quality (upcycling) or equivalent to the original materials (recycling). The aim is to reduce the extraction of new resources and the volume of waste going into landfills.
(5) The sharing and collaboration economy encourages consumers and producers to share resources, knowledge, and infrastructure to optimize product use and reduce environmental impact.
(6) Use of renewable energy and resource efficiency by replacing fossil energy sources with renewable energy and improving resource use efficiency to reduce greenhouse gas emissions and mitigate environmental impacts.

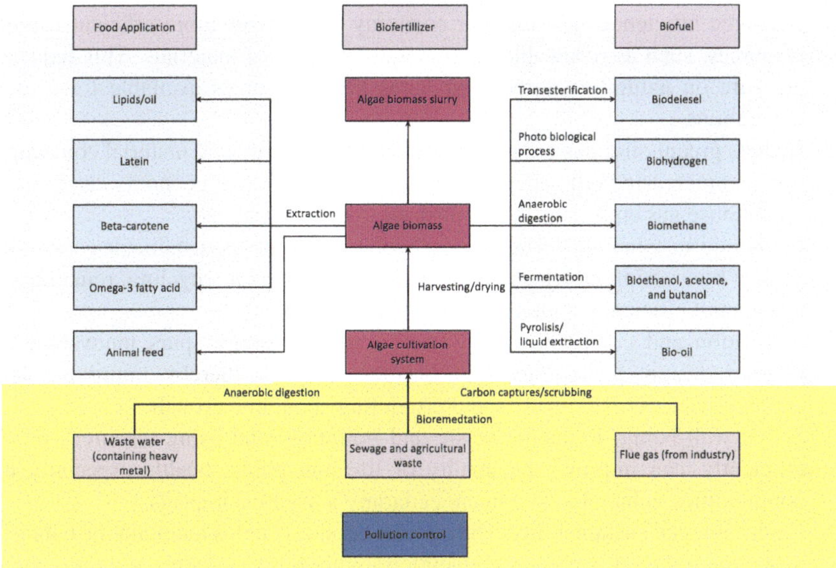

Fig. 17.1 Schematic diagram of the economic benefits of using microalgae

(7) The system approach applies circular economy principles throughout the value chain, including resource management, product design, production, distribution, consumption, and waste management. This approach helps create synergies between different sectors and stakeholders to achieve greater sustainability goals (Taelman et al. 2018), as shown in (Fig. 17.1).

By adopting the circular economy principles, companies, governments, and individuals can reduce their environmental footprint, save costs, and create added value for society and the environment.

17.1.2 The Benefits and Sustainability of the Circular Economy

The circular economy offers many benefits and sustainability that can change how we produce, use, and manage resources (Barros et al. 2021). Here are some of the benefits and sustainability aspects of the circular economy:

(1) Waste and pollution reduction. The redesign of products and processes to reduce waste and pollution from the outset, as well as an emphasis on recycling and upcycling, reduces the amount of waste going into landfills and minimizes pollutant emissions to the environment.

(2) Resource efficiency. The circular economy encourages more efficient use of resources, such as renewable energy and the reuse of materials. This reduces pressure on natural resources and helps keep resources available for future generations.
(3) Reduce greenhouse gas emissions. By reducing energy and material consumption and increasing efficiency in resource use, the circular economy can reduce greenhouse gas emissions that contribute to climate change.
(4) Economic sustainability. The circular economy promotes sustainable economic growth by creating new jobs and industries focusing on recycling, remanufacturing, and efficient resource management.
(5) Innovation and competitiveness. The circular economy spurs innovation in product design, technology, and business models, thereby improving the competitiveness of companies and promoting economic growth.
(6) Social well-being. Reducing waste and pollution, and using resources more efficiently, can improve the quality of life and public health, especially in communities vulnerable to a linear economy's negative impacts.
(7) Environmental sustainability. The circular economy helps maintain the balance of nature and biodiversity by reducing pressure on ecosystems and mitigating the negative impacts of resource extraction, production, and consumption.
(8) System resilience. The circular economy builds resilience in economic and environmental systems by reducing dependence on finite natural resources, reducing risks related to resource prices, and increasing the ability to handle change and disruption.

By applying the circular economy principles, we can create a more sustainable and inclusive system that benefits individuals and the environment (Velenturf and Purnell 2021). These economic, social, and environmental benefits will help ensure long-term sustainability and well-being for future generations.

17.1.3 Circular Economy Applications in the Renewable Energy Industry

The renewable energy industry has great potential to adopt and integrate circular economy principles. Here are some applications of the circular economy in the renewable energy industry:

(1) Design and manufacture of renewable energy equipment. Develop equipment such as solar panels, wind turbines, and energy storage systems that are easily recyclable, use environmentally friendly materials, and have a longer lifespan (Massoud et al. 2023). This may include modularity in design, selection of recyclable materials, and planning for repair, maintenance, and dismantling.
(2) Equipment maintenance and repair. Conduct regular maintenance and repairs on renewable energy equipment to extend its lifespan and reduce the need

for equipment replacement. This includes performance monitoring, preventive maintenance, and timely repair interventions.

(3) Recycling and waste management. Implement efficient recycling and waste management systems for renewable energy equipment that has reached the end of its useful life. For example, old solar panels can be recycled to obtain raw materials such as silicon, metal, and glass, which can then be used to make new panels (Chowdhury et al. 2020).

(4) Integrated energy systems. Integrate various renewable energy sources, such as solar, wind, and hydro, in a flexible and coordinated energy system (Manousakis et al. 2023). This creates synergy and efficiency and ensures a stable and sustainable energy supply.

(5) Service-based business model. Adopt business models emphasizing providing renewable energy services rather than selling energy equipment. These models, such as Energy-as-a-Service (EaaS), encourage manufacturers and service providers to take responsibility for equipment maintenance, repair, and recycling, thus reducing the burden on end-users and promoting resource efficiency (Islam et al. 2022).

(6) Collaboration and sharing economy. Build platforms and networks that enable companies, governments, and communities to share renewable energy knowledge, resources, and infrastructure (Sutherland and Jarrahi 2018). This collaboration could include partnerships for technology development, integrated planning, or shared use of infrastructure such as power grids.

By integrating the principles of the circular economy in the renewable energy industry, we can create an energy system that is more efficient, sustainable, and resilient to resource price fluctuations and changing environmental conditions. This will help accelerate the transition to a greener energy system and support sustainability goals.

17.2 Biodiesel and Microalgae Biomass Conversion Technology

17.2.1 Definition and Characteristics of Biodiesel

Biodiesel is an alternative fuel from renewable feedstocks such as vegetable oils, animal fats, and used cooking oil (Neupane 2023). Biodiesel is a form of bioenergy that has similar properties and characteristics to petroleum-derived diesel, but is more environmentally friendly and, therefore, sustainable. Biodiesel is commonly used as a fuel for diesel engines in the transportation and industrial sectors. Characteristics of biodiesel include:

(1) Renewable feedstock sources. Biodiesel is produced from renewable feedstocks such as vegetable oils (e.g. palm oil, soybean oil, and rapeseed oil) and animal fats. These feedstocks can be sourced locally and replanted after harvesting, which means a more stable and sustainable supply than petroleum.
(2) High biodegradability. Biodiesel is more biodegradable in the environment compared to conventional diesel. This means that biodiesel spills tend to have a lower environmental impact than fossil fuel spills.
(3) Lower greenhouse gas emissions. The use of biodiesel results in lower greenhouse gas emissions, especially carbon dioxide (CO_2), compared to conventional diesel (Xu et al. 2022). This is because the plants used to produce biodiesel absorb CO2 as they grow, which is then released back into the atmosphere when the biodiesel is burned. This cycle results in a better carbon balance.
(4) Good lubricating properties. Biodiesel has better lubricating properties than conventional diesel, which can reduce friction and wear on the engine (Li et al. 2019). This can increase engine life and reduce maintenance costs.
(5) Higher freezing point. Biodiesel has a higher freezing point than conventional diesel, which means that it can agglomerate at higher temperatures (Rao and Chary 2018). This can be a drawback in cold climates, but this problem can be overcome by blending biodiesel with conventional diesel or using additives that lower the freezing point.
(6) Compatibility with diesel engines. Biodiesel is generally compatible with existing diesel engines, although some minor modifications may be required depending on the blend level of biodiesel and conventional diesel used (Brahma et al. 2022). Biodiesel is often used in blended form, such as B5 (5% biodiesel, 95% diesel) or B20 (20% biodiesel, 80% diesel).

17.2.2 Process and Technology for Converting Microalgae Biomass into Biodiesel

Microalgae are an attractive feedstock source for biodiesel production due to their fast growth rate, high lipid content (Kumar et al. 2022), and ability to grow under diverse conditions. The process of converting microalgal biomass to biodiesel involves several main steps, which include microalgae cultivation, biomass separation and drying, lipid extraction, and transesterification. A more detailed description of each step follows.

Microalgae Cultivation

Microalgae cultivation is an important stage in the process of converting microalgae biomass into biodiesel (Zewdie and Ali 2020). This process involves growing microalgae in open (such as ponds) or closed (such as photobioreactors) systems with optimal light, temperature, and nutrient conditions. During cultivation, microalgae produce biomass rich in lipids, which will be used for biodiesel production. In addition, as photosynthetic microorganisms, microalgae can produce proteins and carbohydrates. Lipids produced by microalgae, especially triacylglycerol (TAG), can be converted into biodiesel through a transesterification process. The cultivation process of microalgae in the biodiesel production process needs to pay attention to the following points:

(1) Microalgae strain selection. Microalgae strains that produce high lipid content and have a fast growth rate are selected to improve biodiesel production efficiency (Zhang et al. 2022). In addition, strains capable of growing in a wide range of environmental conditions were favored.
(2) Cultivation conditions. Cultivation conditions such as temperature, light, pH, and nutrients greatly affect the growth and production of lipids by microalgae. Temperature and light intensity must be optimized so that microalgae can grow well and produce lipids in optimal amounts (Christwardana et al. 2022). Nutrients such as nitrogen, phosphorus, and carbon dioxide must also be provided in sufficient quantities.
(3) Cultivation system. There are two main cultivation systems for microalgae biomass production: open systems (ponds or channels) and closed systems (photobioreactors) (Tan et al. 2020). Open systems are cheaper and easier to operate, but susceptible to contamination. Meanwhile, closed systems provide better control over environmental conditions and produce more consistent biomass, but have higher operational costs.
(4) Maintenance and quality control. During the cultivation process, monitoring and quality control are necessary to ensure optimal microalgae growth and high lipid production (Ebhodaghe et al. 2022). Parameters such as biomass concentration, lipid content, and growth rate need to be measured regularly.

Biomass Separation and Drying

Biomass separation and drying are two important steps in the process of converting microalgal biomass into biodiesel. After microalgae are cultivated, the resulting biomass must be separated from the culture medium and dried before the lipid extraction and transesterification process (Ebhodaghe et al. 2022).

(1) Biomass separation or harvesting. After cultivation, microalgae biomass must be harvested to continue the conversion process into biodiesel. The purpose of biomass separation or harvesting is to collect microalgae cells from the culture medium and separate them from water and other solid particles. Harvesting techniques include centrifugation, flotation, filtration, and coagulation-flocculation, with details as follows:

    (a) Centrifugation, is one of the commonly used technologies for harvesting microalgal biomass (Kucmanová and Gerulová 2019). This technique involves separating solid particles (in this case, microalgae cells) from a liquid medium (cultivation medium) by utilizing the centrifugal force generated by the rotation of the centrifuge, in this case, applying it to a mixture of microalgae cells and cultivation medium. The centrifugal force causes the solid particles (microalgae cells) to move to the outside of the centrifuge tube, while the lighter cultivation medium is on the inside. Thus, the microalgae cells will be separated from the cultivation medium, and the resulting biomass will collect at the bottom of the centrifuge tube. The speed and duration of centrifugation affect the efficiency of separating microalgae cells from the cultivation medium. High speeds and longer durations usually result in more efficient separation. However, the optimal speed and duration are highly dependent on the type of microalgae and cultivation conditions. Centrifugation technology can be applied in laboratory as well as industrial scale. For large-scale microalgae biomass production, a continuous centrifuge with high capacity is used to harvest the biomass more efficiently. Centrifugation offers several advantages in the microalgae harvesting process, such as high efficiency, separation speed, and high biomass concentration. This technique also produces dry biomass that is easier for further processing, such as lipid extraction or conversion to biodiesel (Patel et al. 2018). One of the main disadvantages of centrifugation technology is the high energy consumption, which can increase operational costs. In addition, large-scale use of centrifuges may require significant initial investment in equipment and infrastructure. Despite these drawbacks, centrifugation remains a popular technology for harvesting microalgal biomass due to its high efficiency and ability to produce dry biomass ready for conversion to biodiesel or other products.

(b) Flotation, is an alternative technology used to harvest microalgal biomass. This technique involves separating microalgal cells from the cultivation medium through a flotation process (Kucmanová and Gerulová 2019). Flotation is commonly used to overcome some of the limitations of other harvesting technologies (e.g., centrifugation and filtration) by utilizing the density difference between microalgal cells and the cultivation medium. In the flotation process, gas (usually air) is injected into the cultivation medium, forming bubbles that will attach to the microalgae cells. The microalgae cells attached to the bubbles will float to the surface of the liquid, allowing the separation and collection of biomass. One of the commonly used flotation techniques in microalgae harvesting is dissolved air bubble-based flotation (DAF). In the DAF process, water saturated with air at high pressure is injected into the cultivation medium, which produces fine air bubbles that will attach to microalgae cells and float them to the surface. Flotation offers several advantages over other harvesting methods, such as lower energy costs, the ability to harvest biomass with low cell concentrations, and the potential for large-scale processing. This technique is also effective in overcoming solid particle contamination that may be present in the cultivation medium. Although flotation has some advantages, the technique also has some disadvantages. The separation efficiency in flotation may be lower compared to methods such as centrifugation, especially for microalgae species that do not float easily. In addition, flotation may be less effective for producing dry biomass ready for further conversion processes, such as lipid extraction or conversion to biodiesel. Overall, flotation is an attractive technology for harvesting microalgal biomass, especially when energy costs and the ability to overcome solid particle contamination are important considerations (Branyikova et al. 2018). However, the separation efficiency and quality of the biomass produced should be taken into account before adopting this technique for large-scale applications.

(c) Filtration, is a microalga harvesting technology that involves separating microalgae cells from the cultivation medium through a filtration process. This technique utilizes filtration media or membranes with pore sizes designed to retain the microalgae cells while the liquid medium passes through the filter. Filtration utilizes the size difference between microalgae cells and cultivation medium particles (Kucmanová and Gerulová 2019). In the filtration process, the cultivation medium containing microalgae cells is pumped or flows through a filtration medium or membrane. Microalgae cells are retained by the filter because their size is larger than the pore size of the filter, while allowing the cultivation medium to pass through the filter. Various types of filters are used in the microalgae harvesting process, including, among others, fabric, sand, and microfiltration or ultrafiltration membranes. The selection of a suitable filtration medium largely depends on the microalgae cell size and process requirements. Filtration offers several advantages in microalgae harvesting processes, such as relatively lower energy costs, ease of operation, and the ability to produce biomass with lower water content. In addition, filtration can be applied at various scales, ranging from laboratory to industrial. Some disadvantages of filtration technology include the potential for filter clogging, which can reduce harvesting efficiency and require periodic cleaning or replacement of filters. In addition, filtration may not be suitable for microalgae with very small cell sizes or those that tend to form aggregates (Kucmanová and Gerulová 2019), as this may lead to faster clogging. Overall, filtration is an attractive and effective microalga harvesting technology, especially for situations where energy costs and ease of operation are important considerations. However, harvesting efficiency and filter maintenance should be taken into account before adopting this technique for large-scale applications.

(d) Coagulation-flocculation is a microalga harvesting technology that involves agglomerating microalgae cells into larger flocs (Fig. 17.2), which are then easily separated from the cultivation medium (Delgado-Plaza et al. 2019).

This process involves the addition of coagulants and flocculants, which modify the surface properties of the microalgae cells and result in clumping. The coagulation-flocculation process requires two stages. First, a coagulant (such as aluminum or iron salts) is added to the cultivation medium to agglomerate the dispersed microalgae particles. Then, flocculants (such as organic polymers or polyelectrolytes) are added to increase the size of the formed flocs, which will be more easily separated from the cultivation medium. Coagulation-flocculation efficiency depends on several factors,

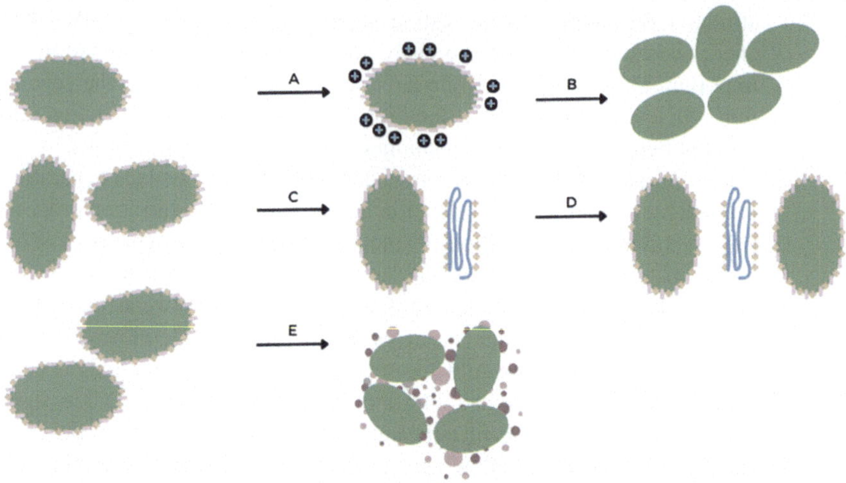

Fig. 17.2 Basic mechanism of flocculation

such as the type and concentration of coagulant and flocculant, pH, and mixing conditions. In many cases, optimization of coagulant and flocculant dosage and process conditions is required to achieve high harvesting efficiency. Coagulation-flocculation offers several advantages in microalgae harvesting processes, such as lower energy costs compared to centrifugation, the ability to treat biomass in low cell concentrations, and the potential for large-scale application (Branyikova et al. 2018). Some disadvantages of coagulation-flocculation technology include chemical consumption that may increase operational costs and have environmental impacts, as well as the possibility of lower biomass quality due to coagulant and flocculant contamination. In addition, separation of the resulting flocs may require additional steps through sedimentation, filtration, or centrifugation. Overall, coagulation-flocculation is an attractive and potentially effective microalgae harvesting technology, especially for situations where energy cost is an important consideration and biomass is in low concentration. However, chemical consumption and the quality of the biomass produced should be taken into account before adopting this technique for large-scale applications.

(2) Biomass drying. After the biomass is successfully separated from the culture medium, the process continues with drying to reduce the moisture content in the biomass. Drying is important to improve lipid extraction efficiency and reduce biomass transportation costs (Razzak et al. 2022). Some commonly used drying methods include:

(a) Convection drying, is one of the commonly used drying techniques to reduce the moisture content in a variety of materials, including harvested microalgal biomass, thereby improving product stability and facilitating use, storage and transportation. Convection drying involves the transfer of heat from a heat source (the dryer) to the microalgal biomass by convection of hot air flowing around the biomass particles. Drying can also be done under direct sunlight (Delgado-Plaza et al. 2019). While this method is simple and inexpensive, the process is slow and dependent on weather conditions. Several steps in the convection drying process using a drying device start with the preparation of the biomass. Once the microalgal biomass is harvested, usually through centrifugation, flotation, or flocculation, it needs to be prepared for the drying process. This involves surface drying or initial mechanical drying, such as the use of filter presses or centrifugation, to reduce the initial moisture content.

The prepared microalgae biomass is then arranged in thin layers or in special racks or containers that allow hot air to flow through them. Arranging the biomass in thin layers helps to improve drying efficiency by ensuring good contact between the hot air and the biomass. A heat source is used to heat the air, which will flow through the microalgae biomass. One can use different heat sources, including an electric heater, gas, or renewable energy, such as solar power. The temperature used in the convection drying process usually ranges from 40 to 80 °C, depending on the nature of the biomass and the needs of the application.

The hot air generated by the heat source is then circulated through the microalgal biomass using a fan or ventilation system (Schmid et al. 2022). This hot air flow creates heat transfer from the air to the biomass through a convection process, causing the water in the biomass to evaporate and be carried away by the air flow. The water vapor is removed from the drying system, usually using a condenser or vapor trapping system. This prevents moisture from getting back into the biomass. It will help to optimize the drying process. Once the moisture content in the microalgae biomass reaches the desired level, the dried product can be stored in sealed containers or packaged for further applications (e.g., biofuel production, animal feed, or nutritional supplements).

(b) Vacuum drying is a drying technique used to reduce the moisture content in various materials, including harvested microalgal biomass (Bheda et al. 2019). The process involves lowering the pressure inside the vacuum drying chamber, thereby reducing the boiling point of water and allowing drying at lower temperatures compared to convection drying. Vacuum drying has several advantages, such as reducing thermal damage to the biomass and improving the quality of the final product.

Once the microalgal biomass has been harvested, usually through centrifugation, flotation, or flocculation, it needs to be prepared for the drying process. This involves surface drying or initial mechanical drying, such as the use of filter presses or centrifugation, to reduce the initial moisture content. The prepared microalgae biomass is then loaded into a vacuum drying chamber. This chamber should be designed to ensure good contact between the biomass and the heating surface, as well as allowing the flow of water vapor generated from the drying process.

The pressure inside the vacuum drying chamber is then lowered, usually using a vacuum pump. This pressure drop reduces the boiling point of water, allowing the water in the biomass to evaporate at a lower temperature. The microalgae biomass in the vacuum chamber is then heated using a heat source, such as an electric heater or steam heater. The temperature used in the vacuum drying process is usually lower compared to convection drying, ranging from 30 to 60 °C, depending on the nature of the biomass and the needs of the application.

The water vapor generated from the drying process is then removed from the vacuum drying chamber via a vacuum pump. This prevents moisture from re-entering the biomass and helps optimize the drying process. Once the moisture content in the microalgae biomass reaches the desired level, the dried product can be stored in sealed containers or packaged for further applications.

The main advantage of the vacuum drying method is that it dries at a lower temperature, which can reduce thermal damage to the biomass and maintain the quality of the final product (Yue et al. 2023). In addition, this process has better energy efficiency and is faster than convection drying, although higher operational costs.

(c) Freeze drying or lyophilization is another drying technique used to reduce the moisture content in various materials, including harvested microalgal biomass (Schmid et al. 2022), which involves freezing the biomass, followed by drying under vacuum to remove the water contained in the biomass through a sublimation process. Freeze-drying has several advantages, such as maintaining the structure and functional properties of the biomass and improving the quality of the final product.

Once the microalgal biomass has been harvested, usually through centrifugation, flotation, or flocculation, it needs to be prepared for the drying process. This involves surface drying or initial mechanical drying, such as the use of filter presses or centrifugation, to reduce the initial moisture content. The prepared microalgal biomass is then frozen, usually by exposing it to very low temperatures (Martins et al. 2022) (− 40 °C to − 80 °C) in a freezer or using rapid cooling such as liquid nitrogen. This freezing turns the water in the biomass into ice.

The frozen microalgae biomass is then placed in a freeze-drying chamber (lyophilizer), which is designed to allow sublimation of ice water from the biomass under vacuum. The pressure inside the freeze-drying chamber is then lowered, typically using a vacuum pump. This pressure drop creates conditions that allow the ice water in the biomass to evaporate directly into water vapor through the sublimation process. During the freeze drying process, gentle heating is applied to the biomass to aid the sublimation process. The heat source can be electric heaters or steam heaters placed around the freeze drying chamber. The temperature used in this process is usually very low (a few degrees above the freezing point of the biomass) to avoid damage to the biomass.

The water vapor generated from the sublimation process is then removed from the freeze-drying chamber via a vacuum pump. This prevents moisture from re-entering the biomass. Once the moisture content in the microalgae biomass reaches the desired level, the dried product can be stored in sealed containers or packaged for further applications.

Lipid Extraction

Lipids are extracted from dried microalgal biomass using organic solvents such as hexane, ethanol, or acetone (Zhou et al. 2022). This process produces microalgal oil rich in triglycerides, which are the main components required for biodiesel production. Lipid extraction from microalgal biomass is a crucial step in the process of converting microalgal biomass into biodiesel. At this stage, the lipids contained in the microalgae cells are extracted and then converted into biodiesel through the transesterification process. Some common methods used to extract lipids from microalgae biomass include:

(1) **Mechanical extraction**, is one of the approaches used to extract lipids from microalgal biomass (Zhou et al. 2022). The aim of mechanical extraction is to break the microalgal cell wall and release the lipids contained within. This process generally involves several techniques, such as high-pressure homogenization, ultrasonic extraction, and extraction using milling equipment such as an expeller press. Although these methods are simple, their efficiency is limited and may not be effective for extracting lipids from hard or difficult-to-break microalgal cells.

(a) Mechanical extraction by high-pressure homogenization begins with pumping a suspension of microalgal biomass through a narrow slit at high pressure. This process generates high shear forces and pressure, which causes the microalgal cell walls to break and release the lipids. High-pressure homogenization is often used to treat relatively small amounts of microalgal biomass.

(b) Ultrasonic mechanical extraction, uses high-frequency sound waves to induce cavitation in the liquid containing the microalgal biomass. The cavitation results in imploding bubbles, which in turn create sufficient pressure and shear forces to break the microalgae cell walls and release the lipids. This technique has the advantage of higher energy efficiency compared to high-pressure homogenization, but may not be suitable for larger industrial scales.

(c) Expeller press is a mechanical extraction method that uses mechanical pressure to squeeze microalgal biomass and release lipids. The microalgal biomass is placed in a cylinder containing a serrated disk or a rotating screw. As the screw rotates, the biomass is forced through the narrow gap between the cylinder wall and the screw, thus generating enough pressure to break the microalgae cell wall and release the lipids. Expeller presses are typically used on a larger industrial scale and have the advantage of lower cost compared to other mechanical extraction methods.

Mechanical extraction methods have the advantage of reducing the use of chemical solvents, which can reduce environmental impact and operational costs (Bitwell et al. 2023). However, the efficiency of lipid extraction using mechanical methods is often lower than that of solvent extraction methods. Therefore, in some cases, mechanical extraction methods might be combined with solvent extraction methods to improve the overall lipid extraction efficiency.

(2) **Solvent extraction** is a common technique used to extract lipids from microalgal biomass. In this method, organic solvents are used to dissolve and separate lipids from other components of the microalgal biomass. Some commonly used solvents in this extraction process include hexane, ethanol, acetone, and chloroform. Solvent extraction is generally more efficient than mechanical extraction, but the use of toxic chemicals and high energy consumption are considerations on an industrial scale (Kapadia et al. 2022).

Before extraction, the microalgal biomass must be dried to reduce the moisture content. Drying can be done using methods such as convection drying, vacuum drying, or freeze drying. The dried microalgae biomass is then mixed with a suitable organic solvent. The biomass to solvent ratio, temperature, and extraction time may vary depending on the type of microalgae and solvent used.

During the extraction process, the lipids contained in the microalgae biomass will dissolve in the organic solvent. This process is usually carried out in a closed container with constant stirring to ensure good contact between biomass and solvent. After the extraction process is complete, the mixture of biomass and solvent will form two separate phases: a solid phase containing the remaining microalgal biomass and a liquid phase containing the lipids dissolved in the solvent. The liquid phase containing lipids can be separated from the solid phase using methods such as centrifugation, decantation, or filtration. The extracted lipids are then separated from the solvent by evaporating the solvent, usually using a device such as a rotary evaporator or vacuum dryer. Once the solvent has evaporated, the remaining lipids can be collected. In some cases, the evaporated solvent can be condensed and recycled for reuse in the extraction process.

Solvent extraction methods have the advantage of higher lipid extraction efficiency compared to mechanical extraction methods (Saini et al. 2021). However, the use of organic solvents can increase operational costs and have a higher environmental impact. Therefore, research is currently being conducted to develop alternative extraction methods that are more environmentally friendly and economically efficient.

(3) Soxhlet extraction is a classic technique that has been widely used to extract lipids from various types of biomass, including microalgae biomass. This method uses a Soxhlet apparatus containing a condenser, heating walls, and an extraction chamber with a cellulose fiber-based thimble that serves as a holder for the biomass sample.

Prior to extraction, the microalgal biomass must be dried to reduce the moisture content. Drying can be done using methods such as convection drying, vacuum drying, or freeze drying. The dried microalgal biomass is then placed in a thimble, which will later be placed in the Soxhlet extraction chamber. A suitable organic solvent, such as hexane, ethanol, or chloroform, is added to a flat-walled flask placed over a heating source. This flask is connected to the Soxhlet extraction chamber containing the biomass. When the flask is heated, the solvent evaporates and rises to the top of the extraction chamber, where a condenser condenses the solvent vapor.

The condensed solvent will drip back into the extraction chamber, dissolving lipids from the microalgal biomass in the thimble (Yeong Hwang et al. 2021). This process will occur repeatedly, so that the solvent saturated with lipids will flow back into the flat-walled flask. The extraction process takes place automatically and continuously, allowing the solvent to efficiently wash the biomass and extract the maximum amount of lipids. Once the extraction process is complete, the flat-walled flask containing the lipid and solvent mixture will be removed from the Soxhlet device. The extracted lipids can be separated from the solvent by evaporating the solvent, usually using a device such as a rotary evaporator or vacuum dryer. Once the solvent has evaporated, the remaining lipids can be collected.

The Soxhlet extraction method has the advantages of high lipid extraction efficiency and a relatively simple process (Figler et al. 2019). However, this process requires a more prolonged period and uses larger amounts of organic solvents, which can increase operational costs and environmental impact. Therefore, alternative extraction methods that are more environmentally friendly and economically efficient are being developed to extract lipids from microalgal biomass.

(4) Enzymatic extraction, is a method that uses enzymes to break the microalgae cell wall and release the lipids contained within. This process is considered more environmentally friendly and selective compared to solvent or mechanical extraction methods. Prior to extraction, the microalgal biomass must be dried to reduce the moisture content. Drying can be done using methods such as convection drying, vacuum drying, or freeze drying.

The dried microalgal biomass is then mixed with the appropriate enzyme solution. Enzymes commonly used in enzymatic extraction include cellulase, hemicellulase, protease, and lipase (Raveendran et al. 2018). These enzymes work by breaking microalgal cell wall components such as cellulose, hemicellulose, and proteins, which in turn will release lipids trapped inside the cells. The biomass and enzyme mixture is incubated at a specific temperature and conditions (e.g., pH and incubation time) that are optimal for enzyme activity. During the incubation process, the enzymes will break the microalgae cell wall and release the lipids.

After incubation, the extracted lipids must be separated from the biomass and enzyme mixture. This separation can be done by centrifugation, filtration, or decantation methods. In some cases, additional solvent extraction may be required to efficiently separate the lipids from the mixture. The extracted lipids then need to be purified to remove residual enzymes and other contaminants. This purification process may involve steps such as precipitation, filtration, or chromatography.

The enzymatic extraction method has several advantages over other extraction methods (Handayani et al. 2018). First, this process is more environmentally friendly as it reduces the use of organic solvents and energy. Second, enzymatic extraction tends to be more selective, resulting in lipids with higher purity. However, this method can have a higher cost due to the use of enzymes and may require a longer time to achieve optimal extraction efficiency. Therefore, research is underway to develop more efficient and economical enzymatic extraction methods.

(5) Supercritical extraction, is an advanced method that uses supercritical fluid as a solvent to extract lipids from microalgal biomass. Supercritical fluids are substances that are above their critical pressure and temperature, which gives them unique properties between gases and liquids. The most commonly used supercritical fluid in the extraction of lipids from microalgal biomass is supercritical carbon dioxide ($SCCO_2$).

Prior to extraction, the microalgal biomass must be dried to reduce the moisture content. Drying can be done using methods such as convection drying, vacuum drying, or freeze drying. The dried microalgal biomass is fed into a high-pressure extraction reactor. A supercritical fluid, such as carbon dioxide, is then injected into the reactor under appropriate pressure and temperature. The pressure and temperature conditions will determine the solvent properties of the supercritical fluid and the lipid extraction efficiency.

During the extraction process, the supercritical fluid will dissolve the lipids present in the microalgal biomass (Zhou et al. 2022). Due to its unique properties, the supercritical fluid can penetrate the microalgae cell wall and dissolve the lipids with high efficiency. The extraction process usually lasts for several hours to a day, depending on the operating conditions and the type of microalgae biomass. After the extraction process is complete, the supercritical fluid containing lipids is expanded to a lower pressure and temperature. This expansion will result in the fluid changing from supercritical phase to gas phase, which results in the lipids separating from the fluid and precipitating as solids or liquids. The separated lipids can then be collected from the extraction reactor. After lipid separation, the supercritical fluid that has turned into the gas phase can be condensed and recycled back to the extraction system, thus reducing the cost and environmental impact associated with solvent use.

Supercritical extraction methods have several advantages compared to conventional extraction methods, such as high extraction efficiency, better product purity, and the use of more environmentally friendly solvents. However, this method has higher operational costs due to the high-pressure equipment required and greater energy consumption.

Transesterification

Transesterification is a key step in the process of converting microalgal biomass into biodiesel (Akubude et al. 2019). After lipids are extracted from microalgae biomass, the triacylglycerol (TAG) compounds contained in the lipids need to be converted into methyl esters (biodiesel) through a transesterification chemical reaction.

Transesterification is the reaction between triacylglycerol (TAG) and an alcohol (usually methanol or ethanol) with the help of a catalyst (acid, base, or enzyme). This reaction produces methyl esters (biodiesel) and glycerol as a by-product. The general reaction can be formulated as follows:

$$\text{TAG} + 3\ \text{Methanol} \rightarrow 3\ \text{Methyl Esters(Biodiesel)} + \text{Glycerol}$$

Catalysts are needed to increase the reaction speed and produce biodiesel with good quality. Catalysts used in the transesterification reaction include:

(a) Base catalysts. Base catalysts such as potassium hydroxide (KOH) or sodium hydroxide (NaOH) are commonly used in the transesterification process because the reaction proceeds faster than with acid catalysts. However, base catalysts are not effective if the lipids contain high amounts of free fatty acids (FFA), as soap will form and reduce the quality of the biodiesel.
(b) Acid catalysts. Acid catalysts such as sulfuric acid (H_2SO_4) or hydrochloric acid (HCl) are used if the lipids contain high FFA. Acid catalysts can overcome soap formation, but the transesterification reaction proceeds more slowly than using base catalysts.
(c) Enzymatic catalyst. Lipase is an enzyme that can be used as a catalyst in transesterification reactions. The advantages of using enzymes as catalysts are that they are environmentally friendly and tolerant of high FFA content. However, the cost of enzymes and slower reaction speed are considerations in industrial scale applications.

After the transesterification reaction is complete, the resulting product is a mixture of methyl esters (biodiesel), glycerol, and catalyst (Anisah et al. 2019). The separation of the methyl ester (biodiesel), glycerol, and catalyst mixture involves physical processes that separate the components based on differences in physical properties, such as density and solubility. Sedimentation, decantation, and centrifugation are commonly used separation methods. All three methods have similarities in that they utilize density differences between components and do not require specialized equipment, except centrifugation requires centrifuge equipment. The differences between the three are:

(a) Sedimentation, is a process of separating a mixture based on the difference in density between solid particles and a liquid where the solid particles sink to the bottom of the container as a precipitate or sediment. In the context of the separation of methyl esters, glycerol and catalysts, the speed of sedimentation depends on the difference in density and particle size. A relatively slow process as it relies on the earth's gravitational force. Less efficient for the separation of components with very similar densities or particles with very small sizes.

(b) Decantation, is a method that allows lighter or clear liquids to be separated from heavier sediment or solid particles by pouring the liquid into another container. This method is often used after sedimentation to further separate the mixture. A process that relies on the user's skill to pour the liquid without disturbing the sediment. It may be less efficient if the sediment is easily disturbed or solid particles easily re-dissolve in the liquid.

(c) Centrifugation is a separation method that uses centrifugal force to separate solid particles from a liquid. A device called a centrifuge is used to speed up the separation process. The separation of a mixture of methyl esters, glycerol, and catalyst can be done more quickly and efficiently using the centrifugation method. Centrifugation requires specialized equipment, the centrifuge. The process is faster because it uses centrifugal force to separate the components.

Various factors need to be considered when choosing a separation method (sedimentation, decantation, or centrifugation) to separate a mixture of methyl esters (biodiesel), glycerol, and catalyst (Ampairojanawong et al. 2023). Some factors to consider include:

(a) Component density. The density difference between the components affects the efficiency of the separation method. If the density difference is large enough, methods such as sedimentation and decantation may be quite effective. However, if the density difference is small, centrifugation may be preferable as it can separate components with very similar densities.

(b) Particle size. Particle size also affects the efficiency of the separation method. If the solid particles are large, sedimentation may be quite effective. However, if the particles are very small or in colloidal form, centrifugation is more suitable as it can separate particles of very small size.

(c) Process time. If the speed of the separation process is a priority, centrifugation may be the best choice as it can separate components quickly compared to sedimentation and decantation. Sedimentation and decantation take longer as they depend on the earth's gravitational force.

(d) Cost and equipment. Sedimentation and decantation do not require specialized equipment and tend to be less expensive. However, centrifugation requires a centrifuge, which can be a higher initial investment. Consider budget and equipment availability when choosing the most suitable separation method.

(e) Separation quality. If the main objective is to obtain a highly efficient separation and high product quality, centrifugation may be preferred as it can produce a better separation. Sedimentation and decantation may be less efficient in some cases, especially if the precipitate is easily disturbed or the solid particles easily re-dissolve in the liquid.

Overall, the specific requirements of the separation process and the above factors are important to consider first when selecting the most suitable method for separating a mixture of methyl esters (biodiesel), glycerol, and catalyst (Ampairojanawong et al. 2023).

Purification of Biodiesel

The biodiesel purification process is an important step in the process of converting microalgae biomass into biodiesel. After the transesterification reaction and product separation, the resulting biodiesel may still contain traces of glycerol, catalyst, water, and other impurities. Purification is necessary to improve the quality of biodiesel and meet the standard specifications set by regulatory agencies. Several common methods are used in the biodiesel purification process:

(1) Water washing. This method involves washing the biodiesel with water to remove residual catalyst, glycerol, and other impurities. The biodiesel is mixed with water, and after some time, the water and biodiesel will separate due to the difference in density. Water containing impurities will settle at the bottom, while clean biodiesel will be on top. Water washing may need to be repeated several times to achieve the desired level of purity.
(2) Acid washing. This method involves washing the biodiesel with a dilute acid solution, such as citric acid or phosphoric acid, to neutralize the residual base catalyst and remove impurities. After washing, the acid solution will separate from the biodiesel, and the impurities will be carried away by the acid solution.
(3) Adsorption. This method uses adsorbents, such as silica gel, magnesium silicate, or activated carbon, to bind and remove impurities from biodiesel. The biodiesel is mixed with the adsorbent, and after some time, the adsorbent containing the impurities will separate from the biodiesel through filtration or centrifugation.
(4) Distillation. This method involves heating the biodiesel under vacuum pressure to vaporize and separate lighter components, such as water and volatile impurities. Vacuum distillation allows the separation of components at lower temperatures, reducing the risk of thermal damage to the biodiesel.
(5) Fractional crystallization. This method uses the principle of different melting points of components in the biodiesel mixture to separate and remove impurities. The biodiesel is cooled to a certain melting point (Samsuri et al. 2017), where unwanted components will crystallize and can be separated through filtration or centrifugation.
(6) Once the refining process is complete, the resulting biodiesel must meet set quality standards, such as viscosity, flash point, moisture content, and free fatty acid (FFA) content. Effective biodiesel purification will lead to improved performance, stability, and lower emissions when used as an alternative fuel.

Biodiesel Quality Standard

Quality standards for microalgae-based biodiesel, similar to biodiesel from other sources, are determined by various quality parameters important for performance, efficiency, and emissions. Some key parameters in biodiesel quality standards include viscosity, flash point, moisture content, and free fatty acid (FFA) content. The following describes each of these parameters:

(1) Biodiesel viscosity, refers to the viscosity of a liquid, which affects the flow of fuel through storage systems, pipes, and fuel injectors (Hoang 2021). Biodiesel

with the right viscosity will flow smoothly through the system and provide efficient combustion. Too high viscosity can cause problems in the fuel system and incomplete combustion. International standards such as ASTM D6751 (USA) and EN 14,214 (Europe) set biodiesel viscosity limits between 1.9 and 6.0 mm^2/s at 40 °C.

(2) The flash point of biodiesel, is the temperature at which fuel vapors can ignite when exposed to a fire source. This parameter is important for the safe storage and transportation of biodiesel. A higher flash point indicates a lower risk of fire. ASTM D6751 and EN 14,214 set the minimum flash point of biodiesel at 93 °C (200 °F) to prevent accidental fires.

(3) Moisture content in biodiesel affects fuel performance, oxidation stability, and fuel system corrosion. High moisture content can lead to the growth of microorganisms in the fuel storage system and increase the risk of corrosion of metal components. Therefore, moisture content should be kept at a low level. ASTM D6751 sets the maximum moisture content at 500 ppm, while EN 14,214 sets a stricter limit at 200 ppm.

(4) The free fatty acid (FFA) content of biodiesel indicates the amount of unesterified fatty acids that can potentially cause corrosion in the fuel system. Biodiesel with a high FFA content may require further treatment, such as esterification, to reduce the FFA content. ASTM D6751 sets the maximum FFA limit at 0.05% (500 ppm) by weight, while EN 14,214 sets a stricter limit at 0.02% (200 ppm) by weight.

The technology and process of converting microalgal biomass into biodiesel is still in the research and development stage. Some of the challenges that must be overcome include high production costs, low lipid extraction efficiency (Saini et al., 2021), and large-scale technologies that have yet to be fully developed. However, with further research and innovation, microalgae could become a sustainable and environmentally friendly feedstock source for biodiesel production.

17.2.3 *The Potential of Microalgae as a Renewable Energy Source*

Microalgae have great potential as a renewable energy feedstock for the following detailed reasons:

(1) Fast growth rate. Microalgae can reproduce quickly; some species can double their biomass within hours to days. This high growth rate enables the production of large amounts of biomass in a short period of time.

(2) High lipid content. Many microalgal species have a high lipid content, which can reach 20–50% of the dry weight of the biomass (Udayan et al. 2022). These lipids can be converted into biodiesel or renewable jet fuel through processes such as transesterification or hydro-processing. Obviously, efficiency can be increased accordingly.

(3) Light and land conversion efficiency. Microalgae have more efficient photosynthetic abilities than terrestrial plants; thus, they can convert solar energy into biomass more efficiently. In addition, microalgae can be grown in open or closed systems, allowing for more efficient land use compared to land crops such as soybean or oil palm.
(4) Use of renewable resources. Microalgae can grow by utilizing renewable resources such as wastewater, CO_2 from industrial emissions, and nutrients that land plants cannot normally utilize. This can reduce pressure on natural resources and help reduce environmental pollution.
(5) Does not compete with food production. Microalgae can be grown on land that is unsuitable for conventional agriculture, such as wastelands, deserts, or rocky areas. As such, microalgae do not compete with food production and can help reduce pressure on existing agricultural land.
(6) Ability to grow in diverse conditions. Microalgae can grow in various environmental conditions, from freshwater to saltwater (Figler et al. 2019) and in a wide temperature range. This makes microalgae more tolerant to climate change and allows them to be grown in various geographical regions.
(7) Use in other energy applications. Besides biodiesel, microalgae also have the potential to be used in the production of bioethanol, biogas, and renewable jet fuel. This diversity can help diversify renewable energy sources and reduce dependence on fossil fuels.

Although microalgae offer great potential as a renewable energy feedstock, there are still several challenges that need to be overcome before this technology can be commercially applied. Some of these challenges include high production costs, large-scale technologies that have yet to be fully developed, and low lipid extraction efficiency.

17.3 Wastewater Management in Biodiesel Production

17.3.1 Sources and Characteristics of Wastewater from Biodiesel Production

Biodiesel production involves several processes that generate wastewater and requires special treatment (Fig. 17.3). The main source of wastewater comes from the biodiesel production process, which includes the following processes:

(1) Oil extraction. The process of extracting oil from feedstocks such as grains, animal fats, and microalgae involves the use of organic solvents such as hexane, ethanol, or acetone. The wastewater generated from these processes generally has a high organic solvent content and may contain feedstock residues.

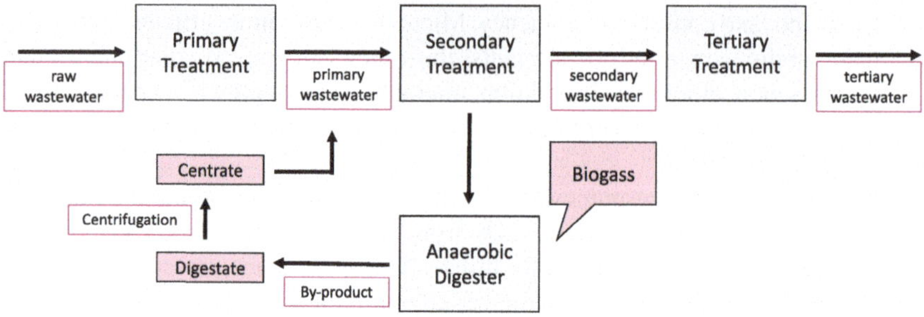

Fig. 17.3 Special treatment of liquid waste

(2) Transesterification. The transesterification process involves the reaction of oil (triglycerides) with alcohol (usually methanol) in the presence of a catalyst (alkali metal or acid). This process produces biodiesel (methyl ester) and glycerol as by-products. The resulting wastewater may contain residual methanol, catalyst, and glycerol, as well as free fatty acids and solids.
(3) Purification of biodiesel. After the transesterification process, biodiesel needs to be separated from glycerol and then purified to remove catalyst, water, and solvent residues. This purification process often involves the use of large amounts of water, so the resulting wastewater may contain residual glycerol, catalyst, and solids (Miyuranga et al. 2022).
(4) Biodiesel washing. Biodiesel washing is an additional step used to remove any remaining contaminants from the final product. This process involves washing the biodiesel with water or water-containing additives. The wastewater produced from this washing is rich in contaminants removed from the biodiesel, such as catalyst residue, glycerol, and solids.

Wastewater from biodiesel production generally has certain characteristics that can indicate the manufacturing process and the materials used. The following are some general characteristics of the wastewater produced (Postaue et al. 2022):

(1) Oil and fat content. Residual oils and fats from raw materials, such as vegetable oils or animal fats, can be found in wastewater. These contents need to be removed before the wastewater can be discharged or treated further.
(2) Glycerol content. Glycerol is a by-product of the transesterification process. This glycerol needs to be separated from the biodiesel and can be found in wastewater.
(3) pH. Wastewater from biodiesel production is usually alkaline due to the use of alkaline catalysts such as sodium hydroxide (NaOH) or potassium hydroxide (KOH). Therefore, the pH of the wastewater will tend to be high and needs to be neutralized before discharge or treatment.
(4) Metal content. Metals such as sodium, potassium, and calcium, which are components of catalysts, can be found in wastewater. The concentration of these metals needs to be reduced to meet discharge or treatment requirements.

(5) Methanol content. During the transesterification process, methanol is used as a reagent. Unreacted methanol that is not separated from the final product can be found in wastewater.
(6) Suspended solids content. The biodiesel production process can generate suspended solids, such as oil or grease particles, which can be found in wastewater.
(7) Organic matter content. Wastewater from biodiesel production may contain unwanted organic substances, such as free fatty acids (FFA) and other organic compounds.

Given these characteristics, wastewater from biodiesel production requires special treatment before it can be discharged or reused. Processes such as coagulation, flocculation, flotation, filtration, and biodegradation can be used to treat wastewater to meet applicable environmental standards (Amenorfenyo et al. 2019); Rinanti et al. 2017).

17.3.2 Conventional Wastewater Treatment Methods and Technologies

Conventional wastewater treatment in the context of biodiesel production involves a range of methods and technologies designed to reduce pollutants and contaminants generated during the production process. Wastewater from biodiesel production can contain chemicals, oils and fats, and suspended solids that need to be removed before being discharged into the environment. The following are some conventional wastewater treatment methods and technologies commonly used in biodiesel production:

(1) Pre-treatment, is the initial stage in wastewater treatment that aims to reduce the content of coarse solids, oil, and grease before the wastewater is treated further in the wastewater treatment system. Pre-treatment helps protect wastewater treatment equipment from damage and clogging that may be caused by coarse particles and non-degradable materials. Some common pre-treatment methods used in microalgae feedstock biodiesel production are:

    (a) Skimming system, used to remove oil, grease, and solids floating on the surface of wastewater. Skimming can be done mechanically using a surface skimmer or by gravity in a skimming tank, where the wastewater is allowed to settle, and the floating oil and grease layer will naturally separate from the wastewater. The separated oil and grease layer can then be removed and further treated.
    (b) Grinding, is a mechanical process used to reduce the particle size of coarse solids in wastewater, thus preventing clogging of wastewater treatment equipment. Grinding can be done using a specially designed grinder to

crush coarse into smaller particles. This process also facilitates more efficient separation and treatment of solids at subsequent stages of wastewater treatment.

(c) Screening, aims to separate coarse solids and particles suspended in wastewater. Screening can be done using various filter types, such as mechanical, sand, or sieve filters. The filter will capture coarse solids and suspended particles, while the filtered water will proceed to the next stage of treatment.

In the context of microalgae-based biodiesel production, wastewater pre-treatment is essential to ensure the efficiency and sustainability of the wastewater treatment process (Krishnamoorthy et al. 2022). Pre-treatment helps reduce the load of coarse solids, oil, and grease on wastewater treatment equipment, and reduces the risk of damage and clogging that may affect the overall performance of the wastewater treatment system.

(2) Primary treatment, in the conventional wastewater treatment commonly used in microalgae-based biodiesel production, aims to remove most of the suspended solids and dissolved organic material. This process involves the physical separation of solids and organic matter from wastewater through sedimentation or coagulation-flocculation. Some of the methods used in primary treatment are:

(a) Sedimentation, is the process of separating suspended solids from wastewater by gravity. In this stage, wastewater is passed through a sedimentation tank (also known as primary clarifier) where the velocity of the water flow is controlled to allow suspended solids to settle to the bottom of the tank. The settled solids, also known as sludge, can then be removed from the bottom of the tank using sludge collection equipment. Water that has passed the sedimentation stage will contain a lower amount of suspended solids and is ready to proceed to the secondary treatment stage.
(b) Coagulation-Flocculation, is a chemical process used to separate suspended and dissolved particles in wastewater that are difficult to remove through sedimentation involves the addition of coagulant chemicals, such as aluminum sulfate or organic polymers, which will react with suspended particles and form larger solid clumps called flocs. Flocculation is the process of combining those flocs into larger aggregates that are easier to settle or filter.
(c) After the coagulation-flocculation process, the wastewater is then passed through a sedimentation tank or filtration unit to separate the flocs from the treated water. As a result, the water that has gone through the primary treatment stage will have lower suspended solids and organic material content and is ready to proceed to the secondary treatment stage.

In the context of microalgae-based biodiesel production, primary treatment of wastewater helps reduce the suspended solids and organic matter load before further treatment. This increases the efficiency of wastewater treatment and helps produce cleaner water that can be reused in the biodiesel production process or discharged into the environment with minimal environmental impact.

(3) In conventional wastewater treatment commonly used in microalgae feedstock biodiesel production, secondary treatment aims to remove most of the remaining organic matter, including suspended and dissolved solids, after the primary treatment stage (Amenorfenyo et al. 2019). This process involves the utilization of microorganisms to decompose organic matter and convert it into gas, water, and microorganism cells. Some of the methods used in secondary treatment are:

   (a) Activated Sludge Process. The sludge activation process involves the use of aerobic microorganisms that grow in a mixture of sludge and wastewater. In this process, wastewater that has gone through primary treatment is mixed with activated sludge rich in aerobic microorganisms, and then flowed into an aeration tank. In the aeration tank, air or oxygen is pumped into the mixture to maintain sufficient oxygen levels for microorganisms to decompose the organic matter. After this process, the mixture is flowed into a secondary sedimentation tank, where the activated sludge is separated from the treated water. A portion of the activated sludge will be returned to the aeration tank to maintain the microorganism population, while the rest will be further treated or disposed of.
   (b) Suspended Growth Systems. Suspended sludge treatment systems involve the use of microorganisms that grow in the form of aggregates or flocs suspended in wastewater. Examples of suspended sludge treatment systems include intermittent batch treatment reactors (SBRs) and installed total oxidation systems (TFOTs). These processes are similar to sludge activation but use different operating systems and tank configurations.
   (c) Attached Sludge Treatment Systems (Attached Growth Systems). Attached sludge treatment systems involve the use of microorganisms growing on solid surfaces, such as filter media, stone, or plastic, in wastewater treatment systems. Examples of attached sludge treatment systems include trickling filters and attached biofilm reactors. In this process, wastewater that has gone through primary treatment is passed through filter media inhabited by aerobic microorganisms. Organic matter in the wastewater is decomposed by the microorganisms, which is then precipitated or filtered from the treated water.

   After secondary treatment, the wastewater will have a much lower organic matter content and is ready to proceed to the tertiary treatment stage (if required) to remove certain nutrients or pollutants before discharge or reuse.

(4) Tertiary treatment is an advanced stage that aims to remove certain pollutants or nutrients that are not removed efficiently in the primary and secondary treatment stages (Geremia et al. 2021). Tertiary treatment includes various physical, chemical, and biological methods that help produce water of very high quality and safe for discharge into the environment or reuse. Some of the methods used in tertiary treatment include:

(a) Filtration, is a physical process used to remove fine suspended particles remaining in wastewater after the secondary treatment stage. Filtration can be done using various types of filter media, such as sand, ceramic, or activated carbon. Filtration can also help remove pathogens, heavy metals, and certain organic chemicals from wastewater.
(b) Adsorption, is the process of binding specific pollutant or nutrient molecules to the surface of a solid material, such as activated carbon or zeolite. Adsorption is used to remove pollutants such as heavy metals, dissolved organic compounds, and nutrients such as phosphate and ammonia from wastewater.
(c) Denitrification, is a biological process used to remove nitrogen from wastewater in the form of nitrate. The process involves the use of denitrifying microorganisms that decompose nitrate into nitrogen gas, which is then released into the atmosphere. Denitrification is usually carried out in anaerobic or anoxic treatment systems.
(d) Phosphorus removal is usually achieved through a chemical process that involves the addition of coagulants such as aluminum sulfate, ferric chloride, or organic polymers. These coagulants react with phosphates in wastewater to form solid compounds that can be precipitated or filtered.

In the context of microalgae-based biodiesel production (Fig. 17.4), tertiary treatment of wastewater helps produce water that is very clean and safe for reuse in the biodiesel production process or be discharged into the environment with minimal environmental impact. Tertiary treatment also ensures that the wastewater meets water quality standards set by environmental regulations and local or national water management authorities.

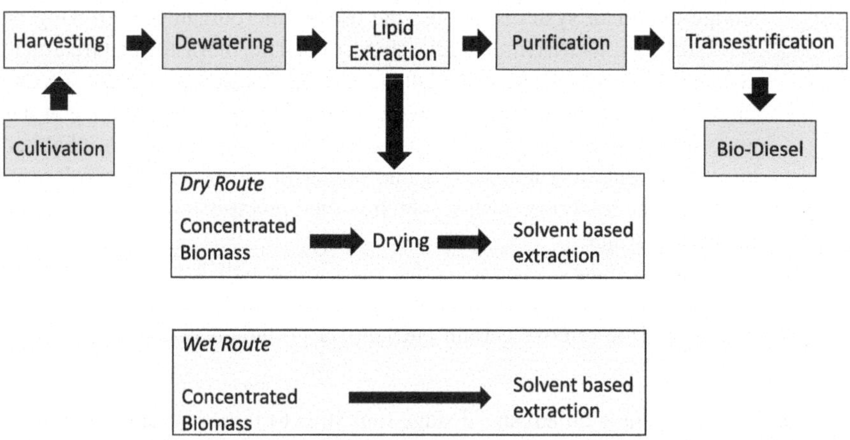

Fig. 17.4 Conversion of microalgae to biodiesel

(5) Disinfection, is the final stage in wastewater treatment that aims to remove or reduce pathogens such as bacteria, viruses, and protozoa. The disinfection process includes the use of chemicals such as chlorine, ozone, or hydrogen peroxide, or physical methods such as ultraviolet light or heating (Ghernaout and Elboughdiri 2020)

(a) Chlorination is the process of using chlorine or chlorine compounds, such as sodium hypochlorite or chlorine gas, to kill or inhibit the growth of pathogens, such as bacteria, viruses, and protozoa. Chlorine is very effective in destroying the cell walls of microorganisms and stopping the enzyme activity of microorganisms. However, there are some disadvantages in using chlorine in wastewater treatment, namely:

- After chlorination, residual chlorine still present in the treated water can become an environmental problem. Residual chlorine can be toxic to aquatic life and affect aquatic ecosystems.
- Chlorine can react with organic matter present in wastewater to form toxic trihalomethane (THM) compounds and potentially harmful halogen organic compounds. These THMs and halogen organic compounds can have negative effects on water quality and the environment as well as human health.
- The use of chlorine in wastewater treatment incorporates chemicals in the system, which may be undesirable in some applications or environments.

To overcome some shortcomings in the use of chlorine, several disinfection alternatives, such as ozonation, ultraviolet disinfection, and heating, can be used. However, chlorination remains a common and economical disinfection method in wastewater treatment, including in microalgae feedstock biodiesel production. In these applications, it is important to monitor and control residual chlorine levels in the treated water and take measures to reduce the formation of harmful compounds such as THMs.

(b) Ozonization, is a disinfection process that uses ozone (O3), a highly reactive form of oxygen, to kill pathogens such as bacteria, viruses, and protozoa. Ozone is very effective in destroying the cell walls of microorganisms and stopping the enzyme activity of pathogenic microorganisms (Xue et al. 2023). Some of the advantages of using ozone in wastewater treatment are:

- It leaves no chemical residue in the treated water and naturally breaks down into oxygen (O_2). This means that no additional contaminants are generated during the disinfection process, which is a major advantage compared to chlorine-based disinfection methods.
- Ozone is highly effective in destroying various types of pathogens, including bacteria, viruses, protozoa, and spores. In addition, ozone

is also effective in oxidizing water-soluble organic compounds, such as pesticides and phenols, and removing unwanted odors and tastes.
- The disinfection process takes place relatively quickly. The contact time required to achieve an effective level of disinfection is usually shorter than other disinfection methods such as chlorination or ultraviolet disinfection.

However, there are some disadvantages in the use of ozone in wastewater treatment, namely:

- Ozonization requires specialized equipment and higher operating costs compared to chlorine-based disinfection methods. This includes the cost of generating ozone and maintaining the necessary equipment.
- While ozone leaves no chemical residue in treated water, it still involves the use of chemicals in the disinfection process, which may be undesirable in some applications or environments.

Although ozonation has some drawbacks, its use in wastewater treatment, including in microalgae feedstock biodiesel production, can provide high disinfection rates with lower environmental impact compared to chlorine-based disinfection methods.

(c) Hydrogen peroxide (H_2O_2) is a chemical that can be used in conventional wastewater treatment commonly used in microalgae feedstock biodiesel production. Hydrogen peroxide has strong oxidative properties and can be used as a disinfecting agent to remove or reduce the number of pathogens in wastewater that has passed through primary, secondary, and tertiary treatment stages. The use of hydrogen peroxide helps produce water that is safe to be reused in the biodiesel production process or discharged into the environment with minimal environmental impact. Some advantages of using hydrogen peroxide in wastewater treatment:

- Once reacted, hydrogen peroxide leaves no harmful chemical residue as it breaks down into water and oxygen, both of which are environmentally safe and do not contaminate the treated water.
- Hydrogen peroxide shows high effectiveness in destroying various types of pathogens, including bacteria, viruses, and protozoa. In addition, hydrogen peroxide is also effective in oxidizing water-soluble organic compounds, such as pesticides and phenols, and removing unwanted odors and tastes.
- Hydrogen peroxide is highly biodegradable, so it breaks down into water and oxygen and poses no long-term environmental risks.

However, there are some disadvantages to the use of hydrogen peroxide in wastewater treatment, namely:

- Hydrogen peroxide is a strong oxidizer and can cause corrosion of wastewater treatment equipment. Therefore, it is necessary to use

corrosion-resistant equipment and compatible materials to reduce the risk of damage to the wastewater treatment system.
- Hydrogen peroxide tends to be unstable and decompose easily, especially when exposed to sunlight or high humidity. Therefore, proper storage and protection from sunlight and moisture are necessary to maintain the stability of hydrogen peroxide during the wastewater treatment process.
- Although the cost of hydrogen peroxide itself may be lower compared to ozone, for example, operating costs associated with equipment maintenance can be higher, so it is necessary to consider the cost of storage, and the use of hydrogen peroxide in large-scale applications.

Although the use of hydrogen peroxide in wastewater treatment has some disadvantages, it is an effective and environmentally friendly alternative for wastewater disinfection, including in microalgae feedstock biodiesel production (Rafa et al. 2021). In these applications, it is important to consider factors such as operating costs, hydrogen peroxide stability, and protection of equipment against corrosion.

(d) The use of ultraviolet (UV) light is an effective physical method for conventional wastewater disinfection in microalgae-based biodiesel production. UV light can inactivate pathogens, including bacteria, viruses, and protozoa, without leaving chemical residues or producing harmful byproducts. The process involves lighting wastewater that has passed through primary, secondary, and tertiary treatment stages with UV light, which damages the DNA and RNA of microorganisms and stops their reproduction. Some of the advantages of using UV light in wastewater treatment are:

- The UV light method does not involve chemicals and leaves no chemical residue in the treated water. This means that no additional contaminants are generated during the disinfection process, which is a major advantage compared to chlorine-based or ozonation disinfection methods.
- UV light is highly effective in inactivating various types of pathogens, including chlorine-resistant bacteria. In addition, UV light can also reduce the amount of water-soluble organic compounds, such as pesticides and phenols, and remove unwanted odors and tastes.
- Disinfection with UV light is a relatively fast process. The contact time required to achieve effective disinfection levels is usually shorter than other disinfection methods such as chlorination or ozonation.

However, there are some disadvantages in using UV light in wastewater treatment, namely:

- The effectiveness of UV light is highly dependent on the transparency of the wastewater. If the water has a high suspended solids content, the UV light may not be able to penetrate deeply enough to effectively inactivate pathogens.

- While the cost of UV light equipment may be lower than chemical-based disinfection methods, the maintenance and energy costs of running a UV light system are important considerations, especially in large-scale applications.
- UV lamps have a limited lifespan and require periodic replacement. This can increase the operation and maintenance costs of UV light disinfection systems.

Although the use of UV light in wastewater treatment has some disadvantages, it is an effective and environmentally friendly alternative for wastewater disinfection, including in microalgae feedstock biodiesel production. In these applications, it is important to consider factors such as water transparency, operating costs, and maintenance of the UV light system. In the context of biodiesel production, wastewater treatment processes need to be optimized to ensure compliance with environmental regulations and to reduce the negative impact of wastewater discharge on ecosystems and water sources.

17.3.3 Advantages and Challenges of Using Microalgae in Wastewater Treatment

Microalgae are producers or photosynthetic microorganisms that can be used to treat wastewater while producing lipid-rich biomass, which can then be converted into biodiesel (Udayan et al. 2022). Several advantages of using microalgae in wastewater treatment are associated with biodiesel production technology:

(1) Wastewater treatment and biodiesel production in one process. Wastewater treatment and biodiesel production in a single process using microalgae offers several advantages that make it an attractive and sustainable solution. Here are some of the advantages of using microalgae in wastewater treatment:

    (a) Wastewater treatment and biomass production. Microalgae can clean wastewater by removing excess nutrients (such as nitrogen and phosphorus) and organic contaminants through their metabolic processes. During this process, microalgae grow and produce lipid-rich biomass, which can be used as a feedstock to produce biodiesel.
    (b) Reduction of greenhouse gas emissions. Microalgae absorb CO_2 from the environment during photosynthesis, thereby reducing greenhouse gas emissions. In addition, biodiesel produced from microalgae produces lower emissions compared to fossil fuels, thus contributing to climate change mitigation.
    (c) Nutrient recycling. In an integrated system, nutrients present in wastewater can be used as a nutrient source for microalgae, thereby reducing the need for additional nutrient inputs and aiding in sustainable nutrient management.

(d) More efficient land use. Microalgae can grow on land that is unproductive or unsuitable for conventional agriculture, such as swampy or contaminated areas. This enables more efficient land use and reduces competition between land for food and renewable energy production.
(e) Product diversification potential. In addition to biodiesel, microalgae can also produce a variety of high-value products, such as proteins, pigments, and omega-3 fatty acids. This opens opportunities for product diversification and additional revenue, which can improve the competitiveness and sustainability of the microalgae-based biodiesel industry.
(f) Lower water consumption. Microalgae can grow in wastewater, brackish water, or saltwater, thus reducing pressure on increasingly scarce freshwater sources.
(g) Increased sustainability. By integrating wastewater treatment and biodiesel production using microalgae, the industry can create a more sustainable system, reducing environmental impact and making efficient use of existing resources.

(2) Energy efficiency. Utilization of microalgae can improve energy efficiency in wastewater treatment because (Plöhn et al. 2021; Permatasari and Rinanti 2018):

(a) Microalgae can use solar energy for growth and biomass production through photosynthesis. This process has the potential to produce more energy per unit area than conventional bioenergy crops, such as oil palm or soybean.
(b) Efficient photosynthesis. Microalgae have efficient photosynthetic capabilities, which allow them to absorb solar energy and convert it into chemical energy in the form of biomass. During this process, microalgae absorb nutrients and contaminants from wastewater, resulting in cleaner water and biomass that can be used to produce renewable energy such as biodiesel.
(c) Biodiesel produced from microalgae is a renewable energy source that is more environmentally friendly than fossil fuels. By replacing fossil fuels with biodiesel, greenhouse gas emissions and fossil energy consumption can be reduced, thereby improving overall energy efficiency.
(d) An integrated system that combines wastewater treatment and biodiesel production using microalgae can create a more efficient energy cycle. For example, methane gas generated during the anaerobic process of wastewater treatment can be collected and used as an energy source for biodiesel production or other energy needs.
(e) Microalgae can grow in a variety of environmental conditions, including wastewater, brackish water, or saltwater. By using less water resources and utilizing the nutrients present in wastewater, microalgae can reduce the energy consumption required for growth and maintenance.
(f) The use of microalgae in wastewater treatment can reduce energy requirements in conventional treatment processes, such as aeration or intensive chemical processes. This can reduce operating costs and energy required to treat wastewater.

(3) Carbon dioxide sequestration. Carbon dioxide (CO_2) sequestration by microalgae is one of the main advantages of using microalgae in wastewater treatment (Molazadeh et al. 2019), especially in the context of climate change mitigation and reduction of greenhouse gas emissions. Here are some important points related to CO_2 absorption by microalgae in wastewater treatment:

   (a) Photosynthesis and CO_2 sequestration. Microalgae, like plants, utilize photosynthesis to produce energy and biomass. During the photosynthesis process, microalgae absorb CO_2 from the environment and convert it into chemical energy and biomass. This helps reduce the concentration of CO_2 in the atmosphere and contributes to the reduction of greenhouse gas emissions.
   (b) Climate change mitigation. Reduction of CO_2 emissions through sequestration by microalgae can help in climate change mitigation. Biodiesel produced from microalgae, as a substitute for fossil fuels, also produces lower emissions, thus contributing to efforts to reduce the impact of climate change.
   (c) Increased productivity. The absorption of CO_2 by microalgae can increase the productivity and growth of microalgae, which in turn increases the efficiency of wastewater treatment and biomass production. The biomass produced can be used to produce biodiesel and other high-value products.
   (d) Utilization of industrial emissions. Microalgae can be grown in integrated systems that use CO_2 emissions from industries, such as power plants or cement factories, to enhance microalgae growth and CO_2 sequestration efficiency. This can reduce the environmental impact of industrial emissions and create a more sustainable system.
   (e) Technology development potential. 2Research and development of microalgae technology for CO_2 sequestration could lead to a more efficient and economical system to reduce greenhouse gas emissions and create a more sustainable energy system.

The utilization of microalgae in wastewater treatment related to biodiesel production technology faces serious challenges related to cost, scalability, and microalgae species selection (Zewdie and Ali 2020).

(1) Cost. One of the main challenges in using microalgae for wastewater treatment and biodiesel production is the high cost of cultivation, harvesting, and lipid extraction. This can include costs for construction and maintenance of photobioreactor systems or open culture ponds, as well as lipid separation and conversion technologies.
(2) Scalability. Developing efficient and economical processes for large-scale cultivation of microalgae remains a challenge. Factors such as optimal environmental conditions, resistance to contamination, and high productivity need to be addressed to achieve commercial scale production.

(3) Microalgae species selection. Selecting the right species from thousands of microalgae species for wastewater treatment and biodiesel production is challenging. The ideal species should be resistant to wastewater conditions, have a fast growth rate, and produce high lipid content for efficient biodiesel conversion.

These constraints must be overcome to utilize the full potential of microalgae in wastewater treatment and biodiesel production technologies. Continued research and development will help optimize processes and reduce associated costs, thus making the use of microalgae in these technologies more economically and environmentally viable.

17.4 Integration of Microalgae in a Circular Economy System

17.4.1 Absorption of Nutrients by Microalgae from Wastewater

Microalgae are photosynthetic organisms that can absorb nutrients from wastewater, such as nitrogen and phosphorus, as well as carbon dioxide from the atmosphere or other emission sources (Al-Jabri et al. 2021). The integration of microalgae in circular economy systems can help reduce the environmental impact of wastewater and energy production and create value-added products. Here are some aspects of nutrient uptake by microalgae from wastewater related to the integration of microalgae in circular economic systems:

(1) Microalgae can be used in wastewater treatment systems to remove nutrients such as nitrogen (N) and phosphorus (P) that can cause eutrophication if released into the environment. The process involves the uptake of nutrients by microalgae, which are then used for growth and biomass production. In the context of a circular economy, this reduces the need for synthetic chemicals or more intensive wastewater treatment processes.
(2) The biomass generated from nutrient uptake by microalgae can be used for various purposes, such as the production of biodiesel, animal feed, fertilizer, or high-value chemicals. In a circular economy system, this biomass can be used to replace fossil raw materials or non-renewable natural resources.
(3) During photosynthesis, microalgae absorb carbon dioxide (CO_2) from the atmosphere or other emission sources, such as industrial exhaust gases. This CO_2 sequestration can help reduce greenhouse gas emissions and contribute to climate change mitigation. In a circular economy system, it creates added value by reducing the environmental impact of CO_2 emissions.

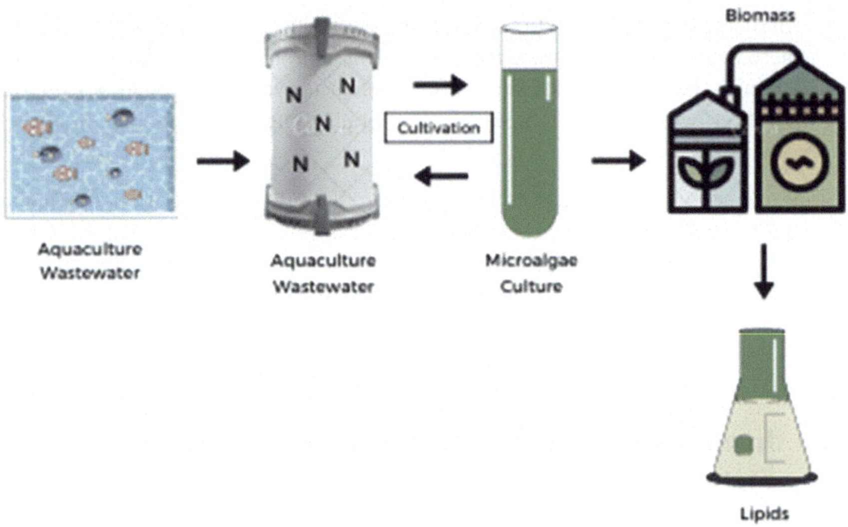

Fig. 17.5 Wastewater treatment process with microalgae

(4) After wastewater treatment and biomass production, the nutrients contained in the microalgal biomass can be recovered and used as fertilizer or other value-added products (Geremia et al. 2021). In a circular economy system, this increases the efficiency of resource use and reduces the need for synthetic fertilizers or minerals extracted from the environment.

(5) The use of microalgae in wastewater treatment and biomass production, as well as nutrient recovery, can help create a circular economy system through more efficient, sustainable, and environmentally friendly production and consumption processes. The process is illustrated in (Fig. 17.5).

17.4.2 Microalgal Biomass Growth and Lipid Extraction for Biodiesel Production

Microalgae offer potential as an alternative source for biodiesel production through biomass growth and lipid extraction. In the context of a circular economic system, the integration of microalgae in biodiesel production can improve resource use efficiency and reduce the environmental impact of energy production.

(1) Microalgae biomass growth. Microalgae grow quickly and can produce large amounts of biomass in a short period of time. They require nutrients, such as nitrogen and phosphorus, and CO_2 for growth. In a circular economy system, microalgae can be grown on wastewater or CO_2 emission sources (e.g., industrial exhaust gas), helping to reduce pollutants and reduce greenhouse gas emissions.

(2) Lipid production. Some species of microalgae produce large amounts of lipids that can be converted into biodiesel through the transesterification process. These lipids are usually stored in the cells in the form of triacylglycerol (TAG). In a circular economy, lipid production by microalgae can replace conventional feedstock sources for biodiesel, such as palm oil or soybean, which are often associated with land use change and greenhouse gas emissions.
(3) Lipid extraction. After the growth of microalgal biomass, lipids must be extracted from the cells for the biodiesel production process. Commonly used lipid extraction methods include solvent extraction, mechanical extraction, or micro-wave or ultrasonic-based technologies. In a circular economy system, the selection of efficient and environmentally friendly extraction methods will be important to reduce the energy consumption and environmental impact of the biodiesel production process.
(4) Conversion of Lipids into Biodiesel. The extracted lipids are then converted into biodiesel through a transesterification process, which involves a chemical reaction between lipids and alcohol (usually methanol) in the presence of a catalyst. The resulting biodiesel can be used as an alternative fuel that is more environmentally friendly than fossil fuels.

During the biodiesel production process, by-products such as glycerol and cellular biomass remaining after lipid extraction can be further processed to produce value-added products, such as fertilizer, animal feed, or industrial chemicals. In a circular economy system, such by-products or Value-Added Products can improve resource use efficiency and reduce waste disposal (Saadaoui et al. 2021; Rinanti and Purwadi 2018).

17.4.3 Utilization of Remaining Microalgae Biomass for Value-Added Products (Animal Feed, Fertilizer, Biogas)

Utilization of remaining microalgae biomass for value-added products such as animal feed, fertilizer, and biogas is an important concept in the integration of microalgae in the circular economy system. Microalgae are microscopic organisms that can grow rapidly and thus can be raw material to produce biofuels, nutrients and food (Fig. 17.6), animal feed, and chemicals.

However, after the extraction of the main products such as oil or protein, there is still a large amount of microalgal biomass remaining. Utilizing this leftover biomass for value-added products can help integrate microalgae in the circular economy system, as follows:

(1) Animal Feed. The remaining microalgal biomass contains protein, fatty acids and other nutrients that are beneficial for animal feed. The protein contained in microalgal biomass can be a sustainable alternative source for animal feed,

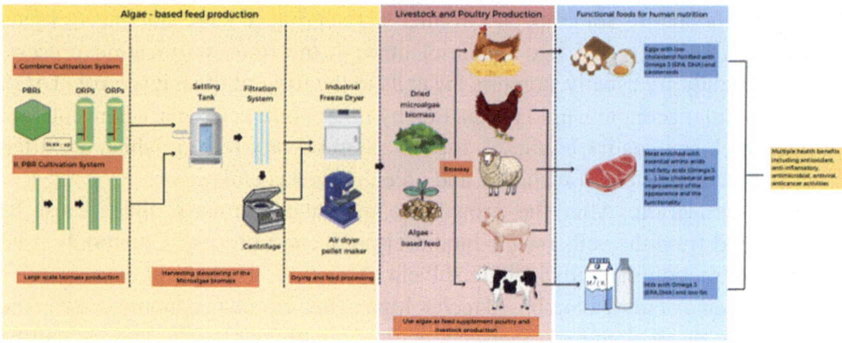

Fig. 17.6 Various use of microalgae

replacing conventional protein sources such as soybean and fish. In addition, the use of microalgal biomass in animal feed can improve the quality and health of livestock and reduce greenhouse gas emissions from the livestock industry (Saadaoui et al. 2021).

(2) Fertilizer. The remaining microalgae biomass can also be used as an organic fertilizer that is rich in nutrients. Microalgae contains nitrogen, phosphorus, and potassium, which are key nutrients required by plants. Using microalgae biomass as fertilizer can improve soil fertility, support healthy plant growth, and reduce dependence on chemical fertilizers that are not environmentally friendly.

(3) Biogas. The remaining microalgal biomass can be utilized to produce biogas through anaerobic processes, such as anaerobic digestion or gasification. The resulting biogas, which mainly consists of methane and carbon dioxide, can be used as a renewable energy source to generate electricity and heat. Biogas production from microalgal biomass can help reduce greenhouse gas emissions and replace fossil fuels in the energy system.

Integration of microalgae in circular economy systems through utilization of residual biomass for value-added products helps to create a more efficient and sustainable production cycle (Sarma et al. 2021) as in (Fig. 17.7). By reducing waste, maximizing resource utilization, and producing value-added products, microalgae can make important contributions to environmental sustainability and the global economy.

17.4.4 *Treatment and Recycling of Water After Treatment with Microalgae*

Water treatment and recycling after wastewater has been treated by microalgae is important and needs to be implemented in the context of microalgae integration in circular economy systems for the following reasons:

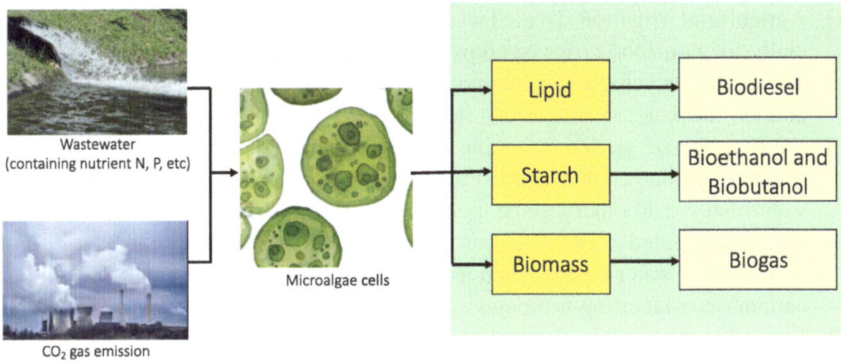

Fig. 17.7 Biogas production using microalgae

(1) Conserving water resources. Water is a limited natural resource that is essential for life. Utilizing microalgae to treat wastewater and then recycling the treated water can save water resources and reduce the consumption of clean water. This will improve water use efficiency and reduce pressure on limited water resources.
(2) Reducing environmental pollution. Microalgae can be utilized to treat wastewater by removing excess nutrients such as nitrogen and phosphorus, as well as heavy metals and other contaminants. This process helps reduce environmental pollution due to the release of untreated or inadequately treated wastewater into the environment. Recycled water after treatment with microalgae will reduce negative impacts on aquatic ecosystems and water quality.
(3) Supporting microalgae growth. Wastewater that has been treated by utilizing microalgae is rich in nutrients and thus becomes recycled water that can be used to support further microalgae growth. As such, the recycled water can help to maintain the availability of nutrients necessary for the sustainable growth of microalgae, while reducing the need for external nutrient inputs.
(4) Minimizing water treatment costs. Water that has been treated with microalgae or recycled water can reduce wastewater disposal costs and the costs associated with using new water sources.
(5) Creating a sustainable cycle. The integration of microalgae in circular economic systems involves closing the loop in the production and consumption process. Water treatment and recycling after wastewater treatment with microalgae is an important part of creating a sustainable process and reducing negative impacts on the environment.

Recycling treated wastewater has great potential in reducing clean water consumption, reducing the environmental impact of wastewater discharge, and utilizing previously unused resources. Some potential utilizations of treated wastewater recycling (Bauer and Wagner 2022) are:

(1) Agricultural irrigation. Treated wastewater can be used for crop irrigation, especially for non-food crops or crops that are not directly consumed by humans. This use of recycled water can reduce pressure on clean water sources and help address the water shortages that many regions are facing.
(2) Landscape and garden irrigation. Treated wastewater can also be used for watering gardens, public parks, and other green spaces. This can save clean water that would otherwise be used for these purposes.
(3) Industry. Treated wastewater can be used in industrial processes that do not require high water quality, such as in cooling systems, equipment cleaning, or certain manufacturing processes.
(4) Groundwater recharge. Treated wastewater can be used to recharge groundwater through infiltration or direct injection into aquifers. This can help maintain groundwater balance and reduce the risk of land subsidence or aquifer depletion.
(5) Use in construction. Treated wastewater can be used as a concrete admixture or to control dust at construction sites and reduce the need for clean water (Mojapelo et al. 2021).
(6) Toilet flushing: Treated wastewater can be used to flush toilets, saving drinking water and reducing the use of clean water.
(7) In some cases, recycled water can also be used as drinking water, especially when the quality of the treated water meets safety standards and has undergone a rigorous advanced treatment and disinfection process.

The recycling potential of treated wastewater is highly dependent on the quality of the treated water and the local policies and regulations in place. To maximize the utilization of treated wastewater, there needs to be a collaboration between government, industry, and communities in developing wastewater treatment technologies, regulating the use of recycled water, and raising public awareness about the importance of sustainable use of water resources. Microalgae-based treatment and recycling of treated water is an implementation of the circular economy system to achieve more efficient and sustainable water management, reduce environmental pollution, and optimize the utilization of natural resources.

17.5 Design and Operation of Microalgae-Based Systems

The design and operation of microalgae-based systems need to be considered in wastewater treatment, especially concerning the circular economy system of biodiesel production, for the following reasons (Srimongkol et al. 2022):

(1) Resource efficiency. Proper design and operation of microalgae-based systems ensure the efficient use of resources, such as water, nutrients, and energy. In a circular economy system, the goal is to reduce resource consumption and wastage through recycling and reuse of materials.

(2) Operation and investment costs. Good design and operation of microalgae-based systems reduce operating and investment costs in wastewater treatment and biodiesel production. In a circular economy system, minimizing costs throughout the product's life cycle is crucial to remain competitive in the market.
(3) Product quality. Effective design and operation of microalgae-based systems yield high-quality products, such as lipid-rich microalgae biomass for biodiesel production. The circular economy system emphasizes the importance of creating high-quality, durable products and services that can enhance value and competitiveness in the market.
(4) Environmental impact reduction. The eco-friendly design and operation of microalgae-based systems reduce the environmental impacts of wastewater treatment and biodiesel production, such as greenhouse gas emissions and eutrophication. Circular economy systems aim to reduce environmental impacts and promote sustainability.
(5) Process synergy. Integrated design and operation of microalgae-based systems enable synergy between wastewater treatment and biodiesel production processes, such as nutrient and energy recovery from wastewater to support microalgae growth. Circular economy systems encourage collaboration and integration between various processes and sectors to create added value and reduce resource wastage.

Further discussions on the design and operation of the microalgae-based systems in wastewater treatment related to the circular economy concept in biodiesel production will be as follows.

17.5.1 Closed (Photobioreactor) and Open (Pond) Systems for Microalgae Cultivation

Microalgae cultivation systems can be classified into two main categories: open systems (ponds) and closed systems (photobioreactors). Both systems have their advantages and disadvantages, which will influence the decision regarding the most suitable system for a specific application (Wimmerova et al. 2023). Open Systems (i.e., ponds) for microalgae cultivation involve the use of ponds that are open to the atmosphere. These ponds can be natural or artificial and generally have a shallower depth. Commonly used open systems in microalgae cultivation include:

(1) Raceway Pool. This oval-shaped pond has a rotating water flow that follows the pond's shape. The raceway pond is equipped with stirrers to ensure circulation and even nutrient distribution while keeping the microalgae exposed to sunlight.
(2) Multipurpose Ponds. These ponds are often used for dual purposes, such as microalgae cultivation and wastewater treatment. They are usually larger and deeper than raceway ponds.

The advantages of an open system include:

(1) Lower initial costs for construction and equipment.
(2) Easier scalability.
(3) Relatively simple maintenance.

The disadvantages of an open system include:

(1) Contamination from foreign organisms (e.g., bacteria, fungi, protozoa, or other algae) may disrupt microalgae growth.
(2) Variability in environmental conditions (e.g., temperature, humidity, wind speed) can affect microalgae growth and quality.
(3) Higher water usage and potential water loss through evaporation.
(4) Lower light capture efficiency.

Closed systems for microalgae cultivation utilize photobioreactors (PBRs), which are controlled and isolated systems from the surrounding environment. PBRs are typically made of transparent materials such as glass or plastic and are designed to optimize microalgae growth conditions. Some common types of photobioreactors used in microalgae cultivation include:

(1) Tube Photobioreactor (Tubular PBR). This type of photobioreactor consists of a series of interconnected transparent tubes where microalgae are cultured. The tubes can be arranged horizontally, vertically, or spirally. Although this system provides excellent lighting and efficient environmental control, it comes with higher investment and maintenance costs than open systems.
(2) Flat Plate Photobioreactor (Flate Plate PBR). Flat plate photobioreactors have transparent plates on which microalgae are cultured. These plates can be placed horizontally or vertically and are generally easier to assemble and maintain than tubular PBR. However, they may be less efficient in terms of light capture and cooling compared to tubular PBR.
(3) Column Photobioreactor (Column PBR). This photobioreactor involves a transparent column filled with microalgae cultures. The column can be a bubble column, where air or gas is passed through the microalgae culture to create circulation and even lighting, or an airlift column, which uses the density difference between the bubble-rich and the bubble-poor zones to create circulation. Column photobioreactors have a simple design and relatively lower investment costs than tube or flat plate photobioreactors, but the light capture efficiency may be lower.
(4) Panel Photobioreactor (Panel PBR). These photobioreactors consist of transparent open or enclosed panels where microalgae are cultured. They can be placed vertically, horizontally, or at an angle and are generally easier to assemble and maintain than tube photobioreactors. The light capture efficiency in panel PBRs is better than flat plate PBRs but may be less efficient than tubular PBRs.

The advantages of a closed system (photobioreactor) for microalgae cultivation (Al-Dailami et al. 2022):

(1) Better environmental control. Photobioreactors allow better control over environmental conditions, such as temperature, pH, nutrient concentration, and light intensity. This enables the setting of optimal conditions for the specific growth of microalgae, ultimately enhancing productivity and product quality.
(2) Lower risk of contamination. The risk of contamination from foreign organisms (e.g., bacteria, fungi, protozoa, or other algae) in closed systems is much lower compared to open systems. This is very important, especially when cultured microalgae are used for applications with high hygiene standards, such as food products, cosmetics, or pharmaceuticals.
(3) Higher light capture efficiency. Photobioreactors are designed to maximize light capture by microalgae. This usually results in higher light capture efficiency than open systems, ultimately improving microalgae productivity.
(4) More Efficient Water Use. Closed systems usually utilize water more efficiently than open systems. This is because closed systems reduce water loss through evaporation and maintain more stable water conditions, thus reducing the need for frequent water replacement.

Disadvantages of the closed system (photobioreactor) for microalgae cultivation are (Al-Dailami et al. 2022):

(1) Higher investment and operational costs. Photobioreactors usually have higher initial investment costs than open systems due to the more advanced equipment and technology. In addition, the operational and maintenance costs for closed systems also tend to be higher, mainly because of greater energy consumption to control environmental conditions and keep the system running efficiently.
(2) Difficulties in Scalability. Closed systems, such as photobioreactors, may be more challenging to scale than open systems. This is because increasing the size and complexity of closed systems often results in higher costs and greater technical challenges.
(3) More complicated maintenance. Photobioreactors generally require more complex maintenance than open systems related to equipment cleaning, sanitization, monitoring, and setup.

17.5.2 Operational Parameters and Factors Affecting the Efficiency of Wastewater Treatment

In the context of a circular economy for biodiesel production using microalgae, wastewater treatment plays a crucial role in supporting microalgae growth and reducing environmental impacts. Several operational parameters and factors influence the efficiency of wastewater treatment, which must be carefully considered to achieve optimal results.

(1) Nutrient concentration. Wastewater contains rich nutrients such as nitrogen and phosphorus, which are essential for microalgae growth. Maintaining proper

nutrient concentration in wastewater affects the growth rate of microalgae, leading to improved nutrient absorption efficiency and contaminant reduction.
(2) Temperature. Temperature is an important factor affecting the growth rate and metabolism of microalgae. Each microalgae species has an optimal temperature range for efficient growth. Therefore, regulating wastewater temperature within the appropriate range enhances the efficiency of wastewater treatment.
(3) pH. The pH level influences nutrient availability and the biochemical activity of microalgae. Most microalgae thrive in neutral to slightly alkaline pH (Yu et al. 2022). Controlling wastewater pH within the optimal range for the microalgae species enhances wastewater treatment efficiency.
(4) Aeration and agitation. Adequate mixing and oxygen supply are essential for microalgae growth and wastewater treatment. Proper aeration and agitation ensure even nutrient distribution, CO_2 absorption, oxygenation, and preventing microalgae settling. Optimizing these parameters leads to high wastewater treatment efficiency.
(5) Light intensity. Microalgae require light for photosynthesis, a fundamental process for growth and metabolism. Providing sufficient light intensity and appropriate lighting duration significantly impacts the growth rate of microalgae and wastewater treatment efficiency.
(6) Microalgae Species Selection. Some microalgae species are more effective in treating wastewater than others. Selecting suitable microalgae species that thrive under wastewater conditions and produce high lipids for biodiesel production will increase the overall system efficiency (Srimongkol et al. 2022). The lipid content in microalgae is strongly influenced by growth conditions such as nutrient deficiency, light intensity, and environmental stress. Under specific stress conditions, such as nutrient deficiencies (especially nitrogen) or increased light intensity, the lipid content in microalgae can significantly increase. Some examples of microalgae with high lipids can be seen in (Table 17.1) (Srimongkol et al. 2022).

In a circular economy, the efficiency of wastewater treatment using microalgae is essential to reduce environmental impact, optimize resource utilization, and establish a sustainable process. It is imperative to optimize operational parameters and consider factors affecting wastewater treatment efficiency to effectively support the circular economy concept (Yue et al. 2023).

17.5.3 Strategy for Optimization of Biodiesel Production and By-Products

Strategies for optimizing the production of biodiesel and by-products by utilizing microalgae in a circular economy are as follows:

(1) Selecting microalgae species with high lipid content as potential raw materials for biodiesel production. Additionally, select microalgae capable of thriving in

Table 17.1 Types of microalgae with lipid content

Microalgae	Lipid content (%) per g dry weight (%)
Chlorella vulgaris	4–58
Chlorella protothecoides	15–55
Scenedesmus obliquus	11–45
Nannochloropsis gaditana	20–50
Neochloris oleoabundans	20–54
Botryococcus braunii	25–75
Dunaliella tertiolecta	15–30
Isochrysis galbana	15–35
Phaeodactylum tricornutum	20–40
Haematococcus pluvialis	25–50
Tetraselmis suecica	10–25
Spirulina platensis	4–16

diverse environmental conditions. Some species may also yield valuable by-products, such as pigments, proteins, and polysaccharides.

(2) Optimizing growth conditions such as temperature, pH, light intensity, and nutrient concentration promotes proper microalgal growth and lipid accumulation (Udayan et al. 2022). Certain stress conditions (e.g., nutrient or light deficiency) can stimulate higher lipid production and desired by-products.

(3) Utilizing nutrient and CO_2-rich wastewater from industrial emissions as a source of nutrients and carbon for microalgae growth. This practice reduces input costs, minimizes environmental impacts, and aligns with the circular economy concept.

(4) Developing energy-efficient technologies to separate microalgal cells from media and extract lipids. Methods such as centrifugation, flotation, and flocculation can be used for cell separation, while solvent extraction, supercritical extraction, and electrocoagulation can be used for lipid extraction.

(5) Optimizing the biodiesel conversion processes, such as transesterification, using efficient and environmentally friendly catalysts to enhance the overall sustainability of the process.

(6) Identify and develop high-value applications for by-products generated during biodiesel production, such as proteins, pigments, and polysaccharides. These valuable by-products can be used in various industries, including food, feed, pharmaceutical, cosmetic, and agricultural, supporting the circular economy concept.

(7) Ensure that the biodiesel and by-product production processes using microalgae can be economically upscaled and integrated with existing industrial processes. This scalability allows for broader implementation of the technology on a larger scale.

By optimizing these strategies, biodiesel and by-product production using microalgae can become integral to the circular economy. This approach fosters efficient resource utilization, reduces environmental impacts, and generates economic value.

17.6 Case Studies and Best Practices

17.6.1 Examples of Circular Economy Applications in Biodiesel Production Using Microalgae

Some examples of circular economy applications in biodiesel production using microalgae that have been carried out in European countries (Casanova et al. 2023), in several projects are shown in (Table 17.2).

Some examples of circular economy applications in biodiesel production using microalgae have been conducted in Asian countries (Casanova et al. 2023):

(1) Sinopec-Algae Biofuel Project (2011–2016) in China. Sinopec, a Chinese energy company, in collaboration with Tsinghua University and the Environmental Research Institute is developing biodiesel production technology from microalgae. The project incorporates wastewater for microalgae cultivation and integrates various processes to produce biodiesel and value-added products such as pigments and polysaccharides.

(2) Bharat-OUCRU Project (2014–2019) in India. Bharat Renewable Energy Limited and Oil and Natural Gas Corporation Limited (OUCRU) collaborate on producing biodiesel from microalgae. The project utilizes industrial and agricultural waste as nutrients for microalgae cultivation and CO_2 from industrial emissions as a carbon source. In this project, the solid waste generated is also utilized as organic fertilizer or animal feed.

(3) Korea Institute of Energy Research (KIER) Algae Biofuel Project in South Korea. KIER has been researching biodiesel production from microalgae, employing wastewater for microalgae cultivation and CO_2 from industrial emissions as a carbon source. The project also integrates biodiesel production with creating value-added products such as proteins, pigments, and omega-3 fatty acids.

(4) Project of Algaetech International Sdn. Bhd. In Malaysia. Algaetech International, a Malaysian company focused on microalgae technology, developed a biodiesel production project from microalgae. In addition, the company is also developing other value-added products, such as pigments, proteins, and nutraceuticals from microalgae. The waste generated in this project is also used as organic fertilizer or animal feed, promoting the circular economy concept.

(5) Taipower-Green Energy Development Corporation (GEDC) project in Taiwan. Taipower collaborates with GEDC to produce biodiesel from microalgae. The

Table 17.2 Examples of circular economy application projects in biodiesel production in Europe

Project title	Duration	Funding	Partners	Focus
BIOFAT	2011–2016	European Union	10 partners from 6 European countries (Italy, Portugal, France, Germany, Slovakia, and Belgium)	• Biodiesel and value-added products from microalgae • Utilization of industrial and agricultural wastewater as a source of nutrients for microalgae cultivation and CO_2 from industrial emissions as a carbon source
SPLASH	2012–2016	European Union	20 partners from 9 European countries (Netherlands, France, Germany, Belgium, Spain, England, Denmark, Austria, and Italy)	• Polymers and high-value chemicals from microalgae and biomass • Combine biodiesel production processes and value-added products to create a more efficient and sustainable circular economy
PUFAChain	2013–2017	European Union	11 partners from 8 European countries (Norway, Germany, France, Belgium, Spain, Italy, Austria, and Greece)	• Polyunsaturated fatty acids (PUFA) from microalgae and biomass • Utilization of waste and renewable resources to produce high-value products and promote a circular economy
ALG-AD	2018–2021	Interreg Program Europe North-West	Partners from England, Belgium, and France	• High-value products (animal protein and pigments) from microalgae and anaerobic digestate waste • Reducing greenhouse gas emissions, saving resources, and creating value-added products from waste from the products they produce

project utilizes wastewater and CO_2 from power plant emissions as inputs for microalgae cultivation. Besides biodiesel, the project also yields other value-added products such as proteins, pigments, and omega-3 fatty acids.

Some examples of circular economy applications in biodiesel production using microalgae have been conducted in American countries (Casanova et al. 2023):

(1) Sapphire Energy project. Sapphire Energy, a US biotechnology company, focused on developing renewable fuels from microalgae. The project employs

wastewater and CO_2 from industrial emissions as inputs for microalgae cultivation. It also produces other value-added products such as proteins, pigments, and omega-3 fatty acids. The waste generated is used as organic fertilizer or animal feed, supporting the circular economy concept.

(2) Solazyme Project. Solazyme, another US biotechnology company, develops technologies to produce biodiesel and high-value chemical products from microalgae. The project utilizes industrial waste and renewable resources to create renewable fuels and value-added products such as proteins, pigments, and nutraceuticals.
(3) Algenol Biofuels project. Algenol Biofuels, a US company, focused on producing renewable fuels from microalgae and developing technologies to produce ethanol, biodiesel, and high-value chemical products. The project incorporates wastewater and CO_2 from industrial emissions as inputs for microalgae cultivation and integrates multiple processes to produce fuels and other value-added products.
(4) Pond Technologies project. Pond Technologies, a Canadian company, develops technologies to produce renewable fuels and high-value products from microalgae. The project collaborates with industry to produce biodiesel and other products from microalgae, utilizing wastewater and CO_2 from industrial emissions as inputs for microalgae cultivation and integrating various processes to produce fuels and other value-added products.

Although research on microalgae and biodiesel production in Africa may be less advanced than in some other regions (Casanova et al. 2023), here are several efforts have been made in African countries to implement circular economy principles in biodiesel production using microalgae:

(1) The SANBI (South African National Biodiversity Institute) project in South Africa. This project involves research and development of biodiesel production from microalgae grown in local waters. It utilizes wastewater and CO_2 from industrial emissions as inputs for microalgae cultivation. Besides biodiesel, the project yields other value-added products like proteins, pigments, and omega-3 fatty acids. The generated waste is also used as organic fertilizer or animal feed, supporting the circular economy concept.
(2) AlgaePARC-Morocco project. AlgaePARC-Morocco is a collaborative project between Wageningen University & Research in the Netherlands and Université Mohammed VI Polytechnique in Morocco. This project focuses on research and technology development to produce biodiesel and other value-added products from local microalgae. It involves using wastewater and CO_2 from industrial emissions as inputs for microalgae cultivation, and integrating various processes to produce fuels and other value-added products.
(3) The University of Ilorin Project. This research initiative by the University of Ilorin, Nigeria, focuses on biodiesel production from local microalgae. It utilizes wastewater and CO_2 from industrial emissions as inputs for microalgae cultivation, integrating various processes to produce fuel and other value-added products such as proteins, pigments, and omega-3 fatty acids.

(4) Jomo Kenyatta University of Agriculture and Technology (JKUAT) project. JKUAT in Kenya has been conducting research on biodiesel production from microalgae. The project involves utilizing wastewater and CO_2 from industrial emissions as inputs for microalgae cultivation and integrating various processes to produce fuel and other value-added products such as proteins, pigments, and omega-3 fatty acids.

17.6.2 Lessons Learned from Case Studies and Best Practices

Based on the lessons learned from the case studies and best practices mentioned above, here are some key points regarding wastewater utilization using microalgae to implement the circular economy for biodiesel production (Geremia et al. 2021):

(1) Utilizing wastewater and CO_2 from industrial emissions as inputs for microalgae cultivation can reduce the environmental impact of water pollution and greenhouse gas emissions while utilizing existing resources, thus reducing pressure on natural resources.
(2) Selecting the appropriate microalgae species is crucial to improve wastewater treatment efficiency and biodiesel production. Fast-growing species with high lipid content are preferred in this context.
(3) Optimizing cultivation conditions such as temperature, light intensity, and nutrient concentration is essential to enhance microalgae productivity and wastewater treatment efficiency.
(4) Efficient and effective cultivation technologies, such as closed photobioreactors or open pond systems, can reduce operational costs and increase biodiesel production efficiency.
(5) Producing value-added products other than biodiesel, such as proteins, pigments, and omega-3 fatty acids, can enhance the overall economic value of the system.
(6) Integrated waste management within biodiesel production systems, such as using solid waste as organic fertilizer or animal feed, can reduce environmental impacts and create additional economic value.
(7) Collaboration between industry, academia, and government in developing and implementing technologies for biodiesel production using microalgae is essential to achieve a sustainable circular economy.

Applying the principle of circular economy in biodiesel production using microalgae is expected to create a more efficient, sustainable, and environmentally friendly system, ultimately providing economic, social, and environmental benefits.

17.6.3 Potential for Scale-Up and Commercialization of Microalgae-Based Systems

The potential for scaling up and commercializing microalgae-based systems in the context of a circular economy for biodiesel production is enormous. Here are some aspects to consider in achieving commercial scale (Bošnjaković and Sinaga 2020):

(1) Further research is needed to optimize microalgae cultivation technologies and methods and improve the conversion efficiency of microalgae biomass into biodiesel. Research on new microalgae species with higher potential for biodiesel production is also important.
(2) Developing innovative technologies, such as more efficient and environmentally friendly photobioreactors and open pond systems, will increase microalgae productivity and reduce operational costs.
(3) Reducing the production cost of microalgae is essential to achieve price competitiveness with fossil fuels. Cost reduction can be achieved through process optimization, improved energy efficiency, and utilization of renewable resources and waste.
(4) Strategic partnerships, including collaboration between industry, government, and research institutions, are expected to facilitate technology transfer, capacity building, and the necessary investments for scaling up and commercializing microalgae-based systems.
(5) Policy support and incentives from the government, such as favorable taxes, subsidies, and regulations promoting renewable fuel use, will influence the adoption and growth of microalgae-based systems in biodiesel production.
(6) Increasing public awareness and understanding of the economic, environmental, and social benefits of microalgae-based systems will help drive demand and adoption of this technology.
(7) Developing markets for value-added products generated from microalgae-based systems, such as proteins, pigments, and nutraceuticals, will enhance the economic competitiveness of this system.

By considering these aspects, microalgae-based systems have significant potential to be integrated into the circular economy and achieve commercial scale in biodiesel production. However, achieving this will require continuous research and development efforts and support from various stakeholders to address existing challenges and create an enabling environment for the growth of this sector.

17.7 Policy, Regulation, and Support

17.7.1 Policies and Regulations that Support the Development of Circular Economy in Biodiesel Production

Policies and regulations that support the development of circular economy in biodiesel production by wastewater management using microalgae are essential to encourage the adoption and growth of this technology (Zewdie and Ali 2020). Policies and regulations that can support the development of microalgae-based systems in the context of circular economy are as follows:

(1) The government can set strict CO_2 emission and wastewater quality standards, encouraging industries to seek environmentally friendly wastewater treatment technologies, such as microalgae-based systems.
(2) The government can provide tax incentives and subsidies to companies investing in microalgae-based technologies or using biodiesel produced from microalgae can stimulate the growth of this industry.
(3) The government can establish targets and mandates for using renewable fuels, such as biodiesel, in the transportation and industrial sectors. This will increase the demand for microalgae-based biodiesel and drive investment in this sector.
(4) The government can support research and development in microalgae technology by providing funding, facilities, and partnerships between research institutions and industries.
(5) The government can develop waste management policies that support the utilization of wastewater as a nutrient source for microalgae cultivation. This may include providing incentives for companies implementing this technology.
(6) The government could develop quality and certification standards for microalgae-based biodiesel, ensuring the product meets certain quality requirements and supporting its adoption in the market.
(7) The government can support education and public awareness programs about the economic, environmental, and social benefits of microalgae-based technologies and circular economy.

By implementing policies and regulations that support the development of circular economy in biodiesel production by wastewater management using microalgae, the government can create an enabling environment for the growth of this industry. A combination of incentives, mandates, and research and development support will drive innovation and investment in microalgae-based technologies, ultimately contributing to the goals of a circular and sustainable economy.

17.7.2 Government Incentives and Support for Research and Development of Green Technologies

Here are some examples of government incentives and support for research and development of green technologies in the context of the circular economy in biodiesel production by managing wastewater using microalgae (Culaba et al. 2020):

(1) The government can provide funding and grants for research and development in microalgae-based technologies, including microalgae cultivation, biomass conversion into biodiesel, and wastewater treatment technologies.
(2) The government can provide tax incentives or rebates for companies investing in microalgae-based technologies or adopting circular economy practices in biodiesel production.
(3) The government can partner with the private sector to develop and commercialize microalgae-based technologies, such as photobioreactors and open pond systems.
(4) The government can support training and capacity-building programs to improve knowledge and skills in microalgae-based technologies and wastewater management.
(5) The government can provide infrastructure support, such as research facilities and laboratories, to facilitate the development of microalgae-based technologies.
(6) The government can establish targets and mandates for using renewable fuels, such as biodiesel, to drive investment and research in microalgae-based technologies.
(7) The government can support international collaboration in the research and development of microalgae-based technologies through knowledge exchange, joint research, and exchange programs.
(8) The government can recognize and reward companies and researchers contributing to developing microalgae-based technologies and circular economy in biodiesel production.

By providing appropriate incentives and support, the government can encourage innovation and investment in microalgae-based technologies, enabling the development of a sustainable circular economy in biodiesel production by wastewater management using microalgae.

17.7.3 The Role of Industry and Society in Adopting the Circular Economy Concept

The role of industry and society in adopting the circular economy concept is crucial in the context of biodiesel production using microalgae and wastewater management. Here are some strategies that industry and society can follow:

(1) Collaboration between industry and research institutions to develop and optimize microalgae-based technologies and identify the best microalgae species for biodiesel production and wastewater treatment.
(2) Increase public awareness and understanding of the economic, environmental, and social benefits of microalgae-based technologies and circular economy concepts. This may include information campaigns, seminars, and training programs.
(3) Industries should invest in innovation and development of microalgae-based technology to improve efficiency and reduce biodiesel production costs.
(4) Industries should adopt sustainable business practices by reducing resource consumption, efficient waste management, and utilizing renewable energy.
(5) Industry can adopt corporate policies that support the use of microalgae-based technologies and the circular economy, such as prioritizing sourcing raw materials and energy from sustainable sources.
(6) Society can support the market for microalgae-based biodiesel products and value-added by-products, such as animal feed and fertilizer, by choosing these products and encouraging companies to adopt microalgae-based technologies.
(7) Communities can participate in local initiatives that support the development of microalgae-based technologies and the circular economy, such as collaborative projects involving universities, government, and industries.
(8) Communities and industries can collectively urge the government to adopt policies that support the development of microalgae-based technologies and the circular economy in biodiesel production.

By following these strategies, industry and society can play a significant role in adopting the circular economy concept in biodiesel production by wastewater management using microalgae. Collaboration among various stakeholders will drive innovation and investment in this technology and help create an enabling environment for the sector's growth.

17.8 Research, Innovation, and Education

17.8.1 Recent Research in Circular Economy and Wastewater Management Using Microalgae

Recent research in circular economy and wastewater management using microalgae (Vaz et al. 2023):

(1) Advanced selection and genetic engineering of microalgae for improved nutrient uptake efficiency and lipid production.
(2) Discovery and development of more efficient and energy-saving photobioreactor or pond systems.
(3) Research on new and efficient methods for lipid extraction and conversion into biodiesel.

17.8.2 Technology and Process Innovations Supporting Circular Economy in Biodiesel Production

Technological and process innovations supporting circular economy in biodiesel production (Wynn and Jones 2022; Rinanti and Purwadi 2019):

(1) Real-time monitoring and control technology to optimize microalgae cultivation and wastewater treatment. Here are some reasons why this technology is important and how it supports the circular economy:

   (a) Efficient wastewater treatment. Real-time monitoring and control enable rapid adjustments to environmental conditions that may affect the performance of microalgae-based wastewater treatment systems. By optimizing operational conditions, this technology helps improve wastewater treatment efficiency and reduces environmental impacts from pollutants.
   (b) Efficient resource utilization. Real-time monitoring and control technology enable adjustment of operational parameters, such as temperature, light, and nutrient concentration, to enhance microalgae productivity and ensure efficient resource utilization. This contributes to circular economy principles of reducing resource consumption and minimizing waste.
   (c) Increased biodiesel yield. Real-time monitoring and control technology aid in optimizing microalgae cultivation conditions to maximize lipid production, thereby increasing biodiesel yield. This supports the creation of efficient and environmentally friendly renewable energy, in line with the circular economy goals.
   (d) Reduce production costs. By monitoring and controlling the process in real time, companies can reduce the operational and maintenance costs of microalgae cultivation and wastewater treatment systems. This cost reduction will enhance the economic competitiveness of microalgae-based biodiesel and encourage wider-scale adoption of this technology.
   (e) System integration. Real-time monitoring and control technologies enable better integration between biodiesel production, wastewater treatment, and utilization of value-added by-products (such as animal feed and fertilizer). This integration supports the circular economy concept by creating a complementary system and maximizing the value of each component.
   (f) Adaptability and flexibility. The ability to monitor and control processes in real-time enables operators to respond quickly to external changes or potential issues that may arise. This ensures the system remains flexible and adaptive, a crucial quality supporting a dynamic and sustainable circular economy.

(2) Integration of Internet of Things (IoT) and Artificial Intelligence (AI) technologies in microalgae-based biodiesel production systems. The integration of the Internet of Things (IoT) and Artificial Intelligence (AI) technologies in microalgae-based biodiesel production systems is an important technological

and process innovation supporting the circular economy in biodiesel production (Dębowski et al. 2023). Here are some reasons why these technologies are important and how they support the circular economy:

(a) Real-time monitoring and control. IoT enables real-time data collection from various sensors and devices placed throughout the microalgae cultivation and wastewater treatment system. AI can analyze this data and make timely decisions regarding adjustments to operational parameters, such as temperature, light, and nutrient concentration. This enhances system efficiency and reduces resource consumption.
(b) Prediction and planning. AI can analyze historical data and identify patterns that may affect system performance. By predicting changes in environmental conditions or production needs, AI can assist operators in planning and anticipating changes in system operations, which supports the circular economy by increasing efficiency and reducing waste.
(c) Process optimization. AI can be used to optimize processes in microalgae-based biodiesel production systems, such as cultivation, lipid extraction, and conversion of lipids to biodiesel. By optimizing these processes, AI helps increase biodiesel yields and reduce production costs, which supports the circular economy.
(d) Energy management. IoT and AI can be used to manage energy consumption in microalgae-based biodiesel production systems, such as optimizing lighting and energy use in photobioreactor systems. Efficient energy management reduces resource consumption and greenhouse gas emissions, aligning with circular economy principles.
(e) Utilization of by-product. IoT and AI can be used to monitor and optimize the utilization of by-products generated from microalgae-based biodiesel production systems, such as animal feed and fertilizer. By optimizing the utilization of these by-products, these technologies support the circular economy by creating added value from waste and reducing environmental impact.
(f) Scalability and flexibility. The integration of IoT and AI allows microalgae-based biodiesel production systems to be more easily scaled and adapted to changing needs. The ability to adapt to market-changing conditions and demands is key to creating a sustainable and dynamic system within the context of a circular economy.

(3) Development of recycling and nutrient recovery technologies from microalgae biomass residues. The development of recycling and nutrient recovery technologies from microalgae biomass residues is an important technological and process innovation to support the circular economy in biodiesel production for the following reasons:

(a) Waste reduction. Recycling and nutrient recovery from microalgae biomass residues reduce the waste generated during biodiesel production. Waste reduction is one of the key principles of the circular economy, aiming to minimize the environmental impact of industrial processes.

(b) Resource efficiency. The nutrient recovery process allows previously considered waste resources (such as phosphorus and nitrogen) to be reused in the microalgae production processes. This enhances resource utilization efficiency and reduces raw material consumption, aligning with circular economy goals.

(c) Production costs reduction. Companies can reduce raw material and waste disposal costs by implementing nutrient recycling and recovery technologies. These cost reductions can improve the economic competitiveness of microalgae-based biodiesel and encourage wider adoption of this technology.

(d) Support for sustainable agriculture. Nutrients recovered from microalgae biomass residues can be used as fertilizer in agriculture. Using these fertilizers can help reduce the environmental impact of conventional agriculture and support the transition to more sustainable agricultural systems.

(e) Support for related industries. Nutrient recycling and recovery technologies can create value-added by-products (such as fertilizer and animal feed) that can be used in related industries. This creates additional business opportunities and strengthens relationships between complementary industries, an essential aspect of the circular economy.

(f) Research and innovation. The development of nutrient recycling and recovery technologies promotes research and innovation in biotechnology and environmental technologies. These innovations can help address the environmental and economic challenges faced by the biodiesel industry and support the transition to a more circular and sustainable economic system.

Overall, the development of recycling and nutrient recovery technologies from microalgae biomass residues can contribute to creating a more efficient, environmentally friendly, and economically viable biodiesel production system, supporting the principles of the circular economy (de Carvalho et al. 2022).

Education and training programs in circular economy and wastewater management using microalgae:

(1) Curriculum and educational programs emphasizing circular economy concepts, biodiesel production, and wastewater management.
(2) Training and certification for professionals working in the biodiesel industry and wastewater management.
(3) Development of educational materials to increase public awareness of circular economy and the benefits of microalgae utilization.

Collaboration between academia, industry, and government:

(1) Collaborative research initiatives involving universities, research centers, and industries focusing on circular economy and wastewater management using microalgae.

(2) Government support for research and development of innovations in circular economy and biodiesel production.
(3) Public–private partnerships for the development and implementation of green technologies.

Case studies and examples of educational applications:

(1) Successful and innovative education program that teaches circular economy concepts and wastewater management using microalgae.
(2) Implementation of educational technologies, such as virtual and augmented reality to enrich the learning experience on circular economy and biodiesel production.
(3) The influence of education and public awareness on the adoption of environmentally friendly technologies.

References

Akubude VC, Nwaigwe KN, Dintwa E (2019) Production of biodiesel from microalgae via nanocatalyzed transesterification process: a review. Mater Sci Energy Technol 2(2):216–225. https://doi.org/10.1016/j.mset.2018.12.006

Al-Dailami A, Koji I, Ahmad I, Goto M (2022) Potential of photobioreactors (PBRs) in cultivation of microalgae. J Adv Res Appl Sci Eng Technol 27(1):32–44. https://doi.org/10.37934/araset.27.1.3244

Al-Jabri H, Das P, Khan S, Thaher M, Abdulquadir M (2021) Treatment of wastewaters by microalgae and the potential applications of the produced biomass—a review. Water (Switzerland) 13(1). MDPI AG. https://doi.org/10.3390/w13010027

Amenorfenyo DK, Huang X, Zhang Y, Zeng Q, Zhang N, Ren J, Huang Q (2019) Microalgae brewery wastewater treatment: potentials, benefits and the challenges. Int J Environ Res Public Health 16(11). https://doi.org/10.3390/ijerph16111910

Ampairojanawong R, Boripun A, Ruankon S, Suwanasri T, Cheenkachorn K, Kangsadan T (2023) Separation process of biodiesel-product mixture from crude glycerol and other contaminants using electrically driven separation technique with ac high voltage. Electrochem 4(1):123–144. https://doi.org/10.3390/electrochem4010011

Anisah PM, Suwandi, Agustian E (2019) Effect of transesterification on the result of waste cooking oil conversion to biodiesel. J Phys: Conf Ser 1170(1). https://doi.org/10.1088/1742-6596/1170/1/012067

Barros MV, Salvador R, do Prado GF, de Francisco AC, Piekarski CM (2021) Circular economy as a driver to sustainable businesses. In: Cleaner environmental systems, vol 2. Elsevier Ltd. https://doi.org/10.1016/j.cesys.2020.100006

Bauer S, Wagner M (2022) Possibilities and challenges of wastewater reuse—planning aspects and realized examples. Water (Switzerland) 14(10). https://doi.org/10.3390/w14101619

Bheda B, Shinde M, Ghadge R, Thorat B (2019) Drying of algae by various drying methods. https://doi.org/10.4995/ids2018.2018.7761

Bitwell C, Indra SS, Luke C, Kakoma MK (2023) A review of modern and conventional extraction techniques and their applications for extracting phytochemicals from plants. In: Scientific African, vol 19. Elsevier B.V. https://doi.org/10.1016/j.sciaf.2023.e01585

Bošnjaković M, Sinaga N (2020) The perspective of large-scale production of algae biodiesel. Appl Sci (Switz) 10(22):1–26. MDPI AG. https://doi.org/10.3390/app10228181

Brahma S, Nath B, Basumatary B, Das B, Saikia P, Patir K, Basumatary S (2022) Biodiesel production from mixed oils: a sustainable approach towards industrial biofuel production. Chem Eng J Adv 10. Elsevier B.V. https://doi.org/10.1016/j.ceja.2022.100284

Branyikova I, Prochazkova G, Potocar T, Jezkova Z, Branyik T (2018) Harvesting of microalgae by flocculation. Fermentation 4(4). MDPI AG. https://doi.org/10.3390/fermentation4040093

Casanova LM, Mendes LBB, de Corrêa TS, da Silva RB, Joao RR, Macrae A, Vermelho AB (2023) Development of microalgae biodiesel: current status and perspectives. Microorganisms 11(1). MDPI. https://doi.org/10.3390/microorganisms11010034

Chowdhury MS, Rahman KS, Chowdhury T, Nuthammachot N, Techato K, Akhtaruzzaman M, Tiong SK, Sopian K, Amin N (2020) An overview of solar photovoltaic panels' end-of-life material recycling. Energy Strategy Rev 27. Elsevier Ltd. https://doi.org/10.1016/j.esr.2019.100431

Christwardana M, Hadiyanto H, Pratiwi WZ (2022) Optimization of light intensity and color temperature in the cultivation of chlorella vulgaris culture using the surface response method. J Bioresources Environ Sci 1(2):33–41. https://doi.org/10.14710/jbes.2022.14410

Culaba AB, Ubando AT, Ching PML, Chen WH, Chang JS (2020) Biofuel from microalgae: sustainable pathways. Sustain (switz) 12(19):1–19. https://doi.org/10.3390/su12198009

de Carvalho JC, Molina-Aulestia DT, Martinez-Burgos WJ, Karp SG, Manzoki MC, Medeiros ABP, Rodrigues C, Scapini T, de Vandenberghe LPS, Vieira S, Woiciechowski AL, Soccol VT, Soccol CR (2022) Agro-industrial wastewaters for algal biomass production, bio-based products, and biofuels in a circular bioeconomy. Fermentation 8(12). MDPI. https://doi.org/10.3390/fermentation8120728

Dębowski M, Świca I, Kazimierowicz J, Zieliński M (2023) Large scale microalgae biofuel technology—development perspectives in light of the barriers and limitations. Energies 16(1). https://doi.org/10.3390/en16010081

Delgado-Plaza E, Peralta-Jaramillo J, Quilambaqui M, Gonzalez O, Reinoso-Tigre J, Arevalo A, Arancibia M, Paucar M, Velázquez-Martí B (2019) Thermal evaluation of a hybrid dryer with solar and geothermal energy for agroindustry application. Appl Sci (Switz) 9(19). https://doi.org/10.3390/app9194079

Ebhodaghe SO, Imanah OE, Ndibe H (2022) Biofuels from microalgae biomass: a review of conversion processes and procedures. Arab J Chem 15(2). Elsevier B.V. https://doi.org/10.1016/j.arabjc.2021.103591

Figler A, B-Béres V, Dobronoki D, Márton K, Nagy SA, Bácsi I (2019) Salt tolerance and desalination abilities of nine common green microalgae isolates. Water (Switz) 11(12). https://doi.org/10.3390/w11122527

Geremia E, Ripa M, Catone CM, Ulgiati S (2021) A review about microalgae wastewater treatment for bioremediation and biomass production—a new challenge for europe. Environments—MDPI 8(12). MDPI. https://doi.org/10.3390/environments8120136

Ghernaout D, Elboughdiri N (2020) On the other side of viruses in the background of water disinfection. Oalib 07(05):1–29. https://doi.org/10.4236/oalib.1106374

Handayani D, Amalia R, Endy Yulianto M, Hartati I, Murni (2018) Determination of influential factors during enzymatic extraction of ginger oil using immobile isolated cow rumen enzymes. Int J Technol 9(3):455–463. https://doi.org/10.14716/ijtech.v9i3.489

Hoang AT (2021) Prediction of the density and viscosity of biodiesel and the influence of biodiesel properties on a diesel engine fuel supply system. J Mar Eng Technol 20(5):299–311. https://doi.org/10.1080/20464177.2018.1532734

Islam MT, Iyer-Raniga U, Trewick S (2022) Recycling perspectives of circular business models: a review. Recycling 7(5). MDPI. https://doi.org/10.3390/recycling7050079

Kapadia P, Newell AS, Cunningham J, Roberts MR, Hardy JG (2022) Extraction of high-value chemicals from plants for technical and medical applications. Int J Mol Sci 23(18). MDPI. https://doi.org/10.3390/ijms231810334

Krishnamoorthy A, Rodriguez C, Durrant A (2022) Sustainable approaches to microalgal pretreatment techniques for biodiesel production: a review. Sustain (Switz) 14(16). MDPI. https://doi.org/10.3390/su14169953

Kucmanová A, Gerulová K (2019) Microalgae harvesting: a review. Res Pap Fac Mater Sci Technol Slovak Univ Technol 27(44):129–143. https://doi.org/10.2478/rput-2019-0014

Kumar L, Anand R, Shah MP, Bharadvaja N (2022) Microalgae biodiesel: a sustainable source of energy, unit operations, technological challenges, and solutions. J Hazard Mater Adv 8:100145. https://doi.org/10.1016/j.hazadv.2022.100145

Li F, Liu Z, Ni Z, Wang H (2019) Effect of biodiesel components on its lubrication performance. J Market Res 8(5):3681–3687. https://doi.org/10.1016/j.jmrt.2019.06.011

Manousakis NM, Karagiannopoulos PS, Tsekouras GJ, Kanellos FD (2023) Integration of renewable energy and electric vehicles in power systems: a review. Processes 11(5). MDPI. https://doi.org/10.3390/pr11051544

Martins PL, Reis A, Duarte LC, Carvalheiro F (2022) Effective fractionation of microalgae biomass as an initial step for its utilization as a bioenergy feedstock. Energy Convers Manage: X 16. https://doi.org/10.1016/j.ecmx.2022.100317

Massoud M, Vega G, Subburaj A, Partheepan J (2023) Review on recycling energy resources and sustainability. Heliyon 9(4). Elsevier Ltd. https://doi.org/10.1016/j.heliyon.2023.e15107

Miyuranga KAV, Arachchige USPR, Jayasinghe RA, Samarakoon G (2022) Purification of residual glycerol from biodiesel production as a value-added raw material for glycerolysis of free fatty acids in waste cooking oil. Energies 15(23). https://doi.org/10.3390/en15238856

Mojapelo KS, Kupolati WK, Ndambuki JM, Sadiku ER, Ibrahim ID (2021) Utilization of wastewater sludge for lightweight concrete and the use of wastewater as curing medium. Case Stud Constr Mater 15. https://doi.org/10.1016/j.cscm.2021.e00667

Molazadeh M, Ahmadzadeh H, Pourianfar HR, Lyon S, Rampelotto PH (2019) The use of microalgae for coupling wastewater treatment with CO_2 biofixation. Front Bioeng Biotechnol 7(MAR). Frontiers Media S.A. https://doi.org/10.3389/fbioe.2019.00042

Morseletto P (2020) Targets for a circular economy. Resour, Conserv Recycl 153. https://doi.org/10.1016/j.resconrec.2019.104553

Neupane D (2023) Biofuels from renewable sources, a potential option for biodiesel production. Bioengineering 10(1). MDPI. https://doi.org/10.3390/bioengineering10010029

Patel A, Mikes F, Matsakas L (2018) An overview of current pretreatment methods used to improve lipid extraction from oleaginous microorganisms. Molecules 23(7). MDPI AG. https://doi.org/10.3390/molecules23071562

Permatasari R, Rinanti A (2018) Treating domestic effluent wastewater treatment by aerobic biofilter with bioballs medium. IOP Conf Ser Earth Environ Sci 106(1).

Plöhn M, Spain O, Sirin S, Silva M, Escudero-Oñate C, Ferrando-Climent L, Allahverdiyeva Y, Funk C (2021) Wastewater treatment by microalgae. Physiol Plant 173(2):568–578. https://doi.org/10.1111/ppl.13427

Postaue N, Fonseca JM, Bergamasco R, da Silva C (2022) Impact of biodiesel production on wastewater generation. Engenharia Sanit e Ambiental 27(2):235–244. ABES—Associacao Brasileira de Engenharia Sanitaria e Ambiental. https://doi.org/10.1590/S1413-415220210086

Rafa N, Ahmed SF, Badruddin IA, Mofijur M, Kamangar S (2021) Strategies to produce cost-effective third-generation biofuel from microalgae. Front Energy Res 9. Frontiers Media S.A. https://doi.org/10.3389/fenrg.2021.749968

Rao PV, Chary DP (2018) Characteristics comparison of biodiesel-diesel blend (B20) fuel with alcohol additives. Int J Adv Eng Res Sci 5(8):128–132. https://doi.org/10.22161/ijaers.5.8.17

Raveendran S, Parameswaran B, Ummalyma SB, Abraham A, Mathew AK, Madhavan A, Rebello S, Pandey A (2018) Applications of microbial enzymes in food industry. Food Technol Biotechnol 56(1):16–30. University of Zagreb. https://doi.org/10.17113/ftb.56.01.18.5491

Razzak SA, Lucky RA, Hossain MM, deLasa H (2022) Valorization of microalgae biomass to biofuel production: a review. Energy Nexus 7. Elsevier Ltd. https://doi.org/10.1016/j.nexus.2022.100139

Rinanti A, Fachrul MF, Hadisoebroto R, Silalahi MDS (2017) Improving biosorption of Cu(II)-ion on artificial wastewater by immobilized biosorbent of tropical microalgae. Int J Geomate 13(36): 6–10.

Rinanti, A, Purwadi R (2018) Harvesting of freshwater microalgae biomass by Scenedesmus sp. as bioflocculant. IOP Conf Ser Earth Environ Sci 106(1).

Rinanti, A, Purwadi, R (2019) Increasing carbohydrate and lipid productivity in tropical microalgae biomass as a sustainable biofuel feed stock. Energy Procedia 158:1215–1222.

Saadaoui I, Rasheed R, Aguilar A, Cherif M, Al Jabri H, Sayadi S, Manning SR (2021) Microalgal-based feed: promising alternative feedstocks for livestock and poultry production. J Anim Sci Biotechnol 12(1). BioMed Central Ltd. https://doi.org/10.1186/s40104-021-00593-z

Saini RK, Prasad P, Shang X, Keum YS (2021) Advances in lipid extraction methods—a review. Int J Mol Sci 22(24). MDPI. https://doi.org/10.3390/ijms222413643

Samsuri S, Aini Amran N, Jia Zheng L, Muhaimin Mohd Bakri M (2017) Effect of coolant temperature and cooling time on fractional crystallization of biodiesel and glycerol. Malays J Fundam Appl Sci 13(4)

Sarma S, Sharma S, Rudakiya D, Upadhyay J, Rathod V, Patel A, Narra M (2021) Valorization of microalgae biomass into bioproducts promoting circular bioeconomy: a holistic approach of bioremediation and biorefinery. 3 Biotech 11(8). Springer Science and Business Media Deutschland GmbH. https://doi.org/10.1007/s13205-021-02911-8

Schmid B, Navalho S, Schulze PSC, Van De Walle S, Van Royen G, Schüler LM, Maia IB, Bastos CRV, Baune MC, Januschewski E, Coelho A, Pereira H, Varela J, Navalho J, Rodrigues AMC (2022) Drying microalgae using an industrial solar dryer: a biomass quality assessment. Foods 11(13). https://doi.org/10.3390/foods11131873

Srimongkol P, Sangtanoo P, Songserm P, Watsuntorn W, Karnchanatat A (2022) Microalgae-based wastewater treatment for developing economic and environmental sustainability: current status and future prospects. Front Bioeng Biotechnol 10. Frontiers Media S.A. https://doi.org/10.3389/fbioe.2022.904046

Sutherland W, Jarrahi MH (2018) The sharing economy and digital platforms: a review and research agenda. Int J Inf Manage 43:328–341. Elsevier Ltd. https://doi.org/10.1016/j.ijinfomgt.2018.07.004

Taelman SE, Tonini D, Wandl A, Dewulf J (2018) A holistic sustainability framework for waste management in European cities: concept development. Sustain (Switz) 10(7). https://doi.org/10.3390/su10072184

Tan JS, Lee SY, Chew KW, Lam MK, Lim JW, Ho SH, Show PL (2020) A review on microalgae cultivation and harvesting, and their biomass extraction processing using ionic liquids. Bioengineered 11(1):116–129. https://doi.org/10.1080/21655979.2020.1711626

Udayan A, Pandey AK, Sirohi R, Sreekumar N, Sang BI, Sim SJ, Kim SH, Pandey A (2022) Production of microalgae with high lipid content and their potential as sources of nutraceuticals. Phytochem Rev. Springer Science and Business Media B.V. https://doi.org/10.1007/s11101-021-09784-y

Vaz SA, Badenes SM, Pinheiro HM, Martins RC (2023) Recent reports on domestic wastewater treatment using microalgae cultivation: towards a circular economy. Environ Technol Innovation 30. https://doi.org/10.1016/j.eti.2023.103107

Velenturf APM, Purnell P (2021) Principles for a sustainable circular economy. Sustain Prod Consumption 27:1437–1457. Elsevier B.V. https://doi.org/10.1016/j.spc.2021.02.018

Wimmerova L, Keken Z, Solcova O, Vavrova K (2023) A comparative analysis of environmental impacts of operational phases of three selected microalgal cultivation systems. Sustain (Switz) 15(1). https://doi.org/10.3390/su15010769

Wynn M, Jones P (2022) Digital technology deployment and the circular economy. Sustain (Switz) 14(15). https://doi.org/10.3390/su14159077

Xu H, Ou L, Li Y, Hawkins TR, Wang M (2022) Life cycle greenhouse gas emissions of biodiesel and renewable diesel production in the United States. Environ Sci Technol 56(12):7512–7521. https://doi.org/10.1021/acs.est.2c00289

Xue W, Macleod J, Blaxland J (2023) The use of ozone technology to control microorganism growth, enhance food safety and extend shelf life: a promising food decontamination technology. Foods 12(4). MDPI. https://doi.org/10.3390/foods12040814
Yeong Hwang T, Mee Kin C, Ling Shing W (2021) Extraction solvents in microalgal lipid extraction for biofuel production: a review. Malays J Anal Sci 25
Yu H, Kim J, Rhee C, Shin J, Shin SG, Lee C (2022) Effects of different ph control strategies on microalgae cultivation and nutrient removal from anaerobic digestion effluent. Microorganisms 10(2). https://doi.org/10.3390/microorganisms10020357
Yue Y, Zhang Q, Wan F, Ma G, Zang Z, Xu Y, Jiang C, Huang X (2023) Effects of different drying methods on the drying characteristics and quality of codonopsis pilosulae slices. Foods 12(6). https://doi.org/10.3390/foods12061323
Zewdie DT, Ali AY (2020) Cultivation of microalgae for biofuel production: coupling with sugarcane-processing factories. Energy, Sustain Soc 10(1). https://doi.org/10.1186/s13705-020-00262-5
Zhang S, Zhang L, Xu G, Li F, Li X (2022) A review on biodiesel production from microalgae: influencing parameters and recent advanced technologies. Front Microbiol 13. Frontiers Media S.A. https://doi.org/10.3389/fmicb.2022.970028
Zhou J, Wang M, Saraiva JA, Martins AP, Pinto CA, Prieto MA, Simal-Gandara J, Cao H, Xiao J, Barba FJ (2022) Extraction of lipids from microalgae using classical and innovative approaches. Food Chem 384. https://doi.org/10.1016/j.foodchem.2022.132236

Chapter 18
Analyzing Techno-economic Feasibility on Advanced Technologies in Biorefineries

Luísa Pereira Pinheiro, Arthur Santos Longati, Andrew Milli Elias, Thais Suzane Milessi, and Andreza Aparecida Longati

Abstract In an era marked by growing environmental concerns, resource scarcity, and the need for sustainable energy solutions, biorefineries have emerged as promising sustainable platforms for converting biomass into a variety of valuable products. Biorefineries can be classified according to their raw materials (and their generations), motivation (production of bioenergy or bioproducts), conversion process and platforms (intermediate products). Despite several studies in the literature and the development of technologies that add value to biomass and its products, studies of techno-economic analysis of these processes are necessary to evaluate the technical and economic feasibility of the process, generate valuable information for process improvements, in order to promote the development of a sustainable bioeconomy. This chapter provides a brief overview of conversion of the technologies used for conversion in biorefineries and identifies in the literature the most recent works of techno-economic analysis of biorefineries. A wide range of works showed the feasibility of different biorefineries, showing that these facilities are promising and play an important role in the world context.

L. P. Pinheiro
Doctoral Program in Electrical Engineering, Federal University of Itajubá, Itajubá, Minas Gerais 37500-903, Brazil

A. S. Longati
Graduate Programs Mechanical Engineering, Federal University of São João del-Rei, Praça Frei Orlando, 170, Centro, São João del-Rei, Minas Gerais CEP.: 36307-352, Brazil

A. M. Elias
Embrapa Instrumentação, São Carlos 13560-970, Brazil

T. S. Milessi · A. A. Longati (✉)
Chemical Engineering Department, Federal University of São Carlos, São Carlos, São Carlos 13565-905, Brazil
e-mail: andrezalongati@gmail.com

T. S. Milessi
Graduate Program in Chemical Engineering (PPGEQ), Federal University of São Carlos, São Carlos, São Carlos 13565-905, Brazil

T. A. Kurniawan and A. Anouzla (eds.), *Algae as a Natural Solution for Challenges in Water-Food-Energy Nexus*, Environmental Science and Engineering,
https://doi.org/10.1007/978-981-97-2371-3_18

18.1 Introduction

Environmental concerns such as climate change, global warming, waste management, pollution and depletion of natural resources have become the focus of research and development teams around the world. These environmental issues are mostly a consequence of anthropogenic activities with extensive exploitation of fossil fuel resources to produce energy, chemicals and synthetic materials (IEA 2022a). The process of urbanization, society development, industrialization and advancement in technology has increase energy consumption over the years, using the fossil fuel as the main source of primary energy. The use of fossil fuels is responsible for the greenhouse gases (GHG) emissions (including CO_2, CH_4, SO_2, NO_x and other particulate matters). As a result, natural, non-renewable resources are depleted and high levels of GHGs are released, causing significant environmental impacts such as rising global temperature (Shukla 2020; Karmaker et al. 2020). According to the IEA, fossil fuels were responsible for more than 80% of the world's total energy supply in 2020 (IEA 2021), while CO2 emissions from these energy sources represent 31.4 Gt worldwide (IEA 2022b).

World environmental conferences have emerged with the goal of bringing together leaders from different countries to discuss, expand debates and generate solutions on environmental impacts, such as the Kyoto Protocol (aimed at limiting and reducing emissions of GHG emissions) (Zhumadilova et al. 2023), ECO-92 (introduced the concept of sustainable development) (Nilander et al. 2023), and the Paris Agreement (to contain and minimize the increase in global warming) (Tavares et al. 2023). A policy of net-zero emissions by 2050 provides a pathway to stabilize the rise in the global average temperature at 1.5 °C, while at the same time providing universal access to modern energy by 2030. This will be a pivotal moment in energy history, reversing the rise of fossil fuels while continuing to grow the global economy(IEA 2022a).

A wide range of strategies are available to reduce GHG emissions and optimize waste management and environmental pollution such as energy efficiency, renewable energy, electrification of transport, and advanced technologies (Buckman et al. 2018; IEA 2022a). In this context, the concept of the circular economy (CE) has emerged to improve the world's sustainability (Yang et al. 2023), defined as the production and consumption economic model that promotes the responsible and circular use of resources (Moraga et al. 2019). In a linear economy (the traditional model), resources are extracted, products are manufactured, used, and then discarded as waste. In contrast, the CE promotes a closed-loop system which products and materials are reused, shared, repaired, remanufactured, and recycled, thereby extending their life and reducing the need for new resource extraction (Wang et al. 2022; Knäble et al. 2022; Rodrigues Dias et al. 2023). In this sense, the development of biorefineries is critical to address global sustainability challenges and provide a more secure, sustainable and affordable energy system (Kokkinos et al. 2020), by producing energy and chemicals from biomass in a sustainable way. As a result, these facilities have an important role to play in the transition from a linear to a circular bio-economy.

Significant advancements in biotechnology and process engineering have led to the development of advanced technologies that enhance the efficiency and product diversity of biorefineries (Yang et al. 2013; Gavrilescu 2014). Many biorefineries are at the proof-of-concept stage, where laboratory-scale studies have been conducted to estimate biochemical and biofuel potential. However, the successful implementation of these technologies to move these facilities from the laboratory and small-scale testing to commercial development and deployment heavily relies on their techno-economic feasibility, which involves carry out their techno-economic analysis (TEA), a tool commonly used to determine the economic feasibility of the early-stage processes. This work provides an overview of advanced conversion technologies of biomass in biorefineries and presents a wide range of studies that perform TEA on these facilities.

18.2 Advanced Technologies in Biorefineries

According to International Energy Agency in the IEA Bioenergy Task 42, biorefinery can be defined as "sustainable processing of biomass into a spectrum of marketable products (food, feed, materials, chemicals) and energy (fuels, power, heat)" (IEA Bioenergy 2009a). Therefore, biorefinery is a facility that integrates various biomass conversion processes to produce a wide range of products and bio-based materials (Gavrilescu 2014; Ng et al. 2017). This concept is similar to a conventional petroleum refinery, but instead of processing crude oil to produce fuels and chemicals, a biorefinery uses renewable biomass sources, such as agricultural residues, forestry by-products, energy crops, and organic waste, as feedstock. As the biorefinery is a complex system that encompasses several technologies with the main objective of the production of biofuels for the replacement of fossil fuels used in the transport sector. However, with the development of biorefinery technologies, several high-value products began to be developed with the aim of improving market competitiveness, increasing process sustainability, developing countries, diversifying corporate portfolios, creating opportunities and research (Hingsamer and Jungmeier 2018).

Biorefinery products are divided into two groups based on their motivation: energy-driven and product-driven. The production of biofuels and bioenergy is the main goal of "energy-driven" biorefineries. In this case, biomass is processed to produce products with low commercial value and in high quantities (Low Value High Volume—LVHV), mainly for the production of biofuels (bioethanol, biodiesel, synthetic biofuels, biogas), heat, cooling and electricity. In addition, the residues from the processes of this plant are sold or used in the production of bioproducts, adding value to the bioproducts and generating economic and environmental benefits (Budzianowski 2017; Yang et al. 2020b). On the other hand, "product-driven" biorefineries focus on the production of products with high added value and in small quantities (High Value Low Volume—HVLV) such as biochemicals, bionutrients and biomaterials (Budzianowski 2017), and the residues are used for energy production

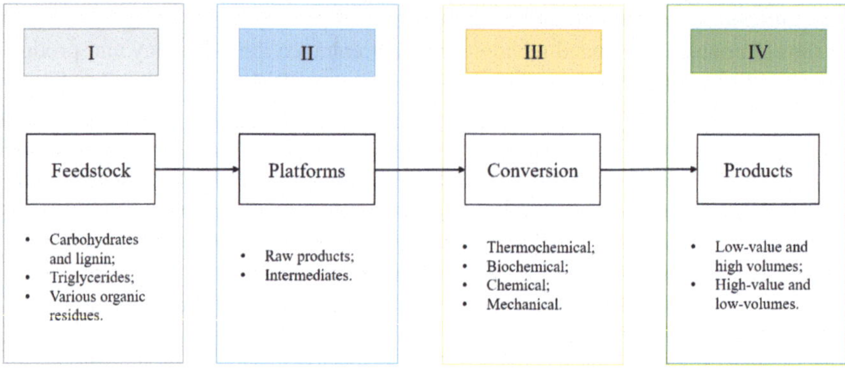

Fig. 18.1 A schematic structure of the biomass conversion processes. *Source* Authors

(as internal consumption of the plant) or feed production (Hingsamer and Jungmeier 2018). The selection and combination of the products to be integrated in the plant is generally based on a technical, economic and environmental analysis, which seeks to find the plant with the greatest profitability and sustainability. Given the challenge for renewable biofuels to achieve cost-competitiveness with relatively cheap fossil fuels, co-production of high-value products can help the biorefinery achieve viability (Elias et al. 2021a).

As shown in IEA Bioenergy Task 42 (IEA Bioenergy 2009b), the classification of biorefineries, based on the schematic representation of full "biomass to end product" chains, consists of four main features (Fig. 18.1): (I) feedstocks, (II) platforms, (III) conversion processes, and (IV) products.

18.3 Feedstock

Biomass is a renewable feedstock derived from organic matter of plant or animal origin, representing the fourth most used primary energy source in the world (IEA 2022b). This feedstock contains a wide variety of materials that differ in consistency, availability and chemical composition (Dahmen et al. 2009; Reck and Graedel 2012; Yin et al. 2021).

The bio-based resources, also known as biomass feedstocks, available for use in biorefineries are derived from agriculture, forest, industry, wet waste, and aquaculture (Maroneze et al. 2019). Agricultural biomass feedstock is defined as all organic matter that is produced as a by-product of the harvesting and processing of agricultural crops (Mohammed et al. 2018). The main agricultural biomasses are sugarcane, sugarcane bagasse, sugarcane straw, corn straw, rice straw, wheat straw, rice husks, corn stalks and straw, banana and pineapple leaves (Panthapulakkal and Sain 2015; Adhikari et al. 2018). Forest biomass feedstock refers to the by-products of forest harvesting, such as residues after logging (thinning, wood cutting or pulp), land clearing or non-tradable activities (branches, crowns, twigs) (Belyakov 2019). Industrial biomass feedstock comes from industrial activities such as manufactories, mining, milling (Vignesh et al. 2021), construction, energy and waste/water services (Millati et al. 2019), including vinasse, bagasse (de Castro et al. 2018), sludge, wood waste, solvents, used oils, muds, tailings, acids, textile waste (Millati et al. 2019). Wet waste biomass feedstock include the generation of animal manure, commercial, institutional and residential residues, such as urban solid waste, biosolids rich in organic matter, plant residues, animal feces, wastewater, sewage sludge (Adhikari et al. 2018; Dehghani et al. 2021). Aquaculture feedstock corresponds to uneaten feed, feces production and dead animals (Dauda et al. 2020). However, the most commonly used classification for biomass is by chemical composition (Takkellapati et al. 2018).

The biomass chemical composition is highly variable, which has an impact on the type of processing technology required to convert a particular biomass feedstock into biofuels and high-value products (Hayes 2013; Gent et al. 2017). The biomass chemical composition can be grouped in three categories: carbohydrates and lignin, triglycerides, and various organic residues (Cherubini 2010). The most traditional and established processes of current biorefineries are based on the utilization of carbohydrates, such as sucrose (composed of glucose and fructose) derived from sugar crops (mainly sugar beet and sugar cane), glucose obtained after hydrolysis of starch (found in wheat, corn, barley, rye, triticale...) or triglycerides derived from vegetable/animal fats, waxes and oils (Nguyen et al. 2017).

Carbohydrates are molecules composed of carbon, hydrogen and oxygen. The simplest carbohydrates, known as monosaccharides, are main represented by glucose, fructose, galactose, mannose, xylose and arabinose. Plants for the processing of monosaccharide biomass are the ones where the processing technologies are the simplest in terms of complexity. Generally, sugarcane and sugar beet are the major sources of sucrose (composed of glucose and fructose) production in the world (Rajaeifar et al. 2019). Polysaccharides consist of several connected monosaccharides, such as starch, cellulose, hemicellulose, and glucose (Nabben and Glatz 2015). Starch is the energy storage form of glucose in plants, composed by amylose and amylopectin. These molecules need to be broken down into smaller chains (monosaccharides or disaccharides) before to being fermented (Höfer 2015) to bioethanol, or used as a substrate for chemical reactions or enzymatic hydrolysis, which represents an opportunity to produce a wide variety of chemical products (Höfer 2015; Chaturvedi et al. 2020). Cellulose consists of glucose molecules linked by glycosidic bonds (Blanco and Blanco 2017). Unlike starch, cellulose is more difficult to digest (Nabben and Glatz 2015) and release the glucose monomers (Cherubini 2010). Hemicellulose is a branched and amorphous heteropolysaccharide composed of xylans, glucomannans, arabinans, galactans and glucans (Yang et al. 2020a). Lignin is a biopolymer composed hydroxylated and methoxylated phenylpropanoid units (Kramer et al. 2020), the largest non-carbohydrated fraction of lignocellulosic (Blanco and Blanco 2017).

Triglycerides (triesters) are molecules formed by a glycerol molecule linked to three fatty acid molecules (Bayly 2014). The main sources of triglycerides are vegetable and animal oils and fats, such as soybean, coconut, olive, palm, sunflower, rapeseed, waste edible oil (from food industry, commercial services, households) and aquatic biomass (Cherubini 2010; Jong et al. 2020). Various organic residues include organic compounds derived from agricultural, industrial and urban activities, such as proteins, manure, municipal solid waste (MSW), sewage slugde, residues from food industries are also source of triglycerides (Chen et al. 2017).

As it was described, the chemical composition of biomass is highly variable, and this has an impact on the type of processing technology that is required to convert a given biomass into biofuels or high-value products (Hayes 2013; Gent et al. 2017). Therefore, the chemical properties of biomass feedstocks directed to which conversion technology will be targeted (Hayes 2013; Adams et al. 2018). There are many aspects to consider when selecting a biomass conversion process to adopt as an industrial process, including chemical composition, type, availability and quantity, seasonality of supply, environmental standards, governmental policies, economic conditions, and project-specific factors (Hayes 2013; Adams et al. 2018).

18.4 Platforms

Platforms refer to the intermediate products between the raw material and the final products of the biorefinery, in other words, they are the products that connect different processes and systems within the biorefinery that are transformed into other intermediate products (a new raw material) (Trokanas et al. 2015; Hingsamer and Jungmeier 2018). Within the biorefinery, the feedstocks are typically fractionated into a number of key intermediates. These intermediate products can be a single compound or a mixture of compounds of varying purity. Several feedstocks can lead to the same platform and a biorefinery can include several platforms (Jong et al. 2020).

As mentioned above, the chemical composition of biomass is highly variable; therefore, in order to obtain some components of biomass (intermediate products), many biomasses must first be prepared by pre-treatment before going through the conversion processes. In lignocellulosic materials, the preparation process is known as pretreatment with the main objective of breaking the recalcitrant structure of the biomass to make cellulose and hemicellulose available for processing and converting the carbohydrate polymers into fermentable sugars. The pretreatments are classified into four groups: physical, chemical, biological and physochemical.

In physical pretreatments the biomass undergoes a physical process without changing its composition (Nunes et al. 2018). This process increases the accessible areas and volume of pores with the reduction of biomass size, crystallinity and polymerization of cellulose fibers, which facilitates the future application of biomass (Ani 2016; Shah et al. 2022). Physical pretreatment types include drying (reducing the amount of water in the biomass), size reduction (milling, grinding, chipping), thermal processes (microwave and infrared radiation, pneumatic dryers, conductive and convective dryers) (Nunes et al. 2018). This method is found described in several works of literature (Vásquez-Parra et al. 2013; Adiletta et al. 2016; Xu et al. 2020; Sun et al. 2021b; Shen et al. 2022).

The chemical pretreatment of biomass uses organic or inorganic chemicals compounds to disrupt the structure of the biomass (Basso 2020). According to Amin et al. (2017), the most used chemicals are sulfuric acid, acetic acid, hydrochloric acid, sodium hydroxide, lime, aqueous ammonia, hydrogen peroxide and potassium hydroxide. This pretreatment was reported in several works of literature (Lemões et al. 2018; Zheng et al. 2018; Hamada et al. 2020; Castro et al. 2020; Jallouli et al. 2021; Swaminathan et al. 2021; Perez et al. 2022; Vannarath and Thalla 2022).

The physicochemical pretreatments combine chemical reactions and physical changes during the pretreatment process. The main physical-chemical pretreatments are: steam explosion, ammonia fiber explosion (AFEX) and liquid hot water (Jędrzejczyk et al. 2019). This work can be found in a wide range of applications in literature (Cao et al. 2013; Imman et al. 2015; Ao et al. 2020; Liu et al. 2021; Xu et al. 2022; Shi et al. 2022; Mokomele et al. 2022; Akizuki et al. 2023; Wang et al. 2023a).

Biological pretreatments use microorganisms or enzymes as catalysts to alter structure of the biomass (through the deconstruction or alteration of lignin structures). This pretreatment is a slower process, cannot easily control and has several factors that affect the pre-treatment (pH, temperature, incubation time, consortium and competition of microorganisms, source of carbon, nitrogen, aeration, organic and inorganic compounds) when compared with others methods (Brémond et al. 2018; Sharma et al. 2019). The discussion and applications of this pretreatment can be found in several works (Fdez.-Güelfo et al. 2011; Castoldi et al. 2014; Sun et al. 2021a; Li et al. 2021; AP et al. 2021; Bhushan et al. 2023; Rouabhia et al. 2023).

18.5 Conversion Technologies

There are several biomass conversion processes. In addition to the composition of the feedstock used biorefineries, the technological pathway used to convert the biomass depends on the desired product and technological advances and improvements in process performance of the pathway. Biorefinery conversion processes can be divided into four categories: biochemical, chemical, mechanical/physical, and thermochemical. Thermochemical conversion is the method that raises the temperature of the biomass to degrade (chemically change) the biomass structure (Wang 2018). Thermochemical conversion includes combustion, pyrolysis, gasification and liquefaction.

Biochemical, or biological, conversion consists of the degradation of biomass by microorganisms. This conversion is characterized as a cost-effective method for the production of fuels, chemicals and materials, which operates at lower temperatures and lower reactions rates (Demirbas 2009; Gnanasekaran et al. 2023). Biochemical conversion encompasses anaerobic digestion, fermentation and photosynthesis (hydrogen production).

Chemical conversion consists of the chemical transformation of the molecular structure by reaction with other substances (Cherubini 2010). In other words, biomass undergoes chemical changes during processing in the presence of a catalyst. Chemical conversion is classified into hydrolysis, transesterification, Fisher-Tropsch synthesis, and steam reforming (IEA Bioenergy 2009a).

Mechanical conversion corresponds to the process that does not alter the state or composition of the biomass. It is commonly used in the biorefinery as the first biomass conversion process, operating with the reduction, standardization, extraction and separation of the biomass (Cherubini 2010; Ng et al. 2017). The mechanical conversion process is mainly present in biorefineries using forest feedstocks. However, in combination with other processes, the main intermediates produced in biorefineries are organic fibers and lignin (Baldoni et al. 2021).

18.6 Bioenergy/Bioproducts

As previously described, biorefinery products are divided into two categories: energy products and material products (Fig. 18.2). According to Cherubini (2010), energy products are products used to provide energy such as heat, electricity and transportation services. Material products correspond to products that are used for non-energy purposes due to their physical or chemical properties.

The main energy products produced in a biorefinery are solid, liquid and gaseous biofuels. Solid biofuels are fuels that are produced from organic materials, biomass, and municipal solid waste. These fuels are obtained from various sources, such

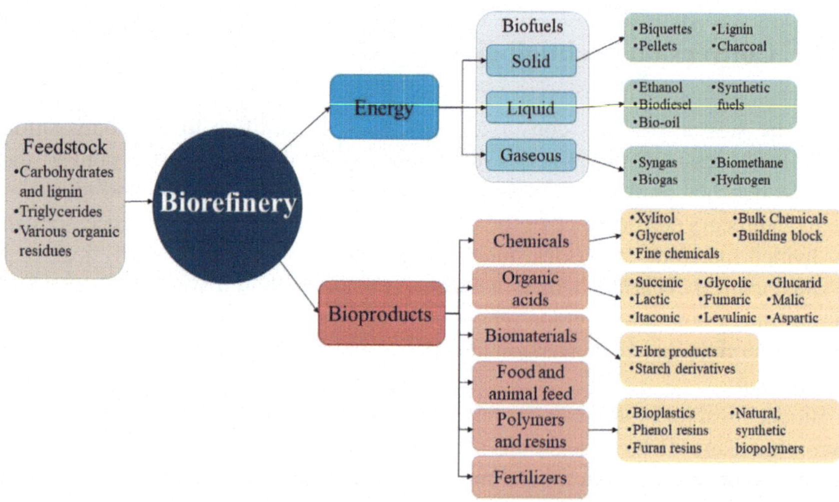

Fig. 18.2 Energy and bioproducts of biorefinery

as agricultural and forest residues and animal remains (Kiehbadroudinezhad et al. 2023). Therefore, solid biofuels are: briquettes, pellets, lignin, charcoal (Cherubini 2010). According to (Angulo-Mosquera et al. 2021), briquettes and pellets are the most widely produced solid biofuel. They are produced by physical processes to generate thermal and/or electrical energy (Islas et al. 2019).

The most commonly produced liquid biofuels are bioethanol, biodiesel, bio-oil and synthetic fuels (Fischer-Tropsch). Liquid biofuels are popular because of their unique characteristics: they are safe to transport, relatively inexpensive, stable in storage, have a high energy-mass ratio, are highly combustible, and have lower greenhouse gas emissions than fossil fuels (Kiehbadroudinezhad et al. 2023). Biodiesel and bioethanol are the world's most widely used liquid biofuels in the transportation, produced through chemical (transesterification) and biochemical (fermentation) conversion. By using these biofuels to replace fossil fuels (blended or not with fossil fuels) it is possible to reduce GHG emissions, pollution and global warming (Carneiro et al. 2017; Wang et al. 2023b). Brazilian bioethanol produced in sugarcane biorefineries is an example of the production of energy product. Its production is an important economic activity for the country, since the energy produced from sugarcane represents about 15.4% of the national energy supply. Additionally, in these facilities are generated sugarcane bagasse to be burnt in boiler in the cogeneration system, which are responsible for 7.6% of the national electric matrix (EPE 2023).

The major gaseous biofuels derived from biomass include syngas, biogas, methane and hydrogen (Yadav et al. 2023). Among then, the most common are biogas and syngas. Biogas, the product of anaerobic digestion (biochemical conversion) of organic waste, is an energy product with a wide range of applications. In the biorefinery context, this biofuel is commonly used in cogeneration systems, in combustion engines for energy generation (thermal and electrical) (Volpi et al. 2021), and in processes to maximize the energy content of the biofuel with the production of biomethane (Haldar et al. 2023).

Material products in biorefineries, bioproducts, are an environmentally friendly alternative to fossil-based products. According to Cherubini (2010), bioproducts are divided into chemicals, organic acids, polymers and resins, biomaterials, food and animal feed, and fertilizers. Some examples of potential biochemicals are fumaric acid, ethylene, glycolic acid, acetoin, bioethanol, methanol (Wagemann 2012), sorbitol, lactic acid, 2,5-furandicarboxylic acid, succinic acid, levulinic acid, 3-hydroxypropionic acid (3-HP), 3-hydroxypropionaldehyde (3-HPA), sorbitol, xylitol, isoprene (Takkellapati et al. 2018), pigments, vitamins, antioxidants (Chew et al. 2017).

One example of biochemical production is biopolymer production. According to Philippini et al. (2020), the bioethanol produced in the first- or second-generation process in the biorefinery is dehydrated and polymerized to form ethylene (an essential chemical in the petrochemical industry and a precursor to polyethylene). Another example is microalgae. Microalgae biorefineries produce high-value products such as pigments, proteins, carbohydrates, vitamins, and antioxidants with multiple applications in the pharmaceutical, cosmetics, and nutrition industries that are difficult to synthesize by chemical methods (Khavari et al. 2021).

For sustainable biorefineries, the goal is to maximize the use of the feedstock, which leads to a wide range of products that can be derived from a feedstock and the plethora of possible pathways that can lead to these products. The studies of the literature indicate that the feasibility of a biorefinery industry can be potentially higher when byproducts are produced than that of a traditional industry (focused on optimizing the production of only a single compound from the feedstock) (Kiskini et al. 2016). Therefore, it is evident that an economic analysis is required for the design of a biorefining process. A brief description of the literature studies carrying out technical and economic analysis in in the context of biorefinery is presented as follows.

18.7 Techno-economic Feasibility of Biorefineries

The increasing use of biofuels and bioproducts (as substitutes for fossil fuels and products) is driving the development of advanced processes to efficiently process biomass in biorefineries. Several benefits can be gained from the development and implementation of biorefineries. The main question is what are the advantages of a biorefinery as a multi-product system compared to conventional bioenergy systems (e.g. single-product systems). These advantages might be in terms of economic, environmental, technological, and social perspectives. Therefore, for the continued development and investment in research in biorefinery areas, it is essential to conduct a TEA of biorefineries to the key factors that affect their profitability and to propose strategies to improve their technical and economic performance. In this sense, the application of TEA (a tool used as a decision support for evaluation of the economic performance, Fig. 18.3) in biorefineries is important to overcome its economic feasibility issues and for the development and improvement of new technologies and the processing of raw materials (Manfren et al. 2021). Through TEA it is possible to obtain a compression of the system, observe the limitations and bottlenecks, the performance of processes, in addition to evaluating the economic and technical feasibility for the implementation.

18.7.1 Methods for the Techno-economic Assessment

Conducting the techno-economic evaluation involves the widespread utilization of process system engineering techniques, including modeling and simulation, to comprehensively analyze and enhance the formulation of biorefinery processes (Giordano et al. 2020). The pivotal phase of process formulation can be efficiently aided by software tools that aid in the selection and arrangement of equipment, as well as the establishment of operational parameters necessary to achieve production objectives.

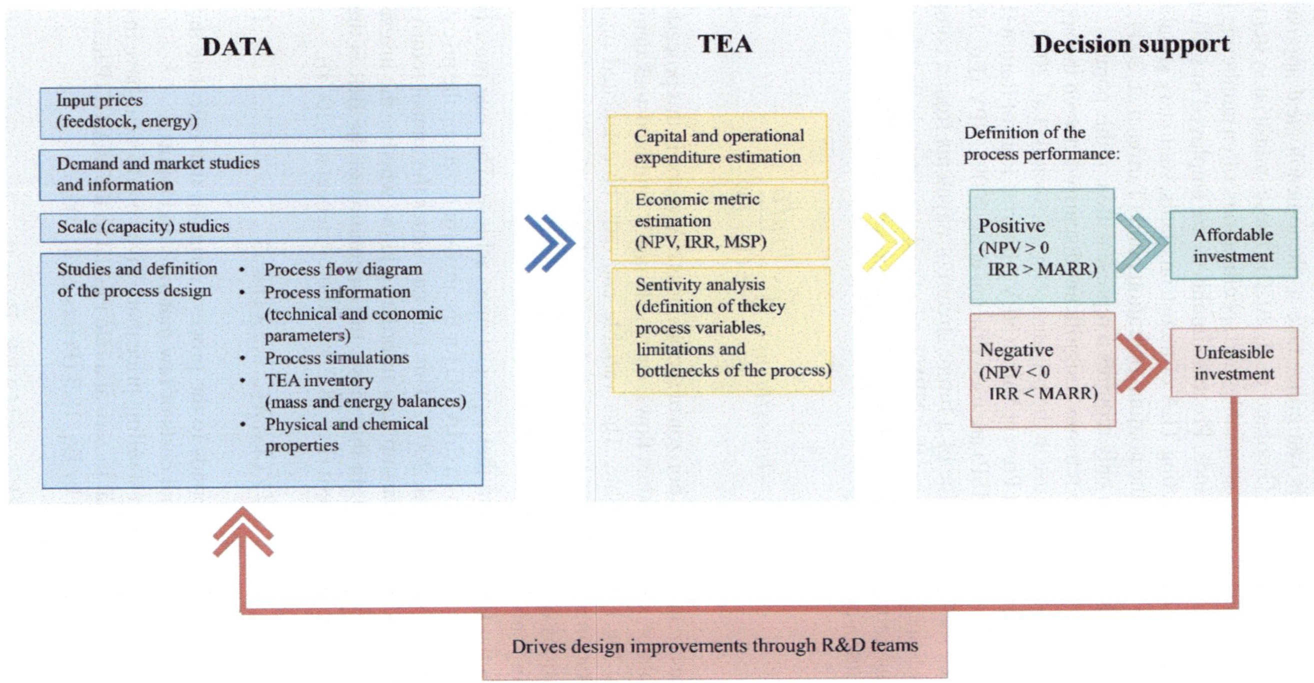

Fig. 18.3 Diagram of the technical–economic evaluation (TEA) of a biorefinery as a decision support tool for a project

This involves constructing mathematical models, primarily centered on mass and energy balances, to describe the behavior of unit operations, (bio)reactors, and various processing units. These models can either take a phenomenological approach or treat the units as black boxes (Giordano et al. 2020). These industrial systems can encompass a multitude of process variables and specifications, often numbering in the tens or even hundreds of thousands. Process simulators are employed to solve these complex models. When conducting TEA of industrial plants, stationary simulations are typically used to represent the nominal operation point (Turton et al. 2009). This is because solving the problem only requires a representation of this point.

Despite its usefulness, there remains a significant disparity between the application of TEA in research and its use in the private sector (Goswami et al. 2022). Most usually, metrics such as the Net Present Value (NPV), Internal Rate of Return (IRR), and Minimum Selling Price (MSP) are used (Turton et al. 2009). NPV (Eq. 18.1) is use to calculate the current value of a future stream of payments from a company, project, or investment. This means that NPV is used to determine whether the anticipated financial gains of a project will outweigh the present-day investment (Peters et al. 2003; Turton et al. 2009).

$$NPV = -Investment + \sum_{i=1}^{N} \frac{Profit}{(1+r)^i} \tag{18.1}$$

The tax rate is r, and the time horizon of the business is N years. In this process, data from the simulator, such as product flows, energy demands, and feedstock demands, are input into the *Profit* variable. This variable can only be calculated once the process simulation converges to a solution. Investment costs are determined based on the dimensions of the equipment, which are either provided by the simulator or calculated using spreadsheets.

The IRR is the rate of return that a project must achieve. This rate must be the minimum acceptable rate of return (MARR) for the investor to recoup the investment made in the project (Peters et al. 2003). This rate calculates the annual return of an investment in relation to the start-up date in order to know whether the investment has appreciated or depreciated. In this sense, the calculation of the IRR is the rate that makes the NPV equal to zero (Eq. 18.2, based on Peters et al. (2003)).

$$NPV(X_1, X_2, \ldots, X_3) = 0 \tag{18.2}$$

where $(X_1, X_2, \ldots, X_i)$ corresponds to the process variables that calculate the size and cost of equipment, flows and costs of raw materials and inputs.

The MSP represents the lowest selling price that a given product can have in order to be profitable, with NPV equal to zero at a rate r equal to the MARR (Peters et al. 2003). The MSP is determined by Eq. 18.3 (Peters et al. 2003):

$$NPV(r = MARR, P_i) = 0 \tag{18.3}$$

However, private companies, evaluating potential investments, typically focus on simpler profit-related indicators that do not take into account the time value of money (Scown et al. 2021). These indicators include revenues, gross margin, return on investment (ROI, %), and payback period (in years). When it comes to policy-making, the key question is whether a new renewable fuel or product can be sustainable in the long run (Goswami et al. 2022). Additionally, it is important to determine the level of economic incentives required to make production profitable soon (Scown et al. 2021).

18.7.2 Tecno-economic Feasibility of Biorefineries

Several examples of techno-economic performance assessments for bioproducts generation are available in the literature. However, it is difficult to compare the results because different approaches have been used in different research papers. The recent available data on TEAs have been summarized in Table 18.1 as follows.

The sugar platform is one of the most widely used platforms in the world for both biochemical and biofuel production. This type of platform uses six-carbon (glucose, fructose and galactose), five-carbon (xylose, pentose and ribose) and 12-carbon (lactose, sucrose and maltose) sugars derived from sugar crops (sugar cane, sugar beet and sweet sorghum), starch hydrolysis (corn, grain and wheat) or hydrolysis of lignocellulosic crops and residues (switchgrass, stover, bagasse, straw) (Jong et al. 2020). This type of platform, with a mature, well-stabilized technology, and commercially consolidated processes with high productivity and profitability, is mainly used in biorefineries for the production of alcohols, organic acids, lipids, alkenes, hydrocarbons and other chemicals (E4tech et al. 2015). Among the products generated in a sugar biorefinery, bioethanol is the most widely produced biofuel in the world, with the United States (from corn) (Alternative Fuels Data Center 2022), Brazil (sugarcane and corn) (Gonçalves et al. 2023), China (corn and cassava) (Zhou 2020), India (sugarcane, rice and wheat) (Government of India 2022), and Canada (corn and wheat) (Government of Canada 2020) as the largest producers.

The most recent works published in the literature use the 1G structure mainly as a base scenario for processes that integrate different technologies. Usually, theses biorefineries are high profitable (Morakile et al. 2022) and generate large amounts of residues (such as sugarcane bagasse and straw in sugarcane biorefineries cases), which can be used as raw material for a wide range of bioproducts of high value. The integration of different processes (LVHV and HVLV) in the 1G structure is promising, since this strategy allows share part of the infrastructure where 1G bioethanol production takes place (for instance concentration, fermentation, distillation, storage and cogeneration facilities). In addition, the process resemblance can be verified inseveral areas such as agricultural, logistics, geographic, industrial, streams, equipment, administrative, and commercial fields (Longati et al. 2020).

Dias et al. (2012) reported the production of bioethanol from sugarcane in several 1G, 2G and 1G2G scenarios. Considering the 1G production, the authors estimated

Table 18.1 The main publications in the literature on TEA of biorefineries

Feedstock	Products	Comments	Reference
Seaweed (*Saccharina latissima*)	2G Bioethanol	MSP of 2.9–7.5 \$/gal, with ≥ 80% yield ≥ 20% solids loading. Feedstock costs about \$50/T	Konda et al. (2015)
Seaweed (*Laminaria/Gracilaria*)	2G Bioethanol	MSP of 2.39 and 2.85 \$/gal, respectively, with EH% solid loading 17.5–20. Feedstock costs about 64.6 and \$26/T	Fasahati et al. (2017)
Sugarcane juice	1G Bioethanol	MSP of 471.3 \$/m^3 (1.78 \$/gal), traditional production of ethanol in Brazil	Longati et al. (2018)
Sugarcane bagassse	1G2G Bioethanol	MSP of 566.07 \$/m^3 (2.14 \$/gal)	Longati et al. (2018)
Sugarcane bagasse	1G2G Bioethanol	MSP of 615.74 \$/m^3 (2.33 \$/gal)	Longati et al. (2018)
Corn stover	Bioethanol	MSP of 0.55 \$/L (2.08 \$/gal)	González-Contreras et al. (2023)
Sugarcane juice and bagasse	1G2G Biethanol	MSP of 499.08 \$/m^3 (1.89 \$/gal)	Elias et al. (2021a)
Sugarcane juice	1G Bioethanol	MSP of 435.39 \$/m^3 (1.65 \$/gal)	Elias et al. (2021a)
Sugarcane straw	2G Bioethanol	MSP of 623.84 USD/m^3 (2.36 \$/gal)	Pratto et al. (2020)
Sugarcane lignocellulosic biomass	Bioethanol	MSP of 639 US\$/m^3 (2.41 \$/gal)	Taylor-de-Lima et al. (2016)
Sugar lignocellulosic biomass (bagasse and trash)	Bioethanol	MSP of 0.58 \$/L (2.20 \$/gal)	Farzad et al. (2017)
Lipid cane	Biodiesel	MSP of 0.59 \$/L	Huang et al. (2016)
Soybean	Biodiesel	MSP of 1.08 \$/L	Huang et al. (2016)
Used cooking oil	Biodiesel	MSP of 0.72 \$/L	Elgharbawy and Ali (2022)
Used cooking oil	Biodiesel	MSP of 0.68 \$/L	Elgharbawy and Ali (2022)
Seaweed (*Scenedesmus* sp.)	Biodiesel	MSP of 1.83 \$/L (2.03 \$/kg)	Mustapha et al. (2022)
Seaweed (*Nannochloropsis gaditana*)	Protein	MSP of 4.38 \$/kg	Slegers et al. (2020)

(continued)

Table 18.1 (continued)

Feedstock	Products	Comments	Reference
Seaweed (*Isochrysis galbana*)	Protein	MSP of 1.61 $/kg	Slegers et al. (2020)
Seaweed (*N. gaditana*)	Lipid	MSP of 7.76 $/kg	Slegers et al. (2020)
Seaweed (*N. gaditana*)	Lipid	MSP of 7.26 $/kg	Slegers et al. (2020)
Corn stover	Lactic acid	MSP of 0.523 $/kg. Feedstock, enzyme, and non-enzyme conversions are major contributors to MSP, respectively	Liu et al. (2015)
Sugarcane bagasse hemicellulose hydrolysate	Glutamic acid	MSP of 2.32 $/kg. Downstream processing is a major contributor, accounting for 17 of the total investments	Özüdoğru et al. (2019)
Wheat straw and distillery waste	Microbial Oil	MSP of 14 $/kg for a microbial oil plant processing capacity of 100 T/year	Parsons et al. (2019)
Sugarcane bagasse	Polyhydroxybutyrate (PHB)	MSP of 2.06 $/kg. Fermentation and downstream process are responsible for 15% of the total installed equipment cost	Nieder-Heitmann et al. (2019)
Sugarcane bagasse	Sophorolipid biosurfactant	MSP of 19.53 US$/kg using acid pretreatment	(Elias et al. 2021b)
Sugarcane bagasse	Sophorolipid biosurfactant	MSP of 19.53 US$/kg using hydrothermal pretreatment	(Elias et al. 2021b)
Sugarcane lignocellulosic biomass	Xylitol	MSP of 3000 $/t	Özüdoğru et al. (2019)
Microalgae (cyanobacteria Arthrospira platensis)	DSFP (dried spirulina food product)—food additive	MSP of 138.7 €/kg for plant capactiry of 1000 kg spirulina/day	Kingsley et al. (2023)
Corn stover	Ethyl lactate	MSP of 1.49 $/L	González-Contreras et al. (2023)
Mollusc waste	Calcium carbonate	MSP of 12.6 €/kg	Andreola et al. (2023)
Mollusc waste	Hydrolysates	MSP of 12.5 €/kg	Andreola et al. (2023)
Fish waste	Hydrolysates	MSP of 259.6 €/kg	Andreola et al. (2023)

(continued)

Table 18.1 (continued)

Feedstock	Products	Comments	Reference
Fish waste	Co-composted biochar	MSP of 1.82 €/kg	Andreola et al. (2023)
Organic fraction municipal solid waste	Succinic acid	MSP of 1.58 US$/kg	Ioannidou et al. (2023)
Organic fraction municipal solid waste	Lactic acid	MSP of 2.14 US$/kg	Ioannidou et al. (2023)
Miscanthus	80% Xylooligosaccharides	MSP of 3.43 $/kg – 7.50 $/kg	Lan et al. (2021)
Miscanthus	95% Xylooligosaccharides	MSP of 4.84 $/kg – 10.64 $/kg	Lan et al. (2021)
Brewers' spent grains	Xylooligosaccharides	MSP of 2.51 $/kg	Swart et al. (2021)
Brewers' spent grains	Xylooligosaccharides	MSP of 2.18 $/kg	Swart et al. (2021)
Brewers' spent grains	Xylooligosaccharides	MSP of 5.50 US$/kg	Van Heerden (2023)
Sugarcane juice	Jet fuel	MSP of 2.2 US$/L (8.34 $/gal)	Diederichs et al. (2016)
Jatropha	Jet fuel	MSP of 3.82 $/gal	Tao et al. (2017b)
Castor bean	Jet fuel	MSP of 9.43 $/gal	Tao et al. (2017b)
Camelina	Jet fuel	MSP of 10.98 $/gal	Tao et al. (2017b)
Sugar lignocellulosic biomass (bagasse and trash)	Butadiene	MSP of 2645 $/t	Farzad et al. (2017)
Microalgae	Butanol	MSP of 8.54 $/kg	Figueroa-Torres and Theodoropoulos (2023)
Microalgae	Diesel	MSP of 2.36 $/kg	Figueroa-Torres and Theodoropoulos (2023)

that 82 l of anhydrous bioethanol were produced per ton of sugarcane at a cost of 0.37 US$ per liter of bioethanol (34% cheaper than the average bioethanol price of 0.6 US$/L). Moncada et al. (2013) simulated a Colombian 1G biorefinery producing sugar, bioethanol, and electricity as a base scenario for the TEA. The biorefinery produced 26.19 liters of bioethanol, 0.11 tons of sugar and 550.76 MJ of electricity per ton of sugarcane. The MSP was estimated at US$ 1.24/L, US$ 0.42/kg, US$ 0.1/kWh for bioethanol, sugar and electricity, respectively. Junqueira et al. (2017) developed the study of the 1G and 2G bioethanol production from sugarcane to compare with other scenarios in the three-time horizons: short (2016–2020), medium (2021–2025), and long (2026–2030) terms. As a result, the biorefinery achieved bioethanol production in the base scenario (short-term) of 85.2 l per ton of sugarcane at a price of R$1.34/L, with an IRR of 12%. For short, medium and long-terms scenarios, the authors estimated the MSP of US$ 0.50/L, US$ 0.37/L and US$ 0.33/

L for the 1G scenario, and US$ 0.64/L, US$ 0.33/L and US$ 0.24/L for 2G scenario, respectively.

Mendes et al. (2017) performed a TEA of 1G bioethanol production from six different hybrid sugarcane, verifying the impact of each feedstock on the IRR. The scenarios accessing the use of sugarcane H89 as feedstock obtained the highest IRR (17.8%) with a production of 85.5 liters of bioethanol per ton of sugarcane and bioethanol price of US$ 0.60/L. The production of bioethanol from sugarcane molasses corresponded to 125 million liters per day (corresponding to 26.19 l of bioethanol per ton of sugarcane), at MPS of US$ 1.24/L. Longati et al. (2018) modeled a 1G biorefinery producing bioethanol and electricity as a base scenario for a comparative TEA. This 1G scenario produces 89 l of bioethanol per ton of sugarcane at a MSP of US$ 0.47/L (the same value obtained in the 10-year average price of bioethanol) with an IRR of 11%.

Tao et al. (2017a) reported a TEA of jet fuel production from bioethanol based on corn stover and grains as feedstocks. As a result, the biorefinery recorded ethanol production of 427.8 l per ton of dry corn kernels at an MSP of $1.98/gal ($0.52/L) and 299.05 l per ton of dry corn at an MSP of $2.75/gal ($0.73/L). For jet fuel, the biorefinery registered a production per ton of dry biomass of 227 l for corn grain and 143.85 l for corn stover at a respective MSP of $4.20/gal ($1.11/L) and $6.14/gal ($1.62/L) at a 10% IRR. Bulkan et al. (2020) performed a TEA of the integration of 2G technology into a 1G grain-based (wheat) bioethanol plant. The 1G plant has the capacity to process 150.4 tons of grain per year and produce 41.35 thousand tons of bioethanol per year with MSP of USD 1.16 per kilogram of bioethanol using the interest rate of 6%. Gerrior et al. (2022) presented a techno-economic and environmental analysis of the integration of 1G corn bioethanol production with an anaerobic digestion system and algae cultivation. The 1G biorefinery produced 94.51 liters of bioethanol per ton of corn at MSP of $5.46/L.

The oil platform in a biorefinery convers fatty acids and triglycerides from different biomasses with multiple applications in chemical industry (from the production of biofuels to the manufacture of cosmetics, pharmaceuticals and use in the food industry) (Cherubini 2010; Jong et al. 2020). These oils are used for biodiesel production by the transesterification process (Bardhan et al. 2022). In recent years, biodiesel production has increased worldwide, and several countries have adopted the addition of biodiesel to diesel as a way to reduce GHG emissions (IEA Bioenergy 2021). Although biodiesel production is an alternative for reducing GHG emissions, economic feasibility studies are needed to evaluate all costs associated with its production. Al-Saadi et al. (2022) conducted a TEA of three different biodiesel production processes. The first scenario was based on a conventional process, the second with a reactive coupling to produce glycerol carbonate, and the third with a reactive coupling to produce solketal. The NPV of the 1G process was $65 million, while the NPVs of second and third scenarios were $631 million and $128 million, respectively. The price of biodiesel, glycerol carbonate, and solketal were $1031 per ton, $15980 per ton, and $3000 per ton, respectively. Huang et al. (2016) modeled a TEA of biodiesel production from lipid cane and soybean. The two biorefineries produce 96 million liters of biodiesel per year. For lipid cane, the MSP of biodiesel was \$0.59/L (52% less than the average price of biodiesel) with an IRR of 24%. For soybean, the MSP of biodiesel was \$1.08/L (11% less than the average market price) with an IRR of 15%.

The lignocellulosic biorefineries used as biomass feedstock energy crops, agro-industrial residues, forestry materials, and organic materials from municipal solid waste (Patel and Shah 2021). This biomass is the most abundant in the world and has great potential for the production of high-value products, chemicals, energy and fuels (Patel and Shah 2021). The expectation is that the integration of lignocellulosic biorefineries into traditional technologies fully utilize biomass, minimize waste, optimize and share processes, and expand sustainable development (Yogalakshmi et al. 2023). Many integration processes can be found in literature, such as the work of Seufitelli et al. (2022) which assessed the production of jet fuel, formic acid and xylitol from dried poplar wood, Morakile et al. (2022) with the assessment of the levulinic acid and xylitol production from molasses (NPV of \$184.2 million at an IRR of 27.59%), Ou et al. (2021) with the TEA of the sugar production from wheat and the co-production of xylitol (from hemicellulose hydrolysate) and polyol (from lignin) (MSP of \$342/t), Pereira et al. (2014) carry out the TEA of butanol from sugarcane (MSP of US\$1.65/kg with an IRR of 13.3%), Longati et al. (2019) with TEA of 1G2G bioethanol production from sugarcane (Scenario 2B: MSP of USD 0.47 /L, NPV of 4.63 million USD at an IRR 13.6%, Scenario 3B: MSP of USD 0.47/L, NPV of—7.13 million USD at an IRR 6.1%), and the study of Ladakis et al. (2022) assessing the production of succinic acid from the organic fraction of municipal solid waste (MSP of \$2.39/kg and NVP equal to \$ 207.1 million).

With the advancement of metabolic engineering and biotechnology, several microorganisms and plants have been developed as a promising strategy to increase productivity, reduce costs, improve efficiency (Wuest et al. 2011) and facilitate the construction of high-performance industrial strains (Teng et al. 2020). Energy cane is an example, this type of sugarcane has been developed to increase the potential for bioelectricity generation and the production of second-generation bioethanol and biodiesel (Thammasittirong et al. 2023). Compared to regular sugarcane, energy cane has greater resistance to pests/diseases and adverse environmental conditions, greater efficiency in the use of water and nutrients (Barbosa et al. 2019; Thammasittirong et al. 2023), higher fiber content and greater ability to accumulate triglycerides (Kumar et al. 2021). Kumar et al. (2021) conducted a TEA of two co-production scenarios for bioethanol and biodiesel from energy cane. In the first scenario, bioethanol and biodiesel are produced from the sugars present in sugarcane juice and the accumulated triglycerides. While the second scenario deals with the production of bioethanol and biodiesel from sugarcane juice and the hydrolysis of bagasse carbohydrates and accumulated triglycerides. As a result, the biorefinery in the first scenario had the best economic performance when the feedstock had 7.7% of accumulated triglycerides, producing biodiesel at \$0.69/L and bioethanol at \$0.35/L (both fuels 31% below the 10-year average market price) with an IRR of 7.84%.

Microalgae have been also extensively studied for the production of biofuels, food supplements, cosmetics, lubricants, and other valuable products (Sivaramakrishnan et al. 2022). Figueroa-Torres and Theodoropoulos (2023) conducted a TEA of a microalgae biorefinery for the production of biofuels. The biorefinery processes 425.24 tons of microalgae per year, producing 29 tons of biobutanol per year with an MSP of $8.34/L in the algae-to-butanol scenario and 22 tons of biodiesel per year with an MSP of $2.48/L in the algae-to-biodiesel scenario. Thomassen et al. (2016) performed the TEA of four algal biorefinery scenarios. The best economic results were found in the scenario assessing the plant producing 1476 tons of fertilizer, 141 tons of b-carotene, and an NPV of 37 million EUR (representing an economically viable scenario). Llano et al. (2023) studied the microalgae biorefinery for the production of bioethanol and of butanol. Both scenarios achieved a positive economic performance, with the bioethanol microalgae biorefinery achieving an NPV of $27.66 million with an IRR of 46%, while the butanol biorefinery achieved an NPV of $16.67 million with an IRR of 37%.

18.8 Feasibility of Biorefineries from Environmental and Social Perspectives

For the implementation of the biorefinery in a sustainable context, it is imperative that the facility pursue a balance between the tripod of the sustainability, i. e., its economic, social and environmental performance. Therefore, for the sustainable implementation of biorefineries, its facilities should be less harmful to the environment, beneficial to society and profitable to the company (Shen et al. 2010). In this sense, the biorefinery is considered a key tool for sustainable development (Radermacher 2016), since the essence of these facilities is the conversion of renewable raw materials, wastes and by-products into marketable biobased products such as bioenergy and bioproducts (Annevelink et al. 2022a) with concomitant benefits to society.

Sustainability analysis in the context of biorefinery is not an easy task as this analysis takes into account the influence of economic, environmental and social aspects from the initial design phase to a more developed process to ensure efficiency and effectiveness in production capacity (Arias et al. 2023). According to Lindorfer et al. (2019), biorefineries are responsible for several direct and indirect impacts on the environment, which requires a systematic assessment of impacts. Life cycle assessment (LCA) is one of the most widely used environmental analysis methods in the world and is used to compare biorefinery processes with petroleum refineries and internally (evaluating different biorefineries configurations), to investigate the sustainability of biofuel and biomaterial production in a biorefinery (Khoshnevisan et al. 2018), and to provide comprehensive information on the environmental impacts of a process, product, or system (Patel and Panwar 2023). The LCA includes several assessment modules from raw material production to product end-of-life, including information on input consumption, products and by-products produced, GHG emissions studied, and energy balance of the biorefinery (Lindorfer et al. 2019). The main barriers for environmental aspects, according to Annevelink et al. (2022a) are the lack of government policies for biorefineries, recycling processes, expensive assessment of environmental sustainability with LCA, sustainability requirements for bio-based products are more stringent than for fossil bio-based products.

Among the pillars of sustainability, social analysis in the context of biorefineries is still a challenge for researchers and project developers. This is due to the lack of appropriate methodologies. In this sense, the issues include the assessment of job creation, food supply and health, indicators of public and geopolitical conditions, inclusion of minorities, dignified working conditions, contribution to community development, transparency, among others. other social impact factors. (Aristizábal-Marulanda et al. 2020). In this context, the social life cycle assessment (S-LCA) is a tool that assesses the potential positive and negative effects of a product or process on society throughout its entire life cycle (Yang et al. 2020c). According to Annevelink et al. (2022a), the main social barriers to biorefineries are: low public awareness of biorefinery products, confusing terminology, perception of lower quality of biobased products, perception of higher price of bioproducts.

18.9 Development and Trends of Biorefineries

Growing global concerns about climate change and food security have focused attention on developing new production models that reduce dependence on fossil fuels, expand the use of renewable resources, reduce GHG emissions, and offer a more sustainable and greener world (Leong et al. 2021). In this context, the development of biorefineries has increased in recent years to address global sustainability challenges, such as the Sustainable Development Goals (SDG) (Solarte-Toro and Cardona Alzate 2021), commitments made by countries at international conferences, and incentives for energy transition, circular economy and bioeconomy (Solarte-Toro et al. 2022).

During the elaboration, creation and implementation of biorefinery, the main barriers to their development are related to technical, economic (Arias et al. 2023), environmental and social aspects. According to Annevelink et al. (2022b), barriers to feedstocks (from the technical perspective) include the complex nature, seasonality, quality loss and degradation of biomass, and uncertainties about the true cost of processing secondary feedstocks (such as waste and recycled materials). From lignocellulosic biomass conversion processes, which include pretreatment, primary and secondary refining, purification, reuse and recovery of by-products, the main technical barriers are biochemical processes with low product concentration and production of multiple secondary products (making downstream processes expensive and energy intensive), limited integration of the primary and secondary conversion process. These technical issues result in high cost investments (economic barriers) (Annevelink et al. 2022b). In terms of platforms and products, the main barriers cited by Annevelink et al. (2022b) are lack of cost competitiveness of products with the same functionality, low diversification of intermediates into new products, some bio-based products have low quality and low performance compared to fossil-based products, difficulty in demonstrating the performance of bio-based products and other new functionalities.

Despite these barriers, the development of biorefineries has increased in the last decade due to numerous government initiatives for biofuels, bioproducts, and biochemicals by fully utilizing resources and extending their useful life (Arias et al. 2023). The development of biorefineries is the key to an integrated and optimized production of biomass processing, combining routes of chemical, biochemical, physical and thermochemical conversions, aiming at the full utilization of available resources, with greater efficiency and the lowest environmental impact (Calvo-Flores and Martin-Martinez 2022). According to Levidow (2015), the development of biorefineries can offer several socio-economic benefits, such as promoting rural development, improving the economic conditions of certain regions through the production of biofuels, and promoting employment opportunities.

The World Bioenergy Association (WBA) has estimated the global techno-economic potential of biomass for energy at 150EJ per year by 2030, mainly from agricultural, forestry and organic wastes (Nakada et al. 2014). Increasing the potential of biorefineries to use these types of biomasses for the production of bioproducts, driven by research and development of biorefinery technologies and processes. These technological advances can expand the market and attract new investment (Fortune Business Insights 2023). The biorefinery market is estimated to reach $210 billion by the end of 2035, with a compound annual growth rate (CAGR) of approximately 8% between 2023 and 2035 (Reasearch Nester 2023). According to (Reasearch Nester 2023), the drivers of this growth are related to the growing demand for renewable fuels, technological advances in the biorefining process, initiatives to reduce carbon emissions into the atmosphere, increased demand and consumption of electricity, and the importance of using energy from waste in bioenergy. Among the growing biorefineries, the microalgae biorefinery has gained a lot of attention because of its specific characteristics for the production of biofuels, proteins, lipids, cellulose, electricity and various other high-value chemicals.

18.10 Conclusions

Market demand for biorefinery products is being driven by several key factors that stem from the need for sustainable and environmentally friendly alternatives to traditional fossil fuel- based products. In this sense, TEA is an important tool for assessing the viability of a process, identifying possible bottlenecks, generating information on the performance of the process, supporting decision making, and developing new technologies and processes. In literature can be found a wide range of TEA applied in biorefineries showing that these facilities are promising and play an important role in the world context.

Acknowledgements Authors are grateful to São Paulo Research Foundation (FAPESP) for the grant #2016/10636-8, #2019/15851-2, #2020/15450-5, #2022/10900-8, and to Brazilian National Council for Scientific and Technological Development (CNPq) for the grants #407716/2021-1 and

#152289/2022-4. This study was financed in part by the Coordenação de Aperfeiçoamento de Pessoal de Nível Superior, Brasil (CAPES, Finance Code 001).

References

Adams P, Bridgwater T, Lea-Langton A et al (2018) Biomass conversion technologies. Greenhouse Gas Balances Bioenergy Syst 107–139. https://doi.org/10.1016/B978-0-08-101036-5.00008-2

Adhikari S, Nam H, Chakraborty JP (2018) Conversion of solid wastes to fuels and chemicals through pyrolysis. In: Waste biorefinery: potential and perspectives. Elsevier, Amsterdam, pp 239–263

Adiletta G, Russo P, Senadeera W, Di Matteo M (2016) Drying characteristics and quality of grape under physical pretreatment. J Food Eng 172:9–18. https://doi.org/10.1016/j.jfoodeng.2015.06.031

Akizuki S, Suzuki H, Fujiwara M, Toda T (2023) Impacts of steam explosion pretreatment on semi-continuous anaerobic digestion of lignin-rich submerged macrophyte. J Clean Prod 385. https://doi.org/10.1016/j.jclepro.2022.135377

Al-Saadi LS, Eze VC, Harvey AP (2022) Techno-economic analysis of processes for biodiesel production with integrated co-production of higher added value products from glycerol. Biofuels 13:489–496. https://doi.org/10.1080/17597269.2020.1767495

Alternative Fuels Data Center (2022) Ethanol Feedstocks. In: U.S. Department of Energy. https://afdc.energy.gov/fuels/ethanol_feedstocks.html

Amin FR, Khalid H, Zhang H et al (2017) Pretreatment methods of lignocellulosic biomass for anaerobic digestion. AMB Express 7

Andreola C, González-Camejo J, Tambone F et al (2023) Techno-economic assessment of biorefinery scenarios based on mollusc and fish residuals. Waste Manage 166:294–304. https://doi.org/10.1016/j.wasman.2023.05.014

Angulo-Mosquera LS, Alvarado-Alvarado AA, Rivas-Arrieta MJ et al (2021) Production of solid biofuels from organic waste in developing countries: a review from sustainability and economic feasibility perspectives. Sci Total Environ 795. https://doi.org/10.1016/j.scitotenv.2021.148816

Ani FN (2016) Utilization of bioresources as fuels and energy generation. Electr Renew Energy Syst 140–155. https://doi.org/10.1016/B978-0-12-804448-3.00008-6

Annevelink B, de Jong E, Mandl M (2022a) Task 42 Biorefining in a circular economy: final task report triennium 2019–2021

Annevelink B, Garcia Chavez L, van Ree R et al (2022b) Global biorefinery status report 2022

Ao T, Luo Y, Chen Y, et al (2020) Towards zero waste: a valorization route of washing separation and liquid hot water consecutive pretreatment to achieve solid vinasse based biorefinery. J Clean Prod 248. https://doi.org/10.1016/j.jclepro.2019.119253

AP Y, Farghali M, Mohamed IMA et al (2021) Potential of biogas production from the anaerobic digestion of *Sargassum fulvellum* macroalgae: Influences of mechanical, chemical, and biological pretreatments. Biochem Eng J 175. https://doi.org/10.1016/j.bej.2021.108140

Arias A, Feijoo G, Moreira MT (2023) Biorefineries as a driver for sustainability: key aspects, actual development and future prospects. J Clean Prod 418. https://doi.org/10.1016/j.jclepro.2023.137925

Aristizábal-Marulanda V, Solarte-Toro JC, Cardona Alzate CA (2020) Economic and social assessment of biorefineries: the case of Coffee Cut-Stems (CCS) in Colombia. Bioresour Technol Rep 9. https://doi.org/10.1016/j.biteb.2020.100397

Baldoni E, Reumerman P, Parisi C et al (2021) Chemical and material driven biorefineries in the EU and beyond. Luxembourg

Barbosa GV de S, Dos Santos JM, Diniz CA et al (2019) Energy cane breeding. In: Sugarcane biorefinery, technology and perspectives. Elsevier, Amsterdam, pp 103–116

Bardhan P, Deka A, Bhattacharya SS, et al (2022) Economical aspect in biomass to biofuel production. Value-Chain Biofuels Fundam Technol Standard 395–427. https://doi.org/10.1016/B978-0-12-824388-6.00003-8

Basso TP (2020) Biomass pretreatment and characterization: a review. In: Basso TO, Luiz Carlos Basso (eds) Biotechnological applications of biomass. IntechOpen, pp 1–17

Bayly GR (2014) Lipids and disorders of lipoprotein metabolism. In: Clinical biochemistry: metabolic and clinical aspects, 3rd edn, pp 702–736. https://doi.org/10.1016/B978-0-7020-5140-1.00037-7

Belyakov N (2019) Bioenergy. In: Sustainable power generation. Elsevier, Amsterdam, pp 461–474

Bhushan S, Jayakrishnan U, Shree B et al (2023) Biological pretreatment for algal biomass feedstock for biofuel production. J Environ Chem Eng 11. https://doi.org/10.1016/j.jece.2023.109870

Blanco A, Blanco G (2017) Carbohydrates. In: Medical Biochemistry. Elsevier Inc, pp 73–97. https://doi.org/10.1016/B978-0-12-803550-4/00004-5

Brémond U, de Buyer R, Steyer JP et al (2018) Biological pretreatments of biomass for improving biogas production: an overview from lab scale to full-scale. Renew Sustain Energy Rev 90:583–604

Buckman G, Ekelund D, Bygrave S et al (2018) Canberra—renewable Energy Leader. In: Urban energy transition. Elsevier, Amsterdam, pp 419–444

Budzianowski WM (2017) High-value low-volume bioproducts coupled to bioenergies with potential to enhance business development of sustainable biorefineries. Renew Sustain Energy Rev 70:793–804. https://doi.org/10.1016/j.rser.2016.11.260

Bulkan G, Ferreira JA, Rajendran K, Taherzadeh MJ (2020) Techno-economic analysis of bioethanol plant by-product valorization: exploring market opportunities with protein-rich fungal biomass production. Fermentation 6. https://doi.org/10.3390/fermentation6040099

Calvo-Flores FG, Martin-Martinez FJ (2022) Biorefineries: achievements and challenges for a bio-based economy. Front Chem 10. https://doi.org/10.3389/fchem.2022.973417

Cao GL, Xia XF, Zhao L, et al (2013) Development of AFEX-based consolidated bioprocessing on wheat straw for biohydrogen production using anaerobic microflora. Int J Hydrogen Energy 15653–15659

Carneiro MLNM, Pradelle F, Braga SL et al (2017) Potential of biofuels from algae: comparison with fossil fuels, ethanol and biodiesel in Europe and Brazil through life cycle assessment (LCA). Renew Sustain Energy Rev 73:632–653

Castoldi R, Bracht A, de Morais GR et al (2014) Biological pretreatment of *Eucalyptus grandis* sawdust with white-rot fungi: study of degradation patterns and saccharification kinetics. Chem Eng J 258:240–246. https://doi.org/10.1016/j.cej.2014.07.090

Castro JDS, da Silva EGP, Virgens CF (2020) Evaluation of models to predict the influence of chemical pretreatment on the peels of *Nephelium lappaceum* L. based on pyrolysis kinetic parameters obtained using a combined Fraser-Suzuki function and Friedman's isoconversional method. J Anal Appl Pyrolysis 149. https://doi.org/10.1016/j.jaap.2020.104827

Chaturvedi T, Torres AI, Stephanopoulos G et al (2020) Developing process designs for biorefineries-definitions, categories, and unit operations. Energies (Basel) 13. https://doi.org/10.3390/en13061493

Chen WS, Strik DPBTB, Buisman CJN, Kroeze C (2017) Production of caproic acid from mixed organic waste: an environmental life cycle perspective. Environ Sci Technol 51:7159–7168. https://doi.org/10.1021/acs.est.6b06220

Cherubini F (2010) The biorefinery concept: using biomass instead of oil for producing energy and chemicals. Energy Convers Manag 51:1412–1421. https://doi.org/10.1016/j.enconman.2010.01.015

Chew KW, Yap JY, Show PL et al (2017) Microalgae biorefinery: high value products perspectives. Bioresour Technol 229:53–62. https://doi.org/10.1016/j.biortech.2017.01.006

Dahmen N, Dinjus E, Kruse A (2009) Fuels—hydrogen production. In: Biomass: thermochemical processes. Encyclopedia of electrochemical power sources, pp 259–267. https://doi.org/10.1016/B978-044452745-5.00313-0

Dauda AB, Ajadi A, Tola-Fabunmi AS, Akinwole AO (2020) Waste production in aquaculture: Parte 1—sources, components and managements in different culture systems. In: Global Seafood. https://www.globalseafood.org/advocate/waste-in-aquaculture-part-1/. Accessed 21 Aug 2023

de Castro REN, Alves RMB, Hawkes A, Nascimento CAO (2018) Open sugarcane process simulation platform. Comput Aided Chem Eng 44:1819–1824. https://doi.org/10.1016/B978-0-444-64241-7.50298-6

Dehghani MH, Omrani GA, Karri RR (2021) Solid waste—sources, toxicity, and their consequences to human health. In: Soft computing techniques in solid waste and wastewater management, pp 205–213. https://doi.org/10.1016/B978-0-12-824463-0.00013-6

Demirbas MF (2009) Biorefineries for biofuel upgrading: a critical review. Appl Energy 86. https://doi.org/10.1016/j.apenergy.2009.04.043

Dias MOS, Junqueira TL, Jesus CDF et al (2012) Improving second generation ethanol production through optimization of first generation production process from sugarcane. Energy 43:246–252. https://doi.org/10.1016/j.energy.2012.04.034

Diederichs GW, Ali Mandegari M, Farzad S, Görgens JF (2016) Techno-economic comparison of biojet fuel production from lignocellulose, vegetable oil and sugar cane juice. Bioresour Technol 216:331–339. https://doi.org/10.1016/j.biortech.2016.05.090

E4tech, RE-CORD, WUR (2015) From the sugar platform to biofuels and biochemicals. London

Elgharbawy AS, Ali RM (2022) Techno-economic assessment of the biodiesel production using natural minerals rocks as a heterogeneous catalyst via conventional and ultrasonic techniques. Renew Energy 191:161–175. https://doi.org/10.1016/j.renene.2022.04.020

Elias AM, Longati AA, de Campos Giordano R, Furlan FF (2021a) Retro-techno-economic- environmental analysis improves the operation efficiency of 1G-2G bioethanol and bioelectricity facilities. Appl Energy 282. https://doi.org/10.1016/j.apenergy.2020.116133

Elias AM, Longati AA, Ellamla HR et al (2021b) Techno-economic-environmental analysis of sophorolipid biosurfactant production from sugarcane bagasse. Ind Eng Chem Res 60:9833–9850. https://doi.org/10.1021/acs.iecr.1c00069

EPE (2023) Brazilian energy balance 2023. Rio de Janeiro

Farzad S, Mandegari MA, Görgens JF (2017) Integrated techno-economic and environmental analysis of butadiene production from biomass. Bioresour Technol 239:37–48. https://doi.org/10.1016/j.biortech.2017.04.130

Fasahati P, Saffron CM, Woo HC, Liu JJ (2017) Potential of brown algae for sustainable electricity production through anaerobic digestion. Energy Convers Manag 135:297–307. https://doi.org/10.1016/j.enconman.2016.12.084

Fdez.-Güelfo LA, Álvarez-Gallego C, Sales Márquez D, Romero García LI (2011) Biological pretreatment applied to industrial organic fraction of municipal solid wastes (OFMSW): effect on anaerobic digestion. Chem Eng J 172:321–325. https://doi.org/10.1016/j.cej.2011.06.010

Figueroa-Torres GM, Theodoropoulos C (2023) Techno-economic analysis of a microalgae-based biorefinery network for biofuels and value-added products. Bioresour Technol Rep 101524. https://doi.org/10.1016/j.biteb.2023.101524

Fortune Business Insights (2023) Biorefinery Market Size Expected to Rise $80.26 billion at CAGR 8.6% by 2029. In: GlobeNewswire. https://www.globenewswire.com/news-release/2023/07/18/2706264/0/en/Biorefinery-Market-Size-Expected-to-Rise-80-26-billion-at-CAGR-8-6-by-2029.html. Accessed 26 Aug 2023

Gavrilescu M (2014) Biorefinery systems: an overview. In: Bioenergy research: advances and applications. Elsevier, Amsterdam, pp 219–241

Gent S, Twedt M, Gerometta C, Almberg E (2017) Introduction to thermochemical conversion processes. In: Theoretical and applied aspects of biomass torrefaction, pp 1–16. https://doi.org/10.1016/B978-0-12-809483-9.00001-4

Gerrior D, Delsoz Bahri K, Kermanshahi-pour A et al (2022) Life cycle assessment and techno-economic analysis of a novel closed loop corn ethanol biorefinery. Sustain Prod Consum 30:359–376. https://doi.org/10.1016/j.spc.2021.12.007

Giordano RC, Elias AM, Furlan FF (2020) Virtual biorefineries: computational tools for bridging the gap toward a low carbon economy. In: Recent advances in bioconversion of lignocellulose to biofuels and value added chemicals within the biorefinery concept. Elsevier, Amsterdam, pp 233–257

Gnanasekaran L, Priya AK, Thanigaivel S et al (2023) The conversion of biomass to fuels via cutting-edge technologies: explorations from natural utilization systems. Fuel 331. https://doi.org/10.1016/j.fuel.2022.125668

Gonçalves F de O, Firmani Perna R, Savioli Lopes E, et al (2023) Strategies to ensure fuel security in brazil considering a forecast of ethanol production. Biomass 3:1–17. https://doi.org/10.3390/biomass3010001

González-Contreras M, Hernández-Escoto H, Aguilar-Garnica E (2023) A comprehensive analysis of bioethanol and ethyl lactate joint production in second-generation biorefinery: simulation, techno-economic, and profitability assessments. Bioresour Technol 385. https://doi.org/10.1016/j.biortech.2023.129470

Goswami L, Kayalvizhi R, Dikshit PK et al (2022) A critical review on prospects of bio-refinery products from second and third generation biomasses. Chem Eng J 448. https://doi.org/10.1016/j.cej.2022.137677

Government of Canada (2020) Ethanol. In: Government of Canada. https://natural-resources.canada.ca/energy-efficiency/transportation-alternative-fuels/alternative-fuels/biofuels/ethanol/3493

Government of India (2022) Ethanol Growth Story

Haldar D, Bhattacharjee N, Shabbirahmed AM et al (2023) Purification of biogas for methane enrichment using biomass-based adsorbents: a review. Biomass Bioenergy 173. https://doi.org/10.1016/j.biombioe.2023.106804

Hamada A, Kita Y, Murakami K et al (2020) Enhancement of transduction efficiency using Adeno-associated viral vectors by chemical pretreatment to mice bladder urothelium. J Virol Methods 279. https://doi.org/10.1016/j.jviromet.2020.113854

Hayes DJM (2013) Biomass composition and its relevance to biorefining. In: The role of catalysis for the sustainable production of bio-fuels and bio-chemicals, pp 27–65. https://doi.org/10.1016/B978-0-444-56330-9.00002-4

Hingsamer M, Jungmeier G (2018) Biorefineries. In: The role of bioenergy in the emerging bioeconomy: resources, technologies, sustainability and policy. Elsevier, Amsterdam, pp 179–222

Höfer R (2015) Sugar- and Starch-based biorefineries. In: Industrial biorefineries and white biotechnology, pp 157–235. https://doi.org/10.1016/B978-0-444-63453-5.00005-7

Huang H, Long S, Singh V (2016) Techno-economic analysis of biodiesel and ethanol co- production from lipid-producing sugarcane. Biofuels Bioprod Biorefin 10:299–315. https://doi.org/10.1002/bbb.1640

IEA (2022a) Energy Efficiency 2022. Paris

IEA (2021) Key World Energy Statistics 2021. Paris

IEA (2022b) Energy statistics data browser. In: IEA. https://www.iea.org/data-and-statistics/data-tools/energy-statistics-data-browser. Accessed 9 Dec 2022

IEA Bioenergy (2009a) IEA Bioenergy Task 42 Biorefinery

IEA Bioenergy (2009b) Biorefineries: adding value to the sustainable utilisation of biomass. Paris

IEA Bioenergy (2021) Implementation of bioenergy in Brazil - 2021 update

Imman S, Arnthong J, Burapatana V et al (2015) Influence of alkaline catalyst addition on compressed liquid hot water pretreatment of rice straw. Chem Eng J 278:85–91. https://doi.org/10.1016/j.cej.2014.12.032

Ioannidou SM, López-Gómez JP, Venus J et al (2023) Techno-economic evaluation and life cycle assessment for sustainable alternative biorefinery concepts using the organic fraction of municipal solid waste. Green Chem 25:4482–4500. https://doi.org/10.1039/D3GC00244F

Islas J, Manzini F, Masera O, Vargas V (2019) Solid Biomass to heat and power. In: The role of bioenergy in the emerging bioeconomy: resources, technologies, sustainability and policy, pp 145–177. https://doi.org/10.1016/B978-0-12-813056-8.00004-2

Jallouli W, Keskes S, Guidara W, et al (2021) Acidic pretreatment as a chemical approach for enhanced *Photorhabdus temperata* bioinsecticide production from industrial wastewater. J Environ Manage 278. https://doi.org/10.1016/j.jenvman.2020.111476

Jędrzejczyk M, Soszka E, Czapnik M et al (2019) Physical and chemical pretreatment of lignocellulosic biomass. In: Second and third generation of feedstocks: the evolution of biofuels. Elsevier, Amsterdam, pp 143–196

Jong E, Stichnothe H, Bell G et al (2020) Bio-based chemicals: a 2020 update

Junqueira TL, Chagas MF, Gouveia VLR et al (2017) Techno-economic analysis and climate change impacts of sugarcane biorefineries considering different time horizons. Biotechnol Biofuels 10:1–12. https://doi.org/10.1186/s13068-017-0722-3

Karmaker AK, Rahman MM, Hossain MA, Ahmed MR (2020) Exploration and corrective measures of greenhouse gas emission from fossil fuel power stations for Bangladesh. J Clean Prod 244. https://doi.org/10.1016/j.jclepro.2019.118645

Khavari F, Saidijam M, Taheri M, Nouri F (2021) Microalgae: therapeutic potentials and applications. Mol Biol Rep 48:4757–4765. https://doi.org/10.1007/s11033-021-06422-w

Khoshnevisan B, Rafiee S, Tabatabaei M et al (2018) Life cycle assessment of castor-based biorefinery: a well to wheel LCA. Int J Life Cycle Assess 23:1788–1805. https://doi.org/10.1007/s11367-017-1383-y

Kiehbadroudinezhad M, Merabet A, Ghenai C, et al (2023) The role of biofuels for sustainable MicrogridsF: a path towards carbon neutrality and the green economy. Heliyon 9. https://doi.org/10.1016/j.heliyon.2023.e13407

Kingsley PR, Braud L, Kumar Mediboyina M et al (2023) Prospects for commercial microalgal biorefineries: integrated pilot demonstrations and process simulations based techno-economic assessment of single and multi-product value chains. Algal Res 74:103190. https://doi.org/10.1016/j.algal.2023.103190

Kiskini A, Zondervan E, Wierenga PA et al (2016) Using product driven process synthesis in the biorefinery. Comput Chem Eng 91:257–268. https://doi.org/10.1016/j.compchemeng.2016.03.030

Yogalakshmi KN, Mohamed MU, Kavitha S et al (2023) Lignocellulosic biorefinery technologies: a perception into recent advances in biomass fractionation, biorefineries, economic hurdles and market outlook. Fermentation 9. https://doi.org/10.3390/fermentation9030238

Knäble D, de Quevedo PE, Pérez-Cornejo C, Baumgärtler T (2022) The impact of the circular economy on sustainable development: a European panel data approach. Sustain Prod Consum 34:233–243. https://doi.org/10.1016/j.spc.2022.09.016

Kokkinos K, Karayannis V, Moustakas K (2020) Circular bio-economy via energy transition supported by Fuzzy Cognitive Map modeling towards sustainable low-carbon environment. Sci Total Environ 721. https://doi.org/10.1016/j.scitotenv.2020.137754

Konda NVSNM, Singh S, Simmons BA, Klein-Marcuschamer D (2015) An Investigation on the economic feasibility of macroalgae as a potential feedstock for biorefineries. Bioenergy Res 8:1046–1056. https://doi.org/10.1007/s12155-015-9594-1

Kramer CAC, Da Silva ARL, De Carvalho LS (2020) Influence of phenylpropanoid units of lignin and its oxidized derivatives on the stability and βo4 binding properties: DFT and QTAIM approach. Org Biomol Chem 18:5897–5905. https://doi.org/10.1039/d0ob01171a

Kumar D, Long SP, Arora A, Singh V (2021) Techno-economic feasibility analysis of engineered energycane-based biorefinery co-producing biodiesel and ethanol. GCB Bioenergy 13:1498–1514. https://doi.org/10.1111/gcbb.12871

Ladakis D, Stylianou E, Ioannidou SM, et al (2022) Biorefinery development, techno- economic evaluation and environmental impact analysis for the conversion of the organic fraction of municipal solid waste into succinic acid and value-added fractions. Bioresour Technol 354. https://doi.org/10.1016/j.biortech.2022.127172

Lan K, Xu Y, Kim H, et al (2021) Techno-economic analysis of producing xylo-oligosaccharides and cellulose microfibers from lignocellulosic biomass. Bioresour Technol 340. https://doi.org/10.1016/j.biortech.2021.125726

Lemões JS, Lemons e Silva CF, Avila SPF et al (2018) Chemical pretreatment of *Arundo donax* L. for second-generation ethanol production. Electron J Biotechnol 31:67–74. https://doi.org/10.1016/j.ejbt.2017.10.011

Leong HY, Chang CK, Khoo KS et al (2021) Waste biorefinery towards a sustainable circular bioeconomy: a solution to global issues. Biotechnol Biofuels 14. https://doi.org/10.1186/s13068-021-01939-5

Levidow L (2015) Eco-efficient biorefineries: techno-fix for resource constraints? Économie rurale 1–23. https://doi.org/10.4000/economierurale.4729

Li R, Li L, Zhang Z et al (2021) Limiting factors of heavy metals removal during anaerobic biological pretreatment of municipal solid waste landfill leachate. J Hazard Mater 416. https://doi.org/10.1016/j.jhazmat.2021.126081

Lindorfer J, Rosenfeld D, Annevelink B, Mandl M (2019) Technical, economic and environmental assessment of biorefinery concepts: developing a practical approach for characterisation

Liu CY, Sun YY, Jia YQ et al (2021) Effect of steam explosion pretreatment on the structure and bioactivity of *Ampelopsis grossedentata* polysaccharides. Int J Biol Macromol 185:194–205. https://doi.org/10.1016/j.ijbiomac.2021.06.002

Liu G, Sun J, Zhang J et al (2015) High titer l-lactic acid production from corn stover with minimum wastewater generation and techno-economic evaluation based on Aspen plus modeling. Bioresour Technol 198:803–810. https://doi.org/10.1016/j.biortech.2015.09.098

Llano T, Arce C, Gallart LE, et al (2023) Techno-economic analysis of macroalgae biorefineries: a comparison between ethanol and butanol facilities. Fermentation 9. https://doi.org/10.3390/fermentation9040340

Longati AA, Batista G, Cruz AJG (2020) Brazilian integrated sugarcane-soybean biorefinery: trends and opportunities. Curr Opin Green Sustain Chem 26. https://doi.org/10.1016/j.cogsc.2020.100400

Longati AA, Lino ARA, Giordano RC et al (2018) Defining research and development process targets through retro-techno-economic analysis: the sugarcane biorefinery case. Bioresour Technol 263:1–9. https://doi.org/10.1016/j.biortech.2018.04.102

Longati AA, Lino ARA, Giordano RC et al (2019) Biogas production from anaerobic digestion of Vinasse in sugarcane biorefinery: a techno-economic and environmental analysis. Waste Biomass Valorization 11:4573–4591. https://doi.org/10.1007/s12649-019- 00811-w

Manfren M, Nastasi B, Tronchin L et al (2021) Techno-economic analysis and energy modelling as a key enablers for smart energy services and technologies in buildings. Renew Sustain Energy Rev 150. https://doi.org/10.1016/j.rser.2021.111490

Maroneze MM, Queiroz Zepka L, Jacob-Lopes E (2019) Introductory chapter: renewable resources and biorefineries. In: Renewable resources and biorefineries. IntechOpen

Mendes FM, Dias MOS, Ferraz A et al (2017) Techno-economic impacts of varied compositional profiles of sugarcane experimental hybrids on a biorefinery producing sugar, ethanol and electricity. Chem Eng Res Des 125:72–78. https://doi.org/10.1016/j.cherd.2017.06.023

Millati R, Cahyono RB, Ariyanto T et al (2019) Agricultural, industrial, municipal, and forest wastes: an overview. In: Sustainable resource recovery and zero waste approaches, pp 1–22. https://doi.org/10.1016/B978-0-444-64200-4.00001-3

Mohammed NI, Kabbashi N, Alade A (2018) Significance of agricultural residues in sustainable biofuel development. In: Agricultural waste and residues. InTech

Mokomele T, da Costa Sousa L, Colbert A et al (2022) Coupling AFEX and steam-exploded sugarcane residue pellets with a room temperature CIIII-activation step lowered enzyme dosage requirements for sugar conversion. Chem Eng J 446. https://doi.org/10.1016/j.cej.2022.137117

Moncada J, El-Halwagi MM, Cardona CA (2013) Techno-economic analysis for a sugarcane biorefinery: Colombian case. Bioresour Technol 135:533–543. https://doi.org/10.1016/j.biortech.2012.08.137

Moraga G, Huysveld S, Mathieux F et al (2019) Circular economy indicators: what do they measure? Resour Conserv Recycl 146:452–461. https://doi.org/10.1016/j.resconrec.2019.03.045
Morakile T, Mandegari M, Farzad S, Görgens JF (2022) Comparative techno-economic assessment of sugarcane biorefineries producing glutamic acid, levulinic acid and xylitol from sugarcane. Ind Crops Prod 184. https://doi.org/10.1016/j.indcrop.2022.115053
Mustapha SI, Bux F, Isa YM (2022) Techno-economic analysis of biodiesel production over lipid extracted algae derived catalyst. Biofuels 13:663–674. https://doi.org/10.1080/17597269.2021.1878577
Nabben M, Glatz JFC (2015) Basics in metabolically relevant biochemistry. In: The scientist's guide to cardiac metabolism. Elsevier, Amsterdam, pp 7–18
Nakada S, Saygin D, Gielen D (2014) Global bioenergy supply and demand projections for the year 2030
Ng DKS, Ng KS, Ng RTL (2017) Integrated biorefineries. In: Encyclopedia of sustainable technologies. Elsevier, Amsterdam, pp 299–314
Nguyen QA, Yang J, Bae HJ (2017) Bioethanol production from individual and mixed agricultural biomass residues. Ind Crops Prod 95:718–725. https://doi.org/10.1016/j.indcrop.2016.11.040
Nieder-Heitmann M, Haigh K, Görgens JF (2019) Process design and economic evaluation of integrated, multi-product biorefineries for the co-production of bio-energy, succinic acid, and polyhydroxybutyrate (PHB) from sugarcane bagasse and trash lignocelluloses. Biofuels Bioprod Biorefin 13:599–617. https://doi.org/10.1002/bbb.1972
Nilander RA, Targa M dos S, Catelani C de S, Balduino AR (2023) Proposal for Analysis of Geoenvironmental Sustainability Indicators for the Municipality of Barcarena in the State of Pará. International Journal of Advanced Engineering Research and Science 10:075–083. https://doi.org/10.22161/ijaers.102.7
Nunes LJR, De Oliveira Matias JC, Da Silva Catalão JP (2018) Physical pretreatment of biomass. In: Torrefaction of biomass for energy applications. Elsevier, Amsterdam, pp 45–88
Ou L, Dou C, Yu JH et al (2021) Techno-economic analysis of sugar production from lignocellulosic biomass with utilization of hemicellulose and lignin for high-value co- products. Biofuels Bioprod Biorefin 15:404–415. https://doi.org/10.1002/bbb.2170
Özüdoğru HMR, Nieder-Heitmann M, Haigh KF, Görgens JF (2019) Techno-economic analysis of product biorefineries utilizing sugarcane lignocelluloses: Xylitol, citric acid and glutamic acid scenarios annexed to sugar mills with electricity co-production. Ind Crops Prod 133:259–268. https://doi.org/10.1016/j.indcrop.2019.03.015
Panthapulakkal S, Sain M (2015) The use of wheat straw fibres as reinforcements in composites. In: Biofiber reinforcements in composite materials. Elsevier, Amsterdam, pp 423–453
Parsons S, Abeln F, McManus MC, Chuck CJ (2019) Techno-economic analysis (TEA) of microbial oil production from waste resources as part of a biorefinery concept: assessment at multiple scales under uncertainty. J Chem Technol Biotechnol 94:701–711. https://doi.org/10.1002/jctb.5811
Patel A, Shah AR (2021) Integrated lignocellulosic biorefinery: gateway for production of second generation ethanol and value added products. J Bioresour Bioprod 6:108–128. https://doi.org/10.1016/j.jobab.2021.02.001
Patel MR, Panwar NL (2023) Biochar from agricultural crop residues: environmental, production, and life cycle assessment overview. Resour Conserv Recycl Adv 19. https://doi.org/10.1016/j.rcradv.2023.200173
Pereira LG, Dias MOS, Junqueira TL et al (2014) Butanol production in a sugarcane biorefinery using ethanol as feedstock. Part II: Integration to a second generation sugarcane distillery. Chem Eng Res Des 92:1452–1462. https://doi.org/10.1016/j.cherd.2014.04.032
Perez CL, Pereira LPR da C, Milessi TS et al (2022) Towards a practical industrial 2G ethanol production process based on immobilized recombinant *S. cerevisiae*: medium and strain selection for robust integrated fixed-bed reactor operation. Renew Energy 185:363–375. https://doi.org/10.1016/j.renene.2021.12.044
Peters MS, Timmerhaus KD, West RE (2003) Plant design and economics for chemical engineers, 5th edn. McGraw-Hill, New York

Philippini RR, Martiniano SE, Ingle AP et al (2020) Agroindustrial byproducts for the generation of biobased products: alternatives for sustainable biorefineries. Front Energy Res 8. https://doi.org/10.3389/fenrg.2020.00152

Pratto B, dos Santos-Rocha MSR, Longati AA et al (2020) Experimental optimization and techno-economic analysis of bioethanol production by simultaneous saccharification and fermentation process using sugarcane straw. Bioresour Technol 297:122494. https://doi.org/10.1016/j.biortech.2019.122494

Radermacher K (2016) Environmental and safety issues of polymers and polymeric material in the printing industry. In: Printing on polymers: fundamentals and applications, pp 397–415. https://doi.org/10.1016/B978-0-323-37468-2.00025-7

Rajaeifar MA, Sadeghzadeh Hemayati S, Tabatabaei M et al (2019) A review on beet sugar industry with a focus on implementation of waste-to-energy strategy for power supply. Renew Sustain Energy Rev 103:423–442. https://doi.org/10.1016/j.rser.2018.12.056

Research Nester (2023) Global biorefinert market size, forecast, and tren highlights over 2023–2035. In: Research Nester. researchnester.com/reports/biorefinery-market/4883. Accessed 26 Aug 2023

Reck BK, Graedel TE (2012) Challenges in metal recycling. Science 337:690–695

Rodrigues Dias R, Deprá MC, Ragagnin de Menezes C et al (2023) The high-value product, bio-waste, and eco-friendly energy as the tripod of the microalgae biorefinery: connecting the dots. Sustainability 15:11494. https://doi.org/10.3390/su151511494

Rouabhia A, Álvarez-Gallego CJ, Fdez-Güelfo LA (2023) Effect of passively aerated biological pretreatment on different biomasses with diverse lignocellulosic fiber profiles. J Environ Manage 342. https://doi.org/10.1016/j.jenvman.2023.118332

Scown CD, Baral NR, Yang M et al (2021) Technoeconomic analysis for biofuels and bioproducts. Curr Opin Biotechnol 67:58–64. https://doi.org/10.1016/j.copbio.2021.01.002

Seufitelli GVS, El-Husseini H, Pascoli DU et al (2022) Techno-economic analysis of an integrated biorefinery to convert poplar into jet fuel, xylitol, and formic acid. Biotechnol Biofuels Bioprod 15. https://doi.org/10.1186/s13068-022-02246-3

Shah AA, Seehar TH, Sharma K, Toor SS (2022) Biomass pretreatment technologies. In: Hydrocarbon biorefinery: sustainable processing of biomass for hydrocarbon biofuels, pp 203–228. https://doi.org/10.1016/B978-0-12-823306-1.00014-5

Sharma HK, Xu C, Qin W (2019) Biological pretreatment of lignocellulosic biomass for biofuels and bioproducts: an overview. Waste Biomass Valorization 10:235–251

Shen L yin, Tam VWY, Tam L, Ji Y bo (2010) Project feasibility study: the key to successful implementation of sustainable and socially responsible construction management practice. J Clean Prod 18:254–259. https://doi.org/10.1016/j.jclepro.2009.10.014

Shen MY, Huang ZH, Kuo YT et al (2022) Improvement of gaseous bioenergy production from spent coffee grounds Co-digestion with pulp wastewater by physical/chemical pretreatments. Int J Hydrogen Energy 47:40664–40671. https://doi.org/10.1016/j.ijhydene.2022.10.041

Shi Y, Zhang S, Xu J et al (2022) Effect of hydrothermal and hydrothermal oxidation pretreatment on the physicochemical properties of biofuel pellet. J Anal Appl Pyrolysis 165. https://doi.org/10.1016/j.jaap.2022.105566

Shukla SK (2020) Renewable biofuels and their by-products for automotive applications. In: Encyclopedia of renewable and sustainable materials, vol 1–5. Elsevier, pp 180–186

Sivaramakrishnan R, Suresh S, Kanwal S et al (2022) Microalgal biorefinery concepts' developments for biofuel and bioproducts: current perspective and bottlenecks. Int J Mol Sci 23. https://doi.org/10.3390/ijms23052623

Slegers PM, Olivieri G, Breitmayer E et al (2020) Design of value chains for microalgal biorefinery at industrial scale: process integration and techno-economic analysis. Front Bioeng Biotechnol 8. https://doi.org/10.3389/fbioe.2020.550758

Solarte-Toro JC, Cardona Alzate CA (2021) Biorefineries as the base for accomplishing the sustainable development goals (SDGs) and the transition to bioeconomy: Technical aspects, challenges and perspectives. Bioresour Technol 340. https://doi.org/10.1016/j.biortech.2021.125626

Solarte-Toro JC, Laghezza M, Fiore S, et al (2022) Review of the impact of socio-economic conditions on the development and implementation of biorefineries. Fuel 328. https://doi.org/10.1016/j.fuel.2022.125169

Sun E, Zhang Y, Yong C, et al (2021a) Biological fermentation pretreatment accelerated the depolymerization of straw fiber and its mechanical properties as raw material for mulch film. J Clean Prod 284. https://doi.org/10.1016/j.jclepro.2020.124688

Sun Y, Liu Y, Chen J et al (2021b) Physical pretreatment of petroleum refinery wastewater instead of chemicals addition for collaborative removal of oil and suspended solids. J Clean Prod 278. https://doi.org/10.1016/j.jclepro.2020.123821

Swaminathan AV, Molitor MS, Burrington KJ et al (2021) A study of various chemical pretreatments to fractionate lipids from whey protein phospholipid concentrate. J Dairy Sci 104:12249–12262. https://doi.org/10.3168/jds.2021-20563

Swart LJ, Petersen AM, Bedzo OKK, Görgens JF (2021) Techno-economic analysis of the valorization of brewers spent grains: production of xylitol and xylo-oligosaccharides. J Chem Technol Biotechnol 96:1632–1644. https://doi.org/10.1002/jctb.6683

Takkellapati S, Li T, Gonzalez MA (2018) An overview of biorefinery-derived platform chemicals from a cellulose and hemicellulose biorefinery. Clean Technol Environ Policy 20:1615–1630. https://doi.org/10.1007/s10098-018-1568-5

Tao L, Markham JN, Haq Z, Biddy MJ (2017a) Techno-economic analysis for upgrading the biomass-derived ethanol-to-jet blendstocks. Green Chem 19:1082–1101. https://doi.org/10.1039/c6gc02800d

Tao L, Milbrandt A, Zhang Y, Wang WC (2017b) Techno-economic and resource analysis of hydroprocessed renewable jet fuel. Biotechnol Biofuels 10. https://doi.org/10.1186/s13068-017-0945-3

Tavares P da S, Acosta R, Nobre P et al (2023) Water balance components and climate extremes over Brazil under 1.5 °C and 2.0 °C of global warming scenarios. Reg Environ Change 23. https://doi.org/10.1007/s10113-023-02042-1

Taylor-de-Lima RLN, da Silva AJG, Legey LFL, Szklo A (2016) Evaluation of economic feasibility under uncertainty of thermochemical route for ethanol production in Brazil. Energy 150:363–376

Teng M, Chen S, Yin X, Zhang G (2020) System metabolic engineering strategies for cell factories construction. Syst Synth Metabol Eng 125–151. https://doi.org/10.1016/B978-0-12-821753-5.00006-X

Thammasittirong SNR, Chatwachirawong P, Khemdee K, Thammasittirong A (2023) Evaluating the potential of newly developed energy cane clones for first- and second- generation ethanol production. Fermentation 9. https://doi.org/10.3390/fermentation9030267

Thomassen G, Egiguren Vila U, Van Dael M et al (2016) A techno-economic assessment of an algal-based biorefinery. Clean Technol Environ Policy 18:1849–1862. https://doi.org/10.1007/s10098-016-1159-2

Trokanas N, Bussemaker M, Velliou E et al (2015) BiOnto: an ontology for biomass and biorefining technologies. In: computer aided chemical engineering. Elsevier B.V., Amsterdam, pp 959–964

Turton R, Bailie RC, Whiting WB, Shaeiwitz JA (2009) Analysis, synthesis and design of chemical processes, 3rd edn. Pearson Education Inc., Boston

Van Heerden C (2023) Techno-economic analysis of 1,3-propanediol, sorbitol, itaconic acid, and xylooligosaccharides production from sugarcane-based feedstocks. Thesis (Master of Chemical Enginering), Faculty of Engineering at Stellenbosh Universty

Vannarath A, Thalla AK (2022) Effects of chemical pretreatments on material solubilization of *Areca catechu* L. *husk*: digestion, biodegradability, and kinetic studies for biogas yield. J Environ Manage 316. https://doi.org/10.1016/j.jenvman.2022.115322

Vásquez-Parra JE, Ochoa-Martínez CI, Bustos-Parra M (2013) Effect of chemical and physical pretreatments on the convective drying of cape gooseberry fruits (*Physalis peruviana*). J Food Eng 119:648–654. https://doi.org/10.1016/j.jfoodeng.2013.06.037

Vignesh KS, Rajadesingu S, Arunachalam KD (2021) Challenges, issues, and problems with zero-waste tools. In: Concepts of advanced zero waste tools: present and emerging waste management practices, pp 69–90. https://doi.org/10.1016/B978-0-12-822183-9.00004-0
Volpi MPC, Fuess LT, Moraes BS (2021) Anaerobic co-digestion of residues in 1G2G sugarcane biorefineries for enhanced electricity and biomethane production. Bioresour Technol 330. https://doi.org/10.1016/j.biortech.2021.124999
Wagemann K (2012) Biorefineries—prerequisite for the realization of a future bioeconomy. In: DGMK conference: reducing the carbon footprint of fuels and petrochemicals. DGMK, Berlin, pp 1–11
Wang K, Costanza-van den Belt M, Heath G et al (2022) Circular economy as a climate strategy: current knowledge and calls-to-action. Washington, DC
Wang S, Wang W, Xu Y et al (2023a) Enhancing bonding synergy and mechanical response of metal/composite hybrid joints through physicochemical surface pretreatment. J Mater Process Technol 315. https://doi.org/10.1016/j.jmatprotec.2023.117923
Wang X, Zhang Y, Xia C et al (2023b) A review on optimistic biorefinery products: biofuel and bioproducts from algae biomass. Fuel 338. https://doi.org/10.1016/j.fuel.2022.127378
Wang Z (2018) Energy and air pollution. In: Comprehensive energy systems. Elsevier, Amsterdam, pp 909–949
Wuest DM, Hou S, Lee KH (2011) Metabolic engineering. In: Moo-Young M (ed) Comprehensive biotechnology, 2nd edn. Elsevier, Amsterdam, pp 617–628
Xu C, Yuan Q, Zhao S, et al (2020) Effects of pretreatments on physical and chemical characteristics of wheat straw used as a maintenance-free compressed green roof substrate material. J Clean Prod 277. https://doi.org/10.1016/j.jclepro.2020.123381
Xu F, Ma Z, Wang X et al (2022) From liquid hot water pretreatment solution to lignin-based hydrophobic deep eutectic solvent for highly efficient extraction of Cr (VI). Int J Biol Macromol 208:883–889. https://doi.org/10.1016/j.ijbiomac.2022.03.190
Yadav A, Sharma V, Tsai ML et al (2023) Development of lignocellulosic biorefineries for the sustainable production of biofuels: towards circular bioeconomy. Bioresour Technol 381
Yang J, An X, Liu L et al (2020a) Cellulose, hemicellulose, lignin, and their derivatives as multi-components of bio-based feedstocks for 3D printing. Carbohydr Polym 250
Yang M, Baral NR, Simmons BA, et al (2020b) Accumulation of high-value bioproducts in planta can improve the economics of advanced biofuels. PNAS 117:8639–8648. www.pnas.org/cgi/doi/10.1073/pnas.2000053117
Yang M, Chen L, Wang J et al (2023) Circular economy strategies for combating climate change and other environmental issues. Environ Chem Lett 21:55–80
Yang S, Ma K, Liu Z et al (2020c) Development and applicability of life cycle impact assessment methodologies. In: Life cycle sustainability assessment for decision-making: methodologies and case studies, pp 95–124. https://doi.org/10.1016/B978-0-12-818355-7.00005-1
Yang S-T, El-Enshasy HA, Thongchul N (eds) (2013) Bioprocessing technologies in biorefinery for sustainable production of fuels, chemicals, and polymers, 1st edn. Wiley, New York
Yin C, Deng X, Yu Z et al (2021) Auto-classification of biomass through characterization of their pyrolysis behaviors using thermogravimetric analysis with support vector machine algorithm: case study for tobacco. Biotechnol Biofuels 14. https://doi.org/10.1186/s13068-021-01942-w
Zheng A, Zhao K, Li L et al (2018) Quantitative comparison of different chemical retreatment methods on chemical structure and pyrolysis characteristics of corncobs. J Energy Inst 91:676–682. https://doi.org/10.1016/j.joei.2017.06.002
Zhou Y (2020) Two birds, one stone: China's portential for zero waste and nationwide ethanol. In: The International Council on Clean Transportation
Zhumadilova A, Zhigitova S, Turalina M (2023) The impact of greenhouse gases on climate change. Sci Horiz 26. https://doi.org/10.48077/scihor6.2023.97

Chapter 19
Microalgal Biofuels in North America Advances and the Way Forward

Lizet Rodríguez-Machín and Luis Ernesto Arteaga-Pérez

Abstract Algae biomass has become an important raw material in the US for the production of third generation biofuels. US and Canada stand for the 30% of the world's algae biofuel production. The objective of this chapter is to present an up-to-date review of major advances, challenges, and drivers for microalgae biofuels production in North America. Bioethanol and biodiesel are the main biofuels produced and marketed from microalgae in North America. Bioethanol from microalgae can be obtained by hydrolysis followed by fermentation, but also by "dark" fermentation, as well as by genetically engineered photofermentation, while biodiesel is obtained by transesterification of the extracted lipids. It can be concluded that the market potential, drivers, and opportunities for microalgae-derived bio-based products in North America are substantial. The advancements in technology, strategic initiatives, and favorable policies contribute to the growth and adoption of microalgae-based biofuels. However, addressing challenges in cultivation, production costs, and refining technologies will be key to unlocking the full potential of microalgae as a sustainable and renewable resource for energy, food, feed, and other valuable products. Continued research, innovation, and collaborations among industries, government organizations, and research institutes will pave the way for a thriving microalgae industry in the future.

L. Rodríguez-Machín (✉)
Universidad Central "Marta Abreu" de Las Villas, Santa Clara, Villa Clara, Cuba
e-mail: lizetr@uclv.edu.cu

L. E. Arteaga-Pérez
Laboratory of Thermal and Catalytic Processes (LPTC-UBB), Wood Engineering Department, University of Bío-Bío, Concepción, Chile
e-mail: larteaga@ubiobio.cl

T. A. Kurniawan and A. Anouzla (eds.), *Algae as a Natural Solution for Challenges in Water-Food-Energy Nexus*, Environmental Science and Engineering,
https://doi.org/10.1007/978-981-97-2371-3_19

19.1 Introduction

The United States (US) has made remarkable strides in reducing its reliance on fossil fuels by embracing renewable energy sources like geothermal, wind, solar, and biomass. In this endeavour, various generations of biofuels have emerged, each with its unique characteristics, but all striving to minimize environmental impacts and meet the crescent energy demands (Jabłońska-Trypuć et al. 2023).

First-generation biofuels, produced from feedstocks such as starch, corn, rapeseed, sugar beet, sunflower, wheat, and barley, have faced criticism due to concerns over the food-versus-fuel debate. This controversy arises from the competition between utilizing these crops for biofuel production or as food sources, a dilemma that poses ethical challenges. Additionally, the profitability of first-generation biofuel production has been relatively low as compared to other products obtained from the same sources of biomass (Ullah et al. 2015). To address the limitations associated to the first generation biofuels, the scientific and industrial communities have increased their efforts on developing sustainable ways for producing second-generation biofuels. The second generation biofuels are those derived from forest biomass, agricultural residues, demolition wood, straw, and grass. Although they offer advantages over first-generation biofuels, such as reduced competition with food production, they also come with drawbacks such as potential biodiversity loss and increased nitrogen emissions from the use of nitrogenous fertilizers during crops cultivation (Sinitsyn and Sinitsyna 2021). Finally, the third-generation biofuels have emerged as a promising solution to overcome the challenges faced by their predecessors. These biofuels are derived from various types of algae, harnessing their immense potential for producing several goods. Microalgae, are the most abundant organisms on Earth, can thrive in diverse habitats, including seawater, salty water, fresh water, and wastewater owing their remarkable adaptability (Abo et al. 2019; John et al. 2011).

Microalgae are rich in essential components such as lipids (8–15% of dry mass in nutrition-rich medium with no starvation period), carbohydrates (20–40% of dry mass), and proteins (30–50% of dry mass). This composition makes microalgae ideal for energy and fuels production, such as alcohols, biodiesel, methane, and hydrogen (Casanova et al. 2022; Jabłońska-Trypuć et al. 2023). Moreover, microalgae contain various complex organic molecules, albeit in smaller quantities, such as vitamins, terpenes, and carotenoids (Casanova et al. 2022). One of the key advantages of microalgae is their rapid growth rate and their ability to thrive without requiring fertile land, freshwater, herbicides, or pesticides, as compared to traditional crops (Abo et al. 2019; Khan et al. 2018; Tan et al. 2020). Furthermore, microalgae cultivation does not disrupt the human and animal food chain, and it is not constrained by seasonal conditions, offering year-round production potential (Abo et al. 2019; Rodríguez-Palacio et al. 2022; Tan et al. 2020). Microalgae cultivation holds significant potential in mitigating the impact of greenhouse gas emissions and global warming by effectively sequestering atmospheric CO_2 and treating wastewater. However, there are certain challenges that need to be addressed in the development of microalgae cultivation. One such challenge is the limited biomass production and the small size of cells when grown in a liquid medium, which escalates the cost of the collection process (Tan et al. 2020).

To optimize microalgae-derived biofuel production, genetically modified algae have been introduced as feedstock. Genetic engineering techniques have been employed to enhance the productivity of specific algae species, offering a promising avenue for biofuel optimization. Nonetheless, the issue of waste disposal remains a significant concern in this context (Jabłońska-Trypuć et al. 2023).

In the US, the algal collection comprises approximately 300,000 specimens, encompassing both natural and genetically modified organisms (Smithsonian 2020). Canada, on the other hand, boasts a collection of over 40,000 algae species (Pankratz et al. 2017), while the global estimate surpasses one million species (Abo et al. 2019). US and Canada produced 30% of the world's algae biofuel (Muthuraman and Kasianantham 2023).

This chapter aims to provide an updated review of the technology, recent advances, challenges, and driving factors in microalgae biofuel production in North America (Canada and the US). By exploring these facets, we provide valuable insights into the current state of the field and its potential for sustainable energy generation in the region.

19.2 Microalgae Cultivation Systems in North America

Microalgae cultivation has the potential to revolutionize biofuel production, with impressive yields of up to 10,000 gallons per acre per year, in stark contrast to the mere 60 gallons per acre obtained from corn crops. This significant difference highlights the value of transitioning from crop-based biofuels to microalgae-based biofuels, as it allows for a substantial reduction in required land area for cultivation (Abo et al. 2019).

In North America, various cultivation methods are employed for microalgae production, including open raceway ponds, closed photobioreactors, hybrid systems, and to a lesser extent, fermenters under heterotrophic and mixotrophic conditions in fresh and sea water mediums (Casanova et al. 2022; Romagnoli et al. 2020; United States Department of Energy 2020).

19.2.1 Open Pond Cultivation

Open pond cultivation represents the oldest and most straightforward method of large-scale microalgae cultivation. It is widely employed in the industry due to its relatively lower construction, maintenance, and operational costs, as well as its low energy demands and scalability (Tan et al. 2020). However, open pond systems do not facilitate complex operations or additional benefits (Sarker and Kaparaju 2023). Among open pond cultures, raceway ponds stand out as highly energy-efficient, their cheaper construction, and their easier scalability. Therefore, they are consequently one of the most prevalent types employed (Casanova et al. 2022; Tan et al. 2020).

In the US, several start-up companies emerged in the 1950s, pioneering the production of *Arthrospira platensis* (Spirulina). Cyanotech Corp. currently holds a prominent position as the main producer (Velazquez-Lucio et al. 2018). Additionally, the production of *Dunaliella salina*, a valuable source of β-carotene, commenced in the early to mid-1980s (Belay 1997; Sarker and Kaparaju 2023). Notably, the Sapphire Energy Columbus Algal Biomass Farm in the US successfully demonstrated the continuous cultivation of microalgae (*Arthrospira sp.*) in large-scale open raceway ponds, generating 520 metric tons of algal biomass (dry weight) over a 2-year period (White and Ryan 2015).

In recent years, researchers in the US have focused on investigating the potential competition for land between open pond microalgae culture and terrestrial land dedicated to feedstock supply systems. Combined nutrient remediation and algal biomass exploration for biofuels are expected to accelerate the commercialization of algae open ponds, offering a cost advantage of approximately 70% compared to conventional activated sludge-based treatments. Industry giants such as ExxonMobil and Synthetic Genomics are also investing in genetic engineering strategies to significantly increase algal lipid production, aiming to reach a production rate of 10,000 barrels of biofuel per day by 2025 (Behera and Paramasivan 2022).

However, open pond raceway systems are not ideally suited for the Canadian context due to the climatic conditions. These systems can only operate for a limited period of four to six months annually, and there have been no reported research attempts to experimentally quantify or model their application in Canada to date (Pankratz et al. 2017). Calipatria, US have the world's largest raceway pond for the production of Spirulina and Spirulina-based products (Chhandama et al. 2021).

In summary, North America utilizes diverse microalgae cultivation systems, with open pond raceway ponds being a prominent choice. The advantages of scalability, cost- effectiveness, and energy efficiency make them a promising option for large-scale microalgae biofuel production. However, specific regional considerations, such as climate, need to be considered when implementing these systems.

The development of large-scale microalgae production facilities traces back to the late 1940s and early 1950s when the Stanford Research Institute and the Carnegie Institute initiated the first large-scale testing of open ponds for microalgae food production in the US. Subsequently, in the 1970s and 1980s, commercial-scale microalgae production commenced in countries such as the US, Australia, India, and Israel (Agrawal et al. 2020; Medipally et al. 2015). Among these countries, the US has emerged as a leader in spatially explicit analyses, identifying optimal areas for locating microalgae production farms (Correa et al. 2021).

19.2.2 Cultivation of Microalgae in Photobioreactors

Photobioreactors (PBRs) are closed artificial cultivation systems that provide an optimal environment for the growth of selected microalgae strains. Compared to open ponds, PBRs offer greater control over culture variables such as lighting, temperature, medium composition, and continuous mixing, resulting in higher biomass productivity (Jabłońska-Trypuć et al. 2023).

PBRs are typically constructed using transparent materials such as acrylic or glass. Various designs are employed, including tubular, plate, helical, horizontal, foil vertical columns, as well as tubular bioreactors, flat plate bioreactors, and fermenter-type bioreactors (Abo et al. 2019; Jabłońska-Trypuć et al. 2023). Studies on outdoor algae PBRs were conducted by researchers in the US and Germany, leading to the establishment of several new companies in the 1950s.

Key factors for successful microalgae cultivation in PBRs include adequate lighting, temperature, medium composition, and a bioreactor design that enables continuous mixing (Jabłońska-Trypuć et al. 2023).

The advantages of PBRs include (Adeniyi et al. 2018; Casanova et al. 2022; Jabłońska-Trypuć et al. 2023):

- Continuous monitoring of growth parameters and culture conditions (e.g., temperature, pH, mixing, CO_2, and O_2).
- Independence from climatic conditions.
- Reduced land requirement.
- No dependency on natural sunlight.
- Decreased CO_2 and water losses.
- Accommodation of higher cell concentrations.
- Enhanced volumetric productivity.
- Improved heat and nutrient dissipation.
- Lower susceptibility to contamination by predatory organisms.
- Better CO_2 utilization and maximum light exposure.

However, PBRs also have some disadvantages, including bio-fouling, overheating, benthic algae growth, cleaning challenges, and limitations on dissolved oxygen build-up that can hinder growth (Narala et al. 2016). Another significant drawback is the higher capital and operational cost and energy consumption compared to open ponds making PBRs less viable for large-scale commercial applications, particularly for wastewater treatment using microalgae cultivation. Additionally, achieving efficient and uniform light penetration into PBRs with operating volumes of 50-100L or higher is challenging (Abo et al. 2019; Casanova et al. 2022; Narala et al. 2016).

The use of computer-based simulations to investigate critical conditions in PBRs has been relatively limited. Over the past two decades, only one simulation study was conducted by German researchers in the US (Sarker and Kaparaju 2023). In 2013, Bioprocess Algae, with a $6.4 million funding from the US Department of Energy, studied the integrated production of microalgae biomass on a commercial scale, focusing on accelerated lipid production and the conversion of lipids into various hydrocarbons for the development of military biofuels (Abo et al. 2019; United States Department of Energy 2020).

In evaluating biomass and lipid productivity, *Nannochloropsis oculata* and Nannochloropsis salina were studied by Quinn et al. (2012) at a large-scale outdoor microalgal culture facility operated by Solix BioSystems in Fort Collins, Colorado, US. The facility utilized advanced photobioreactors (PBRs) and provided valuable data for understanding the productivity potential of these microalgal species (Quinn et al. 2012). Various microalgal species, including *Chlorella minutissima, C.zofingiensis, Nannochloropsis, Scenedesmus obtusus, Thalassiosira sp., Fistulifera sp., Chlorella sp.,* and *C. sorokiniana,* have been cultivated under diverse climatic conditions in Brazil and the US. These cultivation efforts have utilized different types of facilities, including outdoor PBRs, aiming to explore the bioenergy potential of these microalgae (Menegazzo et al. 2022).

The establishment and operation of commercial-scale microalgae production plants not only contribute to the advancement of biofuel technologies but also provide valuable insights into the scalability and economic viability of microalgae-based biofuel production on a larger scale.

19.2.3 Hybrid Systems for Microalgae Cultivation

Hybrid systems, which combine open ponds and closed photobioreactors (PBRs), have emerged as a promising approach for microalgae cultivation, aiming to capitalize on the strengths of both systems. The utilization of a hybrid system involves the strategic use of each cultivation method to optimize productivity, biomass quality, and overall profitability (Ganesan et al. 2020).

One common approach within hybrid systems is to utilize PBRs for the initial growth stage of microalgae. PBRs provide a controlled environment that enables high-density cultivation with minimal contamination. Following this stage, the biomass produced in the PBRs is transferred to open ponds. By subjecting the biomass to stress conditions in the open pond, such as nutrient deprivation or high light intensity, lipids accumulation can be promoted (Casanova et al. 2022).

It is worth noting that this hybrid system, known as a batch method, may incur higher costs and may not be suitable for large-scale biofuel production due to its limited efficiency (Casanova et al. 2022; Medipally et al. 2015). However, there have been studies suggesting economic viability in certain scenarios. For instance, researchers from the US and the United Kingdom found that using 25 m^3 PBRs to provide a continuous and consistent inoculum for culture in 400 m^2 open ponds resulted in more stable cultivation compared to long-term open pond culture alone, mitigating the risk of biological system crashes (Huntley et al. 2015).

The economic feasibility of hybrid systems remains a complex consideration. It is challenging to predict their viability, as it depends on various factors such as the specific costs associated with implementing hybrid solutions and the market availability of microalgae-based products (Sarker and Kaparaju 2023). Further research and economic analysis are needed to better understand the potential advantages and limitations of hybrid systems and their suitability for different applications within the microalgae biofuel industry.

19.3 Status of Microalgae Biofuels Production in North America

Microalgae-derived biofuels are gaining increasing attention in North America as a promising solution for sustainable energy development and to overcome relevant environmental challenges. Considering the energy demands for transportation expected in the 2050s, it is estimated that approximately 70% of the total potential microalgae production areas worldwide would be located in US, China, Russia, Germany, Canada, Japan, UK, India, Brazil, France, South Korea, Mexico, Indonesia, Iran, and Saudi Arabia (Correa et al. 2021).

The versatility of microalgae allows for their utilization in diverse applications, ranging from energy production to the production of high-value food and biomaterials. For example, besides biofuels like bioethanol and biodiesel, microalgae can also be used as a source for food, and biomaterials (Fig. 19.1). Bioethanol, for example, can be produced through the fermentation of sugars obtained from microalgal biomass, while biodiesel is typically derived from the lipid content of microalgae. Moreover, microalgal biomass also holds potential as a source for the production of biohydrogen. Biohydrogen production from microalgae involves harnessing their ability to undergo certain metabolic processes, such as anaerobic fermentation, to generate hydrogen gas. This represents a promising avenue for renewable energy production and has garnered attention for its potential as a clean and sustainable energy source.

Overall, microalgae biofuels production in North America is progressing towards the development of sustainable and efficient processes for harnessing the potential of these microorganisms. Ongoing research and development efforts aim to optimize cultivation techniques, improve lipid productivity, enhance conversion processes, and address economic and environmental considerations. With continued advancements, microalgae biofuels have the potential to play a significant role in the transition towards a more sustainable and renewable energy future in North America and beyond.

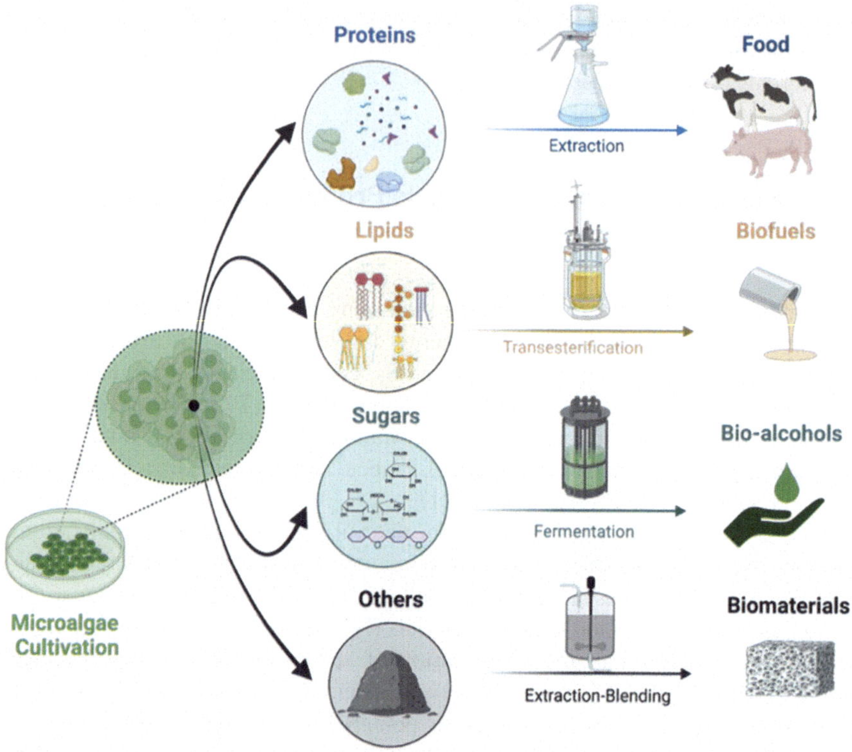

Fig. 19.1 Main products derived from microalgae and their associated production processes

19.3.1 Main Biofuels Produced from Microalgae in North America

Bioethanol and biodiesel are the main biofuels produced and marketed from microalgae in North America. Although, the biohydrogen is also produced but in a lesser extent.

Bioethanol

Bioethanol is becoming the most widely used and popular biofuel due to advances in its production technology and its ability to replace gasoline for the transportation sector (Jabłońska-Trypuć et al. 2023; Jambo et al. 2016). Microalgae can produce between 46,000 and 140,000 L of ethanol per hectare compared to other types of first and second generation biomasses (Muthuraman and Kasianantham 2023).

North America (US, 57% and Canada, 2%) is the world leader in bioethanol production from the three generations of biomass. Brown algae have garnered most of the attention in North American countries and therefore the key drawbacks are mainly related to this group of algae (Kouhgardi et al. 2023). In the US, bioethanol production reached 15.4 billion gallons in 2022, with corn and sorghum being the main feedstocks (USDA 2023). Bioethanol is the most consumed non-petroleum transportation fuel in the country, amounting to 13.9 billion gallons (Johnson et al. 2021; USDA 2023).

The high-octane rating of bioethanol improves engine performance, and it is available in various gasoline-ethanol blends, such as E10, E15, and E85, each catering to different niches and markets. These blends contribute to improving air quality, enhancing energy security, optimizing engine performance, and achieving climate change-related goals (Johnson et al. 2021; Velazquez-Lucio et al. 2018).

Bioethanol production from microalgae biomass utilizes various techniques such as traditional (involving hydrolysis) and "dark" fermentation, as well as photofermentation based on genetic engineering (de Farias Silva and Bertucco 2016; Jabłońska-Trypuć et al. 2023; John et al. 2011). Advances in bioethanol production technology have led to its increasing popularity as a biofuel. Ongoing efforts are focused on improving efficiency and reducing production costs, making third generation bioethanol from microalgae an ecologically and economically viable solution (Jabłońska-Trypuć et al. 2023).

Microalgae species like Porphyridium, Chlorella, Dunaliella, Chlamydomonas, Scenedesmus, and Spirulina are attractive raw materials for bioethanol production due to their high starch, cellulose, and glycogen content, exceeding 50% of their dry weight (Agrawal et al. 2020).

Despite Canada's challenging mandates and funding support for biodiesel production, its technologies in general have been largely overlooked. This may severely limit future Canadian research on biofuels and the development of technologies to address carbon emissions in the future (Scaife et al. 2015).

Biodiesel

Biodiesel is composed of alkyl esters of long-chain fatty acids and can be obtained by transesterification of microalgae oils' triglycerides using alcohols under controlled conditions of temperature, pressure, reaction time, and catalyst. Common solvent extraction methods for lipid extraction include Folch and Bligh-Dryer (Almomani et al. 2023).

The world's main producers and consumers of biodiesel are the US, Brazil and Indonesia. Its growing demand for green energy alternatives and public policies has driven the increase in commercial biodiesel obtained mainly from oilseed crops, such as soybeans, corn, sunflower and oil palm. A 70% growth in the demand for biodiesel as a transportation fuel is projected by 2040. Nevertheless, microalgae biomass is emerging as a source of biodiesel due to its high lipid content, higher growth rates and productivity compared to conventional crops (Casanova et al. 2022). In the EU, biodiesel represents 82% of the total production of biofuels (Mata et al. 2010).

Microalgae species like *Schizochytrium limacium* and *Botryococcus braunii* (UTEX-2441; 48.85%) are known for their ability to produce high levels of total fatty acids, making them suitable for biodiesel production through direct transesterification. *Chlorella* sp. also exhibits a favorable fatty acid profile and holds potential for biodiesel production (Adeniyi et al. 2018; Casanova et al. 2022; da Rosa et al. 2023). Moreover, *Chlorella* sp. biomass has high protein content and nutrient value, making it a valuable resource for wastewater treatment, particularly in the remediation of heavy metals like iron (Adeniyi et al. 2018).

The dry method of biodiesel production from microalgae requires significant energy input for lipid extraction and transesterification, potentially leading to decreased methyl ester yields. Alternatively, the wet route, which involves approximately 80% moisture, offers a promising alternative by enabling acid-catalyzed in situ transesterification of highly saturated microalgae, reducing energy requirements (Abo et al. 2019).

The US leads microalgae research in the world focused on the development of valuable co-products from microalgae in addition to biodiesel production and on carbon capture technologies to improve microalgae biomass yields (Patnaik and Mallick 2021). The US Environmental protection agency confirmed the report by many researchers that biodiesel from microalgae meets the requirement Renewable Fuel Standard 2007 by reducing 50% of greenhouse gas emissions as compared to petroleum diesel (Patnaik and Mallick 2021).

US companies ExxonMobil and Synthetic Genomics teamed up to generate biodiesel from genetically modified microalgae to increase the biosynthesis of target molecules (Casanova et al. 2022).

19.4 First Steps in the Market for Bio-Based Derived Products from Microalgae

In the early twenty-first century, countries such as the US, Brazil, Canada, China, India, Japan, and Europe demonstrated a strong commitment to developing domestic markets for biofuels. They implemented strategic plans to utilize these biofuels (Medipally et al. 2015).

North America has emerged as a leader in the microalgae biofuel market, capturing a 30% share, with a significant number of start-ups and substantial support from the US government in funding the renewable energy sector. Solazyme Inc., a US company, made the pioneering move of delivering 100% microalgae biofuel to the US Navy for certification. Other major players in microalgal biofuel production in the US include Sapphire Energy, Solix Biofuels, Algenol Biotech, Genifuels, Blue Marble Productions, and ExxonMobil. For instance, Solix Biofuels currently produces approximately 3,000 gallons of biofuel from algae on a daily basis (Banu et al. 2020).

Key firms like Sapphire Energy Inc. (San Diego, CA, USA), Algenol (Fort Myers, FL, USA), Solazyme Inc. (South San Francisco, CA, USA), Mera Pharmaceuticals Inc. (Kailua-Kona, HI, USA), and Solix Algadrients (Fort Collins, CO, USA) are at the forefront of global collaboration with industries, government organizations, and research institutes to drive the commercialization of cost-effective and sustainable algal technologies (Bhatia et al. 2020).

The commercialization of viable microalgae cultivation processes has fostered collaborations between China and other countries such as the US, Taiwan, Pakistan, and the UK. These collaborations aim to technoeconomically validate integrated processes involving microalgae cultivation and simultaneous resource reuse (Behera and Paramasivan 2022).

19.4.1 Drivers/Opportunities

Microalgae for biofuels serve as a tool to combat climate change, with studies providing practical considerations for national and regional policies in North America, Australia, and other countries in the Asia Pacific region (Moshood et al. 2021; Show 2022). Microalgae products are already well-established in the North American and Australian markets, serving various applications. The major drivers and opportunities in the microalgae market for biofuels are as follows (DATA BRIDGE MARKET RESEARCH 2022; Sarker and Kaparaju 2023):

- Technological advances and scientific innovation facilitating the development of effective and novel microalgae products.
- Increased funding in advanced microalgae technologies and cultivation methods.
- Growing strategic initiatives by market players.
- Increasing considerations of microalgae's practical benefits in national and regional policies.
- Utilization as a means to combat climate change and promote renewable resources in the domains of food, feed, and fuel.

In the case of ethanol, the Renewable Fuel Standard (RFS) program is widely regarded as the primary driver of market growth in the US (Newes et al. 2022).

North America is expected to dominate the global microalgae market due to increasing consumer awareness of the health benefits associated with microalgae products, diverse industry coverage, rising demand for microalgae products, and significant research and development activities in the nutraceutical industry. Data Bridge Market Research predicts that the microalgae market will reach US$ 1394.95 million by 2030 (DATA BRIDGE MARKET RESEARCH 2022).

19.4.2 Constraints/Challenges

Although microalgae offer numerous advantages, such as the ability to convert atmospheric CO_2 into valuable products like carbohydrates, lipids, and bioactive metabolites, there are challenges associated with their production. Some of the key restraints and challenges in the microalgae market are as follows (DATA BRIDGE MARKET RESEARCH 2022):

- High production costs of microalgae derivatives.
- Complexities involved in microalgae production.
- Stringent regulations pertaining to microalgae.

Different microalgal species, such as Chlorella minutissima, C. zofingiensis, Nannochloropsis, S. obtusus, Thalassiosira sp., Fistulifera sp., Chlorella sp., and C. sorokiniana, have demonstrated bioenergy potential for biofuel production in outdoor photobioreactors at laboratory, pilot, and industrial scales. However, the challenge lies in identifying suitable microalgal species for specific locations (Sarker and Kaparaju 2023).

Recent shocks in the microalgae market

The transition from fossil fuel-based systems to bioenergy and bioproducts is gaining traction due to factors such as population growth and the subsequent rise in energy demand. Global warming, climate change, the COVID-19 pandemic, and events like the Russo-Ukrainian war have also contributed to the urgency of implementing sustainable development plans. Environmental sustainability in biomass production is crucial to unlock the potential benefits of bioenergy products (Aghbashlo et al. 2022).

Although highly industrialized countries like the US still heavily rely on fossil fuels as their primary energy source, there has been a significant reduction in their consumption due to the development of alternative renewable energy sources (Jabłońska-Trypuć et al. 2023).

19.5 Perspectives

19.5.1 Artificial Cultivation

In order to cultivate new species of microalgae effectively, it is important to conduct indoor experiments under near-ideal conditions to determine their optimal culture parameters. The selected algal strain should not only be a competitive feedstock for value-added chemicals but also be able to adapt to diurnal changes when cultivation is done outdoors (Sarker and Kaparaju 2023).

Artificial cultivation of microalgae presents several challenges, including overheating, bio-fouling, oxygen accumulation, scaling up difficulties, cell damage caused by shear stress, and deterioration of materials used for the photostage. These challenges contribute to operating costs and need to be addressed for successful large-scale cultivation (Adeniyi et al. 2018).

19.5.2 Ethanol Production Future

Future research on ethanol production from microalgae biomass offers various promising aspects. These include increasing the carbohydrate content of microalgae, isolating and selecting potential algal strains, optimizing their growth in different environments, selecting appropriate yeast strains for improved fermentation, and identifying biochemical triggers, pretreatment methods, and enzymatic methods to enhance ethanol yield from algae (Agrawal et al. 2020).

19.5.3 Microalgal Biorefinery Technologies

Circular open pond systems for cultivating Chlorella sp. have faced limitations in expansion beyond 10,000 m^2 due to issues related to homogeneous mixing. Open pond setups also face challenges such as improper light distribution, seasonal fluctuations, and cross-contamination. Indoor closed photobioreactors (PBRs) offer better control over cultivation conditions but are associated with high capital installation costs, significant operating expenses (especially due to artificial lighting), and maintenance requirements, which hinder their widespread industrial implementation (Behera and Paramasivan 2022).

Tax credits have been implemented in the US to incentivize the use of biofuels in ground transportation. For instance, a blender tax credit ranging from $1.25 to $1.75 per gallon of sustainable aviation fuels was passed by the US Senate to support the acceptance of these fuels in the market. This measure aims to address the economic challenge of producing 3 billion gallons of sustainable aviation fuel in 2030 and 35 billion gallons in 2050 (Grimme 2023).

Other routes for thermochemical conversion of microalgae are in study despite pathways have been largely investigated for lignocellulosic biomass. These routes can be dry-processes (pyrolysis) and wet-processes (hydrothermal liquefaction) (Chiaramonti et al. 2017).

The future of the use of microalgae goes towards the use of an integrated biorefinery plan for the production of biofuels and value-added products including extraction processes, alcohol fermentation, pyrolysis or hydrothermal liquefaction, anaerobic digestion and gasification to obtain bioactive materials and biofuels (Kouhgardi et al. 2023).

19.6 Conclusions

The prospects of microalgae-derived bio-based products in North America are promising, driven by the increasing demand for renewable energy sources, the need to combat climate change, and the pursuit of sustainable development plans. The market potential of microalgae-derived products is vast, with various types and applications distributed across different regions.

In the market for bio-based products from microalgae, North America stands as a leader, with significant investments, numerous start-ups, and government support in the renewable energy sector. Companies like Solazyme Inc., Sapphire Energy, Solix Biofuels, Algenol Biotech, and Genifuels are major players in the production of algal biofuels, contributing to the reduction of greenhouse gas emissions and the transition from fossil fuels to bioenergy.

The drivers and opportunities for microalgae-based biofuels in North America include technological advances, increased funding, strategic initiatives by market players, and practical considerations in national and regional policies. The market is expected to grow due to rising awareness of the health benefits of microalgae products and their application in various industries. North America is projected to dominate the global microalgae market, reaching a value of US$ 1394.95 million by 2030.

However, there are challenges and restraints that need to be addressed. These include high production costs, complications in microalgae production, stringent regulations, and the identification of suitable microalgae species for specific locations. Overcoming these challenges requires further research and development, optimization of cultivation techniques, and advancements in artificial cultivation methods.

Artificial cultivation of microalgae presents its own set of challenges, such as overheating, bio-fouling, scaling up difficulties, and high operating costs. Research efforts are needed to improve cultivation efficiency and optimize growth conditions for new species of microalgae. Additionally, the production of ethanol from microalgae biomass holds promising potential, with a focus on increasing carbohydrate content, selecting optimal algal strains, and enhancing fermentation processes.

Microalgal biorefinery technologies offer a pathway for the production of bio-based products from microalgae. However, limitations in open pond systems and the high costs associated with indoor closed photobioreactors pose challenges to large-scale industrial implementation. Overcoming these limitations and developing cost-effective and sustainable biorefinery technologies is crucial for the future of microalgae-based industries.

In conclusion, the market potential, drivers, and opportunities for microalgae-derived bio-based products in North America are substantial. The advancements in technology, strategic initiatives, and favorable policies contribute to the growth and adoption of microalgae-based biofuels. However, addressing challenges in cultivation, production costs, and refining technologies will be key to unlocking the full potential of microalgae as a sustainable and renewable resource for energy, food, feed, and other valuable products. Continued research, innovation, and collaborations among industries, government organizations, and research institutes will pave the way for a thriving microalgae industry in the future.

Acknowledgements The authors acknowledge gratefully to the authors of the publications and reports used in this chapter.

References

Abo BO, Odey EA, Bakayoko M, Kalakodio L (2019) Microalgae to biofuels production: a review on cultivation, application and renewable energy 34(1):91–99. https://doi.org/10.1515/reveh-2018-0052

Adeniyi OM, Azimov U, Burluka A (2018) Algae biofuel: current status and future applications. Renew Sustain Energy Rev 90:316–335

Aghbashlo M, Hosseinzadeh-Bandbafha H, Shahbeik H, Tabatabaei M (2022) The role of sustainability assessment tools in realizing bioenergy and bioproduct systems. Biofuel Res J 9(3):1697–1706. https://doi.org/10.18331/brj2022.9.3.5

Agrawal K, Bhatt A, Bhardwaj N, Kumar B, Verma P (2020) Algal biomass: potential renewable feedstock for biofuels production–part I. Biofuel Prod Technol: Crit Anal Sustain 203–237

Almomani F, Hosseinzadeh-Bandbafha H, Aghbashlo M, Omar A, Joo S-W, Vasseghian Y, Karimi-Maleh H, Lam SS, Tabatabaei M, Rezania S (2023) Comprehensive insights into conversion of microalgae to feed, food, and biofuels: current status and key challenges towards implementation of sustainable biorefineries. Chem Eng J 455:140588

Banu JR, Kavitha S, Gunasekaran M, Kumar G (2020) Microalgae based biorefinery promoting circular bioeconomy-techno economic and life-cycle analysis. Biores Technol 302:122822

Behera B, Paramasivan B (2022) Research trends and market opportunities of microalgal biorefinery technologies from circular bioeconomy perspectives. Bioresour Technol 127038

Belay A (1997) Mass culture of *Spirulina* outdoors—the earthrise farms experience. Spirulina Platensis 1(1):131–158

Bhatia L, Bachheti RK, Garlapati VK, Chandel AK (2020) Third-generation biorefineries: a sustainable platform for food, clean energy, and nutraceuticals production. Biomass Convers Biorefinery 1–16

Casanova LM, Mendes LBB, Corrêa TdS, da Silva RB, Joao RR, Macrae A, Vermelho AB (2022) Development of microalgae biodiesel: current status and perspectives. Microorganisms 11(1):34

Correa DF, Beyer HL, Possingham HP, Fargione JE, Hill JD, Schenk PM (2021) Microalgal biofuel production at national scales: reducing conflicts with agricultural lands and biodiversity within countries. Energy 215:119033

Chhandama MVL, Satyan KB, Changmai B, Vanlalveni C, Rokhum SL (2021) Microalgae as a feedstock for the production of biodiesel: a review. Bioresour Technol Rep 15:100771

Chiaramonti D, Prussi M, Buffi M, Rizzo AM, Pari L (2017) Review and experimental study on pyrolysis and hydrothermal liquefaction of microalgae for biofuel production. Appl Energy 185:963–972

da Rosa MDH, Alves CJ, dos Santos FN, de Souza AO, Zavareze EdR, Pinto E, Noseda MD, Ramos D, de Pereira CMP (2023) Macroalgae and microalgae biomass as feedstock for products applied to bioenergy and food industry: a brief review. Energies 16(4):1820

DATA BRIDGE MARKET RESEARCH (2022) Global microalgae market—industry trends and forecast to 2030. https://www.databridgemarketresearch.com/reports/global-microalgae-market

de Farias Silva CE, Bertucco A (2016) Bioethanol from microalgae and cyanobacteria: a review and technological outlook. Process Biochem 51(11):1833–1842

Ganesan R, Manigandan S, Samuel MS, Shanmuganathan R, Brindhadevi K, Lan Chi NT, Duc PA, Pugazhendhi A (2020) A review on prospective production of biofuel from microalgae. Biotechnol Rep (Amst) 27:e00509. https://doi.org/10.1016/j.btre.2020.e00509

Grimme W (2023) The introduction of sustainable aviation fuels—a discussion of challenges, options and alternatives. Aerospace 10(3):218

Huntley ME, Johnson ZI, Brown SL, Sills DL, Gerber L, Archibald I, Machesky SC, Granados J, Beal C, Greene CH (2015) Demonstrated large-scale production of marine microalgae for fuels and feed. Algal Res 10:249–265

Jabłońska-Trypuć A, Wołejko E, Mahmudova ED, Głowacka A, Sokołowska G, Wydro U (2023) Using algae for biofuel production: a review. Energies 16(4)

Jambo SA, Abdulla R, Azhar SHM, Marbawi H, Gansau JA, Ravindra P (2016) A review on third generation bioethanol feedstock. Renew Sustain Energy Rev 65:756–769
John RP, Anisha G, Nampoothiri KM, Pandey A (2011) Micro and macroalgal biomass: a renewable source for bioethanol. Biores Technol 102(1):186–193
Johnson C, Moriarty K, Alleman T, Santini D (2021) History of ethanol fuel adoption in the United States: policy, economics, and logistics. Golden, CO: National Renewable Energy Laboratory. NREL/TP-5400-76260. https://www.nrel.gov/docs/fy22osti/76260.pdf. Accessed 23 May 2023
Khan MI, Shin JH, Kim JD (2018) The promising future of microalgae: current status, challenges, and optimization of a sustainable and renewable industry for biofuels, feed, and other products. Microb Cell Fact 17(1):36. https://doi.org/10.1186/s12934-018-0879-x
Kouhgardi E, Zendehboudi S, Mohammadzadeh O, Lohi A, Chatzis I (2023) Current status and future prospects of biofuel production from brown algae in North America: progress and challenges. Renew Sustain Energy Rev 172:113012
Mata TM, Martins AA, Caetano NS (2010) Microalgae for biodiesel production and other applications: a review. Renew Sustain Energy Rev 14(1):217–232
Medipally SR, Yusoff FM, Banerjee S, Shariff M (2015) Microalgae as sustainable renewable energy feedstock for biofuel production. Biomed Res Int 2015:519513. https://doi.org/10.1155/2015/519513
Menegazzo ML, Nascimento VM, Hestekin CN, Hestekin JA, Fonseca GG (2022) Evaluation of *Chlorella sorokiniana* cultivated in outdoor photobioreactors for biodiesel production. Biofuels 13(4):483–488. https://doi.org/10.1080/17597269.2020.1763094
Moshood TD, Nawanir G, Mahmud F (2021) Microalgae biofuels production: a systematic review on socioeconomic prospects of microalgae biofuels and policy implications. Environ Challenges 5:100207
Muthuraman VS, Kasianantham N (2023) Valorization opportunities and adaptability assessment of algae based biofuels for futuristic sustainability. Process Safety and Environmental Protection
Narala RR, Garg S, Sharma KK, Thomas-Hall SR, Deme M, Li Y, Schenk PM (2016) Comparison of microalgae cultivation in photobioreactor, open raceway pond, and a two-stage hybrid system. Front Energy Res 4. https://doi.org/10.3389/fenrg.2016.00029
Newes E, Clark CM, Vimmerstedt L, Peterson S, Burkholder D, Korotney D, Inman D (2022) Ethanol production in the United States: the roles of policy, price, and demand. Energy Policy 161:112713
Pankratz S, Oyedun AO, Zhang X, Kumar A (2017) Algae production platforms for Canada's northern climate. Renew Sustain Energy Rev 80:109–120
Patnaik R, Mallick N (2021) Microalgal biodiesel production: realizing the sustainability index. Front Bioeng Biotechnol 9:620777
Quinn JC, Yates T, Douglas N, Weyer K, Butler J, Bradley TH, Lammers PJ (2012) Nannochloropsis production metrics in a scalable outdoor photobioreactor for commercial applications. Bioresour Technol 117:164–171. https://doi.org/10.1016/j.biortech.2012.04.073
Rodríguez-Palacio MC, Cabrera-Cruz RBE, Rolón-Aguilar JC, Tobías-Jaramillo R, Martínez-Hernández M, Lozano-Ramírez C (2022) The cultivation of five microalgae species and their potential for biodiesel production. Energy, Sustain Soc 12(1):10. https://doi.org/10.1186/s13705-022-00337-5
Romagnoli F, Ievina B, Arachchige W, Rayan A, Perera P, Ferrari D (2020) Novel stacked modular open raceway ponds for microalgae biomass cultivation in biogas plants: preliminary design and modelling. Environ Climate Technol 24:1–19. https://doi.org/10.2478/rtuect-2020-0050
Sarker NK, Kaparaju P (2023) A critical review on the status and progress of microalgae cultivation in outdoor photobioreactors conducted over 35 years (1986–2021). Energies 16(7):3105
Scaife MA, Merkx-Jacques A, Woodhall DL, Armenta RE (2015) Algal biofuels in Canada: status and potential. Renew Sustain Energy Rev 44:620–642
Show PL (2022) Global market and economic analysis of microalgae technology: status and perspectives. Bioresour Technol 127329

Sinitsyn AP, Sinitsyna OA (2021) Bioconversion of renewable plant biomass. second-generation biofuels: raw materials, biomass pretreatment, enzymes, processes, and cost analysis. Biochem Mosc 86:S166–S195

Smithsonian (2020) Algae research/department of botany, national museum of natural history, Smithsonian institution. Retrieved 11 Aprill 2020. https://naturalhistory.si.edu/research/botany/research/algae

Tan JS, Lee SY, Chew KW, Lam MK, Lim JW, Ho SH, Show PL (2020) A review on microalgae cultivation and harvesting, and their biomass extraction processing using ionic liquids. Bioengineered 11(1):116–129. https://doi.org/10.1080/21655979.2020.1711626

Ullah K, Ahmad M, Sofia SVK, Lu P, Harvey AP, Zafar M, Sultana S (2015) Assessing the potential of algal biomass opportunities for bioenergy industry: a review. Fuel 143:414–423

United States Department of Energy (2020) Federal activities report on the bioeconomy: Algae. U.S. Department of Energy. DOE/EE-2009, Washington D.C

USDA (2023) Economic research service using data from U.S. Department of energy, energy information administration; U.S. Department of Commerice, Bureau of the Census; and USDA, National Agricultural Statistics Service. https://www.ers.usda.gov/data-products/u-s-bioenergy-statistics/. Accessed 23 May 2023

Velazquez-Lucio J, Rodríguez-Jasso RM, Colla LM, Sáenz-Galindo A, Cervantes- Cisneros DE, Aguilar CN, Fernandes BD, Ruiz HA (2018) Microalgal biomass pretreatment for bioethanol production: a review. Biofuel Res J 5(1):780–791. https://doi.org/10.18331/brj2018.5.1.5

White RL, Ryan RA (2015) Long-term cultivation of algae in open-raceway ponds: lessons from the field. Ind Biotechnol 11(4):213–220

Chapter 20
Algae as a Sustainable Source for Energy Storage Technologies

Astri Rinanti, Lutfia Rahmiyati, Melati Ferianita Fachrul, Sarah Aphirta, Sheilla Megagupita Putri Marendra, and Nadia Savira

Abstract This chapter provides a thorough overview of the potential of algae as a renewable energy source and energy storage technology, which includes algae biology, aquaculture technology, conversion processes, energy storage applications, and economic, social and environmental impacts. Algae is a potential raw material for sustainable renewable energy because of its ability to grow rapidly and produce high-energy biomass. As photosynthetic organisms, algae play an important role in the ecosystem and can be developed as an alternative energy source that is environmentally friendly. The discussion begins by explaining the technology and methods of algae cultivation, as well as the process of converting algae biomass into various forms of energy, such as bioethanol, biobutanol, biogas, syngas, bio-oil, biochar, and biodiesel. Energy storage technologies, such as batteries, capacitors, flywheels, and hydro pumps, can be integrated with algal biomass to create more efficient and sustainable renewable energy systems. Relevant application and case studies are also presented to provide a deeper understanding of the potential of algae in the energy sector. The economic, social, and environmental impacts of using algae as a renewable energy source need to be considered. An analysis of the life cycle, production costs, and market implications of algae-based renewable energy is presented to provide

---

A. Rinanti (✉) · L. Rahmiyati · M. F. Fachrul · S. Aphirta · S. M. P. Marendra · N. Savira
Environmental Engineering Department, Faculty of Landscape Architecture and Environmental Technology, Universitas Trisakti, Jakarta, Indonesia
e-mail: astririnanti@trisakti.ac.id

L. Rahmiyati
e-mail: lutfiarahmiyati@trisakti.ac.id

M. F. Fachrul
e-mail: melati@trisakti.ac.id

S. Aphirta
e-mail: aphirtasarah@trisakti.ac.id

S. M. P. Marendra
e-mail: sheilla@trisakti.ac.id

N. Savira
e-mail: nadia082001900051@std.trisakti.ac.id

T. A. Kurniawan and A. Anouzla (eds.), *Algae as a Natural Solution for Challenges in Water-Food-Energy Nexus*, Environmental Science and Engineering,
https://doi.org/10.1007/978-981-97-2371-3_20

a thorough picture of the sustainability of this technology. In addition, the role of government regulations and policies in the development and adoption of algae-based technologies is discussed. Although algae offer great potential as a renewable energy source, there are still some challenges that need to be overcome, such as technological barriers, efficiency, and production costs. It is hoped that the latest innovations and research to address these challenges will provide visions and future projections for the algae-based renewable energy industry.

20.1 Introduction

20.1.1 The Role of Algae in the Ecosystem

Algae are a group of photosynthetic eukaryotic organisms that inhabit aquatic environments—both freshwater and marine ecosystems—as well as moist terrestrial areas. Algae can exist as single-celled or multicellular organisms, often possessing relatively simple body structures. Several types of algae have been utilized in various industrial applications, such as the production of chemicals, pharmaceuticals, cosmetics, food, and renewable energy (Abusweireh et al. 2023; Babich et al. 2022; Boukid and Castellari 2023). Generally, algae can be regarded as highly diverse organisms with numerous benefits for humans and the environment.

Several key roles of algae as photosynthetic organisms in aquatic ecosystems are listed below:

(1) Primary producers: Algae serve as primary producers in aquatic food chains (Nagappan et al. 2021). Through photosynthesis, they convert solar energy into chemical energy, which is utilized by other organisms in the ecosystem as a source of energy (Visca et al. 2017). Algae produce biomass that serves as food for primary consumers, such as zooplankton and certain species of small fish.
(2) Carbon absorption: Algae absorb carbon dioxide (CO_2) from the atmosphere and water through photosynthesis, contributing to the mitigation of greenhouse gas emissions and climate change impacts. Their photosynthetic activity also yields oxygen, which is indispensable for aquatic organisms, underscoring their vital part in maintaining the oxygen cycle within aquatic ecosystems (Iglina et al. 2022; Onyeaka et al. 2021).
(3) Nutrient cycling: Algae help regulate water quality by absorbing excess nutrients, e.g., phosphates and nitrogen. In some cases, algae can also assist in reducing pollutants like heavy metals from aquatic environments (Nguyen et al. 2022; Siwi et al. 2017).
(4) Habitat and protection: Macroalgae, such as seaweed and kelp, play a crucial role in providing habitat and shelter for various aquatic species. Macroalgae create complex, layered structures in water, offering a home to diverse creatures. For instance, kelp forests and seagrass beds provide habitat for fish, crustaceans, molluscs, and numerous invertebrate species. These ecosystems are

often referred to as "underwater forests" that provide abundant protection and food resources for many organisms.

Macroalgae provide protection for the species residing within them, particularly from predators. Animals like crabs, shrimp, and molluscs can hide among the leaves and branches of seaweed or aquatic moss, making them less visible to predators. This also applies to fish larvae and juveniles, which often seek refuge among macroalgae to grow and develop before reaching adulthood.

Many aquatic species utilize macroalgae as a medium for egg-laying and reproduction. For example, certain fish species lay their eggs on seaweed leaves, while molluscs and crustaceans attach their eggs to algal stems (Dharmono et al. 2022). Macroalgae provide a safe shelter for the eggs and larvae to grow and develop. They also serve as a food source for various aquatic organisms. Some species, such as marine herbivores, directly consume algae (Yin et al. 2019), while others feed on detritus (dead and organic particles) produced by macroalgae (Yin et al. 2019). Additionally, macroalgae provide habitat for numerous microorganisms and small invertebrates, which, in turn, serve as food for other aquatic species.

Macroalgae promote species diversity in aquatic ecosystems by providing suitable habitats for new or migratory species. For instance, fish seeking spawning grounds or refuge can find macroalgae as an ideal site, thus eventually enhancing overall ecosystem diversity and health. Moreover, macroalgae contribute to nutrient cycling within aquatic ecosystems by absorbing dissolved nutrients, such as nitrogen and phosphorus. This helps maintain water quality and supports the growth and reproduction of other aquatic species (Nguyen et al. 2022).

(5) Microecosystem foundation: microscopic algae, such as diatoms and dinoflagellates, constitute phytoplankton—the foundation of aquatic ecosystems. Phytoplankton are photosynthetic organisms that float or swim freely in the water column, serving as the cornerstone of aquatic ecosystems, with various vital roles as follows:

   (a) As primary producers, phytoplankton convert solar energy into chemical energy through photosynthesis. They produce biomass that serves as a food source for primary consumers like zooplankton, which in turn, become food for higher-level consumers like fish and marine mammals.
   (b) As oxygen providers, phytoplankton perform photosynthesis which yields oxygen as a byproduct. This oxygen becomes the main source of dissolved oxygen in water, essential for organisms like fish, amphibians, and invertebrates.
   (c) Phytoplankton play a critical role in aquatic food chains, serving as a food source for primary consumers like zooplankton. Zooplankton then sustains higher-level consumers such as small fish, which become prey for larger fish, seabirds, marine mammals, and other predators. Hence, phytoplankton support life throughout the aquatic ecosystem.

(d) Phytoplankton play a pivotal role in the water's nutrient cycle by absorbing dissolved nutrients like nitrogen and phosphorus. They utilize these nutrients for growth and reproduction, and when they die, these nutrients are released back into the environment through decomposition, sustaining nutrient availability for other organisms.
(e) Phytoplankton are crucial components of the global water system. They participate in the global carbon cycle by absorbing carbon dioxide from the atmosphere through photosynthesis and producing the oxygen essential for life on Earth. They also contribute to biomass production, supporting life in water bodies, including fisheries and the fishing industry.

(6) Erosion and Sedimentation. Seaweeds and water mosses are examples of macroalgae that grow in large and structured forms, playing a significant role in mitigating erosion and sedimentation in aquatic ecosystems by retaining soil particles and other materials that could contaminate the water. These macroalgae possess intricate structures, creating a physical network capable of capturing soil particles and other materials in the water column. As a result, macroalgae help reduce water flow velocity and slow down particle movement, effectively decreasing erosion and sedimentation in the ecosystem (Cotas et al. 2023).

Seaweeds and water mosses grow on substrates, e.g., rocks, sand, or mud at the bottom of water bodies. By attaching to such substrates, these organisms contribute to maintaining the stability and integrity of the aquatic bed, decreasing erosion caused by currents, waves, and human activities. Coastal seaweeds, particularly seagrass or kelp, can form dense coverings that protect shorelines from wave-induced erosion. The roots and rhizomes of seagrass aid in binding soil particles and sand, promoting coastal stability and mitigating erosion. Macroalgae (like seaweed and water moss) also function as natural filters, absorbing and storing nutrients, heavy metals, and other pollutants in the water column (Arumugam et al. 2018). Through absorption and accumulation of these matters, macroalgae help lower particle and pollutant concentrations in the water, hence minimizing sedimentation and pollution. They create complex ecosystems, such as seagrass meadows and kelp forests, which provide habitat for various aquatic organisms. These ecosystems enhance the stability of the aquatic bed and damp erosion by offering protection from waves and currents while retaining particles and materials in the water column.

Coastal seaweed and water moss can absorb wave energy, subsequently diminishing water flow velocity and coastal erosion (Sultana et al. 2023). For instance, coastal kelp forests serve as natural shields protecting shorelines from wave-induced erosion. Thus, macroalgae like seaweed and water moss play a vital role in mitigating erosion and sedimentation in aquatic ecosystems through particle trapping, substrate stabilization, and coastal protection.

(7) Water quality indicators: The presence and abundance of algae in water bodies can serve as indicators of water quality, reflecting the ecological and chemical conditions as well as changes occurring in the ecosystem. Various types of algae, especially phytoplankton like diatoms and dinoflagellates, are sensitive

to environmental changes and pollutants. Therefore, the presence, abundance, and species composition of algae can provide valuable insights into the overall condition of the water. Algae are used as bioindicators of water quality in the following ways:

(a) Nutrient indicator: Algae are highly sensitive to nutrient levels in water, particularly phosphates and nitrogen. Increased nutrient levels often lead to excessive algal growth or "blooms," which can result in water quality deterioration and environmental issues—such as eutrophication and hypoxia (Gökçe 2016)
(b) Pollution Indicator: Some types of algae can absorb or accumulate pollutants like heavy metals, pesticides, and organic compounds. The presence and abundance of algae resistant to specific pollutants can indicate pollution in the aquatic ecosystem (Rinanti et al. 2021).
(c) Climate change indicator: Algae are highly sensitive to temperature and light conditions, making fluctuations in their abundance or distribution indicative of climate changes or other environmental situations impacting water bodies.
(d) pH Indicator: Some algae species have specific preferences for water acidity levels. The presence of algae more tolerant to low or high pH levels can indicate changes in water acidity balance.
(e) Dissolved oxygen indicator: Algae influence the concentration of dissolved oxygen in water through photosynthesis and respiration. The existence of algae producing significant amounts of oxygen can represent a healthy water condition, while less algal abundance may indicate issues with dissolved oxygen or poor water conditions.
(f) Biological indicator: Algae species diversity and community composition in water bodies can depict the overall ecosystem health. For example, the presence of certain algae or healthy algal groups can signify good water conditions, whereas decreased diversity or changes in algal community composition may suggest disturbances or stress in the ecosystem.

By monitoring alterations in algal communities, scientists and water resource managers can detect water quality problems and take necessary actions to protect and restore ecosystem health. Hence, algae play a crucial role in maintaining the balance and sustainability of aquatic ecosystems.

20.1.2 History and Development of Renewable Energy

The development of algae-based renewable energy has a long and fascinating history. Algae have been recognized as a source of energy since ancient times, but intensive research and development on utilizing algae as a renewable energy source began in the twentieth century. The following is the historical timeline and advancement of algae-based:

(1) In ancient times, algae (particularly seaweed) were used as food and animal feed in diverse civilizations worldwide. Although algae were not explicitly utilized as an energy source during that period, their diverse applications in daily life showcased their potential and versatility. Seaweed has been consumed as food by humans since ancient times. In Asia, principally in Japan, Korea, and China, it has become an integral part of daily diets and featured in various dishes like sushi, soups, and salads. Seaweed is rich in nutrients, such as vitamins, minerals, fibre, and essential fatty acids like omega-3. Algae were also utilized as fertilizers and animal feed in agriculture: Dried seaweed was used as a nutrient-rich organic fertilizer, enhancing soil fertility and crop yields; Moreover, protein and nutrient-rich algae were used as animal feed, particularly for cattle, sheep, and poultry.

Several ancient civilizations utilized algae, notably seaweeds, for medicinal and healing purposes. As an illustration, algae were incorporated in traditional Chinese medicine to treat various ailments and digestive problems. Although the industrial use of algae was not fully developed in ancient times, some early applications were noted. For example, seaweed extracts containing polysaccharides like agar and alginate were used in food thickening and preservation. Algae were not recognized as an energy source during ancient times, as the knowledge of processes like fermentation and biomass conversion into energy was not fully comprehended. Nevertheless, the use of algae in diverse applications demonstrated their versatility and potential as valuable natural resources. Technological improvements and deeper understanding of algae and their biochemical mechanisms have enabled their utilization as a renewable energy source in modern times (Salami et al. 2021).

(2) In the twentieth century, around the 1950s to 1960s, initial research on the utilization of algae as an energy source commenced, especially in the United States. During the 1970s, the energy crisis caused by the OPEC oil embargo prompted governments and researchers to seek alternatives for sustainable and eco-friendly energy sources. Algae emerged as an attractive option due to their rapid growth rate and ability to produce large amounts of biomass.

(3) In 1978, the United States Department of Energy (DOE) launched the Aquatic Species Program (ASP), operated by the National Renewable Energy Laboratory (NREL). The program aimed to develop biodiesel production technology from microalgae. The program lasted for nearly 20 years and yielded extensive knowledge about algae biology, cultivation technology, and the process of converting algal oil into biodiesel. Although the ASP was discontinued in 1996 due to funding constraints, research on algae as an energy source continued to advance. Several companies and universities in the United States invested resources in developing biofuel production technology from algae, including ExxonMobil, Synthetic Genomics, and Arizona State University. Algae's utilization as an energy source in the United States includes biodiesel, biogas, and bioethanol production.

(4) Since the early twenty-first century, research and development of algae-based renewable energy have rapidly progressed. Various technologies and processes

have been developed to produce various types of fuels from algae, including biodiesel, bioethanol, biogas, and biohydrogen (Salami et al. 2021). Algae are also utilized to generate high-value carbon-based products, such as pigments, omega-3 fatty acids, and protein.

(5) Technological advancements in genetics, biotechnology, and process engineering have allowed increased efficiency and cost reduction in algae-based renewable energy production. Innovations like genetic engineering and selection of highly productive algae strains have resulted in algae strains with higher productivity and lipid content, consequently amplifying fuel production efficiency.

20.1.3 Why Algae Are Important for Renewable Energy?

Algae possess immense potential as a renewable energy source for several compelling reasons as follows:

(1) Rapid growth and high productivity: Algae exhibit a remarkably fast growth rate and can generate a substantial biomass in a relatively short time. It means that algae can yield more energy per unit area compared to other renewable energy crops such as corn or sugarcane (Sawant et al. 2018).

(2) Non-competitive with agricultural land: Algae can thrive in water bodies such as lakes, ponds, or even wastewater, avoiding competition with agricultural land used for food production. This aspect is crucial for global food security and efficient land utilization (Amir et al. 2021).

(3) Carbon dioxide sequestration: Algae absorb carbon dioxide (CO_2) from the atmosphere or industrial emissions as part of the photosynthesis process. Thus, utilizing algae as a renewable energy source can aid in reducing greenhouse gas emissions and contribute to climate change mitigation efforts (Dewi et al. 2014; Iglina et al. 2022; Onyeaka et al. 2021).

(4) High lipid content: Some algae species boast high lipid content, making them ideal for biodiesel production. The biodiesel derived from algae shares similar properties with fossil diesel and can be used in conventional diesel engines without modification (Morales et al. 2021; Udayan et al. 2022)

(5) Fuel production flexibility: Algae can be processed into different types of fuels, including biodiesel, bioethanol, biogas, and biohydrogen. This flexibility allows for various energy needs to be met and promotes the diversification of renewable energy sources (Salami et al. 2021).

(6) Waste and nutrient treatment: Algae can be employed to treat excess waste and nutrients from industries and agriculture. By absorbing nutrients such as nitrogen and phosphorus, algae assist in decreasing water pollution and eutrophication while generating biomass for energy production (Nguyen et al. 2022).

(7) High-value product generation: Beyond energy, algae can yield various high-value products, including omega-3 fatty acids, pigments, and protein. This

heightens the economic value of algae and renders algae-based renewable energy production more financially feasible and sustainable.

Due to these reasons, algae are considered a prominent and promising renewable energy source. However, there are challenges to overcome for achieving efficient and economically viable algae-based energy production on a commercial scale, such as cost reduction and efficiency improvement in manufacturing and processing technologies.

20.2 Biology and Physiology of Microalgae as Energy Raw Materials

20.2.1 Cell Structure and Algal Metabolism

The cellular structure and anatomy of algae vary depending on their types and species. However, several common characteristics are shared by most algae, as depicted in (Fig. 20.1) and elaborated below:

(1) Algal cell structure: Each algal cell typically consists of various components, including the cell nucleus, cytoplasm, and cell membrane. Additional structures such as vacuoles, chloroplasts, and stigma may also be present.
(2) Cell nucleus: This central structure in the algal cell contains genetic material, such as DNA and RNA.
(3) Cytoplasm: The fluid found within the algal cell where various metabolic processes occur.
(4) Cell membrane: The outer layer that encloses the algal cell and regulates the movement of substances in and out of the cell.

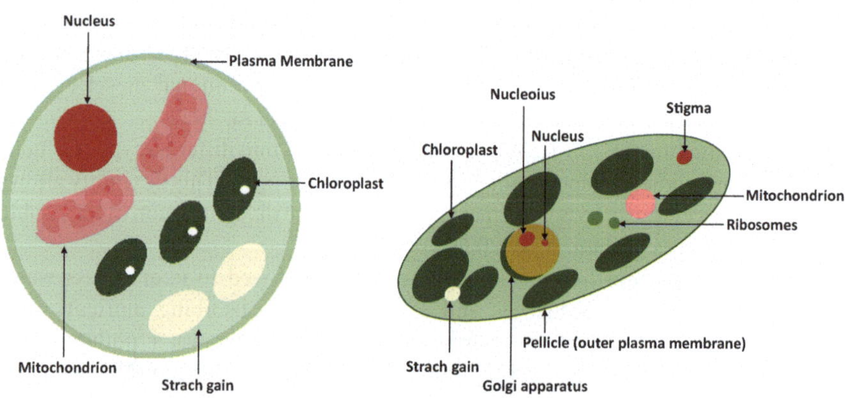

Fig. 20.1 Algal cell structure

(5) Chloroplasts: Organelles that enable algae to conduct photosynthesis. Chloroplasts contain photosynthetic pigments like chlorophyll and carotenoids.
(6) Vacuoles: Bag-shaped organelles within the algal cell responsible for storing reserve substances and maintaining osmotic pressure within the cell.
(7) Stigma: Grain-like structures within the algal cell that play a role in directing algal movement towards light.
(8) Algal cell anatomy: More complex types of algae, such as classes *Rhodophyceae*, *Chlorophyceae*, and *Phaeophyceae*, possess a more organized body structure and specialized tissues, such as leaves, stems, and roots. These body structures comprise several components, including:
(9) Thallus: The body structure consisting of algal cells without specific organs or tissues.
(10) Stipe: The stem-like part that connects the thallus to the substrate where the algae adhere.
(11) Holdfast: A specialized structure that serves as an anchor for the algae to attach to a stronger substrate.
(12) Bladder: A sac-like structure found in certain algal species that functions as a buoyancy aid to keep the algae floating at the water's surface.

The structure and anatomy of algae cells vary depending on their types and species. However, most algae cells share common features with other plant cells, such as chloroplasts, cell membranes, and nuclei. Some algae types also exhibit more intricate body structures and specialized tissues, including thallus, stipe, holdfast, and bladder.

The metabolism of algae cells encompasses a series of biochemical processes that occur within the cell. These processes aim to acquire energy and maintain vital cell functions. Algal cell metabolism consists of two stages, namely anabolism and catabolism, as explained below:

(1)(**Anabolism**The formation of complex compounds from simple molecules. Processes within this stage include photosynthesis and the synthesis of other organic compounds such as proteins and carbohydrates. Several anabolic phases occurred in algae cells are as follows:

   (a) **Photosynthesis**: The conversion of light energy into chemical energy in the form of sugars. During photosynthesis, chloroplasts within the algae cells transform light energy into chemical energy and use CO_2 from the environment to create organic compounds like glucose.
   (b) **Protein synthesis**: The formation of protein compounds from amino acids. During this process, algae cells require nutrients like nitrogen and sulfur.
   (c) **Carbohydrate synthesis**: The formation of carbohydrate compounds from CO_2 and water. This process requires the energy generated during photosynthesis.

(2) **Catabolism**: The breakdown of complex compounds into simpler molecules, releasing energy that powers the algae cells. Processes within this stage include respiration and fermentation. Several catabolic phases occurred in algae cells are as follows;

(a) **Respiration**: The breakdown of organic compounds into CO_2 and H_2O with the aid of oxygen, producing energy in the form of ATP.
(b) **Fermentation**: The breakdown of organic compounds without the aid of oxygen, generating energy in the form of ATP (Catalanotti et al. 2013).

Accordingly, algal cell metabolism encompasses a series of biochemical processes comprising anabolism and catabolism. Anabolism is the stage of synthesizing complex compounds (e.g., proteins, carbohydrates, and glucose) from simple molecules. On the other hand, catabolism is the stage of breaking down complex compounds into simpler molecules like CO_2 and H_2O, yielding energy in the form of ATP.

20.2.2 The Process of Photosynthesis and Biomass Production

Photosynthesis is a fundamental mechanism used by algae and other green plants to generate energy and biomass (Vecchi et al. 2020). It is a process that converts solar energy into chemical energy stored in molecules of glucose and other organic compounds. It involves the conversion of carbon dioxide (CO_2) and water (H_2O) into simple sugars and oxygen, with light energy serving as the trigger. Algal biomass refers to the total dry weight of living algal organisms, composed of organic molecules such as proteins, carbohydrates, and lipids. The formation or production of algal biomass is closely related to photosynthesis (Fig. 20.2) as it provides the energy and raw materials needed for algal growth and reproduction.

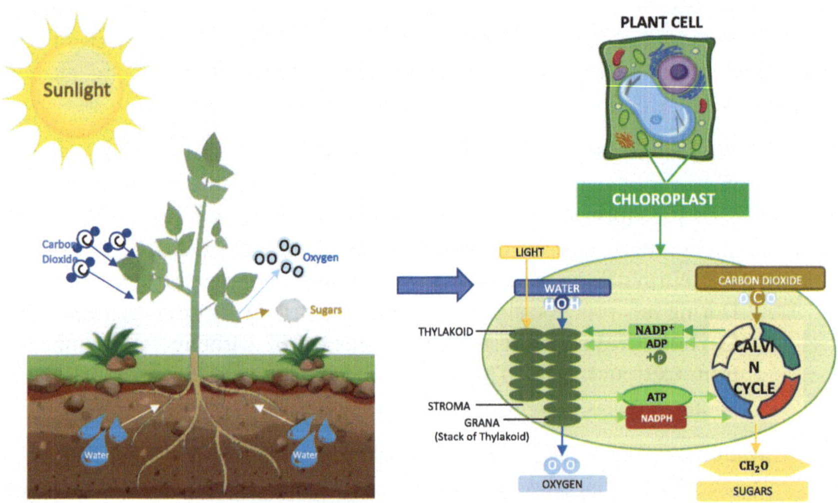

Fig. 20.2 Photosynthesis process in plants

The process of photosynthesis leading to the formation of algal biomass includes several stages as follows:

(1) **Absorption of light energy:** Algae contain photosynthetic pigments like chlorophyll that capture solar energy. These pigments convert light energy into chemical energy used in the process of photosynthesis.

(2) **Carbon fixation:** Algal carbon fixation is the process of "capturing" or absorbing CO_2 from the air and converting it into organic compounds through photosynthesis. Through the Calvin cycle, also known as the Calvin-Benson-Bassham (CBB) cycle, algae absorb CO_2 from the air to produce organic compounds such as glucose, which are used for growth and biomass production. The Calvin cycle consists of three main phases:

    (a) **Carbon fixation:** In this stage, CO_2 from the environment is bound to ribulose-1,5-bisphosphate (RuBP), the CO_2 acceptor. This reaction is catalyzed by the enzyme Rubisco (ribulose-1,5-bisphosphate carboxylase/oxygenase), the most abundant enzyme on Earth. The result of this reaction is two molecules of 3-phosphoglycerate (3-PGA), a three-carbon compound.

    (b) **Reduction:** The reduction phase involves converting 3-PGA into triose phosphates (simple sugars) through several chemical reactions. First, 3-PGA is reduced to 1,3-bisphosphoglycerate (1,3-BPG) by the enzyme phosphoglycerate kinase, using ATP. Next, 1,3-BPG is reduced to glyceraldehyde 3-phosphate (G3P) by the enzyme NADPH-glyceraldehyde-3-phosphate dehydrogenase (GAPDH), using NADPH. In this process, one molecule of G3P is exported from the cycle for glucose and other organic compound synthesis, while the other G3P molecule remains within the cycle.

    (c) **Regeneration:** The regeneration phase involves a series of reactions that convert the remaining G3P molecules back into RuBP, allowing the Calvin cycle to continue. This process incorporates several enzymes (e.g., transketolase, aldolase, and phosphoribulokinase) and consumes ATP. Once RuBP is regenerated, the Calvin cycle can repeat, enabling the fixation of more CO_2 and biomass production.

    Through carbon fixation via the Calvin cycle, microalgae convert CO_2 from the air into organic compounds necessary for growth and biomass production. This process is essential for maintaining the carbon balance in the atmosphere and holds potential for use in climate change mitigation technologies, such as carbon capture and storage (CCS) or environmentally friendly bioenergy production.

(3) **Biomolecule synthesis:** Simple sugars produced through photosynthesis serve as raw materials for synthesizing more complex biomolecules, such as polysaccharides, proteins, and lipids. Simple sugars like glucose, fructose, and galactose are monosaccharides and act as the building blocks of more complex

biomolecules like polysaccharides, proteins, and lipids. Here's how simple sugars are synthesized into more complex biomolecules:

(a) **Polysaccharides**: These are complex carbohydrates composed of long chains of connected monosaccharides. Glucose, for example, can be synthesized into polysaccharides like amylum and glycogen through a process called glycosylation. In glycosylation, enzymes catalyze the formation of glycosidic bonds between monosaccharides, resulting in longer polysaccharides. Amylum and glycogen serve as energy storage forms in plants and animals, respectively.

(b) **Proteins**: Proteins are polymers composed of amino acids as monomers. Protein synthesis involves several steps, including DNA transcription to RNA, RNA translation to polypeptides, and polypeptide folding into functional protein structures. Glucose and other simple sugars can be used as an energy source for these processes. Moreover, glucose can be converted into amino acids through gluconeogenesis and the citric acid cycle (Krebs cycle). In gluconeogenesis, glucose is converted into pyruvate, which can then enter the citric acid cycle. In the citric acid cycle, pyruvate is converted into amino acid precursors such as oxaloacetate and α-ketoglutarate. The produced amino acids can then be used for protein synthesis.

(c) **Lipids**: Lipids include molecules such as triglycerides, phospholipids, and sterols. Glucose can be used as an energy source in lipid synthesis or as a substrate to produce lipid precursors such as glycerol and fatty acids. This process involves several metabolic steps, including glycolysis, the citric acid cycle, and fatty acid synthesis. In glycolysis, glucose is broken down into pyruvate, which can then be converted into acetyl-CoA. Acetyl-CoA is a critical substrate in fatty acid synthesis, which occurs in the cytoplasm through a reaction catalyzed by the enzyme fatty acid synthase complex (FAS). Once fatty acids are formed, they can combine with glycerol to form triglycerides, the primary form of stored fat. Additionally, fatty acids can be used to synthesize phospholipids and sterols like cholesterol.

In summary, simple sugars like glucose play a crucial role in the synthesis of complex biomolecules such as polysaccharides, proteins, and lipids. These biomolecules are essential for building algal cell structures and serve as a source of energy for algal growth and reproduction.

(1) **Growth and reproduction:** Through biomolecule synthesis, algae undergo growth and reproduction, leading to an overall increase in algal biomass. Microalgae employ various methods of reproduction depending on species and environmental conditions. Reproduction in microalgae can be categorized into two main types, namely asexual and sexual reproduction, as elaborated below:

(a) **Asexual reproduction:** a process of microalgal proliferation without involving genetic material exchange. In asexual reproduction, parent cells divide into two or more identical offspring genetically. Several methods of asexual reproduction in microalgae include:

- Binary fission: The most common asexual reproduction method in microalgae. During binary fission, parent cells divide into two equally sized daughter cells. This process involves DNA replication, septum formation, and offspring cell separation.
- Some microalgae species undergo sporulation, forming asexual spores called zoospores or autospores. These spores can withstand harsh environmental conditions and remain dormant until conditions become more favorable. Once conditions improve, these spores germinate and form new algal cells.
- Fragmentation: In some filamentous algae species, filaments can break into fragments that then develop into new individuals.

(b) **Sexual Reproduction:** Sexual reproduction in microalgae comprises the exchange of genetic material between two different individuals. This process results in offspring with unique combinations of parental characteristics. Several methods of sexual reproduction in microalgae include:

- Isogamy: In isogamy, two gametes of identical size and shape (isogametes) fuse to form a zygospore. The zygospore then grows and develops into a new individual.
- Anisogamy: In anisogamy, two gametes of different sizes and shapes (anisogametes) fuse to form a zygospore. This process is similar to isogamy but involves morphologically different gametes.
- Oogamy: In oogamy, smaller, motile male gametes (sperm) fuse with larger, non-motile female gametes (oosphere) to form a zygospore. The zygospore then develops into a new individual. Sexual reproduction usually occurs under unfavorable environmental conditions or when resources are limited. This process results in genetic variation that can enhance the adaptability of microalgae to environmental changes.

In conclusion, photosynthesis plays a vital role in the formation of algal biomass by providing the energy and raw materials necessary for algal growth and reproduction. Improving photosynthetic efficiency can have a positive impact on algal biomass production, which is essential for various industrial applications such as bioenergy, animal feed, and chemical raw materials (Vecchi et al. 2020).

20.2.3 Environmental Factors Affecting Algae Growth

The growth of microalgae in their natural environment is influenced by various physical, chemical, and biological factors. Here are some key aspects that affect microalgae growth:

(1) Physical factors, including:

(a) Sunlight: Solar energy is the primary source of energy for microalgal photosynthesis. The intensity and duration of light exposure affect the rate of photosynthesis and microalgae growth (Metsoviti et al. 2020).
(b) Temperature: Environmental temperature affects the rate of metabolism and enzyme activity within microalgae cells. Each microalgae species has an optimal temperature range for its growth, and temperatures outside this range can reduce growth rate (Novosel et al. 2022).
(c) Mixing and water movement: This aspect influences nutrient uptake and light distribution. Proper mixing enhances nutrient absorption and light exposure, supporting microalgae growth (Chowdury et al. 2020).

(2) Chemical factors, including:

(a) Nutrients: Microalgae require nutrients such as nitrogen, phosphorus, and sulfur for growth and reproduction. The availability of these nutrients in the natural environment affects microalgae growth rate (Chowdury et al. 2020).
(b) Carbon dioxide (CO_2): is the primary substrate for photosynthesis and microalgae growth. The concentration of CO_2 in the natural environment affects the rate of photosynthesis and microalgae growth (Udayan et al. 2022).
(c) pH: Environmental pH balance affects enzyme activity and nutrient uptake by microalgae. Each microalgal species has an optimal pH range for its growth (Latsos et al. 2021).
(d) Salinity: Salt content in water affects microalgae growth. Some microalgal species thrive better in higher salinity, while others are more tolerant of low salinity levels (Latsos et al. 2021).

(3) Biological factors, including:

(a) Microalgae species and strains: Different microalgae species and strains have varying requirements and tolerances to physical and chemical factors, affecting their growth rate.
(b) Interactions with other microorganisms: Interactions between microalgae and other microorganisms like bacteria, protozoa, and viruses can influence microalgae growth. Some of these interactions can be mutualistic, while others can be competitive or parasitic.
(c) Predation: Microalgae serve as a food source for various organisms such as zooplankton and filter feeders. High predation rates can affect the growth of microalgal populations in their natural environment.

20.3 Technology and Methods of Algae Cultivation

The cultivation method of algae is closely linked to the potential of algae as a renewable resource for energy storage technologies. Algae, especially microalgae, offer several advantages as a raw material source for energy storage, such as biogas, biodiesel, bioethanol, and biohydrogen. The connection between algae cultivation technology and the potential of algae in energy storage can be explained as follows:

(1) (Biomass productivity: In algae cultivation, one of the main goals is to increase biomass productivity. High algal biomass will increase the amount of raw material available for energy conversion processes, such as fermentation, transesterification, or gasification (Mahmood et al. 2022).
(2) Lipid and biochemical content: Optimal algae cultivation can increase lipid, carbohydrate, and protein content in algae, directly impacting the potential of algae as a renewable resource (Visca et al. 2017). For instance, algae with high lipid concentrations are more suitable for biodiesel production, while algae with high carbohydrate content are more satisfactory for bioethanol or biogas production (Kumar et al. 2022).
(3) Environmental control and cultivation conditions: Algae cultivation allows for the control of environmental conditions, such as temperature, light intensity, pH, and nutrients. Optimizing these conditions can promote algae growth and the production of renewable energy raw materials. Algae cultivation can also reduce competition with agricultural land and freshwater resources, which are vital for sustainable renewable energy (Mahmood et al. 2022).
(4) Breeding and genetic engineering: Algae cultivation technology involves the use of breeding and genetic engineering techniques to produce algae strains that are more efficient in producing biomass or specific biochemical components. These optimized algae strains can elevate the potential of algae as a renewable resource for energy storage.

By optimizing algae cultivation method, we can improve the potential of algae as a renewable resource for energy storage technologies. This will help decrease dependence on fossil fuels and foster the development of environmentally friendly and sustainable energy technologies.

20.3.1 Algae Cultivation Techniques (Photobioreactor, Open System, Closed System)

Microalgae cultivation refers to the process of cultivating or nurturing microalgae using optimal nutrient, light sources, and environmental conditions to support their growth and reproduction. In microalgae cultivation, there are two primary systems used to support their growth and reproduction: open systems and closed systems.

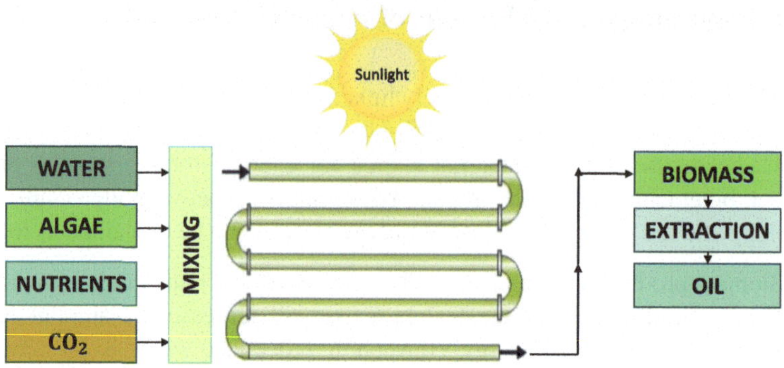

Fig. 20.3 Photobioreactor system for growing algae. Diagram shows the constituents of a tubular photobioreactor

Algae Cultivation in Closed System Photobioreactor

Cultivating microalgae in a closed system refers to a cultivation method where microalgae are grown in containers isolated from the external environment, such as photobioreactors (PBR) or tube bioreactors (Fig. 20.3) (Benner et al. 2022). Closed systems offer several advantages over open systems, such as better control over environmental conditions, protection from contamination, and higher photosynthetic efficiency. Here is how to perform microalgae cultivation in a closed system:

(1) Selecting a suitable photobioreactor or closed system: There are various types of photobioreactors, such as tubular PBRs, flat-plate PBRs, and hybrid PBRs that can be adjusted according to the cultivation needs and objectives.
(2) Preparing the inoculum: The inoculum is a collection of microalgae ready to be added to the culture medium through the inoculation process. The inoculum usually comes from existing algal cultures or isolated from natural sources. Microalgae isolated from the natural environment can be propagated on a laboratory or pilot scale to prepare sufficient inoculum under healthy and contamination-free conditions. The steps for isolating microalgae from the environment include:

    (a) Comprehensive sampling: Samples of water or substrate containing microalgae from the environment (e.g., lakes, ponds, or coastal waters) should be taken from various depths and sampling points to increase the likelihood of finding diverse microalgal species.
    (b) Initial separation: Methods like centrifugation or filtration are employed to separate microalgae from solid particles and other contaminants in the sample. The aim is to concentrate microalgae and facilitate the subsequent isolation process.
    (c) Inoculation: Microalgae are administered into suitable artificial culture media, such as Walne, f/2, or BG-11. The culture is then incubated under appropriate light and temperature conditions for microalgae growth.

(d) Colony isolation: Subsequent to the microalgae growth in the culture media, single colonies are isolated from the culture media using serial dilution or agar streaking methods to obtain pure algal colonies consisting of a single species to avoid cross-contamination.
e) Purification. After acquiring untainted microalgal colonies, further purification can be done, such as single-cell isolation or cryopreservation, to ensure the authenticity and stability of the isolated microalgal species.
(f) Identification and characterization: Identification of the isolated microalgal species can be done using morphological and/or genetic techniques, such as microscopy, pigment analysis, or DNA sequencing. Characterization of microalgal species, including growth, photosynthesis, and biochemical composition testing, is necessary to understand their important properties for further desired applications.
(g) Pre-cultivation. Microalgae are then ready to be cultivated on a laboratory scale using Erlenmeyer flasks, test tubes, or small-scale photobioreactors. The objective is to optimize growth conditions and prepare adequate inoculum for cultivation in large-scale photobioreactors.
(h) Cultivation in photobioreactor: Subsequently, microalgae are all set to be inoculated into the prepared photobioreactor, which already contains the fitting culture media (Dewi et al. 2014).

(3) Preparing the culture media: The culture media should be tailored to the needs of the microalgal species to be cultivated. Culture media are mixtures of nutrients and chemicals required for the growth and development of microorganisms, including microalgae. Culture media can be divided into two main categories, namely natural and artificial culture media, as follows:

(a) **Natural culture media**: These are media or compounds that naturally contain the nutrients and chemical components required for microalgal growth, e.g., seawater, freshwater, and nutrient-rich wastewater (Silva et al. 2014). In natural culture media, microalgae rely on existing nutrients in the environment, such as nitrogen, phosphorus, and trace elements. The advantages of natural culture media include lower costs and simpler processes since the natural nutrients are readily available in the environment. However, it often lack control and may contain contaminants or other microorganisms that can interfere with microalgal growth.
(b) **Artifical culture media**: These are mixtures of nutrients and chemicals prepared artificially to promote microalgal growth. They are typically made by adding essential nutrients like carbon sources (e.g., CO_2 or sugar), nitrogen sources (e.g., nitrate or ammonium), phosphorus sources (e.g., phosphate), as well as trace elements and vitamins required for optimal microalgal growth. Various artificial culture media have been developed for microalgal cultivation, such as Walne media, f/2 media, PHM (Provasoli Haematococcus Media), and BG-11 media. Each medium has a different composition to support the growth of different microalgal species. The advantages of artificial culture media include better control over nutrient

and environmental composition, which can improve microalgal growth efficiency and lessen contamination. However, artificial culture media are often more expensive and require more careful preparation compared to natural culture media.

The choice of culture media in microalgal cultivation depends on the research or production objectives, the microalgal species used, and the environmental conditions. Natural culture media may be more suitable for large-scale applications with low costs, while artificial culture media can be used for scientific studies and applications that require tighter control over nutrients and the environment.

(4) Inoculation: In the context of cultivating microalgae in an artificial medium, inoculation refers to the process of adding microalgal inoculum to the prepared culture medium in the photobioreactor to initiate algal growth and development. Inoculation is a crucial step in microalgal cultivation to ensure that the desired microalgae are present in the culture system and can grow under the provided conditions. Before inoculation, the inoculum needs to be checked for quality, density, and the absence of unwanted contaminants, such as other microorganisms or chemicals (Bani et al. 2021).
(5) Environmental parameter control: In closed systems, environmental parameters such as temperature, pH, light intensity, and gas concentrations (CO_2 and O_2) can be well regulated. Each parameter can be adjusted to the needs of the microalgal species being cultivated to optimize growth and biomass or product production.
(6) Stirring and aeration: Closed systems are usually equipped with stirring and aeration mechanisms to keep microalgae evenly dispersed and maintain optimal oxygen and CO_2 concentrations. This also aids in nutrient absorption and more efficient light exposure.
(7) Monitoring and maintenance: These measures need to carried out periodically, including providing additional nutrients when required. The photobioreactor should be kept clean to lower the risk of potential contamination that could interfere with microalgal growth.
(8) Harvesting: Once the microalgae reach the desired density, harvesting can be performed using methods such as centrifugation, filtration, or flocculation. In closed systems, harvesting is usually easier and more efficient compared to open system.
(9) Biomass processing: After harvesting, the microalgal biomass can undergo further treatment to generate products such as biodiesel, proteins, pigments, animal feed, and more.

Algae Cultivation in Open Pond

Cultivating microalgae in open ponds is one of the most commonly used methods for large-scale microalgae production. Open ponds are typically designed as pools or channels made of soil, concrete, or plastic. The following considerations need to be taken into account when cultivating microalgae in open ponds:

(1) Pond preparation: The location of the open pond should be carefully chosen to ensure good access to water sources, nutrients, and a source of energy such as sunlight. Adequate drainage and water supply systems should be in place. Before cultivation, the pond needs to be cleaned and sterilized to minimize the risk of contamination.
(2) Inoculum preparation: Desired microalgae strains should be cultivated on a laboratory or pilot scale to prepare sufficient inoculum. The inoculum must be in a healthy condition and free from contamination.
(3) Natural cultural media preparation: Natural culture media such as wastewater containing essential nutrients (e.g., nitrogen, phosphorus, sulphur) and trace elements (e.g., iron, manganese, and zinc) need to be prepared.
(4) Inoculation: The prepared inoculum should be introduced into the open pond to mix with the natural culture media. An appropriate inoculum concentration typically ranges from 0.1 to 0.5 g/L (dry weight of microalgae per liter of water).
(5) Agitation and aeration: Regular and periodic agitation in the open pond ensures even distribution of microalgae and proper access to sunlight and nutrients. Aeration can be achieved by activating air pumps or CO_2 injection systems to sustain optimal oxygen and CO_2 concentrations.
(6) Environmental parameter control: Environmental aspects such as temperature, pH, and salinity should be monitored periodically to maintain optimal growth conditions. The ideal temperature and pH vary depending on the cultivated microalgae species.
(7) Maintenance and fertilization: Additional nutrients should be added as needed to support microalgae growth. Fertilization can be done in batches or continuously, depending on the microalgae's nutrient requirements.
(8) Harvesting: Once the microalgae reach the desired density, harvesting can be performed using methods such as centrifugation, filtration, or flocculation. Gradual harvesting is preferred to minimize disruptions to the open pond conditions and ensure sustainable microalgae growth.
(9) Biomass processing: After harvesting, the microalgal biomass can undergo further treatment to yield products such as biodiesel, proteins, pigments, or polysaccharides.

20.3.2 Optimization of Cultivation Conditions (Nutrients, Light, Temperature, pH)

Some principal parameters in microalgae cultivation include light intensity, temperature, pH, gas concentration (CO_2 and O_2), nutrients, and agitation or aeration. Optimizing these factors is paramount to achieve optimal microalgae growth and desired biomass or product generation. Here is how to optimize these parameters:

(1) Light intensity: Light intensity affects photosynthesis efficiency and microalgae growth rate. To optimize it, adjust the light source to provide sufficient and

uniform intensity. In closed systems, use LED or fluorescent lights with appropriate spectra. In open ponds, ensure sufficient sunlight exposure.

(2) Temperature: Optimal temperature for microalgae growth varies among species. Too high or too low temperatures can inhibit growth and cell metabolism. Monitor temperature regularly and adjust it within the optimal range for the cultivated species, typically ranging from 20 to 30 °C (Novosel et al. 2022).

(3) pH: pH affects nutrient availability and enzyme stability within microalgae cells. The optimal pH varies among species but generally falls within the range of 7–9. Monitor pH periodically and adjust it with suitable acid or base augmentations if necessary (Latsos et al. 2021).

(4) Gas concentration (CO_2 and O_2): CO_2 concentration influences photosynthesis rate, while O_2 concentration affects cell respiration. To optimize gas concentration, use a well-designed aeration system to maintain optimal CO_2 and O_2 levels. In closed systems, CO_2 supply can be directly controlled.

(5) Nutrients: Microalgae require nutrients such as nitrogen, phosphorus, sulfur, and trace elements like iron, manganese, and zinc. Nutrient availability determines growth rate and microalgae biomass composition. Use culture media containing the required nutrients and add supplementary nutrients during cultivation if needed.

(6) Agitation or aeration: Agitation and aeration affect nutrient uptake, light exposure, and gas concentration. To optimize agitation and aeration, use suitable pumps, impellers, or air bubble systems to create sufficient and uniform flow in the cultivation system.

Optimizing these parameters can be achieved through systematic experimentation and regular monitoring of cultivation conditions. The use of statistical approaches, such as response surface analysis or the Taguchi method, can also help identify the optimal parameter combinations for achieving satisfactory microalgae growth. Both statistical methods are useful in reducing the number of required experiments and obtaining maximum information about the effects of different parameters on the desired outcomes, such as productive microalgae growth.

(1) Response Surface Analysis (RSM) is a statistical method used to model and analyze the relationship between a response variable (output) and several independent variables (inputs). In the context of microalgae cultivation, response variables may include growth rate, biomass productivity, or secondary metabolite production, while independent variables include light intensity, temperature, pH, gas concentration, nutrients, and agitation or aeration. RSM involves the following steps:

    (a) Designing experiments, comprising different combinations of independent variables using factorial design, Box-Behnken design, or Central Composite design.

    (b) Collecting data through experiments according to the designed experimental setup and recording the results of the response variables.

(c) Building a mathematical model from the collected data by estimating regression coefficients using multiple regression analysis.
(d) Performing parameter optimization. The developed mathematical model from the previous step serves as the basis for identifying the optimal parameter combinations that will yield the desired response.

(2) **Taguchi Method** is a simpler approach to optimize a process using orthogonal experimental designs. This method focuses on reducing process variability and improving product quality by identifying the optimal parameter combinations. The steps in the Taguchi method include:

(a) Designing experiments by selecting an appropriate orthogonal experimental design based on the number of independent variables and desired levels.
(b) Collecting data through experiments according to the orthogonal experimental design and recording the results of the response variables.
(c) Analyzing the results using analysis of variance (ANOVA) to evaluate the influence of independent variables on the response variable and identify the most significant factors.
(d) Performing parameter optimization based on the analysis results and the influence of factors on the response variable. Dengan menggunakan pendekatan statistik seperti RSM atau metode Taguchi, kombinasi parameter optimal untuk pertumbuhan mikroalga yang baik dapat diidentifikasi secara efisien, sehingga memungkinkan peningkatan produktivitas dan efisiensi dalam proses kultivasi mikroalga.

By using statistical approaches such as RSM or the Taguchi method, the optimal parameter combinations for good microalgae growth can be efficiently identified, enabling increased productivity and efficiency in the microalgae cultivation process.

20.3.3 Algal Biomass Separation and Purification Techniques

The objective of algal biomass separation and purification techniques is to separate microalgae biomass from the culture medium and generate high-quality products for various applications. Microalgae have the potential to yield a wide range of high-value products, such as proteins, pigments (e.g., chlorophyll and carotenoids), omega-3 fatty acids, and bioenergy sources like biodiesel, biogas, and bioethanol (Salami et al. 2021).

The process of separating and purifying algal biomass involves the following critical steps:

(a) Biomass Concentration: In this stage, microalgae biomass is concentrated from the liquid culture by decreasing the volume of water. Some commonly used techniques for algal biomass concentration are centrifugation, flotation, filtration, and flocculation.
(b) Drying: After biomass concentration, the remaining water in the algal biomass needs to be removed to reduce transportation and storage costs. Drying also improves product stability and facilitates further processing. Commonly used drying techniques include spray drying, drum drying, freeze drying, and air drying.
(c) Extraction: The extraction process is used to separate specific components from algal biomass, such as proteins, pigments, or lipids, depending on the desired product. Commonly used extraction techniques include solvent extraction, supercritical extraction, enzymatic extraction, and mechanical extraction.
(d) Purification: After extraction, the resulting products often contain contaminants or unwanted components. Purification processes are necessary to enhance the quality and purity of the products. Commonly used purification techniques include chromatography, precipitation, crystallization, and ultrafiltration.

By employing algal biomass separation and purification techniques, high-value products obtained from microalgae can be utilized in various industrial applications, such as food, animal feed, nutraceuticals, pharmaceuticals, cosmetics, and bioenergy.

20.4 The Process of Converting Algae Biomass into Energy

20.4.1 Fermentation (Production of Bioethanol, Biobutanol, Biogas)

Production of Bioethanol

The conversion of algal biomass into bioethanol through fermentation involves several essential phases, commonly comprising pre-processing, hydrolysis, fermentation, and separation or purification of bioethanol. The steps are explained as follows (Fig. 20.4):

(1) Pre-processing: In this stage, the collected algal biomass undergoes a drying process to reduce its water content. Drying can be achieved through methods such as solar drying, oven drying, or spray drying. Subsequently, the algal biomass is crushed into smaller particles to boost enzyme accessibility during the hydrolysis stage.
(2) Hydrolysis: The purpose of hydrolysis is to break down polymers (such as cellulose, hemicellulose, and starch) in the algal biomass into simpler sugar monomers, such as glucose and fructose. Hydrolysis can be performed through enzymatic, chemical, or combined methods. Frequently used enzymes in the enzymatic hydrolysis process include cellulase, hemicellulase, and amylase.

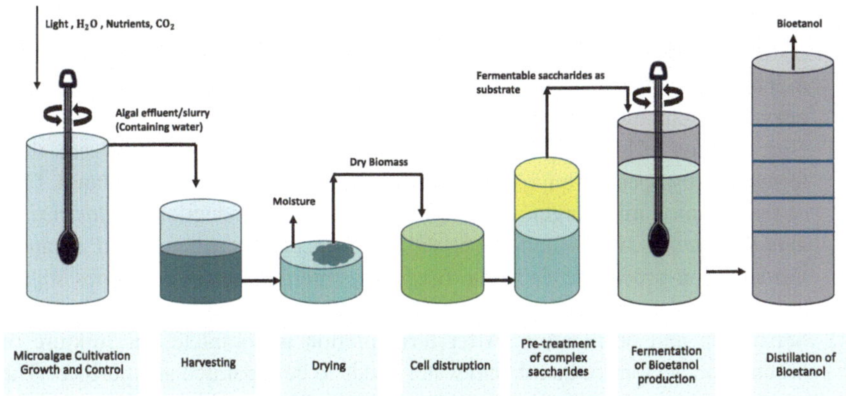

Fig. 20.4 Schematic representation of bioethanol production from microalgae feedstocks

(3) Fermentation: The simple sugars produced during the hydrolysis are then fermented by microorganisms, such as *Saccharomyces cerevisiae* (yeast), to yield ethanol and carbon dioxide. This process can be carried out aerobically or anaerobically, depending on the type of microorganism used. Fermentation typically takes several days to around one week, depending on the conditions and type of microorganism.

(4) Separation and purification: After fermentation is complete, the resulting mixture of ethanol and water needs to be separated and purified. The typical separation process is distillation, where ethanol and water are separated based on their difference in boiling points. Following the distillation, ethanol may still contain a small amount of water and other impurities. Further purification processes, such as dehydration using molecular sieves or adsorbents, may be required to produce ethanol with the desired quality for specific applications.

The bioethanol produced from algae can be used as an environmentally friendly alternative fuel, either in its pure form or as a blend with fossil fuels, such as gasoline. The use of bioethanol has the potential to mitigate greenhouse gas emissions and decrease dependence on fossil fuels.

Production of Biobutanol

The conversion of algal biomass into biobutanol through fermentation involves several primary steps that are similar to the bioethanol production process, consisting of pre-processing, hydrolysis, fermentation, and separation or purification of biobutanol, as follows:

(1) Pre-processing: In this stage, the collected algal biomass undergoes a drying process to reduce its water content. Subsequently, the algal biomass is crushed into smaller particles to enhance enzyme accessibility during the hydrolysis stage.

(2) Hydrolysis: The purpose of hydrolysis is to break down polymers (such as cellulose, hemicellulose, and starch) in the algal biomass into simpler sugar monomers, such as glucose and fructose. Hydrolysis can be performed through enzymatic, chemical, or combined methods.
(3) Fermentation: The simple sugars yielded during the hydrolysis are then fermented by specific microorganisms capable of generating biobutanol. One of the common microorganisms used for biobutanol production is *Clostridium acetobutylicum*, which undergoes the ABE (acetone, butanol, and ethanol) fermentation procedure. In this process, the microorganism consumes sugars and produces acetone, biobutanol, and ethanol as the main products.
(4) Separation and purification: After fermentation is complete, the mixture of biobutanol, acetone, ethanol, and water needs to be separated and purified. The common separation process used is distillation, where biobutanol, acetone, and ethanol are separated based on their difference in boiling points. Further purification, such as extraction with solvents or azeotropic distillation processes, may be required to produce biobutanol with the desired quality for specific applications.

Biobutanol generated from algae has several advantages over bioethanol, such as higher energy density, better compatibility with existing fossil fuel infrastructure, and lower hygroscopicity. The use of biobutanol as an alternative fuel has the potential to diminish greenhouse gas emissions and dependence on fossil fuels. However, biobutanol production through algal fermentation is still in the research and development stage, and several technical and economic challenges need to be addressed before this technology can be commercially applied.

Production of Biogas

The conversion of algal biomass into biogas through fermentation involves several important steps, including pre-processing, anaerobic fermentation, and biogas purification; each of which is described as follows (Fig. 20.5):

(1) Pre-processing: During this stage, the collected algal biomass undergoes drying (if necessary) to reduce its moisture content. Subsequently, the algal biomass is fragmented into smaller particles to enhance microorganism accessibility during anaerobic fermentation.
(2) Anaerobic fermentation: This process converts algal biomass into biogas—primarily composed of methane and carbon dioxide—by anaerobic microorganisms in an oxygen-free environment. Anaerobic fermentation involves several stages of reactions, comprising hydrolysis, acidogenesis, acetogenesis, and methanogenesis (Han et al. 2019).

    (a) Hydrolysis: Large polymers in algal biomass, such as proteins, lipids, and polysaccharides, are broken down into smaller monomers by extracellular enzymes produced by hydrolytic microorganisms.
    (b) Acidogenesis: The monomers resulting from hydrolysis are further converted into volatile organic acids, such as acetic acid, propionic acid,

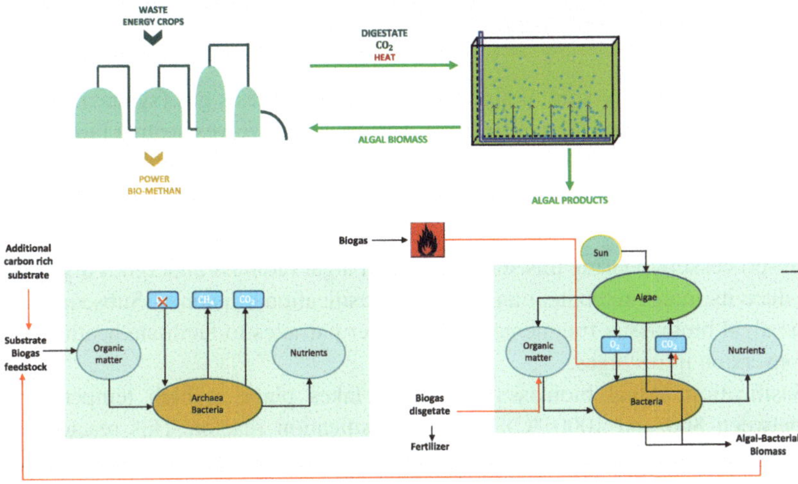

Fig. 20.5 Biogas from algae

and butyric acid, as well as gases like hydrogen and carbon dioxide by acidogenic microorganisms.

(c) Acetogenesis: In this stage, volatile organic acids and hydrogen are converted into acetic acid, hydrogen, and carbon dioxide by acetogenic microorganisms.

(d) Methanogenesis: Finally, methanogens—microorganisms that produce methane—convert acetic acid, hydrogen, and carbon dioxide into methane and carbon dioxide.

(3) Biogas purification: The biogas generated from algal anaerobic fermentation typically contains methane (CH_4), carbon dioxide (CO_2), as well as small amounts of contaminants like hydrogen sulfide (H_2S) and nitrogen (N_2). To improve biogas quality, further purification is required. Biogas purification processes involve the removal of carbon dioxide, hydrogen sulfide, and other contaminants using methods such as water scrubbing, chemical absorption, or membrane technologies.

Biogas yielded from algae can serve as an environmentally-friendly alternative fuel, either in its pure form or as a blend with natural gas. Its utilization has the potential to lessen greenhouse gas emissions and dependence on fossil fuels. Moreover, the nutrient-rich byproduct from anaerobic fermentation can be utilized as organic fertilizer.

20.4.2 Gasification (Syngas Production)

Gasification is the conversion of algal biomass into synthesis gas (syngas), rich in hydrogen (H_2) and carbon monoxide (CO). This process involves heating biomass in an environment with limited oxygen supply, preventing complete combustion (Verma and Mishra 2020). The conversion of algal biomass through gasification unfolds as follows:

(1) Pre-processing: During this stage, collected algal biomass undergoes drying to reduce its moisture content and enhance gasification efficiency. Subsequently, the algal biomass is fragmented into smaller particles to facilitate heating and conversion into syngas.
(2) Gasification: Algal biomass gasification takes place at high temperatures (between 800 and 1000 °C) within a gasification reactor. This reactor can be of various designs, such as downdraft, updraft, fluidized bed, or plasma gasification, depending on specific objectives. During gasification, the algal biomass is heated in an environment with limited oxygen supply, typically using air, pure oxygen, or steam as the gasification agent. This heating induces an endothermic chemical reaction that breaks down the biomass into syngas, comprising hydrogen, carbon monoxide, methane, and small amounts of other gases like nitrogen and carbon dioxide.
(3) Syngas purification: Syngas produced from the gasification may contain contaminants such as tar, solid particles, and acidic gases (such as H_2S and CO_2). Before syngas can be utilized as a feedstock for chemical or fuel production, a purification process is required to remove these contaminants. Syngas purification involves various technologies, such as wet scrubbing, dry scrubbing, or adsorption using media like activated carbon or zeolite.
(4) Syngas utilization: The syngas yielded from algal biomass gasification can be used as a feedstock to generate various products, such as hydrogen, methanol, or synthetic fuels through the Fischer–Tropsch process. Additionally, syngas can be utilized as fuel in gas turbines or internal combustion engines to generate electricity.

Gasification of algal biomass offers several advantages, including more efficient conversion compared to anaerobic fermentation and flexibility in generating diverse end products (Ebhodaghe et al. 2022). However, technical and economic challenges, such as the cost of algal biomass production and processing, need to be addressed before algal gasification can be implemented on a large commercial scale.

20.4.3 Lipid Extraction (Biodiesel Production)

The conversion of algal biomass into biodiesel through lipid extraction involves several essential steps (Fig. 20.6), consisting of pre-processing, lipid extraction, lipid conversion into biodiesel, and biodiesel purification, described as follows:

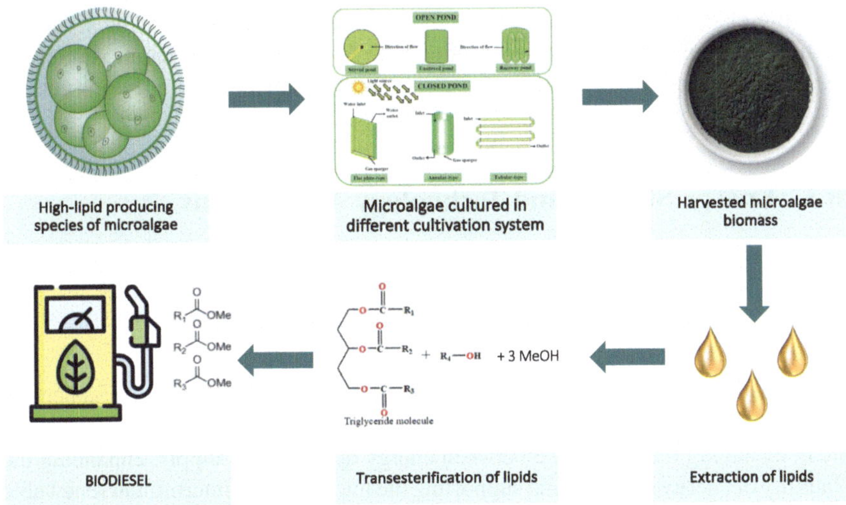

Fig. 20.6 Biodiesel production from microalgae

(1) Pre-processing: During this stage, collected algal biomass undergoes drying to reduce its moisture content. Subsequently, the algal biomass is fragmented into smaller particles to enhance lipid extraction efficiency.
(2) Lipid extraction: The goal of the extraction process is to separate lipids (fats) from the algal biomass. Lipid extraction can be carried out using organic solvents such as hexane, ethanol, or dichloromethane. Alternative methods include supercritical extraction using CO_2, ultrasound-assisted extraction, or microwave-assisted extraction (Sultana et al. 2023). After the extraction process, lipids are separated from the algal biomass, and the solvent is removed, typically through evaporation.
(3) Lipid conversion into biodiesel: The obtained lipids from the algal biomass are then converted into biodiesel through the transesterification process. In this phase, lipids are mixed with alcohol (usually methanol or ethanol) and a catalyst (usually sodium or potassium hydroxide). The reaction between lipids and alcohol produces methyl (or ethyl) esters, known as biodiesel, and glycerol as a byproduct.
(4) Biodiesel purification: After the transesterification process, the resulting mixture of biodiesel and glycerol needs to be separated and purified. Separation is usually accomplished through decantation, centrifugation, or membrane separation. Following the separation, biodiesel may still contain small amounts of contaminants like alcohol, water, and catalyst. Further purification processes, such as washing with water or drying using adsorbents, may be necessary to produce biodiesel with the desired quality for specific applications.

Biodiesel generated from algae can serve as an environmentally-friendly alternative fuel, either in its pure form or blended with fossil diesel fuel. Its utilization has the potential to reduce greenhouse gas emissions and dependence on fossil fuels.

20.5 Energy Storage and Technology Applications

20.5.1 Concept and Types of Energy Storage Technologies (Batteries, Capacitors, Flywheels, Hydro Pumps)

The concept of energy storage involves converting energy from easily generated forms into forms that can be stored and used as needed. Energy storage plays a vital role in managing the imbalance between energy demand and supply, enhancing the reliability of energy systems, and supporting the integration of intermittent renewable energy sources. Various energy storage technologies have been developed and can be classified into several categories based on how energy is stored and released. Some types of energy storage technologies include:

(1) Mechanical energy storage

  (a) Potential energy storage, for example, Pumped Hydro Energy Storage (PHES), where water is pumped to a higher reservoir during times of excess energy supply and released through turbines during periods of increased energy demand.
  (b) Kinetic energy storage, such as a flywheel, where electrical energy is used to rotate a heavy wheel at high speed, and the kinetic energy stored in the wheel can be converted back into electricity through a generator when needed (Sultana et al. 2023).

(2) Electrical energy storage

  (a) Capacitive energy storage, exemplified by capacitors, where electrical energy is stored in the electric field between two conductive plates separated by an insulator.
  (b) Inductive energy storage, like Superconducting Magnetic Energy Storage (SMES), where electrical energy is stored in the magnetic field generated by electric current flowing through superconductor coils.

(3) Electrochemical energy storage

  (a) Batteries, which store energy in the form of chemical energy through reversible redox reactions. Common types of batteries used for energy storage include lithium-ion, nickel–cadmium, and lead-acid batteries.
  (b) Fuel cells, which produce electrical energy through electrochemical reactions between a fuel, such as hydrogen, and an oxidant like oxygen.

(4) Thermal energy storage, which involves storing energy in the form of heat or cold. Examples include hot water storage tanks, where thermal energy from sources like solar thermal panels or heating systems is stored for later use.
(5) Chemical energy storage, where electrical or thermal energy is converted into chemical substances that are stored and then reconverted into energy when required.

20.5.2 Integration of Algal Biomass with Energy Storage Technologies

Integration of algal biomass with energy storage technologies can optimize the use of renewable energy sources and lower the environmental impact of energy systems (Beal et al. 2018). Several instances of this integration include:

(1) Production of biogas and chemical energy storage: Biogas is converted into a form of energy that can be chemically stored and later utilized as needed. One of the methods to achieve this objective is by converting biogas into a more easily storable fuel, such as hydrogen or synthetic methane, which can then be stored and used within a chemical energy storage system.
    (a) Conversion of biogas into hydrogen: Biogas—primarily composed of methane (CH_4) and carbon dioxide (CO_2)—can be transformed into hydrogen through a process called Steam Methane Reforming (SMR). In this process, methane reacts with steam at high temperatures in the presence of a catalyst, producing hydrogen and carbon dioxide. The generated hydrogen can be stored in either gaseous or liquid form at low temperatures and pressures. This hydrogen can then be used as a fuel within chemical energy storage systems, such as hydrogen batteries or fuel cells.
    (b) Conversion of biogas into synthetic methane (power-to-gas): One approach to integrate biogas production with chemical energy storage is through the concept of power-to-gas. In this process, hydrogen generated from water electrolysis or steam methane reforming is combined with carbon dioxide from biogas to produce synthetic methane. Synthetic methane can be stored in existing natural gas infrastructure and used as fuel for electricity generation, heating, or transportation. This approach enables large-scale and long-term energy storage.
    (c) Chemical energy storage system: Energy stored in the form of hydrogen or synthetic methane can be utilized within chemical energy storage systems, such as batteries or fuel cells, to generate electricity as needed. Chemical batteries like lithium-ion batteries or redox flow batteries can store chemical energy in the form of electric charge and release it as electricity when required. Hydrogen or synthetic methane fuel cells can directly convert chemical energy into electrical energy through electrochemical reactions, providing higher efficiency and longer operating times.

(2) Production of biodiesel and chemical energy storage: Microalgae biomass or other raw materials are converted into biodiesel, which can then be stored and used as an efficient and environmentally friendly energy source. Biodiesel is a liquid fuel that can be used as an alternative or blended with conventional diesel fuel. The integration of biodiesel production with chemical energy storage proceeds as follows:

   (a) Biodiesel production: Biodiesel is produced from microalgae oil or other feedstock through a process called transesterification. In this process, triglycerides in the oil react with alcohol (such as methanol) in the presence of a catalyst to produce methyl ester (biodiesel) and glycerol as a byproduct. The biodiesel produced has physical and chemical properties similar to diesel fuel, enabling easy storage and utilization within chemical energy storage systems (Mahmood et al. 2022)
   (b) Chemical energy storage in the form of biodiesel: Biodiesel serves as a form of chemical energy storage because the energy contained in its chemical bonds can be utilized when biodiesel is consumed as fuel. Biodiesel can be stored in conventional fuel storage tanks and used in diesel engines without significant modifications. Moreover, biodiesel offers the advantage of lower greenhouse gas emissions and better biodegradability compared to conventional diesel fuel.
   (c) Applications in energy storage systems: Stored biodiesel can be used as fuel in various applications, such as electricity generation, heating, and transportation. In the context of energy storage, biodiesel can serve as a backup energy source or be used to generate electricity when there is an increase in electricity demand or when renewable energy supply is insufficient. Additionally, biodiesel can be utilized in hybrid renewable energy storage systems, combining renewable energy sources like solar, wind, or hydro with chemical fuel-based energy storage systems to achieve higher reliability and efficiency (Dash et al. 2023).

(3) Production of syngas and chemical energy storage. The production of syngas (synthesis gas) from microalgae biomass involves converting this biomass into a synthetic gas consisting of hydrogen (H_2), carbon monoxide (CO), small amounts of methane (CH_4), and carbon dioxide (CO_2) through the gasification process. In other words, syngas, or synthesis gas, is a gas mixture rich in hydrogen (H_2) and carbon monoxide (CO) produced from the gasification of microalgae biomass or other carbon feedstocks. The integration of syngas production with chemical energy storage proceeds as follows:

   (a) Syngas production: Syngas is generated through gasification, where microalgae biomass or other carbon feedstocks are subjected to high-temperature treatment in a limited oxidation atmosphere. This process yields syngas, which can be used as fuel or as a raw material in other chemical processes, such as the production of methanol, ammonia, or Fischer–Tropsch fuels.

(b) Chemical energy storage in the form of syngas: Syngas can be stored as chemical energy because the energy contained in its chemical bonds can be utilized when syngas is consumed as fuel or as a raw material in other chemical processes. Syngas storage can be achieved in pressurized tanks at appropriate temperature and pressure or by converting syngas into more easily storable and transportable forms, such as synthetic methane, methanol, or ammonia.

(c) Applications in energy storage systems: Stored syngas can be used as fuel in various applications, such as electricity generation, heating, and transportation. In the context of energy storage, syngas can serve as a backup energy source or be used to generate electricity during periods of increased demand or when renewable energy supply is insufficient. Syngas can also be utilized in hybrid renewable energy storage systems, combining renewable energy sources like solar, wind, or hydro with chemical fuel-based energy storage systems to achieve higher reliability and efficiency.

(4) Integration with thermal energy storage: Thermal energy yielded from the conversion of microalgae biomass (e.g., direct combustion, gasification, or fermentation) can be stored in thermal energy storage media (e.g., water, molten salts, or phase change materials). The stored thermal energy can be used to produce steam and generate electricity through steam turbines or for other heating and cooling applications. The thermal energy storage media are described as follows:

(a) Water is a commonly used thermal energy storage medium. Thermal energy generated from the conversion of microalgae biomass is transferred to water, which is then stored in insulated tanks. This hot water can be used for heating applications (Pagkalos et al. 2020; Zakeri et al. 2022).

(b) Molten salts are mixtures of salts with relatively low melting points. Molten salts can absorb and store thermal energy in the form of latent heat when they melt and release this energy when they solidify. Molten salts have higher thermal energy storage capacity compared to water and can store energy at higher temperatures.

(c) Phase change materials (PCMs) are substances that store and release thermal energy during a phase change (e.g., melting and solidification). PCMs can be tailored to store thermal energy at specific temperatures, and they have a higher thermal energy storage capacity compared to water or molten salts.

(5) Integration with electrical energy storage. Electrical energy produced from the conversion of microalgae biomass (via biogas or biodiesel combustion in internal combustion engines or gas turbines) can be stored in electrical energy storage systems, such as batteries or flywheels. The stored electrical energy can be used to address fluctuations in energy supply and demand and to enhance the reliability of the energy system.

(a) Batteries are electrical energy storage devices that convert chemical energy into electrical energy. Electrical energy generated from the conversion of microalgae biomass, like biogas or biodiesel combustion, can be stored in batteries for later use. Batteries offer several advantages, such as the ability to store energy for extended periods, ease of installation, and flexibility in terms of storage capacity. Commonly used batteries in energy storage applications include lithium-ion batteries lead-acid batteries, nickel–cadmium batteries, redox flow batteries.

(b) Flywheels are electrical energy storage devices that store energy in the form of kinetic energy (Fig. 20.7). Electrical energy generated from the conversion of microalgae biomass can be stored in a rapidly rotating flywheel. When electrical energy is needed, the flywheel slows down, and the kinetic energy stored in the flywheel is converted back into electrical energy through a generator. Flywheels are efficient short-term energy storage solutions with longer lifespans compared to batteries. However, flywheel energy storage capacity is lower than that of batteries, and the installation and maintenance costs of flywheels can be higher.

Integrating algae biomass with energy storage technologies can help create a more flexible, sustainable, and environmentally friendly energy system. Combining bioenergy production with various types of energy storage (chemical energy, thermal energy, electrical energy) offers flexible and efficient solutions to manage fluctuations in renewable energy generation and meet varying energy demands. Furthermore, this

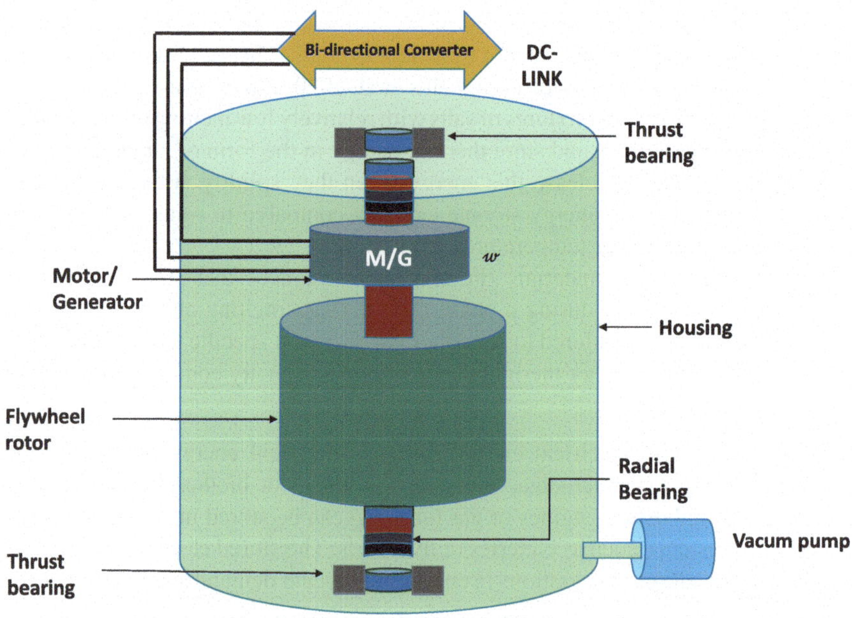

Fig. 20.7 Flywheel energy storage system

technology can contribute to mitigating greenhouse gas emissions and transitioning to a sustainable energy system. However, certain technical and economic challenges need to be addressed before this integration can be commercially applied on a large scale.

20.5.3 Application and Case Studies of Algae-Based Energy Storage Technologies

While algae-based energy storage applications are still in the research and development phase, several case studies have shown the potential of this technology in various sectors. Some applications and case studies related to algae-based energy storage technologies are as follows:

Microgrid and Isolated Energy Systems

Integrating algae biomass with energy storage technologies can be highly beneficial in microgrid scenarios or isolated energy systems, particularly in remote or island regions. A microgrid is an independent electrical network that can operate separately from the main grid, combining various renewable energy sources and energy storage systems (Santos et al. 2022). In remote or island regions, the availability of fossil fuel sources can be a challenge. Algae biomass serves as a local energy source that can be cultivated and converted into biogas, biodiesel, or bioethanol. By using local energy sources, the microgrid system can decrease dependence on imported fossil fuels and transportation costs.

Integrating algae biomass with other renewable energy sources, such as solar, wind, and hydro, in the microgrid system enhances source diversification. With a variety of energy sources, the system becomes more resilient to fluctuations in energy supply and can overcome disruptions in specific energy sources. Energy storage technologies like batteries and flywheels allow energy generated from algae biomass conversion to be stored and used as needed. This is crucial in microgrid systems, considering that energy supply and demand may vary over time. Energy storage also helps manage fluctuations in renewable energy supply, such as solar and wind.

Using algae biomass as an energy source reduces greenhouse gas emissions and enhances the sustainability of the energy system. In the microgrid system, converting algae biomass into biogas, biodiesel, or bioethanol results in lower emissions compared to burning fossil fuels (Jeswani et al. 2020; Osman et al. 2021). In some studies, algae biomass is used to yield biogas, which is then utilized to generate electricity using a gas generator. In certain cases, algae biomass can be cultivated using nutrient-rich wastewater, diminishing water pollution and recovering resources that would otherwise go to waste. This process can also aid in wastewater treatment and decrease wastewater treatment costs.

Integration with Renewable Energy Systems

In several studies, algae biomass is combined with other renewable energy technologies, such as solar and wind energy, to create hybrid energy systems. In these systems, energy from various sources like solar, wind, and algae biomass is combined to enhance the reliability and efficiency of energy supply.

In hybrid energy systems, nutrient-rich wastewater can be used for algae growth. This not only helps reduce water pollution but also cuts down on the nutrient input costs for algae growth. Renewable energy sources like solar and wind tend to be fluctuating and weather-dependent. In hybrid energy systems, algae biomass can be converted into biogas or biodiesel, serving as a backup energy source when energy from renewable sources is unavailable. This enhances the energy system's reliability and reduces dependence on fossil fuel sources. Hybrid energy systems involving algae biomass and other renewable energy sources can overall improve energy efficiency. In some cases, heat produced from algae biomass conversion can be used for water or air heating, which can further be utilized in heating or cooling systems.

Use of Algae Biogas in Public Transportation

An interesting case study is the integration of biogas generated from algae in public transportation systems. In Göttingen, Germany, a successful project yielded biogas from algae grown in photobioreactors, which was then used to power public buses. This showcases the potential of this technology to lower greenhouse gas emissions and dependency on fossil fuels in the transportation sector.

Biogas produced from algae may contain contaminants such as carbon dioxide (CO_2), hydrogen sulfide (H_2S), and water vapor (Dash et al. 2023). Prior to being used as fuel, biogas needs to be purified to decrease the content of these contaminants. Purification processes comprise absorption, adsorption, membrane, and absorption technologies (Dash et al. 2023; Paglini et al. 2022). Once purified, the generated methane can be used as fuel in public vehicles, such as buses, trucks, or taxis. Some vehicles may require modifications to their fuel systems and engines to use biogas as an alternative fuel. In certain cases, biogas can be used in the form of compressed natural gas (CNG) or liquefied natural gas (LNG), which is easier to store and transport (Prussi et al. 2021; Shah et al. 2017).

The use of biogas in public transportation systems requires the development of suitable refueling infrastructure, including refueling stations equipped with apparatus to store, fill, and measure the biogas used in vehicles.

The use of biogas yielded from algae in public transportation offers several advantages, e.g., mitigating greenhouse gas emissions, improving energy sustainability, and potential cost savings. Algae biogas is a renewable energy source that can decrease reliance on fossil fuels and contribute to achieving greenhouse gas emission reduction targets.

Integration with the Agricultral Sector

In some case studies, algae biomass is utilized to generate energy and valuable byproducts in the context of industrial and waste treatment facilities. For instance, algae can be grown using CO_2 discharge from industrial processes, helping lower greenhouse gas emissions. The yielded algae biomass can be converted into energy and stored in energy storage systems, which can be used to meet energy needs in these facilities.

In the agricultural sector, biogas from algae can be utilized for electricity generation, heating, and cooling. For example, farmers can use algae biomass growing in ponds or photobioreactors to produce biogas through anaerobic processes. This biogas can then be stored in energy storage systems, such as batteries or flywheels and used to operate farm equipment, irrigation pumps, or heating systems for grain drying. Additionally, the residual algae biomass from biogas extraction can be incorporated as organic fertilizer for crops (Sarwer et al. 2022).

Algae can be combined with other renewable energy technologies, such as solar panels or wind turbines, to create hybrid energy systems that provide more stable and sustainable energy for agricultural operations (Morales et al. 2019). Energy storage systems enable farmers to optimize energy consumption and decrease dependence on the electrical grid. Agricultural wastewater, such as effluents from washing and processing agricultural produce, is rich in nutrients and can be used as a growth medium for algae. By integrating algae cultivation systems with agricultural wastewater management, farmers can diminish environmental pollution loads and simultaneously produce energy that can be stored and used for agricultural purposes.

Case studies and applications of algae-based energy storage technologies in the agricultural sector are still limited, but the potential it offers indicates that integrating this innovation can contribute to energy sustainability and efficiency in agricultural systems. Although algae-based energy storage technologies have shown promising potential in various applications, there are still several technical and economic challenges that need to be addressed before widespread adoption of this technology can be realized.

20.6 Economic, Social and Environmental Studies

20.6.1 Life Cycle Analysis and Environmental Impact Assessment

Life Cycle Analysis (LCA) and Environmental Impact Assessment (EIA) are two crucial tools used to evaluate the environmental impact and sustainability of algae-based energy storage technologies (Chowdury et al. 2020; Villacreses-Freire et al. 2022). LCA enables an overall assessment of the environmental impacts associated with all stages of the life cycle of a product or technology, from raw material extraction to final disposal. On the other hand, EIA is a systematic process used to identify, predict, and evaluate the environmental impacts of a proposed project or activity. Several aspects that need to be taken into account in LCA and EIA related to algae-based energy storage technologies are as follows:

(1) Resources and raw materials: The assessment should consider the resources used in algae cultivation, e.g., water, land, nutrients, and energy. LCA should comprehend the efficiency of resource utilization and the potential environmental impacts associated with their use, such as water eutrophication and energy consumption.
(2) Algae cultivation and harvesting: The environmental impact of algae cultivation and harvesting, including greenhouse gas emissions, water usage, and land utilization, should be evaluated in LCA. EIA should consider the potential consequences of algae cultivation activities on the local ecosystem and water quality.
(3) Conversion of algae biomass into energy: LCA should encompass emissions and energy consumption associated with the conversion of algae biomass into various energy products, such as biodiesel, biogas, or syngas. EIA should assess the impact of energy conversion facilities on air quality and the surrounding environment.
(4) Energy utilization and storage: The environmental impact of using and storing energy generated from algae biomass should be considered in LCA. EIA should assess the impact of energy storage infrastructure, such as tanks or battery systems, on the surrounding environment.
(5) End-of-life disposal and recycling: LCA should assess the environmental impact of disposing residues and byproducts from algae-based energy storage technologies. EIA should evaluate the potential impact of disposing these products on water, soil, and air quality.

By conducting LCA and EIA related to algae-based energy storage technologies, we can identify areas where environmental impacts can be minimized and improve the method to achieve greater sustainability. Moreover, the results of LCA and EIA can aid decision-making and policy planning concerning the development and implementation of algae-based energy storage technologies.

20.6.2 Cost of Production and Market of Renewable Energy from Algae

Renewable energy from algae is one of the environmentally friendly and promising sources that can replace fossil fuels. It has been explained that algae can be converted into various energy products such as biodiesel, bioethanol, biogas, and biohydrogen. Algae-based energy storage technology entails production costs and the microalgae-based renewable energy market.

Production Costs

The production costs of renewable energy from algae include raw material, technology, operational, and maintenance costs. Several factors influencing production costs are as follows:

(1) Raw materials: Algae can be obtained from natural water bodies or cultivated in laboratories. The cost of raw materials includes the price of algae, substrates, and nutrients required for algae growth.
(2) Technology: The cost of technology comprises equipment used for cultivation, algae cell separation, lipid extraction, energy conversion, and energy storage. More efficient and effective technologies will reduce production costs.
(3) Operational: Operational costs include labor, energy, and chemicals used in the production and energy storage.
(4) Maintenance: Maintenance costs include equipment maintenance and repair, as well as the management of waste generated from the production process.

Market of Renewable Energy from Algae

The market of algae-based renewable energy continues to grow as awareness of the negative impacts of fossil fuels on the environment and human health increases. Factors affecting this market are as follows:

(1) Government policies: Regulations, incentives, and government support for renewable energy development will influence the algae energy market.
(2) Demand: The demand for renewable energy from algae is influenced by changes in fossil fuel prices, environmental awareness, and increasing energy needs.
(3) Research and development: Studies and advancement of more efficient and effective technologies in the algae production and energy storage will boost the competitiveness of this energy in the market.
(4) Infrastructure: Adequate infrastructure for the distribution and storage of renewable energy from algae will support market growth.

20.6.3 Associated Social and Employment Implications

The implementation of algae-based energy storage technologies has several major social and employment impacts. It can create new jobs in various sectors, such as research and development, production, construction, maintenance of infrastructure, as well as the operation and maintenance of energy storage systems. This will increase employment opportunities, especially in areas with abundant algae resources.

The use of algae-based renewable energy as an alternative to fossil fuels can mitigate greenhouse gas emissions and air pollution (Onyeaka et al. 2021), leading to positive impacts on public health and quality of life. Furthermore, providing cleaner and renewable energy will improve energy resilience and decrease dependence on imported fossil fuels. As the algae-based energy industry grows, the demand for skilled workers will also increase, driving opportunities for education and training in related fields, such as biotechnology, chemical engineering, and environmental engineering. Improved access to high-quality education and training will help people take advantage of job opportunities created by this industry.

The implementation of algae-based energy storage technology will affect related industries, such as agriculture, fisheries, and the chemical industry. For instance, algae can be cultivated alongside food crops or used to process industrial and agricultural waste, which will help reduce environmental impacts and create additional economic value (Amir et al. 2021). The development and implementation of algae-based energy storage technology must consider social and environmental justice issues, such as equitable access to renewable energy, local community participation in decision-making, and protection of indigenous rights. The development of algae-based energy storage technology will require collaboration and partnerships between governments, industries, and research institutions at the international level. International cooperation can accelerate research and development, promote technology and knowledge transfer, and enhance the adoption and efficiency of this technology worldwide.

20.6.4 Government Regulations and Policies Regarding Algae-Based Renewable Energy

Government regulations and policies play a crucial role in the development and implementation of algae-based renewable energy. Several aspects commonly covered in government regulations and policies are as follows:

(1) Fiscal incentives: Governments often provide fiscal incentives such as taxes, subsidies, and tax credits to support research, development, and investment in algae-based renewable energy. These incentives aim to decrease production costs and increase the competitiveness of renewable energy in the market.
(2) Renewable energy standards: Some countries have adopted Renewable Portfolio Standards (RPS) or biofuel blending mandates that require energy companies or fuel users to incorporate a portion of energy from renewable sources, including algae. These standards aim to promote the adoption of renewable energy and lower greenhouse gas emissions.
(3) Environmental regulations: Governments also issue environmental regulations that affect the algae-based renewable energy industry, such as emission permits, air and water quality standards, and waste management requirements. These regulations aim to ensure that the production and use of algae-based energy do not have negative impacts on the environment and human health.
(4) Research and development: Governments often support research and development of algae-based renewable energy through funding, research facilities, and partnerships between research institutions, industries, and the government. The goal is to accelerate the development of more efficient, effective, and environmentally friendly technologies.
(5) Promotion and education: Governments also play a role in promoting algae-based renewable energy through public awareness campaigns, education

programs, and training. Promotion and education aim to raise public understanding of the benefits and potential of algae-based renewable energy and develop the skills necessary to support this industry.

(6) International cooperation: Governments need to collaborate with other countries, international organizations, and donor agencies to develop and promote algae-based renewable energy. International cooperation includes knowledge and technology exchange, joint funding for research and development projects, and the establishment of global policies and standards.

20.7 Challenges and Future of Algae-Based Renewable Energy

The challenges and barriers to adopting algae as a renewable energy source can be categorized into technical, economic, social, and environmental issues, as elaborated below.

20.7.1 Barriers to Technology Development and Adoption

Technical Challenges and Obstacles

Although algae-based renewable energy offers many benefits, there are several technical challenges and barriers that need to be addressed before this technology can be widely adopted, including:

(1) Production efficiency: The generation of renewable energy from algae still requires significant amounts of energy and chemicals, such as nutrients and carbon dioxide. Algae also require specific growth conditions and precise temperature control to achieve optimal production rates (Mahmood et al. 2022). Therefore, further research is needed to improve the production efficiency of renewable energy from algae.

(2) Supply Stability and Availability: The supply of renewable energy from algae is still not stable and is dependent on environmental factors such as weather and climate. This makes conventional energy sources more stable and reliable for meeting energy needs.

(3) Technological advancement: Production and conversion technologies for renewable energy from algae are still in the development phase, requiring further research and development of more efficient and effective technologies, including the cultivation systems, processing and purification technologies, and energy storage and distribution systems. In contrast, production technologies for conventional energy sources are already well-established and efficient, making conventional energy production cheaper and more efficient than renewable energy from algae.

(4) High Production Costs: The production costs of renewable energy from algae are still higher compared to fossil fuels. This makes renewable energy from algae less appealing in the market, requiring efforts to lower production costs through increased production efficiency and the use of more effective technologies.
(5) Scaling issues: The production of renewable energy from algae is still limited to small and restricted scales, necessitating efforts to elevate production scale to meet larger energy demands.
(6) Transportation and logistics issues: Transportation and logistics are significant challenges in distributing renewable energy from algae, especially if algae resources are located far from production centers. This requires the development of better infrastructure and transportation networks to ensure an effective and efficient distribution of renewable energy from algae.
(7) Waste management: The production of renewable energy from algae generates waste that needs proper disposal to avoid environmental pollution. Hence, effective waste management technologies and strategies are needed to minimize environmental impacts.
(8) Regulatory challenges: Inconsistent or unclear regulations and policies can hinder the development of algae-based renewable energy. Thus, cooperation with the government as well as clear and consistent regulations are needed to support the development of algae-based renewable energy.

Economic Challenges and Obstacles

The development of renewable energy from algae also faces economic challenges and obstacles in competing with conventional energy sources. Here are some economic issues that need to be addressed in adopting algae-based renewable energy:

(1) High production costs: The generation of renewable energy from algae still requires significant costs, including production, transportation, and distribution expenses. These high production costs make renewable energy from algae less competitive in the market compared to conventional energy sources.
(2) Limited production scale: The production of renewable energy from algae is still limited to small and restricted scales. The lack of production scale increases production costs and hinders the large-scale adoption of renewable energy from algae.
(3) Market uncertainty: The market for renewable energy from algae is not yet stable and can be affected by changes in fossil fuel prices. This market uncertainty drives investors reluctant to invest in the development of algae-based renewable energy.
(4) Inadequate infrastructure: Insufficient infrastructure, such as inadequate distribution networks and production facilities, can obstruct the development of renewable energy from algae. It can lead to increased production costs and delay the advancement of renewable energy technology from algae.

To tackle these economic challenges and obstacles, support from the government, industry, and society is essential to promote and develop renewable energy from algae.

This aid can include fiscal incentives, education and training programs, investment in research and development, as well as the establishment of more efficient and effective infrastructure and distribution networks. Additionally, efforts should be made to improve production efficiency and lessen the production costs of algae-based renewable energy through research and the development of more efficient technologies.

Environmental Challenges and Obstacles

The development of renewable energy from algae also faces environmental challenges and obstacles in competing with conventional energy sources. Some of the environmental hindrances that need to be addressed in adopting renewable energy from algae are as follows:

(1) Impact on water quality: The production of renewable energy from algae requires water and nutrients. Excessive withdrawal of water and nutrients from the environment can decrease water quality and trigger the growth of harmful wild algae, bringing negative impacts on the environment and society (Silaen et al. 2022).
(2) Impact on biodiversity: Extracting nutrients and water from the environment for algae-based renewable energy production can affect biodiversity in water bodies. The use of wild algae for renewable energy production can also impact marine biodiversity (Correa et al. 2017).
(3) Greenhouse gas emissions: The production and transportation processes of renewable energy from algae can result in greenhouse gas emissions if not managed properly (Arora et al. 2021; Sunaryo et al. 2023).
(4) Waste management: The production of renewable energy from algae generates waste that needs to be appropriately handled to avoid environmental pollution. This residue is typically nutrient-rich, so meticulous management is pivotal to prevent algal blooms that can disrupt the balance of aquatic ecosystems (Ali et al. 2020).
(5) Sustainable resource utilization: The development of renewable energy from algae must consider sustainable resource utilization, as it should not cause environmental degradation or unsustainable resource use.

To overcome these environmental challenges and obstacles, strict environmental management and monitoring are compulsory during the production of renewable energy from algae. Besides, there is a need for technological innovation of more environmentally friendly technologies, as well as the establishment of policies and regulations that promote sustainable and eco-friendly production of renewable energy from algae. Community engagement and participation are also requisite in decision-making and sustainable environmental management.

Social Challenges and Obstacles

In addition to technical challenges, the development of renewable energy from algae also faces social challenges and obstacles in competing with conventional energy source, including:

(1) Community and local resistance: Some communities and local groups may reject the development of renewable energy from algae due to environmental concerns, aesthetic reasons, or economic interests. Such opposition can slow down or even halt the development of renewable energy from algae in a particular area.
(2) Uncertainty in regulations and policies: Ambiguity in government regulations and policies regarding renewable energy from algae can be a barrier to adopting this technology, affecting investments, research and development, and decisions concerning the use of algae-based renewable energy.
(3) Lack of public understanding and awareness: Insufficient public understanding and awareness of the benefits of renewable energy from algae can hinder the adoption of this technology. Public education and campaign efforts can help accelerate its development and adoption.
(4) Imbalance in benefit distribution: The inequality in the distribution of benefits can influence public acceptance of renewable energy from algae. For instance, communities may not accept this innovation if its benefits are only experienced by a small group of individuals or companies.
(5) Social and environmental justice: The development of renewable energy from algae should consider social and environmental justice issues, such as the rights of indigenous communities, equitable access to renewable energy, and environmental protection. This can help reduce negative impacts and enhance positive benefits from the development of algae-based renewable energy.

To address these social challenges and obstacles, support from the government, industry, and society is needed to promote and develop renewable energy from algae in an inclusive and sustainable manner. This aid can include educational and training programs, community dialogues and participation, investment in social and environmental development, as well as the development of comprehensive and sustainable policies.

20.7.2 Innovation and Current Research

Recent advancements in renewable energy technology from algae encompass various aspects, ranging from improved energy conversion efficiency to cost reduction. Here are some principal developments and innovations that have the potential to transform the renewable energy sector using algae:

- Algae genetics and engineering: Advances in biotechnology and genetics allow researchers to optimize algae strains that are more efficient in producing biomass or desired fuels. Genetic engineering can also help lower contamination and enhance algae's tolerance to varying environmental conditions (Sreenikethanam et al. 2022).
- More efficient photobioreactors: Photobioreactors are technologies that enable algae growth in controlled environments. Innovations in photobioreactor design and materials can increase the efficiency of algae biomass production, reduce

costs, and enable larger industrial-scale operations (Fernández del Olmo et al. 2021).

- More effective biomass conversion methods: Research is ongoing to develop more effective algae biomass conversion approaches, such as hydrothermal processes, electrochemistry, and enzymatic methods (Han et al. 2019). Innovations in conversion technology can improve the quality and quantity of energy generated from algae.
- Integration of algae cultivation with other systems, such as wastewater treatment or carbon dioxide (CO_2) sequestration from factories, can enhance overall process sustainability and efficiency. This integration also helps mitigate greenhouse gas emissions and utilize previously unused resources.
- Utilization of byproducts such as proteins, pigments, and polysaccharides from algae conversion processes can create added value and help lessen the production costs of algae-based energy (Chiong et al. 2016).
- Collaboration between the public and private sectors, as well as international cooperation, can accelerate research and development of renewable energy technology from algae. These partnerships can also help address technical, economic, and environmental challenges related to this sector.

Future projections indicate that renewable energy from algae has the potential to become a significant energy source, particularly in replacing fossil fuels and diminishing greenhouse gas emissions. Furthermore, innovations in algae-based renewable energy technology can create new opportunities in the fields of energy, environment, and economy.

20.7.3 Strategies to Increase Efficiency and Reduce Production Costs

To address the technical challenges and barriers in adopting renewable energy from algae as a power source, it is necessary to develop strategies that enhance efficiency and decrease production costs. Several schemes that can be employed to achieve these goals in utilizing algae for energy storage technology are as follows:

(1) Advancement of production technology: One of the key strategies to increase production efficiency and reduce the costs of renewable energy from algae is the innovation of production technology. Utilizing more sophisticated production methods can boost productivity and efficiency, consequently lowering production costs.
(2) Development of processing technology: Another strategy to improve production efficiency and decrease production costs of algae-based renewable energy is the development of more efficient processing technology, as it can elevate the quality and quantity of production, leading to cost abatement.

(3) Optimal nutrient utilization: Increasing the appropriate use of nutrients can enhance productivity and efficiency of renewable energy from algae, resulting in cost reductions.
(4) Augmenting nutrient availability: Improved nutrient availability can elevate productivity and efficiency as well as decrease the costs of algae-based renewable energy.
(5) Efficient resource utilization: Employing more efficient resource management is another approach to boost production efficiency and lower costs of renewable energy from algae.
(6) Infrastructure and distribution network development: Establishing more efficient infrastructure and distribution networks can improve productivity and efficiency, ultimately leading to cost minimization.

To implement these strategies, support from the government, industry, and society is indispensable in promoting and developing renewable energy from algae.

20.7.4 Vision and Future Projection for the Algae-Based Renewable Energy Industry

From the discussions in the previous sections, algae possess significant potential as a future source of renewable energy and energy storage. Various technologies have been developed, ranging from cultivation processes, cultivation techniques, conversion technologies, to energy storage applications. The development of algae-based renewable energy industries can contribute to reducing dependence on fossil energy sources and mitigating negative impacts on the environment. Therefore, a grand vision for the renewable energy industry is necessary, where the sustainability and success of algae-based industries heavily rely on strong government investment, innovation, collaboration, and policy support. This vision can accelerate the growth of algae-based industries and create a multiplier effect on various other sectors, ultimately enhancing national economic growth. The use of algae as a renewable energy source presents several promising future visions and projections, including:

(1) Greenhouse gas emission mitigation: Algae can potentially lower greenhouse gas emissions by absorbing CO_2 from the atmosphere during photosynthesis.
(2) Large-scale algae-based fuels: Algae-based energy industries can develop large-scale production for algae-based fuels. The future projection for algae-based renewable energy industries includes creating systems capable of producing fuels like biodiesel or bioethanol from algae more efficiently, enabling them to compete with fossil fuels and reducing dependence on conventional fuels.
(3) Utilization as food and nutrition source: Algae can also be used as a nutritious food source, containing proteins, vitamins, minerals, and omega-3 fatty acids beneficial to human health. Developing algae production as safe and highly nutritious food is a future projection for achieving food security.

(4) Algae usage in industrial processes: Algae can be employed in various industrial processes to generate high-value products such as pigments, proteins, and antioxidants. This usage can expand the economic base of the algae energy industry and raise production efficiency.
(5) Technological innovation: Algae-based energy industries will continue to evolve. Future projections include developing more efficient photobioreactors and cultivation methods to increase algae harvesting yields. Additionally, processing and extraction technologies will be continuously improved to maximize the value of algae products.
(6) Collaboration between industry and research: To accelerate the development of algae-based energy industries, collaboration between industrial sectors and researchers is fundamental. Such partnership will have a positive impact through knowledge and technology exchange, addressing technical challenges, and expediting the adoption of algae-based solutions across various sectors. Some countries have begun recognizing the significant potential of the algae industries, such as the European Union (EU), which has conducted studies on algae production and its applications in Europe. Though still in the initiation phase, the EU also incorporates the algae sector in its economic strategy, namely the EU Blue Bioeconomy.
(7) Implementation of advanced biofuel technology: Algae-based energy industries also have the potential to adopt more advanced biofuel technologies such as bio-hydrogen, biobutanol, and methane yielded by algae. The use of these technologies can provide more sustainable and efficient alternatives to meet global energy demand.

References

Abusweireh RS, Rajamohan N, Sonne C, Vasseghian Y (2023) Algae biogas production focusing on operating conditions and conversion mechanisms—a review. Heliyon 9(7):e17757. https://doi.org/10.1016/j.heliyon.2023.e17757

Ali J, Wang L, Waseem H, Song B, Djellabi R, Pan G (2020) Turning harmful algal biomass to electricity by microbial fuel cell: a sustainable approach for waste management. Environ Pollut 266. https://doi.org/10.1016/j.envpol.2020.115373

Amir ARG, Soeprobowati TR, Hariyati R, Jumari, Murningsih, Khotimperwati L, Tarwotjo U (2021) Culturing microalgae from nature: simple experiment during pandemic covid-19. E3S Web of Confe 317. https://doi.org/10.1051/e3sconf/202131701102

Arora P, Chance RR, Hendrix H, Realff MJ, Thomas VM, Yuan Y (2021) Greenhouse gas impact of algal bio-crude production for a range of CO_2 supply scenarios. Appl Sci (Switzerland), 11(24). https://doi.org/10.3390/app112411931

Arumugam N, Chelliapan S, Kamyab H, Thirugnana S, Othman N, Nasri NS (2018) Treatment of wastewater using seaweed: a review. In: Int J Environ Res Public Health 15(12). https://doi.org/10.3390/ijerph15122851

Babich O, Sukhikh S, Larina V, Kalashnikova O, Kashirskikh E, Prosekov A, Noskova S, Ivanova S, Fendri I, Smaoui S, Abdelkafi S, Michaud P, Dolganyuk V (2022) Algae: Study of edible and

biologically active fractions, their properties and applications. Plants 11(6):1–30). https://doi.org/10.3390/plants11060780
Bani A, Fernandez FGA, D'Imporzano G, Parati K, Adani F (2021) Influence of photobioreactor set-up on the survival of microalgae inoculum. Bioresour Technol 320. https://doi.org/10.1016/j.biortech.2020.124408
Beal CM, Archibald I, Huntley ME, Greene CH, Johnson ZI (2018) Integrating algae with bioenergy carbon capture and storage (ABECCS) increases sustainability. Earth's Future 6(3):524–542. https://doi.org/10.1002/2017EF000704
Benner P, Meier L, Pfeffer A, Krüger K, Oropeza Vargas JE, Weuster-Botz D (2022) Lab-scale photobioreactor systems: principles, applications, and scalability. Bioprocess Biosyst Eng 45(5):791–813. https://doi.org/10.1007/s00449-022-02711-1
Boukid F, Castellari M (2023) Algae as nutritional and functional food sources. Foods 12(1). https://doi.org/10.3390/foods12010122
Catalanotti C, Yang W, Posewitz MC, Grossman AR (2013) Fermentation metabolism and its evolution in algae. Front. Plant Sci 4. https://doi.org/10.3389/fpls.2013.00150
Chiong T, Acquah C, Lau SY, Khor EH, Danquah MK (2016) Microalgal-based protein by-products: extraction, purification, and applications. In: Protein byproducts: transformation from environmental burden into value-added products. Elsevier, Amsterdam, pp 213–234. https://doi.org/10.1016/B978-0-12-802391-4.00012-4
Chowdury KH, Nahar N, Deb UK (2020) The growth factors involved in microalgae cultivation for biofuel production: a review. Comput Water Energy Environ Eng 09(04):185–215. https://doi.org/10.4236/cweee.2020.94012
Correa DF, Beyer HL, Possingham HP, Thomas-Hall SR, Schenk PM (2017) Biodiversity impacts of bioenergy production: microalgae vs. first generation biofuels. Renew Sustain Energy Rev 74:1131–1146. Elsevier, Amsterdam. https://doi.org/10.1016/j.rser.2017.02.068
Cotas J, Gomes L, Pacheco D, Pereira L (2023) Ecosystem services provided by seaweeds. Hydrobiology 2(1):75–96. https://doi.org/10.3390/hydrobiology2010006
Dash RL, Mohanty B, Hota PK (2023) Energy, economic and environmental (3E) evaluation of a hybrid wind/biodiesel generator/tidal energy system using different energy storage devices for sustainable power supply to an Indian archipelago. Renew Energy Focus 44:357–372. https://doi.org/10.1016/j.ref.2023.01.004
Dharmono, Mahrudin, Irianti R, Fajeriadi H (2022) Aquatic plants as niche for lay eggs and raising juveniles by freshwater fish in three swamp habitats in South Kalimantan, Indonesia. Biodiversitas 23(3):1520–1526. https://doi.org/10.13057/biodiv/d230341
Dewi, K, Rinanti, A, Astuti, DI, Halomoan, N (2014) Preliminary study on biomitigation green house gas carbon dioxide in closed system bubble photobioreactor: Relationship among the mass transfer rate and CO2 removal efficiency in high level of CO2. J Teknologi 69(6):93–9.
Ebhodaghe SO, Imanah OE, Ndibe H (2022) Biofuels from microalgae biomass: a review of conversion processes and procedures. Arab J Chem 15(2). https://doi.org/10.1016/j.arabjc.2021.103591
Fernández del Olmo P, Acién FG, Fernández-Sevilla JM (2021) Analysis of productivity in raceway photobioreactor using computational fluid dynamics particle tracking coupled to a dynamic photosynthesis model. Biores Technol 334(125226):1–11. https://doi.org/10.1016/j.biortech.2021.125226
Gökçe D (2016) Algae as an indicator of water quality. In: Algae—organisms for imminent biotechnology. InTech. https://doi.org/10.5772/62916
Han Y, Hoekman SK, Cui Z, Jena U, Das P (2019) Hydrothermal liquefaction of marine microalgae biomass using co-solvents. Algal Res 38. https://doi.org/10.1016/j.algal.2019.101421
Iglina T, Iglin P, Pashchenko D (2022) Industrial CO_2 capture by algae: a review and recent advances. Sustainability (Switzerland) 14(7). https://doi.org/10.3390/su14073801
Jeswani HK, Chilvers A, Azapagic A (2020) Environmental sustainability of biofuels: a review: environmental sustainability of biofuels. Proc R Soc A Math Phys Eng Sci 476(2243). https://doi.org/10.1098/rspa.2020.0351

Kumar L, Anand R, Shah MP, Bharadvaja N (2022) Microalgae biodiesel: a sustainable source of energy, unit operations, technological challenges, and solutions. J Hazard Mater Adv 8:100145. https://doi.org/10.1016/j.hazadv.2022.100145

Latsos C, Bakratsas G, Moerdijk T, van Houcke J, Timmermans KR (2021) Effect of salinity and pH on growth, phycoerythrin, and non-volatile umami taste active compound concentration of Rhodomonas salina using a D-optimal design approach. J Appl Phycol 33(6):3591–3602. https://doi.org/10.1007/s10811-021-02547-4

Mahmood T, Hussain N, Shahbaz A, Mulla SI, Iqbal HMN, Bilal M (2022) Sustainable production of biofuels from the algae-derived biomass. Bioprocess Biosyst Eng. https://doi.org/10.1007/s00449-022-02796-8

Metsoviti MN, Papapolymerou G, Karapanagiotidis IT, Katsoulas N (2020) Effect of light intensity and quality on growth rate and composition of *Chlorella vulgaris*. Plants 9(1). https://doi.org/10.3390/plants9010031

Morales M, Aflalo C, Bernard O (2021). Microalgal lipids: A review of lipids potential and quantification for 95 phytoplankton species. Biomass Bioenergy 150(206108):1–25. https://doi.org/10.1016/j.biombioe.2021.106108

Morales M, Hélias A, Bernard O (2019) Optimal integration of microalgae production with photovoltaic panels: environmental impacts and energy balance. Biotechnol Biofuels 12(239):1–17. https://doi.org/10.1186/s13068-019-1579-4

Nagappan S, Das P, AbdulQuadir M, Thaher M, Khan S, Mahata C, Al-Jabri H, Vatland AK, Kumar G (2021) Potential of microalgae as a sustainable feed ingredient for aquaculture. J Biotechnol 341:1–20. https://doi.org/10.1016/j.jbiotec.2021.09.003

Nguyen LN, Aditya L, Vu HP, Johir AH, Bennar L, Ralph P, Hoang NB, Zdarta J, Nghiem LD (2022) Nutrient removal by algae-based wastewater treatment. Curr Pollut Rep 8(4):369–383. https://doi.org/10.1007/s40726-022-00230-x

Novosel N, Mišić Radić T, Zemla J, Lekka M, Čačković A, Kasum D, Legović T, Žutinić P, Gligora Udovič M, Ivošević DeNardis N (2022) Temperature-induced response in algal cell surface properties and behaviour: an experimental approach. J Appl Phycol 34(1):243–259. https://doi.org/10.1007/s10811-021-02591-0

Onyeaka H, Miri T, Obileke KC, Hart A, Anumudu C, Al-Sharify ZT (2021) Minimizing carbon footprint via microalgae as a biological capture. In: Carbon capture science and technology, vol. 1. Elsevier, Amsterdam. https://doi.org/10.1016/j.ccst.2021.100007

Osman AI, Mehta N, Elgarahy AM, Al-Hinai A, Al-Muhtaseb AH, Rooney DW (2021) Conversion of biomass to biofuels and life cycle assessment: a review. Environ Chem Lett 19(6):4075–4118. https://doi.org/10.1007/s10311-021-01273-0

Pagkalos C, Dogkas G, Koukou MK, Konstantaras J, Lymperis K, Vrachopoulos MG (2020) Evaluation of water and paraffin PCM as storage media for use in thermal energy storage applications: a numerical approach. Int J Thermofluids 1–2(100006):1–7. https://doi.org/10.1016/j.ijft.2019.100006

Paglini R, Gandiglio M, Lanzini A (2022) Technologies for deep biogas purification and use in zero-emission fuel cells systems. Energies 15(10):1–30. https://doi.org/10.3390/en15103551

Prussi M, Julea A, Lonza L, Thiel C (2021) Biomethane as alternative fuel for the EU road sector: analysis of existing and planned infrastructure. Energ Strat Rev 33(100612):1–9. https://doi.org/10.1016/j.esr.2020.100612

Rinanti A, Fachrul MF, Hadisoebroto R, Desty S, Rahmadhania R, Widyaningrum DA, Saad NA (2021) Heavy metal pollutant sorption in aquatic environment by microalgae consortium. Indones J Urban Environ Technol 51–71. https://doi.org/10.25105/urbanenvirotech.v5i1.10747

Salami R, Kordi M, Bolouri P, Delangiz N, Asgari Lajayer B (2021) Algae-based biorefinery as a sustainable renewable resource. Circ Econ Sustain 1(4):1349–1365. https://doi.org/10.1007/s43615-021-00088-z

de Santos DCLP, Ihler JB, Soares LO, Alves YA, Souza CG, Boloy RAM (2022). Global trends and opportunities in hybrid microgrid systems using renewable energies. Int J Energy Econ Policy 12(4):263–273. https://doi.org/10.32479/ijeep.13175

Sarwer A, Hamed SM, Osman AI, Jamil F, Al-Muhtaseb AH, Alhajeri NS, Rooney DW (2022) Algal biomass valorization for biofuel production and carbon sequestration: a review. Environ Chem Lett 20(5):2797–2851. https://doi.org/10.1007/s10311-022-01458-1

Sawant SS, Gajbhiye BD, Mathpati CS, Pandit R, Lali AM (2018) Microalgae as sustainable energy and its cultivation. IOP Conf Ser Mater Sci Eng 360(1). https://doi.org/10.1088/1757-899X/360/1/012025

Shah MS, Halder PK, Shamsuzzaman ASM, Hossain MS, Pal SK, Sarker E (2017) Perspectives of biogas conversion into Bio-CNG for automobile fuel in Bangladesh. J Renew Energy 2017:1–14. https://doi.org/10.1155/2017/4385295

Silaen S, Olfan W, Barat B (2022) The effect of water quality on growth of microalgaes *Chlorella pyrenoidosa.* J Appl Geospatial Inform 6(2):638–640. http://jurnal.polibatam.ac.id/index.php/JAGI

Silva LHS, Huszar VLM, Marinho MM, Rangel LM, Brasil J, Domingues CD, Branco CC, Roland F (2014) Drivers of phytoplankton, bacterioplankton, and zooplankton carbon biomass in tropical hydroelectric reservoirs. Limnologica 48:1–10. https://doi.org/10.1016/j.limno.2014.04.004

Siwi, WP, Rinanti, A, Silalahi, MDS, Hadisoebroto, R, Fachrul, MF (2017) Effect of immobilized biosorbents on the heavy metals (Cu 2+) biosorption with variations of temperature and initial concentration of waste. IOP Conf Ser Earth Environ Sci 106(1).

Sreenikethanam A, Raj S, J, Rajesh Banu J, Gugulothu P, Bajhaiya AK (2022) Genetic engineering of microalgae for secondary metabolite production: recent developments, challenges, and future prospects. Front Bioeng Biotechnol 10:1–14. https://doi.org/10.3389/fbioe.2022.836056

Sultana F, Wahab MA, Nahiduzzaman M, Mohiuddin M, Iqbal MZ, Shakil A, Mamun AA, Khan MSR, Wong LL, Asaduzzaman M (2023) Seaweed farming for food and nutritional security, climate change mitigation and adaptation, and women empowerment: a review. Aquacult Fish 8(5):463–480. https://doi.org/10.1016/j.aaf.2022.09.001

Sunaryo, T, Radyati, MRN, Utha, MA, Rinanti, A, Minarti, A (2023) Research development on carbon capture, utilization and storage as a climate change mitigation technology. Environmental Research, Engineering and Management 79(1):37–55.

Udayan A, Pandey AK, Sirohi R, Sreekumar N, Sang BI, Sim SJ, Kim SH, Pandey A (2022) Production of microalgae with high lipid content and their potential as sources of nutraceuticals. In: Phytochemistry reviews. Springer Science and Business Media B.V. https://doi.org/10.1007/s11101-021-09784-y

Vecchi V, Barera S, Bassi R, Dall'osto L (2020) Potential and challenges of improving photosynthesis in algae. Plants 9(1). https://doi.org/10.3390/plants9010067

Verma M, Mishra V (2020) An introduction to algal biofuels, pp 1–34. https://doi.org/10.1007/978-981-15-7190-9_1

Villacreses-Freire D, Ketzer F, Rösch C (2022) Advanced metabolic engineering approaches and renewable energy to improve environmental benefits of algal biofuels: LCA of large-scale biobutanol production with *Cyanobacteria synechocystis* pcc6803. Bioenergy Research 15(3):1515–1530. https://doi.org/10.1007/s12155-021-10323-y

Visca A, Di Caprio F, Spinelli R, Altimari P, Cicci A, Iaquaniello G, Toro L, Pagnanelli F (2017) Microalgae cultivation for lipids and carbohydrates production. Chem Eng Trans 57:127–132. https://doi.org/10.3303/CET1757022

Yin X, Wang J, Yang M, Xie X (2019) Evaluation of macroalgal detritus as food source for juvenile Manila clam, *Ruditapes philippinarum*: effects on growth, amino acid content and fatty acid composition. Aquac Res 50(12):3579–3588. https://doi.org/10.1111/are.14313

Zakeri B, Hunt JD, Laldjebaev M, Krey V, Vinca A, Parkinson S, Riahi K (2022) Role of energy storage in energy and water security in Central Asia. J Energy Stor 50(104587):1–14. https://doi.org/10.1016/j.est.2022.104587

Part IV
Microalgae and Cyanobacteria for Food Applications

Chapter 21
Food and Feed Preparation Using Algae

John N. Idenyi and Jonathan C. Eya

Abstract Algae have gained prominence as a sustainable source of food for man and feed for animals because of their high nutritional content, presenting a potential solution to global food shortages. Algae possess a variety of high-value chemicals, such as essential lipids (polyunsaturated fatty acids, omega-3 and omega-6), proteins, carbohydrates, and vitamin and minerals comparable to ingredients found in other sources such as marine products. Moreover, algae pigments have antioxidant properties and are also natural colorants, besides providing numerous functional foods or other health benefits. Among other benefits, algae supplements can enhance growth, lower feeding requirements, improve immunity, improve disease resistance, provide antimicrobial properties, and enrich animal products with natural active ingredients, among others. Recent advances in algal biotechnology have transformed algae into an effective "warehouse of cells" used in the food industry. This accelerated the growth of the algae market in food and feed. This chapter began by providing a general overview of algae and their potential. Next, algae nutritional compositions were presented as well as potential uses in human foods and animal feeds. Final steps included the identification of problems and prospects, as well as the conclusions.

Keyword Algal products · Animal feeds · Functional foods · Omega-3 fatty acids

21.1 Introduction

In recent years, the world's population has grown rapidly, increasing food demand. A growing population of urban dwellers and older demographics have substantially contributed to the increase in demand for dietary proteins. However, multiple challenges have slowed global efforts towards eradicating hunger and malnutrition by 2030 (FAO, The State of World Fisheries and Aquaculture 2022). The combined

J. N. Idenyi (✉) · J. C. Eya
Department of Biology/Gus R. Douglass Institute, West Virginia State University, Institute, WV, USA
e-mail: john.idenyi@wvstateu.edu

T. A. Kurniawan and A. Anouzla (eds.), *Algae as a Natural Solution for Challenges in Water-Food-Energy Nexus*, Environmental Science and Engineering,
https://doi.org/10.1007/978-981-97-2371-3_21

impact of the catastrophic effects of COVID-19, ecological degradation, climate change, biodiversity decline, and humanitarian emergencies have exacerbated global food insecurity, with approximately 811 million people experiencing hunger and 3 billion unable to access nutritious food sources (FAO, The State of World Fisheries and Aquaculture 2022).

Algae technology has emerged as a promising solution to tackle current food crises, offering the potential for an alternative and sustainable source of food and feed (Alam et al. 2020; Becker 2013). It is projected that algae will transform the food and feed productions as a result of their enormous availability, fast growth rate, and impressive nutritional profile (Wells 2017). There are many types of algae that can be found in marine environments, including microalgae, macroalgae (seaweeds), and cyanobacteria, which thrive in freshwater, marine, and even extreme aquatic environments (Adesalu et al. 2022; Pereira 2021). These organisms convert sunlight and carbon dioxide into valuable organic compounds through the process of photosynthesis, which reduces environmental impact, improves resource efficiency, and enhances the nutritional value of food (Sidari and Tofalo 2019).

A key characteristic that distinguishes algae is their nutritional composition. Several algae species contain proteins, carbohydrates, lipids, vitamins, minerals, and bioactive compounds (Idenyi 2021). Particularly, Spirulina and Chlorella contain high levels of protein, exceeding those found in conventional sources such as soybeans. Furthermore, a variety of essential amino acids can be derived from these microorganisms, making them an ideal protein source for vegetarians, vegans, and people with special dietary needs.

Besides proteins, algae contain large amounts of carbohydrates, including simple sugars and complex polysaccharides like agar and carrageenan (Adesalu et al. 2022). Carbohydrates contribute to the desirable texture of algae-based foods as well as being a source of energy (Borowitzka 2013). Algae contain dietary fiber, which may benefit digestion and cholesterol control. Algae contain omega-3 fatty acids, including eicosapentaenoic acid (EPA) and docosahexaenoic acid (DHA), which are essential for the development of the brain, the health of the cardiovascular system, and the management of inflammation. In comparison with traditional sources such as fish oil, algae provide a sustainable, non-animal source of omega-3 fatty acids (Charoensiddhi et al. 2018; Sidari and Tofalo 2019).

In addition to being food, algae have many other applications. A sustainable alternative to animal feeds, algae-based feeds are becoming increasingly popular in aquaculture and livestock (Idenyi et al. 2022; Sidari and Tofalo 2019). A superior feed ingredient than traditional feed ingredients, algae reduces reliance on land and water resources while ensuring livestock and fish health and productivity (Idenyi et al. 2022). The inclusion of algae in nutrition may contribute to the preservation of the environment, improve the health of humans and animals, and provide solutions to global food security issues (Idenyi et al. 2022). Researchers and industry experts, in fact the general public, are being encouraged to take an interest in these high-value compounds derived from algae, with the aim of marketing them as a viable source of food and feed. Moreover, this chapter presents an overview of the various applications of microalgae in human nutrition and animal feed, including the latest advancements in the food and feed industries as they pertain to algae.

21.2 Algal-Derived Functional Foods

Functional foods are foods that provide health benefits that extend beyond nutritional needs (Domínguez 2013). They can improve health as well as reduce the risk of disease and health complications (Reyes et al. 2021). Nutraceuticals, bioactive foods, novel foods, designed foods, health foods, medicinal foods, and therapeutic foods are some terms used to describe functional ingredients (Alam et al. 2020; Bidlack 1998; Domínguez 2013; Guo 2013; Sankaran and Mouly 2007). Generally, functional foods have health benefits associated with the presence of biologically active compounds or chemical substances that do not commonly occur in conventional foods (Biesalski et al. 2009). Some of the functional compounds found in algae include minerals, vitamins, polyphenols, prebiotics, proteins, amino acids, carotenoids, and antioxidant compounds, all of which may have certain physiological effects on human health (Donn et al. 2022; Hasler 2002; Idenyi et al. 2022). In studies on functional foods, a broad range of issues are addressed, including the impact that these foods have on the gut microbiota, transit time, nutrient bioavailability, endocrine and immune systems, antioxidants, and redox systems, reducing cardiovascular disease risk and insulin resistance, improving infant and fetal nutrition, reducing the risk of cancer and insulin resistance, reducing environmental carcinogenesis and toxicity, and improving mental, behavioral, and cognitive functions (Guo 2013; Idenyi et al. 2022). Functional foods derived from algae are proving to be valuable sources for human consumption (Fig. 21.1). Currently, algae are gaining attention and being evaluated as functional food ingredients source, which highlights their significant contribution to a healthy, balanced diet.

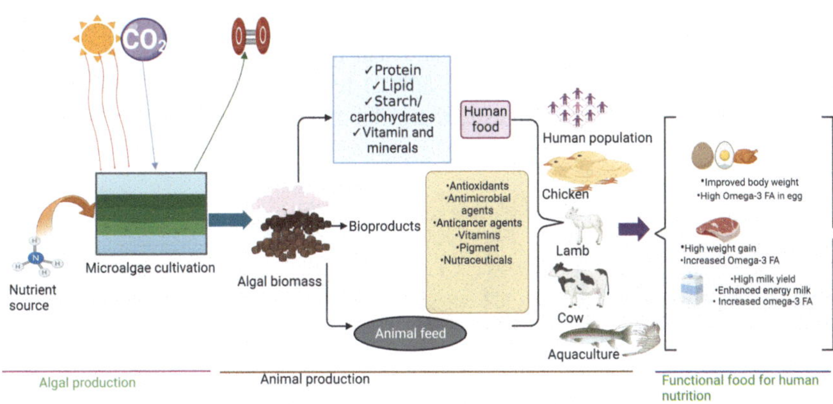

Fig. 21.1 Algae as source of nutrition for human and animals

21.3 Algal Protein

Food production. (i) Algal-based meat alternatives. Due to its high protein content, amino acid profile, and textured properties, algae protein has been utilized as an ingredient in plant-based meat alternatives. Adding algae protein to food substances can enhance their nutritional value and improve their sensory properties (Boukid and Castellari 2021). It has been shown in research that adding algal protein to algal-based meat formulations can enhance texture, juiciness, and general acceptability (Barros et al. 2021). (ii) Bakery and Snacks (Fig. 21.2): Algal protein can be used in the manufacture of bakery products and snacks to improve their nutritional content (Colgrave et al. 2021; Kadam and Prabhasankar 2010). The addition of algal protein powders, isolates, or concentrates can enhance the nutritional content of bread, cookies, crackers, and other baked goods (Kadam and Prabhasankar 2010; Mu et al. 2019). Moreover, algae proteins have functional properties, such as water retention and emulsification, which have been shown to significantly enhance the texture and shelf life of baked foods (Plaza et al. 2009). (iii) Dairy alternatives. Milk, yogurt, and cheese can be made with algae protein as an alternative to dairy products. Algal protein isolates and hydrolysates can be used to manufacture dairy analogs that are similar to traditional dairy products in terms of texture and composition (Raheem et al. 2018) (Fig. 21.2). There are many potential advantages to using algae protein-based dairy alternatives, including the availability of high-quality protein, essential fatty acids, and bioactive ingredients (Colgrave et al. 2021; Spolaore et al. 2006).

Feed Production: (i) Aquaculture: The use of algal protein in aquatic feeds has been shown to improve the growth and health of a variety of aquatic animals. The high protein content, essential amino acid profile, and beneficial fatty acids of algae including Chlorella, Spirulina, and Nannochloropsis are commonly incorporated into fish, shrimp, and mollusk nutrition (Colgrave et al. 2021; Kent et al. 2015). In aquaculture species, the inclusion of algae protein can enhance feed conversion efficiency, growth rates, immune response, and general performance in aquaculture species (Shah 2018). (ii) Livestock and Poultry: The use of algae protein in livestock and poultry feeds can provide a more sustainable alternative to conventional protein sources such as soybean meal and fish meal. Algal protein has an ideal amino acid profile, and its inclusion in animal feeds has been shown to improve growth performance, nutrient digestion, and feed conversion (Eilam et al. 2023).

Algae cake   Algae bread   Algae cookies   Algae pie

Fig. 21.2 Baked foods derived from algae

In addition, algal protein possesses high-value products that provide animals with health benefits (Sutherland et al. 2021).

21.4 Algal Lipid

Algal oils or lipids. Their potential applications in the food and feed industries have gained significant attention in recent years. Eicosapentaenoic acid (EPA) and docosahexaenoic acid (DHA) are omega-3 fatty acids that are essential to human health and are plentiful in algae lipids (Chen et al. 2023). Algal lipids are an excellent source of omega-3 fatty acids that are both sustainable and vegan, making them an attractive alternative to fish-derived omega-3 oils (Chen et al. 2023). They can be added to a wide range of food products in order to enhance their dietary properties. A number of foods, such as bread, yogurt, milk, and margarine, have been enriched with omega-3 fatty acids through the use of algae oils (Ganesan et al. 2014). As a valuable source of omega-3 fatty acids, these enriched products can provide benefits to vegetarians and vegans.

Algal lipids are a common ingredient in dietary supplements and nutraceuticals. There are various ways in which they may be encapsulated and formulated, including softgels, capsules, and powders. Due to their omega-3 fatty acid content, these products are very popular, and they are highly valued due to the potential health benefits they provide, including improved cardiovascular health, cognitive function, and vision (Brenna and Carlson 2014).

Moreover, algal lipids are used in the manufacture of aquafeed, which is a feed that is specifically designed for aquatic animals, including fish and shrimp (Ahmad et al. 2022). The majority of aquafeed do not contain the essential fatty acids EPA and DHA, which are essential for the growth and development of aquatic species. However, algal lipids, which are rich in omega-3 fatty acids, can be used in the formulation of aquafeed as a reliable and sustainable source of these essential nutrients (Tocher 2015).

Other than aquafeed, algal lipids are also used as a nutrition source for domestic animals, such as poultry, swine, and cattle. Omega-3 enriched animal products, including meat, eggs, and milk, can be obtained by incorporating algal lipids into the diet of these animals (Lum et al. 2013) (Fig. 21.1). Thus, higher omega-3 fatty acid levels can contribute to the production of healthier animals.

21.5 Algal Carbohydrates

The potential applications of algal carbohydrates in various industries, including food and feed production, have attracted a great deal of attention in recent years. It is well known that algae are abundant sources of carbohydrates, which can be utilized as valuable ingredients for a wide range of functional and nutritional purposes.

1. Algal carbohydrates in food production

   i. Alginate: Alginate is a polysaccharide produced by brown algae that is commonly used in a variety of foods as a thickening, stabilizing, and gelling ingredient (Qin et al. 2018; Raus et al. 2021). Products such as ice cream, yogurt, dressings, and sauces contain this additive. Films and coatings derived from alginate can also be used to preserve and extend the shelf life of food products (Raus et al. 2021). Alginate has been used in wound dressing and has demonstrated remarkable efficacy in promoting wound healing and regeneration (Varaprasad et al. 2020). Considering that alginate is widely used in food preparation for human consumption, it is regarded as safe for biomedical applications as well (Raus et al. 2021). Moreover, alginate is biodegradable and dissolves slowly in the body when cross-linking agents within the material interact with monovalent cations in the body (Raus et al. 2021) (Fig. 21.3).
   ii. Carrageenan: Carrageenan, which is a red algae product, has also become a popular ingredient in food products (Rupert et al. 2022). It is used in dairy products, processed meats, desserts, and beverages as a stabilizer, thickener, and emulsifier. Carrageenan enhances food products' texture, flavor, and sensory characteristics (Jönsson et al. 2020; Rupert et al. 2022; Sipahigil and Dortunc 2001) (Fig. 21.3).
   iii. Agar: Agar is a gelling agent derived from red algae that is used in the food industry to enhance taste and texture (Fig. 21.3). A variety of confectionery, desserts, jellies, and bakery products are commonly prepared with it. Agar also acts as a fat substitute and contributes to low-fat or reduced-calorie food formulations (Jin et al. 2021).

2. Algal carbohydrates in feed production.

The use of microalgae in livestock and aquaculture has been recognized as a potential feed ingredient for various livestock and aquatic species due to their rich carbohydrates content (Idenyi 2021). Macroalgae, also known to as seaweeds, are rich

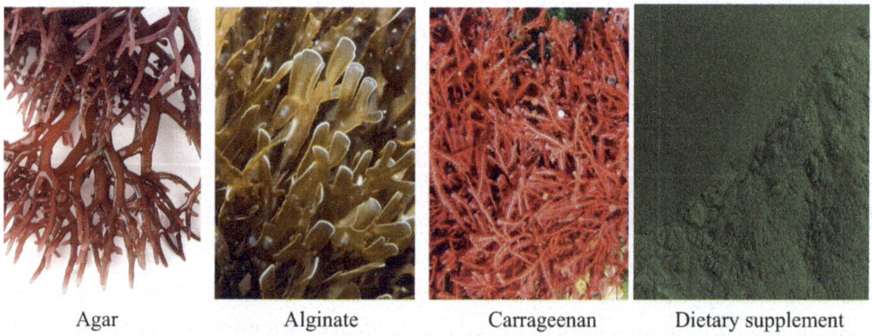

Fig. 21.3 Superfoods from algae

in complex carbohydrates such as laminarin and fucoidan, as well as a variety of other useful compounds. The potential benefits of these additives for livestock, poultry, and aquaculture species have been examined (Lopez-Santamarina et al. 2020). Fiber and prebiotics: Algal carbohydrates, such as alginate and laminarin, can promote the growth of beneficial gut microbes in animals (Fig. 21.3). Macroalgae produce these polysaccharides which may be used as prebiotics to improve animal digestive health (O'Sullivan et al. 2010). They are non-digestible compounds specifically fermented by the gut microbiota, stimulating growth and activity of beneficial organisms in the digestive tract, thus promoting nutrient utilization, boost immune function, and overall health of the host (O'Sullivan et al. 2010). In addition to their unique properties, algae carbohydrate is a promising candidate for the development of sustainable and functional ingredients.

21.6 Algal Pigments

Besides the well-known chlorophylls, the primary pigments responsible for photosynthetic activity, microalgae produce a large variety of other pigments. These microalgae also contain carotenoids and phycobiliproteins. The food industry benefits significantly from these natural pigments because they are both colorant agents and antioxidants (Sun et al. 2023). An example of such a pigment is phycocyanin, which is produced primarily by *A. platensis*. Several applications of this compound in the food industry exist, most notably in fermented milk, milkshakes, and ice cream (Dineshbabu et al. 2019). It is also used as an ingredient in jellies, coated soft candies, soft drinks, health drinks, sour milk, green tea, confectioneries, desserts, and sweet cake decorations (Luzardo-Ocampo et al. 2021).

The blue phycocyanin extracted from the red microalga *Porphyridium aerugineum* is another notable pigment. This substance is used in beverages such as Pepsi® and Bacardi Breezer® because of its resistance to pH changes and exposure to light (Sun et al. 2023). Phycocyanin is present in Linablue®, a blue food colorant derived from Arthrospira. It is used in beverages, frozen desserts, macarons, as well as candy-coated chocolates (Sidari and Tofalo 2019). The coloring properties and antioxidant properties of microalgal pigments like phycocyanin have led to their increasing popularity (Sun et al. 2023). The use of these compounds in a wide range of food products reflects the increasing recognition of their potential in the food industry, reflected in the number of ongoing research studies and patents.

Beta-carotene, a pigment known to be an antioxidant and a precursor to vitamin A, is one of the phytochemicals found in *Dunaliella salina* microalgae. Considering its role as a provitamin A compound, beta-carotene plays a significant role in the human diet. In the human body, it is converted into vitamin A, a vital nutrient necessary to maintain healthy vision, support the immune system, and promote overall growth and development (Pourkarimi et al. 2020). Moreover, beta-carotene is an antioxidant that protects cells from free radical damage, thereby lowering the risk of chronic disease (Pourkarimi et al. 2020).

Astaxanthin, a red pigment found in *Haematococcus pluvialis* microalgae, is employed as a natural colorant in salmon, shrimp, and other seafood products (Ambati et al. 2014). It plays an essential role in fish nutrition by acting as a potent antioxidant, protecting against oxidative stress and cellular damage caused by free radicals (Lim et al. 2018). When astaxanthin is included in fish diets, it enhances pigmentation, resulting in the vibrant red or pink color of their flesh (rainbow trout), supplying a high-quality product (Rahman 2016). The addition of astaxanthin to rainbow trout diets increases growth hormone levels, enhances nutrient absorption, and enhances metabolism, resulting in faster and more efficient growth (Zhang et al. 2023). Astaxanthin has immune-modulatory properties that enhance immune cell activity, increase antibody production, and improve fish and rainbow trout's ability to fight infection and disease (Jafari et al. 2021).

21.7 Algal Vitamins and Minerals

Biomass derived from microalgae is an excellent source of many vitamins (Fig. 21.4) and minerals and could be used to enhance the nutritional value of human and animal diets (Fabregas and Herrero 1990; Matanjun 2009). The vitamins and minerals function as cofactors in biological pathways and are ultimately responsible for a number of cellular functions, such as enhancing vision and immunity, promoting growth and development, and restoring damaged cells (Mahata et al. 2022). Additionally, deficiency of certain vitamins can lead to a variety of diseases, such as rickets, scurvy, and beriberi (Mahata et al. 2022). Microalgae species Spirulina, Scenedesmus, and Chlorella are considered to be good sources of abundant amounts of vitamins A and B complex (B1 (thiamine), B2 (riboflavin), B3 (niacin), B6 (pyridoxine), and B12 (cobalamin)), which are essential for energy metabolism and various physiological functions (Jesus Raposo et al. 2013). Also, they are critical to brain development, cellular metabolism, and the prevention of infection (Christaki et al. 2011; Kusmayadi et al. 2021). Furthermore, *Dunaliella tertiolecta* is a good source of vitamin B2, vitamin B12, and vitamin E, as is *Tetraselmis suecica*, which synthesizes vitamin C and vitamin B complex (Carballo-Cárdenas et al. 2003; Koyande et al. 2019).

Among the major bioactive compounds found in marine microalgae is vitamin E, which has been found to be beneficial in the treatment of various diseases including eye disease, cancer, skin diseases, and high blood pressure (Hernández-Carmona 2009; Kusmayadi et al. 2021). This substance functions as a natural antioxidant and can maintain male fertility (Rengaraj and Hong 2015).

A variety of minerals are critical to the growth and maintenance of the body, including calcium, phosphate, potassium, magnesium, sodium, iron, zinc, and copper (Madeira et al. 2017; Mahata et al. 2022). An inadequate potassium diet can lead to convulsions, while a deficiency of calcium and phosphate results in skeleton deformities. Similarly, magnesium, iron, copper, and zinc play significant roles in the movement of muscles, the formation of skeletons, and the prevention of fatigue (Mahata et al. 2022). Fortunately, microalgae contain these minerals, which could

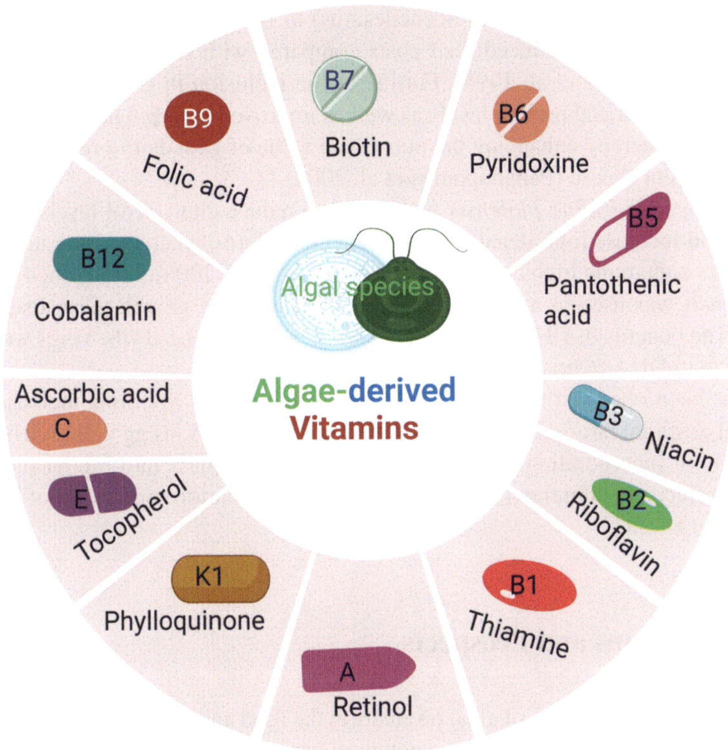

Fig. 21.4 A variety of vitamins found in algal biomass

be used in the diet of humans and animals (Dineshbabu et al. 2019). There are a number of promising mineral sources for fish diets, including *Isochrysis galbana*, *Chlorella stigmatophora*, *Tetraselmis suecica*, and *Dunaliella tertiolecta* (Fábregas and Herrero 1986). Meanwhile, marine *Spirulina platensis* has been found to be a valuable source of minerals for animal feed (Alagawany et al. 2021).

21.8 Algae Biomass for Ruminants and Non-Ruminants' Nutrition

Ruminant and non-ruminant (monogastric) animals consume algae as part of their diets to enhance the nutritional value of their products (Christaki et al. 2011) (Fig. 21.1). The dietary inclusion of microalgae such as *Schizochytrium* sp. has been demonstrated to increase the level of valuable polyunsaturated fatty acids such as DHA and conjugated linoleic acid in milk fat while reducing saturated fatty acids in dairy cows (Franklin et al. 1999). It was also found that feeding microalgae suspensions (e.g., Chlorella and Scenedesmus) to fattening calves improved crude fiber digestibility and reduced feed costs compared with feeding sesame seed oil to them (Chowdhury et al. 1995). Further, algae inclusion in goat feed resulted in increased milk fat and protein levels as well as improved milk polyunsaturated fatty acid content, thereby enhancing the nutritional value of goat dairy products such as yogurt and feta cheese (Papadopoulos et al. 2002).

Microalgae *Spirulina platensis* was found to reduce cholesterol levels in rabbit serum, and increase high-density lipoprotein cholesterol, when incorporated into a rabbit diet containing high cholesterol levels (Colla et al. 2008). For weaned piglets, *S. platensis* was used as a protein supplement to improve growth (Grinstead et al. 2000). The benefits of DHA in pork were significantly enhanced when pigs were fed algae rich in DHA (Christaki et al. 2011; Sardi et al. 2006).

It has been observed that feeding chickens Chlorella or Spirulina increased the levels of essential fatty acids such as linolenic acid and DHA in egg yolks (Grigorova et al. 2006). Incorporating 0.8% *Chlorella vulgaris* biomass into broiler diets has positive effects on immune responses related to inflammation and growth of broilers (Roques et al. 2022).

21.9 Problems and Prospects

The widespread utilization of algal products in the food and feed industries is subject to certain challenges despite various potential beneficial applications.

1. Scaling and achieving efficiency of production. Despite algae's ability to grow rapidly, the cultivation process needs to be optimized to ensure high production yields and cost efficiency. It is possible to enhance control over cultivation parameters, maximize productivity, and reduce the risk of contamination through the development of advanced cultivation systems, such as photobioreactors and closed-loop systems.
2. Standardization and quality control of algae-based products are crucial to their acceptance in the feed and food industries. In order to achieve uniform product characteristics, it is essential to establish standard practices regarding cultivation, harvesting, and processing. Moreover, robust quality control measures should be in place throughout the production chain for the detection of toxins, heavy metals, and microbial contaminants, in order to meet regulatory requirements and ensure consumer safety.
3. There is a direct relationship between consumers' acceptance of algae-based products and their taste, texture, and palatability. Algae species must overcome the inherent seaweed-like flavors and textures in order to be more widely popular. Innovation in product development, including formulation techniques, flavor masking, and texture modification, can help create a wide range of appealing and consumer-friendly algae-based food and feed products.

4. Awareness and acceptance of algae-based foods and feeds. These products are relatively new to the market, and consumer awareness and acceptance are still in early stages. To ensure the successful adoption of algae-based products, it is crucial to educate consumers about their nutritional benefits, sustainability aspects, and versatility. It is possible to build trust and create demand for algae-based products by developing effective marketing strategies, providing transparent labeling, and working closely with chefs, specialists in nutrition and food science.

The following are future prospects and strategies for the of development algal-based ingredients.

1. The application of genetic engineering and strain selection techniques can significantly improve the desirable traits of algae, such as growth rate, nutritional content, and stress tolerance (Moellering and Benning 2010). Algae strains optimized for food and feed production can be developed through genetic manipulation and breeding programs.
2. To enhance the sustainability of algae-based systems, we recommend integrating algae cultivation with other industries such as aquaculture and wastewater treatment. Using waste streams and co-cultivating algae with other organisms can improve nutrient recycling and resource utilization, resulting in more efficient and sustainable practices.
3. Innovations in processing technologies, such as techniques for extraction, drying, and fractionation, can be explored to boost the yield of high-quality algae-derived ingredients. The development of value-added products with improved functionality, stability, and sensory properties is possible through advances in these areas.
4. Establishing clear regulatory frameworks and standards is essential for the successful integration of algae-based foods and feeds into the market. Developing appropriate guidelines, ensuring product safety, and fostering consumer confidence require collaboration between industry stakeholders, policymakers, and regulatory agencies.

21.10 Conclusions

A remarkable benefit of algae, particularly microalgae, resides in their capability to produce assorted compounds such as proteins, carbohydrates, lipids, and carotenoids, depending on species. The food, feed, and nutraceutical industries have already benefited from the use of some microalgae species, such as *Chlorella* sp., *Arthrospira platensis*, *Haematococcus pluvialis*, *Dunaliella salina*, and Nannochloropsis sp. They have the potential to produce more than just their current uses. They can produce substantial quantities of protein, carbohydrates, carotenoids and lipids, including omega-6 and omega-3. Microalgae synthesize compounds that have great potential in a variety of markets. Their protein and lipid content makes them suitable for use in human foods, animal feeds and aquaculture. In addition, these microalgae can be found to have applications in the medicine and cosmetic industry due to their ability to produce beneficial compounds. Finally, microalgae's ability to produce different useful chemicals makes them quite valuable. Algal technology has significant potential for expansion across a wide range of industries, including ruminants and non-ruminants' nutrition. In the future, as science and technology advance, algae will likely be used more widely in a variety of markets.

References

Adesalu TA et al (2022) Emerging eco-and bio-technologies in the use of algal biomass for biofuels and high-value products. Fermentation and algal biotechnologies for the food, beverage and other bioproduct industries. CRC Press, pp 213–232

Ahmad A, Hassan SW, Banat F (2022) An overview of microalgae biomass as a sustainable aquaculture feed ingredient: food security and circular economy. Bioengineered 13(4):9521–9547. https://doi.org/10.1080/21655979.2022.2061148

Alagawany M et al (2021) Nutritional applications of species of Spirulina and Chlorella in farmed fish: a review. Aquaculture 542:736841. https://doi.org/10.1016/j.aquaculture.2021.736841

Alam MA, Xu J-L, Wang Z (2020) Microalgae biotechnology for food, health and high value products. Springer

Ambati RR et al (2014) Astaxanthin: sources, extraction, stability, biological activities and its commercial applications—a review. Mar Drugs 12(1):128–152. https://doi.org/10.3390/md12010128

Barros JC et al (2021) Healthy beef burgers: effect of animal fat replacement by algal and wheat germ oil emulsions. Meat Sci 173:108396. https://doi.org/10.1016/j.meatsci.2020.108396

Becker EW (2013) Microalgae for human and animal nutrition. Handbook of microalgal culture: applied phycology and biotechnology, pp 461–503 https://doi.org/10.1002/9781118567166.ch25

Bidlack WW (1999) Functional foods: biochemical and processing aspects. In: Mazza G (ed) Technomic Publishing Co., Inc., Lancaster, PA, Taylor & Francis, p 437

Biesalski H-K et al (2009) Bioactive compounds: safety and efficacy. Nutrition 25(11–12):1206–1211. https://doi.org/10.1016/j.nut.2009.06.014

Borowitzka MA (2013) High-value products from microalgae—their development and commercialization. J Appl Phycol 25:743–756. https://link.springer.com/article/10.1007/s10811-013-9983-9

Boukid F, Castellari M (2021) Food and beverages containing algae and derived ingredients launched in the market from 2015 to 2019: a front-of-pack labeling perspective with a special focus on Spain. Foods 10(1):173. https://doi.org/10.3390/foods10010173

Brenna JT, Carlson SE (2014) Docosahexaenoic acid and human brain development: evidence that a dietary supply is needed for optimal development. J Hum Evol 77:99–106. https://doi.org/10.1016/j.jhevol.2014.02.017

Carballo-Cárdenas EC et al (2003) Vitamin E (α-tocopherol) production by the marine microalgae Dunaliella tertiolecta and Tetraselmis suecica in batch cultivation. Biomol Eng 20(4–6):139–147. https://doi.org/10.1016/S1389-0344(03)00040-6

Charoensiddhi S et al (2018) Process and economic feasibility for the production of functional food from the brown alga Ecklonia radiata. Algal Res 29:80–91. https://doi.org/10.1016/j.algal.2017.11.022

Chen W et al (2023) Microalgal polyunsaturated fatty acids: hotspots and production techniques. Front Bioeng Biotechnol 11:1146881. https://doi.org/10.3389/fbioe.2023.1146881

Chowdhury S et al (1995) Study on the use of algae as a substitute for oil cake for growing calves. Livest Res Rural Dev 6(3):1995

Christaki E, Florou-Paneri P, Bonos E (2011) Microalgae: a novel ingredient in nutrition. Int J Food Sci Nutr 62(8):794–799. https://doi.org/10.3109/09637486.2011.582460

Colgrave ML et al (2021) Perspectives on future protein production. J Agric Food Chem 69(50):15076–15083. https://doi.org/10.1021/acs.jafc.1c05989

Colla LM, Muccillo-Baisch AL, Costa JAV (2008) Spirulina platensis effects on the levels of total cholesterol, HDL and triacylglycerols in rabbits fed with a hypercholesterolemic diet. Braz Arch Biol Technol 51:405–411. https://doi.org/10.1590/S1516-89132008000200022

de Jesus Raposo MF, de Morais RMSC, de Morais AMMB (2013) Health applications of bioactive compounds from marine microalgae. Life Sci 93(15):479–486. https://doi.org/10.1016/j.lfs.2013.08.002

Dineshbabu G et al (2019) Microalgae–nutritious, sustainable aqua-and animal feed source. J Funct Foods 62:103545. https://doi.org/10.1016/j.jff.2019.103545

Domínguez H (2013) Algae as a source of biologically active ingredients for the formulation of functional foods and nutraceuticals. In: Functional ingredients from algae for foods and nutraceuticals. Elsevier, pp 1–19. https://doi.org/10.1533/9780857098689.1

Donn P et al (2022) Functional foods based on the recovery of bioactive ingredients from food and algae by-products by emerging extraction technologies and 3D printing. Food Biosci 49:101853. https://doi.org/10.1016/j.fbio.2022.101853

Eilam Y et al (2023) Microalgae—sustainable source for alternative proteins and functional ingredients promoting gut and liver health. Glob Challenges 2200177. https://doi.org/10.1002/gch2.202200177

Fabregas J, Herrero C (1990) Vitamin content of four marine microalgae. Potential use as source of vitamins in nutrition. J Ind Microbiol Biotechnol 5(4):259–263. https://doi.org/10.1007/BF01569683

Fábregas J, Herrero C (1986) Marine microalgae as a potential source of minerals in fish diets. Aquaculture 51(3–4):237–243. https://doi.org/10.1016/0044-8486(86)90315-7

FAO, The State of World Fisheries and Aquaculture (2022) Towards blue transformation. FAO Rome. https://doi.org/10.4060/cc0461en

Franklin ST et al (1999) Dietary marine algae (*Schizochytrium sp.*) increases concentrations of conjugated linoleic, docosahexaenoic and transvaccenic acids in milk of dairy cows. J Nutr 129(11):2048–2054. https://doi.org/10.1093/jn/129.11.2048

Ganesan B, Brothersen C, McMahon DJ (2014) Fortification of foods with omega-3 polyunsaturated fatty acids. Crit Rev Food Sci Nutr 54(1):98–114. https://doi.org/10.1080/10408398.2011.578221

Grigorova S et al (2006) The effect of biomass from green algae of Chlorella genus on the biochemical characteristics of table eggs. J Central Eur Agric 7(1):111–116

Grinstead G et al (2000) Effects of Spirulina platensis on growth performance of weanling pigs. Anim Feed Sci Technol 83(3–4):237–247. https://doi.org/10.1016/S0377-8401(99)00130-3

Guo M (2013) Functional foods: principles and technology. Elsevier

Hasler CM (2002) Functional foods: benefits, concerns and challenges—a position paper from the American council on science and health. J Nutr 132(12):3772–3781. https://doi.org/10.1093/jn/132.12.3772

Hernández-Carmona G et al (2009) Monthly variation in the chemical composition of Eisenia arborea JE Areschoug. J Appl Phycol 21:607–616. https://link.springer.com/article/10.1007/s10811-009-9454-5

Idenyi JN et al (2021) Characterization of strains of Chlorella from Abakaliki, Nigeria, for the production of high-value products under variable temperatures. J Appl Phycol 33:275–285. https://link.springer.com/article/10.1007/s10811-020-02313-y

Idenyi JN et al (2022) Aquaculture sustainability through alternative dietary ingredients: microalgal value-added products. Eng Microbiol 100049. https://doi.org/10.1016/j.engmic.2022.100049

Jafari Z et al (2021) Nanotechnology-abetted astaxanthin formulations in multimodel therapeutic and biomedical applications. J Med Chem 65(1):2–36. https://doi.org/10.1021/acs.jmedchem.1c01144
Jin M et al (2021) Polysaccharide-based biomaterials in tissue engineering: a review. Tissue Eng Part B Rev 27(6):604–626. https://doi.org/10.1089/ten.teb.2020.0208
Jönsson M et al (2020) Extraction and modification of macroalgal polysaccharides for current and next-generation applications. Molecules 25(4):930
Kadam S, Prabhasankar P (2010) Marine foods as functional ingredients in bakery and pasta products. Food Res Int 43(8):1975–1980. https://doi.org/10.1016/j.foodres.2010.06.007
Kent M et al (2015) Nutritional evaluation of Australian microalgae as potential human health supplements. PLoS ONE 10(2):e0118985. https://doi.org/10.1371/journal.pone.0118985
Koyande AK et al (2019) Microalgae: a potential alternative to health supplementation for humans. Food Sci Human Wellness 8(1):16–24. https://doi.org/10.1016/j.fshw.2019.03.001
Kusmayadi A et al (2021) Microalgae as sustainable food and feed sources for animals and humans–biotechnological and environmental aspects. Chemosphere 271:129800. https://doi.org/10.1016/j.chemosphere.2021.129800
Lim KC et al (2018) Astaxanthin as feed supplement in aquatic animals. Rev Aquac 10(3):738–773. https://doi.org/10.1111/raq.12200
Lopez-Santamarina A et al (2020) Potential use of marine seaweeds as prebiotics: a review. Molecules 25(4):1004. https://doi.org/10.3390/molecules25041004
Lum KK, Kim J, Lei XG (2013) Dual potential of microalgae as a sustainable biofuel feedstock and animal feed. J Anim Sci Biotechnol 4:1–7. https://link.springer.com/article/10.1186/2049-1891-4-53
Luzardo-Ocampo I et al (2021) Technological applications of natural colorants in food systems: a review. Foods 10(3):634. https://doi.org/10.3390/foods10030634
Madeira MS et al (2017) Microalgae as feed ingredients for livestock production and meat quality: a review. Livest Sci 205:111–121. https://doi.org/10.1016/j.livsci.2017.09.020
Mahata C et al (2022) The potential of marine microalgae for the production of food, feed, and fuel (3F). Fermentation 8(7):316. https://doi.org/10.3390/fermentation8070316
Matanjun P et al (2009) Nutrient content of tropical edible seaweeds, Eucheuma cottonii, Caulerpa lentillifera and Sargassum polycystum. J Appl Phycol 21:75–80. https://link.springer.com/article/10.1007/s10811-008-9326-4
Moellering ER, Benning C (2010) RNA interference silencing of a major lipid droplet protein affects lipid droplet size in Chlamydomonas reinhardtii. Eukaryot Cell 9(1):97–106. https://doi.org/10.1128/ec.00203-09
Mu N et al (2019) Recent advances in microalgal bioactives for food, feed, and healthcare products: commercial potential, market space, and sustainability. Compr Rev Food Sci Food Saf 18(6):1882–1897. https://doi.org/10.1111/1541-4337.12500
O'Sullivan L et al (2010) Prebiotics from marine macroalgae for human and animal health applications. Mar Drugs 8(7):2038–2064. https://doi.org/10.3390/md8072038
Papadopoulos G et al (2002) Effects of dietary supplements of algae, containing polyunsaturated fatty acids, on milk yield and the composition of milk products in dairy ewes. J Dairy Res 69(3):357–365
Pereira L (2021) Macroalgae. Encyclopedia 1(1):177–188. https://doi.org/10.3390/encyclopedia1010017
Plaza M et al (2009) Innovative natural functional ingredients from microalgae. J Agric Food Chem 57(16):7159–7170. https://doi.org/10.1021/jf901070g
Pourkarimi S et al (2020) Factors affecting production of beta-carotene from Dunaliella salina microalgae. Biocatal Agric Biotechnol 29:101771. https://doi.org/10.1016/j.bcab.2020.101771
Qin Y et al (2018) Applications of alginate as a functional food ingredient. In: Biopolymers for food design. Elsevier, pp 409–429. https://doi.org/10.1016/B978-0-12-811449-0.00013-X
Raheem A et al (2018) A review on sustainable microalgae based biofuel and bioenergy production: recent developments. J Clean Prod 181:42–59. https://doi.org/10.1016/j.jclepro.2018.01.125

Rahman MM et al (2016) Effects of dietary inclusion of astaxanthin on growth, muscle pigmentation and antioxidant capacity of juvenile rainbow trout (Oncorhynchus mykiss). Prev Nutr Food Sci 21(3):281. https://doi.org/10.3746/pnf.2016.21.3.281

Raus RA, Nawawi WMFW, Nasaruddin RR (2021) Alginate and alginate composites for biomedical applications. Asian J Pharm Sci 16(3):280–306. https://doi.org/10.1016/j.ajps.2020.10.001

Rengaraj D, Hong YH (2015) Effects of dietary vitamin E on fertility functions in poultry species. Int J Mol Sci 16(5):9910–9921. https://doi.org/10.3390/ijms16059910

Reyes LI et al (2021) Actions in global nutrition initiatives to promote sustainable healthy diets. Glob Food Sec 31:100585. https://doi.org/10.1016/j.gfs.2021.100585

Roques S et al (2022) Effect of feeding 0.8% dried powdered Chlorella vulgaris biomass on growth performance, immune response, and intestinal morphology during grower phase in broiler chickens. Animals 12(9):1114. https://doi.org/10.3390/ani12091114

Rupert R et al (2022) Carrageenan from Kappaphycus alvarezii (Rhodophyta, Solieriaceae): metabolism, structure, production, and application. Front Plant Sci 13:859635. https://doi.org/10.3389/fpls.2022.859635

Sankaran JK, Mouly VS (2007) Managing innovation in an emerging sector: the case of marine-based nutraceuticals. R&D Manage 37(4):329–344. https://doi.org/10.1111/j.1467-9310.2007.00479.x

Sardi L et al (2006) Effects of a dietary supplement of DHA-rich marine algae on Italian heavy pig production parameters. Livest Sci 103(1–2):95–103

Shah MR et al (2018) Microalgae in aquafeeds for a sustainable aquaculture industry. J Appl Phycol 30:197–213. https://link.springer.com/article/10.1007/s10811-017-1234-z

Sidari R, Tofalo R (2019) A comprehensive overview on microalgal-fortified/based food and beverages. Food Rev Intl 35(8):778–805. https://doi.org/10.1080/87559129.2019.1608557

Sipahigil O, Dortunc B (2001) Preparation and in vitro evaluation of verapamil HCl and ibuprofen containing carrageenan beads. Int J Pharm 228(1–2):119–128. https://doi.org/10.1016/S0378-5173(01)00814-6

Spolaore P et al (2006) Commercial applications of microalgae. J Biosci Bioeng 101(2):87–96. https://doi.org/10.1263/jbb.101.87

Sun H et al (2023) Microalgae-derived pigments for the food industry. Mar Drugs 21(2):82. https://doi.org/10.3390/md21020082

Sutherland DL et al (2021) How microalgal biotechnology can assist with the UN sustainable development goals for natural resource management. Curr Res Environ Sustain 3:100050. https://doi.org/10.1016/j.crsust.2021.100050

Tocher DR (2015) Omega-3 long-chain polyunsaturated fatty acids and aquaculture in perspective. Aquaculture 449:94–107. https://doi.org/10.1016/j.aquaculture.2015.01.010

Varaprasad K et al (2020) Alginate-based composite materials for wound dressing application: a mini review. Carbohyd Polym 236:116025. https://doi.org/10.1016/j.carbpol.2020.116025

Wells ML et al (2017) Algae as nutritional and functional food sources: revisiting our understanding. J Appl Phycol 29:949–982. https://link.springer.com/article/10.1007/S10811-016-0974-5

Zhang J et al (2023) Effects of natural and synthetic astaxanthin on growth, body color, and transcriptome and metabolome profiles in the leopard coralgrouper (Plectropomus leopardus). Animals 13(7):1252. https://doi.org/10.3390/ani13071252

Chapter 22
Algae-Based Food Technologies

Lena-Sophie Bischoff

Abstract Microalgae are known to produce and accumulate desirable components for human consumption such as vitamins, fatty acids, protein and minerals. With the ability to grow certain species heterotrophically as well as autotrophically in different cultivation systems, such as raceways, open ponds, or closed tubular/flat plate/column bioreactors, there are a variety of cultivation options depending on space and regional limitations. Open systems are mainly used for large scale biomass production as they are cost effective and low maintenance, but due to their design they are more susceptible to contamination by other algae, grazers or bacteria. Closed, or heterotrophic systems are typically high-maintenance and therefore more expensive but allow for more controlled cultivation parameters and the production of pure, high-value compounds. As macro- and micronutrients vary greatly depending on the given cultivation conditions, such as media components, light or temperature, the ability to grow microalgae in different cultivation systems helps to establish a sustainable and nutrient-rich production of algal biomass depending on the given environment and product.

22.1 Algae-Based Food Technologies

The development of sustainable and nutritious food resources is an increasingly important issue considering the challenges posed by climate change, a growing population, and the increasing inaccessibility of arable land. According to the Food and Agriculture Organization (FAO), food production has to increase by 60 % till 2050 in order to feed a population of 9.3 billion people (United Nations 2012; Ciani et al. 2021). In 2020, about 112 teragrams of nitrogen fertilizer (56 % of all fertilizers) are used in agriculture, increasing to about 232 teragrams in 2050 (Bodirsky et al. 2014; International Fertilizer Industry Association 2023). This excessive and currently rather ineffective usage in traditional agriculture leads to multiple environmental

L.-S. Bischoff (✉)
Universitätsstr. 25, 33615 Bielefeld, Germany
e-mail: lena-sophie.bischoff@uni-bielefeld.de; lenabisch@gmail.com

T. A. Kurniawan and A. Anouzla (eds.), *Algae as a Natural Solution for Challenges in Water-Food-Energy Nexus*, Environmental Science and Engineering,
https://doi.org/10.1007/978-981-97-2371-3_22

problems, such as the pollution of farming land, air and drinking water (Bodirsky et al. 2014; van Grinsven et al. 2013). Bringing up not only environmental pollution but also direct impacts on human health like the development of cancer and asthma (Sutton and Ayyappan 2013; Bodirsky et al. 2014). To minimize the environmental and health impact of, for example fertilizers, there is a need to find food production technologies which use nutrients more efficiently. This could be achieved by taking up less space, preventing leakage and using organisms with a high biomass output compared to their nutrient input. To achieve global food security, production not only has to become more efficient, but also accessibility must be ensured.

In rural, poor and harsh environments where traditional farming is not efficient new farming approaches are needed (Sutton and Ayyappan 2013; United Nations 2012). In Africa in particular, it is predicted that only 458 to 629 m^2 of arable land per person will be available by 2100 (Ullmann and Grimm 2021). If novel technologies such as algae cultivation become cheaper in the future, they could be used locally, even in environments where soil quality is not suitable for commercial plant cultivation (Ullmann and Grimm 2021).

In comparison to other plants microalgae have a higher photosynthetic efficiency and are more effective in their uptake and utilization of micro- and macronutrients (Ciani et al. 2021).

Compared to conventional agriculture, the resources used for bioreactor cultivation, such as nitrogen and phosphorus, remain completely or mostly within the cultivation system and are available to the algae. Due to this, sparce and valuable nutrients like phosphorus can be used more efficiently without being lost to the environment. Compared to especially closed system reactors, conventional plant cultivation leads to a greater loss of nutrients due to the open environment. Nutrients escaping the cultivation system into the environment then also can cause ocean acidification, nutrient pollution of water and the destruction of the natural soil and water composition (Ciani et al. 2021; Bodirsky et al. 2014). Algae cultivation does not directly compete with conventional agriculture, as bioreactor systems do not rely on fertile soil. Especially in urban areas where space is limited bioreactor systems can yield biomass where conventional agriculture would not be possible. Autotrophic cultivation could also be used to directly utilize CO_2 from industrial processes and factories (Matassa et al. 2016; Pikaar et al. 2018). With this approach cities could not only reduce their environmental impact by lowering the CO_2 levels but also use the carbon to growth a reliable food source independent from otherwise less optimal growth conditions for traditional agriculture.

When it comes to future sustainable industrial scale cultivation, the energy costs should be minimized. Studies on the energy input revealed that current mass production of algae (autotrophic and heterotrophic) has a higher environmental impact than most traditional plant-based foods and protein sources (Smetana et al. 2017; Ullmann and Grimm 2021). Only when compared to the commercial production of pork and beef, *C. vulgaris* had a smaller energy and resource demand (Smetana et al. 2017). The studies conclude that the energy and resource costs come mainly from the stirring, lighting and heat required for cultivation and glucose needed for heterotrophic growth.

Processing also has a high impact, as heat is usually used to dry algae biomass (Ho and Redan 2022). Therefore, it is crucial to find more efficient species as well as optimize processing and cultivation. A more efficient nutrient source for the algae could for example be achieved by using hydrolyzed food waste, reducing the environmental impact of cultivation by 4.5 times (Smetana et al. 2017). New cultivation systems that enable high-density cultivation would also improve the amount of biomass relative to the energy input.

Algae and cyanobacteria can be used in a variety of products. In general, there are two main forms of algae products: Supplements and food products with algae as an additive. There are products that use dry biomass in the form of powder or tablets, but also applications where fresh whole cells are used, such as pasta or drinks. Valuable ingredients such as omega-3-fatty acids or vitamins also get sold in isolated form. Supplements are sold to improve bodily functions, such as supporting the immune system or helping to regulate sugar levels (Matos et al. 2017). Other products are not directly marketed as health supplements and include algae as a high protein source or to provide vitamins in products like bread, noodles, or drinks. With algae products being able to be vegan and organic they are suitable for a broad consumer market.

Due to the diversity of microalgae species, different amounts of macro- and micronutrients can be obtained from them. Even within a species, amounts do vary depending on the cultivation conditions, such as temperature, light availability, or culture media (Bernal et al. 2011; Phillips and Williams 2011; Sandgruber et al. 2021). In outdoor cultivation seasonal changes may lead to a difference in the nutrients profile due to the cells having direct contact with the environment. Microalgae and cyanobacteria accumulate around 40–60 % of their dry weight in protein, making them comparable to other commercially used high-protein vegetables such as soybeans (Phillips and Williams 2011). Due to protein being predominantly composed of nitrogen, the amount of nitrogen available to the cells in cultivation is proportional to the amount of protein produced (Phillips and Williams 2011). If the goal is to produce high quantities of protein an increased nitrogen concentration to the media is helpful. Testing protein contents of commercially available algae powders of 10 different species showed a range of 0.1 to 43.4 % dry weight for pure protein content (Ciani et al. 2021; Sandgruber et al. 2021). These differences show the impact of the cultivation conditions on the quantity of compounds.

In terms of protein quality, *Chlorella* and *Arthrospira* do have a balanced amino acid profile for human consumption, comparable to protein sources such as eggs and soybeans (Becker 2007; Caporgno and Mathys 2018). For human consumption the quality of the protein is also determined by the bioavailability of the compounds in the algae. Treatments to improve accessibility are variable depending on the algae or cyanobacteria used. In general, cyanobacteria are naturally better digestible than most algae (Yamaguchi 1996). Due to the lack of a cell wall *Dunaliella* species are also easier to digest (Caporgno and Mathys 2018). Compounds like a rigid cell wall hinder the accessibility of proteins and nutrients from inside the cell. But the algae can be treated enzymatically or with pH alternating treatments (Caporgno and Mathys 2018; Phillips and Williams 2011). This disrupts the cell wall and leads to a higher accessibility of the inner cell compounds to the consumer.

Another big sector is the production of vitamins with microalgae, as they are able to produce a variety of vitamins, like vitamin B, C, D, E, but also due to the fact that they can produce vitamin D and K, which otherwise are found only in small quantities in higher plants (Ampofo and Abbey 2022; Del Mondo et al. 2020). Microalgae can also be used to produce vitamin precursors, for example carotenes, which the human body can convert into vitamin A (Ampofo and Abbey 2022). As already stated before, the quantity of vitamins is highly variable depending on the growth conditions. For example, a 2020 study by Ljubic et al. showed that vitamin D levels in *Nannochloropsis oceanica* were 1 $\mu g\ g^{-1}$ of their dry weight when cells were exposed to 36 $kJ/m^2/day$ UV-B radiation in comparison to $< 0.004\ \mu g\ g^{-1}$ dry weight with no exposure to UV-B. Due to the higher concentration of vitamins in algae compared to other plants, supplements such as powders can be taken in smaller amounts relative to their biomass, making them more convenient for consumption (Fu et al. 2017). Although the amount of vitamins is promising, more studies on bioaccessibility need to be conducted (Ljubic et al. 2020).

Polyunsaturated fatty acids (PUFAs) are an interesting compound for production in algae as they cannot be synthesized by the human body and must be obtained by food (Ampofo and Abbey 2022; Chen et al. 2023). Algae produce a variety of PUFAs such as linolenic acid, ARA, EPA, docosapentaenoic acid or DHA, with algae-derived DHA being already established on the market (Chen et al. 2021, 2023). The main genera used for DHA production are *Crypthecodinium, Schizochytrium* and *Ulkenia sp.*, but there are also other promising genera such as *Monodus* or *Nannochloropsis* (Ampofo and Abbey 2022; Sahin et al. 2018), thus enabling a wide spectrum of possible species and systems useable for cultivation. Yields are dependent on the substrate used, particularly the carbon and nitrogen sources (Sahin et al. 2018). Studies by Sahin et al. in 2018 showed that high biomass production does not correspond to a high DHA yield, as cultures with glycerol and ethanol were having up to 71.26 % lower biomass but higher yields with an increase from 29.94 to 40.04 % in total lipid content. In large-scale cultivation the addition of ethanol not throughout the whole cultivation but at specific times could increase the quantity of DHA without decimating biomass production at such a high rate. Because of this time point specific intervention controllable reactor setups may be best for cultivation.

Algae are not only great for providing energy and nutrition to the consumer, but also contain compounds that reduce the likelihood of chronic diseases (Matos et al. 2017; Villaró et al. 2021). Algae products have been shown to support health by providing antioxidants, antitumor, anticancer, and anti-inflammatory agents to the consumer (Fu et al. 2017; Sandgruber et al. 2021). For health applications secondary metabolites, such as pigments (carotenoids, phenolics etc.), mycosporine-like amino acids and phytosterols are of most interest (Sidari and Tofalo 2019; Ampofo and Abbey 2022). They are produced by the algae to defend themselves against environmental stress. Examples of metabolites that have a positive effect on human health are listed in Table 22.1.

The production of active metabolites in algae can be more efficient compared to conventional higher plant resources. For example, the production of lutein in *C. protothecoides* and *Muriellopsis sp.* is 150 and 18 times higher when compared to the

Table 22.1 Algae species, the secondary metabolite produced by them and the heath impact depending on a given concentration

Species	Metabolite	Concentration	Health impact	Reference
Dunaliella salina, Chlorella sorokiniana, Chlorella protothecoides, Muriellopsis sp.	Lutein	0.4–0.8 %, 5.21 mg/g	Antioxidative, anticancer, cataract and macular degeneration protection	Ampofo and Abbey (2022); Fu et al. (2017); Lin et al. (2015)
Heterochlorella luteoviridis, Dunaliella salina	β-carotene	0.19 mg/g, 0.01–15.0 g/L	Anticancer, antioxidative, antihypertensive, neuroprotective, protection against macular degeneration, anticholesterol	Sui et al. (2021); Morowvat and Ghasemi (2016)
Chlorella zofingiensis, Haematococcus pluvialis	Astaxanthin	1.5 %, 1.95–2.75 %	Antibacterial, anticancer, anti-inflammatory, antioxidative, neuroprotective, antimicrobial	Butler and Golan (2020); Wen et al. (2020)
Arthrospira platensis, Anabaena sp., Cyanidium caldarium, Oscillatoria quadripunctulata, Spirulina platensis	Phycocyanin	54.65 mg/g, 8.3 %, 4–53 mg/g, 27.43 mg/g, 6.5–152 mg/g	Neuroprotective, anti-inflammatory, antioxidative, hepatoprotective, antiaging, antitumor	Butler and Golan (2020); Wen et al. (2020)
Dunaliella salina, Heterochlorella luteoviridis	Zeaxanthin	6 mg/g, 0.244 mg/g	Antidiabetic, supporting eye health	Butler and Golan (2020); Wen et al. (2020)

extraction from marigold (Lin et al. 2015). These values also show the wide range in between species, hence making certain species more favorable compared to others.

Most microalgae can be cultivated with carbon from different sources. Autotrophic cultures use CO_2 and light as the primary energy source and can be used by all species of microalgae, while heterotrophic cultivation relies on the ability of the organism to convert and digest sugar molecules as the carbon source (Ruiz et al. 2022). Since autotrophic cultivation is dependent on light, cell density is an important factor in growth limitation (Jareonsin and Pumas 2021). In biotechnological applications, this means that for high-density cultivation, reactors such as flat panels or sloping cascades can be used more efficiently than large suspension ponds, as they provide greater exposure to light (Pandey et al. 2015). Heterotrophic cultivation, on the other hand, does not require light and allows for a more controllable environment.

Culture systems can also be open or closed. The most commercially used open system, the raceway pond, consists of a circular reservoir, open to the air and a paddle wheel to stir the cells. They are mainly used for biofuel production, but can also be used for food appliances when using different downstream processing (Diaz et al. 2022; Khan et al. 2018). To date, open cultivation systems are widely used for commercial-scale biomass production because their main advantage is cost effectiveness. A raceway pond system with the same volume as a closed photobioreactor can be fifteen times cheaper to construct and twenty times cheaper in terms of maintenance costs (Li et al. 2011), although this may vary depending on the organism used. In general, the costs are lower because they require less energy for resuspension and fewer regulatory elements (Narala et al. 2016; Richardson et al. 2012), thus making them an affordable cultivation method. Due to their open design, cells are in direct contact with the environment. Because of that, cultures are more susceptible to contamination by other microalgae, bacteria or grazers (Cho et al. 2017; Narala et al. 2016). When compared to closed systems, this may cause an inconsistent nutritional value, heath concerns and in general a less productive system (Mata et al. 2010; Narala et al. 2016). Regular testing should therefore be carried out.

A major hurdle to high-density cultivation is light penetration through the media. In open systems such as circular and raceway ponds, the reservoir is typically made of a non-transparent material, meaning that light only penetrates the surface of the culture, whereas in closed bioreactors, light can come from multiple directions. Light penetration is further hindered by self-shading of the cells at an increased depth of the reservoir (Slade and Bauen 2013; Sheehan et al. 1998). Therefor circular and raceway pond systems, typically consisting of a depth of 0.15–0.45 m produce lower cell densities compared to sloped open bioreactors with a depth of approximately 1 cm (Murthy 2011; Ravishankar and Ambati 2020; Slade and Bauen 2013).

Closed cultivation systems are generally less prone to contamination than open systems, as cells are isolated from the external environment (Diaz et al. 2022; Narala et al. 2016). Parameters like light exposure, temperature, pH, CO_2 levels etc. can be controlled more reliably (Mata et al. 2010; Ruiz et al. 2022), thus enabling the production of high valuable, pure products. Due to the higher production and initial costs closed bioreactors are usually not economic for lower quality products (Molina Grima et al. 1999; Narala et al. 2016; Ruiz et al. 2022). Typical closed systems are flat panel reactors or horizontal and vertical tube reactors (Ravishankar and Ambati 2020). Natural light can be used as the light source when grown outdoors or in a greenhouse, whereas indoor cultivation relies on an artificial light source. In comparison to open photobioreactor systems such as open raceway ponds, closed systems enable a more controlled environment and usually a more sufficient light availability due to for example a tubular designs and more controllable parameters (Benner et al. 2022). Closed autotrophic systems are usually more expensive than open systems because of energy costs for cultivation and biofilm removal in closed systems (Benner et al. 2022; Zakova et al. 2019).

Cultivation systems can increase in diameter, length, or height depending on the system to be used for larger cultivations, but because cells in photobioreactors depend on light, scale-up is more challenging compared to reactors that do not require light (Benner et al. 2022; Ugwu et al. 2008). Other factors that make scale-up more complex are CO_2 distribution and temperature in a larger system.

A major challenge and cost factor in especially complex photobioreactors like tubular bioreactors is the formation of biofilm and the removal of it.

In heterotrophic cultivation, cells do not require photosynthetic conditions to grow, so higher cell densities can be achieved compared to photobioreactors (Chen 1996; Chen and Chen 2006), resulting in high yields with low space requirements. Higher cell density also reduces harvesting and downstream processing costs, as there is less media to separate relative to the cell mass. In the case of heterotrophic cultivation, there is an additional cost of the reduced carbon and usually a more complex cultivation system, which is why this cultivation method is only used for high-value products such as omega-6 fatty acids or vitamins (Diaz et al. 2022; Morales-Sánchez et al. 2016).

When it comes to challenges during the cultivation process, contaminants in the form of microorganisms can be directly or indirectly harmful to the biomass used for consumption because the organism itself, or a metabolite produced by it is not safe for consumption. Contamination also leads to variation in the nutritional value of the product, making products from different locations or batches more variable and difficult to market (Ciani et al. 2021; Sandgruber et al. 2021).

Another factor that hinders productivity is biofilm formation. It is a major limitation to the productivity of photobioreactors, due to light limitation, chemical changes, and contamination issues (Benner et al. 2022; Zakova et al. 2019). Studies presented by Huang et al. in 2016 showed that a biomass density of 40 g m^{-2} was sufficient to prevent light penetration through the biofilm, resulting in lower cell densities. Studies by Schnurr et al. in 2014 further demonstrated the effect of shading, as a biofilm of 150 μm reduced light intensity from 100 to 20 μmol m^{-2} s^{-1}. An increasing biofilm can also lead to cell death and fouling of cells deeper within the layer, as these algae do have a limited supply of nutrients and CO_2 (Zeriouh et al. 2017). According to Zeriouh et al. this can also lead to the release of chemicals that promote the loss of pigmentation to other cells in suspension.

Biofilm formation can also increase the risk of contamination, as it can lead to the development of dead zones with an anaerobic environment, which are suitable for microbes like bacteria (Muhammad et al. 2020; Zeriouh et al. 2017). As bacteria biofilms can also produce extracellular polymeric substances, cleaning biofilm removal can be harder depending on the method used for cleaning.

To hinder biofilm formation the material of the cultivation vessel can be modified, so microorganisms cannot adhere to the walls. This can be achieved through the coating of the vessel with polymers such as methacryloyloxyethyl phosphorylcholine (Yin et al. 2021). Other ways to hinder or remove biofilms are presented in Table 22.2.

In conclusion, the main target to further establish algae in the food industry and as a sustainable food source will be to reduce the cost of cultivation and the initial cost of the system. This can be achieved by researching new strains and species of algae that are more efficient. In addition, novel low-tech applications could help establish algae cultivation in poor regions. Accessibility of compounds should be tested, especially in whole cell supplements. A more economical design compared to high nutrient accessibility could be a way to ensure nutrition in the future.

Table 22.2 Treatments and their advantages/disadvantages for the removal of algae biofilm (Nawar et al. 2017; Zakova et al. 2019)

Treatment	Principle	Advantages	Disadvantages
Mechanical	Abrasive removal due to shear stress from inert solid materials	Effective, no chemical residue	Scratching, dead zones may not be accessible, interruption of cultivation, addition of particles
Physical/chemical/ enzymatic	Ultrasound, ozone, oxidating agents or enzymatic detergents	Can reach less accessible locations	Interruption of cultivation, cost inefficient, chemical residue may be left
Optimizing hydrodynamics	Hydrodynamic conditions to prevent attachment	No cultivation interruption, non-scratching	Higher energy demand, stress induction due to turbulence

References

Ampofo J, Abbey (2022) Microalgae: bioactive composition, health benefits, safety and prospects as potential high-value ingredients for the functional food industry. Foods 11(12):S.1744. https://doi.org/10.3390/foods11121744

Becker EW (2007) Micro-algae as a source of protein. Biotechnol Adv 25 (2):S.207–210. https://doi.org/10.1016/j.biotechadv.2006.11.002

Benner P, Meier L, Pfeffer A, Krüger K, Oropeza V, José E, Weuster-Botz D (2022) Lab-scale photobioreactor systems: principles, applications, and scalability. Bioprocess Biosyst Eng 45(5):S.791–813. https://doi.org/10.1007/s00449-022-02711-1

Bernal J, Mendiola JA, Ibáñez E, Cifuentes A (2011) Advanced analysis of nutraceuticals. J Pharm Biomed Anal 55(4):S.758–774. https://doi.org/10.1016/j.jpba.2010.11.033

Bodirsky BL, Popp A, Lotze-Campen H, Dietrich JP, Rolinski S, Weindl I et al (2014) Reactive nitrogen requirements to feed the world in 2050 and potential to mitigate nitrogen pollution. Nat Commun 5:S.3858. https://doi.org/10.1038/ncomms4858

Butler T, Golan Y (2020) Astaxanthin production from microalgae. In: Microalgae biotechnology for food, health and high value products. Springer, Singapore, pp S.175–242

Caporgno MP, Mathys A (2018) Trends in microalgae incorporation into innovative food products with potential health benefits. Front Nutr 5:S.58. https://doi.org/10.3389/fnut.2018.00058

Chen G-Q, Chen F (2006) Growing phototrophic cells without light. Biotechnol Lett 28(9):S.607–616. https://doi.org/10.1007/s10529-006-0025-4

Chen W, Wang J, Ren Y, Chen H, He C, Wang Q (2021) Optimized production and enrichment of α-linolenic acid by Scenedesmus sp. HSJ296. Algal Res 60:S.102505. https://doi.org/10.1016/j.algal.2021.102505

Chen W, Li T, Du S, Chen H, Wang Q (2023) Microalgal polyunsaturated fatty acids: hotspots and production techniques. Front Bioeng Biotechnol 11:S.1146881. https://doi.org/10.3389/fbioe.2023.1146881

Chen F (1996) High cell density culture of microalgae in heterotrophic growth. Trends Biotechnol 14(11):S.421–426. https://doi.org/10.1016/0167-7799(96)10060-3

Cho D-H, Choi J-W, Kang Z, Kim B-H, Oh H-M, Kim H-S, Ramanan R (2017) Microalgal diversity fosters stable biomass productivity in open ponds treating wastewater. Sci Rep 7(1):S.1979. https://doi.org/10.1038/s41598-017-02139-8

Ciani M, Lippolis A, Fava F, Rodolfi L, Niccolai A, Tredici MR (2021) Microbes: food for the future. Foods (Basel, Switz) 10(5). https://doi.org/10.3390/foods10050971

Del Mondo A, Smerilli A, Sané E, Sansone C, Brunet C (2020) Challenging microalgal vitamins for human health. Microb Cell Fact 19(1):S.201. https://doi.org/10.1186/s12934-020-01459-1

Diaz CJ, Douglas KJ, Kang K, Kolarik AL, Malinovski R, Torres-Tiji Y et al (2022) Developing algae as a sustainable food source. Front Nutr 9:S.1029841. https://doi.org/10.3389/fnut.2022.1029841

Fu W, Nelson DR, Yi Z, Xu M, Khraiwesh B, Jijakli K et al (2017) Bioactive compounds from microalgae: current development and prospects. Stud Nat Prod Chem 54:199–225. https://doi.org/10.1016/B978-0-444-63929-5.00006-1

Ho KKHY, Redan BW (2022) Impact of thermal processing on the nutrients, phytochemicals, and metal contaminants in edible algae. Crit Rev Food Sci Nutr 62(2):S.508–526. https://doi.org/10.1080/10408398.2020.1821598

International Fertilizer Industry Association (2023) Global consumption of agricultural fertilizer from 1965 to 2020, by nutrient (in million metric tons). Hg. v. International Fertilizer Industry Association. Statista Research Department. Available online: https://www.statista.com/statistics/438967/fertilizer-consumption-globally-by-nutrient/

Jareonsin S, Pumas C (2021) Advantages of heterotrophic microalgae as a host for phytochemicals production. Front Bioeng Biotechnol 9:S.628597. https://doi.org/10.3389/fbioe.2021.628597

Khan MI, Shin JH, Kim JD (2018) The promising future of microalgae: current status, challenges, and optimization of a sustainable and renewable industry for biofuels, feed, and other products. Microbial Cell Fact 17(1):S.36. https://doi.org/10.1186/s12934-018-0879-x

Li J, Zhu D, Niu J, Shen S, Wang G (2011) An economic assessment of astaxanthin production by large scale cultivation of Haematococcus pluvialis. Biotechnol Adv 29(6):S.568–574. https://doi.org/10.1016/j.biotechadv.2011.04.001

Lin J-H, Lee D-J, Chang J-S (2015) Lutein production from biomass: marigold flowers versus microalgae. Biores Technol 184:421–428. https://doi.org/10.1016/j.biortech.2014.09.099

Ljubic A, Jacobsen C, Holdt SL, Jakobsen J (2020) Microalgae nannochloropsis oceanica as a future new natural source of vitamin D3. Food Chem 320:S.126627. https://doi.org/10.1016/j.foodchem.2020.126627

Mata TM, Martins AA, Caetano NS (2010) Microalgae for biodiesel production and other applications: a review. Renew Sustain Energy Rev 14(1):S.217–232. https://doi.org/10.1016/j.rser.2009.07.020

Matassa S, Verstraete W, Pikaar I, Boon N (2016) Autotrophic nitrogen assimilation and carbon capture for microbial protein production by a novel enrichment of hydrogen-oxidizing bacteria. Water Res 101:137–146. https://doi.org/10.1016/j.watres.2016.05.077

Matos J, Cardoso C, Bandarra NM, Afonso C (2017) Microalgae as healthy ingredients for functional food: a review. Food Funct 8(8):S.2672–2685. https://doi.org/10.1039/c7fo00409e

Molina Grima E, Fernández FGA, Garcıa Camacho F, Chisti Y (1999) Photobioreactors: light regime, mass transfer, and scaleup. J Biotechnol 70(1–3):S.231–247. https://doi.org/10.1016/S0168-1656(99)00078-4

Morales-Sánchez D, Kyndt J, Ogden K, Martinez A (2016) Toward an understanding of lipid and starch accumulation in microalgae: a proteomic study of Neochloris oleoabundans cultivated under N-limited heterotrophic conditions. Algal Res 20:22–34. https://doi.org/10.1016/j.algal.2016.09.006

Morowvat MH, Ghasemi Y (2016) Culture medium optimization for enhanced β-carotene and biomass production by Dunaliella salina in mixotrophic culture. Biocatal Agr Biotechnol 7:S.217–223. https://doi.org/10.1016/j.bcab.2016.06.008

Muhammad MH, Idris AL, Fan X, Guo Y, Yu Y, Jin X et al (2020) Beyond risk: bacterial biofilms and their regulating approaches. Front Microbiol 11:S.928. https://doi.org/10.3389/fmicb.2020.00928

Murthy GS (2011) Chapter 18—overview and assessment of algal biofuels production technologies. In: Pandey A, Larroche C, Ricke SC, Dussap C-G, Gnansounou E (eds) Biofuels. Alternative feedstocks and conversion processes. Academic Press, Amsterdam, S.415–437

Narala RR, Garg S, Sharma KK, Thomas-Hall SR, Deme M, Li Y, Schenk PM (2016) Comparison of microalgae cultivation in photobioreactor, open raceway pond, and a two-stage hybrid system. Front Energy Res 4. https://doi.org/10.3389/fenrg.2016.00029

Nawar A, Khoja AH, Akbar N, Ansari AA, Qayyum M, Ali E (2017) Physical abrasion method using submerged spike balls to remove algal biofilm from photobioreactors. BMC Res Notes 10(1):S.666. https://doi.org/10.1186/s13104-017-2995-9

Pandey A, Höfer R, Larroche C, Taherzadeh M, Nampoothiri KM (2015) Industrial biorefineries and white biotechnology. Elsevier, Amsterdam

Phillips GO, Williams PA (2011) Handbook of food proteins. Elsevier Science, San Diego, CA, USA

Pikaar I, de Vrieze J, Rabaey K, Herrero M, Smith P, Verstraete W (2018) Carbon emission avoidance and capture by producing in-reactor microbial biomass based food, feed and slow release fertilizer: potentials and limitations. Sci Total Environ 644:S.1525–1530. https://doi.org/10.1016/j.scitotenv.2018.07.089

Ravishankar GA, Ambati RR (eds) (2020) Handbook of algal technologies and phytochemicals. Volume I: food, health and nutraceutical applications. CRC Press, Boca Raton, FL. Available online: unter https://www.taylorfrancis.com/books/9780429054242

Richardson JW, Johnson MD, Outlaw JL (2012) Economic comparison of open pond raceways to photo bio-reactors for profitable production of algae for transportation fuels in the Southwest. Algal Res 1(1):S.93–100. https://doi.org/10.1016/j.algal.2012.04.001

Ruiz J, Wijffels RH, Dominguez M, Barbosa MJ (2022) Heterotrophic vs autotrophic production of microalgae: bringing some light into the everlasting cost controversy. Algal Res 64:S.102698. https://doi.org/10.1016/j.algal.2022.102698

Sahin D, Tas E, Altindag UH (2018) Enhancement of docosahexaenoic acid (DHA) production from Schizochytrium sp. S31 using different growth medium conditions. AMB Express 8(1):S.7. https://doi.org/10.1186/s13568-018-0540-4

Sandgruber F, Gielsdorf A, Baur AC, Schenz B, Müller SM, Schwerdtle T et al (2021) Variability in macro- and micronutrients of 15 commercially available microalgae powders. Marine Drugs 19(6):S.310. https://doi.org/10.3390/md19060310

Sheehan J, Dunahay T, Benemann J, Roessler P (1998) Look back at the U.S. Department of energy's aquatic species program: biodiesel from algae; close-out Report (National Renewable Energy Lab. (NREL), Golden, CO (United States), NREL/TP-580-24190). Available online: https://www.osti.gov/biblio/15003040

Sidari R, Tofalo R (2019) A comprehensive overview on microalgal-fortified/based food and beverages. Food Rev Int 35(8):S.778–805. https://doi.org/10.1080/87559129.2019.1608557

Slade R, Bauen A (2013) Micro-algae cultivation for biofuels: cost, energy balance, environmental impacts and future prospects. Biomass Bioenerg 53:29–38. https://doi.org/10.1016/j.biombioe.2012.12.019

Smetana S, Sandmann M, Rohn S, Pleissner D, Heinz V (2017) Autotrophic and heterotrophic microalgae and cyanobacteria cultivation for food and feed: life cycle assessment. Bioresour Technol 245(Pt A):S.162–170. https://doi.org/10.1016/j.biortech.2017.08.113

Sui Y, Mazzucchi L, Acharya P, Xu Y, Morgan G, Harvey PJ (2021) A comparison of β-carotene, phytoene and amino acids production in Dunaliella salina DF 15 (CCAP 19/41) and Dunaliella salina CCAP 19/30 using different light wavelengths. Foods 10(11). https://doi.org/10.3390/foods10112824

Sutton MA, Ayyappan S (2013) Our nutrient world. The challenge to produce more food and energy with less pollution. [Global overview on nutrient management]. Centre for Ecology et Hydrology, Edinburgh

Ugwu CU, Aoyagi H, Uchiyama H (2008) Photobioreactors for mass cultivation of algae. Bioresour Technol 99(10):S.4021–4028. https://doi.org/10.1016/j.biortech.2007.01.046

Ullmann J, Grimm D (2021) Algae and their potential for a future bioeconomy, landless food production, and the socio-economic impact of an algae industry. Org Agr 11(2):S.261–267. https://doi.org/10.1007/s13165-020-00337-9

United Nations (2012) Feeding the world sustainably|United Nations. Available online: https://www.un.org/en/chronicle/article/feeding-world-sustainably

van Grinsven Hans JM, Holland M, Jacobsen BH, Klimont Z, Sutton MA, Jaap WW (2013) Costs and benefits of nitrogen for Europe and implications for mitigation. Environ Sci Technol 47(8):S.3571–3579. https://doi.org/10.1021/es303804g

Villaró S, Viñas I, Lafarga T (2021) Consumer acceptance and attitudes toward microalgae and microalgal-derived products as food. In: Acien G, Lafarga T, Acien G, Lafarga T (eds) Cultured microalgae for the food industry: current and potential applications. Academic Press, S.367–385

Wen X, Wang Z, Ding Y, Geng Y, Li Y (2020) Enhancing the production of astaxanthin by mixotrophic cultivation of Haematococcus pluvialis in open raceway ponds. Aquacult Int 28(2):S.625–638. https://doi.org/10.1007/s10499-019-00483-2

Yamaguchi K (1996) Recent advances in microalgal bioscience in Japan, with special reference to utilization of biomass and metabolites: a review. J Appl Phycol 8(6):S.487–502. https://doi.org/10.1007/BF02186327

Yin W, Xu S, Wang Y, Zhang Y, Chou S-H, Galperin MY, He J (2021) Ways to control harmful biofilms: prevention, inhibition, and eradication. Crit Rev Microbiol 47(1):S.57–78. https://doi.org/10.1080/1040841X.2020.1842325

Zakova T, Jirout T, Kratky L, Belohlav V (2019) Hydrodynamics as a tool to remove biofilm in tubular photobioreactor. Available online: https://www.semanticscholar.org/paper/Hydrodynamics-as-a-Tool-to-Remove-Biofilm-in-Varbanov-Walmsley/d999f40881141fda84714e3f7cbd033afac0bbce

Zeriouh O, Reinoso-Moreno JV, López-Rosales L, Del Cerón-García MC, Sánchez-Mirón A, García-Camacho F, Molina-Grima E (2017) Biofouling in photobioreactors for marine microalgae. Crit Rev Biotechnol 37(8):S.1006–1023. https://doi.org/10.1080/07388551.2017.1299681

Chapter 23
Bioactivity and Biofunctionality Characterization of Algal Biomass

J. Echave, P. Barciela, A. Perez-Vázquez, S. Seyyedi-Mansour, P. Donn, L. Cassani, M. A. Prieto, J. Simal-Gándara, Paz Otero, and M. Fraga-Corral

Abstract Algae are considered a valuable and varied source of natural compounds and ingredients which presence is highly demanded in the market by consumers. Algae have attracted the attention of industrial sectors for their cost-effective and sustainable production. They are photosynthetic organisms capable of sequestering huge amounts of CO_2 from the atmosphere. In addition, algae present a rich chemical composition in terms of macro- and micronutrients but also in minerals and secondary metabolites. Among the macronutrients they stand out for their content in polysaccharides and proteins, two ingredients with a wide range of organoleptic but also biological properties with applications in food industry. Regarding micronutrients, algae are a relevant source of vitamins and minerals of key importance in nutraceuticals, cosmetics and pharmaceutical sectors. Finally, algae are also a reliable source of secondary metabolites, of which phenolic compounds represent vast family of compounds associated to several bioactivities, namely antioxidant, anti-inflammatory, anti-microbial, anti-cancer, anti-diabetic/anti-obesity, hypotensive or neuroprotective. Algae are highly valuable resource associated to a sustainable productivity. Apart from being used as healthy food ingredients they also represent an innovative, natural and environmentally friendly matrix to obtain a wide variety of molecules with industrial and innovative applications, especially when applied for the development of functional products.

Keywords Algae · Applications · Bioactivities · Macroalgae · Sustainable · Functional products

---

J. Echave · P. Barciela · A. Perez-Vázquez · S. Seyyedi-Mansour · P. Donn · L. Cassani · M. A. Prieto · J. Simal-Gándara · P. Otero (✉) · M. Fraga-Corral (✉)
Nutrition and Bromatology Group, Department of Analytical Chemistry and Food Science, Faculty of Science, Universidade de Vigo, 32004 Ourense, Spain
e-mail: paz.otero@uvigo.es

M. Fraga-Corral
e-mail: mfraga@uvigo.es

T. A. Kurniawan and A. Anouzla (eds.), *Algae as a Natural Solution for Challenges in Water-Food-Energy Nexus*, Environmental Science and Engineering,
https://doi.org/10.1007/978-981-97-2371-3_23

23.1 Introduction

In recent years, extensive research has highlighted the exceptional properties of seaweeds, positioning them as valuable and sustainable reservoirs of distinctive bioactive compounds with biological activities and essential nutritional components for human consumption. These discoveries hold substantial economic prospects in food, pharmaceuticals, and cosmetics. Also, algae have garnered significant attention due to their remarkable productivity, vast diversity, and the continuous discovery of novel bioactivities, along with advancements in much more efficient and environmentally friendly cultivation methods. Consequently, algae have become a subject of immense interest in scientific investigations and explorations (Belghit et al. 2017; Ordóñez 2012; Stengel and Connan 2015).

Algae, a diverse group of aquatic organisms, are classified into macroalgae and microalgae, depending on their size. Macroalgae are classified into three main groups: brown algae (Phaeophyta), red algae (Rhodophyta), and green algae (Chlorophyta). Each group has distinct characteristics and ecological roles within marine ecosystems. They have long been recognized as prominent natural sources for profitable commercial production (Bleakley and Hayes 2017; Yaamini et al. 2023).

The global population of brown algae comprises an estimated 1500–2000 species, with a significant number of these species being utilized for human consumption. Furthermore, brown algae of various types are extensively employed as animal feed. Brown algae are characterized by the pigment fucoxanthin, giving them their distinctive olive-brown color. Unlike other algae, brown algae exist only in multicellular form. Green algae, comprising around 8000 species, encompass a wide range of unicellular and multicellular forms. Their pigmentation primarily arises from chlorophyll a and b, alongside other pigments like β-carotene and xanthophylls. Red algae represent a substantial group of multicellular macroalgae, encompassing approximately 6000 species. Red algae are distinguished by the occurrence of phycobilins, which are accountable for imparting their characteristic red coloring. Their high protein content and delightful taste caused these species to be extensively consumed in Asia and Western countries, establishing them as one of the most popular seaweed choices (Bleakley and Hayes 2017).

Macroalgae biomass can be obtained from natural harvesting or cultivated sources, which are likely to differ depending on their origin; (1) Harvesting: (a) Macroalgae harvesting involves the manual collection, which is widely regarded as the most environmentally friendly approach; (b) various machine-assisted methods are employed to gather macroalgae from the beach and nearby waters. Harvesting from wild stocks, compared to seaweed aquaculture, targets a greater diversity of species and yields a higher production volume (Schultz-Zehden and Matczak 2012). The global production derived from wild algal stocks has remained relatively stable, at approximately 1.1 million tons per year since 1969. However, aquaculture played a pivotal role in supporting global seaweed production. In 1969, the world's total seaweed production was evenly divided between harvesting and wild farming, amounting to 2.2 million tons. Half a century later, while wild harvesting remained at 1.1 million tons in 2019, cultivation surged to 34.7 million tons, accounting for 97% of the global seaweed production. (2) Cultivation (Fig. 23.1): Many macroalgae species are amenable to cultivation, and the equipment required can range from simple and low-tech to sophisticated and expensive. There are two main methods for cultivating macroalgae. First, they can be grown in tanks or ponds on land classified as circular, raceway and unstirred ponds. Each of them possesses distinct and unique features and choices of them related to algal species, economics and climate conditions. Despite their cost-effectiveness, these methods present several drawbacks, including water evaporation, CO_2 diffusion into the environment, limited light utilization by cells, and land requirements. Circular and raceway ponds are the oldest and most usual open cultivation systems for algal biomass production, respectively. Second, and more commonly, they are cultivated in open-sea coastal or offshore environments. In offshore production, opportunities for optimizing the growth and composition of algae are limited (Kumar 2015; Mehariya et al. 2021; Stengel and Connan 2015). The most common method employed in macroalgae cultivation is long-line structures, owing to its simplicity and cost-effectiveness. Additionally, other simple cultivation can be carried out from small boats by cutting the algae from the ropes or pulling the ropes onto the boat. For large-scale macroalgae cultivation spanning over 1000 ha, the cultivation system should be designed to be simple and cost-efficient (Schultz-Zehden and Matczak 2012). The potential of seaweed aquaculture to meet the increasing demand of the processing industry for traceable, high-quality, and predictable biomass yields is widely recognized. It is acknowledged that a shift from wild seaweed harvesting to aquaculture is necessary to satisfy the growing demand while mitigating the risk of overexploitation of natural seaweed resources. Moreover, there is an increasing interest in implementing eco-friendly farming practices such as Integrated Multi-trophic Aquaculture (IMTA) systems where macroalgae play a pivotal role. IMTA are currently promoted at both European and national levels (Alexander et al. 2016; Buck et al. 2018; Kleitou et al. 2018). The concept of IMTA revolves around the co-cultivation of species from different trophic levels (two or more) (Chopin et al. 2001; Neori et al. 2004; Troell et al. 2009). IMTA offers numerous benefits compared to monoculture systems. These benefits include the potential for multiple harvests, enhanced profitability within a specific area, and reduced production costs by efficiently sharing resources such as boats and labor. Furthermore, the cultivation of macroalgae in nutrient-rich environments as part of an IMTA setup can provide additional nutritional advantages compared to standalone seaweed farms. The quality of seaweed cultivated in recirculating aquaculture systems may be adversely impacted due to increased levels of nitrogen and phosphate in the water, which can enhance the absorption of metals, especially harmful elements (Wan et al. 2019). This indicates that there is still a need to acquire further knowledge regarding the innovative IMTA methods.

Therefore, since algae represent a sustainable source of natural and environmentally friendly biomolecules, this chapter aims to investigate the composition of the most abundant nutrients of seaweeds, to evaluate their bioactivity, as well as the potential of their application in a wide range of industrial sectors, from agriculture, healthcare and cosmetics to pharmaceutical uses.

Fig. 23.1 Production systems mainly used for seaweeds production

23.2 Chemical Composition of Seaweeds

Algae are considered a valuable and varied source of natural, cost-effective, and sustainable ingredients, as well as healthy food due to its chemical composition in terms of macro- and micronutrients, also taking into consideration their secondary metabolites (Table 23.1). Regarding the macronutrients, polysaccharides and proteins stand out as the most prominent components both in quantity and in biological properties. Regarding micronutrients, algae are a relevant source of vitamins and minerals, especially I, K, Ca or Mg. Last but not least, seaweeds are a reliable source of secondary metabolites, such as terpenes, or phenolic compounds, of which kelps have been found to be rich source of.

23.2.1 Macronutrients

Being plant organisms, seaweeds contain high levels of water (80–90%) and about 15–25% of organic components on its fresh weight. However, since fresh seaweeds are highly odorous and perishable, these are often dried and marketed this way (Afonso et al. 2022). Among these, carbohydrate and ash contents are generally higher in seaweeds in comparison to terrestrial plants, and among seaweeds, brown seaweeds tend to display a higher content of these components (Taboada et al. 2013).

Table 23.1 Chemical composition of macroalgae

Species	Macronutrients	Amount (% or g/100 g dw)	Micronutrients	Amount (% or g/100 g dw)	Secondary metabolites	Amount (% or g/100 g dw)	References
Rhodophyceae							
Chondrus crispus	C	50–57			TPC	0.3–0.4	Nunes et al. (2017) and Tibbetts et al. (2016)
	DF	46			Carotenoids	0.2	
	L	2–4					
	P	6–18					
	A	19–23					
	C						
	DF						
	L						
	P						
	A						
Gracilaria gracilis	C	52					Ferreira et al. (2021)
	DF						
	L	1					
	P	19					
	A	26					
Palmaria palmata	C	31–59			TPC	0.4	Tibbetts et al. (2016)
	DF						
	L	5–13					
	P	15–31					
	A	9–23					

(continued)

Table 23.1 (continued)

Species	Macronutrients	Amount (% or g/100 g dw)	Micronutrients	Amount (% or g/100 g dw)	Secondary metabolites	Amount (% or g/100 g dw)	References
Porphyra purpurea	C	21					Taboada et al. (2013)
	DF	22					
	L	1					
	P	33					
	A	21					
Chlorophyceae							
Cladophora rupestris	C	39					Olsson et al. (2020)
	DF						
	L	2					
	P	18	Ca	0.007			
	A	20	Fe	0.4			
Codium tomentosum (Rodrigues et al. 2015)	C						
	DF						
	L						
	P		Ca				
	A		Fe				
Enteromorpha intestinalis	C	36					Olsson et al. (2020)
	DF						
	L	1					
	P	9	Ca	0.8			
	A	31	Fe	0.08			

(continued)

Table 23.1 (continued)

Species	Macronutrients	Amount (% or g/100 g dw)	Micronutrients	Amount (% or g/100 g dw)	Secondary metabolites	Amount (% or g/100 g dw)	References
Ulva lactuca	C	34					Olsson et al. (2020) and Tabarsa et al. (2012)
	DF	5					
	L	2					
	P	9–10	Ca	2.7			
	A	18–32	Fe	0.1			
Ulva rigida	C	45	Folic acid	0.1			Ferreira et al. (2021) and Taboada et al. (2010)
	DF		Vit C	0.9			
	L	1	Tocopherols	0.1			
	P	22	Mg	2.9			
	A	30	Fe	0.2			
Phaeophyceae							
Ascophyllum nodosum	C	31–56	Mannitol	8.8	TPC	5.8	Olsson et al. (2020) and Tibbetts et al. (2016)
	DF						
	L	2–8					
	P	5–8	Ca	1.2			
	A	20–24	Mg	0.8			
Fucus vesiculosus	C	53	Mannitol	8.1	TPC	4.6	Olsson et al. (2020) and Tibbetts et al. (2016)
	DF						
	L	8					
	P	12	Ca				
	A	22	Mg				

(continued)

Table 23.1 (continued)

Species	Macronutrients	Amount (% or g/100 g dw)	Micronutrients	Amount (% or g/100 g dw)	Secondary metabolites	Amount (% or g/100 g dw)	References
Macrocystis pyrifera	C	75					Ortiz et al. (2009)
	DF						
	L	0.7					
	P	13					
	A	10					
Saccharina japonica	C	53			TPC		Lopez-Santamarina et al. (2023)
	DF	37					
	L	< 1					
	P	7	Ca	1.6			
	A	29	Fe	0.04			
Saccharina latissima	C	59			TPC	1.1	Tibbetts et al. (2016)
	DF						
	L	5					
	P	8					
	A	24					
Undaria pinnatifida	C	21			TPC		Tibbetts et al. (2016)
	DF	30					
	L	2					
	P	16					
	A	28					

Abbreviations *C* Carbohydrates, *L* Lipids, *P* Proteins, *A* Ash, *DF* Dietary fiber

In general terms, lipid content in seaweeds is generally very low, with values oscillating between < 1 and 4% dw (Cherry et al. 2019). Nonetheless, among these, brown algae generally display higher proportions of unsaturated fatty acids such as oleic (C18:1) or linoleic (C18:2), which may represent up to 20–40% of total fatty acids (Ortiz et al. 2009).

Protein content is much more variable among the different seaweed classes, with the general consensus being that the highest protein content is found in red (12–35%), followed by green (10–23%), and lastly brown (5–16%) (Echave et al. 2022). Seaweed protein content is also highly variable depending on species, climatological/environmental conditions, and seasonality. For example, the brown seaweed *Ascophyllum nodosum* harvested from Northern Portugal contained 6.9% dw of protein during spring, but increased to 9.4% in autumn (Vieira et al. 2018). However, when this same seaweed was harvested in October from Northern Norway its protein content was of just 3% dw (Biancarosa et al. 2017). Seaweed protein is usually calculated by quantifying the elemental nitrogen present and applying a factor of 6.25. However, in recent years this has been repeatedly disputed as biasing towards an overestimation of seaweed actual protein content. To amend this, the factor has been corrected in some cases closer to 5, or by quantifying the total amino acid composition (Shuuluka et al. 2013). Regarding the mentioned *A. nodosum* seaweed collected from Northern Norway, by calculating the protein content by N × 6.25, its protein reached 4.5% dw, but by measuring its amino acids, this was lowered to 3% dw. For other species analyzed in this study, the alteration was more notable, like *Porphyra dioica* shifting from 31 to 20.6% dw (Biancarosa et al. 2017). Nevertheless, it has been observed that seaweed protein accounts for significant nutritional quality, with a proportion of essential amino acids ranging from 30 to 40% of its total protein content (Echave et al. 2022).

23.2.2 Micronutrients

Seaweeds are rich in various micronutrients. Foremost, as marine plant organisms, they contain high levels of Na, Cl and K.

23.2.3 Secondary Metabolites

A wide variety of secondary metabolites with bioactive potential can be found in seaweeds. Among these, unique structural polysaccharides may be found in diverse seaweed groups, specifically, fucoidan and alginate in brown seaweeds, agar and carrageenan in red seaweeds and ulvan in certain green seaweeds (mostly *Ulva* sp.). Additionally, storage polysaccharides such as Floridean starch (red seaweeds) and laminarin (brown seaweeds), are also present in seaweed and studied for a variety of applications (Otero et al. 2021). In regard to their bioactive properties, which include antioxidant, anti-inflammatory or antibacterial activities among others, sulfated polysaccharides (ulvan, carrageenan, fucoidan) have been found to be the most bioactive of these, with a positive correlation to their sulfation degree.

Phenolic compounds, molecules that have at least a phenolic ring in their structure, and found in many terrestrial plants, have been thoroughly studied for their antioxidant capacity, and can also be found in seaweeds. Seaweeds may contain phenolic acids and various flavonoids. For example, brown and red seaweeds have been reported to contain significant levels of rutin (> 2 mg/g dw) or the flavonol hesperidin (> 10 mg/g dw), among other phenolic compounds, such as gallic acid. On the other hand, brown seaweeds are known to possess the unique phlorotannins, complex phloroglucinol macromolecules with very high antioxidant activity (Park et al. 2023). Additionally, seaweeds exhibit mycosporine-like amino acids. These are compact nitrogenous secondary compounds with a molecular weight lower than 400 Da that display potent antioxidant properties and the ability to absorb ultraviolet (UV) radiation. Consequently, these molecules are valuable for shielding cells against oxidative reactions and harm triggered by solar UV exposure. Seaweeds and microalgae are plentiful reservoirs of these substances, predominantly found in red seaweeds (Sun et al. 2020a).

Finally, seaweed pigments are greatly diverse in nature and are increasingly drawing attention due to their varied range of bioactive properties and potential applications. For instance, not only seaweeds contain chlorophylls, predominant in green seaweeds, but also phycobiliproteins in red seaweeds and various xanthophylls in brown seaweeds (Echave et al. 2022; Pereira et al. 2021). The fact that phycobiliproteins are red seaweed's pigments may help to explain their significantly higher protein content. On the other hand, brown seaweeds display significant levels of various interesting xanthophyll's that are more uncommon in terrestrial plants, such as zeaxanthin, b-cryptoxanthin and specifically fucoxanthin, the latter being unique to brown algae (Pereira et al. 2021).

Seaweed also synthesize other secondary metabolites of interest with diverse potential or current applications, such as various alkaloids, terpenes, terpenoids, indoles or alcohol sugars (i.e. mannitol) (Soares et al. 2021).

23.3 Bioactive Properties of Seaweed

Different classes of algae have been tested for their potential bioactivities. Most of the current knowledge has been developed through the assessment of their in vivo and in vitro capacities, while very scarce data has been collected from clinical trials. Among the tested bioactivities the antioxidant, anti-inflammatory, anti-cancer, antidiabetic and anti-obesity, antimicrobial, antihypertensive or neuroprotective are described below (Table 23.2).

Table 23.2 Biocompounds obtained from seaweeds with biological activities tested under different assays (in vitro, in vivo, clinical trial)

Species	Biocompounds	Assay	Bioactivity	Results	References
Rhodophyta					
Pyropia yezoensis	Phycobiliproteins	ORAC	Antioxidant	432.763 mM TE/mg	Ulagesan et al. (2022)
		TEAC		64.42 mM TE/mg	
		DPPH radical scavenging activity		58.99%	
Palmaria palmata	Protein hydrolyzed by chymotrypsin	ORAC	Antioxidant	37.84 μmol TE/mg	Beaulieu et al. (2016)
		FRAP		42.86 μmol TE/mg	
		DPPH radical scavenging activity		17.99%	
Palmaria palmata	Peptide fraction (SDITRPGGQM)	ORAC	Antioxidant	152.43 nmol TE/μmol	Harnedy et al. (2017)
		FRAP		21.23 nmol TE/μmol	
Palmaria palmata	Renin inhibitory peptide protein hydrolyzed by papain (IRLIIVLMPILMA)	Renin inhibitory bioassay of hydrolysates	Antihypertensive	58.97%	Fitzgerald et al. (2012)
Palmaria palmata	Tridecapeptide hydrolysate (IRLIIVLMPILMA)	Renin inhibitory assay using spontaneously hypertensive rats	Antihypertensive	3.34 μM (IC_{50})	Fitzgerald et al. (2014)
Iyengaria stellata (Moheimanian et al. 2022)	Fucoxanthin	Inhibitory assay against α-glucosidase	Antidiabetic	0.33 μg/mL (IC_{50})	
Gracilariopsis lemaneiformis (Zhang et al. 2019)	Hydrophobic and/or aromatic amino acids	DPPH radical scavenging activity	Antioxidant	72%	

(continued)

Table 23.2 (continued)

Species	Biocompounds	Assay	Bioactivity	Results	References
Gracilariopsis lemaneiformis	Protein hydrolyzed by α-chymotrypsin	DPPH radical scavenging activity	Antioxidant	71.75%	Zhang et al. (2019)
Gracilariopsis lemaneiformis	Purified peptide Gln-Val-Glu-Tyr (QVEY)	ACE-inhibition ability	Antihypertensive	474.36 μM (IC_{50})	Cao et al. (2017)
Solieria chordalis	Chymotrypsin hydrolysate (< 10 kDa)	ORAC	Antioxidant	903.64 μmol TE/g	Bondu et al. (2015)
		FRAP		19.12 μmol TE/g	
		DPPH radical scavenging activity		11.86 μg (IC_{50})	
		Copper-chelating activity by assay of β-carotene oxidation		97.5%	
Phorphyra columbina	Protein hydrolyzed enriched in Asp, Ala and Glu	DPPH radical scavenging activity	Antioxidant	81.1%	Cian et al. (2012)
		$ABTS^+$ radical cation decolorization assay		61.4%	
Porphyra yezoensis	Pepsin hydrolysate	In vitro anticoagulant assay using human plasma	Anticoagulant	0.3 μM (IC_{50})	Indumathi and Mehta (2016)
Chlorophyta					
Ulva lactuca	Protein hydrolyzates produced using papain, protease, and α-chymotrypsin	Fe^{2+} chelating activity	Antioxidant	46.68%	Hamit et al. (2022)
Ulva lactuca	Trypsin hydrolysate (> 10 kDa)	FRAP	Antioxidant	29.72 μmol TE/g	Bondu et al. (2015)

(continued)

Table 23.2 (continued)

Species	Biocompounds	Assay	Bioactivity	Results	References
		DPPH radical scavenging activity		24.70 μg (IC_{50})	
Ulva lactuca	PG peptide	In silico methods using Biopep enzyme action tool	Antithrombotic activity hypolipidemic	–	Amin et al. (2022)
	GP peptide				
	EF peptide				
Ulva intestinalis	Phe-Gly-Met-Pro-Leu-Asp-Arg (FGMPLDR)	ACE-inhibition ability	Antihypertensive	219.35–236.85 μM (IC_{50})	Sun et al. (2019)
Ulva lactuca (Amin et al. 2022)	EF peptide	In silico methods	Antidiabetic	Inhibition against dipeptidyl-peptidase-IV (DPP-IV)	
Ulva spp. (Cian et al. 2018)	Hydrolysate	In vitro immodulation assays	Anti-inflammatory	Ability to stimulate the production of the anti-inflammatory cytokine IL-10	
Auxenochlorella pyrenoidosa (Zhu et al. 2022)	Peptide (AGWACLVG)	DPPH radical scavenging activity	Antioxidant	68.88 μM	
		ABTS assay		6.20 mmol TE/g	
Phorphyra columbina (Cian et al. 2012)	Hydrolyzed peptide	DPPH radical scavenging activity	Antioxidant	81%	
		ABTS assay		61%	
		β-carotene oxidation		97%	
Chaetomorpha linum (Qin et al. 2022)	Sulfated polysaccharides	Islet amyloid polypeptide aggregation inhibition assay	Antidiabetic	Prevention of the toxicity that polypeptide aggregates exert on pancreatic β-cells	

(continued)

Table 23.2 (continued)

Species	Biocompounds	Assay	Bioactivity	Results	References
Halimeda macroloba (Chin et al. 2015)	Protein hydrolyzates	Inhibitory assays against α-glucosidase	Antidiabetic	6.388 mg/mL (IC_{50})	
Caulerpa cupressoides	Lectin	In vivo nociceptive assay	Neuroprotective	–	da Conceição Rivanor et al. (2018)
Sargassum crassifolium	Protein hydrolyzates produced using papain, protease, and α-chymotrypsin	Fe^{2+} chelating activity	Antioxidant	15.63%	Hamit et al. (2022)
Sargassum maclurei	Protein hydrolysates (RWDISQPY)	ACE-inhibition ability	Antihypertensive	72.24 μM (IC_{50})	Zheng et al. (2020)
Sargassum hystrix	Polyphenols and phlorotannins	Inhibitory assay against α-amylase and α-glucosidase	Antidiabetic	0.53–0.61 mg/mL (IC_{50})	
Sargassum wightii (Raji et al. 2023)	Fucoxanthin	In vitro assay in diabetic and hypertensive rats	Antidiabetic	Restoration of the activity of several liver enzymes	
Ascophyllum nodosum (Frazzini et al. 2022)	Phlorotannins	DPPH radical scavenging activity	Antioxidant	43%	
Iyengaria stellata (Moheimanian et al. 2022)	Fucoxanthin	Inhibitory assay against α-glucosidase	Antidiabetic	0.33 μg/mL (IC_{50})	
Ishige foliacea (Lee et al. 2016)	Octaphlorethol A (phlorotannin)	Tested in vivo in diabetic C57BL/Ks mice	Antidiabetic	Beneficial effects on the blood glucose level regulation and serum insulin levels	
Ecklonia cava (Ahn et al. 2007)	Eckol (phlorotannin)	DPPH radical scavenging activity	Antioxidant	93%	

(continued)

Table 23.2 (continued)

Species	Biocompounds	Assay	Bioactivity	Results	References
Fucus vesiculosus (Bogolitsyn et al. 2019)	Oligomeric phlorotannins	DPPH radical scavenging activity	Antioxidant	502–905 mg of ascorbic acid	
Myagropsis myagroides (Heo et al. 2010)	Fucoxanthin	Enzymatic assays	Anti-inflammatory	Inhibition of iNOS and COX-2	
Saccharina longicruris	Protein hydrolysate fraction	Liquid growth inhibition assays using an optimized critical microdilution method	Antibacterial	0.19 μmax	Beaulieu et al. (2015)
Auxenochlorella pyrenoidosa	Protein hydrolyzed by papain (AGWACLVG)	DPPH radical scavenging activity	Antioxidant	68.88 μM	Zhu et al. (2022)
		$ABTS^+$ radical cation decolorization assay		6.20 mmol TE/g	
Undaria pinnatifida	AIYK	ACE inhibition assay	Antihypertensive	213 μM (IC_{50})	Suetsuna and Nakano (2000)
	YKYY			64 μM (IC_{50})	
	KFYG			90 μM (IC_{50})	
	YNKL			21 μM (IC_{50})	

Abbreviations ORAC Oxygen radical absorbance capacity, *TEA* Trolox equivalent antioxidant capacity, *DPPH* Diphenyl-2-picrylhydrazyl, *FRAP* Ferric-reducing antioxidant power, *ACE* Angiotensin I-converting enzyme, *ABTS*, *iNOS* Nitric oxide synthase, *COX-2* Cyclooxygenase-2

23.3.1 Antioxidant

Oxidative stress is linked to the interaction of an overabundance of reactive oxygen species (ROS) with cell components such as DNA or proteins, among others, leading to the disruption of normal cellular metabolism (Echave et al. 2022). Antioxidants play a key role in protecting against cellular damage induced by free radicals and are essential for maintaining an optimal health and well-being (Sen and Chakraborty 2011). Seaweed peptides represent an important fraction in several algae groups such as in red algae where they may represent a third part of the total weight. Marine peptides have been reported to act as antioxidant agents by direct metal chelation, ROS scavenging and inhibition of lipid peroxidation cascade (O'Connor et al. 2022). Seaweed bioactive peptides contain 3–40 amino acids that are usually inactive inside the parent protein sequence but may be released by gastrointestinal digestion, food processes or fermentation (Pangestuti and Kim 2015). Compared to larger peptides, hydrolysates of lower molecular weight have more amino acid groups ready to donate electrons to radicals and when tested with antioxidant assays such as DPPH, demonstrate stronger ability (Yu et al. 2016). For example, bioactive peptides from aqueous extracts of *Ulva lactuca* and *Sargassum crassifolium*, exhibited higher antioxidant activity than crude proteins. In another study, the smaller molecular size of hydrophobic and/or aromatic amino acids of α-chymotrypsin hydrolysates from *Gracilariopsis lemaneiformis* was concluded to positively contribute to the antioxidant potential with 72% DPPH radical scavenging activity. The peptide AGWACLVG extracted from *Auxenochlorella pyrenoidosa* achieved high antioxidant performance in ABTS (6.20 mmol TE/g) and DPPH (68.88 μM) assays. It was suggested to be suitable as biological ingredient in food formulation or further industrial purposes (Zhu et al. 2022). Similarly, another hydrolyzed peptide from *Phorphyra columbina* displayed remarkable antioxidant data (61%, 81% and 97% for $ABTS^+$, DPPH and β-carotene oxidation, respectively) associated to its low molecular weight (Cian et al. 2012).

Notwithstanding the reported results, structure–activity relationship of bioactive peptides has not yet been fully unraveled. However, peptide length, abundance of aromatic (Tyr, His, Trp and Phe) and hydrophobic (Val, Leu, Ile, Pro, Ala, Met and Gly) amino acids, and the structure of C- and N-terminal residues seem to dictate their antioxidant potential (Harnedy et al. 2017). The lower molecular weight of the peptides (about 340 Da) and the presence of free amino acids seem to be responsible for this improved bioactivity (Hamit et al. 2022) since smaller peptides may easily cross intestinal barrier and interact with free radicals, which could terminate the radical chain (Zhang et al. 2019). As can be observed in Table 23.2 and Fig. 23.2, in vitro antioxidant assays are the most generally described in the use and application of bioactive peptides from seaweeds.

Regarding polysaccharides, fucoidans from *Turbinaria decurrens* were demonstrated to enhance the activity of several antioxidant enzymes such as superoxide

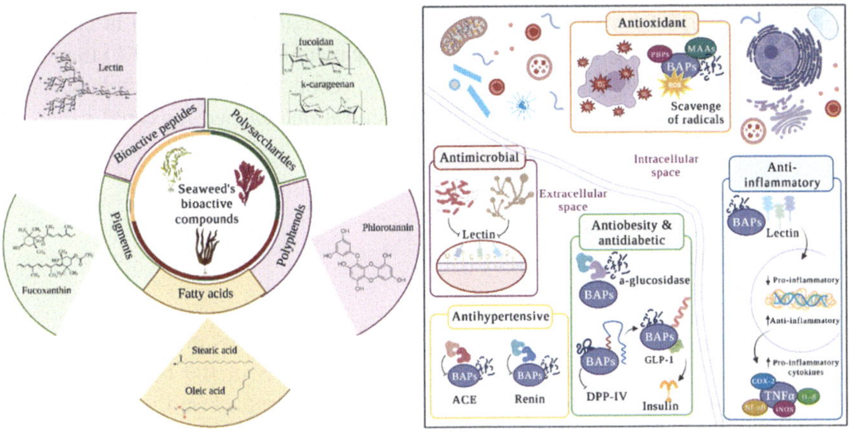

Fig. 23.2 Main described bioactive properties of biocompounds extracted from seaweeds

dismutase (SOD), catalase (CAT) or glutathione peroxidase (GP) among others in IC-21 macrophages (Manikandan et al. 2020). Regarding seaweed secondary metabolites pigments have been repeatedly reported to possess antioxidant activity. Among them, carotenoids, isoprene derivatives have been recognized as antioxidants, even for the algae itself under conditions of UV radiation excess. Indeed, both chlorophylls and carotenoids, such as astaxanthin, zeaxanthin, β-carotene, or fucoxanthin, may mitigate UV harmful effects (Ariede et al. 2017). For instance, astaxanthin, has stronger antioxidant potential than α-tocopherol and was demonstrated to prevent UV damage in skin and to decrease age-induced macular degeneration. β-carotene acts as provitamin A which directly confers antioxidant properties (Anyanwu et al. 2018). Fucoxanthin, especially present in brown algae, is capable of minimize oxidative damage induced by ROS, reduces the activity of tyrosinase and the expression of matrix metalloproteases (MMP), related to cancerous processes. Thus, fucoxanthin administration has been reported to inhibit or minimize melanogenesis processes (Couteau and Coiffard 2016; Wang et al. 2017). Another relevant group of secondary metabolites in seaweed are phlorotannins mainly present in brown algae. Phlorotannins have been repeatedly proved to possess antioxidant properties. For instance, among the three purified phlorotannins from *Ecklonia cava*, eckol showed the best outcomes (93% DPPH) although phloroglucinol and dieckol also demonstrate to prevent DNA damage mediated by excess of ROS when tested in mice T-cells lymphoma (Ahn et al. 2007). Similarly, oligomeric phlorotannins from artic *Fucus vesiculosus* that contained 3–8 phloroglucinol units (374–994 Da) showed radical scavenging activity equivalent to that of 502–905 mg of ascorbic acid (Bogolitsyn et al. 2019).

Apart from macroalgae, also microalgae have shown antioxidant properties which may get maximized through the mixture of algae extracts. That was the analyzed case for *Ascophyllum nodosum* and *Lithothamnium calcareum* which in combination provided a higher inhibition of the radical scavenging activity (49%) than individually at 50% of concentration (43% and 2%, respectively). Nevertheless, *Ascophyllum nodosum* when tested at 100% concentration reached maximal values of 58% inhibition (Frazzini et al. 2022).

23.3.2 Anti-inflammatory

Inflammation is a complex defensive strategy that co-evolves in high-level organisms in reaction to stresses that disrupt the homeostasis of the body. An excess and uncontrolled inflammatory condition may have an impact on the pathogenesis of a wide range of illnesses (Saraswati et al. 2019). Anti-inflammatory trials are often carried out using cell cultures to evaluate the production or inhibition of pro-inflammatory mediators (Echave et al. 2021). Regarding the protein fraction of algae, hydrolysates from *Ulva* spp. exerted in vitro immunomodulation consistent with an anti-inflammatory effect through Toll-like innate immunity receptors (TLRs). Results revealed the capacity of peptides to stimulate the production of the anti-inflammatory cytokine IL-10 in both basal and boosted conditions, in splenocytes, macrophages and splenic lymphocytes. Peptides appear to have a basic mechanism of action source of strong anti-inflammatory activity (Cian et al. 2018). In a further study, protein hydrolysates extracted from *P. columbina* displayed considerable immunomodulatory properties in vitro in macrophages and lymphocytes T (primary splenocytes). Particularly, they induced IL-10 synthesis in both cell lines as well as the production of other anti-inflammatory cytokines. This peptide from *P. columbia* presented a high degree of hydrolysis and high content in Asp, Ala and Glu. These findings could be key to handle inflammatory and oxidative stress states (Cian et al. 2012).

Regarding some of the secondary metabolites of algae, fucoxanthin has been widely proved to possess anti-inflammatory potential. For instance, fucoxanthin extracted from *Myagropsis myagroides*, inhibit the nitric oxide synthase (iNOS), cyclooxygenase 2 (COX-2), tumor necrosis factor-α (TNF-α) and few interleukins (IL-1β and IL-6) (Heo et al. 2010). Another study reported that apart from inhibiting the previous pro-inflammatory mediators, fucoxanthin was capable of inhibiting the mRNA expressions of the plasminogen activator inhibitor-1 (PAI-1) in lipopolysaccharide (LPS) stimulated Raw 264.7 macrophages (Wang et al. 2018). Similarly, algae polysaccharides were demonstrated to have inhibiting capacity over some of the key pro-inflammatory pathways when using LPS stimulation on different cell models. For instance, sulfated polysaccharides obtained from the green algae *Caulerpa lentillifera*, was found to inhibit IL-1β and TNF-α expression in the HT29 cell line (Sun et al. 2020b). Fucoidans from *Sargassum swartzii* showed inhibition patterns for iNOS, COX-2, NO, IL-1β, IL-6 and TNF-α as well as for Nuclear Factor κB (NF-κB) and mitogen-activated protein kinase (MAPK) pathways in RAW 264.7

macrophages (Jayawardena et al. 2020). Whereas fucoidans from *Turbinaria decurrens* inhibit the expression of IL-1β, COX-2, MMP-9 together with the enhancement of the activity of several antioxidant enzymes (SOD, CAT, GP, etc.) in IC-21 macrophages (Manikandan et al. 2020). In a atherosclerosis context, a sulfated polysaccharide, named after GNP, extracted from *Gelidium crinale* restored some inflammatory-related pathways such as MAPK, NF-κB or m-TOR. Besides GNP prevent cellular mechanisms such as apoptosis, invasion and migration (Zheng et al. 2023).

Also, non-identified molecules present in algae extracts have been showed to have anti-inflammatory activity. That is the case of the green algae *Enteromorpha compressa* extracts that were demonstrated to possess an anti-inflammatory capacity comparable to the salicylic acid. Specifically, methanolic extracts showed higher capacity for the inhibition of hemolysis, protein denaturation, the activity of the proteinase or the lipoxygenase (Pradhan et al. 2022).

23.3.3 Anti-microbial

Seaweeds amino acids with short or large peptide chains have also been proven to have antimicrobial growth inhibitory effects based on their amphiphilic nature. These peptides interfere with cellular processes, leading to the development of pores and thereby triggering membrane disruption and cell breakdown (Afzal et al. 2023). Proteins extracted from *Saccharina longicruris* and hydrolyzed with trypsin enzyme showed antibacterial potential. In fact, the hydrolyzed protein fraction (> 10 kDa) demonstrated activity against *Staphylococcus aureus* with a marked reduction in the maximum specific growth rate from concentrations of 0.31–2.5 mg/mL (Beaulieu et al. 2015). Besides, marine microbial systems have been pointed out that they may be based on efficacious peptide resources with unique pharmacological characteristics, such as broad-spectrum cytostatic and bactericidal activity, immunosuppressants, antivirals, antimalarials, and insecticides (De Mandal et al. 2021).

Polysaccharides found in *Fucus vesiculosus* were demonstrated to inhibit the growth of gram-positive and gram-negative bacterial strains. Fucoidans fractions at concentrations of 12.5 and 25.0 g/kg showed antibacterial activity against *Bacillus subtilis* and *Listeria innocua* (Gram-positive) and *Escherichia coli* and *Pseudomonas fluorescens* (Gram-negative) (Cabral et al. 2021). Similarly, sulfated polysaccharides obtained from *F. vesiculosus* showed antibacterial activity against few dental plaque gram-positive bacteria. Specifically, the fucoidan fraction named after F85 was highly effective against *Streptococcus mutans* KCTC 5458 with a minimum inhibitory concentration (MIC) between 125 and 1000 μg/mL and *Listeria monocytogenes* KCTC 13,064 (MIC 250 μg/mL). But this fucoidan fraction also have MICs of 500 μg/mL against *Staphylococcus aureus* KCTC 3881 and the lactic acid bacteria *Lactobacillus acidophilus* KCTC 3164 and *Streptococcus thermophilus* KCTC 3658. Besides, this fucoidan fraction was capable of suppressing the biofilm formation and planktonic cell growths of *S. mutans* and *S. sobrinus* when applied at concentrations higher than 250 μg/mL (Jun et al. 2018). Another work where polysaccharide from the green algae *Chaetomorpha aerea* and the red algae *Agardhiella subulata* and *Hypnea cornuta* were tested, displayed efficacy against the biofilm formation induced by *Pseudomonas aeruginosa* ATCC 27853 and *Staphylococcus aureus* ATCC 29213 (Zammuto et al. 2022).

Regarding fatty acids capacity to inhibit bacterial growth, several fractions obtained from the green algae *Ulva rigida* were found possess strong antibacterial properties. Indeed, MICs of these fatty fractions ranged between 10 and 250 μg/ml against *Staphylococcus aureus* ATCC 25923 and *Enterococcus faecalis* ATCC 29212. The identified fatty acids were oleic, linoleic, palmitic, and stearic acids (Ismail et al. 2018).

Phlorotannins, a family of molecules with phloroglucinol as common element, belong to the chemical group of the polyphenolic compounds which are well-known to possess a wide variety of bioactivities, such as antimicrobial. In a study six phlorotannins, namely dieckol, dioxinodehydroeckol, eckol, fucofuroeckol-A, phlorofucofuroeckol-A and 7-phloroeckol A extracted from *Eisenia bicyclis* were tested against *Candida albicans*. Among them, the phlorofucofuroeckol-A and fucofuroeckol-A were the most effective, although just the latter was fully tested. Fucofuroeckol-A was demonstrated to have stronger antifungal capacity against resistant strains of *C. albicans* than the control, fluconazole. Indeed, the phlorotannin was suggested to create a synergy with fluconazole that reinforces its antifungal capacity and even to revert the azole-resistance (Kim et al. 2018).

Methanolic extracts obtained from *Ascophyllum nodosum*, *Chlorella vulgaris*, *Lithothamnium calcareum*, *Schizochytrium* spp. and *Arthrospira platensis* displayed inhibition growth capacity against the O138 strain of *E. coli* when diluted at 1:4 (v/v). Moreover, *L. calcareum* was effective at a dilution 1:8 (v/v) while *A. nodosum* presented antibacterial capacity in further dilutions (up to 1:16, v/v) (Frazzini et al. 2022). Few extracts with no chemical identified profile obtained from the green algae *E. compressa* were demonstrated to possess an anti-bacterial capacity comparable to the two positive controls norfloxacin and ciprofloxacin. The methanolic and ethanolic extracts showed higher antimicrobial capacity against the gram negative bacteria *Pseudomonas aeruginosa* and two gram positive strains *Staphylococcus aureus* and *Streptococcus pneumoniae* (Pradhan et al. 2022).

23.3.4 Antidiabetic and Anti-obesity

Obesity is a current threat for the human health in the current occidental society, being associated with diabetes type 2, metabolic syndrome, and different cardiovascular diseases (Wan-Loy and Siew-Moi 2016). Diabetes is also a problematic disease that causes death and disability worldwide, affects people of any country, age group, and sex (GBD 2021 Diabetes Collaborators 2023). The disorders previously mentioned can be minimized by the inhibition of the activity of some enzymes and regulation systems, by the ingestion of different compounds, including seaweed peptides (Lafarga et al. 2020). In this way, antithrombotic peptides have been used for the prevention of blood coagulation (Lafarga et al. 2020). NMEKGSSSVVSS-RMKQ is a peptide obtained by the hydrolysis of *Porphyra yezoensis* as an anti-coagulant. Authors suggested that the dose-dependent behavior of the anticoagulant activity of this peptide was achieved by its interaction with the clotting factors that involved in the coagulation process (Indumathi and Mehta 2016; Lafarga et al. 2020). The antidiabetic potential of four Malaysian seaweeds (*Halimeda macroloba*, *Padina sulcata*, *Sargassum binderi* and *Turbinaria conoides*) was also found. For this study, the water crude extracts of the different seaweeds were tested, so the activity found could be related not only to the peptide fraction. After running an inhibitory assay against α-lucosidase, results showed IC_{50} values of 6.388 mg/mL for the water extract of *H. macroloba*, while the water extracts of the brown seaweeds showed a potent dipeptidyl-peptidase-IV (DPP-IV) activity (Chin et al. 2015). This enzyme is involved in the rapid metabolism of incretins (e.g., glucagon-like peptides) (Page 2008). Finally, all the brown seaweed tested showed a stimulation of the secretion of glucose-dependent insulinotropic polypeptide (GIP) and glucagon-like peptide-1 (GLP-1) from pGIP neo STC-1 cells in vitro, while *H. macroloba* stimulated GLP-1 secretion but not GIP secretion (Chin et al. 2015). *Ulva lactuca* hydrolysates were also studied by in silico methods, showing inhibitory activity against DPP-IV (Amin et al. 2022). In this way, DPP-IV inhibitors can be used as therapeutic approach of type 2 diabetes, since they prolong the action of incretin compounds (Page 2008). Furthermore, the bioactive peptides PG and GP identified in the *Ulva lactuca* study showed antithrombotic activity by activating the anticoagulation and depolymerization system, and promoting the recovery of enzymatic fibrinolytic activity (Amin et al. 2022). This study also identified the hypolipidemic activity of EF peptide, being able to inhibit the expression of a sterol regulatory element binding protein-1c and its lipogenic target genes by AMP-activated protein kinase (AMPK) activation.

Regarding the group of the phlorotannins, octaphlorethol A extracted from the brown algae *Ishige foliacea* was tested in vivo in diabetic C57BL/Ks mice. Outcomes revealed that the administration of octaphlorethol A have positive effects on the regulation of blood glucose levels, glucose tolerance and serum insulin levels. The molecular mechanisms regulated by octaphlorethol A are similar to those previously observed for peptides. It included an increased the activation of AMPK and the expression of glucose transporter 4 (GLUT4) protein together with the inhibition of the expression levels of hepatic gluconeogenesis enzymes (Lee et al. 2016).

Other studies just analyzed the capacity of algal extracts to inhibit α-amylase and α-glucosidase activity. That is the case of the work that tested polyphenols and phlorotannins obtained from *Sargassum hystrix* and *Eucheuma denticulatum*. According to the authors, results pointed to the polyphenolic compounds of *S. hystrix* as the most promising antidiabetic ones with IC_{50} values for α-amylase and α-glucosidase ranging from 0.56 to 0.59 mg/mL (acarbose values range: 0.53–0.61 mg/mL). Even though the inhibitory capacity of the polyphenolic molecules from *E. denticulatum* was lower (IC_{50} between 0.86 and 1.92 mg/mL) they still displayed activity, specially phlorotannins (Husni et al. 2018). Similarly, phenolic extracts from

Padina tetrasomatica showed maximum inhibition percentages for α-amylase (27%) and α-glucosidase (39%) when tested at a concentration of 25 μg. Some the phenolic compounds identified were quercetin and hydroxyl-ferulic acid derivatives, luteolin, or rosmarinic acid, among others (Naveen et al. 2021).

Regarding pigments, fucoxanthin was identifies as the major responsible for the antidiabetic capacity of several brown algae. In fact, aqueous methanolic extracts of *Colpomenia sinuosa*, *Sargassum acinaciforme*, *Iyengaria stellata*, *Sirophysalis trinodis*, and *Polycladia myrica* showed α-glucosidase inhibitory capacities. The most active extract was that from *I. stellata* (IC_{50} = 0.33 μg/mL) followed by *C. sinuosa* (IC_{50} = 3.5 μg/mL). However, nearly all the obtained extracts showed higher α-glucosidase inhibitory activity than the positive control, acarbose (IC_{50} = 160 μg/mL). These results underline the antidiabetic potential of brown algae, and more specifically of the fucoxanthin (Moheimanian et al. 2022). Other, recently published work also pointed to fucoxanthin purified from *Sargassum wightii* as antidiabetic molecule. Fucoxanthin was tested in diabetic and hypertense-induced Sprague Dawley rats. Fucoxanthin restored the activity of several hepatic gluconeogenesis-related enzymes and the lipidic and glycogen levels while improved the oxidative status of the rats (Raji et al. 2023).

The group of polysaccharides has also been found to possess antidiabetic capacities. Indeed, sulfated polysaccharides obtained from *Chaetomorpha linum* was revealed as antidiabetic. They were capable of inhibiting the islet amyloid polypeptide aggregation. This roles is crucial to prevent the toxicity that the polypeptide aggregates exert over the pancreatic β-cells (Qin et al. 2022).

Unidentified compounds extracted using methanol from the green algae *E. compressa* were proved to have an antidiabetic potential comparable to acarbose. These extracts showed IC_{50} values for α-amylase and α-glucosidase activities of 3.67 and 3.76 μg/mL whereas acarbose IC_{50} values were 2.85 and 3.19 μg/mL, respectively (Pradhan et al. 2022).

23.3.5 Antihypertensive

Hypertension is a common pathology characterized by the increment in the blood pressure, usually above 140/90 mmHg (Muslin 2012). This disease can be treated by inhibiting renin, angiotensin I-converting enzyme (ACE-I), and other enzymes that are involved in the renin-angiotensin-aldosterone system (RAAS) (Mercier et al. 2014). ACE-I participates in the regulation of blood pressure in the renin-angiotensin system (Suetsuna and Nakano 2000). In this way, an inhibitory activity of ACE-I was achieved by using hydrolysates of two microalgae, *Spirulina platensis* and *Polysiphonia urceolata* (He et al. 2007). Thus, blood vessels were relaxed and widen with hydrolysates concentrations ranging between 0.1 and 1.0, depending on the enzymatic treatment (He et al. 2007). Moreover, four tetra-peptides were found in the peptic digest of *Undaria pinnatifida* (AIYK, YKYY, KFYG, and YNKL), showing ACE-I IC_{50} values of 213, 64, 90, and 21 μM, respectively. After the administration of these peptides in hypertensive rats, these results show a significant decreased in the blood pressure, suggesting the positive effect of the incorporation of the peptides in the diet for the control of blood pressure diseases (Suetsuna and Nakano 2000). In other study, *Pyropia columbina* proteins obtained by hydrolysis were tested ACE-I inhibition, showing a IC_{50} value of 1.2 g/L by uncompetitive mechanism (Cian et al. 2015).

Hydrolysates of *Ulva lactuca* were also studied by using *in-silico* methods to determine both availability and abundance of peptides in this seaweed. Results showed a high number of bioactive peptides with ACE inhibitory activity, including EF, RW, VW, GF, AH, GH, GL, KY, PL, RL, VF and DY peptides (Amin et al. 2022). Moreover, a tridecapeptide hydrolysate (which amino acid sequence is IRLI-IVLMPILMA) was obtained from *Palmaria palmata* by using papain enzyme. The hypotensive effect of this peptide was assessed achieving a renin inhibitory IC_{50} value of 3.34 mM. Moreover, a simulation of gastrointestinal (GI) digestion was performed, showing how the tridecapeptide is cleaved and activated to the dipeptide IR when it travels through the GI tract. Finally, both IRLIIVLMPILMA and IR showed a reduction of spontaneously hypertensive rats blood pressure after their administration during 24 h (Fitzgerald et al. 2014). Antihypertensive activity, second only to the antioxidant property, would be the most well-reported concerning the potential of seaweed-derived peptides, in Table 23.2 and Fig. 23.2 we have gathered several examples of this activity principally tried and tested in the ability to inhibit ACE-I.

Much less information was found for other compounds. Nevertheless, polysaccharides, pigments and phenolic compounds have been found to also exert antihypertensive properties. For instance, a previously mentioned sulfated polysaccharide (GNP) extracted from the red algae *G. crinale* inhibited not just ACE but also the low-density lipoprotein receptor-1 (LOX-1) involved in the development of endothelial damage that promotes the formation and rupture of atherosclerotic plaques. The response of GNP against ACE is to act as antagonist since it competes for biding the same active site in a dose-dependent manner. Besides, GNP was demonstrated to reduce the systolic blood pressure (Zheng et al. 2023). Similar results were observed for the methanolic extracts from *Spirogyra* sp. were isolated and purified to contain a high proportion of gallic acid. This gallic acid rich extracts inhibited ACE-I and reduced the systolic blood pressure (Kang et al. 2015). Finally, identical mechanisms were found for the fucoxanthin purified from *S. wightii* and tested in Sprague Dawley rats. Fucoxanthin inhibited ACE-I and reduced blood pressure (Raji et al. 2023).

23.3.6 Other Bioactivities: Antimicrobial, and Neuroprotective Effect

The neuroprotective activity is defined as all the strategist and mechanisms that shield the central nervous system from neuronal damage, whether it is produced by chronic or acute neurodegenerative diseases (Elufioye et al. 2017). In this way, some bioactive peptides have been classified as neuropeptides. YR peptide present in *Ulva lactuca* hydrolysates is an example, since it is involved in the pain relief mechanism (Amin et al. 2022). Moreover, a lectin fraction extracted from *Caulerpa cupressoides* inhibited the inflammatory nociception in the temporomandibular joint (TMJ) of rats. TMJ is considered as a degenerative disorder since when the inflammation in the joint occurs, nociceptors become sensitive producing hyperalgesia and increasing the excitability of the primary afferent neurons. Results of the study showed how lectin from *C. cupressoides* was able to inhibit the nociceptive response induced by 1.5% Capsaicin. Therefore, *C. cupressoides* lectin has an antinociceptive effect that is produced by acting directly on the primary nociceptive neurons (da Conceição Rivanor et al. 2018). Regarding, other major compounds present in algae, lectins, carbohydrate-binding proteins, displayed several bioactivities, including the antidepressant. Lectins extracted from *Solieria filiformis* were purified and administered to LPS treated Swiss mice. Main changes related with the antidepressant capacity of lectins included their restoral capacity against the oxidative and nitrosative damage observed in the prefrontal cortex, hippocampus, and striatum parts of the animals' brains (Abreu et al. 2022).

In summary, algal extracts have been widely and repeatedly demonstrate to possess positive health effects. Nevertheless, research involving the aforementioned biological properties is limited, which suggests that this is an interesting and innovative area that could be further explored and may lead to encouraging findings.

23.4 Industrial Application

Algae represent cost-effective and sustainable sources of natural compounds and ingredients highly demanded in markets. Therefore, seaweeds, both as foodstuff and as matrix to recover bio-compounds, have attracted the attention of industrial sectors. Indeed, polysaccharides and proteins are well-regarded ingredients in the food industry for their wide range of organoleptic and biological properties. Vitamins and minerals are key micronutrients in nutraceuticals, cosmetics, and pharmaceutical sectors as well as their content in phenolic compounds. Hence, algae are a highly valuable resource that can be used as ingredients, but also as matrix to obtain a wide variety of molecules with industrial and innovative applications, especially when applied for the development of nutraceuticals, functional foods, food additives, cosmetics, and pharmaceutical products (Fig. 23.2).

23.4.1 Food Industry

Among the multiple applications that seaweed extracts may have in the food industry, they have been repeatedly utilized in meat, dairy, bakery, pasta, and juice industries.

Among the most employed species those belonging to brown algae have been found to possess several applications (Tagliapietra and Clerici 2023). For instance, *Himanthalia elongata*, *Undaria pinnatifida*, *Laminaria japonica*, and *Porphyra umbilicalis* have been used for their capacity to improve water-binding, emulsion stability, nutritional, mineral, and phytochemical profile in meat industries. Other species such as *Fucus vesiculosus* was used in the juice industry for its bacteriostatic and bactericidal effects against *Salmonella typhimurium* and *Listeria monocytogenes.* Besides, the antioxidant activity of most of the brown algae extracts, such as those from *Fucus vesiculosus*, *Ulva* spp. and *Gracilaria* spp. have prompt their use to prevent oxidation and improve product shelf-life, such as for sausages (Tagliapietra and Clerici 2023). These properties have been exploited in the manufacturing of edible eco-friendly food packaging (Mishra 2022). Another application of algae is their utilization as natural food colorants with low side effects (low toxicity, carcinogenicity, and allergenicity) in comparison with synthetic pigments. The asserted biological activities of seaweed pigments have potentiated their use as nutraceuticals, antioxidants, food additives, and dietary supplements. Indeed, pigments such as chlorophylls, β-carotene, astaxanthin, lutein, fucoxanthin, phycocyanin, allophycocyanin or phycoerythrin have been applied in the manufacturing of beverages, juice, dairy products, confectioneries, cakes, chewing gums, ice creams, baked foods, candies, and sweets (Sun et al. 2023). In this sense, *Chlorella vulgaris*, *Monoraphidium dybowskii*, *Scenedesmus dimorphus*, *Chlamydomonas reinhardtii*, *Pavlova lutheri*, *Saccharina latissimi*, *Sargassum fusifiorme*, *Gracilaria* sp. or *Corallina* sp. are exploited as sources of chlorophylls for the green color they provide. *Dunaliella salina*, *D. bardawil*, *Grateloupia filicina*, *S. fusiforme*, *Laminaria digitata*, *Palmaria palmata* or *Ulva* sp. were reported to produce β-carotene which is applied for their ability to provide red orange coloration. For the extraction of lutein that provides yellow orange coloration *Chlorella protothecoides*, *Muriellopsis* sp., *Scenedesmus almerienses*, *D. salina*, *D. tertiolecta* or *Amphiroa rigida* have been cultured. Fucoxanthin, with yellow brown coloration, has been extracted from *Undaria pinnatifida*, *Sargassum* sp., *Fucus* sp., *Palmaria palmata*, *Cylindrotheca closterium*, *Phaeodactylum tricornutum*, *Isochrysis galbana*, *Mallomonas* sp., *Nitzschia laevis*, *Odontella aurita* or *Chaetoceros* sp., among others. Finally, *Spirulina platensis*, *Geitlerinema* sp., *Porphyridium* sp. or *Gracilariopsis lemaneiformis* were described to generate phycobiliproteins with red coloration (Cikoš et al. 2022; Sun et al. 2023). Apart from the biocompounds that may be extracted and purified from seaweeds, some algae itself has been considered an important nutrient source. That is the case of *Arthrospira* (*Spirulina*) which has been classified as safe foodstuff by the U.S. Food and Drug Administration (FDA). *Spirulina* is used as dietary supplement and functional ingredient. It contains 65% of proteins, and is also a source of minerals, vitamins, polyunsaturated fatty acids, and pigments (chlorophyll and β-carotene, phycocyanin) (Marles et al. 2011). So, algae represent a valuable source of natural ingredients in food industries, to produce healthy foods with beneficial properties.

23.4.2 Cosmetics

Nowadays, the application of seaweed in cosmetic industries is highly developed and exploited by many firms since the biological properties of algae or their extracted bioactive compounds are considering beneficial. Depending on the algae species and their respective compounds, different products or applications can be developed from macro and micro-algae in cosmetic industries (Thiyagarasaiyar et al. 2020). In general terms, they have been used in the manufacturing of many products of beauty such as body milk, creams, soap, and hair products. The main properties of algal extracts included anti-aging, anti-wrinkling, skin whitening, skin sensitizer, antioxidant or antimicrobial but they have been also used as thickening agent (Thiyagarasaiyar et al. 2020). Apart from the associated benefits, seaweeds represent natural sources of compounds, so they are considered eco-friendly. Among the biomolecules used in cosmetics phenolic compounds, fatty acids, pigments, polysaccharides and vitamins stand out. For instance, species belonging to Rhodophyta has been used to extract fatty acids (unsaturated: Ω-3 such as linolenic acid, Ω-9 such as oleic acid and saturated: palmitic acid) and vitamins A, D2, K1 and E. Among the associated properties they have acted as emollient, moisturizing, hair nutritive, skin-soothing, anti-aging, and anti-inflammatory (Lourenço-Lopes et al. 2020). Regarding Phaeophyta algae some of the most utilized molecules include fucoxanthin and its ester derivates and polyunsaturated fatty acids. They are recognized to provide moisturizing, anti-aging, anti-wrinkle, antioxidant and smoothing properties; besides they offer UV protection (Lourenço-Lopes et al. 2020). Finally, from the class of the Chlorophyta unsaturated fatty acids (Ω-3 linolenic acid, Ω-6 linoleic acid and Ω-9 oleic acid) and liposoluble vitamins (D2, K1 and E) were extracted. These molecules possess several bioactivities such as antioxidant, anti-inflammatory, promoters of collagen synthesis, skin elasticity boosters, anti-wrinkle, emollient or moisturizing (Lourenço-Lopes et al. 2020). Regarding the major applications of microalgae in cosmetics, they have been used as matrix to recover pigments, fatty acids, polysaccharides, and vitamins. Thus, some pigments like phycocyanin extracted from *Spirulina*, cantaxanthin, and fucoxanthin from *Isochrysis*, chlorophyll-a, chlorophyll-b, and β-carotene from *Dunaliella Salina*, *Chlorella vulgaris*, *Tetraselmiss uecica*, and *Botryococcus braunii* are all used as antioxidants. These microalgae are also the source of many compounds of interest for cosmetical industries like fatty acids (linolenic, mistiric, oleic, stearic, palmitic, and palmitoleic acid), polysaccharides, vitamin E, and metabolites (phycocyanobilin, phycoerythrobilin, β-cryptoxanthin). It is set that the advances in research concerning algae have enabled the cosmetic industry to create new products made from natural ingredients with more functionality, thus preserving the environment and improving consumer health (Pereira et al. 2020).

23.4.3 Pharmaceutical

The research related to the production of new drugs made with bioactive ingredients of natural sources with less secondary effects have increased the interest of pharmaceuticals industries toward natural matrixes like seaweeds. The main pharmaceutical applications of algae are as anti-cancer, anti-bacterial, antiviral, antihypertensive, antidiabetic, anti-obesity, antihypertensive, anti-inflammatory, and anticoagulant. Thus, to understand the mechanism of action of the biological active compounds extracted from algae, some in vitro and in vivo studies have been achieved on different classes of algae (Lomartire and Gonçalves 2022). In the research fields of pharmaceutical application, the macroalgae *Sargassum fusiforme* (Phaeophyceae) is amongst the most important. Its use has a previous long trajectory since it was applied in traditional medicine for the treatment of diseases such as tumors, goiter, neck mass, and swellings (Dai et al. 2022). Currently, some scientific studies have shown that fucoidans extracted from *Sargassum fusiforme* can be explored in pharmaceutical industries for its anti-diabetic and anti-obesity properties. Also, the polysaccharides of this algae have antidiabetic and neuroprotective capacity, and their extract demonstrated potential in alleviating postmenopausal symptoms, thus proving the importance of *Sargassum fusiforme* for pharmaceutical industries. The pigment fucoxanthin was demonstrated through in vitro and in vivo studies to be effective against distinctive types of cancer and so it may be applied as antitumoral compound. For example, fucoxanthin extracted from *Laminaria japonica* presented antitumoral activity against lung cancer while fucoxanthin extracted from *Colpomenia sinuosa*, and *Sargassum prismaticum*, displayed antitumoral activity on breast cancer, hepatic cancer and colon cancer cellular lines. So, seaweeds and their biologically active components found diverse applications for pharmaceutical industries. However, even though many studies have demonstrated in vitro and in vivo biological activities of seaweeds, many of them more toxicological and clinical studies are still needed to guarantee consumer safety.

23.5 Conclusions

The overpopulation reached over the last decades is expected to continue growing up to 10 billion people by 2050. The increasing population has associated an increment in the demand of food, among other products. The current models of production require alternative sources of ingredients to fulfill the high rates of consumers' demands. In this sense, macroalgae and microalgae represent two natural sources of biocompounds. They are considered to be environmentally friendly since they are capable of sequestering CO_2 from the atmosphere. Besides, they offer several harvesting approaches. Macroalgae can be collected ashore which reduce the mortality of the fauna living in the sand, has no economic costs associated and do not need to be cultured. However, this method does not permit standardize the outcomes. For this end, culturing macroalgae is key. Among the culturing systems, the one based on IMTA allows a better reutilization of biological wastes along the trophic chain levels implied. Regarding microalgae bioreactors seems the most profit option although freshwater of marine blooms may be considered for their exploitation, however at this point it has to be considered the probable presence of phycotoxins and their associated risks.

Independently from the harvesting technique adopted, algae have been demonstrated to offer a rich and variated chemical profile with several nutritional but also biological properties associates such as colorant, antioxidant, anti-inflammatory or antimicrobial, among others. These biomolecules have been demonstrated to have huge industrial potential. Besides, algae bioactivities are still under study, and it is expected to disclose additional bioproperties. Therefore, algae provide innovative ingredients obtained from natural and sustainable sources that may offer an alternative to the current commercialized options with higher carbon footprint.

Acknowledgements The research leading to these results was supported by MICINN supporting the Ramón y Cajal's grant for M. A. Prieto (RYC-2017-22891) and for Paz Otero (RYC2022-036690-I). This research was funded by Xunta de Galicia that supports the program EXCELENCIA-ED431F (1 December 2020) and the program Grupos de Referencia Competitiva (GRUPO AA1-GRC 2018), and the post-doctoral grant of M. Fraga-Corral (ED481B-2019/096) and L. Cassani (ED481B-2021/152). The research leading to these results was also supported by the Ibero-American Program on Science and Technology (CYTED—AQUA-CIBUS, P317RT0003), to the Bio Based Industries Joint Under-taking (JU) under grant agreement Nono. 888003 UP4HEALTH Project (H2020-BBI-JTI-2019). The JU receives support from the European Union's Horizon 2020 research and innovation program and the Bio Based Industries Consortium. The project SYSTEMIC Knowledge hub on Nutrition and Food Security, has received funding from national research funding parties in Belgium (FWO), France (INRA), Germany (BLE), Italy (MIPAAF), Latvia (IZM), Norway (RCN), Portugal (FCT), and Spain (AEI) in a joint action of JPI HDHL, JPI-OCEANS and FACCE-JPI launched in 2019 under the ERA-NET ERA-HDHL (n° 696295).

References

Abreu TM, Corpe FP, Teles FB, da Conceição Rivanor RL, de Sousa CNS, da Silva Medeiros I, de Queiroz INL, Figueira-Mansur J, Mota ÉF, Mohana-Borges R, Macedo DS, de Vasconcelos SMM, Júnior JERH, Benevides NMB (2022) Lectin isolated from the red marine alga *Solieria filiformis* (Kützing) P.W. Gabrielson: secondary structure and antidepressant-like effect in mice submitted to the lipopolysaccharide-induced inflammatory model of depression. Algal Res 65:102715. https://doi.org/10.1016/j.algal.2022.102715

Afonso C, Julião DR, Pinto E, Almeida A, Ferreira IMPLVO, Bandarra NM, Cardoso C (2022) The effect of drying process on undervalued brown and red seaweed species: elemental composition. J Appl Phycol 34(3):1749–1761. https://doi.org/10.1007/s10811-022-02741-y

Alexander KA, Freeman S, Potts T (2016) Navigating uncertain waters: European public perceptions of integrated multi trophic aquaculture (IMTA). Environ Sci Policy, 61:230–237. https://doi.org/10.1016/j.envsci.2016.04.020

Afzal S, Yadav AK, Poonia AK, Choure K, Yadav AN, Pandey A (2023) Antimicrobial therapeutics isolated from algal source: retrospect and prospect. Biologia 78(2):291–305. https://doi.org/10.1007/s11756-022-01207-3

Ahn G-N, Kim K-N, Cha S-H, Song C-B, Lee J, Heo M-S, Yeo I-K, Lee N-H, Jee Y-H, Kim J-S, Heu M-S, Jeon Y-J (2007) Antioxidant activities of phlorotannins purified from *Ecklonia cava* on free radical scavenging using ESR and H_2O_2-mediated DNA damage. Eur Food Res Technol 226(1):71–79. https://doi.org/10.1007/s00217-006-0510-y

Amin MA, Chondra U, Mostafa E, Alam MM (2022) Green seaweed *Ulva lactuca*, a potential source of bioactive peptides revealed by in silico analysis. Inform Med Unlocked 33. https://doi.org/10.1016/j.imu.2022.101099

Anyanwu RC, Rodriguez C, Durrant A, Olabi AG (2018) Micro-macroalgae properties and applications. In: Reference module in materials science and materials engineering, pp 1–28. https://doi.org/10.1016/B978-0-12-803581-8.09259-6

Ariede MB, Candido TM, Jacome ALM, Velasco MVR, de Carvalho JCM, Baby AR (2017) Cosmetic attributes of algae—a review. Algal Res 25:483–487. https://doi.org/10.1016/j.algal.2017.05.019

Beaulieu L, Bondu S, Doiron K, Rioux LE, Turgeon SL (2015) Characterization of antibacterial activity from protein hydrolysates of the macroalga *Saccharina longicruris* and identification of peptides implied in bioactivity. J Funct Foods 17:685–697. https://doi.org/10.1016/j.jff.2015.06.026

Beaulieu L, Sirois M, Tamigneaux É (2016) Evaluation of the in vitro biological activity of protein hydrolysates of the edible red alga, *Palmaria palmata* (dulse) harvested from the Gaspe coast and cultivated in tanks. J Appl Phycol 28(5):3101–3115. https://doi.org/10.1007/s10811-016-0850-3

Belghit I, Rasinger JD, Heesch S, Biancarosa I, Liland N, Torstensen B, Waagbø R, Lock E-J, Bruckner CG (2017) In-depth metabolic profiling of marine macroalgae confirms strong biochemical differences between brown, red and green algae. Algal Res 26:240–249. https://doi.org/10.1016/j.algal.2017.08.001

Biancarosa I, Espe M, Bruckner CG, Heesch S, Liland N, Waagbø R, Torstensen B, Lock EJ (2017) Amino acid composition, protein content, and nitrogen-to-protein conversion factors of 21 seaweed species from Norwegian waters. J Appl Phycol 29(2):1001–1009. https://doi.org/10.1007/s10811-016-0984-3

Bleakley S, Hayes M (2017) Algal proteins: extraction, application, and challenges concerning production. Foods 6(5):33. https://doi.org/10.3390/foods6050033

Bogolitsyn K, Dobrodeeva L, Druzhinina A, Ovchinnikov D, Parshina A, Shulgina E (2019) Biological activity of a polyphenolic complex of Arctic brown algae. J Appl Phycol 31(5):3341–3348. https://doi.org/10.1007/s10811-019-01840-7

Bondu S, Bonnet C, Gaubert J, Deslandes É, Turgeon SL, Beaulieu L (2015) Bioassay-guided fractionation approach for determination of protein precursors of proteolytic bioactive metabolites from macroalgae. J Appl Phycol 27(5):2059–2074. https://doi.org/10.1007/s10811-014-0425-0

Buck BH, Troell MF, Krause G, Angel DL, Grote B, Chopin T (2018) State of the art and challenges for offshore integrated multi-trophic aquaculture (IMTA). Front Mar Sci 5:165. https://doi.org/10.3389/fmars.2018.00165

Cabral EM, Mondala JRM, Oliveira M, Przyborska J, Fitzpatrick S, Rai DK, Sivagnanam SP, Garcia-Vaquero M, O'Shea D, Devereux M, Tiwari BK, Curtin J (2021) Influence of molecular weight fractionation on the antimicrobial and anticancer properties of a fucoidan rich-extract from the macroalgae *Fucus vesiculosus*. Int J Biol Macromol 186:994–1002. https://doi.org/10.1016/j.ijbiomac.2021.06.182

Cao D, Lv X, Xu X, Yu H, Sun X, Xu N (2017) Purification and identification of a novel ACE inhibitory peptide from marine alga *Gracilariopsis lemaneiformis* protein hydrolysate. Eur Food Res Technol 243(10):1829–1837. https://doi.org/10.1007/s00217-017-2886-2

Cherry P, O'Hara C, Magee PJ, Mcsorley EM, Allsopp PJ (2019) Risks and benefits of consuming edible seaweeds. Nutr Rev 77(5):307–329. https://doi.org/10.1093/nutrit/nuy066

Chin YX, Lim PE, Maggs CA, Phang SM, Sharifuddin Y, Green BD (2015) Anti-diabetic potential of selected Malaysian seaweeds. J Appl Phycol 27(5):2137–2148. https://doi.org/10.1007/S10811-014-0462-8

Chopin T, Buschmann AH, Halling C, Troell M, Kautsky N, Neori A, Kraemer GP, Zertuche-Gonzalez JA, Yarish C, Neefus C (2001) Integrating seaweeds intomarine aquaculture systems: a key towards sustainability. J Phycol 37:975–986

Cian RE, López-Posadas R, Drago SR, Sánchez De Medina F, Martínez-Augustin O (2012) A *Porphyra columbina* hydrolysate upregulates IL-10 production in rat macrophages and lymphocytes through an NF-κB, and p38 and JNK dependent mechanism. Food Chem 134(4):1982–1990. https://doi.org/10.1016/j.foodchem.2012.03.134

Cian RE, Garzón AG, Ancona DB, Guerrero LC, Drago SR (2015) Hydrolyzates from *Pyropia columbina* seaweed have antiplatelet aggregation, antioxidant and ACE I inhibitory peptides which maintain bioactivity after simulated gastrointestinal digestion. LWT 64(2):881–888. https://doi.org/10.1016/j.lwt.2015.06.043

Cian RE, Hernández-Chirlaque C, Gámez-Belmonte R, Drago SR, Sánchez de Medina F, Martínez-Augustin O (2018) Green alga *Ulva* spp. hydrolysates and their peptide fractions regulate cytokine production in splenic macrophages and lymphocytes involving the TLR4-NFκB/MAPK pathways. Mar Drugs 16(7). https://doi.org/10.3390/md16070235

Cikoš A-M, Šubarić D, Roje M, Babić J, Jerković I, Jokić S (2022) Recent advances on macroalgal pigments and their biological activities (2016–2021). Algal Res 65:102748. https://doi.org/10.1016/j.algal.2022.102748

Couteau C, Coiffard L (2016) Seaweed application in cosmetics. In: Seaweed in health and disease prevention. Elsevier Inc. https://doi.org/10.1016/B978-0-12-802772-1.00014-2

da Conceição Rivanor RL, Do Val DR, Ribeiro NA, Silveira FD, de Assis EL, Franco ÁX, Vieira LV, de Queiroz INL, Chaves HV, Bezerra MM, Benevides NMB (2018) A lectin fraction from green seaweed *Caulerpa cupressoides* inhibits inflammatory nociception in the temporomandibular joint of rats dependent from peripheral mechanisms. Int J Biol Macromol 115:331–340. https://doi.org/10.1016/j.ijbiomac.2018.04.065

Dai N, Wang Q, Xu B, Chen H (2022) Remarkable natural biological resource of algae for medical applications. Front Mar Sci 9:1–16. https://doi.org/10.3389/fmars.2022.912924

De Mandal S, Panda AK, Murugan C, Xu X, Senthil Kumar N, Jin F (2021) Antimicrobial peptides: novel source and biological function with a special focus on entomopathogenic nematode/bacterium symbiotic complex. Front Microbiol 12:1–16. https://doi.org/10.3389/fmicb.2021.555022

Echave J, Fraga-Corral M, Garcia-Perez P, Popović-Djordjević J, Avdović EH, Radulović M, Xiao J, Prieto MA, Simal-Gandara J (2021) Seaweed protein hydrolysates and bioactive peptides: extraction, purification, and applications. Mar Drugs 19(9):1–22. https://doi.org/10.3390/md19090500

Echave J, Otero P, Garcia-Oliveira P, Munekata PES, Pateiro M, Lorenzo JM, Simal-Gandara J, Prieto MA (2022) Seaweed-derived proteins and peptides: promising marine bioactives. Antioxidants 11(1):1–26. https://doi.org/10.3390/antiox11010176

Elufioye TO, Berida TI, Habtemariam S (2017) Plants-derived neuroprotective agents: cutting the cycle of cell death through multiple mechanisms. Evid-Based Complement Altern Med 2017. Hindawi Limited. https://doi.org/10.1155/2017/3574012

Ferdouse F, Holdt SL, Smith R, Murua P, Yang Z (2018) The global status of seaweed production, trade and utilization. Food Agriculture Organization of the United Nations

Ferreira M, Teixeira C, Abreu H, Silva J, Costas B, Kiron V, Valente LMP (2021) Nutritional value, antimicrobial and antioxidant activities of micro- and macroalgae, single or blended, unravel their potential use for aquafeeds. J Appl Phycol 33(6):3507–3518. https://doi.org/10.1007/s10811-021-02549-2

Fitzgerald C, Mora-Soler L, Gallagher E, O'Connor P, Prieto J, Soler-Vila A, Hayes M (2012) Isolation and characterization of bioactive pro-peptides with in vitro renin inhibitory activities from the macroalga *Palmaria palmata*. J Agric Food Chem 60(30):7421–7427. https://doi.org/10.1021/jf301361c

Fitzgerald C, Aluko RE, Hossain M, Rai DK, Hayes M (2014) Potential of a renin inhibitory peptide from the red seaweed *Palmaria palmata* as a functional food ingredient following confirmation and characterization of a hypotensive effect in spontaneously hypertensive rats. J Agric Food Chem 62(33):8352–8356. https://doi.org/10.1021/jf500983n

Frazzini S, Scaglia E, Dell'Anno M, Reggi S, Panseri S, Giromini C, Lanzoni D, Sgoifo Rossi CA, Rossi L (2022) Antioxidant and antimicrobial activity of algal and cyanobacterial extracts: an in vitro study. Antioxidants 11(5). https://doi.org/10.3390/antiox11050992

GBD 2021 Diabetes Collaborators (2023) Global, regional, and national burden of diabetes from 1990 to 2021, with projections of prevalence to 2050: a systematic analysis for the global burden of disease study 2021. Lancet (London, England). https://doi.org/10.1016/S0140-6736(23)01301-6

Hamit CS, Nam KC, Abeyrathne EDNS (2022) Bioactive peptide production and determination of functional properties using crude water extracts of *Ulva lactuca* (Chlorophyta) and *Sargassum crassifolium* (Phaeophyceae) collected from Sri Lanka. Appl Phycol 3(1):73–82. https://doi.org/10.1080/26388081.2022.2063758

Harnedy PA, O'Keeffe MB, FitzGerald RJ (2017) Fractionation and identification of antioxidant peptides from an enzymatically hydrolysed *Palmaria palmata* protein isolate. Food Res Int 100(1):416–422. https://doi.org/10.1016/j.foodres.2017.07.037

He H-L, Chen X-L, Wu H, Sun C-Y, Zhang Y-Z, Zhou B-C (2007) High throughput and rapid screening of marine protein hydrolysates enriched in peptides with angiotensin-I-converting enzyme inhibitory activity by capillary electrophoresis. Biores Technol 98:3499–3505. https://doi.org/10.1016/j.biortech.2006.11.036

Heo SJ, Yoon WJ, Kim KN, Ahn GN, Kang SM, Kang DH, Affan A, Oh C, Jung WK, Jeon YJ (2010) Evaluation of anti-inflammatory effect of fucoxanthin isolated from brown algae in lipopolysaccharide-stimulated RAW 264.7 macrophages. Food Chem Toxicol 48(8–9):2045–2051. https://doi.org/10.1016/j.fct.2010.05.003

Husni A, Pratiwi T, Ustadi, Samudra AG, Nugroho AE (2018) In vitro antidiabetic activity of *Sargassum hystrix* and *Eucheuma denticulatum* from Yogyakarta Beach of Indonesia: antidiabetic activity of *S. hystrix* and *E. denticulatum*. Proc Pak Acad Sci B Life Environ Sci 55(3 SE-Research Articles):1–8. http://www.ppaspk.org/index.php/PPAS-B/article/view/68

Indumathi P, Mehta A (2016) A novel anticoagulant peptide from the Nori hydrolysate. J Funct Foods 20:606–617. https://doi.org/10.1016/j.jff.2015.11.016

Ismail A, Ktari L, Ben Redjem Romdhane Y, Aoun B, Sadok S, Boudabous A, El Bour M (2018) Antimicrobial fatty acids from green alga *Ulva rigida* (Chlorophyta). BioMed Res Int 2018:3069595. https://doi.org/10.1155/2018/3069595

Jayawardena TU, Sanjeewa KKA, Nagahawatta DP, Lee H-G, Lu Y-A, Vaas APJP, Abeytunga DTU, Nanayakkara CM, Lee D-S, Jeon Y-J (2020) Anti-inflammatory effects of sulfated polysaccharide from *Sargassum swartzii* in macrophages via blocking TLR/NF-Kb signal transduction. Mar Drugs 18(12). https://doi.org/10.3390/md18120601

Jun J-Y, Jung M-J, Jeong I-H, Yamazaki K, Kawai Y, Kim B-M (2018) Antimicrobial and antibiofilm activities of sulfated polysaccharides from marine algae against dental plaque bacteria. Mar Drugs 16(9). https://doi.org/10.3390/md16090301

Kang N, Lee J-H, Lee W, Ko J-Y, Kim E-A, Kim J-S, Heu M-S, Kim GH, Jeon Y-J (2015) Gallic acid isolated from *Spirogyra* sp. improves cardiovascular disease through a vasorelaxant and antihypertensive effect. Environ Toxicol Pharmacol 39(2):764–772. https://doi.org/10.1016/j.etap.2015.02.006

Kim K-H, Yu D, Eom S-H, Kim H-J, Kim D-H, Song H-S, Kim D-M, Kim Y-M (2018) Fucofuroeckol-A from edible marine alga *Eisenia bicyclis* to restore antifungal activity of fluconazole against fluconazole-resistant *Candida albicans*. J Appl Phycol 30(1):605–609. https://doi.org/10.1007/s10811-017-1232-1

Kleitou P, Kletou D, David J (2018) Is Europe ready for integrated multi-trophic aquaculture? A survey on the perspectives of European farmers and scientists with IMTA experience. Aquac 490:136–148. https://doi.org/10.1016/j.aquaculture.2018.02.035

Kumar S (2015) GM algae for biofuel production: biosafety and risk assessment. Collect Biosaf Rev 9:52–75

Lafarga T, Acién-Fernández FG, Garcia-Vaquero M (2020) Bioactive peptides and carbohydrates from seaweed for food applications: natural occurrence, isolation, purification, and identification. Algal Res 48:101909. https://doi.org/10.1016/J.ALGAL.2020.101909

Lee S-H, Ko S-C, Kang M-C, Lee DH, Jeon Y-J (2016) Octaphlorethol A, a marine algae product, exhibits antidiabetic effects in type 2 diabetic mice by activating AMP-activated protein kinase and upregulating the expression of glucose transporter 4. Food Chem Toxicol 91:58–64. https://doi.org/10.1016/j.fct.2016.02.022

Lomartire S, Gonçalves AMM (2022) An overview of potential seaweed-derived bioactive compounds for pharmaceutical applications. Mar Drugs 20(2). https://doi.org/10.3390/md20020141

Lopez-Santamarina A, Sinisterra-Loaiza L, Mondragón-Portocarrero A, Ortiz-Viedma J, Cardelle-Cobas A, Abuín CMF, Cepeda A (2023) Potential prebiotic effect of two Atlantic whole brown seaweeds, *Saccharina japonica* and *Undaria pinnatifida*, using in vitro simulation of distal colonic fermentation. Front Nutr 10:1–12. https://doi.org/10.3389/fnut.2023.1170392

Lourenço-Lopes C, Fraga-Corral M, Jimenez-Lopez C, Pereira AG, Garcia-Oliveira P, Carpena M, Prieto MA, Simal-Gandara J (2020) Metabolites from macroalgae and its applications in the cosmetic industry: a circular economy approach. Resources 9(9). https://doi.org/10.3390/resources9090101

Manikandan R, Parimalanandhini D, Mahalakshmi K, Beulaja M, Arumugam M, Janarthanan S, Palanisamy S, You S, Prabhu NM (2020) Studies on isolation, characterization of fucoidan from brown algae *Turbinaria decurrens* and evaluation of it's in vivo and in vitro anti-inflammatory activities. Int J Biol Macromol 160:1263–1276. https://doi.org/10.1016/j.ijbiomac.2020.05.152

Marles RJ, Barrett ML, Barnes J, Chavez ML, Gardiner P, Ko R, Mahady GB, Dog TL, Sarma ND, Giancaspro GI, Sharaf M, Griffiths J (2011) United States pharmacopeia safety evaluation of *Spirulina*. Crit Rev Food Sci Nutr 51(7):593–604. https://doi.org/10.1080/10408391003721719

Mehariya S, Goswami RK, Verma P, Lavecchia R, Zuorro A (2021) Integrated approach for wastewater treatment and biofuel production in microalgae biorefineries. Energies 14:2282. https://doi.org/10.3390/en14082282

Mercier K, Smith H, Biederman J (2014) Renin-angiotensin-aldosterone system inhibition: overview of the therapeutic use of angiotensin-converting enzyme inhibitors, angiotensin receptor blockers, mineralocorticoid receptor antagonists, and direct renin inhibitors. Prim Care Clin Off Pract 41(4):765–778. https://doi.org/10.1016/J.POP.2014.08.002

Mishra A (2022) Algal functional foods and nutraceuticals: benefits, opportunities, and challenges. https://doi.org/10.2174/97898150518721220101

Moheimanian N, Mirkhani H, Sohrabipour J, Jassbi AR (2022) Inhibitory potential of six brown algae from the Persian Gulf on α-glucosidase and in vivo antidiabetic effect of *Sirophysalis trinodis*. Iran J Med Sci 47(5):484–493. https://doi.org/10.30476/IJMS.2021.91258.2245

Muslin AJ (2012) The pathophysiology of heart failure. In: Muscle: fundamental biology and mechanisms of disease, pp 1–2, 523–535. https://doi.org/10.1016/B978-0-12-381510-1.00037-5

Naveen J, Baskaran R, Baskaran V (2021) Profiling of bioactives and in vitro evaluation of antioxidant and antidiabetic property of polyphenols of marine algae *Padina tetrastromatica*. Algal Res 55:102250. https://doi.org/10.1016/j.algal.2021.102250

Neori A, Chopin T, Troell M, Buschmann AH, Kraemer GP, Halling C, Shpigel M, Yarish C (2004) Integrated aquaculture: rationale, evolution and state of the artemphasizing seaweed biofiltration in modern mariculture. Aquac 231:361–391

Nunes N, Ferraz S, Valente S, Barreto MC, Pinheiro de Carvalho MAA (2017) Biochemical composition, nutritional value, and antioxidant properties of seven seaweed species from the Madeira Archipelago. J Appl Phycol 29(5):2427–2437. https://doi.org/10.1007/s10811-017-1074-x

O'Connor J, Garcia-Vaquero M, Meaney S, Tiwari BK (2022) Bioactive peptides from algae: traditional and novel generation strategies, structure-function relationships, and bioinformatics as predictive tools for bioactivity. Mar Drugs 20(5). https://doi.org/10.3390/md20050317

Olsson J, Toth GB, Albers E (2020) Biochemical composition of red, green and brown seaweeds on the Swedish west coast. J Appl Phycol 32(5):3305–3317. https://doi.org/10.1007/s10811-020-02145-w

Ordóñez EG (2012) Evaluación nutricional y propiedades biológicas de algas marinas comestibles. Estudios in vitro e in vivo (Doctoral dissertation, Universidad Complutense de Madrid)

Ortiz J, Uquiche E, Robert P, Romero N, Quitral V, Llantén C (2009) Functional and nutritional value of the Chilean seaweeds *Codium fragile*, *Gracilaria chilensis* and *Macrocystis pyrifera*. Eur J Lipid Sci Technol 111(4):320–327. https://doi.org/10.1002/ejlt.200800140

Otero P, Carpena M, Garcia-Oliveira P, Echave J, Soria-Lopez A, Garcia-Perez P, Fraga-Corral M, Cao H, Nie S, Xiao J, Simal-Gandara J, Prieto MA (2021) Seaweed polysaccharides: emerging extraction technologies, chemical modifications and bioactive properties. Crit Rev Food Sci Nutr 1–29. https://doi.org/10.1080/10408398.2021.1969534

Page RCL (2008) Insulin, other hypoglycemic drugs, and glucagon. In: Side effects of drugs annual, vol 30. Elsevier, pp 494–506. https://doi.org/10.1016/S0378-6080(08)00042-1

Pangestuti R, Kim S (2015) Seaweed proteins, peptides, and amino acids. In: Seaweed sustainability. Elsevier Inc., pp 125–140. https://doi.org/10.1016/B978-0-12-418697-2/00006-4

Park E, Yu H, Lim JH, Choi JH, Park KJ, Lee J (2023) Seaweed metabolomics: a review on its nutrients, bioactive compounds and changes in climate change. Food Res Int 163:112221. https://doi.org/10.1016/j.foodres.2022.112221

Pereira AG, Jimenez-Lopez C, Fraga M, Lourenço-Lopes C, García-Oliveira P, Lorenzo JM, Perez-Lamela C, Prieto MA, Simal-Gandara J (2020) Extraction, properties and applications of bioactive compounds obtained from microalgae. Curr Pharm Des 26(16):1929–1950. https://doi.org/10.2174/1381612826666200403172206

Pereira AG, Otero P, Echave J, Carreira-Casais A, Chamorro F, Collazo N, Jaboui A, Lourenço-Lopes C, Simal-Gandara J, Prieto MA (2021) Xanthophylls from the sea: algae as source of bioactive carotenoids. Mar Drugs 19(4):1–31. https://doi.org/10.3390/md19040188

Pradhan B, Patra S, Dash SR, Satapathy Y, Nayak S, Mandal AK, Jena M (2022) In vitro antidiabetic, anti-inflammatory and antibacterial activity of marine alga *Enteromorpha compressa* collected from Chilika lagoon, Odisha, India. Vegetos 35(3):614–621. https://doi.org/10.1007/s42535-022-00359-6

Qin L, Yang Y, Hao J, He X, Liu S, Chu X, Mao W (2022) Antidiabetic-activity sulfated polysaccharide from *Chaetomorpha linum*: characteristics of its structure and effects on oxidative stress and mitochondrial function. Int J Biol Macromol 207:333–345. https://doi.org/10.1016/j.ijbiomac.2022.02.129

Raji V, Loganathan C, Ramesh T, Thayumanavan P (2023) Dual antidiabetic and antihypertensive activity of fucoxanthin isolated from *Sargassum wightii* Greville in in vivo rat model. Food Sci Hum Wellness 12(5):1693–1700. https://doi.org/10.1016/j.fshw.2023.02.037

Rodrigues D, Freitas AC, Pereira L, Rocha-Santos TA, Vasconcelos MW, Roriz M, ... & Duarte AC (2015) Chemical composition of red, brown and green macroalgae from Buarcos bay in Central West Coast of Portugal. Food Chem 183:197–207. https://doi.org/10.1016/j.foodchem.2015.03.057

Saraswati, Giriwono PE, Iskandriati D, Tan CP, Andarwulan N (2019) Sargassum seaweed as a source of anti-inflammatory substances and the potential insight of the tropical species: a review. Mar Drugs 17(10):1–35. https://doi.org/10.3390/md17100590

Schultz-Zehden A, Matczak M (2012) Submariner compendium: an assessment of innovative and sustainable uses of Baltic marine resources. Maritime Institute in Gdansk: Gdansk. ISBN 978-83-62438-14-3.260

Sen S, Chakraborty R (2011) The role of antioxidants in human health. In: ACS symposium series, vol 1083, pp 1–37. https://doi.org/10.1021/bk-2011-1083.ch001

Shuuluka D, Bolton JJ, Anderson RJ (2013) Protein content, amino acid composition and nitrogen-to-protein conversion factors of *Ulva rigida* and *Ulva capensis* from natural populations and *Ulva lactuca* from an aquaculture system, South Africa. J Appl Phycol 25(2):677–685. https://doi.org/10.1007/s10811-012-9902-5

Soares C, Paíga P, Marques M, Neto T, Carvalho AP, Paiva A, Simões P, Costa L, Bernardo A, Fernández N, Pinto E, Almeida A, Bronze MR, Delerue-Matos C, Grosso C (2021) Multi-step subcritical water extracts of *Fucus vesiculosus* L. and *Codium tomentosum* Stackhouse: composition, health-benefits and safety. Processes 9(5):893. https://doi.org/10.3390/pr9050893

Stengel DB, Connan S (2015) Marine algae: a source of biomass for biotechnological applications. In: Stengel D, Connan S (eds) Natural products from marine algae. Methods in molecular biology, vol 1308. Humana Press, New York, NY. https://doi.org/10.1007/978-1-4939-2684-8_1

Suetsuna K, Nakano T (2000) Identification of an antihypertensive peptide from peptic digest of wakame (*Undaria pinnatifida*). J Nutr Biochem 11(9):450–454. https://doi.org/10.1016/S0955-2863(00)00110-8

Sun S, Xu X, Sun X, Zhang X, Chen X, Xu N (2019) Preparation and identification of ACE inhibitory peptides from the marine macroalga *Ulva intestinalis*. Mar Drugs 17(3):1–17. https://doi.org/10.3390/md17030179

Sun Y, Zhang N, Zhou J, Dong S, Zhang X, Guo L, Guo G (2020a) Distribution, contents, and types of mycosporine-like amino acids (MAAs) in marine macroalgae and a database for MAAs based on these characteristics. Mar Drugs 18(1):43. https://doi.org/10.3390/md18010043

Sun Y, Liu Z, Song S, Zhu B, Zhao L, Jiang J, Liu N, Wang J, Chen X (2020b) Anti-inflammatory activity and structural identification of a sulfated polysaccharide CLGP4 from *Caulerpa lentillifera*. Int J Biol Macromol 146:931–938. https://doi.org/10.1016/j.ijbiomac.2019.09.216

Sun H, Wang Y, He Y, Liu B, Mou H, Chen F, Yang S (2023) Microalgae-derived pigments for the food industry. Mar Drugs 21(2). https://doi.org/10.3390/md21020082

Tabarsa M, Rezaei M, Ramezanpour Z, Waaland JR (2012) Chemical compositions of the marine algae *Gracilaria salicornia* (Rhodophyta) and *Ulva lactuca* (Chlorophyta) as a potential food source. J Sci Food Agric 92(12):2500–2506. https://doi.org/10.1002/jsfa.5659

Taboada C, Millán R, Míguez I (2010) Composition, nutritional aspects and effect on serum parameters of marine algae *Ulva rigida*. J Sci Food Agric 90(3):445–449. https://doi.org/10.1002/jsfa.3836

Taboada C, Millan R, Miguez I (2013) Evaluation of marine algae *Undaria pinnatifida* and *Porphyra purpurea* as a food supplement: composition, nutritional value and effect of intake on intestinal, hepatic and renal enzyme activities in rats. J Sci Food Agric 93(8):1863–1868. https://doi.org/10.1002/jsfa.5981

Tagliapietra BL, Clerici MTPS (2023) Brown algae and their multiple applications as functional ingredient in food production. Food Res Int 167:112655. https://doi.org/10.1016/j.foodres.2023.112655

Thiyagarasaiyar K, Goh B-H, Jeon Y-J, Yow Y-Y (2020) Algae metabolites in cosmeceutical: an overview of current applications and challenges. Mar Drugs 18(6). https://doi.org/10.3390/md18060323

Tibbetts SM, Milley JE, Lall SP (2016) Nutritional quality of some wild and cultivated seaweeds: nutrient composition, total phenolic content and in vitro digestibility. J Appl Phycol 28(6):3575–3585. https://doi.org/10.1007/s10811-016-0863-y

Troell M, Joyce A, Chopin T, Neori A, Buschmann AH, Fang JG (2009) Ecological engineering in aquaculture—potential for integrated multitrophic aquaculture (IMTA) in marine offshore systems. Aquac 297(1–4):1–9

Ulagesan S, Eom T, Nam TJ, Choi YH (2022) Antioxidant and chemoprotective peptides from simulated gastrointestinal digested (SGID) protein hydrolysate of *Pyropia yezoensis* against acetaminophen-induced HepG2 cells. Bioprocess Biosyst Eng 45(10):1645–1660. https://doi.org/10.1007/s00449-022-02770-4

Vieira EF, Soares C, Machado S, Correia M, Ramalhosa MJ, Oliva-teles MT, Paula Carvalho A, Domingues VF, Antunes F, Oliveira TAC, Morais S, Delerue-Matos C (2018) Seaweeds from the Portuguese coast as a source of proteinaceous material: total and free amino acid composition profile. Food Chem 269:264–275. https://doi.org/10.1016/j.foodchem.2018.06.145

Wan AH, Davies SJ, Soler-Vila A, Fitzgerald R, Johnson MP (2019) Macroalgae as a sustainable aquafeed ingredient. Rev Aquac 11(3):458–492. https://doi.org/10.1111/raq.12241

Wan-Loy C, Siew-Moi P (2016) Marine algae as a potential source for anti-obesity agents. Mar Drugs 14(12):222. https://doi.org/10.3390/MD14120222

Wang HMD, Li XC, Lee DJ, Chang JS (2017) Potential biomedical applications of marine algae. Biores Technol 244:1407–1415

Wang L, Park YJ, Jeon YJ, Ryu BM (2018) Bioactivities of the edible brown seaweed, *Undaria pinnatifida*: a review. Aquaculture 495:873–880. https://doi.org/10.1016/j.aquaculture.2018.06.079

Yaamini M, Venil CK, Dufossé L (2023) Valuable bio-products from algae and their commercial applications. In: Algal metabolites (pp 25–45). Apple Academic Press. https://doi.org/10.1201/9781003346081-3

Yu J, Hu Y, Xue M, Dun Y, Li S, Peng N, Liang Y, Zhao S (2016) Purification and identification of antioxidant peptides from enzymatic hydrolysate of *Spirulina platensis*. J Microbiol Biotechnol 26(7):1216–1223. https://doi.org/10.4014/jmb.1601.01033

Zammuto V, Rizzo MG, Spanò A, Spagnuolo D, Di Martino A, Morabito M, Manghisi A, Genovese G, Guglielmino S, Calabrese G, Capparucci F, Gervasi C, Nicolò MS, Gugliandolo C (2022) Effects of crude polysaccharides from marine macroalgae on the adhesion and biofilm formation of *Pseudomonas aeruginosa* and *Staphylococcus aureus*. Algal Res 63:102646. https://doi.org/10.1016/j.algal.2022.102646

Zhang X, Cao D, Sun X, Sun S, Xu N (2019) Preparation and identification of antioxidant peptides from protein hydrolysate of marine alga *Gracilariopsis lemaneiformis*. J Appl Phycol 31(4):2585–2596. https://doi.org/10.1007/s10811-019-1746-9

Zheng Y, Zhang Y, San S (2020) Efficacy of a novel ACE-inhibitory peptide from *Sargassum maclurei* in hypertension and reduction of intracellular endothelin-1. Nutrients 12(3):1–14. https://doi.org/10.3390/nu12030653

Zheng H, Pei Y, Zhou C, Hong P, Qian Z-J (2023) Amelioration of atherosclerosis in ox-LDL induced HUVEC by sulfated polysaccharides from *Gelidium crinale* with antihypertensive activity. Int J Biol Macromol 228:671–680. https://doi.org/10.1016/j.ijbiomac.2022.12.245

Zhu Z, Chen Y, Jia N, Zhang W, Hou H, Xue C, Wang Y (2022) Identification of three novel antioxidative peptides from *Auxenochlorella pyrenoidosa* protein hydrolysates based on a peptidomics strategy. Food Chem 375:131849. https://doi.org/10.1016/j.foodchem.2021.131849

Chapter 24
Biochemical, Techno-Functional and Sensory Properties of Food Prepared with Algae

Hang-kin Kong

Abstract Asians have a long history of consuming seaweeds (macroalgae), such as kombu, wakame, hiziki and nori. They are served with soups, noodles, salads, and sushi to enhance the umami in the cuisines. In terms of nutrition, the addition of seaweeds increases the dietary fiber and mineral contents of the dishes. Moreover, seaweeds are rich in different hydrocolloids, like agar, alginate and carrageenan which have been authorized as thickeners, stabilizers and emulsifiers (E400–E407a) in the European Union, and widely used in the food industries. Apart from macroalgae, microalgae have recently become popular nutritional supplements, food ingredients and additives because of their richness in proteins, essential amino acids (EAAs), polyunsaturated fatty acids (PUFAs), antioxidants, pigments, and antimicrobial compounds, etc. Since both macroalgae and microalgae in general produce broad ranges of metabolites, numerous research have been made to incorporate algal biomass or refined algal metabolites into different food products to improve their nutritional, biochemical, techno-functional and sensory properties. In this review, the applications of different algal biomasses and metabolites in various food products will be summarized. Moreover, the impact of algal ingredients on the sensory profiles of the food products will also be reviewed. The main aim is to shed light on the future development of the applications of algal biomasses and metabolites in the food industries through an extensive review of pertinent literature.

Keywords Macroalgae · Microalgae · Food · Nutritional supplements · Food additives · Sensory acceptability

---

H. Kong (✉)
Department of Food Science and Nutrition, Research Institute of Future Food, The Hong Kong Polytechnic University, Hung Hom, Kowloon, Hong Kong SAR, China
e-mail: hnag-kin.kong@polyu.edu.hk

T. A. Kurniawan and A. Anouzla (eds.), *Algae as a Natural Solution for Challenges in Water-Food-Energy Nexus*, Environmental Science and Engineering,
https://doi.org/10.1007/978-981-97-2371-3_24

24.1 Introduction

Kombu (edible species in *Laminariaceae*), wakame (*Undaria pinnatifida*), hiziki (*Sargassum fusiforme*) and nori (edible species in *Pyropia* and *Porphyra*) are the seaweeds (macroalgae) that have, for centuries, been commonly consumed in China, Japan and Korea. They are not only a medicinal substances that have been used in food, but also a traditional Chinese medicine to treat illnesses such as scrofula, goiter, arteriosclerosis, high blood pressure and chronic bronchitis (Liu et al. 2012; Tseng 1944). In early nineteenth century Europe, ashes of seaweeds were prescribed to patients with goiter due to their high iodine content (Kelly 1961). Macroalgae are good sources of different hydrocolloids, including agar, alginate, and carrageenan, which have been used as stabilizers, emulsifiers, thickening and gelling agents in the food industries. These hydrocolloids have been registered in the European Union (with the Europe numbers E400–E407a) and considered safe for human consumption. Unlike macroalgae, which have long been commonly trusted, the consumption of microalgae came under the spotlight two decades ago. While microalgae are not good sources of hydrocolloids, they are generally rich in proteins, polyunsaturated fatty acids (PUFAs), essential amino acids, vitamins, antioxidants, phenolic compounds, and pigments, etc. These properties make microalgae potentially suitable as alterative protein sources, health supplements and food additives (Ampofo and Abbey 2022). For this reason, many researches have been conducted to extract proteins, lipids and metabolites from different microalgae, namely species in the family of *Chlorella*, *Dunaliella*, *Haematococcus*, *Isochrysis*, *Schizochytrium*, and *Spirulina* to develop various food items. Meanwhile, the safety of microalgae consumption has been scrutinized extensively. Recently, the European Food Safety Authority has acknowledged *Schizochytrium* sp. as a novel food, since there is no concern regarding its toxicity (EFSA Panel on Nutrition et al. 2020). The United States Food and Drug Administration also recognized that a few microalgal species and their extracted metabolites as safe for human consumption, including *Haematococcus pluvialis*, *Schizochytrium* sp., and *Ulkenia* sp. (Hosseinkhani et al. 2022). As there is more and more evidence supporting the safety of microalgae, consumption, more new food products with microalgal based ingredients are expected to be developed, registered, and go to market in the near future. As a consequence, the market value of algae-based food and nutraceutical products is expected to grow continuously (Credence Research 2022).

Taste, aroma, appearance, texture, nutritional value, and shelf life are key considerations in food product development. Traditionally, dried macroalgae are used in Asian countries to prepare broth, soup, salad, sushi, and crispy seaweed snacks. The addition of macroalgae aims primarily to give an umami taste and enrich the aroma and flavor of the cuisines and food products (Mouritsen et al. 2018). Depending on the cooking recipe, seaweeds could also provide a chewy or crunchy texture to the food items (Mouritsen et al. 2018). With regard to nutrition, the additions of macroalgae significantly increase the amounts of dietary fiber, minerals, and vitamins in the dishes and food, items as macroalgae are in generally rich in these nutrients

(Rajapakse and Kim 2011). It is clear that the addition of macroalgae can significantly alter different sensory properties and nutritional values of cuisines simultaneously. Furthermore, same as in traditional cuisines, using either algal biomasses or extracts to manufacture food products might alter not just their sensory and nutritional properties, but also their techno-functional and biochemical properties. For example, the addition of *Spirulina* biomass to pasta (20 g of biomass per 100 g of flour) only improved the nutritional values of the pasta, but also modified its texture (Rodríguez De Marco et al. 2014). Nevertheless, while adding algal ingredients might significantly improve certain properties of the food products, it might concurrently affect the sensory profiles, which might weaken customer acceptance due to the unexpected and undesired taste, aroma, color and texture of the food products (Matos et al. 2022). As such, reduced the sensory acceptability is known to be one of the major obstacles that impede the commercialization of food products. Through summarizing pertinent literature, this review attempts to conduct an in depth discussion on recent developments in the manufacturing of food products using algal biomasses or extracts. Details concerning the researches on the development of food products using algal biomasses or extracts are summarized in Table 24.1.

24.2 Altering the Properties of Food Items with Algal Biomass

Algal biomass in dry powder form is commonly added to food products to archive certain proposes, since this is the simplest and cheapest way to manufacture a food product with algal ingredients. Depending on the cooking recipes and manufacturing processes, dry algal powder is often used to partially replace certain other ingredients in the original recipes, as seen for example in its partial replacement of flour in certain bakery and pastry products (Quitral et al. 2022). Supplementation is another common method to incorporate algal biomasses into different food products, like supplementing *Spirulina platensis* and *Chlorella vulgaris* into probiotic fermented milks (Beheshtipour et al. 2013). No matter is replacement or supplementation, it is expected that the nutritional, sensory, biochemical and techno-functional properties of the food products would be altered or should even be improved. Turning bakery goods, pastry goods, meat products, dairy products and extruded snacks green or yellow were one of the most common and obvious effects after adding algal biomass to these food items (García-Segovia et al. 2017; Ozyurt et al. 2015). Other effects of incorporating algal biomass on the properties of various food items were summarized and discussed in the following paragraphs.

24.2.1 Bakery and Pastry Goods

Algal biomass was commonly used to partially replace flour in bakery and pastry goods. It is expected that the textures and techno-functional properties of the dough

Table 24.1 Summary of food products with algal ingredients

Food products	Algal species	Algal ingredients and amount used	Key modified properties	Sensory acceptability	References
Meat products					
Chicken fillet	*Halimeda opuntia*, *Actinotrichia fragilis*, and *Turbinaria turbinata*	Lyophilized ethanolic extract: 4 and 6% (w/w)	Extract from *H. opuntia* reduced viability of *Staphylococcus aureus* in fillets after storage	Fillets with 4 and 6 % of *H. opuntia* extracts had significant higher overall acceptability than controls	Hamad et al. (2023)
Low-fat pork patties	*Schizochytrium* sp.	Emulsion gel with 20% algal oil: 10 and 20 g of gel/100 g of patties	Reduced fat content Increased moisture Increased PUFA content Increased lipid oxidation after 15-day storage	Reduced acceptability at high % due to unpleasant color and taste	Cirstea Lazar et al. (2023)
Calf fillet	*Spirulina platensis* and *Chlorella vulgaris*	Biomass mixed with chitosan and natamycin: 1, 1.5 and 2% (v/v)	Reduced viability microbe after stored for 28 days Increased antioxidant capacity	Fillets with 1% of *S. platensis* and 2% of *C. vulgaris* had significant higher overall acceptability than controls	Shafiei and Mostaghim (2022)
Low-fat pork burger	*Undaria* sp.	Biomass: 3 g/100 g	Increased water and oil holding capacity Increased hardness and chewiness Increased greenness Increased antioxidant capacity Increased SFA, MUFA and PUFA	N/A	Nagai et al. (2022)

(continued)

Table 24.1 (continued)

Food products	Algal species	Algal ingredients and amount used	Key modified properties	Sensory acceptability	References
Beef sausage	*Bryopsis plumosa*	Extracted polysaccharide: 0.05, 0.125, and 0.25% (w/w)	Reduced lipid and myoglobin oxidation Reduced microbial load during storage	N/A	Ghariani et al. (2022)
Minced beef meat	*Porphyridium cruentum*	Sulphated exopolysaccharide: 0.5, 1 and 2% (v/w)	Reduced protein and lipid oxidation Inhibited microflora proliferation	N/A	Ben Hlima et al. (2021)
Beef sausage	*Falkenbergia rufolanosa*	Extracted polysaccharide: 0.05, 0.125, and 0.25% (w/w)	Reduced lipid and myoglobin oxidation Reduced microbial load during storage Increased hardness and chewiness when increased %	Had similar acceptability as control	Feki et al. (2021)
Frankfurt sausages	*Ulva* spp., *Gracilaria* spp., *Fucus vesiculosus*	Powder mixture	Increased mineral contents Reduced Na/K ratio Increased antibacterial activities	N/A	Marcal et al. (2021)

(continued)

Table 24.1 (continued)

Food products	Algal species	Algal ingredients and amount used	Key modified properties	Sensory acceptability	References
Frankfurters	*Porphyra umbilicalis*, *Palmaria palmata*, *Himanthalia elongata* and *Undaria pinnatifida*	Air-dried biomass: 1% (w/w)	Reduced NaCl Increased protein content and moisture Reduced hardness and chewiness	Frankfurter with *H. elongata* had the highest acceptability among them, but lower than control Frankfurter with *P. palmata* had the lowest acceptability due to seaweed flavor	Vilar et al. (2020)
Tunisian beef sausage	*Chaetomorpha linum*	Polysaccharide: 0.05, 0.125 and 0.25% (w/w)	Reduced myoglobulin and lipid oxidation Inhibited microbial growth during storage	N/A	Hamzaoui et al. (2020)
Fish-surimi restructured products	*Ulva intestinalis*	Biomass: 27.7 g/kg Sulphated polysaccharide: 5 g/kg	Increased moisture content Reduced cooking loss Biomass increased greenness of surimi Reduced lipid oxidation Sulphated polysaccharide reduced hardness, and chewiness	Overall acceptability of surimi with sulphated polysaccharide were higher than controls after storage for 6 months	Jannat-Alipour et al. (2019)
Fresh pork sausages	*Chlorella* sp. and *Spirulina* sp.	Protein: 1% (w/w)	Increased greenness Reduced hardness and chewiness Increased essential amino acid content	N/A	Marti-Quijal et al. (2019a)

(continued)

Table 24.1 (continued)

Food products	Algal species	Algal ingredients and amount used	Key modified properties	Sensory acceptability	References
Pork liver pâté, turkey patties, fresh pork sausages, cooked turkey breast and marinated turkey breast	*Chlorella* sp. and *Spirulina* sp.	Protein Pork liver pâté: 5% Turkey patty: 1% Fresh pork sausage: 1% Cooked turkey breast: 2.67% brine Marinated turkey breast: 5% brine	Increased total amino acid content	Reduced sensory acceptability	Zamuz et al. (2019)
Canned fish burgers	*Chlorella minutissima*, *Picochlorum* sp. and *Isochrysis galbana*	Biomass: 0.5, 1, and 1.5% (w/v)	Increased swelling ability, water and oil holding capacity Increased total dietary fiber, greenness and antioxidant capacity Burger with *Picochlorum* increased hardness and chewiness	Reduced acceptability at 1.5% due to undesirable off-flavors	Ben Atitallah et al. (2019a)
Turkey burgers	*Chlorella* sp., *Spirulina* sp.	Protein: 1%	Increased greenness Reduced adhesiveness Increased essential amino acid content	N/A	Marti-Quijal et al. (2019b)

(continued)

Table 24.1 (continued)

Food products	Algal species	Algal ingredients and amount used	Key modified properties	Sensory acceptability	References
Rainbow trout fillet	*Spirulina platensis*, *Chlorella vulgaris*	Alcoholic extract: 0.05 and 0.1%	Increase antioxidant capacity Reduced lipid oxidation	Had similar acceptability as control: fillet with extract had higher acceptabilities than control after storage	Takyar et al. (2019)
Fish burger	*Chlorella minutissima*, *Isochrysis galbana*, and *Picochlorum* sp.	Biomass: 0.5, 1 and 1.5% (w/v)	Increased swelling ability Increased water and oil retention capacities Increased dietary fiber content, pigment and antioxidant capacity at 1% Increased hardness and chewiness	Reduced sensory acceptability at 1.5% due to off-flavors	Ben Atitallah et al. (2019b)
Fish burger	*Cystoseira compressa* and *Jania adhaerens*	Biomass: 0.5, 1 and 1.5% (w/v)	Increased dietary fiber content at 1% Increased water and oil holding capacity at 1% Increased swelling at 1% Increased antioxidant capacity	Reduced acceptability at 1.5% due to poor taste	Hentati et al. (2019)

(continued)

Table 24.1 (continued)

Food products	Algal species	Algal ingredients and amount used	Key modified properties	Sensory acceptability	References
Fish burger	*Spirulina platensis*	Biomass: 0.5, 1, and 1.5% (w/v)	Increased swelling ability Increased water and oil holding capacity Increased dietary fiber content Increased hardness and chewiness Increased antioxidant capacity Increased greenness	Reduced acceptability when increasing % due to unpleasant color and taste	Barkallah et al. (2019)
Raw ground pork meat	*Haematococcus pluvialis*	Extract with 5% astaxanthin: 0.15, 0.3, 0.45 g/kg	Increased redness Reduced lipid oxidation	Increased acceptability due to pleasant color at high %	Pogorzelska et al. (2018)
Beef patties	*Chlorella* sp. and *Spirulina* sp.	Protein: 1% (w/w)	Increased greenness Increase essential amino acid content	N/A	Zugcic et al. (2018)
Turkey breasts	*Chlorella* sp., *Spirulina* sp.	Isolated protein in brine: 2.67%	Increased greenness Reduced hardness and cohesiveness Had similar essential amino acid content as control	N/A	Marti-Quijal et al. (2018)
Chorizo sausage	*Spirulina* and *Chlorella*	Protein: 3% (w/w)	Increased protein content Increased greenness Increased hardness	N/A	Thirumdas et al. (2018)

(continued)

Table 24.1 (continued)

Food products	Algal species	Algal ingredients and amount used	Key modified properties	Sensory acceptability	References
Pork liver pate	*Ascophyllum nodosum*, *Fucus vesiculosus*, *Bifurcaria bifurcata*	Extract: 500 mg/kg	No significant difference observed	N/A	Agregan et al. (2018)
Chicken rotti	*Chlorella* sp. and *Spirulina* sp.	Protein: 1% (w/w)	Increased total amino acid and lipid content Reduced hardness and chewiness	Reduced sensory acceptability	Parniakov et al. (2018)
Chinese-style pork sausage	*Spirulina platensis*	Polysaccharide: 0.1, 0.25, and 0.5% (w/w)	Decreased lipid oxidation Increase antioxidant capacity Maintained redness of sausages	Increased sensory acceptability compared to control	Luo et al. (2017)
Low-fat beef patties	DHASCO® oil, obtained from *Crypthecodinium cohnii*	Emulsion gel with 1% algal oil: 9 g of gel/ 100 g of beef	Reduced fat content Increased moisture Increased DHA content Reduced lipid oxidation	Reduced acceptability	Alejandre et al. (2017)
Turkey meat sausage	*Cystoseira barbata*	Aqueous extract: 0.01–0.4%	Reduced sodium nitrite content Reduced lipid oxidation Increased antioxidant capacity	N/A	Sellimi et al. (2017a)
Turkey meat sausages	*Cystoseira barbata*	Fucoxanthin: 0.01, 0.02 and 0.04%	Increased redness and yellowness Reduced lipid oxidation after storage	N/A	Sellimi et al. (2017b)

(continued)

Table 24.1 (continued)

Food products	Algal species	Algal ingredients and amount used	Key modified properties	Sensory acceptability	References
Ground pork meat product	Not specified	Marine algae calcium powder: 0.5% (w/w)	Reduced hardness and chewiness Increased cooking loss	N/A	Bae et al. (2017)
White puddings	Not specified	Seaweed wakame: 3.3%	Reduced NaCl content Increased greenness Increased hardness and chewiness	Increased spiciness and juiciness Reduced acceptability due to unpleasant color and flavor	Fellendorf et al. (2016b)
Black puddings	Not specified	Seaweed wakame: 3.3%	Reduced NaCl content	Increased spiciness and saltiness Had the similar acceptability as control	Fellendorf et al. (2016a)
Lower-salt cooked ham	*Palmaria palmata*	Aqueous extract: 1 and 1.2%	Reduced NaCl content in cooked ham by 30%	Ham with extract had higher taste intensity, saltiness and acceptability than control	Barbieri et al. (2016)
Restructured pork	*Porphyra umbilicalis*, *Undaria pinnatifida*, and *Himanthalia elongata*	Biomass: 5 g/100 g meat	Pork with *H. elongata* inhibited α-glucosidase activity Pork with *P. umbilicalis*/*U. pinnatifida* reduced glucose diffusion	N/A	Garcimartin et al. (2015)

(continued)

Table 24.1 (continued)

Food products	Algal species	Algal ingredients and amount used	Key modified properties	Sensory acceptability	References
Reduced-fat Bologna sausage	DHASCO® oil, obtained from *Crypthecodinium cohnii*	Emulsion with mixture linseed and algal oils in 1:1 ratio: 5% Lyophilized aqueous extract from *Melissa officinalis*: 200 ppm	Reduced fat content Increased ALA and DHA contents Reduced lipid oxidation	Had the same acceptability as control	Berasategi et al. (2014)
Minced pork patties	*Laminaria digitata*	Extract with laminarin (9.3%) and fucoidan (7.8%): 0.01, 0.1 and 0.5% (w/w)	Reduced lipid oxidation at 0.5% in cooked patties Increased lipid oxidation in fresh patties Increased greenness	Reduced acceptability at high % Patties with 0.01% had similar acceptability as control	Moroney et al. (2013)
Beef patties	*Himanthalia elongata*	Blanched biomass: 10, 20, 30, and 40%	Reduced cooking loss Reduced lipid oxidation Inhibited microbial growth Increased dietary fiber, total phenolic content and antioxidant capacity	Patties with 40% biomass had the highest overall acceptability due to better texture and taste	Cox and Abu-Ghannam (2013)
Reduced fat pork patties	*Laminaria japonica*	Biomass: 1, 3, and 5% (w/w)	Increased moisture and dietary fiber Reduced protein, and fat content Reduced hardness Increased greenness	Patties with 1 and 3% had higher overall acceptability than control	Choi et al. (2012)

(continued)

Table 24.1 (continued)

Food products	Algal species	Algal ingredients and amount used	Key modified properties	Sensory acceptability	References
Low-salt restructured poultry steaks	*Himanthalia elongata*	Biomass: 3% (w/w)	Increased dietary fibre Reduced cooking loss Increased yellowness Increased viability of lactic acid bacteria	Reduced sensory acceptability due to unexpected taste	Cofrades et al. (2011)
Fermented sausage	DHASCO® oil, obtained from *Crypthecodinium cohnii*	Mixture of linseed oil and algae oil (3:2): 62.5 g/kg	Increased PUFA content	Had similar acceptability as the control	de Ciriano et al. (2010a)
Pork meat batters	*Himanthalia elongata*	Biomass: 3.4%	Increased hardness and chewiness Enhanced the water and oil retention capacity	N/A	Fernandez-Martin et al. (2009)
Low-fat frankfurters	*Himanthalia elongata*	Biomass: 5%	Reduced processing loss Increased hardness and chewiness Increased greenness	Reduced acceptability due to off-flavors	Lopez-Lopez et al. (2009)
Low-fat frankfurters	Not specified	Carrageenan: 0.41, 0.5, 0.57 and 0.7% Replaced fat	Increased hardness, gumminess and chewiness Reduced cooking loss	Appearance, texture, and taste of frankfurter with carrageenan had higher acceptability than control	Cierach et al. (2009)

(continued)

Table 24.1 (continued)

Food products	Algal species	Algal ingredients and amount used	Key modified properties	Sensory acceptability	References
Fish cutlet	*Kappaphycus alvarezii*	Biomass: 5, 7.5, 10, 12.5, and 15% (w/w)	Increased moisture and dietary fiber Decreased protein and fat content Increased hardness and firmness	Cutlet with up to 10% biomass had similar acceptability than control, except for texture Reduced acceptability above 10%	Senthil et al. (2005)
Low-fat ground pork patties	Not specified	Carrageenan: 0.25, 0.5 and 0.75%	Increased moisture content Reduced cooking loss Reduced calories	Patties with 0.5% carrageenan had similar acceptability as control	Kumar and Sharma (2004)
Bakery products					
Cookie	*Ulva intestinalis*	Biomass: 1, 2.5, 5, and 10% (w/w) Fragmented: 1, 2.5, 5 and 10% (w/w)	Increased protein, and lipid contents at 2.5% (powdered) and 5% (fragmented)	Reduced acceptability when increasing % due to fishy odor	Mohibbullah et al. (2023)
Rice-based dough	*Kappaphycus alvarezii*	Dried biomass: 2, 5 and 10% (w/w)	Increased hardness, cohesiveness, and springiness Decreased the adhesiveness of the dough at 5 and 10% Increased dark color with increases in %	N/A	Lopes et al. (2023)

(continued)

Table 24.1 (continued)

Food products	Algal species	Algal ingredients and amount used	Key modified properties	Sensory acceptability	References
Bread, cake, and cookies	*Ulva lactuca* and *Gracilaria corticata*	Biomass 2 g in 180 g bread 2 g in 345 g cake 2 g in 242 g cookie	Increased protein and mineral contents	Breads with *G. corticata*, cakes with *U. lactuca*, and cookies with *U. lactuca* had sensory attributes as controls	Turuk and Banerjee (2023)
Cracker	*Kappaphycus alvarezii*	Biomass: 1, 2.5, 5, 7.5, and 10% (w/w)	Increased fiber contents with increase in % Reduced fat content Increased greenness Decreased hardness with increases in %	N/A	Aganduk et al. (2023)
Pasta	*Fucus vesiculosus*	Biomass: 1, 5.5, and 10% (w/w)	Increased in total fiber at 10% Disrupted the gluten–starch matrix at 10% Darker in color when increasing %	All were negatively rated Decreased acceptability when increasing %	Ribeiro et al. (2022)
Pasta	*Spirulina* sp.	Biomass: 2, 5, 7.5, 10, 12.5 and 15% (w/w)	Increased protein, fat, total phenols, flavonoids, iron and calcium contents when increasing % Increased antioxidant capacity Reduced firmness	Pasta with 12.5 and 15% had similar acceptability as control, but others did not	Koli et al. (2022)

(continued)

Table 24.1 (continued)

Food products	Algal species	Algal ingredients and amount used	Key modified properties	Sensory acceptability	References
Wheat bread	*Tetraselmis chuii*, *Chlorella vulgaris*, and *Nannochloropsis gaditana*	Biomass/ ethanol-treated biomass: 4% (w/w)	Increased protein content Bread with *T. chuii* had higher mineral content especially the Ca Breads with biomass had higher total phenolic content and antioxidant capacities, but not ethanol treated biomass Increased crumb firmness Reduced bread volumes Breads with biomass had higher green tonality	Bread with ethanol-treated *C. vulgaris* had similar sensory acceptability as control	Qazi et al. (2022)
Wheat bread	*Tetraselmis chuii*	Biomass/ ethanol-treated biomass: 4, 8, 12, and 16% (w/w)	Increased crumb firmness when increasing % biomass Breads with 4% ethanol-treated biomass had similar color and firmness to control	N/A	Qazi et al. (2021b)

(continued)

Table 24.1 (continued)

Food products	Algal species	Algal ingredients and amount used	Key modified properties	Sensory acceptability	References
Cracker	*Undaria pinnarifida*, and *Himanthalia elongata*	Biomass: 5, 10, 15, and 20% w/w	Incorporation of *U. pinnarifida* slowed down starch digestion Increased hardness at 5 and 10% Crackers with 15 and 20% had similar fracturability as controls	All had reduced sensory acceptability Increased fishiness and salty attributes Reduced the sweetness attribute	Wilkin et al. (2021)
Wheat bread	*Spirulina platensis* and Kelp	Biomass: 2, and 4% (w/w)	Increased thiamine (vitamin B1) and riboflavin (vitamin B2) contents at 4%	N/A	Chochkov et al. (2021)
Bread	*Spirulina platensis*	Biomass: 2, 4, 6, and 8%	Increased total phenolic contents and antioxidant capacity	Reduced acceptability when increasing % Breads with 2% had similar acceptability as control after 7-day storage	Saharan and Jood (2021)
Wheat bread	*Microchloropsis gaditana*, *Tetraselmis chuii*, and *Chlorella vulgaris*	Biomass/ethanol-treated biomass: 12% (w/w)	Increased essential amino acid content Improved protein content and quality Decreased in vitro protein digestibility	N/A	Qazi et al. (2021a)

(continued)

Table 24.1 (continued)

Food products	Algal species	Algal ingredients and amount used	Key modified properties	Sensory acceptability	References
Gluten-free bread	*Nannochloropsis gaditana* L2 and *Chlamydomonas* sp. EL5	Dried biomass: 1, and 3% (w/w)	Increased protein, lipid, Fe, and Ca contents Increased green tonalities	Had higher acceptance than control	Khemiri et al. (2020)
Cookie	*Sargassum fulvellum*, *Enteromorpha linza*, *Codium fragile*, and *Hizikia fusiforme*	Biomass: 5%	Increased solvent absorption capacity Increased moisture content Increased hardness	Reduced sensory acceptability compared to control	Oh et al. (2020)
Wheat bread	*Spirulina platensis* and Kelp	Biomass: 2, and 4% (w/w)	4% had high content of phenolic compounds and antioxidant capacity	N/A	Zlateva et al. (2020)
Gluten-free bread	*Tetraselmis chuii*	Biomass: 1, 2, and 4% (w/w)	Increased total phenolic content and antioxidant capacity Increased firmness Increased greenness	Reduced acceptability when increasing % due to fishy flavors	Nunes et al. (2020)
Gluten-free bread	*Chlorella sorokiniana*	Biomass powder, 2.5, and 5% (w/w)	Increased protein, fatty acids and carotenoid contents Darker color when increasing %	All had bitter after taste and mate herb aroma Reduced sensory acceptability compared to control	Diprat et al. (2020)
Wheat bread	*Spirulina platensis*	Biomass: 2, and 4% (w/w)	Stored softness and elasticity for a longer time	N/A	Zlateva and Chochkov (2019)

(continued)

Table 24.1 (continued)

Food products	Algal species	Algal ingredients and amount used	Key modified properties	Sensory acceptability	References
Wheat cracker	*Arthrospira platensis* F&M-C256, *Chlorella vulgaris* Allma, *Tetraselmis suecica* F&M-M33, and *Phaeodactylum tricornutum* F&M-M40	Biomass: 2, and 6%	Increased protein and total phenolic contents Increased antioxidant capacity Increased green tonality	Reduced acceptance at all % compared to control Had fishy off-flavors	Batista et al. (2019)
Cookie	*Bifurcaria bifurcate*	Biomass: 3, 6, and 9% (w/w)	Increased total phenolic content and antioxidant capacity Increase hardness when increased %	N/A	Arufe et al. (2019a)
Cookie	*Chlorella* sp. (Abca-17)	Microalgal meal: 3, 6, 9, and 12% (w/w)	Increased moisture content and hardness Increased greenness	3% had similar acceptance to control, but the rest had reduced acceptance	Sahni et al. (2019)
Cracker and bread	*Nannochloropsis* sp. and *Tetraselmis* sp.	Biomass Bread: 1, 2 and 3% (w/w) Cracker: 1.25, 2.25, 3.75% (w/w)	Increased phenolic content and antioxidant capacity Increased green tonality	Reduced sensory acceptance of bread (control > 1% of *Nannochloropsis* sp. > 2% of *Tetraselmis* sp.) Cracker with 1% of *Nannochloropsis* sp. had a higher overall acceptance than control	Lafarga et al. (2019b)

(continued)

Table 24.1 (continued)

Food products	Algal species	Algal ingredients and amount used	Key modified properties	Sensory acceptability	References
Breadsticks	*Chlorella vulgaris* and *Arthrospira platensis*	Biomass: 1.5% (w/w)	Decreased hardness, and crispiness Increased mineral content, especially Fe and Se Increased green tone	N/A	Uribe-Wandurraga et al. (2019)
Crostini	*Arthrospira platensis*	Biomass: 2, 6 and 10% (w/w)	Increased free amino acid, protein, and phenolic content Increased antioxidant capacity Increased greenness when increasing %	Decreased acceptability when increasing % 2% had a better appreciation among all %	Niccolai et al. (2019)
Cake	*Eucheuma cottonii*	Biomass: 5, 10, 15, and 20% (w/w)	Increased total dietary fiber Increased chewiness Less fragile	Increase in % resulted in decrease in sensory acceptance 5% had the best score but lower than control	Huang and Yang (2019)
Cookies	*Bifurcaria bifurcata, Fucus vesiculosus* and *Ascophyllum nodosum*	Biomass: 3, 6, and 9% (w/w)	Increased total polyphenolic contents and antioxidant capacities of doughs before baking	N/A	Arufe et al. (2019b)
Wheat bread	*Fucus vesiculosus*	Biomass: 2, 4, 6, and 8% (w/w)	Increased greenness Increased bread density	N/A	Arufe et al. (2018)

(continued)

Table 24.1 (continued)

Food products	Algal species	Algal ingredients and amount used	Key modified properties	Sensory acceptability	References
Biscuits	*Caulerpa racemose*	Biomass: 1, 5, and 10% (w/w)	Increased water and oil absorption capacity Increased protein, fiber, and phenolic content Increased antioxidant capacity	Sensory acceptability declined when increasing % 1% had a similar but lower score than control	Kumar et al. (2018)
Wheat bread	*Chlorella vulgaris*	Biomass: 1, 2, 3, 4, and 5% (w/w)	Positive effect on the rheology of dough and its viscoelastic feature (1–3%) Poor elasticity at high %	N/A	Graça et al. (2018)
Wheat bread	*Isochrysis galbana*, *Tetraselmis suecica*, *Scenedesmus almeriensis* and *Nannochloropsis gaditana*	Biomass: 1.5% (w/w)	Increased green tonalities	N/A	García-Segovia et al. (2017)
Gluten-free bread	*Ascophyllum nodosum*	Biomass: 2, 4, 6, 8, and 10% (w/w)	Increased the elasticity of crumb Reduced chewiness and firmness Increased total phenol content and antiradical potential Darker color when increasing %	Reduced sensory acceptance when increasing %	Różyło et al. (2017)

(continued)

Table 24.1 (continued)

Food products	Algal species	Algal ingredients and amount used	Key modified properties	Sensory acceptability	References
Cookie	*Haematococcus pluvialis*	Biomass: 5, 10, and 15%	Increased total phenolic content Increased antioxidant capacity Reduced glycaemic response	Increased green color in all samples with all %	Hossain et al. (2017)
Cookie	*Arthrospira platensis* F&M-C256, *Chlorella vulgaris* Allma, *Tetraselmis suecica* F&M-M33 and *Phaeodactylum tricornutum* F&M-M40	Biomass: 2, and 6% (w/w)	Increased protein and total phenolic contents Increased antioxidant capacity Increased dark green tonality	2% *A. platensis* had a pleasant taste but others had strong fishy flavors	Batista et al. (2017)
Instant fried noodles	*Eucheuma cottonii*, *Spirulina platensis*, and *Chlorella vulgaris*	Mixtures of 3 extracts: Ec 5%, Sp 2.5–15%, Cv 2.5–15% (w/w)	Increased protein, fat and fiber contents Decreased water absorption during cooking Increased cooking loss	Reduced sensory acceptability Noodles with 5% of *E. cottonii* + 5% of *S. platensis* had the highest score among other combination, but still lower than control	Kumoro et al. (2016)
Bread	*Cladophora* spp. and *Ulva* spp.	Biomass: 2.5, 5 and 7.5% (w/w)	Increased protein and fiber content Reduced lipid content	No significant difference among different % No control included	Menezes et al. (2015)

(continued)

Table 24.1 (continued)

Food products	Algal species	Algal ingredients and amount used	Key modified properties	Sensory acceptability	References
Pasta	*Spirulina platensis*	Biomass: 5, 10 and 15% (w/w)	Increased greenness when increasing %	Strange flavors was detected at 15% 10% formulation was preferred for its flavors and appearance	Ozyurt et al. (2015)
Bread	*Kappaphycus alvarezii*	Biomass: 2, 4, 6, and 8% (w/w)	Increased water absorption, firmness and yellowness when increased %	N/A	Mamat et al. (2014)
Bread	*Palmaria palmata*	Protein hydrolysate: 4% (w/w)	Increased hardness Reduced cohesiveness and bread volume Increased renin inhibitory activity	Reduced sensory acceptability	Fitzgerald et al. (2014)
Pasta	*Spirulina platensis*	Biomass: 5, 10, and 20% (w/w)	Increased total phenolic content Increased antioxidant capacity Increased water absorption at high % Increased chewiness and firmness Increased protein content Increased starch digestibility	N/A	Rodríguez De Marco et al. (2014)

(continued)

Table 24.1 (continued)

Food products	Algal species	Algal ingredients and amount used	Key modified properties	Sensory acceptability	References
Pasta	*Isochrysis galbana*, and *Diacronema vlkianum*	Biomass: 0.5, 1, and 2% (w/w)	Increased PUFA content Reducing cooking loss, swelling index, and water absorption	Low sensory acceptance than control Increased fishy taste when increasing %	Fradique et al. (2013)
Doughnut	*Spirulina platensis*	Biomass: 2.59 and 5.41% (w/w)	Increased protein, lipid and fiber contents	No significant change in acceptability	Rabelo et al. (2013)
Bread	*Ascophyllum nodosum*	Biomass: 1, 2, 3, and 4% (w/w)	Reduced energy intake No effect in glycemic index	Reduced sensory acceptability	Hall et al. (2012)
Pasta	*Arthrospira platensis*	Biomass: 1, 2, and 3% (w/w)	Increased protein content Increased swelling index Increased firmness Increased greenness Increased antioxidant capacity	All had higher taste and overall acceptability scores 2% had the highest score	Zouari et al. (2011)
Bread	*Ascophyllum nodosum*	Biomass: 5, 10, 15 and 20 g/400 g loaves	Increased dietary fiber	Sensory acceptability decreased at all concentrations compared to control 5 g/400 g has the highest acceptability among all concentrations	Hall et al. (2010)

(continued)

Table 24.1 (continued)

Food products	Algal species	Algal ingredients and amount used	Key modified properties	Sensory acceptability	References
Pasta	*Undaria pinnatifida*	Biomass: 5, 10, 20, 30% (w/w)	Increased protein and total phenolic contents Increased antioxidant capacity Increased amino acid and PUFA contents at 10%	Pasta with 5 and 10% had similar acceptability as control Reduced acceptability when increasing % due to seaweed flavors	Prabhasankar et al. (2009)
Chinese fresh egg noodles	*Monostroma nitidum*	Biomass: 4, 6, and 8% (w/w)	Increased fiber contents Reduced cooking loss Reduced springiness and extensibility when increasing %	Color and texture of noodles with algal biomass were found to be different from control	Chang and Wu (2008)
White pan bread	*Schizochytrium* sp., and *Crypthecodinium cohnii*	Algal oil: 0.325 and 0.65% Microencapsulated algal oil powder: 1.6 and 3.2%	Breads with microencapsulated algal oil powder had the poorest crumb texture, and lowest bread density Increased PUFA content	Bread with 0.325% algal oil had a similar acceptability as control Bread with 1.6% microencapsulated algal oil powder had higher acceptability than control	Serna-Saldivar et al. (2006)
Wheat bread	*Undaria pinnatifida*, and *Chondrus crispus*	Biomass: amount not specified	Decreased the digestion rate of bread starch Lower glycemic index Dairy product	N/A	Goni et al. (2002)

(continued)

Table 24.1 (continued)

Food products	Algal species	Algal ingredients and amount used	Key modified properties	Sensory acceptability	References
Catalan fresh cheese	*Ericaria selaginoides*	Crude extract: 14, 28, and 70 mg/g Non-polar subfraction: 5, 10, and 20 mg/g Mid-polar subfraction: 10, 25 and 50 mg/g	Inhibited growth of *Bacillus cereus* in the cheese with the extracts	N/A	Rubino et al. (2023)
Ayran drink	*Chondrus crispus*, *Chlorella vulgaris* and 1:1 mixture	Biomass: 0.25% (w/w)	Reduced time required for fermentation Increased viability of yogurt bacteria Increased protein and fat content	Reduced acceptability compared to control	Pehlivan et al. (2023)
Cheese spread	*Osmundea pinnatifida*	Aqueous extract (Viscozyme L): 3% (w/w)	Increased viability of *Lactobacillus acidophilus* and *Bifidobacterium animalis* Increased antioxidant capacity	N/A	Faustino et al. (2023)
Yogurt	*Undaria pinnatifida*	Fucoxanthin/ encapsulated fucoxanthin: 8 μg/mL	Yogurt with encapsulated fucoxanthin Increased antioxidant capacity and viability of *Lactobacillus*	Yogurt with fucoxanthin had lowest acceptability after 7 days storage	Wang et al. (2023)

(continued)

Table 24.1 (continued)

Food products	Algal species	Algal ingredients and amount used	Key modified properties	Sensory acceptability	References
White cheese	*Spirulina platensis*	Crude extract/crude extract nanoliposome: 3 mg/g	Inhibited growths of *Staphylococcus aureus* and *Listeria monocytogenes*	N/A	Zanganeh et al. (2022)
Dairy curd	*Nannochloropsis salina*	Biomass: 14.2% (w/w) Ruptured biomass: 7.5% (w/w)	Only curd with biomass increased EPA content Ruptured biomass impaired rennet gelation and disrupted casein microstructure in curd	N/A	Munoz-Tebar et al. (2022)
Kefir	*Arthrospira platensis*, *Chlorella vulgaris*	Biomass: 0.5 and 1% (w/v)	Increased total phenolic content Increased antioxidant capacity Kefir with *C. vulgaris* increased viability of *Lactobacillus acidophilus* and *Bifidobacterium lactis*	N/A	Ilikkan and Bagdat (2022)
Vegan kefir	*Spirulina platensis*	Biomass: 0.25 and 0.5%	Increased viability of *Lactobacilli* and *Lactococci* Increased total phenolic content	N/A	Atik et al. (2021)

(continued)

Table 24.1 (continued)

Food products	Algal species	Algal ingredients and amount used	Key modified properties	Sensory acceptability	References
Goat cheese	*Schizochytrium limacinum*	Biomass: 15% (w/w)	Increased caprylic fatty acid, capric fatty acid, total n-3 PUFA, DHA and rumenic acid	N/A	Bodnar et al. (2020)
Ice cream	*Nannochloropsis oculata*, *Porphyridium cruentum* and *Diacronema vlkianum*	Spray-dried biomass: 0.1, 0.2 and 0.3 g/100 g	Altered color of ice cream Ice cream with *P. cruentum* had higher consistency while others reduced consistency Increased phenolic content	Reduced acceptability when increasing %, except for *P. cruentum*	Durmaz et al. (2020)
Ice cream	*Spirulina platensis* LEB-52	Phycocyanin: 0.25 mg/mL	Increased blueness Increased antioxidant capacity after in vitro simulated digestion	N/A	Campos Assumpcao de Amarante et al. (2020)
Yogurt	*Aurantiochytrium* sp.	Biomass: 2% (w/w)	Increased DHA content	N/A	Paulo et al. (2020)
Feta-type cheese	*Arthrospira platensis*	Biomass: 0.5, 1 and 1.5% (w/w)	Increased viability of *Lactobacillus casei* Increased protein content Increased greenness and yellowness	Lower acceptability at higher algal concentration	Golmakani et al. (2019)

(continued)

Table 24.1 (continued)

Food products	Algal species	Algal ingredients and amount used	Key modified properties	Sensory acceptability	References
Smoothie	*Ulva* spp., *Laminaria ochroleuca*, *Undaria pinnatifida*, *Himanthalia elongata*, *Palmaria palmata*, *Chondrus crispus*, *Porphyra* spp., *Spirulina*, and *Chlorella*	Biomass: 2.2% (w/w)	Smoothies with *Spirulina* and *Chlorella* had higher vitamin B12 contents Smoothies with *L. ochroleuca*, *U. pinnatifida* and *H. elongata* had higher fucose contents	Smoothies with *C. crispus* and *Chlorella* had the lowest acceptability due to strong marine odor and flavor Smoothies with *Spirulina* and *L. ochroleuca* had the highest acceptability similar to control	Castillejo et al. (2018)
Camembert-type cheese	*Palmaria palmata*, and *Saccharina longicruris*	Biomass: 2% (w/w)	Cheese with *S. longicruris* showed a higher ACE-inhibitory activity	N/A	Hell et al. (2018)
Cheese	*Chlorella vulgaris*	Biomass: 2, 4 and 6% (w/w) Slurry: 4%	Increased mineral content Increased antioxidant capacity	Decreased acceptability when increasing % All % had lower acceptability than control	Tohamy et al. (2018)
Semi-hard cheese	*Himanthalia elongata*, *Laminaria ochroleuca*, *Porphyra umbilicalis*, *Ulva lactuca*, and *Undaria pinnatifida*	Biomass: 10 g/kg of curd	Increased antioxidant capacity in cheese with *H. elongata* Altered color of the cheese Reduced fracturability of the cheese	Additions of algal odor and flavor	del Olmo et al. (2018)

(continued)

Table 24.1 (continued)

Food products	Algal species	Algal ingredients and amount used	Key modified properties	Sensory acceptability	References
Yogurt, and quark	*Himanthalia elongata*; *Porphyra umbilicalis*; *Saccharina latissimi*; *Ulva lactuca* and *Undaria pinnatifida*	Biomass: 0.25, 0.5, 0.75, and 1% (w/w)	N/A	Texture and flavor quality decreased with increasing % Yogurts and quarks with *H. elongata* or *S. latissimi* had less seaweed odor and flavor compared to other algae Yogurts and quarks with 0.5% of *U. lactuca* had the lowest flavor quality	Nunez and Picon (2017)
Yogurt	*Spirulina*	Biomass: 0.25, 0.5, 0.75 and 1% (w/v)	Reduced firmness at higher % Increased dietary fiber, protein, Fe and Ca content Increased antioxidant capacity	Reduced acceptability when increasing %	Barkallah et al. (2017)
Yogurt	*Pavlova lutheri*	Extract: 0.25 and 0.5% (w/v)	Increased total n-3 PUFA content Increased green tonality	Reduced acceptability due to undesirable flavor and fishy taste	Robertson et al. (2016)
Shake	*Spirulina* sp. LEB 18	Biomass: 750 mg/100 g	No significant change in protein Reduced shelf life	Had high acceptance as the control	Santos et al. (2016)

(continued)

Table 24.1 (continued)

Food products	Algal species	Algal ingredients and amount used	Key modified properties	Sensory acceptability	References
Milk	*Ascophyllum nodosum* and *Fucus vesiculosus*	Aqueous/ethanolic extracts: 0.25 and 0.5% (w/w)	Increased greenness and yellowness Milk with FV ethanolic extract had less lipid oxidation Increased antioxidant capacity	Reduced acceptability when increasing %	O'Sullivan et al. (2014)
Yogurt	Not specified	Algal oil emulsion: 1.84 g/100 g	Increase PUFA content	Fishy off-flavors were detected in yogurt with algal oil Acceptance of yogurt with the emulsion was similar to the control	Chee et al. (2005)
Snacks and others					
Couscous	*Chlorella vulgaris*	Biomass: 6% (w/w)	Lower glycemic index than control Increased total phenolic content	N/A	Khemiri et al. (2022)
Pear-cantaloupe juice	*Spirulina* sp.	Biomass: 0.3, and 2% (w/v)	Increased total phenolic, flavonoid and mineral content, especially potassium and phosphorus Increased antioxidant capacity	Juice with 0.3% had the optimal sensory profile	Hassanzadeh et al. (2022)

(continued)

Table 24.1 (continued)

Food products	Algal species	Algal ingredients and amount used	Key modified properties	Sensory acceptability	References
Snack bar	*Spirulina* sp. LEB 18	Biomass: 2 and 6%	Increased protein content Increased greenness Reduced hardness at 6% Increased moisture at 6%	Had similar sensory profiles and acceptability as control	Lucas et al. (2020)
Broccoli soups	*Spirulina* sp., *Chlorella* sp., and *Tetraselmis* sp.	Biomass: 0.5, 1, 1.5 and 2% (w/v)	Increased total phenolic content Increased antioxidant capacity	Reducing acceptability when increasing % Soup with 0.5% of *Tetraselmis* sp. had the highest acceptability but still lower than control	Lafarga et al. (2019a)
Extruded corn snack	*Sirophysalis trinodis*, and *Polycladia myrica*	Aqueous extract: 2 and 4% (w/w) Aqueous concentrate: 2% (w/w)	Increased total phenolic contents at high % Increased antioxidant activities	Had similar acceptability as controls	Etemadian et al. (2018)
Extruded snack	*Spirulina* sp. LEB 18	Dried biomass: 2.6% (w/w)	Increased protein, lipid, and mineral contents Enhanced greenness	Snacks with microalgal biomass had similar sensory profiles and acceptability as controls, except appearance	Lucas et al. (2018)

(continued)

Table 24.1 (continued)

Food products	Algal species	Algal ingredients and amount used	Key modified properties	Sensory acceptability	References
Extruded snack	*Spirulina* sp. LEB 18	Biomass: 0.4, 1.0, 1.8, 2.6, and 3.2% microstructure	Increased hardness Increased water absorption index Increased bulk density Increased protein content Increased greenness	N/A	Lucas et al. (2017)
Corn extrudate	*Spirulina*	Biomass: 2, 4, 6, and 8%	Increased protein, fiber, β-carotene contents Increased greenness and yellowness Reduced expansion, hardness and crispness	Extrudate with 2% had similar sensory attributes and acceptability as control Acceptability decreased as the % increased	Tańska et al. (2017)
Extruded maize product	*Porphyra columbina*	Biomass: 1.7, 3.5 and 5.2%	Increased total phenolic content Increased antioxidant capacity Increased greenness	N/A	Cian et al. (2014)

would be altered. In general, replacing flour with macroalgal biomass in the bakery items would increase the hardness of the items. The hardness of cookies increased when the percentage of *Bifurcaria bifurcate* in the cookie dough increased from 3 to 9% (Arufe et al. 2019a). Same as the cookies, the firmness of the breads increased after adding *Kappaphycus alvarezii* (2–8%) in a does depend manner (Mamat et al. 2014). Cakes with 5–20% of *Eucheuma cottonii* were found to be chewier in texture (Huang and Yang 2019). The hardness of the rice-based doughs with 5% or 10% of *K. alvarezii* were found to increase, while the adhesiveness of the doughs was found to reduce (Lopes et al. 2023). Properties of wheat bread dough with 2–8% of *Fucus vesiculosus* were found to be modified, resulting in high density of the wheat bread (Arufe et al. 2018). A few research have showed that the hardness of the bakery and pastry goods might not be directly proportional to the amount of algal biomass added to the doughs and batters. The hardness of the crackers with *Undaria pinnarifida* or *Himanthalia elongata* increased when the doughs contained 5 or 10% of the algal biomass (Wilkin et al. 2021). Once the amount raised to 15 or 20%, the crackers with the algal biomass would have similar textures and fracturabilities as the ones prepared according to the original receipts. Decreases in hardness have been observed in crackers and gluten free breads prepared from *K. alvarezii* (1–10%) and *Ascophyllum nodosum* (2–10%) in a dose dependent manner respectively (Aganduk et al. 2023; Różyło et al. 2017). Recent study on the texture of pasta with 10% of *F. vesiculosus* has revealed that the incorporation of the algal biomass increased the total fibre content and disrupted gluten-starch matrix, reducing the hardness and firmness of the pasta (Ribeiro et al. 2022). Just like macroalgae, incorporating microalgal biomass, like *C. vulgaris* (1.5%) and *S. platensis* (1.5%) into breadsticks also resulted in decreases in hardness and crispiness of the breadsticks (Uribe-Wandurraga et al. 2019).

Capacity of absorbing water and oil of bakery goods and pastry goods might be modified by replacing flour with algal biomass perhaps due to the increase in dietary fibre content. The solvent absorption capacities of cookies with 5% of macroalgal biomass, like *Sargassum fulvellum*, *Enteromorpha linza*, *Codium fragile*, or *Hizikia fusiforme* increased meanwhile the hardness of these cookies also increased (Oh et al. 2020). Biscuits with 1–10% of *Caulerpa racemose* and breads with 2–8% of *K. alvarezii* were found to have higher capacities for absorbing water and oil (Kumar et al. 2018; Mamat et al. 2014). The capacity for water absorption of pasta could increase by incorporating 1–20% of microalgal biomass of *S. platensis*, while the texture of the pasta became chewier and firmer (Rodríguez De Marco et al. 2014; Zouari et al. 2011).

The changes in textures as well as the capacities of water and oil absorption of the bakery and pastry goods after incorporating algal biomass seemed to be related to the algal dietary fibres. Besides dietary fibre, both macroalgal and microalgal biomass were well-known to contain other nutrients, so the nutritional values of bakery and pastry goods supplementing with algal biomass should be fortified. Baking breads, cakes, and cookies with doughs contained around 0.5–1% of *Ulva lactuca* or *Gracilaria corticata* could enrich protein and mineral contents, especially calcium and magnesium in these bakery goods (Turuk and Banerjee 2023). Cookies prepared with 1–10% of *Ulva intestinalis* had higher protein and lipid contents than cookies prepared with original receipt (Mohibbullah et al. 2023). Microalga *S. platensis* was also commonly added to enhance the nutritional values of bakery and pastry goods. Crostini with 2–10% of *S. platensis* had higher protein and free amino acid contents (Niccolai et al. 2019). Doughnuts prepared from around 2.5 to 5.5% of *S. platensis* had higher protein, and lipid contents (Rabelo et al. 2013). Another attempt has found out that supplementing pasta with 2–15% of *Spirulina* sp. could increase protein, lipid and mineral contents, especially calcium and iron in the pasta (Koli et al. 2022). Amounts of iron and selenium of breadsticks supplemented with 1.5% of *C. vulgaris* or *S. platensis* were found to be increased (Uribe-Wandurraga et al. 2019). Wheat bread prepared from 2 or 4% of *S. platensis* and Kelp was found to contain high levels of vitamin B1 and B2 (Chochkov et al. 2021). Protein, fatty acid, and carotenoid contents of gluten-free breads could be enhanced by supplementing with 2.5 or 5% of *Chlorella sorokiniana* (Diprat et al. 2020). Supplementing pasta with 0.5–2% of *Isochrysis galbana* or *Diacronema vlkianum* could significantly increase PUFA content of the pasta (Fradique et al. 2013).

Not just improving the nutrition values, incorporating algal biomass with high phenolic contents might increase antioxidant capacities of the bakery and pastry goods. The addition of *B. bifurcate* (3–9% of final weight), *F. vesiculosus* (2–8% of final weight), and *A. nodosum* (3–9% of final weight) in cookie doughs increased total phenolic contents and antioxidant capacities before baking (Arufe et al. 2019a, b). Total phenolic content and antioxidant capacity of gluten-free breads and cookies increased significantly after supplemented with *Tetraselmis chuii* (1–4% of final weight) and *H. pluvialis* (5–15% of final weight) respectively (Hossain et al. 2017; Nunes et al. 2020). Crackers and breads prepared from around 1 to 4% of *Nannochloropsis* sp. or *Tetraselmis* sp. were found to have high total phenolic contents and antioxidant capacities (Lafarga et al. 2019b). Despite the examples just highlighted, most of other studies mentioned in previous paragraphs have found that incorporating algal biomass into the bakery and pastry goods could increase total phenolic contents and antioxidant capacities of the goods, which might extend their shell lives (Saharan and Jood 2021).

Last but not least, the incorporations of algal biomass might modify digestion rates of starch and glycaemic indexes of the bakery and pastry goods. The rate of starch digestion of wheat breads with *U. pinnatifida* or *Chondrus crispus* were found to be slower than those prepared according to the original receipts, resulting in lower glycaemic indexes of the wheat breads (Goni et al. 2002). Adding 5–20% of *U. pinnarifida* or *H. elongata* to crackers slowed down starch digestion (Wilkin et al. 2021). The glycaemic responses of the cookies with 5–15% of *H. pluvialis* were found to be decreased (Hossain et al. 2017). However, contradictory findings have been reported that the breads with biomass of *A. nodosum* only reduced energy intakes of the human subjects, but had no effect on the blood glucose levels after meals (Hall et al. 2012). Another study reported that pasta with 5–20% of *S. platensis* indeed had a higher starch digestibility (Rodríguez De Marco et al. 2014). This might be due to the disruption of protein-starch matrix, making starch more ready for digestion.

Unfortunately, sensory acceptability of nearly all of the bakery and pastry goods were lower than those prepared according to the original receipts. Most of the time, the sensory acceptability declined as the amount of the algal biomass increased to above 5–8% of final weight in the bakery and pastry goods (Huang and Yang 2019; Kumar et al. 2018; Niccolai et al. 2019; Ribeiro et al. 2022; Różyło et al. 2017; Saharan and Jood 2021). Low sensory acceptances were mainly due to unpleasant fishy off-flavour in the bakery and pastry goods prepared from the algal biomass (Batista et al. 2017, 2019; Diprat et al. 2020; Fradique et al. 2013; Mohibbullah et al. 2023; Nunes et al. 2020; Ozyurt et al. 2015; Prabhasankar et al. 2009; Wilkin et al. 2021). Just a few bakery and pastry goods had better sensory attributes after incorporating with the algal biomass, like pasta prepared from 1 to 3% of *S. platensis* and Gluten-free breads prepared from 1 to 3% of *Nannochloropsis gaditana* or *Chlamydomonas* sp. (Khemiri et al. 2020; Zouari et al. 2011). Crackers contained 1% of *Nannochloropsis* sp. had better sensory profiles than the ones with 1% of *Tetraselmis* sp. and the ones prepared according to the original receipts (Lafarga et al. 2019b).

24.2.2 Meat Products

Just like the bakery and pastry goods, algal biomass was be used to partially replace meat, fat, or other ingredients, like soy protein in patties, sausages and burgers (Espinosa-Ramírez et al. 2023; Wang et al. 2022). Obviously, replacing animal fat with algal biomass significantly reduced the fat contents of the meat products, like low-fat pork patties prepared from 1 to 5% of *Laminaria japonica* (Choi et al. 2012). Meanwhile, the PUFA contents of the meat products could be increased by the supplementations of algal biomass, like adding 3% of *Undaria* sp. to low-fat pork burgers (Nagai et al. 2022).

When replacing animal fats with algal biomass, dietary fibre contents of the meat products would significantly increase, modifying both nutritional values and techno-functional properties of the meat products. The texture of the low-fat frankfurters with 5% of *H. elongata* was found to be harder than the ones prepared using the original ingredients (Lopez-Lopez et al. 2009). Moreover, increases in dietary fibre contents would not just increase hardness, but also increase the capacities of water and oil absorption of the meat products. Fish cutlets mixed with 5–15% of *K. alvarezii* had higher water and oil capacities as well as firmer textures (Senthil et al. 2005). Incorporating either macroalgal biomass, like *Cystoseira compressa* (0.5–1.5%) and *Jania adhaerens* (0.5–1.5%) or microalgal biomass, like *Chlorella minutissima* (0.5–1.5%), *I. galbana* (0.5–1.5%), *Picochlorum* sp. (0.5–1.5%) and *S. platensis* (0.5–1.5%) into fish burgers could increase the capacities on water and oil absorptions as well as chewiness of the fish burgers (Barkallah et al. 2019; Ben Atitallah et al. 2019a, b; Hentati et al. 2019). Besides the fish burgers, other studies have attempted to mix pork with 3.4% of *H. elongata* to make meat batters (Fernandez-Martin et al. 2009). Chewiness as well as capacities of water and oil absorption of the pork meat batters have been enhanced.

Apart from dietary fibre, various algal biomasses are rich in phenolic metabolites which might act as natural antioxidants in various meat products. Fish burgers contained microalgal biomass, like *C. minutissima* (0.5–1.5%), *I. galbana* (0.5–1.5%), *Picochlorum* sp. (0.5–1.5%) and *S. platensis* (0.5–1.5%) had higher antioxidant capacities that the original fish burgers (Barkallah et al. 2019; Ben Atitallah et al. 2019a, b). Beef patties prepared from 10 to 40% of *H. elongata* were found to have high total phenolic contents and antioxidant capacities which prevented lipid oxidation in the beef patties during storage (Cox and Abu-Ghannam 2013). Moreover, proliferations of microbes in the beef patties during storage were found to be inhibited due to the incorporations of *H. elongata*. Another study has demonstrated that viabilities of microbes of calf fillets could be reduced by coating the fillets with 1–2% of *S. platensis* or *C. vulgaris* (Shafiei and Mostaghim 2022).

As algae contains various minerals in high levels, so algal biomass could serve as seasonings in meat products. Wakame (3.3%) have been used to replace salt (sodium chloride) in black and white puddings (Fellendorf et al. 2016a, b). Frankfurters seasoned with *Porphyra umbilicalis*, *Palmaria palmata*, *H. elongata* or *U. pinnatifida* (1% of final weight) were found to have less salt contents however textures of the frankfurters became less chewy (Vilar et al. 2020). Incorporating mixture of *Ulva* sp., *Gracilaria* sp. and *F. vesiculosus* into frankfurters not just reduced their salt contents but also increased other mineral contents in the frankfurters, and improving antimicrobial activities of the frankfurters (Marcal et al. 2021).

Although using algal biomass to make various meat products could fortified their nutritional values, the additions of the algal biomass have negative impacts on the sensory profiles of the meat products. In brief, increases in algal biomass to around 1.5% of final weight in the meat products introduced unexpected seaweed off-flavours which reduced acceptances of the panellists (Barkallah et al. 2019; Ben Atitallah et al. 2019a, b; Hentati et al. 2019). Fish cutlets with more 10% of *K. alvarezii* would had poor sensory profiles (Senthil et al. 2005). Only limited meat products contained algal biomass had better sensory attributes than the original receipts, including the calf fillets coated with *S. platensis* or *C. vulgaris* (1–2% of final weight) as well as the low-fat pork patties with 1–3% of *L. japonica* (Choi et al. 2012; Shafiei and Mostaghim 2022). Obviously, the natural flavour of the algae played a critical role in the sensory acceptability of the meat products. The frankfurter with 1% of *P. palmata* had a significant lower acceptability than the one with 1% of *H. elongata* (Vilar et al. 2020).

24.2.3 Dairy Products

Different from meat products, bakery goods and pastry items, dairy products were supplemented with algal biomass for enhancing their nutrition values, such as increasing PUFA contents in yogurts, and goat cheese by adding *Spirulina* sp. (2% of final weight) and *Schizochytrium limacinum* (15% of final weight) respectively (Bodnar et al. 2020; Paulo et al. 2020). Moreover, vitamin B12 contents of smoothies could be fortified by supplementing with *Spirulina* sp. or *Chlorella* sp. (2.2% of final weight) (Castillejo et al. 2018). Furthermore, the supplementations of 0.5–1%

of *S. platensis* or *C. vulgaris* could increase total phenolic contents, subsequently increasing antioxidant capacities of kefirs (Ilikkan and Bagdat 2022). Along with the increases in antioxidant capacities, cheese with 2–6% of *C. vulgaris* and yogurt with 0.25–1% of *Spirulina* sp. were found to enriched with mineral contents, especially iron and calcium (Barkallah et al. 2017; Tohamy et al. 2018).

Apparently, nutrients of algal biomass not only increased the nutritional values of the dairy products, but also promoted growths of probiotics, facilitating fermentations of the dairy products. The viability of *Lactobacillus* and *Lactococci* could be enhanced by incorporating 0.5–1.5% of *S. platensis* into feta-type cheese and vegan kefirs (Atik et al. 2021; Golmakani et al. 2019). In another study on kefirs, the addition of 0.5–1% of *C. vulgaris* improved the viabilities of *Lactobacillus acidophilus* and *Bifidobacterium lactis* in the kefirs (Ilikkan and Bagdat 2022). By adding both *C. crispus* and *C. vulgaris* (0.25% of final volume) into Aryan drinks was found to shorten time required for fermentation and increased viabilities of yogurt bacteria (Pehlivan et al. 2023).

Using algal biomass to make dairy products might alter the textures of the dairy products as well. The firmness of yogurt was found to be reduced after adding *S. platensis* (0.25–1% of final weight) in a dose dependent manner (Barkallah et al. 2017). The consistency of ice cream could be modified by adding *Nannochloropsis oculate*, *Porphyridium cruentum* or *Diacronema vlkianum* (0.1–0.3% of final weight) (Durmaz et al. 2020). The fracturabilities of semi-hard cheese reduced once adding *H. elongata*, *Laminaria ochroleuca*, *P. umbilicalis*, *U. lactuca* or *U. pinnatifida* (1% of final weight) (del Olmo et al. 2018). Subsequent analysis has revealed that supplementing dairy curds with microalgal biomass would impair rennet gelation and disrupt casein microstructure, which altering the textures of the curds (Munoz-Tebar et al. 2022).

Similar to bakery goods, pastry goods and meat products, incorporating algal biomass resulted in decreases in sensory acceptability of the dairy products. Higher the amount of algal biomass incorporated over around 0.5% of final weight or volume leaded to lower sensory attributes of the dairy products (Barkallah et al. 2017; Golmakani et al. 2019; Tohamy et al. 2018). Undesired seaweed odour and flavour was the key reason that lowered the acceptability of the dairy products comparing to the ones prepared based on the original receipts (del Olmo et al. 2018). Among different trials, smoothies with 2.2% of *Spirulina* sp. or *L. ochroleuca*, yogurts with 0.25–1% of *H. elongata* or *Sachharina latissimi*, as well as ice cream with 0.1–0.3% of *P. cruentum* got relatively high sensory ratings in the sensory evaluations (Castillejo et al. 2018; Durmaz et al. 2020; Nunez and Picon 2017).

24.2.4 Snacks and Others

Publics in general consider snacks as junk food since they have low nutritional values. A few researches have attempted to supplement extruded snacks with algal biomass to enhance their nutritional values. Only *Spirulina* sp. and *Porphyra columbina* have been used to make extruded snacks. The extruded snacks with 2–8% of the algal biomass had higher protein, lipid, mineral, dietary fibre, and carotene contents (Lucas et al. 2018; Tańska et al. 2017). Moreover, the total phenolic content of the extruded maize snack could be fortified by adding *P. columbina* up to around 1.7–5.2% of the final weights, as a consequence increasing the antioxidant capacities of the snack (Cian et al. 2014). Although the nutritional values and antioxidant capacities were enhanced, the incorporations of algal biomass did reduce expansion indices, as a result the hardness and densities of the snacks increased (Lucas et al. 2017; Tańska et al. 2017). Sensory evaluations of the extruded snacks revealed that the snacks with around 2% of algal biomass had similar sensory profiles and acceptability as the controls (Lucas et al. 2018, 2020). While further increases in the amount of algal biomass would reduce the acceptability, mainly due to appearance and texture (Tańska et al. 2017).

Apart from the extruded snacks, algal biomass was supplemented to beverages for enhancing their nutritional values. The amounts of total phenolic compounds, flavonoids, and minerals as well as the antioxidant capacity of pear-cantaloupe juice were found to be increased by supplementing with *Spirulina* sp. (0.3% of final weight) which had high acceptability in the sensory evaluation (Hassanzadeh et al. 2022). The additions of *Spirulina* sp., *Chlorella* sp., or *Tetraselmis* sp. (0.5–2% of final volume) into broccoli soups leaded to increases in total phenolic content and antioxidant capacity with decreases in the sensory acceptability, except the broccoli soups with 0.5% of *Tetraselmis* sp. (Lafarga et al. 2019a). Last but not least, mixing *C. vulgaris* (6% of final weight) with couscous enhanced the total phenolic content and lowered the glycaemic index of the couscous (Khemiri et al. 2022).

24.2.5 Common Effects of Algal Biomass on the Properties of Different Food Items

Incorporating algal biomass into bakery goods, pastry goods, meat products, dairy products, snacks and other food items did alter their nutritional values, biochemical properties, techno-functional properties, and sensory attributes. By reviewing the effects of algal biomass on the properties of multiple types of food items side by side, a few effects were commonly observed among these food items fortified with the algal biomass. On top of the changes in nutritional values and colour, one of the noticeable common effects of incorporating algal biomass enhanced the total phenolic contents, as a result improving the antioxidant capacities and nutritional values of the food items. The increases in the total phenolic contents of these food items were obvious because other ingredients in these food products contain relatively low amounts of phenolic compounds, except for milk (Sik et al. 2023). As algal biomass in general were rich in phenolic compounds (Jimenez-Lopez et al. 2021), so the increases in total phenolic contents and antioxidant capacities would be obvious and significant even small amounts of algal biomass were added to the bakery goods, pastry goods, meat products, dairy products and snacks. In perspectives of manufacturing, increasing the antioxidant capacities of the food items should be always welcomed as the shelf lives of the food items could be extended.

Another noticeable effect was that the incorporations of algal biomass modified textures of the food items, especially the bakery goods, pastry goods, and cheeses by disrupting their microstructures. Since customers have their preferences on the textures of the food items, so the changes in textures of the food items contained algal biomass might be undesirable, like reducing the crispness of the extruded snacks made from *Spirulina* sp. (Tańska et al. 2017). The third noticeable effect was improving the capacities of water and oil absorptions of the food items, perhaps due to increasing the amounts of dietary fibre in these food items. The capacities of water and oil absorption is directly correlated to water activities in the food items. Increasing water activity shortens shelf life of meat products, bakery goods, and pastry goods (Barcenilla et al. 2022; Smith et al. 2004). Thus, algal-fortified meat products with higher water absorption capacities would likely had longer shelf lives (Hentati et al. 2019). In addition, the changes of capacities on water and oil absorption might alter the spreading factors and moisture contents in bakery goods as well as swelling indexes in the pastry goods (Kumar et al. 2018; Oh et al. 2020; Zouari et al. 2011).

Although certain properties of the food items could be modified by algal biomass, people have low acceptances to the sensory profiles of algal biomass fortified food items, due to unpleasant and undesired seaweed off-flavour, especially in the dairy products. The sensory acceptability of the dairy products dropped even just around 0.5% of algal biomass being added. The sensory acceptability of the meat products and snacks contained algal biomass seemed to be higher. Meat products and snacks usually have stronger savoury flavours than bakery goods, pastry goods and dairy products. Thus, it is believed that the strong savoury flavours partially covered up the undesired seaweed off-flavour in the meat products and snacks with algal biomass. To improve sensory acceptability of food items with algal ingredients, one of the potential solutions is to remove metabolites that cause unpleasant off-flavour of algal biomass before adding to food items. Purifying refined algal metabolites for food productions would be an alternative solution to avoid introducing unpleasant algal off-flavour to the food items, which decreasing their sensory attributes. Therefore, incorporating algal extracts would be a potential strategy to improve certain properties of a food item without drawbacks.

24.3 Altering Properties of Food Items with Algal Extracts

Agar, alginate and carrageenan are polysaccharides extracted from different species of macroalgae and have been used as emulsion stabilizers, thickeners and gelling agents in food industries for decades (Chen et al. 2021; Qin et al. 2018; Udo et al. 2023). Their safeties have been evaluated by the expert panel on food additives and flavourings from European Food Safety Authority and assigned with Europe number E400 to E407a accordingly. Macroalgae in *Gelidium* and *Gracilaria* genera are the dominant industrial sources for extracting agar which is commonly used as a gelling agent in bakery goods (Rioux and Turgeon 2015). Alginate extracted from macroalgae in *Laminaria* genus is usually used as a thickener in ice cream, soup, sauces and beverages (Rioux and Turgeon 2015). Carrageenan is often used as a stabilizer and thickener in dairy products as carrageenan was found to interact with casein and stabilize casein micelles in the dairy products (Spagnuolo et al. 2005). Carrageenan is mostly extracted from *K. alvarezii* and *E. cottonii.* In addition, microalgae, like *H. pluvialis, Dunaliella salina,* etc., could be an excellent source for food colorants due to rich in various carotenoids, like β-carotene, astaxanthin, fucoxanthin, etc. (Generalic Mekinic et al. 2023). Thermal stable phycocyanin from *S. platensis* could be a natural blue food colorant (Jespersen et al. 2005). Apart from traditional usages of refined algal polysaccharides and pigments, recent developments on the applications of algal extracts in food industries were summarized and discussed in the below sections.

24.3.1 Bakery and Pastry Goods

The key drawbacks of incorporating algal biomass into food items were introducing unpleasant fishy and earthy taste as well as unappealing colour due to presences of volatile unsaturated fatty acid derived aldehydes, ketones and alcohols as well as various pigments in the algal biomass (Coleman et al. 2022; García-Vaquero et al. 2021). To reduce the unpleasant taste and unappealing colour, a few research have attempted to treat the biomass of *T. chuii* with ethanol before the treated algal biomass was used to partially replace flour in wheat breads (Qazi et al. 2021b). The wheat breads with the ethanol treated algal biomass (4% of final weight) had similar appearances and tastes as the ones prepared according to the original receipts (Qazi et al. 2022). However, their textures were harder than the original receipts due to high dietary fibre contents. The incorporation of ethanol treated *T. chuii, C. vulgaris* or *Microchloropsis gaditana* (12% of final weight) significantly enhanced protein qualities and essential amino acid contents of the wheat breads (Qazi et al. 2021a). However, the treatments with ethanol significantly reduced the total phenolic contents of the treated algal biomasses, decreasing the antioxidant capacities of the wheat breads prepared from the treated algal biomasses (Qazi et al. 2022). Although certain nutritional properties have been diminished after the treatments with ethanol, the treatments successfully decreased the unpleasant tasty and unappealing colour of the algal biomasses, making the wheat breads with 4% of the ethanol treated *C. vulgaris* had similar sensory acceptability as the controls (Qazi et al. 2022).

Instead of removing the undesired taste and colour of the algal biomass, metabolites of interest could be extracted from the algal biomass for improving certain properties of food items. A previous study has attempted to improve the functionality of the bread by adding protein hydrolysate of *Palmaria plamata* (4% of final weight) which could inhibit activity of renin (Fitzgerald et al. 2014). After baking, the renin inhibitory activity still remained, but the texture and flavour of the bread were altered due to incorporating with the protein hydrolysate, resulting in low sensory acceptability (Fitzgerald et al. 2014). Protein, fat and fibre contents of instant fried noodles were found to be fortified by adding different of extracts from *E. cottonii*, *S. platensis* and *C. vulgaris* (Kumoro et al. 2016). In another study, algal oil (0.3–0.65% of final weight) or microencapsulated algal oil in powder form (1.6–3.2% of final weight) were added to white pan breads to fortify their PUFA contents instead of adding algal biomass (Serna-Saldivar et al. 2006). Although the texture and density of the bread have been affected by adding the microencapsulated algal oil, the bread indeed had a higher sensory acceptability than the original breads (Serna-Saldivar et al. 2006). Obviously, adding partially purified or highly refined algal metabolites could be a potential solution to enhance nutritional values or certain techno-functional properties of the food items without introducing algal unpleasant taste and unappealing colour. However, this approach hinders by chemical natures of algal metabolites. If the chemical natures of the algal metabolites of interest were similar to those undesired volatile unsaturated fatty acid derived aldehydes, ketones and alcohols as well as pigments in the algal biomass, it would be a challenge to purify the desired metabolites without the contaminations.

24.3.2 Meat Products

As mentioned, carrageenan from macroalgae has been used in the food industry as a binding agent for decades. Not just used as a binding agent, carrageenan (0.25–0.75% of final weight) has been used to partially replace fat in the meat products to produce low-fat frankfurters and ground pork patties (Cierach et al. 2009; Kumar and Sharma 2004). Calories of these low-fat products were reduced as expected, meanwhile cooking losses were also reduced. Even though the low-fat meat products were harder in texture due to carrageenan, panellists showed higher acceptance to them than those prepared according to the original receipts (Cierach et al. 2009). Other than carrageenan, algae are known to contain other polysaccharides, like laminarin and fucoidan. These polysaccharides extracted from *Laminaria digitata* were used to prepare minced pork patties (Moroney et al. 2013). With laminarin or fucoidan, the patties appeared to be greener in colour than controls. The algal polysaccharides only reduced lipid oxidation of the patties during cooking. The acceptability of the patties with the algal polysaccharides reduced as the amount of the algal polysaccharides increased (Moroney et al. 2013).

Apart from the well-known polysaccharides, no matter macroalgae or microalgae, they are still rich in other types of polysaccharides which might also be used as ingredients for manufacturing meat products. Crude sulphated polysaccharides extracted from *U. intestinalis* (0.5% of final weight) has been used to partially replace wheat starch and bread flour as a binding agent for restructured fish surimi (Jannat-Alipour et al. 2019). The additions of algal sulphated polysaccharides reduced hardness and chewiness of the surimi as well as cooking loss. Moreover, the surimi prepared from the *U. intestinalis* polysaccharides gained a higher acceptability than the surimi with 2.7% of *U. intestinalis* biomass as well as the ones prepared according to the original receipt after 6 months storage due to better texture (Jannat-Alipour et al. 2019). This implied that the sulphated polysaccharide might prolong the shelf life of the surimi. Furthermore, acceptances on surimi prepared from biomass of *U. intestinalis* were lower than the one with just the sulphated polysaccharide since the surimi contained the algal biomass were poor in texture and had an unpleasant green colour (Jannat-Alipour et al. 2019).

Other researches on investigating properties of algal polysaccharides revealed that the incorporations of algal polysaccharides could enhance the antioxidant capacities of various meat products. The polysaccharides extracted from *Bryposis plumosa* (0.05–0.25% of final weight), *Falkenbergia rufolanosa* (0.05–0.25% of final weight), *P. cruentum* (0.5–2% of final weight), *S. plantensis* (0.1–0.5% of final weight) and *Chaetomorpha linum* (0.05–0.25% of final weight) could reduce myoglobulin and lipid oxidations in beef sausages, minced beef meat, and pork sausages (Ben Hlima et al. 2021; Feki et al. 2021; Ghariani et al. 2022; Hamzaoui et al. 2020; Luo et al. 2017). The finding indicated that various algal polysaccharides might replace antioxidants like vitamin C in the original receipts of various meat products to prolong their shelf lives. Further investigations unveiled that the algal polysaccharides also exerted antimicrobial activities which inhibiting microbial growth during storages (Ben Hlima et al. 2021; Feki et al. 2021; Ghariani et al. 2022; Hamzaoui et al. 2020). Although the textures of these meat products were always herder and chewier than the ones based on the original receipts due to the incorporations of the algal polysaccharides, these products had similar or even higher acceptability than the original receipts (Feki et al. 2021; Luo et al. 2017).

Not just algal polysaccharides exerted both antioxidant and antimicrobial activities, aqueous extracts of algae might also prevent oxidations of nutrients in meat products. The crude aqueous extract from *Cystoseira barbata* (0.01–0.04% of final weight) could reduce lipid oxidation in turkey meat sausages (Sellimi et al. 2017a). Fucoxanthin was found to be one of the key ingredients in the extract of *C. barbata* that reduced lipid oxidation in the sausages (Sellimi et al. 2017b). Both the extract and fucoxanthin from *C. barbata* (0.01–0.04% of final weight) could serve as potential preservatives to replace sodium nitrite in processed meat products. The extract from *H. pluvialis* which contained around 5% of astaxanthin was found to reduce lipid oxidation and improve the appearance of the ground pork meat (Pogorzelska et al. 2018). Lipid oxidations of rainbow trout fillets were reduced after marinated with alcoholic extracts from *S. platensis* and *C. vulgaris* (Takyar et al. 2019). The marinated fillet had a higher acceptability than control after storage which indicated that the microalgal alcoholic extracts could act as preservatives to extend the shelf life of the fillet. By adding the ethanolic extract from *Halimeda opuntia* (4–6% of final weight), the growth of *Staphylococcus aureus* in chicken fillets was found to be inhibited without sacrificing sensory attributes of the fillets (Hamad et al. 2023). Algae are also rich in minerals which make their aqueous extracts as potential flavourings in meat products. The aqueous extract from *P. palmata* was used to partially replace salt in ham (Barbieri et al. 2016). Although the ham with around 1–1.2% of the aqueous algal extract had less salt content, the panellists had a higher acceptance the ham due to strong flavour.

As algae in general are rich in PUFA, a few research have supplemented sausages with around 2% of algal oil extracted from *Crypthecodinium cohnii* (de Ciriano et al. 2010a). The supplemented sausages were found to be rich in PUFA and also had similar acceptability as the sausages prepared according to the original receipt. Subsequent researches have attempted to partially replace animal fat in beef and pork patties with emulsion gel (9–20%) containing different concentrations of algal oil extracted from *C. cohnii* and *Schizochytrium* sp. (Alejandre et al. 2017; Cirstea Lazar et al. 2023). The total fat contents of these patties were reduced and the lipid profiles also became healthier as the PUFA contents increased. Meanwhile, their shelf lives have been extended due to decreases in lipid oxidation. Unfortunately, the algal oil still contained pigments as well as unsaturated fatty acid derived aldehydes, ketones and alcohols generated by oxidations. Thus, the acceptability of these patties were found to be low due to unpleasant appearances and tastes (Alejandre et al. 2017; Cirstea Lazar et al. 2023). Obviously, partially replacing animal fat with emulsions of algal oil significantly improved the nutritional values of the meat products. However, more antioxidants or preservatives might be required to prevent the oxidations of the algal oil. Instead of using antioxidants and preservatives from other origins, the aqueous extract from *Melissa officinalis* was used as an antioxidant (200 ppm) in low-fat Bologna sausages with emulsions containing oil from *C. cohnii* (Berasategi et al. 2014). ALA and DHA contents of the low-fat sausages not just increased, but the lipid oxidation was also reduced. Less algal lipid oxidation resulted in high sensory acceptability of the low-fat sausages similar as the original receipts.

24.3.3 Dairy Products

Just like incorporating algal biomass into cheeses and kefirs, adding the aqueous extract from *Osmundea pinnatifida* (3% of final weight) could increase the viabilities of probiotics, like *L. acidophilus* and *Bifidobacterium animalis* as well as antioxidant capacities of the cheese spreads (Faustino et al. 2023). Surprisingly, extracts from a few species of macroalgae and microalgae were found to inhibit growths of pathogens in dairy product. The non-polar extract from *Ericaria selaginoides* showed an excellent inhibition on the growths of *Bacillus cereus* in Catalan fresh cheese (Rubino et al. 2023). The crude extract from *S. platensis*, not matter alone (0.3% of final weight) or encapsulated in nanoliposomes (0.3% of final weight) were able to inhibit the growth of *S. aureus*, and *Listeria monocytogenes* in white cheese (Zanganeh et al. 2022). The shelf lives of the dairy products might be prolonged not just due to the antimicrobial effects of the algal extracts, but might also due to the antioxidant capacities of the algal extracts. Milks with 5% of ethanolic extract from *F. vesiculosus* showed a significant higher antioxidant capacity, and less lipid oxidation which might prolong the shelf life of the milk (O'Sullivan et al. 2014). However, the milk with the ethanolic algal extract was found to be less acceptable due to fishy taste from the algal extract. Refined algal pigments, like fucoxanthin from *U. pinnatifida* and phycocyanin from *S. platensis* were also used to enhance the antioxidant capacities of yogurts and ice creams respectively (Campos Assumpcao de Amarante et al. 2020; Wang et al. 2023). Although these algal pigments indeed increased the antioxidant capacities of the dairy products, the yogurts with around 8 ppm of fucoxanthin were found to be less acceptable to panellists (Wang et al. 2023). Yogurt supplemented with around 1.8% of algal oil emulsion for fortifying its PUFA content were also found to have fishy off-flavour (Chee et al. 2005). Although the properties of the dairy products could be improved by adding small amounts of algal extracts, the dairy products still had unacceptable sensory profiles.

24.3.4 Snacks and Others

As mentioned, making extruded snacks using algal biomass might alter the expansion indexes, reducing the crispness of the snacks. Moreover, adding algal biomass would affect the colours of the extruded snacks as well. Instead of using algal biomass, aqueous extracts from *Sirophysalis trinodis* and *Polycladia myrica* were added to extruded corn snacks (2–4% of final weight) to increase total phenolic contents and antioxidant capacities of the snacks (Etemadian et al. 2018). The modified snacks indeed had similar acceptability as the ones prepared as usual.

24.3.5 Are Algal Extracts Better Choice for Making Food Products?

As mentioned, incorporating algal biomass into food items always reduced their sensory acceptability compared to the ones prepared according to the original receipts. This was due to the algal biomass naturally contain some metabolites which caused unpleasant off-flavour in the food items. Thus, the key proposes of using the algal extracts to make food products was to improve certain properties of the food products without adding those undesired algal metabolites and sacrificing the sensory profiles of the food products. Obviously, incorporating various algal extracts did fortify different properties of the bakery goods, pastry goods, meat products, dairy products and extruded snacks. However, the sensory profiles of these food items with algal extracts did not entirely gain high acceptances in the sensory evaluations. The dairy products with around 0.5–1.8% of the algal extracts still had low sensory acceptability compared to the ones prepared according to the original receipts (Chee et al. 2005; O'Sullivan et al. 2014; Robertson et al. 2016). In contrast, wheat breads with around 0.3% of algal oil or 4% of ethanol treated algal biomass had sensory profiles similar to the original wheat breads (Qazi et al. 2022; Serna-Saldivar et al. 2006). Among different meat products, chicken fillets with 4–6% of ethanolic extracts from

H. opuntia, low-fat frankfurters with 0.4–0.7% of carrageenan, surimi with 0.5% of sulphated polysaccharides from *U. intestinalis*, ground pork meat with around 2000 ppm of astaxanthin, and Chinese-style pork sausages with 0.1–0.5% of polysaccharides from *S. platensis* were found to have higher sensory acceptability than the original products (Cierach et al. 2009; Hamad et al. 2023; Jannat-Alipour et al. 2019; Luo et al. 2017; Pogorzelska et al. 2018). Thus, it seemed that algal extracts would be suitable ingredients or additives to improve both functional properties and sensory profiles of meat products.

In general, an algal extract or metabolite was expected to improve a single aspect of a food items. Adding astaxanthin or fucoxanthin to pork meat and turkey meat sausages aimed to reduce lipid oxidations in the meat products respectively (Pogorzelska et al. 2018; Sellimi et al. 2017b). The aqueous extract from *P. palmata* acted as a flavouring to replace salt in the ham (Barbieri et al. 2016). The PUFA content of sausage could be fortified by adding algal oil (de Ciriano et al. 2010b). Among different algal extracts and metabolites, sulphated polysaccharides or crude polysaccharides extracted from both macroalgae and microalgae appeared to exert both antioxidant and antimicrobial activities (Ben Hlima et al. 2021; Feki et al. 2021; Ghariani et al. 2022; Hamzaoui et al. 2020). Meanwhile, adding these algal polysaccharide extract could also modify the texture of the meat products (Feki et al. 2021; Jannat-Alipour et al. 2019). Thus, the algal polysaccharide extracts could be further developed into multiple functional food additives, which might serve as preservatives and texturizers simultaneously.

24.4 Commercially Food Items with Algal Ingredients and Prospectives

Numerous researches have been conducted to extract algal metabolites for food productions and develop novel food items with algal ingredients. In the market, refined agar, alginate, and carrageenan as well as encapsulated algal biomass supplements have been commercialized for decades. In the past decade, a few companies have started to manufacture algal oil and alternative protein, like Kuehnle AgroSystems Inc. (https://www.kuehnleagro.com/kuehnle-agrosystems-alternative-protein) and Bioriginal (https://www.bioriginal.com/products/algalpure-dha-algal-oil/). The algal oil and protein are mainly used as supplements to increase the nutritional values of food items and animal feeds. Although algal phenolic compounds and sulphated polysaccharides were potential antioxidants, antimicrobial agents, preservatives or even texturizer for food productions, none of them have been successfully commercialized yet.

Now a day, just a few novel meat alternative and beverage with algal ingredients are available in markets, like algal-based bacon, algal-based tuna and *Spirulina*-based soda (Matos et al. 2022). The algal-based bacon alternative is made of proteins extracted from red seaweed manufactured by Umaro Foods Inc. (https://www.umarofoods.com/). The algal-based tuna manufactured by Hooked contains algal protein as one of the main ingredients (https://hookedfoods.com/product/toonish/). The *Spirulina*-based soda is manufactured by FUL Water (https://fulfoods.com/collections/shop-all). These meat alternatives and beverage products have already been sold across United States and Europe.

These meat alternatives and beverage not just have high sensory profiles and nutritional values, but also are thought to be more eco-friendly than conventional products since cultivations of algae were considered as efficient strategies for carbon capture and storage (Onyeaka et al. 2021). Thus, both macroalgae and microalgae were thought to be sustainable sources for food ingredients and food additives to combat against climate change and food insecurity (Diaz et al. 2022). Thus, it is expected that food industry, academia and investors would put more efforts to keep on developing novel algal sustainable food ingredients and additives. In the near future, it will not be surprised that our food products contain multiple different algal ingredients and additives.

References

Aganduk AA, Matanjun P, Tan TS, Khor BH (2023) Proximate and physical analyses of crackers incorporated with red seaweed, *Kappaphycus alvarezii*. J Appl Phycol. https://doi.org/10.1007/s10811-023-03022-y

Agregan R, Franco D, Carballo J, Tomasevic I, Barba FJ, Gomez B, Muchenje V, Lorenzo JM (2018) Shelf life study of healthy pork liver pate with added seaweed extracts from *Ascophyllum nodosum*, *Fucus vesiculosus* and *Bifurcaria bifurcata*. Food Res Int 112:400–411. https://doi.org/10.1016/j.foodres.2018.06.063

Alejandre M, Passarini D, Astiasaran I, Ansorena D (2017) The effect of low-fat beef patties formulated with a low-energy fat analogue enriched in long-chain polyunsaturated fatty acids on lipid oxidation and sensory attributes. Meat Sci 134:7–13. https://doi.org/10.1016/j.meatsci.2017.07.009

Ampofo J, Abbey L (2022) Microalgae: bioactive composition, health benefits, safety and prospects as potential high-value ingredients for the functional food industry. Foods 11(12). https://doi.org/10.3390/foods11121744

Arufe S, Della Valle G, Chiron H, Chenlo F, Sineiro J, Moreira R (2018) Effect of brown seaweed powder on physical and textural properties of wheat bread. Eur Food Res Technol 244(1):1–10. https://doi.org/10.1007/s00217-017-2929-8

Arufe S, Chenlo F, Sineiro J, Moreira R (2019a) Effect of brown seaweed addition and starch gelatinization on gluten-free chestnut flour doughs and cookies. J Food Meas Charact 13(4):2571–2580. https://doi.org/10.1007/s11694-019-00177-6

Arufe S, Sineiro J, Moreira R (2019b) Determination of thermal transitions of gluten-free chestnut flour doughs enriched with brown seaweed powders and antioxidant properties of baked cookies. Heliyon 5(6):e01805. https://doi.org/10.1016/j.heliyon.2019.e01805

Atik DS, Gurbuz B, Boluk E, Palabiyik I (2021) Development of vegan kefir fortified with *Spirulina platensis*. Food Biosci 42:101050. https://doi.org/10.1016/j.fbio.2021.101050

Bae SM, Cho MG, Jeong JY (2017) Effects of various calcium powders as replacers for synthetic phosphate on the quality properties of ground pork meat products. Korean J Food Sci Anim Resour 37(3):456–463. https://doi.org/10.5851/kosfa.2017.37.3.456

Barbieri G, Barbieri G, Bergamaschi M, Francheschini M, Berizi E (2016) Reduction of NaCl in cooked ham by modification of the cooking process and addition of seaweed extract (*Palmaria palmata*). LWT 73:700–706. https://doi.org/10.1016/j.lwt.2016.06.057

Barcenilla C, Alvarez-Ordonez A, Lopez M, Alvseike O, Prieto M (2022) Microbiological safety and shelf-life of low-salt meat products—a review. Foods 11(15). https://doi.org/10.3390/foods11152331

Barkallah M, Dammak M, Louati I, Hentati F, Hadrich B, Mechichi T, Ayadi MA, Fendri I, Attia H, Abdelkafi S (2017) Effect of *Spirulina platensis* fortification on physicochemical, textural, antioxidant and sensory properties of yogurt during fermentation and storage. LWT 84:323–330. https://doi.org/10.1016/j.lwt.2017.05.071

Barkallah M, Ben Atitallah A, Hentati F, Dammak M, Hadrich B, Fendri I, Ayadi MA, Michaud P, Abdelkafi S (2019) Effect of *Spirulina platensis* biomass with high polysaccharides content on quality attributes of common carp (*Cyprinus carpio*) and common barbel (*Barbus barbus*) fish burgers. Appl Sci 9(11)

Batista AP, Niccolai A, Fradinho P, Fragoso S, Bursic I, Rodolfi L, Biondi N, Tredici MR, Sousa I, Raymundo A (2017) Microalgae biomass as an alternative ingredient in cookies: sensory, physical and chemical properties, antioxidant activity and in vitro digestibility. Algal Res 26:161–171. https://doi.org/10.1016/j.algal.2017.07.017

Batista AP, Niccolai A, Bursic I, Sousa I, Raymundo A, Rodolfi L, Biondi N, Tredici MR (2019) Microalgae as functional ingredients in savory food products: application to wheat crackers. Foods 8(12). https://doi.org/10.3390/foods8120611

Beheshtipour H, Mortazavian AM, Mohammadi R, Sohrabvandi S, Khosravi-Darani K (2013) Supplementation of *Spirulina platensis* and *Chlorella vulgaris* algae into probiotic fermented milks. Compr Rev Food Sci Food Saf 12(2):144–154. https://doi.org/10.1111/1541-4337.12004

Ben Atitallah A, Barkallah M, Hentati F, Dammak M, Ben Hlima H, Fendri I, Attia H, Michaud P, Abdelkafi S (2019a) Physicochemical, textural, antioxidant and sensory characteristics of microalgae-fortified canned fish burgers prepared from minced flesh of common barbel (*Barbus barbus*). Food Biosci 30:100417. https://doi.org/10.1016/j.fbio.2019.100417

Ben Atitallah A, Hentati F, Dammak M, Hadrich B, Fendri I, Ayadi MA, Michaud P, Abdelkafi S, Barkallah M (2019b) Effect of microalgae incorporation on quality characteristics and functional and antioxidant capacities of ready-to-eat fish burgers made from common carp (*Cyprinus carpio*). Appl Sci Basel 9(9):1830. https://doi.org/10.3390/app9091830

Ben Hlima H, Smaoui S, Barkallah M, Elhadef K, Tounsi L, Michaud P, Fendri I, Abdelkafi S (2021) Sulfated exopolysaccharides from *Porphyridium cruentum*: a useful strategy to extend the shelf life of minced beef meat. Int J Biol Macromol 193(Pt B):1215–1225. https://doi.org/10.1016/j.ijbiomac.2021.10.161

Berasategi I, Navarro-Blasco I, Calvo MI, Cavero RY, Astiasaran I, Ansorena D (2014) Healthy reduced-fat Bologna sausages enriched in ALA and DHA and stabilized with *Melissa officinalis* extract. Meat Sci 96(3):1185–1190. https://doi.org/10.1016/j.meatsci.2013.10.023

Bodnar A, Egerszegi I, Klecska E, Poti P, Pajor F (2020) Effect of *Schizochytrium limacinum* marine alga supplementation on fatty acid profile of goat cheese. Bulg J Agric Sci 26(3):646–651. WOS: 000547563100021

Campos Assumpcao de Amarante M, Cavalcante Braga AR, Sala L, Juliano Kalil S (2020) Colour stability and antioxidant activity of C-phycocyanin-added ice creams after in vitro digestion. Food Res Int 137:109602. https://doi.org/10.1016/j.foodres.2020.109602

Castillejo N, Martinez-Hernandez GB, Goffi V, Gomez PA, Aguayo E, Artes F, Artes-Hernandez F (2018) Natural vitamin B12 and fucose supplementation of green smoothies with edible algae and related quality changes during their shelf life. J Sci Food Agric 98(6):2411–2421. https://doi.org/10.1002/jsfa.8733

Chang HC, Wu LC (2008) Texture and quality properties of Chinese fresh egg noodles formulated with green seaweed (*Monostroma nitidum*) powder. J Food Sci 73(8):S398–S404. https://doi.org/10.1111/j.1750-3841.2008.00912.x

Chee CP, Gallaher JJ, Djordjevic D, Faraji H, McClements DJ, Decker EA, Hollender R, Peterson DG, Roberts RF, Coupland JN (2005) Chemical and sensory analysis of strawberry flavoured yogurt supplemented with an algae oil emulsion. J Dairy Res 72(3):311–316. https://doi.org/10.1017/S0022029905001068

Chen X, Fu X, Huang L, Xu J, Gao X (2021) Agar oligosaccharides: a review of preparation, structures, bioactivities and application. Carbohydr Polym 265:118076. https://doi.org/10.1016/j.carbpol.2021.118076

Chochkov R, Zlateva D, Stefanova D (2021) Effect of *Spirulina platensis* and Kelp algae on the content of thiamine and riboflavin in wheat bread. Ukr Food J 10(1):145–157. WOS: 000656057100011

Choi YS, Choi JH, Han DJ, Kim HY, Kim HW, Lee MA, Chung HJ, Kim CJ (2012) Effects of *Laminaria japonica* on the physico-chemical and sensory characteristics of reduced-fat pork patties. Meat Sci 91(1):1–7. https://doi.org/10.1016/j.meatsci.2011.11.011

Cian RE, Caballero MS, Sabbag N, González RJ, Drago SR (2014) Bio-accessibility of bioactive compounds (ACE inhibitors and antioxidants) from extruded maize products added with a red seaweed *Porphyra columbina*. LWT Food Sci Technol 55(1):51–58. https://doi.org/10.1016/j.lwt.2013.08.011

Cierach M, Modzelewska-Kapitula M, Szacilo K (2009) The influence of carrageenan on the properties of low-fat frankfurters. Meat Sci 82(3):295–299. https://doi.org/10.1016/j.meatsci.2009.01.025

Cirstea Lazar N, Nour V, Boruzi AI (2023) Effects of pork backfat replacement with emulsion gels formulated with a mixture of olive, chia and algae oils on the quality attributes of pork patties. Foods 12(3). https://doi.org/10.3390/foods12030519

Cofrades S, López-López I, Ruiz-Capillas C, Triki M, Jiménez-Colmenero F (2011) Quality characteristics of low-salt restructured poultry with microbial transglutaminase and seaweed. Meat Sci 87(4):373–380. https://doi.org/10.1016/j.meatsci.2010.11.014

Coleman B, Van Poucke C, Dewitte B, Ruttens A, Moerdijk-Poortvliet T, Latsos C, De Reu K, Blommaert L, Duquenne B, Timmermans K, van Houcke J, Muylaert K, Robbens J (2022) Potential of microalgae as flavoring agents for plant-based seafood alternatives. Future Foods 5:100139. https://doi.org/10.1016/j.fufo.2022.100139

Cox S, Abu-Ghannam N (2013) Enhancement of the phytochemical and fibre content of beef patties with *Himanthalia elongata* seaweed. Int J Food Sci Technol 48(11):2239–2249. https://doi.org/10.1111/ijfs.12210

Credence Research (2022) Algae products market by type (lipids, carrageenan, carotenoids, algae protein, alginates, other types), by application (food & beverage, nutraceutical & diet supplement, animal feed, personal care, pharmaceutical, other applications), by source (brown algae, blue-green algae, green algae, red algae, other sources), by form (solid, liquid)—growth, share, opportunities & competitive analysis, 2016–2028. https://www.credenceresearch.com/report/algae-products-market#summary

de Ciriano MGI, Larequi E, Rehecho S, Calvo MI, Cavero RY, Navarro-Blasco I, Astiasaran I, Ansorena D (2010a) Selenium, iodine, omega-3 PUFA and natural antioxidant from *Melissa officinalis* L.: a combination of components from healthier dry fermented sausages formulation. Meat Sci 85(2):274–279. https://doi.org/10.1016/j.meatsci.2010.01.012

de Ciriano MGI, Rehecho S, Calvo MI, Cavero RY, Navarro I, Astiasaran I, Ansorena D (2010b) Effect of lyophilized water extracts of *Melissa officinalis* on the stability of algae and linseed oil-in-water emulsion to be used as a functional ingredient in meat products. Meat Sci 85(2):373–377. https://doi.org/10.1016/j.meatsci.2010.01.007

del Olmo A, Picon A, Nuñez M (2018) Cheese supplementation with five species of edible seaweeds: effect on microbiota, antioxidant activity, colour, texture and sensory characteristics. Int Dairy J 84:36–45. https://doi.org/10.1016/j.idairyj.2018.04.004

Diaz CJ, Douglas KJ, Kang K, Kolarik AL, Malinovski R, Torres-Tiji Y, Molino JV, Badary A, Mayfield SP (2022) Developing algae as a sustainable food source. Front Nutr 9:1029841. https://doi.org/10.3389/fnut.2022.1029841

Diprat AB, Thys RCS, Rodrigues E, Rech R (2020) *Chlorella sorokiniana*: a new alternative source of carotenoids and proteins for gluten-free bread. LWT Food Sci Technol 134:109974. https://doi.org/10.1016/j.lwt.2020.109974

Durmaz Y, Kilicli M, Toker OS, Konar N, Palabiyik I, Tamtürk F (2020) Using spray-dried microalgae in ice cream formulation as a natural colorant: effect on physicochemical and functional properties. Algal Res 47:101811. https://doi.org/10.1016/j.algal.2020.101811

EFSA Panel on Nutrition, Novel Foods and Food Allergens (NDA), Turck D, Castenmiller J, De Henauw S, Hirsch-Ernst KI, Kearney J, Maciuk A, Mangelsdorf I, McArdle HJ, Naska A, Pelaez C, Pentieva K, Siani A, Thies F, Tsabouri S, Vinceti M, Cubadda F, Engel KH, Frenzel T, Heinonen M, Marchelli R, Neuhauser-Berthold M, Poulsen M, Sanz Y, Schlatter JR, van Loveren H, Ferreira L, Knutsen HK (2020) Safety of *Schizochytrium* sp. oil as a novel food pursuant to regulation (EU) 2015/2283. EFSA J 18(10):e06242. https://doi.org/10.2903/j.efsa.2020.6242

Espinosa-Ramírez J, Mondragón-Portocarrero AC, Rodríguez JA, Lorenzo JM, Santos EM (2023) Algae as a potential source of protein meat alternatives [review]. Front Nutr 10. https://doi.org/10.3389/fnut.2023.1254300

Etemadian Y, Shabanpour B, Ramzanpour Z, Shaviklo AR, Kordjazi M (2018) Evolution of *Sirophysalis trinodis* and *Polycladia myrica* water extract properties on corn snacks treated with them during 3 months of storage at ambient temperature. J Food Process Preserv 42(4):e13579. https://doi.org/10.1111/jfpp.13579

Faustino M, Machado D, Rodrigues D, Andrade JC, Freitas AC, Gomes AM (2023) Design and characterization of a cheese spread incorporating *Osmundea pinnatifida* extract. Foods 12(3):611. https://doi.org/10.3390/foods12030611

Feki A, Sellem I, Hamzaoui A, Ben Amar W, Mellouli L, Zariat A, Nasri M, Ben Amara I (2021) Effect of the incorporation of polysaccharide from *Falkenbergia rufolanosa* on beef sausages for quality and shelf life improvement. LWT Food Sci Technol 143:111139. https://doi.org/10.1016/j.lwt.2021.111139

Fellendorf S, O'Sullivan MG, Kerry JP (2016a) Impact of ingredient replacers on the physicochemical properties and sensory quality of reduced salt and fat black puddings. Meat Sci 113:17–25. https://doi.org/10.1016/j.meatsci.2015.11.006

Fellendorf S, O'Sullivan MG, Kerry JP (2016b) Effect of using ingredient replacers on the physicochemical properties and sensory quality of low-salt and low-fat white puddings. Eur Food Res Technol 242(12):2105–2118. https://doi.org/10.1007/s00217-016-2707-z

Fernandez-Martin F, Lopez-Lopez I, Cofrades S, Colmenero FJ (2009) Influence of adding Sea Spaghetti seaweed and replacing the animal fat with olive oil or a konjac gel on pork meat batter gelation. Potential protein/alginate association. Meat Sci 83(2):209–217. https://doi.org/10.1016/j.meatsci.2009.04.020

Fitzgerald C, Gallagher E, Doran L, Auty M, Prieto J, Hayes M (2014) Increasing the health benefits of bread: assessment of the physical and sensory qualities of bread formulated using a renin inhibitory *Palmaria palmata* protein hydrolysate. LWT Food Sci Technol 56(2):398–405. https://doi.org/10.1016/j.lwt.2013.11.031

Fradique M, Batista AP, Nunes MC, Gouveia L, Bandarra NM, Raymundo A (2013) *Isochrysis galbana* and *Diacronema vlkianum* biomass incorporation in pasta products as PUFA's source. LWT Food Sci Technol 50(1):312–319. https://doi.org/10.1016/j.lwt.2012.05.006

García-Segovia P, Pagán-Moreno MJ, Lara IF, Martínez-Monzó J (2017) Effect of microalgae incorporation on physicochemical and textural properties in wheat bread formulation. Food Sci Technol Int 23(5):437–447. https://doi.org/10.1177/1082013217700259

García-Vaquero M, Brunton N, Lafarga T (2021) Microalgae as a source of pigments for food applications. In: Lafarga T, Acién G (eds) Cultured microalgae for the food industry. Academic Press, pp 177–198. https://doi.org/10.1016/B978-0-12-821080-2.00014-9

Garcimartin A, Benedi J, Bastida S, Sanchez-Muniz FJ (2015) Aqueous extracts and suspensions of restructured pork formulated with *Undaria pinnatifida*, *Himanthalia elongata* and *Porphyra umbilicalis* distinctly affect the in vitro alpha-glucosidase activity and glucose diffusion. LWT Food Sci Technol 64(2):720–726. https://doi.org/10.1016/j.lwt.2015.06.050

Generalic Mekinic I, Simat V, Rathod NB, Hamed I, Cagalj M (2023) Algal carotenoids: chemistry, sources, and application. Foods 12(14). https://doi.org/10.3390/foods12142768

Ghariani M, Ben Saad H, Hamzaoui A, Ajela M, Hilali A, Ben Amara I (2022) Effect of the incorporation of polysaccharides from green alga *Bryopsis plumosa* on beef sausages quality. Cell Mol Biol 68(10):30–37. https://doi.org/10.14715/cmb/2022.68.10.5

Golmakani M-T, Soleimanian-Zad S, Alavi N, Nazari E, Eskandari MH (2019) Effect of *Spirulina* (*Arthrospira platensis*) powder on probiotic bacteriologically acidified feta-type cheese. J Appl Phycol 31(2):1085–1094. https://doi.org/10.1007/s10811-018-1611-2

Goni I, Valdivieso L, Gudiel-Urbano M (2002) Capacity of edible seaweeds to modify in vitro starch digestibility of wheat bread. Nahrung-Food 46(1):18–20. https://doi.org/10.1002/1521-3803(20020101)46:1<18::Aid-Food18>3.0.Co;2-C

Graça C, Fradinho P, Sousa I, Raymundo A (2018) Impact of *Chlorella vulgaris* on the rheology of wheat flour dough and bread texture. LWT 89:466–474. https://doi.org/10.1016/j.lwt.2017.11.024

Hall AC, Fairclough AC, Mahadevan K, Paxman JR (2010) Seaweed (*Ascophyllum nodosum*) enriched bread is acceptable to consumers. Proc Nutr Soc 69(OCE5):E352. https://doi.org/10.1017/S0029665110002132

Hall AC, Fairclough AC, Mahadevan K, Paxman JR (2012) *Ascophyllum nodosum* enriched bread reduces subsequent energy intake with no effect on post-prandial glucose and cholesterol in healthy, overweight males. A pilot study. Appetite 58(1):379–386. https://doi.org/10.1016/j.appet.2011.11.002

Hamad G, Amer A, Kirrella G, Mehany T, Elfayoumy RA, Elsabagh R, Elghazaly EM, Esatbeyoglu T, Taha A, Zeitoun A (2023) Evaluation of the prevalence of *Staphylococcus aureus* in chicken fillets and its bio-control using different seaweed extracts. Foods 12(1):20. https://doi.org/10.3390/foods12010020

Hamzaoui A, Ghariani M, Sellem I, Hamdi M, Feki A, Jaballi I, Nasri M, Amara IB (2020) Extraction, characterization and biological properties of polysaccharide derived from green seaweed "*Chaetomorpha linum*" and its potential application in Tunisian beef sausages. Int J Biol Macromol 148:1156–1168. https://doi.org/10.1016/j.ijbiomac.2020.01.009

Hassanzadeh H, Ghanbarzadeh B, Galali Y, Bagheri H (2022) The physicochemical properties of the *Spirulina*-wheat germ-enriched high-protein functional beverage based on pear-cantaloupe juice. Food Sci Nutr 10(11):3651–3661. https://doi.org/10.1002/fsn3.2963

Hell A, Labrie S, Beaulieu L (2018) Effect of seaweed flakes addition on the development of bioactivities in functional Camembert-type cheese. Int J Food Sci Technol 53(4):1054–1064. https://doi.org/10.1111/ijfs.13681

Hentati F, Barkallah M, Ben Atitallah A, Dammak M, Louati I, Pierre G, Fendri I, Attia H, Michaud P, Abdelkafi S (2019) Quality characteristics and functional and antioxidant capacities of algae-fortified fish burgers prepared from common barbel (*Barbus barbus*). Biomed Res Int 2019:2907542. https://doi.org/10.1155/2019/2907542

Hossain A, Brennan MA, Mason SL, Guo X, Zeng XA, Brennan CS (2017) The effect of astaxanthin-rich microalgae "*Haematococcus pluvialis*" and wholemeal flours incorporation in improving the physical and functional properties of cookies. Foods 6(8). https://doi.org/10.3390/foods6080057

Hosseinkhani N, McCauley JI, Ralph PJ (2022) Key challenges for the commercial expansion of ingredients from algae into human food products. Algal Res 64:102696. https://doi.org/10.1016/j.algal.2022.102696

Huang M, Yang H (2019) Eucheuma powder as a partial flour replacement and its effect on the properties of sponge cake. LWT 110:262–268. https://doi.org/10.1016/j.lwt.2019.04.087

Ilikkan K, Bagdat ES (2022) The effect of kefir enrichment with *Arthrospira platensis* and *Chlorella vulgaris* on kefir microbiota, antioxidant, and physicochemical properties. J Appl Phycol. https://doi.org/10.1007/s10811-022-02892-y

Jannat-Alipour H, Rezaei M, Shabanpour B, Tabarsa M, Rafipour F (2019) Addition of seaweed powder and sulphated polysaccharide on shelf_life extension of functional fish surimi restructured product. J Food Sci Technol 56(8):3777–3789. https://doi.org/10.1007/s13197-019-03846-y

Jespersen L, Strømdahl LD, Olsen K, Skibsted LH (2005) Heat and light stability of three natural blue colorants for use in confectionery and beverages. Eur Food Res Technol 220(3):261–266. https://doi.org/10.1007/s00217-004-1062-7

Jimenez-Lopez C, Pereira AG, Lourenço-Lopes C, Garcia-Oliveira P, Cassani L, Fraga-Corral M, Prieto MA, Simal-Gandara J (2021) Main bioactive phenolic compounds in marine algae and their mechanisms of action supporting potential health benefits. Food Chem 341:128262. https://doi.org/10.1016/j.foodchem.2020.128262

Kelly FC (1961) Iodine in medicine and pharmacy since its discovery—1811–1961. Proc R Soc Med 54(10):831–836. https://www.ncbi.nlm.nih.gov/pubmed/19994130

Khemiri S, Khelifi N, Nunes MC, Ferreira A, Gouveia L, Smaali I, Raymundo A (2020) Microalgae biomass as an additional ingredient of gluten-free bread: dough rheology, texture quality and nutritional properties. Algal Res 50:101998. https://doi.org/10.1016/j.algal.2020.101998

Khemiri S, Nunes MC, Raymundo A, Smaali I (2022) In vitro starch digestibility and estimation of glycemic index in algae-based couscous. Int J Food Sci Technol 57(11):7245–7253. https://doi.org/10.1111/ijfs.16073

Koli DK, Rudra SG, Bhowmik A, Pabbi S (2022) Nutritional, functional, textural and sensory evaluation of *Spirulina* enriched green pasta: a potential dietary and health supplement. Foods 11(7). https://doi.org/10.3390/foods11070979

Kumar M, Sharma BD (2004) The storage stability and textural, physico-chemical and sensory quality of low-fat ground pork patties with carrageenan as fat replacer. Int J Food Sci Technol 39(1):31–42. https://doi.org/10.1111/j.1365-2621.2004.00743.x

Kumar A, Krishnamoorthy E, Devi HM, Uchoi D, Tejpal CS, Ninan G, Zynudheen AA (2018) Influence of sea grapes (*Caulerpa racemosa*) supplementation on physical, functional, and antioxidant properties of semi-sweet biscuits. J Appl Phycol 30(2):1393–1403. https://doi.org/10.1007/s10811-017-1310-4

Kumoro AC, Johnny D, Alfilovita D (2016) Incorporation of microalgae and seaweed in instant fried wheat noodles manufacturing: nutrition and culinary properties study. Int Food Res J 23(2):715–722. WOS: 000423020500036

Lafarga T, Acién-Fernández FG, Castellari M, Villaró S, Bobo G, Aguiló-Aguayo I (2019a) Effect of microalgae incorporation on the physicochemical, nutritional, and sensorial properties of an innovative broccoli soup. LWT 111:167–174. https://doi.org/10.1016/j.lwt.2019.05.037

Lafarga T, Mayre E, Echeverria G, Viñas I, Villaró S, Acién-Fernández FG, Castellari M, Aguiló-Aguayo I (2019b) Potential of the microalgae *Nannochloropsis* and *Tetraselmis* for being used as innovative ingredients in baked goods. LWT 115:108439. https://doi.org/10.1016/j.lwt.2019.108439

Liu L, Heinrich M, Myers S, Dworjanyn SA (2012) Towards a better understanding of medicinal uses of the brown seaweed *Sargassum* in traditional Chinese medicine: a phytochemical and pharmacological review. J Ethnopharmacol 142(3):591–619. https://doi.org/10.1016/j.jep.2012.05.046

Lopes JS, Simao RD, Mendes GD, Demarco M, de Moraes JO, Hayashi L, Laurindo JB, Tribuzi G (2023) *Kappaphycus alvarezii* flours as an ingredient for seaweed-enriched, rice-based, snacks: raw algae pretreatment and physical properties of the dough and snacks. Int J Food Sci Technol 58(5):2448–2457. https://doi.org/10.1111/ijfs.16383

Lopez-Lopez I, Cofrades S, Jimenez-Colmenero F (2009) Low-fat frankfurters enriched with n−3 PUFA and edible seaweed: effects of olive oil and chilled storage on physicochemical, sensory and microbial characteristics. Meat Sci 83(1):148–154. https://doi.org/10.1016/j.meatsci.2009.04.014

Lucas BF, de Morais MG, Santos TD, Costa JAV (2017) Effect of *Spirulina* addition on the physicochemical and structural properties of extruded snacks. Food Sci Technol 37

Lucas BF, de Morais MG, Santos TD, Costa JAV (2018) *Spirulina* for snack enrichment: nutritional, physical and sensory evaluations. LWT 90:270–276. https://doi.org/10.1016/j.lwt.2017.12.032

Lucas BF, de Rosa APC, de Carvalho LF, de Morais MG, Santos TD, Costa JAV (2020) Snack bars enriched with *Spirulina* for schoolchildren nutrition. Food Sci Technol 40

Luo A, Feng J, Hu B, Lv J, Chen CO, Xie S (2017) Polysaccharides in *Spirulina platensis* improve antioxidant capacity of Chinese-style sausage. J Food Sci 82(11):2591–2597. https://doi.org/10.1111/1750-3841.13946

Mamat H, Matanjun P, Ibrahim S, Md. Amin SF, Abdul Hamid M, Rameli AS (2014) The effect of seaweed composite flour on the textural properties of dough and bread. J Appl Phycol 26(2):1057–1062. https://doi.org/10.1007/s10811-013-0082-8

Marcal C, Pinto CA, Silva AMS, Monteiro C, Saraiva JA, Cardoso SM (2021) Macroalgae-fortified sausages: nutritional and quality aspects influenced by non-thermal high-pressure processing. Foods 10(2). https://doi.org/10.3390/foods10020209

Marti-Quijal FJ, Zamuz S, Galvez F, Roohinejad S, Tiwari BK, Gomez B, Barba FJ, Lorenzo JM (2018) Replacement of soy protein with other legumes or algae in turkey breast formulation: changes in physicochemical and technological properties. J Food Process Preserv 42(12):e13845. https://doi.org/10.1111/jfpp.13845

Marti-Quijal FJ, Zamuz S, Tomašević I, Gómez B, Rocchetti G, Lucini L, Remize F, Barba FJ, Lorenzo JM (2019a) Influence of different sources of vegetable, whey and microalgae proteins on the physicochemical properties and amino acid profile of fresh pork sausages. LWT 110:316–323. https://doi.org/10.1016/j.lwt.2019.04.097

Marti-Quijal FJ, Zamuz S, Tomasevic I, Rocchetti G, Lucini L, Marszalek K, Barba FJ, Lorenzo JM (2019b) A chemometric approach to evaluate the impact of pulses, *Chlorella* and *Spirulina* on proximate composition, amino acid, and physicochemical properties of turkey burgers. J Sci Food Agric 99(7):3672–3680. https://doi.org/10.1002/jsfa.9595

Matos ÂP, Novelli E, Tribuzi G (2022) Use of algae as food ingredient: sensory acceptance and commercial products [mini review]. Front Food Sci Technol 2. https://doi.org/10.3389/frfst.2022.989801

Menezes BS, Coelho MS, Meza SLR, Salas-Mellado M, Souza MRAZ (2015) Macroalgal biomass as an additional ingredient of bread. Int Food Res J 22(2):812–817. WOS: 000422938700050

Mohibbullah M, Amin A, Talha MA, Baten MA, Rana MM, Sabuz AA, Newaz AW, Choi JS (2023) Physicochemical and nutritional characteristics of cookies prepared with untapped seaweed *Ulva intestinalis*: an approach to value addition as a functional food. Foods 12(1). https://doi.org/10.3390/foods12010205

Moroney NC, O'Grady MN, O'Doherty JV, Kerry JP (2013) Effect of a brown seaweed (*Laminaria digitata*) extract containing laminarin and fucoidan on the quality and shelf-life of fresh and cooked minced pork patties. Meat Sci 94(3):304–311. https://doi.org/10.1016/j.meatsci.2013.02.010

Mouritsen OG, Rhatigan P, Pérez-Lloréns JL (2018) World cuisine of seaweeds: science meets gastronomy. Int J Gastron Food Sci 14:55–56. https://doi.org/10.1016/j.ijgfs.2018.09.002

Munoz-Tebar N, Ong L, Gamlath CJ, Yatipanthalawa BS, Ashokkumar M, Gras SL, Berruga MI, Martin GJO (2022) Nutrient enrichment of dairy curd by incorporation of whole and ruptured microalgal cells (*Nannochloropsis salina*). Innov Food Sci Emerg Technol 82:103211. https://doi.org/10.1016/j.ifset.2022.103211

Nagai NF, Lorenzo JM, Ranalli N, Pérez-Álvarez JÁ, Sepulveda N, Domínguez R, Santos EM, Teixeira A, Andrés SC (2022) Use of seaweed powder (*Undaria* sp.) as a functional ingredient in low-fat pork burgers. Algal Res 67:102862. https://doi.org/10.1016/j.algal.2022.102862

Niccolai A, Venturi M, Galli V, Pini N, Rodolfi L, Biondi N, D'Ottavio M, Batista AP, Raymundo A, Granchi L, Tredici MR (2019) Development of new microalgae-based sourdough "crostini": functional effects of *Arthrospira platensis* (*Spirulina*) addition. Sci Rep 9(1):19433. https://doi.org/10.1038/s41598-019-55840-1

Nunes MC, Fernandes I, Vasco I, Sousa I, Raymundo A (2020) *Tetraselmis chuii* as a sustainable and healthy ingredient to produce gluten-free bread: impact on structure, colour and bioactivity. Foods 9(5). https://doi.org/10.3390/foods9050579

Nunez M, Picon A (2017) Seaweeds in yogurt and quark supplementation: influence of five dehydrated edible seaweeds on sensory characteristics. Int J Food Sci Technol 52(2):431–438. https://doi.org/10.1111/ijfs.13298

O'Sullivan AM, O'Callaghan YC, O'Grady MN, Waldron DS, Smyth TJ, O'Brien NM, Kerry JP (2014) An examination of the potential of seaweed extracts as functional ingredients in milk. Int J Dairy Technol 67(2):182–193. https://doi.org/10.1111/1471-0307.12121

Oh H, Lee P, Kim SY, Kim YS (2020) Preparation of cookies with various native seaweeds found on the Korean Coast. J Aquat Food Prod Technol 29(2):167–174. https://doi.org/10.1080/10498850.2019.1707925

Onyeaka H, Miri T, Obileke K, Hart A, Anumudu C, Al-Sharify ZT (2021) Minimizing carbon footprint via microalgae as a biological capture. Carbon Capture Sci Technol 1:100007. https://doi.org/10.1016/j.ccst.2021.100007

Ozyurt G, Uslu L, Yuvka I, Gokdogan S, Atci G, Ak B, Isik O (2015) Evaluation of the cooking quality characteristics of pasta enriched with *Spirulina platensis*. J Food Qual 38(4):268–272. https://doi.org/10.1111/jfq.12142

Parniakov O, Toepfl S, Barba FJ, Granato D, Zamuz S, Galvez F, Lorenzo JM (2018) Impact of the soy protein replacement by legumes and algae based proteins on the quality of chicken rotti. J Food Sci Technol 55(7):2552–2559. https://doi.org/10.1007/s13197-018-3175-1

Paulo MC, Marques J, Cardoso C, Coutinho J, Gomes R, Gomes-Bispo A, Afonso C, Bandarra NM (2020) The development of a novel functional food: bioactive lipids in yogurts enriched with *Aurantiochytrium* sp. biomass. Food Funct 11(11):9721–9728. https://doi.org/10.1039/d0fo01884h

Pehlivan AD, Yadel I, Kilic N, Ozturk HI (2023) The incorporation of *Chlorella vulgaris* and *Chondrus crispus* algae in the production of functional ayran drinks: effects on physicochemical, microbiological, and sensory characteristics. J Food Meas Charact 17(3):3019–3032. https://doi.org/10.1007/s11694-023-01840-9

Pogorzelska E, Godziszewska J, Brodowska M, Wierzbicka A (2018) Antioxidant potential of *Haematococcus pluvialis* extract rich in astaxanthin on colour and oxidative stability of raw ground pork meat during refrigerated storage. Meat Sci 135:54–61. https://doi.org/10.1016/j.meatsci.2017.09.002

Prabhasankar P, Ganesan P, Bhaskar N, Hirose A, Stephen N, Gowda LR, Hosokawa M, Miyashita K (2009) Edible Japanese seaweed, wakame (*Undaria pinnatifida*) as an ingredient in pasta: chemical, functional and structural evaluation. Food Chem 115(2):501–508. https://doi.org/10.1016/j.foodchem.2008.12.047

Qazi WM, Ballance S, Kousoulaki K, Uhlen AK, Kleinegris DMM, Skjanes K, Rieder A (2021a) Protein enrichment of wheat bread with microalgae: *Microchloropsis gaditana*, *Tetraselmis chuii* and *Chlorella vulgaris*. Foods 10(12). https://doi.org/10.3390/foods10123078

Qazi WM, Ballance S, Uhlen AK, Kousoulaki K, Haugen JE, Rieder A (2021b) Protein enrichment of wheat bread with the marine green microalgae *Tetraselmis chuii*—impact on dough rheology and bread quality. LWT Food Sci Technol 143:111115. https://doi.org/10.1016/j.lwt.2021.111115

Qazi MW, de Sousa IG, Nunes MC, Raymundo A (2022) Improving the nutritional, structural, and sensory properties of gluten-free bread with different species of microalgae. Foods 11(3). https://doi.org/10.3390/foods11030397

Qin Y, Jiang J, Zhao L, Zhang J, Wang F (2018) Applications of alginate as a functional food ingredient. In: Grumezescu AM, Holban AM (eds) Biopolymers for food design. Academic Press, pp 409–429. https://doi.org/10.1016/B978-0-12-811449-0.00013-X

Quitral V, Sepúlveda M, Gamero-Vega G, Jiménez P (2022) Seaweeds in bakery and farinaceous foods: a mini-review. Int J Gastron Food Sci 28:100403. https://doi.org/10.1016/j.ijgfs.2021.100403

Rabelo SF, Lemes AC, Takeuchi KP, Frata MT, de Carvalho JCM, Danesi EDG (2013) Development of cassava doughnuts enriched with *Spirulina platensis* biomass. Braz J Food Technol 16

Rajapakse N, Kim S-K (2011) Nutritional and digestive health benefits of seaweed. In: Kim S-K (ed) Advances in food and nutrition research, vol 64. Academic Press, pp 17–28. https://doi.org/10.1016/B978-0-12-387669-0.00002-8

Ribeiro AR, Madeira T, Botelho G, Martins D, Ferreira RM, Silva AMS, Cardoso SM, Costa R (2022) Brown algae *Fucus vesiculosus* in pasta: effects on textural quality, cooking properties, and sensorial traits. Foods 11(11). https://doi.org/10.3390/foods11111561

Rioux L-E, Turgeon SL (2015) Seaweed carbohydrates. In: Tiwari BK, Troy DJ (eds) Seaweed sustainability. Academic Press, pp 141–192. https://doi.org/10.1016/B978-0-12-418697-2.00007-6

Robertson RC, Gracia Mateo MR, O'Grady MN, Guihéneuf F, Stengel DB, Ross RP, Fitzgerald GF, Kerry JP, Stanton C (2016) An assessment of the techno-functional and sensory properties of yoghurt fortified with a lipid extract from the microalga *Pavlova lutheri*. Innov Food Sci Emerg Technol 37:237–246. https://doi.org/10.1016/j.ifset.2016.03.017

Rodríguez De Marco E, Steffolani ME, Martínez CS, León AE (2014) Effects of *Spirulina* biomass on the technological and nutritional quality of bread wheat pasta. LWT Food Sci Technol 58(1):102–108. https://doi.org/10.1016/j.lwt.2014.02.054

Różyło R, Hameed Hassoon W, Gawlik-Dziki U, Siastała M, Dziki D (2017) Study on the physical and antioxidant properties of gluten-free bread with brown algae. CyTA J Food 15(2):196–203. https://doi.org/10.1080/19476337.2016.1236839

Rubino S, Aymerich T, Peteiro C, Bover-Cid S, Hortos M (2023) Challenge test in Catalan "mató" fresh cheese to assess the antimicrobial activity of *Ericaria selaginoides* extracts against *Bacillus cereus*. Appl Sci Basel 13(4):2207. https://doi.org/10.3390/app13042207

Saharan V, Jood S (2021) Effect of storage on *Spirulina platensis* powder supplemented breads. J Food Sci Technol 58(3):978–984. https://doi.org/10.1007/s13197-020-04612-1

Sahni P, Sharma S, Singh B (2019) Evaluation and quality assessment of defatted microalgae meal of *Chlorella* as an alternative food ingredient in cookies. Nutr Food Sci 49(2):221–231. https://doi.org/10.1108/Nfs-06-2018-0171

Santos TD, de Freitas BCB, Moreira JB, Zanfonato K, Costa JAV (2016) Development of powdered food with the addition of *Spirulina* for food supplementation of the elderly population. Innov Food Sci Emerg Technol 37:216–220. https://doi.org/10.1016/j.ifset.2016.07.016

Sellimi S, Benslima A, Ksouda G, Montero VB, Hajji M, Nasri M (2017a) Safer and healthier reduced nitrites turkey meat sausages using lyophilized *Cystoseira barbata* seaweed extract. J Complement Integr Med 15(1). https://doi.org/10.1515/jcim-2017-0061

Sellimi S, Ksouda G, Benslima A, Nasri R, Rinaudo M, Nasri M, Hajji M (2017b) Enhancing colour and oxidative stabilities of reduced-nitrite turkey meat sausages during refrigerated storage using fucoxanthin purified from the Tunisian seaweed *Cystoseira barbata*. Food Chem Toxicol 107(Pt B):620–629. https://doi.org/10.1016/j.fct.2017.04.001

Senthil A, Mamatha BS, Mahadevaswamy M (2005) Effect of using seaweed (eucheuma) powder on the quality of fish cutlet. Int J Food Sci Nutr 56(5):327–335. https://doi.org/10.1080/09637480500224205

Serna-Saldivar SO, Zorrilla R, De la Parra C, Stagnitti G, Abril R (2006) Effect of DHA containing oils and powders on baking performance and quality of white pan bread. Plant Foods Hum Nutr 61(3):121–129. https://doi.org/10.1007/s11130-006-0009-5

Shafiei R, Mostaghim T (2022) Improving shelf life of calf fillet in refrigerated storage using edible coating based on chitosan/natamycin containing *Spirulina platensis* and *Chlorella vulgaris* microalgae. J Food Meas Charact 16(1):145–161. https://doi.org/10.1007/s11694-021-01153-9

Sik B, Buzás H, Kapcsándi V, Lakatos E, Daróczi F, Székelyhidi R (2023) Antioxidant and polyphenol content of different milk and dairy products. J King Saud Univ Sci 35(7):102839. https://doi.org/10.1016/j.jksus.2023.102839

Smith JP, Daifas DP, El-Khoury W, Koukoutsis J, El-Khoury A (2004) Shelf life and safety concerns of bakery products—a review. Crit Rev Food Sci Nutr 44(1):19–55. https://doi.org/10.1080/10408690490263774

Spagnuolo PA, Dalgleish DG, Goff HD, Morris ER (2005) Kappa-carrageenan interactions in systems containing casein micelles and polysaccharide stabilizers. Food Hydrocolloids 19(3):371–377. https://doi.org/10.1016/j.foodhyd.2004.10.003

Takyar MBT, Khajavi SH, Safari R (2019) Evaluation of antioxidant properties of *Chlorella vulgaris* and *Spirulina platensis* and their application in order to extend the shelf life of rainbow trout (*Oncorhynchus mykiss*) fillets during refrigerated storage. LWT Food Sci Technol 100:244–249. https://doi.org/10.1016/j.lwt.2018.10.079

Tańska M, Konopka I, Ruszkowska M (2017) Sensory, physico-chemical and water sorption properties of corn extrudates enriched with *Spirulina*. Plant Foods Hum Nutr 72(3):250–257. https://doi.org/10.1007/s11130-017-0628-z

Thirumdas R, Brncic M, Brncic SR, Barba FJ, Galvez F, Zamuz S, Lacomba R, Lorenzo JM (2018) Evaluating the impact of vegetal and microalgae protein sources on proximate composition, amino acid profile, and physicochemical properties of fermented Spanish "chorizo" sausages. J Food Process Preserv 42(11):e13817. https://doi.org/10.1111/jfpp.13817

Tohamy MM, Ali MA, Shaaban HA, Mohamad AG, Hasanain AM (2018) Production of functional spreadable processed cheese using *Chlorella vulgaris*. Acta Sci Pol Technol Aliment 17(4):347–358. https://doi.org/10.17306/J.AFS.0589

Tseng CK (1944) Utilization of seaweeds. Sci Mon 59(1):37–46. http://www.jstor.org/stable/18503

Turuk AS, Banerjee K (2023) Blending seaweed into bakery products. J Appl Phycol. https://doi.org/10.1007/s10811-023-02982-5

Udo T, Mummaleti G, Mohan A, Singh RK, Kong F (2023) Current and emerging applications of carrageenan in the food industry. Food Res Int 173:113369. https://doi.org/10.1016/j.foodres.2023.113369

Uribe-Wandurraga ZN, Igual M, Garcia-Segovia P, Martinez-Monzo J (2019) Effect of microalgae addition on mineral content, colour and mechanical properties of breadsticks. Food Funct 10(8):4685–4692. https://doi.org/10.1039/c9fo00286c

Vilar EG, Ouyang H, O'Sullivan MG, Kerry JP, Hamill RM, O'Grady MN, Mohammed HO, Kilcawley KN (2020) Effect of salt reduction and inclusion of 1% edible seaweeds on the chemical, sensory and volatile component profile of reformulated frankfurters. Meat Sci 161:108001. https://doi.org/10.1016/j.meatsci.2019.108001

Wang M, Zhou J, Tavares J, Pinto CA, Saraiva JA, Prieto MA, Cao H, Xiao J, Simal-Gandara J, Barba FJ (2022) Applications of algae to obtain healthier meat products: a critical review on nutrients, acceptability and quality. Crit Rev Food Sci Nutr 1–18. https://doi.org/10.1080/10408398.2022.2054939

Wang C, Wang E, Bai Y, Lu Y, Qi H (2023) Encapsulated fucoxanthin improves the structure and functional properties of fermented yogurt during cold storage. Food Chem 419:136076. https://doi.org/10.1016/j.foodchem.2023.136076

Wilkin JD, Ross K, Alric T, Hooper M, Grigor JV, Chu BS (2021) Optimisation of concentration of *Undaria pinnarifida* (Wakame) and *Himanthalia elongata* (Sea Spaghetti) varieties to effect digestibility, texture and consumer attribute preference. J Aquat Food Prod Technol 30(8):932–943. https://doi.org/10.1080/10498850.2021.1958114

Zamuz S, Purrinos L, Galvez F, Zdolec N, Muchenje V, Barba FJ, Lorenzo JM (2019) Influence of the addition of different origin sources of protein on meat products sensory acceptance. J Food Process Preserv 43(5):e13940. https://doi.org/10.1111/jfpp.13940

Zanganeh E, Mirzaei H, Jafari SM, Javadi A, Afshar Mogaddam MR (2022) *Spirulina platensis* extract nanoliposomes: preparation, characterization, and application to white cheese. J AOAC Int 105(3):827–834. https://doi.org/10.1093/jaoacint/qsab162

Zlateva D, Chochkov R (2019) Effect of *Spirulina platensis* on the crumb firming of wheat bread during storage. Ukr Food J 8(4):851–860. https://doi.org/10.24263/2304-974x-2019-8-4-15

Zlateva D, Ivanova P, Chochkov R, Stefanova D (2020) Effect of *Spirulina platensis* and Kelp on the antioxidant activity of wheat bread. Ukr Food J 9(3):636–650. https://doi.org/10.24263/2304-974x-2020-9-3-12

Zouari N, Abid M, Fakhfakh N, Ayadi MA, Zorgui L, Ayadi M, Attia H (2011) Blue-green algae (*Arthrospira platensis*) as an ingredient in pasta: free radical scavenging activity, sensory and cooking characteristics evaluation. Int J Food Sci Nutr 62(8):811–813. https://doi.org/10.3109/09637486.2011.582461

Zugcic T, Abdelkebir R, Barba FJ, Rezek-Jambrak A, Galvez F, Zamuz S, Granato D, Lorenzo JM (2018) Effects of pulses and microalgal proteins on quality traits of beef patties. J Food Sci Technol 55(11):4544–4553. https://doi.org/10.1007/s13197-018-3390-9

Chapter 25
Safety, Toxicological and Allergenic Aspects of Using Algae for Food

Christine Kyarimpa, Tom Omute, Caroline K. Nakiguli, Alice V. Khanakwa, Christopher Angiro, Ivan Kahwa, Fortunate Ahumuza, and Timothy Omara

Abstract Consumption of algae has been historically practiced, especially in East Asia and the Pacific region cultures. However, sporadic events and empirical studies have suggested that some compounds could be triggering intoxications, allergic reactions and mortalities in humans who consumed algae. This chapter is an effort to explore with in-depth attention the safety, toxicological and allergic reactions following human consumption of algae. Based on retrieved literature, it is clear that toxicities and allergies from ingestion of algae are not a rarity, and to date, at least seventy (70) illnesses, six (6) allergic reactions and fourteen (14) mortalities have

C. Kyarimpa
Department of Chemistry, Kyambogo University, P.O. Box 1, Kampala, Uganda

T. Omute
Department of Biochemistry, Lira University, P.O. Box 1035, Lira, Uganda

C. K. Nakiguli
Department of Chemistry, Mbarara University of Science and Technology, P.O. Box 1410, Mbarara, Uganda

A. V. Khanakwa
Department of Environmental Health and Disease Prevention, Lira University, P.O. Box 1035, Lira, Uganda

C. Angiro
Centre for Water, Environment and Development – School of Water, Energy and Environment, Cranfield University, College Road, Cranfield MK43 0AL, UK

I. Kahwa
Pharm-BioTechnology and Traditional Medicine Centre of Excellence, Mbarara University of Science and Technology, P.O. Box 1410, Mbarara, Uganda

F. Ahumuza · T. Omara (✉)
Food Safety Laboratories (Chemistry Division), Testing Department, Uganda National Bureau of Standards, P.O. Box 6329, Kampala, Uganda
e-mail: prof.timo2018@gmail.com

T. Omara
Department of Chemistry, College of Natural Sciences, Makerere University, P.O. Box 7062 Kampala, Uganda

T. A. Kurniawan and A. Anouzla (eds.), *Algae as a Natural Solution for Challenges in Water-Food-Energy Nexus*, Environmental Science and Engineering,
https://doi.org/10.1007/978-981-97-2371-3_25

been reported globally. Toxicities and mortalities from intake of edible algae has been reported in species of *Gracilaria, Caulerpa* and *Acanthophora* genus, and are associated with their bioaccumulation of contaminants such as excess iodine, heavy metals, cyanotoxins or toxic inherent compounds such as caulerpenyne, manauealides A and C, prostaglandin E_2, polycavernosides, aplysiatoxins and their derivatives. Allergenicity has been reported in *Arthrospira, Chlorella, Chondrus, Eucheuma, Gigartina* and *Palmaria* species, with the sulfated polysaccharide (carrageenan) and the photosynthetic pigment (C-phycocyanin) being the implicated allergens. These allergic reactions are mediated through activation of innate immune pathways of inflammation that trigger NF-kB activation, modification of gut microbiota and thickness of mucus barrier. We contend that appropriate labelling of algae-derived food products, public education, proper cleaning of fresh algae before consumption and profiling of toxic and allergenic algal species and compounds could aid in reducing intoxications and allergic reactions from algae used in food and food products. Future studies should consider examining edible algae for contaminants of emerging concern such as microplastics, cyanotoxins, emerging per- and polyfluoroalkyl substances, pharmaceutical residues and personal care products.

Keywords Arsenic · *Caulerpa* toxins · Carrageenan · *Hizikia fusiforme* · Iodine · Polycavernosides

25.1 Global Overview of Algae and Allied Products Consumption

Edible micro- and macroalgae (especially seaweeds; SWDs) are a class of pigmented autotrophic organisms with undifferentiated roots, stems, leaves and total absence of vascular bundles. They are consumed, used as medicinal and nutraceutical products or incorporated into foods (as additives, extracts and sources of functional ingredients) (Debbarma et al. 2022; Ibrahim et al. 2022; Lomartire et al. 2021; Mendes et al. 2022; Michalak and Chojnacka 2018).

Consumption of algae is historic, and has been part of habitual diets in many East Asian cultures (Preeprame et al. 2001). Algae consumption can be traced to as early as 500 BCE (14,000 years ago) according to records on American, Asian and European culinary recipes (dietary staples) and complimentary therapies (Boukid and Castellari 2022; Dillehay et al. 2008; Hallsson 1961; Haroun et al. 2019; Mouritsen et al. 2013; Nisizawa et al. 1987; Pereira 2016). For instance, *Sargassum horneri* is a savoury algae (seaweed) that is popularly consumed in Japan (Preeprame et al. 2001). Similarly, *Grateloupia turuturu* (a red seaweed) on the Atlantic Coast of Europe forms part of the Japanese and Korean oriental diet (da Costa et al. 2021; Pacheco et al. 2020). In Great Britain and Ireland, the seaweed (*Porphyra umbilicalis*) is a popular ingredient of laverbread (Harford 2023). The red algae "*ogonori*" (*Gracilaria verrucosa*) is widely eaten in form of a Japanese salad, *Sashimi* (Higa and Kuniyoshi 2000).

Commercialization of algae farming has surged in an effort to address global food security and this is evidenced by various algae-based functional foods and dietary supplements in form of snacks, condiments, pasta and salads on the market (European Food Safety Authority et al. 2023; Zhang et al. 2022a, b). According to the Algae Products Global Market Report 2023, algal products market soared from to US $12.1 billion in 2022 to US $13.2 billion in 2023, representing a compound annual growth rate (CAGR) of 8.9%. This is anticipated to approach $17.89 billion by 2027, growing at a CAGR of 7.9% (Cision 2023). These values account for foods, bioplastics, industrial lubricants, green chemicals, and cosmetics. Dietary intake (37%) is the major use of algae (mostly red and brown SWDs), followed by pharmaceutical and nutraceutical applications at 8.1% (FAO 2018; FAO and WHO 2022; Transparency Market Research 2022).

The renewed interest in algae is due to the expanding view of macroalgae as healthy foods i.e., 'super-foods' and their antimicrobial, neuroprotective, anti-allergic, anti-oxidant, antitumor, anti-fungal, chemoprotective, anti-obesity, antidiabetic and hepatoprotective properties (Aakre et al. 2021; Cotas et al. 2021; Ganesan et al. 2019; Shannon and Abu-Ghannam 2019). Despite the exponential growth in algae consumption and their use as perfect functional ingredients, there is brooding body of knowledge that their ingestion and inclusion in food products present food safety risks (Filippini et al. 2021; Wu et al. 2022). The food safety hazards in algae and algal products are of physical, microbiological and chemical nature, with the latter being more associated with illnesses, allergies and mortalities. The following subsections will discuss the toxicities and allergic reactions associated with consumption of algae and algae-derived products, the constitutive compounds and their supposed mechanisms of action.

25.2 Poisoning and Mortality Incidences Due to Consumption of Edible Algae and Their Products

To date, illnesses, toxicities and mortalities from ingestion of algae is traced to mainly three genera namely; *Acanthophora*, *Caulerpa* and *Gracilaria*. Majority of the cases were recorded in countries of the Pacific Rim (Guam, California and Japan), arguably due to the voluminous consumption of SWDs in this region (Marshall and Vogt 1998) (Table 25.1). The largest number of incidences pertaining to algal poisoning comes from the genus *Gracilaria* (represented by four species, one of which now fall under genus *Polycavernosa*). Again, this is explained by the wide consumption of raw species from this genus in traditional cuisines of Asia and the Pacific Rim (Cheney 2016; Marshall and Vogt 1998). Two poisoning cases have been associated with the Japanese favorite brown algae *Nemacystus decipiens* (*ito-mozuku*) (Fusetani and

Hashimoto 1981) which is common in Okinawa prefecture (Nishitsuji et al. 2019)—and *Cladosiphon okamuranus*—which is endemic to the Ryukyu Islands (Sato et al. 2023). Taken together, seventy (70) illnesses and fourteen (14) fatality cases due to intake of algae have been reported globally (Table 25.1).

25.3 Chemicals Implicated in the Toxicity of Edible Algae and Their Mechanisms of Action

The toxicity of algae has been ascribed to both biosynthesized compounds and xenobiotics such as heavy metals (arsenic), excess iodine absorbed, carrageenans and cyanotoxins (Desideri et al. 2016; Vellinga et al. 2022). These are discussed in the following with other potential contaminants that can induce toxicity and illnesses.

25.3.1 Toxins in Caulerpa *Algae*

The first incidence of poisoning from ingestion of algae was described by Doty and Aguilar-Santos (1966) and their chemical analysis pointed to the presence of a sphingosine derivative (caulerpicin, **1**) from *Caulerpa racemosa* (Fig. 25.1). Further efforts of Doty and Aguilar-Santos (1970) led to the identification of another potentially toxic orange-red pigment (caulerpin, **2**) in *C. racemosa* and *C. lamourouxii.* Their study led to the discovery that there was potential trophic transfer and biomagnification of the toxins (**1** and **2**) by sea slug (*Oxynoe panamensis*) which feeds on *Caulerpa* species (Doty and Aguilar-Santos 1970).

Another unusual *Caulerpa* toxin (caulerpenyne, **3**) was isolated from the Mediterranean *C. prolifera* (Amico et al. 1978) and *C. taxifolia* (Marić et al. 2017). It has since been isolated from nearly all (other nine) *Caulerpa* species native to the tropical Pacific and the Caribbean Sea. This acetylenic sesquiterpenoid was appreciated to possess strong ichthyo-toxicity and antigrazer activities (Paul et al. 1987), cytoxicity (Lemée et al. 1993) as well as inhibitory effects on uptake transporters of organic anions (Oatp1d1) and cations (Oct1) (Marić et al. 2017). Caulerpinic acid (**4**), also common in the *Caulerpa* genus, is known to be as mosquitocidal as the bisindole alkaloid caulerpin (Alarif et al. 2010). Therefore, these secondary metabolites isolated from *Caulerpa* genus (**1–4**) are evolutionarily-developed chemical defense mechanisms against various herbivores, epiphytes and other competitors (Marić et al. 2017). This may explain the fact that poisoning from this genus occurs during the rainy season when toxin concentrations increase and in part reflects the influence of seasons on the biosynthesis of these metabolites (Lemée et al. 1993). For example, a potential allelopathic effect of caulerpenyne (**3**) from *C. racemosa* var. *cylindracea* was cited, and this could be fundamental in the species' out competition of other

Table 25.1 Synopsis of poisoning, illnesses and mortality events arising from consumption of edible algae based on published literature

Algae	Report on incidence	Year	Country	Implicated compounds	Reference(s)
Acanthophora specifera	15 illnesses and 3 mortalities recorded	2003	Philippines	Polycavernoside A, C and C2	Yotsu-Yamashita et al. (2004, 2007)
	12 illnesses and 3 mortalities recorded	2002			
Gracilaria edulis	9 illnesses and 2 mortalities recorded	2002	Philippines	Polycavernoside A, Polycavernoside D	Haddock and Cruz (1991), Yotsu-Yamashita et al. (1993, 2004) and Navarro et al. (2015)
Gracilaria coronopifolia	Vomiting, diarrhea, burning sensation of the mouth and throat and generalized myalgia. Seven (7) illnesses	1994	Hawaii	Aplysiatoxins and debromoaplysiatoxin, manauealides A–C	Marshall and Vogt (1998) and Nagai et al. (1996, 1997)
Gracilaria verrucosa	Nausea, hypotension and death of a woman due to hypotensive shock 14 h following ingestion of 20–30 g of steeped algae. Husband recovered	1993	Japan	Prostaglandin E	Fusetani and Hashimoto (1984) and Noguchi et al. (1994)
Polycavernosa tsudai (previously *Gracilaria edulis*)	13 suffered illnesses and 3 died	1991	Guam (USA)	Polycavernoside A and B (LD_{99} = 200–400 μg/kg for both)	Haddock and Cruz (1991)

(continued)

Table 25.1 (continued)

Algae	Report on incidence	Year	Country	Implicated compounds	Reference(s)
Gracilaria chorda	Six victims suffered from nausea, vomiting, stomach pain, diarrhea and 1 died	1982	Japan	Prostaglandin E	Fusetani and Hashimoto (1984) and Noguchi et al. (1994)
	Four victims suffered from nausea, vomiting, stomach pain and diarrhea; 1 died	1980			
Cladosiphon okamuranus and *Nemacystus decipiens*	2 illnesses	1967	Japan	Diethyl peroxides	Fusetani and Hashimoto (1981)
Caulerpa racemosa	Not reported	1966	Philippines	Caulerpin, caulerpicin	Doty and Aguilar-Santos (1966, 1970)

plants with macrophytes such as the seagrass: *Cymodocea nodosa* (Raniello et al. 2007).

Other potentially toxic compounds have been elucidated in *Caulerpa* species. For example, 10,11-epoxy-caulerpenyne along with oxytoxin 1, taxifolial A, B, C and D has been reported in *C. taxifolia* (Guerriero et al. 1992; Lemée et al. 1993). Though taxifolial A and D are not toxic to both mice and mammalian cells, 10,11-epoxy-caulerpenyne was lethal to mice at 75 mg/kg and had $IC_{50} = 11$ μg/mL against mammalian cells (Lemée et al. 1993).

The mechanism of toxicity of most *Caulerpa* toxins are not clearly understood. Caulerpicin (**1**) has been implicated in the manifestation of toxic symptoms such as mild anaesthetizing sensation, numbness of the tongue and cold sensation in the feet and fingers (Doty and Aguilar-Santos 1966). Although discussed controversially, caulerpin (**2**) is known to induce mild anesthetic action, trigger difficulties in breathing, sedation, and loss of balance which is similar to the toxic syndromes observed in victims intoxicated by ciguatoxins i.e., ciguatera fish poisoning (Kase et al. 2020). Both **1** and **2** were initially indicated to be toxic to mice and rats (Doty and Aguilar-Santos 1970), but this toxicity is not supported by acute toxicity studies of Vidal et al. (1984) in mice.

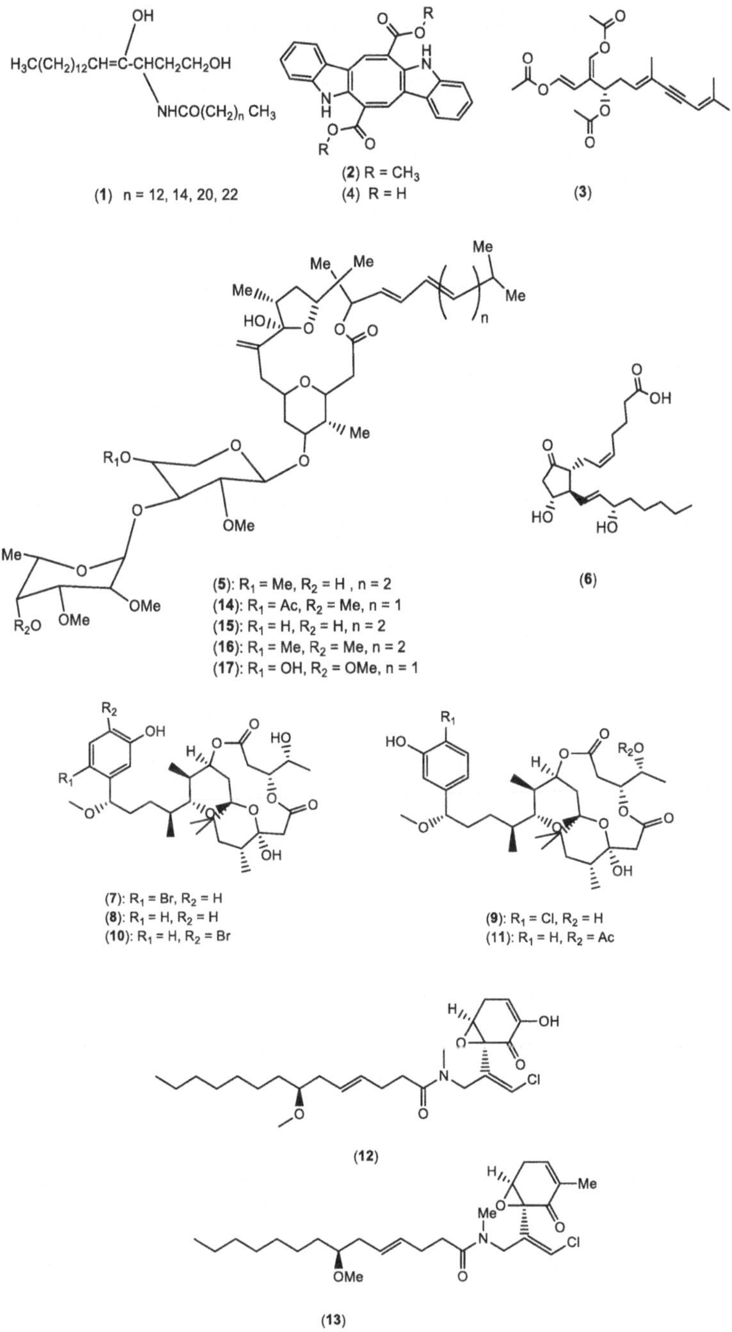

Fig. 25.1 Chemical structure of compounds implicated in illnesses, toxicities and mortalities from consumption of edible algae

Recently, the bioaccumulation and biochemical perturbations due to caulerpin (upto 50 μg/g) from *C. racemosa* in stomach contents of the white seabream (*Diplodus sargus*) was reported (Felline et al. 2012). At a molecular level, caulerpin induced alterations in fish antioxidant defenses (e.g., glutathione reductase and glutathione levels), augmented activities of cytochrome P450 (through triggering the activity of 7-etoxyresorufin O-deethylase) (Felline et al. 2012), glutathione S-transferases and acyl CoA oxidase, gene transcription for peroxisome proliferator-activated receptor alpha, cytochrome P4501A and vitellogenin 1 (Gorbi et al. 2014). Because caulerpin, a bisindolic alkaloid, possess a molecular structure comparable to indole alkaloids (i.e., possesses two ^{1}H indole moieties joined by an eight-membered cyclooctatetraene ring) (Zhan et al. 2010), lower levels of acetylcholinesterase reported in fish with caulerpin suggest that it may be largely neurotoxic (Felline et al. 2012). Alteration of lipid metabolism (De Pascali et al. 2015; Del Coco et al. 2018) has also been cited in fish following caulerpin ingestion.

The most potent of the *Caulerpa* toxins (**3**) is known to negatively affect P-glycoprotein-ATPase and inhibits overall protein phosphorylation (Pesando et al. 1996). It also directly modified the electrical properties of touch mechanosensory cells of leech (*Hirudo medicinalis*), leading to partially irreversible afterhyperpolarization mediated by the activity of Na^+/K^+-ATPase and to some extent a calcium-dependent potassium current (Brunelli et al. 2000; Mozzachiodi et al. 2001). Caulerpenyne is also an inhibitor of human 5-lipoxygenase (Richter et al. 2014). To date, only caulerpenyne (**3**) is considered to be the major toxin in *Caulerpa* SWDs (Guerriero et al. 1992; Higa and Kuniyoshi 2000), and its concentration necessary to induce toxicity in human hematopoietic progenitors, melanocytes and keratinocytes ($IC_{50} = 6$–$24\ \mu M$) is high (Parent-Massin et al. 1996).

25.3.2 Toxins in Acanthophora *Algae*

By percentage, the *Acanthophora specifera* intoxication which occurred in Philippines in 2002 and then again in 2003 represents the highest number of illnesses and fatalities in algal poisoning history (Table 25.1). However, no satisfactory empirical data exists on the causative compounds partly due to absence of adequate samples that could be analyzed at that time (Yotsu-Yamashita et al. 2004). Nevertheless, the symptoms associated with these events were in complete agreement with those due to ingestion of *G. edulis* in 2002. Based on a later argument by the same research group (Yotsu-Yamashita et al. 2007), polycavernoside A (**5**) (initially identified in *G. edulis*) was characterized in *A. specifera.* This discovery made it plausible to cite that the *Acanthorphora* outbreaks recorded in Philippines mirrored *G. edulis* outbreaks in the same country, and could have been triggered by one or more blue-green algae (Cheney 2016). Blue-green algae (also called cyanobacteria) are known to produce hepatotoxic, neurotoxic and immunotoxic secondary metabolites which are potentially lethal to humans (Omara et al. 2023).

25.3.3 Toxins in Gracilaria *Algae*

In *Gracilaria* species, the major toxins incriminated in algae toxicity are prostaglandins (Fusetani and Hashimoto 1984; Noguchi et al. 1994). Specifically, prostaglandin E_2 (**6**) has been isolated from *G. verrucosa* and *G. lichenoides* (Gregson et al. 1979). Also known as dinoprostone, prostaglandin E_2 (PGE_2) is a naturally-occurring prostaglandin established to induce physiological reactions in humans such as nausea, diarrhea, hypertension and bleeding (Noguchi et al. 1994). Moreover, the concentration of bioavailable and toxic PGE_2 levels seem to come from unsaturated fatty acids (suspectedly arachidonic acid) (Änggård and Oliw 1981; Gregson et al. 1979; Noguchi et al. 1994), a fact that has been proven over the years by other research groups (Ricciotti and FitzGerald 2011). PGE_2 along with PGF2a was found in *G. lichenoides* from Australia, and it occurs that the concentrations always found in the intact algae are safe (Gregson et al. 1979). However, toxicity events involving PGE_2 where those in which the victims consumed freshly harvested algae (and at least one of the victims steeped it in fresh water overnight). It is argued that such a treatment might have washed out the inherent inhibitor of prostaglandin synthesis, which could have induced PGE_2 generation to toxic concentrations (Fusetani and Hashimoto 1984).

The 1994 successive poisoning incidences by *G. coronopifolia* in islands of Maui, Hawaii and Oahu presented a new symptom (burning sensation of the throat and mouth), suggesting that it involved some other unknown toxins. In an attempt to arrive at the causative agents, Nagai et al. (1996) identified aplysiatoxin (**7**) and debromoaplysiatoxin (**8**) from the original algal collection. In addition to these, manauealides A–C (**9**–**11**) (Nagai et al. 1997), malyngamides M (**12**) and N (**13**) were subsequently isolated and characterised (Kan et al. 1998). Manauealides—just as aplysiatoxin and debromoaplysiatoxin—are known to induce diarrhea in mice (Nagai et al. 1997). On the other hand, malyngamides elicited low cytotoxic effects on mouse neuroblastoma cells (Kan et al. 1998).

Another toxicity incidence in USA detailed by Haddock and Cruz (1991) pointed to the presence of glycosidic macrolides: polycavernoside A (**5**) and B (**14**) in *Polycavernosa tsudai* (previously *Gracilaria edulis*) as the cause of illnesses and deaths. Indeed, polycavernoside A has a glycosidic macrolide with an unusual tricyclic aglycon stemming from a partially dehydrated and unsaturated 3,5,7,13,15-pentahydroxy-9,10-dioxotricosanoic acid (Blakemore and White 2005). In addition to **14**, three novel potentially toxic glycosidie macrolides (polycavernoside A2, A3 and B2; **15**–**17**) has been implicated in the poisoning incident which ensued ingestion of the red algae *Polycavernosa tsudai* (Yotsu-Yamashita et al. 1995). Polycavernosides are toxic but the mechanism of their toxicities are not well elucidated (Louzao et al. 2014).

25.3.4 *Toxins in* Cladosiphon, Sphaerotrichia *and* Nemacystus *Algae*

Two cases of algae poisoning, the so called—"*mozuku*" poisoning—was reported to be from brown algae: *Cladosiphon okamuranus* and *Nemacystus decipiens* (Fusetani and Hashimoto 1981). Owing to the absence of adequate samples, the team examined another edible brown algae (*Sphaerotrichia divaricata*) and characterized diethyl peroxides in fat-soluble fractions that were lethal to mice when hot, suggesting that illnesses associated with these species arise following steeping (Fusetani and Hashimoto 1981). In this context, there exists knowledge gaps that need to be unveiled to understand the mechanisms and responsible compounds for poisonings recorded in some of these algae species.

25.3.5 *Bioaccumulated Potentially Toxic Metals, Metalloids and Organic Contaminants*

Unlike in fresh algae, concerns on the safety of algal products stretch beyond inherent toxic algal metabolites but relates to bioaccumulation and translocation of potentially toxic elements (PTEs) as well as other organic contaminants in algae tissues (Vellinga et al. 2022). Table 25.2 shows a summary of reports on PTEs, and conclusions made on the levels obtained in some edible algae.

At a glance, the concentration of PTEs reported in edible algae appears to be safe. However, very high levels of inorganic and total arsenic (up to 117 and 149 mg/kg dw, respectively) has been reported in *Hizikia fusiforme* (*H. fusiforme*; synonym: *Sargassum fusiforme*) (Almela et al. 2006). Such arsenic levels are far beyond the compliance limit(s) for this metalloid in food products (3 mg/kg or less) in USA and France. In the same study (Almela et al. 2006), unacceptably high levels of Cd in some species of edible *Porphyra* (0.089–3.19 mg/kg) and *H. fusiforme* SWDs (0.511–1.53 mg/kg) were recorded. Very high levels of inorganic arsenic (32–70 and 67–96 mg/kg) and total As (18–124 and 103–147 mg/kg) has also been reported in *H. fusiforme* from UK (Rose et al. 2007) and Spain (Besada et al. 2009).

In regards to organic contaminants, SWDs (*Fucus distichus*, *F. spiralis* and *Nereocystis luetkeana*) sampled from the Salish Sea of Western United States and Canada were found to contain Cd (1.15–7.91 mg/kg), Cr (below detection limit (BDL)—5.49 mg/kg), Hg (BDL—0.25 mg/kg), Pb (BDL—13.4 mg/kg), persistent organic pollutants and polychlorinated biphenyls (0.139–3.12 μg/kg) at levels that surpassed either US EPA noncancer-based screening levels or international compliance limits (Hahn et al. 2022). It should be noted that the effects of PTEs maybe cumulative, and therefore even at very low levels, the overall health effects maybe not negligible. It should be reiterated that both inorganic and organic arsenic are carcinogenic, but the former is more readily absorbed into the body. For example, *H. fusiforme* was

Table 25.2 Concentration of potentially toxic elements (PTEs) reported in some edible algae and implications for public health protection

Area (country)	Algae	PTEs (mg/kg)[a]	Daily intake level or hazard index	Implications for public health	References
Salish Sea (United States and Canada)	*Fucus distichus*, *F. spiralis* and *Nereocystis luetkeana*	Cd (1.15–7.91), Cr (BDL—5.49), Hg (BDL—0.25), Pb (BDL—13.4)	Not calculated	High health risk	Hahn et al. (2022)
Italy	*Ascophyllum*, *Himanthalia*, *Laminaria*, *Palmaria*, *Porphyra*, *Saccharina*, *Ulva* and *Undaria* species	Cd (0.02–0.93), Cr (0.09–0.30), Ni (0.24–0.51), Hg (BDL—0.01), Zn (1.58–18.12), total As (0.62–6.74), inorganic As (0.12–0.18)	Hazard index was less than 1	Low health risk	Filippini et al. (2021)
Zhejiang (China)	Red and brown seaweeds	As (0.185–71), Cd (0.002–6.4), Cr (0.089–35.7), Cu (BDL—39.2), Hg (BDL—0.391), Mn (0.404–407), Ni (BDL—8.82), Pb (BDL—6.96)	Hazard index was less than 1	Low health risk	Chen et al. (2018)

(continued)

Table 25.2 (continued)

Area (country)	Algae	PTEs (mg/kg)[a]	Daily intake level or hazard index	Implications for public health	References
Bodø, Trondheim, Trondheimsfjord (Norway), Pleubian (France)	*Alaria esculenta, Palmaria palmata, Saccharina latissima*	Total As (7.192–99.112), Cd (0.024–2.622), Hg (0.0009–0.3139), Pb (0.0301–0.7126)	Hazard index was less than 1	Low health risk	Roleda et al. (2019)
Seoul (South Korea)	Laver, brown seaweeds, Kelp and Sea lettuce (n = 426)	Total As (BDL—88.8), Hg (0.001–0.050), Pb (BDL—2.7), Cd (BDL—2.9)	0.11, 0.65, and 0.45 μg/kg body weight/week for total Hg, Pb and Cd, respectively	Low health risk	Hwang et al. (2010)
Auckland and Nelson (New Zealand)	*Macrocystis pyrifera, Undaria pinnatifida, Porphyra* and *Ecklonia radiata, Ulva stenophylla, Durvillaea antarctica, Hormosira banksii*	Total As (3.8–97), inorganic As (0.1–1.5), Fe (13.7–1227), Zn (10.74–45.03), Hg (0.01–0.17), Pb (0.20–0.41)	Tolerable intake levels were not exceeded	Low health risk	Smith et al. (2010)

(continued)

Table 25.2 (continued)

Area (country)	Algae	PTEs (mg/kg)[a]	Daily intake level or hazard index	Implications for public health	References
Spain	*Chondrus crispus*, *Eisenia bicyclis*, *Gelidium* species, *Himanthalia elongata*, *Hizikia fusiforme*, *Laminaria* species, *Porphyra umbilicales*, *Undaria pinnatifida*, and *Ulva rigida*	Cd (0.025–4.82), Pb (< 0.008–1.35), Hg (0.001–0.057), Cu (0.410–7.70), inorganic As (67–96), total As (103–147)	High arsenic levels in *Hizikia fusiforme*	High risk	Besada et al. (2009)
London and its environs (UK)	Hijiki, Nori, Kombu, Wakame, Arame	Total As (18–124), inorganic As (67–96)	0.57 mg/kg/ body weight	High risk. UK Food Standards Agency advised not consuming products with *Hizikia fusiforme*	Rose et al. (2007)

(continued)

Table 25.2 (continued)

Area (country)	Algae	PTEs (mg/kg)[a]	Daily intake level or hazard index	Implications for public health	References
Spain	Various seaweeds of Japanese, Korean, Spanish and Chilean origins[b]	Cd (< 0.003–3.55), total As (0.031–149), inorganic As (< 0.014–117), Pb (< 0.050–12.1)	Limit of total As per national regulatory limit	High health risk, especially due to high As levels in *Hizikia fusiforme*	Almela et al. (2006)

Note BDL below method detection limit

[a] PTEs: *Pb* lead, *As* arsenic, *Ni* nickel, *Mn* manganese, *Zn* zinc, *Cd* cadmium, *Cr* chromium, *Hg* mercury, *Fe* iron

[b] Species included: *Enteromorpha* species, *Ulva pertusa, Porphyra tenera, P. umbilicalis, Palmaria palmata, Palmaria* species, *Rhodymenia palmata, Chondrus crispus, Laminaria* species, *L. japonica, L. digitata, Eisenia bicyclis, Undaria pinnatifida, Hizikia fusiforme, Fucus vesiculosus, Himanthalia elongata, Durvillaea antartica* (Almela et al. 2006)

indicated to have led to arsenic urinary excretion levels similar to those in individuals affected by arsenic poisoning when a volunteer consumed the equivalent of eight servings of commercial *H. fusiforme* food (Nakajima et al. 2006). An incidence of a 54-year-old woman from California who presented with symptoms of arsenic poisoning after taking two to four tablets of a kelp (*Laminaria*) supplement per day for several months has been published (Amster et al. 2007). Later, McGuffin and Dentali (2007) argued that the case in question should have emanated from intake of excess iodine, which is known to share toxicity symptoms with arsenic. Nevertheless, similar elevation of inorganic arsenic levels have recently been appreciated among Japanese children and expectant mothers who consumed *H. fusiforme* (Mise et al. 2019).

In addition to other organic contaminants, the accumulation of microplastics on edible SWDs such as nori (0.9–3.0 items/g dw) (Li et al. 2020), *Gracilaria fisheri* and *Caulerpa lentillifera* (16.46–181.73 particles per 100 g wet weight) (Klomjit et al. 2021) has joined algae literature. Microplastics are known to act as binders, concentrators and vectors of toxic contaminants and pathogenic microorganisms (Liu et al. 2021a, b; Tumwesigye et al. 2023). There is therefore need for future studies to examine the relationship between microplastic levels in algae and heavy metal contents as well as other organic contaminants. Further research efforts should consider examining edible algae for contaminants of emerging concern such as microplastics (Gutow et al. 2016), cyanotoxins, emerging per- and polyfluoroalkyl substances (e.g., ADONA and GenX), current use pesticides, active pharmaceutical ingredients and personal care products (Hahn et al. 2022).

25.3.6 Absorbed Iodine

The iodine content of algae (SWDs) is often high (Desideri et al. 2016), and this makes them a veritable source of iodine for tackling dietary iodine insufficiency among individuals (Aakre et al. 2021; Bouga and Combet 2015; Ficheux et al. 2023). Across the three taxonomic categories of green (Chlorophyta), red (Rhodophyta), and brown (Phaeophyceae) SWDs, the latter (specifically species of *Laminaria* genus and *Laminariale* genus i.e., kelp) may absorb iodine up to 100,000 times higher than that in seawater in which they are growing (Milinovic et al. 2021; Yeh et al. 2014). A survey of 224 products with SWDs in the United Kingdom by Bouga and Combet (2015) showed that 26 (11.6%) of the products could translate into iodine intakes higher than the European tolerable adult upper level of 600 μg/day. This study suggested that there is scope to improve product labelling of algae-containing products to promote their safe consumption. Similar alarming levels of iodine exceeding tolerable limits (30–62,400 μg and upto 6770 mg/kg) were recently found in commercial macroalgae-based foods and supplements in Norway (Aakre et al. 2021) and edible SWDs marketed in Italy (Filippini et al. 2021).

Although dietary intake of iodine is instrumental for good human health, low and very high intake are known to increase risks of diseases such as goiter, thyroiditis, hypothyroidism, hyperthyroidism and papillary thyroid cancer (European Food Safety Authority et al. 2023; Winder et al. 2022). Michikawa et al. (2012), for instance, investigated the relationship between SWDs consumption and the supposed risk of thyroid cancer in Japanese women, and concluded that consumption of SWDs by postmenopausal women exposed them to higher risks of papillary carcinoma. Maternal consumption of SWDs is a common practice in some Asian cultures as a natural way to promote breast milk supply (Emder and Jack 2011). A report from Korea found high daily intakes as well as breast milk iodine levels in women who took large volumes of *Undaria pinnatifida* (kelp) soup during their early postpartum (Rhee et al. 2011).

Most preceding epidemiological and experimental studies of algae ingestion did not find such an association, or gave controversial and diametrically opposite conclusions regarding effect of excess iodine intake from edible algae on papillary thyroid cancer incidences. That notwithstanding, it has been confirmed that iodine nutrition influences the overall prevalence, distribution of histological types, and clinicopathological aggressiveness of papillary thyroid cancer (Zhang et al. 2022a, b).

As a matter of fact, huge economic losses have been associated with iodine poisoning in SWDs. A Bonsoy soy milk manufacturer in Australia which incorporated powder of the seaweed (*Laminaria* species) was ordered to compensate nearly 500 victims with US $25 million. The victims had suffered various health consequences of consuming soy with excessive levels of iodine (The Sydney Morning Herald 2014).

25.3.7 Carrageenan

Carrageenans (CGNs) is a group of indigestible sulphated polygalactans (kappa, iota, and lambda; chemically differentiable by the repetition of their sulphation units) obtained from edible red SWDs: *Chondrus* (the original commercial source), *Eucheuma*, and *Kappaphycus* (Guan et al. 2017; McKim 2014). They are made up of D-galactose residues linked by β-1,4 and α-1,3 galactose-galactose bond (Borsani et al. 2021). The CGNs content of SWDs vary (22–88% dw), contingent on the species and ecophysiological variables (Asni and Najamuddin 2021; Rupert et al. 2022).

Native (undegraded) CGNs have the "generally recognised as safe" food additives status as per the Food and Drug Administration of the United States, European Economic Community and the WHO Joint Expert Committee on Food Additives list (Liu et al. 2021a, b). CGNs typically possess no known nutritive values, but is used as a food additive (E-407) i.e., thickener, gelling agent, emulsifier or stabilizer (Borsani et al. 2021). However, studies dating as far back as 1980 have illustrated that CGNs may induce inflammatory reactions and toxic effects on colon microbiota, thereby contributing to gastrointestinal disorders such as ulcers and colon cancer (David et al. 2018; Venkatesan et al. 2015). While this specific group of compounds are implicated in some toxic effects of algal products, a recent communication by the European Food Safety Authority (EFSA Panel on Food Additives and Nutrient Sources Added to Food et al. 2018) has refuted the correlation of CGNs and cancer, which may be due to inappropriate additive application (e.g., using very high concentrations), generalization of in vitro outcomes to possible in vivo health effects as well as utilization of lower molecular weight (degraded) CGNs other than the approved undegraded form (poligeenan). This is supported by a recent observation (Alli et al. 2022) where CGNs did not significantly decrease cell and tissue viability of human salivary gland cells and reconstructed human oral epithelium when compared to sodium dodecyl sulphate. Potential allergenicity of CGNs is discussed in the later part of this chapter but its definite toxicity in humans is an inconclusive debate and appears to be rather concentration-dependent.

25.4 Incidences of Allergies from Consumption of Edible Algae and Their Products

Allergenicity in food science refers to the tendency of a food or food product to induce adverse immunologic (immunoglobulin E)-mediated response(s) or side effect(s), usually in genetically predisposed (atopic) individuals (Waserman et al. 2018). Allergenicity is among the overriding concerns impeding effective utilization of algae as food in sensitive individuals (FAO and WHO 2022; James et al. 2023). Although algal biomass and extracts are useful food ingredients which could suffice as nutrient-rich, sustainable and healthy foods, allergenicity of algae has retarded their acceptability in some parts of the world. The first concern pertains to the palatability of algae in terms of sensory attributes and allergen content (James et al. 2023).

Algae allergenicity has been encountered following consumption of products containing *Spirulina* (*Arthrospira*), *Chlorella*, *Chondrus*, *Palmaria*, *Eucheuma* and *Gigartina* species. The complete range of reactions varied from mild simple allergic reactions to incidences of anaphylaxis (Table 25.3). One of the rare cases involving *Chlorella vulgaris* allergenicity was recorded following consumption of a powdered supplement based on this alga (Abasszade et al. 2020). The causative compounds inducing allergic reactions have not been identified in all the reported incidences. The sulfated polysaccharide (carrageenan) and the photosynthetic pigment (C-phycocyanin) are the major allergens that has so far identified in algae and algal products (Kular et al. 2018; Petrus et al. 2010). Other cases involving algae allergenicity has been reported in a randomized clinical trial (Vojdani and Vojdani 2015) and cross-sectional studies (Tiberg et al. 1990, 1995; Tiberg and Einarsson 1989).

Table 25.3 Incidences of allergic reactions from consumption of algae

Country	Algae species	Symptoms	Responsible allergen	References
Australia	*Chlorella vulgaris*	A 75-year-old male presented with wheeze, severe hypotension, urticarial rash, diarrhoea and elevated blood tryptase levels, suggesting anaphylaxis	Not reported	Abasszade et al. (2020)
UK	Nori, along with *Chondrus crispus*, and *Palmaria palmata*	A twenty-seven-year-old male with abdominal discomfort and high body temperature	Not reported	Thomas et al. (2018)
Canada	*Eucheuma*, *Chondrus*, *Gigartina* SWDs	A ten-month-old infant developed lip angioedema after eating fruit cake icing	Carrageenan	Kular et al. (2018)
The Netherlands	*Spirulina platensis* (tablet)	Seventeen-year-old male developed anaphylaxis after ingesting a *Spirulina*	Not reported	Le et al. (2014)
France	*Arthrospira platensis*	Fourteen-year-old adolescent presented with urticaria, labial oedema and asthma six hours of taking 5 *Spirulina* tablets	C-phycocyanin	Petrus et al. (2010)
South Korea	*Chlorella* species	Eleven-year-old male who took *Chlorella* tablets for 3 months had anemia, increased levels of creatinine, hypokalemia, hypo-uricemia, hypophosphatemia, proteinuria, hypergammaglobulinemia, leukocyturia and glucosuria	Not reported	Yim et al. (2007)

25.5 Mechanisms of Allergenicity of Compounds in Edible Algae

As shown in Table 25.3, carrageenan and C-phycocyanin are the major allergens identified in algae and algal products. Carrageenan has been labelled as a pseudo-latex allergy implicated in triggering anaphylaxis during barium enema (Tarlo et al. 1995). The underlying cause of carrageenan allergenicity is believed to be due to triggering of an immune response i.e., it mediates its allergenic effect through activation of innate immune pathways of inflammation encompassing pathways of NF-kB activation with a central role in transcriptional activation of the interleukin-8 gene, modification of gut microbiota constitution and thickness of mucus barrier (Bhattacharyya et al. 2008; Borthakur et al. 2007; Fahoum et al. 2017).

The mechanism of allergy induction by C-phycocyanin is not reported in open literature but it should be emphasized that appropriate labelling of algae products should be encouraged to reduce incidences of allergies from ingestion of algal products in sensitive individuals.

25.6 Future Perspectives

Consumption of algae and algal products have markedly increased globally but research to date indicates that there are potential health risks that could arise from their unregulated consumption. Several illnesses, allergic reactions and deaths so far reported have specifically been in the algae genera of *Acanthophora*, *Caulerpa* and *Gracilaria* genus. However, it should be expected that such toxicities may occur in other edible algae as well. Future research efforts should consider examining edible algae for contaminants of emerging concern such as microplastics, cyanotoxins, emerging per- and polyfluoroalkyl substances (e.g., ADONA and GenX), current use pesticides, active pharmaceutical ingredients and personal care products. Moreover, appropriate labelling of algae-derived food products (for example with their allergen or iodine contents), wider public education, cleaning of fresh algae before consumption and profiling of toxic and allergenic algal species and compounds could reduce intoxications and allergic reactions from algae consumption.

References

Aakre I, Solli DD, Markhus MW, Mæhre HK, Dahl L, Henjum S et al (2021) Commercially available kelp and seaweed products—valuable iodine source or risk of excess intake? Food Nutr Res 65. https://doi.org/10.29219/fnr.v29265.27584

Abasszade JH, Rama Raj P, Tinson AJ (2020) Anaphylaxis following enteral exposure to *Chlorella vulgaris*. BMJ Case Rep 13:e237054

Alarif WM, Abou-Elnaga ZS, Ayyad SEN, Al-Lihaibi SS (2010) Insecticidal metabolites from the green alga *Caulerpa racemosa*. CLEAN Soil Air Water 38:548–557

Alli BY, Upadhyay A, Zhang Y, Nicolau B, Tran SD (2022) In vitro toxicity evaluation of carrageenan on cells and tissues of the oral cavity. Mar Drugs 20:502

Almela C, Clemente MJ, Vélez D, Montoro R (2006) Total arsenic, inorganic arsenic, lead and cadmium contents in edible seaweed sold in Spain. Food Chem Toxicol 44:1901–1908

Amico V, Oriente G, Piattelli M, Tringali C, Fattorusso E, Magno S, Mayol L (1978) Caulerpenyne, an unusual sequiterpenoid from the green alga *Caulerpa prolifera*. Tetrahedron Lett 19(38):3593–3596

Amster E, Tiwary A, Schenker MB (2007) Case report: potential arsenic toxicosis secondary to herbal kelp supplement. Environ Health Perspect 115:606–608

Änggård E, Oliw E (1981) Formation and metabolism of prostaglandins in the kidney. Kidney Int 19:771–780

Asni A, Najamuddin (2021) Analysis on carrageenan content of seaweed *Kappaphycus alvarezii* at different water condition in Bantaeng District. IOP Conf Ser Earth Environ Sci 860:012069

Besada V, Andrade JM, Schultze F, González JJ (2009) Heavy metals in edible seaweeds commercialised for human consumption. J Mar Syst 75:305–313

Bhattacharyya S, Dudeja PK, Tobacman JK (2008) Carrageenan-induced NFκB activation depends on distinct pathways mediated by reactive oxygen species and Hsp27 or by Bcl10. Biochim Biophys Acta 1780:973–982

Blakemore PR, White JD (2005) The synthesis of polycavernoside A. An example of conformationally guided macrolactonization. In: Strategies and tactics in organic synthesis, vol 6, pp 173–210

Borsani B, De Santis R, Perico V, Penagini F, Pendezza E, Dilillo D et al (2021) The role of carrageenan in inflammatory bowel diseases and allergic reactions: where do we stand? Nutrients 13:3402

Borthakur A, Bhattacharyya S, Dudeja PK, Tobacman JK (2007) Carrageenan induces interleukin-8 production through distinct Bcl10 pathway in normal human colonic epithelial cells. Am J Physiol 292:G829–G838

Bouga M, Combet E (2015) Emergence of seaweed and seaweed-containing foods in the UK: focus on labeling, iodine content, toxicity and nutrition. Foods 4:240–253

Boukid F, Castellari M (2022) Algae as nutritional and functional food sources. Foods 12(1):122

Brunelli M, Garcia-Gil M, Mozzachiodi R, Roberto M, Scuri R, Traina G, Zaccardi ML (2000) Neurotoxic effects of caulerpenyne. Prog Neuro-Psychopharmacol Biol Psychiatry 24:939–954

Chen Q, Pan XD, Huang BF, Han JL (2018) Distribution of metals and metalloids in dried seaweeds and health risk to population in southeastern China. Sci Rep 8:3578

Cheney D (2016) Toxic and harmful seaweeds. In: Seaweed in health and disease prevention, pp 407–421

Cision (2023) Algae products global market report 2023. Accessed 04 June 2023. https://www.prnewswire.com/news-releases/algae-products-global-market-report-2023-301772768.html

Cotas J, Pacheco D, Gonçalves AM, Silva P, Carvalho LG, Pereira L (2021) Seaweeds' nutraceutical and biomedical potential in cancer therapy: a concise review. J Cancer Metastasis Treat 7:13

da Costa E, Melo T, Reis M, Domingues P, Calado R, Abreu MH, Domingues MR (2021) Polar lipids composition, antioxidant and anti-inflammatory activities of the Atlantic red seaweed *Grateloupia turuturu*. Mar Drugs 19(8):414

David S, Levi CS, Fahoum L, Ungar Y, Meyron-Holtz EG, Shpigelman A, Lesmes U (2018) Revisiting the carrageenan controversy: do we really understand the digestive fate and safety of carrageenan in our foods? Food Funct 9(3):1344–1352

De Pascali SA, Del Coco L, Felline S, Mollo E, Terlizzi A, Fanizzi FP (2015) ^{1}H NMR spectroscopy and MVA analysis of *Diplodus sargus* eating the exotic pest *Caulerpa cylindracea*. Mar Drugs 13:3550–3566

Debbarma J, Viji P, Rao BM, Ravishankar CN (2022) Seaweeds: potential applications of the aquatic vegetables to augment nutritional composition, texture, and health benefits of food and food products. In: Ranga Rao A, Ravishankar GA (eds) Sustainable global resources of seaweeds, vol 2. Springer, Cham, pp 3–54

Del Coco L, Felline S, Girelli CR, Angilè F, Magliozzi L, Almada F et al (2018) ^{1}H NMR spectroscopy and MVA to evaluate the effects of caulerpin-based diet on *Diplodus sargus* lipid profiles. Mar Drugs 16:390

Desideri D, Cantaluppi C, Ceccotto F, Meli MA, Roselli C, Feduzi L (2016) Essential and toxic elements in seaweeds for human consumption. J Toxicol Environ Health 79:112–122

Dillehay TD, Ramírez C, Pino M, Collins MB, Rossen J, Pino-Navarro JD (2008) Monte Verde: seaweed, food, medicine, and the peopling of South America. Science 320:784–786

Doty M, Aguilar-Santos G (1966) Caulerpicin, a toxic constituent of Caulerpa. Nature 211:990

Doty M, Aguilar-Santos G (1970) Transfer of toxic algal substances in marine food chains. Pac Sci 24:351–355

EFSA Panel on Food Additives and Nutrient Sources Added to Food, Younes M, Aggett P, Aguilar F, Crebelli R, Filipič M et al (2018) Re-evaluation of carrageenan (E 407) and processed Eucheuma seaweed (E 407a) as food additives. EFSA J 16:e05238

Emder PJ, Jack MM (2011) Iodine-induced neonatal hypothyroidism secondary to maternal seaweed consumption: a common practice in some Asian cultures to promote breast milk supply. J Paediatrics Child Health 47:750–752

European Food Safety Authority (EFSA), Dujardin B, Ferreira de Sousa R, Gómez Ruiz JÁ (2023) Dietary exposure to heavy metals and iodine intake via consumption of seaweeds and halophytes in the European population. EFSA J 21:e07798

Fahoum L, Moscovici A, David S, Shaoul R, Rozen G, Meyron-Holtz EG, Lesmes U (2017) Digestive fate of dietary carrageenan: evidence of interference with digestive proteolysis and disruption of gut epithelial function. Mol Nutr Food Res 61:3

FAO (2018) The global status of seaweed production, trade and utilization, vol 124. Globefish Research Programme, Rome, 120 pp

FAO and WHO (2022) Risk assessment of food allergens: part 1: review and validation of Codex alimentarius priority allergen list through risk assessment: meeting report. Food Agricultural Organisation/World Health Organisation, Rome. https://www.who.int/publications/i/item/9789240042391. Accessed 15 July 2023

Felline S, Caricato R, Cutignano A, Gorbi S, Lionetto MG, Mollo E et al (2012) Subtle effects of biological invasions: cellular and physiological responses of fish eating the exotic pest *Caulerpa racemosa*. PLoS ONE 7:e38763

Ficheux AS, Boniou B, Durand G, Garrec RL, Pierre O, Roudot AC (2023) Dietary exposure and risk assessment to trace elements and iodine in seaweeds. J Trace Element Med Biol 78:127187

Filippini M, Baldisserotto A, Menotta S, Fedrizzi G, Rubini S, Gigliotti D et al (2021) Heavy metals and potential risks in edible seaweed on the market in Italy. Chemosphere 263:127983

Fusetani N, Hashimoto K (1981) Diethyl peroxides probably responsible for Mozuku poisoning. Nippon Suisan Gakkaishi 47:1059–1063

Fusetani N, Hashimoto K (1984) Prostaglandin E2: a candidate for causative agent of "ogonori [marine algae *Gracileria verrucosa*]" poisoning. Bull Jpn Soc Sci Fish 50:465–469

Ganesan AR, Tiwari U, Rajauria G (2019) Seaweed nutraceuticals and their therapeutic role in disease prevention. Food Sci Hum Wellness 8(3):252–263

Gorbi S, Giuliani ME, Pittura L, d'Errico G, Terlizzi A, Felline S et al (2014) Could molecular effects of *Caulerpa racemosa* metabolites modulate the impact on fish populations of *Diplodus sargus*? Mar Environ Res 96:2–11

Gregson RP, Marwood JF, Quinn RJ (1979) The occurrence of prostaglandins PGE2 and PGF2α in a plant-the red alga *Gracilaria lichenoides*. Tetrahedron Lett 20(46):4505–4550

Guan J, Li L, Mao S (2017) Applications of carrageenan in advanced drug delivery. In: Seaweed polysaccharides. Elsevier, pp 283–303

Guerriero A, Meinesz A, D'Ambrosio M, Pietra F (1992) Isolation of toxic and potentially toxic sesqui- and monoterpenes from the tropical green seaweed *Caulerpa taxifolia* which has invaded the region of Cap Martin and Monaco. Helv Chim Acta 75:689–695

Gutow L, Eckerlebe A, Giménez L, Saborowski R (2016) Experimental evaluation of seaweeds as a vector for microplastics into marine food webs. Environ Sci Technol 50:915–923
Haddock R, Cruz O (1991) Foodborne intoxication associated with seaweed. Lancet 338:195–196
Hahn JL, Van Alstyne KL, Gaydos JK, Wallis LK, West JE et al (2022) Chemical contaminant levels in edible seaweeds of the Salish Sea and implications for their consumption. PLoS ONE 17:e0269269
Hallsson S (1961) The uses of seaweeds in Iceland. In: Proceedings of the fourth international seaweed symposium, Biarritz, France, 18–24 Sept 1961
Harford R (2023) Laver. https://www.eatweeds.co.uk/laver-porphyra. Accessed 29 July 2023
Haroun R, Gil-Rodríguez MC, Neto AI, Machín-Sánchez M, Viera-Rodríguez MA (2019) A review of current uses and potential biotechnological applications of seaweeds from the Macaronesian region (Central-East Atlantic Ocean). J Appl Phycol 31:3777–3790
Higa T, Kuniyoshi M (2000) Toxins associated with medicinal and edible seaweeds. Toxin Revs 19:119–137
Hwang YO, Park SG, Park GY, Choi SM, Kim MY (2010) Total arsenic, mercury, lead, and cadmium contents in edible dried seaweed in Korea. Food Addit Contam 3:7–13
Ibrahim RYM, Hammad HBI, Gaafar AA, Saber AA (2022) The possible role of the seaweed *Sargassum vulgare* as a promising functional food ingredient minimizing aspartame-associated toxicity in rats. Int J Environ Health Res 32(4):752–771
James CA, Welham S, Rose P (2023) Edible algae allergenicity—a short report. J Appl Phycol 35:339–352
Kan Y, Fujita T, Nagai H, Sakamoto B, Hokama Y (1998) Malyngamides M and N from the Hawaiian Red Alga *Gracilaria coronopifolia*. J Nat Prod 61:152–155
Kase AGO, Calumpong H, Rupidara A (2020) Secondary metabolites of some varieties of *Caulerpa* species. IOP Conf Ser Mater Sci Eng 823:012041
Klomjit A, Sutthacheep M, Yeemin T (2021) Occurrence of microplastics in edible seaweeds from aquaculture. Ramkhamhaeng Int J Sci Technol 4:38–44
Kular H, Dean J, Cook V (2018) A case of carrageenan allergy in a pediatric patient. Ann Allergy Asthma Immunol 121:S119
Le T, Knulst A, Röckmann H (2014) Anaphylaxis to *Spirulina* confirmed by skin prick test with ingredients of *Spirulina* tablets. Food Chem Toxicol 74:309–310
Lemée R, Pesando D, Durand-Clément M, Dubreuil A, Meinesz AAG, Pietra F (1993) Preliminary survey of toxicity of the green alga *C. taxifolia* introduced into Mediterranean. J Appl Phycol 5:485–493
Li Q, Feng Z, Zhang T, Ma C, Shi H (2020) Microplastics in the commercial seaweed nori. J Hazard Mater 388:122060
Liu F, Hou P, Zhang H, Tang Q, Xue C, Li RW (2021a) Food-grade carrageenans and their implications in health and disease. Comp Rev Food Sci Food Safe 20:3918–3936
Liu S, Shi J, Wang J, Dai Y, Li H, Li J et al (2021b) Interactions between microplastics and heavy metals in aquatic environments: a review. Front Microbiol 12:652520
Lomartire S, Marques JC, Gonçalves AMM (2021) An overview to the health benefits of seaweeds consumption. Mar Drugs 19(6):341
Louzao MC, Vilariño N, Yotsu-Yamashita M (2014) Polycavernosides and other scarce new toxins. In: Seafood and freshwater toxins, 3rd edn. CRC Press, 16 pp
Marić P, Ahel M, Senta I, Terzić S, Mikac I, Žuljević A, Smital T (2017) Effect-directed analysis reveals inhibition of zebrafish uptake transporter Oatp1d1 by caulerpenyne, a major secondary metabolite from the invasive marine alga *Caulerpa taxifolia*. Chemosphere 174:643–654
Marshall K, Vogt R (1998) Illness associated with eating seaweed, Hawaii, 1994. West J Med 169:293–295
McGuffin M, Dentali S (2007) Safe use of herbal kelp supplements. Environ Health Perspect 115:A575–A577
McKim JM (2014) Food additive carrageenan: part I: a critical review of carrageenan in vitro studies, potential pitfalls, and implications for human health and safety. Crit Rev Toxicol 44:211–243

Mendes MC, Navalho S, Ferreira A, Paulino C, Figueiredo D, Silva D et al (2022) Algae as food in Europe: an overview of species diversity and their application. Foods 11:1871
Michalak I, Chojnacka K (2018) Seaweeds as a component of the human diet. In: Chojnacka K, Wieczorek P, Schroeder G, Michalak I (eds) Algae biomass: characteristics and applications. Developments in applied phycology, vol 8. Springer, Cham, pp 57–71
Michikawa T, Inoue M, Shimazu T, Sawada N, Iwasaki M, Sasazuki S et al (2012) Seaweed consumption and the risk of thyroid cancer in women: the Japan Public Health Center-based prospective study. Eur J Cancer Prevent 21:254–260
Milinovic J, Rodrigues C, Diniz M, Noronha JP (2021) Determination of total iodine content in edible seaweeds: application of inductively coupled plasma-atomic emission spectroscopy. Algal Res 53:102149
Mise N, Ohtsu M, Ikegami A, Mizuno A, Cui X, Kobayashi Y et al (2019) Hijiki seaweed consumption elevates levels of inorganic arsenic intake in Japanese children and pregnant women. Food Addit Contam 36:84–95
Mouritsen OG, Dawczynski C, Duelund L, Jahreis G, Vetter W, Schröder M (2013) On the human consumption of the red seaweed dulse (*Palmaria palmata* (L.) Weber & Mohr). J Appl Phycol 25:1777–1791
Mozzachiodi R, Scuri R, Roberto M, Brunelli M (2001) Caulerpenyne, a toxin from the seaweed *Caulerpa taxifolia*, depresses afterhyperpolarization in invertebrate neurons. Neuroscience 107:519–526
Nagai H, Yasumoto T, Hokama Y (1996) Aplysiatoxin and debromoaplysiatoxin as the causative agents of a red alga *Gracilaria coronopifolia* poisoning in Hawaii. Toxicon 37:753–761
Nagai H, Yasumoto T, Hokama Y (1997) Manauealides, some of the causative agents of a red alga *Gracilaria coronopifolia* poisoning in Hawaii. J Nat Prod 60:925–928
Nakajima Y, Endo Y, Inoue Y, Yamanaka K, Kato K, Wanibuchi H, Endo G (2006) Ingestion of Hijiki seaweed and risk of arsenic poisoning. Appl Organom Chem 20:557–564
Navarro G, Cummings S, Lee J, Moss N, Glukhov E, Valeriote FA, Gerwick WH (2015) Isolation of polycavernoside D from a marine cyanobacterium. Environ Sci Technol Lett 2:166–170
Nishitsuji K, Arimoto A, Higa Y, Mekaru M, Kawamitsu M, Satoh N, Shoguchi E (2019) Draft genome of the brown alga, *Nemacystus decipiens*, Onna-1 strain: fusion of genes involved in the sulfated fucan biosynthesis pathway. Sci Rep 9:4607
Nisizawa K, Noda H, Kikuchi H, Watanable T (1987) The main seaweed foods in Japan. Hydrobiologia 151(152):5–29
Noguchi T, Matsu T, Miyazawa K, Asakawa M, Iijima N, Shida Y et al (1994) Poisoning by the red alga "Ogonori" (*Gracilaria verrucsa*) on the Nojima Coast, Yokohama, Kanagawa Prefecture, Japan. Toxicon 32:1533–1538
Omara T, Nagawa C, Kyarimpa C, Böhmdorfer S, Rosenau T, Lugasi SO et al (2023) Lacustrine cyanobacteria, algal blooms and cyanotoxins in East Africa: implications for human and ecological health protection. Phycology 3:147–167
Pacheco D, Araújo GS, Cotas J, Gaspar R, Neto JM, Pereira L (2020) Invasive seaweeds in the Iberian Peninsula: a contribution for food supply. Mar Drugs 18:560
Parent-Massin D, Fournier V, Amade P, Lemée R, Durand-Clément M, Delescluse C, Pesando D (1996) Evaluation of the toxicological risk to humans of Caulerpenyne using human hematopoietic progenitors, melanocytes, and keratinocytes in culture. J Toxicol Environ Health 47:47–59
Paul VJ, Littler MM, Littler DS, Fenical W (1987) Evidence for chemical defense in tropical green alga *Caulerpa ashmeadii* (Caulerpaceae: Chlorophyta): isolation of new bioactive sesquiterpenoids. J Chem Ecol 13:1171–1185
Pereira L (2016) Edible seaweeds of the world. CRC Press, Boca Raton, FL
Pesando D, Lemée R, Ferrua C, Amade P, Girard JP (1996) Effects of caulerpenyne, the major toxin from *Caulerpa taxifolia* on mechanisms related to sea urchin egg cleavage. Aquat Toxicol 35:139–155

Petrus M, Culerrier R, Campistron M, Barre A, Rouge P (2010) First case report of anaphylaxis to *Spirulina*: identification of phycocyanin as responsible allergen. Allergy 65:924–925

Preeprame S, Hayashi K, Lee JB, Sankawa U, Hayashi T (2001) A novel antivirally active fucan sulfate derived from an edible brown alga, *Sargassum horneri*. Chem Pharm Bull 49:484–485

Raniello R, Mollo E, Lorenti M, Gavagnin M, Buia MC (2007) Phytotoxic activity of caulerpenyne from the Mediterranean invasive variety of *Caulerpa racemosa*: a potential allelochemical. Biol Invas 9:361–368

Rhee SS, Braverman LE, Pino S, He X, Pearce EN (2011) High iodine content of Korean seaweed soup: a health risk for lactating women and their infants? Thyroid 21:927–928

Ricciotti E, FitzGerald GA (2011) Prostaglandins and inflammation. Arterioscler Thromb Vas Biol 31:986–1000

Richter P, Schubert G, Schaible AM, Cavas L, Werz O, Pohnert G (2014) Caulerpenyne and related bis-enol esters are novel-type inhibitors of human 5-lipoxygenase. ChemMedChem 9:1655–1659

Roleda MY, Marfaing H, Desnica N, Jónsdóttir R, Skjermo J, Rebours C, Nitschke U (2019) Variations in polyphenol and heavy metal contents of wild-harvested and cultivated seaweed bulk biomass: health risk assessment and implication for food applications. Food Control 95:121–134

Rose M, Lewis J, Langford N, Baxter M, Origgi S, Barber M et al (2007) Arsenic in seaweed—forms, concentration and dietary exposure. Food Chem Toxicol 45:1263–1267

Rupert R, Rodrigues KF, Thien VY, Yong WTL (2022) Carrageenan from *Kappaphycus alvarezii* (Rhodophyta, Solieriaceae): metabolism, structure, production, and application. Front Plant Sci 13:859635

Sato Y, Inomata E, Nagoe H, Ito M, Konishi T, Fujimura H et al (2023) Nutrient uptake characteristics of *Cladosiphon okamuranus* (Phaeophyceae) from the Ryukyu Islands of Japan. J Appl Phycol 35:265–275

Shannon E, Abu-Ghannam N (2019) Seaweeds as nutraceuticals for health and nutrition. Phycologia 58:563–577

Smith J, Summers G, Wong R (2010) Nutrient and heavy metal content of edible seaweeds in New Zealand. New Zealand J Crop Hort Sci 38:19–28

Tarlo SM, Dolovich J, Listgarten C (1995) Anaphylaxis to carrageenan: a pseudo-latex allergy. J Allergy Clin Immunol 95:933–936

The Sydney Morning Herald (2014) Bonsoy to pay soy milk victims $25m in record class action. https://www.smh.com.au/business/bonsoy-to-pay-soy-milk-victims-25m-in-record-class-action-20141124-11spe4.html. Accessed 01 Aug 2023

Thomas I, Chean Siew LQ, Watts TJ, Haque R (2018) Seaweed allergy. J Allergy Clin Immunol 7:714–715

Tiberg E, Einarsson R (1989) Variability of allergenicity in eight strains of the green algal genus *Chlorella*. Int Arch Allergy Immunol 90:301–306

Tiberg E, Rolfsen W, Einarsson R, Dreborg S (1990) Detection of *Chlorella*-specific IgE in mould-sensitized children. Allergy 45:481–486

Tiberg E, Dreborg S, Björkstén B (1995) Allergy to green algae (*Chlorella*) among children. J Allergy Clin Immunol 96:257–259

Transparency Market Research (2022) Algae market (type: microalgae and macroalgae: and production method: harvesting and cultivation {open raceway pond and photobioreactors})-global industry analysis, size, share, growth, trends, and forecast, 2022–2031. https://www.transparencymarketresearch.com/algae-market.html. Accessed 01 Aug 2023

Tumwesigye E, Nnadozie CF, Akamagwuna FC, Noundou XS, Nyakairu GW, Odume ON (2023) Microplastics as vectors of chemical contaminants and biological agents in freshwater ecosystems: current knowledge status and future perspectives. Environ Pollut 330:121829

Vellinga RE, Sam M, Verhagen H, Jakobsen LS, Ravn-Haren G, Sugimoto M et al (2022) Increasing seaweed consumption in the Netherlands and Portugal and the consequences for the intake of iodine, sodium, and exposure to chemical contaminants: a risk-benefit study. Front Nutr 8:792923

Venkatesan J, Lowe B, Anil S, Manivasagan P, Kheraif AAA, Kang KH, Kim SK (2015) Seaweed polysaccharides and their potential biomedical applications. Starch-Stärke 67:381–390

Vidal JP, Laurent D, Kabore SA, Rechencq E, Boucard M, Girard JP et al (1984) Caulerpin, Caulerpicin, *Caulerpa scalpelliformis*: comparative acute toxicity study. Bot Mar 27:533–537

Vojdani A, Vojdani C (2015) Immune reactivities against gums. Altern Ther Health Med 21:60–68

Waserman S, Bégin P, Watson W (2018) IgE-mediated food allergy. Allergy Asthma Clin Immunol 14:55

Winder M, Kosztyła Z, Boral A, Kocełak P, Chudek J (2022) The impact of iodine concentration disorders on health and cancer. Nutrients 14:2209

Wu G, Zhuang D, Chew KW, Ling TC, Khoo KS, Van Quyen D et al (2022) Current status and future trends in removal, control, and mitigation of algae food safety risks for human consumption. Molecules 27:6633

Yeh TS, Hung NH, Lin TC (2014) Analysis of iodine content in seaweed by GC-ECD and estimation of iodine intake. J Food Drug Anal 22:189–196

Yim H, Yoo K, Seo W, Won N, Hong Y, Lee J (2007) Acute tubulointerstitial nephritis following ingestion of *Chlorella* tablets. Pediatr Nephrol 22:887–888

Yotsu-Yamashita M, Haddock R, Yasumoto T (1993) Polycavernoside A: a novel glycosidic macrolide from the red alga *Polycavernosa tsudai* (*Gracilaria edulis*). J Am Chem Soc 115:1147–1148

Yotsu-Yamashita M, Seki T, Paul V, Naoki H, Yasumoto T (1995) Four new analogs of polycavernoside A. Tetrahedron Lett 36:5563–5566

Yotsu-Yamashita M, Yasumoto T, Yamada S, Bajarias F, Formeloza M, Romero M, Fukuyo Y (2004) Identification of polycavernoside A as the causative agent of the fatal food poisoning resulting from ingestion of the red alga *Gracilaria edulis* in the Philippines. Chem Res Toxicol 17:1265–1271

Yotsu-Yamashita M, Abe K, Seki T, Fujiwara K, Yasumoto T (2007) Polycavernoside C and C2, the new analogs of the human lethal toxin polycavernoside A, from the red alga, *Gracilaria edulis*. Tetrahedron Lett 48:2255–2259

Zhan Z, Yu Q, Wang Z, Shan W (2010) Indole alkaloids from *Ervatamia hainanensis* with potent acetylcholinesterase inhibition activities. Bioorg Med Chem Lett 20:6185–6187

Zhang L, Wei L, Huang Y, Wen Y, Chu Y, Zhao C (2022a) Global seaweed farming and processing in the past 20 years. Food Prod Process Nutr 4:23

Zhang X, Zhang F, Li Q, Feng C, Teng W (2022b) Iodine nutrition and papillary thyroid cancer. Front Nutr 9:1022650

Chapter 26
Microalgae and Cyanobacteria Are Potential Sources of Food in the Future

Aria Babakhani and Fatemeh Zahmatkesh

Abstract In light of the rapid growth of the global population, increasing the productivity of agricultural systems has become imperative. Microalgae and cyanobacteria offer promising solutions because of their unique properties, making them ideal for a sustainable agriculture and energy production. These organisms can help address the global demand for food, animal feed, biofuels, and chemicals when integrated with traditional crops. This abstract explores the potential advantages of incorporating microalgae and cyanobacteria into agricultural systems and emphasizes the need for more research. Microalgae and cyanobacteria have been identified as promising sources of biomass, which can be combined with traditional crops to meet the increasing global demand for food, animal feed, biofuels, and chemicals. These organisms have the potential to make a significant contribution to sustainable agriculture and energy production due to their unique properties. Sustainable culture of microalgae and cyanobacteria can be achieved by improving the effectiveness of nutrient application and recycling waste streams as nutrients. However, addressing the challenges of contaminated biomass production and the limited bioavailability of certain nutrients from waste streams is crucial. The sustainability and efficiency of microalgae and cyanobacterial production can be greatly improved by using these strategies.

Keywords Microalgae · Cyanobacteria · Nutritional compounds · Functional food · Health

A. Babakhani (✉) · F. Zahmatkesh
Fisheries Department, Faculty of Natural Resources, University of Guilan, Sowmeh Sara, Iran
e-mail: Babakhani@guilan.ac.ir

T. A. Kurniawan and A. Anouzla (eds.), *Algae as a Natural Solution for Challenges in Water-Food-Energy Nexus*, Environmental Science and Engineering,
https://doi.org/10.1007/978-981-97-2371-3_26

26.1 Introduction

26.1.1 Microalgae and Cyanobacteria: Architects of Diverse Ecosystems

Microalgae and cyanobacteria, two types of microorganisms, play a crucial role in shaping and sustaining many ecosystems. They may be found in a wide range of environments, including aquatic habitats and desert surfaces, and often engage in mutually beneficial interactions with other organisms. Within this group, there are remarkable organisms known as cyanobacteria that contribute to the whole color palette with their distinctive green–blue shades. The resilience exhibited by these organisms is remarkable, as they are able to flourish in a wide range of conditions. They are capable of withstanding extreme temperature variations, adapting to different levels of sunlight intensity, effectively managing pH levels, and successfully navigating through changing degrees of salinity (Barsanti et al. 2008).

Microalgae and cyanobacteria hold significant promise in addressing pressing challenges related to food security and sustainability. With the global population projected to reach 9.7 billion by 2050, traditional food production systems are likely to face constraints in meeting the escalating demand for protein-rich nourishment (FAO 2017). Moreover, the reverberations of the COVID-19 pandemic have left an indelible mark on the intricate web of food production, distribution, and consumption. As a grim consequence, it is projected that a staggering 130 million individuals will bear the brunt of hunger a stark reminder of the urgent need to address the critical challenges facing our global food systems (Kakaei et al. 2022).

In this context, the cultivation of microorganisms as alternative protein sources emerges as a viable solution to alleviate food scarcity. In an effort to explore this predisposition, firms in the food and nutraceutical industries are increasingly focusing on harnessing the potential of natural high-value compounds, indicating a growing interest in their large-scale manufacturing. In the realm of pigments used in food and health applications, there has been a noticeable movement in market interest towards organically sourced compounds produced by biological synthesis. Currently, the majority of food pigments available in the commercial market are produced using chemical synthesis. However, this technique is typically accompanied by safety issues due to the molecular structures and organic processes involved in their production. In stark contrast, natural pigments exhibit a notable absence of toxicity or detrimental impacts on the human body, thereby assuming an extra function as nutritional enhancers that possess a range of biological advantages. Although some plant-based pigments have been successfully incorporated into commercial products, the unique appeal of microalgal pigments lies in their remarkable capacity to significantly enhance various activities (De Mendonca et al. 2021; Udayan et al. 2022).

When compared to their synthetic counterparts, natural pigments exhibit a notable absence of toxicity or adverse effects on the human body. Instead, they serve as nutritional enhancers, imbued with an array of biologic al activities (Kakaei et al.

2022). While certain plant-derived pigments have found their way into commercial markets, the vigor of microalgal pigments has been found to amplify manifold. In the role of a pivotal food constituent, the allure of microalgal pigments beckons manufacturers and consumers alike. Projections indicate that by the year 2025, the market value of microalgal pigments is poised to ascend to USD 452.4 million, bolstered by a Compound Annual Growth Rate (CAGR) of 4%. It is foreseen that the realm of microalgae products will burgeon into a substantial enterprise, carving a substantial niche in the commercial landscape (Patel et al. 2022).

In an effort to explore this propensity, food and nutraceutical producers are increasingly focusing on harnessing the potential of naturally occurring substances with significant value, indicating a shift towards their large-scale manufacturing. In the realm of pigments utilized in both food and health sectors, there has been a noticeable shift in market focus towards naturally sourced chemicals produced through biological synthesis. Currently, the majority of food pigments available in the commercial market are produced using chemical synthesis. However, this technique is frequently associated with safety issues due to the molecular structures and organic processes involved in their production. In stark contrast, natural pigments exhibit a notable absence of toxicity or detrimental impacts on the human body, thereby assuming an extra function as nutritional enhancers that possess a diverse range of biological advantages. Although some plant-based pigments have been incorporated into commercial products, the appeal of micro-algal pigments is unique due to their capacity to significantly enhance various activities (Sun et al. 2023).

Cyanobacteria and microalgae, are found in a wide range of habitats, including freshwater, marine, terrestrial, and severe conditions such as hot springs and desert rocks. Due to their exceptional plasticity, they play a crucial role in the generation of biomass in situations where sustainability is of utmost importance. According to Camacho et al. (2019), there exists a symbiotic relationship between these entities, wherein they provide economic growth while ensuring the preservation of agricultural resources. This relationship also promotes the efficient reuse of water and mitigates the release of harmful greenhouse gases. Moreover, the commercial value of cyanobacteria and microalgae is further amplified by their revolutionary potential. The adaptability of these microorganisms is evidenced by their ability to produce a wide range of products, including biofuels, cosmetics, and renewable chemicals (Camacho et al. 2019). In an intriguing turn of events, they assume significant roles in supporting the nutritional needs of both animals and humans, acting as a valuable source of vital proteins, lipids, and phytochemicals (Ferrazzano et al. 2020).

During the process of photosynthesis, microalgae harness solar energy and combine it with several vital nutrients such as carbon, nitrogen, phosphorus, sulfur, potassium, iron. This enables them to synthesize biomass molecules and undergo cellular multiplication. Microalgae necessitate precise amounts of these fundamental components in order to generate biomass (Markou et al. 2014). The utilization of cyanobacteria could enable the extraction and recovery of metals from Martian rocks, demonstrating their adaptability to extreme environments and their potential significance in future space exploration and resource utilization endeavors (Cockell et al. 2021). The primary objective of this study article is to investigate the viability of microalgae and cyanobacteria as sustainable food sources. Furthermore, the study seeks to examine the nutritional composition, bioactive constituents, and potential physiological advantages associated with them.

26.2 Microalgae

Microalgae are oxygen-evolving, eukaryotic, unicellular, photosynthetic microorganisms with cells that are generally several micrometers in size. According to the results reported by researchers, the categorization of organisms within the algal kingdom is determined by the presence of chlorophyll a (Chl a), in addition absence of features such as stems, roots, leaves, and flowers (Lee 1999). The previous defining criterion serves as a comprehensive representation of the fundamental characteristics that define an alga amidst the complex range of living organisms. Exploring the intricate profiles of microalgae through molecular analyses, a contemporary approach has emerged that involves identifying the genetic signatures of microalgae through DNA sequencing. This process involves examining the conserved domains of DNA, such as the Ribulose Bisphosphate Carboxylase Large subunit gene (rbcL), along with the 16S and 18S ribosomal RNA (rRNA) genes. These molecular breadcrumbs provide insights into the genetic makeup of prokaryotic and eukaryotic microalgae, shedding light on the complexities of their existence (Ballesteros et al. 2021; Hadi et al. 2016). Microalgae, categorized into distinct groups based on eukaryotic and prokaryotic divisions, offer a fascinating array of diversity. Among the eukaryotic classes are *Heterokontophyceae, Cryptophyceae, Dinophyceae, Chlorophyceae, Haptophyceae, Phodophyceae, Bacillariphyceae, Eustigmatophyceae,* and more. Prominent species such as *Skeletonema*, *Thalassiosira*, *Isochrysis*, *Nannochloropsis*, and *Chlorella* find their place within these classes. The prokaryotic realm encompasses *Prochlorophyceae* and *Cyanophyceae*, housing widely used species such as *Spirulina*, *Dunaliella*, *Arthrospira*, and *Chlamydomonas*. This intricate taxonomy underscores the richness of microalgal diversity, shaping their significance in various applications, making them a captivating subject of study in our exploration (Becker 2013). Microalgae, by harnessing CO2 fixation, have the capacity to convert solar energy into chemical energy. Their efficiency is more than tenfold greater than that of terrestrial plants. Microalgae serve as a natural reservoir of biologically active compounds with high dynamism (de Morais et al. 2015). Microalgae have extensive commercial and industrial applications in fields such as metallurgy, chemistry, pharmaceuticals, energy, food, paper, glass, and textiles (Mehta et al. 2018; Nedaei and Shokrkar 2022). They provide various valuable products including different types of fats, unsaturated fatty acids, pigments, natural colorants, sugars, antioxidants, biopolymers, peptides, and livestock feed materials (Vassilev and Vassileva 2016). Despite challenges in their production process, such as scale-up limitations and carbon dioxide sequestration issues, the accumulation of high-value compounds within microalgae makes them a promising resource for a sustainable and optimized future (Panis and Carreon 2016). Certainly, the diversity of microalgae is astonishing, encompassing species with significant evolutionary differences (de Vargas et al. 2015). Despite their minuscule size, spanning from 0.2 to 200 μm, these organisms possess the capacity to play a role in mitigating climate change through carbon fixation (Ratnapuram et al. 2018; Vu et al. 2018). Remarkably, the biological capture and sequestration of carbon by microalgae have garnered acknowledgment as a paramount and efficient method for carbon sequestration (Alami et al. 2021; Moreira and Pires 2016).

The species *Chlorella* and *Spirulina* have achieved global cultivation as functional foods, with their respective industries valued at $397 million in 2018 and forecasted to escalate to $850 million by 2026. Algae consumption represents a burgeoning trend, attributed to expanding populations, limited arable land availability, and an increasing shift toward Veganism in support of sustainable development and carbon emission reduction. The majority of algae belong to the category of autotrophic or mixotrophic organisms, relying on light and uncomplicated nutrients for their flourishing (Ruzik 2023). Chlorella has notable attributes as a hepatoprotective and hypocholesterolemic agent, offering potential benefits in instances of starvation and methionine intoxication. It has been observed to significantly decrease blood glucose levels and improve hemoglobin concentrations. Furthermore, Chlorella possesses an active immunostimulatory known as β-1,3-glucan, which plays a role in reducing blood lipids and functions as a robust scavenger of free radicals. Spirulina, a well-known microalga sometimes referred to as a "superfood," is classified under the blue-green photoautotrophic genus of unicellular microalgae. Microalgae function as an extraordinary natural repository of key nutrients, including vitamins A, B1, B2, and B12, necessary fatty acids, and valuable pigments, such as xanthophyll and carotenoids. Due to the inability of the human body to manufacture these beneficial components, significant attention has been directed toward the cultivation of Spirulina to exploit its manifold advantages (Ruzik 2023; Sathasivam et al. 2019).

26.3 Cyanophyta

Cyanobacteria possess a significant evolutionary lineage extending over around 3.5 billion years in the past (Whitton 2012). Prokaryotic organisms, which encompass a wide array of species, exhibit remarkable diversity and inhabit a multitude of environments (Smith and Thanh Doan 1999). Cyanobacteria, as live photosynthetic organisms (Pagels et al. 2019), possess the ability to synthesize chlorophyll-a (Whitton 2012). They are found in soil habitats, where they interact with plants, algae, and bacteria (Smith and Thanh Doan 1999). The majority of cyanobacteria are capable of synthesizing the pigment phycocyanin, which imparts a bluish hue to their thick colonies, hence leading to their designation as "blue-green algae".

Cyanobacteria possess the ability to engage in the process of photosynthesis as well as nitrogen fixation. The aforementioned capabilities, in conjunction with their remarkable capacity to acclimate to diverse soil compositions, confer upon them the potential to flourish in virtually all habitats they occupy (Mishra and Pabbi 2004). Although they possess a microscopic cell size, these organisms become observable to the unaided human eye during periods of blooming.

These microorganisms are predominantly single-celled and sometimes exist in simple multicellular forms. They exhibit an extremely rapid growth rate and can thrive in challenging environmental conditions, including saline waters and arid regions. They acquire nutrients from diverse sources, such as wastewater, and have the ability to absorb carbon dioxide from emissions produced by fossil fuel power plants. Additionally, they can generate valuable byproducts. Utilizing cyanobacteria offers multiple advantages due to their ability to flourish in these conditions and their potential for various applications (Hu et al. 2008).

Cyanobacteria require light, carbon dioxide, nitrogen, phosphorus, and potassium for their growth. In a relatively short period, they produce various metabolic products, including different types of lipids, carbon hydrates, and proteins, in noticeable quantities (Mata et al. 2010).

Cyanobacteria exhibit important and practical features in biotechnology. They are capable of producing molecular hydrogen and various bioactive compounds, including anti-cancer agents like microcystin. These attributes highlight their significance in microbial biotechnology. Moreover, protein-rich materials can also be derived from these microorganisms, particularly Spirulina, which can serve as a source of food.

Cyanobacteria possess several advantages over other microorganisms, including their larger cell volume, high attachment tendency, and simple nutritional requirements, which make them easily cultivable in large-scale laboratory settings (Abdel-Aty et al. 2013). The overall quality of life on Earth is closely linked to the quality of the environment. In earlier times, we believed that the Earth and its resources were limitless, but the current state of available resources indicates that we have used them carelessly and without considering their finite nature. Currently, polluted lands pose a serious threat to human health, and extensive international efforts have been made to address this issue.

Many cyanobacteria have applications in pharmaceuticals and agriculture. They are used in areas such as drug production, toxins, endotoxins, cytotoxins, hepatotoxins, and more8. Some of these microorganisms produce toxic and highly dangerous substances, including neurotoxins. The toxicity of cyanobacteria has become a globally significant issue. The desired temperature for many cyanobacteria is generally a few degrees higher than that of most eukaryotic algae. Some cyanobacteria have a high tolerance for ultraviolet radiation and are among the most successful organisms in hypersaline environments (Whitton 1992). The growth of cyanobacteria is limited by the availability of phosphorus (Mishra and Pabbi 2004). Most cyanobacteria thrive in alkaline pH conditions. Interactions with calcareous rock are a characteristic feature of certain cyanobacteria (Whitton 2012).

Indeed, while around 2700 cyanobacteria species have been identified and described, prediction models indicate that the actual number in nature could range between 2000 and 8000 species (Nabout et al. 2013). Many of these species have been investigated and recognized for their potential applications in diverse industries such as agriculture, energy, food, and pharmaceuticals. These cyanobacteria exhibit capabilities like oil production, atmospheric nitrogen fixation, and high concentrations of vitamins, minerals, and proteins, making them valuable candidates for various applications (Singh et al. 2020).

26.4 Use of Microalgae and Cyanobacteria in the Food

Indeed, microalgae and cyanobacteria are prolific producers of a diverse range of metabolites. These biologically active compounds hold significant potential for agricultural applications, serving as bio-fertilizers, bio stimulants, and bio-pesticides. Among these metabolites, phenolic compounds, terpenoids, free fatty acids (FFAs), polysaccharides, carotenoids, and phytohormones stand out due to their ability to promote plant growth. Their unique biological activities make them valuable tools for enhancing agricultural productivity and sustainability (Gonçalves 2021; Kusvuran and Kusvuran 2019; Pan et al. 2019; Singh et al. 2017). According to Uribe-Wandurraga et al. (2021), it is possible to commercially exploit microalgae derivatives in various forms such as food supplements, including powder and oil, as well as ingredients or additives. The utilization of microalgae biomass in the production of innovative food and ingredients for beverages, bread, and dough formulations has been a subject of significant progress, as highlighted in studies conducted by (Niccolai et al. 2020; Ververis et al. 2020). According to Canelli et al. (2020), microalgae exhibit potential as a viable reservoir of bio accessible proteins. Chlorella exhibits a substantial concentration of carbohydrates. Likewise, *Nannochloropsis oculata* and *Phaeodactylum tricornutum* serve as notable sources of omega 3 fatty acids (Niccolai et al. 2019).

Exactly, phenolic compounds are organic compounds characterized by an aromatic hydrocarbon structure bonded to a hydroxyl group. They can be categorized into two main groups: phenols and polyphenols (Gonçalves 2021). Phenols are further divided into subgroups such as phenolic acids, flavonoids, and phenylpropanoids. These diverse compounds play important roles in various biological activities due to their unique chemical structures and properties (Pan et al. 2019).

Microalgae have garnered considerable interest as a prospective food source owing to its high nutritional content and environmentally friendly growth techniques. The microscopic organisms possess a high concentration of vital nutrients, including proteins, beneficial fats, vitamins, minerals, and antioxidants. The potential of microalgae such as Spirulina and Chlorella as dietary supplements has been investigated, primarily due to their high protein content, which surpasses that of the conventional plant source.

Microalgae oil offers a more efficient energy yield per hectare compared to land-based oil seeds like sunflower and rapeseed, presenting a promising avenue for biofuel production (Basha and Shrikanya 2023). Additionally, microalgae's consumption of atmospheric CO_2 results in environmentally friendly biodiesel production, emitting lower levels of carbon dioxide compared to conventional petroleum-based fuels (Palma et al. 2023).

Furthermore, the cultivation of microalgae necessitates a smaller allocation of land and water resources in comparison to conventional crops, rendering them a viable and sustainable alternative for the production of food. These substances can be employed in diverse formats, encompassing supplementary forms, powdered compositions, and even integration into comestible items such as snacks, beverages, and plant-based meat substitutes. Nevertheless, there are several obstacles that must be overcome in order to successfully include microalgae into widely adopted dietary practices. These hurdles encompass factors such as flavor, digestibility, and the ability to scale up production. Absolutely, microalgae and cyanobacteria offer a wide range of biotechnological applications across various industries. These applications can be categorized into closed and open systems, each with distinct benefits.

26.4.1 Microalgae and Cyanobacteria: Harnessing High-Value Compounds

In closed systems, like photobioreactors, microalgae can be cultivated under tightly controlled conditions. This precision allows for the production of high-value compounds such as food additives, nutraceuticals, pharmaceuticals, and recombinant proteins. These systems provide the ability to regulate factors like pH, temperature, CO_2 supply, light exposure, nutrient availability, broth collection, oxygen levels, and sterility, all contributing to the production of pure and valuable compounds (Gonçalves 2021). On the other hand, open systems involve cultivating microalgae in natural environments, like ponds or open tanks. While these systems are less controlled, they offer advantages in terms of scalability and cost-effectiveness. Open systems are often used for bulk production of biofuels and biomass for animal feed or biofertilizers. Both closed and open systems have their unique advantages and are employed based on the specific goals and requirements of the applications. The versatility of microalgae and cyanobacteria in producing a variety of valuable compounds makes them promising candidates for advancing biotechnological innovations in numerous industries (Gonçalves 2021). Certainly, cyanobacteria exhibit a remarkable ecological significance, boasting an estimated total wet biomass of approximately one billion metric tons (10^{15} g). This colossal biomass underscores their pivotal role as one of the most paramount life forms on Earth. Their adeptness in photosynthetic activities not only shapes biogeochemical cycles but also profoundly influences the intricate dynamics of ecosystems across the globe (Hachicha et al. 2022).

Microalgae and Cyanobacteria are renowned for their abundant reservoirs of vegetable proteins, serving as essential functional ingredients with multifaceted applications across various industries (Ibañez and Cifuentes 2013).

Certainly, Microalgae and Cyanobacteria demonstrate a remarkable capacity to synthesize proteins, accounting for a considerable portion of their dry mass, which can range from 6% to an impressive 80%, depending on the specific species (Becker 2007; Ibañez and Cifuentes 2013). Noteworthy examples encompass Chlorella vulgaris of the Chlorophyta lineage (with protein content spanning 51–58% of dry weight), Arthrospira maxima hailing from Cyanobacteria (exhibiting a protein content range of 60–71% of dry weight), Isochrysis galbana within Haptophyta (showcasing protein content of 50–56% of dry weight), and Porphyridium purpureum from Rhodophyta (showcasing protein content of 28–39% of dry weight) (Ibañez and Cifuentes 2013). Additionally, the integration of bioactive peptides adds an extra dimension of value, endowing these microorganisms with compounds that hold promise not only in augmenting nutrition but also in therapeutic applications, elevating their significance (Ibañez and Cifuentes 2013).

26.4.2 Microalgae: Space Travel Nutrition and Antioxidant Potential:

Chlorella's proteins encompass all essential amino acids, rendering it an ideal food source for long-duration space travel. To ensure sustenance during extended space missions, scientists have ingeniously designed an ecological cycle using Chlorella. These microalgae, with their exceptional attributes, are considered the most valuable biotic substance on Earth. They serve as the cornerstone of the food chain and possess an extraordinary capacity for rapid reproduction, positioning them at the forefront of biological productivity.

Cyanobacterial carotenoids, including canthaxanthin, beta-carotene, nostoxanthin, and zeaxanthin, have been harnessed to meet the escalating demand for food supplements, animal feed, food additives, and colorants. Among these, Spirulina, a cyanobacterial species renowned for its substantial content of riboflavin, thiamine, β-carotene, and vitamin B12, stands as a notable illustration. These value-added supplements are commercially accessible in a range of formats, encompassing capsules, granules, and tablets, catering to a diverse array of consumer preferences and nutritional needs (Ayesha Algade et al. 2021).

Certainly, let's delve into a more sophisticated explanation. Astaxanthin, a keto-carotenoid of significance, has emerged as a formidable antioxidant, outclassing conventional contenders like vitamin C, vitamin A, and other carotenoids. Its exceptional capability lies in its adeptness at shielding human cellular structures from oxidative harm stemming from photooxidation (Singh et al. 2017). This feature ignites contemplation about its potential role in disease management, with a noteworthy mention of HIV (Singh et al., 2017). These nutritional supplements, distinctively, harness the entirety of cyanobacterial biomass, offering a departure from the pharmaceutical-oriented extraction processes commonly observed (Chronakis and Madsen 2011). Undoubtedly, certain microalgae species has taken center stage as notable contributors to the human food supply. Aphanizomenon flos-aquae, *Chlorella* sp., *Dunaliella Salina*, *Dunaliella tertiolecta*, and *Spirulina plantensis* emerge as frontrunners in this regard. Their selection is rooted in their impressive protein content and elevated nutritive worth. These microalgae, through meticulous cultivation and harvesting methods, have become renowned sources of sustenance, heralding a pivotal role in meeting dietary requirements and promoting overall human well-being (de Oliveira Rangel-Yagui et al. 2004; Soletto et al., 2005).

26.4.3 Polysaccharides from Microalgae: Diverse Applications and Health Benefits

Polymeric structures known as polysaccharides, which are carbohydratebased, find extensive use in the food industry. Currently, researchers are ac- tively investigating novel natural antioxidants, leveraging microalgae like Chlorella and Spirulina due to their high content of polysaccharides (Abbaspour et al. 2020). These polysaccharides not only contribute to bolstering the immune system but also enhance enzymatic activity within nuclear cells and facilitate DNA repair. Particularly noteworthy are the polysaccharides extracted from Spirulina platensis and Chlorella pyr- enoidosa, which showcase remarkable antioxidant and anti-tumor proper- ties, alongside their ability to modulate the immune system and efficiently neutralize superoxide and hydroxyl radicals (Gómez et al. 2023).

Amoral and Imulina, derived respectively from Chlorella pyrenoidosa and Spirulina platensis, exhibit superior anti-cancer effects when compared to fungal polysaccharides like schizophyllan, lentinan, and krestin. Recently, microalgal polysaccharides have been recognized as potential substitutes in industries spanning nutrition, cosmetics, pharmaceuticals, and more (Nowruzi et al. 2020). These polysaccharides from microalgae have also demonstrated effective antiviral activity against the herpes simplex virus, further solidifying their versatile applications (Nowruzi et al. 2023).

26.5 Nutritional Benefits and Applications of Chlorella and Spirulina

The active components of Chlorella, including its immune system-stimulating properties, free radical scavenging, and blood lipid-reducing effects, have been recognized. Prepared soup from the microalgae species Chlorella sp. has been shown to improve energy, weight, and overall health in patients with leprosy as a dietary supplement. Japanese researchers have created various food products such as green tea, soup, noodles, bread, rolls, cookies, ice cream, and soy sauce using Chlorella as a powdered ingredient. The microalgae *Scenedesmus* sp. is also recognized as a nutritious substance, although its commercial production is limited. Its extract is used in regular foods like desserts, fruit pudding, yogurt, noodles, and soup. Chlorella and Spirulina, in particular, are rich in protein content (more than 60% of dry weight) and can serve as sources of essential amino acids such as lysine, leucine, isoleucine, tryptophan, and valine. The lipid content in algal biomass varies across different divisions of algae and depends on ecological factors during growth. High lipid content, reaching up to 9% of dry weight, has been observed in certain algal species like Chlorella, Crysiophyta, and Bacillariophyceae according to Liu et al. (2021).

According to the U.S. Food and Drug Administration (FDA) and the European Food Safety Authority (EFSA), food products based on Spirulina sp. are classified as healthy foods. Clinical studies have shown that consuming Spirulina can lead to reduced cholesterol levels, protect against certain types of cancers, enhanced immune response, increased levels of gut lactobacillus (probiotics), and protection against radiation (using sunscreen lotion). *sp. Spirulina* has been consumed as a staple food by people in regions like Mexico (Aztecs) and Chad (Africa) for an extended period. Many food products, such as instant soups, puddings, cake mix, and biscuits, can be enriched with microalgae or their bioactive compounds. Similarly, the protein content of bread increased due to the addition of dried Spirulina platensis. Additionally, the concentration of 11 amino acids, particularly essential ones like training, methionine, leucine, and isoleucine, significantly increased (Gantar and Svirčev 2008).

26.6 Bioactive Compounds from Cyanobacteria

Potential and Applications Cyanobacteria are highly regarded for their newfound potential in producing novel bioactive compounds with significant biochemical activity. Many cyanobacteria, such as Spirulina, Anabaena, Nostoc, and synthesize various secondary metabolites and bioactive compounds. The presence of enzymes like peptide synthetase is common in cyanobacteria and contributes to the biosynthesis of certain active compounds, such as Microcystis (Liu et al. 2021). For instance, Scenedesmus and Chlorella spp., including the quadricauda species, can produce polysaccharides that act as protective agents against oxidative stress (Kultschar and Llewellyn 2018). Some cyanobacteria like *Phormidium tenue* and *Lyngbya lagerheimii* generate active sulfolipids coupled with various fatty acids. Cryptophytic, a natural product initially isolated from blue-green algae Nostoc, demonstrates potent anti-tumor activity both in vitro and in vivo (Guedes et al. 2011).

Furthermore, in the algal culture medium of *Chlorella vulgaris*, a type of glycoprotein called arabinogalactan was discovered to possess protective properties against tumor metastasis and suppression of chemotherapy-induced immune suppression in mice. Several cyclic and linear peptides derived from cyanobacteria exhibit protease inhibition qualities and are utilized in treating conditions such as strokes and coronary artery blockage (Liu et al. 2014). For example, aruginozines isolated from Microcystis demonstrate inhibitory properties against enzymes such as thrombin, plasmin, and trypsin. Other derived peptides, like micropeptin, microcystyleide, cyanopeptolin, oscillapeptin, and nostocyclin, also function as inhibitors of enzymes like trypsin, plasmin, thrombin, and chymotrypsin (Encarnação et al. 2015).

26.7 Future

Microalgae present a potentially viable solution for tackling the complex issues arising from the increasing global demands for water, food, and energy, as well as mitigating the adverse impacts of climate change (Ferreira de Oliveira and Bragotto 2022). These purposes include the production of biomass, natural compounds with added value, and functional foods (Liu et al. 2021). With the increasing scarcity of natural resources, the significance of optimizing resource utilization becomes of utmost importance. The protection of biodiversity and the preservation of ecosystems are of utmost importance, as emphasized by (Bianchi et al. 2014; Pascoal et al. 2021). Microalgae possess significant biological and economic significance owing to their diverse array of global uses, encompassing several sectors such as the food industry, medicines, immunological stimulants, biofuels, cosmetic compounds, and agriculture (Liu et al. 2021; Nowruzi et al. 2023). The ability of microalgae to offer sustainable solutions is in line with the pressing necessity to reconcile human advancement with the preservation of the environment. The inclusion of microalgae enriched with carotenoids or secondary metabolites extracted from microalgae in the diets of humans and animals has been shown to enhance their nutritional composition (Liu et al. 2021). Hence, it is advisable to implement quality control measures for dietary supplements sourced from diverse microalgae species in order to safeguard customer well-being (Nowruzi et al. 2020).

26.8 Conclusion

Microalgae and cyanobacteria are commonly found in various habitats, including water bodies, desert crusts, and in symbiotic relationships with other organisms. These organisms possess the ability to withstand a diverse range of environmental circumstances, including extreme temperatures, intense or mild lighting, varying pH levels, and high salinity. There is a prevailing belief that a reduction or complete elimination of prominent risk factors, namely smoking, radiation exposure, alcohol consumption, and infections, would lead to a notable 30% decline in mortality rates associated with cancer.

The biotechnological capabilities of microalgae have been explored extensively in recent years, mostly due to the discovery of various synthetic chemicals produced by these microbes. The extensive range of microalgae species and their abundance biochemical constituents, such as unsaturated fatty acids, carotenoids, phycobilins, polysaccharides, vitamins, sterols, and various other bioactive compounds, has resulted in their cultivation for commercial purposes.

It is widely believed that microalgae are valuable reservoirs of bioactive compounds that have potential health-promoting properties and exhibit inhibitory effects against a range of illnesses. As a result, there has been a notable rise in the widespread cultivation and commercialization of microalgae in the past several decades, mostly driven by their utilization in aquaculture and as a supplement in animal feed. In contemporary times, the utilization of dietary supplements sourced from microalgae has been prevalent in augmenting the nutritional value of particular dietary regimens. Furthermore, microalgae have been shown to have therapeutic potential in alleviating various ailments that affect both human and animal populations. Nevertheless, it is crucial to acknowledge that cyanobacteria have the capability to generate certain poisons that can potentially be encountered by individuals through the ingestion of algal products.

References

Abbaspour S, Nowruzi B, Hamdi SMM (2020) Optimizing the Cultivation conditions of *Fischerella* sp. SH.A Cyanobacterium for maximizing polysaccharide production with antibacterial activity. Biol J Microorg 9(34):23–53. https://doi.org/10.22108/bjm.2020.120846.1261

Abdel-Aty AM, Ammar NS, Ghafar HHA, Ali RK (2013) Biosorption of cadmium and lead from aqueous solution by fresh water alga *Anabaena sphaerica* biomass. J Adv Res 4(4):367–374

Alami AH, Alasad S, Ali M, Alshamsi M (2021) Investigating algae for CO_2 capture and accumulation and simultaneous production of biomass for biodiesel production. Sci Total Environ 759:143529

Ayesha Algade A, Kweku Amoako Atta D-J, Gabriel Komla A (2021) Industrial applications of cyanobacteria. In: Wael NH (ed) Cyanobacteria. IntechOpen. https://doi.org/10.5772/intechopen.98859

Ballesteros I, Terán P, Guamán-Burneo C, González N, Cruz A, Castillejo P (2021) DNA barcoding approach to characterize microalgae isolated from freshwater systems in Ecuador. Neotrop Biodiv 7(1):170–183

Barsanti L, Coltelli P, Evangelista V, Frassanito AM, Passarelli V, Vesentini N, Gualtieri P (2008) Oddities and curiosities in the algal world. In: Evangelista V, Barsanti L, Frassanito AM, Passarelli V, Gualtieri P (2008) Algal toxins: nature, occurrence, effect and detection. Dordrecht.

Basha F, Shrikanya Rao KVL (2023) A review on third-generation biofuels from marine diatoms. GSC Biol Pharmaceut Sci 24(1):373–379

Becker EW (2007) Micro-algae as a source of protein. Biotechnol Adv 25(2):207–210

Becker EW (2013) Microalgae for aquaculture: nutritional aspects. In: Handbook of microalgal culture: applied phycology and biotechnology, pp 671–691

Bianchi M, Chopin F, Farme T, Franz N, Fuentevilla C, Garibaldi L, Laurenti A (2014) FAO: the state of world fisheries and aquaculture. Food and Agriculture Organization of the United Nations, Rome, Italy, pp 1–230

Camacho F, Macedo A, Malcata F (2019) Potential industrial applications and commercialization of microalgae in the functional food and feed industries: a short review. Mar Drugs 17(6):312. https://www.mdpi.com/1660-3397/17/6/312

Canelli G, Tarnutzer C, Carpine R, Neutsch L, Bolten CJ, Dionisi F, Mathys A (2020) Biochemical and nutritional evaluation of Chlorella and Auxenochlorella biomasses relevant for food application. Front Nutr 7:565996

Chronakis IS, Madsen M (2011) Algal proteins. In: Handbook of food proteins. Elsevier, pp 353–394

Cockell CS, Santomartino R, Finster K, Waajen AC, Nicholson N, Loudon CM, Eades LJ, Moeller R, Rettberg P, Fuchs FM, Van Houdt R, Leys N, Coninx I, Hatton J, Parmitano L, Krause J, Koehler A, Caplin N, Zuijderduijn L et al (2021) Microbially-enhanced vanadium mining and bioremediation under micro- and mars gravity on the international space station. Front Microbiol 12:641387. https://doi.org/10.3389/fmicb.2021.641387

De Mendonca HV, Assemany P, Abreu M, Couto E, Maciel AM, Duarte RL, dos Santos MGB, Reis A (2021) Microalgae in a global world: new solutions for old problems? Renew Energy 165:842–862

de Morais MG, Vaz Bda S, de Morais EG, Costa JA (2015) Biologically active metabolites synthesized by microalgae. Biomed Res Int 2015:835761. https://doi.org/10.1155/2015/835761

de Oliveira Rangel-Yagui C, Danesi EDG, De Carvalho JCM, Sato S (2004) Chlorophyll production from *Spirulina platensis*: cultivation with urea addition by fed-batch process. Biores Technol 92(2):133–141

de Vargas C, Audic S, Henry N, Decelle J, Mahé F, Logares R, Lara E, Berney C, Le Bescot N, Probert I, Carmichael M, Poulain J, Romac S, Colin S, Aury JM, Bittner L, Chaffron S, Dunthorn M, Engelen S et al (2015) Ocean plankton. Eukaryotic plankton diversity in the sunlit ocean. Science 348(6237):1261605. https://doi.org/10.1126/science.1261605

Encarnação T, Pais AA, Campos MG, Burrows HD (2015) Cyanobacteria and microalgae: a renewable source of bioactive compounds and other chemicals. Sci Prog 98(2):145–168

Ferrazzano GF, Papa C, Pollio A, Ingenito A, Sangianantoni G, Cantile T (2020) Cyanobacteria and microalgae as sources of functional foods to improve human general and oral health. Molecules 25(21):5164. https://www.mdpi.com/1420-3049/25/21/5164

Ferreira de Oliveira AP, Bragotto APA (2022) Microalgae-based products: food and public health. Fut Foods 6:100157. https://doi.org/10.1016/j.fufo.2022.100157

Gantar M, Svirčev Z (2008) Microalgae and cyanobacteria: food for thought 1. J Phycol 44(2):260–268

Gómez PI, Mayorga J, Flaig D, Castro-Varela P, Jaupi A, Ulloa PA, Soto-Bartierra J, Henríquez V, Rojas V (2023) Looking beyond Arthrospira: comparison of antioxidant and anti-inflammatory properties of ten cyanobacteria strains. Algal Res 103182

Gonçalves AL (2021) The use of microalgae and cyanobacteria in the improvement of agricultural practices: a review on their biofertilising, biostimulating and biopesticide roles. Appl Sci 11(2):871. https://www.mdpi.com/2076-3417/11/2/871

Guedes AC, Amaro HM, Malcata FX (2011) Microalgae as sources of carotenoids. Mar Drugs 9(4):625–644

Hachicha R, Elleuch F, Ben Hlima H, Dubessay P, de Baynast H, Delattre C, Pierre G, Hachicha R, Abdelkafi S, Michaud P, Fendri I (2022) Biomolecules from microalgae and cyanobacteria: applications and market survey. Appl Sci 12(4):1924. https://www.mdpi.com/2076-3417/12/4/1924

Hadi SI, Santana H, Brunale PP, Gomes TG, Oliveira MD, Matthiensen A, Oliveira ME, Silva FC, Brasil BS (2016) DNA barcoding green microalgae isolated from neotropical inland waters. PLoS ONE 11(2):e0149284

Hu Q, Sommerfeld M, Jarvis E, Ghirardi M, Posewitz M, Seibert M, Darzins A (2008) Microalgal triacylglycerols as feedstocks for biofuel production: perspectives and advances. Plant J 54(4):621–639

Ibañez E, Cifuentes A (2013) Benefits of using algae as natural sources of functional ingredients. J Sci Food Agric 93(4):703–709

Kakaei H, Nourmoradi H, Bakhtiyari S, Jalilian M, Mirzaei A (2022) Effect of COVID-19 on food security, hunger, and food crisis. In: Dehghani MH, Karri RR, Roy S (eds) COVID-19 and the

sustainable development goals. Elsevier, Amsterdam, pp 3–29. https://doi.org/10.1016/B978-0-323-91307-2.00005-5
Kultschar B, Llewellyn C (2018) Secondary metabolites in cyanobacteria. Second Metabol Sour Appl 64
Kusvuran A, Kusvuran S (2019) Using of microbial fertilizer as biostimulant alleviates damage from drought stress in guar (*Cyamopsis tetragonoloba* (L.) Taub.) seedlings. Int Lett Nat Sci 76
Lee RE (1999) Phycology. Cambridge University Press, Cambridge
Liu D, Liberton M, Hendry JI, Aminian-Dehkordi J, Maranas CD, Pakrasi HB (2021) Engineering biology approaches for food and nutrient production by cyanobacteria. Curr Opin Biotechnol 67:1–6. https://doi.org/10.1016/j.copbio.2020.09.011
Liu L, Jokela J, Wahlsten M, Nowruzi B, Permi P, Zhang YZ, Xhaard H, Fewer DP, Sivonen K (2014). Nostosins, trypsin inhibitors isolated from the terrestrial cyanobacterium *Nostoc* sp. strain FSN. J Nat Prod 77(8):1784–1790
Markou G, Vandamme D, Muylaert K (2014) Microalgal and cyanobacterial cultivation: the supply of nutrients. Water Res 65:186–202. https://doi.org/10.1016/j.watres.2014.07.025
Mata TM, Martins AA, Caetano NS (2010) Microalgae for biodiesel production and other applications: a review. Renew Sustain Energy Rev 14(1):217–232
Mehta P, Singh D, Saxena R, Rani R, Gupta RP, Puri SK, Mathur AS (2018) High-value coproducts from algae—an innovational way to deal with advance algal industry. Waste Wealth, pp 343–363
Mishra U, Pabbi S (2004) Cyanobacteria: a potential biofertilizer for rice. Resonance 9:6–10
Moreira D, Pires JC (2016) Atmospheric CO_2 capture by algae: negative carbon dioxide emission path. Biores Technol 215:371–379
Nabout JC, da Silva Rocha B, Carneiro FM, Sant'Anna CL (2013) How many species of Cyanobacteria are there? Using a discovery curve to predict the species number. Biodiv Conserv 22:2907–2918
Nedaei L, Shokrkar H (2022) A review on the extraction of chlorophyll and carotenoids from microalgae. Iran Chem Eng J 21(123):45–58. https://doi.org/10.22034/ijche.2021.299781.1134
Niccolai A, Bažec K, Rodolfi L, Biondi N, Zlatić E, Jamnik P, Tredici MR (2020) Lactic acid fermentation of *Arthrospira platensis* (Spirulina) in a vegetal soybean drink for developing new functional lactose-free beverages. Front Microbiol 11:560684
Niccolai A, Zittelli GC, Rodolfi L, Biondi N, Tredici MR (2019) Microalgae of interest as food source: biochemical composition and digestibility. Algal Res 42:101617
Nowruzi B, Adelkhani M, Anvar A (2023) A review of the cytotoxic, antioxidant and antimicrobial activity of nanoparticles biosynthesized from cyanobacteria. J Compar Pathobiol 20(1):4039–4052
Nowruzi B, Sarvari G, Blanco S (2020) Applications of cyanobacteria in biomedicine. In: Handbook of algal science, technology and medicine. Elsevier, Amsterdam, pp 441–453
Pagels F, Guedes AC, Amaro HM, Kijjoa A, Vasconcelos V (2019) Phycobiliproteins from cyanobacteria: chemistry and biotechnological applications. Biotechnol Adv 37(3):422–443
Palma A, Cossu M, Deligios P, Ledda L, Tiloca M, Sassu M, D'Aquino S (2023) Organic versus conventional globe artichoke: Influence of cropping system and harvest date on physiological activity, physicochemical parameters, and bioactive compounds. Sci Hortic 321:112304
Pan S, Jeevanandam J, Danquah MK (2019) Benefits of algal extracts in sustainable agriculture. In: Grand challenges in algae biotechnology, pp 501–534
Panis G, Carreon JR (2016) Commercial astaxanthin production derived by green alga *Haematococcus pluvialis*: a microalgae process model and a techno-economic assessment all through production line. Algal Res 18:175–190. https://doi.org/10.1016/j.algal.2016.06.007
Pascoal PV, Ribeiro DM, Cereijo CR, Santana H, Nascimento RC, Steindorf AS, Calsing LC, Formighieri EF, Brasil BS (2021) Biochemical and phylogenetic characterization of the wastewater tolerant *Chlamydomonas biconvexa* Embrapal LBA40 strain cultivated in palm oil mill effluent. PLoS ONE 16(4):e0249089

Patel AK, Albarico FPJB, Perumal PK, Vadrale AP, Nian CT, Chau HTB, Anwar C, ud din Wani HM, Pal A, Saini R (2022) Algae as an emerging source of bioactive pigments. Bioresour Technol 351:126910

Ratnapuram HP, Vutukuru SS, Yadavalli R (2018) Mixotrophic transition induced lipid productivity in *Chlorella pyrenoidosa* under stress conditions for biodiesel production. Heliyon 4(1):e00496. https://doi.org/10.1016/j.heliyon.2017.e00496

Ruzik L (2023) Microalgae with active biological metal-nanoparticles as a novel food. Biosynthesis, characterization and bioavailability investigation—review. Trends Food Sci Technol 139:104127. https://doi.org/10.1016/j.tifs.2023.104127

Sathasivam R, Radhakrishnan R, Hashem A, Abd_Allah EF (2019) Microalgae metabolites: a rich source for food and medicine. Saudi J Biol Sci 26(4):709–722. https://doi.org/10.1016/j.sjbs.2017.11.003

Singh R, Parihar P, Singh M, Bajguz A, Kumar J, Singh S, Singh VP, Prasad SM (2017) Uncovering potential applications of cyanobacteria and algal metabolites in biology, agriculture and medicine: current status and future prospects. Front Microbiol 8:515

Singh SP, Ansari MI, Pandey B, Srivastava JK, Yadav TP, Rani H., Parveen, A., Mala, J., & Singh, A. K. (2020). Recent trends and advancement toward phyto-mediated fabrication of noble metallic nanomaterials: focus on silver, gold, platinum, and palladium. Nanomater Environ Biotechnol 87–105

Smith GD, Thanh Doan N (1999) Cyanobacterial metabolites with bioactivity against photosynthesis in cyanobacteria, algae and higher plants. J Appl Phycol 11:337–344

Soletto D, Binaghi L, Lodi A, Carvalho J, Converti A (2005) Batch and fed-batch cultivations of *Spirulina platensis* using ammonium sulphate and urea as nitrogen sources. Aquaculture 243(1–4):217–224

Sun H, Wang Y, He Y, Liu B, Mou H, Chen F, Yang S (2023) Microalgae-derived pigments for the food industry. Mar Drugs 21(2): 82. https://www.mdpi.com/1660-3397/21/2/82

Udayan A, Pandey AK, Sirohi R, Sreekumar N, Sang B-I, Sim SJ, Kim SH, Pandey A (2022) Production of microalgae with high lipid content and their potential as sources of nutraceuticals. Phytochem Rev 1–28

Uribe-Wandurraga ZN, Igual M, Reino-Moyón J, García-Segovia P, Martínez-Monzó J (2021) Effect of microalgae (*Arthrospira platensis* and *Chlorella vulgaris*) addition on 3D printed cookies. Food Biophys 16:27–39

Vassilev SV, Vassileva CG (2016) Composition, properties and challenges of algae biomass for biofuel application: an overview. Fuel 181:1–33. https://doi.org/10.1016/j.fuel.2016.04.106

Ververis E, Ackerl R, Azzollini D, Colombo PA, de Sesmaisons A, Dumas C, Fernandez-Dumont A, da Costa LF, Germini A, Goumperis T (2020) Novel foods in the European Union: scientific requirements and challenges of the risk assessment process by the European Food Safety Authority. Food Res Int 137:109515

Vu CHT, Lee H-G, Chang YK, Oh H-M (2018) Axenic cultures for microalgal biotechnology: establishment, assessment, maintenance, and applications. Biotechnol Adv 36(2):380–396

Whitton BA (1992) Diversity, ecology, and taxonomy of the cyanobacteria. In: Photosynthetic prokaryotes. Springer, Berlin, pp 1–51

Whitton BA (2012) Ecology of cyanobacteria II: their diversity in space and time. Springer Science & Business Media, Berlin

Chapter 27
Enhanced Production of Carotenoids from Microalgae: A Study of Anti-obesity Potential in *C. elegans*

Elamathi Vimali, Johnson Prasanth, Kalimuthu Meena, Nagamalai Sakthi Vignesh, Velmurugan Ajithkumar, Balasubramaniem Ashokkumar, and Perumal Varalakshmi

Abstract Secondary metabolites from algae have a huge impact on human health and are being used to produce many value-added products by Industries. Generally, the pigments (Astaxanthin, Beta-carotene, Lutein, Fucoxanthin and Phycobilin) extracted from micro and macroalgae are become important source for manufacturing diverse products in nutraceuticals, food pharmaceuticals, aquaculture and textile industries. Herein, the carotenoid content from *Scenedesmus* sp. IS3 was attained and carotenoid accumulation was highly influenced by various stress factors like nitrogen starvation, high salinity (3%) and Methyl Jasmonate. In order to optimize the carotenoid production using various factors, the Response Surface Methodology (RSM) using Central Composite Design (CCD) was employed. Further, the modern techniques like Ultrasound and Microwave-Assisted Extraction (UAE and MAE) were used to extract the carotenoids. Then, the extracted carotenoids were evaluated for their anti-obesity potential using *Caenorhabditis elegans* for assessing the fat deposition.

Keywords Pigment · Microalgae · Ultrasound assisted extraction · Microwave assisted extraction · *C. elegans*

E. Vimali · J. Prasanth · K. Meena · N. S. Vignesh · V. Ajithkumar · P. Varalakshmi (✉)
Department of Molecular Microbiology, School of Biotechnology, Madurai Kamaraj University, Madurai, Tamil Nadu 625021, India
e-mail: vara5277@gmail.com; pvlakshmi.biotech@mkuniversity.ac.in

B. Ashokkumar
Department of Genetic Engineering, School of Biotechnology, Madurai Kamaraj University, Madurai, Tamil Nadu 625021, India

T. A. Kurniawan and A. Anouzla (eds.), *Algae as a Natural Solution for Challenges in Water-Food-Energy Nexus*, Environmental Science and Engineering,
https://doi.org/10.1007/978-981-97-2371-3_27

27.1 Introduction

Obesity is one of the common diseases that could be formed by the deposition of excess fat globules in the liver and adipose tissue. Obesity becomes a major root cause for other diseases like cardiac diseases, diabeties type II, hypertension, and some type of cancers. Further, it was estimated to reduce BMI level upto 5% by 2030, along with a decrease in medical expenditure for obese related disorders of about €495 million in next 20 years (Keaver et al. 2013; Yao et al. 2021). The carotenoids are known by their nutraceutical effects and the global market value in 2019 was about $1.5 billion and it will reach upto $2.0 billion by 2026 (Ashokkumar et al. 2022). Moreover, the complete global value of three commercially important carotenoids such as astaxanthin, beta carotene and lutein had reached about 60% (Charalampia et al. 2017). In general, these carotenoids are produced by cyanobacteria, fungi, algae and archaea bacteria. On contrary, it cannot be naturally synthesized by human metabolism but it could be consumed from the diet only. Carotenoids are specifically known for its photoharvesting function that leads to ROS formation (D'Alessandro and Antoniosi Filho 2016). Furthermore, it has been well reported that the microalgal carotenoids has anti-inflammatory, anti-oxidant and anti-cancer properties (Guedes et al. 2011; Ciccone et al. 2013). Among the different carotenoid producers, microalgae are one of the copious producer (Lafarga et al. 2020) with plausible advantages such as highpigment production, easy purification steps, economically feasible and no requirement of land for cultivation etc.

Generally, the carotenoid production from microalgae has performed using various stress conditions especially depleting the nitrogen, phosphorous and certain trace elements. So far, there are several reports dealt with the production of carotenoids (lutein, β-carotene and astaxanthin) in the absence of nitrogen (Shi et al. 2020; Chen et al. 2015). The microalgae grown at high salt concentration becomes a major factor that influencing the increased carotenoid yield. For example, the report dealt with beta carotene production from *Dunaliella salina* was increased from 4 to 9% salinity with a 30-fold increase yield than the control (Coesel et al. 2008). Recently, it was reported that the astaxanthin production from *Haematococcus pluvialis* yielded about 5.01 mg/g that is seems to be 42% higher than the control at 0.2 g/L sodium chloride (Ding et al. 2019). Therefore, choosing the suitable stress factors for high carotenoid accumulation without losing its nutraceutical property is mandatory.

In order to decode the anti-obesity potential of astaxanthin and lutein from microalgae, *Caenorhabditis elegans* was used as an animal model. In general, *C. elegans* are well-known for its shorter generation time, easy to handle and genome similarity of about 60–80% that is related to human diseases and metabolic pathways (Kaletta and Hengartner 2006). Thus, this study is highly intended to investigate the suitable stress factors (Sodium nitrate, Sodium chloride, and Methyl jasmonate) for carotenoid accumulation by applying modern extraction methods to improve the yield % of astaxanthin and lutein. Further, the anti-obesity potential at various concentrations (25 μg and 50 μg/mL) of astaxanthin and lutein were evaluated using Nile red Imaging.

27.2 Materials and Methods

27.2.1 Sample Collection

The freshwater microalgal sample was collected from Melayakudi, Paramakudi, Tamil Nadu, India. The collected water sample was undergone for serial dilution followed by spread plate technique. Further, the unknown isolate IS3 was enriched with 100 mL of BG11 medium in 250 mL Erlemmeyer flask and it was maintained at light intensity (1500 lx), with an alternative light and dark illumination (12 h: 12 h) and temperature was maintained at 24 ± 2 °C (Angelaalincy et al. 2023).

27.2.2 Molecular Identification of Unknown Isolate

The genomic DNA was extracted from an unknown isolate IS3 using CTAB method (Karpagam et al. 2015). Further, the DNA (3 μL: 50 ng/μL) amplification was done using ITS1 forward primer (5′-GAGCATGTCTGCCTCAGC-3′) and reverse primer of ITS2 (5′-GGTAGCCTTGCCTGAGC-3′) (2 μL each) with 10 X reaction buffer (2.5 μL), dNTP's (0.5 μL), Taq Polymerase (0.5 μL) and MilliQ (8 μL) by Biorad T100™ Thermal cycler. The amplification was done by following conditions such as initial denaturation at 94 °C for 5 min followed by denaturation at 94 °C for 30 s, annealing at 60 °C for 1 min, extension at 72 °C for 1 min and final extension at 72 °C for 10 min. Further, the PCR products were purified and sequenced by Sanger dideoxy method (Karpagam et al. 2015).

27.2.3 Growth Kinetics of Scenedesmus *sp. IS3*

In order to examine the growth pattern of *Scenedesmus* sp. IS3, chlorophyll estimation was performed. Microalgal cells were collected after 20 days of incubation with 12 h light and 12 h dark condition at 24 ± 2 °C and 1 mL of 100% methanol was added to the collected samples in 1.5 mL eppendorf and freeze thawed at 55 °C until the algal pellet become colourless. Then, the chlorophyll content (μg/mL) of *Scenedesmus* sp. IS3 was measured by following equation (Pushpalatha et al. 2021).

$$\text{Chlorophyll a }(C_a) = 15.65 \cdot OD_{666} - 7.340 \cdot OD_{653} \tag{27.1}$$

$$\text{Chlorophyll b }(C_b) = 27.05 \cdot OD_{653} - 11.21 \cdot OD_{666} \tag{27.2}$$

$$\text{Total carotenoids} = \text{Eq. }27.1 + \text{Eq. }27.2 \tag{27.3}$$

MAE

The lyophilized microalgal powder (100 mg) was taken in 15 mL falcon tube and added with 5 mL of methanol: dichloromethane (75:25). The experiment was done by applying 800 W of microwave for 1 min.

27.2.8 Characterization of Carotenoids

27.2.8.1 Thin Layer Chromatography (TLC)

The extracted microalgal pigment was qualitatively analyzed by TLC using silica gel 60 F_{254} plate (20 cm X 20 cm Merck, Darmstadt, Germany plate). 10 μL of pigments extracted from microalgae was loaded along with standards (astaxanthin, beta-carotene, and lutein) onto the TLC plate. Then, the plate was kept in a developing chamber containing a solvent mixer of hexane: diethyl ether: acetic acid (70:30:1) used as a mobile phase. Further, a developing solution was prepared by mixing of copper sulphate solution with phosphoric acid (8%) and slowly sprayed onto the TLC plate (Pushpalatha et al. 2021).

27.2.8.2 High Performance Liquid Chromatography

To investigate each carotenoid, the concentrated pigments obtained from RSM was diluted in HPLC grade methanol (1 mL) with a certain concentration to attain separate peaks for different carotenoids. Thus, the HPLC analysis was performed by SPD-20A, Shimadzu model, which has Supelco-LC-18 column (25 cm × 4.6 mm) and PAD detector (5 μm). Further, the solvent mixture such as 0.2% of methanol (450 mL), ammonium acetate (25 mL), butylated hydroxyl toluene (50 mg), acetonitrile (455 mL), isopropanol (20 mL) and diisopropylamine (0.2 mL) used for mobile phase. The conditions used for HPLC quantification were as follows: 1.9 mL/min flow rate, injection volume (20 μL) and detected at 448. Each carotenoid was identified as individual peaks and compared with the retention time of standards, viz. lutein, β-carotene, astaxanthin and zeaxanthin. Moreover, various carotenoids were quantified by biomass weight, standard concentration, both sample and standard peak area, standard purity and dilution factors (Rodriguez-Amaya and Kimura 2004; Karpagam et al. 2018).

27.2.9 *Anti-obesity Studies*

27.2.9.1 *C. elegans* Maintenance

Caenorhabditis elegans was maintained in nematode growth medium (NGM) at 20 °C and fed with *E. coli* OP50 as a food source.

27.2.9.2 AGE Synchronization of Worms

The adult gravid worms were collected using M9 buffer and washed thrice until the bacterial debris were removed. The collected worms were bleached with sodium hypochlorite solution for an efficient disruption. The worms were added with 1 mL of bleaching solution and vortexed vigorously for 5–7 min. To this content, M9 buffer was added after vigorous vortexing to neutralize the worms immediately. This contents was centrifuged at 3000 rpm for 2–3 min and the pellet containing eggs were used for further studies.

27.2.9.3 In-Vivo Analysis for Anti-obesity Studies

The AGE synchronised worms were collected and added on 12 well plate containing S complete media. In this study, the complete medium contains cholesterol about 3 μg/mL as control, high fat diet (cholesterol about 15 μg/mL) to induce the obesity in worm further, astaxanthin and lutein (25 and 50 μg/mL) was added separately along with 10 μL of *E. coli* OP50 and the plates were maintained at 20 °C for two days.

27.2.9.4 Nile Red O Staining

After 2 days of incubation, the worms were washed off using M9 buffer until the bacterial debris was removed. Then, the worms were added with freshly prepared Nile red (6 μL of working stock in 1 mL of 40% isopropanol) and kept in an incubator for 30 min under dark condition. Then, the imaging was analyzed using two different filters (Brightfield and Fluor155 Red and green) under Leica Microsystem to visualize the fat deposition in *C. elegans*.

27.3 Results and Discussion

27.3.1 Isolation and Molecular Identification of Microalgae

The unicellular, multinucleate, ovoid, lunate shaped unknown freshwater isolate IS3 was screened and characterized by sequencing. The IS3 isolate was shown similarity to the previously reported *Scenedesmus* sp. KGU-Y002 and *Scenedesmus* sp. KGU-D002 (Aburai et al. 2013). The obtained sequence was undergone for BLAST analysis to know the edges of ITS1-5.8S-ITS2 with previously available sequencing in NCBI database. Further, the phylogenetic tree was constructed based on the Neighbor-joining tree method with 1000 bootstrap replications by ribosomal sequence using Mega 6.0 software. This method witnessed that the obtained sequence belongs to the Order *Sphaeropleales* and Genera *Scenedesmus* sp. and the identified *Scenedesmus* sp. was shown the query coverage about 98 and 100% identity (Fig. 27.1). The SEM micrograph of *Scenedesmus* sp. IS3 was depicted in Fig. 27.2 (Silva et al. 2020).

27.3.2 Chlorophyll Estimation of Scenedesmus *sp. IS3*

The microalgal growth rate is an important feature for producing various bio active products. In this study, the *Scenedesmus* sp. IS3 growth rate was analyzed by its chlorophyll content and the growth rate was quantified by absorbance at 660 nm using spectrophotometer for 0–24 days of cultivation. The experimental results evidently showed that the maximum growth was increased on 22nd day and then started to decline (Fig. 27.3). The doubling time of *Scenedesmus* sp. IS3 was estimated as 16.98 h that was comparatively lower than the previous report on microalgal consortia of *Scenedesmus obliquus* showing 22.2 h (Gatamaneni Loganathan et al. 2020).

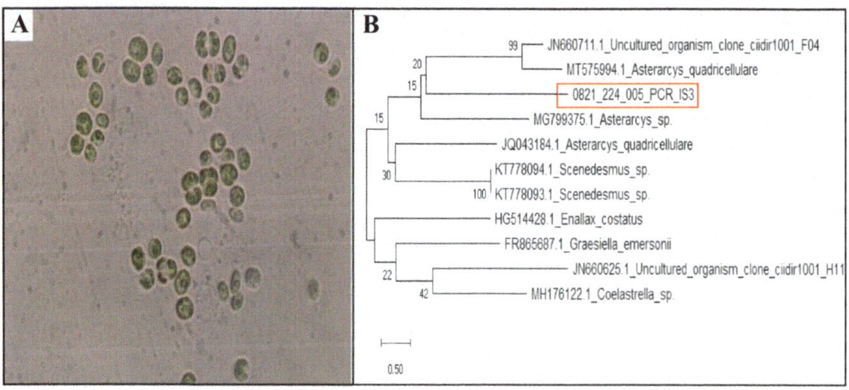

Fig. 27.1 **A** Microscopic view and **B** phylogenetic tree analysis of *Scenedesmus* sp. IS3

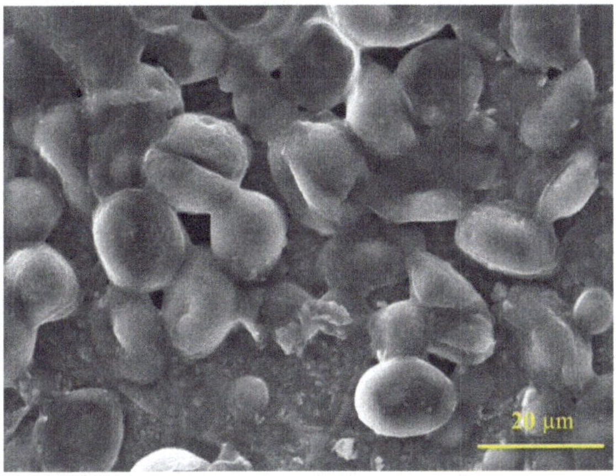

Fig. 27.2 SEM micrograph of *Scenedesmus* sp. IS3

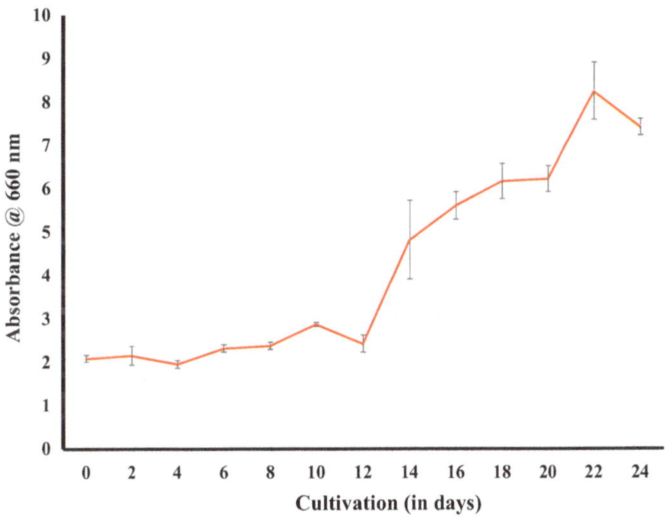

Fig. 27.3 Growth rate of *Scenedesmus* sp. IS3 grown in BG11 media at 24 ± 2 °C for 25 days (12 h light: 12 h dark)

27.3.3 Optimization of Pigment Accumulation

Central composite design (CCD) is an optimizing tool used to standardize the conditions employed for pigment accumulation in *Scenedesmus* sp. IS3 using Design Expert software. In this report, the PC, PP (mg/L/day) and BP (mg/L/day) were analyzed. Generally, microalgae need a simplest form of energy supplements such as phosphate, nitrate, carbondioxide (CO_2) for their growth (Mc Gee et al. 2020). From the experimental study, the actual results were similar to the predicted values (Table 27.2) which implied that the design is in line with a linear equation of critical

Table 27.2 Experimental design of CCD design using three different variables ($NaNO_3$, NaCl, and MeJA) and responses such as BP, PP and PC.

Std	Run	Factors			Actual value			Predicted value		
		$NaNO_3$ (g/L)	NaCl (g/L)	MeJA (mg/L)	BP (mg/L/day)	PC (%)	PP (mg/L/day)	BP (mg/L/day)	PC (%)	PP (mg/L/day)
1	7	0.6	1	0.25	20.84	31.30	13.17	21.4	55.14	24.0916
2	4	0.7	1	0.25	23.28	23.97	10.59	28.37	27.06	26.2212
3	6	0.6	3	0.25	4.606	45.35	4.424	7.18	57.39	8.7624
4	14	0.7	3	0.25	11.52	27.92	6.538	11.56	36.74	10.8922
5	10	0.6	1	0.75	20.80	35.08	14.62	23.36	41.2	23.2328
6	13	0.7	1	0.75	23.15	68.05	31.21	23.17	70.95	25.3626
7	11	0.6	3	0.75	11.71	21.30	5.153	9.21	33.15	7.9038
8	1	0.7	3	0.75	4.38	**79.24**	7.095	6.42	70.34	10.0334
9	15	0.57	2	0.5	16.94	65.97	22.63	16.29	41.16	15.2716
10	8	0.73	2	0.5	22.84	45.12	20.53	19.82	48.82	18.8534
11	5	0.65	0.32	0.5	31.07	57.87	**35.84**	27.43	43.68	29.939
12	9	0.65	3.68	0.5	1.46	52.02	1.2	1.42	45.05	4.186
13	3	0.65	2	0.07	21.55	56.82	24.97	19.47	49.55	17.801
14	16	0.65	2	0.09	21.86	63.31	26.58	20.29	49.1	17.7666
15	2	0.65	2	0.5	23.46	50.04	23.48	27.87	48.71	17.0626
16	12	0.65	2	0.5	**31.66**	43.35	26.4	27.87	48.71	17.0626

factors. The three-dimensional graph revealed that the relation between each variable factors ($NaNO_3$, NaCl and MeJA) involved in influencing carotenoid accumulation (Fig. 27.4). Among the experimental trials, Run 1 achieved higher PC (79.24%) which is 1.77 folds higher than the control (28.59%).

27.3.3.1 Effect of Sodium Nitrate

In general, the nitrogen source is the most common stress factor that encompasses about 1–10% of microalgal DCW (Wijffels et al. 2010; Minhas et al. 2016). Among the varies nitrogen sources such as urea, ammonia, sodium nitrate and peptone etc Thus, in this study, 0.7 g/L of nitrogen positively increased the PC but not PP in Run 1. Priyanka et al. (2020) reported that the supplement of nitrogen source (910 mg/L) enhanced the zeaxanthin yield. Moreover, the sodium nitrate plays a crucial role in microalgae growth and its cellular activities. Commonly, nitrogen source in the media is actively stimulate the microalgae during photosynthesis for procuring energy resources.

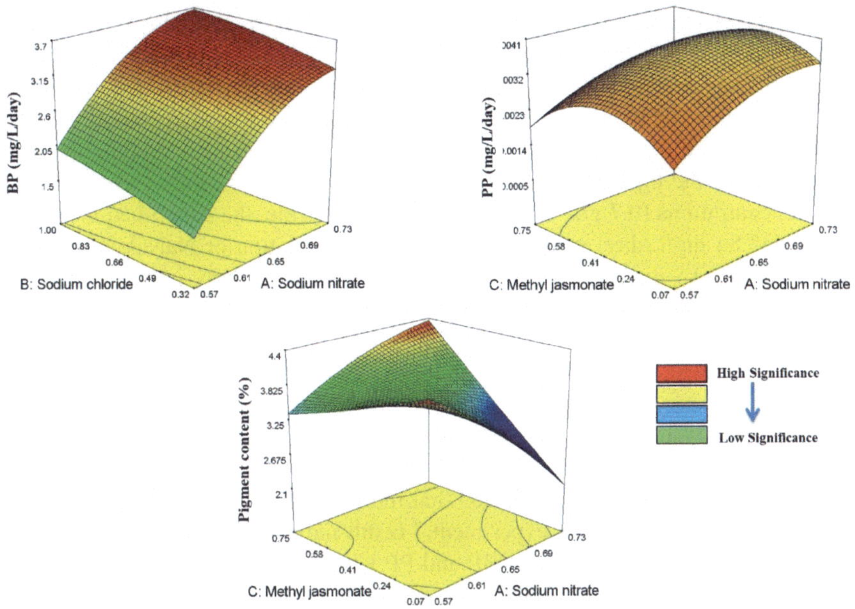

Fig. 27.4 3D counter-plot representation of three different variables and its interaction between MeJA, NaCl and $NaNO_3$

27.3.3.2 Effect of Salinity

Salinity is another important factor used to accumulate the pigment in microalgae (Subramanian and Gurunathan 2020). The mechanism behind utilizing high salinity leads to unwinding of phycobilisomes present in the thylakoid membrane to the photosynthesis reaction centre and produce pigments (Begum et al. 2016). The increased concentration of salinity reduces the growth of microalgae, thus the NaCl (3 g/L) given in Run 1 and Run 6 had decreased the BP (4.38 and 4.60 mg/L/day) and PP (4.24 and 7.07 mg/L/day) than the predicted values. Meanwhile, the interaction between NaCl and $NaNO_3$ improves the biomass productivity of about 31.66 mg/L/day (Fig. 27.3). Earlier, the canthaxanthin was highly increased while employing different salt concentrations (50–400 mM) (Del Campo et al. 2004). The increased concentration of salinity is enhanced the photosynthesis followed by enriched the production of ATP and NADH which gives rise to primary and secondary metabolites production (Sedjati et al. 2019).

27.3.3.3 Effect of MeJA

MeJA is an oxylipid molecule that plays a vital role in pigment accumulation (Udayan et al. 2020; Mc Gee et al. 2021). In this report, a increased concentration of MeJA

(0.75 mg/L) yielded increased PC (79.24%) (Table 27.2) but reduced the BP (4.38 mg/L/day) and PP (7.095 mg/L/day) drastically. The effect of MeJA is dependent upon the species and dose specific. In one of the earlier reports, Jasmonic acid increased the astaxanthin gene expression (Gao et al. 2012), and it was also improved the beta carotene ketolase gene expression (Lu et al. 2010) in *H. pluvialis.* Herein, the Run 1 was showed the higher PC (79.24%) while the $NaNO_3$ and MeJA employed in high concentrations (0.7 g/L and 0.75 mg/L). Meanwhile, the Run 5 had increased the PP (35.84 mg/L/day) when the moderate concentration of $NaNO_3$ and MeJA (0.65 g/L and 0.5 mg/L) was applied. These results are clearly indicated that the strong interaction between $NaNO_3$ and MeJA that highly influenced in BP and PP (Fig. 27.3).

27.3.3.4 Analysis of Variance (ANOVA)

The experimental results are disclosed that the increase of BP and PP (mg/L/day) in Runs 1, 5, 12 and 13. The experimental result indicated that the model used in this study was more significant in BP and PP with *P*-values of 0.0164 and 0.0427 but not in PC (Table 27.3). Further, the adjusted R^2 was found to be > 0.05 thus the improvised model is significant. All these experiments were done in triplicates. Then, the total carotenoid was calculated and it is further validation was done in an outdoor cultivation using 5 L media with optimized parameters of RSM.

27.3.3.5 Quantification of Carotenoids from RSM

The carotenoid extraction from RSM design were analyzed by both TLC and HPLC. Initially, TLC was done for all the 16 experimental runs and it clearly depicted that the Run 1 and Run 13 yielded all the three carotenoids Lutein, Astaxanthin and Beta carotene (Fig. 27.5). Thus, the carotenoids from Run 1 and 13 were quantitatively analyzed by HPLC and compared with the control which already reported from our laboratory (Pusphalatha et al., 2021) (Fig. 27.6 and Table 27.4) Among the carotenoids, lutein becomes more valuable due to its nutraceutical properties that cannot be synthesized by human (Zheng et al. 2022; Xie et al. 2021). Thus, this study exhibited higher lutein content in Run 1 (2.164 mg/g) than the control (1.981 mg/g) under different stress factors employed. Earlier, the lutein was extracted from *Auxenochlorella protothecoides* SAG 211-7a was about 1.303 mg/L under optimised stress factors such as sucrose, EDTA, yeast extract and $MgSO_4$ which was lower than our report (Shinde and Lele 2010). *Scenedesmus* sp. FSP3 had accumulated 2.30 mg/L/day lutein and it was extracted by using two-stage method that was higher in our report (Li et al. 2022). It was also stated that *Chlorella saccharophila* was produced 2.26 mg/g of lutein using solvent extraction which was likely similar to our study (Paliwal et al. 2022). Besides, the astaxanthin was also extracted from the foresaid microalgae since it has good nutraceuticals potential which profoundly present in fish and crustaceans etc (Niizawa et al. 2018). The media optimization studies using RSM

Table 27.3 Analysis of variance for quadratic model dictated that the significance of F-value (> 0.05)

ANOVA for RSM-quadratic model															
Analysis of variance table [partial sum of squares—type III]															
Source	BP (mg/L/day)					PP (mg/L/day)					PC (%)				
	Sum of squares	df	Mean square	F-value	*P*-value	Sum of squares	df	Mean square	F-value	*P*-value	Sum of squares	df	Mean square	F-value	*P*-value
Model	1115.85	9	123.98	6.56	**0.016***	2.049	3	06.829	3.71	**0.0427***	2076.29	9	230.70	0.62	0.7506
A-$NaNO_3$	122.68	1	122.68	6.50	0.0436	38,713.0	1	38,713	0.21	0.6549	0.20	1	0.20	5.256	0.9825
B-NaCl	16.59	1	16.59	0.88	0.3848	2.004	1	2.004	10.88	0.0064	6.63	1	6.63	0.018	0.8981
C-MeJA	54.36	1	54.36	2.88	0.1407	5964.84	1	5964.84	0.032	0.8602	1548.70	1	1548.7	4.17	0.0873
Lack of fit	79.70	5	15.94	0.47	0.7939	2.200	11	2.000	18.82	0.1781	2207.94	5	441.59	19.73	0.1692
	Significant					Significant					Non-significant				

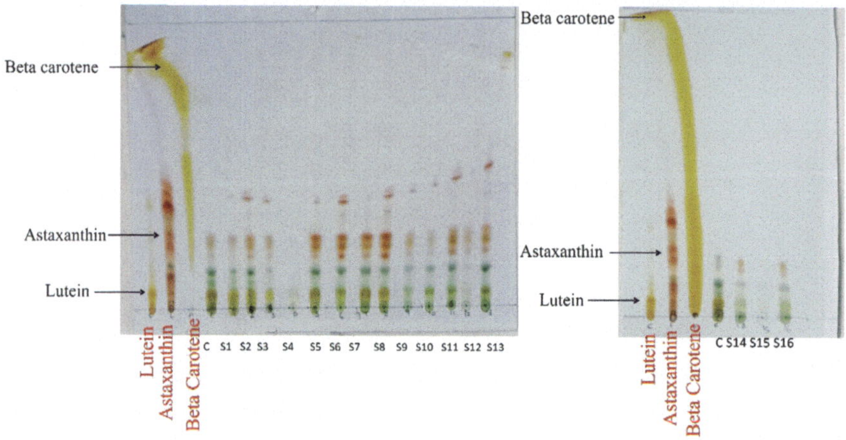

Fig. 27.5 Qualitative analysis of carotenoids from *Scenedesmus* sp. IS3 experimental runs from 1 to 16 in TLC

revealed that *Scenedesmus* sp. IS3 grown in optimized condition of Run 1 produced 1.981 mg/g DCW of astaxanthin. Recently, the overexpression of gene related to beta carotene production in *Phaeodactylum tricornutum* increased the yield 4.34 mg/g (Cen et al. 2022). Herein, the optimised condition has facilitated the beta carotene yield in *Scenedesmus* sp. IS3 upto 4 to five fold higher in Run 13 (0.125 mg/g) and Run 1 (0.162 mg/g) which was higher than the control (0.033 mg/g). In comparison with β-carotene, astaxanthin and lutein production, the freshwater microalgae *Scenedesmus* sp. IS3 yielded more lutein (2.14 and 2.164 mg/g) in RSM optimized conditions. In one of the previous studies, the beta carotene from *Coelastrella* sp. M60 was about 0.02 mg/g DCW grown under different stress condition (nitrogen deprivation, 3% salinity and TiO_2) which was comparatively lower than this report (0.125 and 0.162 mg/g in Run 1 and 13) (Pushpalatha et al. 2021).

27.3.4 Effect of Modern Methods in Extraction

The carotenoid accumulation in *Scenedesmus* sp.IS3 cultivated in 5L of optimised media conditions such as $NaNO_3$, NaCl and MeJA of Run 1 from CCD design in an outdoor condition. After 70 days of cultivation, the accumulated pigment was extracted using three different methods such as conventional, MAE and UAE. Commonly, the different solvents such as acetone, hexane, chloroform, and diethyl ether were used to extract the value-added products by maceration and Soxhlet method of extraction. But this method required lengthy processing time, extensive solvent usage and lower yield % are some noticeable disadvantages (Hadiyanto and Suttrisnorhadi 2016).

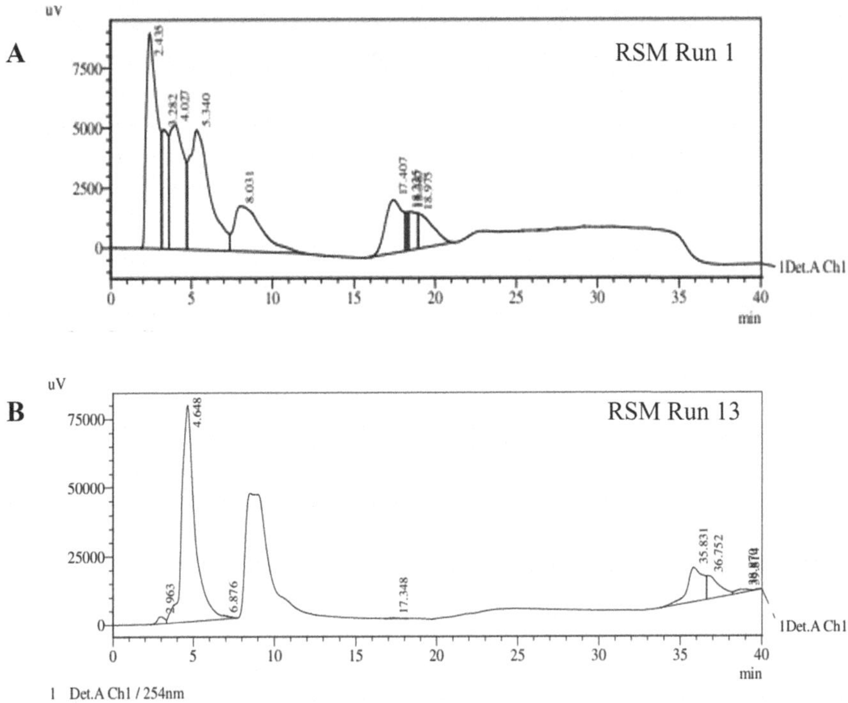

Fig. 27.6 Carotenoid profiling of *Scenedesmus* sp. IS3 obtained from RSM optimised media conditions of **A** Run 1 and **B** Run 13

Table 27.4 Profiling of carotenoids from Run 1 and Run 13

Treated conditions	Carotenoid content (mg/g DCW)		
	β-carotene	Lutein	Astaxanthin
Control	0.033	0.915	1.981
Run 13	0.125	2.14	–
Run 1	0.162	2.164	1.275

27.3.4.1 UAE and MAE

To overcome the aforesaid drawbacks, the modern methods like UAE and MAE were used for extraction. In MAE process, the pigment led to regulated heat transfer in a synchronic mode that causes a proper heat transfer towards the sample. Moreover, the pigments can able to withstand upto 60 °C, especially the pro- vitamin carotenoids such as β-carotene, α-carotene, and β-cryptoxanthin (Fratianni et al. 2010; Pasquet et al. 2011).

The UAE for carotenoid extraction becomes feasible, economical, and time consuming. In one of the earlier reports, ultrasound was used to extract the lutein from

Chlorella vulgaris by applying 35 kHz of ultrasound power, followed with an extraction duration of 162 min for 35.6 mg/L of Solid: Liquid ratio and yielded 3.36 mg/g of lutein (Deenu et al. 2013). This disruption not showing any effect on the reliability, structural and functional properties of the value-added product. In another study on applying ultrasound upto 400 W for 7.5 min in *Nannochloropsis* sp. improved the yield of chlorophyll upto 59% (Parniakov et al. 2015). In another case, the lutein from *Chlorella pyrenoidosa* was treated with HCl (15.63%) and extracted by employing ultrasound of about 35 kHz and yielded about 3.5 mg/g (Arun et al. 2017) and in another report, the ultrasound power (1000 W) using subcritical CO_2 extraction and yielded 124.01 mg/100 g of lutein from *Chlorella pyrenoidosa* (Fan et al. 2015). Gayathri et al. (2018) addressed that the lutein extraction from marine microalgae *Chlorella salina* with the following conditions: Ultrasound frequency about 35 kHz for 30 min at 40 °C and achieved 2.92 ± 0.40 mg/g dry weight. While discussing about beta carotene extraction using UAE method, the microwave power 167 W was applied upto 30 min and extracted 47% of β-carotene in *Spirulina plantesis* (Dey and Rathod 2013). Jaeschke et al. (2017) suggested that *Heterochlorella luteoviridis* with following conditions of 40–80% of ultrasound intensity and obtained 80% yield while employing 50% of UAE to extract β-carotene. In another study, the 4.66 mg of β-carotene was obtained from *Spirulina platensis* by applying 20 kHz with 130 W at various time intervals of 2, 4, 6, 8, 10 and 12 min (Bachchhav et al. 2020). Recently, the microalgal isolate *Tetradesmus obliquus* SGM19 used to extract β-carotene 0.67 mg/g DCW with the conditions of 33 kHz and 31.5% of yield was obtained (Singh et al. 2019). These reports revealed that the efficiency of ultrasound power was improved the carotenoid yield within less extraction time, higher yield % and also low solvent utilization that paves the way to an efficient extraction and becomes an alternative to commercially used solvent methods.

MAE is another important method to extract carotenoids from microalgae. Earlier, the beta carotene was extracted from *H. pluvialis* by applying the conditions of 141 W for 5 min 30 s and achieved 0.59% DCW of astaxanthin (Zhao et al. 2009). On contrary, the microwave conditions1200 W output (60%) at 75 °C for 5 min and achieved 74% astaxanthin (Ruen-ngam et al. 2010). In another report, 50 W of 56 °C upto 3–5 min was applied to *Dunaliella tertiolecta, Cylindrotheca closterium* gave 0.12% of astaxanthin and 0.42% of fucoxanthin was obtained (Pasquet et al. 2011).

In another report, *Arthrospira platensis* exposed to different temperatures 70 and 60 °C yielded β-carotene about 0.063% (Esquivel-Hernández et al. 2016) and 0.014% of DCW (Esquivel-Hernández et al. 2017) with the similar microwave 400 W for 15 min. In this study, after the pigment accumulation in large-scale cultivation the *Scenedesmus* sp. IS3 was collected and used to extract carotenoids by using conventional, UAE and MAE (Fig. 27.7). The quantitative analysis of total carotenoids obtained from MAE was 29.3 mg/g DCW and UAE was 69.9 mg/g DCW in this report.

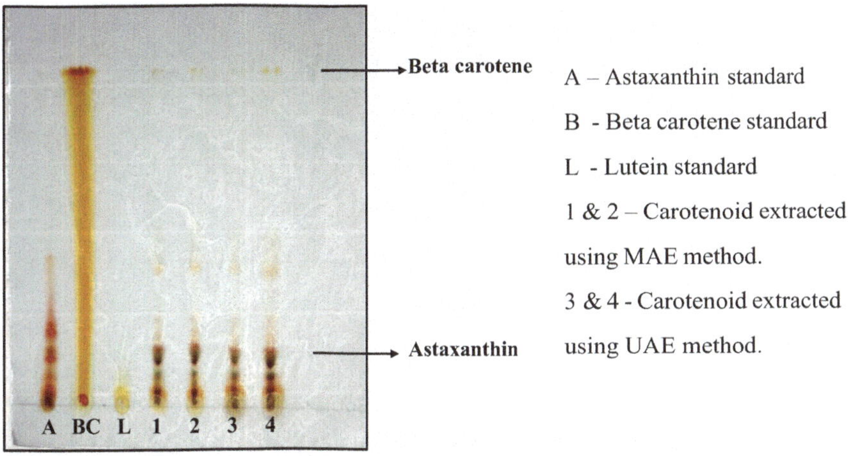

Fig. 27.7 Qualitative analysis of carotenoids extracted from *Scenedesmus* sp. IS3 on 50th day using the modern methods for extraction (UAE and MAE)

27.3.5 Effect of Carotenoids in Anti-obesity

Obesity has been considered as global pandemic since 1997 by World Health Organization (WHO) which leads to cause different severe diseases like cancer, diabetics, hypertension and cardio vascular diseases. Thus, there is need for suitable alternative to arrest the lipid depots in adipose tissue, liver and other organs. Generally, there are few reports dealt about the carotenoid function against obesity. In one of the early reports, the astaxanthin was exposed to obesed mice at different concentrations such as 1.2, 6 and 30 mg/kg for 60 days and the study clearly evidenced that the weight, triglyceride of liver, plasma triglyceride and blood cholesterol was highly reduced (Ikeuchi et al. 2007). Meantime, the astaxanthin was given to a young adult at different concentration of 5 and 20 mg for three weeks and found that decreased plasma content. Moreover, astaxanthin acted as barrier for mitochondria to avoid the damage and lessen the oxidative stress related to non-alcoholic fatty liver disease (Baburina et al. 2019). The astaxanthin treatment had shown improved resistance glucose metabolism and insulin resistance in mice (Bhuvaneswari and Anuradha 2012) and which activated the insulin signalling pathway by inducing the post-insulin receptor signalling, induced TNFα signalling pathway, lesser the oxidative stress and proinflammatory cytokines (Ishiki et al. 2013; Zhang et al. 2014). Recently, there was an another report dealt about the carotenoid extraction from *Dunaliella salina* was tested in high fat diet rats and its effects against obese related cardiac diseases had led to cause reduction in adiponectin and serum glycogen level (El-Baz et al. 2020). With this supporting information, in this study the carotenoids extracted from *Scenedesmus* sp. IS3 were subjected to *C. elegans* at two different concentrations (25 and 50 μg/mL) employed to high fat diet as cholesterol. The lutein and astaxanthin were the two different carotenoids obtained from *Scenedesmus* sp. IS3 exposed to *C.*

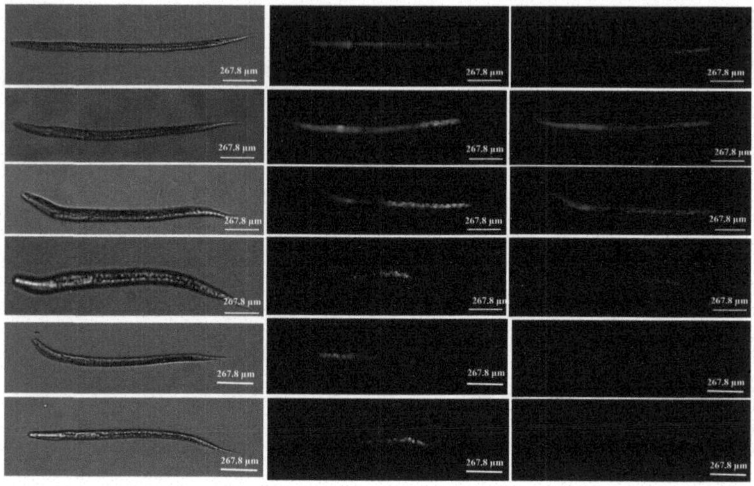

Fig. 27.8 Nile red imaging of lipid accumulation upon exposure to Astaxanthin and Lutein (25 and 50 µg/mL) in *C. elegans*

elegans for two days. After the post treatment, the fluorescence imaging was done to analyse the lipid droplets in both qualitatively and quantitatively using ImageJ software. Commonly, *C. elegans* are readily store their fatty acid granules in the intestine and hypodermal cells in which intestine contains small granules or vesicles which are structurally or functionally active. Meantime, some granules become self-fluorescein and the lysosome-based organelles like GLO-3, FUS-1 and PGP-2 are fluorescence while treated with Nile-red (Hermann et al. 2005; Nunes et al. 2005; Schroeder et al. 2007; Treusch et al. 2004). This insight gives an understanding of lysosome associated organelles reacted with Nile red and auto-fluorescence. Thus, the same pattern of lipid variation was found in our study compared with worms grown in complete medium and high fat diet (Fig. 27.8).

27.4 Conclusion

This chapter is completely dealt about the carotenoids from microalga and its nutraceutical potential against obesity. Further, the optimization studies pave the way to improvise the yield of carotenoids by employing suitable factors with an appropriate concentration. Following this, the modern extraction methods also involved in extracting the carotenoids within short duration and becomes economical that would enroute for commercialization in near future.

Acknowledgements Authors thankfully acknowledge RUSA-MKU (013/RUSA/MKU/2020-2021), AquAgri, New Delhi, SERB (EEQ/2021/000251), DST-PURSE, DST-FIST, UGC-SAP,

UGC-UPE and Central Instrumentation facility of Madurai Kamaraj University, Madurai for providing the instrumentation facilities for SEM and Imaging.

References

Aburai N, Ohkubo S, Miyashita H, Abe K (2013) Composition of carotenoids and identification of aerial microalgae isolated from the surface of rocks in mountainous districts of Japan. Algal Res 2(3):237–243. https://doi.org/10.1016/j.algal.2013.03.001

Angelaalincy M, Nishtha P, Ajithkumar V, Ashokkumar B, Moorthy IMG, Brindhadevi K, Chi NTL, Pugazhendhi A, Varalakshmi P (2023) Phycoremediation of Arsenic and biodiesel production using green microalgae *Coelastrella* sp. M60—an integrated approach. Fuel 333:126427. https://doi.org/10.1016/j.fuel.2022.126427

Arun J, Avinash U, Arun Krishna B, Pandimadevi M, Gopinath KP (2017) Ultrasound assisted enhanced extraction of lutein (β, ε-carotene-3, 3′-diol) from microalga (*Chlorella pyrenoidosa*) grown in wastewater: optimization through response surface methodology. Glob NEST J 19(4):574–583

Ashokkumar V, Flora G, Sevanan M, Sripriya R, Chen WH, Park JH, Kumar G (2022) Technological advances in the production of carotenoids and their applications—a critical review. Biores Technol 128215. https://doi.org/10.1016/j.biortech.2023.129057

Baburina Y, Krestinin R, Odinokova I, Sotnikova L, Kruglov A, Krestinina O (2019) Astaxanthin inhibits mitochondrial permeability transition pore opening in rat heart mitochondria. Antioxidants 8(12):576. https://doi.org/10.3390/antiox8120576

Bachchhav MB, Kulkarni MV, Ingale AG (2020) Process-intensified extraction of phycocyanin followed by β-carotene from *Spirulina platensis* using ultrasound-assisted extraction. Sep Sci Technol 55(5):932–944. https://doi.org/10.1080/01496395.2019.1580293

Begum H, Yusoff FM, Banerjee S, Khatoon H, Shariff M (2016) Availability and utilization of pigments from microalgae. Crit Rev Food Sci Nutr 56(13):2209–2222. https://doi.org/10.1080/10408398.2013.764841

Bhuvaneswari S, Anuradha CV (2012) Astaxanthin prevents loss of insulin signaling and improves glucose metabolism in liver of insulin resistant mice. Can J Physiol Pharmacol 90(11):1544–1552. https://doi.org/10.1139/y2012-119

Cen SY, Li DW, Huang XL, Huang D, Balamurugan S, Liu WJ, Zheng JW, Yang WD, Li HY (2022) Crucial carotenogenic genes elevate hyperaccumulation of both fucoxanthin and β-carotene in *Phaeodactylum tricornutum*. Algal Res 64:102691. https://doi.org/10.1016/j.algal.2022.102691

Charalampia D, Antonios KE, Constantina N, Haralabos KC (2017) Current trends and emerging technologies in biopigment production processes: industrial food and health applications. Int J Hortic Agric Food Sci 1:33–46. ISSN: 2456-8635

Chen G, Wang B, Han D, Sommerfeld M, Lu Y, Chen F, Hu Q (2015) Molecular mechanisms of the coordination between astaxanthin and fatty acid biosynthesis in *Haematococcus pluvialis* (Chlorophyceae). Plant J 81(1):95–107. https://doi.org/10.1111/tpj.12713

Ciccone MM, Cortese F, Gesualdo M, Carbonara S, Zito A, Ricci G, De Pascalis F, Scicchitano P, Riccioni G (2013) Dietary intake of carotenoids and their antioxidant and anti-inflammatory effects in cardiovascular care. Mediat Inflamm 2013. https://doi.org/10.1155/2013/782137

Coesel SN, Baumgartner AC, Teles LM, Ramos AA, Henriques NM, Cancela L, Varela JCS (2008) Nutrient limitation is the main regulatory factor for carotenoid accumulation and for Psy and Pds steady state transcript levels in *Dunaliella salina* (Chlorophyta) exposed to high light and salt stress. Mar Biotechnol 10:602–611. https://doi.org/10.1007/s10126-008-9100-2

D'Alessandro EB, Antoniosi Filho NR (2016) Concepts and studies on lipid and pigments of microalgae: a review. Renew Sustain Energy Rev 58:832–841. https://doi.org/10.1016/j.rser.2015.12.162

Deenu A, Naruenartwongsakul S, Kim SM (2013) Optimization and economic evaluation of ultrasound extraction of lutein from *Chlorella vulgaris*. Biotechnol Bioprocess Eng 18:1151–1162. https://doi.org/10.1007/s12257-013-0213-8

Del Campo JA, Rodriguez H, Moreno J, Vargas MA, Rivas J, Guerrero MG (2004) Accumulation of astaxanthin and lutein in *Chlorella zofingiensis* (Chlorophyta). Appl Microbiol Biotechnol 64:848–854. https://doi.org/10.1007/s00253-003-1510-5

Dey S, Rathod VK (2013) Ultrasound assisted extraction of β-carotene from *Spirulina platensis*. Ultrason Sonochem 20(1):271–276. https://doi.org/10.1016/j.ultsonch.2012.05.010

Ding W, Cui J, Zhao Y, Han B, Li T, Zhao P, Xu JW, Yu X (2019) Enhancing *Haematococcus pluvialis* biomass and γ-aminobutyric acid accumulation by two-step cultivation and salt supplementation. Biores Technol 285:121334. https://doi.org/10.1016/j.biortech.2019.121334

El-Baz FK, Aly HF, Abd-Alla HI (2020) The ameliorating effect of carotenoid rich fraction extracted from *Dunaliella salina* microalga against inflammation-associated cardiac dysfunction in obese rats. Toxicol Rep 7:118–124. https://doi.org/10.1016/j.toxrep.2019.12.008

Esquivel-Hernández DA, López VH, Rodríguez-Rodríguez J, Alemán-Nava GS, Cuéllar-Bermúdez SP, Rostro-Alanis M, Parra-Saldívar R (2016) Supercritical carbon dioxide and microwave-assisted extraction of functional lipophilic compounds from *Arthrospira platensis*. Int J Mol Sci 17(5):658. https://doi.org/10.3390/ijms17050658

Esquivel-Hernández DA, Rodríguez-Rodríguez J, Rostro-Alanis M, Cuéllar-Bermúdez SP, Mancera-Andrade EI, Núñez-Echevarría JE, García-Pérez JS, Chandra R, Parra-Saldívar R (2017) Advancement of green process through microwave-assisted extraction of bioactive metabolites from *Arthrospira platensis* and bioactivity evaluation. Biores Technol 224:618–629. https://doi.org/10.1016/j.biortech.2016.10.061

Fan XD, Hou Y, Huang XX, Qiu TQ, Jiang JG (2015) Ultrasound-enhanced subcritical CO_2 extraction of lutein from *Chlorella pyrenoidosa*. J Agric Food Chem 63(18):4597–4605. https://doi.org/10.1021/acs.jafc.5b00461

Fratianni A, Cinquanta L, Panfili G (2010) Degradation of carotenoids in orange juice during microwave heating. LWT-Food Sci Technol 43(6):867–871. https://doi.org/10.1016/j.lwt.2010.01.011

Gao Z, Meng C, Zhang X, Xu D, Zhao Y, Wang Y, Lv H, Yang L, Chen L, Ye N (2012) Differential expression of carotenogenic genes, associated changes on astaxanthin production and photosynthesis features induced by JA in *H. pluvialis*. https://doi.org/10.1371/journal.pone.0042243

Gatamaneni Loganathan B, Orsat V, Lefsrud M, Wu BS (2020) A comprehensive study on the effect of light quality imparted by light-emitting diodes (LEDs) on the physiological and biochemical properties of the microalgal consortia of *Chlorella variabilis* and *Scenedesmus obliquus* cultivated in dairy wastewater. Bioprocess Biosyst Eng 43(8):1445–1455. https://doi.org/10.1007/s00449-020-02338-0

Gayathri S, Rajasree Radhika SR, Suman TY, Aranganathan L (2018) Ultrasound-assisted microextraction of β, ε-carotene-3, 3′-diol (lutein) from marine microalgae *Chlorella salina*: effect of different extraction parameters. Biomass Convers Bioref 8:791–797. https://doi.org/10.1007/s13399-018-0331-9

Guedes AC, Amaro HM, Malcata FX (2011) Microalgae as sources of carotenoids. Mar Drugs 9:625–644. https://doi.org/10.3390/md9040625

Hadiyanto H, Suttrisnorhadi S (2016) Response surface optimization of ultrasound assisted extraction (UAE) of phycocyanin from microalgae *Spirulina platensis*. Emir J Food Agric 227–234. https://doi.org/10.9755/ejfa.2015-05-193

Hermann GJ, Schroeder LK, Hieb CA, Kershner AM, Rabbitts BM, Fonarev P, Grant BD, Priess JR (2005) Genetic analysis of lysosomal trafficking in *Caenorhabditis elegans*. Mol Biol Cell 16:3273–3288. https://doi.org/10.1091/mbc.e05-01-0060

Ikeuchi M, Koyama T, Takahashi J, Yazawa K (2007) Effects of astaxanthin in obese mice fed a high-fat diet. Biosci Biotechnol Biochem 71(4):893–899. https://doi.org/10.1271/bbb.60521

Ishiki M, Nishida Y, Ishibashi H, Wada T, Fujisaka S, Takikawa A, Urakaze M, Sasaoka T, Usui I, Tobe K (2013) Impact of divergent effects of astaxanthin on insulin signaling in L6 cells. Endocrinology 154(8):2600–2612. https://doi.org/10.1210/en.2012-2198

Jaeschke DP, Rech R, Marczak LDF, Mercali GD (2017) Ultrasound as an alternative technology to extract carotenoids and lipids from *Heterochlorella luteoviridis*. Biores Technol 224:753–757. https://doi.org/10.1016/j.biortech.2016.11.107

Kaletta T, Hengartner MO (2006) Finding function in novel targets: *C. elegans* as a model organism. Nat Rev Drug Discov 5(5):387–399. https://doi.org/10.1038/nrd2031

Karpagam R, Raj KJ, Ashokkumar B, Varalakshmi P (2015) Characterization and fatty acid profiling in two fresh water microalgae for biodiesel production: lipid enhancement methods and media optimization using response surface methodology. Biores Technol 188:177–184. https://doi.org/10.1016/j.biortech.2015.01.053

Karpagam R, Jawaharraj K, Ashokkumar B, Sridhar J, Varalakshmi P (2018) Unraveling the lipid and pigment biosynthesis in *Coelastrella* sp. M-60: genomics-enabled transcript profling. Algal Res 29:277–289. https://doi.org/10.1016/j.algal.2017.11.031

Keaver L, Webber L, Dee A, Shiely F, Marsh T, Balanda K, Perry I (2013) Application of the UK foresight obesity model in Ireland: the health and economic consequences of projected obesity trends in Ireland. https://doi.org/10.1371/journal.pone.0079827

Lafarga T, Clemente I, Garcia-Vaquero M (2020) Carotenoids from microalgae. In: Carotenoids: properties, processing and applications. Academic Press, pp 149–187. https://doi.org/10.1016/B978-0-12-817067-0.00005-1

Li J, Zhao X, Chang JS, Miao X (2022) A two-stage culture strategy for *Scenedesmus* sp. FSP3 for CO_2 fixation and the simultaneous production of lutein under light and salt stress. Molecules 27(21):7497. https://doi.org/10.3390/molecules27217497

Lu Y, Jiang P, Liu S, Gan Q, Cui H, Qin S (2010) Methyl jasmonate- or gibberellins A3-induced astaxanthin accumulation is associated with up-regulation of transcription of β-carotene ketolase genes (bkts) in microalga *Haematococcus pluvialis*. Biores Technol 101(16):6468–6474. https://doi.org/10.1016/j.biortech.2010.03.072

Mc Gee D, Archer L, Fleming GT, Gillespie E, Touzet N (2020) The effect of nutrient and phytohormone supplementation on the growth, pigment yields and biochemical composition of newly isolated microalgae. Process Biochem 92:61–68. https://doi.org/10.1016/j.procbio.2020.03.001

Mc Gee D, Archer L, Parkes R, Fleming GT, Santos HM, Touzet N (2021) The role of methyl jasmonate in enhancing biomass yields and bioactive metabolites in *Stauroneis* sp. (Bacillariophyceae) revealed by proteome and biochemical profiling. J Proteomics 249:104381. https://doi.org/10.1016/j.jprot.2021.104381

Minhas AK, Hodgson P, Barrow CJ, Adholeya A (2016) A review on the assessment of stress conditions for simultaneous production of microalgal lipids and carotenoids. Front Microbiol 7:546. https://doi.org/10.3389/fmicb.2016.00546

Niizawa I, Espinaco BY, Leonardi JR, Heinrich JM, Sihufe GA (2018) Enhancement of astaxanthin production from *Haematococcus pluvialis* under autotrophic growth conditions by a sequential stress strategy. Prep Biochem Biotechnol 48(6):528–534. https://doi.org/10.1080/10826068.2018.1466159

Nunes F, Wolf M, Hartmann J, Paul RJ (2005) The ABC transporter PGP-2 from *Caenorhabditis elegans* is expressed in the sensory neuron pair AWA and contributes to lysosome formation and lipid storage within the intestine. Biochem Biophys Res Commun 338:862–871. https://doi.org/10.1016/j.bbrc.2005.10.023

Paliwal C, Rehmanji M, Shaikh KM, Zafar SU, Jutur PP (2022) Green extraction processing of lutein from *Chlorella saccharophila* in water-based ionic liquids as a sustainable innovation in algal biorefineries. Algal Res 66:102809. https://doi.org/10.1016/j.algal.2022.102809

Parniakov O, Apicella E, Koubaa M, Barba FJ, Grimi N, Lebovka N, Pataro G, Ferrari G, Vorobiev E (2015) Ultrasound-assisted green solvent extraction of high-added value compounds from microalgae *Nannochloropsis* spp. Biores Technol 198:262–267. https://doi.org/10.1016/j.biortech.2015.09.020

Pasquet V, Chérouvrier JR, Farhat F, Thiéry V, Piot JM, Bérard JB, Kaas R, Serive B, Patrice T, Cadoret JP, Picot L (2011) Study on the microalgal pigments extraction process: performance of microwave assisted extraction. Process Biochem 46(1):59–67. https://doi.org/10.1016/j.procbio.2010.07.009

Priyanka S, Kirubagaran R, Leema JM (2020) Optimization of algal culture medium for zeaxanthin production by *Dunaliella tertiolecta*: an RSM based approach. Curr Sci 119(12):1997

Pushpalatha S, Sangeetha R, Ariraman S, Ashokkumar B, Varalakshmi P (2021) Photocatalyst (TiO_2) as an enhancer: an attempt to enhance the production of carotenoids and lipids with the combined oxidative stresses in *Coelastrella* sp. M60. Clean Technol Environ Policy 23:41–53. https://doi.org/10.1007/s10098-020-01879-y

Rodriguez-Amaya DB, Kimura M (2004) HarvestPlus handbook for carotenoid analysis. International Food Policy Research Institute (IFPRI), International Center for Tropical Agriculture (CIAT), Washington, DC, Cali. http://ebrary.ifpri.org/cdm/singleitem/collection/p15738coll2/id/125148/rec/25

Ruen-ngam D, Shotipruk A, Pavasant P (2010) Comparison of extraction methods for recovery of astaxanthin from *Haematococcus pluvialis*. Sep Sci Technol 46(1):64–70. https://doi.org/10.1080/01496395.2010.493546

Schroeder LK, Kremer S, Kramer MJ, Currie E, Kwan E, Watts JL, Lawrenson AL, Hermann GJ (2007) Function of the *Caenorhabditis elegans* ABC transporter PGP-2 in the biogenesis of a lysosome-related fat storage organelle. Mol Biol Cell 18:995–1008. https://doi.org/10.1091/mbc.e06-08-0685

Sedjati S, Santosa GW, Yudiati E, Supriyantini E, Ridlo A, Kimberly FD (2019) Chlorophyll and carotenoid content of *Dunaliella salina* at various salinity stress and harvesting time. IOP Conf Ser Earth Environ Sci 246(1):012025. https://doi.org/10.1088/1755-1315/246/1/012025

Shi TQ, Wang LR, Zhang ZX, Sun XM, Huang H (2020) Stresses as first-line tools for enhancing lipid and carotenoid production in microalgae. Front Bioeng Biotechnol 8:610. https://doi.org/10.3389/fbioe.2020.00610

Shinde S, Lele SS (2010) Statistical media optimization for lutein production from microalgae *Auxenochlorella protothecoides* SAG 211-7A. http://hdl.handle.net/10044/1/17821

Silva A, Coimbra RN, Escapa C, Figueiredo SA, Freitas OM, Otero M (2020) Green microalgae *Scenedesmus obliquus* utilization for the adsorptive removal of nonsteroidal anti-inflammatory drugs (NSAIDs) from water samples. Int J Environ Res Public Health 17(10):3707. https://doi.org/10.3390/ijerph17103707

Singh N, Roy K, Goyal A, Moholkar VS (2019) Investigations in ultrasonic enhancement of β-carotene production by isolated microalgal strain *Tetradesmus obliquus* SGM19. Ultrason Sonochem 58:104697. https://doi.org/10.1016/j.ultsonch.2019.104697

Subramanian P, Gurunathan J (2020) Differential production of pigments by halophilic bacteria under the effect of salt and evaluation of their antioxidant activity. Appl Biochem Biotechnol 190:391–409. https://doi.org/10.1007/s12010-019-03107-w

Treusch S, Knuth S, Slaugenhaupt SA, Goldin E, Grant BD, Fares H (2004) *Caenorhabditis elegans* functional orthologue of human protein h-mucolipin-1 is required for lysosome biogenesis. Proc Natl Acad Sci USA 101:4483–4488. https://doi.org/10.1073/pnas.0400709101

Udayan A, Sabapathy H, Arumugam M (2020) Stress hormones mediated lipid accumulation and modulation of specific fatty acids in *Nannochloropsis oceanica* CASA CC201. Biores Technol 310:123437. https://doi.org/10.1016/j.biortech.2020.123437

Wijffels RH, Barbosa MJ, Eppink MH (2010) Microalgae for the production of bulk chemicals and biofuels. Biofuels Bioprod Bioref Innov Sustain Econ 4(3):287–295. https://doi.org/10.1002/bbb.215

Xie Y, Xiong X, Chen S (2021) Challenges and potential in increasing lutein content in microalgae. Microorganisms 9(5):1068. https://doi.org/10.3390/microorganisms9051068

Yao N, Yan S, Guo Y, Wang H, Li X, Wang L, Hu W, Li B, Cui W (2021) The association between carotenoids and subjects with overweight or obesity: a systematic review and meta-analysis. Food Funct 12(11):4768–4782. https://doi.org/10.1039/D1FO00004G

Zhang J, Sun Z, Sun P, Chen T, Chen F (2014) Microalgal carotenoids: beneficial effects and potential in human health. Food Funct 5(3):413–425. https://doi.org/10.1039/C3FO60607D

Zhao L, Chen G, Zhao G, Hu X (2009) Optimization of microwave-assisted extraction of astaxanthin from *Haematococcus pluvialis* by response surface methodology and antioxidant activities of the extracts. Sep Sci Technol 44(1):243–262. https://doi.org/10.1080/01496390802282321

Zheng H, Wang Y, Li S, Nagarajan D, Varjani S, Lee DJ, Chang JS (2022) Recent advances in lutein production from microalgae. Renew Sustain Energy Rev 153:111795. https://doi.org/10.1016/j.rser.2021.111795

Part V
Algal Roles in Climate Change Mitigation

Chapter 28
The Role of Brown Algae in Global Warming Mitigation

Fayaz A. Malla, Afaan A. Malla, Suhaib A. Bandh, Nazir A. Sofi, and Mukhtar Ahmed

Abstract There is a critical need to pay attention to and implement effective mitigation methods for the global warming crisis. In this chapter, we review many marine photosynthetic organisms known as brown algae and their potential role in slowing climate change. The amazing carbon absorption capacities and potential to offset greenhouse gas emissions of brown algae have attracted a lot of attention. This chapter takes a look at the planet's current predicament due to global warming. The ecological importance of brown algae is discussed afterward, with special attention paid to their remarkable characteristics including strong carbon fixation capacities and fast growth rates. Furthermore, the chapter investigates the many ways in which brown algae can help reduce greenhouse gas emissions. These methods include carbon sequestration via photosynthesis and the use of harvested biomass for carbon storage. The possible difficulties and restrictions of using brown algae on a massive scale as a mitigation strategy are also discussed. Challenges lie in areas such as technological feasibility, financial viability, and potential effects on marine ecosystems. In addition, the chapter covers the latest findings and developing technology for maximizing carbon sequestration in brown algae during its culture, harvesting, and application. This chapter's overarching goal is to provide a more nuanced picture of brown algae's possible involvement in mitigating global warming and to highlight the need for additional study and development in this area. To counteract global

F. A. Malla (✉) · S. A. Bandh
Department of Environmental Science, Higher Education Department, Govt. of Jammu and Kashmir, Jammu 190001, India
e-mail: nami.fayaz@gmail.com

A. A. Malla
Department of Environmental Science, Government Degree College Baramulla, (Indira Gandhi National Open University), Baramulla 193101, India

N. A. Sofi
Department of Agriculture Research Information System, Sher-e-Kashmir University of Agricultural Sciences and Technology, Shalimar Campus, Srinagar, Jammu and Kashmir 190025, India

M. Ahmed
Meteorological Centre, India Meteorological Department, Rambagh, Srinagar 190015, India

T. A. Kurniawan and A. Anouzla (eds.), *Algae as a Natural Solution for Challenges in Water-Food-Energy Nexus*, Environmental Science and Engineering,
https://doi.org/10.1007/978-981-97-2371-3_28

warming and secure a better future for our world, harnessing the power of brown algae could pave the way for sustainable and environmentally friendly solutions.

Keywords Brown algae · Climate change · Marine ecosystems · Carbon dioxide

28.1 Introduction to Brown Algae

The Phaeophyta, or brown algae, are an exceptionally varied group of marine organisms that play a crucial role in marine ecosystems (Hakim and Patel 2020). They are a remarkable and sophisticated species of multicellular photosynthetic creatures that thrive in the cold and temperate oceans. Brown algae have captured the interest of scientists, marine biologists, and ecologists due to their distinctive properties and ecological significance. With an emphasis on their taxonomy, morphology, life cycle, ecological significance, and economic importance, this introductory section tries to provide a comprehensive review of the brown algae (Behera et al. 2022).

28.1.1 Taxonomy and Classification

Chromista, which also includes diatoms and similar microbes, is the kingdom that brown algae call home. The name Phaeophyta comes from the Greek terms for "brown" and "plant," phaeo and phyton, respectively. There are around 2000 species in this group, the smallest of which are filamentous and the largest of which can grow to over 60 m in length. Pigmentation, cellular structure, and reproductive traits are the primary criteria for categorizing brown algae (Wynne 1981).

28.1.2 Morphology and Structure

Brown algal morphology displays a rich diversity of shapes and structures. While chlorophyll an is the primary pigment in the chloroplasts of red and green algae, chlorophyll c and a third pigment called fucoxanthin are present in brown algae. This fucoxanthin is responsible for coloring these organisms a distinctive brown. Moreover, brown algae have cell walls that are intricately made of cellulose, alginic acid, and other polysaccharides (Charrier et al. 2012).

A thallus describes the body structure of brown algae, which can be greatly separated into several areas. The thallus is attached to the ground by a holdfast structure at its base. The blades, which resemble leaves, are supported by the stipe, a stalk-like structure that rises from the holdfast. Light energy for photosynthesis is mostly captured by the blades, which are the primary photosynthetic organs (Ihua et al. 2020).

28.1.3 Life Cycle

Brown algae have a convoluted life cycle that includes sexual and asexual reproduction. One distinctive aspect of their life cycle is the periodic reversal of their generational status. Meiosis occurs during the diploid sporophyte phase, resulting in haploid spores. The haploid gametophytes that develop from these spores are responsible for making both male and female gametes. After the male and female gametes have fused to create a zygote, the resulting diploid sporophyte matures and the life cycle is complete (Ningrum and Chasani 2021).

28.1.4 Ecological Significance

Brown algae are extremely important to marine food webs. They play a crucial role in the creation of organic matter through photosynthesis as primary producers. The kelp forests that brown algae create are home to a diverse array of marine life, from fish and invertebrates to various types of algae. Forests like these also aid in stabilizing coastal sediments, protecting against erosion caused by waves (Dobrinčić et al. 2020).

In addition, brown algae play a role in the recycling of nutrients in marine ecosystems. Large quantities of nutrients, such as nitrogen and phosphorus, can be accumulated and stored by these organisms, and then released back into the water column upon decomposition. The coastal ecosystems' productivity and biodiversity benefit from the nutrient recycling.

28.1.5 Economic Importance

The direct and indirect contributions of brown algae to the economy are substantial. As a commercial crop, they are used to make alginates, a class of polysaccharides with numerous applications in the culinary, pharmaceutical, and cosmetics sectors. Bioactive chemicals found in brown algae may one day be used in pharmaceutical research (Charoensiddhi et al. 2018).

Brown algae are useful for the economy in ways other than their immediate worth. As a result, ecotourism and leisure options like scuba diving, snorkeling, and exploring natural regions flourish in coastal locations where they are found. Divers and other fans of the great outdoors often make a point of visiting kelp forests (King 2019).

In addition, brown algae help control global warming. Their strong capacity to absorb carbon dioxide (CO_2) is one way in which they reduce the impact of emissions of greenhouse gases. Brown algae absorb atmospheric carbon dioxide (CO_2) during photosynthesis and use it as a metabolic fuel. By lowering atmospheric CO_2

levels, this carbon sequestration method helps fight against climate change and ocean acidification (Saha and Pradhan 2023).

Brown algae are also a part of the bioremediation process for cleaning up marine ecosystems that have been polluted. They can help clean polluted water along polluted coastlines by accumulating heavy metals and other toxins in their bodies. Brown algae perform this ecological service, which aids in the upkeep of thriving marine habitats (El-Sheekh et al. 2021).

There may be potential financial prospects associated with brown algae research as well. The creation of biofuels, in particular, has piqued the interest of scientists who are investigating the potential of brown algae as a renewable energy source. Brown algae are a possible alternative to fossil fuels due to their rapid growth rates and the fact that they can be cultivated in industrial seaweed farms (Thanigaivel et al. 2022).

Further, studies are being conducted to find ways to use brown algae's beneficial components. These compounds are useful for the creation of medicines, nutraceuticals, and cosmeceuticals because of their antibacterial, antioxidant, and anti-inflammatory characteristics. The discovery and extraction of these beneficial chemicals from brown algae could give rise to whole new economic sectors (Lourenço-Lopes et al. 2020).

28.2 Brown Algae and Carbon Sequestration

Emissions of carbon dioxide (CO_2) and its potential influence on global warming have recently emerged as a major topic of discussion. To lessen these impacts, researchers are looking at different methods of "carbon sequestration," which involve removing CO_2 from the air and depositing it in underground or aboveground storage facilities. The diverse group of marine organisms known as brown algae has been getting a lot of attention in recent years for what could be an important role in carbon sequestration (Kouhgardi et al. 2023). This chapter presents the scientific consensus on the mechanics, efficacy, and potential future implications of the link between brown algae and carbon sequestration.

28.2.1 Brown Algae and Carbon Sequestration Mechanisms

Brown algae can efficiently capture atmospheric carbon dioxide and incorporate it into their biomass due to their special properties. Brown algae synthesize organic molecules from sunlight, carbon dioxide, and water via photosynthesis, with oxygen as a byproduct. Plants are able to take up atmospheric CO_2 and turn it into carbon-based compounds, primarily carbohydrates, through a process called photosynthesis (Chan et al. 2022).

In addition, brown algae can develop and accumulate biomass at an astounding rate. They grow quickly and can colonize large areas, creating what are called "kelp forests" in the water. These woods have a lot of ground for photosynthetic processes and absorbing carbon dioxide. Brown algae are effective carbon sinks due to their rapid growth and their capacity to store carbon in their tissues (Pedersen et al. 2021).

It has recently come to light that brown algae may help with carbon sequestration initiatives. Their potential to store carbon over extended periods of time and strong rates of growth make them attractive candidates for reducing the effects of climate change. However, there are a number of obstacles that need to be overcome, including the need to scale up production, ensure economic sustainability, and handle potential environmental concerns.

28.2.2 *Effectiveness of Brown Algae in Carbon Sequestration*

Research suggests that brown algae could be very efficient in sequestering carbon. Scientific investigations of kelp forests have revealed their remarkable capacity to absorb and store carbon dioxide. Because of their fast growth rates, brown algae are able to rapidly absorb CO_2 from the water.

Brown algae can store a lot of carbon, and that carbon might be sequestered for a long time. The carbon-rich biomass of brown algae, which sinks to the ocean floor as the algae die or are harvested, can be retained for long periods of time (Collins et al. 2022). Marine carbon burial aids in the long-term reduction of atmospheric CO_2 concentrations.

Brown algae not only help mitigate ocean acidification but also serve as a reservoir for carbon (Campbell et al. 2017). Brown algae are important because they absorb CO_2, keeping saltwater from becoming too acidic and harming marine ecosystems. Brown algae indirectly aid the health and survival of a wide range of marine creatures by reducing the effects of ocean acidification (Campbell et al. 2017).

28.2.3 *Implications and Challenges*

Important implications for measures to mitigate climate change stem from brown algae's ability for carbon sequestration (Yong et al. 2022). Brown algae, if managed properly, could be a useful tool for lowering atmospheric CO_2 levels, fighting global warming, and softening the blow of climate change on coastal ecosystems.

However, numerous obstacles and factors must be overcome before brown algae can reach its full carbon sequestration potential. To begin, substantial carbon sequestration would demand for the widespread cultivation and expert administration of brown algae farms. To maximize carbon uptake and biomass output, it is crucial to optimize cultivation techniques including nutrient availability, water quality, and farm design (Sarwer et al. 2022).

Further research is also required to determine the cost-effectiveness and scalability of carbon sequestration systems based on brown algae (Lourenço-Lopes et al. 2020). Key considerations include the refinement of efficient harvesting methods, the creation of low-cost growing strategies, and the discovery of novel uses for the produced biomass.

Further, it is important to assess the results of growing brown algae on a massive scale. Sustainable and ecologically friendly methods require in-depth research and management of potential ecological repercussions such changes in local biodiversity, nutrient cycle, and interactions with native species.

28.2.4 Future Perspectives

Further study and technological developments are required to fully realize the potential of brown algae in carbon sequestration, but their future is bright. Some potential outcomes are listed in Fig. 28.1.

Enhanced Cultivation Techniques: To maximize carbon uptake and biomass production, scientists are looking at the best ways to cultivate brown algae (Collins et al. 2022). Researching how growth rates and carbon sequestration effectiveness are affected by factors including nutrition availability, light intensity, and temperature is a part of this. Increases in scalability and productivity could result from developments in growing techniques like offshore farms or integrated multi-trophic aquaculture systems.

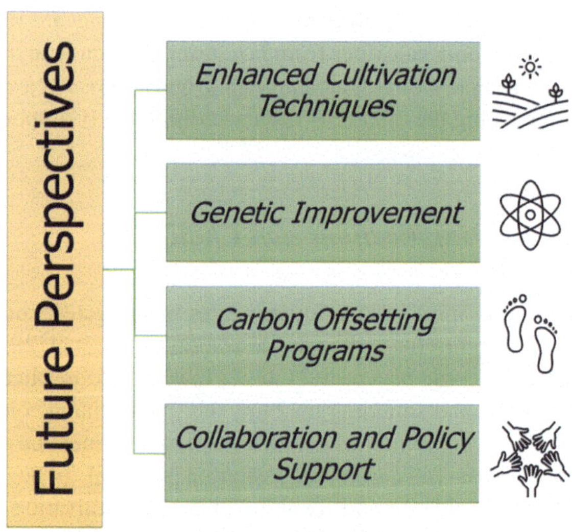

Fig. 28.1 Future perspectives for carbon sequestration

Genetic Improvement: The ability of brown algae to sequester carbon could be improved by genetic engineering and selective breeding (Kouhgardi et al. 2023). Scientists may be able to improve strains' carbon uptake rates and biomass accumulation by discovering and altering genes responsible for carbon fixing and growth.

Carbon Offsetting Programs: Potentially included in carbon offsetting schemes are brown algae-based carbon sequestration initiatives (Roy et al. 2023). Individuals and businesses can make amends for their carbon footprints through these programs by funding initiatives that reduce atmospheric CO_2 levels by the same amount. Economic incentives for the development of brown algae farms could come from the creation of carbon credits that could be exchanged on carbon markets (Sydney et al. 2019).

Collaboration and Policy Support: For brown algae-based carbon sequestration programs to move forward, scientists, policymakers, and industrial stakeholders must work together (Cai et al. 2021). Policy support, research funding, and advocacy for sustainable practices are all areas in which governments and international organizations can play a crucial role.

To fully utilize brown algae for carbon sequestration, it will be necessary to invest in research, technological innovation, and sustainable practices. We can get closer to meeting global emission reduction targets and building a sustainable future by using brown algae-based carbon sequestration in extensive climate change mitigation initiatives (Roy et al. 2023).

28.3 Brown Algae and Ocean Acidification

The effects of ocean acidification, caused by rising amounts of carbon dioxide (CO_2) in the atmosphere, are becoming increasingly urgent, particularly for marine ecosystems. The pH of the ocean drops as CO_2 combines with water molecules to generate carbonic acid (Stallinga 2018). Although this change in pH can be harmful to calcifying creatures, new studies suggest that some marine animals, including brown algae, may help to lessen the effects of ocean acidification (Wahl et al. 2018).

28.3.1 Brown Algae and Ocean Acidification Responses

Many different kinds of adaptations of brown algae have been shown in response to ocean acidification (Porzio et al. 2020). They are not completely safe from it, but research suggests they may be more resistant than other calcifying species.

Physiological Responses: When CO_2 levels rise, brown algae can adapt by changing their internal activities (Raddatz et al. 2017). They have been shown to boost their photosynthetic rates in response to higher CO_2 concentrations, leading to increased growth and production. It is hypothesized that the availability of extra carbon for photosynthesis is what prompted this reaction. Brown algae may also display altered metabolic pathways and ion control mechanisms, all in an effort to preserve cellular homeostasis in an acidic environment (Zou et al. 2019).

Calcification and Morphology: Brown algae aren't calcifiers, meaning they don't need to build skeletons out of calcium carbonate to thrive (Moore et al. 2022). As a result, it's possible they won't be adversely affected by ocean acidification's reduction in carbonate ion availability. However, some research indicates that brown algae's calcified features, such as the composition and density of their cell walls, can undergo modifications in response to acidified environments.

Interactive Effects: Other environmental conditions, such as temperature and nutrient availability, may affect how brown algae react to ocean acidification. Some brown algae species may benefit from increased growth and photosynthetic rates in the presence of increasing CO_2 and higher temperatures, according to studies (Zhang et al. 2020). However, further research is needed to understand the interacting impacts of numerous stressors on the physiology and ecology of brown algae.

28.3.2 Ecological Implications

Complex and situationally dependent, the ecological ramifications of brown algae reactions to ocean acidification are still being explored. Changes in the amount and distribution of brown algae can have far-reaching consequences for coastal ecosystems, despite the fact that the algae themselves may be rather resilient (Brown et al. 2018).

Habitat Modification: Coastal ecosystems rely heavily on brown algae for habitat and structural complexity (Bringloe et al. 2020). The loss of these organisms or a reduction in their numbers as a result of ocean acidification could have repercussions for the fish, invertebrates, and other algae that rely on them.

Trophic Interactions: In many coastal environments, brown algae serve as the foundation of food webs. Herbivores and predators that rely on brown algae as a food supply may be adversely affected if their productivity and nutritional quality decreases in acidified environments (Raddatz et al. 2017).

Biodiversity and Community Composition: The competitive dynamics between algae species are susceptible to changes in environmental conditions brought on by ocean acidification (Horwitz et al. 2017). Under more acidic conditions, some types of brown algae may flourish. Changes in the makeup of coastal communities can have far-reaching effects on biodiversity and ecosystem function.

28.3.3 Management and Conservation Considerations

Protecting brown algae and maximizing its potential advantages in the face of ocean acidification calls for careful management and conservation (Albright and Cooley 2019). Some essential factors to think about:

Marine Protected Areas (MPAs): The creation of MPAs is an effective way to protect brown algae ecosystems. The protection of vital ecosystems and the restoration of degraded regions are facilitated by the establishment of protected zones in which fishing, mining, and other destructive activities are restricted or forbidden. The effects of ocean acidification on brown algae and their biological communities can be better understood with the help of study and monitoring made possible by MPAs.

Sustainable Harvesting Practices: To protect the longevity of brown algae populations, responsible harvesting methods must be put into place. This includes arranging seasonal or rotational harvesting to give populations time to recover and setting proper catch limits. Guidelines for sustainable harvesting must be developed and implemented with input from local communities, fishermen, and industry stakeholders.

Reduction of Local Stressors: Protecting brown algae ecosystems requires taking action against threats like pollution and habitat loss. It is important to safeguard the health of coastal ecosystems and keep water quality high by taking steps to limit nutrient runoff, sedimentation, and chemical pollution. It is also crucial to encourage prudent coastal development practices and lessen the negative effects of coastal infrastructure on brown algae ecosystems.

Climate Change Mitigation: In order to prevent further ocean acidification and protect brown algae ecosystems, it is crucial to take steps to limit emissions of greenhouse gases. Taking action to increase renewable energy, energy efficiency, and environmentally responsible land use can help lessen the effects of climate change. Global efforts to counteract ocean acidification require widespread support for international agreements and activities aimed at lowering CO_2 emissions, such as the Paris Agreement.

Integrated Coastal Management: The conservation and sustainable use of coastal resources can be approached in a comprehensive and coordinated manner by using integrated coastal management concepts. This necessitates taking into account the interplay and interdependence of a wide range of players, industries, and ecological systems. To address the complex difficulties faced by ocean acidification and to maintain brown algae habitats, integrated coastal management plans should involve scientific research, stakeholder participation, and adaptive management measures.

Public Awareness and Education: Brown algae and ocean acidification are two issues that need more attention in order to rally public support for conservation measures. Understanding the ecological relevance of brown algae and the importance of conserving them can be improved by public education campaigns, community outreach programs, and educational efforts in schools and colleges. Participation in citizen science programs can increase both data quality and the public's sense of ownership and responsibility for coastal ecosystems.

Scientific research, policy interventions, stakeholder collaboration, and public awareness are all essential components of an effective strategy for the management and conservation of brown algae in the context of ocean acidification. Brown algae habitats can be safeguarded to preserve their important role in coastal ecosystems through the use of sustainable practices, the reduction of local stressors, the mitigation of climate change, and the promotion of ecosystem-based management. These actions are crucial for the long-term health of our marine ecosystems.

28.4 Brown Algae and Biofuel Production

There is a growing interest in investigating alternative biofuel solutions in light of the pressing need to reduce greenhouse gas emissions and meet the rising demand for renewable energy (Antar et al. 2021). The numerous sea organisms known collectively as brown algae have recently attracted interest as a possible feedstock for biofuel generation (Suparmaniam et al. 2019). The biochemical makeup, cultivation methods, and biofuel conversion mechanisms of brown algae are discussed in this article from a scientific point of view.

28.4.1 Biochemical Composition of Brown Algae

Brown algae are ideally suited for the production of biofuels due to their distinctive metabolic features (Marella et al. 2020). They have a high concentration of fermentable sugar-producing carbs, including laminarin and mannitol. Brown algae also have a high concentration of lipids, such as fatty acids, which can be used to produce biodiesel (Fawzy and Gomaa 2020). Diverse species of brown algae have diverse carbohydrate and lipid compositions, allowing for the selection of strains with favorable properties for biofuel production (Debnath et al. 2021).

28.4.2 Cultivation Techniques for Brown Algae

Efficient methods of growing brown algae are needed to keep up with the growing demand for biofuel (Kouhgardi et al. 2023). Many different types of farming, such as ocean farming, coastal farming, and multi-trophic aquaculture systems, are being researched.

Open Ocean Farming: Brown algae can be farmed in the open water in offshore places. This technique performs well in regions with plenty of sunshine and water that is rich in nutrients. Algae can be grown in the water column when they are anchored to floating structures or wires. Selecting the right site for open ocean farming is essential for achieving optimum growth conditions while minimizing environmental damage.

Coastal Farms: Brown algae is grown in coastal farms in near-shore regions like bays and estuaries. This is because brown algae thrives in protected environments with easy access to nutrient-rich waters. Algae can be grown in coastal farms that are supported by ropes, nets, or other structures. Sustainable farming along the coast requires diligent attention to management details like water quality monitoring and the prevention of the introduction of unwanted species.

Integrated Multi-Trophic Aquaculture (IMTA) Systems: Synergistically growing various species, such as brown algae, is at the heart of integrated multi-trophic aquaculture (IMTA) systems. Waste from other aquaculture species, such finfish or shellfish, can be a rich source of nutrients for brown algae development in IMTA systems. The environmental implications of aquaculture can be minimized while resource utilization is increased through this coordinated strategy.

28.4.3 Conversion Processes for Biofuel Production

The conversion of brown algae biomass into biofuels involves several steps, including extraction, pretreatment, and conversion processes.

Biomass Extraction: Biomass extraction from harvested or grown algae is the initial stage in brown algae biofuel manufacturing. Algae can have their carbohydrates and lipids extracted using a variety of techniques, such as mechanical pressing, solvent extraction, or enzymatic hydrolysis.

Pretreatment: The effectiveness of the succeeding conversion procedures can be improved by employing pretreatment treatments. The complex structure of brown algal biomass can be broken down through physical, chemical, or biological processes to allow for the release of fermentable sugars or lipids. Efficient, cost-effective, and environmentally friendly pretreatment conditions are best.

Conversion Processes: The extracted carbohydrates and lipids can be converted into biofuels through different processes.

(a) *Fermentation*: Bioethanol can be made by fermenting the sugars extracted from brown algae with yeast or bacteria. The process of fermentation can only occur under certain temperature, pH, and nutrient availability conditions.
(b) *Transesterification*: Transesterification allows the lipids isolated from brown algae to be processed into biodiesel. The fatty acid methyl esters (FAME) that are the backbone of biodiesel are created by reacting the lipids with an alcohol,

often methanol or ethanol, in the presence of a catalyst (Nisar et al. 2021). Different catalysts, such as alkaline or acid catalysts, and reaction conditions are required for transesterification to occur.

Upgrading and Refining: Biofuel products may be upgraded and refined further after conversion operations to ensure they are of sufficient quality for commercial use (Kesieme et al. 2019). Biofuels may require additional processing such as purification, impurity removal, and property tweaking to reach their full potential.

28.4.4 Challenges and Future Perspectives

While brown algae hold promise as a feedstock for biofuel production, several challenges need to be addressed for large-scale implementation:

Cultivation Efficiency: Improving yields of biomass while decreasing expenses necessitates fine-tuning farming methods. Maximizing biomass yield requires better strain selection, optimized growth conditions, and more effective harvesting methods.

Conversion Efficiency: It is essential to improve the extraction, pretreatment, and conversion procedures used in the biomass conversion process. New and improved methods of conversion need to be developed, and existing processes should be optimized to increase yields while decreasing energy use.

Sustainability and Environmental Impacts: Brown algae cultivation and extraction for biofuel generation must be carried out in an environmentally responsible manner (Kouhgardi et al. 2023). Water conservation, chemical avoidance, and mitigation of ecological threats including fertilizer runoff and the introduction of non-native species are all part of this process.

Techno-Economic Viability: Production of biofuels from brown algae raises important questions about its commercial viability. To lower production costs, increase overall process efficiency, and create effective downstream processing and refining procedures, continued R&D is required (Kouhgardi et al. 2023).

Integrated Approaches: The sustainability and economic viability of brown algae culture can be improved by investigating integrated approaches that combine biofuel production with other value-added products, such as biofertilizers, animal feed, or bioplastics (Arora et al. 2021).

Due to their distinct biochemical composition and abundance in marine environments, brown algae are a suitable feedstock for biofuel production. Key factors for effective biofuel production include efficient farming methods, optimal conversion processes, and sustainable practices (Zhu et al. 2017). In order to overcome obstacles and realize the full potential of brown algae as a renewable and ecologically benign source of biofuels, continued study, innovation, and collaboration among scientists, engineers, politicians, and industry stakeholders is essential. Brown algae-based biofuels can help reduce greenhouse gas emissions and meet the global demand for renewable energy with additional technological breakthroughs and the adoption of sustainable practices (Shuba and Kifle 2018).

28.5 Potential of Brown Algae in Combating Climate Change

The earth faces serious threats from climate change, which calls for immediate action to curb emissions of greenhouse gases and lessen the effects of this phenomenon (Shen et al. 2020). The unique features and biological responsibilities of brown algae, a varied group of marine organisms, make them a promising tool in the fight against climate change. In order to combat climate change, brown algae have the ability to sequester carbon, cycle nutrients, and restore ecosystems, all of which are discussed in this article from a scientific point of view (Kritzberg et al. 2020).

28.5.1 Carbon Sequestration

Carbon sequestration, the process by which carbon dioxide (CO_2) is trapped and stored, with the goal of reducing atmospheric CO_2 levels, is greatly aided by brown algae (Tarafdar et al. 2023). The photosynthesis of brown algae allows it to rapidly multiply and take up large amounts of carbon dioxide from the ocean in its immediate environment. Carbon dioxide is taken out of the air and stored permanently as biomass once it has been transformed from CO_2 (Daval 2018).

Coastal Blue Carbon Habitats: Kelp forests and seagrass meadows are two examples of coastal blue carbon habitats that feature prominently in the presence of brown algae (Duarte et al. 2022). In terms of carbon sequestration, these ecosystems are second only to terrestrial forests. Because of their high biomass, the brown algae found in these environments act as a major carbon sink.

Carbon Storage in Sediments: Long-term carbon storage can be aided by the shipment of brown algal biomass offshore or its deposition in coastal sediments. Brown algae produce carbon-rich organic materials that can be stored in sediments for hundreds of years, if not millennia (Huxham et al. 2018). Burial or sedimentary storage is one method for removing carbon from the carbon cycle and cutting back on CO_2 in the air.

28.5.2 Nutrient Cycling and Ecosystem Services

Brown algae contribute to climate change mitigation through carbon sequestration and other ecosystem services they provide.

Nutrient Uptake: Nitrogen and phosphorus are just two of the nutrients that brown algae can take in and retain. Brown algae can help mitigate eutrophication—a process linked to the emission of greenhouse gases like nitrous oxide (N_2O)—by efficiently absorbing the nutrients that contribute to it (Kumar et al. 2019). Brown algae's ability to assimilate nutrients aids in the upkeep of nutrient equilibrium and the enhancement of water quality in coastal habitats.

Enhanced Primary Productivity: Through photosynthesis, which uses sunlight as an energy source, brown algae increase the overall productivity of coastal habitats. Increased primary productivity from their rapid development and effective carbon fixation capacity benefits several marine species by maintaining complex food webs and providing suitable habitat. Brown algae indirectly aid in carbon sequestration and atmospheric CO_2 reduction by increasing primary productivity (Raven 2017).

28.5.3 Ecosystem Restoration and Resilience

In addition to aiding in ecosystem restoration, brown algae can make coastal habitats more resistant to the effects of climate change.

Habitat Structure and Coastal Protection: Brown algae, especially kelp species, construct elaborate three-dimensional structures that provide shelter for many different kinds of marine life (Rueda et al. 2021). Numerous species benefit from these structures since they offer protection, nourishment, and spawning areas. Brown algae also physically help to disperse wave energy and stabilize sediments, which protects coastlines from erosion and storms (James et al. 2021).

Coastal Acidification Mitigation: Brown algae may help reduce the effects of ocean acidification, one of the many threats to coastal ecosystems posed by climate change (Hall-Spencer and Harvey 2019). By collecting CO_2 through photosynthesis, brown algae can alter the chemistry of the saltwater around them, increasing the pH and mitigating the consequences of acidification (Wahl et al. 2018). Because of this "canopy effect," coastal ecosystems are able to withstand natural disasters and remain healthy.

Brown algae has enormous and varied potential for use in climate change mitigation efforts. Greenhouse gas emissions can be reduced, coastal resilience can be strengthened, and overall adaptation techniques can benefit from brown algae because of their ability to sequester carbon, cycle nutrients, and restore ecosystems (Roberts et al. 2017). However, appropriate conservation measures, sustainable management methods, and research are needed to unlock brown algae's full potential. Recognizing and using brown algae's potential allows us to take advantage of their biological services to combat climate change and build a more sustainable future for our world (Bibi et al. 2017).

28.6 Conclusion

Brown algae are a marine algae species well-known for their capacity to take in atmospheric carbon dioxide. In terms of carbon sequestration, they are so efficient that they can even compete with terrestrial forests. Because of this, brown algae may be used to help slow climate change. Several strategies exist for employing brown algae in the fight against climate change. One option is to use massive aquaculture farms to cultivate them. The open sea or the shore would provide ideal locations for these farms due to the abundance of both sunlight and water. Algae would be able to absorb carbon dioxide from the air and hold onto it for millennia if collected and utilized to make biofuels or bioproducts, or if buried at the bottom of the ocean. Brown algae can also be used to construct artificial reefs, which is something that can help with the issue of global warming. These reefs have the potential to halt beach erosion while also serving as a haven for marine life. In addition to lowering atmospheric greenhouse gas levels, the algae would keep soaking up carbon dioxide from the air. We can assist mitigate the effects of climate change by increasing their numbers in the ocean through large-scale aquaculture farms or by constructing artificial reefs. The direct and indirect contributions of brown algae to the economy are substantial. As a commercial crop, they are used to make alginates, a class of polysaccharides with numerous applications in the culinary, pharmaceutical, and cosmetics sectors. Bioactive chemicals found in brown algae may one day be used in pharmaceutical research. The development of biofuels is only one area where research into brown algae could lead to lucrative prospects in the future. Brown algae are a possible alternative to fossil fuels due to their rapid growth rates and the fact that they can be cultivated in industrial seaweed farms. Brown algae have bioactive substances with potential medical, nutritional, and therapeutic uses that are now being investigated. Because of their ability to take in and store large volumes of carbon dioxide from the water, brown algae show great promise as a carbon sequestration tool. Growing the crop on a larger scale, making it economically viable, and minimizing any negative effects on the environment are all obstacles that must be overcome.

References

Albright R, Cooley S (2019) A review of interventions proposed to abate impacts of ocean acidification on coral reefs. Reg Stud Mar Sci 29:100612. https://doi.org/10.1016/j.rsma.2019.100612

Antar M, Lyu D, Nazari M, Shah A, Zhou X, Smith DL (2021) Biomass for a sustainable bioeconomy: an overview of world biomass production and utilization. Renew Sustain Energy Rev 139:110691. https://doi.org/10.1016/j.rser.2020.110691

Arora K, Kaur P, Kumar P, Singh A, Patel SKS, Li X, Yang Y, Bhatia SK, Kulshrestha S (2021) Valorization of wastewater resources into biofuel and value-added products using microalgal system. Front Energy Res 9:646571. https://doi.org/10.3389/fenrg.2021.646571

Behera DP, Ingle KN, Mathew DE, Dhimmar A, Sahastrabudhe H, Sahu SK, Krishnan MG, Shinde PB, Ganesan M, Mantri VA (2022) Epiphytism, diseases and grazing in seaweed aquaculture: a comprehensive review. Rev Aquac 14(3):1345–1370. https://doi.org/10.1111/raq.12653

Bibi R, Ahmad Z, Imran M, Hussain S, Ditta A, Mahmood S, Khalid A (2017) Algal bioethanol production technology: a trend towards sustainable development. Renew Sustain Energy Rev 71:976–985. https://doi.org/10.1016/j.rser.2016.12.126

Bringloe TT, Starko S, Wade RM, Vieira C, Kawai H, De Clerck O, Cock JM, Coelho SM, Destombe C, Valero M, Neiva J, Pearson GA, Faugeron S, Serrão EA, Verbruggen H (2020) Phylogeny and evolution of the brown algae. Crit Rev Plant Sci 39(4):281–321. https://doi.org/10.1080/07352689.2020.1787679

Brown NEM, Milazzo M, Rastrick SPS, Hall-Spencer JM, Therriault TW, Harley CDG (2018) Natural acidification changes the timing and rate of succession, alters community structure, and increases homogeneity in marine biofouling communities. Glob Change Biol 24(1):e112–e127. https://doi.org/10.1111/gcb.13856

Cai J et al (2021) Seaweeds and microalgae: an overview for unlocking their potential in global aquaculture development. FAO Fish Aquac Circ 1229

Campbell JE, Sneed J, Johnston L, Paul V (2017) Effects of ocean acidification and contact with the brown alga *Stypopodium zonale* on the settlement and early survival of the coral *Porites astreoides*. Mar Ecol Prog Ser 577:67–77. https://doi.org/10.3354/meps12249

Chan SS, Low SS, Chew KW, Ling TC, Rinklebe J, Juan JC, Ng EP, Show PL (2022) Prospects and environmental sustainability of phyconanotechnology: a review on algae-mediated metal nanoparticles synthesis and mechanism. Environ Res 212(A):113140. https://doi.org/10.1016/j.envres.2022.113140

Charoensiddhi S, Lorbeer AJ, Franco CMM, Su P, Conlon MA, Zhang W (2018) Process and economic feasibility for the production of functional food from the brown alga *Ecklonia radiata*. Algal Res 29:80–91. https://doi.org/10.1016/j.algal.2017.11.022

Charrier B, Le Bail A, de Reviers B (2012) Plant Proteus: brown algal morphological plasticity and underlying developmental mechanisms. Trends Plant Sci 17(8):468–477. https://doi.org/10.1016/j.tplants.2012.03.003

Collins N, Kumar Mediboyina M, Cerca M, Vance C, Murphy F (2022) Economic and environmental sustainability analysis of seaweed farming: monetizing carbon offsets of a brown algae cultivation system in Ireland. Biores Technol 346:126637. https://doi.org/10.1016/j.biortech.2021.126637

Daval D (2018) Carbon dioxide sequestration through silicate degradation and carbon mineralisation: promises and uncertainties. npj Mater Degrad 2(1):11. https://doi.org/10.1038/s41529-018-0035-4

Debnath C, Bandyopadhyay TK, Bhunia B, Mishra U, Narayanasamy S, Muthuraj M (2021) Microalgae: sustainable resource of carbohydrates in third-generation biofuel production. Renew Sustain Energy Rev 150:111464. https://doi.org/10.1016/j.rser.2021.111464

Dobrinčić A, Balbino S, Zorić Z, Pedisić S, Bursać Kovačević D, Elez Garofulić I, Dragović-Uzelac V (2020) Advanced technologies for the extraction of marine brown algal polysaccharides. Mar Drugs 18(3):168. https://doi.org/10.3390/md18030168

Duarte CM, Gattuso J, Hancke K, Gundersen H, Filbee-Dexter K, Pedersen MF, Middelburg JJ, Burrows M, Krumhansl KA, Wernberg T, Moore P, Pessarrodona A, Ørberg SB, Pinto IS, Assis J, Queirós A, Smale DA, Bekkby T, Serrão E, Field R (2022) Global estimates of the extent and production of macroalgal forests. Glob Ecol Biogeogr 31(7):1422–1439. https://doi.org/10.1111/geb.13515

El-Sheekh MM, El-Kassas HY, Shams El-Din NG, Eissa DI, El-Sherbiny BA (2021) Green synthesis, characterization applications of iron oxide nanoparticles for antialgal and wastewater bioremediation using three brown algae. Int J Phytorem 23(14):1538–1552. https://doi.org/10.1080/15226514.2021.1915957

Fawzy MA, Gomaa M (2020) Pretreated fucoidan and alginate from a brown seaweed as a substantial carbon source for promoting biomass, lipid, biochemical constituents and biodiesel quality of *Dunaliella salina*. Renew Energy 157:246–255. https://doi.org/10.1016/j.renene.2020.05.065

Hakim MM, Patel IC (2020) A review on phytoconstituents of marine brown algae. Future J Pharm Sci 6:1–11

Hall-Spencer JM, Harvey BP (2019) Ocean acidification impacts on coastal ecosystem services due to habitat degradation. Emerg Top Life Sci 3(2):197–206. https://doi.org/10.1042/ETLS20180117

Horwitz R, Hoogenboom MO, Fine M (2017) Spatial competition dynamics between reef corals under ocean acidification. Sci Rep 7(1):1

Huxham M, Whitlock D, Githaiga M, Dencer-Brown A (2018) Carbon in the coastal seascape: how interactions between mangrove forests, seagrass meadows and tidal marshes influence carbon storage. Curr For Rep 4(2):101–110. https://doi.org/10.1007/s40725-018-0077-4

Ihua MW, FitzGerald JA, Guihéneuf F, Jackson SA, Claesson MJ, Stengel DB, Dobson ADW (2020) Diversity of bacteria populations associated with different thallus regions of the brown alga *Laminaria digitata*. PLoS ONE 15(11):e0242675. https://doi.org/10.1371/journal.pone.0242675

James RK, Lynch A, Herman PMJ, van Katwijk MM, van Tussenbroek BI, Dijkstra HA, van Westen RM, van der Boog CG, Klees R, Pietrzak JD, Slobbe C, Bouma TJ (2021) Tropical biogeomorphic seagrass landscapes for coastal protection: persistence and wave attenuation during major storms events. Ecosystems 24(2):301–318. https://doi.org/10.1007/s10021-020-00519-2

Kesieme U, Pazouki K, Murphy A, Chrysanthou A (2019) Biofuel as an alternative shipping fuel: technological, environmental and economic assessment. Sustain Energy Fuels 3(4):899–909. https://doi.org/10.1039/C8SE00466H

King AH (2019) Brown seaweed extracts (alginates). In: Food hydrocolloids. CRC Press, pp 115–188

Kouhgardi E, Zendehboudi S, Mohammadzadeh O, Lohi A, Chatzis I (2023) Current status and future prospects of biofuel production from brown algae in North America: progress and challenges. Renew Sustain Energy Rev 172:113012. https://doi.org/10.1016/j.rser.2022.113012

Kritzberg ES, Hasselquist EM, Škerlep M, Löfgren S, Olsson O, Stadmark J, Valinia S, Hansson LA, Laudon H (2020) Browning of freshwaters: consequences to ecosystem services, underlying drivers, and potential mitigation measures. Ambio 49(2):375–390. https://doi.org/10.1007/s13280-019-01227-5

Kumar R, Kumar R, Prakash O (2019) The impact of chemical fertilizers on our environment and ecosystem. Chief, 35th edn, p 69

Lourenço-Lopes C, Garcia-Oliveira P, Carpena M, Fraga-Corral M, Jimenez-Lopez C, Pereira AG, Prieto MA, Simal-Gandara J (2020) Scientific approaches on extraction, purification and stability for the commercialization of fucoxanthin recovered from brown algae. Foods 9(8):1113. https://doi.org/10.3390/foods9081113

Marella TK, López-Pacheco IY, Parra-Saldívar R, Dixit S, Tiwari A (2020) Wealth from waste: diatoms as tools for phycoremediation of wastewater and for obtaining value from the biomass. Sci Total Environ 724:137960. https://doi.org/10.1016/j.scitotenv.2020.137960

Moore D, Heilweck M, Petros P (2022) What should be done. In: Aquaculture: ocean blue carbon meets UN-SDGS. Springer International Publishing, pp 217–242

Ningrum AM, Chasani AR (2021) Numerical phenetic and phylogenetic relationships in silico among brown seaweeds (Phaeophyceae) from Gunungkidul, Yogyakarta, Indonesia. Biodiversitas J Biol Divers 22(6). https://doi.org/10.13057/biodiv/d220607

Nisar S, Hanif MA, Rashid U, Hanif A, Akhtar MN, Ngamcharussrivichai C (2021) Trends in widely used catalysts for fatty acid methyl esters (FAME) production: a review. Catalysts 11(9):1085. https://doi.org/10.3390/catal11091085

Pedersen MF, Filbee-Dexter K, Frisk N, Sárossy Z, Wernberg T (2021) Carbon sequestration potential increased by incomplete anaerobic decomposition of kelp detritus. Mar Ecol Prog Ser 660:53–67. https://doi.org/10.3354/meps13613

Porzio L, Arena C, Lorenti M, De Maio A, Buia MC (2020) Long-term response of *Dictyota dichotoma* var. *intricata* (C. Agardh) Greville (Phaeophyceae) to ocean acidification: insights from high pCO_2 vents. Sci Total Environ 731:138896. https://doi.org/10.1016/j.scitotenv.2020.138896

Raddatz S, Guy-Haim T, Rilov G, Wahl M (2017) Future warming and acidification effects on anti-fouling and anti-herbivory traits of the brown alga *Fucus vesiculosus* (Phaeophyceae). J Phycol 53(1):44–58. https://doi.org/10.1111/jpy.12473

Raven JA (2017) The possible roles of algae in restricting the increase in atmospheric CO_2 and global temperature. Eur J Phycol 52(4):506–522. https://doi.org/10.1080/09670262.2017.1362593

Roberts CM, O'Leary BC, McCauley DJ, Cury PM, Duarte CM, Lubchenco J, Pauly D, Sáenz-Arroyo A, Sumaila UR, Wilson RW, Worm B, Castilla JC (2017) Marine reserves can mitigate and promote adaptation to climate change. Proc Natl Acad Sci USA 114(24):6167–6175. https://doi.org/10.1073/pnas.1701262114

Roy P, Mohanty AK, Misra M (2023) Prospects of carbon capture, utilization and storage for mitigating climate change. Environ Sci Adv 2(3):409–423. https://doi.org/10.1039/D2VA00236A

Rueda JL et al (2021) Benthic fauna of littoral and deep-sea habitats of the Alboran Sea: a hotspot of biodiversity. In: Alboran Sea—ecosystems and marine resources. Springer International Publishing, pp 285–358

Saha C, Pradhan S (2023) Microorganisms as effective CO_2 assimilator for biofuel production. In: Biofuel extraction teçhniques, pp 495–522

Sarwer A, Hamed SM, Osman AI, Jamil F, Al-Muhtaseb AH, Alhajeri NS, Rooney DW (2022) Algal biomass valorization for biofuel production and carbon sequestration: a review. Environ Chem Lett 20(5):2797–2851. https://doi.org/10.1007/s10311-022-01458-1

Shen M et al (2020) (Micro)plastic crisis: un-ignorable contribution to global greenhouse gas emissions and climate change. J Clean Prod 254:120138

Shuba ES, Kifle D (2018) Microalgae to biofuels: 'promising' alternative and renewable energy [review]. Renew Sustain Energy Rev 81:743–755. https://doi.org/10.1016/j.rser.2017.08.042

Stallinga P (2018) Carbon dioxide and ocean acidification. Eur Sci J

Suparmaniam U, Lam MK, Uemura Y, Lim JW, Lee KT, Shuit SH (2019) Insights into the microalgae cultivation technology and harvesting process for biofuel production: a review. Renew Sustain Energy Rev 115:109361. https://doi.org/10.1016/j.rser.2019.109361

Sydney EB et al (2019) Potential carbon fixation of industrially important microalgae. In: Biofuels from algae. Elsevier, pp 67–88

Tarafdar A, Sowmya G, Yogeshwari K, Rattu G, Negi T, Awasthi MK, Hoang A, Sindhu R, Sirohi R (2023) Environmental pollution mitigation through utilization of carbon dioxide by microalgae. Environ Pollut 328:121623. https://doi.org/10.1016/j.envpol.2023.121623

Thanigaivel S, Priya AK, Dutta K, Rajendran S, Vasseghian Y (2022) Engineering strategies and opportunities of next generation biofuel from microalgae: a perspective review on the potential bioenergy feedstock. Fuel 312:122827. https://doi.org/10.1016/j.fuel.2021.122827

Wahl M, Schneider Covachã S, Saderne V, Hiebenthal C, Müller JD, Pansch C, Sawall Y (2018) Macroalgae may mitigate ocean acidification effects on mussel calcification by increasing pH and its fluctuations. Limnol Oceanogr 63(1):3–21. https://doi.org/10.1002/lno.10608

Wynne MJ (1981) Phaeophyta: morphology and classification. In: Biology of seaweeds, pp 52–85

Yong WTL, Thien VY, Rupert R, Rodrigues KF (2022) Seaweed: a potential climate change solution. Renew Sustain Energy Rev 159:112222. https://doi.org/10.1016/j.rser.2022.112222

Zhang X, Xu D, Guan Z, Wang S, Zhang Y, Wang W, Zhang X, Fan X, Li F, Ye N (2020) Elevated CO_2 concentrations promote growth and photosynthesis of the brown alga *Saccharina japonica*. J Appl Phycol 32(3):1949–1959. https://doi.org/10.1007/s10811-020-02108-1

Zhu L, Nugroho YK, Shakeel SR, Li Z, Martinkauppi B, Hiltunen E (2017) Using microalgae to produce liquid transportation biodiesel: what is next? Renew Sustain Energy Rev 78:391–400. https://doi.org/10.1016/j.rser.2017.04.089

Zou P, Lu X, Zhao H, Yuan Y, Meng L, Zhang C, Li Y (2019) Polysaccharides derived from the brown algae *Lessonia nigrescens* enhance salt stress tolerance to wheat seedlings by enhancing the antioxidant system and modulating intracellular ion concentration. Front Plant Sci 10:48. https://doi.org/10.3389/fpls.2019.00048

Chapter 29
Algae-Based Bioenergy Production as a Carbon Mitigation Technology

Santosh Kumar and Makarand M. Ghangrekar

Abstract The looming threat of global warming from various anthropogenic interventions has become increasingly visible globally. The atmospheric carbon dioxide (CO_2) concentration, the most prevalent greenhouse gas, crossed 410 ppm for the first time in 2017. Alarmingly, this value further elevated to an all-time high of 420 ppm in January 2023. Even though extensive research has been conducted on developing efficient carbon capture and storage technologies, most are expensive, have a short working life, and impose high environmental impacts. On the other hand, algal cultivation is a green alternative for CO_2 sequestration into biomass with derived applications, such as feedstock for producing biofuel, biogas, enzymes, antioxidants, vitamins, biofertilizers, and other value-added products. Additionally, algae-based bioelectrochemical systems (BESs), like microbial carbon-capture cells and biophotovoltaics, are considered more advanced technologies owing to their multidimensional usages, including CO_2 sequestration, bioenergy production, wastewater treatment, and valuable product recovery. Therefore, the current chapter comprehensively describes algae-based carbon capture, recent advancements in algal cultivation, and associated roadblocks. Furthermore, this chapter also highlights algae-mediated CO_2 capture through BESs, process optimisation, and challenges in the large-scale implementation of these systems.

Keywords Algae · Carbon capture and storage · Greenhouse gas · Global warming · Microbial carbon-capture cells

S. Kumar · M. M. Ghangrekar (✉)
P. K. Sinha Centre for Bioenergy and Renewables, Indian Institute of Technology Kharagpur, Kharagpur 721302, India
e-mail: ghangrekar@civil.iitkgp.ac.in

M. M. Ghangrekar
Department of Civil Engineering, Indian Institute of Technology Kharagpur, Kharagpur 721302, India

T. A. Kurniawan and A. Anouzla (eds.), *Algae as a Natural Solution for Challenges in Water-Food-Energy Nexus*, Environmental Science and Engineering,
https://doi.org/10.1007/978-981-97-2371-3_29

29.1 Introduction

Climate change poses a critical threat to the environment, significantly endangering the survival of diverse plants and animals including humans. The exacerbation of climate change has been attributed to the persistent escalating emission of greenhouse gas (GHG), including carbon dioxide (CO_2), methane (CH_4), water vapor, nitrous oxide (N_2O), and fluorinated gases (Kweku et al. 2018). Among all the GHGs, CO_2 alone accounts for > 65% of total emissions (Malyan 2016) and it is the main GHG affecting the climate. The increasing concentration of atmospheric CO_2 also alters the aquatic ecosystem by raising acidity (Prasad et al. 2021). Thus, it is of great importance at present to formulate a suitable approach for mitigating or maintaining the level of atmospheric CO_2. The CO_2 concentration can be controlled by implementing a strategy involving a decline in fossil fuel utilisation coupled with an augmented adoption of renewable energy sources and implementing carbon capture and storage (CCS) technologies. Moreover, shifting toward bioenergy production might be a big move to reduce total CO_2 in the atmosphere.

The quest for sustainable and environmentally-friendly methods for bioenergy production and carbon mitigation has garnered considerable interest recently, owing to global warming and rising pollution rates. In this regard, algae-based bioenergy production has surfaced as a promising avenue for mitigating carbon emissions with concurrent wastewater treatment within the spectrum of feasible technologies (Vo Hoang Nhat et al. 2018; Kumar et al. 2023). Algae, commonly called "green gold," encompassing a wide range of photosynthetic organisms, fulfil a pivotal function in carbon cycling and provide a natural mechanism for carbon sequestration. Algae can sequester and retain atmospheric CO_2 via photosynthesis, transforming it into organic biomass (Onyeaka et al. 2021). The distinctive attribute of algae renders it an appealing contender for mitigating carbon emissions and reducing atmospheric CO_2 concentration. Moreover, algal biomass has been identified for its elevated lipids, carbohydrates, and protein levels, rendering it a highly suitable raw material for generating green fuels, including biodiesel, bioethanol, and biogas (Sarwer et al. 2022). Furthermore, biofuels derived from algae have the potential to function as environmentally sustainable substitutes for fossil fuels, leading to a decrease in greenhouse gas emissions and a reduced reliance on non-renewable energy resources.

Algal-based bioenergy production was started in the 1970s by the U.S. Department of Energy (DOE) with the aim of wastewater treatment via algal cultivation and biomethane production from the algal biomass (Mahmood et al. 2023). Further, the Aquatic Species Program (ASP), which was financially supported by the U.S. Department of Energy's Office of Fuels Development (DOE-OFD), was established with the primary objective of assessing the viability of utilizing microalgae with high lipid content for the production of biodiesel (Pienkos and Darzins 2009). This program specifically focused on cultivating microalgae biomass in ponds and harnessing the emitted CO_2 from coal-fired power plants as a carbon source. Several microalgal species were examined and studied under different physicochemical conditions, such as temperature, pH, and salinity. However, the expected outcomes still needed to be attained. The ASP accomplished two significant achievements. Firstly, it established a centre for the cultivation of microalgae culture collections. Secondly, it successfully implemented a pilot-scale microalgae cultivation raceway pond in New Mexico.

Many businesses and academic institutions are attempting to make the algae-based bioenergy system acceptable to the general public while the research process is still ongoing. Concurrently, a more sophisticated approach, commonly referred to as the algal bioelectrochemical system (A-BES), has been explored to investigate the potential for generating bioelectricity as an additional outcome. The A-BES, including microbial carbon-capture cells (MCCs) and biophotovoltaics (BPVs), are widely recognized for their enhanced efficiency in carbon mitigation, wastewater treatment, and biomass production for biofuel purposes. This chapter aims to serve readers with an understanding of the progress made in both traditional and contemporary methods of algal cultivation for biochemical conversion technologies, seeking to recover valuable biochemicals and bioenergy. The present chapter also discusses the utilisation of the BES system and its associated limitations to facilitate the preparation of this emerging technology for widespread deployment.

29.2 Carbon Dioxide Emission and Its Threat to the Environment

The escalating CO_2 emissions constitute a paramount environmental menace, primarily responsible for driving global climate change and inflicting adverse impacts on ecosystems, as illustrated in Fig. 29.1 (Ramirez-Corredores et al. 2023). Industrial revolution and urbanization skyrocketed the atmospheric CO_2 to unprecedented levels owing to burning fossil fuels, cutting down trees, and other industrial operations (Chandra Voumik and Sultana 2022). Being a potent greenhouse gas, an elevated concentration of atmospheric CO_2 captures heat, magnifying the inherent greenhouse effect and propelling a surge in global temperatures. The events, including heatwaves, droughts, storms, and floods, detrimentally affect ecosystems, agriculture, and human settlements (Clarke et al. 2022). Additionally, as CO_2 levels rise, ocean acidification occurs, endangering marine life, particularly coral reefs and shellfish. Indirectly, such disruption of marine ecosystems risks food security for communities reliant on seafood (Doney et al. 2020).

To combat these challenges several international agreements, such as the Paris Agreement (United Nations Climate Change Conference in Paris, France, on 12 December 2015), have been implemented to mitigate global warming and reduce greenhouse gas emissions (Branco-Vieira et al. 2022). Moreover, implementing renewable energy sources, enhancing energy efficiency, and advocating for sustainable land-use practices are imperative measures in mitigating carbon dioxide emissions. Nevertheless, climate change mitigation necessitates a collaborative endeavour, including enacting legislation by governmental bodies, adopting sustainable practices by industries, and the deliberate decision-making of people to prioritise

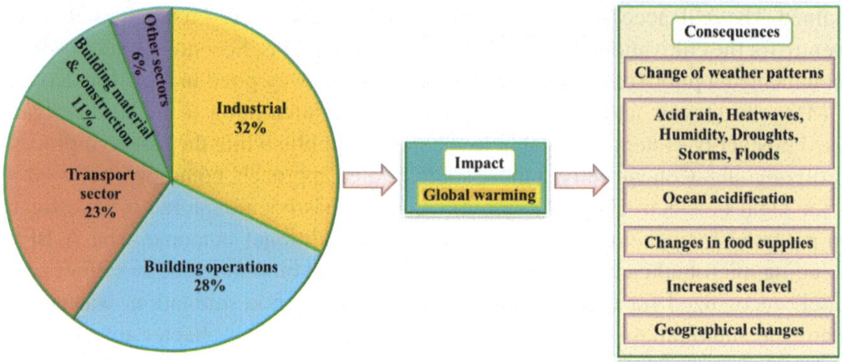

Fig. 29.1 Global CO_2 emissions by different sectors and its impacts and consequences on the environment

ecological considerations in their day-to-day activities. The potential consequences of not acting are severe since uncontrolled carbon dioxide emissions would persistently worsen climate change, posing a threat to the equilibrium of ecosystems and the welfare of current and future populations. Hence, it is essential to promptly and consistently undertake measures to mitigate CO_2 emissions to ensure our planet's long-term viability and liveability.

29.3 Algae for CO_2 Mitigation

The CO_2 mitigation is a comprehensive approach aimed at managing CO_2 emissions and regulating its atmospheric concentration. This involves adopting practices such as utilization of renewable energy sources, engaging in reforestation efforts, and implementing advanced CCS technologies to curb CO_2 emissions. In this regard, algae-based CO_2 bio-mitigation has emerged as a potent practice that simultaneously facilitates wastewater treatment and bioenergy and biochemical production (Nguyen et al. 2022; Saravanan et al. 2022).

Microalgae possess a specific cellular mechanism, referred to as the carbon concentration mechanism (CCM), which enables them to ingest CO_2. As per the CCM, the pyrenoid, a distinct cellular organelle, facilitates an elevation in the concentration of CO_2 around the thylakoid membrane inside the algal cell. The elevation in CO_2 levels in the vicinity of the thylakoid membrane serves as a stimulus for enhancing the functionality of ribulose-1,5-bisphosphate carboxylase/oxygenase (referred to as Rubisco), a crucial enzyme involved in photosynthesis, namely in the assimilation of carbon (Kumar et al. 2023). Microalgae are a kind of photosynthetic microorganisms that use CO_2 as their main source of carbon, outpacing terrestrial plants in their rate of CO_2 fixation by a factor of 100 (Farooq 2022). Theoretically, the production of 1 kg of algal biomass results in the capture and storage of about

1.88 kg of CO_2 (Adamczyk et al. 2016). In addition, it is noteworthy that algae have a remarkable ability to sequester CO_2, at a rate of 6.24 kg/m^3 day (Nagendranatha Reddy et al. 2019). Their significant contribution to oxygen synthesis is also remarkable, accounting for nearly 50% of the world's generation of oxygen (Chapman 2013).

The rate of CO_2 fixation depends on various factors including algal strain, CO_2 concentration in the cultivation medium, and physicochemical parameters (pH, temperature, light intensity, salinity, nutrient content, type of cultivation system). A connection between carbon capture rates and algal biomass productivity is shown by the data presented in Table 29.1. The results indicate that there is a positive correlation between greater carbon sequestration and enhanced algal biomass production. Additionally, several research have examined the possibilities of employing algae culture to capture flue gas pollutants including NO_X and SO_X (Yen et al. 2015). Therefore, the cultivation of algae for CO_2 mitigation arises as one of the most ecologically sustainable strategies to combat global warming.

29.3.1 Algal Strain Selection and Cultivation Methods for CO_2 Mitigation

Algae cultivation may be done without having a large environmental effect, unlike certain bioenergy feedstocks that compete with food production and cause deforestation (Mathimani and Pugazhendhi 2019). Algae-based bioenergy generation minimizes the change of natural ecosystems and eases the burden on freshwater resources by employing non-arable land or wastewater. However, algal cultivation requires optimal CO_2 level, generally recommended in the range of 0.038–10% (v/v) (Gabrielyan et al. 2022). In contrast, certain algal strain can thrive in elevated CO_2 concentrations (10–15%), albeit at the cost of reduced growth rates, carbon fixation, and productivity compared to optimal CO_2 concentrations (Paul et al. 2020). Some strains, such as *Chlorella* sp. ZY-1 and *Chlorella* sp. KR-1 have undergone evolutionary adaptations that enable them to thrive even in extremely high CO_2 concentrations of up to 70% (Sung 1999; Yue and Chen 2005). Notably, *Chlorella* sp. T-1 can withstand a remarkable 100% CO_2 environment (Maeda et al. 1995). Additionally, with a stepwise adaptation technique, *Chlorococcum littorale* has demonstrated the ability to tolerate 60% of CO_2 and continue growing (Viswanaathan et al. 2022). Thus, algal strain selection is necessary for efficient CO_2 mitigation and biomass production. Consequently, harnessing algae photosynthesis to reduce atmospheric CO_2 directly is considered advantageous and environmentally safe.

After strain selection, the next step is selecting a cultivation system for large-scale algal biomass production. Generally, two types of algal cultivation systems are popular: open pond and photobioreactor (PBR) cultivation systems (Gupta et al. 2015). Open pond cultivation is the oldest and most cost-effective method for large-scale algal cultivation, widely used in the industry owing to its affordability in

Table 29.1 Algal CO_2 fixation rate and biomass production at optimal CO_2 concentration

Algal strain	CO_2 fixation rate (mg/L/d)	CO_2 concentration (%)	Biomass productivity (mg/L/d)	References
Tetraselmis suecica	111.26	5	65.96	(Kassim and Meng 2017)
Chlorella sp.	95.73	5	59.33	
	96.89	15	56.22	
Chlorella vulgaris	835.7	Not reported	444.5	(Li et al. 2019)
Scenedesmus obliquus CNW-N	390.2	20	201.4	(Ho et al. 2010)
Scenedesmus obliquus AS-6-1	290.2		150.7	
Chlorella kessleri	Not reported	6	87	(Morais and Costa 2007)
Chlorella sorokiniana GS03	660	5	360	(Jin et al. 2021)
Heynigia riparia SX01	710	15	390	
Botryococcus braunii	496.98	Not reported	207.3	(Sydney et al.2010)
Spirulina platensis	318.61		145.3	
Dunaliella tertiolecta	272.4		143.3	
Chlorella vulgaris	251.64		129.3	
Scenedesmus obliquus	288	10	144	(Tang et al. 2011)
Chlorella pyrenoidosa	260		155	
Chlorella vulgaris	223.68	NR	121.57	(Hu et al. 2016)
	887.8		426.7	
Nannochloropsis oculata NCTU-3	Not reported	15	372	(Chiu et al. 2009)
		2	480	
		10	398	
		5	441	

construction, maintenance, and operation (Borowitzka and Moheimani 2013). Open ponds can be natural water bodies like lakes or artificial ones such as circular and raceway ponds, and sometimes tanks are also used for algae culture. However, natural water sources can yield lower algal concentrations, necessitating efficient harvesting techniques.

Challenges associated with open pond systems include environmental factors like rainwater runoff affecting salinity and pH, bank erosion causing leakage, and water turbidity impacting productivity (Jerney and Spilling 2018). Contamination by protozoa and bacteria is a concern, which can be mitigated by cultivating microalgae capable of surviving extreme conditions (Lam et al. 2018). Circular ponds, characterized by circular-shaped tanks, are an artificial option; however, their size limitations, energy usage, and construction costs are drawbacks. Raceway ponds, with closed-loop channels and paddlewheels for recirculation, are energy-efficient and highly regarded for commercial application.

The PBR is a closed bioreactor system for algal cultivation, preventing direct material exchange between the culture and the environment (Gupta et al. 2015). It offers advantages over open pond systems, including compact size, efficient land usage, contamination-free culture conditions, and high nutrient and metabolic efficiency (Gupta et al. 2015). However, its practical scalability is limited due to design flaws and associated high capital and operating costs (Chen et al. 2011). Tubular PBR consist of long, transparent tubes oriented in various ways to capture sunlight, however, they suffer from poor mass transfer due to concentration differences along the tubes.

Vertical column PBR use a vertical cylindrical tubing and sparger for efficient gas–liquid mass transfer, they lack adequate lighting and face high construction costs (Ugwu et al. 2008). Flat-plate PBR, with rectangular compartments and airlift systems for mixing, offer high photosynthetic efficiency however, can stress algal cells (Ugwu et al. 2008). Innovative designs like twin-layer and plastic sheets have been introduced to enhance efficiency.

29.3.2 Algae in Biorefinery

Algae, classified as photosynthetic organisms, possess the remarkable capability to harness solar energy and convert it into a diverse array of valuable products, thus playing a critical role as primary producers within ecosystems. The algae shoulder the vital responsibility of contributing to nearly half of the byproducts generated during the photosynthesis process. Consequently, substantial strides have been made in the field of algal bioprocesses over the past few decades. These advancements primarily focus on harnessing solar energy for the sustainable production of essential commodities, including carbohydrates, biofuels, pharmaceuticals, and innovative waste remediation techniques (Laurens et al. 2017).

The biorefinery concept came from the petroleum refineries, where multiple fuels and chemicals are extracted from crude oil as the initial feedstock. Similarly, an algal biorefinery is the legitimate processing of algal biomass into a wide array of commercially viable products and energy (Trivedi et al. 2015). Algae are increasingly being recognized as a multifaceted and environmentally-friendly resource within biorefinery. They offer solution to many persistent issues, such as ensuring energy stability, promoting environmental sustainability, and fostering the establishment of a circular economy (Trivedi et al. 2015). The algal biomass has a high essential components content, including proteins, carbohydrates, and lipids, making it a promising

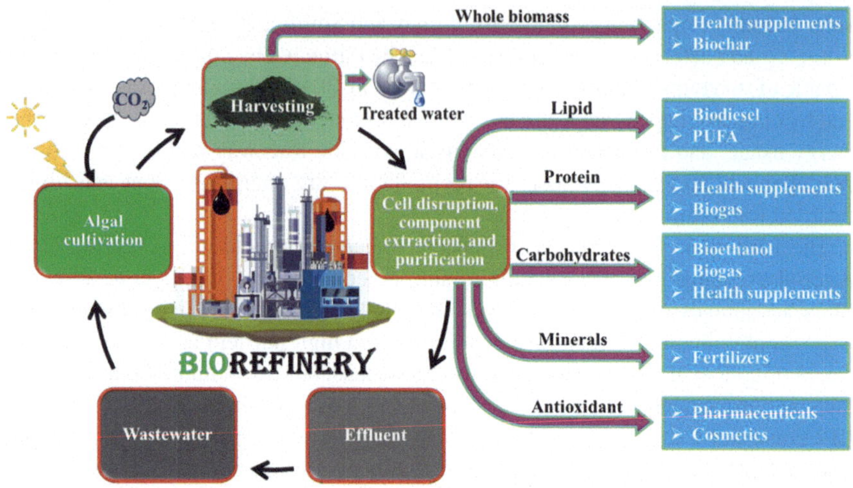

Fig. 29.2 Schematic blueprint of algal biorefinery

source material for producing algae biofuels and biochemicals (Yadav et al. 2022). However, commercial algal fuel production is challenging owing to the substantial energy and cost requirements associated with algal cultivation and harvesting (Yadav et al. 2022).

The occurrence of bottleneck issues is a prevalent challenge encountered during the pilot-scale processing of algae for fuel or chemical synthesis. The implementation of algal biorefinery idea has the potential to mitigate energy consumption and minimize expenses. In terms of the recovery of value-added bioproducts such as biopolymers, biochemicals, biofertilizers, and biofuels (as shown in Fig. 29.2), this approach proves to be more cost-effective when compared to traditional methods. Furthermore, this approach will reduce the total production costs, facilitating the establishment of a sustainable algae-based green economy.

For an instant, an investigation on an integrated algae-based biorefinery for biodiesel production demonstrate a remarkable cost reduction, dropping from $2.42/kg to $0.48/kg to additional extraction of astaxanthin and poly-hydroxybutyrate (García Prieto et al. 2017). Similarly, in a biorefinery experimental approach, astaxanthin was extracted from the biomass of microalga *Haematococcus pluvialis*, the residual was fermented to produce ethanol, again the fermented residual was anaerobically digested to produce biomethane. In this process 264.8 mL biomethane/g volatile solid was obtained, which was 2.9-fold greater than methane yield from raw microalgal biomass (Hosseini et al. 2020). In conclusion, in this approach of astaxanthin extraction followed by the production of ethanol and biogas from *H. pluvialis* biomass, yielding 45.8 g of astaxanthin per kg of unprocessed biomass and producing 70,953 kJ of energy (Hosseini et al. 2020). Thus, biorefinery approach could be the only hope to make algal system viable to obtain biodiesel and other concurrent product.

29.4 Algae-Based Bioelectrochemical System for CO_2 Mitigation and Energy Production

The algae-based bioelectrochemical system (A-BES) represents an advanced iteration of the conventional algal biorefinery, enabling the synchronized achievement of surplus power generation and high-concentration chemical oxygen demand (COD) wastewater treatment, in addition to all the standard algal biorefinery outcomes. What sets this system apart is its integration with a bioelectrochemical component, often consisting of MCC or BPV, also known as photomicrobial fuel cells or microbial solar cells. The A-BES facilitates the generation of electricity through the metabolic activities of the algae and/or other associated microorganisms, effectively turning CO_2 into a source of renewable energy. Moreover, A-BES can be implemented in various environments, including wastewater treatment plants, industrial facilities, and even in remote locations, contributing to localized CO_2 mitigation efforts and energy production.

29.4.1 Microbial Carbon-Capture Cells

The MCC is an upgraded iteration of the Microbial Fuel Cell (MFC) in which algal cultivation in the cathodic chamber provides O_2 as terminal electron acceptors (TEA) for cathodic reduction to support power generation (Jadhav et al. 2019). Moreover, the algal cultivation in the cathodic chamber of the reactor capture CO_2 during photosynthesis utilizing nutrient (nitrates and phosphate) present in the wastewater for their growth, thus treating wastewater (Ghangrekar et al. 2022). The algal biomass is used in biorefinery to produce biofuel and many other valuable products. Thus, MCC process serves to aid in the cathodic reduction reaction and the subsequent generation of bioenergy, as seen in Fig. 29.3.

In the anodic chamber of MCC, anaerobic bacteria facilitate the oxidation of organic pollutants found in wastewater, resulting in the production of CO_2 and the release of e^- and H^+ (Jadhav et al. 2017). Cultivating microalgae in the cathodic chamber facilitates CO_2 capture, nutrient (nitrate and phosphate) removal, and generating O_2 as an electron acceptor (Das et al. 2019). The H^+ migrate via the proton exchange membrane (PEM) towards the cathode, whilst the e^- are transported from the anode to the cathode through an external circuit. At the cathode, O_2 function as TEA and undergo reduction by e^- and H^+ to form H_2O, resulting in the simultaneous production of bioelectricity (Ghangrekar et al. 2022; Kumar et al. 2023). The comprehensive biochemical reactions taking place in the anodic and cathodic chambers of MCC are represented by Eqs. (29.1)–(29.4) (Saba et al. 2017; Jadhav et al. 2017):

At anode:

$$CH_3COO^- + 2H_2O \rightarrow 2CO_2 + 7H^+ + 8e^- \qquad (29.1)$$

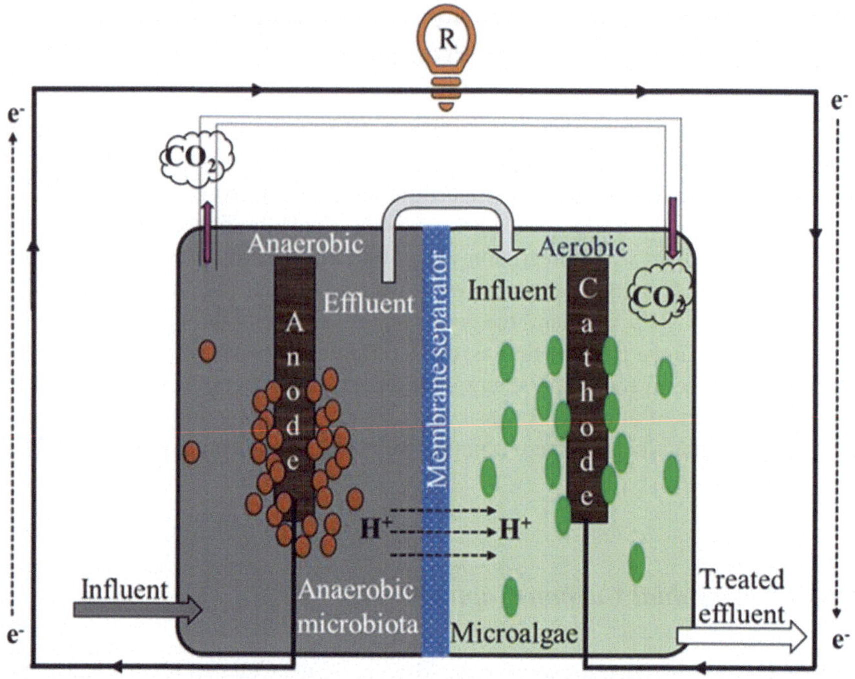

Fig. 29.3 A brief schematic representation of microbial carbon-capture cell

At cathode:

$$2O_2 + 8e^- + 8H^+ \rightarrow 4H_2O \text{ (At the cathode surface)} \tag{29.2}$$

$$nCO_2 + nH_2O \rightarrow (CH_2O)_n + nO_2 \text{ (Light-dependent (photosynthesis))} \tag{29.3}$$

$$C_2H_4O_2 + 2O_2 \rightarrow 2CO_2 + 2H_2O \text{ (Light independent (respiration))} \tag{29.4}$$

Currently, the research in the field of MCC has primarily concentrated on boosting performance (in terms of power generation and biomass yields) and optimising byproduct synthesis through enhancements in cathodic configurations and operational parameters. The power generation in a MCC is directly related to the rate of oxidation–reduction reaction (ORR). The ORR is a fundamental step for power generation in MFCs (Dange et al. 2022). The rate of ORR is regulated by factors such as electrode material, PEM, the efficiency of electron transfer from the microbial community to the cathode surface, cathode surface area, and the rate of algal growth that regulates O_2 concentration in the cathodic chamber (Arun et al. 2020; Dange et al. 2022). Thus, the algal growth rate is correlated not only to the rate of carbon capture, however, also related indirectly to power output by the MCC.

Low algal growth rate and biomass yield are one of the major limitations of MCC that hampers the biorefinery aspects. Algal growth can be improved by the modification of cathodic chamber of MCC. For instance, in a comparative investigation of a conventional MCC and an air-lift-type microbial carbon-capture cell (ALMCC), the ALMCC was reported to produce power generation of 972.5 mW/m^3 (46.47% higher than MCC) and a 9.98-folds higher CO_2 fixation rate than MCC (Hu et al. 2015). In the same investigation, the reported COD, ammonium nitrogen, and phosphorus removal by the ALMCC were 86.69%, 70.52%, and 69.24%, respectively; however, for MCC it was 82.45%, 60.29%, and 65.39%, respectively (Hu et al. 2015). Apart of reactor configuration the biomass yield of algae can be improved by the selection of robust algal strain, few past investigations reported in this regard are shown in Table 29.2.

The algal growth in the cathodic chamber of MCC also depends on the other parameters, such as light intensity, pH, temperature, and nutrient content in the culture medium. Thus, future research direction to make the MCC technology acceptable for practical application should be in the direction of optimisation of reactor configuration, anodic microbial and cathodic algal strain selection, electrode material selection and modification, possible application of nanomaterials as catalyst to improve ORR kinetics and microbial growth, and optimisation of physicochemical growth parameters.

29.4.2 *Biophotovoltaics*

Biophotovoltaics, often abbreviated as BPVs, represents an emerging discipline within the MFC domain that harnesses the power of photosynthetic microorganisms, such as algae, offering a biological solution for renewable energy generation (Tschörtner et al. 2019). The algae leverage their photosynthetic capabilities to split water molecules into H^+, e^-, and O_2, as illustrated in Fig. 29.4 (Wey et al. 2019). In a manner akin to MFCs, the generated H^+ ions through PEM migrate to the cathode. At the same time, electrons move to the anode and subsequently flow through an external circuit from the anode to the cathode. At the cathode, O_2 acts as TEA, being reduced by e^- and H^+ ions to H_2O to produce bioelectricity Fig. 29.4 (Zhu et al. 2023).

The overall BPV process can be categorized into two phases: charging and discharging. During the charging phases, the photosystem absorbs solar photons, leading to the generation of excited electrons. These electrons are then stored in high-energy intermediates, such as plastoquinol (PQH_2), NADPH, or organic molecules that arise from carbon fixation (McCormick et al. 2015; Saper et al. 2018). Subsequently, the electrons that have been captured are transferred from the interior of the cell to the external electrical circuit, resulting in the generation of a current. This step is often referred to as the discharging phase (McCormick et al. 2015).

Algae, functioning as natural bio-electrical catalysts, exhibit a remarkable capacity for self-sustenance and energy production by harnessing the power of

Table 29.2 Efficiency of MCC under different reactor configuration and algal strain

Algal strain	MCC configuration		COD removal efficiency (%)	Biomass production (mg/L/d)	Power generation (W/m^3)	References
Anabaena sparged	Dual-chambered flat plate mediator-less MFCs separated by an anion exchange membrane		NR	NR	57.8#	(Pandit et al. 2012)
Chlorella vulgaris	Double-chamber		NR	NR	5.6	(Wang et al. 2010)
Chlorella vulgaris	Double-chamber	Suspended algal culture	84.2	NR	1.325	(Zhou et al. 2012)
		Immobilized algal culture	84.8	NR	2.485	
Chlorella sorokiniana	Double-chamber, with a coconut shell-based proton exchange membrane		65.97 ± 0.83	272.2	3.2	(Neethu et al. 2018)
Chlorella pyrenoidosa	Double-chamber, having a low-cost clayware separator		87.3 ± 2.6	66.4 ± 4.7	6.4	(Jadhav et al. 2017)
Anabaena ambigua			82.4 ± 2.1	50.4 ± 3.9	4.29	
Mixed culture microalgae	Double-chamber		82 ± 2	830	7.12	(Neethu et al. 2020)
Scenedesmus dimorphus	Double-chamber		72	75.75 ± 9.2	0.38	(Mehrotra et al. 2021)
Chlorella vulgaris	Double-chamber		Ammonia nitrogen: 85.6; TN: 70.2; TOC: 93.9	14.7×10^5@	3.72	(Zhang et al. 2019)
Chlorella vulgaris	Double-chamber, immobilized algal culture		92.1	NR	2.573	(He et al. 2014)

Note MCC microbial carbon-capture cell; *COD* chemical oxygen demand; *NR* not reported; # mW/m^2; @ cells/mL d; *TN* total nitrogen; *TOC* total organic carbon

water and sunlight. In BPV, oxygen is liberated at the anode and released into the surrounding environment, while the water produced at the cathode can either be collected for various uses or evaporate harmlessly. Its minimal environmental impact sets this technique apart, as it does not yield hazardous chemical waste. The versatility of BPV technology opens the door to many applications, including water desalination and the co-production of valuable chemicals like formic acid (Soni et al. 2016; Anam et al. 2021). Formic acid, a product of BPV technology, holds immense promise as a sustainable fuel source (Soni et al. 2016). Additionally, this technology paves the way for hydrogen generation, contributing to the growing repertoire of

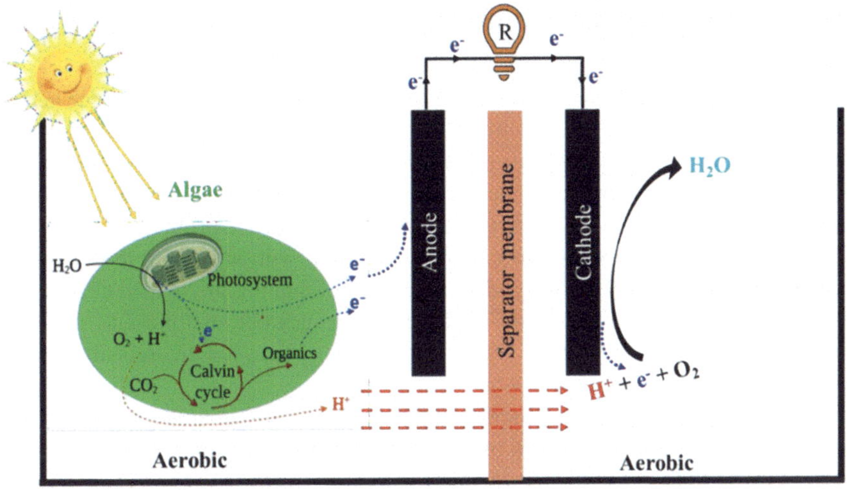

Fig. 29.4 A brief schematic representation of biophotovoltaics

clean energy solution (Soni et al. 2016). Thus, BPV can be a potential system to promote algal biorefinery.

The BPV systems have traditionally relied on immobilising photosynthetic cells on anodes or using external electron mediators to enable electron transfer. However, these approaches face limitations due to the low electron transfer capacity of cyanobacteria and microalgae, as well as a need for an understanding of intrinsic electron exchange mechanisms. As a result, power outputs in mediator-less BPV systems have been far below the theoretical maximum of 7 W/m^2 (McCormick et al. 2015; Sawa et al. 2017). While synthetic mediators can enhance electron transmission, they do not efficiently accept electrons from the photosynthetic chain and may introduce toxicity risks, increasing overall costs. Consequently, BPV technology faces practical challenges that hinder its widespread use. Overcoming these limitations will be crucial for realising the full potential of BPV systems as sustainable energy sources and reducing their dependence on exogenous mediators while enhancing their efficiency and safety.

Despite the potential improvements offered by abiotic engineering techniques, such as electrode modification and device optimization in enhancing the electrochemical interaction between live cells and electrodes, the power densities of BPV remain limited owing to the relatively low exoelectrogenic activity shown by algae. To overcome the low exoelectrogenic activity shown by algae, genetic modification of algal strains could be one of the possible approaches; however, no such investigation is still being reported in the literature as per the author's knowledge. In addition, the reactor modification, electrode and PEM material selection, and other physiochemical parameters optimisation are still required to upscale this system for biorefinery aspects.

29.5 Conclusion and Future Outlook

This chapter extensively explores the multidimensional applications of algae across a multitude of domains, ranging from biofuel production and power generation to carbon capture and storage, wastewater treatment, and the recovery of valuable products. It highlights how crucial algae are to solving today's problems (such as global warming, wastewater treatment, renewable energy source) and promoting innovation in a variety of industries. The increasing fascination with algae and their byproducts is being handled by skilfully identifying and evaluating the related challenges. Significantly, the chapter not only identifies challenges but also presents practical solutions, identifying the need of bridging the disparity between theoretical potential and practical application in real-world scenarios.

Algae have recently gained attention as a viable and sustainable resource for the production of renewable biofuels, medicinal chemicals, and important food components, due to their resemblance to microscopic bio-factories. The scope of their responsibilities includes carbon sequestration, the treatment of industrial effluent, and participation in the flourishing cosmeceutical business, which is fuelled by the growing customer preference for natural goods. Moreover, algae play a significant role in the synthesis of bioactive chemicals that exhibit a wide range of characteristics. In order to fully comprehend their potential, it is imperative to develop cost-effective and ecologically sustainable methods for algae production, since this is crucial for the successful commercialization of biofuels. Researchers advocate the incorporation of nanotechnology within microalgae biorefinery methodologies as a means to augment economic efficiency. They highlight the necessity of investigating, advancing, and consistently implementing this approach in real-world scenarios, with the aim of ushering in a transformative era of biofuels and microalgae products of significant value.

References

Adamczyk M, Lasek J, Skawińska A (2016) CO_2 biofixation and growth kinetics of *Chlorella vulgaris* and *Nannochloropsis gaditana*. Appl Biochem Biotechnol 179:1248–1261. https://doi.org/10.1007/s12010-016-2062-3

Anam M, Gomes HI, Rivers G et al (2021) Evaluation of photoanode materials used in biophotovoltaic systems for renewable energy generation. Sustain Energy Fuels 5:4209–4232. https://doi.org/10.1039/D1SE00396H

Arun S, Sinharoy A, Pakshirajan K, Lens PNL (2020) Algae based microbial fuel cells for wastewater treatment and recovery of value-added products. Renew Sustain Energy Rev 132:110041. https://doi.org/10.1016/j.rser.2020.110041

Borowitzka MA, Moheimani NR (2013) Open pond culture systems. Algae for biofuels and energy. Springer, The Netherlands, pp 133–152. https://doi.org/10.1007/978-94-007-5479-9_8

Branco-Vieira M, Lopes MPC, Caetano N (2022) Algae-based bioenergy production aligns with the Paris agreement goals as a carbon mitigation technology. Energy Rep 8:482–488. https://doi.org/10.1016/j.egyr.2022.01.081

Chandra Voumik L, Sultana T (2022) Impact of urbanization, industrialization, electrification and renewable energy on the environment in BRICS: fresh evidence from novel CS-ARDL model. Heliyon 8:e11457. https://doi.org/10.1016/j.heliyon.2022.e11457

Chapman RL (2013) Algae: the world's most important "plants"—an introduction. Mitig Adapt Strateg Glob Chang 18:5–12. https://doi.org/10.1007/s11027-010-9255-9

Chen C-Y, Yeh K-L, Aisyah R et al (2011) Cultivation, photobioreactor design and harvesting of microalgae for biodiesel production: a critical review. Bioresour Technol 102:71–81. https://doi.org/10.1016/j.biortech.2010.06.159

Chiu S-Y, Kao C-Y, Tsai M-T et al (2009) Lipid accumulation and CO_2 utilization of *Nannochloropsis oculata* in response to CO_2 aeration. Bioresour Technol 100:833–838. https://doi.org/10.1016/j.biortech.2008.06.061

Clarke B, Otto F, Stuart-Smith R, Harrington L (2022) Extreme weather impacts of climate change: an attribution perspective. Environ Res Clim 1:012001. https://doi.org/10.1088/2752-5295/ac6e7d

Dange P, Savla N, Pandit S et al (2022) A comprehensive review on oxygen reduction reaction in microbial fuel cells. J Renew Mater 10:665–697. https://doi.org/10.32604/jrm.2022.015806

Das S, Das S, Das I, Ghangrekar MM (2019) Application of bioelectrochemical systems for carbon dioxide sequestration and concomitant valuable recovery: a review. Mater Sci Energy Technol 2:687–696. https://doi.org/10.1016/j.mset.2019.08.003

de Morais MG, Costa JAV (2007) Isolation and selection of microalgae from coal fired thermoelectric power plant for biofixation of carbon dioxide. Energy Convers Manag 48:2169–2173. https://doi.org/10.1016/j.enconman.2006.12.011

Doney SC, Busch DS, Cooley SR, Kroeker KJ (2020) The impacts of ocean acidification on marine ecosystems and reliant human communities. Annu Rev Environ Resour 45:83–112. https://doi.org/10.1146/annurev-environ-012320-083019

Farooq W (2022) Maximizing energy content and CO_2 bio-fixation efficiency of an indigenous isolated microalga *Parachlorella kessleri* HY-6 through nutrient optimization and water recycling during cultivation. Front Bioeng Biotechnol 9. https://doi.org/10.3389/fbioe.2021.804608

Gabrielyan DA, Gabel BV, Sinetova MA et al (2022) Optimization of CO_2 supply for the intensive cultivation of *Chlorella sorokiniana* IPPAS C-1 in the laboratory and pilot-scale flat-panel photobioreactors. Life 12:1469. https://doi.org/10.3390/life12101469

García Prieto CV, Ramos FD, Estrada V et al (2017) Optimization of an integrated algae-based biorefinery for the production of biodiesel, astaxanthin and PHB. Energy 139:1159–1172. https://doi.org/10.1016/j.energy.2017.08.036

Ghangrekar MM, Kumar S, Ahmad A, Das S (2022) Concomitant bioenergy production and wastewater treatment employing microbial electrochemical technologies. In: Biofuels and bioenergy. Elsevier, pp 359–385. https://doi.org/10.1016/B978-0-323-85269-2.00015-0

Gupta PL, Lee S-M, Choi H-J (2015) A mini review: photobioreactors for large scale algal cultivation. World J Microbiol Biotechnol 31.1409–1417. https://doi.org/10.1007/s11274-015-1892-4

He H, Zhou M, Yang J et al (2014) Simultaneous wastewater treatment, electricity generation and biomass production by an immobilized photosynthetic algal microbial fuel cell. Bioprocess Biosyst Eng 37:873–880. https://doi.org/10.1007/s00449-013-1058-4

Ho S-H, Chen C-Y, Yeh K-L et al (2010) Characterization of photosynthetic carbon dioxide fixation ability of indigenous *Scenedesmus obliquus* isolates. Biochem Eng J 53:57–62. https://doi.org/10.1016/j.bej.2010.09.006

Hosseini A, Jazini M, Mahdieh M, Karimi K (2020) Efficient superantioxidant and biofuel production from microalga *Haematococcus pluvialis* via a biorefinery approach. Bioresour Technol 306:123100. https://doi.org/10.1016/j.biortech.2020.123100

Hu X, Liu B, Zhou J et al (2015) CO_2 fixation, lipid production, and power generation by a novel air-lift-type microbial carbon capture cell system. Environ Sci Technol 49:10710–10717. https://doi.org/10.1021/acs.est.5b02211

Hu X, Zhou J, Liu B (2016) Effect of algal species and light intensity on the performance of an air-lift-type microbial carbon capture cell with an algae-assisted cathode. RSC Adv 6:25094–25100. https://doi.org/10.1039/c5ra26299b

Jadhav DA, Jain SC, Ghangrekar MM (2017) Simultaneous wastewater treatment, algal biomass production and electricity generation in clayware microbial carbon capture cells. Appl Biochem Biotechnol 183:1076–1092. https://doi.org/10.1007/s12010-017-2485-5

Jadhav DA, Neethu B, Ghangrekar MM (2019) Microbial carbon capture cell: advanced bio-electrochemical system for wastewater treatment, electricity generation and algal biomass production. Appl Microalgae Wastewater Treat 317–338. https://doi.org/10.1007/978-3-030-13909-4_14

Jerney J, Spilling K (2018) Large scale cultivation of microalgae: open and closed systems, pp 1–8. https://doi.org/10.1007/7651_2018_130

Jin X, Gong S, Chen Z et al (2021) Potential microalgal strains for converting flue gas CO_2 into biomass. J Appl Phycol 33:47–55. https://doi.org/10.1007/s10811-020-02147-8

Kassim MA, Meng TK (2017) Carbon dioxide (CO_2) biofixation by microalgae and its potential for biorefinery and biofuel production. Sci Total Environ 584–585:1121–1129. https://doi.org/10.1016/j.scitotenv.2017.01.172

Kumar S, Priyadarshini M, Ahmad A, Ghangrekar MM (2023) Advanced biological and non-biological technologies for carbon sequestration, wastewater treatment, and concurrent valuable recovery: a review. J CO_2 Util 68:102372. https://doi.org/10.1016/j.jcou.2022.102372

Kweku D, Bismark O, Maxwell A et al (2018) Greenhouse effect: greenhouse gases and their impact on global warming. J Sci Res Rep 17:1–9. https://doi.org/10.9734/JSRR/2017/39630

Lam TP, Lee T-M, Chen C-Y, Chang J-S (2018) Strategies to control biological contaminants during microalgal cultivation in open ponds. Bioresour Technol 252:180–187. https://doi.org/10.1016/j.biortech.2017.12.088

Laurens LML, Markham J, Templeton DW et al (2017) Development of algae biorefinery concepts for biofuels and bioproducts; a perspective on process-compatible products and their impact on cost-reduction. Energy Environ Sci 10:1716–1738. https://doi.org/10.1039/C7EE01306J

Li M, Zhou M, Tan C, Tian X (2019) Enhancement of CO_2 biofixation and bioenergy generation using a novel airlift type photosynthetic microbial fuel cell. Bioresour Technol 272:501–509. https://doi.org/10.1016/j.biortech.2018.10.078

Maeda K, Owada M, Kimura N et al (1995) CO_2 fixation from the flue gas on coal-fired thermal power plant by microalgae. Energy Convers Manag 36:717–720. https://doi.org/10.1016/0196-8904(95)00105-M

Mahmood T, Hussain N, Shahbaz A et al (2023) Sustainable production of biofuels from the algae-derived biomass. Bioprocess Biosyst Eng 46:1077–1097. https://doi.org/10.1007/s00449-022-02796-8

Malyan S (2016) Nitrification inhibitors: a perspective tool to mitigate greenhouse gas emission from rice soils. Curr World Environ 11:423–428. https://doi.org/10.12944/CWE.11.2.10

Mathimani T, Pugazhendhi A (2019) Utilization of algae for biofuel, bio-products and bio-remediation. Biocatal Agric Biotechnol 17:326–330. https://doi.org/10.1016/j.bcab.2018.12.007

McCormick AJ, Bombelli P, Bradley RW et al (2015) Biophotovoltaics: oxygenic photosynthetic organisms in the world of bioelectrochemical systems. Energy Environ Sci 8:1092–1109. https://doi.org/10.1039/C4EE03875D

Mehrotra S, Kiran Kumar V, Man Mohan K et al (2021) Bioelectrogenesis from ceramic membrane-based algal-microbial fuel cells treating dairy industry wastewater. Sustain Energy Technol Assess 48:101653. https://doi.org/10.1016/j.seta.2021.101653

Nagendranatha Reddy C, Nguyen HTH, Noori MT, Min B (2019) Potential applications of algae in the cathode of microbial fuel cells for enhanced electricity generation with simultaneous nutrient removal and algae biorefinery: current status and future perspectives. Bioresour Technol 292:122010. https://doi.org/10.1016/j.biortech.2019.122010

Neethu B, Bhowmick GD, Ghangrekar MM (2018) Enhancement of bioelectricity generation and algal productivity in microbial carbon-capture cell using low cost coconut shell as membrane separator. Biochem Eng J 133:205–213. https://doi.org/10.1016/j.bej.2018.02.014

Neethu B, Tholia V, Ghangrekar MM (2020) Optimizing performance of a microbial carbon-capture cell using Box-Behnken design. Process Biochem 95:99–107. https://doi.org/10.1016/j.procbio.2020.05.018

Nguyen LN, Aditya L, Vu HP et al (2022) Nutrient removal by algae-based wastewater treatment. Curr Pollut Reports 8:369–383. https://doi.org/10.1007/s40726-022-00230-x

Onyeaka H, Miri T, Obileke K et al (2021) Minimizing carbon footprint via microalgae as a biological capture. Carbon Capture Sci Technol 1:100007. https://doi.org/10.1016/j.ccst.2021.100007

Pandit S, Nayak BK, Das D (2012) Microbial carbon capture cell using cyanobacteria for simultaneous power generation, carbon dioxide sequestration and wastewater treatment. Bioresour Technol 107:97–102. https://doi.org/10.1016/j.biortech.2011.12.067

Paul V, Chandra Shekharaiah PS, Kushwaha S et al (2020) Role of algae in CO_2 sequestration addressing climate change: a review, pp 257–265. https://doi.org/10.1007/978-981-32-9578-0_23

Pienkos PT, Darzins A (2009) The promise and challenges of microalgal-derived biofuels. Biofuels, Bioprod Biorefining 3:431–440. https://doi.org/10.1002/bbb.159

Prasad R, Gupta SK, Shabnam N et al (2021) Role of microalgae in global CO_2 sequestration: physiological mechanism, recent development, challenges, and future prospective. Sustainability 13:13061. https://doi.org/10.3390/su132313061

Ramirez-Corredores MM, Goldwasser MR, Falabella de Sousa Aguiar E (2023) Carbon dioxide and climate change. In: Hybrid nanomaterials for sustainable applications. Elsevier, pp 1–14

Saba B, Christy AD, Yu Z, Co AC (2017) Sustainable power generation from bacterio-algal microbial fuel cells (MFCs): an overview. Renew Sustain Energy Rev 73:75–84. https://doi.org/10.1016/j.rser.2017.01.115

Saper G, Kallmann D, Conzuelo F et al (2018) Live cyanobacteria produce photocurrent and hydrogen using both the respiratory and photosynthetic systems. Nat Commun 9:2168. https://doi.org/10.1038/s41467-018-04613-x

Saravanan A, Deivayanai VC, Senthil Kumar P et al (2022) CO_2 bio-mitigation using genetically modified algae and biofuel production towards a carbon net-zero society. Bioresour Technol 363:127982. https://doi.org/10.1016/j.biortech.2022.127982

Sarwer A, Hamed SM, Osman AI et al (2022) Algal biomass valorization for biofuel production and carbon sequestration: a review. Environ Chem Lett 20:2797–2851. https://doi.org/10.1007/s10311-022-01458-1

Sawa M, Fantuzzi A, Bombelli P et al (2017) Electricity generation from digitally printed cyanobacteria. Nat Commun 8:1327. https://doi.org/10.1038/s41467-017-01084-4

Soni RA, Sudhakar K, Rana RS (2016) Biophotovoltaics and biohydrogen through artificial photosynthesis: an overview. Int J Environ Sustain Dev 15:313. https://doi.org/10.1504/IJESD.2016.077391

Sung K (1999) CO_2 fixation by *Chlorella* sp. KR-1 and its cultural characteristics. Bioresour Technol 68:269–273. https://doi.org/10.1016/S0960-8524(98)00152-7

Sydney EB, Sturm W, de Carvalho JC et al (2010) Potential carbon dioxide fixation by industrially important microalgae. Bioresour Technol 101:5892–5896. https://doi.org/10.1016/j.biortech.2010.02.088

Tang D, Han W, Li P et al (2011) CO_2 biofixation and fatty acid composition of *Scenedesmus obliquus* and *Chlorella pyrenoidosa* in response to different CO_2 levels. Bioresour Technol 102:3071–3076. https://doi.org/10.1016/j.biortech.2010.10.047

Trivedi J, Aila M, Bangwal DP et al (2015) Algae based biorefinery—how to make sense? Renew Sustain Energy Rev 47:295–307. https://doi.org/10.1016/j.rser.2015.03.052

Tschörtner J, Lai B, Krömer JO (2019) Biophotovoltaics: green power generation from sunlight and water. Front Microbiol 10. https://doi.org/10.3389/fmicb.2019.00866

Ugwu CU, Aoyagi H, Uchiyama H (2008) Photobioreactors for mass cultivation of algae. Bioresour Technol 99:4021–4028. https://doi.org/10.1016/j.biortech.2007.01.046

Viswanaathan S, Perumal PK, Sundaram S (2022) Integrated approach for carbon sequestration and wastewater treatment using algal-bacterial consortia: opportunities and challenges. Sustainability 14:1075. https://doi.org/10.3390/su14031075

Vo Hoang Nhat P, Ngo HH, Guo WS et al (2018) Can algae-based technologies be an affordable green process for biofuel production and wastewater remediation? Bioresour Technol 256:491–501. https://doi.org/10.1016/j.biortech.2018.02.031

Wang X, Feng Y, Liu J et al (2010) Sequestration of CO_2 discharged from anode by algal cathode in microbial carbon capture cells (MCCs). Biosens Bioelectron 25:2639–2643. https://doi.org/10.1016/j.bios.2010.04.036

Wey LT, Bombelli P, Chen X et al (2019) The development of biophotovoltaic systems for power generation and biological analysis. ChemElectroChem 6:5375–5386. https://doi.org/10.1002/celc.201900997

Yadav K, Vasistha S, Nawkarkar P et al (2022) Algal biorefinery culminating multiple value-added products: recent advances, emerging trends, opportunities, and challenges. 3 Biotech 12:244. https://doi.org/10.1007/s13205-022-03288-y

Yen H-W, Ho S-H, Chen C-Y, Chang J-S (2015) CO_2, NO_x and SO_x removal from flue gas via microalgae cultivation: a critical review. Biotechnol J 10:829–839. https://doi.org/10.1002/biot.201400707

Yue L, Chen W (2005) Isolation and determination of cultural characteristics of a new highly CO_2 tolerant fresh water microalgae. Energy Convers Manag 46:1868–1876. https://doi.org/10.1016/j.enconman.2004.10.010

Zhang Y, Zhao Y, Zhou M (2019) A photosynthetic algal microbial fuel cell for treating swine wastewater. Environ Sci Pollut Res 26:6182–6190. https://doi.org/10.1007/s11356-018-3960-4

Zhou M, He H, Jin T, Wang H (2012) Power generation enhancement in novel microbial carbon capture cells with immobilized *Chlorella vulgaris*. J Power Sources 214:216–219. https://doi.org/10.1016/j.jpowsour.2012.04.043

Zhu H, Wang H, Zhang Y, Li Y (2023) Biophotovoltaics: recent advances and perspectives. Biotechnol Adv 64:108101. https://doi.org/10.1016/j.biotechadv.2023.108101

Chapter 30
Algae at Nexus of Eutrophication and Climate Change

Rukhsana Kausar

Abstract Algae, the plural form of alga, refers to an unconventional assembly of photosynthetic organisms that generally exhibit simpler morphologies compared to members of the plant kingdom, encompassing bryophytes, pteridophytes, and seed plants. While some algae exist as unicellular entities, others manifest as larger, multicellular organisms commonly known as "seaweeds."

30.1 Algae

Algae, the plural form of alga, refers to an unconventional assembly of photosynthetic organisms that generally exhibit simpler morphologies compared to members of the plant kingdom, encompassing bryophytes, pteridophytes, and seed plants. While some algae exist as unicellular entities, others manifest as larger, multicellular organisms commonly known as "seaweeds." The classification of algae primarily relies on distinctive plastid accessory pigments, with various groups identified by color, such as red, green, and brown. Contemporary taxonomic frameworks incorporate additional criteria like cell wall components, reproductive structures, and unique food storage molecules. Given their prevalence as primary producers in aquatic ecosystems, algae hold significant ecological importance.

30.2 Eutrophication

The word "Eutrophic" means well nourished (From Greek). Eutrophication is a result of aquatic condition with a high level of nutrient concentrations like Phosphorous and Nitrogen which causes enriched algal growth that may deteriorate water quality

R. Kausar (✉)
Department of Environmental Science, International Islamic University Islamabad, Islamabad, Pakistan
e-mail: rukhsana.tariq@iiu.edu.pk

T. A. Kurniawan and A. Anouzla (eds.), *Algae as a Natural Solution for Challenges in Water-Food-Energy Nexus*, Environmental Science and Engineering,
https://doi.org/10.1007/978-981-97-2371-3_30

in the aquatic ecosystems. (Bali Gueddari 2019; Sonarghare et al. 2019) Eutrophication usually happens as a part of normal natural process with the aging of an aquatic ecosystem when gradual increase happens the concentration of Phosphorous Nitrogen such as a lake. The productivity as wel as fertility of the system naturally increases as the amount of organic material that can be broken down into nutrient increases.

Eutrophication is recognized as a significant force influencing the modification and transformation of coastal areas globally. Many regions have initiated measures to reduce nutrient inputs in response to this phenomenon. However, the variability of nutrient concentrations in coastal waters over time and space poses a logistical challenge when assessing the impacts of these eutrophication mitigation efforts (Salo et al. 2022).

Additionally, factors influencing eutrophication extend beyond nutrient reduction measures. Other growth factors, such as the availability of limiting elements like sunlight, carbon dioxide, and nutrient fertilizers, contribute to excessive plant growth and increased requirements for photosynthesis (Songhare et al. 2019; Zhang et al. 2023).

Anthropogenic activities play a pivotal role in elevating nutrient levels, particularly nitrate and phosphate, in water bodies. This continuous influx surpasses the self-purifying capacity of these aquatic systems, leading to cultural eutrophication. The resultant acceleration of eutrophication beyond natural levels induces undesirable changes in the ecological system, marking a departure from its natural state (Sonarghare et al. 2020; Knight 2021).

The "ocean economy," constituting approximately 3%–5% of the worldwide GDP, with assets valued at $24 trillion (equivalent to around $74,000 per person in the US), has garnered attention. Prominent ocean advocates such as Al Gore, Michelle Bachelet, and Marc Benioff discussed this matter during the World Economic Forum (Prasad and Prasad 2019).

The greening or eutrophication of the Earth's lakes is projected to lead to a 30 to 90 percent rise in methane emissions into the atmosphere within the next century, according to researchers. This elevated methane release corresponds to 18–33% of the yearly carbon dioxide emissions resulting from the combustion of fossil fuels (Science news, University of Minnesota, March 2019).

Eutrophication, now acknowledged as a significant factor, contributes to changes in habitats and the expansion of harmful algal bloom species both in terms of geography and time. Cultural eutrophication, may be considered as a major global pollution issue resulting as a consequence of population growth and food production including all forms of agriculture and farming practices and power generation and utilization. By throwing sewage and runoff from agricultural and over populated areas to the oceans has also resulted in significant landscape alterations. The substantial surge in the use of chemical fertilizers and further escalation in the upcoming decades has contributed to eutrophication. While both nitrogen and phosphorus are concerns in this context, nitrogen garners more attention due to its role in limiting primary production in estuaries and coastal waters, along with its higher global application from synthetic fertilizers.

Global rises in population and food production, leading to eutrophication, exert substantial environmental effects, affecting the microorganisms in adjacent waters. Eutrophication is now acknowledged as a key factor driving shifts in habitats and the geographic and temporal expansion of harmful algal bloom (HAB) species.

The correlation between climate changes, nutrient changes, and the global increase in eutrophication and algal blooms is indisputable. While there are dissenting voices, the prevailing consensus is that the proliferation of Algae is significantly linked to nutrient pollution and climate shifts. The human activities, including primarily nutrient pollution and the effect of climate change, may be regarded as the primary contributors, to this expansion. Climate projections even indicate an escalation in the occurrence of events like red tides, exemplified by the persistent occurrences in Florida in recent years.

In essence, the eutrophication-induced transformation of lakes and its collaboration with human activities, such as nutrient pollution and climate change, are instrumental drivers in the global spread of harmful algal blooms.

30.3 Algae at Nexus of Eutrophication and Climate Change

Elevated water temperatures and increased nutrient concentrations serve as stimuli for the growth of algae, resulting in water eutrophication and the development of algal blooms. The rise in phosphorous and nitrogen concentrations in lakes, rivers, and estuaries leads to an increased dominance of cyanobacteria.

30.3.1 Effect of Temperature

In the context of climate change, temperature plays a crucial role. Cyanobacteria, often known as "blue-green algae" due to their color, thrive in various water types and utilize sunlight for sustenance. They exhibit rapid growth in warm, nutrient-rich environments. Water bodies with frequent cyanobacteria blooms exhibit characteristics that can negatively impact water quality and the surrounding environment, particularly when water temperatures exceed 25 °C (Hudnell 2008; Havens 2008) Moreover, higher temperatures can potentially trigger earlier and prolonged periods of algal blooms, directly influenced by the warming environment. These warmer temperatures also affect nutrient loadings from soil and sediment, influencing the trophic status of water (Dale et al. 2006). Additionally, as surface water warms, decreased viscosity leads to increased nutrient diffusion towards the surface, causing larger phytoplankton to sink and leading to a higher abundance of cyanobacteria (Peperzak 2003; Reynolds 2006)

In summary, the escalation of air temperature contributes to increased water temperature, exacerbating water quality issues by accelerating the eutrophication process in water bodies, posing environmental and health-related concerns (Nazari-Sharabian et al. 2018)

30.3.2 Precipitation

In regions expecting heightened precipitation, the likelihood of intense extreme precipitation events is conceivable, potentially causing increased erosion and re-suspension of sediments. This, in turn, results in elevated concentrations of sediments and nutrients in receiving water bodies (Whitehead et al. 2009). Additionally, these extreme events could amplify contaminant discharge, impacting non-point pollution by mobilizing pollutants across land and increasing nutrient concentrations in receiving water bodies, ultimately leading to water quality degradation.

Conversely, reduced precipitation poses a risk of eutrophication by diminishing minimum flows. In such instances, the decreased water volume for pollutant dilution could lead to heightened contaminant concentrations, contributing to oxygenation, as manifested by a decrease in dissolved oxygen concentration (DO) and an increase in biochemical oxygen demand (BOD). Consequently, the eutrophication risk, particularly in water bodies with limited re-aeration capacity, is heightened. Therefore, in the context of climate change conditions and the alteration of regional precipitation patterns, water bodies are susceptible to increased nutrient loads, ultimately culminating in the deterioration of water quality.

30.3.3 Wind

Climate change will impact the wind, leading to direct and indirect consequences on water resources (Reynolds 2006). The direct effects involve the blowing of algae from the water surface to lakeshores or riverbanks, causing algal blooms and altering environmental conditions. Indirectly, wind-induced disturbances can circulate water, mixing different layers of the water column, facilitating the blending of nutrients, and expediting the release of nutrients from sediments. Additionally, as air temperature rises, the wind mixes warmer upper layers of water with colder lower layers, accelerating the volatilization, migration, and transformation of pollutants (Xia et al. 2012).

30.3.4 Solar Radiation

Global warming and solar radiation are interconnected phenomena (Frey et al. 2011). Serving as a crucial energy source, solar radiation plays a pivotal role in photosynthesis across various ecosystems, influencing the growth of phytoplankton and other aquatic species. Consequently, the efficiency of photosynthesis relies on the temporal and spatial variations of solar radiation. Adequate sunlight, combined with nutrient presence, creates favorable conditions for the growth of algae and phytoplankton, leading to water eutrophication (Ihnken et al. 2010).

Solar radiation's impact extends to a wide array of living organisms by permeating aquatic systems and serving as the energy source for plant photosynthesis. In instances where plants lack sufficient sunlight, they absorb oxygen from the water, resulting in a depletion of dissolved oxygen (DO). Under anaerobic conditions, phosphorus release from sediments can contribute to water eutrophication (NEP China 2012; Nesa et al. 2012). Algae distribution is also influenced by the intensity of solar radiation at different depths. However, increased sunlight does not necessarily lead to a proportional increase in algae growth, as algae exhibit a maximum growth rate, and beyond this threshold, the growth rate diminishes (Craig et al. 2014).

The scientific consensus is evident: we must not solely decrease carbon emissions to prevent irreversible climate change impacts and constrain global warming. Additionally, it's imperative to extract carbon from the atmosphere to align with the specified 1.5 °C target of the Paris Climate Agreement. Surprisingly, algae, often observed on pond surfaces and linked to detrimental algal blooms causing marine life loss, might provide a significant solution for tackling climate change. Similar to trees, algae employ carbon dioxide for photosynthesis, effectively capturing CO_2 during its growth.

30.4 Algae as a Carbon Sequestration Tool

Projections concerning climate change notably indicate that global warming will surpass 1.5 °C by around 2050. Simultaneously, the world's population is expected to grow, intensifying demands for essential resources. This exacerbates nutrient pollution issues. Patricia M. Glibert has highlighted the intricate impacts of eutrophication and climate on harmful algae, affecting both harmful and non-harmful species at a physiological level.

In a future ocean scenario, non-harmful diatoms may face disproportionate stress, while mixotrophic organisms could gain advantages due to changes in nutrient composition, temperature, stratification, and oceanic pH.

The shifting pattern of climate change and nutrient dynamics is mainly contributing to the global expansion of eutrophication and the broader presence of harmful algal blooms (HABs). These blooms are occurring more abundantly, persisting longer, and appearing in new and diverse locations across freshwater and marine environments. They bring forth a range of consequences and toxicities in the oceans.

30.5 Algae at Eutrophication and Climate Change Nexus.

The research articles of Tina solo and Sonja Salovius (July 22) has discussed the significant repercussions of eutrophication on aquatic ecosystems, specifically focusing on the Baltic Sea, and sheds light on the intricate dynamics of harmful algal blooms (HABs). It delves into several critical aspects related to how excessive nutrient input, predominantly nitrogen and phosphorus from terrestrial sources, can have detrimental effects on water quality, biodiversity, and the delicate ecological equilibrium. It also explores how environmental factors like stratification and local geographical features influence the distribution and accumulation of algal species those are more harmful. The main Impacts on environment can be summarized as follows:

30.5.1 Impact of Eutrophication

Eutrophication, triggered by an oversupply of nutrients, particularly nitrogen and phosphorus, originating from sources on land, leads to heightened growth of various types of algae, including micro and macro algae. This excessive proliferation can obstruct sunlight penetration and disrupt the habitats of larger aquatic plants (macrophytes). The result is diminished light availability, which impacts the ability of predators to catch prey.

30.5.2 Hypoxia and Dead Zones

The decomposition of organic matter resulting from the escalated algal growth can deplete oxygen levels in the water, culminating in hypoxia, a condition marked by insufficient oxygen supply. This is detrimental to organisms dwelling on the seafloor, often causing mortality and disturbing the aquatic ecosystem's balance.

30.5.3 Eutrophication in the Baltic Sea

An example of vulnerability of the Baltic Sea to eutrophication, is a pressing environmental concern. Although measures have been taken to mitigate nutrient inputs, challenges persist. The situation could exacerbate due to amplified precipitation and the influence of climate change.

30.5.4 Understanding Harmful Algal Blooms (HABs)

The text explores the underlying factors driving the formation of HABs. Elements such as imbalanced nutrient ratios, prevailing environmental conditions, interactions with other organisms, and the role of physical dynamics collectively contribute to fostering or curtailing the growth of harmful algal species. Furthermore, the text points out that these HABs tend to aggregate in specific regions, often where light and nutrient conditions intersect.

30.5.5 Distorted Nutrient Ratios

Modifications in nutrient usage patterns, notably a pronounced increase in nitrogen-based fertilizers compared to phosphorus, can lead to skewed nutrient loads within aquatic ecosystems. This imbalance can pave the way for altered ecological dynamics and the prevalence of specific algal species.

30.5.6 Interaction with Physical Factors

The presence of physical aspects like water stratification and coastal features significantly influences the dispersion and accumulation of harmful algal species. These species can amass in thin layers at locations where light and nutrient conditions converge. So there is an intricate interplay between nutrient pollution, physical elements, and ecological responses in aquatic environments, with a specific focus on the phenomenon of harmful algal blooms. A intricate interconnection While findings from the IPCC in 2018 and the US National Climate Assessments (Reidmiller et al. 2018) can deduce, by employing constructs of numerous models depict that upto 2050 considerable warming will transpire, while predictions concerning shifts in nutrient dynamics are far less definitive.

Not only do uncertainties exist within climate models, as each model inherently harbors its individual bias or uncertainty, but the uncertainties related to scenarios are even more substantial (e.g., Freer et al. 2018), especially regarding nutrient changes. IPCC has devised Millennium Ecosystem Assessment (MA) to foresee future eutrophication using a range of assumptions in the Four distinct scenarios, grounded in storylines devised by the IPCC. These studies has translated into alterations in primary anthropogenic drivers, economic growth, demography, and agricultural output (Alcamo et al. 2006).

Despite the observable trend of nutrient utilization and discharge into aquatic systems having seemingly surged over the past few decades, it is expected that the amount of nutrients might be causing less pollution in some areas. For example declines may be expected in Europe, and stabilization could be projected in North America and Australia. The scope of regional nutrient reductions, which falls under the purview of local management decisions, showcases substantial variability. The industrialized nations has initiated a decrease in the mid of 1980 and to mid of 1990s, in phosphorus use, that is achieved by its removal from detergents. They have made more improvements in sewage treatment processes, particularly more efficient in P (Phosphorous) removal than nitrogen (Litke 1999; Van Drecht et al. 2009).

In the United States, where fertilizer and manure application is monitored at the state level, it is evident that the proportional use of nitrogen (N) and phosphorus (P) varies significantly among states. Moreover, many states have reduced fertilizer application from 2007 to 2011 compared to the period between 2002 and 2006. Interestingly, more states have implemented measures to control P usage rather than N. While some states have increased the application of either N or P, a larger number have witnessed an increase in N use relative to P. China also demonstrates a significant investment in environmental concerns, including nutrient reduction strategies (Glibert et al. 2013; 2014a; Sutton et al. 2013; Bouwman et al. 2017).

The impact of a warmer and more extreme climate, characterized by high temperatures, extends to the habitat for algal blooms (HABs). Rising temperatures are expected to favor species with higher temperature preferences for growth while negatively affecting those favoring lower temperatures. Diatoms, categorized as "Cold-water flora," typically thrive in lower temperature conditions, while many cyanobacteria and dinoflagellates, including HAB species, flourish in warmer environments. However, the growth of these species is influenced by various environmental factors, not solely temperature optimization.

Temperature not only promotes the growth of certain algae, phytoplankton, and dinoflagellates but also has the capacity to induce changes in their chemical composition. Elevated levels of CO_2 can lead to alterations in cellular chemical composition, resulting in heightened carbon-to-nutrient ratios within phytoplankton. These changes have cascading effects on trophic efficiency, food quality, and ultimately, the structure of food webs. (Urabe et al. 2003; Schoo et al. 2013).

In oceanic contexts, warm and nutrient-depleted waters often harbor small cells, while larger cells are associated with cooler, more nutrient-rich waters. Changes in precipitation patterns, such as increased drought occurrences and heightened storm events, may influence the expansion of harmful algal species. Extreme precipitation events and subsequent floods induce modifications in coastal ecosystems, fostering conditions conducive to harmful algal blooms due to the substantial influx of runoff.

The combination of amplified freshwater input and elevated temperatures heightens the likelihood and suitable habitats for specific types of HABs. Anticipated climate shifts are expected to transform seasonality, leading to extended springs and autumns as well as warmer summers in some regions. The impact of increased CO_2 levels on algae growth varies among different species, and changes in pH due to elevated CO_2 levels can alter HAB habitats. pH decreases can favor cyanobacteria over eukaryotic algae, leading to shifts in HAB composition.

The prediction that smaller cells are favored in a warming world is contingent on nutrient conditions, and the complex interactions of pCO_2, temperature, and nutrient supply complicate the understanding of pCO_2 effects on HAB toxicity. Efforts to describe the impacts of eutrophication and climate change, either separately or in combination, on HABs have primarily focused on HAB species. (e.g., Davis et al. 2009; Fu et al. 2012).

References

Bali M, Gueddari M (2019) Removal of phosphorus from secondary effluents using infiltration–percolation process. Appl Water Sci 9(54):1–8. https://doi.org/10.1007/s13201-019-0945-5

Blaas H, Kroeze C (2014) Possible future effects of large-scale algae cultivation for biofuels on coastal eutrophication in Europe, Affiliations expand, PMID: 25058933. https://doi.org/10.1016/j.scitotenv.2014.06.131. Green algae as bio indicators for long-term nutrient pollution along a coastal eutrophication gradient

Craig EW, Zepp RG, Lucas RM, Madronich S, Austin TA, Ballare CL, Norval M, Sulzberger B, Bais AF, McKenzie RL, Robinson SA, Hader DP, Paul ND, Bornman JF (2014) Solar ultraviolet radiation in a changing climate. Nat Clim Chang 4:434–441

Dale B, Edwards M, Reid PC (2006) Climate change and harmful algal blooms. In: Ecology of harmful algae, ecological studies (Analysis and synthesis), vol 189. Springer, Heidelberg, pp 367–378

Frey K, Perovich DK, Light B (2011) The spatial distribution of solar radiation under a melting Arctic sea ice cover. Geophys Res Lett 38(22):1–6

Glibert PM (2020) Harmful algae at the complex nexus of eutrophication and climate change

Havens KE (2008) Cyanobacteria blooms: effects on aquatic ecosystems. Adv Exp Med Biol 619:733–747

Hudnell HK (2008) Cyanobacteria harmful algal blooms: state of the science and research needs. Springer, Heidelberg

Ihnken S, Eggert A, Beardall J (2010) Exposure times in rapid light curves affect photosynthetic parameters in algae. Aquat Bot 93(3):185–194

Nazari-Sharabian N, Sajjad A, Karakouzian M (2018) Climate change and eutrophication: a short review. Eng Technol Appl Sci Res 8(6):3668–3672

NEP (2012) National "12th Five-Year Plan" for environmental protection. China Environmental Science Press, Beijing, China

Nesa B, Baird AH, Harii S, Yakovleva I, Hidaka M (2012) Algal symbionts increase DNA damage in coral planulae exposed to sunlight. Zool Stud 51(1):12–17

Paerl HW, Otten TG (2013) Blooms bite the hand that feeds them. Science 342(6157):433–434

Paerl HW, Valdes LM, Joyner AR, Peierls BL, Piehler MF, Riggs SR, Christian RR, Eby LA, Crowder LB, Ramus JS, Clesceri EJ, Buzzelli CP, Luettich RA (2006) Ecological response to hurricane events in the Pamlico sound system, NC and implications for assessment and management in a regime of increased frequency. Estuaries Coasts 29(6):1033–1045

Peperzak L (2003) Climate change and harmful algal blooms in the North Sea. Acta Oecologica 24(1):139–144

Rodgers EM (2021) Adding climate change to the mix: responses of aquatic ectotherms to the combined effects of eutrophication and warming, Published: 27 October 2021. https://doi.org/10.1098/rsbl.2021.0442

Reynolds CS (2006) Ecology of phytoplankton (Ecology, biodiversity and conservation). Cambridge University Press, Cambridge

Salo T, Salovius-Laurén S (2020) Ecological indicators, v.140, p 109034. ISSN: 1470-160X

Sonarghare PC, Masram SC, Sonparote UR, Khaparde KP, Kharkate SK (2020) Causes and effects of Eutrophication on aquatic life (a review). Int J Env Rehab Conserv XI(SP2):213–218

Tong Y, Sun J, Uddin M (2021) Perspectives and challenges of applying the water-food-energy nexus approach to lake eutrophication modelling. Water Security 14(7749):100095. https://doi.org/10.1016/j.wasec.2021.100095

Valdemarsen T, Quintana CO, Flindt MR, Kristensen E (2015) Organic N and P in eutrophic fjord sediments—rates of mineralization and consequences for internal nutrient loading. Biogeosciences 12:1765–1779

Whitehead PG, Wilby RL, Battarbee RW (2009) A review of the potential impacts of climate change on surface water quality. Hydrol Sci J 54(1):101–123

Xia R, Chen Z, Zhou Y (2012) Impact assessment of climate change on algal blooms by a parametric modeling study in Han River. J Resources Ecol 3(3):209–219

Zhang H, Chen J, Haffner G (2023) Plateau Lake water quality and eutrophication: status and challenges. Water (Basel) 15(337):1–6. https://doi.org/10.3390/w15020337

Zhang H, Lin Y, He Y, Chen J (2023) Plateau Lake water quality and Eutrophication : status and challenges. Water (Basal) 15(337):1–6

Chapter 31
Algae for Environmental Sustainability: Trends and Future Outlook

Mariany Costa Deprá, Rosangela Rodrigues Dias, Leila Queiroz Zepka, and Eduardo Jacob-Lopes

Abstract Microalgae are a potential renewable source that can contribute to sustainable development to provide and design a sustainable living environment for modern society. However, although these biofactories hold great promise for sustainable production, the transition of technological maturity as well as its consolidation on an industrial scale presents numerous challenges that still need to be overcome. Thus, a promising strategy that can serve as a springboard to leverage its advancement is the use of innovative technologies, such as the use of resources linked to information technology. In view of this, this chapter aims to emphasize the advancement of microalgal biotechnology in environmental management through the incorporation of technological innovations such as artificial intelligence. In addition, the main challenges inherent in its consolidation process towards a greener industry were examined, as well as competition with other sources of food additives, social acceptance, cost viability, environmental compensation, safety, and compatibility.

Keywords Microalgae · Technological innovations · Environmental indicators · Sustainability approaches

31.1 Introduction

The demand and dependence enhancement on raw materials from natural reserves, accompanied by large amounts of pollutants released into the environment, has seriously threatened the ecological balance and human survival (Zhao et al. 2023). This widespread problem has intensified the scarcity of natural resources beyond enhancing climate change, putting future generations at risk (Oliveira et al. 2022). Thus, in order to avoid the depletion of conventional non-renewable sources,

M. C. Deprá (✉) · R. R. Dias · L. Q. Zepka · E. Jacob-Lopes
Bioprocess Intensification Group, Federal University of Santa Maria, UFSM, Roraima Avenue 1000, Santa Maria, RS 97105-900, Brazil
e-mail: marianydepra@gmail.com

T. A. Kurniawan and A. Anouzla (eds.), *Algae as a Natural Solution for Challenges in Water-Food-Energy Nexus*, Environmental Science and Engineering,
https://doi.org/10.1007/978-981-97-2371-3_31

the world and modern society were pressured to explore and develop economically efficient, socially equitable, and environmentally rational strategies to achieve sustainable development (Kholssi et al. 2023).

Among the proposed solutions, biotechnological process development, from which products with high added value can be obtained, using non-valuable or unexplored industrial flows, has become a primary objective in sustainability and circular economy context (Barbosa et al. 2023). It is believed that the various peculiarities of innovation in biotechnology and biomass for generation of cleaner goods, services, and or energy indicate prosperous insights in the face of sustainability transitions (Losacker et al. 2023). Therefore, given the current competitiveness of biotechnological bioresources, the relevance of microalgae as sustainable cellular biofactories has emerged as a silver bullet for modern society.

In fact, it is known that, since the dawn of humanity, man has used microalgae as a food bioresource. However, their constant evolution and adaptation over billions of years caused the cellular mechanisms of microalgae to present diversity and complexity that allow them a range of applications. Since then, research on the development of microalgae and its contribution to economic production has been carried out, enabling various products and processes to be obtained and exploited through its renewable raw material (Calijuri et al. 2022).

However, even though the scientific and industrial community has worked exhaustively on its technological development, the optimistic values of productivity and process control based on microalgae are still challenges to be overcome when considering the large-scale production and dissemination of bioproducts in the microalgae market. Furthermore, the high energy requirements, the use of nonrenewable resources, as well as high sensitivity to environmental disturbances put the real sustainability referred to this biotechnology to the test. As a result, its consolidation in markets and the transfer of technologies remain a difficult undertaking to achieve (Bröring and Vanacker 2022). Thus, comprehensive analyzes focusing on bioprocesses, environmental sustainability, and economical cultivation are essential to establish the current status and identify knowledge gaps and opportunities in microalgae biotechnology.

At the same time, intelligent systems enabled by artificial intelligence (AI) have been recognized as promising strategies to face the challenges of sustainable development since they make it possible to analyze interconnected databases on a large scale to develop joint actions aimed at preserving the environment, economic sector and social (Vinuesa et al. 2020). Thus, when integrated into microalgae-based processes and products, this computational technology allows the identification and screening of microalgae species and strains, monitoring and management of microalgae cultivation, and microalgae conversion (Teng et al. 2020). In this way, artificial intelligence (AI) algorithms provide guidance to decision-makers as they offer innovative strategies to assess, predict and regulate sustainability uncertainties in microalgal biotechnology (Oruganti et al. 2023).

Against this background, this chapter aims to emphasize the advancement of microalgae biotechnology in environmental management through technological innovation incorporation as artificial intelligence algorithms. In addition, the main challenges inherent to its consolidation process towards a greener industry were examined, considering its competition with other sources of food additives, social acceptance, cost viability, environmental compensation, safety, and compatibility.

31.2 Microalgae as Multi-functional Options in Sustainable Modern Society

Today, biomanufacturing and environmental sustainability are areas of intense research in modern society. The production of chemical bioproducts, biomaterials, and bioenergy is based on cleaner industrial practices, which would ensure social well-being concomitant with global economic growth (Whitford et al. 2022; Akaniro et al. 2023).

Microalgae offer fascinating highlights to enable them as elective raw materials with multiple functions in your biomanufacturing process. These cellular biofactories are seen as a key cog in the machinery of how the planetary ecosystem can maintain its dynamic balance. Therefore, exploring biomass and bioprocesses can be a potential alternative to meet environmental ambitions and achieve sustainable development goals (Bhattacharya and Goswami 2020).

Food is essential to sustain life. However, recent estimates indicate that 1 in 9 people in the world has symptoms that indicate malnutrition, the most important factor being protein malnutrition (Almomani et al. 2023). On the other hand, conventional sources of protein, more precisely animal protein, have faced serious criticism regarding their mode of production. This is because the livestock system has been referred to as responsible for more than 15% of the total emissions of the food production chain (~ 30% of global emissions), representing about 7100 $MtCO_{2\text{-eq}}$/yr (mega tonne of carbon dioxide equivalent per year) (Deprá et al. 2022). Thus, since there is a need to expand protein production to values greater than 50% to meet the growing global population (~ 10 billion people by 2050), debates about the excessive use of natural resources, moral issues related to animal welfare, in addition to potential problems in human health, represent an unprecedented threat. Current agricultural approaches are being significantly affected by climate change, and inefficient food distribution systems are unable to cover current needs and deal with increased nutritional demands for the future (Helmy et al. 2023).

Influenced by this paradigm, microalgae have been placed as a potential food solution for this sustainable food crisis. However, for these microorganisms to be considered as new potential food sources, crucial factors such as their composition and nutritional content needed to be investigated. Pioneering studies report that species such as *Arthrospira platensis* have up to 70% of their biomass as protein content. Other strains, such as *Chlamydomonas reinhardtii*, *Auxenochlorella protothecoides*, *Chlorella vulgaris*, *Dunaliella bardawil*, and *Euglena gracilis* have approximately 40% protein content, which is characterized by being relatively high when compared to conventional vegetable sources such as soy (~ 38%) and vegetable protein sources incorporated into meat analogs such as lentil and chickpea protein (~ 20–30%) (Alonso-Miravalles et al. 2022). Nevertheless, microalgae not only have a high protein content, but their composition is much richer in essential amino acids compared to common plant proteins (Torres-Tiji et al. 2020). In addition, microalgal biomass is rich in a variety of bioactive compounds seen as health-promoting elements, such as polyunsaturated fatty acids (EPA and DHA), photosynthetic pigments (chlorophylls, carotenoids, and phycobiliproteins), polysaccharides, vitamins, polyphenols, and phytosterols (Chen et al. 2022).

Concomitantly, most microalgal cells require high loads of carbon dioxide (CO_2) and solar energy as resources for bioconversion into biomass. Thus, the cultivation of microalgae has been presented as a highly promising biological method for capturing CO_2 from greenhouse gas emissions. It is estimated that about 1.8 kg of CO_2 is incorporated into the microalgal biomass; this element, in turn, is stored in the biomass in the form of carbohydrates and lipids, which can be directed to a wide range of products, such as bioenergy and biofuels. Thus, carbon fixed in biomass can not only play a fundamental role in the face of its biological functions in the biochemical apparatus of these microorganisms but can also be an integrated alternative in the face of mitigating environmental burdens. At the level of knowledge, studies indicate that certain species of microalgae such as *Chlorella vulgaris* can achieve a flue gas decarburization rate of approximately 50% through cultivation (Yadav et al. 2020).

On the other hand, microalgae can also benefit from other forms of impurities. It is well known that pollution from commercial sectors is mostly released into water currents. These generated wastewaters have a variety of contaminants, including minerals, and heavy metals, in addition to a high load of organic matter—depending on the type of company, negatively resulting in fresh and saltwater ecosystems (Kandasamy et al. 2023). However, the bioremediation of wastewater through microalgal biotechnology has inevitably taken center stage in scientific research lines and platforms for waste treatment. In fact, microalgae proved to be excellent bioremovers of pollutants from industrial and urban effluents. This cellular capacity is attributed to its metabolic ability to survive in the face of various adverse environmental conditions, in addition to its potential to bioconvert organic matter into bioproducts of commercial interest, such as biofuels, biofertilizers or even chemical products with high added value (Bhatt et al. 2022). In this way, integrating effluents as a basic resource for microalgae cultures, allows these microorganisms to use pollutants, which until then were considered a problem, in nutrients to carry out their cellular processes and generate useful biomass (Ahmad et al. 2022). Therefore, the cultivation of microalgae in wastewater can be advantageous and viable, given the environmental problems caused by a significant amount of wastewater comprising poisonous and harmful compounds, helping as small and fundamental building blocks to build a sustainable future.

As seen above, the multiple functions performed by microalgae can be assumed as promising tools to encourage the race for sustainability. This is because among the 17-goal sustainable development (SDG) structural pyramid, microalgae processes and products can be part of 5 of the main objectives, namely: SDG-2 zero hunger, SDG-6 clean water and sanitation, SDG-14 life below water, ODS-7 affordable and clean energy, and SDG-13 climate action. However, it is clear that there are still heterogeneous bottlenecks that must be overcome to consolidate the biotechnological routes of microalgae and make them a robust instrument to achieve sustainability. Among them, factors such as monitoring of cultivation systems (pH, temperature, salinity, turbidity, and nutrient availability), design and projection of processing plants, in addition to meticulous selection and mapping of species are just a few gaps that need to be filled. Thus, recognizing the weaknesses of these systems and bringing to light new opportunities for improvement, through the evolution and wide dissemination of artificial intelligence and the automation of systems, is essential to reach a new level of sustainability through processes and products based on microalgae.

31.3 Closed Loop in Microalgae Biotechnology—How Artificial Intelligence Can on a Stepping Stone to Achieving Sustainability?

Biotechnological approaches involve a sequence of complex, multiscalar, and high-dimensional biconversions processed over time and space, resulting from the interaction of many process parameters, which are generally difficult to control (Cheng et al. 2023).

Likewise, the bottlenecks of microalgal biotechnology are present in all operating units. This is because current strategies for the reuse and adaptation of conventional industrial plants for application in biological processes are generally insufficient. In addition, the equipment responsible for the core of the processes, namely the photobioreactors and fermenters, are extrapolations of equipment used in chemical processes, or still need to be built from scratch, resulting in arduous and costly research and development projects (Deprá et al. 2021). As a result, microalgal biotechnology has been questioned given its real viability, since the processes have high demands for fossil resources and unsatisfactory cellular yields, causing an inefficient cost-benefit ratio for sustainable industrial facilities. Thus, the starting point for having a closed loop of success in microalgal biotechnology is the successful control and management of bioprocesses (Aniza et al. 2023).

Faced with this scenario, artificial intelligence emerges today with its ability to break down traditional barriers and allow the creation of intelligent systems that can perform complex tasks that would normally require excellent human intelligence, such as fast learning, problem-solving, and decision-making (Holzinger et al. 2023). Among its aspects, artificial intelligence has two segments, although related, but distinct, such as (i) machine learning (ML) and (ii) deep learning (DL). Machine learning is characterized by encompassing the training of digital computers to perform functions without explicit instructions, applying patterns and insights from data. In contrast, deep learning uses artificial neural networks with many spheres to learn and make decisions. It is particularly proficient for tasks involving the analysis of large data such as imaging and genetic sequencing (Ahmad et al. 2023; Himmel et al. 2023).

Furthermore, artificial intelligence models have been recognized as some of the most accurate and at the same time the simplest of these mathematical models to predict process output (Zhang et al. 2023). Thus, artificial intelligence has been incorporated into microalgal biotechnology as a strategy to assist in the progress of these systems.

Therefore, since the origin of these facilities occurs in the selection of robust cell lines, bioengineering applied to microalgal biotechnology has deepened its investigations in front of the optimization of lineage selections and supply of nutrients using artificial intelligence as a promising resource. Indeed, recent scientific advances in gene editing along with the possibility of modifying metabolic pathways to provide higher biomass yields have become a reality. However, the status quo of understanding and applying these techniques is still limited, as the methodology is generally long, complex, and costly (Teng et al. 2020).

Thus, the application of artificial intelligence algorithms like machine learning (ML) has also been suggested to analyze transcriptome data and cell structure data from many microalgae datasets available to predict the connection between genes and pathways/networks to reveal new mechanisms of gene regulation. This makes it possible to test and validate the best biosynthetically sorting pathways in stable microalgae for enhanced alternative protein production, increasing their potential as a sustainable food strategy (Helmy et al. 2023). On the other hand, classification techniques for microalgal strains can also be performed by the convolutional neural network (CNN), which comprises deep layers such as fully-connected Layer, ReLU, convolutional, and pooling (Chong et al. 2023).

Concurrently, recent technological advances in the downstream stages of microalgae-based processes have also enabled consistent monitoring of microalgal culture systems. The planning of controlled cultivation systems such as photobioreactors, improved cultivation techniques, as a variety of sensors were incorporated into their equipment, making it possible to change the cultivation conditions as necessary. However, the implementation of artificial intelligence has resulted in progress in elucidating kinetic parameters through the use of machine learning. In addition, studies show that artificial intelligence was more successful than expected in identifying the best time to harvest microalgal cells when compared to traditional methods (Peter et al. 2023). Likewise, the artificial backpropagation neural network (BPANN) has been used to validate its ability to accurately predict the flocculation performance of microalgae cells (Suparmaniam et al. 2022). These upgrades are important resources in promoting less energy-intensive biotech pathways. This is because, in practice, greater process control allows for higher cell productivity rates, resulting in higher yields of biomass/bioproducts. Besides, the higher cell density decreases energy demands for cell harvesting, resulting in more profitable and, consequently, more sustainable industrial processes.

Other application models of artificial intelligence in microalgal biotechnology also extend to environmental applications. As an example, Kushwaha et al. (2023) developed a new computational model, used to predict CO_2 fixation by microalgae, based on the hybrid approach of the adaptive neuro-fuzzy inference system (ANFIS) and genetic algorithm (GA). Under the same bias, fuzzy-type modeling was also used to determine and optimize the best formula for the production of biofuels using microalgal fuel cells (Sayed et al. 2023).

Given the above, it is important to emphasize that scientific signs of progress to fill in the gaps and circumvent the bottlenecks hitherto faced by microalgal biotechnology are in progress in the broad fields of process bioengineering. However, when advances in microalgae biotechnology are combined with advances in artificial intelligence, unprecedented new potential solutions become available. This could open the door to a more consolidated and efficient future for biotechnology and microalgae-based products.

31.4 Conclusion

The need to produce more sustainable bioresources to replace finite resources is obvious to everyone. Thus, the complex and diverse nature of microalgae has been recognized as the silver bullet of the new era of sustainability. However, these microorganisms require a large amount of data and robust interdisciplinary knowledge in order to circumvent and clarify bottlenecks that involve species selection, cultivation parameters, reactor design, to the most complex conversion technologies to consolidate their applications. Thus, the incorporation of artificial intelligence tools may be a necessity for microalgal biotechnology. This is because the use of models and algorithms has already shown to be promising in terms of reducing the time and effort required to optimize process yields. Therefore, the use of these resources can be a springboard for microalgal biotechnology to reach its more ecological, autonomous, and economically viable production goals.

References

Ahmad A, Banat F, Alsafar H, Hasan SW (2022) Algae biotechnology for industrial wastewater treatment, bioenergy production, and high-value bioproducts. Sci Total Environ 806:150585. https://doi.org/10.1016/j.scitotenv.2021.150585

Ahmad S, Shakeel I, Mehfuz S, Ahmad J (2023) Deep learning models for cloud, edge, fog, and IoT computing paradigms: survey, recent advances, and future directions. Comput Sci Rev 49:100568. https://doi.org/10.1016/j.cosrev.2023.100568

Akaniro IR, Chibuike IV, Onwujekwe EC, Gbadamosi FA, Enyi DO, Onwe ON (2023) Penicillium species as chassis for biomanufacturing and environmental sustainability in the modern era: progress, challenges, and future perspective. Fungal Biol Rev 46:100326. https://doi.org/10.1016/j.fbr.2023.100326

Almomani F, Hosseinzadeh-Bandbafha H, Aghbashlo M, Omar A, Joo SW, Vasseghian Y, Karimi-Maleh H, Lam SS, Tabatabaei M, Rezania S (2023) Comprehensive insights into conversion of microalgae to feed, food, and biofuels: current status and key challenges towards implementation of sustainable biorefineries. Chem Eng J 455:140588. https://doi.org/10.1016/j.cej.2022.140588

Alonso-Miravalles L, Barone G, Waldron D, Bez J, Joehnke MS, Petersen IL, Zannini E, Arendt EK, O'Mahony JA (2022) Formulation, pilot-scale preparation, physicochemical characterization and digestibility of a lentil protein-based model infant formula powder. J Sci Food Agric 102(12):5044–5054. https://doi.org/10.1002/jsfa.11199

Aniza R, Chen WH, Pétrissans A, Hoang AT, Ashokkumar V, Petrissans M (2023) A review of biowaste remediation and valorization for environmental sustainability: artificial intelligence approach. Environ Pollut 121363. https://doi.org/10.1016/j.envpol.2023.121363

Barbosa MJ, Janssen M, Südfeld C, D'Adamo S, Wijffels RH (2023) Hypes, hopes, and the way forward for microalgal biotechnology. Trends Biotechnol. https://doi.org/10.1016/j.tibtech.2022.12.017

Bhatt P, Bhandari G, Turco RF, Aminikhoei Z, Bhatt K, Simsek H (2022) Algae in wastewater treatment, mechanism, and application of biomass for production of value-added product. Environ Pollut 309:119688. https://doi.org/10.1016/j.envpol.2022.119688

Bhattacharya M, Goswami S (2020) Microalgae—a green multi-product biorefinery for future industrial prospects. Biocatal Agric Biotechnol 25:101580. https://doi.org/10.1016/j.bcab.2020.101580

Bröring S, Vanacker A (2022) Designing business models for the bioeconomy: what are the major challenges? EFB Bioecon J 100032. https://doi.org/10.1016/j.bioeco.2022.100032

Calijuri ML, Silva TA, Magalhães IB, de Paula Pereira ASA, Marangon BB, de Assis LR, Lorentz JF (2022) Bioproducts from microalgae biomass: technology, sustainability, challenges and opportunities. Chemosphere 135508. https://doi.org/10.1016/j.chemosphere.2022.135508

Chen C, Tang T, Shi Q, Zhou Z, Fan J (2022) The potential and challenge of microalgae as promising future food sources. Trends Food Sci Technol. https://doi.org/10.1016/j.tifs.2022.06.016

Cheng Y, Bi X, Xu Y, Liu Y, Li J, Du G, Lv X, Liu L (2023) Artificial intelligence technologies in bioprocess: opportunities and challenges. Biores Technol 128451. https://doi.org/10.1016/j.biortech.2022.128451

Chong JWR, Khoo KS, Chew KW, Vo DVN, Balakrishnan D, Banat F, Munawaroh HS, Iwamoto K, Show PL (2023) Microalgae identification: future of image processing and digital algorithm. Biores Technol 369:128418. https://doi.org/10.1016/j.biortech.2022.128418

Deprá MC, Severo IA, Dias RR, Zepka LQ, Jacob-Lopes E (2021) Photobioreactor design for microalgae culture. In: Microalgae. Academic Press, pp 35–61. https://doi.org/10.1016/B978-0-12-821218-9.00002-5

Deprá MC, Dias RR, Sartori RB, de Menezes CR, Zepka LQ, Jacob-Lopes E (2022) Nexus on animal proteins and the climate change: the plant-based proteins are part of the solution? Food Bioprod Process 133:119–131. https://doi.org/10.1016/j.fbp.2022.03.006

Helmy M, Elhalis H, Yan L, Chow Y, Selvarajoo K (2023) Perspective: multi-omics and machine learning help unleash the alternative food potential of microalgae. Adv Nutr. https://doi.org/10.1016/j.advnut.2022.11.002

Himmel A, Matschek J, Kok R, Morabito B, Nguyen HH, Findeisen R (2023) Machine learning for process control of (bio)chemical processes. arXiv preprint. https://doi.org/10.48550/arXiv.2301.06073

Holzinger A, Keiblinger K, Holub P, Zatloukal K, Müller H (2023) AI for life: trends in artificial intelligence for biotechnology. New Biotechnol 74:16–24. https://doi.org/10.1016/j.nbt.2023.02.001

Kandasamy S, Narayanan M, Raja R, Devarayan K, Kavitha R (2023) The current state of algae in wastewater treatment and energy conversion: a critical review. Curr Opin Environ Sci Health 100469. https://doi.org/10.1016/j.coesh.2023.100469

Kholssi R, Riouchi O, Douhri H, Debdoubi A (2023) Microalgae as a sustainable energy source: growth and lipids production of *Nitzschia* sp., *Nannochloropsis* sp., and *Tetraselmis* sp. from Mediterranean seawater. Biocatal Agric Biotechnol 50:102720. https://doi.org/10.1016/j.bcab.2023.102720

Kushwaha OS, Uthayakumar H, Kumaresan K (2023) Modeling of carbon dioxide fixation by microalgae using hybrid artificial intelligence (AI) and fuzzy logic (FL) methods and optimization by genetic algorithm (GA). Environ Sci Pollut Res 30(10):24927–24948. https://doi.org/10.1007/s11356-022-19683-0

Losacker S, Heiden S, Liefner I, Lucas H (2023) Rethinking bioeconomy innovation in sustainability transitions. Technol Soc 102291. https://doi.org/10.1016/j.techsoc.2023.102291

Oliveira CYB, Jacob A, Nader C, Oliveira CDL, Matos ÂP, Araújo ES, Shabnam N, Ashok B, Gálvez AO (2022) An overview on microalgae as renewable resources for meeting sustainable development goals. J Environ Manage 320:115897. https://doi.org/10.1016/j.jenvman.2022.115897

Oruganti RK, Biji AP, Lanuyanger T, Show PL, Sriariyanun M, Upadhyayula VK, Gadhamshetty V, Bhattacharyya D (2023) Artificial intelligence and machine learning tools for high-performance microalgal wastewater treatment and algal biorefinery: a critical review. Sci Total Environ 876:162797. https://doi.org/10.1016/j.scitotenv.2023.162797

Peter AP, Chew KW, Pandey A, Lau SY, Rajendran S, Ting HY, Munawaroh HS, Van Phuong N, Show PL (2023) Artificial intelligence model for monitoring biomass growth in semi-batch *Chlorella vulgaris* cultivation. Fuel 333:126438. https://doi.org/10.1016/j.fuel.2022.126438

Sayed ET, Rezk H, Abdelkareem MA, Olabi AG (2023) Artificial neural network based modelling and optimization of microalgae microbial fuel cell. Int J Hydrogen Energy. https://doi.org/10.1016/j.ijhydene.2022.12.081

Suparmaniam U, Shaik NB, Lam MK, Lim JW, Uemura Y, Shuit SH, Show PL, Tan IS, Lee KT (2022) Valorization of fish bone waste as novel bioflocculant for rapid microalgae harvesting: experimental evaluation and modelling using back propagation artificial neural network. J Water Process Eng 47:102808. https://doi.org/10.1016/j.jwpe.2022.102808

Teng SY, Yew GY, Sukačová K, Show PL, Máša V, Chang JS (2020) Microalgae with artificial intelligence: a digitalized perspective on genetics, systems and products. Biotechnol Adv 44:107631. https://doi.org/10.1016/j.biotechadv.2020.107631

Torres-Tiji Y, Fields FJ, Mayfield SP (2020) Microalgae as a future food source. Biotechnol Adv 41:107536. https://doi.org/10.1016/j.biotechadv.2020.107536

Vinuesa R, Azizpour H, Leite I, Balaam M, Dignum V, Domisch S, Felländer A, Langhans SD, Tegmark M, Fuso Nerini F (2020) The role of artificial intelligence in achieving the sustainable development goals. Nat Commun 11(1):233. https://doi.org/10.1038/s41467-019-14108-y

Whitford B, Jones D, Kinnane S (2022) Biomanufacturing design: reducing the environmental burden. Curr Opin Biotechnol 76:102717. https://doi.org/10.1016/j.copbio.2022.102717

Yadav G, Dubey BK, Sen R (2020) A comparative life cycle assessment of microalgae production by CO_2 sequestration from flue gas in outdoor raceway ponds under batch and semi-continuous regime. J Clean Prod 258:120703. https://doi.org/10.1016/j.jclepro.2020.120703

Zhang G, Liu J, Pan X, Abed AM, Le BN, Ali HE, Ge Y (2023) Latest avenues and approaches for biohydrogen generation from algal towards sustainable energy optimization: recent innovations, artificial intelligence, challenges, and future perspectives. Int J Hydrogen Energy. https://doi.org/10.1016/j.ijhydene.2022.10.224

Zhao X, Meng X, Dang B, Zhang T, Shi W, Hou N, Yan Q, Li C (2023) Succession dynamics of microbial communities responding to the exogenous microalgae ZM-5 and analysis of the environmental sustainability of a constructed wetland system. Biores Technol 128642. https://doi.org/10.1016/j.biortech.2023.128642

Chapter 32
Algae as Nature-Based Solutions for Climate Change Adaptation

Caroline Samberger

Abstract This chapter explores the use of algae as nature-based solutions for climate change adaptation. It first touches on a few definitions about climate change adaptation versus mitigation, nature-based solutions and biomimicry and explains how algae-based measures can be beneficial as adaptation actions. Historical uses of algae in civilisations are also briefly mentioned to highlight the fact that algae are not a newly discovered nature-based solution but were already used thousands of years ago in food or construction applications. The various components and benefits of algae are then discussed along with how they are currently used to promote climate adaptation in the different areas of the economy i.e. built environment, food production, new energy systems and non-fossil-fuel materials such as bio-plastics. Lastly, the chapter explores the risks and reasons why algae as nature-based solutions have not yet been implemented more extensively and the reasons for the existing barriers to implementing algae technologies in some areas of our current economy.

Keywords Algae · Climate change · Adaptation

Abbreviations

C	Carbon
CO_2	Carbon dioxide
COD	Chemical Oxygen Demand
GHG	Green House Gas
IPCC	Intergovernmental Panel for Climate Change
N	Nitrogen
NBS	Nature Based Solutions
P	Phosphorus

C. Samberger (✉)
Stantec Ltd.—Wastewater, Energy and Environmental Services, Buckingham Court, Kingsmead Business Park, Frederick Place, London Rd, High Wycombe, London HP11 1JU, UK
e-mail: Caroline.Samberger@stantec.com

T. A. Kurniawan and A. Anouzla (eds.), *Algae as a Natural Solution for Challenges in Water-Food-Energy Nexus*, Environmental Science and Engineering,
https://doi.org/10.1007/978-981-97-2371-3_32

32.1 Introduction

Due to climate change, the world is already experiencing more frequent extreme weather events such as cyclones or storms, more intense rainfalls and floodings, extreme temperature variations, and some shifts in seasons with multiple heat waves and/or freezing temperatures. Therefore, the longer we put off our efforts to adapt and become more resilient to climate change, the more difficult and expensive it will be to respond (UNFCC, Accessed Oct 2023).

32.1.1 Climate Change Adaptation versus Climate Change Mitigation

The Intergovernmental Panel for Climate Change (IPCC) defines climate change adaptation as "the process of adjustment to ecological, social or economic systems in response to actual or expected climate and its effects" (UNCC 2023).

Countries and communities need to develop adaptation solutions to respond to climate change impacts, lessen and counteract potential damages and benefit from opportunities associated with climate change (UNFCC, Accessed Oct 2023). Communities' ability to adapt to climate change can be improved by reducing their exposure and sensitivity to the negative effects of climate change (Austin 2021).

However, there is no "one-size-fits-all" approach to adaptation because it can be accomplished in a variety of ways and effective solutions depend on the specifics of the projects they are applied to, their location, and the local communities they affect (UNFCC, Accessed Oct 2023).

Making sure we manage our food, water, and other natural resources wisely, diversifying crops that can withstand more extreme climatic conditions, assisting communities in lowering their risk from rising sea levels and increased flooding, and ensuring infrastructure can withstand more extreme weather are all examples of adaptation strategies (WWF 2023). Hence solutions for climate adaptation can include planting drought-tolerant crops for food and feed production, erecting flood barriers along coastlines, installing early warning systems for severe weather, or even changing governmental policies and regulations. Many nations have already begun to change their societies and economies to become more resilient. However, more work is required to manage climate change risks economically, both now and in the future (UNFCC, Accessed Oct 2023).

In contrast to mitigation, which aims to lessen the causes of these effects, adaptation focuses on reducing or coping with the effects of climate change on life and populations by gaining benefits from them. Although mitigation and adaptation strategies are two complementary approaches to combating climate change, there are times when they overlap (as shown in Fig. 32.1). For example, the use of algae in construction and green infrastructure may both contribute to mitigation (through CO_2 absorption) and adaptation (through thermal insulation and heat wave resilience).

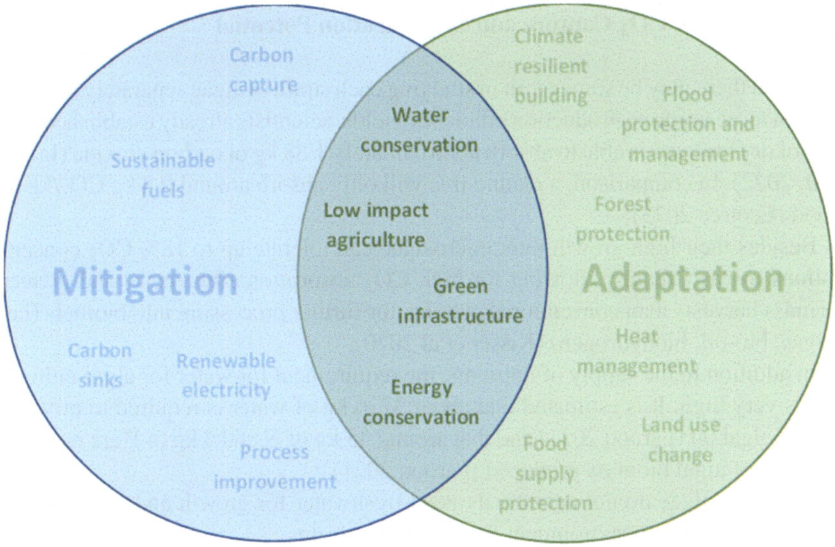

Fig. 32.1 Overlap of climate change mitigation and climate change adaptation

In this chapter, Section 32.1 covers the various characteristics and attributes of algae, after which Section 32.2 discusses the potential for climate adaptation using algae in different application examples. Finally, Section 32.3 briefly discusses the obstacles to algae exploitation for climate change adaptation.

Although the roles of adaptation and mitigation strategies can occasionally overlap, this chapter only discusses how algae might be used in adaptation to climate change. It does not go into great detail about how algae might be used for climate change mitigation.

32.1.2 Algae power

Algae can absorb CO_2 from atmosphere and nitrogen and phosphorus from water to grow, are not dependent on agricultural cultivable land, do not require fertilisers or freshwater and have been estimated to include between 30,000 and more than 1 million species according to (Guiry 2012).

However, of these many species, (Siedenburg 2022) states that only few are relatively well-known, and many others would probably deserve more investigation into their multiple economic, environmental and health benefits.

32.1.2.1 Algae CO_2 Capture and Sequestration Potential

Although there may be some merit in studying each strain of algae separately to come up with more accurate production efficiency yields, scientists already established that 1 kg of dry biomass is able to absorb approximately 1.88 kg of carbon dioxide (Iglina et al. 2022). In comparison, a mature tree will only absorb around 0.8 kg CO_2/kg of wood (Ecotree 2023).

Besides their high growth rate, microalgae can tolerate up to 18% CO_2 concentrations in gas streams, allowing for high CO_2 absorption efficiencies and greater biomass harvests than conventional cultures for further processing into biofuels (i.e. biogas, bio-oil, biohydrogen) (Kisser et al 2020).

In addition to the supply of nutrients, the requirement for water for algae cultivation is very high. It is estimated that about 3726 kg of water is required to produce 1 kg of algal oil (Farooq 2021) and that around 45 kg of N and 4 kg of P are required for 1 ton of algal biomass produced (Farooq 2021).

However, algae do not necessarily need freshwater for growth and can thrive in wastewater or saline environment depending on strains.

Additionally, harvested algal biomass presents various other properties and economic opportunities:

- several microalgae can synthesize large quantities of proteins (around 46–71% of their dry matter) (Chen et al. 2022; Colla et al. 2020)
- algae are also a great source of carbohydrates (8–17% of dry matter) (Chen et al. 2022; Colla et al. 2020)
- algae can synthetise a large proportion of lipids (2–22% of dry matter) (Colla et al. 2020).

Proteins, carbohydrates and lipids proportions vary depending on the algae strain. However, their high protein content means that they are more likely to be considered as an alternative source of proteins for food and feed. Additionally, all 3 components contained in algae biomass are precursors to various other potential products such as biofuels, bioplastics, fertilizers, pigments or even pharmaceutical and nutraceutical products.

Therefore, you can eat, wear or build from algae biomass, which offers considerable climate change adaptation potential.

This chapter explores the various current applications of algae as nature-based solutions.

32.1.3 Nature-Based Solutions and Biomimicry

The idea of finding inspiration in natural processes to address problems in human design is known as biomimicry. "Bioinspired design" generally refers to engineering and design methodologies that leverage biology as a source of inspiration, such as biomimicry.

Biomimicry is the foundation of Nature Based Solutions (NBS) and involves working with and enhancing nature to help address societal challenges (Veron et al. 2013; Proksch 2012). They include a wide range of engineered solutions, such as the incorporation of green infrastructure in urban areas, protecting and managing natural ecosystems, and the application of ecosystem-based principles to agriculture practices (Seddon et al. 2020).

Based on their properties and characteristics, algae implemented as nature-based solutions have therefore the potential to significantly contribute to tackling both climate mitigation and adaptation at relatively low-cost while delivering multiple additional benefits for people and nature.

32.2 Algae as Nature-Based Solutions

32.2.1 Algae in the Built Environment

32.2.1.1 Impacts of Climate Change on the Built Environment

Many cities and villages around the world are already experiencing multiple impacts from climate change such as loss of coastal area, wetlands or forests, repeated flooding from heavy rains and more frequent and intense heat waves (USEPA, Climate Change Impacts on the Built Environment | US EPA 2022).

As the built environment interacts with and depends on other sectors of the economy, such as water, transportation, and energy (FAO 2014), all these resources compete with each other, which is sometimes referred to as the energy-food-water-land nexus (Samberger 2022a, b). Therefore, climate impacts on these economic sectors also affect the built environment (USEPA, Climate Change Impacts on the Built Environment | US EPA 2022) with:

- more urban flooding, especially in urbanized areas with a lot of paved and engineered surfaces.
- increased heat impacts like unusually hot summers in many cities with urban residents facing more heat related issues.
- increased property damage and damage to historic buildings and landmarks due to coastal flooding and sea level rise.
- reduced air quality due to wildfires also posing risks to people and property.

32.2.1.2 Algae in Architecture and Design

The use of natural resource as construction materials is nothing new. Ancient and new civilizations have been using natural resources around them (plants, organic matter, straw, algae) to turn them into construction and building structures for thousands of years (TIME 2022).

Nowadays, green facades and microalgae structures are the most popular algae applications, usually applied with the primary objective to purify ambient air in cities and usually designed to enhance the aesthetics of urban spaces (Kisser et al. 2020).

The BIQ, a groundbreaking residential structure, was created and constructed from December 2011 to April 2013 for the International Building Exhibition in Hamburg. It is the first structure in the world to have a bioenergy façade, which incorporates living microorganisms into the comprehensive design of the heating system (photo.Synthetica 2017). Algae take in sunlight and convert carbon dioxide into biomass. Extra light that is not absorbed by photosynthesis is converted to heat. The building is able to provide all of its own energy needs thanks to the on-site harvesting of the heat and biomass produced by the solar panels. This is the first system of its kind for urban construction. Growing algae are converted to biomass by the CO_2 added from a micro Combined Heat and Power system powered by biogas. The building reduces CO_2 emissions by up to 6 tonnes annually (Kisser et al. 2020; Elmeligy and Elhassan 2019; Veron et al. 2013; European Union 2013).

Other microalgae cultivation in photo-bioreactors for construction include design studies and temporary constructions such as the following projects (Violano and Cannaviello 2019):

- in Switzerland, Geneva (2014), The Cloud Collective (INHABITAT 2023) installed an algae cultivation system on a highway overpass, to demonstrate the role urban settings could play in the production of food and plants;
- in France, Paris (2017), photo-bioreactors were incorporated in typical Parisian Morris advertising columns by Suez and start-up Fermentalg (UP'Magazine 2017);
- the French Dream Towers by XTU (2019) were built with parts of bio facades micro-algae culture and vegetated facades and the integration of the water cycle with the recycling of rainwater and grey water by phytoremediation in basins (ITSLIQUID 2019).
- in 2021, ecoLogicStudio created an air-purifying micro-algae cultures playground for children The "AirBubble" in the centre of Warsaw (Poland) (Ecologicsstudio 2023).

Urban Cladding

Some companies also innovated on some urban cladding systems based on algae, such as PhotoSynthetica who created a photosynthetic algae building curtain system using light to absorb CO_2 and pollutants from the atmosphere and produce a renewable food resource. Only 2 m^2 of this cladding absorb the same amount of CO_2 as a mature tree (photoSynthetica 2023). This "algae curtain" was displayed in November 2018 at Dublin Climate Innovation Summit.

In June 2017, PhotoSynthetica also completed a prototype project for a future dwelling, the "BIO.tech HUT" in Kazakhstan, which is energetically self-sufficient, carbon neutral and nutritionally circular, and hosts a total of 1600 l of living cultures. The designer claims that in optimal conditions, the system can absorb 2 kg of CO_2/

day which is about the equivalent absorption capacity of 32 large trees, and produce 1.5 kg of O_2. For comparison's sake, an average adult inhales 550 l O_2/day which corresponds to the O_2 production from 8 large trees while they also exhale 1 kg CO_2/day, which takes approximately 16 large trees to adsorb. The building is also claimed to produce on average 1.12 kg of dry algae per day. Green microalgae can contain up to 60% oil in dry weight. Hence the project can produce 672 g of oil per day from which approximately 1 kg of biofuel can be produced, enough to power an average UK home (photo.Synthetica 2017).

Algae also contain up to 60% proteins, meaning that every day, the building could potentially produce 672 g of proteins, which is enough to supply the recommended daily intake of 12 adults. Comparatively, an average cow provides 52 kg of meat-based proteins (but takes 660 days to mature) (photo.Synthetica 2017).

Transparency and Sun Shading

At Milan 2015 exposition on Future Food (Atlas 2015) some innovative solutions of building integrated urban agriculture were on display. The "Urban Algae Folly" (Ecologicstudio 2023) integrated living micro-algal cultures, a built example of architecture's bio-digital future, converting solar energy into biomass and oxygen.

On average, the installation produced 2 kg of oxygen per day, the equivalent oxygen that 3 adults need to survive, which would take 25 large urban trees to produce. Additionally, it also produced the equivalent in proteins of 2 kg of meat per day (enough for 12 adults) with no animals being killed and no methane being released, and capturing CO_2 at a rate of 4 kg per day.

Finally, with light, the microalgae grow rapidly thus increasing the shading potential of the architectural structure and improving temperature comfort in buildings.

Light Sources

"Liquid trees" were installed in late 2021 in Serbia, based on an innovation by scientists from the University of Belgrade (Newsweek 90, 2023).

One LIQUID3 bioreactor can be as efficient at CO_2 absorption as a 10-year-old adult tree, or a full 200 m^2 of lawn (Newsweek 90, 2023). The bioreactor can be built and installed quickly and start processing polluted air far more quickly than growing a full mature tree. Much of the criticisms against the bioreactor are unjustified, as the reactor is not intended to replace trees in cities but rather designed to fit into narrow and congested spaces where growing a tree is impractical. In addition, bioreactors can offer efficiencies that trees and grasses simply cannot compete with. Scientific literature claims that microalgae are capable of removing CO_2 from atmosphere at a rate 10 to 50 times the rate of mature trees (Hackaday 2023) (Newsweek 90, 2023).

From 2015 to 2016, Living Things created an exhibit of custom glass bioreactors designed as household furnishings. The exhibition was featured in various furniture shops (Frier 2016).

An energy-free bio lighting system based on bioluminescence, a natural chemical process present in living organisms (particularly fireflies and algae), has been developed by the French start-up Glowee. Bioluminescence is the term used to describe how some living things, including some marine organisms and glow-worms and fireflies, produce and emit light. Since light pollution is the root of many urban issues, such as artificial nighttime lighting interfering with plant and wildlife ecosystems as well as human health, inspiration from how nature functions (or biomimicry) was used to develop city lighting with no pollution and a very low environmental impact. The Glowee solution is now employed in conferences as a calming lighting solution (Glowee 2022). Glowee works side by side with the municipality of Rambouillet (France) through an innovation public market contract, to experiment the first bioluminescent signage panel (BBC 2022).

Nevertheless, bioluminescent lighting is not a novel concept. In about 350 BC, Aristotle described bioluminescence from glow worms and fireflies as a type of "cold" light used in mines to prevent deadly explosions. Meanwhile, Indian tribes have long employed luminescent fungi to light up dense vegetation (BBC 2022).

However, Glowee is the first company in the world to commercialize these bioluminescent applications and would be in negotiations with 40 cities across Europe such as in Belgium, France, Portugal, and Switzerland.

Despite these innovations, scientists believe there are serious challenges still ahead before bioluminescence can be applied in large-scale commercial deployment (BBC 2022) and bioluminescence most significant potential currently lies in building decoration, and temporary art displays (Veron et al. 2013).

Algae as Thermal Insulation

Since the seventeenth century, the algae gathered on the beaches of the Danish Island of Læsø have been used to build roofs for houses. In Denmark's traditional homes as well as contemporary houses, marine algae have also recently been used at scale for thermal insulation (Violano and Cannaviello 2019).

As another example, fourteen homes were constructed as part of the Formentera Residential Construction project in Spain using algae as insulation material in order to reduce waste and emissions, and the project consequently won a prize in the Low Carbon Category of the 2017 Green Solutions Awards (Idealista 2019).

As part of a significant development to reduce the emissions contribution of insulation material in the construction and automotive sectors, the Posidonia-Dämmstoffe EU-funded project tested using Neptune Grass—seaweed—naturally non-inflammable fibers as a fire-resistant insulating material in 2013 (EU 2016). However, this company does not seem to exist anymore.

Despite these few projects, the application of algae biomass for thermal insulation and building/architectural design seems to remain at prototype/experimental stage.

32.2.1.3 Algae as Bricks

Cement production is carbon-intensive and accounts for around 8% of global CO_2 emissions (TIME 2022).

According to (TIME 2022), Prometheus Materials was created in 2021 and takes microalgae from lakes or ponds to grow them in bioreactors. The algae use ambient air, to absorb the carbon dioxide it contains, as well as sea water and light from LED lamps. A cement-like substance capable of bonding together sand with gravel or stone to make concrete is produced from the algal biomass. The algae-based bricks are targeted to be commercially available in 2023 (Prometheus Materials 2023).

32.2.2 Algae in Agri-food Production

In 2021, the IPCC forecasted that heat waves, droughts, and a rise in rainfall variability would have a negative impact on crop yields and livestock productivity, which would then lead to issues with food availability and nutrient quality.

Consequently, due to their content in proteins and lipids as well as their resilience to extreme weather variations, algae are seen as one of the potential solutions to counteract the impacts of climate change on food and feed availability.

32.2.2.1 History of Algae in Food and Feed

However, eating algae is also nothing new. Coastal populations have used marine plants as food additives, soil fertilizers, fuel, and animal feed since ancient times. Scientists believe that algae were used as food ingredients in very early times thanks to the presence of algae traces in the ashes of prehistoric fireplaces (Ifremer 2004).

According to archaeological findings in Chile, people were already collecting seaweeds for food and medicine 14,000 years ago (Caria Mendes et al. 2022). Another source (Aaronson 2008) states that the earliest archaeological evidence for the consumption of algae and kelp found thus far was discovered in ancient dumping grounds along the coast of Peru from 2500 to 1300 B.C and that these findings indicated that marine algae were employed by the ancient Peruvians to supplement their diets.

32.2.2.2 Impact of Climate Change on Agriculture and Food

The occurrence, frequency, and severity of plant diseases are impacted by climate change. Crop yield declines, increased pest and disease pressure, increased crop water demands, and doubts about the sustainability of some extremely vulnerable crops are all already affecting human agricultural practices (Pathak et al. 2018).

Increased temperatures caused by climate change are also particularly influencing the agriculture and the supply of high-demand luxury foods such as coffee, chocolate, honey, wine, seafood, or fruits such as avocados, strawberries and bananas. And these foods have already been declared to be at risk of extinction by 2050 (Insights 2021).

32.2.2.3 Algae as Promising Future Food Sources

Microalgae production differs from plant or animal production as it does not require arable land as it grows on water and therefore does not compete with agricultural land. Additionally, (Nuwen 2022) states that it takes up 20 to 30 times less land to grow algae than terrestrial agriculture: one hectare of cultivated algae can produce the equivalent of 20 ha of soybeans or 50 ha of corn.

Algae can grow and be processed without necessarily needing fresh water, and in the majority of food-processing operations, algae are even rinsed with sea water. For instance, the production of 1 kg of rice requires 1000 liters of fresh water, 1 kg of chicken requires 400 liters, but 1 kg of algae requires no freshwater at all. Since algae growth is also largely unaffected by regional weather patterns, it is also much less vulnerable to the effects of climate change than other land-based crops (Nuwen 2022).

Therefore, microalgae cultivation is climate resilient as:

- it does not require arable land and can thrive in saline waters (Siedenburg 2022; Lopes Amorim et al. 2021) so does not compete with traditional agriculture (Chia et al. 2020)
- it uses less water/kg of protein than other crops (Habib 2008)
- microalgae strains can be selected or engineered to match varying climatic conditions (Walsh et al. 2015)

As a result, (Sills et al. 2020) came to the conclusion that the positive effects of using algae in place of conventional animal feed come primarily from avoiding the use of land and freshwater to grow crops.

These characteristics are particularly well suited to cope with climatic shocks, so microalgae production offers lots of climate resilience agricultural alternatives to conventional practices (Conversation 2022).

Hence, thanks to their composition, algae can be used in a multitude of food applications.

32.2.2.4 Algae as Pharmaceuticals/nutraceuticals

Due to their bioactive molecules and their effect on human health, algae are an extensively demanded product for use in nutraceuticals (Kumari et al. 2023). Different beneficial pharmaceutical properties have also been attributed to microalgae, such as their capacity to improve health or reduce the risk of disease (Khavari et al. 2021). Studies have shown that bioactive compounds in microalgae may present antioxidant activity, which can be associated with some of their therapeutic properties, including their anticancer and anti-inflammatory properties (Colla et al. 2020). Saide et al. (2021) also suggested that a wide range of conditions can be treated with the nutraceuticals and pharmaceuticals that can be extracted from microalgae. Their biological functions are extremely varied, and they are said to have advantages in slowing down the progression of some cancers and lowering the risks of developing cardiovascular disease, diabetes, or neurodegenerative diseases.

Nutraceutical molecules from algae range from Carotenoids to Astaxanthin (Garcia et al. 2017). Algae also contain essential Vitamins (e.g A, B1, B2, B6, B212, C, E, folic acid). For example, one-half of the adult daily requirement for vitamin A would be fulfilled by the consumption of only 1 g of Spirulina (Colla et al. 2020). They also contain fatty acids, omega-3s and omega-6s (Chen et al. 2022). Climate change induced global warming and ozone depletion

are also leading to increase of sun exposure, for which algae seem to offer skin photoprotection properties (Kumari et al. 2023).

Nevertheless, some of these benefits still need to be confirmed by solid clinical trials (Khavari et al. 2021).

32.2.2.5 Algae as Gelling Agents and Pigments

Producing sustainable food ingredients while promoting human health through balanced nutrition are the two main challenges the world is currently facing (Liao et al. 2021). Food additives are added to food to improve the quality/consistency/shelf life of prepared food, including gelling agents. The primary gelling agents currently in use in the food and beverage industry are animal-based and numerous personal care items, including shampoos, face masks, and other cosmetics, also contain animal jellies. Animal gelatine is obtained by boiling skin, bones, ligaments and tendons of animals, usually cows or pigs (OIPA 2017).

However, algae derived products are becoming more and more widely used in the food industry as thickening, gelling and emulsifying agents. Algae are becoming increasingly popular as "superfoods," replacing animal products with additive ingredients like flavouring agents or natural colorants. 13,090 different foods and drinks with algae ingredients were made available globally between 2015 and 2019 (Boukid and Castellari 2021). Foods are frequently coloured and made thicker by the addition of algae and numerous food items, including dairy, desserts, yogurt, baby formula, and ice cream, now contain algae.

Algae can also be used as pigments since they contain chlorophylls or carotenoids (Chen et al. 2022).

32.2.2.6 Algae for Aquaculture and Animal Feed

Since animal-derived foods, such as meat and dairy, are important components of diets in many cultures around the world as a source of protein, livestock production is essential to the human food system. Securing sufficient, nutrient-rich livestock feed in a sustainable manner is a major challenge for humanity. However, it has been calculated that 14.5% of all current greenhouse gas emissions come from livestock (Gerber 2013).

According to (Cordeiro, et al. 2022), climate change can negatively impact feed supplies by endangering the quantity and quality of pastures and feed crops, endangering livestock production by leading to animal illnesses and deaths. As a result, feed insecurity is an emerging societal and economic concern linked to climate change.

Meanwhile, the steps taken to secure feed can increase greenhouse gas emissions and thus aggravate climate change, notably where this involves converting forests to pastures or cropland for feed crops (Siedenburg 2022).

Microalgae could play an interesting role in resolving sustainable human nutrition challenges since swine are some of the main competitors with the human food chain, mostly because their diets are typically composed of two main ingredients: corn and soybean. (Valente et al. 2021). Moreover, corn and soy, which need a lot of land to grow, are the primary conventional feeds for poultry.

Microalgae might be a good substitute for plant proteins like soy because of its high protein content. And in order to reduce the amount of soy in poultry diets, algae such as spirulina have the potential to replace a significant portion of popular plant-based protein sources in animal feed (Valente et al. 2021).

The increase in global population has also led to a rise in the demand for fish products, which has prompted significant growth in aquaculture over the past ten years. The main sources of proteins and oils for the diets of fish and shrimp have historically been fish meat and fish oil. Aquaculture, however, has shifted to alternative sources of proteins for fish feed, such as algae feed, as a result of the serious problems the world's oceans are facing due to overfishing and the disruption of marine ecosystems caused by climate change.

Novel feed such as algae-based feed offers options to improve the sustainability of livestock production. And algae minimum use at present has shown higher productivity and better nutritional quality of products (Siedenburg 2022).

32.2.3 Algae in Cosmetics

People use cosmetics extensively all over the world for aesthetic reasons and to shield their skin from weather damage. Since it was established that synthetic cosmetics had negative side effects and a low absorption rate due to the large molecules of the chemicals, consumers have now started to prefer natural cosmetics. As one of the richest marine resources that are considered safe and have very low toxicity effects on humans, marine algae have also attracted a lot of attention in the cosmetics and cosmeceutical industries. It is claimed that they have medicinal or drug-like benefits. (Thiyagarasaiyar et al. 2020). The active ingredients extracted from algal biomass can be used as additives, stabilizers or preservatives of cosmetic products.

32.2.4 Algae as Biofertilizers and Bio Stimulants for Crop Support

World agriculture depends heavily on chemical fertilizers, which has led to soil degradation, aggravating the vulnerability of crops to climate change by reducing the water-holding capacity of soils and erosion control. Chemical fertilizers manufacturing is also a major greenhouse gas emitter.

Algae-based biofertilizers and bio-stimulants are natural alternatives that can increase crop productivity, while making it more sustainable and resilient to climate stress and variability (Siedenburg 2022; Garcia-Gonzalez and Sommerfeld 2016; Van Oosten et al. 2017). They are attracting growing attention from both researchers and farmers. At present, however, these products remain largely unexploited in the agriculture sector (Siedenburg 2022; Ronga et al. 2019).

32.2.5 Algae as Biochar Feedstocks

Microalgae can also be used for land application as biochar, or charred biomass to improve soil fertility and enhance soil's capacity to hold water (Conversation 2022). Such effects could help crops cope with climate change impacts like extreme weather events such as torrential rainfalls. A key benefit of biochar is its stability with its carbon components and climate change benefits due to a residence time in soil of hundreds to thousands of years (Siedenburg 2022). Soils treated with biochar by the Amerindians in Brazil many centuries ago were found to have a productivity twice as high as the neighbouring lands.

32.2.6 Algae as Plastics

32.2.6.1 Climate Change and the Future of Fossil Fuel Plastics

Climate change, meanwhile, also impacts plastic pollution. Extreme temperatures can cause plastics to break apart more rapidly, which accelerates the creation of micro and nano plastics. More extreme weather events can promote the spread of plastic pollutants and microparticles around the world (UK-Natural-History-Museum 2022). These particles have been linked with a variety of health impacts on animals and the food chain.

Polyhydroxylalkanoates (PHA) and polylactic acid (PLA) are two of the most notable bioplastics produced by algal metabolism. Algae biomass can be directly moulded into bioplastic beads or sheets, along with algal metabolites, with the potential to replace conventional plastics (Sreenikethanam and Bajhaiya 2021).

Microalgae are well known for their ease of cultivation, ability to grow in wastewater, and ability to survive in harsh conditions, but producing a marketable product faces many challenges from development to commercialization due to high harvesting costs and insufficient quantities for industrial production. As a result, only 1% of plastics are currently biodegradable, with tests only being done in laboratories or on small scales.

32.2.7 Algae for Energy Systems and Biofuels

The majority of biofuels produced today are derived from food crops, competing for land and water with food and feed supply for human consumption. The mass production of biofuel requires more arable agricultural lands, which results in reduced lands for the production of food for humans and animals. This is thought to be the cause of the global food shortage, particularly in developing nations. As a result, these first and second generation biofuels pose significant economic, environmental, and political concerns (Alam et al. 2015). Since algae may be cultivated in saltwater and do not compete with food or feed crops for arable land, they offer a sustainable alternative to these first and second generation biofuels. These are referred to as third-generation biofuels. Algal biomass feedstock has not, however, yet demonstrated itself to be a financially viable source of biofuel or biofuel production.

32.2.8 Wastewater Treatment and Water Conservation

32.2.8.1 Wastewater Treatment

Algae can potentially also replace conventional wastewater treatment systems for small communities as N, P and C are essential nutrients for algal growth and wastewater is an ideal growth media (Samberger 2022b).

The use of algae for wastewater reclamation has been studied since the 1960s and have been applied to the treatment of wastewater due to their capacity to absorb and utilize organic and inorganic carbon, nitrogen, and phosphorus, thereby reducing N, P, and chemical oxygen demand (COD) in the wastewater while simultaneously producing valuable biomass (Plohn, et al. 2021).

Microalgae are not yet used at industrial scale in the treatment of wastewater. Although many microalgal strains have the potential to remove pollutants from wastewater, only a very small proportion seems to be currently used, most likely due to their rapid growth rate, low production cost and high tolerance to extreme environmental conditions but only on side streams treatment.

Furthermore, their high adaptability to different environments allows them to grow in a wide variety of conditions and waters, including municipal, industrial, and agricultural wastewater.

However, current biomass harvesting techniques in wastewater treatment are expensive, and although microalgae have potential to contribute to a circular bioeconomy (Samberger 2022b), further research and development are needed to over-come the current challenges (Plohn, et al. 2021).

In order to overcome the current global water scarcity and the expensive operation and maintenance costs of wastewater treatment plants, the INCOVER project was financed by the EU with a view to move wastewater treatment from being primarily a sanitation technology to being a bio-product recovery industry and a recycled water supplier. INCOVER plants will generate benefits from wastewater through bio-production of bioplastics and organic acids via microalgae/bacteria and yeast biotechnology (Incover 2019–2023).

32.2.8.2 Flood Protection

Heavy rainfall increases pollutant runoff and sedimentation on rivers, lakes, streams (USEPA, Climate Adaptation and Source Water Impacts 2023)

Water quality can be degraded, adequate stormwater management can be prevented, and storage capacity can be reduced by increased erosion and sedimentation. Increased saltwater infiltration into source waters can be a result of drought and sea level rise. The quality and quantity of source water that is readily available may subsequently be lowered as a result of saltwater intrusion, whether it occurs in groundwater or surface water (USEPA, Climate Adaptation and Source Water Impacts 2023).

Climate change also increases the risk of surface water flooding and sewer flooding.

It is especially crucial to protect and restore the foreshore vegetation since extreme weather events are occurring more frequently. Many concerned nations, however, do not have the resources or infrastructure necessary to implement engineering solutions for beach protection. Because of this, many coasts continue to erode and vanish, making them more susceptible to flooding and endangering coastal structures and tourism. Therefore, it is crucial that communities and nations understand how a vegetated foreshore ecosystem can stop coastal erosion and thereby increase these areas' resistance to storm surges and flooding (James et al. 2019).

Algae culture can potentially contribute some of the ecosystem functions by creating coastal habitats that natural kelp forests and macroalgal beds support (Duarte et al. 2017) with benefits to eco-systems such as:

- shoreline protection via dissipation of wave energy
- avoid ocean deoxygenation with warming
- high daytime pH in seaweed to the benefit of calcifiers (such as molluscs), serve a role in protecting them from projected ocean acidification.

32.3 Major Risks and Obstacles to Algae Commercialisation as Nature Based Solutions

Despite all the benefits from algae discussed in this chapter, implementation of algae solutions for climate change adaptation remains limited.

Algae present good opportunities for climate change adaptation such as applications in the built environment in energy efficient buildings or insulation materials. However, the use of algae of algae in this urban sector is still limited.

Perhaps the more advanced application of algae in the economy is for its food applications as nutraceutical and food supplements, which still represents a niche market with only a few algae strains been exploited.

There remain various barriers to the use of algae as alternative to conventional food system, which pertain mainly to product safety as well as commercialization at industrial scale.

Overall, the economic feasibility of large-scale biomass production still remains a major obstacle to algae commercialization (Miranda 2022) and climate adaptation techniques due to some or all of the following barriers:

- few algae species being well known and designated as generally safe for human consumption (Siedenburg 2022)
- contamination with biotic pathogens (Siedenburg 2022) especially for production in open-air systems (Garcia et al. 2017) and microbiological contaminants (Caria-Mendes 2022)
- consumers' perception and acceptance (Siedenburg 2022; Garcia et al. 2017; Caria Mendes et al. 2022)
- affordability (Siedenburg 2022) and cost-effective biomass production (Garcia et al. 2017)
- safety and regulatory issues on toxicity (Garcia et al. 2017), allergens (Caria Mendes et al. 2022) and heavy metals in case of excessive consumption.

More research into the benefits and the safe application of algae into the various sectors of the economy are required in the coming years.

References

Aaronson S (2008) II.C.1 - algae important vegetable supplements. In: Aaronson (ed) The Cambridge World history of food. Cambridge University Press

Alam F, Mobin S, Chowdhury H (2015) Third generation biofuel from Algae. Procedia Eng 105(2015):763–768

Atlas N (2015) Urban Algae Canopy will generate a 4-hectare forest's worth of oxygen. Retrieved from https://newatlas.com/urban-algae-canopy-milan-expo/37480/

Austin WCF-L (2021) COP26 Universities network briefing—April 2021. Retrieved from https://www.naturebasedsolutionsinitiative.org/wp-content/uploads/2021/08/Media_790171_smxx.pdf

BBC (2022) The French town where the lighting is alive. Retrieved from https://www.bbc.com/future/article/20220407-the-living-lights-that-could-reduce-energy-use#:~:text=In%202019%2C%20Rambouillet%20city%20hall,a%20transformation%20across%20the%20country

Boukid F, Castellari M (2021) Food and beverages containing algae and derived ingredients launched in the market from 2015 to 2019: a front-of-pack labeling perspective with a special focus on Spain. Foods

Caria Mendes M, Navalho S, Ferreira A, Paulino C, Figueiredo D, Silva D, Speranza L et al (2022) Algae as food in Europe: an overview of species diversity and their application. Foods 11:1871. https://doi.org/10.3390/foods11131871

Caria-Mendes (2022) Algae as food in Europe: an overview of species diversity and their application. Foods

Chen C, Tang T, Shi Q, Zhou Z, Fan J (2022) The potential and challenge of microalgae as promisingfuture food sources. Trends Food Sic Technol. https://doi.org/10.1016/j.tifs.2022.06.016

Chia S, Macharia J, Diiro G, Kassie M, Ekesi S, van Loon J, Tanga C et al (2020) 21. Smallholder farmers' knowledge and willingness to pay for insect-based feeds in Kenya. PLOS ONE

Colla E, Lupatini Menegotto A, Lahis Kalschne D, Aparecida da Silva-Buzanello R, Canan C, Drunkler D (2020) Microalgae: a new and promising source of food. Handb Algal Sci Technol Med, pp 507–518 – Chapter 32

Conversation T (2022) Climate change threatens food but microscopic algae offer answers. Retrieved from https://theconversation.com/climate-change-threatens-food-but-microscopic-algae-offer-answers-189735

Cordeiro M, Mengistu G, Pogue S, Legesse G, Gunte K, Taylor A, McAllister T (2022) Assessing feed security for beef production within livestockintensive regions. Agric Syst

Duarte CM, Wu J, Xiao X, Bruhn A, Krause-Jensen D (2017) Can Seaweed farming play a role in climate change mitigation and adaptation?—Can seaweed farming play a role in climate change mitigation and adaptation? Frontiers

Ecologicsstudio (2023) Air Bubble. Retrieved from https://www.ecologicstudio.com/projects/air bubble-playground-and-exhibition

Ecologicstudio (2023) Urban algae folly. Retrieved from https://www.ecologicstudio.com/projects/urban-algae-folly-2-0

Ecotree (2023) How much CO_2 does a tree absorb? Retrieved from https://ecotree.green/en/how-much-co2-does-a-tree-absorb

Elmeligy D, Elhassan Z (2019) The Bio-adaptive algae contribution for sustainable architecture. IOPScience

EU (2016) Environmentally friendly, fire-resistant insulation material produced from seagrass. Retrieved from https://ec.europa.eu/regional_policy/en/projects/germany/environmentally-friendly-fire-resistant-insulation-material-produced-from-seagrass

European Union (2013) Retrieved from https://2021.prizes.new-european-bauhaus.eu/index.php/node/269469

FAO (2014) The Water-Energy-Food Nexus—A new approach in support of food security and sustainable agriculture. Retrieved from https://www.fao.org/3/bl496e/bl496e.pdf

Farooq W (2021) Sustainable production of microalgae biomass for biofuel and chemicals through recycling of water and nutrient within the biorefinery context: A review. GCb-Bioenergy

Frier (2016) Living Things. Retrieved from http://www.ethanfrier.com/living-things/

Garcia J, de Vicente M, Galan B (2017) Microalgae, old sustainable food and fashion nutraceuticals. Microbial Biotechnol

Garcia-Gonzalez J, Sommerfeld M (2016) Biofertilizer and biostimulant properties of the microalga Acutodesmus dimorphus. J Appl Phycol

Gerber (2013) Tackling climate change through livestock—A global assessment of emissions and mitigation opportunities. Retrieved from http://www.fao.org/3/a-i3437e.pdf

Glowee (2022) Glopolis—integrate bioluminescence in your development projects. Retrieved from https://en.glowee.com/glowpolis

Guiry MD (2012) How many species of algae are there? J Phycol 48(5)
Habib (2008) A review on the culture, production and use of Spirulina as a food for humans and feeds for domestic animals and fish. In: FAO fisheries and aquaculture circular No. 1034. United Nations Food and Agriculture Organization
Hackaday (2023) The liquid trees of belgrade: the facts behind the furore. Retrieved from https://hackaday.com/2023/04/04/the-liquid-trees-of-belgrade-the-facts-behind-the-furore/
Idealista (2019) These ecological houses in Formentera will use seaweed as a thermal insulator. Retrieved from https://www.idealista.com/en/news/lifestyle-spain/2019/10/08/7025-these-ecological-houses-formentera-will-use-seaweed-thermal
Ifremer (2004) Algae used as food. Retrieved from https://aquaculture.ifremer.fr/en/Info.-Card/Algae-sector/Algae-used-as-food#:~:text=The%20Wakame%20(%20Undaria%20pinnatifida)%20is,sashimis%20(raw%20fish%20preparation).&text=Undaria%2C%20highly%20appreciated%2C%20is%20cooked,also%20consumed%20grilled%20
Iglina T, Iglin P, Pashchenko D (2022) Industrial CO_2 Capture by Algae: a review and recent advances. Sustainability 14:3801. https://doi.org/10.3390/su14073801
Incover (2019-2023) Incover—project summary. Retrieved from https://incover-project.eu/about
INHABITAT (2023) Retrieved from https://inhabitat.com/overpass-algae-garden-turns-co2-emissions-into-combustible-biomass-in-switzerland/
Insights T (2021) 8 foods going extinct due to climate change. Retrieved from https://topicinsights.com/sustainability/foods-going-extinct-due-to-climate-change/#:~:text=Increased%20temperatures%20in%20the%20atmosphere,extinction%20in%20the%20near%20future
ITSLIQUID. (2019) French Dream Towers by XTU. Retrieved from https://www.itsliquid.com/french-dream-towers-by-xtu.html
James R, Silva R, van Tussenbroek B, Escudero-Castillo M, Marino-Tapia I, Dijkstra H, Bouma Tjeerd J (2019) Maintaining tropical beaches with seagrass and Algae: a promising alternative to engineering solutions. BioScience
Khavari F, Saidijam M, Taheri M, Nouri F (2021) Microalgae: therapeutic potentials and applications. Mol Biol R
Kisser J et al (2020) A review of nature-based solutions for resource recovery in cities. Blue-Green Syst 2(1)
Kumari A, Garima, Bharadvaja N (2023) A comprehensive review on algal nutraceuticals as prospective therapeutic agent for different diseases. Biotech
Liao Y-C, Chang C-C, Nagarajan D, Chen C-Y, Chang J-S (2021) Algae-derived hydrocolloids in foods: applications and health-related issues. Bioengineered
Lopes Amorim M, Soares J, Dos Reis Coimbra J, de Oliveira Leite M, Teixeire Albino L, Aredes Martins M (2021) Microalgae proteins: production, separation, isolation, quantification and application in food and feed. Crit Rev Food Sci Nutr 61(12)
Miranda (2022) Trends on CO_2 capture with microalgae: a bibliometric analysis. Molecules
Newsweek 90 (2023, 03 04) New "Liquid Trees" divide the internet. Retrieved from https://www.newsweek.com/viral-post-liquid-tree-algae-carbon-dioxide-1792214#:~:text=The%20%22liquid%20trees%22%20actually%20emerged,(CO2)%20from%20city%20air
Nuwen (2022) Why algae have become a must for the food challenge? Retrieved from https://www.nuwen.com/en/food/why-algae-have-become-a-must-for-the-food-challenge/
OIPA (2017) Do you know what gelatin is made of? Retrieved from https://www.oipa.org/international/vegan-gelatin/
Van Oosten M, Pepe O, De Pascale S, Silletti S, Maggio A (2017) The role of biostimulants and bioeffectors as alleviators of abiotic stress in crop plants. Chem Biol Technol Agric 4:5. https://doi.org/10.1186/s40538-017-0089-5
Pathak T, Maskey M, Dahlberg J, Kearns F, Bali K, Zaccaria D (2018) Climate change trends and impacts on California agriculture: a detailed review. Agronomy 8:25. https://doi.org/10.3390/agronomy8030025
photo.Synthetica (2017) Bio.Tech HUT. Retrieved from https://www.photosynthetica.co.uk/copy-of-application-interior-1

photoSynthetica (2023) Algae curtain. Retrieved from https://www.photosynthetica.co.uk/cladding
Plohn M, Spain O, Sirin S, Silva M, Escudero-Onate C, Ferrando-Climent L, Funk C (2021) Wastewater treatment by microalgae. Physiol Plant
Proksch G (2012) Growing sustainability—Integrating Algae cultivation into the built environment. Edinb Architectural Res J Edinb Sch Architect Edinb UK 33
Prometheus Materials (2023) Retrieved from https://prometheusmaterials.com/home/
Ronga D, Biazzi E, Parati K, Carminati D, Carminati E, Tava A (2019) Microalgal biostimulants and biofertilisers in crop productions. Agronomy
Saide A, Martinez K, Ianora A, Lauritano C (2021) Unlocking the health potential of microalgae as sustainable sources of bioactive compounds. Mol Sci
Samberger C (2022a) Food-water-energy nexus for (big) children and Scepticals. Am J Biomed Sci Res
Samberger, C (2022b) The role of water circularity in the food-water-energy nexus and climate change mitigation. Energy Nexus
Seddon N, Chausson A, Berry P, Girardin CA, Smith A, Turner B (2020) Understanding the value and limits of nature-based solutions to climate change and other global challenges. Phil Trans R Soc B375:20190120. Philos Trans R Soc B. https://doi.org/10.1098/rstb.2019.0120
Siedenburg J (2022) Could microalgae offer promising options for climate action via their agri-food applications? Frontiers
Sills D, Gerber van Doren L, Beal C, Raynor E (2020) The effect of functional unit and co-product handling methods on life cycle assessment of an algal biorefinery
Sreenikethanam A,Bajhaiya A (2021) Algae based bio-plastics: future of green economy. https://doi.org/10.5772/intechopen.100981. In Biernat, *Biorefineries*
Thiyagarasaiyar K, Goh B-H, Jeon Y-J, Yow Y-Y (2020) Algae metabolites in cosmeceutical: an overview of current applications and challenges. Mar Drugs
TIME (2022) *Cutting* construction's climate impact with Algae bricks | TIME. Retrieved from https://time.com/6192603/algae-plant-buildings-carbon/
UK-Natural-History-Museum (2022) The solutions to climate change and plastic pollution are linked. Retrieved from https://www.nhm.ac.uk/discover/news/2022/july/solutions-climate-change-plastic-pollution-linked.html
UNCC (2023) Adaptation and resilience. Retrieved from https://unfccc.int/topics/adaptation-and-resilience/the-big-picture/introduction#:~:text=Adaptation%20refers%20to%20adjustments%20in,opportunities%20associated%20with%20climate%20change
UNFCC 2023. (Accessed Oct 2023). Retrieved from https://unfccc.int/topics/adaptation-and-resilience/the-big-picture/introduction
UP'Magazine (2017) Pollution guzzling algae have entered Paris. Retrieved from https://up-magazine.info/en/le-vivant/innovations-vertes/6870-des-algues-devoreuses-de-pollution-sont-entrees-dans-paris/
USEPA (2022) Climate change impacts on the built environment | *US EPA*. Retrieved from https://www.epa.gov/climateimpacts/climate-change-impacts-built-environment
USEPA (2023) Climate adaptation and source water impacts. Retrieved from https://www.epa.gov/arc-x/climate-adaptation-and-source-water-impacts
USEPA. (n.d.). Climate change impacts on the built environment | *US EPA*. Retrieved from https://www.epa.gov/climateimpacts/climate-change-impacts-built-environment
Valente L, Cabrita A, Maia M, Valente I, Engrola S., Fonseca A, Bengala Freire J et al (2021) Chapter 9—Microalgae as feed ingredients for livestock production and aquaculture. In: Microalgae. Academic Press
Veron A, Morales Ibarria M, Lopez E (2013) Bioluminescent algae and possible implications in architectural design. In: PLEA2013—29th conference, sustainable architecture for a renewable future, Munich, Germany 10–12 Sept 2013. Munich
Violano A, Cannaviello M (2019) Green-algae resilient architecture

Walsh B, Rydzak F, Palazzo A, Kraxner F, Herrero M, Schenk P, Obersteiner M et al (2015) New feed sources key to ambitious climate targets. Carbon Balance Manag

WWF (2023) Retrieved from What's the difference between climate change mitigation and adaptation? | Stories | WWF: https://www.worldwildlife.org/stories/what-s-the-difference-between-climate-change-mitigation-and-adaptation

Chapter 33
Climate Change: Deducing the Importance of Algae as a Significant Tool for Mitigation of the Eminent Threat of Climate Induced Changes of Environment

Rohan Kr Biswas and Avik Kumar Choudhury

Abstract In the past few decades, "Climate Change" has emerged as one of the biggest concerns for scientists, ecologists, and environmentalists as well as for the common people. The main reason of changing climate is global warming. The increase of global temperature triggers different phenomenon like melting of glaciers and rise in sea level, increase in sea surface temperature and ocean acidification, loss of biodiversity and decreased ecosystem functioning. Population growth further adds to anthropogenic pollution and nutrient enrichment of habitats. These phenomena negatively affect both nature and the human population. The most affecting cause of global warming is the increase in Green House Gases. The Green House Gases increase mainly due to burning of fossil fuels. Over exploitation of fossil fuel have negative impacts on environment as well as on economy. Algae not only reduce atmospheric GHGs by capturing carbon di oxide, but also used to make green and sustainable energy products, which have the potential to replace fossil fuel and contribute towards carbon reduction. In this chapter, importance of algae as a potential tool for mitigation of climate change is discussed with an insight into their potentiality in contributing towards carbon sequestration and sustainable blue economy.

Keywords Climate change · Carbon emission · Greenhouse gases (GHGs) · Sustainable products · Carbon mitigation

---

R. K. Biswas · A. K. Choudhury (✉)
Phycology Ecology and Remote Sensing Lab, Department of Botany, Ramakrishna Mission Vivekananda Centenary College, Rahara, Kolkata, West Bengal 700118, India
e-mail: avik.choudhury@rkmvccrahara.org

T. A. Kurniawan and A. Anouzla (eds.), *Algae as a Natural Solution for Challenges in Water-Food-Energy Nexus*, Environmental Science and Engineering,
https://doi.org/10.1007/978-981-97-2371-3_33

33.1 Introduction

One of the most concerned topics in recent times is "Climate change." As opined by Paul et al. (2020), "Climate Change" is the long-term change of climatic condition at a regional or global scale (Paul et al. 2020). Historically, the Earth has evolved for the past 450 billion years and is still evolving. As any system evolves it undergoes changes and so did the Earth from a completely nitrogen enriched anoxygenic environment to oxygenic environment. Nearly 400 billion years ago life forms came into existence as evidenced from Urey and Miller's Experiment through formation of "ceocervates" (Miller and Urey 1959).

However, the emergence of life forms like cyanobacteria, photosynthetic prokaryotes and photosynthetic protists completely altered the anoxygenic harsh environment of primitive Earth to a more suited oxygenic environment that paved the way for the biodiversity "boom" on the face of the Earth. Such changes were for the benefit of biological population with evolution of human race from the primates. However, as human population continued to evolve, exploration of newer habitats for settlements started to impact biodiversity.

Furthermore, at the turn of the century as urbanization and industrialization became the call of the hour, unplanned and merciless exploitation of natural resources and deforestation took a heavy toll on natural biodiversity inclusive of both floral and faunal populations. At the beginning of evolution as human population was at the mercy of nature, they were highly concerned about the wellbeing of the habitats and ecosystems around them as sources of food and livelihood. As sciences progressed, emergence of alternates to natural resources made human population less concerned about the environment that led to more intense exploitation of natural biodiversity and resources around the world. Moreover, industrialization and population growth led to higher anthropogenic carbon footprints on the Earth that eventually enhanced the buildup of greenhouse gases.

33.2 The Greenhouse Effect and Climate Change

The root cause of this change in climate is the increase of Green House Gases (GHGs) (Paul et al. 2020). A gas that can absorb infra-red (IR) radiation, can be considered as GHG like Carbon dioxide (CO_2), Ozone (O_3), Methane (CH_4), Water vapour (H_2O). The Earth receives Ultraviolet (UV) and Visible (Vis) radiation from the Sun and it emits InfraRed (IR) radiation to the atmosphere. These GHGs are important as they trap the IR radiation, which is emitted from the Earth and reradiate it to the surface of the Earth (Fig. 33.1) (Kweku et al. 2018). This is very important for maintaining the Earth's atmospheric temperature. This phenomenon is generally called "primary greenhouse effect." The most dominant GHG for the primary greenhouse effect is H_2O, not the other ones mentioned above. So, if the greenhouse effect is good and important for us, why all the blaming it for our destruction. It is not the primary greenhouse effect which the scientists are talking about, it is the "Secondary greenhouse effect" that is a cause of concern for the present day as opined by environmentalist and climate change activists (Tuckett 2019). Due to the industrial revolution and different anthropogenic activities, the concentration of GHGs increased and due to that more IR radiation is trapped that has led to global warming (Kweku et al. 2018). The most dominant GHG for secondary greenhouse effect is CO_2, which increased by almost 43% in last 270 years (Tuckett 2019). Among all GHGs, CO_2 contributes half of the total share of GHGs (Gutsche et al. 2021) (Fig. 33.2). The main reason of increasing CO_2 in the atmosphere is the fossil fuel burning (Kumar et al. 2010). Fossil fuel based economic growth is not only harmful for the climate on a global scenario, economically it is dangerous too (Nel and Cooper 2009). It creates a shortage in fuel storage mainly for the developing countries (Asadian et al. 2018). More the CO_2 concentration in the atmosphere, more IR radiation is absorbed and the temperature of Earth's atmosphere will rise (Tuckett 2019). From current concentration, if the CO_2 concentration gets doubled, the rise of temperature will be 2 °C (Paul et al. 2020). A prediction was made by Intergovernmental Panel on Climate Change (IPCC), where it was estimated to be a global temperature rise of 1–6 °C within 2100 (Asadian et al. 2018).

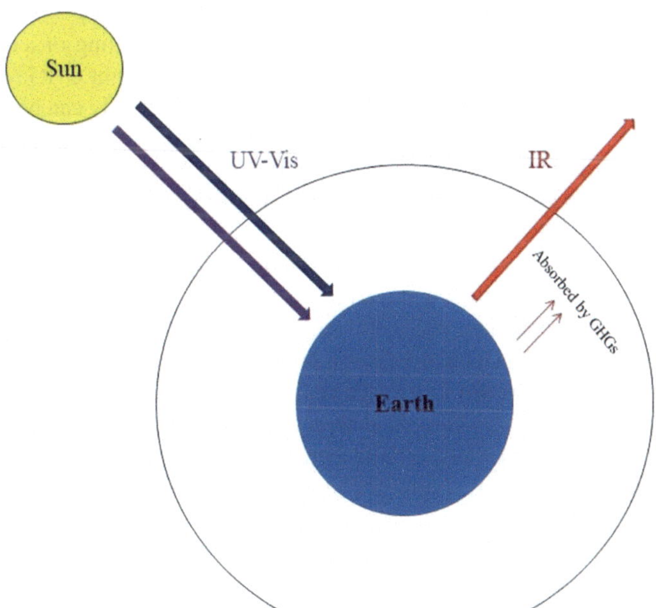

Fig. 33.1 Earth receives Ultraviolet (UV) and Visible (Vis) radiation from the Sun and it emits IR radiation to the atmosphere, which are absorbed by GHGs (Tuckett 2019)

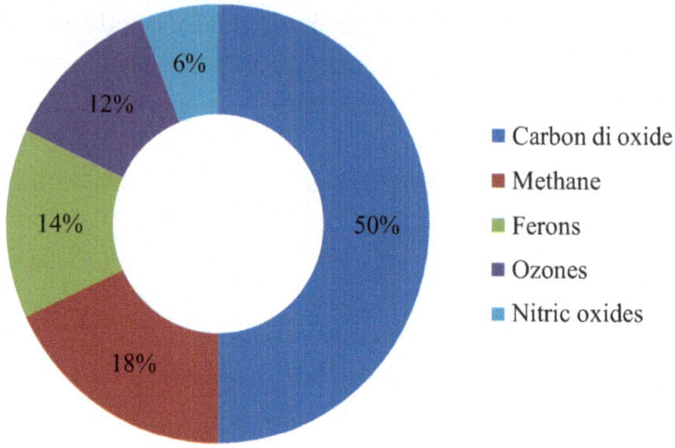

Fig. 33.2 Percentage share of greenhouse gases in Earth's atmosphere (Gutsche et al. 2021)

As the temperature is rising due to global warming, different effects are seen in nature. According to NASA's Jet Propulsion Laboratory, Pasadena, California, the frequency of extremely high clouds responsible for heavy rainfall and storm increased due to warming of the atmosphere (https://www.jpl.nasa.gov/). The increasing concentration of GHGs is responsible for increased ocean heat content (OHC), which mainly affects the euphotic zone (Trenberth et al. 2018). The increased sea surface temperature (SST) acts as a driving force for storms and it changes the climate (Vecchi and Soden 2007). The sea-ice of Arctic is melting due to increase of heat and this phenomenon itself enhance the greenhouse effect. As the melting of ice from the sea is enhanced, sunlight reflection is decreased and an important GHG Methane is released (Ming et al. 2014). Increased nutrients and metal toxicity can be correlated with melting of glaciers and increased CO_2 levels in the ocean, which is the reason for decrease in pH and increase in ocean acidification (Asadian et al. 2018). Ocean acidification is a reason for limited bioavailability of nutrients in marine systems (Riebesell and Tortell 2011).

Precipitation reports and different drought indices show a clear increase in aridity and its positive relation with global warming (Dai 2013). Since the impact of global warming is more severe on the poles, the depletion of polar ice may account for a rise in sea level. Such increase in water level is another consequence of global warming (Shukla et al. 2017). The sea is a major sink for CO_2 and the solubility of CO_2 is much higher in seawater as compared to atmospheric air. Global warming causes increase in the expanse of water volume thereby allowing even higher solubilization of CO_2 in seawater. As CO_2 solubilizes, it enhances the level of carbonic acid that contributes to a rise in H^+ concentrations which is the primary reason for ocean acidification (Six et al. 2013).

$$H_2O + CO_2 \rightarrow H^+ + HCO_3{}^-;\ HCO_3{}^- \rightarrow H^+ + CO_3{}^{2-}$$

This changes the carbonate: bicarbonate ratio with higher residence time of bicarbonate in the water column reduces the availability of assimilable carbonate for members like coccolithophores. Biological diversity is also negatively affected due to global warming (Bálint et al. 2011). Increasing temperature also hampers the agricultural sectors and food production (Singh et al. 2014). Increase the levels of GHGs in atmosphere modifies marine's parameters such as, marine pH, salinity and temperature (Harley et al. 2006).

33.3 Role of Algae in Capturing CO_2

CO_2 sequestration technologies can be divided into two categories—physical and biological (Berberoglu et al. 2009). Photosynthesis is the most sustainable method for biological CO_2 sequestration (Kumar et al. 2010). Algae are photosynthetic organisms (Lee 2018), having both unicellular and multicellular members (Falkowski and Raven 1997; Lee 2018). Algae are the main producer community of natural aquatic environment inclusive of both lotic and lentic habitats (Lee 2018). Microalgae produce more than 50% of global oxygen (O_2) by using CO_2 from the atmosphere (Sahoo et al. 2012). In comparison with higher plant groups, CO_2 fixation by algal groups is more beneficial as the rate of growth of algae is higher, the CO_2 concentrating mechanism is more efficient and CO_2 fixation ability of algae is higher (Maeda et al. 1995; Metting 1996). Microalgae grow in diverse habitats; it can grow in freshwater, brackish water or in saline water which makes it a very important candidate for global CO_2 sequestration (Cantrell et al. 2008). Algae even grow in wastewater (Ashokkumar and Rengasamy 2012). Clearly the role of algae in capturing CO_2 is very important in this era when CO_2 is rising. Thus, algae can be considered as a potential tool for CO_2 sequestration. There are some specific algal species present which are more tolerant in higher CO_2 concentration. *Scenedesmus obliquus* is found to be most productive when the CO_2 concentration is 6% (De Morais and Costa 2007). Though, there are many algal species that are tolerant to higher concentration of CO_2, the growth rate, productivity and capacity of CO_2 fixation is compromised. *Chlorella* sp. ZY-1, *Chlorella* sp. KR-1 and *Chlorella* sp. T-1 can tolerate 70%, 70% and 100% CO_2 concentration respectively (Kodama 1993; Sung et al. 1999). *Chlorococcum littorale* can tolerate 60% CO_2 concentration (Sasaki et al. 1998). Clearly, microalgal photosynthesis process is one of the most economic, sustainable and environment friendly process of CO_2 fixation (Mukherjee and Moroney 2011). The best CO_2 mitigation strategy for algal cultivation is the use of flue gas. Thermophilic algal strains are used in the mitigation of flue gas which is directly incorporated from coal power stations at temperature from 60 to 100 °C. Cooling cost can be cut if the thermophilic strains are used such as *Cyanidium caldarium* and *Synechococcus elongatus* (Miyairi 1995).

33.4 Algae as Bioindicator of Climate Change

Though algae have a significance role in CO_2 capture and help in mitigation of climate change especially global warming, it also affects the algal members in terms of growth and survival. It is necessary to determine the responses of algae when affected by climate change (Walther et al. 2002; Merilä 2012). It should be kept in mind that, "Climate Change" is a simultaneous change of various parameters like pH, temperature, CO_2 concentration etc. and these changes affects physiology, morphology, behaviour and genetics of algae (Williams et al. 2015; Poloczanska et al. 2013). Due to the change of physical and chemical properties of water the dynamics of algal community changes and assessment of algal community provide us with an idea about changing environment (Nash et al. 2015; Baker et al. 2008; Kordas et al. 2015). Algae are now considered as a potent bioindicator for changing climatic conditions (Arts et al. 2015). Phytoplankton is important bioindicators because they are free floating with a short life cycle and remain undisturbed (Hays et al. 2005). While assessing climate change by observing algae, the change in biomass, changes in capacity of photosynthesis, blooms, water colour are determined. The physiology and reproduction of algae, concentration of stress enzymes and defense mechanisms are altered due to the stress from climatic change and these characteristics are observed and analyzed to get the idea about the changing climate (Torres et al. 2008). Diatoms are one of the potent bioindicators, especially in higher altitudes. Diatoms have siliceous cell wall whose morphology can be altered due to different changes in the physical or chemical parameters. By observing the frustules, the siliceous cell wall of diatom, these changes can be assessed (Desrosiers et al. 2013). Diatom is excellent when it comes to indicate the metal toxicity (Biswas and Choudhury 2021). Due to presence of silica in their cell wall diatom is an excellent tool for micropaleontology (Espinosa et al. 2012). Nitrogen (N) and/or Phosphorus regulates the primary productivity of phytoplankton community and changes in concentration of N and/or P in water bodies can hamper the process; even in freshwater system, the reduced diatom concentration is an indication for changes in the P concentration (Woundie 2006). Different physiological responses of algae to different environmental conditions are shown in Table 33.1.

33.5 Low-Carbon Bioenergy Derived from Algae

One of the main problems in consuming fossil fuel are reducing storage and burning that cause pollution. For both the problems the solution is to find an alternative source which is carbon neutral or low in carbon. Producing biofuel from the crops is problematic as it has low capacity of carbon acquisition and need land and water (Valentine et al. 2012). Algae derived biofuel can be an alternative as it has many advantages like high production rate and low cost with high photosynthetic yield (Chiaramonti et al. 2017). The photosynthetic yield can be enhanced in algal cultures

Table 33.1 Different physiological responses of algae to different environmental conditions (as modified from Asadian et al. 2018)

Physiological response	Environmental conditions				
	Increase of temperature	Ocean acidification	Increase of salinity	Enhancement of stratification	Decreased light penetration
Reproduction	Increased	Decreased	Decreased	Decreased	Decreased
Photosynthesis	Increased	Increased or not change	Decreased	Decreased	Decreased
Biofixation of nutrients	Increased	Decreased	Decreased	Decreased	No change
Calcification	Decreased	Decreased	Decreased	Decreased	Decreased

by genetic engineering (Asadian et al. 2018). There are generally four pathways well known for formation of biofuels from algal cultures:

1. Biodiesel production from Triacylglycerol transesterification,
2. Fermentation of algal carbohydrates to produce bio-alcohol,
3. Biogas production from anaerobic digestion,
4. Process of gasification (Craggs et al. 2011).

33.5.1 Biodiesel

Triacylglycerol is the main component for biodiesel production. Algae can store a good amount of Triacylglycerol which make these organisms a potent candidate for biodiesel production. Genetic engineering techniques are used to enhance the yield (Huang et al. 2010; Talebi et al. 2014). Algal members like *Chlorella vulgaris* (Zheng et al. 2011), *Dunaliella salina* (Liu et al. 2013), *Nannochloropsis* sp. (Loong and Idris 2017), *Botryococcus braunii* (Cook et al. 2017) are used for biodiesel production.

33.5.2 Bioalcohol

Algae can produce high amount of carbohydrate due to higher photosynthetic rate and they can convert it into fermentable sugars without energy-consuming pretreatment stage, making it a good candidate for bioalcohol production (Demirbas and Demirbas 2011; Jambo et al. 2016).

33.5.3 Biogas

Production of algae-based biogas is more efficient than production of biogas from crops (Sialve et al. 2009). After extraction of lipids and fermented sugars, the algal biomass is used to produce the biogas (Asadian et al. 2018).

33.5.4 Production of Hydrogen

Hydrogen gas is considered as one of the alternatives of fossil fuel. Different algal members like *Chlamydomonas reinhardtii, Chlorella fusca* produce hydrogen gas when present in anaerobic condition. *Chlamydomonas reinhardtii* produce hydrogen, instead of oxygen, in the photosynthesis process when it is present in a sulphur free environment (Melis et al. 2000). Clearly these algal members are potential candidates for hydrogen production in a large scale.

33.6 Conclusion

In this era of climate change as carbon footprints continue to increase due to rise in carbon emission and global warming, the severity of climatic changes can be multifarious. To reduce such impacts, carbon sequestration through different means is of utmost necessity. Plants are the main natural sequester of carbon footprints through photosynthesis with production of organic carbon to be used as natural food sources. However, algae are more efficient in carbon mitigation through production of a diverse array of biomolecules having significant economic importance. The mechanism of carbon fixation in algae ranges from C1 to C4 types which are most diverse among different plant groups. So, carbon mitigation through production of important biomolecules with commercial importance is most diverse in algal forms. Thus, carbon mitigation through algae will be the future direction for successful "blue economy" and sustenance of future environment of the Earth.

Acknowledgements Mr. Rohan Kr Biswas acknowledges the Human Resource Development Group, Council of Scientific and Industrial Research (CSIR-HRDG), for providing CSIR-Junior Research Fellowship.

References

Arts MT, Hixson SM, Mercieca S (2015) Climate warming reduces essential fatty acid production in algae

Asadian M et al (2018) Algal communities: an answer to global climate change. Clean (Weinh) 46(10):1800032. https://doi.org/10.1002/clen.201800032
Ashokkumar V, Rengasamy R (2012) Mass culture of *Botryococcus braunii* Kutz. under open raceway pond for biofuel production. Bioresour Technol 104:394–399. https://doi.org/10.1016/j.biortech.2011.10.093
Baker AC, Glynn PW, Riegl B (2008) Climate change and coral reef bleaching: an ecological assessment of long-term impacts, recovery trends and future outlook. Estuar Coast Shelf Sci 80(4):435–471. https://doi.org/10.1016/j.ecss.2008.09.003
Bálint M et al (2011) Cryptic biodiversity loss linked to global climate change. Nat Clim Change 1(6):313–318. https://doi.org/10.1038/nclimate1191
Berberoglu H, Gomez PS, Pilon L (2009) Radiation characteristics of *Botryococcus braunii*, *Chlorococcum littorale*, and *Chlorella* sp. used for CO_2 fixation and biofuel production. J Quant Spectrosc Radiat Transf 110(17):1879–1893. https://doi.org/10.1016/j.jqsrt.2009.04.005
Biswas RK, Choudhury AK (2021) Diatoms: miniscule biological entities with immense importance in synthesis of targeted novel bioparticles and biomonitoring. J Biosci 46:102. https://doi.org/10.1007/s12038-021-00222-x
Cantrell KB, Ducey T, Ro KS (2008) Livestock waste-to-bioenergy generation opportunities. Bioresour Technol 99(17):7941–7953. https://doi.org/10.1016/j.biortech.2008.02.061
Chiaramonti D et al (2017) Review and experimental study on pyrolysis and hydrothermal liquefaction of microalgae for biofuel production. Appl Energy 185:963–972. https://doi.org/10.1016/j.apenergy.2015.12.001
Cook C et al (2017) Third-generation biofuels from the microalga, *Botryococcus braunii*. Biomass Bioenergy. https://doi.org/10.1002/9781118350553.ch10
Craggs RJ et al (2011) Algal biofuels from wastewater treatment high rate algal ponds. Water Sci Technol 63(4):660–665. https://doi.org/10.2166/wst.2011.100
Dai A (2013) Increasing drought under global warming in observations and models. Nat Clim Change 3(1):52–58. https://doi.org/10.1038/nclimate1633
De Morais MG, Costa JAV (2007) Carbon dioxide fixation by *Chlorella kessleri*, *C. vulgaris*, *Scenedesmus obliquus* and *Spirulina* sp. cultivated in flasks and vertical tubular photobioreactors. Biotechnol Lett 29:1349–1352. https://doi.org/10.1007/s10529-007-9394-6
Demirbas A, Demirbas MF (2011) Importance of algae oil as a source of biodiesel. Energy Convers Manag 52(1):163–170. https://doi.org/10.1016/j.enconman.2010.06.055
Desrosiers C, Leflaive J, Eulin A (2013) Bioindicators in marine waters: benthic diatoms as a tool to assess water quality from eutrophic to oligotrophic coastal ecosystems. Ecol Indic 32:25–34. https://doi.org/10.1016/j.ecolind.2013.02.021
Espinosa LF, Pantoja S, Rullkötter J (2012) Application of gas chromatography in a study of marine biogeochemistry. In: Salih B (ed) Gas chromatography—biochemicals, narcotics and essential oils, p 97. https://doi.org/10.5772/32318
Falkowski PG, Raven JA (1997) Aquatic photosynthesis. Blackwater Science, London
Gutsche J, Muślewski Ł, Dzioba A et al (2021) Identification and analysis of factors influencing climate change in terms of CO_2 emissions. MATEC Web Conf 332:01002. https://doi.org/10.1051/matecconf/202133201002
Harley CD et al (2006) The impacts of climate change in coastal marine systems. Ecol Lett 9(2):228–241. https://doi.org/10.1111/j.1461-0248.2005.00871.x
Hays GC, Richardson AJ, Robinson C (2005) Climate change and marine plankton. Trends Ecol Evol 20(6):337–344. https://doi.org/10.1016/j.tree.2005.03.004
Huang G, Chen F, Wei D, Zhang X, Chen G (2010) Biodiesel production by microalgal biotechnology. Appl Energy 87(1):38–46. https://doi.org/10.1016/j.apenergy.2009.06.016
Jambo SA, Abdulla R, Azhar SHM et al (2016) A review on third generation bioethanol feedstock. Renew Sustain Energy Rev 65:756–769. https://doi.org/10.1016/j.rser.2016.07.064
Kodama M (1993) A new species of highly CO_2-tolerant, fast-growing marine microalga suitable for high-density culture. J Mar Biotechnol 1:21–25

Kordas RL et al (2015) Intertidal community responses to field-based experimental warming. Oikos 124(7):888–898. https://doi.org/10.1111/oik.00806

Kumar A et al (2010) Enhanced CO_2 fixation and biofuel production via microalgae: recent developments and future directions. Trends Biotechnol 28(7):371–380. https://doi.org/10.1016/j.tibtech.2010.04.004

Kweku DW, Bismark O, Maxwell A (2018) Greenhouse effect: greenhouse gases and their impact on global warming. J Sci Res 17(6):1–9. https://doi.org/10.9734/JSRR/2017/39630

Lee RE (2018) Phycology. Cambridge University Press. https://doi.org/10.1017/9781316407219

Liu J, Mukherjee J, Hawkes JJ et al (2013) Optimization of lipid production for algal biodiesel in nitrogen stressed cells of *Dunaliella salina* using FTIR analysis. J Chem Technol Biotechnol 88(10):1807–1814. https://doi.org/10.1002/jctb.4027

Loong TC, Idris A (2017) One step transesterification of biodiesel production using simultaneous cooling and microwave heating. J Clean Prod 146:57–62. https://doi.org/10.1016/j.jclepro.2016.03.155

Maeda K et al (1995) CO_2 fixation from the flue gas on coal-fired thermal power plant by microalgae. Energy Convers Manag 6(36):717–720. https://doi.org/10.1016/0196-8904(95)00105-M

Melis A, Zhang L, Forestier M (2000) Sustained photobiological hydrogen gas production upon reversible inactivation of oxygen evolution in the green alga *Chlamydomonas reinhardtii*. Plant Physiol 122(1):127–136. https://doi.org/10.1104/pp.122.1.127

Merilä J (2012) Evolution in response to climate change: in pursuit of the missing evidence. BioEssays 34(9):811–818. https://doi.org/10.1002/bies.201200054

Metting FB (1996) Biodiversity and application of microalgae. J Ind Microbiol 17:477–489. https://doi.org/10.1007/BF01574779

Miller SL, Urey HC (1959) Organic compound synthesis on the primitive earth. Science 130(3370):245–251. https://doi.org/10.1126/science.130.3370.245

Ming T, Liu W, Caillol S (2014) Fighting global warming by climate engineering: is the earth radiation management and the solar radiation management any option for fighting climate change? Renew Sustain Energy Rev 31:792–834. https://doi.org/10.1016/j.rser.2013.12.032

Miyairi S (1995) CO_2 assimilation in a thermophilic cyanobacterium. Energy Convers Manag 36(6–9):763–766. https://doi.org/10.1016/0196-8904(95)00116-U

Mukherjee B, Moroney JV (2011) Algal carbon dioxide concentrating mechanisms. https://doi.org/10.1002/9780470015902.a0000314.pub4

Nash MC, Uthicke S, Negri AP (2015) Ocean acidification does not affect magnesium composition or dolomite formation in living crustose coralline algae, *Porolithon onkodes* in an experimental system. Biogeosciences 12(17):5247–5260. https://doi.org/10.5194/bg-12-5247-2015

Nel WP, Cooper CJ (2009) Implications of fossil fuel constraints on economic growth and global warming. Energy Policy 37(1):166–180. https://doi.org/10.1016/j.enpol.2008.08.013

Paul V et al (2020) Role of algae in CO_2 sequestration addressing climate change: a review. Renew Energy Clim Change. In: Proceedings of REC 2019, pp 257–265. https://doi.org/10.1007/978-981-32-9578-0_23

Poloczanska ES et al (2013) Global imprint of climate change on marine life. Nat Clim Change 3(10):919–925. https://doi.org/10.1038/nclimate1958

Riebesell U, Tortell PD (2011) Effects of ocean acidification on pelagic organisms and ecosystems. Ocean Acidif. https://doi.org/10.1093/oso/9780199591091.003.0011

Sahoo D, Elangbam G, Devi SS (2012) Using algae for carbon dioxide capture and bio-fuel production to combat climate change. Phykos 42(1):32–38

Sasaki T, Kurano N, Miyachi S (1998) Cloning and characterization of high-CO_2-specific cDNAs from a marine microalga, *Chlorococcum littorale*, and effect of CO_2 concentration and iron deficiency on the gene expression. Plant Cell Physiol 39(2):131–138. https://doi.org/10.1093/oxfordjournals.pcp.a029349

Shukla JB, Verma M, Misra AK (2017) Effect of global warming on sea level rise: a modeling study. Ecol Complex 32:99–110. https://doi.org/10.1016/j.ecocom.2017.10.007

Sialve B, Bernet N, Bernard O (2009) Anaerobic digestion of microalgae as a necessary step to make microalgal biodiesel sustainable. Biotechnol Adv 27(4):409–416. https://doi.org/10.1016/j.biotechadv.2009.03.001
Singh SK, Dixit K, Sundaram S (2014) Algal-based CO_2 sequestration technology and global scenario of carbon credit market: a review. Am J Eng Res 3(4):35–37
Six KD et al (2013) Global warming amplified by reduced sulphur fluxes as a result of ocean acidification. Nat Clim Change 3(11):975–978. https://doi.org/10.1038/nclimate1981
Sung KD, Lee JS, Shin CS et al (1999) CO_2 fixation by *Chlorella* sp. KR-1 and its cultural characteristics. Bioresour Technol 68(3):269–273. https://doi.org/10.1016/S0960-8524(98)00152-7
Talebi AF et al (2014) Manipulation of carbon flux into fatty acid biosynthesis pathway in *Dunaliella salina* using AccD and ME genes to enhance lipid content and to improve produced biodiesel quality. Biofuel Res J 1(3):91–97
Torres MA et al (2008) Biochemical biomarkers in algae and marine pollution: a review. Ecotoxicol Environ Saf 71(1):1–15. https://doi.org/10.1016/j.ecoenv.2008.05.009
Trenberth KE et al (2018) Hurricane Harvey links to ocean heat content and climate change adaptation. Earth's Future 6(5):730–744. https://doi.org/10.1029/2018EF000825
Tuckett R (2019) Greenhouse gases. In: Encyclopedia of analytical science. Elsevier, pp 362–372. https://doi.org/10.1016/B978-0-12-409547-2.14031-4
Valentine J et al (2012) Food vs. fuel: the use of land for lignocellulosic 'next generation' energy crops that minimize competition with primary food production. Glob Change Biol Bioenergy 4(1):1–19. https://doi.org/10.1111/j.1757-1707.2011.01111.x
Vecchi GA, Soden BJ (2007) Effect of remote sea surface temperature change on tropical cyclone potential intensity. Nature 450(7172):1066–1070. https://doi.org/10.1038/nature06423
Walther GR et al (2002) Ecological responses to recent climate change. Nature 416(6879):389–395. https://doi.org/10.1038/416389a
Williams CM, Henry HA, Sinclair BJ (2015) Cold truths: how winter drives responses of terrestrial organisms to climate change. Biol Rev 90(1):214–235. https://doi.org/10.1111/brv.12105
Woundie A (2006) Temporal dynamics of biomass and primary production of phytoplankton in relation to some physico chemical factors in Lake Hora, Ethiopia. Doctoral dissertation, M.Sc. thesis, Addis Ababa University, Ethiopia
Zheng H et al (2011) Disruption of *Chlorella vulgaris* cells for the release of biodiesel-producing lipids: a comparison of grinding, ultrasonication, bead milling, enzymatic lysis, and microwaves. Appl Biochem Biotechnol 164:1215–1224. https://doi.org/10.1007/s12010-011-9207-1

Chapter 34
Bioremediation of Microalgae-Based Pesticides

Walter José Martínez-Burgos, Roberta Pozzan, Alexander da Silva Vale, Júlio Cesar de Carvalho, Hissashi Iwamoto, Luciana Porto de Souza Vandenberghe, Maria Clara Manzoki, Thamarys Scapini, Ihana Aguiar Severo, and Carlos Ricardo Soccol

Abstract Pesticides are generally synthetic compounds used in agriculture to combat diseases caused by pests in crops and at the same time increase productivity. Around 2.7 million tons of these compounds are applied around the world each year. Nevertheless, the intensified use of agrotoxics can cause severe impacts in different environmental matrices, mainly in soil and water bodies due to their accumulation and persistence. Pesticides can seriously affect human health, as they can induce hormonal and reproductive changes, liver and kidney damage, immune dysfunction, cognitive and neuromotor disorders and cancer. In addition, these compounds can affect species in different ecosystems. Therefore, effective technologies are needed to remove these pollutants. Currently, the processes for eliminating xenobiotic compounds are focused on the use of biological and environmentally sustainable materials and economic strategies through approaches to bioremediation processes, being microalgae one of the microorganisms with promising results. In this work, the different mechanisms of pesticide removal by microalgae and the different methods that can be used to increase the removal capacity by these microorganisms were addressed.

Keywords Water environment · Environmental contamination · Modern agriculture · Volatile compounds · Microalgae bioremediation · Toxic compounds

W. J. Martínez-Burgos (✉) · A. da Silva Vale · J. C. de Carvalho · H. Iwamoto · L. P. de Souza Vandenberghe · M. C. Manzoki · T. Scapini · C. R. Soccol
Department of Bioprocess Engineering and Biotechnology, Federal University of Paraná, Centro Politécnico, Curitiba, PR 81531-990, Brazil
e-mail: ing.wjmartinez@gmail.com

R. Pozzan
Cellular Toxicology Laboratory, Department of Cellular Biology, Federal University of Paraná, Centro Politécnico, Curitiba, PR, Brazil

I. A. Severo
Sustainable Energy Research and Development Center (NPDEAS), Federal University of Paraná, Curitiba, PR 81531-980, Brazil

T. A. Kurniawan and A. Anouzla (eds.), *Algae as a Natural Solution for Challenges in Water-Food-Energy Nexus*, Environmental Science and Engineering,
https://doi.org/10.1007/978-981-97-2371-3_34

34.1 Introduction

Pesticides are widely used in various industries and include insecticides, herbicides, fungicides, acaricides, rodenticides, nematicides, defoliants, plant growth regulators, etc. Those chemicals occupy an indispensable position in modern agriculture, also in other industries (e.g. sylviculture and animal husbandry), to enhance agricultural production and raise economic gains. Nonetheless, the excessive use of pesticides worldwide has been related to several problems.

In this sense, the main sources of environmental contamination by pesticides are the surface runoff from contaminated agricultural soils, the condensation process of volatile compounds, the transportation, storage and application of pesticides, and the release of non-treated residual waters from industries. As a result, they can be bioavailable in environmental matrices and be ingested by non-target organisms, reaching different trophic levels in the food chain through bioaccumulation and biomagnification processes (Nie et al. 2020). Moreover, drinking water has been identified, even at low-level contamination, as one of the ways for human exposure to these compounds (Goh et al. 2023).

Many efforts have been made to mitigate the disturbances that pesticides have caused to the environment, especially in aquatic systems. Remediation methods of contaminants, including pesticides, can be realized through physical, chemical, and/or biological approaches. Bioremediation methods allow that those contaminants be removed or transformed into compounds, gases and/or molecules with less complexity and harmless. At the same time, the pesticides can be a source of food and energy of some microorganisms, promoting their growth.

Phytoremediation by microalgae have shown many advantages for wastewater treatment and have been very studied. Microalgae are aquatic organisms that have fast growth and are naturally present in both freshwater and oceanic ecosystems. They occupy the base of the food chain and, consequently, are the initial point of trophic transfer processes. Because of this, microalgae are essential for the aquatic ecosystem equilibrium (Goh et al. 2023).

Many studies demonstrated that microalgae can effectively bioaccumulate pesticides. Some species can mineralize or transform completely the pesticides into less toxic compounds, while other species suffer inhibition because of the toxic effects of these metabolites (McLellan et al. 2019). Along this chapter, an overview of the microalgae-based treatment for pesticides bioremediation will be discussed, from the specific removal mechanisms for important pesticides, to the strategies for enhanced biotreatment of wastewaters.

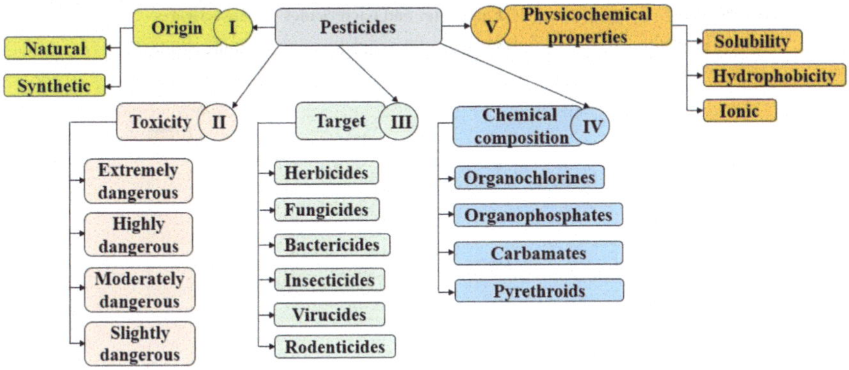

Fig. 34.1 Classification of pesticides. Adapted from: Goh et al. (2023) and Singhal et al. (2021)

34.2 Pesticides: Classification and Market

34.2.1 Classification of Pesticides

Pesticides can be defined as compounds or mixtures of compounds that aim to control some type of pest and, in general, are used to control and protect crops. In this sense, pesticides can be categorized in different groups (Fig. 34.1). Depending on the origin they can be named as natural or synthetic (I), and according to the toxicity (II) they are divided into the following categories: slightly dangerous, moderately dangerous, highly dangerous, and extremely dangerous (Kaur et al. 2019).

Another classification shows that pesticides can be grouped depending on the target or type of organism their action is intended (III) (Goh et al. 2023). For example, insecticides are designed to eliminate insects, fungicides to eliminate fungi, acaricides to mites and herbicides to weeds, among others.

Another classification is depending on the chemical composition (IV). For instance, in the case of insecticides they can be classified into the following groups: organochlorines, organophosphates, carbamates and pyrethroids and pyrethrins (Kaur et al. 2019; Singhal et al. 2021). Finally, with respect to the physicochemical properties of the compounds or active principles (V) these can be soluble, hydrophobic or ionic (Goh et al. 2023).

34.2.2 Pesticide Market

Pesticides in their various forms have been used in agriculture for millennia. Currently, these compounds are generally produced synthetically to target specific organisms (Ritchie et al. 2022). Through Fig. 34.2, it can be seen the production of active ingredients for pesticides in the world is constantly increasing, rising from

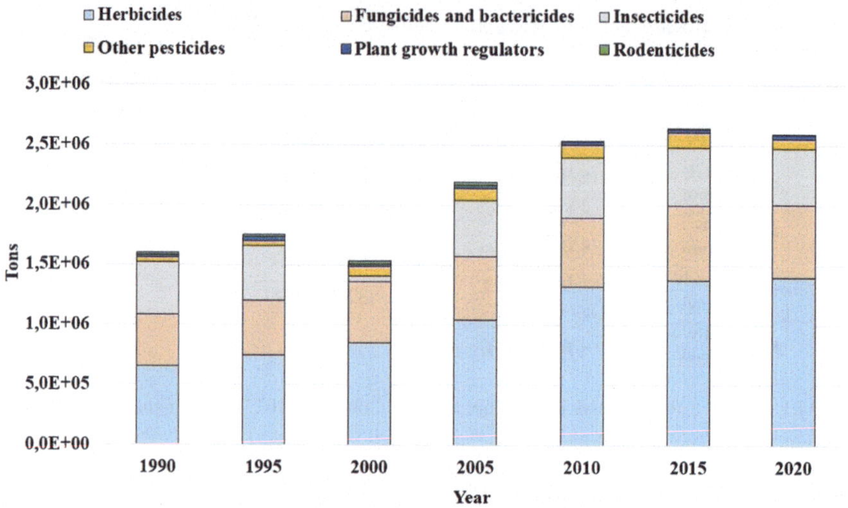

Fig. 34.2 Trends of pesticide production in the world between the years 1990–2020. Adapted from: Ritchie et al. (2022)

1.5 to 2.7 millions of tons (Mt) for the 1990 and 2020, respectively (Ritchie et al. 2022). This is equivalent to around 7.2 Mt of formulated pesticides currently being produced with an approximate value of USD 41.1 billion (FAO 2022). The types of pesticides most produced are herbicides, fungicides, bactericides and insecticides with 55%, 23.4% and 18.2%, respectively. The rise in pesticide production can be explained by the increased demand for agricultural products, whether for human or animal consumption, as well as for energy production, and the significant increase in the amount of pesticide applied per area, which also rised significantly, going from 1.2 to 1.8 kg/ha.

34.3 Environmental Impacts of Pesticides

Contamination and pollution of natural ecosystems by xenobiotics from anthropogenic activities are global problems that have been detected for a long time and have been gaining increasing attention and concern given the growing scarcity of natural resources, such as water (Hassaan and El Nemr 2020). Pesticides are one of the main classes of pollutants that bring great impacts to the environment, since they are used worldwide on a large scale, either by rural producers, or in domestic applications (Kaur et al. 2019).

After being released, pesticides undergo a series of chemodynamic processes, which involve mechanisms such as transport, degradation, and biotic absorption, and which determine their persistence (or half-life) and distribution in the environment (Tudi et al. 2021). In addition to their physical and chemical particularities, pesticides may be influenced by environmental agents, such as pH, temperature, precipitation, and luminosity (Rezende-Teixeira et al. 2022).

Degradation is one of the most relevant chemodynamic processes, since it consists of the chemical transformation of the original compound into other forms and metabolites, which can be even more toxic than the original molecule (Meka and Dubiwak 2021).

Because of their toxicity, necessary to combat the agricultural pests against which they are developed, pesticides have been the subject of several studies that associate their use with risks to human health and numerous environmental impacts (Rezende-Teixeira et al. 2022). Although most pesticides do not lead to lethality in non-target organisms, including humans, at concentrations released into the environment, severe toxic effects can be observed on biota at these doses, as well as on the entire food web of ecosystems exposed to these chemical agents (Raffa and Chiampo 2021; Shefali et al. 2021). Some examples of studies that associate cellular and physiological toxic effects of pesticides in different species are shown in Table 34.1.

Pesticides enter the organisms mainly through ingestion and inhalation, or absorption through the skin and mucous membranes. After overcoming cellular protection barriers, they can reach various intracellular compartments and thus affect the physiology of various systems (Shefali et al. 2021). In mammals, such as humans, contamination by pesticides can lead to serious hepatic and renal problems, since the kidneys and liver are the main organs involved in the detoxification, biotransformation, and excretion of xenobiotics (Rezende-Teixeira et al. 2022). In addition, exposure to this class of xenobiotics has been linked with carcinogenesis, genotoxicity, neurotoxicity and endocrine disruption in humans and several other animal species, constituting a great danger to living beings (Kaur et al. 2019; Sabarwal et al. 2018).

Pesticides reach natural environments mainly through their preparation and application, especially through pulverization. From the moment they are released into the crop or soil, these compounds are absorbed by the target organisms and, after or parallel to this, undergo degradation, transport to groundwater and surface water, volatilization into the atmosphere and/or adsorption by the soil, thus being able to reach and contaminate the three main components of the biosphere, water (hydrosphere), air (atmosphere) and soil (lithosphere) and, consequently, reach non-target organisms, including natural predators of agricultural pests (Meka and Dubiwak 2021; Raffa and Chiampo 2021; Tudi et al. 2021).

Contamination of water resources by pesticides is an eminent and global problem. Surface and underground water sources often contain traces of pesticides and their metabolites, coming directly from spraying, runoff from farmland, and even effluent from agricultural industries. Furthermore, it is important to consider that surface water is associated to atmospheric water and groundwater with the hydrological cycle (Hassaan and El Nemr 2020; Tudi et al. 2021). In this sense, the possible pollution of aquatic ecosystems by waste from pesticides and their metabolites represents serious risks to the biota, both for species that inhabit these ecosystems (e.g. fish, amphibians, zooplankton, microorganisms) and for species that use water for consumption or prey on aquatic organisms (e.g. birds and reptiles) (Shefali et al. 2021).

Table 34.1 Effects of different pesticides on non-target organisms

Pesticide	Classification	Conditions	Main observed effects	Reference
Malathion	Organophosphate insecticide	Exposure of fish *Channa punctatus* to 0.4 mg L^{-1}; assessments at various intervals (1, 4, 8, 12 days)	Morphometric indices decreased with exposure; serum glucose values decreased, total proteins increased; increased levels of creatinine, urea and Ca^{2+} indicate strong renal dysfunction; high antioxidant activity; histopathological damage in kidney and liver	(Bharti and Rasool 2021)
Parathion	Organophosphate insecticide	Evaluation of the effect of repeated doses (3–12 mg^{-1} kg^{-1}. day^{-1} for 5 days) in mice	Dose-dependent effects on sperm quality, lipid peroxidation and DNA integrity; analyses of seminiferous tubules atrophies and spermatogenesis progression interruption	(Urióstegui-Acosta et al. 2020)
Chlorpyrifos	Organophosphate insecticide	Exposure of mangrove crab *Episesarma tetragonum* to 0.0294 and 0.0588 ppm for 7 and 28 days	Cytopathological changes in the gills, hepatopancreas and muscle	(Ravi et al. 2023)
Dichloro-iphenyl-trichloroethane (DDT)	Organochlorine insecticide	Evaluation of wild rats collected in areas sprayed with DDT in South Africa	Significant association between DDT levels and constitutive androstane receptor gene expression levels and related enzymes, such as CYP2b1, in liver tissue	(Motohira et al. 2023)

(continued)

Table 34.1 (continued)

Pesticide	Classification	Conditions	Main observed effects	Reference
Carbaryl	Carbamate insecticide	Exposure of *Apis mellifera* bee larvae to 2 mg L^{-1} of carbaryl (concentration considered as NOAEC—no observed adverse effect concentration) for 4 days	Exposure caused transcriptional disturbances associated with oxidative stress; metabolic homeostasis was disrupted, affecting the metabolism of amino acids and purines and pyrimidines, and the biosynthesis of flavones and flavonoids	(Gao et al. 2022)
Methomyl	Carbamate insecticide	Evaluation of toxic effects in the development stages of *Bufo arabicus*; exposure to 10 ppm for two weeks	Hyperactivity, abnormal swimming, malformation, teratogenicity, and neurodegenerative effects were observed in treated organisms	(Seleem 2019)
Maneb	Fungicide	Exposure of zebrafish (*Danio rerio*) embryos to 0.1, 0.5, 1.0 and 10.0 μM of Maneb for 96 h; dark photokinesis assay in larvae after 7 days of exposure to 0.1, 0.5 and 1.0 μM to assess neurotoxicity	Decreased hatching rate and increased notochord deformity rate at 1.0 and 10.0 μM after 72 and 96 h; reduction of the total body length at 1.0 μM; signs of hypoactivity associated with notochord defects in larvae exposed to 0.5 and 1.0 μM	(Cao et al. 2019)
Mancozeb	Fungicide	Exposure of adult male rats to 150, 300 and 600 mg kg^{-1} of Manzozeb per day for 4 weeks	Abnormal and significant changes in hematological and biochemical parameters, in addition to hepatorenal toxicity and changes in the lipid profile; increase in organ weight of intoxicated rats	(Kotb et al. 2019)

(continued)

Table 34.1 (continued)

Pesticide	Classification	Conditions	Main observed effects	Reference
Glyphosate	Organophosphate herbicide	Exposure of adult *Apis mellifera ligustica* and *Apis cerana cerana* to realistic doses through sucrose solution feeding	Impacts on the immune, digestive and nervous systems, on amino acid and carbohydrate metabolism; affected the growth and development of both species	(Zhao et al. 2020)
Alpha-cypermethrin	Pyrethroid insecticide	Exposure of microalgae *Chlorella* sp. at an EC50 concentration (16.22 ± 0.8 mg L^{-1}) of alpha-cypermethrin for 1 day	Affected proteins related to several important processes, such as photosynthesis, stress response, carbohydrate and lipid metabolism, and amino acid and nucleotide biosynthesis	(Chanu et al. 2023)
Diquat	Bipyridylium herbicide	Exposure to 0.05, 0.11 and 0.14 mg L^{-1} during metamorphosis of the amphibian *Xenopus laevis*	Reduction in the length of the limbs; developmental arrest at the two highest concentrations; histopathological changes in the thyroid gland	(Babalola and van Wyk 2021)

Agricultural soil is a major source of contamination of water resources by pesticides. The coarser the texture of the soil, as in sandy soils and gravel, the greater its infiltration capacity, and thus pesticides are more likely to deposit there or even to reach groundwater, increasing the chances of contamination of natural aquifers. Soils with finer textures, such as clayey soils, have less infiltration capacity, allowing a greater flow of rainwater, for example, which ends up dragging the soil contaminated with pesticides, taking it to nearby surface waters, such as rivers and lakes (Shefali et al. 2021).

In any case, as soil particles have a great capacity to retain pesticides by adsorption, they can act as a pesticide reservoir (Raffa and Chiampo 2021). This can cause damage to microorganisms and beneficial organisms in the soil (e.g. earthworms and bacteria), leading to a loss of soil biodiversity and, consequently, to biologically impoverished of natural soils and to losses of production in later plantings in agricultural areas (Tudi et al. 2021; Rezende-Teixeira et al. 2022).

The retention capacity of pesticides in the soil thus has a great influence on the environmental distribution of this class of xenobiotics. More hydrophobic pesticides, such as organochlorines (e.g. DDT, endosulfan, heptachlor), tend to accumulate in soil and thus be more persistent, which is why they have been banned in many countries. Less persistent pesticides (e.g. carbamates, some organophosphates, fungicides) are more susceptible to being drained and leached into water environments (Tudi et al. 2021).

The volatilization of pesticides can also cause atmospheric pollution, which seriously impacts flora, fauna (e.g. birds and bees) and human health (Rezende-Teixeira et al. 2022). After being volatilized, pesticides and their residues tend to be dispersed and transported over long distances and may even precipitate again in soils and aquatic environments, with rain, for example (Tudi et al. 2021).

Aquatic organisms (e.g. fish) are extremely affected by pesticides because they are directly in contact with these compounds deposited in the water and, therefore, has been the study target of many toxicology studies involving pesticides (Maurya et al. 2019; Martyniuk et al. 2020; Yang et al. 2020). These living beings are exposed to agrotoxics in three main ways: dermal (through the skin); respiratory (through the gills); and orally (through the mouth, either through food or contact with water) (Shefali et al. 2021). Some studies even associated the decrease in the number of fish species with pesticide's toxic effects (Ewing 1999; Scholz et al. 2012).

Pesticides affect living beings directly, causing cellular, physiological, and behavioral changes, or indirectly, resulting in phenomena such as algae and microalgae blooms because of the increase of organic matter in the water. Changes caused by pesticides to organisms involve not only survival mechanisms, but can affect their growth rate, reproductive performance, sexual maturity, spawning capacity and egg growth, fertility rates and other reproductive disorders (Meka and Dubiwak 2021).

In addition, pesticides can alter or inhibit numerous metabolic and enzymatic pathways (e.g. acetylcholinesterase, cytochrome P450, oxidative stress enzymes), leading to cellular damage such as lipid peroxidation, damage to nucleic acids and protein carbonylation, thus affecting the functioning of organelles and important cellular processes (Suliman et al. 2020; Marins et al. 2021; Pérez-Iglesias et al.

2023). All these alterations lead the organism to present severe oxidative stress and affected signaling pathways, causing an enormous imbalance in cellular homeostasis, which, in turn, can trigger serious diseases such as cancer, neurodegeneration, and cardiovascular, respiratory, reproductive, and immunological problems (Buralli et al. 2020; Fucic et al. 2021; Pedroso et al. 2022; Vellingiri et al. 2022).

Pesticides from the organophosphate class, for example, act on insects by inhibiting acetylcholinesterase, an enzyme directly involved in cellular communication necessary for nerve impulses, thus affecting neural transmission in these organisms (Tudi et al. 2021). However, studies point to similar effects in other species, including humans, which represents a real danger to the biota and human health (Duncan et al. 2020; Čadež et al. 2021). Despite having low persistence in the environment, they can cause high acute toxicity in mammals. As most are liposoluble, they are easily absorbed through contact with the skin and mucous membranes, in addition to the respiratory route (Kaur et al. 2019; Meka and Dubiwak 2021).

Carbamic acid from carbamate pesticides acts in a similar way, suppressing the action of cholinesterase enzymes and thus affecting the nervous system of insects. Carbamates strongly affect insects of the Hymenoptera class, consisting of bees and wasps, during their collection of nectar and pollen through contaminated air or plant surfaces. The impact on these organisms, largely responsible for natural pollination, can affect an entire local economy in view of the biological role and ecosystem services performed by them (Rezende-Teixeira et al. 2022). In addition, these pesticides have shown a high endocrine disrupting potential in some species (Raffa and Chiampo 2021).

On the other hand, organochlorines, which constitute the group of pesticides most widely found in natural environments and which have very stable compounds that can remain in the environment for a long time, have been strongly associated with the development of liver, testicular and breast cancers in humans (Tudi et al. 2021). Also liposoluble, they can bioaccumulate in adipose tissue and in mammalian biofluids, such as blood and breast milk, and can cause long-term damage (Meka and Dubiwak 2021). Bioaccumulation is also influenced by the contact time between the organism's absorption pathway and the pesticide, also by the transference of these compounds through the food chain (biomagnification) (Rezende-Teixeira et al. 2022).

Pyrethroids, in turn, affect the sodium channels in the cell membranes of the nervous system, keeping them open, which ends up triggering a neuronal discharge or repetitive excitation and, consequently, impairment of neurological functions (Rezende-Teixeira et al. 2022). In addition, they can cause inhibition of Ca^{2+} and Mg^{2+}-ATPase, resulting in the removal of calcium from nerve endings, releasing neurotransmitters into the postsynaptic cleft (Shefali et al. 2021). Despite their rapid degradation in soil and plants, this class of pesticides has great potential for bioaccumulation, and is highly toxic to fish, amphibians, bees and crustaceans (Rezende-Teixeira et al. 2022).

Herbicides, produced to annihilate weeds, can affect plants in different ways and in different mechanisms, such as photosynthesis, cellular respiration, growth, cell division, or lipid and protein synthesis. Glyphosate, globally the most widely used herbicide, inhibits the enzyme EPSPS (5-enolpyruvoylshikimate 3-phosphate synthase), the main catalyst in the biosynthetic pathway of amino acids. By inhibiting EPSPS, glyphosate affects protein synthesis in plants exposed to this herbicide (Shefali et al. 2021).

Another widely used herbicide is 2,4-D (2,4-dichlorophenoxyacetic acid), of the chlorophenoxy class, which mimics the growth hormone auxin in plants, causing various growth abnormalities, such as curling of leaves or stems, inhibition of shoot and root growth, and even plant necrosis and death (Shefali et al. 2021). These herbicides can cause the same changes in non-target plants that are found in natural environments close to the application sites, which may culminate in their population decline (Kaur et al. 2019). The reduction in the abundance of primary producers in food webs may lead to a decrease in primary consumers and, consequently, in secondary consumers (Hassaan and El Nemr 2020).

Fungicides disrupt energy supply in fungi (through inhibition of the electron transport chain during cellular respiration) and inhibit spore germination of these organisms (Shefali et al. 2021). Some fungicides have low solubility in water and therefore are highly persistent in surface water and soil. Exposure to fungicides, in a chronic or acute way, can cause the development of neurological, reproductive, and endocrine disorders in humans (De Souza et al. 2020).

Thus, assessing whether a pesticide is safe to actually regulate its use requires detailed analyses of its toxic potential, as well as its residues and metabolites. Regulatory authorities set maximum residue levels in food and water, but strong surveillance is indispensable to maintain levels that are safe for ecosystems and human health, especially in developing countries (Meka and Dubiwak 2021).

Finally, it is also necessary to develop policies and mechanisms to mitigate environmental impacts caused by pesticides, such as bioremediation. Since these chemical agents also end up being degraded through chemical and physical reactions, such as photodecomposition, or biological degradation carried out by terrestrial and aquatic microorganisms, such as microalgae, this can be an effective technology to minimize the environmental damages arising from the application of pesticides, as discussed in the next sections of this chapter.

34.4 Microalgae Mechanisms for Pesticide Removal

The processes of pesticide removal by microalgae can occur by pathways of active processes (when external energy is required) or by passive processes (involving physicochemical interaction with the chemical structure of the cell wall), and it is widely accepted that the removal mechanisms involve (mainly) bioadsorption, bioaccumulation and biodegradation (Fig. 34.3) (Avila et al. 2021; Deng et al. 2022; Rempel et al. 2021; Sheng et al. 2022; Xiong et al. 2018). The determination of the preferential mechanisms used for the degradation or transformation of pesticides is associated to the physicochemical properties of the compound, for example, the greater the water solubility of the substance, greater occurrence in the aqueous phase, therefore, the availability for biodegradation (Sheng et al. 2022).

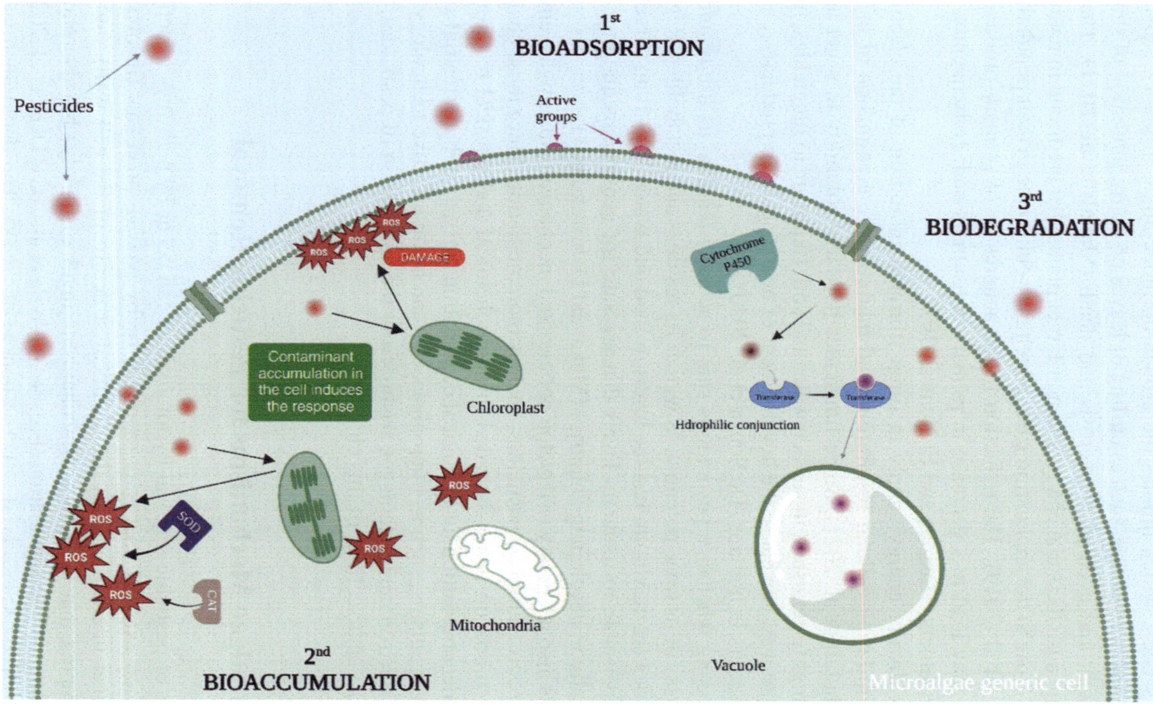

Fig. 34.3 Mechanisms for pesticide removal that can be realized during microalgal metabolism (SOD—superoxide dismutase; CAT—catalase; ROS—reactive oxidative species). Adapted: Nie et al. (2020) and Rempel et al. (2021)

Bioadsorption is an energy-independent (passive) mechanism of pesticide removal since the pollutant molecules interact with functional groups of the organism's cell wall and are retained on the surface, e.g., by ion exchange and surface precipitation. Sorption is very important in pesticide removal processes by microalgae since it does not demand biological activity because it is a passive process, so it can be conducted with live or dead biomass, which makes it a fast and less economically costly process (Avila et al. 2021; Bilal et al. 2018; Bose et al. 2021; Jin et al. 2018). For example, with an 1 h exposure time, it was detected that freeze-dried biomass of *Chlorella vulgaris* exhibited pesticide removal results of greater than 96% for initial concentrations of 2 and 10 $\mu g\ L^{-1}$ (Hussein et al. 2017). Avila et al. (2021) observed that the concentration of chlorpyrifos (an organophosphate pesticide) was reduced 11 times in 2 days of bioprocess.

Regarding bioadsorption, it is important to emphasize that it is a process dependent on the species used, the affinity of the functional groups with the molecule to be retained, the chemical concentration, and the physicochemical properties of the environment (pH, temperature, redox) (Choi and Lee, 2015; Sheng et al. 2022; Sutherland and Ralph, 2019). Strategies to improve the efficiency of the bioabsorption process can be achieved by cell surface modification via the chemical treatment of the biomass (Ali et al. 2018). Furthermore, the bioadsorption process is non-selective, what characterizes a challenge when the medium is contaminated with several contaminants, which can alter the compound's adsorption rate (Sutherland and Ralph 2019; Rempel et al. 2021).

The bioaccumulation mechanism is an energy-dependent process that occurs by transporting the pesticide into the microalgae cell via association with cell wall components and proteins. Molecules that penetrate the cell are bioaccumulated or metabolized, based on the type of biomass (there is variation among microalgae species), characteristics and bioavailability of the accumulated compound, dissolved organic matter, and environmental conditions and this process can be facilitated during the metabolic nutrient's synthesis from the culture medium (Nie et al. 2020; Qiu et al. 2017; Wang et al. 2014). In an experimental setting, this process is characterized by a bioconcentration factor, defined by the proportion between the compound concentration in the organism relative to the concentration in the surrounding environment at a steady state (Wang et al. 2014; Qiu et al. 2017).

During the bioaccumulation process of pesticides inside the cell, there may be a stimulus to the formation of a higher concentration of reactive oxygen species (ROS). Under optimal culture conditions, ROS are essential signaling molecules to maintain cell metabolism, by controlling processes, such as cellular growth, death, and defense. However, when the formation is elevated by the presence of non-favorable environmental conditions—such as an intracellular increase in pesticide concentration—cellular damage can occur, such as DNA's oxidation and lipids from membranes, causing functional metabolic disturbances and even cell death (Nie et al. 2020; Rempel et al. 2021).

Damage by high ROS formation can be reduced by enzymatic pathways since photosynthesizing microorganisms can increase the production of enzymes from the antioxidant system (e.g., glutathione, superoxide dismutase (SOD), catalase (CAT)) during bioaccumulation to preserve the equilibrium between free radicals and peroxides and avoid cell death (Rempel et al. 2021; Sheng et al. 2022; Song et al. 2019). Detection in the culture medium of increased SOD enzyme activity may indicate cell membrane peroxidation due to stress caused by contaminant bioaccumulation. When there is not enough excretion to purge O^{2-}, the equilibrium is not recovered and may lead to peroxidation of the cell membrane (Song et al. 2019).

Bioaccumulation is considered a pace that precedes the mechanism of intracellular contaminant biodegradation (Rempel et al. 2021). The intracellular biodegradation is carried out by an enzyme complex that breaks down primary compounds into molecules of lower structural complexity (Nie et al. 2020; Rempel et al. 2021).

As well as bioaccumulation, in biodegradation enzymes play an essential role and involve several steps that may be particularly different for each microalgae species assessed. Generally, it is accepted that the degradation of contaminants, such as pesticides and drugs, is conducted by microalgae cells through three phases, being: (i) pesticides without functional groups in the structure are activated by different reactions (e.g. reduction, oxidation, hydroxylation) to obtain more hydrophilic compounds, the enzyme cytochrome P450 monooxygenase being responsible for this process; (ii) the activated pesticides (or pesticides that have a functional group) form hydrophilic conjugates with transferase enzymes in the cytosol, e.g. glutathione, glucose, or malonate; (iii) this conjugate is transported to the vacuole through an ATP-binding cassette (Asad et al. 2017; Nie et al. 2020).

Biodegradation can also be conducted extracellularly via the secretion of polysaccharides, enzymes, proteins, and lipids, which can form matrices that remain adhered to the cell, enabling the metabolization of contaminants in the dissolved phase (Rempel et al. 2021). It should be noted that intra and extracellular biodegradation mechanisms may differ within different species and the character of the pesticides evaluated.

In addition to biotic mechanisms, in pesticide removal systems using microalgae, abiotic mechanisms of molecules degradation can be influenced by light (photodegradation), since mixotrophic (light and carbon) and autotrophic (light) cultures are commonly used for these photosynthesizing microorganisms. The photodegradation process can act synergistically with the biological system, improving the total performance of the process (Avila et al. 2021).

It should be noted that when using microalgae biomass, the mechanisms involved in removing pesticides do not occur in isolation. Contaminants in the environment trigger different mechanisms simultaneously, which can cause their biomaccumulation and biodegradation. For example, the application of the microalgae *Scenedesmus obliquus* for removal of the triadimefon (fungicide) resulted in accumulation of the compound in the cell, and the presence of the degradation product (triadimenol) of this pesticide in the medium was also detected, indicating that the biodegradation process also occurred in the medium (Xu and Huang 2017).

34.5 Microalgae Used for Pesticide Removal

Microalgae are of wide diversity, with over 50,000 species classified (Ampofo and Abbey 2022). Since their nutritional requirements are related mainly to mixotrophy, they are interesting candidates for bioremediation of wastewaters. Microalgae show fast growth rates and high productivities, being capable of surviving even extreme environmental conditions (Bhatt et al. 2022). Different species are being studied aiming bioremediation of pesticides, which will be discussed further. According to Goh et al. (2023) the efficiency of pesticide removal using this type of microorganisms can reach up to 95%.

Those microalgae can be produced in open raceways or closed photobioreactors. Concerning the goal of pesticides bioremediation, however, not many studies have been made in large-scale systems. The use of closed photobioreactors may be a better option because allows a better efficiency of light use, higher biomass production and finer control over operational parameters, pollutant losses by volatilization is less prone to happen in closed systems (Avila et al. 2022). It is also known that the remediation of pesticides using microalgae cultivation is related not only to biotic factors, but also to abiotic ones. For example, sunlight and aeration can also have an influence on pesticide bioremediation using microalgae (Hultberg et al. 2016; Matamoros et al. 2015; Mojiri et al. 2022).

One of the most used microalgae genera in pesticides bioremediation—as well as in other kinds of wastewater treatments—is *Chlorella*. It is constituted of unicellular green organisms, which contains approximately 61.6% of proteins, 12.5% of fats and 13.7% of carbohydrates, presenting also high levels of vitamins, minerals, and pigments (Abdel-Latif et al. 2022). This microalgae presents high growth and is utilized for generating a broad range of products, such as biofuels, food supplements, food additives, colorants, and animal feed, or in processes like in the case of biofertilization and sewage treatment (Safi et al. 2014). *Chlorella* cells can be used in bioremediation of pesticides in live of dead forms. It was used for bioremediate carbofuran, dimethoate, atrazine, isoproturon, metolachlore, moline, simazine, pendimenthalin, propanil, pyriproxyfen, among others, especially by adsorption methodologies (Hussein et al. 2017).

The well-known cyanobacteria Spirulina (*Arthospira* genus), which have strong adaptability to extreme environments, also have potential to bioremediate pesticides, such as DDT (Kurashvili et al. 2018). The genus *Scenedesmus* is another one known for its capacity to bioremediate pesticides and as other compounds in wastewaters. It was observed to bioremediate oxamyl, a carbamate, dimethomorph and pyrimethanil, fungicides, and isoproturon, an herbicide, acetamiprid, propanil, dichlorophenols, among others (Dosnon-Olette et al. 2010; Quan et al. 2023). Other microalgae species utilized belong to *Chlamydomonas* and *Dunaliella* genera (Wan et al. 2020; Nasiri et al. 2023). Table 34.2 presents examples of microalgae utilized in pesticides bioremediation, along with the mechanisms they perform in the process and the efficiencies obtained.

Table 34.2 Examples of pesticides bioremediation using different microalgae species

Microalgae	Pesticide	Mechanism	Efficiency	Reference
Chlorella vulgaris	Atrazine, carbofuran, dimethoate, isoproturon, metolachlore, moline, pendimenthalin, propanil, pyriproxyfen, simazine	Bioadsorption	86% (living cells) 96% (lyophilized cells)	(Hussein et al. 2017)
Chlorella vulgaris + biochar	Chlorpyrifos and cypermethrin	Biodegradation and bioadsorption	88.80% (chlorpyrifos) 93.12% (cypermethrin)	(Mojiri et al. 2022)
Chlorella sp. and *Scenedesmus* sp.	Acetamiprid and propanil	Biodegradation	99% (propanil) 71% (acetamiprid)	(Avila et al. 2021)
Scenedesmus sp.	Thiamethoxam	Biodegradation	97.5%	(Quan et al. 2023)
Scenedesmus obliquus and *Scenedesmus quadricauda*	Dimethomorph, pyrimethanil and isoproturon	Bioadsorption and biodegradation	10% (pyrimethanil) 24% (dimethomorph) and 58% (isoproturon)	(Dosnon-Olette et al. 2010)
Spirulina platensis	1,1′-(2,2,2-trichloroethane-1,1-diyl)bis(4-chlorobenzene) (DDT)	Bioadsorption and biotransformation	70%	(Kurashvili et al. 2018)
Chlamydomonas reinhardtii	Trichlorfon (TCF)	Biotransformation	100%	(Wan et al. 2020)
Dunaliella salina	Diazinon, chlorpyrifos, ethion, and profenofos	Biodegradation	70.25% (diazinon) 47.65% (ethion) 53.89% (profenofos) 41.84% (chlorpyrifos)	(Nasiri et al. 2023)

34.5.1 Microalgae Consortia

Microalgae consortia can also be used to bioremediate pesticides. The advantage of utilizing a consortium is that different bioremediation mechanisms can be attained, and a broad range of pesticides can be treated. The microalgae *Anabaena oryzae*, *Nostoc muscorum*, and *Spirulina platensis* were discovered to be capable of biodegrading and utilizing malathion, an organophosphate pesticide, as a source of phosphorus in growth. The three species grew at high concentrations of malathion, and the tenors of proteins and carbohydrates on the biomass were enhanced. *N. muscorum* was highlighted as the species who presented highest removal of malathion (90%) (Ibrahim et al. 2014). In another study, a consortium composed of *Scenedesmus quadricuda*, *Chlorella vulgaris*, and *Spirulina platensis* was utilized for removing the same pesticide, as well as the heavy metals cadmium, lead, and nickel, from urban and agricultural drainage water from Egypt. More than 99% of malathion was removed, and the considerable capacity of the consortium of microorganisms to degrade malathion was attributed to synergistic interactions whereby the overall degradative efficiency was increased (Abdel-Razek et al. 2019).

Besides strictly microalgae consortia, microalgae can be combined with other microorganisms such as bacteria. The synergistic interplay among microalgae, bacteria, and other microorganisms can significantly boost the remediation capacity of these systems (Sheng et al. 2022). Both microalgae and bacteria have demonstrated their effectiveness in removing pesticides, employing distinct mechanisms. Microalgae excel at efficiently accumulating, adsorbing, and transforming pesticides, while bacteria excel at complete mineralization of the pollutants. Partnering these two microorganisms can lead to a remarkable acceleration of the overall detoxification process (Avila et al. 2021; McLellan et al. 2019; Sheng et al. 2022; Xiong et al. 2018).

34.5.2 Microalgae Biomass Destination

The microalgal biomass utilized in pesticides bioremediation processes can be posteriorly processed to generate more aggregated-valued products, such as biodiesel, biochar, and others, enabling the economic feasibility of the process and aligning it with the biorefinery conception (Nie et al. 2020). In biochar production, for example, biomass is submitted to pyrolysis under oxygen-deficient conditions, which renders it a great adsorbent for the elimination of water pollutants (Nie et al. 2020). Anaerobic digestion is also an option for energy recovery, and the result of the possible presence of pesticides and/or its biodegradation products on the digestion performance of the algal biomass can also be assessed (Avila et al. 2021).

34.6 Processes to Increase Pesticide Removal by Microalgae

As already known, various microalgal species remediate different pesticides and other pollutants. In this section, we will discuss the manners of biotechnology and bioprocess that could promote efficiency in contaminant removal. Among these strategies, we can find the microalgae selection and adaptation, the consortium between microalgae and bacteria, and the microalgae immobilization (Ghaffar et al. 2023; Nie et al. 2020). A schematic illustration is presented (Fig. 34.4) for a better comprehension of each strategy and the interactions among them.

34.6.1 Strain Selection and Adaptation Process

To achieve a practical application of microalgae strains with certain characteristics, they may have to be tested and adapted to the target environment (e.g. pesticide-contained effluent). The strain selection process, also known as screening or bioprospection, is a good option because indigenous strains could possibly be tolerant on severe wastewater conditions. For a potential strain selection, the generally performed procedure follows the steps of environment selection, strains purification from the selected environment, the samples are tested at a primary screening for dominant microalgae selection, followed by isolation, second screening to explore the isolated microalgae potential, amplification, and collection (Nie et al. 2020).

The adaptation process can be performed over a melded approach of non-genetic and genetic modification (in the laboratory). At first, the strain is cultivated in the existence of different contaminant concentrations. When the conditions are

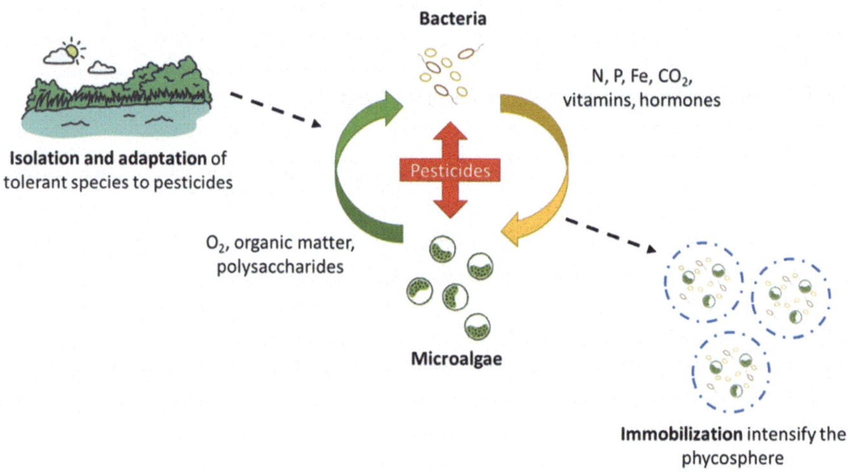

Fig. 34.4 Schematic illustration of the three strategies (*isolation*, *consortium*, and *immobilization*) on enhancement of pesticide removal

altered, microalgae have the tendency to reestablish homeostasis in internal system through an acclimation procedure, following a modified gene expression. In parallel, genetic engineering could enhance the microalgae tolerance and remedial potential to pesticides.

Obtaining a robust microalgae strain demands a long adaptation period. Besides that, another method would be to produce clone library, with a methodology of ultraviolet mutagenesis, and after that, a robust strain could be assessed with culture, in a continuous or semi-continuous way (Ghaffar et al. 2023). Finally, it is important to point out that the treatment procedure must be achieved without unwanted by-products production, that could generate secondary pollution. Fortunately, it can be prevented by employing the intrinsic biological processes of microalgae. As evidence, simultaneous treatment of three pesticides (bentazone, acetamiprid, and propanil) could result in no removal of bentazone, rapid removal of acetamiprid (71%), and propanil removal efficiency raised in > 99% during the treatment because of the biomass acclimation (Avila et al. 2022).

34.6.2 Microalgae and Bacteria Co-cultivation

Microalgae do not live alone in natural water bodies. Microalgae bioremediation studies are executed under aseptic environment at the laboratory; despite, for practical application, the cultivation is frequently coexisted with by bacteria (Nie et al. 2020). In the consortia of microalgae and microbiome, the first are typically more susceptible to organic herbicides and pesticides than the microbial partners. Besides that, microalgal growth, in a stress condition, is supported by bacterial activity, which converts the pesticides into molecules that stimulate growth (like phytohormones) or absorbable metabolites used by microalgae. In the phycosphere (region of microalgal-microbial interaction), bacteria supply microalgae macro and micronutrients, and carbon dioxide. In exchange, microalgae supply oxygen and organic carbon to bacteria. In conclusion, nutritional interactions between microalgae and bacteria aid the maintenance of the phycosphere for survival growth, which can degradate and remove of dangerous pollutants (Ashraf et al. 2023).

34.6.3 Microalgae Immobilization

The immobilization of microalgae implicates the capture of the microalgal cells with a matrix to restrict their mobility whereas keeping their metabolic activity as much time as possible. This technique has been accomplished for several reasons. During wastewater treatment, immobilization can raise survival and metabolism of cells and facilitate the isolation of microalgal biomass subsequent to treatment. Moreover, immobilizing the consortium of bacteria and microalgae can enrich the symbiotic relations between them caused by a more stable phycosphere (Goh et al. 2023).

Naturally occurring biofilms could be observed in the practical process of treatment. The efficacy of this methodology is even more supported by a study about how biofilm establishment enhance the degradation kinetics of diazinon, an organophosphate insecticide. The study shows that removal of diazinon was 99.9% by this type of consortiums in biofilm in comparison with 27% removal by the same species not in biofilm (Tien et al. 2011).

34.7 Photobioreactors Used for Pesticide Bioremediation

For the correct development of bioremediation processes with microalgae, equipment is necessary to facilitate the growth of the microorganisms with the bioremediation of the pollutant. Among the systems that have been used for pesticide bioremediation are open ones such as the raceway pond, circular pond, and unstirred pound and closed systems like tubular photobioreactor, bubble column photobioreactor, and plastic bag photobioreactor (Fig. 34.5) (Arutselvan et al. 2022).

According to Goh et al. (2023) the conventional way to cultivate microalgae is through open systems such as circular ponds or channels (e.g. raceway pond). These systems are simple, with a low operational cost, usually containing one or more paddle wheels that facilitate the homogenization of the medium and the circulation of microorganisms and thus improve their efficiency, except for the unstirred pound that do not contain agitation (Arutselvan et al. 2022). Paddle wheels generally operate at a speed of no more than 55 rpm, and the flow speed in these systems is in the range of 0.10–0.25 m/s (Borowitzka 1999).

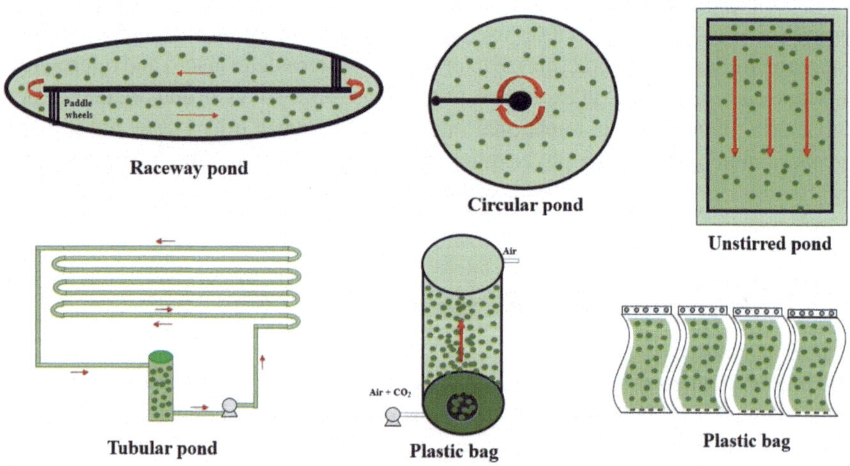

Fig. 34.5 Types of photobioreactors used in pesticide bioremediation. Adapted: Arutselvan et al. (2022) and Singhal et al. (2021)

Open systems are shallow, with approximately a non-maximum depth of 0.5 m to facilitate penetration of light photons into the medium. The main disadvantages of open systems are the lack of control of growth parameters, high susceptibility to contamination and the low cell concentrations that can be produced, approximately 0.25–1 g L^{-1}. However, according to Ruas et al. (2022) in these systems it is also possible to co-cultivate bacteria, so these microorganisms can degrade organic matter while consuming the oxygen produced by microalgae via photosynthesis. Matamoros et al. (2015) reported approximately 75% removal of 2,4-D diazinon and atrazine pesticides in an open pond equipped with paddle wheels using different hydraulic retention time (HRT).

On the other hand, there are closed systems or photobioreactors that are more sophisticated than open systems, in which it is possible to regulate the process parameters and avoid contamination of the microalgal culture by other microorganisms (Arutselvan et al. 2022). Agitation in the case of tubular photobioreactors is carried out by pumps that recirculate the medium. These systems are constructed of transparent materials (e.g. glass) or different types of plastics that allow the passage of light; they can be operated with sunlight and sometimes they also use artificial light. In these systems the productivity is also higher than in open systems, for example in tubular photobioreactors the biomass productivity can hit 40 mg L^{-1} day^{-1} (Christenson and Sims 2011). In general, the diameter of the pipes in these systems is approximately 10 cm (Goh et al. 2023).

In the case of bag photobioreactors, they are also closed systems made mainly with polyethylene films which allow light to pass from one side to the other. Generally, polyethylene bags do not exceed a volume of 100 L and the system is agitated by air flows.

Other closed photobioreactors are the airlift, and in these bioreactors agitation is carried out by small and homogeneous bubbles of air. One of the principal advantages of these systems is that they avoid cell damage during agitation. In general, this system consists of a tube or cylinder and a sprinkler that controls the air flow and the size of the bubbles. In airlift systems flue gases can be used as stirring fluid. According to, Arutselvan et al. (2022) large amounts of microalgal biomass can be produced in airlift systems, in fact Ammar (2016) reported a biomass productivity of 140 mg L^{-1} day^{-1}. Castellanos-Estupiñan et al. (2022) removed organophosphate pesticides in a 2 L closed bioreactor stirred with an airflow of 0.748 L min^{-1}. The reported results reports that *Chlorella* sp., *Scenedesmus* and *Hapalosyphon* sp. removed 100%, 75% and 50% of the chlospyrifos insecticide, respectively. Cheng et al. (2022) also used closed photobioreactors to remove imidacloprid and thiacloprid pesticides using *Scenedesmus* sp.; the elimination rate of these pollutants was 55.54% and 2.71%, respectively.

34.8 Conclusion

In addition to the different applications and use of microalgae in the production of different biomolecules of commercial interest, they appear as a potential alternative for the bioremediation of xenobiotic compounds. Among them, the insecticides, which have immeasurable impacts on the environment and on human and animal health. Therefore, bioremediation appears as a solution for the elimination of these compounds through the use microalgae as important agents in this process. Different methods and/or mechanisms can be used in this context, including biosorption, bioaccumulation and biodegradation of insecticides with efficiencies of up to 100%, depending on the species used.

Optimizations of the bioremediation process using microalgae are being proposed with the inclusion of species improvement processes, immobilization, which enables the continuous use of these biological agents, and their co-cultivation, for example, with bacteria, that help to minimize microalgae susceptibility to organic herbicides and pesticides. However, the allocation of microalgae biomass at the end of the bioremediation process could be a serious problem, but solutions are being proposed such as the production of new products with added value such as biodiesel, biochar and others, allowing the economic viability of the process allied to the concept of biorefinery.

References

Abdel-Latif HMR, El-Ashram S, Sayed AEDH, Alagawany M, Shukry M, Dawood MAO, Kucharczyk D (2022) Elucidating the ameliorative effects of the cyanobacterium Spirulina (*Arthrospira platensis*) and several microalgal species against the negative impacts of contaminants in freshwater fish: a review. Aquaculture 554. https://doi.org/10.1016/j.aquaculture.2022.738155

Abdel-Razek MA, Abozeid AM, Eltholth MM, Abouelenien FA, El-Midany SA, Moustafa NY, Mohamed RA (2019) Bioremediation of a pesticide and selected heavy metals in wastewater from various sources using a consortium of microalgae and cyanobacteria. Slov Vet Res 56:61–73. https://doi.org/10.26873/SVR-744-2019

Ali MEM, Abd El-Aty AM, Badawy MI, Ali RK (2018) Removal of pharmaceutical pollutants from synthetic wastewater using chemically modified biomass of green alga *Scenedesmus obliquus*. Ecotoxicol Environ Saf 151:144–152. https://doi.org/10.1016/j.ecoenv.2018.01.012

Ammar SH (2016) Cultivation of Microalgae for Biomass Production. Al-Khwarizmi Eng J 12(1):90–99

Ampofo J, Abbey L (2022) Microalgae: bioactive composition, health benefits, safety and prospects as potential high-value ingredients for the functional food industry. Foods 11(12). https://doi.org/10.3390/foods11121744

Arutselvan C, Seenivasan HK, Lewis Oscar F, Ramya G, Thuy Lan Chi N, Pugazhendhi A, Thajuddin N (2022) Review on wastewater treatment by microalgae in different cultivation systems and its importance in biodiesel production. Fuel 324:124623. https://doi.org/10.1016/j.fuel.2022.124623a

Asad MAU, Lavoie M, Song H, Jin Y, Fu Z, Qian H (2017) Interaction of chiral herbicides with soil microorganisms, algae and vascular plants. Sci Total Environ 580:1287–1299. https://doi.org/10.1016/j.scitotenv.2016.12.092

Ashraf N, Ahmad F, Lu Y (2023) Synergy between microalgae and microbiome in polluted waters. Trends Microbiol 31(1):9–21. https://doi.org/10.1016/j.tim.2022.06.004

Avila R, Peris A, Eljarrat E, Vicent T, Blánquez P (2021) Biodegradation of hydrophobic pesticides by microalgae: transformation products and impact on algae biochemical methane potential. Sci Total Environ 754:142114. https://doi.org/10.1016/j.scitotenv.2020.142114

Avila R, García-Vara M, López-García E, Postigo C, López de Alda M, Vicent T, Blánquez P (2022) Evaluation of an outdoor pilot-scale tubular photobioreactor for removal of selected pesticides from water. Sci Total Environ 804. https://doi.org/10.1016/j.scitotenv.2021.150040

Babalola OO, van Wyk HJ (2021) Exposure impacts of diquat dibromide herbicide formulation on amphibian larval development. Heliyon 7(4). https://doi.org/10.1016/j.heliyon.2021.e06700

Bharti S, Rasool F (2021) Analysis of the biochemical and histopathological impact of a mild dose of commercial malathion on *Channa punctatus* (Bloch) fish. Toxicol Rep 8:443–455. https://doi.org/10.1016/j.toxrep.2021.02.018

Bhatt P, Bhandari G, Bhatt K, Simsek H (2022) Microalgae-based removal of pollutants from wastewaters: occurrence, toxicity and circular economy. Chemosphere 306. https://doi.org/10.1016/j.chemosphere.2022.135576

Bilal M, Rasheed T, Sosa-Hernández J, Raza A, Nabeel F, Iqbal H (2018) Biosorption: an interplay between marine algae and potentially toxic elements—a review. Mar Drugs 16(2):65. https://doi.org/10.3390/md16020065

Borowitzka M (1999) Commercial production of microalgae: ponds, tanks, and fermenters. Prog Ind Microbiol 35:313–321. https://doi.org/10.1016/S0079-6352(99)80123-4

Bose S, Kumar PS, Vo D-VN, Rajamohan N, Saravanan R (2021) Microbial degradation of recalcitrant pesticides: a review. Environ Chem Lett 19(4):3209–3228. https://doi.org/10.1007/s10311-021-01236-5

Buralli RJ, Dultra AF, Ribeiro H. 2020. Respiratory and allergic effects in children exposed to pesticides—a systematic review. Int J Environ Res Public Health 17(8). https://doi.org/10.3390/ijerph17082740

Čadež T, Kolić D, Šinko G, Kovarik Z (2021) Assessment of four organophosphorus pesticides as inhibitors of human acetylcholinesterase and butyrylcholinesterase. Sci Rep 11(1). https://doi.org/10.1038/s41598-021-00953-9

Cao F, Souders CL, Li P, Pang S, Liang X, Qiu L, Martyniuk CJ (2019) Developmental neurotoxicity of maneb: notochord defects, mitochondrial dysfunction and hypoactivity in zebrafish (*Danio rerio*) embryos and larvae. Ecotoxicol Environ Saf 170:227–237. https://doi.org/10.1016/j.ecoenv.2018.11.110

Castellanos-Estupiñan MA, Carrillo-Botello AM, Rozo-Granados LS, Becerra-Moreno D, García-Martínez JB, Urbina-Suarez NA, López-Barrera GL, Barajas-Solano AF, Bryan SJ, Zuorro A (2022) Removal of nutrients and pesticides from agricultural runoff using microalgae and cyanobacteria. Water (Switz) 14(4):1–11. https://doi.org/10.3390/w14040558

Chanu NK, Mandal MK, Srivastava A, Mishra Y, Chaurasia N (2023) Proteomics reveals damaging effect of alpha-cypermethrin exposure in a non-target freshwater microalga *Chlorella* sp. NC-MKM. Curr Microbiol 80(5). https://doi.org/10.1007/s00284-023-03179-2

Cheng Y, Wang J, Quan L, Li D, Chen Y, Zhang Z, Yang L, Li B, Wu L (2022) Construction of microalgae-bacteria consortium to remove typical neonicotinoids imidacloprid and thiacloprid from municipal wastewater: difference of algae performance, removal effect and product toxicity. Biochem Eng J 187:108634. https://doi.org/10.1016/j.bej.2022.108634

Choi H-J, Lee S-M (2015) Heavy metal removal from acid mine drainage by calcined eggshell and microalgae hybrid system. Environ Sci Pollut Res 22(17):13404–13411. https://doi.org/10.1007/s11356-015-4623-3

Christenson L, Sims R (2011) Production and harvesting of microalgae for wastewater treatment, biofuels, and bioproducts. Biotechnol Adv 29(6):686–702. https://doi.org/10.1016/j.biotechadv.2011.05.015

De Souza RM, Seibert D, Quesada HB, de Jesus BF, Fagundes-Klen MR, Bergamasco R (2020) Occurrence, impacts and general aspects of pesticides in surface water: a review. Process Saf Environ Prot 135:22–37. https://doi.org/10.1016/j.psep.2019.12.035

Deng Z, Zhu J, Yang L, Zhang Z, Li B, Xia L, Wu L (2022) Microalgae fuel cells enhanced biodegradation of imidacloprid by *Chlorella* sp. Biochem Eng J 179:108327. https://doi.org/10.1016/j.bej.2021.108327

Dosnon-Olette R, Trotel-Aziz P, Couderchet M, Eullaffroy P (2010) Fungicides and herbicide removal in Scenedesmus cell suspensions. Chemosphere 79(2):117–123. https://doi.org/10.1016/j.chemosphere.2010.02.005

Duncan WP, Idalino JJS, da Silva AG, Moda RF, da Silva HCM, Matoso DA, Gomes ALS (2020) Acute toxicity of the pesticide trichlorfon and inhibition of acetylcholinesterase in *Colossoma macropomum* (Characiformes: Serrasalmidae). Aquacult Int 28(2):815–830. https://doi.org/10.1007/s10499-019-00497-w

Ewing RD (1999) Dimishing returns: Salmon decline and pesticides

FAO (2022) Pesticides use, pesticides trade and pesticides indicators (1990–2020). fao.org/3/cc0918en/cc0918en.pdf

Fucic A, Duca R, Galea KS, Maric T, Garcia K, Bloom M, Andersen HR, Vena J (2021) Reproductive health risks associated with occupational and environmental exposure to pesticides. Int J Environ Res Public Health 18(12). https://doi.org/10.3390/ijerph18126576

Gao J, Yang Y, Ma S, Liu F, Wang Q, Wang X, Wu Y, Zhang L, Liu Y, Diao Q et al (2022) Combined transcriptome and metabolite profiling analyses provide insights into the chronic toxicity of carbaryl and acetamiprid to *Apis mellifera* larvae. Sci Rep 12(1). https://doi.org/10.1038/s41598-022-21403-0

Ghaffar I, Hussain A, Hasan A, Deepanraj B (2023) Microalgal-induced remediation of wastewaters loaded with organic and inorganic pollutants: an overview. Chemosphere 320:137921. https://doi.org/10.1016/j.chemosphere.2023.137921

Goh PS, Lau WJ, Ismail AF, Samawati Z, Liang YY, Kanakaraju D (2023) Microalgae-enabled wastewater treatment: a sustainable strategy for bioremediation of pesticides. Water (Switz) 15(1). https://doi.org/10.3390/w15010070

Hassaan MA, El Nemr A (2020) Pesticides pollution: Classifications, human health impact, extraction and treatment techniques. Egypt J Aquat Res 46(3):207–220. https://doi.org/10.1016/j.ejar.2020.08.007

Hultberg M, Bodin H, Ardal E, Asp H (2016) Effect of microalgal treatments on pesticides in water. Environ Technol (UK) 37(7):893–898. https://doi.org/10.1080/09593330.2015.1089944

Hussein MH, Abdullah AM, Badr El Din NI, Mishaqa ESI (2017) Biosorption potential of the microchlorophyte chlorella vulgaris for some pesticides. J Fertil Pestic 08(01). https://doi.org/10.4172/2471-2728.1000177

Ibrahim WM, Karam MA, El-Shahat RM, Adway AA (2014) Biodegradation and utilization of organophosphorus pesticide malathion by cyanobacteria. Biomed Res Int. https://doi.org/10.1155/2014/392682

Jin Y, Luan Y, Ning Y, Wang L (2018) Effects and mechanisms of microbial remediation of heavy metals in soil: a critical review. Appl Sci 8(8):1336. https://doi.org/10.3390/app8081336

Kaur R, Mavi GK, Raghav S, Khan I (2019) Pesticides classification and its impact on environment. Int J Curr Microbiol Appl Sci 8(03):1889–1897. https://doi.org/10.20546/ijcmas.2019.803.224

Kotb GAM, Farag AAG, El-Halwagy MEA, Nahas AA, Ziada RM (2019) Subacute influence of mancozeb (80 % WP) on hematological and biomarkers in male rats. Int J Adv Res Biol Sci 6(12):73–80. https://doi.org/10.22192/ijarbs

Kurashvili M, Varazi T, Khatisashvili G, Gigolashvili G, Adamia G, Pruidze M, Gordeziani M, Chokheli L, Japharashvili S, Khuskivadze N (2018) Blue-green alga Spirulina as a tool against water pollution by 1,1′-(2,2,2-trichloroethane-1,1-diyl)bis(4-chlorobenzene) (DDT). Ann Agrar Sci 16(4):405–409. https://doi.org/10.1016/j.aasci.2018.07.005

Marins AT, Severo ES, Cerezer C, Leitemperger JW, Müller TE, Floriano L, Prestes OD, Zanella R, Loro VL (2021) Environmentally relevant pesticides induce biochemical changes in Nile tilapia (*Oreochromis niloticus*). Ecotoxicology 30(4):585–598. https://doi.org/10.1007/s10646-021-02368-8

Martyniuk CJ, Mehinto AC, Denslow ND (2020) Organochlorine pesticides: agrochemicals with potent endocrine-disrupting properties in fish. Mol Cell Endocrinol 507. https://doi.org/10.1016/j.mce.2020.110764

Matamoros V, Gutiérrez R, Ferrer I, García J, Bayona JM (2015) Capability of microalgae-based wastewater treatment systems to remove emerging organic contaminants: a pilot-scale study. J Hazard Mater 288:34–42. https://doi.org/10.1016/j.jhazmat.2015.02.002

Maurya PK, Malik DS, Yadav KK, Gupta N, Kumar S (2019) Haematological and histological changes in fish *Heteropneustes fossilis* exposed to pesticides from industrial waste water. Hum Ecol Risk Assess 25(5):1251–1278. https://doi.org/10.1080/10807039.2018.1482736

McLellan J, Gupta SK, Kumar M (2019) Feasibility of using bacterial-microalgal consortium for the bioremediation of organic pesticides: application constraints and future prospects. Application of microalgae in wastewater treatment. Springer International Publishing, Cham, pp 341–362

Meka W, Dubiwak AD (2021) Environmental Impact of pesticide and its adverse effect on human health: a narrative review. EC Pharmacol Toxicol 9:45–54

Mojiri A, Zhou JL, Nazari VM, Rezania S, Farraji H, Vakili M (2022) Biochar enhanced the performance of microalgae/bacteria consortium for insecticides removal from synthetic wastewater. Process Saf Environ Prot 157:284–296. https://doi.org/10.1016/j.psep.2021.11.012

Motohira K, Yohannes YB, Ikenaka Y, Eguchi A, Nakayama SMM, Wepener V, Smit NJ, Vav Vuren JHJ, Ishizuka M (2023) Investigation of dichlorodiphenyltrichloroethane (DDT) on xenobiotic enzyme disruption and metabolomic bile acid biosynthesis in DDT-sprayed areas using wild rats. J Vet Med Sci 85(2):236–243. https://doi.org/10.1292/jvms.22-0490

Nasiri M, Ahmadzadeh H, Amiri AH (2023) Biodegradation and metabolic fate of organophosphorus pesticides in well water using *Dunaliella salina*. Int J Environ Sci Technol 20(1):981–992. https://doi.org/10.1007/s13762-022-04620-y

Nie J, Sun Y, Zhou Y, Kumar M, Usman M, Li J, Shao J, Wang L, Tsang DCW (2020) Bioremediation of water containing pesticides by microalgae: mechanisms, methods, and prospects for future research. Sci Total Environ 707:136080. https://doi.org/10.1016/j.scitotenv.2019.136080

Pedroso TMA, Benvindo-Souza M, de Araújo Nascimento F, Woch J, dos Reis FG, de Melo e Silva D (2022) Cancer and occupational exposure to pesticides: a bibliometric study of the past 10 years. Environ Sci Pollut Res 29(12):17464–17475. https://doi.org/10.1007/s11356-021-17031-2

Pérez-Iglesias JM, Natale GS, Brodeur JC, Larramendy ML (2023) Realistic scenarios of pesticide exposure alters multiple biomarkers in *Boana pulchella* (Anura) adult frogs. Ecotoxicology 32(3):309–320. https://doi.org/10.1007/s10646-023-02639-6

Qiu Y-W, Zeng EY, Qiu H, Yu K, Cai S (2017) Bioconcentration of polybrominated diphenyl ethers and organochlorine pesticides in algae is an important contaminant route to higher trophic levels. Sci Total Environ 579:1885–1893. https://doi.org/10.1016/j.scitotenv.2016.11.192

Quan L, Cheng Y, Wang J, Chen Y, Li D, Wang S, Li B, Zhang Z, Yang L, Wu L (2023) Efficient removal of thiamethoxam by freshwater microalgae *Scenedesmus* sp. TXH: removal mechanism, metabolic degradation and application. J Environ Manage 332. https://doi.org/10.1016/j.jenvman.2023.117388

Raffa CM, Chiampo F (2021) Bioremediation of agricultural soils polluted with pesticides: a review. Bioengineering 8(7). https://doi.org/10.3390/bioengineering8070092

Ravi R, Athisuyambulingam M, Kanagaraj S, Tresnakova N, Impellitteri F, Viswambaran G, Faggio C (2023) Impact of chlorpyrifos on cytopathological indices in mangrove crab, *Episesarma tetragonum* (Fabricius). Vet Sci 10(53). https://doi.org/10.3390/vetsci10010053

Rempel A, Gutkoski JP, Nazari MT, Biolchi GN, Cavanhi VAF, Treichel H, Colla LM (2021) Current advances in microalgae-based bioremediation and other technologies for emerging contaminants treatment. Sci Total Environ 772:144918. https://doi.org/10.1016/j.scitotenv.2020.144918

Rezende-Teixeira P, Dusi RG, Jimenez PC, Espindola LS, Costa-Lotufo LV (2022) What can we learn from commercial insecticides? Efficacy, toxicity, environmental impacts, and future developments. Environ Pollut 300. https://doi.org/10.1016/j.envpol.2022.118983

Ritchie H, Roser M, Rosado P (2022) Pesticides. Retrieved from: https://ourworldindata.org/pesticides

Ruas G, Lacerda SF, Nantes MA, Serejo ML, Da Silva GHR, Boncz MÁ (2022) CO_2 addition and semicontinuous feed regime in shaded HRAP—pathogen removal performance. Water (Switz) 14(24):1–12. https://doi.org/10.3390/w14244047

Sabarwal A, Kumar K, Singh RP (2018) Hazardous effects of chemical pesticides on human health–cancer and other associated disorders. Environ Toxicol Pharmacol 63:103–114. https://doi.org/10.1016/j.etap.2018.08.018

Safi C, Zebib B, Merah O, Pontalier P-Y, Vaca-Garcia C, Vaca C, Morphology G (2014) Morphology, composition, production, processing and applications of chlorella vulgaris: a review. Renew Sustain Energy Rev 35. https://doi.org/10.1016/j.rser.2014.04.007ï

Scholz NL, Fleishman E, Brown L, Werner I, Johnson ML, Brooks ML, Mitchelmore CL, Schlenk D (2012) A perspective on modern pesticides, pelagic fish declines, and unknown ecological resilience in highly managed ecosystems. Bioscience 62(4):428–434. https://doi.org/10.1525/bio.2012.62.4.13

Seleem AA (2019) Teratogenicity and neurotoxicity effects induced by methomyl insecticide on the developmental stages of *Bufo arabicus*. Neurotoxicol Teratol 72:1–9. https://doi.org/10.1016/j.ntt.2018.12.002

Shefali, Kumar R, Sankhla MS, Kumar R, Sonone SS (2021) Impact of pesticide toxicity in aquatic environment. Biointerface Res Appl Chem 11(3):10131– 10140. https://doi.org/10.33263/BRIAC113.1013110140

Sheng Y, Benmati M, Guendouzi S, Benmati H, Yuan Y, Song J, Xia C, Berkani M (2022) Latest eco-friendly approaches for pesticides decontamination using microorganisms and consortia microalgae: a comprehensive insights, challenges, and perspectives. Chemosphere 308:136183. https://doi.org/10.1016/j.chemosphere.2022.136183

Singhal M, Jadhav S, Sonone SS, Sankhla MS (2021) Microalgae based sustainable bioremediation of water contaminated by pesticides. Biointerface Res Appl Chem 12(1):149–169. https://doi.org/10.33263/briac121.149169

Song C, Wei Y, Qiu Y, Qi Y, Li Y, Kitamura Y (2019) Biodegradability and mechanism of florfenicol via *Chlorella* sp. UTEX1602 and L38: experimental study. Bioresour Technol 272:529–534. https://doi.org/10.1016/j.biortech.2018.10.080

Suliman KA, Shah SSA, Gulfam N, Khisroon M, Zahoor M (2020) Toxicity evaluation of pesticide chlorpyrifos in male Japanese quails (*Coturnix japonica*). Environ Sci Pollut Res 27(20):25353–25362. https://doi.org/10.1007/s11356-020-08953-4

Sutherland DL, Ralph PJ (2019) Microalgal bioremediation of emerging contaminants—opportunities and challenges. Water Res 164:114921. https://doi.org/10.1016/j.watres.2019.114921

Tien C-J, Chuang T-L, Chen CS (2011) The role of naturally occurring river biofilms on the degradation kinetics of diazinon. Clean (Weinh) 39(10):931–938. https://doi.org/10.1002/clen.201000200

Tudi M, Ruan HD, Wang L, Lyu J, Sadler R, Connell D, Chu C, Phung DT (2021) Agriculture development, pesticide application and its impact on the environment. Int J Environ Res Public Health 18(1112):1–24. https://doi.org/10.3390/ijerph18031112

Urióstegui-Acosta M, Tello-Mora P, Solís-Heredia M de J, Ortega-Olvera JM, Piña-Guzmán B, Martín-Tapia D, González-Mariscal L, Quintanilla-Vega B (2020) Methyl parathion causes genetic damage in sperm and disrupts the permeability of the blood-testis barrier by an oxidant mechanism in mice. Toxicology 438. https://doi.org/10.1016/j.tox.2020.152463d

Vellingiri B, Chandrasekhar M, Sri Sabari S, Gopalakrishnan AV, Narayanasamy A, Venkatesan D, Iyer M, Kesari K, Dey A (2022) Neurotoxicity of pesticides—a link to neurodegeneration. Ecotoxicol Environ Saf 243. https://doi.org/10.1016/j.ecoenv.2022.113972

Wan L, Wu Y, Ding H, Zhang W (2020) Toxicity, biodegradation, and metabolic fate of organophosphorus pesticide trichlorfon on the freshwater algae *Chlamydomonas reinhardtii*. J Agric Food Chem 68(6):1645–1653. https://doi.org/10.1021/acs.jafc.9b05765

Wang Y, Wen Y, Li JJ, He J, Qin WC, Su LM, Zhao YH (2014) Investigation on the relationship between bioconcentration factor and distribution coefficient based on class-based compounds: the factors that affect bioconcentration. Environ Toxicol Pharmacol 38(2):388–396. https://doi.org/10.1016/j.etap.2014.07.003

Xiong J-Q, Kurade MB, Jeon B-H (2018) Can microalgae remove pharmaceutical contaminants from water? Trends Biotechnol 36(1):30–44. https://doi.org/10.1016/j.tibtech.2017.09.003

Xu P, Huang L (2017) Stereoselective bioaccumulation, transformation, and toxicity of triadimefon in *Scenedesmus obliquus*. Chirality 29(2):61–69. https://doi.org/10.1002/chir.22671

Yang C, Lim W, Song G (2020) Mediation of oxidative stress toxicity induced by pyrethroid pesticides in fish. Comp Biochem Physiol Part C Toxicol Pharmacol 234. https://doi.org/10.1016/j.cbpc.2020.108758

Zhao H, Li G, Guo D, Wang Y, Liu Q, Gao Z, Wang H, Liu Z, Guo X, Xu B (2020) Transcriptomic and metabolomic landscape of the molecular effects of glyphosate commercial formulation on *Apis mellifera* ligustica and *Apis cerana* cerana. Sci Total Environ 744. https://doi.org/10.1016/j.scitotenv.2020.140819

Correction to: Laboratory and Field Studies of Microalgae in Wastewater Treatment in the Removal of Heavy Metals

Ojeaga Evans Imanah, Blessing Edidiong Akachukwu, and Omolola Valentina Imanah

Correction to: Chapter 8 in: T. A. Kurniawan and A. Anouzla (eds.), *Algae as a Natural Solution for Challenges in Water-Food-Energy Nexus*, Environmental Science and Engineering, https://doi.org/10.1007/978-981-97-2371-3_8

In the original version of the book, the following belated corrections have been incorporated: In Chapter 8 the author name Osemudiamhen Destiny Amienghemhen has been removed as he has not contributed to the volume. The book and the chapter have been updated with the change.

The updated version of this chapter can be found at
https://doi.org/10.1007/978-981-97-2371-3_8

T. A. Kurniawan and A. Anouzla (eds.), *Algae as a Natural Solution for Challenges in Water-Food-Energy Nexus*, Environmental Science and Engineering,
https://doi.org/10.1007/978-981-97-2371-3_35

The manufacturer's authorised representative in the EU is Springer Nature Customer Service Centre GmbH, Europaplatz 3, 69115 Heidelberg, Germany. If you have any concerns regarding our products, please contact ProductSafety@springernature.com

Printed and bound by CPI Group (UK) Ltd, Croydon, CR0 4YY
23/09/2026
02221540-0001